Principles of Development

Principles of Development

THIRD EDITION

Lewis Wolpert

Thomas Jessell

Peter Lawrence

Elliot Meyerowitz

Elizabeth Robertson

Jim Smith

OXFORD
UNIVERSITY PRESS

Great Clarendon Street, Oxford OX2 6DP

Oxford University Press is a department of the University of Oxford.
It furthers the University's objective of excellence in research, scholarship,
and education by publishing worldwide in

Oxford New York

Auckland Cape Town Dar es Salaam Hong Kong Karachi
Kuala Lumpur Madrid Melbourne Mexico City Nairobi
New Delhi Shanghai Taipei Toronto

With offices in

Argentina Austria Brazil Chile Czech Republic France Greece
Guatemala Hungary Italy Japan Poland Portugal Singapore
South Korea Switzerland Thailand Turkey Ukraine Vietnam

Oxford is a registered trade mark of Oxford University Press
in the UK and in certain other countries

Published in the United States
by Oxford University Press Inc., New York

First published by Current Biology Ltd. and Oxford University Press 1998
Second edition 2002

British Library Cataloguing in Publication Data
Data available

Library of Congress Cataloguing in Publication Data
Data available

Typeset by Cepha Imaging Pvt. Ltd., Bangalore, India.
Printed in Great Britain
on acid-free paper by
Bath Press Ltd., Bath, UK

ISBN 0-19-927537-8 978-0-19-927537-3
ISBN 0-19-927536-X (Pbk.) 978-0-19-927536-6 (Pbk.)

1 3 5 7 9 10 8 6 4 2

Preface

Developmental biology is at the core of all biology of multicellular organisms. It deals with the process by which the genes in the fertilized egg control cell behavior in the embryo and so determine the nature of the animal or plant. Evolution operates by changing the development of the organism so that a better-adapted form develops. With the application of advances in cell and molecular biology, the progress in developmental biology in recent years has been remarkable, and an enormous amount of information is now available. In this third edition we have included many recent advances, for example in the understanding of axis formation, neural development, and stem cells, and this new material is complemented by additional illustrations.

Principles of Development is designed for undergraduates and the emphasis is on principles and key concepts. Central to our approach is that development can be best understood by understanding how genes control cell behavior. We have assumed that students have some basic familiarity with cell biology and genetics, but all key concepts, such as the control of gene activity, are explained in the text.

Conscious of the pressures on students, we have tried to make the principles as clear as possible and to provide numerous summaries, in both words and pictures, and have tried to avoid providing too much detail. The illustrations in this book are a special feature and have been carefully designed and chosen to illuminate both experiments and mechanisms.

We have resisted the temptation to cover every aspect of development and have, instead, focused on those systems that best illuminate common principles. Indeed, a theme that runs throughout the book is that universal principles govern the process of development. At all stages, what we included has been guided by what we believe undergraduates should know about development.

We have thus concentrated our attention on vertebrates and *Drosophila,* but not to the exclusion of other organisms, such as the nematode and the sea urchin, where they best illustrate a concept. Because *Drosophila* development is so well understood and has been so influential, in this edition we have started the book with it rather than with vertebrates. An important feature of our book is the inclusion of plant development, which is usually neglected in general textbooks of developmental biology. There have been striking advances in the understanding of plant development in recent times, and some unique and important features have emerged.

A departure from previous editions is the inclusion of the descriptions of the embryology and genetics of our model organisms as parts of the chapters that primarily deal with their early development, rather than placing them together in a separate chapter.

The emphasis in the book is on early development and the laying down of body plans and organ systems, such as limbs and the nervous system, but we also include later aspects of development, including growth and regeneration. The book concludes with a consideration of evolution and development.

In providing further reading, our prime concern has been to guide the student to helpful papers rather than to give credit to all the scientists who have made major contributions: to those whom we have neglected, we apologize.

For this new edition we welcome Elizabeth Robertson as a co-author. Each chapter has also been reviewed by a number of experts (see page xix), to whom we give thanks. I made the initial revisions, which were then deciphered, edited, and incorporated by our editor Eleanor Lawrence, whose expertise and influence pervades the book. The new illustrations were brilliantly drawn or adapted by Matthew McClements, who created the illustrations for the first edition.

We are indebted to Ruth Hughes and Jonathan Crowe at Oxford University Press, for their help and patience throughout the preparation of this new edition.

London *L. W.*
March 2006

About the authors

Lewis Wolpert is Emeritus Professor of Biology as Applied to Medicine, in the Department of Anatomy and Developmental Biology, University College London, London, UK. He is an author of *The Triumph of the Embryo, A Passion for Science, The Unnatural Nature of Science,* and *Six Impossible Things Before Breakfast.*

Thomas Jessell is Professor of Biochemistry and Molecular Biophysics, a member of the Center for Neurobiology and Behaviour, and a Howard Hughes Medical Institute investigator at the Department of Biochemistry and Molecular Biophysics, University of Columbia Medical Center, New York, USA. He is an author of *Principles of Neural Science* and *Essentials of Neural Science and Behaviour.*

Peter Lawrence is in the Cell Biology Division at the Medical Research Council Laboratory of Molecular Biology, Cambridge, UK. He is the author of *The Making of a Fly.*

Elliot Meyerowitz is the George W. Beadle Professor of Biology and Chair of the Division of Biology at the California Institute of Technology, Pasadena, CA, USA.

Elizabeth Robertson is a Wellcome Trust Principal Fellow and Professor in the Division of Medical Sciences at the University of Oxford, Oxford, UK.

Jim Smith is Chairman of the Wellcome Trust/Cancer Research UK Gurdon Institute of Cancer and Developmental Biology, Cambridge, UK.

Eleanor Lawrence is a freelance science writer and editor.

Matthew McClements is an illustrator who specializes in design for scientific, technical, and medical communication.

Summary of contents

Contents

Chapter 1: History and basic concepts

Chapter 2: Development of the *Drosophila* body plan

Chapter 4: Patterning the vertebrate body plan II: the somites and early nervous system

Chapter 5: Development of nematodes, sea urchins, ascidians, and slime molds

Chapter 6: Plant development

Chapter 7: Morphogenesis: change in form in the early embryo

Chapter 8: Cell differentiation and stem cells

Chapter 9: Organogenesis

Chapter 10: Development of the nervous system

Chapter 11: Germ cells, fertilization, and sex

Chapter 12: Growth and post-embryonic development

Chapter 13: Regeneration

Chapter 14: Evolution and development

Text acknowledgements

Many thanks to the following who kindly reviewed various parts of the book.

Chapter 2

David Ish-Horowicz Cancer Research UK, London
Daniel St Johnston Wellcome Trust/Cancer Research UK Gurdon Institute, Cambridge

Chapter 3

Richard Harland University of California, Berkeley
Janet Heasman Cincinnati Children's Hospital Medical Center
Claudio Stern University College London

Chapter 4

Olivier Pourquie Stowers Institute for Medical Research
Derek Stemple Wellcome Trust Sanger Institute
Claudio Stern University College London
David Wilkinson National Institute for Medical Research, London

Chapter 5

Jeff Hardin University of Wisconsin, Madison
Jonathan Hodgkin University of Oxford
David McClay Duke University
Jeff Williams University of Dundee

Chapter 6

Enrico Coen John Innes Centre, Norwich
Caroline Dean John Innes Centre, Norwich
Ottoline Leyser University of York

Chapter 7

Marianne Bronner-Fraser University of California, Irvine
David McClay Duke University

Chapter 8

Nicole Le Douarin Academy of Sciences, Institute of France
Roger Patient University of Nottingham
Fiona Watt Cancer Research UK, London

Chapter 9

Cheryl Tickle University of Dundee

Chapter 10

Andrew Lumsden MRC Centre for Developmental Neurobiology, London
David Wilkinson National Institute for Medical Research, London

Chapter 11

Denise Barlow University of Vienna
Tom Cline University of California, Berkeley
Jonathan Hodgkin University of Oxford
Michael Whitaker University of Newcastle

Chapter 12

Michael Ashburner University of Cambridge
Chris Graham University of Oxford
Tom Kirkwood University of Manchester
Jamshed R. Tata National Institute for Medical Research, London

Chapter 13

Hans Bode University of California, Irvine
Susan Bryant University of California, Irvine
David Gardiner University of California, Irvine
Vernon French University of Edinburgh
Jeremy Brockes University College London

Chapter 14

Michael Coates University of Chicago
John Gerhart University of California, Berkeley
David Fitch New York University

General

Stefan Baumgartner University of Lund
Peter Carlsson University of Goteborg
David Champlin University of Southern Maine
Nigel Finn University of Bergen
Tom Fleming University of Southampton
Danny Huylebroeck University of Leuvan
Kelly Kruger California State University
Peter Olive University of Newcastle
Sebastian Shimeld University of Reading
Frank Slack Yale University
Francois van Herp University of Nijmegen
Tanya Whitfield University of Sheffield
Clive Wilson University of Oxford
Hugh Woodland University of Warwick

Figure acknowledgements

Chapter 1

Fig. 1.17 illustration after Moore, K.L.: *Before we are Born: Basic Embryology and Birth Defects, 2nd edition.* Philadelphia: W.B. Saunders, 1983.

Chapter 2

Fig. 2.1 top photograph reproduced with permission from Turner, F.R., Mahowald, A.P.: **Scanning electron microscopy of *Drosophila* embryogenesis. I. The structure of the egg envelopes and the formation of the cellular blastoderm.** *Dev. Biol.* 1976, **50:** 95–108. Middle photograph reproduced with permission from Turner, F.R., Mahowald, A.P.: **Scanning electron microscopy of *Drosophila* melanogaster embryogenesis. III. Formation of the head and caudal segments.** *Dev. Biol.* 1979, **68:** 96–109.

Fig. 2.4 left photograph reproduced with permission from Turner, F.R., Mahowald, A.P.: **Scanning electron microscopy of *Drosophila* melanogaster embryogenesis. II. Gastrulation and segmentation.** *Dev. Biol.* 1977, **57:** 403–416. Center photograph reproduced with permission from Alberts, B., Bray, D., Lewis, J., Raft, M., Roberts, K., Watson, J.D.: *Molecular Biology of the Cell,* 3rd edition. New York, Garland Publishing, 1994.

Fig. 2.11 photograph reproduced with permission from Griffiths, A.J.H., Miller, J.H., Suzuki, D.T., Lewontin, R.C., Gelbart, W.M.: *An Introduction to Genetic Analysis,* 6th edition. New York: W.H. Freeman & Co., 1996.

Fig. 2.12 photograph reproduced with permission from Griffiths, A.J.H., Miller, J.H., Suzuki, D.T., Lewontin, R.C., Gelbart, W.M.: *An Introduction to Genetic Analysis,* 6th edition. New York: W.H. Freeman & Co., 1996.

Fig. 2.16 illustration after González-Reyes, A., Elliott, H., St Johnston, D.: **Polarization of both major body axes in *Drosophila* by *gurken-torpedo*** signalling. *Nature* 1995, **375:** 654–658.

Fig. 2.20 illustration after St Johnston, D.: **Moving messages: the intracellular localization of mRNAs.** *Nat. Rev. Mol. Cell Biol.* 2005, **6:** 363–375.

Fig. 2.28 illustration after Lawrence, P.: *The Making of a Fly.* Oxford: Blackwell Scientific Publications, 1992.

Fig. 2.32 photograph reproduced with permission from Lawrence, P.: *The Making of a Fly.* Oxford: Blackwell Scientific Publications, 1992.

Fig. 2.33 illustration after Lawrence, P.: *The Making of a Fly.* Oxford: Blackwell Scientific Publications, 1992.

Box 2D illustration after Lawrence, P.: *The Making of a Fly.* Oxford: Blackwell Scientific Publications, 1992.

Fig. 2.38 illustration after Lawrence, P.: *The Making of a Fly.* Oxford: Blackwell Scientific Publications, 1992.

Fig. 2.45 illustration after Lawrence, P.: *The Making of a Fly.* Oxford: Blackwell Scientific Publications, 1992. Photograph reproduced with permission from Bender, W., Akam, A., Karch, F., Beachy, P.A., Peifer, M., Spierer, P., Lewis, E.B., Hogness, D.S.: **Molecular genetics of the bithorax complex in *Drosophila melanogaster.*** *Science* 1983, **221:** 23–29 (image on front cover). © 1983 American Association for the Advancement of Science.

Box 2E illustration after Lawrence, P.A., Casal, J., Struhl, G.: **Cell interactions and planar polarity in the abdominal epidermis of *Drosophila*.** *Development* 2004, **131:** 4651–4664.

Box 2E illustration after Strutt, H., Strutt, D.: **Long-range coordination of planar polarity in *Drosophila*.** *BioEssays* 2005, **27:** 1218–1227.

Chapter 3

Fig. 3.3 top photograph reproduced with permission from Alberts, B., Bray, D., Lewis, J., Raff, M., Roberts, K., Watson, J.D.: *Molecular Biology of the Cell,* 3rd edition. New York: Garland Publishing, 1994.

Fig. 3.5 photographs reproduced with permission from Kessel, R.G., Shih, C.Y.: *Scanning Electron Microscopy in Biology: A Student's Atlas of Biological Organization.* London, Springer-Verlag, 1974. © 1974 Springer-Verlag GmbH & Co. KG.

Fig. 3.6 illustration after Balinsky, B.I.: *An Introduction to Embryology.* Fourth edition. Philadelphia, W.B. Saunders, 1975.

Fig. 3.13 top photograph reproduced with permission from Kispert, A., Ortner, H., Cooke, J., Herrmann, B.G.: **The chick *Brachyury* gene: developmental expression pattern and response to axial induction by localized activin.** *Dev. Biol.* 1995, **168:** 406–415.

Fig. 3.15 illustration adapted, with permission, from Balinsky, B.I.: *An Introduction to Embryology,* 4th edition. Philadelphia, W.B. Saunders, 1975. © 1975 Saunders College Publishing.

Fig. 3.17 illustration after Patten, B.M.: *Early Embryology of the Chick.* New York, Mc Graw-Hill, 1971.

Fig. 3.19 photographs reproduced with permission from Kispert, A., Ortner, H., Cooke, J., Herrmann, B.G.: **The chick *Brachyury* gene: developmental expression pattern and response to axial induction by localized activin.** *Dev. Biol.* 1995, **168:** 406–415.

Fig. 3.20 illustration after Patton, B.M.: **The first heart beat and the beginning of embryonic circulation.** *American Scientist* 1951, **39:** 225–243.

Fig. 3.21 top photograph reproduced with permission from Bloom, T.L.: **The effects of phorbol ester on mouse blastomeres: a role for protein kinase C in compaction?** *Development* 1989, **106:** 159–171 Published by permission of The Company of Biologists Ltd.

Fig. 3.23 illustration after Hogan, B., Beddington, R., Costantini, F., Lacy, E.: *Manipulating the Mouse Embryo: A Laboratory Manual,* 2nd edition. New York: Cold Spring Harbor Laboratory Press, 1994.

Fig. 3.24 illustration adapted, with permission, from McMahon, A.P.: **Mouse development. Winged-helix in axial patterning.** *Curr. Biol.* 1994, **4:** 903–906.

Fig. 3.26 illustration after Kaufman, M.H.: *The Atlas of Mouse Development.* London: Academic Press, 1992.

Fig. 3.39 illustration after Rodriguez, T.A., Srinivas, S., Clements, M.P., Smith, J.C., Beddington, R.S.: **Induction and migration of the anterior visceral endoderm is regulated by the extra-embryonic ectoderm.** *Development* 2005, **132:** 2513–2520.

Fig. 3.61 photograph reproduced with permission from Smith W.C., Harland, R.M.: **Expression cloning of noggin, a new dorsalizing**

factor localized to the Spemann organizer in *Xenopus* embryos. *Cell* 1992, **70:** 829–840. © 1992 Cell Press.

Fig. 3.64 illustration after Beddington, S.P., Robertson, E.J.: **Axis development and early asymmetry in mammals**. *Cell* 1999, **96:** 195–209.

Chapter 4

Fig. 4.3 photograph reproduced with permission from Hausen, P., Riebesell, M.: *The Early Development of Xenopus laevis.* Berlin: Springer-Verlag, 1991.

Fig. 4.7 illustration from Deschamps, J., van Nes, J.: **Developmental regulation of the Hox genes during axial morphogenesis in the mouse**. *Development* 2005, **132:** 2931–2942.

Fig. 4.12 illustration after Burke, A.C., Nelson, C.E., Morgan, B.A., Tabin, C.: **Hox genes and the evolution of vertebrate axial morphology.** *Development* 1995, **121:** 333–346.

Fig. 4.17 illustration after Johnson, R.L., Laufer, E., Riddle, R.D., Tabin, C.: **Ectopic expression of *Sonic hedgehog* alters dorsal-ventral patterning of somites.** *Cell* 1994, **79:** 1165–1173.

Fig. 4.19 illustration from Kiecker, C., Niehrs, C: **The role of prechordal mesendoderm in neural patterning.** *Curr. Opin. Neurobiol.* 2001, **11:** 27–33.

Fig. 4.23 illustration from Kudoh, T. *et. al.*: **Combinatorial Fgf and Bmp signalling patterns the gastrula ectoderm into prospective neural and epidermal domains**. *Development* 2004, **131:** 3581–3592.

Fig. 4.24 illustration after Kintner, C.R., Dodd, J.: **Hensen's node induces neural tissue in *Xenopus* ectoderm. Implications for the action of the organizer in neural induction.** *Development* 1991, **113:** 1495–1505.

Fig. 4.25 adapted from Delfino-Machin, M. *et. al.*: **Specification and maintenance of the spinal cord stem zone**. *Development* 2005, **132:** 4273–4283.

Fig. 4.26 illustration after Mangold, O.: **Über die induktionsfahigkeit der verschiedenen bezirke der neurula von urodelen.** *Naturwissenschaften* 1933, **21:** 761–766.

Fig. 4.27 illustration after Kelly, O.G., Melton, D.A.: **Induction and patterning of the vertebrate nervous system.** *Trends Genet.* 1995, **11:** 273–278.

Fig. 4.28 illustration adapted, with permission, from Lumsden, A.: **Cell lineage restrictions in the chick embryo hindbrain.** *Phil. Trans. R. Soc. Lond. B* 1991, **331:** 281–286.

Fig. 4.29 illustration adapted, with permission, from Lumsden, A.: **Cell lineage restrictions in the chick embryo hindbrain.** *Phil. Trans. R. Soc. Lond. B* 1991, **331:** 281–286.

Fig. 4.31 illustration after Krumlauf, R.: **Hox genes and pattern formation in the branchial region of the vertebrate head.** *Trends Genet.* 1993, **9:** 106–112.

Fig. 4.32 photograph reproduced with permission from Lumsden, A., Krumlauf, R.: **Patterning the vertebrate neuraxis.** *Science* 1996, **274:** 1109–1115 (image on front cover). © 1996 American Association for the Advancement of Science.

Box 4A illustration after Coletta, P.L., Shimeld, S.M., Sharpe, P.T.: The molecular anatomy of Hox gene expression. *J. Anat.* 1994, **184:** 15–22.

Chapter 5

Fig. 5.6 photograph reproduced with permission from Strome, S., Wood, W.B.: **Generation of asymmetry and segregation of germ-line granules in early *C. elegans* embryos.** *Cell* 1983 **35:** 15–25. © 1983 Cell Press.

Fig. 5.7 illustration after Sulston, J.E., Schierenberg, E., White, J.G., Thompson, J.N.: **The embryonic cell lineage of the nematode *C. elegans*.** *Dev. Biol.* 1983, **100**:69–119.

Fig. 5.8 photographs reproduced with permission from Wood W.B: **Evidence from reversal of handedness in *C. elegans* embryos for early cell interactions determining cell fates.** *Nature* 1991, **349:** 536–538. © 1991 Macmillan Magazines Ltd.

Fig. 5.9 illustration after Mello, C.C., Draper, B.W., Priess, J.R.: **The maternal genes *apx-1* and *glp-1* and establishment of dorsal-ventral polarity in the early *C. elegans* embryo.** *Cell* 1994, **77:** 95–106.

Fig. 5.12 illustration after Bürglin, T.R., Ruvkun, G.: **The *Caenorhabditis elegans* homeobox gene cluster.** *Curr. Opin. Gen. Dev.* 1993, **3:** 615–620.

Fig. 5.18 top and middle photographs reproduced with permission from Jim Coffman. Lower photograph reproduced with permission from Oxford Scientific Films.

Fig. 5.24 illustration from Oliveri, P., Davidson, E.H.: **Gene regulatory network controlling embryonic specification in the sea urchin.** *Curr. Opin. Genet. Dev.* 2004, **14:** 351–360.

Fig. 5.26 illustration from Oliveri, P., Davidson, E.H.: **Gene regulatory network controlling embryonic specification in the sea urchin.** *Curr. Opin. Genet. Dev.* 2004, **14:** 351–360.

Fig. 5.28 middle photograph reproduced with permission from Corbo, J.C., Levine, M., Zeller, R.W.: **Characterization of a notochord-specific enhancer from the Brachyury promoter region of the ascidian, *Ciona intestinalis*.** *Development* 1997, **124:** 589–602. Published by permission of The Company of Biologists Ltd. Top photograph reproduced with permission from Shigeki Fujiwara and Naoki Shimozono. Lower photograph reproduced with permission from Andrew Martinez.

Fig. 5.30 illustration from Nishida, H.: **Specification of embryonic axis and mosaic development in ascidians.** *Dev. Dyn.* 2005, **233:** 1177–1193.

Fig. 5.31 illustration after Conklin, E.G.: **The organization and cell lineage of the ascidian egg.** *J. Acad. Nat. Sci. Philadelphia* 1905, **13:** 1–119.

Fig. 5.32 illustration after Nakatani, Y., Yasuo, H. Satoh, N., Nishida, H.: **Basic fibroblast growth factor induces notochord formation and the expression of *As-T*, a *Brachyury* homolog, during ascidian embryogenesis.** *Development* 1996, **122:** 2023–2031.

Fig. 5.33 top photograph reproduced with permission from Early, A.E., Gaskell, M.J., Traynor, D., Williams, J.G.: **Two distinct populations of prestalk cells within the tip of the migratory Dictyostelium slug with differing fates at culmination.** *Development* 1993, **118:** 353–362. Published by permission of The Company of Biologists Ltd. Middle panel, photograph reproduced with permission from Jermyn K., Traynor, D., Williams, J.: **The initiation of basal disc formation in *Dictyostelium discoideum*** is an early event in culmination. *Development* 1996, **122:** 753–760. Published by permission of The Company of Biologists Ltd.

Chapter 6

Fig. 6.4 illustration after Scheres, B., Wolkenfelt, H., Willemsen, V., Terlouw, M., Lawson, E., Dean, C., Weisbeek, P.: **Embryonic origin of the *Arabidopsis* primary root and root meristem initials.** *Development* 1994, **120:** 2475–2487.

Fig. 6.5 illustration after Friml, J., *et al.*: **Efflux-dependent auxin gradients establish the apical-basal axis of *Arabidopsis*.** *Nature* 2003, **426:** 147–153.

Fig. 6.9 illustration after Alberts, B., Bray, D., Lewis, J., Raff, M., Roberts, K., Watson, J.D.: *Molecular Biology of the Cell,* 2nd edition. New York: Garland Publishing, 1989.

Fig. 6.13 illustration after Steeves, T.A., Sussex, I.M.: *Patterning in Plant Development.* Cambridge: Cambridge University Press, 1989.

Fig. 6.15 illustration after McDaniel, C.N., Poethig, R.S.: **Cell lineage patterns in the shoot apical meristem of the germinating maize embryo.** *Planta* 1988, **175:** 13–22.

Fig. 6.16 illustration after Irish, V.F.: **Cell lineage in plant development.** *Curr. Opin. Gen. Devel.* 1991, **1:** 169–173.

Fig. 6.18 top panel, illustration after Poethig, R.S., Sussex, I.M.: **The cellular parameters of leaf development in tobacco: a clonal analysis.** *Planta* 1985, **165:** 170–184. Bottom panel, illustration after Sachs, T.: *Pattern Formation in Plant Tissues.* Cambridge: Cambridge University Press, 1994.

Fig. 6.20 adapted from Heisler, M.G. *et al.*: **Patterns of auxin transport and gene expression during primordium development revealed by live imaging of the *Arabadopsis* inflorescence meristem.** *Curr. Biol.* 2005, **15:** 1899–1911.

Fig. 6.22 illustration after Scheres, B., Wolkenfelt, H., Willemsen, V., Terlouw, M., Lawson, E., Dean, C., Weisbeek, P.: **Embryonic origin of the *Arabidopsis* primary root and root meristem initials.** *Development* 1994, **120**: 2475–2487.

Fig. 6.24 photograph reproduced with permission from Meyerowitz, E.M., Bowman, J.L., Brockman, L.L., Drews, G.N., Jack, T., Sieburth, L.E., Weigel, D.: **A genetic and molecular model for flower development in *Arabidopsis thaliana*.** *Development Suppl.* 1991, 157–167. Published by permission of The Company of Biologists Ltd.

Fig. 6.25 illustration after Coen, E.S., Meyerowitz, E.M.: **The war of the whorls: genetic interactions controlling flower development.** *Nature* 1991, **353:** 31–37.

Fig. 6.26 photographs reproduced with permission from Meyerowitz, E.M., Bowman, J.L., Brockman, L.L., Drews, G.N., Jack, T., Sieburth, L.E., Weigel, D.: **A genetic and molecular model for flower development in *Arabidopsis thaliana*.** *Development Suppl.* 1991, 157–167. Published by permission of The Company of Biologists Ltd. (left panel); center panel from Bowman, J.L., Smyth, D.R., Meyerowitz, E.M.: **Genes directing flower development in *Arabidopsis*.** *Plant Cell* 1989, **1:** 37–52. Published by permission of The American Society of Plant Physiologists.

Fig. 6.29 adapted from Lohmann, J.U., Weigel, D.: Building beauty: the genetic control of floral patterning. *Dev. Cell* 2002, **2:** 135–142.

Fig. 6.30 photograph reproduced with permission from Coen, E.S., Meyerowitz, E.M.: **The war of the whorls: genetic interactions controlling flower development.** *Nature* 1991, **353:** 31–37. © 1991 Macmillan Magazines Ltd.

Fig. 6.31 illustration after Drews, G.N., Goldberg, R.B.: **Genetic control of flower development.** *Trends Genet.* 1989 **5:** 256–261.

Fig. 6.35 illustration from Bl·zquez, M.A.: **The right time and place for making flowers.** *Science* 2005, **309:** 1024–1025.

Box 6C (Fig. 1) illustration after Dennis, E., Bowman, J.L.: **Manipulating floral identity.** *Curr. Biol.* 1993, **3:** 90–93.

Box 6C (Fig. 2) illustration after Meyerowitz, E.M., Bowman, J.L., Brockman, L.L., Drews, G.N., Jack, T., Sieburth, L.E., Weigel, D.: **A genetic and molecular model for flower development in *Arabidopsis thaliana*.** *Development Suppl.* 1991, 157–167.

Chapter 7

Fig. 7.3 photographs reproduced with permission from Steinberg, M.S., Takeichi, M.: **Experimental specification of cell sorting, tissue spreading, and specific spatial patterning by quantitative differences in cadherin expression.** *Proc. Natl Acad. Sci./Dev. Biol* 1994, **91:** 206–209. © 1994 National Academy of Sciences.

Fig. 7.6 illustration after Strome, S.: **Determination of cleavage planes.** *Cell* 1993, **72:** 3–6.

Fig. 7.9 photograph reproduced with permission from Bloom T.L.: **The effects of phorbol ester on mouse blastomeres: a role for protein kinase C in compaction?** *Development* 1989, **106:** 159–171.

Fig. 7.12 illustration after Coucouvanis, E., Martin, G.R.: **Signals for death and survival: a two-step mechanism for cavitation in the vertebrate embryo.** *Cell* 1995, **83:** 279–287.

Fig. 7.16 photograph reproduced with permission from Merrill, J.B., Santos, L.L.: **A scanning electron micrographical overview of cellular and extracellular patterns during blastulation and gastrulation in the sea urchin, *Lytechinus variegatus*.** In *The Cellular and Molecular Biology of Invertebrate Development.* Edited by Sawyer, R.H. and Showman, R.M. University of South Carolina Press, 1985; pp. 3–33.

Fig. 7.18 illustration after Odell, G.M., Oster, G., Alberch, P., Burnside, B.: **The mechanical basis of morphogenesis. I. Epithelial folding and invagination.** *Dev. Biol.* 1981, **85:** 446–462.

Fig. 7.20 photographs reproduced with permission from Leptin, M., Casal, J., Grunewald, B., Reuter, R.: **Mechanisms of early *Drosophila* mesoderm formation.** *Development Suppl.* 1992, 23–31. Published by permission of The Company of Biologists Ltd.

Fig. 7.21 illustration after Bertet, C., Sulak, L., Lecuit, T. **Myosin-dependent junction remodelling controls planar cell intercalation and axis elongation.** *Nature* 2004, **429:** 667–671.

Fig. 7.23 illustration after Balinsky, B.I.: *An Introduction to Embryology,* 4th edition. Philadelphia, W.B. Saunders, 1975.

Fig. 7.24 illustration after Hardin, J., Keller, R.: **The behavior and function of bottle cells during gastrulation of *Xenopus laevis*.** *Development* 1988, **103:** 211–230.

Fig. 7.27 photograph reproduced with permission from Smith, J.C., Cunliffe, V., O'Reilly, M-A.J., Schulte-Merker, S., Umbhauer, M.: ***Xenopus Brachyury*.** *Semin. Dev. Biol.* 1995, **6:** 405–410. © 1995 by permission of the publisher, Academic Press Ltd., London.

Fig. 7.28 illustration after Montero, J.A., Heisenberg, C.P. **Gastrulation dynamics: cells move into focus.** *Trends Cell Biol.* 2004, **14:** 620–627.

Fig. 7.31 illustration after Wallingford, J.B., Fraser, S., Harland, R.M. **Convergent extension: the molecular control of polarized cell movement during embryonic development.** *Dev. Cell* 2002, **2:** 695–706.

Fig. 7.33 illustration after Myers, D.C., Sepich, D.S., Solnica-Krezel, L. **Convergence and extension in vertebrate gastrulae: cell movements according to or in search of identity?** *Trends Genet.* 2002, **18:** 447–455.

Fig. 7.36 illustration after Schoenwolf, G.C., Smith, J.L.: **Mechanisms of neurulation: traditional viewpoint and recent advances.** *Development* 1990 **109:** 243–270.

Fig. 7.41 illustration after Alberts, B., Bray, D., Lewis, J., Raft, M., Roberts, K., Watson, J.D.: *Molecular Biology of the Cell,* 2nd edition. New York: Garland Publishing, 1989.

Fig. 7.44 photographs reproduced with permission from Priess, J.R., Hirsh, D.I.: ***Caenorhabditis elegans morphogenesis:* the role of the cytoskeleton in elongation of the embryo.** *Dev. Biol* 1986, **117:** 156–173. © 1986 Academic Press.

Fig. 7.46 photographs reproduced with permission from Tsuge, T., Tsukaya, H., Uchimaya, H.: **Two independent and polarized processes of cell elongation regulate leaf blade expansion in *Arabidopsis thaliana* (L.) Heynh.** *Development* 1996, **122:** 1589–1600. Published by permission of The Company of Biologists Ltd.

Chapter 8

Fig. 8.3 illustration after Tijian, R.: **Molecular machines that control genes.** *Sci. Am.* 1995, **272:** 54–61.

Fig. 8.6 illustration after Alberts B., Bray, D., Lewis, J., Raff, M., Roberts, K., Watson, J.D.: *Molecular Biology of the Cell,* 2nd edition. New York: Garland Publishing, 1989.

Fig. 8.8 illustration after Alberts, B., Bray, D., Lewis, J., Raft, M., Roberts, K., Watson, J.D.: *Molecular Biology of the Cell,* 2nd edition. New York: Garland Publishing, 1989.

Fig. 8.10 illustration after Alberts, B., Bray, D., Lewis, J., Raft, M., Roberts, K., Watson, J.D.: *Molecular Biology of the Cell,* 2nd edition. New York: Garland Publishing, 1989.

Fig. 8.12 illustration after Kluger, Y., Lian, Z., Zhang, X., Newburger, P.E., Weissman, S.M.: **A panorama of lineage-specific transcription in hematopoiesis.** *BioEssays* 2004, **26:** 1276–1287.

Fig. 8.14 illustration after Metcalf, D.: **Control of granulocytes and macrophages: molecular, cellular, and clinical aspects.** *Science* 1991, **254:** 529–533.

Fig. 8.17 illustration after Crossley, M., Orkin, S.H.: **Regulation of the β-globin locus.** *Curr. Opin. Genet. Dev.* 1993, **3:** 232–237.

Fig. 8.18 illustration after Janeway, C.A., Travers, P.: *Immunobiology: The Immune System in Health and Disease,* 3rd edition. London: Current Biology/Garland, 1997.

Fig. 8.19 illustration after Janeway, C.A., Travers, P.: *Immunobiology: The Immune System in Health and Disease,* 3rd edition. London: Current Biology/Garland Publishing, 1997.

Fig. 8.24 illustration after Taupin, P.: **Adult neurogenesis in the mammalian central nervous system: functionality and potential clinical interest**. *Med. Sci. Monit.* 2005, **11:** 247–252.

Fig. 8.29 illustration after Riedl, S.J., Shi, Y.: **Molecular mechanisms of caspase regulation during apoptosis.** *Nat. Rev. Mol. Cell Biol.* 2004, **5:** 897–907.

Fig. 8.33 illustration after Okada, T.S.: *Transdifferentiation.* Oxford: Clarendon Press, 1992.

Fig. 8.34 illustration after Doupe, A.J., Landis, S.C., Patterson, P.H.: **Environmental influences in the development of neural crest derivatives: glucocorticoids, growth factors, and chromaffin cell plasticity.** *J. Neurosci.* 1985, **5:** 2119–2142.

Chapter 9

Fig. 9.10 photograph reproduced with permission from Cohn, M.J., Izpisúa-Belmonte, J.C., Abud, H., Heath, J.K., Tickle, C.: **Fibroblast growth factors induce additional limb development from the flank of chick embryos.** *Cell* 1995, **80:** 739–746. © 1995 Cell Press.

Fig. 9.18 adapted from Wellik, D.M., Capecchi, M.R.: ***Hox10* and *Hox11* genes are required to globally pattern the mammalian skeleton**. *Science* 2003, **301:** 363–367.

Fig. 9.21 photograph reproduced with permission from Garcia-Martinez, V., Macias, D., Gañan, Y., Garcia-Lobo, J.M., Francia, M.V., Fernandez-Teran, M.A., Hurle, J.M.: **Internucleosomal DNA fragmentation and programmed cell death (apoptosis) in the interdigital tissue of embryonic chick leg bud.** *J. Cell Sci.* 1993, **106:** 201–208. Published by permission of The Company of Biologists Ltd.

Fig. 9.23 illustration after French, V., Daniels, G.: **Pattern formation: the beginning and the end of insect limbs.** *Curr. Biol.* 1994, **4:** 35–37.

Fig. 9.26 photograph reproduced with permission from Nellen, D., Burke, R., Struhl, G., Basler, K.: **Direct and long-range action of a dpp morphogen gradient.** *Cell* 1996, **85:** 357–368. © 1996, Cell Press.

Fig. 9.29 adapted from Crozatier, M., Glise, B., Vincent, A.: **Patterns in evolution: veins of the *Drosophila* wing.** *Trends Genet.* 2004, **20:** 498–505.

Fig. 9.30 photograph reproduced with permission from Zecca, M., Basler, K., Struhl, G.: **Direct and long-range action of a wingless morphogen gradient.** *Cell* 1996, **87:** 833–844. © 1996 Cell Press.

Fig. 9.32 illustration after Bryant, P.J. **The polar coordinate model goes molecular.** *Science* 1993, **259:** 471–472

Fig. 9.47 adapted from Kelly, R.G., Buckingham, M.E.: **The anterior heart-forming field: voyage to the arterial pole of the heart.** *Trends Genet.* 2002, **18:** 210–216.

Box 9B top illustration after Meinhardt, H., Gierer, A.: **Applications of a theory of biological pattern formation based on lateral inhibition.** *J. Cell Sci.* 1974, **15:** 321–346.

Chapter 10

Fig. 10.5 adapted from Knust, E.: **G protein signaling and asymmetric cell division**. *Cell* 2001, **107:** 125–128.

Fig. 10.6 illustration after Jan, Y.N., Jan, L.Y.: **Genes required for specifying cell fates in Drosophila embryonic sensory nervous system.** *Trends Neurosci.* 1990, **13:** 493–498.

Fig. 10.7 illustration after Campuzano, S., Modolell, J.: **Patterning of the *Drosophila* nervous system: the achaete-scute gene complex.** *Trends Genet.* 1992, **8:** 202–208.

Fig. 10.10 illustration after Rakic, P.: **Mode of cell migration to the superficial layers of fetal monkey neocortex.** *J. Comp. Neurol.* 1972, **145:** 61–83.

Fig. 10.11 adapted from Honda, T., Tabata, H., Nakajima, K.: **Cellular and molecular mechanisms of neuronal migration in neocortical development.** *Semin. Cell Dev. Biol.* 2003, **14:** 169–174.

Fig. 10.18 adapted from Dasen, J., Tice, B., Brenner-Morton, S., Jessell, T.: **A Hox regulatory network establishes motor neuron pool identity and target-muscle connectivity.** *Cell* 2005, **123:** 477–491.

Fig. 10.19 illustration after Alberts, B., Bray, D., Lewis, J., Raff, M., Roberts, K., Watson, J.D.: *Molecular Biology of the Cell,* 2nd edition. New York: Garland Publishing, 1989.

Fig. 10.24 illustration after Tessier-Lavigne, M., Placzek, M.: **Target attraction: are developing axons guided by chemotropism?** *Trends Neurosci.* 1991, **14:** 303–310.

Fig. 10.25 photograph reproduced with permission from Serafini, T., Colamarino, S.A., Leonardo, E.D., Wang, H., Beddington, R., Skarnes, W.C., Tessier-Lavigne, M.: **Netrin-1 is required for commissural axon guidance in the developing vertebrate nervous system.** *Cell* 1996, **87:** 1001–1014. © 1996 Cell Press.

Fig. 10.31 illustration after Kandell, E.R., Schwartz, J.H., Jessell, T.M.: *Principles of Neural Science,* 3rd edition. New York: Elsevier Science Publishing Co., Inc., 1991.

Fig. 10.33 illustration after Li, Z., Sheng, M.: **Some assembly required: the development of neuronal synapses**. *Nat. Rev.* 2003, **4**: 833–841.

Fig. 10.35 illustration after Goodman, C.S., Shatz, C.J.: **Developmental mechanisms that generate precise patterns of neuronal connectivity.** *Cell Suppl.* 1993, **72:** 77–98

Fig. 10.36 illustration after Goodman, C.S., Shatz, C.J.: **Developmental mechanisms that generate precise patterns of neuronal connectivity.** *Cell Suppl.* 1993, **72:** 77–98

Fig. 10.37 illustration after Kandell, E.R., Schwartz, J.H., Jessell, T.M.: *Essentials of Neural Science and Behavior.* Norwalk, Connecticut: Appleton & Lange, 1991.

Fig. 10.38 illustration after Goodman, C.S., Shatz, C.J.: **Developmental mechanisms that generate precise patterns of neuronal connectivity.** *Cell Suppl.* 1993, **72:** 77–98

Chapter 11

Fig. 11.3 adapted from Hogan, B.: **Decisions, decisions**. *Nature* 2002, **418**: 282, and Saitou, M., Barton, S. Surani, M.: **A molecular programme for the specification of germ cell fate in mice**. *Nature*, 2002, **418:** 293–300.

Fig. 11.5 illustration after Wylie, C.C., Heasman, J.: **Migration, proliferation, and potency of primordial germ cells.** *Semin. Dev. Biol.* 1993, **4:** 161–170.

Fig. 11.11 illustration after Alberts, B., Bray, D., Lewis, J., Raft, M., Roberts, K., Watson, J.D.: *Molecular Biology of the Cell,* 2nd edition. New York: Garland Publishing, 1989.

Fig. 11.16 illustration after Goodfellow, P.N., Lovell-Badge, R.: ***SRY* and sex determination in mammals.** *Ann. Rev. Genet.* 1993, **27:** 71–92.

Fig. 11.18 illustration after Higgins, S.J., Young, P., Cunha, G.R.: **Induction of functional cytodifferentiation in the epithelium of tissue recombinants II. Instructive induction of Wolffian duct epithelia by neonatal seminal vesicle mesenchyme.** *Development* 1989, **106:** 235–250.

Fig. 11.22 illustration after Cline, T.W.: **The *Drosophila* sex determination signal: how do flies count to two?** *Trends Genet.* 1993 **9:** 385–390.

Fig. 11.26 illustration after Clifford, R., Francis, R., Schedl, T.: **Somatic control of germ cell development.** *Semin. Dev. Biol* 1994, **5:** 21–30.

Chapter 12

Fig. 12.3 illustration after Edgar, B.A., Lehman, D.A., O'Farrell, P.H.: **Transcriptional regulation of *string (cdc25)*: a link between developmental programming and the cell cycle.** *Development* 1994, **120:** 3131–3143.

Fig. 12.5 photograph reproduced with permission from Harrison, R.G.: *Organization and Development of the Embryo.* New Haven: Yale University Press, 1969. © 1969 Yale University Press.

Fig. 12.6 illustration after Gray, H.: *Gray's Anatomy.* Edinburgh: Churchill-Livingstone, 1995.

Fig. 12.9 illustration after Walls, G.A.: **Here today, bone tomorrow.** *Curr. Biol.* 1993, **3:** 687–689.

Fig. 12.10 illustration after Kronenberg, H.M.: **Developmental regulation of the growth plate.** *Nature (Insight)* 2003, **423:** 332–336.

Fig. 12.16 illustration after Tata, J.R.: **Gene expression during metamorphosis: an ideal model for post-embryonic development.** *BioEssays* 1993, **15:** 239–248.

Fig. 12.17 illustration after Tata, J.R.: **Gene expression during metamorphosis: an ideal model for post-embryonic development.** *BioEssays* 1993, **15:** 239–248.

Chapter 13

Fig. 13.11 photographs reproduced with permission from Pecorino, L.T. Entwistle A., Brockes, J.P.: **Activation of a single retinoic acid receptor isoform mediates proximodistal respecification** *Curr. Biol.*1996, **6:** 563–569.

Fig. 13.13 illustration after French, V., Bryant, P.J., Bryant, S.V.: **Pattern regulation in epimorphic fields.** *Science* 1976, **193:** 969–981.

Fig. 13.14 illustration after French, V., Bryant, P.J., Bryant, S.V.: **Pattern regulation in epimorphic fields.** *Science* 1976, **193:** 969–981.

Fig. 13.15 photograph reproduced with permission from Müller, W.A.: **Diacylglycerol-induced multihead formation in *Hydra.*** *Development* 1989, **105:** 309–316. Published by permission of The Company of Biologists Ltd.

Fig. 13.21 photograph reproduced with permission from Müller, W.A.: **Diacylglycerol-induced multihead formation in *Hydra.*** *Development* 1989, **105:** 309–316. Published by permission of The Company of Biologists Ltd.

Chapter 14

Fig. 14.6 illustration after Larsen, W.J.: *Human Embryology.* New York: Churchill Livingstone, 1993.

Fig. 14.7 illustration after Romer, A.S.: *The Vertebrate Body.* Philadelphia: W.B. Saunders, 1949.

Fig. 14.9 photographs reproduced with permission from Sordino, P., van der Hoeven, F., Duboule, D.: **Hox gene expression in teleost fins and the origin of vertebrate digits.** *Nature* 1995, **375:** 678–681. © 1995 Macmillan Magazines Ltd.

Fig. 14.17 illustration after Akam, M.: **Hox genes and the evolution of diverse body plans.** *Phil, Trans. R. Soc. Lond. B* 1995, **349:** 313–319.

Fig. 14.18 illustration after Ferguson, E.L.: **Conservation of dorsa-ventral patterning in arthropods and chordates.** *Curr. Opin. Genet. Dev.* 1996, **6:** 424–431.

Fig. 14.19 illustration after Gerhart, J., Lowe, C., Kirschner, M.: **Hemichordates and the origins of chordates.** *Curr. Opin. Genet. Dev.* 2005, **15:** 461–467.

Fig. 14.20 illustration after Gompel, N. et al.: **Chance caught on the wing: cis-regulatory evolution and the origin of pigment patterns in *Drosophila*.** *Nature*, 2005, **433**: 481–487.

Fig. 14.21 illustration after Gregory, W.K.: *Evolution Emerging.* New York: Macmillan, 1957.

1

History and basic concepts

- The origins of developmental biology
- A conceptual tool kit

The aim of this chapter is to provide a conceptual framework for the study of development. We start with a brief history of the study of embryonic development, which illustrates how some of the key questions in developmental biology were first formulated, and continue with some of the essential principles of development. The big question is how does a single cell—the fertilized egg—give rise to a multicellular organism, in which a multiplicity of different cell types are organized into tissues and organs to make up a three-dimensional body. This question can be studied from many different viewpoints, all of which have to be fitted together to obtain a complete picture of development: which genes are expressed, and when and where; how cells communicate with each other; how a cell's developmental fate is determined, how cells proliferate and differentiate into specialized cell types; and how major changes in body shape are produced. We shall see that an organism's development is ultimately driven by the regulated expression of its genes, determining which proteins are present in which cells and when. In turn, proteins largely determine how a cell behaves. The genes provide a generative program for development, not a blueprint, as their actions are translated into developmental outcomes through cellular behavior such as intercellular signaling, cell proliferation, cell differentiation, and cell movement.

The development of multicellular organisms from a single cell—the fertilized egg—is a brilliant triumph of evolution. During embryonic development, the egg divides to give rise to many millions of cells, which form structures as complex and varied as eyes, arms, heart, and brain. This amazing achievement raises a multitude of questions. How do the cells arising from division of the fertilized egg become different from each other? How do they become organized into structures such as limbs and brains? What controls the behavior of individual cells so that such highly organized patterns emerge? How are the organizing principles of development embedded within the egg, and in particular within the genetic material, DNA? Much of the excitement in developmental biology today comes from our growing understanding of how genes direct these developmental processes, and genetic control is one of the main themes of this book. Thousands of genes are involved in controlling development, but we will focus only on those that have key roles and illustrate general principles.

The development of an embryo from the fertilized egg is known as **embryogenesis**. One of its first tasks is to lay down the overall body plan of the organism, and we shall see that different organisms solve this fundamental

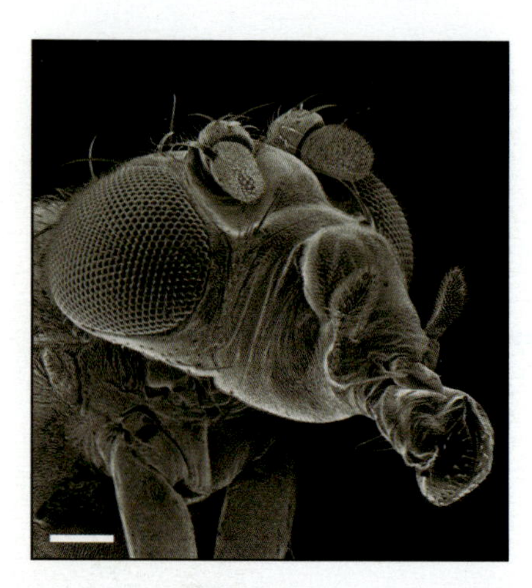

Fig. 1.1 Scanning electron micrograph of the head of an adult *Drosophila melanogaster*. Scale bar = 0.1 mm.

Photograph by D. Scharfe, from Science Photo Library.

Fig. 1.2 Photograph of a lizard, the south eastern five-lined skink, after it has released its tail in defense. This species can deliberately shed its tail as a technique to avoid capture by predators; and then regenerate it. A piece of discarded tail can be seen below the skink.

Photograph from Oxford Scientific Films.

problem in several ways. The focus of this book is mainly on animal development—that of vertebrates such as frogs, birds, fish, and mammals, and of a selection of invertebrates, such as the sea urchin, ascidians, and, above all, the fruit fly *Drosophila melanogaster* (Fig. 1.1) and the nematode worm *Caenorhabditis elegans*. It is in these last two organisms that our understanding of the genetic control of development is most advanced, and the main features of their early development are considered in Chapters 2 and 5, respectively. In Chapter 6, we look briefly at some aspects of plant development, which differs in some respects from that of animals but involves similar principles.

The development of individual organs such as the vertebrate limb, the insect eye, and the nervous system illustrates multicellular organization and tissue differentiation at later stages in embryogenesis, and we consider some of these systems in detail in Chapters 8–10. We also deal with the development of sexual characteristics (Chapter 11). The study of developmental biology, however, goes well beyond the development of the embryo. We also need to understand how some animals can regenerate lost organs (Fig. 1.2, and Chapter 13), and how post-embryonic growth of the organism is controlled, a process that includes metamorphosis and aging (Chapter 12). Taking a wider view, we consider in Chapter 14 how developmental mechanisms have evolved and how they constrain the very process of evolution itself.

One might ask whether it is necessary to cover different organisms and developmental systems in order to understand the basic features of development. The answer at present is yes. Developmental biologists do indeed believe that there are general principles of development that apply to all animals, but that life is too wonderfully diverse to find all the answers in a single organism. As it is, developmental biologists have tended to focus their efforts on a relatively small number of animals, chosen originally because they were convenient to study and amenable to experimental manipulation or genetic analysis. This is why some creatures, such as the frog *Xenopus laevis* (Fig. 1.3), the nematode *Caenorhabditis*, and the fruit fly *Drosophila*, have such a dominant place in developmental biology, and are encountered again and again in this book. Indeed, it is very encouraging that so few systems need to be studied in order to understand animal development. Similarly, the thale-cress, *Arabidopsis thaliana*, can be used as a model plant to consider the basic features of plant development.

One of the most exciting and satisfying aspects of developmental biology is that understanding a developmental process in one organism can help to illuminate similar processes elsewhere, for example in organisms much more like ourselves. Nothing illustrates this more dramatically than the influence that our understanding of *Drosophila* development, and especially of its genetic basis, has had throughout developmental biology. In particular, the identification of genes controlling early embryogenesis in *Drosophila* has led to the discovery of related genes being used in similar ways in the development of mammals and other vertebrates. Such discoveries encourage us to believe in the emergence of general developmental principles.

Frogs have long been a favorite organism for studying development because their eggs are large, and their embryos are robust, easy to grow in a simple culture medium, and relatively easy to experiment on. The South African frog *Xenopus* is the model organism for many aspects of vertebrate development, and the main features of its development (Box 1A, pp. 4–5) serve to illustrate some of the basic stages of development in all animals. The early development of *Xenopus* and the other model vertebrates is discussed in Chapters 3 and 4.

In the rest of this chapter we first look at the history of **embryology**—as the study of developmental biology has been called for most of its existence. The term developmental biology itself is of much more recent origin. We then introduce some key concepts that are used repeatedly in studying and understanding development.

The origins of developmental biology

Many questions in embryology were first posed hundreds, and in some cases thousands, of years ago. Appreciating the history of these ideas helps us to understand why we approach developmental problems in the way that we do today.

1.1 Aristotle first defined the problem of epigenesis and preformation

A scientific approach to explaining development started with Hippocrates in Greece in the 5th century BC. Using the ideas current at the time, he tried to explain development in terms of the principles of heat, wetness, and solidification. About a century later the study of embryology advanced when the Greek philosopher Aristotle formulated a question that was to dominate much thinking about development until the end of the 19th century. Aristotle addressed the problem of how the different parts of the embryo were formed. He considered two possibilities: one was that everything in the embryo was preformed from the very beginning and simply got bigger during development; the other was that new structures arose progressively, a process he termed epigenesis (which means 'upon formation') and that he likened metaphorically to the 'knitting of a net'. Aristotle favored epigenesis and his conjecture was correct.

Aristotle's influence on European thought was enormous and his ideas remained dominant well into the 17th century. The contrary view to epigenesis, namely that the embryo was preformed from the beginning, was championed anew in the late 17th century. Many could not believe that physical or chemical forces could mold a living entity like the embryo. Along with the contemporaneous background of belief in the divine creation of the world and all living things, was the belief that all embryos had existed from the beginning of the world, and that the first embryo of a species must contain all future embryos.

Even the brilliant 17th century Italian embryologist Marcello Malpighi could not free himself from preformationist ideas. While he provided a remarkably accurate description of the development of the chick embryo, he remained convinced, against the evidence of his own observations, that the embryo was already present from the very beginning (Fig. 1.4). He argued that at very early stages the parts were so small that they could not be observed, even with his best microscope. Yet other preformationists believed that the sperm contained the embryo and some even claimed to be able to see a tiny human—an homunculus—in the head of each human sperm (Fig. 1.5).

The preformation/epigenesis issue was the subject of vigorous debate throughout the 18th century. But the problem could not be resolved until one of the great advances in biology had taken place—the recognition that living things, including embryos, were composed of cells.

Fig. 1.3 Photograph of the adult South African claw-toed frog, *Xenopus laevis*. Scale bar = 1 cm.

Photograph courtesy of J. Smith.

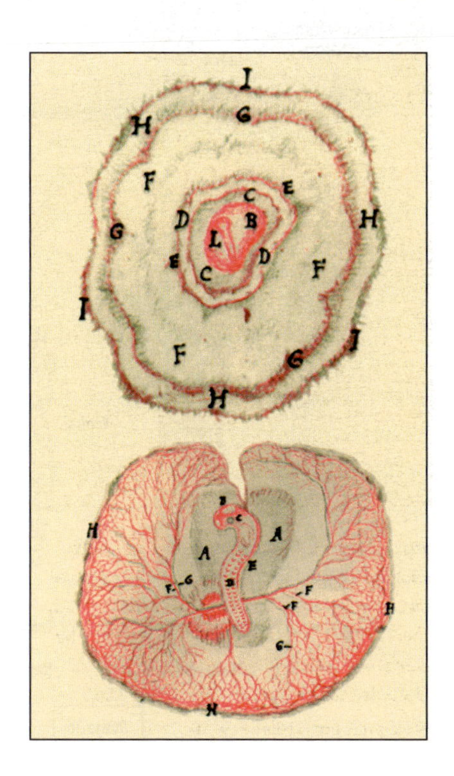

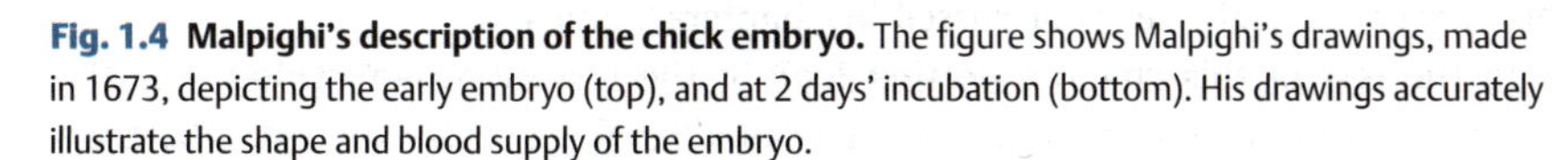

Fig. 1.4 Malpighi's description of the chick embryo. The figure shows Malpighi's drawings, made in 1673, depicting the early embryo (top), and at 2 days' incubation (bottom). His drawings accurately illustrate the shape and blood supply of the embryo.

Reprinted by permission of the President and Council of the Royal Society.

Box 1A Basic stages of *Xenopus laevis* development

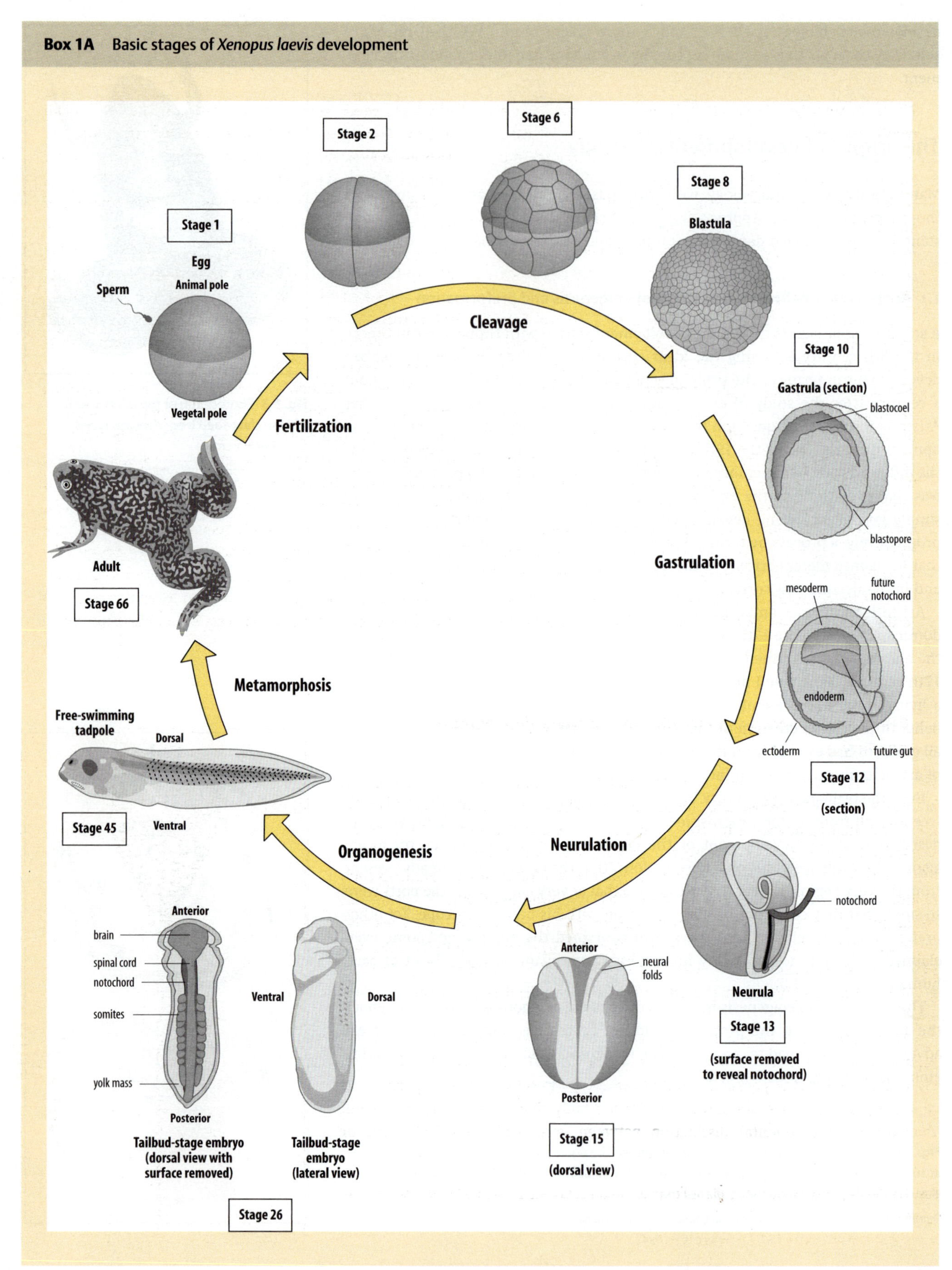

Continued

Box 1A (continued) Basic stages of *Xenopus laevis* development

Although vertebrate development is very varied, there are a number of basic stages that can be illustrated by following the development of the frog *Xenopus laevis*, which is a favorite organism for experimental embryology. The unfertilized egg is a large cell. It has a pigmented upper surface (the **animal pole**) and a lower region (the **vegetal pole**) characterized by an accumulation of yolk granules. So even at the beginning, the egg is not uniform; in subsequent development, cells from the animal half become the anterior (head) end of the embryo.

After fertilization of the egg by a sperm, and the fusion of male and female nuclei, **cleavage** begins. Cleavages are mitotic divisions in which cells do not grow between each division, and so with successive cleavages the cells become smaller. After about 12 division cycles, the embryo, now known as a **blastula**, consists of many small cells surrounding a fluid-filled cavity (the **blastocoel**) above the larger yolky cells. Already, changes have occurred within the cells and they have interacted with each other so that some future tissue types—the **germ layers**—have become partly specified. The future **mesoderm** for example, which gives rise to muscle, cartilage, bone, and other internal organs like the heart, blood, and kidney, is present in the blastula as an equatorial band. Adjacent to it is the future **endoderm**, which gives rise to the gut, lungs, and liver. The animal region will give rise to **ectoderm**, which forms both the epidermis and the nervous system. The future endoderm and mesoderm, which are destined to form internal organs, are still on the surface of the embryo. During the next stage—**gastrulation**—there is a dramatic rearrangement of cells; the endoderm and mesoderm move inside, and the basic body plan of the tadpole is established. Internally, the mesoderm gives rise to a rod-like structure (the **notochord**), which runs from the head to the tail, and lies centrally beneath the future nervous system. On either side of the notochord are segmented blocks of mesoderm called **somites**, which will give rise to the muscles and vertebral column, as well as the dermis of the skin.

Shortly after gastrulation, the ectoderm above the notochord folds to form a tube (the **neural tube**), which gives rise to the brain and spinal cord—a process known as **neurulation**. By this time, other organs, such as limbs, eyes, and gills, are specified at their future locations, but only develop a little later, during **organogenesis**. During organogenesis, specialized cells such as muscle, cartilage, and neurons differentiate. Within 48 hours the embryo has become a feeding tadpole with typical vertebrate features. Because the timing of each stage can vary, depending on environmental conditions, developmental stages in *Xenopus* and other embryos are often denoted by stage numbers rather than by hours of development.

1.2 Cell theory changed the conception of embryonic development and heredity

The cell theory developed between 1820 and 1880 by, among others, the German botanist Matthias Schleiden and the physiologist Theodor Schwann, was one of the most illuminating advances in biology, and had an enormous impact. It was at last recognized that all living organisms consist of cells, which are the basic units of life, and which arise only by division from other cells. Multicellular organisms such as animals and plants could now be viewed as communities of cells. Development could not therefore be based on preformation but must be epigenetic, because during development many new cells are generated by division from the egg, and new types of cells are formed. A crucial step forward in understanding development was the recognition in the 1840s that the egg itself is but a single, albeit specialized, cell.

An important advance was the proposal by the 19th century German biologist August Weismann that the offspring does not inherit its characteristics from the body (the soma) of the parent but only from the **germ cells**—egg and sperm—and that the germ cells are not influenced by the body that bears them. Weismann thus drew a fundamental distinction between germ cells and body cells or

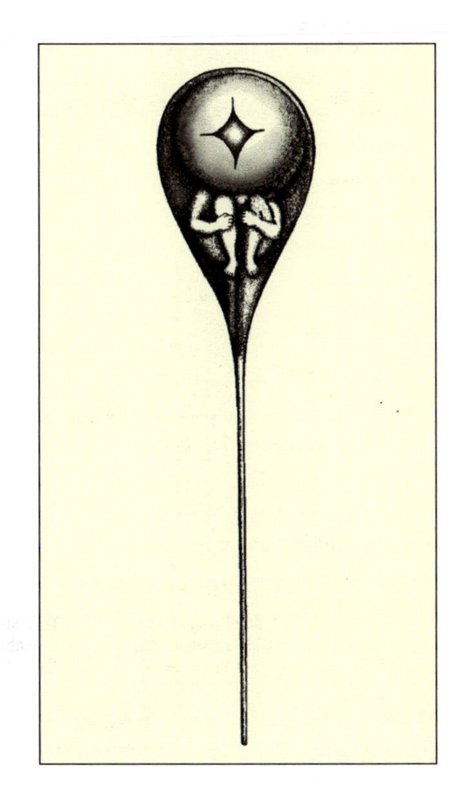

Fig. 1.5 Some preformationists believed that an homunculus was curled up in the head of each sperm.

An imaginative drawing, after Nicholas Hartsoeker (1694).

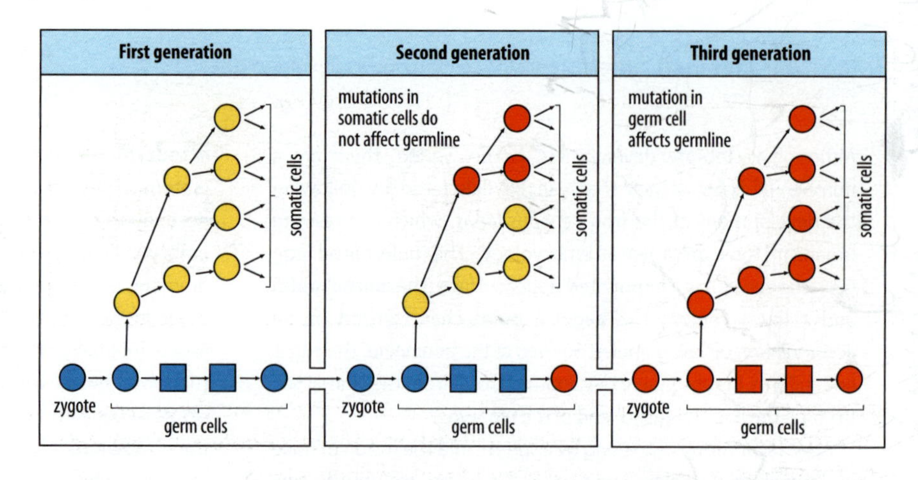

Fig. 1.6 The distinction between germ cells and somatic cells. In each generation germ cells give rise to both somatic cells and germ cells, but inheritance is through the germ cells only. Changes that occur due to mutation in somatic cells can be passed on to their daughter cells but do not affect the germline.

somatic cells (Fig. 1.6). Characteristics acquired by the body during an animal's life cannot be transmitted to the germline. As far as heredity is concerned, the body is merely a carrier of germ cells. As the English novelist and essayist Samuel Butler put it: "A hen is only an egg's way of making another egg."

Work on sea urchin eggs showed that after fertilization the egg contains two nuclei, which eventually fuse; one of these nuclei belongs to the egg while the other comes from the sperm. Fertilization therefore results in an egg carrying a nucleus with contributions from both parents, and it was concluded that the cell nucleus must contain the physical basis of heredity. The climax of this line of research was the eventual demonstration, toward the end of the 19th century, that the chromosomes within the nucleus of the fertilized egg—the **zygote**—are derived in equal numbers from the two parental nuclei, and the recognition that this provided a physical basis for the transmission of genetic characters according to laws developed by the Austrian botanist and monk Gregor Mendel. The constancy of chromosome number from generation to generation in the somatic cells was found to be maintained by a reduction division (**meiosis**) that halved the chromosome number in the germ cells, whereas somatic cells divide by the process of **mitosis**, which maintains chromosome number. The precursors to the germ cells contain two copies of each chromosome, one maternal and one paternal, and are called **diploid**. This number is halved by meiosis during formation of the gametes, so that each germ cell contains only one copy of each chromosome and are called **haploid**. The diploid number is restored at fertilization.

1.3 Two main types of development were originally proposed

Once it was recognized that the cells of the embryo arose by cell division from the zygote, the question emerged of how cells became different from one another. With the increasing emphasis on the role of the nucleus, in the 1880s Weismann put forward a model of development in which the nucleus of the zygote contained a number of special factors or **determinants** (Fig. 1.7). He proposed that while the fertilized egg underwent the rapid cycles of cell division known as cleavage, these determinants would be distributed unequally to the daughter cells and so would control the cells' future development. The fate of each cell was therefore predetermined in the egg by the factors it would receive during cleavage. This type of model was termed 'mosaic', as the egg could be considered to be a mosaic of

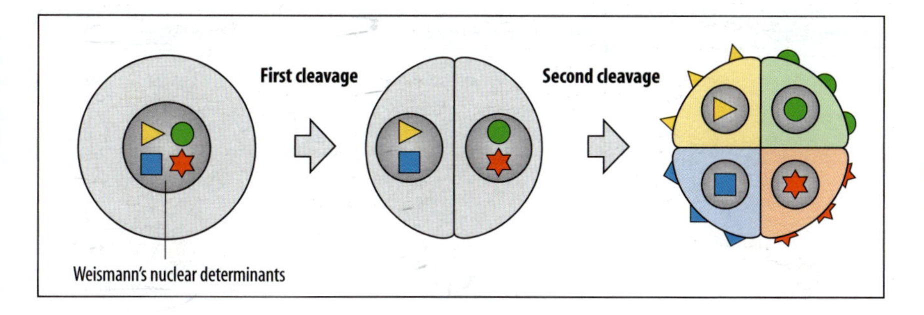

Fig. 1.7 Weismann's theory of nuclear determination. Weismann assumed that there were factors in the nucleus that were distributed asymmetrically to daughter cells during cleavage and directed their future development.

discrete localized determinants. Central to Weismann's theory was the assumption that early cell divisions must make the daughter cells quite different from each other as a result of unequal distribution of nuclear components.

In the late 1880s, initial support for Weismann's ideas came from experiments carried out independently by the German embryologist Wilhelm Roux, who experimented with frog embryos. Having allowed the first cleavage of a fertilized frog egg, Roux destroyed one of the two cells with a hot needle and found that the remaining cell developed into a well-formed half-larva (Fig. 1.8). He concluded that the "development of the frog is based on a mosaic mechanism, the cells having their character and fate determined at each cleavage".

But when Roux's fellow countryman, Hans Driesch, repeated the experiment on sea urchin eggs, he obtained quite a different result (Fig. 1.9). He wrote later: "But things turned out as they were bound to do and not as I expected; there was, typically, a whole gastrula on my dish the next morning, differing only by its small size from a normal one; and this small but whole gastrula developed into a whole and typical larva."

Driesch had completely separated the cells at the two-cell stage and obtained a normal but small larva. That was just the opposite of Roux's result, and was the first clear demonstration of the developmental process known as **regulation**. This is the ability of the embryo to develop normally even when some portions are removed or rearranged. We shall see many examples of regulation throughout the book (an explanation of Roux's experiment, and why he got the result he did, is also given later, in Section 3.7).

1.4 The discovery of induction showed that one group of cells could determine the development of neighboring cells

Although the concept of regulation implied that cells must interact with each other, the central importance of cell–cell interactions in embryonic development was not really established until the discovery of the phenomenon of **induction**, in which one cell, or tissue, directs the development of another, neighboring, cell or tissue.

The importance of induction and other cell–cell interactions in development was proved dramatically in 1924 when Hans Spemann and his assistant Hilde Mangold carried out a famous transplant experiment in amphibian embryos. They showed that a partial second embryo could be induced by grafting one small region of an early newt embryo onto another at the same stage (Fig. 1.10). The grafted tissue was taken from the dorsal lip of the **blastopore**—the slit-like invagination that forms where gastrulation begins on the dorsal surface of the amphibian embryo (see Box 1A, pp. 4–5). This small region they called the **organizer**, since it seemed to be ultimately responsible for controlling the organization of a complete

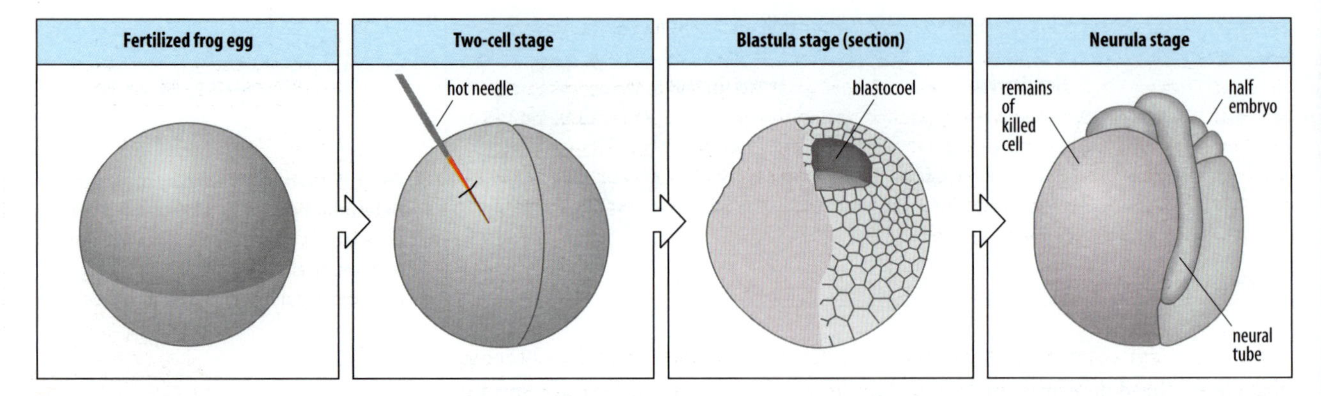

Fig. 1.8 Roux's experiment to investigate Weismann's theory of mosaic development. After the first cleavage of a frog embryo, one of the two cells is killed by pricking it with a hot needle; the other remains undamaged. At the blastula stage the undamaged cell can be seen to have divided as normal into many cells that fill half of the embryo. The development of the blastocoel is also restricted to the undamaged half. In the damaged half of the embryo, no cells appear to have formed. At the neurula stage, the undamaged cell has developed into something resembling half a normal embryo.

embryonic body, and it is commonly known as the **Spemann–Mangold organizer**, or just the **Spemann organizer**. For their discovery, Spemann received the Nobel Prize for Physiology or Medicine in 1935, one of only two ever given for embryological research. Sadly, Hilde Mangold had died earlier, in an accident, and so could not be honored.

1.5 The study of development was stimulated by the coming together of genetics and development

During much of the early part of the twentieth century there was little connection between embryology and genetics. When Mendel's laws were rediscovered in 1900 there was a great surge of interest in mechanisms of inheritance, particularly in relation to evolution, but less so in relation to development. Genetics was seen as the study of the transmission of hereditary elements from generation to generation, whereas embryology was the study of how an individual organism develops and, in particular, how cells in the early embryo became different from each other. Genetics seemed, in this respect, to be irrelevant to development.

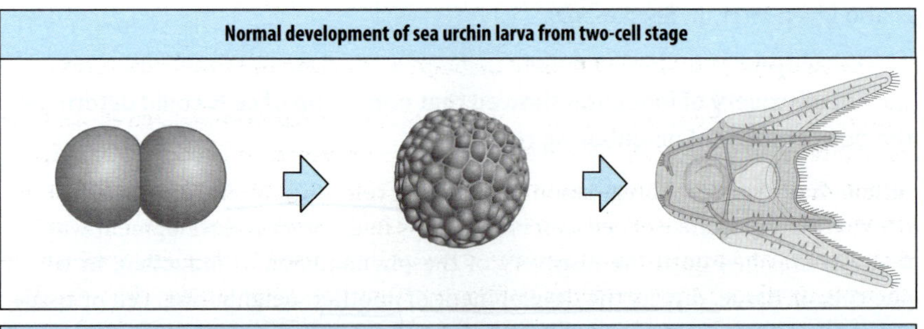

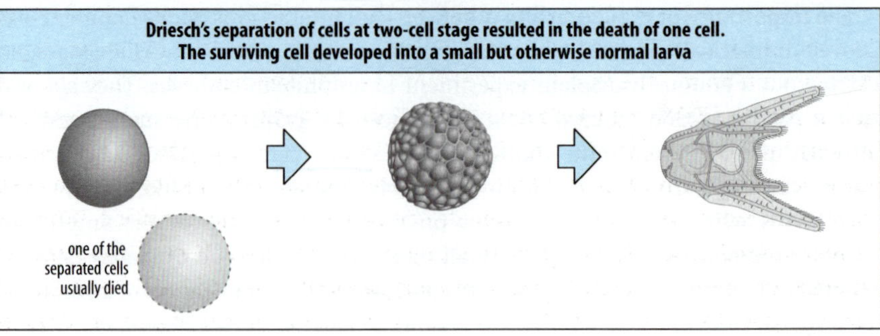

Fig. 1.9 The outcome of Driesch's experiment on sea urchin embryos, which first demonstrated the phenomenon of regulation. After separation of cells at the two-cell stage, the remaining cell develops into a small, but whole, normal larva. This contradicts Roux's earlier finding that if one of the cells of a two-cell frog embryo is damaged, the remaining cell develops into a half-embryo only (see Fig. 1.8).

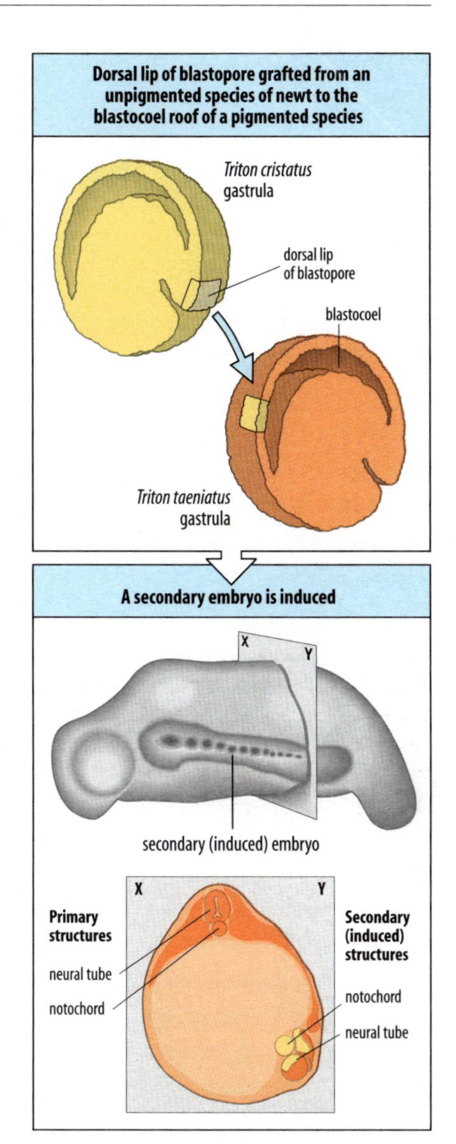

Fig. 1.10 The dramatic demonstration by Spemann and Mangold of induction of a new main body axis by the organizer region in the early amphibian gastrula. A piece of tissue (yellow) from the dorsal lip of the blastopore of a newt (*Triton cristatus*) gastrula is grafted to the opposite side of a gastrula from another, pigmented, newt species (*Triton taeniatus*, pink). The grafted tissue induces a new body axis containing neural tube and somites. The unpigmented graft tissue forms a notochord at its new site (see section in lower panel) but the neural tube and the other structures of the new axis have been induced from the pigmented host tissue. The organizer region discovered by Spemann and Mangold is known as the Spemann organizer.

An important concept that eventually helped to link genetics and embryology was the distinction between **genotype** and **phenotype**. This was first put forward by the Danish botanist Wilhelm Johannsen in 1909. The genetic endowment of an organism—the genetic information it acquires from its parents—is the genotype. Its visible appearance, internal structure, and biochemistry at any stage of development is the phenotype. While the genotype certainly controls development, environmental factors interacting with the genotype influence the phenotype. Despite having identical genotypes, identical twins can develop considerable differences in their phenotypes as they grow up (Fig. 1.11), and these tend to become more evident with age. The problem of development could now be posed in terms of the relationship between genotype and phenotype; how the genetic endowment becomes 'translated' or 'expressed' during development to give rise to a functioning organism.

The coming together of genetics and embryology was a slow and tortuous process. Little progress was made until the nature and function of the genes were much better understood. The discovery in the 1940s that genes encode proteins was a major turning point. As it was already clear that the properties of a cell are determined by the proteins it contains, the fundamental role of genes in development could at last be appreciated. By controlling which proteins were made in a cell, genes could control the changes in cell properties and behavior that occurred during development. A further major advance in the 1960s was the understanding that some genes encode proteins that control the activity of other genes.

Fig. 1.11 The difference between genotype and phenotype. These identical twins have the same genotype because one fertilized egg split into two during development. Their slight difference in appearance is due to nongenetic factors, such as environmental influences.

Photograph courtesy of Josè and Jaime Pascual.

1.6 Development is studied mainly through a selection of model organisms

Although the development of a wide variety of species has been studied at one time or another, a relatively small number of organisms provide most of our knowledge about developmental mechanisms. We can thus regard them as models for understanding the processes involved and they are often called **model organisms**. Sea urchins and amphibians were the main animals used for the first experimental investigations at the beginning of the twentieth century because their developing embryos are both easy to obtain and, in the case of the frog, sufficiently large and robust for relatively easy experimental manipulation, even at quite late stages. Among vertebrates, the frog *X. laevis*, the mouse (*Mus musculus*), the chicken (*Gallus gallus*), and the zebrafish (*Danio rerio*), are the main model organisms now studied. Among invertebrates, the fruit fly *D. melanogaster* and the nematode worm *C. elegans* have been the focus of most attention, because a great deal is known about their developmental genetics and they can also be easily genetically modified. With the advent of modern methods of genetic analysis, there has also been a resurgence of interest in the sea urchin *Strongylocentrotus purpuratus*. For plant developmental biology, the thale-cress *A. thaliana* serves as the main model organism. The life cycles and background details for these model organisms are given in the relevant chapters later in the book.

The reasons for these choices are partly historical—once a certain amount of research has been done on one animal it is more efficient to continue to study it rather than start at the beginning again with another species—and partly a question of ease of study and biological interest. Each species has its advantages and disadvantages as a developmental model. The chick embryo, for example, has long been studied as an example of vertebrate development because fertile eggs are easily available, the embryo withstands experimental microsurgical manipulation very well, and it can be cultured outside the egg. A disadvantage, however, was that little was known about the chick's developmental genetics. However, we know a great deal about the genetics of the mouse, although the mouse is more difficult to study in some ways as development takes place entirely within the mother. Many developmental mutations have been identified in the mouse, and it is also amenable to genetic modification by transgenic techniques. It is also the best experimental model we have for studying mammalian development, including that of humans. The zebrafish is a more recent addition to the select list of vertebrate model systems; it is easy to breed in large numbers, the embryos are transparent and so cell divisions and tissue movements can be followed visually, and it has great potential for genetic investigations.

A major goal of developmental biology is to understand how genes control embryonic development, and to do this one must first identify those genes critically involved in controlling development. This task can be approached in a variety of ways, depending on the organism involved, but the general starting point is the identification of mutations that alter development in some specific and informative way, as described in the following section. Techniques for identifying genes that control development and detecting and manipulating their expression in the organism are described throughout the book, along with techniques for genetic manipulation.

Some of our model organisms are more amenable to conventional genetic analysis than others. Despite its importance in developmental biology, little conventional genetics has been done on *X. laevis*, which has the disadvantage of a **tetraploid** genome (one with four sets of chromosomes in the somatic cells) and a relatively long breeding period, taking 1–2 years to reach sexual maturity. Using modern genetic and bioinformatics methods, however, many developmental genes have been identified in *X. laevis* by direct DNA sequence comparison with known genes in organisms such as *Drosophila* and mice. The closely related frog *Xenopus (Silurana) tropicalis* is a much more attractive organism for genetic analysis; it is diploid and can also be genetically manipulated to produce transgenic organisms. The *X. tropicalis* genome is now being sequenced, which will help identify developmental genes in *X. laevis*. The situation with birds has been rather similar, but here also, direct DNA comparisons with other organisms can be used to identify developmental genes. The DNA sequence of the chicken genome has now been completed, which should be a considerable help in the genetic analysis of development in this species. Sequencing of the genome of the sea urchin *S. purpuratus*, one of the oldest model organisms in developmental biology, is also under way. Complete genome sequences have been available for some years for human, mouse, *Drosophila*, and *Caenorhabditis*.

In general, when an important developmental gene has been identified in one animal, it has proved very rewarding to consider whether a corresponding gene is present and is acting in a developmental capacity in other animals. Such genes are often identified by a sufficient degree of nucleotide sequence similarity to indicate descent from a common ancestral gene. Genes that meet this criterion are known as **homologous genes**. As we shall see in Chapter 4, this approach identified a hitherto unsuspected class of vertebrate genes that specify the regular segmented pattern from head to tail that is represented by the different types of vertebrae at different positions. These genes were identified by their homology with genes that specify the identities of the different body segments in *Drosophila*.

1.7 The first developmental genes were identified as spontaneous mutations

Most of the organisms dealt with in this book are sexually reproducing diploids: their somatic cells contain two copies of each gene, with the exception of those on sex chromosomes. *X. laevis* is an exception, as it is tetraploid; that is, it has four copies of the basic genome in its somatic cells, and two copies in the germ cells. This means that it will have up to four copies of each gene in its somatic cells, which complicates genetic analysis. In a diploid species, one copy, or **allele**, of each gene is contributed by the male parent and the other by the female. For many genes there are several different 'normal' alleles present in the population, which leads to the variation in normal phenotype one sees within any sexually reproducing species. Occasionally, however, a mutation will occur spontaneously in a gene and there will be marked change, usually deleterious, in the phenotype of the organism.

Many of the genes that affect development have been identified by spontaneous mutations that disrupt their function and produce an abnormal phenotype. Mutations are classified broadly according to whether they are dominant or recessive (Fig. 1.12). **Dominant** and **semi-dominant** mutations are those that produce a distinctive phenotype when the mutation is present in just one of the alleles of

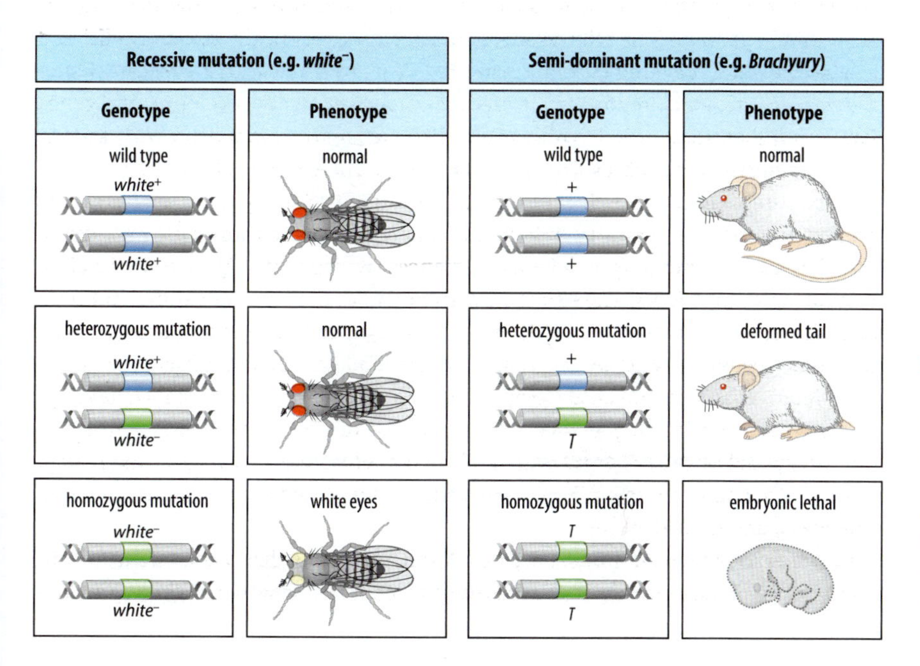

Fig. 1.12 Types of mutations. Left: a mutation is recessive when it only has an effect in the homozygous state; that is, when both copies of the gene carry the mutation. Right: by contrast, a dominant or semi-dominant mutation produces an effect on the phenotype in the heterozygous state; that is, when just one copy of the mutant gene is present. A plus sign denotes wild type, and a minus sign recessive. *T* is the mutant form of the gene *Brachyury*.

a pair; that is, it exerts an effect in the **heterozygous** state. Dominant mutations can be missed as they are often lethal. By contrast, **recessive** mutations such as *white* in *Drosophila*, which produces flies with white eyes rather than the normal red, alter the phenotype only when both alleles of a pair carry the mutation; that is, when they are **homozygous**.

In general, dominant mutations are more easily recognized, particularly if they affect gross anatomy or coloration, provided that they do not cause the early death of the embryo in the heterozygous state. Truly dominant mutations are rare, however. A mutation in the mouse gene *Brachyury* is a classic example of a semi-dominant mutation and was originally identified because mice heterozygous for this mutation (symbolized by *T*) have short tails. When the mutation is homozygous it has a much greater effect, and embryos die at an early stage, indicating that the gene is required for normal embryonic development (Fig. 1.13). Once breeding studies had confirmed that the *Brachyury* trait was due to a single gene, the gene could be mapped to a location on a particular chromosome by classical genetic mapping techniques. Identifying recessive mutations is more laborious, as the heterozygote has a phenotype identical to a normal wild-type animal, and a carefully worked-out breeding program is required to obtain homozygotes. Identifying potentially lethal recessive developmental mutations requires careful observation and analysis in mammals, as the homozygotes may die unnoticed inside the mother.

Very rigorous criteria must be applied to identify those mutations that are affecting a genuine developmental process and not just affecting some vital but routine housekeeping function without which the animal cannot survive. One simple criterion for a developmental mutation is embryonic lethality, but this also catches mutations in genes involved in housekeeping functions. Mutations that produce abnormal patterns of embryonic development are much more promising candidates for true developmental mutations. In later chapters we shall see how large-scale screening for mutations following mutagenesis using chemicals or X-rays has identified many more developmental genes than would ever have been picked up by rare spontaneous mutations.

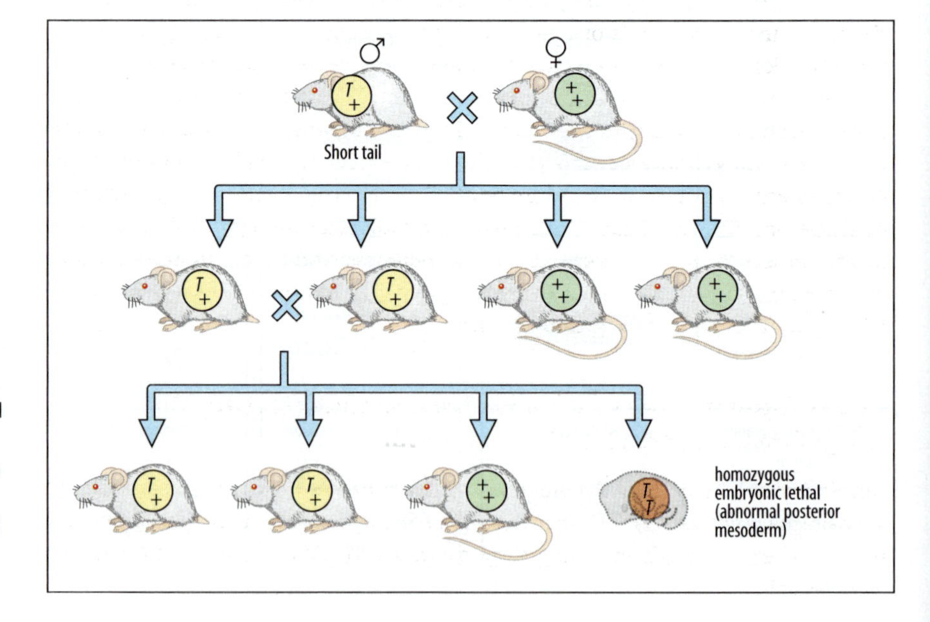

Fig. 1.13 Genetics of the semi-dominant mutation *Brachyury (T)* in the mouse. A male heterozygote carrying the *T* mutation merely has a short tail. When mated with a normal (wild type, ++) female, some of the offspring will also be heterozygotes and have short tails. Mating two heterozygotes together will result in some of the offspring being homozygous (*T/T*) for the mutation, resulting in a severe and lethal developmental abnormality in which the posterior mesoderm does not develop.

Summary

The study of embryonic development started with the Greeks more than 2000 years ago. Aristotle put forward the idea that embryos were not contained completely preformed in miniature within the egg, but that form and structure emerged gradually, as the embryo developed. This idea was challenged in the 17th and 18th centuries by those who believed in preformation, the idea that all embryos that had been, or ever would be, had existed from the beginning of the world. The emergence of the cell theory in the 19th century finally settled the issue in favor of epigenesis, and it was realized that the sperm and egg are single, albeit highly specialized, cells. Some of the earliest experiments showed that very early sea urchin embryos are able to regulate, that is to develop normally even if cells are removed or killed. This established the important principle that development must depend at least in part on communication between the cells of the embryo. Direct evidence for the importance of cell–cell interactions came from the organizer graft experiment carried out by Spemann and Mangold in 1924, showing that the cells of the amphibian organizer region could induce a new partial embryo from host tissue when transplanted into another embryo. The role of the genes in controlling development has only been fully appreciated in the past 30 years and the study of the genetic basis of development has been made much easier in recent times by the techniques of molecular biology.

A conceptual tool kit

Development into a multicellular organism is the most complicated fate a single living cell can undergo; in this lies both the fascination and the challenge of developmental biology. Yet only a few basic principles are needed to start to make sense of developmental processes. The rest of this chapter is devoted to introducing these key concepts. These principles are encountered repeatedly throughout the book, as we look at different organisms and developmental systems, and should be regarded as a conceptual tool kit, essential for embarking on a study of development.

Genes control development by controlling where and when proteins are synthesized, and many thousands of genes are involved. Gene activity sets up intracellular networks of interactions between proteins and genes, and between proteins and proteins, that confer on cells their particular properties. One of these properties is the ability to communicate with and respond to other cells in specified ways. It is these **cell–cell interactions** that determine how the embryo develops; no developmental process can therefore be attributed to the function of a single gene or single protein. The amount of genetic and molecular information on developmental processes is now enormous. In this book, we will be highly selective and describe only those molecular details that give an insight into the mechanisms of development and illustrate general principles.

1.8 Development involves cell division, the emergence of pattern, change in form, cell differentiation, and growth

Development is essentially the emergence of organized structures from an initially very simple group of cells. It is convenient to distinguish five main developmental processes, even though in reality they overlap with and influence one another considerably.

Fig. 1.14 Light micrograph of *Xenopus* eggs after four cell divisions. Scale bar = 1 mm.

Photograph courtesy of J. Slack.

The first is the process known as **cleavage**, which is the division of the fertilized egg into a number of smaller cells (Fig. 1.14). Unlike the cell divisions that take place during cell proliferation and growth of a tissue, there is no increase in cell mass between each cleavage division. The cell cycles during cleavage consist simply of phases of DNA replication, mitosis, and cell division, with no intervening stage of cell growth. We shall discuss the various types of cell cycle in Chapter 12. The cleavage stage of embryogenesis thus rapidly divides the embryo into a number of cells, each containing a copy of the genome.

Pattern formation is the process by which a spatial and temporal pattern of cellular activities is organized within the embryo so that a well ordered structure develops. In the developing arm, for example, pattern formation is the process that enables cells to 'know' whether to make an upper arm or fingers, and where the muscles should form. There is no single universal strategy or mechanism of patterning; rather, it is achieved by a variety of cellular and molecular mechanisms in different organisms and at different stages of development.

Pattern formation initially involves laying down the overall **body plan**—defining the main body axes of the embryo so that the head (anterior) and tail (posterior) ends, and the back (dorsal) and underside (ventral) are specified. Most of the animals we shall deal with in this book have a head at one end and a tail at the other, with the left and right sides of the body being bilaterally symmetrical—that is, a mirror image of each other. In such animals, the main body axis is the **antero-posterior axis**, which runs from the head to the tail. Bilaterally symmetrical animals also have a **dorso-ventral axis**, running from the back to the belly. A striking feature of these axes is that they are almost always at right angles to one another and can thus be thought of as making up a system of coordinates on which any position in the body could be specified (Fig. 1.15). In plants, the main body axis runs from the growing tip (the apex) to the roots and is known as the **apical-basal axis**. Plants also have radial symmetry, with a **radial axis** running from the center of the stem outwards. Even before the body axes become clear, eggs and embryos often show a distinct **polarity**, which in this context means that one end is distinguishable from the other in its structure or properties. Many individual cells in the developing embryo also have polarity, with one end of the cell structurally or functionally different from the other.

The next stage in pattern formation in animal embryos is allocation of cells to the different **germ layers**—the ectoderm, mesoderm, and endoderm (Box 1B, p. 15). During further pattern formation, cells of these germ layers acquire different

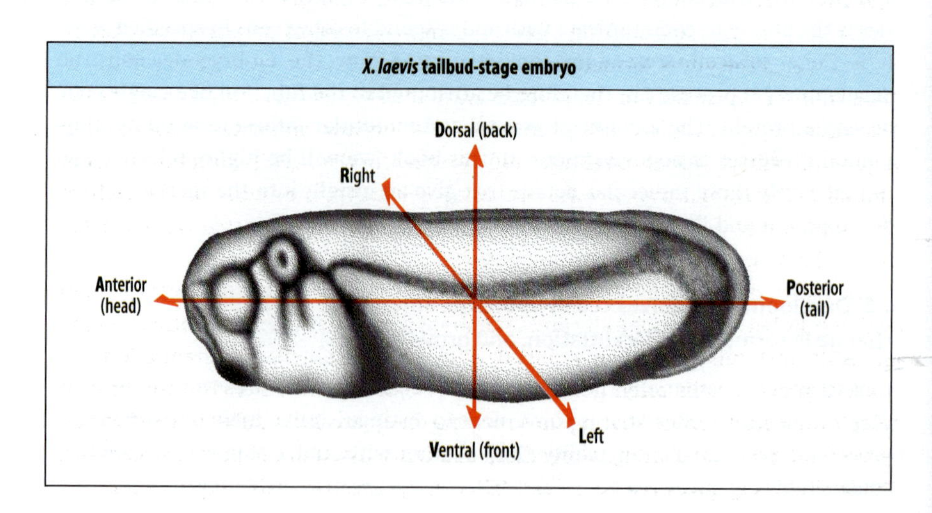

Fig. 1.15 The main axes of a developing embryo. The antero-posterior axis and the dorso-ventral axis are at right angles to one another, as in a coordinate system.

Box 1B Germ layers

The concept of germ layers is useful to distinguish between regions of the early embryo that give rise to quite distinct types of tissues. It applies to both vertebrates and invertebrates. All the animals considered in this book, except for the coelenterate *Hydra*, are triploblasts, with three germ layers: the endoderm, which gives rise to the gut and its derivatives, such as the liver and lungs in vertebrates; the mesoderm, which gives rise to the skeleto-muscular system, connective tissues, and other internal organs such as the kidney and heart; and the ectoderm, which gives rise to the epidermis and nervous system. These are specified early in development. The boundaries between the different layers can be fuzzy and there are notable exceptions. The neural crest in vertebrates, for example, is ectodermal in origin but gives rise both to neural tissue and to some skeletal elements, which would normally be considered mesodermal in origin.

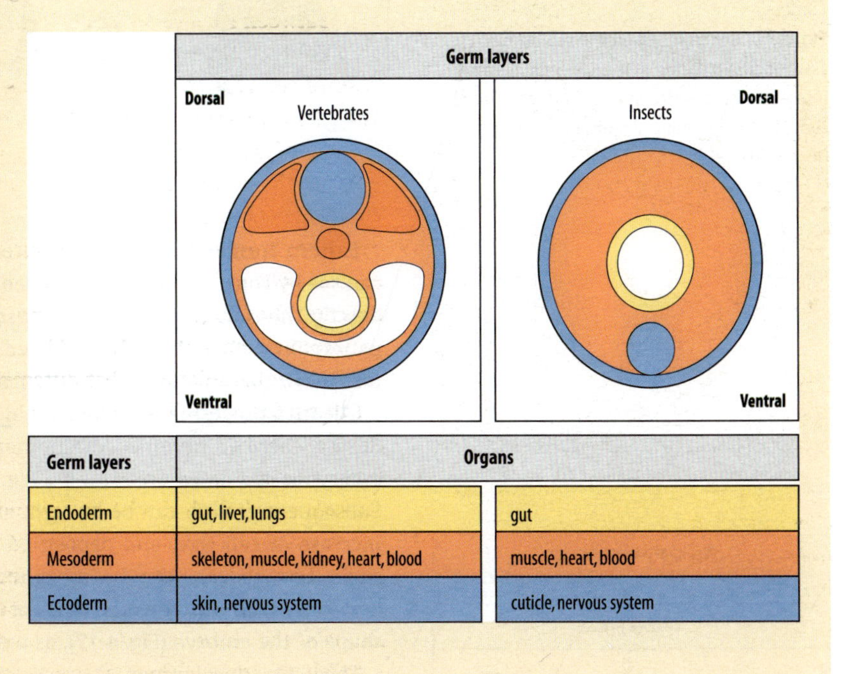

Germ layers	Organs (Vertebrates)	Organs (Insects)
Endoderm	gut, liver, lungs	gut
Mesoderm	skeleton, muscle, kidney, heart, blood	muscle, heart, blood
Ectoderm	skin, nervous system	cuticle, nervous system

identities so that organized spatial patterns of cell differentiation emerge, such as the arrangement of skin, muscle, and cartilage in developing limbs, and the arrangement of neurons in the nervous system. In the earliest stages of pattern formation, differences between cells are not easily detected and probably consist of subtle differences caused by a change in activity of a very few genes.

The third important developmental process is change in form, or **morphogenesis** (Chapter 7). Embryos undergo remarkable changes in three-dimensional form—you need only look at your hands and feet. At certain stages in development, there are characteristic and dramatic changes in form, of which **gastrulation** is the most striking. Almost all animal embryos undergo gastrulation, during which the gut is formed and the main body plan emerges. During gastrulation, cells on the outside of the embryo move inwards and, in animals such as the sea urchin, gastrulation even transforms a hollow spherical blastula into a gastrula with a hole through the middle—the gut (Fig. 1.16). Morphogenesis in animal embryos can also involve extensive cell migration. Most of the cells of the human face, for example, are derived from cells that migrated from a tissue called the neural crest, which originates on the back of the embryo.

The fourth developmental process we must consider here is **cell differentiation**, in which cells become structurally and functionally different from each other, ending up as distinct cell types such as blood, muscle, or skin cells. Differentiation is a gradual process, cells often going through several divisions between the time at which they start differentiating and the time they are fully differentiated (when some cell types stop dividing altogether), and is discussed in Chapter 8. In humans, the fertilized egg gives rise to at least 250 clearly distinguishable types of cell.

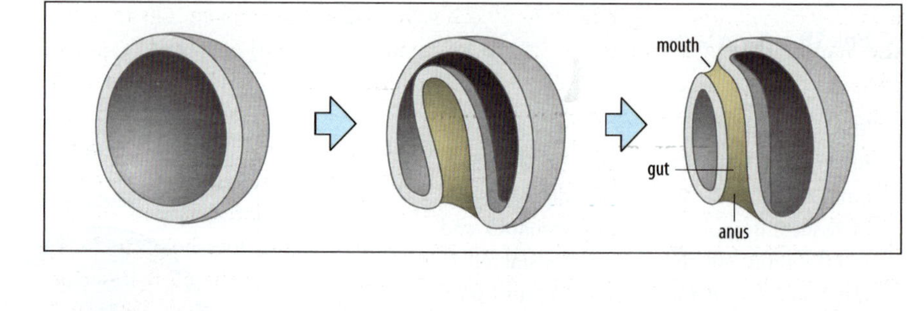

Fig. 1.16 Gastrulation in the sea urchin. Gastrulation transforms the spherical blastula into a structure with a hole through the middle, the gut. The left-hand side of the embryo has been removed.

Pattern formation and cell differentiation are very closely interrelated, as we can see by considering the difference between human arms and legs. Both contain exactly the same types of cell—muscle, cartilage, bone, skin, and so on—yet the pattern in which they are arranged is clearly different. It is essentially pattern formation that makes us different from elephants and chimpanzees.

The fifth process is **growth**—the increase in size. In general there is little growth during early embryonic development and the basic pattern and form of the embryo is laid down on a small scale, always less than a few millimeters in extent. Subsequent growth can be brought about in a variety of ways: cell multiplication, increase in cell size, and deposition of extracellular materials such as bone and shell. Growth can also be morphogenetic in that differences in growth rates between organs, or between parts of the body, can generate changes in the overall shape of the embryo (Fig. 1.17), as we shall see in more detail in Chapter 12.

These five developmental processes are neither independent of each other nor strictly sequential. In very general terms, however, one can think of pattern formation in early development specifying differences between cells that lead to changes in form, cell differentiation, and growth. But in any real developing system there will be many twists and turns in this sequence of events.

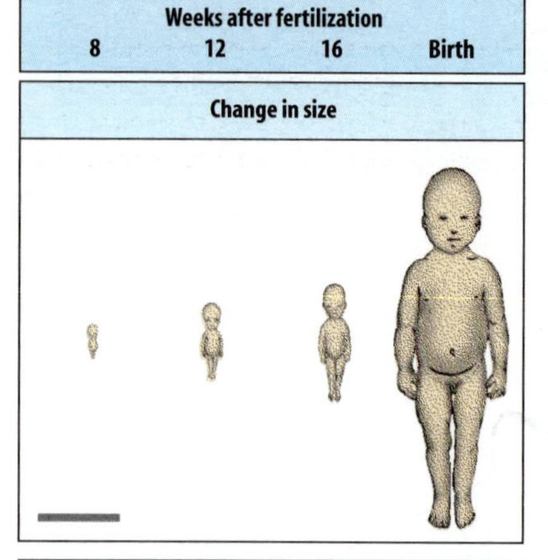

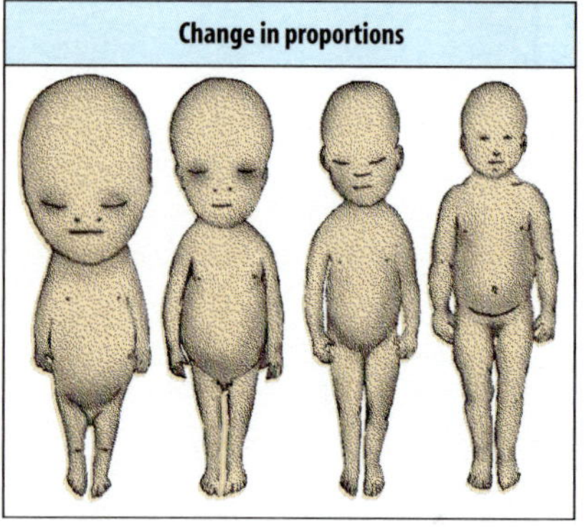

Fig. 1.17 The human embryo changes shape as it grows. From the time the body plan is well established at 8 weeks until birth the embryo increases in length some tenfold (upper panel), while the relative proportion of the head to the rest of the body decreases (lower panel). As a result, the shape of the embryo changes. Scale bar = 10 cm. After Moore, K.L.: 1983.

1.9 Cell behavior provides the link between gene action and developmental processes

Gene expression within cells leads to the synthesis of proteins that specify particular cellular properties and behavior, which in turn determine the course of embryonic development. The past and current pattern of gene activity confers on a cell a certain state, or identity, at any given time, which is reflected in its molecular organization—in particular which proteins are present. As we shall see, embryonic cells and their progeny undergo many changes in state as development progresses. Other categories of cell behavior that will concern us are intercellular communication, also known as **cell–cell signaling**, changes in cell shape and cell movement, cell proliferation, and cell death.

Changing patterns of gene activity during early development are essential for pattern formation. They give cells identities that determine their future behavior and lead eventually to their final differentiation. And, as we saw in the example of induction by the Spemann organizer, the capacity of cells to influence each other's fate by producing and responding to signals is crucial for development. By their movement or change in shape, cells generate the physical forces that bring about morphogenesis (Fig. 1.18). The curvature of a sheet of cells into a tube, as happens in *Xenopus* and other vertebrates during formation of the neural tube (see Box 1A, pp. 4–5), is the result of contractile forces generated by cells changing their shape at certain positions within the cell layer. An important feature of cell surfaces is the presence of adhesive proteins known as cell-adhesion molecules that serve a variety of functions: they hold cells together in tissues, they enable cells to sense the nature of the surrounding extracellular matrix, and they serve to guide migratory cells such as the neural crest cells of vertebrates, which leave the neural tube to form structures elsewhere in the body.

Later in development, growth involves cell proliferation, which can also influence final form, as parts of the body grow at different rates. The death of cells, known as **programmed cell death** or **apoptosis**, is also an intrinsic part of the developmental process; cell death in developing hands and feet helps to form the fingers and toes from a continuous sheet of tissues. We can therefore describe and explain developmental processes in terms of how individual cells and groups of cells behave. Because the final structures generated by development are themselves composed of cells, explanations and descriptions at the cellular level can provide an account of how these adult structures are formed.

Since development can be understood at the cellular level, we can pose the question of how genes control development in a more precise form. We can now ask how genes are controlling cell behavior. The many possible ways in which a cell can behave therefore provide the link between gene activity and the morphology of the adult animal—the final outcome of development. Cell biology provides the means by which the genotype becomes translated into the phenotype.

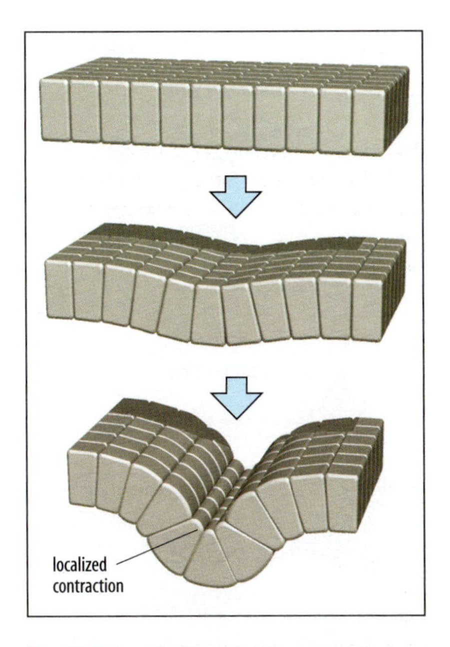

Fig. 1.18 Localized contraction of particular cells can cause a whole sheet of cells to fold. Contraction of a line of cells at their apices due to the contraction of cytoskeletal elements causes a furrow to form in a sheet of epidermis.

1.10 Genes control cell behavior by specifying which proteins are made

What a cell can do is determined very largely by the proteins present within it. The hemoglobin in red blood cells enables them to transport oxygen; skeletal muscle cells are able to contract because they contain contractile structures composed of the proteins myosin, actin, and tropomyosin, and other muscle-specific proteins. All these are very special, or 'luxury', proteins that are not involved in the housekeeping activities that are common to all cells and keep them alive and functioning. Housekeeping activities include the production of energy and the metabolic pathways involved in the breakdown and synthesis of molecules necessary for the life of the cell. Although there are qualitative and quantitative variations in housekeeping proteins in different cells, they are not important players in development. In development we are concerned primarily with those luxury or tissue-specific proteins that make cells different from one another.

Genes control development mainly by specifying which proteins are made in which cells and when. In this sense they are passive participants in development, compared with the proteins for which they code, which are the agents that directly determine cell behavior, including which genes are expressed. To produce a particular protein its gene must be switched on and **transcribed** into **messenger RNA (mRNA)**; the mRNA must then be **translated** into protein. Both these processes are under several layers of control, and translation does not automatically follow transcription. Fig. 1.19 shows the main stages in gene expression at which the production of a protein can be controlled. For example, mRNA may be degraded before it can be exported from the nucleus. Even if it reaches the cytoplasm, its translation may be inhibited there. In the eggs of many animals, preformed mRNA is prevented from being translated until after fertilization. Another factor determining which proteins are produced is **RNA processing**. The initial RNA transcripts of many genes in eukaryotes can be spliced in different ways to give rise to two or more different mRNAs; thus a single gene may be able to produce a number of different proteins with different properties.

Even if a gene has been transcribed and the mRNA translated, the protein may still not be able to function. Many newly synthesized proteins require further

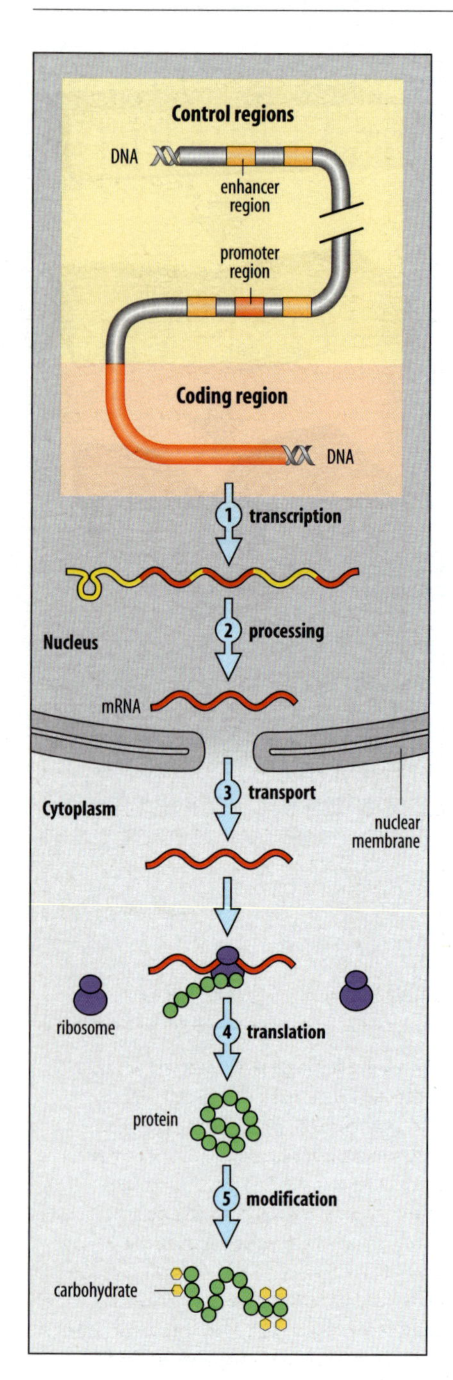

Fig. 1.19 Gene expression and protein synthesis. A protein-coding gene comprises a stretch of DNA that contains a coding region, which contains the instructions for making the protein, and adjacent control regions—promoter and enhancer regions—at which the gene is switched on or off. The promoter region is the site at which RNA polymerase binds and starts transcribing. The enhancer regions may be thousands of base pairs distant from the promoter. Transcription of the gene into RNA (1) may be either stimulated or inhibited by transcription factors that bind to promoter and enhancer regions. The RNA formed by transcription is spliced to remove introns (yellow) and processed within the nucleus (2) to produce mRNA that is exported to the cytoplasm (3) for translation into protein at the ribosomes (4). Control of gene expression and protein synthesis occurs mainly at the level of transcription but can also occur at later stages. For example, mRNA may be degraded before it can be translated. If it is not translated immediately it may be stored in inactive form in the cytoplasm for translation at some later stage. Some proteins require post-translational modification (5) to become biologically active.

post-translational modification before they acquire biological activity. Reversible post-translational modifications such as phosphorylation can also significantly alter protein function. Alternative RNA splicing and post-translational modification together mean that the number of functionally different proteins that can be produced is considerably greater than the number of protein-coding genes would indicate—perhaps as much as 10 times more.

Some genes, such as those for ribosomal RNAs (rRNAs) and transfer RNAs (tRNAs), do not code for proteins; in this case the RNAs themselves are the end products. A recently discovered category of genes is that for **microRNAs** (**miRNAs**), small RNA molecules that inhibit the translation of specific mRNAs (see Chapter 5, Box 5B, p. 197). Some microRNAs are known to be involved in gene regulation in development.

An intriguing question is how many genes out of the total genome are **developmental genes**—that is, genes specifically required for embryonic development. This is not an easy estimate to make. In a few particularly well studied cases we have rough estimates of the minimum number of genes involved in a particular aspect of development. In the early development of *Drosophila* at least 60 genes are directly involved in pattern formation up to the time that the embryo becomes divided into segments. In *Caenorhabditis*, at least 50 genes are needed to specify a small reproductive structure known as the vulva. All these are quite small numbers compared to the thousands of genes that are active at the same time; some of these are essential to development in that they are necessary for maintaining life, but provide no or little information that influences the course of development. Recent studies have shown that many genes change their activity during development, and as the total number of genes in the nematode and *Drosophila* is around 19,000 and 20,000, respectively, the total number of genes involved in development is many thousands. A systematic analysis of nearly 90% of the nematode's genes showed that 1722 genes, at least, were involved in development.

Developmental genes typically code for proteins involved in the regulation of cell behavior—receptors, growth factors, intracellular signaling proteins, and gene-regulatory proteins. Many of these genes, especially those for receptors and signaling molecules, are used throughout an organism's life, but others are active only during embryonic development.

1.11 The expression of developmental genes is under the control of complex control regions

All the somatic cells in an embryo are derived from the fertilized egg by successive rounds of mitotic cell division. Thus, with rare exceptions, they all contain identical genetic information, the same as that in the zygote. The differences between cells must therefore be generated by differences in gene activity that lead to the synthesis of different proteins. Turning the correct genes on or off in the correct cells at the correct time becomes the central issue in development. The genes do not provide a blueprint for development but a set of instructions. Key elements in regulating the readout of these instructions are the control regions located adjacent to most luxury genes and developmental genes. These are acted on by **gene-regulatory proteins**, or **transcription factors**, which switch genes on or off by, respectively, activating or repressing transcription. Some gene-regulatory proteins act by binding directly to the **control regions** (see Fig. 1.19) whereas others interact with transcription factors already bound to the DNA.

Developmental genes are highly regulated to ensure they are switched on only at the right time and place in development. This is a fundamental feature of development. To achieve this, they usually have extensive and complex control regions composed of one or more regulatory modules, known as ***cis*-regulatory modules** (the '*cis*' refers to the fact that the regulatory module is on the same DNA molecule as the gene it controls). Each module contains multiple binding sites for different transcription factors, and the combination of factors that binds determines whether the gene is on or off. On average, a module will have binding sites for four to eight different transcription factors.

Different genes can have the same control module, which usually means that they will be expressed together, or different genes may have modules that contain some, but not all, of the binding sites in common, introducing subtle differences in the timing or location of expression. A gene with more than one regulatory module can be expressed at different times and places during development, depending on which module predominates at any given time in directing gene expression. Thus, an organism's genes are linked in complex interdependent networks of expression through their regulatory modules and the proteins that bind to them. Common examples of such regulation are **positive feedback** and **negative feedback** loops in which a transcription factor respectively promotes or represses the expression of a gene whose product maintains that gene expression (Fig. 1.20). Indeed, the network of interactions between gene-regulatory modules that has been described for early sea urchin development is quite bewildering at first, or even second, sight, as we shall see in Chapter 5.

As all the key steps in development reflect changes in gene activity, one might be tempted to think of development simply in terms of mechanisms for controlling gene expression. But this would be highly misleading. For gene expression is only the first step in a cascade of cellular processes that lead via protein synthesis to changes in cell behavior and so direct the course of embryonic development. To think only in terms of genes is to ignore crucial aspects of cell biology, such as change in cell shape, that may be initiated at several steps removed from gene activity. In fact, there are very few cases where the complete sequence of events from gene expression to altered cell behavior has been worked out. The route leading from gene activity to a structure such as the five-fingered hand may be tortuous.

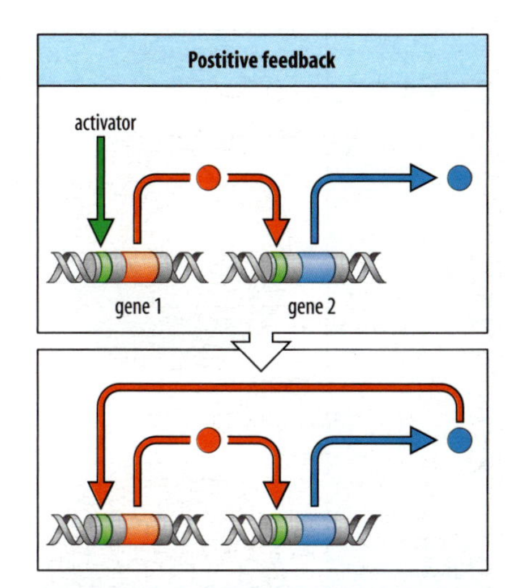

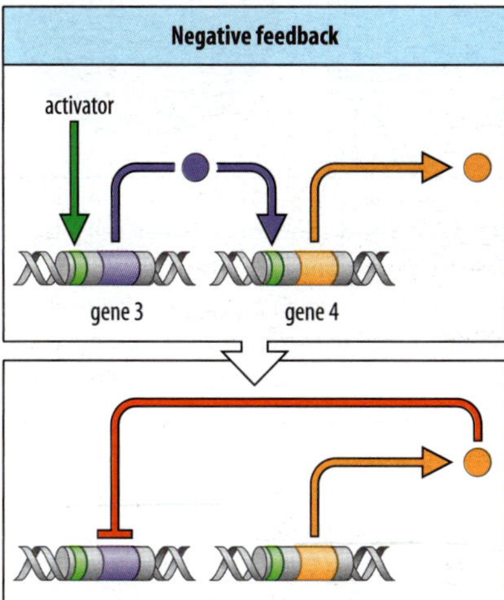

Fig. 1.20 Simple genetic feedback loops. Top: gene 1 is turned on by an activating transcription factor (green); the protein it produces (red) activates gene 2. The protein product of gene 2 (blue) not only acts on further targets, but also activates gene 1, forming a positive feedback loop that will keep genes 1 and 2 switched on even in the absence of the original activator. Bottom: when the product of the gene at the end of the pathway inhibits the first gene, a negative feedback loop is formed. Arrows indicate activation; barred line indicates inhibition.

1.12 Development is progressive and the fate of cells becomes determined at different times

As embryonic development proceeds, the organizational complexity of the embryo becomes vastly increased over that of the fertilized egg. Many different cell types are formed, spatial patterns emerge, and there are major changes in shape. All this occurs more or less gradually, depending on the particular organism. But, in

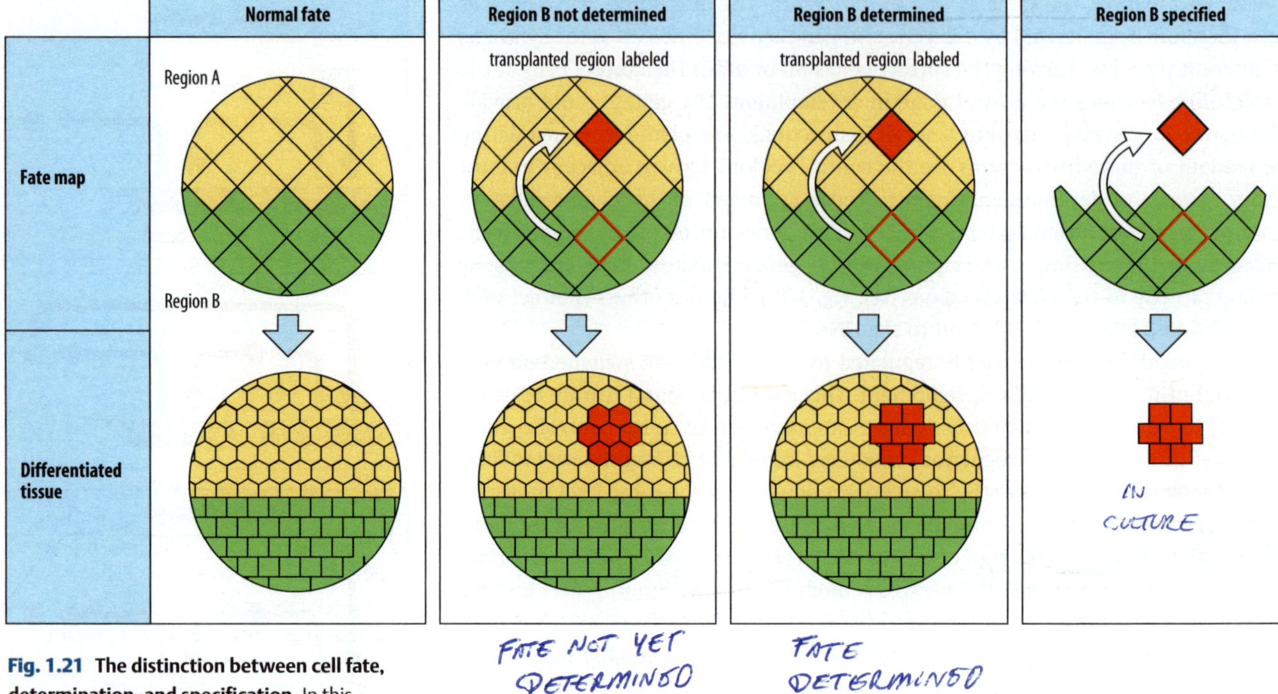

Fig. 1.21 The distinction between cell fate, determination, and specification. In this idealized system, regions A and B differentiate into two different sorts of cells, depicted as hexagons and squares. The fate map (first panel) shows how they would normally develop. If cells from region B are grafted into region A and now develop as A-type cells, the fate of region B has not yet been determined (second panel). By contrast, if region B cells are already determined when they are grafted to region A, they will develop as B cells (third panel). Even if B cells are not determined, they may be specified, in that they will form B cells when cultured in isolation from the rest of the embryo (fourth panel).

general, the embryo is first divided up into a few broad regions, such as the future germ layers (mesoderm, ectoderm, and endoderm). Subsequently, the cells within these regions have their fates more and more finely determined. Mesoderm, for example, becomes differentiated into muscle cells, cartilage cells, bone cells, the fibroblasts of connective tissue, and the cells of the dermis of the skin. **Determination** implies a stable change in the internal state of a cell, and an alteration in the pattern of gene activity is assumed to be the initial step, leading to a change in the proteins produced in the cell.

It is important to understand clearly the distinction between the normal fate of a cell at any particular stage, and its state of determination. The **fate** of a group of cells merely describes what they will normally develop into. By marking cells of the early embryo one can find out, for example, which ectodermal cells will normally give rise to the nervous system, and of those, which to the retina in particular. However, that in no way implies that those cells can only develop into a retina, or are already committed, or determined, to do so.

A group of cells is called **specified** if, when isolated and cultured in the neutral environment of a simple culture medium away from the embryo, they develop more or less according to their normal fate (Fig. 1.21). For example, cells at the animal pole of the amphibian blastula (see Box 1A, pp. 4–5) are specified to form ectoderm, and will form epidermis when isolated. Cells that are specified in this technical sense need not yet be determined, for influences from other cells can change their normal fate; if tissue from the animal pole is put in contact with cells from the vegetal pole, the animal pole tissue will form mesoderm instead of epidermis. At a later stage of development, however, the cells in the animal region have become determined as ectoderm and their fate cannot then be altered. Tests for specification rely on culturing the tissue in a neutral environment lacking any inducing signals, and this is often difficult to achieve.

The state of determination of cells at any developmental stage can be demonstrated by transplantation experiments. At the blastula stage of the amphibian

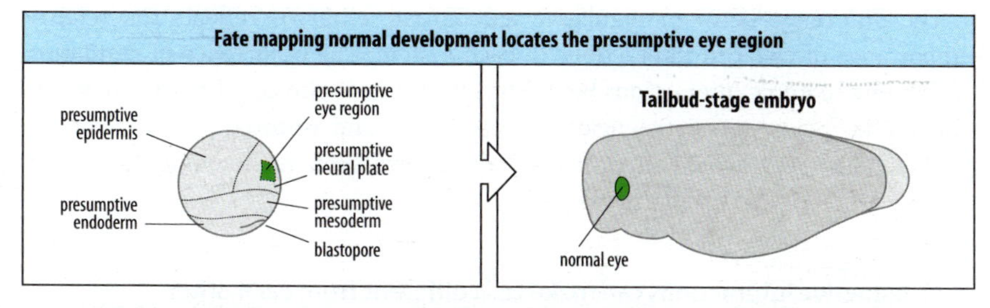

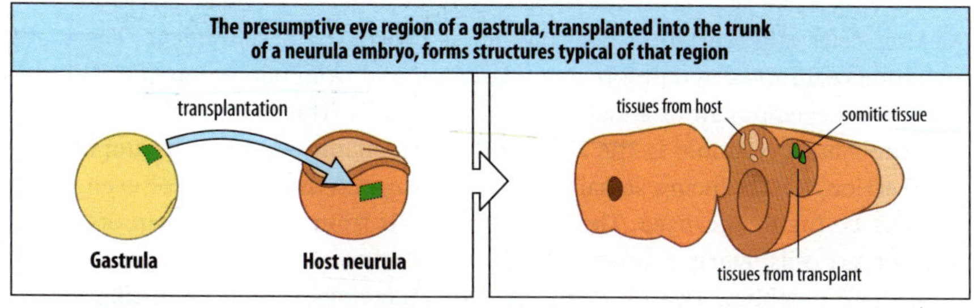

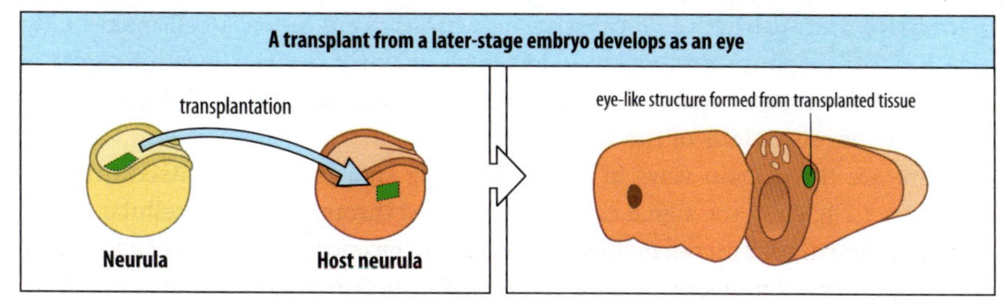

Fig. 1.22 Determination of the eye region with time in amphibian development. If the region of the gastrula that will normally give rise to an eye is grafted into the trunk region of a neurula (middle panel), the graft forms structures typical of its new location, such as notochord and somites. If, however, the eye region from a neurula is grafted into the same site (bottom panel), it develops as an eye-like structure, since at this later stage it has become determined.

embryo, one can graft the ectodermal cells that give rise to the eye into the side of the body and show that the cells develop according to their new position; that is, into mesodermal cells like those of the notochord and somites (Fig. 1.22). At this early stage, their potential for development is much greater than their normal fate. However, if the same operation is done at a later stage, then the future eye region will form structures typical of an eye. At the earlier stage the cells were not yet determined as eye cells, whereas later they had become so.

It is a general feature of development that cells in the early embryo are less narrowly determined than those at later stages; with time, cells become more and more restricted in their developmental potential. We assume that determination involves a change in the genes that are expressed by the cell and that this change fixes or restricts the cell's fate, thus reducing its developmental options.

We have already seen how, even at the two-cell stage, the cells of the sea urchin embryo do not seem to be determined. Each has the potential to generate a whole new larva (see Section 1.3). Embryos like these, where the potential of cells is much greater than that indicated by their normal fate, are termed **regulative**. Vertebrate embryos are among those capable of considerable regulation. In contrast, those embryos where, from a very early stage, the cells can develop only according to their early fate are termed **mosaic**, As we saw in Section 1.3, the term mosaic has a long history. It describes eggs and embryos that develop as if their pattern of future development is laid down very early, even in the egg, as a mosaic of different molecules. The different parts of the embryo then develop quite independently of each other. In such embryos, cell interactions may be quite limited. The demarcation between the regulative and mosaic strategies of development is not always sharp, and in part reflects the time when determination occurs; it occurs much earlier in mosaic systems.

The difference between regulative and mosaic embryos reflects the relative importance of cell–cell interactions in each system. The occurrence of regulation absolutely requires interactions between cells, for how else could normal development take place and deficiencies be recognized and restored? A truly mosaic embryo, however, would in principle not require such interactions. No purely mosaic embryos are known to exist.

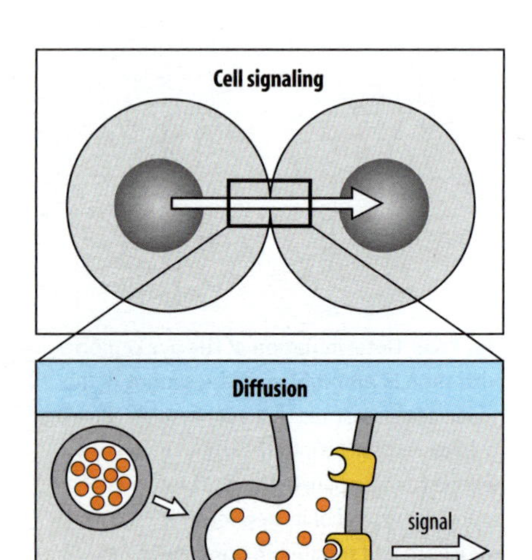

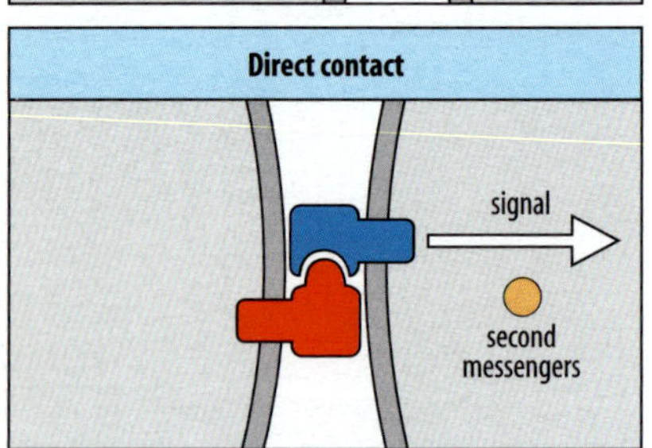

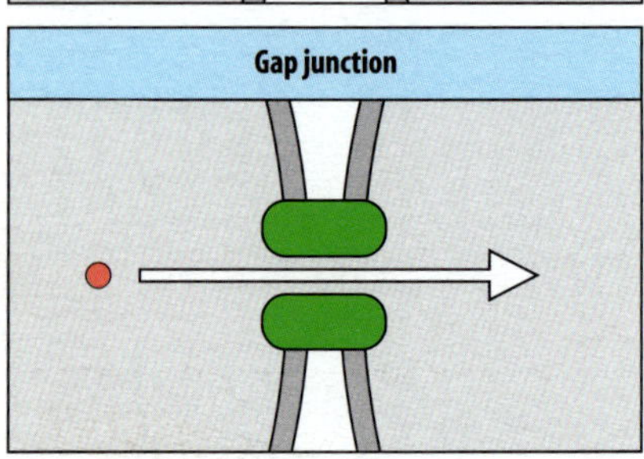

Fig. 1.23 An inducing signal can be transmitted from one cell to another in three main ways. The signal can be a diffusible molecule, which interacts with a receptor on the target cell surface (second panel), or the signal can be produced by direct contact between two complementary proteins at the cell surfaces (third panel). If the signal involves small molecules it may pass directly from cell to cell through gap junctions in the plasma membrane (fourth panel).

1.13 Inductive interactions can make cells different from each other

Making cells different from one another is central to development. There are numerous examples in development where a signal from one group of cells influences the development of an adjacent group of cells. This is known as **induction**, and the classic example is the action of the Spemann organizer in amphibians (see Section 1.4). Inducing signals may be propagated over several or even many cells, or be highly localized. The inducing signals from the amphibian organizer affect many cells, whereas other inducing signals may pass from a single cell to its immediate neighbor. Two different types of inductions should be distinguished: **permissive** and **instructive**. Permissive inductions occur when a cell makes only one kind of response to a signal, and makes it when a given level of signal is reached. By contrast, in an instructive induction, the cells respond differently to different concentrations of the signal.

There are three main ways in which inducing signals may be passed between cells (Fig. 1.23). First, the signal can be transmitted through the extracellular space, usually by means of a secreted diffusible molecule. Second, cells may interact directly with each other by means of molecules located on their surfaces. In both these cases, the signal is generally received by receptor proteins in the cell membrane and is subsequently relayed through intracellular signaling systems to produce the cellular response. Third, the signal may pass from cell to cell directly. In animal cells, this is through gap junctions, which are specialized protein pores in the apposed plasma membranes providing direct channels of communication between the cytoplasms of adjacent cells through which small molecules can pass. Plant cells are connected by strands of cytoplasm called plasmodesmata, through which even quite large molecules such as proteins can pass directly from cell to cell.

In the case of signaling by a diffusible molecule or by direct contact, the signal is received at the cell membrane. If it is to alter gene expression in the nucleus, the signal has to be transmitted from the membrane to the cell's interior. This process is known generally as **signal transduction**, and is carried out by relays of intracellular signaling molecules that are activated when the extracellular signal molecule binds to its receptor. Intracellular signaling proteins and small-molecule second messengers such as cyclic AMP interact with one another to transmit the signal onward in the cell. Activation and deactivation of the pathway components by phosphorylation is an important feature of most signaling pathways. In the case of development, most of the signaling we are concerned with leads to gene activation or repression, but signaling pathways are also used to temporarily alter enzyme activity and metabolic activity within cells and to relay nerve impulses. The variety of signals a cell may be receiving at any one time are integrated by cross-talk between the different signaling pathways to produce an appropriate response.

An important feature of induction is whether or not the responding cell is **competent** to respond to the inducing signal. This competence may depend, for example, on the presence of the appropriate receptor and transducing mechanism, or on the presence of the particular transcription factors needed for gene activation. A cell's competence for a particular response can change with time; for example, the Spemann organizer can induce changes in the cells it affects only during a restricted time window.

In embryos, it seems that small is generally beautiful where signaling and pattern formation are concerned. Whenever a pattern is being specified, the size of the group of cells involved is barely, if ever, greater than 0.5 mm in any direction; that is, some 50 cell diameters. Many patterns are specified on a much smaller scale and involve just tens or a few hundreds of cells. This means that the inducing signals involved in pattern formation reach over distances of the order of only 10 times a cell diameter. The final organism may be very big, but this is almost entirely due to growth of the basic pattern.

1.14 The response to inductive signals depends on the state of the cell

Inductive signals can alter how the induced cells develop. They can thus be regarded as providing the cells with instructions as how to behave. After receiving an inductive signal, a cell usually then develops **autonomously**; that is, without new signals from another cell, for some time. It is important to realize that the response to inductive signals is entirely dependent on the current state of the cell. Not only does it have to be competent to respond, but the number of possible responses is usually very limited. An inductive signal can only select one response from a small number of possible cellular responses. Thus instructive inducing signals would be more accurately called selective signals. A truly instructive signal would be one that provided the cell with entirely new information and capabilities, by providing it with, for example, new genes, which is not thought to occur during development.

The fact that, at any given time, an inductive signal selects one out of several possible responses has several important implications for biological economy. On the one hand, it means that different signals can activate a particular gene at different stages of development: the same gene is often turned on and off repeatedly during development. On the other, the same signal can be used to elicit different responses in different cells. A particular signaling molecule, for example, can act on several types of cell, evoking a characteristic and different response from each, depending on their developmental history. This enables the fixed set of genes and signals to be related to each other in a variety of different combinations, effectively expanding the number of developmentally distinct signals and responses that can be made by cells. As we will see in future chapters, evolution has been lazy with respect to this aspect of development, and the same basic intracellular signaling pathways are used again and again for different purposes.

1.15 Patterning can involve the interpretation of positional information

One general mode of pattern formation can be illustrated by considering the patterning of a simple non-biological model—the French flag (Fig. 1.24). The French flag has a simple pattern: one-third blue, one-third white, and one-third red, along just one axis. Moreover, the flag comes in many sizes but always the same pattern, and thus can be thought of as mimicking the capacity of an embryo to regulate. Given a line of cells, any one of which can be blue, white, or red, and given also that the line of cells can be of variable length, what sort of mechanism is required for the line to develop the pattern of a French flag?

One solution is for the group of cells to acquire **positional information**. That is, each cell acquires an identity, or **positional value**, that is related to its position along the line with respect to the boundaries at either end. After they have

Fig. 1.24 The French flag.

acquired their positional values, the cells interpret this information by differentiating according to their genetic program. Those in the left-hand third of the line will become blue, those in the middle third white, and so on.

Pattern formation using positional information thus implies at least two distinct stages: first the positional value has to be specified with respect to some boundary, and then it has to be interpreted. The separation of these two processes has an important implication: it means that there does not need to be a set relation between the positional values and how they are interpreted. In different circumstances, the same set of positional values could be used to generate the Italian flag or another pattern. How positional values will be interpreted will depend on the particular genetic instructions active in the group of cells and will be influenced by their developmental history.

Cells could have their position specified by various mechanisms. The simplest is based on a gradient of some substance. If the concentration of some chemical decreases from one end of a line of cells to the other, then the concentration of that chemical in any cell along the line effectively specifies the position of the cell with respect to the boundary (Fig. 1.25). A chemical whose concentration varies and which is involved in pattern formation is called a **morphogen**. In the case of the French flag, we assume a source of morphogen at one end and a sink at the other, and that the concentrations of morphogen at both ends are kept constant but are different from each other. Then, as the morphogen diffuses down the line, its concentration at any point effectively provides positional information. If the cells can respond to **threshold concentrations** of the morphogen—for example, above a particular concentration the cells develop as blue, while below this concentration they become white, and at yet another, lower, concentration they become red—the line of cells will develop as a French flag (see Fig. 1.25). Thresholds can represent the amount of morphogen that must bind to receptors to activate an intracellular signaling system, or concentrations of transcription factors required to activate particular genes. The use of threshold concentrations of transcription factors to specify position is most beautifully illustrated in the early development of *Drosophila*, as we shall see in Chapter 2. Other ways of specifying positional information are by direct intercellular interactions and by timing mechanisms, as discussed in Chapter 4 and elsewhere in the book.

The French flag model illustrates two important features of development in the real world. The first is that, even if the length of the line varies, the system regulates and the pattern will still form correctly, given that the boundaries of the system are properly defined by keeping the different concentrations of morphogen constant at either end. The second is that the system could also regenerate the complete original pattern if it were cut in half, provided that the boundary concentrations were re-established. It is therefore truly regulative. We have discussed it

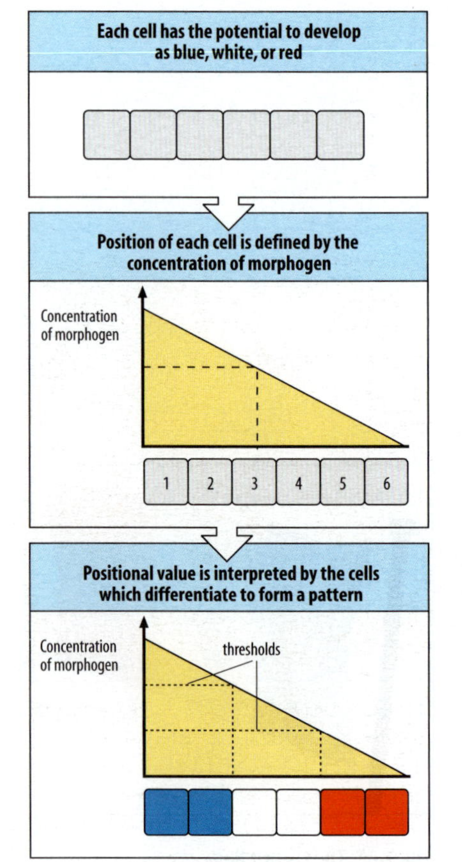

Fig. 1.25 The French flag model of pattern formation. Each cell in a line of cells has the potential to develop as blue, white, or red. The line of cells is exposed to a concentration gradient of some substance and each cell acquires a positional value defined by the concentration at that point. Each cell then interprets the positional value it has acquired and differentiates into blue, white, or red, according to a predetermined genetic program, thus forming the French flag pattern. Substances that can direct the development of cells in this way are known as morphogens. The basic requirements of such a system are that the concentration of substance at either end of the gradient must remain different from each other but constant, thus fixing boundaries to the system. Each cell must also contain the necessary information to interpret the positional values. Interpretation of the positional value is based upon different threshold responses to different concentrations of morphogen.

here as a one-dimensional patterning problem, but the model can easily be extended to provide two-dimensional patterning (Fig. 1.26).

In many cases it is still unclear how positional information is specified—morphogen diffusion is but one of the mechanisms that have been discovered—but the evidence for it is strong. However, we do not know how fine-grained the differences in position are. For example, has every cell in the early embryo a distinct positional value of its own and is it able to make use of this difference?

An important aspect of many patterning processes is the setting up of a polarity in the plane of a sheet of cells. It is as if a direction arrow is present in all the cells in the sheet making one end slightly different from the other, rather like a compass needle pointing north. This type of polarity is known as planar cell polarity and is discussed further in Chapters 2 and 7. One obvious example of planar cell polarity is the hairs on the *Drosophila* wing, which all point in the same direction.

Fig. 1.26 Positional information can be used to generate patterns. A good example, as shown here, is where the people seated in a stadium each have a position defined by their row and seat numbers. Each position has an instruction about which colored card to hold up, and this makes the pattern. If the instructions were changed, a different pattern would be formed, and so positional information can be used to produce an enormous variety of patterns.

1.16 Lateral inhibition can generate spacing patterns

Many structures, such as the feathers on a bird's skin, are found to be more or less regularly spaced with respect to one another. One mechanism that gives rise to such spacing is called **lateral inhibition** (Fig. 1.27). Given a group of cells that all have the potential to differentiate in a particular way, for example as feathers, it is possible to regularly space the cells that will form feathers by a mechanism in which cells that begin to form feathers inhibit the adjacent cells from doing so. This is reminiscent of the spacing of trees in a forest being caused by competition for sunlight and nutrients. In embryos, lateral inhibition is often the result of the differentiating cell secreting an inhibitory molecule that acts locally on the nearest cell neighbors to prevent them developing in a similar way.

1.17 Localization of cytoplasmic determinants and asymmetric cell division can make cells different from each other

Positional specification is just one way in which cells can be given a particular identity. A separate mechanism is based on **cytoplasmic localization** and **asymmetric cell division** (Fig. 1.28). Asymmetric divisions are so called because they result in daughter cells having properties different from each other, independently of any environmental influence. The properties of such cells therefore depend on their **lineage** or line of descent, and not on environmental cues. Although some asymmetric cell divisions are also unequal divisions in that they produce cells of different sizes, this is not usually the most important feature in animals; it is the unequal distribution of cytoplasmic factors that makes the division asymmetric. An alternative way of making the French flag pattern from the egg would be to have chemical differences (representing blue, white, and red) distributed in the egg in the form of determinants that foreshadowed the French flag. When the egg underwent cleavage, these cytoplasmic determinants would become distributed among the cells in a particular way and a French flag would develop. This would require no interactions between the cells, which would have their fates determined from the beginning.

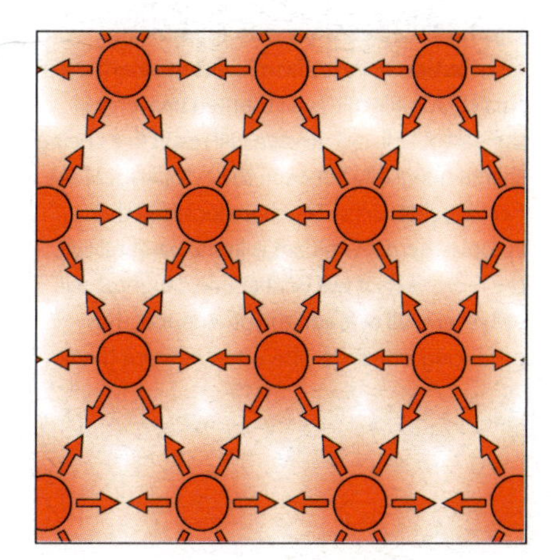

Fig. 1.27 Lateral inhibition can give a spacing pattern. Lateral inhibition occurs when developing structures produce an inhibitor that prevents the formation of any similar structures in the area adjacent to them, and the structures become evenly spaced as a result.

Although such extreme examples of mosaic development are not known in nature, there are well known cases where eggs or cells divide so that some cytoplasmic determinant becomes unequally distributed between the two daughter cells and they develop differently. This happens at the first cleavage of the nematode egg, for example, and defines the antero-posterior axis of the embryo. The germ cells of

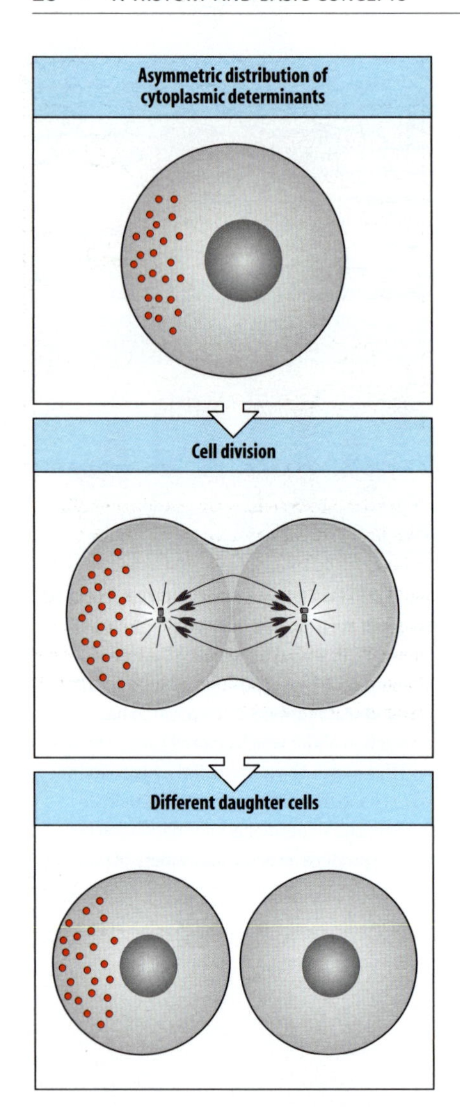

Fig. 1.28 Cell division with asymmetric distribution of cytoplasmic determinants. If a particular molecule is distributed unevenly in the parent cell, cell division will result in its being shared unequally between the cytoplasm of the two daughter cells. The more localized the cytoplasmic determinant is in the parental cell, the more likely it will be that one daughter cell will receive all of it and the other none, thus producing a distinct difference between them.

Drosophila are also specified by cytoplasmic determinants, in this case contained in the cytoplasm located at the posterior end of the egg, and in *Xenopus* a key determinant in the first stages of embryonic development is a protein known as VegT, which is localized in the vegetal region of the fertilized egg. In general, however, as development proceeds, daughter cells become different from each other because of signals from other cells or from their extracellular environment, rather than because of the unequal distribution of cytoplasmic determinants.

A very particular type of asymmetric division is that of **stem cells**, which are cells that are capable of repeated division and that produce daughter cells, one of which remains a stem cell while the other differentiates into a variety of cell types (Fig. 1.29). Stem cells that are capable of giving rise to all the cell types in the body are present in embryos, while in adults, stem cells of apparently more limited potential are responsible for the continual renewal of tissues such as blood, epidermis, and gut epithelium and for the replacement of tissues such as muscle when required. The difference in the daughter cells can be due either to asymmetric distribution of cytoplasmic determinants or to the effects of external signals. The generation of neurons in *Drosophila*, for example, depends on the distribution of cytoplasmic determinants in neuronal stem cells. Stem cells that can give rise to all other types of cells, but not to a complete embryo, are known as **pluripotent**. The fertilized egg, which can give rise to a complete embryo, is described as **totipotent**.

1.18 The embryo contains a generative rather than a descriptive program

All the information for embryonic development is contained within the fertilized egg. So how is this information interpreted to give rise to an embryo? One possibility is that the structure of the organism is somehow encoded as a descriptive program in the genome. Does the DNA contain a full description of the organism to which it will give rise: is it a blueprint for the organism? The answer is no. Instead, the genome contains a program of instructions for making the organism—a generative program—in which the cytoplasmic constituents of eggs and cells are essential players along with the genes.

A descriptive program such as a blueprint or a plan describes an object in some detail, whereas a generative program describes how to make an object. For the same object the programs are very different. Consider origami, the art of paper folding. By folding a piece of paper in various directions it is quite easy to make a paper hat or a bird from a single sheet. To describe in any detail the final form of the paper with the complex relationships between its parts is really very difficult, and not of much help in explaining how to achieve it. Much more useful and easier to formulate are instructions on how to fold the paper. The reason for this is that simple instructions about folding have complex spatial consequences. In development, gene action similarly sets in motion a sequence of events that can bring about profound changes in the embryo. One can thus think of the genetic information in the fertilized egg as equivalent to the folding instructions in origami; both contain a generative program for making a particular structure.

A further distinction can be made between the embryo's genetic program and the developmental program of a particular cell or group of cells. The genetic program refers to the totality of information provided by the genes, whereas a developmental program may refer only to that part of the genetic program that is controlling a particular group of cells. As the embryo develops, different parts acquire their own developmental programs as a result of cell–cell interactions and the activities of selected sets of genes. Each cell in the embryo thus has its own developmental program, which may change as development proceeds.

1.19 The reliability of development is achieved by a variety of means

The development of embryos is remarkably consistent and reliable; you need only look at the similar lengths of your legs, each of which develops quite independently of the other for some 15 years (Chapter 9). Embryonic development needs to be reliable in the sense that the adult organism can function properly. For example, both wings of a bird must be very similar in size and shape if it is to be able to fly satisfactorily. How such reliability is achieved is an important problem in development.

Development needs to be reliable in the face of fluctuations that occur either within the embryo or in the external environment. Internal fluctuations include small changes in the concentrations of molecules, and also changes due, for example, to mutations in genes not directly linked to the development of the organ in question. External factors that could perturb development include temperature and environmental chemicals.

One of the central problems of reliability relates to pattern formation. How are cells specified to behave in a particular way in a particular position? Two ways in which reliability is assured are apparent redundancy of mechanism and negative feedback. **Redundancy** is when there are two or more ways of carrying out a particular process; if one fails for any reason another will still function. It is like having two batteries in your car instead of just one. True redundancy—such as having two identical genes in a haploid genome with identical functions—is rare except for cases such as rRNA, where there can be hundreds of similar genes. Apparent redundancy, on the other hand, where a given process can be specified by several different mechanisms, is probably one of the ways that embryos can achieve such precise and reliable results. It is like being able to draw a straight line with either a ruler or a piece of taut string. This type of situation is not true redundancy, but may give the impression of being so if one mechanism is removed and the outcome is still apparently normal.

Negative feedback also has a role in ensuring consistency; here, the end product of a process inhibits an earlier stage and thus keeps the level of product constant (see, for example, Fig. 1.20). The classic example is in metabolism, where the end product of a biochemical pathway inhibits one of the enzymes that act early in the pathway. Yet another reliability mechanism relates to the complexity of the networks of gene activity that operate in development. There is evidence that these networks are robust and relatively insensitive to small changes in, for example, the rates of the individual processes involved.

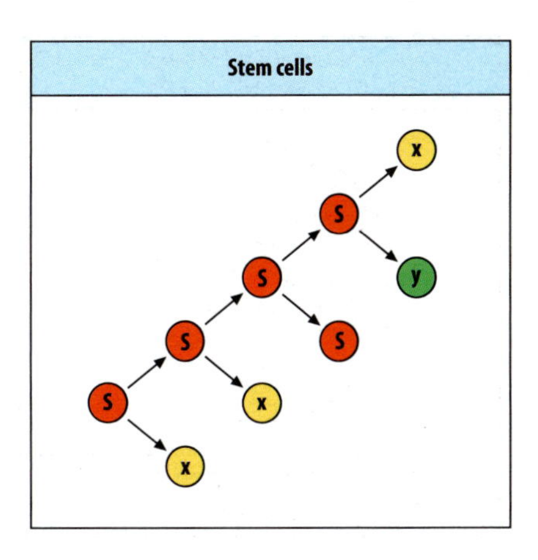

Fig. 1.29 Stem cells. Stem cells (S) are cells that both renew themselves and give rise to differentiated cell types. Thus, a daughter cell arising from stem cell division may either develop into another stem cell or give rise to a quite different type of cell (X). It may also divide to give rise to two stem cells or to two different types of cells (X and Y).

1.20 The complexity of embryonic development is due to the complexity of cells themselves

Cells are, in a way, more complex than the embryo itself. In other words, the network of interactions between proteins and DNA within any individual cell contains many more components and is very much more complex than the interactions between the cells of the developing embryo. However clever you think cells are, they are almost always far cleverer. Each of the basic cell activities involved in development, such as cell division, response to cell signals, and cell movements, are the result of interactions within a population of many different intracellular proteins whose composition varies over time and between different locations in the cell. Cell division for example, is a complex cell-biological program that takes place over a period of time, in a set order of stages, and requires the construction and precise organization of specialized intracellular structures at mitosis.

Any given cell is expressing thousands of different genes at any given time. Much of this gene expression may reflect an intrinsic program of activity, independent of external signals. It is this complexity that determines how cells respond to the signals they receive; how a cell responds to a particular signal depends on its internal state. This state can reflect the cell's developmental history—cells have good memories—and so different cells can respond to the same signal in very different ways. We shall see many examples of the same signals being used over and over again by different cells at different stages of embryonic development, with different biological outcomes.

We have at present only a fragmentary picture of how all the genes and proteins in a cell, let alone a developing embryo, interact with one another. But new technologies are now making it possible to detect the simultaneous activity of hundreds of genes in a given tissue. The discipline of systems biology is also beginning to develop techniques to reconstruct the highly complex signaling networks used by cells. How to interpret this information, and make biological sense of the patterns of gene activity revealed, is a massive task for the future.

Summary

Development results from the coordinated behavior of cells. The major processes involved in development are cell division, pattern formation, morphogenesis or change in form, cell differentiation, cell migration, cell death, and growth. Genes control cell behavior by controlling where and when proteins are synthesized, and thus cell biology provides the link between gene action and developmental processes. During development, cells undergo changes in the genes they express, in their shape, in the signals they produce and respond to, in their rate of proliferation, and in their migratory behavior. All these aspects of cell behavior are controlled largely by the presence of specific proteins; gene activity controls which proteins are made. Since the somatic cells in the embryo generally contain the same genetic information, the changes that occur in development are controlled by the differential activity of selected sets of genes in different groups of cells. The genes' control regions are fundamental to this process. Development is progressive and the fate of cells becomes determined at different times. The potential for development of cells in the early embryo is usually much greater than their normal fate, but this potential becomes more restricted as development proceeds. Inductive interactions, involving signals from one tissue or cell to another, are one of the main ways of changing cell fate and directing development. Asymmetric cell divisions, in which cytoplasmic components are unequally distributed to daughter cells, can also make cells different. One widespread means of pattern generation is through positional information; cells first acquire a positional value with respect to boundaries and then interpret their positional values by behaving in different ways. Developmental signals are more selective than instructive, choosing one or other of the developmental pathways open to the cell at that time. The embryo contains a generative, not a descriptive, program—it is more like the instructions for making a structure by paper folding than a blueprint. A variety of mechanisms, including apparent redundancy and negative feedback, are involved in making development remarkably reliable. The complexity of development lies within the cells.

SUMMARY TO CHAPTER 1

All the information for embryonic development is contained within the fertilized egg—the diploid zygote. The genome of the zygote contains a program of instructions for making the organism. In the working out of this developmental program, the cytoplasmic constituents of the egg and of the cells it gives rise to are essential players along with the genes. Strictly regulated gene activity, by controlling which proteins are synthesized, and where and when, directs a sequence of cellular activities that brings about the profound changes that occur in the embryo during development. The major processes involved in development are cell division, pattern formation, morphogenesis, cell differentiation, cell migration, cell death, and growth. All of these processes can be influenced by communication between the cells of the embryo. Development is progressive, with the fate of cells becoming more precisely specified as development proceeds. Cells in the early embryo usually have a much greater potential for development than is evident from their normal fate, and this enables embryos to develop normally even if cells are removed, added, or transplanted to different positions. Cells' developmental potential becomes much more restricted as development proceeds. Many genes are involved in controlling the complex interactions that occur during development, and reliability is achieved in a variety of ways. Thousands of genes control the development of animals and plants, and a full understanding of the processes involved is far from being achieved. While basic principles and certain developmental systems are quite well understood, there are still many gaps.

FURTHER READING

The origins of developmental biology

Cole, F.J.: *Early Theories of Sexual Generation*. Oxford: Clarendon Press, 1930.

Hamburger, V.: *The Heritage of Experimental Embryology: Hans Spemann and the Organizer*. New York: Oxford University Press, 1988.

Milestones in Development [http://www.nature.com/milestones/development/index.html]

Needham, J.: *A History of Embryology*. Cambridge: Cambridge University Press, 1959.

Sander, K.: **"Mosaic work" and "assimilating effects" in embryogenesis: Wilhelm Roux's conclusions after disabling frog blastomeres**. *Roux's Arch. Dev. Biol.* 1991, **200**: 237–239.

Sander, K.: **Shaking a concept: Hans Driesch and the varied fates of sea urchin blastomeres**. *Roux's Arch. Dev. Biol.* 1992, **201**: 265–267.

Wilson, E.B.: *The Cell in Development and Heredity*. New York: Macmillan, 1896.

Wolpert, L.: **Evolution of the cell theory**. *Phil. Trans. Roy. Soc. Lond. B* 1995, **349**: 227–233.

A conceptual tool kit

Alberts, B., *et al.*: *Essential Cell Biology: An Introduction to the Molecular Biology of the Cell*. 2nd edn. New York: Garland Science, 2003.

Graveley, B.R.: **Alternative splicing: increasing diversity in the proteomic world.** *Trends Genet.* 2001, **17**: 100–107.

Howard, M.L., Davidson, E.H.: ***cis*-Regulatory control circuits in development**. *Dev. Biol.* 2004, **271**: 109–118.

Jordan, J.D., Landau, E.M., Lyengar, R.: **Signaling networks: the origins of cellular multitasking**. *Cell* 2000, **103**: 193–200.

Levine, M., Tjian, R.: **Transcriptional regulation and animal diversity.** *Nature* 2003, **424**: 147–151.

Nelson, W.J.: **Adaptation of core mechanisms to generate polarity.** *Nature* 2003, **422**: 766–774.

Papin, J.A., Hunter, T., Palsson, B.O., Subramanian, S.: **Reconstruction of cellular signalling networks and analysis of their properties**. *Nat. Rev. Mol. Biol.* 2005, **6**: 99–111.

Wolpert, L.: **Do we understand development?** *Science* 1994, **266**: 571–572.

Wolpert, L.: **One hundred years of positional information**. *Trends Genet.* 1996, **12**: 359–364.

Development of the *Drosophila* body plan

2

- *Drosophila* life cycle and overall development
- Setting up the body axes
- Localization of maternal determinants during oogenesis
- Patterning the early embryo
- Activation of the pair-rule genes and the establishment of parasegments
- Segmentation genes and compartments
- Specification of segment identity

The fruit fly *Drosophila melanogaster* is an invertebrate whose early development is better understood than that of any other animal of similar or greater morphological complexity. Its early development exemplifies clearly many of the principles introduced in Chapter 1. The genetic basis of development is particularly well understood in *Drosophila* and many developmental genes in other organisms, particularly vertebrates, have been identified by their homology with *Drosophila* genes. In this chapter we look at the crucial role of maternal gene products in the earliest development of the zygote and how they specify the main axes and regions of the body. We shall see how gradients of maternal morphogens specify the antero-posterior and dorso-ventral pattern of the early embryo, and how *cis*-acting regulatory elements then control the precise spatial expression of the zygote's own genes to divide the embryo into segments. Another set of genes acting along the antero-posterior axis then confers on each segment its unique character, as displayed by the appendages the segment eventually bears in the adult, such as wings, legs, or antennae.

We are much more like flies in our development than you might think. Astonishing discoveries in developmental biology over the past 20 years have revealed that many of the genes that control the development of the fruit fly *Drosophila* are similar to those controlling development in vertebrates, and indeed in many other animals. It seems that once evolution finds a satisfactory way of patterning animal bodies, it tends to use the same mechanisms and molecules over and over again with, of course, some important modifications.

About 13,600 protein-coding genes have been predicted from the genome sequence of *Drosophila*, which is only twice the number in unicellular yeast and fewer than the 19,000 genes of the morphologically simpler nematode *Caenorhabditis*. A recent large-scale analysis of the RNAs transcribed both during development and in adult *Drosophila* has, however, revealed a high incidence of alternative RNA splicing and thousands of as-yet-uncharacterized transcripts, some of which may represent regulatory RNAs such as microRNAs. The *Drosophila* genome has much still to reveal.

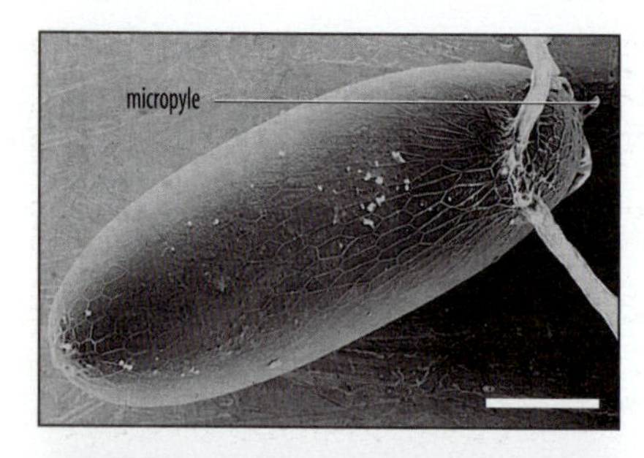

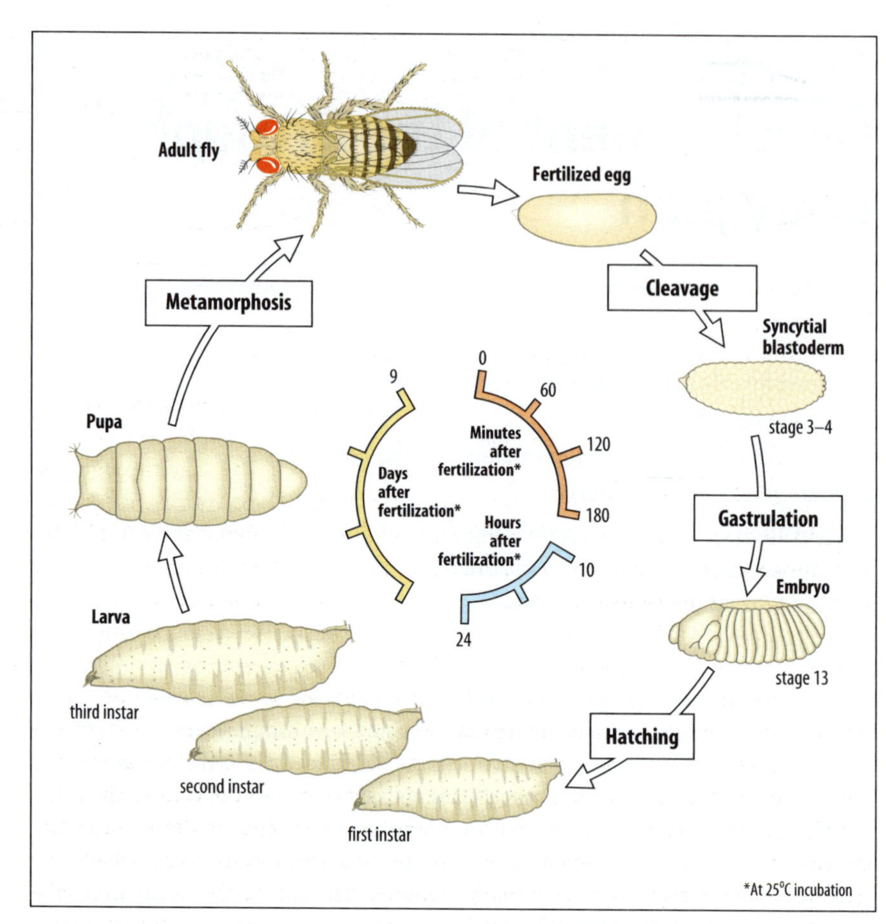

Fig. 2.1 Life cycle of *Drosophila melanogaster*. After cleavage and gastrulation the embryo becomes segmented and hatches out as a feeding larva. The larva grows and goes through two molts (instars), eventually forming a pupa that will metamorphose into the adult fly. The photographs show scanning electron micrographs of: a *Drosophila* egg before fertilization (top). The sperm enters through the micropyle. The dorsal filaments are extra-embryonic structures; a *Drosophila*, second-instar larva (middle); and a *Drosophila* pupa (bottom). Scale bars = 0.1 mm.

Photographs courtesy of F. R. Turner (top, from Turner, F.R., et al.: 1976; middle, from Turner, F.R., et al.: 1979).

The pre-eminent place of *Drosophila* in modern developmental biology was recognized by the award of the 1995 Nobel Prize for Physiology or Medicine for work that led to a fundamental understanding of how genes control development in the fly embryo. This was only the second time that the Nobel Prize had been awarded for work in developmental biology. While insect and vertebrate development may seem to be very different, much has been learned that can be applied to vertebrate development.

The first part of this chapter describes the life cycle of *Drosophila* and its overall development. The remaining parts look at how the basic body plan of the *Drosophila* larva is established up to the stage at which the embryo becomes segmented, and how the segments are patterned and acquire their unique identities. More about *Drosophila* gastrulation, adult organ development, and neural development will be found in Chapters 7, 9, and 10, respectively.

Drosophila life cycle and overall development

The fruit fly *Drosophila melanogaster* is a small dipteran insect, about 3 mm long as an adult, which undergoes embryonic development inside an egg and hatches from the egg as a larva. This then goes through two more larval stages, growing bigger each time, and eventually becomes a pupa, in which metamorphosis into the adult occurs. The life cycle of *Drosophila* is shown in Fig. 2.1.

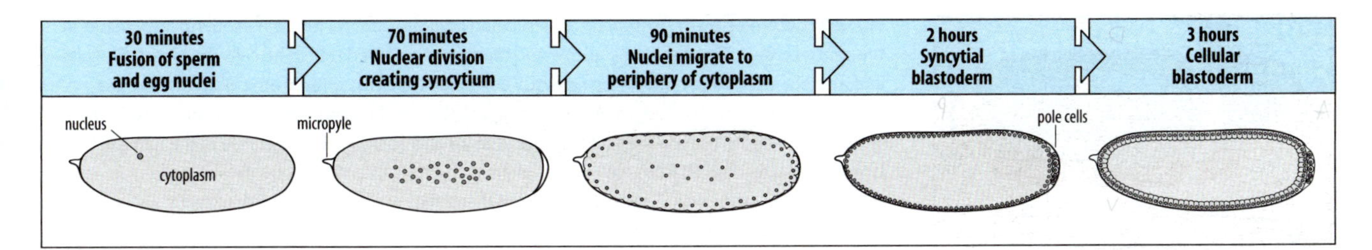

Fig. 2.2 Cleavage of the *Drosophila* embryo. After fusion of the sperm and egg nuclei there is rapid nuclear division but no cell walls form, the result being a syncytium of many nuclei in a common cytoplasm. After the ninth division the nuclei move to the periphery to form the syncytial blastoderm. After about 3 hours, cell walls develop, giving rise to the cellular blastoderm. About 15 pole cells, which will give rise to germ cells, form a separate group at the posterior end of the embryo. Times given are for incubation at 25°C.

2.1 The early *Drosophila* embryo is a multinucleate syncytium

The *Drosophila* egg is sausage-shaped and the future anterior end is easily recognizable by the micropyle, a nipple-shaped structure in the tough external coat surrounding the egg. Sperm enter the anterior end of the egg through the micropyle. After fertilization and fusion of the sperm and egg nuclei, the zygote nucleus undergoes a series of rapid mitotic divisions, one about every 9 minutes but, unlike most animal embryos, there is no cleavage of the cytoplasm. The result after 12 nuclear divisions is a **syncytium** in which around 6000 nuclei are present in a common cytoplasm (Fig. 2.2); the embryo essentially remains a single cell during its early development. After nine divisions the nuclei move to the periphery to form the **syncytial blastoderm**. This is equivalent to the blastula or blastoderm stage of other animals. Shortly afterwards, membranes grow in from the surface to enclose a nucleus and form cells, and the blastoderm becomes truly cellular after about 13 mitoses. Not all the nuclei give rise to the cells of the cellular blastoderm; some 15 or so end up at the posterior end of the embryo and develop into **pole cells**, which later give rise to germ cells from which the gametes—sperm and eggs—eventually develop. Because of the formation of a syncytium, even large molecules such as proteins can diffuse between nuclei during the first 3 hours of development and, as we shall see, this is of great importance to early *Drosophila* development.

In this chapter we will be concerned with the development of the embryo to the stage at which it hatches out of the egg as a larva. This stage naturally takes place inside the opaque egg, and to study embryonic development embryos are freed from the egg case by treating with bleach. The embryos are then placed in a special oil for observation. This can be done at any stage and they will continue to develop and even hatch. The embryo is about 0.5 mm long.

2.2 Cellularization is followed by gastrulation, segmentation, and the formation of the larval nervous system

All the future tissues are derived from the single epithelial layer of the cellular blastoderm. For example, prospective **mesoderm** (see Box 1B, p. 15) is located in the most ventral region, while the future midgut derives from two regions of prospective **endoderm**, one at the anterior and the other at the posterior end of the embryo. Endodermal and mesodermal tissues move to their future positions inside the embryo during **gastrulation**, leaving **ectoderm** as the outer layer (Fig. 2.3). Gastrulation starts at about 3 hours after fertilization when the future mesoderm in the ventral region invaginates to form a furrow along the ventral midline. The mesodermal cells are initially internalized by formation of a tube of mesoderm, which will be described in more detail in Chapter 7. The mesoderm cells then separate from the surface layer of the tube and migrate under the ectoderm to internal locations, where they later give rise to muscle and other connective tissues.

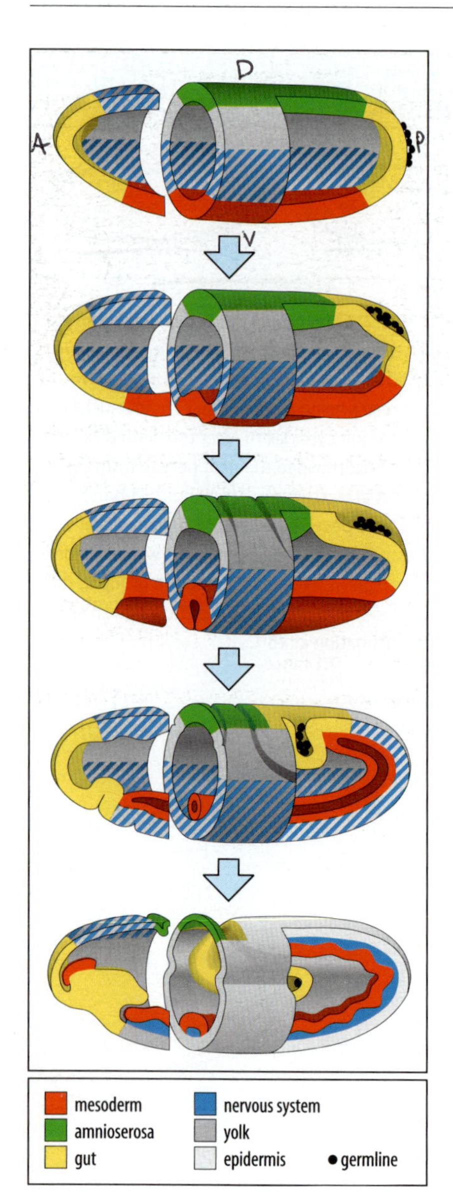

Fig. 2.3 Gastrulation in *Drosophila*. Gastrulation begins when the future mesoderm invaginates in the ventral region, first forming a furrow and then an internalized tube. The cells then leave the tube and migrate internally under the ectoderm. The nervous system comes from cells that leave the surface of the ventral blastoderm and form a layer between the ventral ectoderm and mesoderm. The gut forms from two invaginations at the anterior and posterior end that fuse in the middle. The midgut region is derived from endoderm, whereas the foregut and hindgut are of ectodermal origin. Regions striped blue and gray represent tissue that will give rise to both nervous system and epidermis. The amnioserosa, an extra-embryonic membrane, is discussed in Section 2.5.

In insects, as in all arthropods, the main nerve cord lies ventrally, rather than dorsally as in vertebrates. Shortly after the mesoderm has invaginated, ectodermal cells of the ventral region that will give rise to the nervous system leave the surface individually and form a layer of **neuroblasts** between the mesoderm and the outer ectoderm. At the same time, two tube-like invaginations develop at the sites of the future anterior and posterior midgut. These grow inward and eventually fuse to form the endoderm of the midgut, while ectoderm is dragged inward behind them at each end to form the foregut and the hindgut. The outer ectoderm layer develops into the epidermis. There are no cell divisions during gastrulation, but once it is completed, cells start to divide again. The cells of the epidermis only divide twice more before they secrete a thin cuticle composed largely of protein and the polysaccharide chitin.

Also during gastrulation, the ventral blastoderm or **germ band**, which comprises the main trunk region, undergoes germ-band extension, which drives the posterior trunk regions round the posterior end and onto what was the dorsal side (Fig. 2.4). The germ band later retracts as embryonic development is completed. At the time of germ-band extension the first external signs of **segmentation** can be seen. A series of evenly spaced grooves form more or less at the same time and these demarcate **parasegments**, which later give rise to the **segments** of the larva and adult. Parasegments and segments are out of register, so that a segment is formed by the posterior region of one parasegment and the anterior region of the next. There are 14 parasegments: three contribute to mouthparts of the head, three to the thoracic region and eight to the abdomen.

2.3 After hatching the *Drosophila* larva develops through several larval stages, pupates, and then undergoes morphogenesis to become an adult

The larva (Fig. 2.5) hatches about 24 hours after fertilization, but the different regions of the larval body are well defined several hours before that. The head is a complex structure, largely hidden from view before the larva hatches. A specialized structure associated with the most anterior region of the head is the **acron**. Another specialized structure, called the **telson**, marks the posterior end of the larva. Between the head and telson, three thoracic segments and eight abdominal segments can be distinguished by specializations in the cuticle. On the ventral side of each segment are belts of small tooth-like outgrowths called **denticles**, and other cuticular structures characteristic of each segment. As the larva feeds and grows, it molts, shedding its cuticle. This occurs twice, each stage being called an **instar**. After the third instar, the larva becomes a **pupa**, inside which **metamorphosis** into the adult fly occurs.

The *Drosophila* larva has neither wings nor legs; these and other organs are formed when the larva undergoes hormone-induced metamorphosis in the pupal stage. These structures are, however, already present in the larva as **imaginal discs**, small sheets of prospective epidermal cells derived from the cellular blastoderm

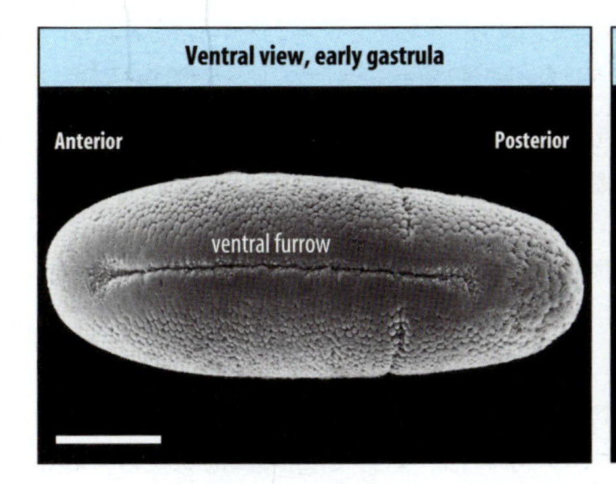

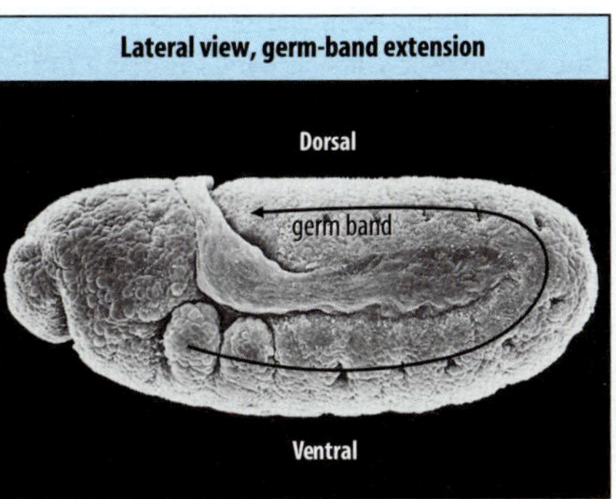

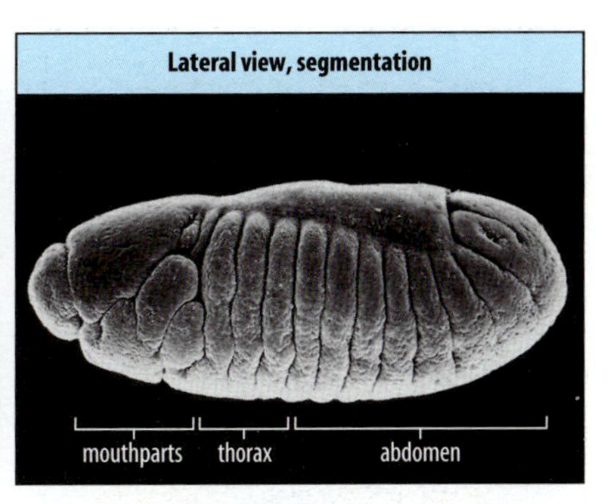

Fig. 2.4 Gastrulation, germ-band extension, and segmentation in the *Drosophila* embryo. Gastrulation involves the future mesoderm moving inside through the ventral furrow. During gastrulation, the ventral blastoderm (the germ band) extends, driving the posterior trunk region onto the dorsal side and segmentation now takes place. Later the germ band shortens. Scale bar = 0.1 mm.

Photographs courtesy of F. Turner (left from Turner, F.R., et al.: 1977; middle from Alberts, B., et al.: 1994).

and usually containing about 40 cells each. These discs grow throughout larval life and form folded sacs of epithelia to accommodate their increase in size. There are imaginal discs for each of the six legs, two wings, and the two halteres (balancing organs), and for the genital apparatus, eyes, antennae, and other adult head structures (Fig. 2.6). At metamorphosis, these develop into the adult organs. We will discuss the development of the imaginal discs in Chapter 9; they provide continuity between the pattern of the larval body and that of the adult, even though metamorphosis intervenes.

2.4 Many developmental mutations have been identified in *Drosophila* through induced mutation and large-scale genetic screening

Valuable though spontaneous mutations have been in the study of development, suitable mutations are rare. Many more developmental genes have been identified by inducing random mutations in a large number of organisms by chemical treatments or irradiation with X-rays, and then screening for mutants of developmental interest. The overall aim is to treat a large enough population so that, in total, a mutation is induced in every gene in the genome. This sort of approach is best used in organisms that breed rapidly and can be obtained and treated conveniently in very large numbers.

Many of the developmental mutations that have led to our present understanding of early *Drosophila* development came from a brilliantly successful screening program that searched the *Drosophila* genome systematically for mutations affecting the patterning of the early embryo. Its success was recognized with the award of a Nobel Prize for Physiology or Medicine to Edward Lewis, Christiane Nüsslein-Volhard, and Eric Wieschaus in 1995.

In this screening program thousands of flies were treated with a chemical mutagen, and were bred and screened according to the strategy described in Box 2A (pp. 36–37). Given the number of progeny involved, it was important to devise a strategy that would reduce the number of flies that had to be examined to find a mutation. So the search was restricted to mutations on just one chromosome at a time. As described in Box 2A, the program also incorporated a means of identifying flies homozygous for the male-derived chromosomes carrying the induced mutations and, most important, also included a way of automatically removing from the population flies that could not be carrying a mutated chromosome.

A phenotypic character of particular value in screening for patterning mutants in *Drosophila* is the stereotyped pattern of small outgrowths, or denticles, on the larval segments, irregularities in which enable mutants to be quickly recognized visually. In this way the key genes involved in patterning the early *Drosophila* embryo were

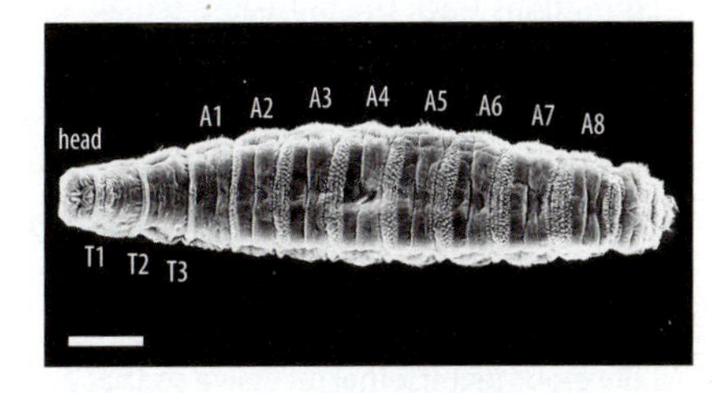

Fig. 2.5 Ventral view of a *Drosophila* larva. T1 to T3 are the thoracic segments and A1 to A8, the abdominal segments. The characteristic pattern of denticles can be seen in the anterior region of each abdominal segment. Scale bar = 0.1 mm.

Photograph courtesy of F.R. Turner.

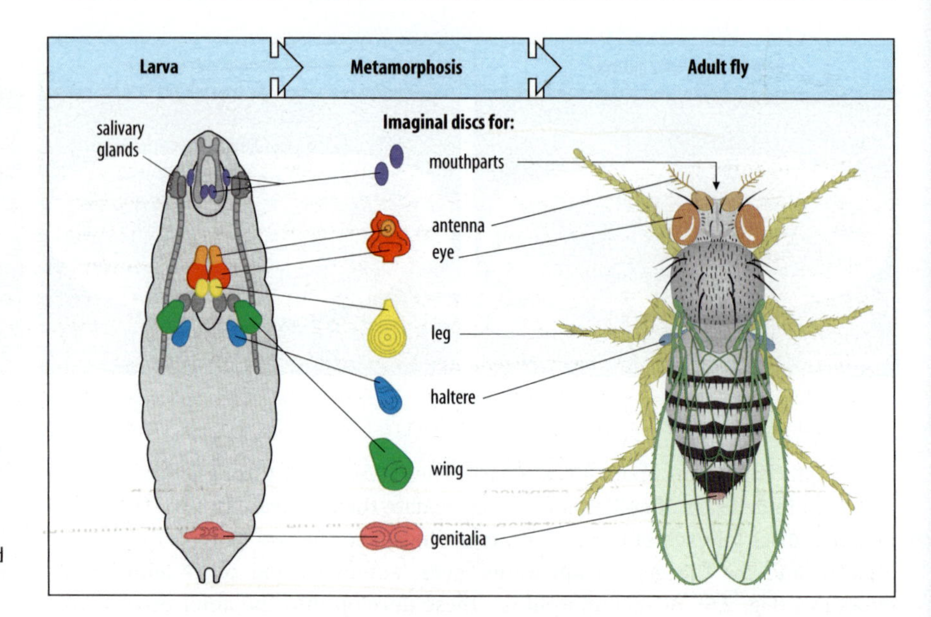

Fig. 2.6 Imaginal discs give rise to adult structures at metamorphosis. The imaginal discs in the *Drosophila* larva are small sheets of epithelial cells; at metamorphosis they give rise to a variety of adult structures. The abdominal cuticle comes from groups of histoblasts located in each larval abdominal segment.

Box 2A Mutagenesis and genetic screening strategy for identifying developmental mutants in *Drosophila*

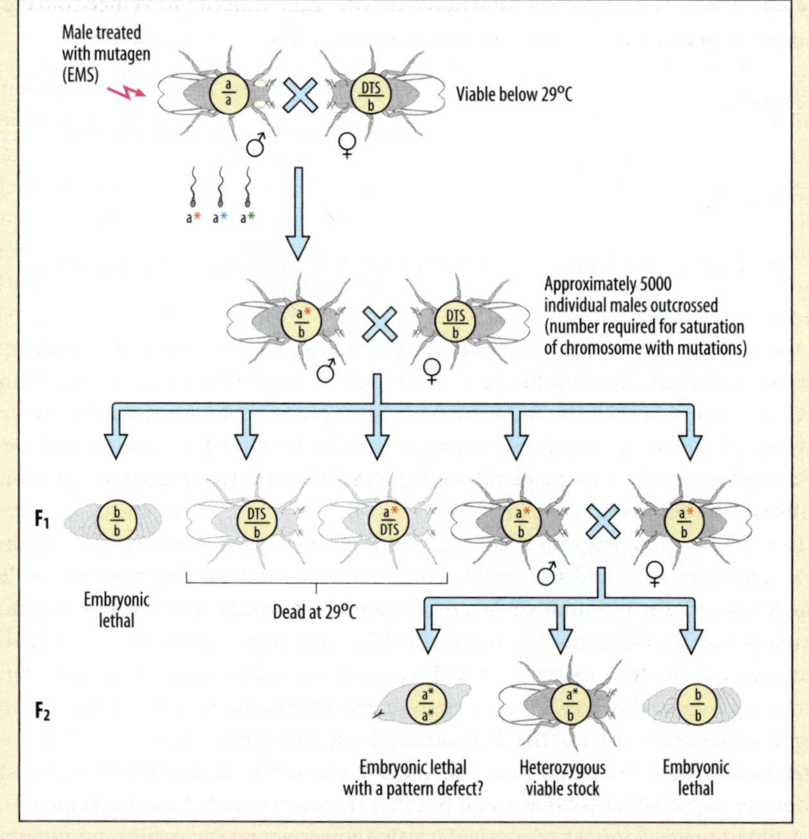

The mutagen ethyl methane sulfonate (EMS) was applied to large numbers of male flies homozygous for a recessive mutation on the selected chromosome. The chromosome marked in this way is designated a in the figure. A recessive mutation is chosen that gives adult flies an easily distinguishable but viable phenotype when homozygous (such as the mutation *white*⁻; see Fig. 1.12).

The treated males, who now produce sperm with a variety of induced mutations on the a chromosomes (a*), were crossed with untreated females that carried different mutations (*DTS* and *b*) on their two a chromosomes, but were otherwise wild type. These mutations track the untreated female-derived chromosomes and automatically eliminate all embryos carrying two female-derived chromosomes in subsequent generations. *DTS* is a dominant temperature-sensitive mutation that causes death of the fly when the incubation temperature is raised to 29°C. *b* is a non-developmental lethal recessive so that any flies homozygous for this female-derived chromosome will die as normal-looking embryos and be automatically eliminated. The female flies also carried a balancer chromosome (not shown) that prevented recombination at meiosis; this is to prevent recombination between male-derived and female-derived chromosomes in females. There is no recombination at meiosis in male *Drosophila*.

Continued

Box 2A (continued) Mutagenesis and genetic screening strategy for identifying developmental mutants in *Drosophila*

To identify the new recessive mutations (a*) caused by the chemical treatment, a large number of the heterozygous males arising from this first cross were again outcrossed to *DTS/b* females. Of the offspring of each individual cross, only a*/*b* flies survive when placed at 29°C; all other combinations die. The surviving siblings were then intercrossed and the offspring of each cross were screened for patterning mutants. There are three possible outcomes: flies homozygous for the induced mutation a* (which will also be homozygous for the original mutation marking chromosome a in males); heterozygous a* flies; and homozygous *b* flies (which die as embryos).

If a* is indeed a patterning mutation which is lethal in the larva, then the culture tube in which the cross is made will contain no adult flies with the original male phenotype–say white-eyed. Therefore tubes containing white-eyed flies can be discarded immediately, as the induced mutation a* they carry must have allowed them to develop to adulthood even when homozygous. If no white-eyed flies are present in a tube, then the homozygous a*/a* embryos may have died or been arrested in their development as a result of abnormal development, and the mutation is of potential interest. The embryos of larvae from this cross can then be examined for pattern defects, as in the example illustrated here. The phenotypically wild-type adults in this tube are heterozygous for a* and are used as breeding stock to study the mutant further. This whole program has to be repeated for each of the four chromosome pairs of *Drosophila*.

first identified. Screening was also successful because the mutations that were inevitably produced in housekeeping genes were mostly rescued by the actions of maternal housekeeping genes—that is, those genes acting in the mother that provide many housekeeping functions to the egg. Indeed, it is necessary to do special screens for **maternal-effect mutations**. These are mutations in genes acting in the mother that are specifically involved in patterning the egg during oogenesis.

Setting up the body axes

2.5 The body axes are set up while the *Drosophila* embryo is still a syncytium

The insect body, like that of vertebrates and most of the other organisms described in this book, is bilaterally symmetrical. Like all animals with bilateral symmetry, the *Drosophila* larva has two distinct and largely independent axes: the antero-posterior and dorso-ventral axes, which are at right angles to each other. These axes are already partly set up in the *Drosophila* egg, and become fully established and patterned in the very early embryo while it is still in the syncytial stage. Along the antero-posterior axis the embryo becomes divided into several broad regions, which will become the head, **thorax**, and **abdomen** of the larva (Fig. 2.7). The thorax and abdomen become divided into segments as the embryo develops, while two regions of endoderm at each end of the embryo invaginate at gastrulation to form the gut (see Fig. 2.3). Each segment has its own unique character in the larva, as revealed by both its external cuticular structures and its internal organization.

The dorso-ventral axis of the embryo becomes divided up into four regions early in embryogenesis: from ventral to dorsal these are the **mesoderm**, which will form muscles and other internal connective tissues; the **ventral ectoderm**, which gives rise to the larval nervous system and is thus also called the neurogenic ectoderm or **neurectoderm**; the dorsal ectoderm, which gives rise to the larval epidermis; and the **amnioserosa**, which gives rise to an extra-embryonic membrane on the dorsal side of the embryo (see Fig. 2.7). Organization along the antero-posterior and dorso-ventral axes of the early embryo develops more or less simultaneously,

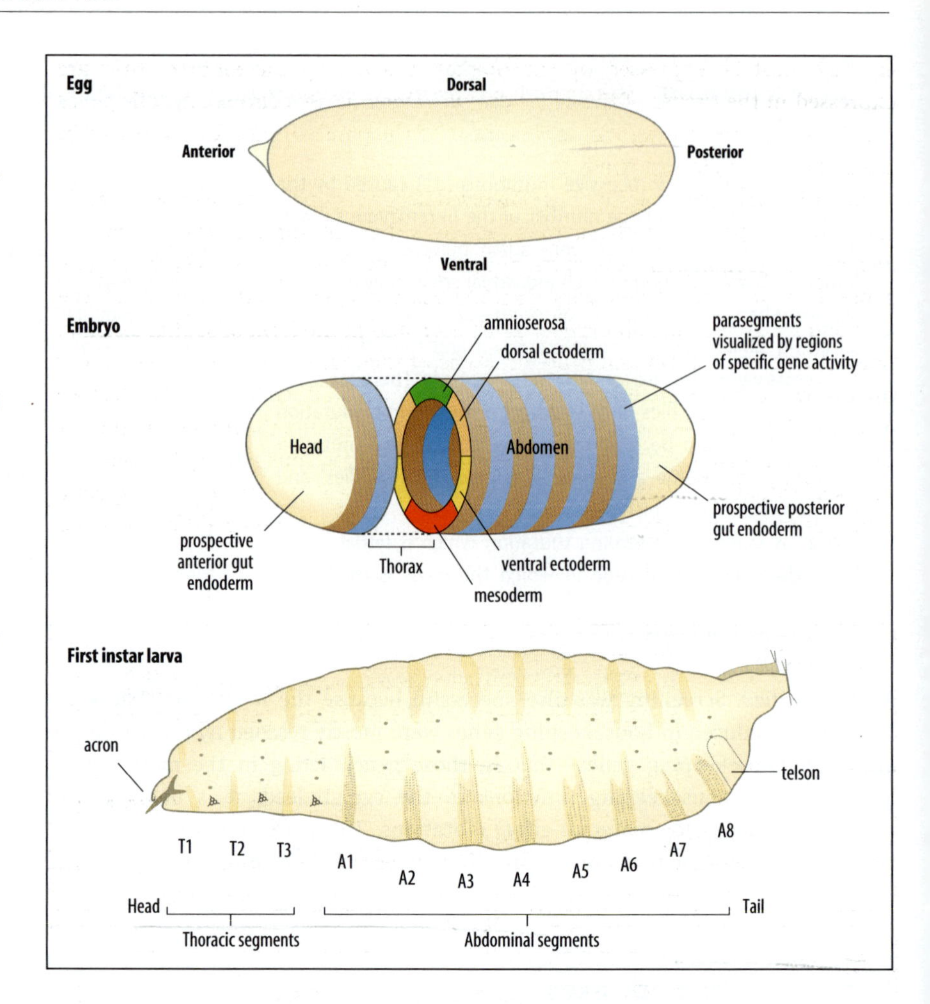

Fig. 2.7 Patterning of the *Drosophila* embryo. The body plan is patterned along two distinct axes. The antero-posterior and dorso-ventral axes are at right angles to each other and are laid down in the egg. In the early embryo, the dorso-ventral axis is divided into four regions: mesoderm (red), ventral ectoderm (yellow), dorsal ectoderm (orange), and amnioserosa (an extra-embryonic membrane; green). The ventral ectoderm gives rise to both ventral epidermis and neural tissue, the dorsal ectoderm to epidermis. The antero-posterior axis becomes divided into different regions that later give rise to the head, thorax, and abdomen. After the initial division into broad body regions, segmentation begins. The future segments can be visualized as transverse stripes by staining for specific gene activity; these stripes demarcate 14 parasegments, 10 of which are marked. The embryo develops into a segmented larva. By the time the larva hatches, the 14 parasegments have been converted into thoracic (T1–T3) and abdominal (A1–A8) segments, which are offset from the parasegments by one half segment. Different segments are distinguished by the patterns of bristles and denticles on the cuticle. Specialized structures, the acron and telson, develop at the head and tail ends, respectively.

but is specified by independent mechanisms and by different sets of genes in each axis.

The early development of *Drosophila* is peculiar to insects, as patterning occurs within a multinucleate syncytial blastoderm (see Fig. 2.2). Only after the beginning of segmentation does the embryo become truly multicellular. At the syncytial stage many proteins, including those that are not normally secreted from cells such as transcription factors, can diffuse throughout the blastoderm and enter other nuclei. Concentration gradients of transcription factors that provide positional information for the nuclei to interpret (see Section 1.15) can thus be set up in the syncytial blastoderm.

Early development is essentially two-dimensional as patterning occurs mainly in the blastoderm. But the larva is a three-dimensional object, with internal structures. This third dimension develops later, at gastrulation, when parts of the surface layer move into the interior to form the gut, the mesodermal structures that will give rise to muscle, and the ectodermally derived nervous system.

2.6 Maternal factors set up the body axes and direct the early stage of *Drosophila* development

The earliest stage of *Drosophila* development is guided by **maternal factors**, the mRNAs and proteins that are synthesized and laid down in the egg by the mother fly. The genes responsible for this maternal contribution are known as **maternal genes** as they must be expressed by the mother and not by the embryo; they are expressed in the tissues of the ovary during oogenesis. By contrast, **zygotic genes** are those required during the development of the embryo; they are expressed in the nuclei of the embryo itself.

About 50 maternal genes in all are involved in setting up the two axes and a basic framework of positional information, which is then interpreted by the embryo's own genetic program. All later patterning, which involves expression of the zygotic genes, is built on this framework (Fig. 2.8). Maternal gene products establish the axes and set up regional differences along each axis in the form of spatial distributions of maternal mRNA and proteins. Some of these are localized at each end of the egg while it is being formed in the ovary, as we shall see later in the chapter. After development begins, maternal RNAs are translated and the maternal proteins then activate zygotic genes in the nuclei along both axes, thus setting the scene for the next round of patterning. The sequential activities of maternal and zygotic genes pattern the embryo in a series of steps. Broad regional differences are

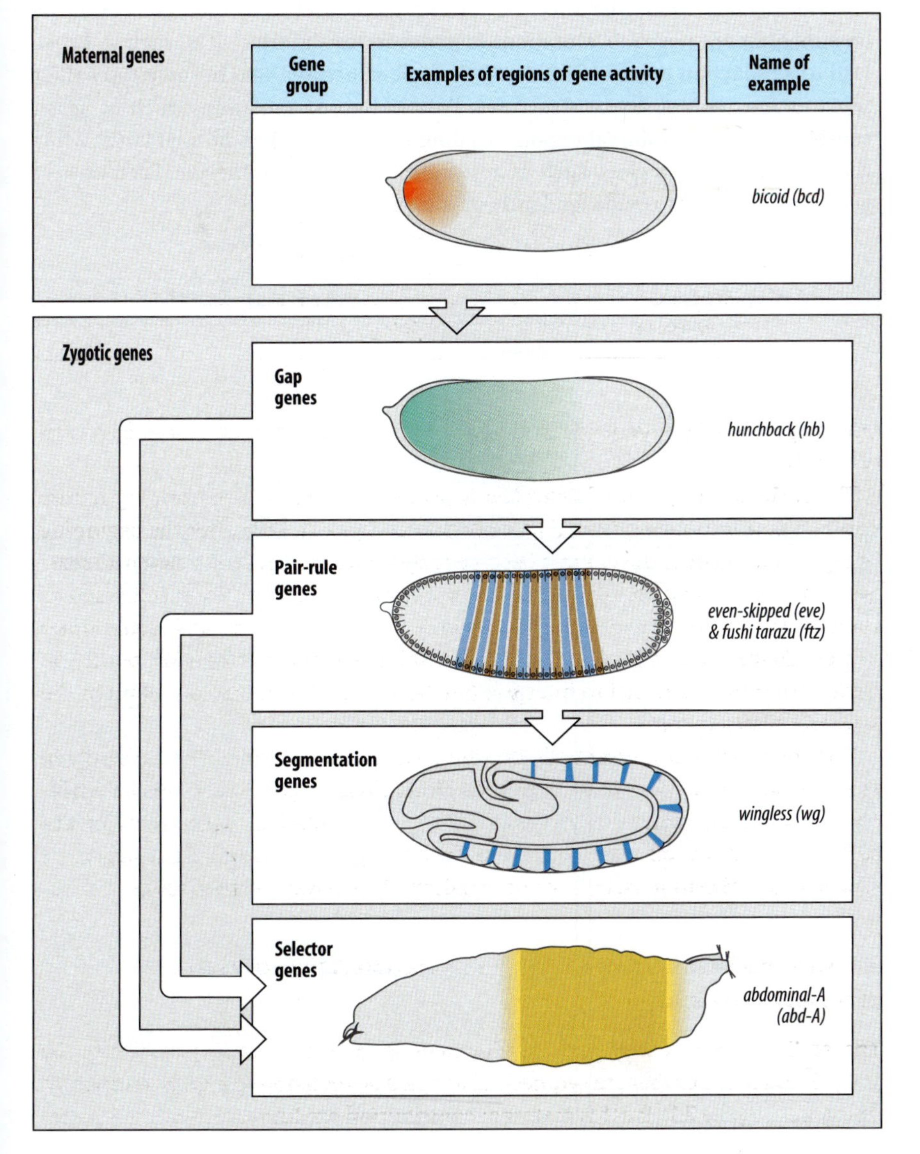

Fig. 2.8 The sequential expression of different sets of genes establishes the body plan along the antero-posterior axis. After fertilization, maternal gene products laid down in the egg, such as *bicoid* mRNA, are translated. They provide positional information which activates the zygotic genes. The four main classes of zygotic genes acting along the antero-posterior axis are the gap genes, the pair-rule genes, the segmentation genes, and the selector, or homeotic, genes. The gap genes define regional differences that result in the expression of a periodic pattern of gene activity by the pair-rule genes, which define the parasegments and foreshadow segmentation. The segmentation genes elaborate the pattern in the segments, and segment identity is determined by the selector genes. The functions of each of these classes of genes are discussed in this chapter.

established first, and these are then refined to produce a larger number of smaller developmental domains, each characterized by a unique profile of zygotic gene activity. Developmental genes act in a strict temporal sequence. They form a hierarchy of gene activity in which the action of one set of genes is essential for another set of genes to be activated, and thus for the next stage of development to occur. We first look at how maternal gene products specify the antero-posterior axis.

2.7 Three classes of maternal genes specify the antero-posterior axis

The expression of maternal genes during egg formation in the mother creates differences in the egg along the antero-posterior axis even before it is fertilized. These differences already distinguish the future head and posterior ends of the adult. Maternal genes are identified by mutations that when present in the mother do not damage her but have effects on the development of her progeny. The roles of the maternal genes can be deduced from the effects of these maternal-effect mutations on the embryo. The mutations fall into three classes: those that affect anterior regions; those that affect posterior regions; and those that affect both the terminal regions (Fig. 2.9). Mutations of genes in the anterior class, such as *bicoid*, lead to a reduction or loss of head and thoracic structures, and in some cases their replacement with posterior structures. Posterior-group mutations, such as *nanos*, cause the loss of abdominal regions, leading to a smaller than normal larva, while those of the terminal class, such as *torso*, affect the acron and telson. Each class of gene acts more or less independently of the others.

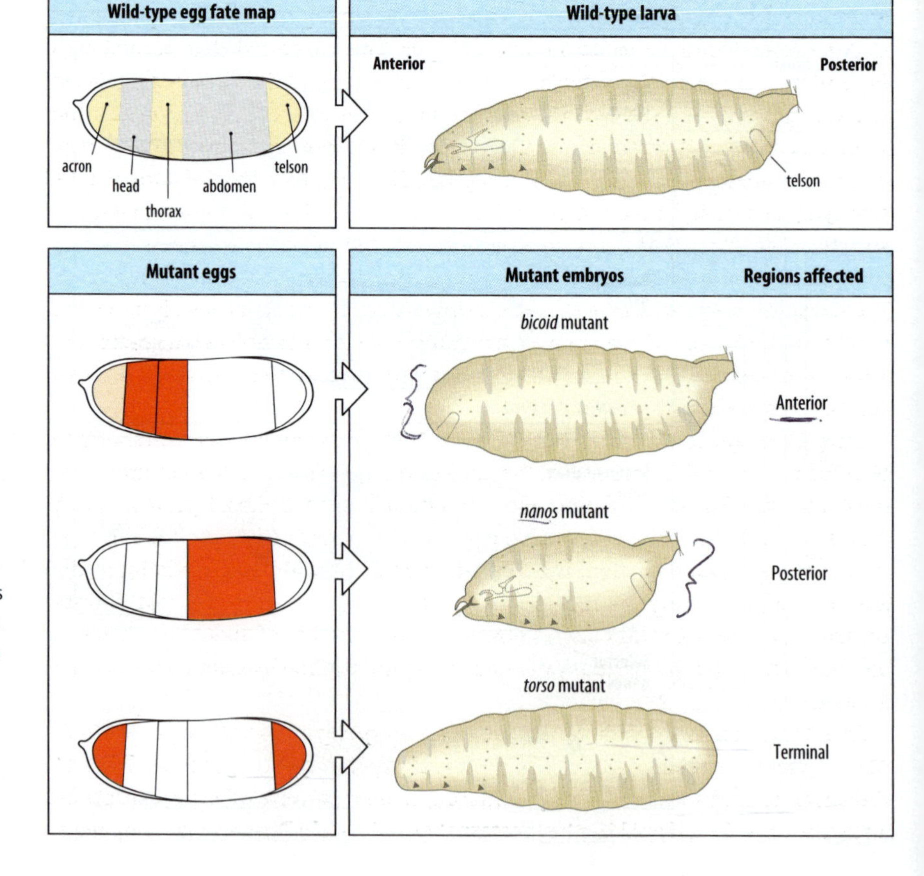

Fig. 2.9 The effects of mutations in the maternal gene system. Mutations in maternal genes lead to deletions and abnormalities in anterior, posterior, or terminal structures. The wild-type fate map shows which regions of the egg give rise to particular regions and structures in the larva. Regions that are affected in mutant eggs and which lead to lost or altered structures in the larva are shaded in red. In *bicoid* mutants there is a partial loss of anterior structures and the appearance of a posterior structure—the telson—at the anterior end. *nanos* mutants lack a large part of the posterior region. *torso* mutants lack both acron and telson.

The idiosyncratic naming of genes in *Drosophila* usually reflects the attempts by the discoverers to describe the mutant phenotype. In this chapter we meet quite a number of gene names; all these are listed, together with their functions where known, in the table at the end of this chapter (p. 85). Of the 50 or so maternal genes, the products of four in particular—*bicoid, nanos, hunchback*, and *caudal*—become distributed along the antero-posterior axis and are crucial in establishing it.

Some maternal genes exert their effect by way of expression in the follicle cells in the mother's ovary, which are derived from the mesoderm of the mother and form a bag that contains the germline-derived oocyte and nurse cells. We shall return to the crucial role of the follicle cells in patterning the oocyte later in this chapter.

2.8 The *bicoid* gene provides an antero-posterior gradient of morphogen

In the unfertilized egg, *bicoid* mRNA is localized at the anterior end. After fertilization it is translated, and the Bicoid protein diffuses from the anterior end and forms a concentration gradient along the antero-posterior axis. This provides the positional information required for further patterning along this axis. Historically, the Bicoid gradient provided the first concrete evidence for the existence of the morphogen gradients that had been postulated to control pattern formation (see Section 1.15).

The role of the *bicoid* gene was first elucidated by a combination of genetic and physical experiments on the *Drosophila* embryo. Female flies lacking *bicoid* gene expression produce embryos that have disrupted anterior segments and thus have no proper head and thorax (see Fig. 2.9). They also have a telson instead of an acron at the head end. In a separate line of investigation into the role of localized cytoplasmic factors in anterior development, normal eggs were pricked at the anterior ends and some cytoplasm allowed to leak out. The embryos that developed bore a striking resemblance to *bicoid* mutant embryos. This suggested that normal eggs have some factor(s) in the cytoplasm at their anterior end that is absent in *bicoid* mutant eggs. And indeed, *bicoid* mutant embryos can be partially rescued, in the sense that they will develop more normally, if anterior cytoplasm of wild-type embryos is injected into their anterior regions (Fig. 2.10). Moreover, if normal anterior cytoplasm is injected into the middle of a fertilized *bicoid* mutant egg, head structures develop at the site of injection and the adjacent segments become thoracic segments, setting up a mirror-image body pattern at the site of injection. The simplest interpretation of these experiments is that the *bicoid* gene is necessary for the establishment of the anterior structures because it establishes a gradient in some substance whose source and highest level are at the anterior end: this substance is the Bicoid protein.

Using *in situ* hybridization (see Box 3D, pp. 112–113) *bicoid* mRNA has been shown to be present in the anterior region of the unfertilized egg. This mRNA is not translated until after fertilization. Staining with an antibody against Bicoid protein shows that it is absent from the unfertilized egg, but that after fertilization the mRNA is translated into Bicoid protein, which forms a gradient with the high point at the anterior end of the egg, at the site of its synthesis (Fig. 2.11). As Bicoid diffuses through the embryo, it also breaks down—it has a half-life of about 30 minutes—and this breakdown is important in establishing the antero-posterior concentration gradient.

Bicoid is a transcription factor and acts as a morphogen, as will be described in more detail in Section 2.18. It switches on particular zygotic genes at different threshold concentrations, thereby initiating a new pattern of gene expression along the axis. Thus, *bicoid* is a key maternal gene in early *Drosophila* development.

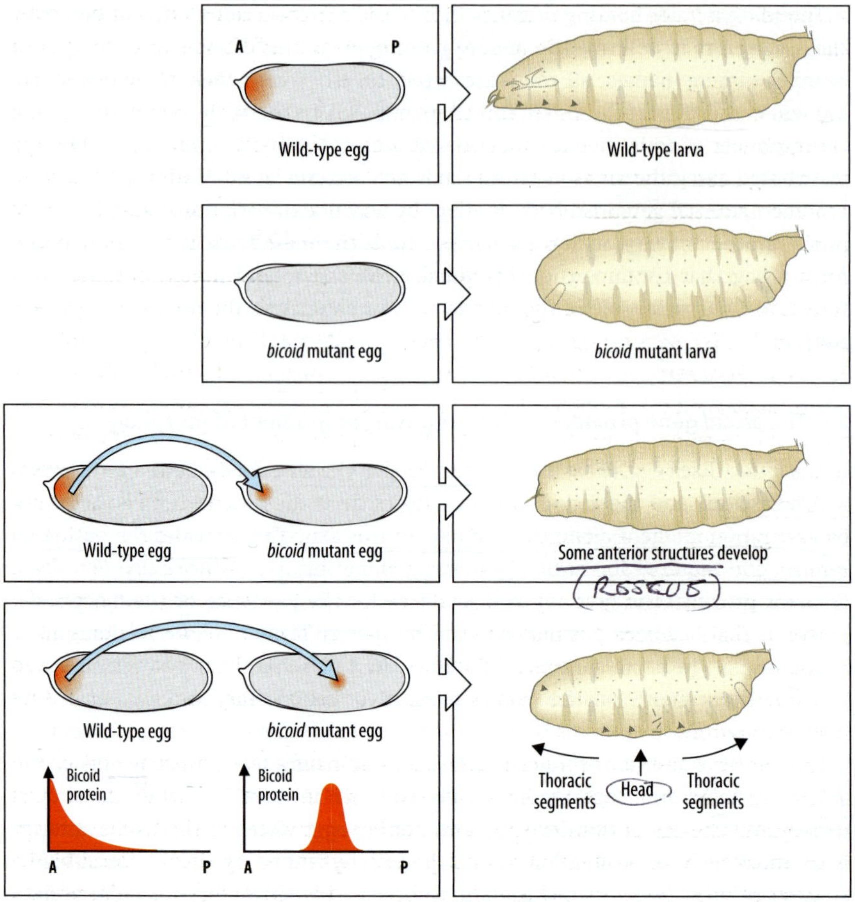

Fig. 2.10 The *bicoid* gene is necessary for the development of anterior structures. Embryos whose mothers lack the *bicoid* gene lack anterior regions (second row). Transfer of anterior cytoplasm from wild-type embryos to *bicoid* mutant embryos causes some anterior structure to develop at the site of injection (third row). If wild-type anterior cytoplasm is transplanted to the middle of a *bicoid* mutant egg or early embryo, head structures develop at the site of injection, flanked on both sides by thoracic-type segments (fourth row). These results can be interpreted in terms of the anterior cytoplasm setting up a gradient of Bicoid protein with the high point at the site of injection (see graphs, bottom left panel). A, anterior; P, posterior.

The other maternal genes of the anterior group are mainly involved in the localization of *bicoid* mRNA to the anterior end of the egg during oogenesis and in the control of its translation.

2.9 The posterior pattern is controlled by the gradients of Nanos and Caudal proteins

For proper patterning along an axis both ends need to be specified, and Bicoid defines only the anterior end of the antero-posterior axis. The posterior end is specified by the actions of at least nine maternal genes—the posterior-group genes. Just as mutations in the *bicoid* gene result in larvae in which head and thoracic regions do not develop normally, mutations in the posterior-group genes result in larvae in which abdominal development is abnormal. These mutant embryos are shorter than normal because there is no abdomen (see Fig. 2.9). One of the actions of the maternal posterior-group genes, for example *oskar*, is to localize *nanos* mRNA at the extreme posterior pole of the unfertilized egg, as well as to specify the posterior **germplasm** in the egg that gives rise to the **germ cells,** the cells that will give rise to sperm and eggs. Like *bicoid* mRNA, *nanos* mRNA is translated after fertilization to give a concentration gradient of Nanos protein, in this case with the highest level at the posterior end of the embryo.

Unlike Bicoid protein, however, Nanos does not act directly as a morphogen to specify the abdominal pattern. It has a quite different role. Its function is to suppress, in a graded way, the translation of the maternal mRNA of another gene, *hunchback*, so that a clear gradient of zygotically expressed Hunchback protein can be subsequently established and act as a morphogen for the next stage of patterning. Maternal *hunchback* mRNA is uniformly distributed in the embryo, but the embryo's own *hunchback* genes are also activated at the anterior end of the embryo by the high concentrations of Bicoid after fertilization. To establish a clear antero-posterior gradient of Hunchback protein, maternal *hunchback* translation has to be prevented, as otherwise there would be too high a concentration of Hunchback protein in the posterior region. The Nanos protein prevents maternal Hunchback translation (Fig. 2.12) by binding to a complex of *hunchback* mRNA and the protein encoded by the posterior-group gene *pumilio*. If maternal Hunchback is completely removed from embryos, then Nanos becomes completely dispensable for antero-posterior patterning, showing that in this context it is needed only to remove the maternal Hunchback protein. Evolution can only work on what is already there, it cannot 'look' at the whole system and redesign it economically. So if a gene causes a difficulty by being expressed in the wrong place, the problem may be solved not by reorganizing the expression pattern, but by bringing in a new function to remove the unwanted protein. This happens in the case of *nanos* and *hunchback*. Hunchback is made maternally so it tends to spread all over the egg, and get to the posterior region where it is not wanted. The special function of *nanos* is to remove that unwanted protein and this is all it does. So if maternal Hunchback is removed by experiment, then Nanos becomes completely unnecessary and the *nanos*$^-$ flies develop normally.

The fourth maternal product crucial to establishing the posterior end of the axis is *caudal* mRNA. Like maternal *hunchback* mRNA, it is uniformly distributed throughout the egg at first. A posterior to anterior gradient of the Caudal protein is established by inhibition of Caudal protein synthesis by Bicoid. Because the concentration of Bicoid is low at the posterior end of the embryo, Caudal protein concentration is highest there. Mutations in the *caudal* gene result in abnormal development of abdominal segments.

Soon after fertilization, therefore, several gradients of maternal proteins have been established along the antero-posterior axis. Two gradients—Bicoid and Hunchback proteins—run in an anterior to posterior direction, while Caudal protein is graded posterior to anterior. We next look at the quite different mechanism that specifies the two termini of the embryo.

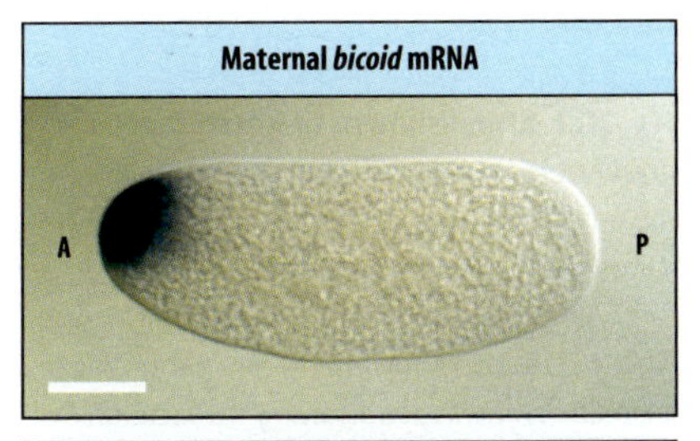

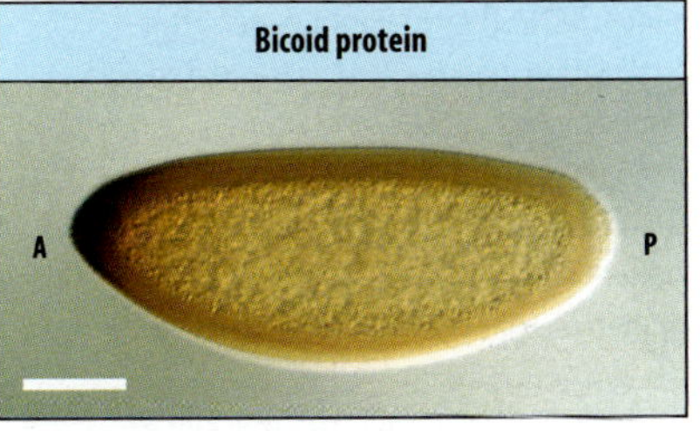

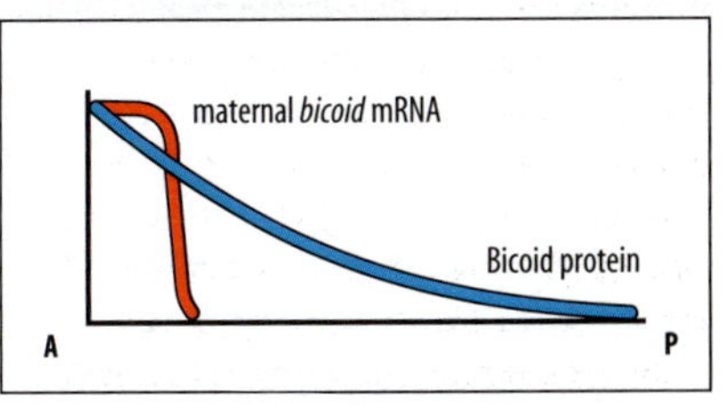

Fig. 2.11 The distribution of the maternal mRNA for *bicoid* in the egg and the gradient of Bicoid after fertilization. Top panel: the mRNA is visualized by *in situ* hybridization. Middle panel: the Bicoid protein is stained with a labeled antibody. Bottom panel: translation of *bicoid* mRNA and diffusion of Bicoid from its site of synthesis produces an antero-posterior gradient of Bicoid in the embryo. Scale bars = 0.1 mm.

Photographs courtesy of R. Lehmann, from Suzuki, D.T., et al.: 1996.

2.10 The anterior and posterior extremities of the embryo are specified by cell-surface receptor activation

A third group of maternal genes specifies the structures at the extreme ends of the antero-posterior axis—the acron and the head region at the anterior end, and the telson and the most posterior abdominal segments at the posterior end. A key gene in this group is *torso*; mutations in *torso* can result in embryos developing neither acron nor telson (see Fig. 2.9). This indicates that the two terminal regions, despite their topographical separation, are not specified independently but use the same pathway.

The terminal regions are specified by an interesting mechanism that involves the localized activation of a receptor protein that is itself present throughout the membrane of the fertilized egg. The activated receptor sends a signal to the adjacent cytoplasm that specifies it as terminal. The receptor is known as Torso, as mutations in the maternal *torso* gene produce embryos lacking terminal regions. After fertilization the maternal *torso* mRNA laid down in the oocyte is

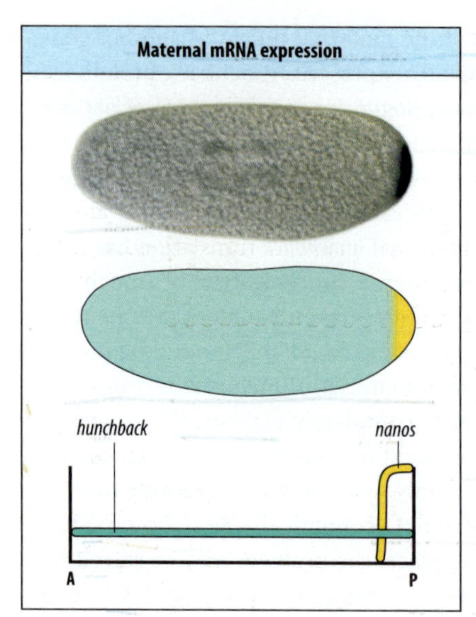

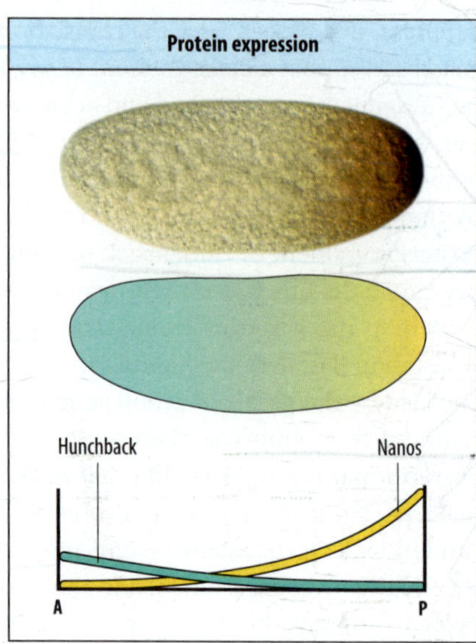

Fig. 2.12 Establishment of a maternal gradient in Hunchback protein. Left panel: in the unfertilized egg, maternal *hunchback* mRNA (turquoise) is present at a relatively low level throughout the egg, whereas *nanos* mRNA (yellow) is located posteriorly. The photograph is an *in situ* hybridization showing the location of *nanos* mRNA (black). Right panel: after fertilization, *nanos* mRNA is translated and Nanos protein blocks translation of *hunchback* mRNA in the posterior regions, giving rise to a shallow antero-posterior gradient in maternal Hunchback protein. The photograph shows the graded distribution of Nanos, detected with a labeled antibody.

Photographs courtesy of R. Lehmann, from Suzuki, D.T., et al.: 1996.

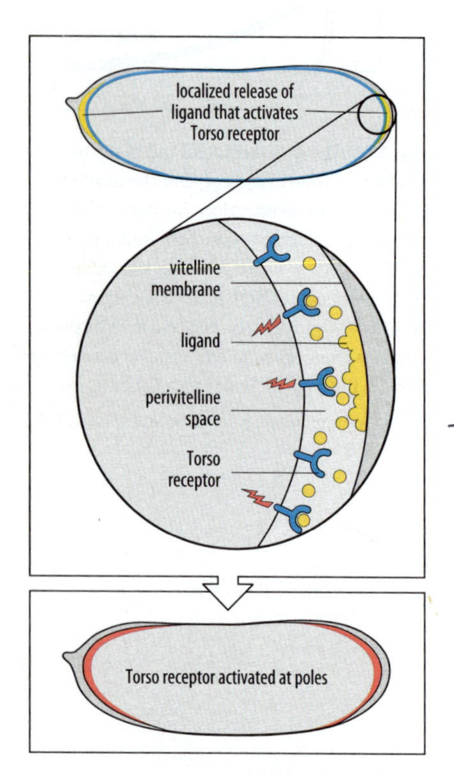

Fig. 2.13 The receptor protein Torso is involved in specifying the terminal regions of the embryo. The receptor protein encoded by the gene *torso* is present throughout the egg plasma membrane. Its ligand is laid down in the vitelline membrane at each end of the egg during oogenesis. After fertilization, the ligand is released and diffuses across the perivitelline space to activate the Torso protein at the ends of the embryo only.

translated and the Torso protein is uniformly distributed throughout the fertilized egg plasma membrane. It is, however, only activated at the ends of the fertilized egg because the protein ligand that stimulates it is only present there.

This ligand is thought to be a fragment of a protein known as Trunk, and to be localized to the poles of the fertilized egg because the protease activity that produces it from Trunk is present only in these two regions. Trunk itself is probably present throughout the perivitelline space. By the time development begins after fertilization, the ligand has been released into the perivitelline space, where it can bind to Torso. The ligand is only present in small quantities and so most becomes bound to Torso at the poles, with little left to diffuse further away. In this way, a localized area of receptor activation is set up at each pole (Fig. 2.13). Stimulation of Torso produces a signal that is transduced across the plasma membrane to the interior of the developing embryo. This signal directs the activation of zygotic genes in nuclei at both poles, thus defining the two extremities of the embryo. The Torso protein is one of a large group of transmembrane receptors known as receptor tyrosine kinases, which have an intrinsic tyrosine protein kinase in the cytoplasmic portion of the receptor. This is activated when the extracellular part of the receptor binds its ligand, and transmits the signal onwards by phosphorylating cytoplasmic proteins.

The ingenious mechanism for setting up a localized area of receptor activation is not confined to determination of the terminal regions of the embryo, but is also used in setting up the dorso-ventral axis, which we consider next.

2.11 The dorso-ventral polarity of the embryo is specified by localization of maternal proteins in the egg vitelline envelope

The dorso-ventral axis is specified by a different set of maternal genes from those that specify the anterior-posterior axis. Like the terminal regions, the ventral end of the axis is determined by the localized production in the vitelline envelope of the ligand for a receptor that is present in the membrane of the fertilized egg. The ligand is a protein fragment produced by processing of a maternal protein

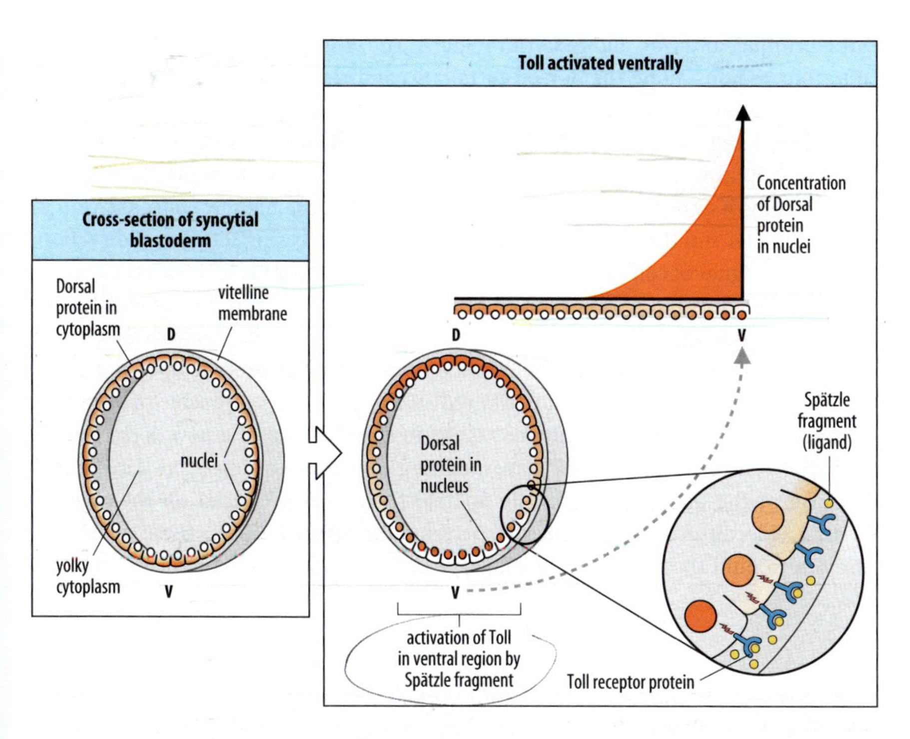

Fig. 2.14 Toll protein activation results in a gradient of intranuclear Dorsal protein along the dorso-ventral axis. Before Toll protein is activated, the Dorsal protein (red) is distributed throughout the peripheral band of cytoplasm. The Toll protein is a receptor that is only activated in the ventral region, by a maternally derived ligand (the Spätzle fragment), which is processed in the perivitelline space after fertilization. The localized activation of Toll results in the entry of Dorsal protein into nearby nuclei. The intranuclear concentration of Dorsal protein is greatest in ventral nuclei, resulting in a ventral to dorsal gradient. D, dorsal; V, ventral.

called Spätzle. After fertilization, Spätzle itself is uniformly distributed throughout the extra-embryonic perivitelline space. Its localized processing is controlled by a small set of maternal genes that are expressed only in the follicle cells that surround the future ventral region—about a third of the total surface of the developing egg. The key gene here is *pipe*, which codes for an enzyme, a heparan sulfate sulfotransferase, that is secreted into the oocyte vitelline envelope by these cells. The subsequent activity of Pipe leads, in some way not yet fully understood, to the localization of a protease activity in the vitelline envelope on the ventral side of the embryo. The processed Spätzle fragment is thus only present in the ventral perivitelline space.

The transmembrane receptor that interacts with the Spätzle fragment is called Toll. *toll* mRNA is laid down in the oocyte but is probably not translated until after fertilization. Toll is present throughout the membrane of the fertilized egg but is only activated in the future ventral region of the embryo, owing to the localization of its ligand there. Toll activation is greatest where the concentration of its ligand is highest, and falls off rapidly, probably due to the limited amount of ligand being mopped up by the receptors. Activation of Toll sends a signal to the adjacent cytoplasm of the embryo. At this stage, the embryo is still a syncytial blastoderm and this signal causes a maternal gene product in the cytoplasm—the Dorsal protein—to enter nearby nuclei (Fig. 2.14). This protein is a transcription factor with a vital role in organizing the dorso-ventral axis, which now becomes cellularized.

2.12 Positional information along the dorso-ventral axis is provided by the Dorsal protein

The initial dorso-ventral organization of the embryo is established at right angles to the antero-posterior axis at about the same time that this axis is being divided into terminal, anterior, and posterior regions. The embryo initially becomes divided into four regions along the dorso-ventral axis and this patterning is controlled by the distribution of the maternal protein Dorsal (see Fig. 2.7).

Unlike Bicoid, Dorsal is uniformly distributed in the egg. Initially it is restricted to the cytoplasm, but under the influence of signals from the ventrally activated Toll proteins it enters nuclei in a graded fashion, with the highest concentration in ventral nuclei and the concentration progressively decreasing in a dorsal direction, as the Toll signal becomes weaker (see Fig. 2.14). Thus there is little or no Dorsal in nuclei in the dorsal regions of the embryo. The role of Toll was first established by the observation that mutant embryos lacking it are strongly **dorsalized**—that is, no ventral structures develop. In these embryos, Dorsal protein does not enter nuclei but remains uniformly distributed in the cytoplasm. Transfer of wild-type cytoplasm into *Toll*-mutant embryos results in specification of a new dorso-ventral axis, the ventral region always corresponding to the site of injection. This occurs because in the absence of Toll, the Spätzle fragments produced on the original ventral side diffuse throughout the perivitelline space because there is no Toll protein to bind them. When the wild-type cytoplasm is injected, the Toll proteins

Box 2B The Toll signaling pathway: a multifunctional pathway

The interaction between the Dorsal and Cactus proteins in the Toll signaling pathway of *Drosophila* is of more than local interest: Dorsal protein is a transcription factor with homology to the Rel/NFκB family of vertebrate transcription factors, which are involved in regulation of gene expression in immune responses, and the Toll signaling pathway is also used in the adult fly in defense against infection. So what might seem at first sight a rather specialized mechanism for confining transcription factors to the cytoplasm until it is time for them to enter the nucleus is likely to be widely used for controlling gene expression and cell differentiation.

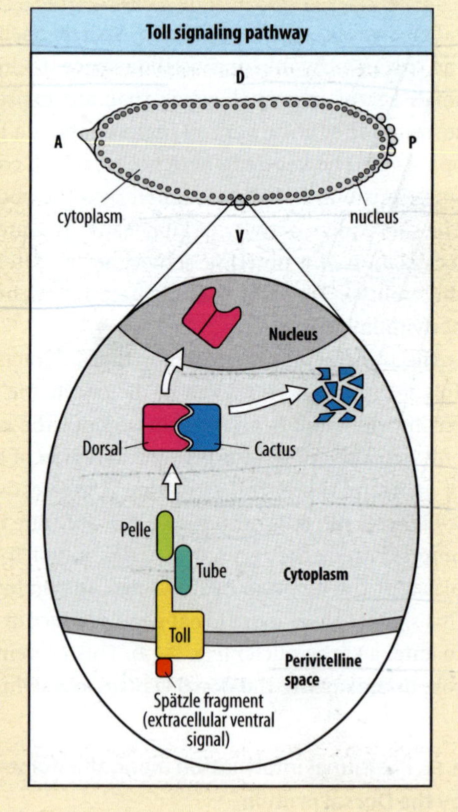

The Toll signaling pathway is thus a good example of a conserved intracellular signaling pathway that is used by multicellular organisms in different contexts, in this case ranging from embryonic development to defense against disease. All members of the Rel/NFκB family are typically held inactive in the cytoplasm until the cell is stimulated through an appropriate receptor. This leads to degradation of the inhibitory protein, which releases the transcription factor. This then enters the nucleus and activates gene transcription (see figure). In the *Drosophila* embryonic Toll pathway, Dorsal is held inactive in the syncytium cytoplasm by the protein Cactus. When Toll is activated by binding the Spätzle fragment, its cytoplasmic domain binds an adaptor protein, Tube, which in turn interacts with and activates the protein kinase Pelle. Activation of Pelle leads, through several more intermediate steps, which have not yet been fully established, to the phosphorylation and degradation of Cactus. This releases Dorsal, which is then free to enter a nucleus.

In adult *Drosophila*, the Toll receptor is stimulated by fungal and bacterial infection and the signaling pathway results in the production of antimicrobial peptides. In humans, Toll-like receptors acting by essentially the same pathway are also involved in innate immunity to microbial infection. MyD88, IRAK, and IκB are the mammalian homologs of Tube, Pelle, and Cactus, respectively, and play the same roles in this pathway. In vertebrates, NFκB is also activated in response to signaling through receptors other than Toll.

it contains enter the membrane at the site of the injection. Spätzle fragments then bind to these receptors, setting in motion the chain of events that defines the ventral region at the site of injection.

In the absence of a signal from Toll protein, Dorsal protein is prevented from entering nuclei by being bound in the cytoplasm to another maternal gene product, the Cactus protein. As a result of Toll activation, Cactus is degraded and no longer binds Dorsal, which is then free to enter the nuclei. The pathway leading from Toll to the activation of Dorsal is shown in Box 2B (opposite). In embryos lacking Cactus protein, almost all of the Dorsal protein is found in the nuclei; there is a very poor concentration gradient and the embryos are **ventralized**—that is, no dorsal structures develop.

Summary

Maternal genes act in the ovary of the mother fly to set up differences in the egg in the form of localized deposits of mRNAs and proteins. After fertilization, maternal mRNAs are translated and provide the embryonic nuclei with positional information in the form of protein gradients or localized protein. Along the antero-posterior axis there is an anterior to posterior gradient of maternal Bicoid protein, which controls patterning of the anterior region. For normal development it is essential that maternal Hunchback protein is absent from the posterior region and its suppression is the function of the posterior to anterior gradient of Nanos. The extremities of the embryo are specified by localized activation of the receptor protein Torso at the poles. The dorso-ventral axis is established by intranuclear localization of the Dorsal protein in a graded manner (ventral to dorsal), as a result of ventrally localized activation of the receptor protein Toll by a fragment of the protein Spätzle.

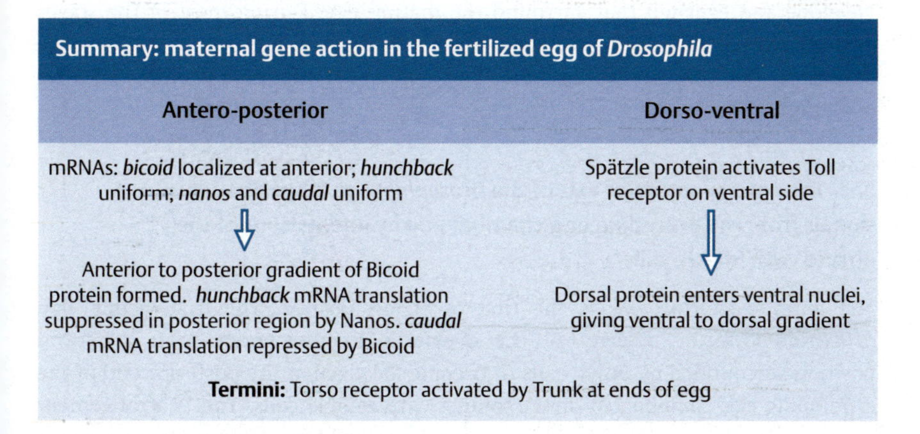

Localization of maternal determinants during oogenesis

Having considered the importance of localized maternal gene products in the egg in setting the basic framework for future development, we now look at how they come to be localized so precisely. When the *Drosophila* egg is released from the ovary it already has a well defined organization. *bicoid* mRNA is located at the anterior end and *nanos* mRNA at the opposite end. Torso-like protein is present in the vitelline envelope at both poles, and other maternal proteins are localized in the

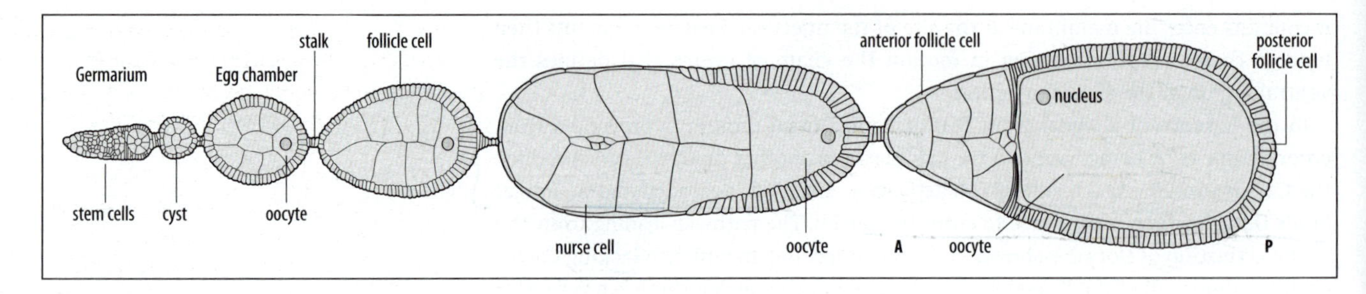

Fig. 2.15 Egg development in *Drosophila*. Oocyte development begins in a germarium, with stem cells at one end. One stem cell will divide four times to give 16 cells with cytoplasmic connections between each other. One of the cells that is connected to four others will become the oocyte, the others will become nurse cells. The nurse cells and oocyte become surrounded by follicle cells and the resulting structure buds off from the germarium as an egg chamber. Successively produced egg chambers are still attached to each other at the poles. The oocyte grows as the nurse cells provide material through the cytoplasmic bridges. The follicle cells have a key role in patterning the oocyte.

ventral vitelline envelope. Numerous other maternal mRNAs, such as those for Caudal, Hunchback, Toll, Torso, Dorsal, and Cactus, are distributed uniformly. How do these maternal mRNAs and proteins get laid down in the egg during **oogenesis**—its period of development in the ovary—and how are they localized in the correct places?

The development of an egg in the *Drosophila* ovary is shown in Fig. 2.15. A diploid germline stem cell in the **germarium** divides asymmetrically to produce another stem cell and a cell called the cystoblast, which undergoes a further four mitotic divisions to give 16 cells with cytoplasmic bridges between them; this group of cells is known as the germline cyst. One of these 16 cells will become the **oocyte**; the other 15 will develop into **nurse cells**, which produce large quantities of proteins and RNAs that are exported into the oocyte through the cytoplasmic bridges. Somatic ovarian cells make up a sheath of **follicle cells** around the nurse cells and oocyte to form the **egg chamber**. The follicle cells have a key role in patterning the egg's axes. During oogenesis, they become subdivided into functionally different populations at various locations around the oocyte; these subpopulations express different genes and thus have differing effects on the parts of the oocyte adjacent to them (Fig. 2.16). Follicle cells also secrete the materials of the vitelline envelope and eggshell that surround the mature egg. During most of the stages of development discussed here the oocyte is arrested in the first prophase stage of meiosis (meiosis will be described in Chapter 11). Meiosis is completed after ovulation.

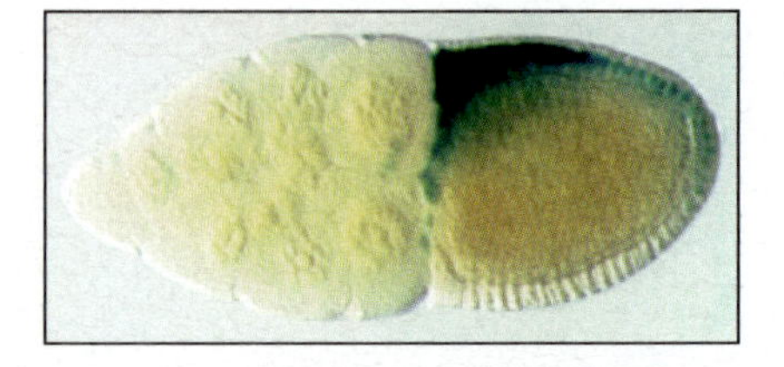

Fig. 2.16 Drosophila oocyte development. A developing *Drosophila* oocyte (right) is shown attached to its 15 nurse cells (left) and surrounded by a monolayer of 700 follicle cells. The oocyte and follicle layer are cooperating at this time to define the future dorso-ventral axis of the egg and embryo, as indicated by the expression of a gene only in the follicle cells overlying the dorsal anterior region of the oocyte (blue staining).

Photograph courtesy of A. Spradling.

2.13 The antero-posterior axis of the *Drosophila* egg is specified by signals from the preceding egg chamber and by interactions of the oocyte with follicle cells

The antero-posterior axis is the first to be established. The first visible sign of antero-posterior polarity in the oocyte is its movement from a central position surrounded by nurse cells to become localized at the posterior end of the developing egg chamber, in direct contact with follicle cells. This rearrangement occurs while the egg chamber is becoming separated from the germarium, and is due to preferential adhesion between the future posterior end of the oocyte and the adjacent posterior follicle cells. This polarity is the result of signaling from the anterior of the older egg chamber to the posterior of the younger chamber (Fig. 2.17). Two signaling pathways are involved and act in relay to convey a signal from the older egg chamber to the younger one. The germline cyst in the older chamber first signals to its anterior follicle cells via a widely used signaling pathway, the Delta–Notch pathway, which will be described in detail later in the book (see Fig. 4.6). This signaling specifies several of the follicle cells to become specialized anterior polar follicle cells. These in turn signal to adjacent follicle cells via

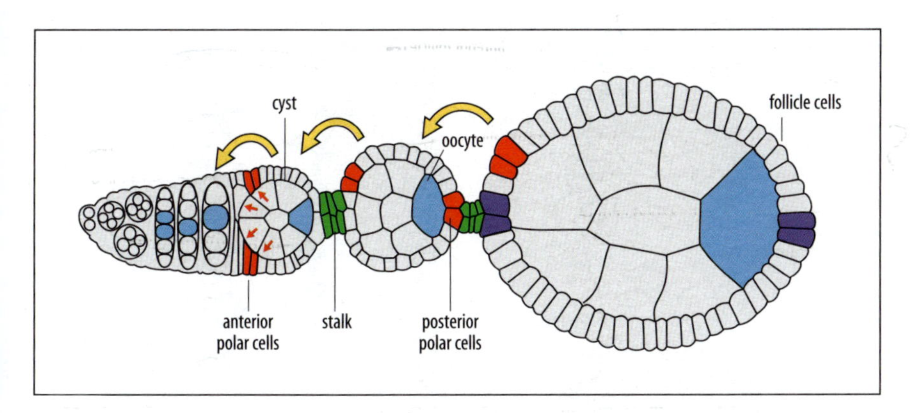

Fig. 2.17 Signals from older to younger egg chambers initially polarize the *Drosophila* oocyte. As a germline cyst buds from the germarium, it signals through the Delta–Notch pathway (small red arrows) to induce the formation of anterior polar follicle cells (red). These in turn signal to the adjacent cells anterior to them and induce them to become stalk (green). The stalk induces the adjacent younger cyst to round up and the oocyte and the posterior follicle cells to produce cadherin, which positions the oocyte at the posterior of the egg chamber. The yellow arrows indicate the overall direction of signaling from older to younger egg chambers.

receptors that stimulate another intracellular signaling pathway called the JAK–STAT pathway, and induce them to form the stalk between the two egg chambers. Signals, as yet unknown, from the stalk cells then cause the younger egg chamber to round up and cause the oocyte and the posterior follicle cells in this chamber to express the adhesion molecule E-cadherin, which anchors the oocyte in a posterior position. (The way in which cadherins and other adhesion molecules work is explained in Box 7A; p. 259.) Thus, an antero-posterior polarity is propagated from one egg chamber to the next. How the very first egg chamber produced gets polarized is, however, not yet understood.

Once the oocyte has become located posteriorly, the next steps in its antero-posterior polarization are mediated by a protein called Gurken, which is a member of the transforming growth factor-α (TGF-α) family of cytokines (examples of the most commonly used growth factor families in *Drosophila* are given in Fig. 2.18). Early in oocyte development *gurken* mRNA is translated at the posterior end of the oocyte, producing a local posterior concentration of the protein, which is secreted

Common intercellular signals used in *Drosophila*		
Family and examples	**Receptors**	**Examples of roles in development**
Hedgehog family		
Hedgehog	Patched	Patterning of insect segments Positional signaling in insect leg and wing discs
Wingless (Wnt) family		
Wingless and other Wnt proteins	Frizzled	Insect segment and imaginal disc specification (Wingless) Other Wnts have roles in development
Delta and Serrate		
Transmembrane signaling proteins	Notch	Roles at many stages in development Specification of oocyte polarity
Transforming growth factor-α (TGF-α) family		
Gurken, Spitz, Vein	EGF receptor (receptor tyrosine kinase) One only in *Drosophila*, known as DER or Torpedo.	Polarization of the oocyte Eye development, wing vein differentiation (see Chapter 9)
Transforming growth factor-β (TGF-β) family		
Decapentaplegic	Receptors act as heterodimers of type I (e.g. Thick veins) and type II (e.g. Punt) subunits Serine/threonine kinases	Patterning of the dorso-ventral axis and imaginal discs
Fibroblast growth factor (FGF) family		
A small number of FGF homologs (e.g. Branchless)	FGF receptors (receptor tyrosine kinases) Two in *Drosophila*, e.g. Breathless	Migration of tracheal cells (see Chapter 9)

Fig. 2.18 Common intercellular signals in *Drosophila*.

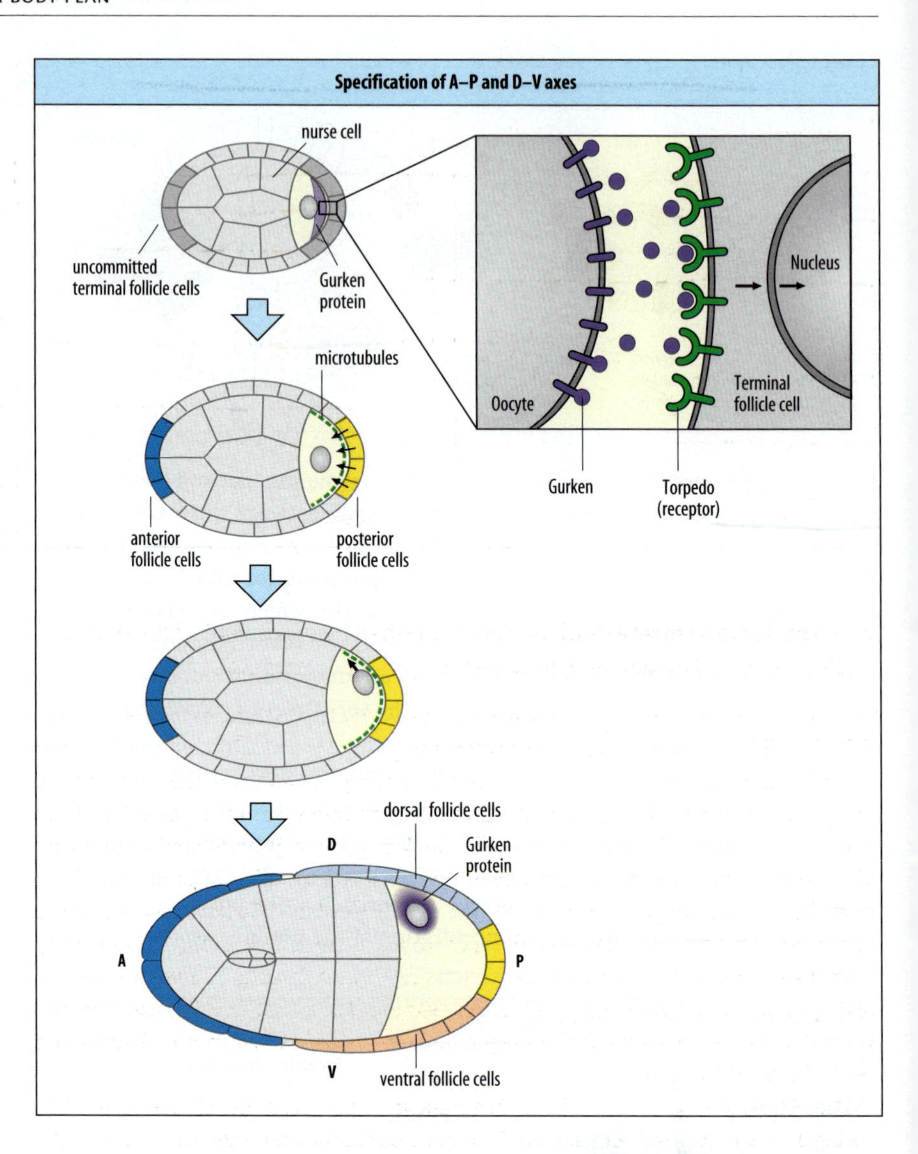

Fig. 2.19 Specification of the antero-posterior and dorso-ventral axes during *Drosophila* oogenesis. The oocyte moves to the posterior end of the egg chamber and comes into contact with the polar follicle cells. It is separated from follicle cells at the anterior end (blue) by the nurse cells. *gurken* mRNA in the oocyte is localized to the posterior. It is translated and Gurken protein is secreted locally. The binding of this protein to the receptor protein Torpedo on the adjacent follicle cells initiates their specification as posterior terminal follicle cells (yellow). These send a signal back to the oocyte that reorganizes the oocyte microtubule cytoskeleton. The oocyte nucleus now moves to an anterior dorsal position and *gurken* mRNA coming into the oocyte from the nurse cells is transported to the region surrounding the nucleus. Translation of the mRNA and local release of Gurken protein specifies the adjacent follicle cells as dorsal follicle cells and that side of the oocyte as the future dorsal side.

across the oocyte membrane. Gurken induces the terminal follicle cells to adopt a posterior fate by locally stimulating the receptor protein Torpedo, which is present on the surface of follicle cells (Fig. 2.19). Torpedo is a receptor tyrosine kinase and is the *Drosophila* equivalent of the epidermal growth factor (EGF) receptor of mammals. In response to Gurken signaling through Torpedo, the posterior follicle cells produce an as yet unidentified signal that induces a repolarization of the oocyte's microtubule cytoskeleton, which localizes *bicoid* mRNA and *oskar* mRNA to their final sites in the egg (Fig. 2.20).

bicoid mRNA is originally made by nurse cells located next to the anterior end of the developing oocyte, and is transferred from them to the egg. The anterior localization of *bicoid* mRNA is thought to involve its transport along microtubules. Similarly, *oskar* mRNA is delivered into the oocyte by nurse cells and moved to the posterior end through its interaction with the microtubule array. One role of *oskar* mRNA and protein is to nucleate the assembly of the **pole plasm** at the posterior end of the egg. This cytoplasm contains the germline determinants that direct the formation of the primordial germ cells at the posterior of the embryo. The localization of *nanos* mRNA to the posterior end (see Section 2.9) depends on the earlier localization of *oskar* mRNA and the assembly of the pole plasm.

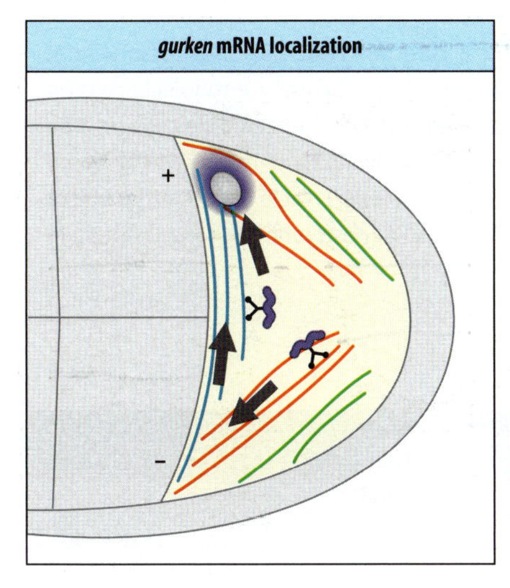

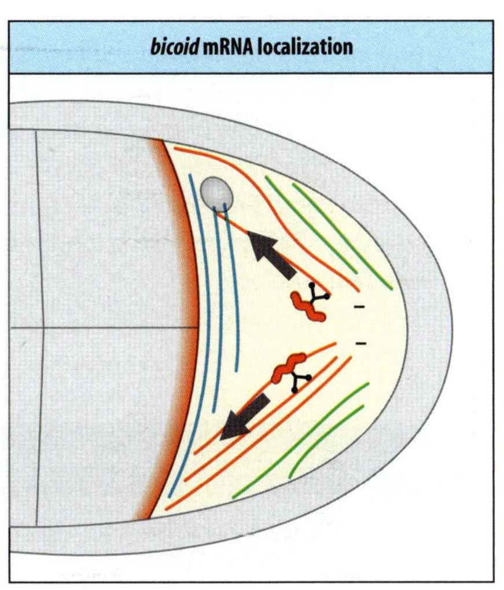

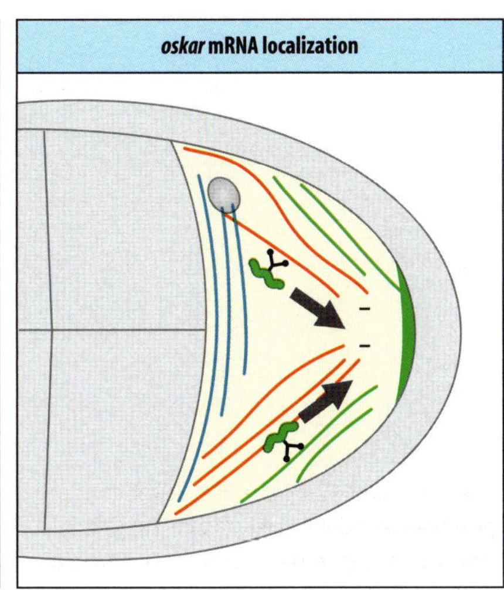

Fig. 2.20 ***bicoid*** **and** ***oskar*** **mRNAs are localized to the anterior and posterior ends of the oocyte respectively.** Localization of maternal mRNAs delivered to the oocyte by the nurse cells is by transport along microtubules. The motor protein dynein transports *gurken* and *bicoid* mRNAs towards the plus ends of the microtubules. *oskar* mRNA is transported towards the minus ends of the microtubules, possibly by the motor protein kinesin. Three different populations of microtubules are thought to be involved in the transport of *gurken*, *bicoid*, and *oskar* mRNAs.

2.14 The dorso-ventral axis of the egg is specified by movement of the oocyte nucleus followed by signaling between oocyte and follicle cells

The setting-up of the egg's dorso-ventral axis involves a further set of oocyte-follicle cell interactions, which occur after the posterior end of the oocyte has been specified and which depend on the previous reorganization of the microtubule array. The oocyte nucleus moves along the microtubules from the posterior of the oocyte to a site on the anterior margin. Gurken protein is expressed at this new site, possibly from mRNA that has been relocated to one side of the nucleus from elsewhere in the oocyte (see Fig. 2.19). The locally produced Gurken acts as a signal to the adjacent follicle cells, specifying them as dorsal follicle cells; the side away from the nucleus thus becomes the ventral region by default. The ventral follicle cells produce proteins, such as Pipe (see Section 2.11), that are deposited in the ventral vitelline envelope of the oocyte and are instrumental in establishing the ventral side of the axis.

The Torso-like protein, which distinguishes the termini of the *Drosophila* embryo, is synthesized and secreted by both posterior and anterior follicle cells, but not by the other follicle cells. It is thus only deposited in the vitelline envelope at both ends of the egg during oogenesis. After fertilization it will act together with other proteins to process Trunk and produce the ligand for Torso (see Section 2.10).

Summary

Drosophila oocytes develop inside individual egg chambers that are successively produced from a germarium, which contains germline stem cells, which give rise to the oocyte and nurse cells, and stem cells that give rise to the somatic follicle cells that surround the oocyte. The nurse cells provide the oocyte with large amounts of mRNAs and proteins, some of which become localized in particular sites. As a result of signals from the adjacent older egg chamber, an oocyte becomes localized posteriorly in its egg chamber as a result of differential cell adhesion to posterior follicle cells. Subsequently the oocyte sends a signal to these follicle cells, which respond with a signal that causes a reorganization of the oocyte cytoskeleton that localizes *bicoid* mRNA to the anterior end and other mRNAs to the posterior end of the oocyte, thus setting up the beginnings of the embryonic antero-posterior axis. The dorso-ventral axis of the oocyte is also initiated by a local signal from the oocyte to follicle cells on the future dorsal side of the egg, thus specifying them as dorsal follicle cells. Follicle cells on the opposite side of the oocyte specify the ventral side of the oocyte by deposition of maternal proteins in the ventral vitelline envelope. Follicle cells at either end of the oocyte similarly specify the termini by localized deposition of maternal protein in the vitelline envelope.

Summary: polarization of axes in the *Drosophila* oocyte

Antero-posterior		Dorso-ventral
oocyte localized at posterior end of follicle by cadherin		
⇩		
oocyte Gurken protein induces posterior follicle cells via Torpedo		
⇩		
posterior signal from follicle cells reorganizes oocyte cytoskeleton	⇨	nucleus moves dorsally
⇩		⇩
bicoid mRNA localized in anterior, *oskar* and other mRNAs in posterior		oocyte Gurken induces dorsal follicle cells
		⇩
		ventral follicle cells deposit ventral proteins in oocyte viteline envelope

Termini: follicle cells at both ends of the egg deposit ligand for the Torso protein in the vitelline envelope

Patterning the early embryo

Understanding in such detail how the main body axes of *Drosophila* are specified was a major achievement, and those who work on other animals like the frog and chick, of which less is known about developmental genetics, are justifiably somewhat envious. We have seen how gradients of Bicoid, Hunchback, and Caudal proteins are established along the antero-posterior axis, and how intranuclear Dorsal protein is graded along the ventral to dorsal axis. This maternally derived framework of positional information is interpreted and elaborated on by zygotic genes to give each region of the embryo an identity. Most of the zygotic genes that are the first to be activated along the antero-posterior and dorso-ventral axes encode transcription factors, which are thus localized along the axes and activate yet more zygotic genes. We first consider the patterning along the dorso-ventral axis, which is somewhat simpler than that along the antero-posterior axis.

2.15 The expression of zygotic genes along the dorso-ventral axis is controlled by Dorsal protein

After Dorsal protein has entered the nuclei, its effects on gene expression divide the dorso-ventral axis into well defined regions. This is also the stage at which the

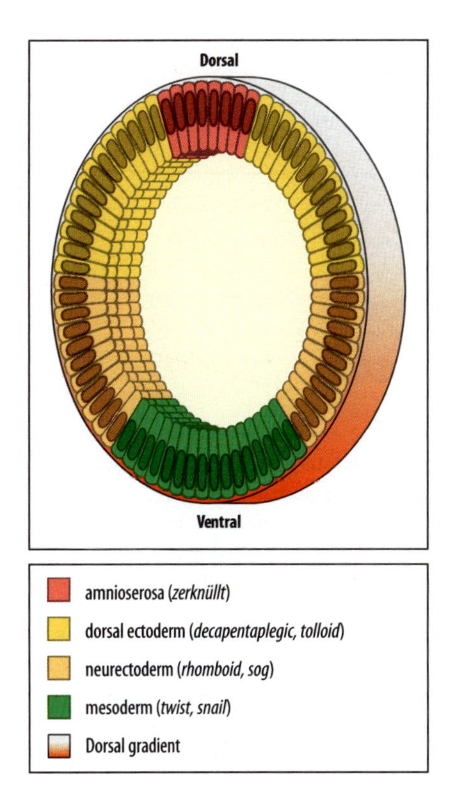

Fig. 2.21 The subdivision of the dorso-ventral axis into different regions by the gradient in intranuclear Dorsal protein. A gradient of intranuclear Dorsal protein is formed with its high point at the ventral midline and little or no Dorsal protein in the dorsal half of the blastoderm. In the dorsal region, the genes *tolloid*, *zerknüllt*, and *decapentaplegic*, which are repressed elsewhere by Dorsal protein, are expressed. In the ventral half of the blastoderm, the Dorsal protein activates the genes *twist*, *snail*, *rhomboid*, and *short gastrulation* (*sog*). *twist* is autoregulatory, maintaining its own expression, and also activates *snail*; Snail protein inhibits *rhomboid* and *sog* expression, repressing these genes in the future mesoderm. *twist* and *snail* require high levels of Dorsal for their activation whereas *rhomboid* and *sog* can still be activated when lower levels of Dorsal are present.

germ layers become distinguished, as Dorsal also acts to specify the ventral-most cells as prospective mesoderm. It is estimated that 30 genes are direct targets of the Dorsal gradient. Going from ventral to dorsal, the main regions are mesoderm, ventral ectoderm, dorsal ectoderm, and prospective amnioserosa. The mesoderm gives rise to internal soft tissues such as muscle and connective tissue; the ventral ectoderm becomes the neurectoderm, which gives rise to all the nervous tissue as well as ventral epidermis; the dorsal ectoderm gives rise only to epidermis. The third germ layer, the endoderm, which is located at either end of the embryo, and which we do not consider here, gives rise to the mid-gut (see Fig. 2.3).

Patterning along the axes poses a problem like that of patterning the French flag (see Section 1.15). Expression of zygotic genes in localized regions along the dorso-ventral axis is initially controlled by the graded concentration of the intranuclear Dorsal protein, which falls off rapidly in the dorsal half of the embryo; little Dorsal protein is found in nuclei above the equator. In the ventral region, Dorsal protein has two main functions—it activates certain genes at specific positions in the ventral region and represses the activity of other genes, which are therefore only expressed in the dorsal region (Fig. 2.21).

In the ventral-most region, where concentrations of intranuclear Dorsal protein are highest, the zygotic genes *twist* and *snail* are activated by Dorsal protein in a strip of nuclei along the ventral middle line of the embryo; soon after this the blastoderm becomes cellular. This ventral strip of cells will form the mesoderm. The expression of *twist* and *snail* is required both for development of the cells as mesoderm and for gastrulation, during which the ventral band of prospective mesoderm cells moves into the interior of the embryo. In the future neurectoderm, the gene *rhomboid* is activated at low levels of Dorsal protein, but is not expressed in more ventral regions because it is repressed there by the Snail protein.

The genes *decapentaplegic*, *tolloid*, and *zerknüllt* are repressed by Dorsal protein and so their activity is confined to the more dorsal regions of the embryo, where there is virtually no Dorsal protein in the nuclei. *zerknüllt* is expressed most dorsally and appears to specify the amnioserosa. *decapentaplegic* is a key gene in the specification of pattern in the dorsal part of the dorso-ventral axis and its role is considered in detail in Section 2.16.

Mutations in the maternal dorso-ventral genes can cause dorsalization or ventralization of the embryo (see Section 2.12). In dorsalized embryos, Dorsal protein is excluded uniformly from the nuclei. This has a number of effects, one of which is that the *decapentaplegic* gene is no longer repressed and is expressed everywhere. In contrast, *twist* and *snail* are not expressed at all in dorsalized embryos, as they need high intranuclear levels of Dorsal protein to be activated. The opposite result is obtained in mutant embryos where the Dorsal protein is present at high concentration in all the nuclei, and the embryos are ventralized; *twist* and *snail* are expressed throughout and *decapentaplegic* is not expressed at all (Fig. 2.22).

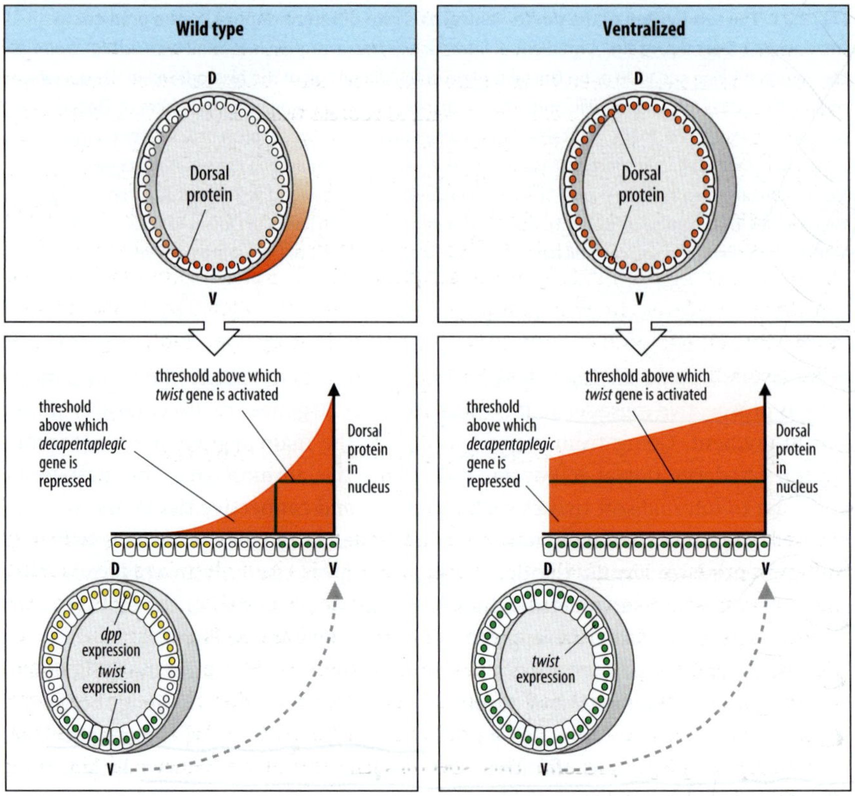

Fig. 2.22 The nuclear gradient in Dorsal protein is interpreted by the activation of other genes, such as *twist* and *decapentaplegic*. Left panels: in normal larvae, the *twist* gene is activated above a certain threshold concentration (green line) of Dorsal protein, whereas above a lower threshold (yellow line), the *decapentaplegic* (*dpp*) gene is repressed. Right panels: in ventralized larvae, the Dorsal protein is present in all nuclei; *twist* is now also expressed everywhere, whereas *decapentaplegic* is not expressed at all, because Dorsal protein is above the threshold level required to repress it everywhere.

Genes whose expression is regulated by the Dorsal protein, such as *twist*, *snail*, and *decapentaplegic*, contain binding sites for the protein in their regulatory DNA that activate or repress gene expression at particular concentrations of the Dorsal protein. This **threshold effect** on gene expression is the result of the integrating function of these binding sites, as discussed in Section 1.11. The ability of genes to respond in a threshold-like manner to varying concentrations of Dorsal protein is due to the presence of both high- and low-affinity binding sites for the protein in their regulatory DNA. In the most ventral part of the embryo (a strip 12–14 cells wide), where the concentration of Dorsal protein is high, the spatial extent of gene expression is delimited by the low-affinity sites, whereas high-affinity sites control expression in slightly more dorsal regions (up to 20 cells from the ventral midline) where there is less Dorsal protein. It is very likely that the threshold response involves cooperativity between the different binding sites; that is, binding at one site makes binding at a nearby site easier and so facilitates further binding. Inhibitory interactions are also involved in delimiting areas of gene expression. For example, the regulatory regions of the *rhomboid* gene contain binding sites for both Dorsal and Snail proteins; Dorsal activates while Snail represses. Snail protein thus represses the expression of *rhomboid* in the ventral region and this helps confine *rhomboid* expression to the neurectoderm.

An approximately twofold difference in the level of Dorsal determines whether an unspecified embryonic cell forms mesoderm or neurectoderm. Five different thresholds in Dorsal concentration pattern the future ventral midline and the neurectoderm. The neurectoderm, for example, is subsequently subdivided into three layers along the dorso-ventral axis, which will form three distinct columns in the future nerve cord (we shall return to this topic in Chapter 10). This subdivision is primarily the result of activation of genes for three different transcription factors at distinct thresholds of the Dorsal gradient, and the pattern is maintained by regulatory interactions between these genes and with other genes, in which those genes expressed in the more ventral regions tend to repress those expressed more dorsally.

The gradient of Dorsal protein is therefore effectively acting as a morphogen gradient along the dorso-ventral axis, activating specific genes at different threshold concentrations, and so defining the dorso-ventral pattern. The regulatory sequences in these genes can be thought of as developmental switches, which when thrown by the binding of transcription factors activate genes and set cells off along new developmental pathways. The Dorsal protein gradient is one solution to the French flag problem. But it is not the whole story—yet another gradient is also involved.

2.16 The Decapentaplegic protein acts as a morphogen to pattern the dorsal region

As with the antero-posterior axis, each end of the dorso-ventral axis is specified by different proteins. The gradient of Dorsal protein, with its high point in the ventral-most region, specifies the initial pattern of zygotic gene activity, and patterns the ventral mesoderm and neurectoderm. But the dorsal region is not similarly specified by a low-level gradient in Dorsal. Indeed, there is little or no Dorsal protein in the nuclei of the dorsal half of the embryo. The more dorsal part of the dorso-ventral pattern is determined by a gradient in the activity of another morphogen, the Decapentaplegic protein. This specifies the dorsal ectoderm and the most dorsal region, the amnioserosa.

Soon after the gradient of intranuclear Dorsal protein has become established the embryo becomes cellular, and transcription factors can no longer diffuse between nuclei. Secreted or transmembrane proteins and their corresponding receptors must now be used to transmit developmental signals between cells. Decapentaplegic is one such secreted signaling protein and is a member of the TGF-β family of cytokines (see Fig. 2.18). As we shall see in Chapter 3, Decapentaplegic is a homolog of bone morphogenetic protein-4 (BMP-4), a TGF-β cytokine in vertebrates that is also involved in patterning the dorso-ventral axis. Decapentaplegic is also involved in many other developmental processes throughout *Drosophila* development, including the patterning of the wing imaginal discs, which is discussed in Chapter 9.

The *decapentaplegic* (*dpp*) gene is expressed throughout the dorsal region where Dorsal protein is not present in the nuclei. Evidence that a gradient in Decapentaplegic protein activity specifies dorsal pattern comes from experiments in which *decapentaplegic* mRNA is introduced into an early wild-type embryo. As more mRNA is introduced and the concentration of Decapentaplegic protein increases above the normal level, the cells along the dorso-ventral axis adopt a more dorsal fate than they would normally. Ventral ectoderm becomes dorsal ectoderm and, at very high concentrations of *decapentaplegic* mRNA all the ectoderm develops as the amnioserosa.

Decapentaplegic is initially produced uniformly throughout the dorsal region just as cellularization is beginning, but within less than an hour its activity appears to be restricted to a dorsal strip of some five to seven cells and is much lower in the adjacent prospective dorsal ectoderm (Fig. 2.23). This sharp peak of Decapentaplegic is not a matter of simple diffusion, however; instead it illustrates how a gradient in activity can be derived from a fairly uniform initial distribution of the protein by interactions with other proteins. In addition, Decapentaplegic activity is also thought to be modified by interactions with different forms of its receptors, further refining the activity gradient.

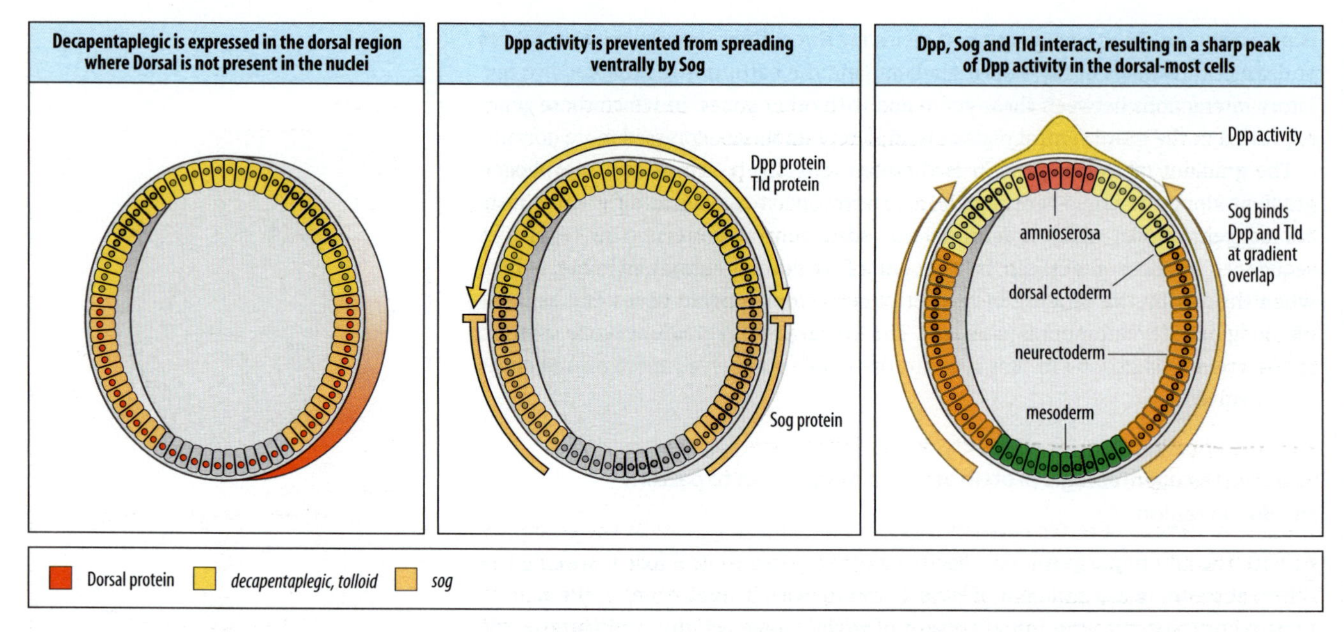

Fig. 2.23 Decapentaplegic protein activity is restricted to the dorsal-most region of the embryo by the antagonistic activity of the Short gastrulation protein. Left panel: the gradient of intranuclear Dorsal protein leads to the expression of the gene *short gastrulation* (*sog*) in the non-neural ventral ectoderm of the early blastoderm. *decapentaplegic* and *tolloid* are expressed throughout the dorsal ectoderm. Middle panel: *sog* encodes a secreted protein, Sog, that forms a ventral to dorsal gradient, while the secreted Decapentaplegic protein (Dpp) is initially present throughout the dorsal region. The protease Tolloid (Tld) is also expressed in the same region as Dpp. Where Sog meets Dpp, it binds it and thus prevents Dpp interacting with its receptors. This prevents Dpp signaling from spreading ventrally into the neurectoderm. Right panel: Dpp bound to Sog is also thought to be carried towards the dorsal region by the developing Sog gradient, thus helping to concentrate its activity in a sharp peak in the dorsal-most region. Tld is also thought to be involved in creating the final sharp dorsal peak of Dpp activity. It binds to the Sog–Dorsal complex and cleaves Sog. This releases Dorsal, which can then bind to its receptors. In the mid-blastoderm embryo the dorsal region has been subdivided by Sog activity into a region of high Dpp signaling, which will become the amnioserosa, and a zone of lower Dpp signaling, which is the dorsal ectoderm. After Ashe and Levine: 1999.

The mechanism that generates the sharp transition between the most dorsal ectoderm cells and those just lateral to it involves the proteins Short gastrulation (Sog), Twisted gastrulation (Tsg), and Tolloid. Sog and Tsg are BMP-related proteins that can bind Decapentaplegic and thus prevent it binding to its receptors, and in this sense are inhibitory. Tolloid is a metalloproteinase that degrades Sog when it is bound to Decapentaplegic, thus releasing Decapentaplegic.

Sog is expressed throughout the neurectoderm and binds Decapentaplegic, preventing its activity spreading into this region. Sog protein in turn is degraded by Tolloid, which is expressed throughout the dorsal region; this sets up a gradient in Sog with a high point in the neurectoderm and a low point at the dorsal midline and ensures that Sog, along with its bound Decapentaplegic, will be shuttled towards the dorsal region. Degradation of Sog or Tsg by Tolloid releases free Decapentaplegic, which helps to generate a gradient of Decapentaplegic activity with its high point in the dorsal-most region (see Fig. 2.23). Another factor proposed to sharpen the gradient is the rapid internalization and degradation of Decapentaplegic when it binds to its receptors. Decapentaplegic activity is also affected by its synergistic action with another member of the TGF-β family, Screw, which forms heterodimers with Decapentaplegic that signal more strongly than homodimers of Decapentaplegic or Screw alone; these are likely to be responsible for most of the Decapentaplegic activity. Experimental observations and mathematical modeling suggest that heterodimers are formed preferentially in the dorsal-most region, which also helps explain the high Decapentaplegic activity here.

Homodimers of Decapentaplegic and of Screw are responsible for lower levels of signaling elsewhere in the prospective dorsal ectoderm. Despite much work on this system however, the contributions of all these components to generating the gradient in Decapentaplegic activity is not completely understood.

As we shall see in Chapter 3, the Decapentaplegic/Sog pairing has an exact counterpart in vertebrates, the BMP-4/Chordin partnership (Chordin is the vertebrate homolog of Sog), which is also involved in patterning the dorso-ventral axis. The nerve cord runs dorsally in vertebrates, however, not ventrally as in insects, and thus the dorso-ventral axis is reversed in vertebrates compared to insects; we shall see in Chapter 3 how this is reflected in a reversal of the pattern of activity of BMP-4 and Chordin.

2.17 The antero-posterior axis is divided up into broad regions by gap-gene expression

We now return to the antero-posterior axis at an earlier stage, while the embryo is still acellular. The **gap genes** are the first zygotic genes to be expressed along the antero-posterior axis, and all code for transcription factors. Their expression is initiated by the antero-posterior gradient of Bicoid protein while the embryo is still essentially a single multinucleate cell. Bicoid primarily activates anterior expression of the gap gene *hunchback*, which in turn is instrumental in switching on the expression of the other gap genes, among which are *giant*, *Krüppel*, and *knirps*, which are expressed in this sequence along the antero-posterior axis (Fig. 2.24). (*giant* is in fact expressed in two bands, one anterior and one posterior, but its posterior expression does not concern us here.)

Gap genes were initially recognized by their mutant phenotypes, in which quite large sections of the body pattern along the antero-posterior axis are missing. Although the mutant phenotype of a gap gene usually shows a gap in the antero-posterior pattern in more-or-less the region in which the gene is normally expressed, there are also more wide-ranging effects. This is because gap-gene expression is also essential for later development along the axis.

As the blastoderm is still acellular at the stage at which the gap genes are expressed, the gap-gene proteins can diffuse away from their site of synthesis. They are short-lived proteins with half-lives of minutes. Their distribution therefore extends only slightly beyond the region in which the gene is expressed, and this typically gives a bell-shaped protein concentration profile. *hunchback* is exceptional as it is expressed over a broad anterior region and so sets up a steep antero-posterior gradient. The control of zygotic *hunchback* expression by Bicoid is best understood and will be considered first.

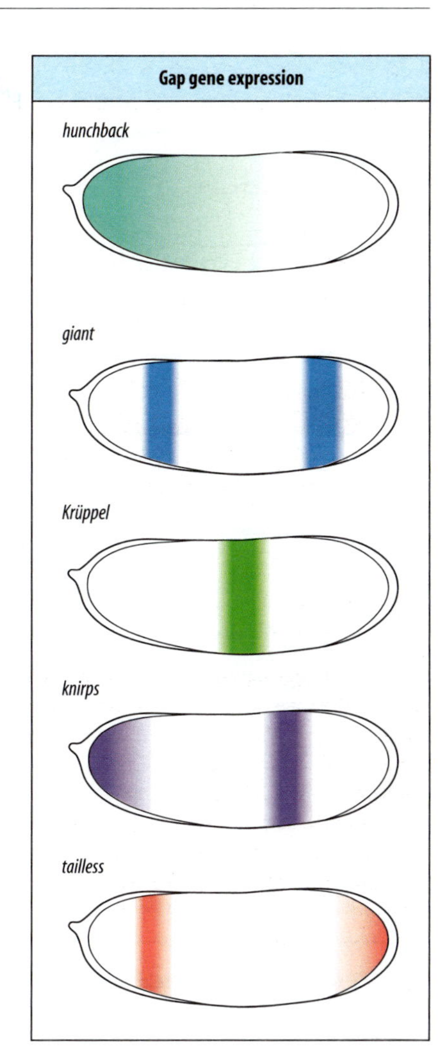

Fig. 2.24 The expression of the gap genes *hunchback*, *Krüppel*, *giant*, *knirps*, and *tailless* in the early *Drosophila* embryo. Gap-gene expression at different points along the antero-posterior axis is controlled by the concentration of Bicoid and Hunchback proteins, together with interactions between the gap genes themselves. The expression pattern of the gap genes provides an aperiodic pattern of transcription factors along the antero-posterior axis, which delimits broad body regions.

2.18 Bicoid protein provides a positional signal for the anterior expression of *hunchback*

Zygotic expression of the *hunchback* gene in normal embryos occurs over most of the anterior half of the embryo. This zygotic expression is superimposed on a low level of maternal *hunchback* mRNA, whose translation is suppressed posteriorly by Nanos (see Section 2.9). This results in a gradient of Hunchback protein in the posterior half of the embryo, running anterior to posterior.

The localized anterior expression of Hunchback is an interpretation of the positional information provided by the Bicoid protein gradient. The *hunchback* gene is switched on only when Bicoid, a transcription factor, is present at a certain threshold concentration. This level is attained only in the anterior third of the embryo,

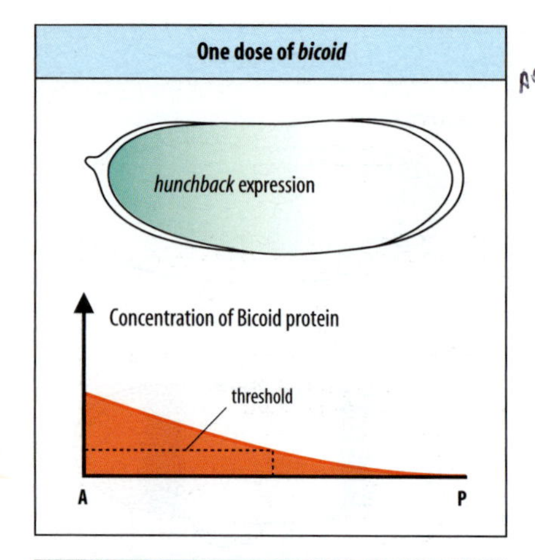

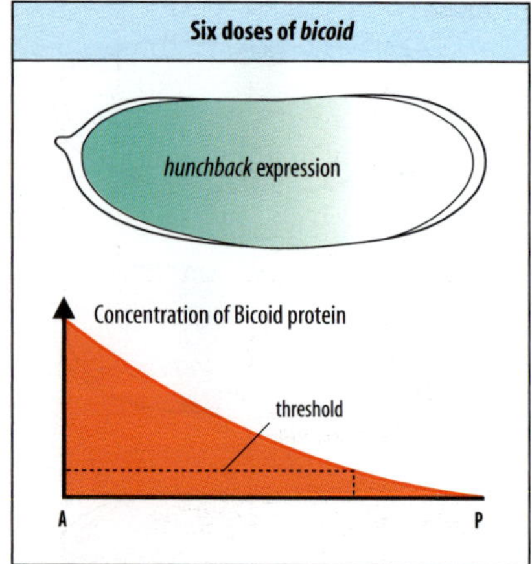

Fig. 2.25 Maternal Bicoid protein controls zygotic *hunchback* expression. If the dose of maternal *bicoid* is increased sixfold, the extent of the Bicoid gradient also increases. The activity of the *hunchback* gene is determined by the threshold concentration of Bicoid, so at the higher dose, its region of expression is extended toward the posterior end because the region in which Bicoid concentration exceeds the threshold level also extends more posteriorly (see graph, bottom panel).

close to the site of Bicoid synthesis, which restricts *hunchback* expression to this region. There is some evidence that maternal Hunchback protein is also necessary for spatial control of zygotic *hunchback* expression.

The relationship between Bicoid concentration and *hunchback* gene expression can be illustrated by looking at how *hunchback* expression changes when the Bicoid concentration gradient is changed by increasing the maternal dosage of the *bicoid* gene (Fig. 2.25). The result is that expression of *hunchback* extends more posteriorly, because the region in which the concentration of the Bicoid protein is above the threshold for *hunchback* gene activation is also extended posteriorly. By calibrating the extension of *hunchback* expression with the maternal dosage of the *bicoid* gene, it can be calculated that a twofold increase in the concentration of Bicoid can switch *hunchback* gene expression from off to on.

The Bicoid protein gradient also provides us with an excellent example of how developmental mechanisms can deal with considerable variation between embryos and still produce the same outcome. Measurements of the Bicoid gradient reveal that it is highly variable in strength at a given point from embryo to embryo, yet the spatial positioning of expression of zygotic *hunchback* remains the same. The mechanism underlying this robustness is not yet understood.

Bicoid is a member of the homeodomain family of transcriptional activators and activates the *hunchback* gene by binding to regulatory sites within the promoter region. Direct evidence of *hunchback* activation by Bicoid has been obtained by gene-transfer experiments (Box 2C, opposite) using a fusion gene constructed from *hunchback* promoter regions and a bacterial reporter gene, *lacZ*, and then introduced into the fly genome. *lacZ* codes for the enzyme β-galactosidase, which is easily made visible by histochemical staining. The extent of *lacZ* expression in embryos carrying this transgene exactly parallels normal *hunchback* expression if the promoter region of the transgene is complete, but not if a large part of it is deleted (Fig. 2.26). The large promoter region required for completely normal gene expression can be whittled down to an essential sequence of 263 base pairs that will still give almost normal activity in this situation. This sequence has several sites at which Bicoid can bind, and it seems that cooperative binding of Bicoid proteins is involved in establishing the threshold response.

The regulatory sequences of genes such as these are yet further examples of developmental switches that direct nuclei along a new developmental pathway. We will encounter many more examples of these transcriptional switches in *Drosophila* early development.

2.19 The gradient in Hunchback protein activates and represses other gap genes

Expression of the other gap genes is localized in bands across the antero-posterior axis (see Fig. 2.23). The Hunchback protein is itself a transcription factor and acts as a morphogen to which the other gap genes respond. The bands of gap-gene expression are delimited by mechanisms that depend on the gene control regions being sensitive to different concentrations of Hunchback protein, and also to other proteins, including Bicoid. Expression of the *Krüppel* gene, for example, is activated by a combination of Bicoid and low levels of Hunchback, but is repressed at high concentrations of Hunchback. Within this concentration window *Krüppel* remains activated (Fig. 2.27, top panel). But below the lower threshold concentration of Hunchback it is not activated. In this way, the gradient in Hunchback protein precisely locates a band of *Krüppel* gene activity near the center of the embryo. Refined spatial localization is brought about by repression of *Krüppel* by other gap-gene proteins.

Box 2C P-element-mediated transformation

Transgenic fruit flies have contributed greatly to *Drosophila* developmental genetics. They are made by inserting a known sequence of DNA into the *Drosophila* chromosomal DNA, using as a carrier a **transposon** that occurs naturally in some strains of *Drosophila*. This transposon is known as a **P element**, and the technique as P-element-mediated transformation (see figure).

P elements can insert at almost any site on a chromosome, and can also hop from one site to another within the germ cells, an action that requires an enzyme called a transposase. As hopping can cause genomic instability, carrier P elements have had their own transposase gene removed. The transposase required to insert the P element initially is instead provided by a helper P element, which cannot itself insert into the host chromosomes and is thus quickly lost from cells. The two elements are injected together into the posterior end of the egg where the germ cells are made.

As well as the gene to be inserted, an additional marker gene, such as the wild-type *white*$^+$ gene, is added to the P element. When *white*$^+$ is the marker, the P element is inserted into flies homozygous for the mutant *white*$^-$ gene (which have white eyes rather than the red eyes of the wild-type *Drosophila*). Red eyes are dominant over white, and so flies in which the P element has become integrated into the chromosome, and is being expressed, can be detected by their red eyes.

In the first generation, all flies have white eyes, as any P element that has integrated is still restricted to the germ cells. But in the second generation, a few flies will have wild-type red eyes, showing that they carry the inserted P element in their somatic cells.

This technique can be used to increase the number of copies of a particular gene, or to introduce a mutated gene that has its control or coding regions altered in a known way, or to introduce new genes. It is also possible to introduce genes that carry a marker coding sequence such as *lacZ* (encoding the bacterial enzyme β-galactosidase), whose expression is detectable by histochemical staining (see Box 3B, p. 99). The P element itself can also be used as a mutagen, as its insertion into a gene usually destroys that gene's function.

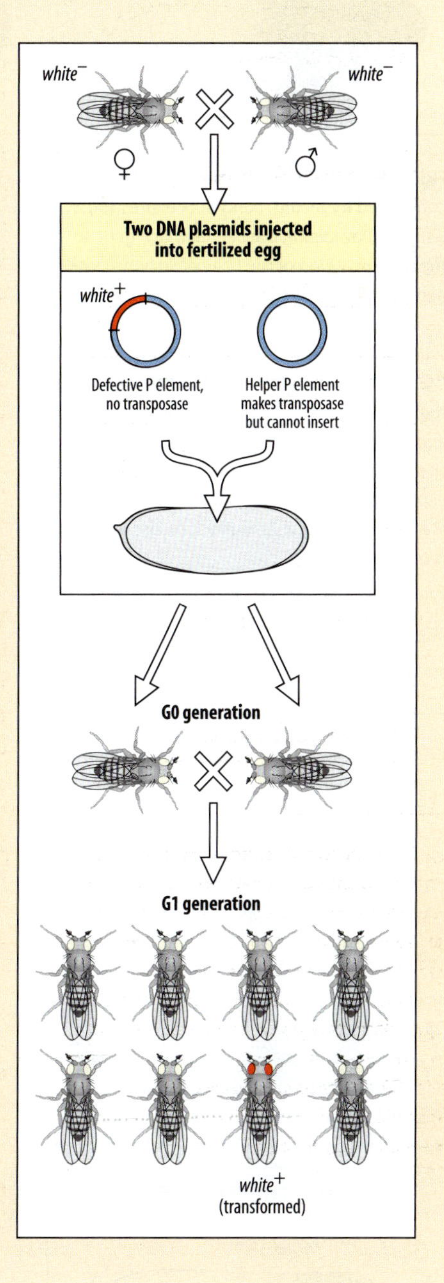

Such relationships were worked out by altering the concentration profile of Hunchback protein systematically, while all other known influences were eliminated or held constant. Increasing the dose of Hunchback protein, for example, results in a posterior shift in its concentration profile, and this results in a posterior shift in the posterior boundary of *Krüppel* expression. In another set of experiments on embryos lacking Bicoid protein, so that only the maternal Hunchback protein gradient is present, the level of Hunchback is such that *Krüppel* is even activated at the anterior end of the embryo (Fig. 2.27, bottom panel).

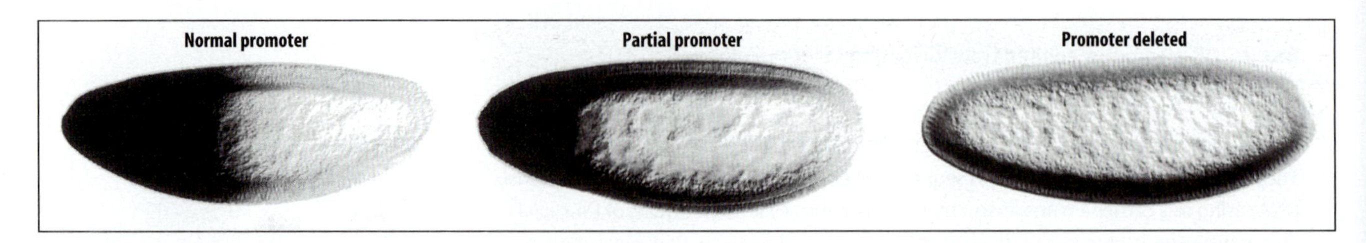

Fig. 2.26 Zygotic *hunchback* expression is controlled by Bicoid. *hunchback* expression is visualized by joining the bacterial *lacZ* gene to the control region of the *Drosophila hunchback* gene and inserting this construct into the fly's genome. With the normal *hunchback* control region, *lacZ* is expressed in the anterior half of the embryo (left); with only a partial control region its expression is more restricted (center); and when the construct lacks a Bicoid-binding site, *lacZ* expression is absent (right). *lacZ* expression is visualized by histochemical staining for the *lacZ* gene product, the enzyme β-galactosidase.

Photographs courtesy of D. Tautz.

The Hunchback protein is also involved in specifying the anterior border of the bands of expression of the gap genes *knirps* and *giant*, again by a mechanism involving thresholds for repression and activation of these genes. At high concentrations of Hunchback, *knirps* is repressed, and this specifies its anterior margin of expression. The posterior margin of the *knirps* band is specified by a similar type of interaction with the product of another gap gene, *tailless*. Where the regions of expression of the gap genes overlap, there is extensive cross-inhibition between them, their proteins all being transcription factors. These interactions are essential to sharpen and stabilize the pattern of gap gene expression. For example, the anterior border of *Krüppel* expression lies four to five nuclei posterior to nuclei that express *giant*, and that anterior border is set by low levels of giant protein.

The antero-posterior axis becomes divided into a number of unique regions on the basis of the overlapping and graded distributions of different transcription factors. This beautifully elegant method of delimiting regions can, however, only work in an embryo such as the acellular blastoderm of *Drosophila*, where the transcription factors are able to diffuse throughout the embryo. This regional distribution of the gap-gene products provides the starting point for the next stage in development—the activation of the pair-rule genes and the beginning of segmentation.

Summary

Gradients of maternally derived transcription factors along the dorso-ventral and antero-posterior axes provide positional information that activates zygotic genes at specific locations along these axes. The dorso-ventral axis becomes divided into four regions: ventral mesoderm, ventral ectoderm (neurectoderm), dorsal ectoderm (dorsal epidermis), and amnioserosa. A ventral to dorsal gradient of maternal Dorsal protein both specifies the ventral mesoderm and defines the dorsal region; a second gradient, of the Decapentaplegic protein, specifies the dorsal ectoderm. Along the antero-posterior axis the maternal gradient in Bicoid protein initiates the activation of zygotic gap genes to specify general body regions. Interactions between the gap genes, all of which code for transcription factors, help to define their borders of expression. Patterning along the dorso-ventral and antero-posterior axes divides the embryo into a number of discrete regions, each characterized by a unique pattern of zygotic gene activity.

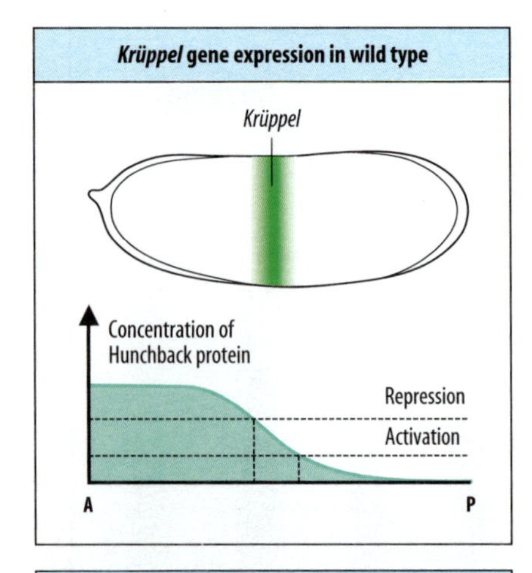

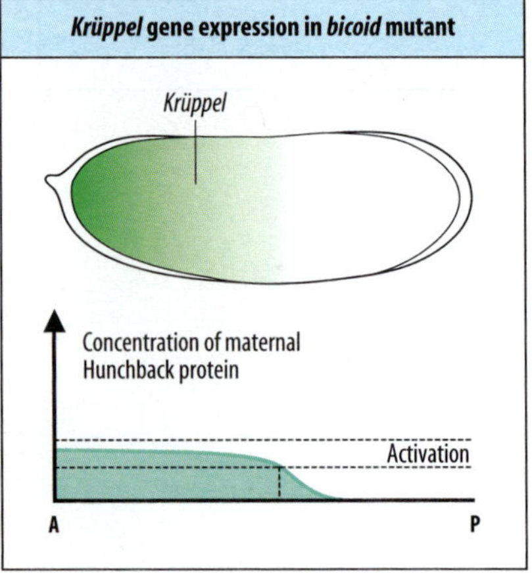

Fig. 2.27 *Krüppel* gene activity is specified by Hunchback protein. Top panel: above a threshold concentration of Hunchback protein, the *Krüppel* gene is repressed; at a lower concentration, above another threshold value, it is activated. Bottom panel: in mutants lacking the *bicoid* gene, and thus also lacking zygotic *hunchback* gene expression, only maternal Hunchback protein is present, which is located at the anterior end of the embryo at a relatively low level. In these mutants, *Krüppel* is activated at the anterior end of the embryo, giving an abnormal pattern.

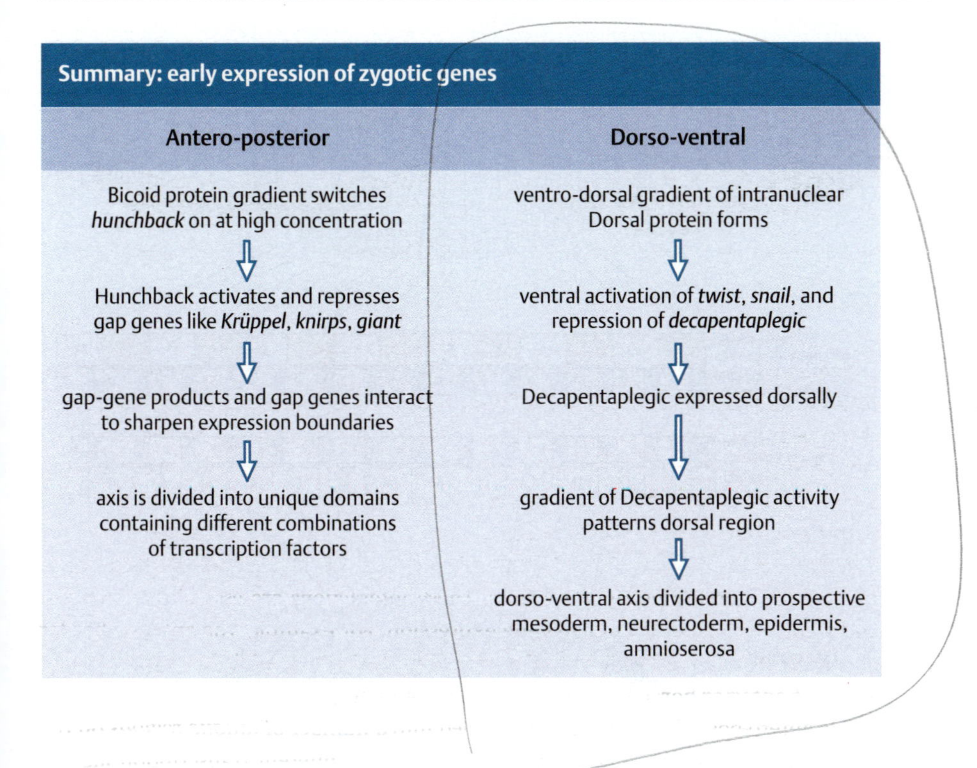

Activation of the pair-rule genes and the establishment of parasegments

The most obvious feature of a *Drosophila* larva is the regular segmentation of the larval cuticle along the antero-posterior axis, each segment carrying cuticular structures that define it as, for example, thorax or abdomen. This pattern of segmentation is carried over into the adult, in which each segment has its own identity. Adult appendages such as wings, halteres, and legs are attached to particular segments (Fig. 2.28), but the discernible segments of the mature larva are not in fact the first units of segmentation along this axis. The basic developmental modules, whose definition we will follow in some detail, are the parasegments, which are specified first and from which the segments derive.

2.20 Parasegments are delimited by expression of pair-rule genes in a periodic pattern

The first visible signs of segmentation in the embryo are transient grooves that appear on the surface of the embryo after gastrulation. These grooves define the parasegments. There are 14 parasegments, which are the fundamental units in the segmentation of the *Drosophila* embryo. Once each parasegment is delimited, it behaves as an independent developmental unit, under the control of a particular set of genes, and in this sense at least the embryo can be thought of as being built up piecemeal of modules. The parasegments are initially similar but each will eventually acquire its own unique identity. The parasegments are out of register with the final segments by about half a segment; each segment is therefore made up of the posterior region of one parasegment and the anterior region of the next (see Fig. 2.28). In the anterior head region the segmental arrangement is lost when some of the anterior parasegments fuse.

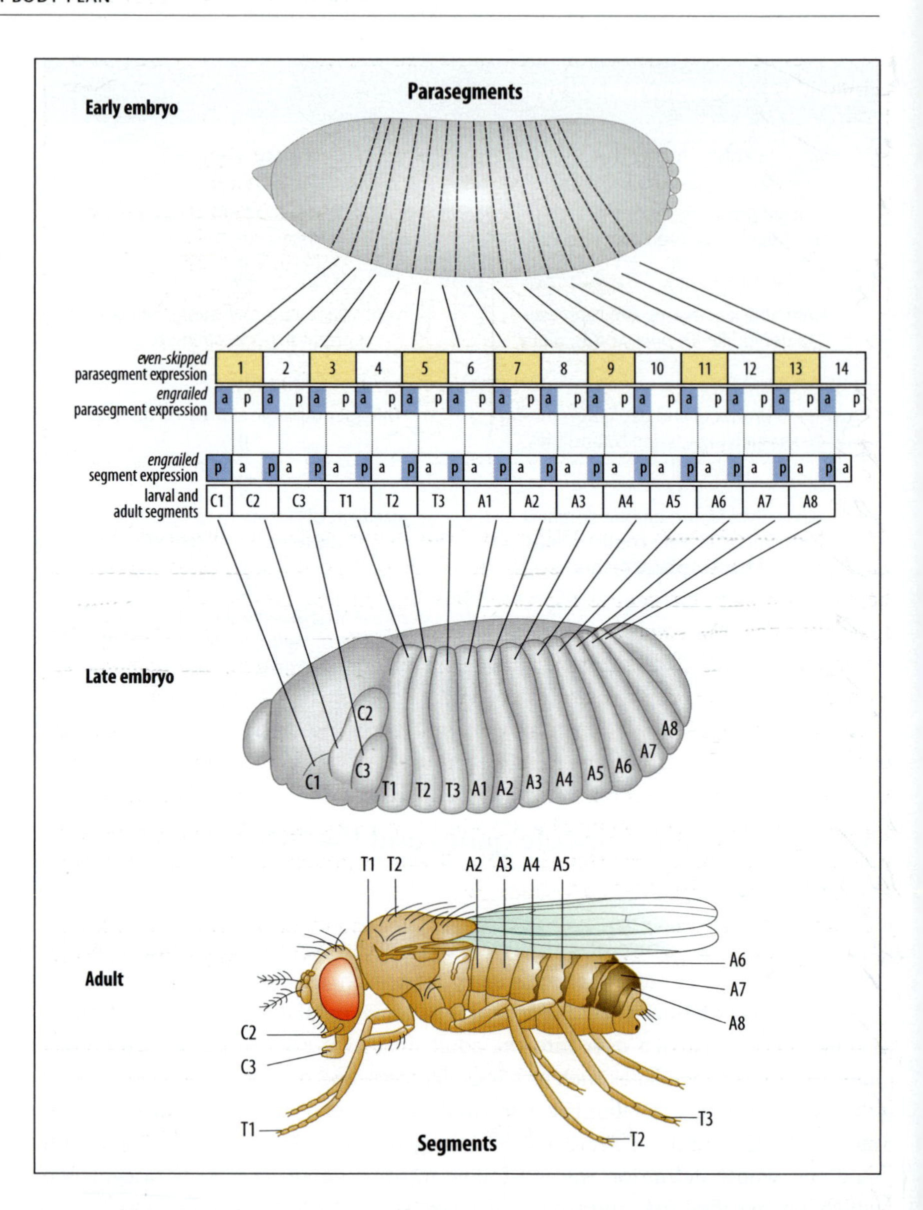

Fig. 2.28 The relationship between parasegments and segments in the early embryo, late embryo, and adult fly. Initially, pair-rule genes are expressed in the embryo as stripes in every second parasegment. *even-skipped* (yellow), for example, is expressed in odd-numbered parasegments. The segmentation selector gene *engrailed* (blue) is expressed in the anterior region of every parasegment, and delimits the anterior margin of each parasegment. Each larval segment is composed of the posterior region of one parasegment and the anterior region of the next. The anterior region of a parasegment becomes the posterior portion of a segment. Segments are thus offset from the original parasegments by about half a segment, and *engrailed* is expressed in the posterior region of each segment. In this figure, a and p refer to the anterior and posterior compartments of segments or parasegments. The segment specification is carried over into the adult and results in particular appendages, such as legs and wings, developing on specific segments only. C1, C2, and C3 represent segments which become fused to form the head region. T, thoracic segments; A, abdominal segments. After Lawrence, P.: 1992.

The parasegments in the thorax and abdomen are delimited by the action of the **pair-rule genes**, each of which is expressed in a series of seven transverse stripes along the embryo, each stripe corresponding to every second parasegment. When pair-rule gene expression is visualized by staining for the pair-rule proteins, a striking zebra-striped embryo is revealed (Fig. 2.29).

The positions of the stripes of pair-rule gene expression are determined by the pattern of gap gene expression—a non-repeating pattern of gap gene activity is converted into repeating stripes of pair-rule gene expression. We now consider how this is achieved.

2.21 Gap-gene activity positions stripes of pair-rule gene expression

Pair-rule genes are expressed in stripes, with a periodicity corresponding to alternate parasegments. Mutations in these genes thus affect alternate segments. Some pair-rule genes (e.g. *even-skipped*) define odd-numbered parasegments, whereas others (e.g. *fushi tarazu*) define even-numbered parasegments. The striped pattern of

Fig. 2.29 The striped patterns of activity of pair-rule genes in the *Drosophila* embryo just before cellularization. Parasegments are delimited by pair-rule gene expression, each pair-rule gene being expressed in alternate parasegments. Expression of the pair-rule genes *even-skipped* (blue) and *fushi tarazu* (brown) is visualized by staining with antibody for their protein products. *even-skipped* is expressed in odd-numbered parasegments, *fushi tarazu* in even-numbered parasegments. Scale bar = 0.1 mm.

Photograph from Lawrence, P.: 1992.

expression of pair-rule genes is present even before cells are formed, while the embryo is still a syncytium, although cellularization occurs soon after expression begins. Each pair-rule gene is expressed in seven stripes, each of which is only a few cells wide. For some genes, such as *even-skipped*, the anterior margin of the stripe corresponds to the anterior boundary of a parasegment; the domains of expression of other pair-rule genes, however, cross parasegment boundaries.

The striped expression pattern appears gradually; the stripe of the *even-skipped* gene is initially fuzzy, but eventually acquires a sharp anterior margin. At first sight this type of patterning would seem to require some underlying periodic process, such as the setting up of a wave-like concentration of a morphogen, with each stripe forming at the crest of a wave. It was surprising, therefore, to discover that each stripe is specified independently.

As an example of how the pair-rule stripes are generated we will look in detail at the expression of the second *even-skipped* stripe (Fig. 2.30). The appearance of this

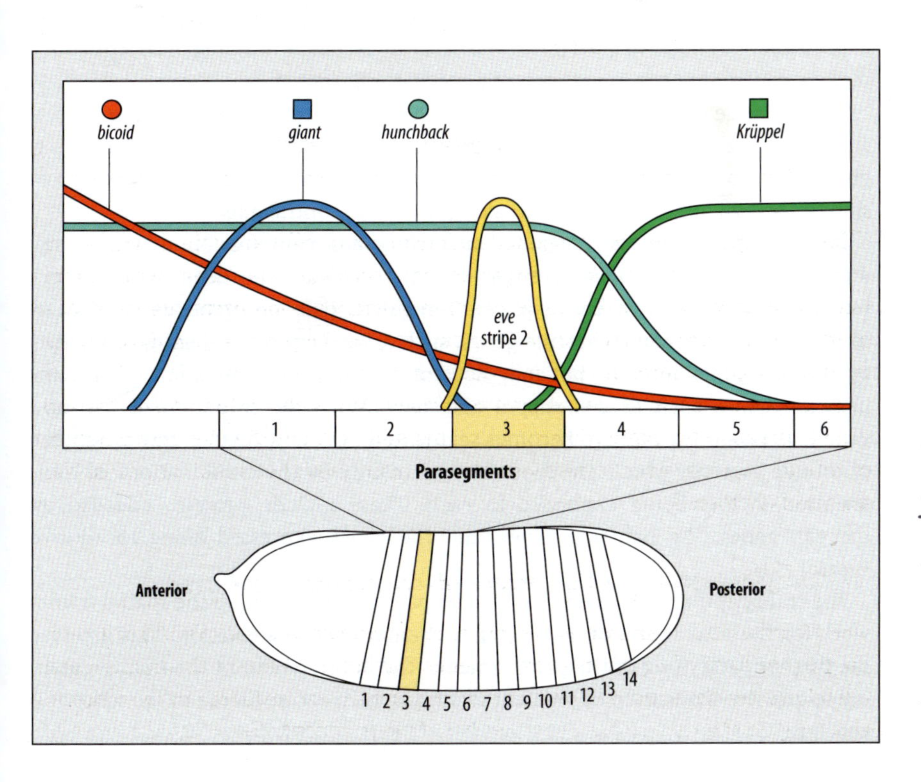

Fig. 2.30 The specification of the second *even-skipped* (*eve*) stripe by gap gene proteins. The different concentrations of transcription factors encoded by the gap genes *hunchback*, *giant*, and *Krüppel* localize *even-skipped*—expressed in a narrow stripe at a particular point along their gradients—in parasegment 3. Bicoid and Hunchback proteins activate the gene in a broad domain, and the anterior and posterior borders are formed through repression by Giant and Krüppel proteins, respectively.

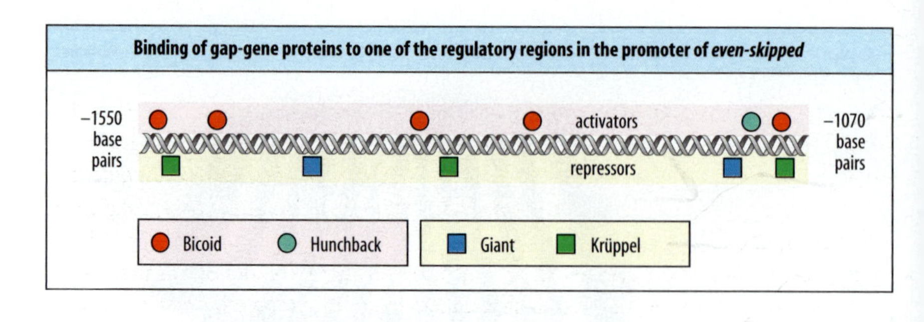

Fig. 2.31 Sites of action of activating and repressing transcription factors in the region of the *even-skipped* promoter involved in expression of the second *even-skipped* stripe. A promoter region of around 500 base pairs, located between 1070 and 1550 base pairs upstream of the transcription start site, directs formation of the second *even-skipped* stripe. Gene expression occurs when the Bicoid and Hunchback transcription factors are present above a threshold concentration, with the Giant and Krüppel proteins acting as repressors where they are above threshold levels. The repressors may act by preventing binding of activators.

stripe depends on the normal expression of *bicoid* and of the three gap genes *hunchback*, *Krüppel*, and *giant* (only the anterior band of *giant* expression is involved in specifying the second *even-skipped* stripe; see Fig. 2.24). Bicoid and Hunchback proteins are required to activate the *even-skipped* gene, but they do not define the boundaries of the stripe. These are defined by Krüppel and Giant proteins, by a mechanism based on repression of *even-skipped*. When concentrations of Krüppel and Giant are above certain threshold levels, *even-skipped* is repressed, even if Bicoid and Hunchback are present. The anterior edge of the stripe is localized at the point of threshold concentration of giant protein, whereas the posterior border is similarly specified by Krüppel protein.

The independent localization of each of the stripes by the gap-gene transcription factors requires that, in each stripe, pair-rule genes respond to different concentrations and combinations of the gap-gene transcription factors. The pair-rule genes thus require complex *cis*-control regions with multiple binding sites for each of the different factors. Examination of the regulatory regions of the *even-skipped* gene reveals seven separate regions, each controlling the localization of a different stripe. Using the *lacZ* reporter gene technique (see Box 2C, p. 59), regulatory regions of around 500 base pairs have been isolated, and each determines the expression of a single stripe (Fig. 2.31). This is an excellent example of *cis*-regulatory regions controlling the expression of a gene at different sites.

The presence of control regions which, when activated, can lead to gene expression in a specific position in the embryo, is a fundamental principle controlling gene action in development. Other examples are provided by the localized expression of the gap genes, and of genes along the dorso-ventral axis.

Each of the regulatory regions on such genes contains binding sites for different transcription factors, some of which activate the gene, while others repress it. In this way, the gap genes regulate pair-rule gene expression in each parasegment. Some pair-rule genes, such as *fushi tarazu*, may not be regulated by the gap genes directly, but may depend on the prior expression of primary pair-rule genes such as *even-skipped* and *hairy*. With the initiation of pair-rule gene expression the embryo becomes segmented; it is now divided into a number of unique regions, which are characterized mainly by the combinations of transcription factors being expressed in each. These include proteins encoded by the gap genes, the pair-rule genes, and the genes expressed along the dorso-ventral axis.

The transcription factors encoded by the pair-rule genes set up the spatial framework for the next round of patterning by transcriptional activation. This involves the further patterning of the parasegments, the development of the final segmentation and the acquisition of segment identity, which is considered in the following sections.

Summary

The activation of the pair-rule genes by the gap genes results in the transformation of the embryonic pattern along the antero-posterior axis from an aperiodic regionalization to a periodic one. The pair-rule genes define 14 parasegments. Each parasegment is defined by narrow stripes of pair-rule gene activity. These stripes are uniquely defined by the local concentration of gap-gene transcription factors acting on the regulatory regions of the pair-rule genes. Each pair-rule gene is expressed in alternate parasegments—some in odd-numbered, others in even-numbered. Most pair-rule genes code for transcription factors.

Summary: pair-rule genes and segmentation

production of local combinations of gap-gene transcription factors

activation of each pair-rule gene in seven transverse stripes along the antero-posterior axis

pair-rule gene expression defines 14 parasegments, each pair-rule gene being expressed in alternate parasegments

Segmentation genes and compartments

The expression of the pair-rule genes defines the anterior boundaries of all 14 parasegments but, like the gap genes, their activity is only temporary. Moreover, at this time the blastoderm becomes cellularized. How, therefore, are the positions of parasegment boundaries fixed, and how do the final segment boundaries of the larval epidermis become established? This is the role of the **segmentation genes**. Unlike the gap genes and pair-rule genes, which all encode transcription factors, the segmentation genes are a diverse group of genes that bear no obvious relation to each other in their protein products or mechanisms of operation.

Segmentation genes are activated in response to pair-rule gene expression. They are each expressed in 14 transverse stripes, one stripe corresponding to each parasegment. During pair-rule gene expression the blastoderm becomes cellularized, so the segmentation genes are acting in a cellular rather than a syncytial environment. One of the segmentation genes to be activated by the pair-rule genes is *engrailed*, which is expressed in the anterior region of every parasegment. *engrailed* is of particular interest as its expression delimits a boundary of cell-lineage restriction. Moreover, *engrailed* is also a **selector gene**, a gene that confers a particular identity on a region or regions by controlling the activity of other genes, and which continues to act for an extended period.

2.22 Expression of the *engrailed* gene delimits a cell-lineage boundary and defines a compartment

The *engrailed* gene has a key role in segmentation and, unlike the pair-rule and gap genes, whose activity is transitory, it is expressed throughout the life of the fly. *engrailed* activity first appears at the time of cellularization as a series of

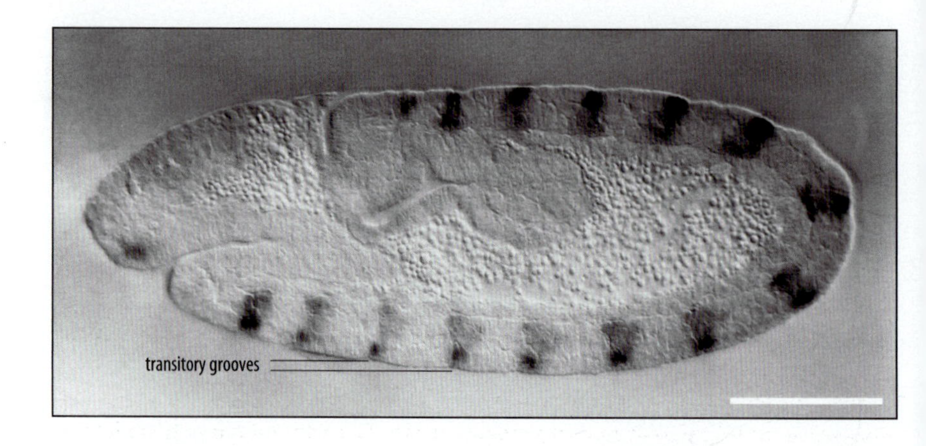

Fig. 2.32 The expression of the *engrailed* gene in a late (stage 11) *Drosophila* embryo. The gene is expressed in the anterior region of each parasegment and the transitory grooves between parasegments can be seen. At this stage of development, the germ band has temporarily extended and is curved over the back of the embryo. Scale bar = 0.1 mm.

Photograph from Lawrence, P.: 1992.

14 transverse stripes. Fig. 2.32 shows expression at the later extended germ-band stage, when part of the ventral blastoderm (the germ band) has extended over the dorsal side of the embryo. *engrailed* is initially expressed in a single line of cells at the anterior margin of each parasegment, which is itself only about three cells wide (Fig. 2.33). It is likely that this periodic pattern of *engrailed* activity is the result of the combinatorial action of transcription factors encoded by pair-rule genes, including *fushi tarazu* and *even-skipped*. Evidence that the pair-rule genes do control *engrailed*

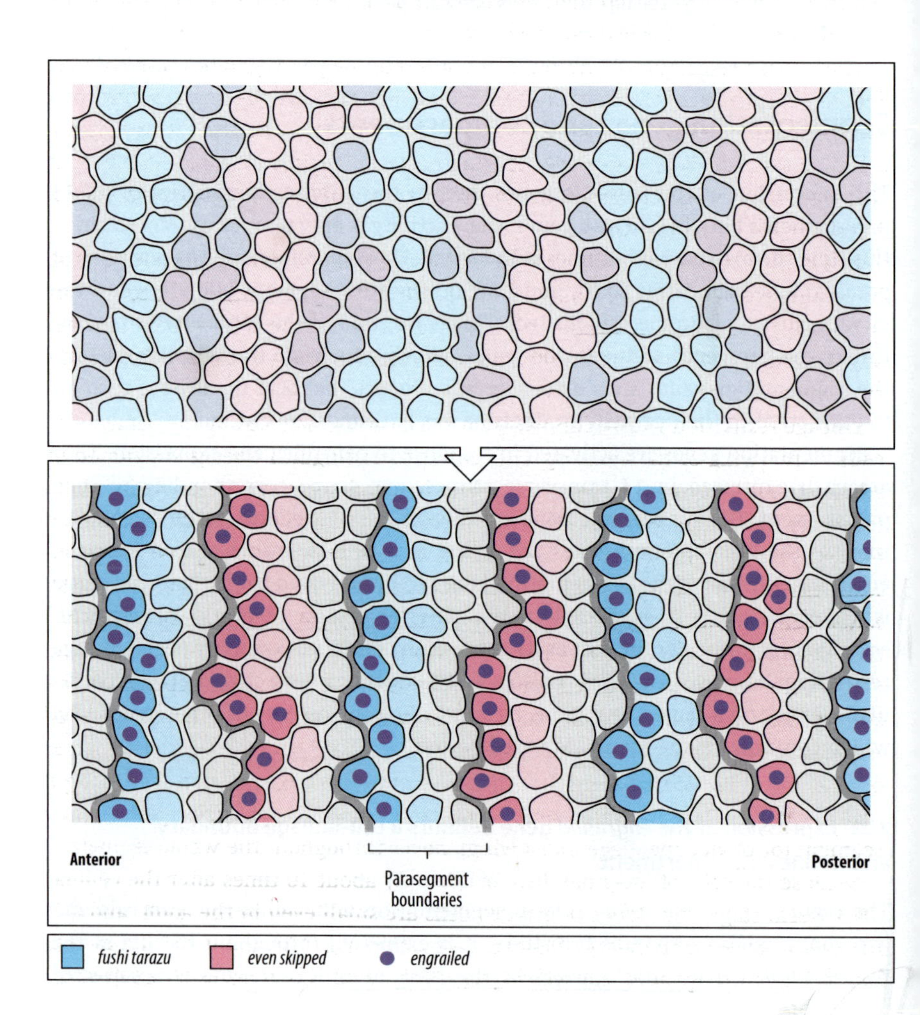

Fig. 2.33 The expression of the pair-rule genes *fushi tarazu* (blue), *even-skipped* (pink), and *engrailed* (purple dots) in parasegments. *engrailed* is expressed at the anterior margin of each stripe and delimits the anterior border of each parasegment. The boundaries of the parasegments become sharper and straighter later on. After Lawrence, P.: 1992.

expression is provided, for example, by embryos carrying mutations in *fushi tarazu*, in which *engrailed* expression is absent only in even-numbered parasegments (in which *fushi tarazu* is normally expressed).

The anterior margin of the parasegment has a very important property—it is a boundary of **cell-lineage restriction**. Cells and their descendants from one parasegment never move into adjacent ones. This implies that the cells in a parasegment are under some common genetic control which both prevents them mixing with their neighbors and controls their later development. Such domains of lineage restriction are known as **compartments**. A compartment can be defined as a region in the embryo, larva, or adult that contains all the descendants of the cells present when the compartment is set up, and no others.

The existence of compartments can be detected by cell-lineage studies. Cell lineage can be followed by marking single cells in the early embryo so that all their descendants (clones) can be identified at later stages of development. One technique is to inject the egg with a harmless fluorescent compound which is incorporated into all the cells of the embryo. In the early embryo, a fine beam of ultraviolet light directed onto a single cell activates the fluorescent compound. All the descendants of that cell will be fluorescent and can therefore be identified. Examination of these clones, together with an analysis of *engrailed* expression, shows that cells at the anterior margin of the parasegment have no descendants that lie on the other side of that margin. The anterior margin is therefore a boundary of lineage restriction; that is, cells and their descendants that are on one or other side of the boundary when it is formed can never cross it.

The lineage restriction of the anterior margin of the parasegment is carried over into the segments of the larva and the adult. But because the anterior part of a parasegment becomes the posterior part of a segment, the lineage restriction within a segment is between anterior and posterior regions. The segment is thus divided into anterior and posterior compartments, with *engrailed* expression defining the posterior compartment.

In a compartment, the cells may all initially be under common genetic control. Imagine a group of cells becoming divided into two separate regional populations by means of signals that turn on different genes in each region. If there is never any mixing of the descendants of the original cells across the boundary the regions are compartments.

Lineage restriction within compartments is clearly illustrated by the behavior of cells in the wing of the adult fly. It is easier to distinguish lineage restriction in adult structures than in embryonic and early larval structures, because in adults there has been considerable cell division since the initial event that determined the compartment in the embryo. Compartments are made visible by making genetically mosaic flies composed of two distinguishable kinds of cells (Box 2D, p. 68). Single cells in the embryonic blastoderm or in larval epidermis are given a distinctive phenotype by X-ray- or laser-induced mitotic recombination. The fate of all the descendants of this marked cell are then followed. Their behavior depends on the stage of development at which the founder nucleus was marked. Descendants of nuclei marked at early stages during cleavage become part of many tissues and organs, but those of nuclei marked at the blastoderm stage or later have a more restricted fate. They are found only in the anterior or the posterior part of each segment (or of an appendage like a wing), never throughout the whole segment.

Because the cells of imaginal discs divide only about 10 times after the cellular blastoderm stage, the clones of marked cells are small even in the adult, and it is not easy to detect a boundary of lineage restriction (Fig. 2.34, top panel). Clone size can be increased by the *Minute* technique, which results in the marked cell dividing

Box 2D Genetic mosaics and mitotic recombination

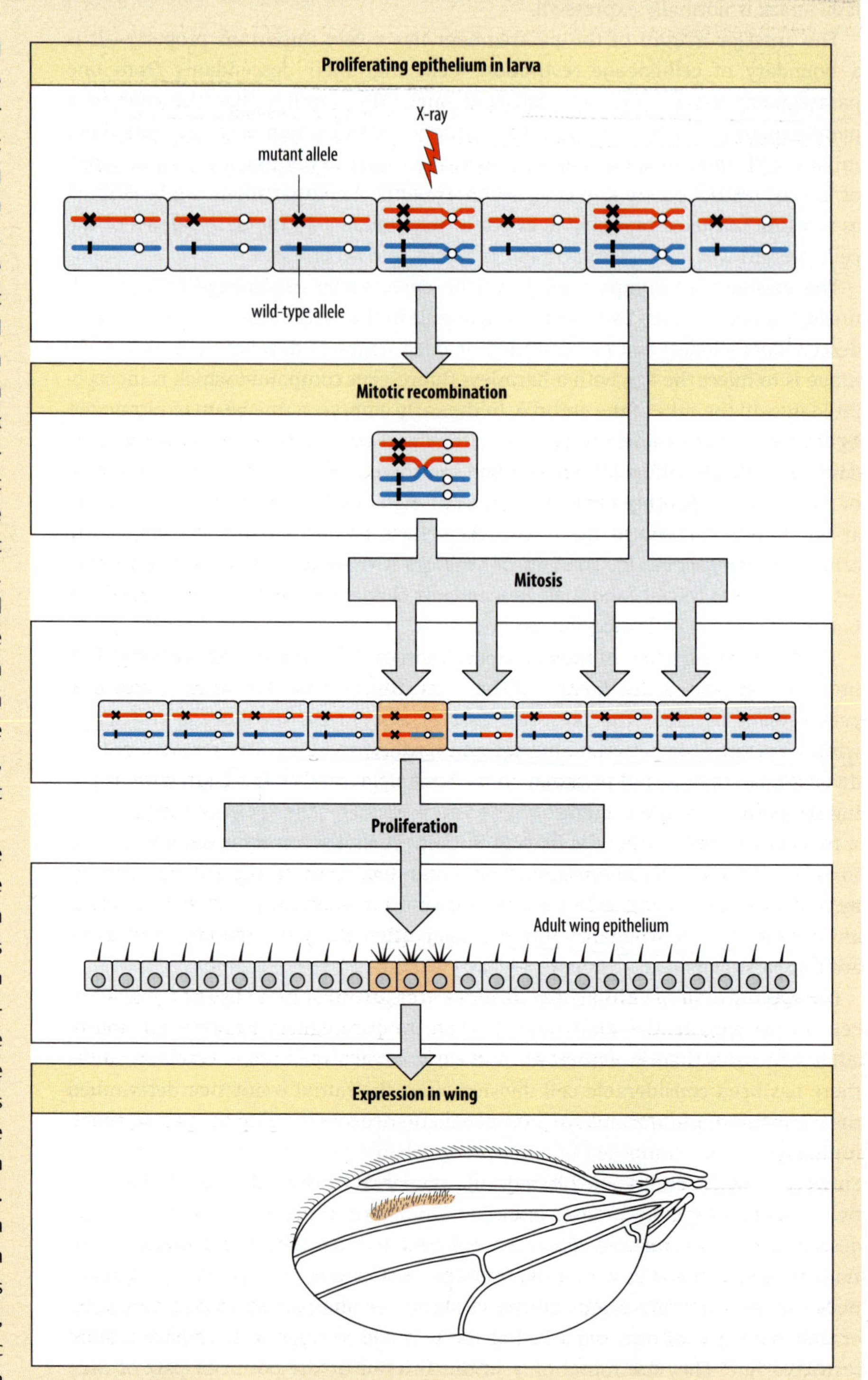

Genetic mosaics are embryos derived from a single genome but in which there is a mixture of cells with rearranged or inactivated genes. In flies, genetic mosaics can be generated by inducing rare mitotic recombination events in the somatic cells of the embryo or larva. The original method was to induce chromosome breaks using X-rays just after the chromosomes have replicated to form two chromatids; this results in the exchange of material between homologous chromosomes as the break is repaired. Today, mitotic recombination is induced in strains of transgenic flies that carry the gene for the yeast recombinase FLP and its target sequence, *FRT*, on their chromosomes. Activation of the recombinase will result in recombination involving the *FRT* sequence. A mitotic recombination event can generate a single cell with a unique genetic constitution that will be inherited by all the cell's descendants, which in flies usually form a coherent patch of tissue.

Easily distinguishable mutations like the recessive *multiple wing hairs* can be used to identify the marked clone; if a cell homozygous for this mutation is generated by mitotic recombination in a heterozygous larva, all of its descendant cells will have multiple hairs (see figure). Marked epidermal clones made by this method are usually small, as there is little cell proliferation after the recombination event. Larger clones can be made using the *Minute* technique. The cells in flies carrying a mutation in the *Minute* gene grow more slowly than those in the wild type. By using flies heterozygous for the *Minute* mutation, clones can be made in which the mitotic recombination event has generated a marked cell that is normal because it has lost the *Minute* mutation, and so is wild type. This normal cell proliferates faster than the slower-growing background and thus large clones of the marked cells are produced (see Fig. 2.34). The mitotic recombination technique has many applications. If clones of marked cells are generated at different stages of development, one can trace the fate of the altered cells and thus see what structures they are able to contribute to. This can provide information on their state of determination or specification at different developmental stages. The technique can also be used to study the localized effects of homozygous mutations that are lethal if present in the homozygous state throughout the whole animal. Illustration after Lawrence, P.: 1992.

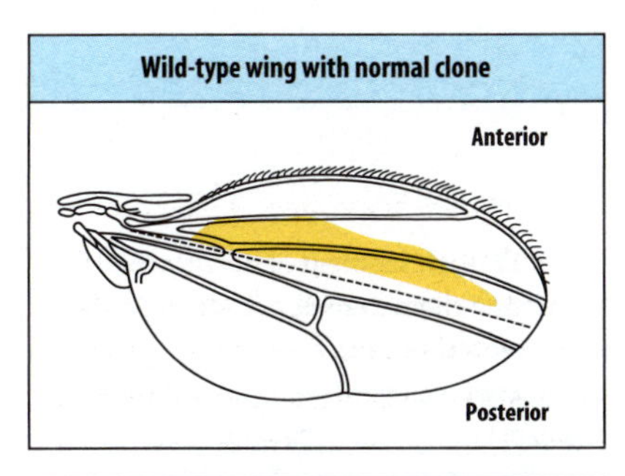

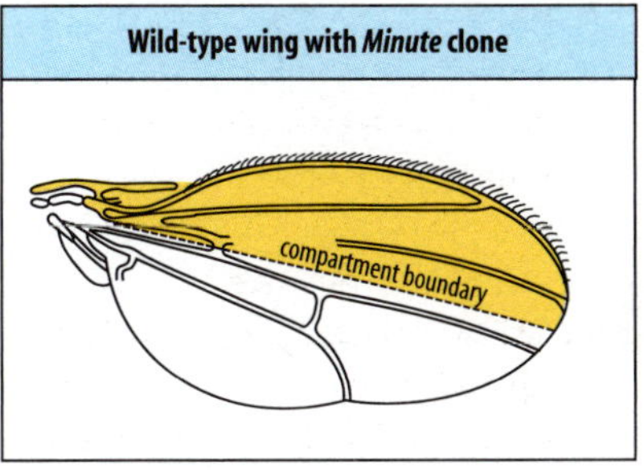

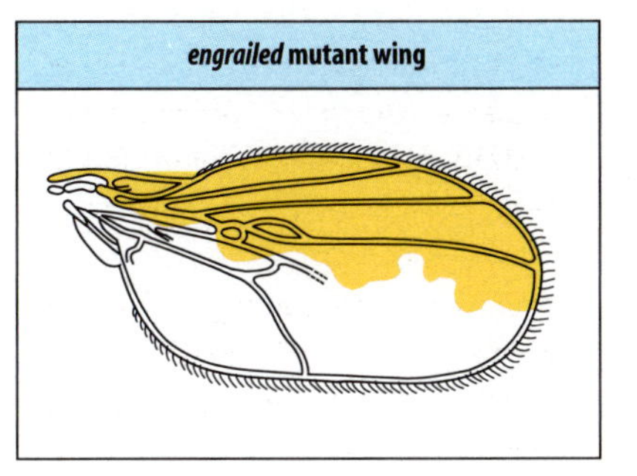

Fig. 2.34 The boundary between anterior and posterior compartments in the wing can be demonstrated by marked cell clones. Top panel: in the wild-type wing, clones marked by mitotic recombination in the embryo, which gives marked cells a phenotype different from other cells of the wing, are too small to demonstrate the compartment boundary. Middle panel: the use of the *Minute* technique (see Box 2D, opposite) produces an increased rate of cell division in the marked cell and gives clones large enough to determine that cells from one compartment do not cross the boundary into an adjoining compartment. Bottom panel: in the wing of an *engrailed* mutant in which the engrailed protein is not produced, there is no posterior compartment or boundary. Clones in the anterior part of the wing cross over into the posterior region and the posterior region is transformed into a more anterior-like structure, bearing anterior-type hairs on its margin. The *engrailed* gene is required for the maintenance of the character of the posterior compartment, and for the formation of the boundary.

many more times than the other cells (see Box 2D, opposite). A single clone of such cells can almost fill either the anterior or posterior part of the wing, and this makes the boundary that the cells never cross more evident: this boundary separates the anterior and posterior compartments (see Fig. 2.34, middle panel). In a normal wing, each compartment is constructed by all the descendants of a set of founding cells. The compartment boundary is remarkably sharp and straight and does not correspond to any structural features in the wing. These experiments also show that the pattern of the wing is in no way dependent on cell lineage. A single marked embryonic cell can give rise to about a twentieth of the cells of the adult wing or, using the *Minute* technique to increase clone size, about half the wing. The lineage of the wing cells in each case is quite different, yet the wing's pattern is quite normal.

The compartment pattern in the adult *Drosophila* wing is carried over from its initial specification in the imaginal discs. When epidermal cells are set aside in the embryo to form the imaginal discs, each disc carries over the compartment pattern of the parasegments from which it arises. A wing disc, for example, develops at the boundary between the two parasegments that contribute to the second thoracic segment. Thus, the wing is divided into an anterior compartment and a posterior compartment, the compartment boundary (the old parasegment boundary) running in a straight line more or less down the middle of the wing.

The specification of cells as the posterior compartment of a segment (the anterior of a parasegment) initially occurs when the parasegments are set up, and is due to the *engrailed* gene. Expression of *engrailed* is required both to confer a 'posterior segment' identity on the cells and to change their surface properties so that they cannot mix with the cells adjacent to them, hence setting up the parasegment boundary.

Direct evidence for the role of *engrailed* comes from the behavior of clones of wing cells in an *engrailed* mutant (see Fig. 2.34, bottom panel). In the absence of normal *engrailed* expression, clones are not confined to anterior or posterior parts of the segment and there is no compartment boundary. Moreover, in *engrailed* mutants, the posterior compartment is partly transformed so that it comes to resemble the pattern of the anterior part of the wing. For example, bristles normally found only at the anterior margin of the wing are also found at the posterior margin.

The *engrailed* gene needs to be expressed continuously throughout larval and pupal stages and into the adult to maintain the character of the posterior compartment of the segment. Thus, *engrailed* is also an example of a selector gene—a gene whose activity is sufficient to cause cells to adopt a particular fate. Selector genes

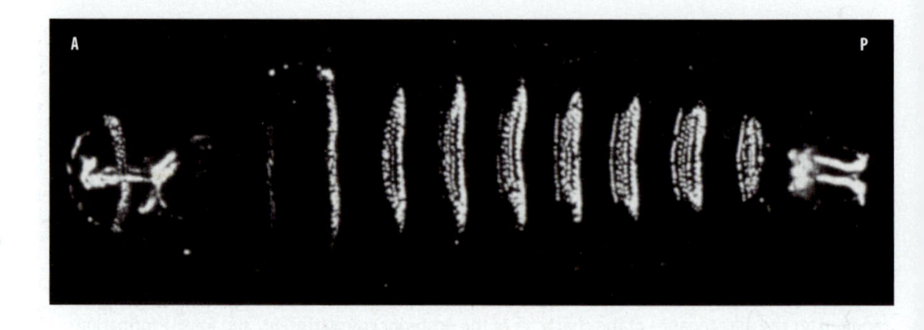

Fig. 2.35 Each segment of the *Drosophila* larva has a characteristic pattern of denticles on its ventral surface. Denticles are confined to the anterior regions of segments, with each segment having its own pattern.

can control the development of a region such as a compartment and, by controlling the activity of other genes, give the region a particular identity.

2.23 Segmentation genes stabilize parasegment boundaries and set up a focus of signaling at the boundary that patterns the segment

Each larval segment has a well defined antero-posterior pattern, which is easily seen on the ventral epidermis of the abdomen: the anterior region of each segment bears denticles while the posterior region is naked (Fig. 2.35). The rows of denticles, of which there are six types, make a distinct pattern. Mutations in segmentation genes often alter the denticle pattern, and this is how such genes were first discovered. For example, a mutation in a segmentation gene known as *wingless* (named after its phenotype in the adult fly) results in the whole of the ventral abdomen being covered in denticles, but in the posterior half of each segment the denticle pattern is reversed. In this mutant, the anterior region of each segment has been duplicated in mirror-image polarity, and the posterior region is lost. Mutations in another segmentation gene, *hedgehog*, give a similar phenotype.

The larval denticle patterns depend on the correct establishment and maintenance of the embryonic parasegment boundaries. Establishment of the parasegment boundary depends on an intercellular signaling circuit being set up between adjacent cells which delimits the boundary between them. This circuit primarily involves the segmentation genes *wingless*, *hedgehog*, and *engrailed*, which are expressed in restricted domains within the parasegment (Fig. 2.36). *engrailed*, as we saw earlier, encodes a transcription factor, while Wingless and Hedgehog are secreted signal proteins that act via receptor proteins on cell surfaces to alter

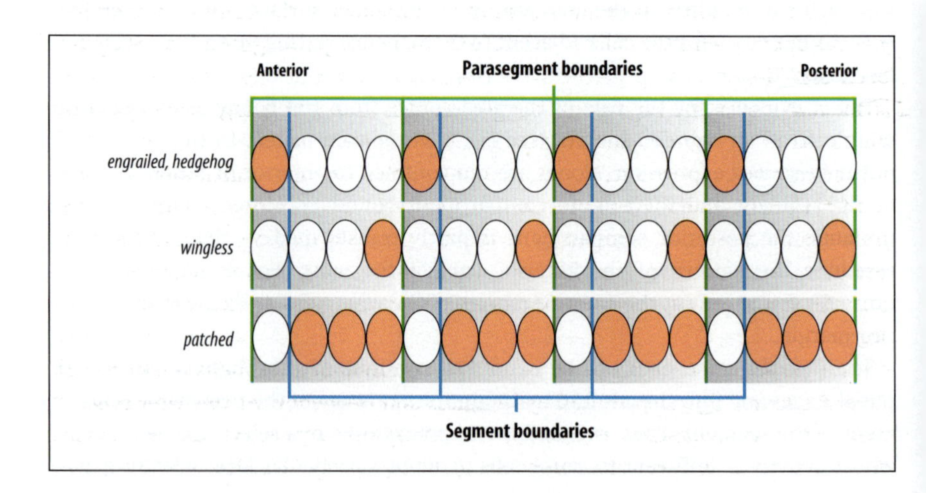

Fig. 2.36 The domains of expression of the segmentation genes. At the cellular blastoderm stage, the *engrailed* gene is expressed at the anterior margin of each parasegment together with *hedgehog*, while *wingless* is expressed at the posterior margin. After gastrulation, *patched* is expressed in all of the cells that express neither *engrailed* nor *hedgehog*. About two cell divisions occur between the time that the parasegments are delimited and the time the embryo hatches.

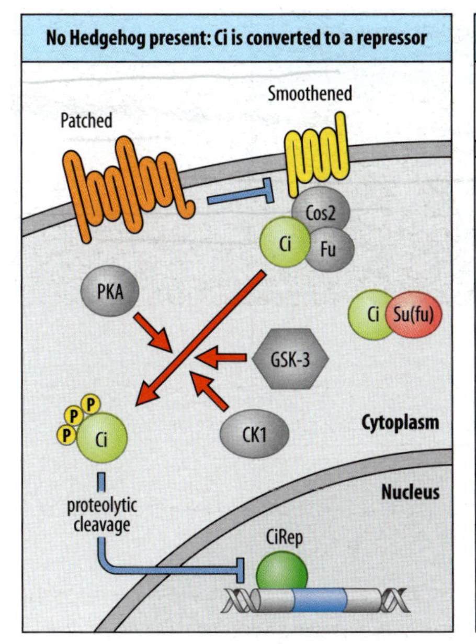

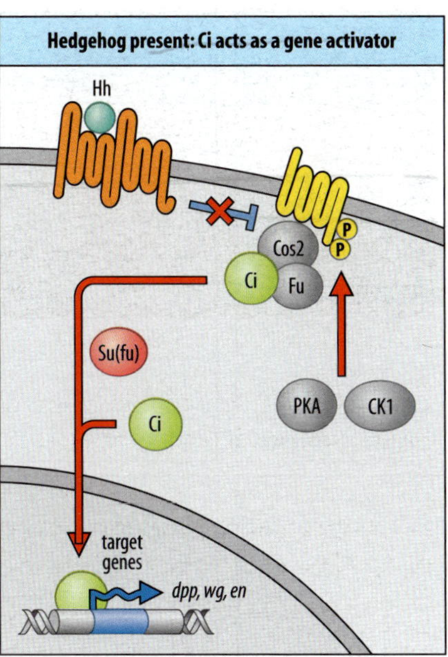

Fig. 2.37 The Hedgehog signaling pathway. Left panel: in the absence of Hedgehog, the membrane protein Patched, which is the receptor for Hedgehog, inhibits the membrane protein Smoothened. In the absence of Smoothened activity, the transcription factor Cubitus interruptus (Ci) is held in the cytoplasm in two protein complexes—one associated with Smoothened and another with the protein Suppressor of fused (Su(fu)). In the absence of Hedgehog, Ci in the Smoothened complex is also phosphorylated by several protein kinases (glycogen synthase kinase (GSK-3), protein kinase A (PKA) and casein kinase 1 (CK1)), which results in the proteolytic cleavage of Ci and formation of the truncated protein CiRep. This enters the nucleus and acts as a repressor of Hedgehog target genes. Right panel: Hedgehog (Hh) binding to Patched lifts the inhibition on Smoothened and blocks production of CiRep. Ci is released from both its complexes in the cytoplasm, enters the cell nucleus, and acts as a gene activator. Genes activated in response to Hedgehog signaling include *wingless* (*wg*), *decapentaplegic* (*dpp*), and *engrailed* (*en*).

patterns of gene expression. Wingless is one of the prototype members of the so-called **Wnt family** of signal molecules, which are key players in development in vertebrates as well as invertebrates and are involved in determining cell fate and cell differentiation in many different aspects of development. **Hedgehog** similarly has homologs in other animals, including vertebrates.

We have seen how the pair-rule genes first delimit the parasegments by inducing the expression of *engrailed* at the anterior margin of each parasegment (see Fig. 2.33). Cells expressing *engrailed* now also express the secreted signal Hedgehog, which activates and maintains expression of the signal protein Wingless in the row of cells immediately anterior to the Engrailed-expressing cells. The general signaling pathway by which Hedgehog acts is shown in Fig. 2.37. The secreted Wingless protein in turn provides a signal that feeds back over the boundary to maintain *hedgehog* and *engrailed* expression, thus stabilizing and maintaining the compartment boundary (Fig. 2.38). At parasegment boundaries and in many of their other developmental functions, Wingless and other Wnt proteins act through the signaling pathway illustrated in Fig. 2.39, which is often called the **canonical Wnt pathway**. A computer model of the Wingless–Hedgehog–Engrailed circuitry shows that it is remarkably robust and resistant to variations in the genetic components governing the circuit's behavior. Some time after the parasegment boundary has been established, a deep groove develops at the posterior edge of each *engrailed* stripe and marks the edge of each segment. This means that *engrailed* expression marks the posterior compartment of the larval segment.

In the embryo, therefore, the continuous epithelium of the abdomen becomes divided up into a succession of anterior and posterior compartments, with a former parasegment boundary forming the boundary between the anterior and posterior compartments of a segment (see Fig. 2.28). Once compartment boundaries are consolidated, signaling from the parasegment boundary sets up the pattern of each segment, leading eventually to the differentiation of the epidermal cells to produce the cuticle pattern evident on the larva. The pattern of denticle belts is specified in response to the signals provided by Wingless and Hedgehog. At this later stage of embryonic development, Hedgehog and Wingless expression are no longer dependent on each other and can be analyzed separately. Wingless protein moves

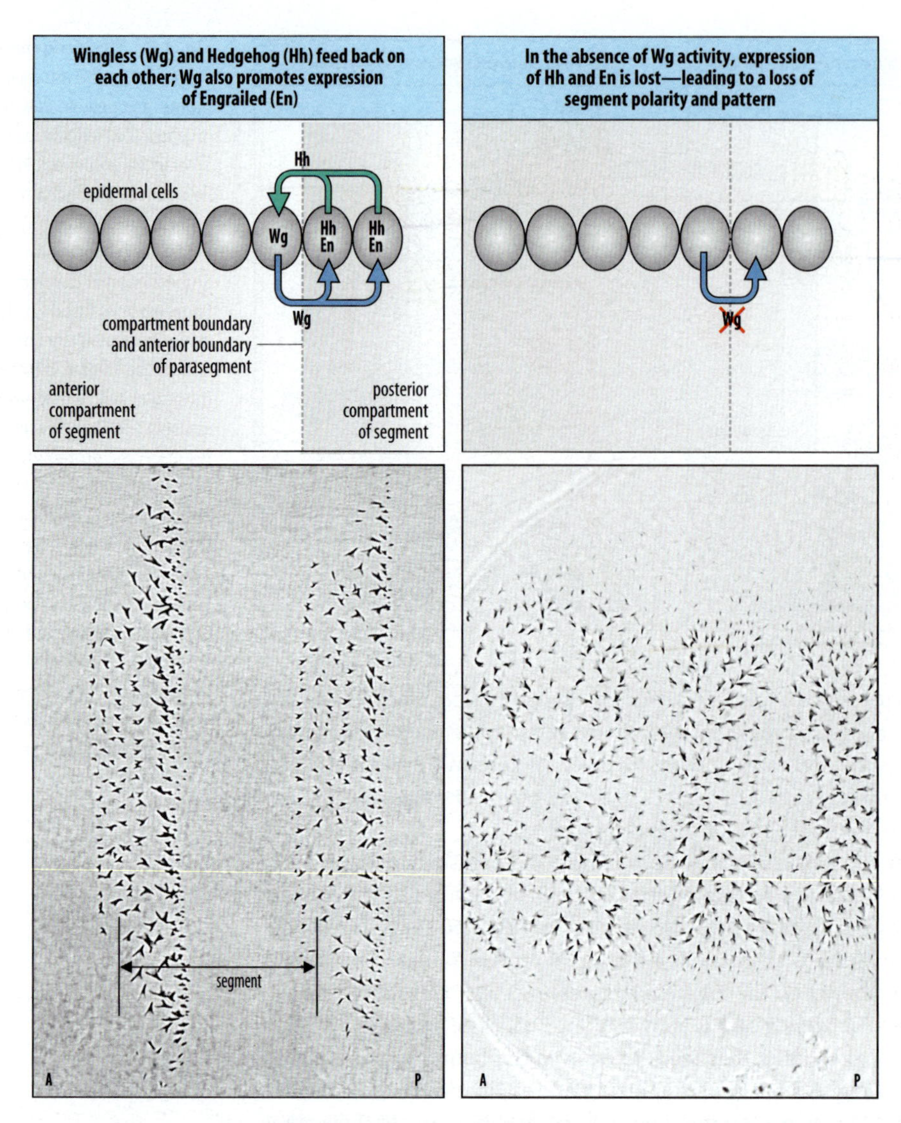

Fig. 2.38 Interactions between *hedgehog*, *wingless*, and *engrailed* genes and proteins at the compartment boundary control denticle pattern. Top left panel: the *engrailed* gene, which encodes a homeodomain transcription factor, is expressed in cells along the anterior margin of the parasegment, which delimits the future boundary of the posterior compartment in the segment; these cells also express the segmentation gene *hedgehog* and secrete the Hedgehog protein. Hedgehog protein activates and maintains expression of the segmentation gene *wingless* in adjacent cells across the compartment boundary. Wingless protein feeds back on *engrailed*-expressing cells to maintain expression of the *engrailed* gene and *hedgehog*. These interactions stabilize and maintain the compartment boundary. Top right panel: in mutants where the *wingless* gene is inactivated and the Wingless protein is absent, neither *hedgehog* nor *engrailed* genes are expressed. This leads to loss of the compartment boundary and the normally well defined pattern of denticles within each abdominal segment. Bottom left panel: in the wild-type larva the denticle bands in the ventral cuticle are confined to the anterior part of the segment and are dependent on the activity of *hedgehog* and *wingless* genes. Bottom right panel: in the *wingless* mutant, denticles are present across the whole ventral surface of the segment in what looks like a mirror-image repeat of the anterior segment pattern.

Photographs from Lawrence, P.: 1992.

posteriorly over the compartment boundary to pattern the posterior compartment of the segment, and also moves anteriorly acts to pattern the anterior part of the segment immediately anterior to the cells in which it is expressed. Wingless moves a shorter distance posteriorly than anteriorly as it is more rapidly degraded in the posterior compartment, and so its effects on patterning extend over a longer range

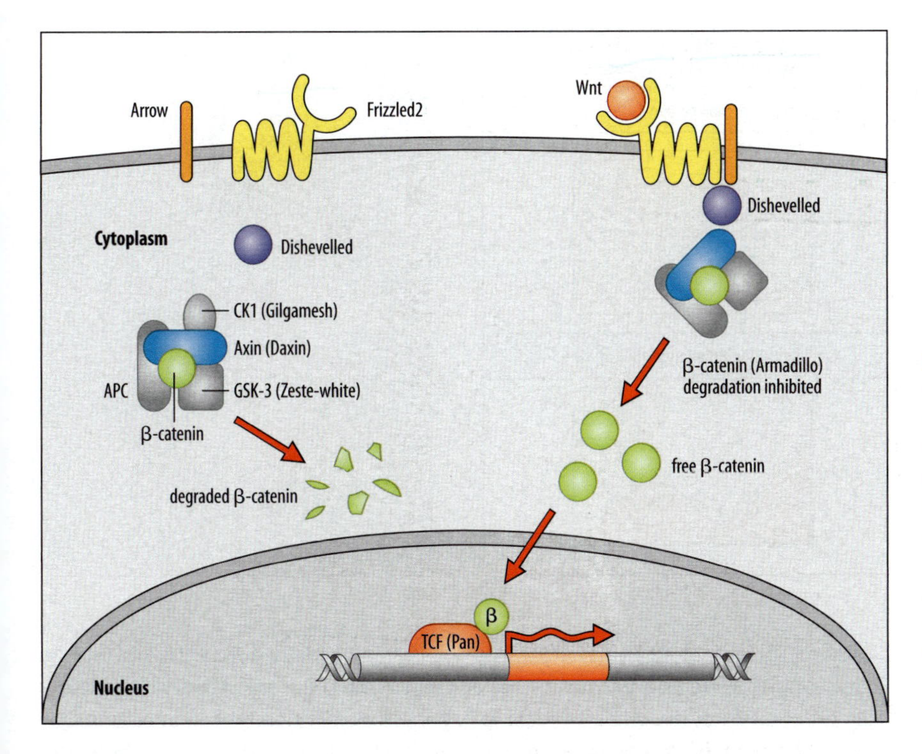

Fig. 2.39 The Wingless signaling pathway in *Drosophila*. When Wingless binds to its receptor, Frizzled, a signal is transduced across the membrane to the intracellular signaling protein Dishevelled. Activated Dishevelled interacts with Daxin (Axin) to prevent a protein complex also containing the protein kinase Zeste-white (GSK-3) and APC from preparing Armadillo (β-catenin) for ubiquitin-mediated degradation. Stabilized β-catenin enters the nucleus where it forms a complex with the transcription factor Pangolin (TCF; Pan) and regulates expression of target genes.

in the anterior compartment. In the anterior compartment, Wingless represses genes required for denticle formation, and the cells that lie immediately anterior to the Wingless-expressing line of cells are specified as epidermal cells that will produce smooth cuticle, while the anterior extent of Wingless signaling delimits the posterior edge of the denticle belts in that segment. In mutants that lack *wingless* function, denticles are produced in the normally smooth regions (see Fig. 2.38).

Hedgehog signaling does not pattern the posterior segment compartment in which Hedgehog is actually expressed, as posterior compartment cells are 'blind' to its action. The Hedgehog signals move anteriorly to maintain Wingless function and also posteriorly to help pattern the anterior compartment of the next segment. The signals provided by Hedgehog and Wingless lead, by a complex set of interactions, to the expression of a number of genes in non-overlapping narrow stripes that specify the larval segmental pattern of smooth cuticle or specific rows of denticles.

Patterning within compartment boundaries is also visible in the pattern of the epidermis on the abdominal segments of the adult fly (Fig. 2.40). The epidermis of each segment is subdivided into a number of types—at least five in the anterior compartment and three in the posterior compartment—which are distinguished by the presence or absence of pigmentation and the positioning of bristles and hairs.

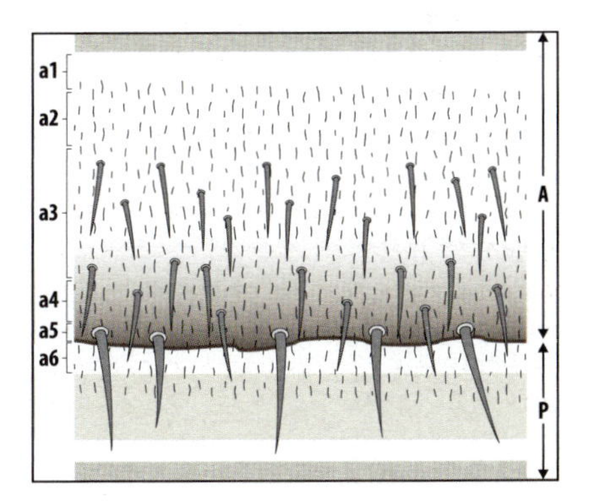

Fig. 2.40 The epidermis of each segment in the abdomen of the adult *Drosophila* is divided into an anterior and a posterior compartment. A number of different regions, a1 to a6, can be distinguished in the anterior compartment, characterized by different bristles, pigmentation, and gene expression.

2.24 Insect epidermal cells become individually polarized in an antero-posterior direction in the plane of the epithelium

The insect epidermis in each segment not only becomes patterned into bands of epidermal cells of different types, but each cell acquires an individual antero-posterior polarity, reflected in the fact that the hairs and bristles on the adult

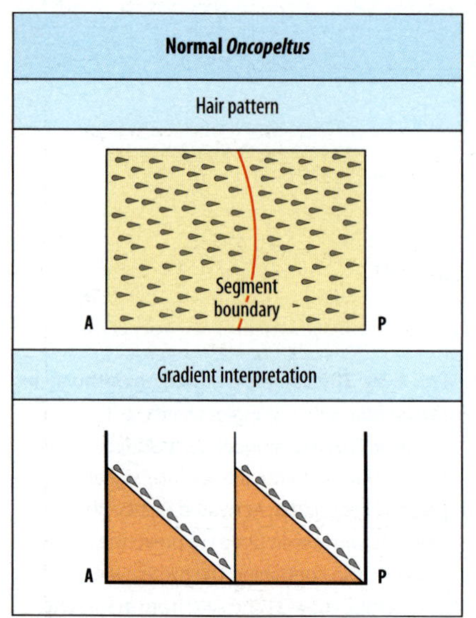

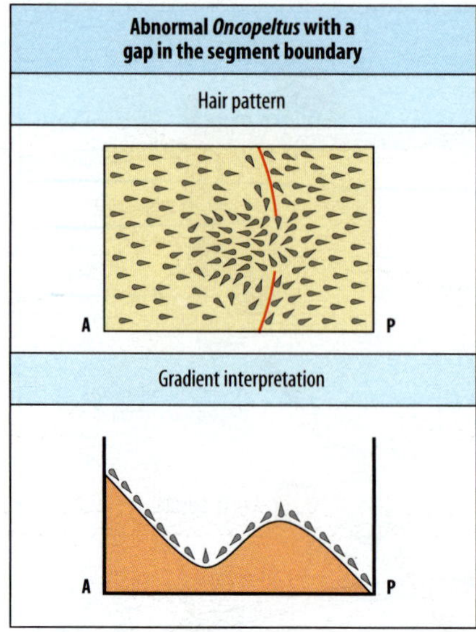

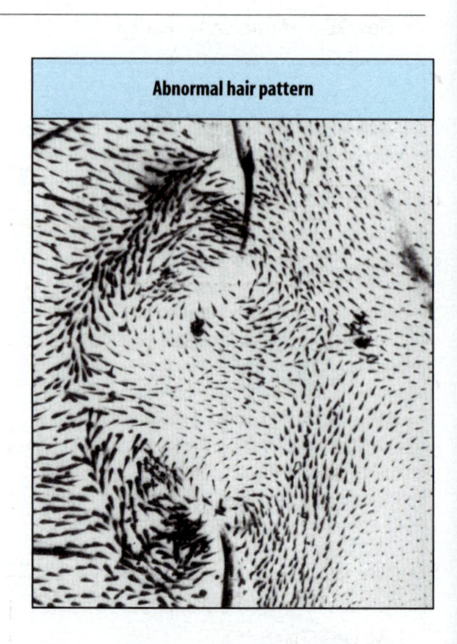

Fig. 2.41 Gradients could specify polarity in the segments of *Oncopeltus*. The hairs on the cuticle point in a posterior direction and this may reflect an underlying gradient in a morphogen that is partly maintained by the segment boundary (left panels). When there is a gap in the boundary (center panels) the sharp discontinuity in morphogen concentration smoothes out, with a resulting local reversal of the gradient and of the direction in which the hairs point. This is illustrated in the photograph (right panel).

Drosophila abdomen all point backwards (see Fig. 2.40). Many different types of cells have polarity, which simply means that one end of the cell is structurally or functionally different from the other. For example, polarization is clearly seen in cells undergoing chemotaxis in response to a gradient of a chemoattractant, where the advancing end of the cell is quite different from the posterior end. This type of cell polarity is called **planar cell polarity**, to distinguish it from the apical–basal polarity often present across an epithelial sheet. Planar cell polarity is also clearly visible in the *Drosophila* wing, which is composed of a thin sheet of epithelium. The wing epidermal cells all bear hairs and these hairs form at the ends of the cells away from the body—the **distal** ends of the cells—and all point away from the body (see Fig. 2.34). Another example is the insect compound eye, where the individual units (ommatidia) in each half of the eye point away from the midline (see Chapter 9), which is the result of the establishment of planar polarity in the developing photoreceptor cells in the eye imaginal disc. In vertebrates, examples of planar polarity occur in the stereocilia in the inner ear and in the scales of fish. Planar cell polarity is also important in morphogenetic cell movements such as the tissue extension that occurs at gastrulation in *Xenopus*, when the spherical frog blastula is being converted into the more elongated gastrula; this will be discussed in Chapter 7.

The mechanism by which planar cell polarity is established is, however, not yet completely understood. Current hypotheses propose the existence of an informational gradient across the tissue that sets the overall direction of polarity, while local signaling between one end of a cell and the other, and between one cell and its neighbor are thought to set the polarity of the individual cell (see Box 2E, opposite). In the case of the adult insect abdominal epidermis, for example, the signaling gradients set up at compartment boundaries have been proposed to have a role in establishing planar cell polarity as well as in patterning the segment.

A natural case where loss of planar cell polarity can be observed is in adults of the plant-feeding bug *Oncopeltus*, which, like *Drosophila*, have hairs covering each segment, all pointing in a posterior direction (Fig. 2.41, left panel). In *Oncopeltus*, some individuals naturally have a gap in the segment boundary; this results in a rather precise pattern of hairs with altered orientation, many of the hairs near the gap pointing in the reverse direction (Fig. 2.41, center and right panels). If the slope of a gradient within each segment gives the hairs their polarity, a gap in a segment

Box 2E Planar cell polarity

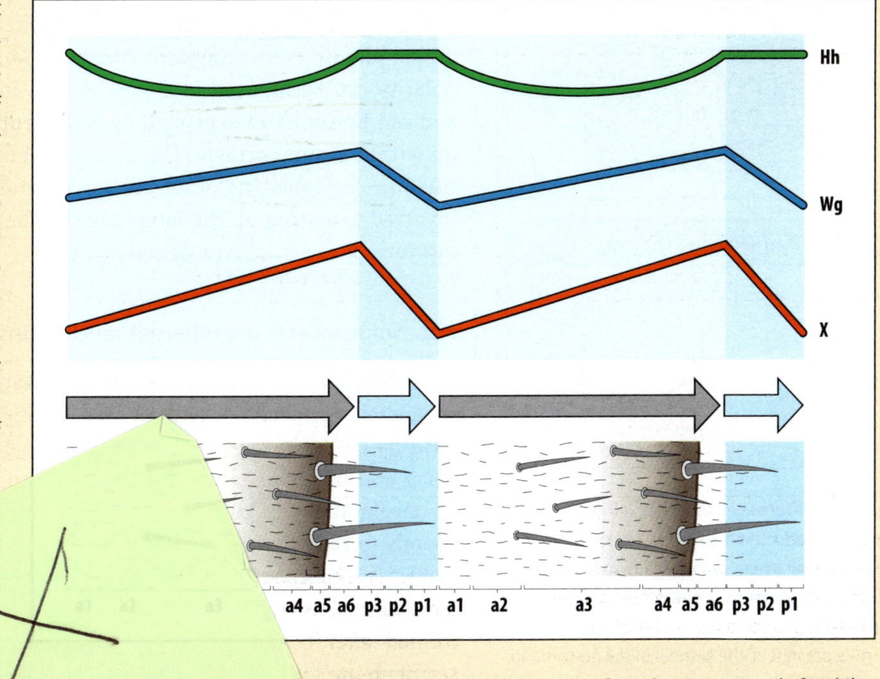

Cells in an epithelium have a distinct apical–basal polarity across the epithelium, but more relevant to developmental patterning is a polarity in the plane of the epithelium, which is known as planar cell polarity. This is visible, for example, in the epidermis of the adult fly, where the abdominal epidermis bears hairs all pointing in the posterior direction. (see figure, right; shows two adult epidermal segments, labeled as in Fig. 2.40; arrows indicate the direction of polarity). Other examples in *Drosophila* are the proximo-distal polarity of the epidermal cells of the wing (see Chapter 9), and the orientation of the ommatidia in the *Drosophila* eye (see Chapter 9).

Planar cell polarity poses two main questions. How is the overall direction of the polarity set throughout the tissue? And how are individual cells polarized? Molecular and genetic studies in *Drosophila* have identified two maingroups of proteins involved in planar polarity, and homologs of these proteins seem to be involved in planar polarity in other organisms, including vertebrates. In the fly abdominal epidermis, for example, one group of proteins appears to generate an informational gradient of activity (X) across the tissue in response to the segment patterning set up by the actions of Hedgehog (Hh) and Wingless (Wg) at the compartment boundaries (see figure). This group includes two cadherins, Dachsous and Fat, that are involved in forming the long-range gradient. Cadherins are transmembrane cell-adhesion molecules that are involved in cell–cell communication and cell polarization through their interactions with the intracellular cytoskeleton (see Box 7A, p. 259). Thus the information gradient may be formed by transfer of polarity information from cell to cell by Fat and Dachsous, rather than by the diffusion of some extracellular molecule. The abdominal hairs in *Drosophila* face in the same direction in both the anterior and posterior compartments, and so cells must read the gradient in one direction in the anterior compartment and in the opposite direction in the posterior compartment (see figure above).

Mutations in the second group of planar polarity proteins result in complete or partial loss of orientation of the abdominal hairs, which point in all directions. This group includes the cell-surface receptor Frizzled. When it acts in planar cell polarity, Frizzled stimulates a different signaling pathway from the canonical Wnt pathway triggered by Wingless in embryonic segment patterning (see Fig. 2.38). The signaling pathway used in planar cell polarity in both insects and vertebrates is known as the non-canonical Wnt pathway (see Fig. 7.33 for the proposed pathway in vertebrates), and in the case of *Drosophila* planar polarity Wingless itself is probably not involved. The signals that activate Frizzled in *Drosophila* planar cell polarity are so far unknown. Instead of stabilizing β-catenin, the pathway leads via Dishevelled to the activation of the small GTPase Rho and the protein kinases Rho-associated kinase (ROK) and Jun kinase (JNK), and typically results in polarization of the cell's internal cytoskeleton, among other effects. In vertebrates, as we shall see in Chapter 7, the same signaling pathway is used to regulate cell polarity and cell movements during gastrulation.

How the pathway is activated in response to the cadherin-mediated gradient is still a matter for debate. One suggestion is that cells align their polarity along the gradient by setting up a corresponding gradient of Frizzled activity. Feedback mechanisms result in cells averaging their level of Frizzled activity between that of their neighbors, and cells form hairs pointing away from the neighbor with higher Frizzled activity (see figure below; the Frizzled gradient is indicated in blue and the arrows indicate the direction of cell polarity). Cell–cell communication between the cells on the boundary of a clone of cells lacking Frizzled (*fz*$^-$) results in an averaged level of Frizzled activity, resulting in polarity reversals on one side of the clone. Other hypotheses depend on the polarized distribution of Frizzled and the proteins it interacts with to different faces of the cell, thus setting the direction of signaling between adjacent cells.

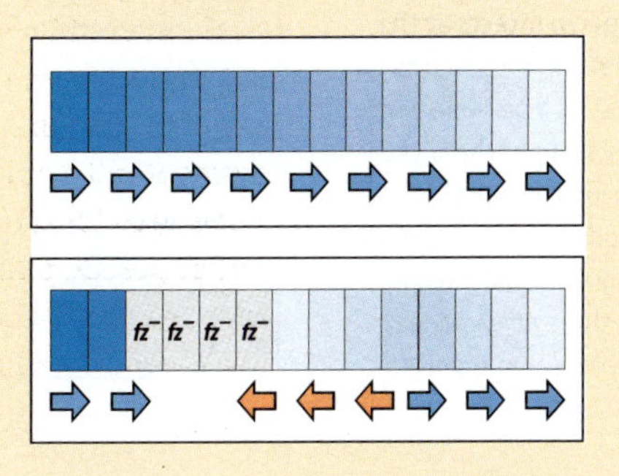

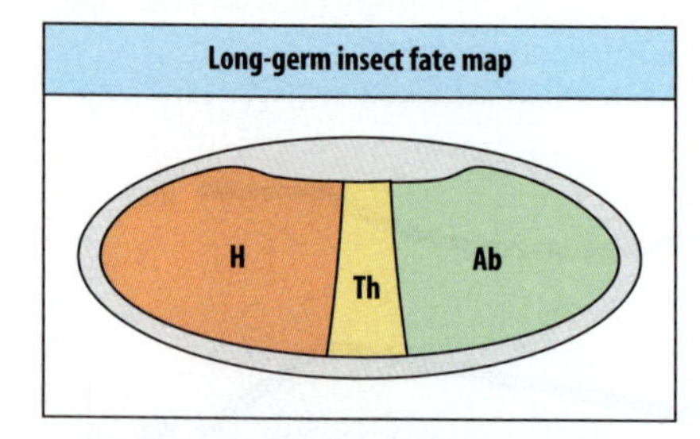

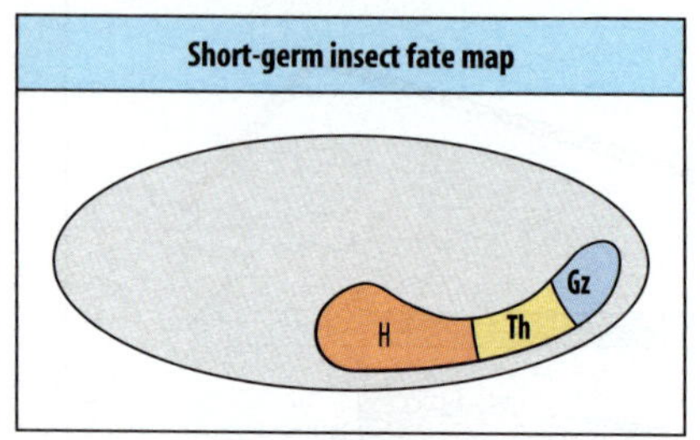

Fig. 2.42 Differences in the development of long-germ and short-germ insects. Top panel: the general fate map of long-germ insects such as *Drosophila* shows that the whole of the body plan—head (H), thorax (Th), and abdomen (Ab)—is present at the time of initial germ-band formation. Bottom panel: in short-germ insects, only the anterior regions of the body plan are present at this embryonic stage. Most of the abdominal segments develop later, after gastrulation, from a posterior growth zone (Gz).

boundary would result in a local change in the gradient, the sharp change in concentration at the normal boundary being smoothed out and forming a local gradient running in the opposite direction. This would result in the hairs in this region pointing in the opposite direction. Mutations that cause similar reversals in polarity can be induced in clones of cells in the *Drosophila* abdominal epidermis, and can be similarly explained by the disruption of an informational gradient set up within a compartment. Genes essential for planar cell polarity have been identified in *Drosophila* and other organisms, and have identified some of the proteins involved in setting up the long-range gradient and in the signaling pathways that interpret it to set individual cell polarity.

2.25 Some insects use different mechanisms for patterning the body plan

Drosophila belongs to an evolutionarily advanced group of insects, in which all the segments are specified more or less at the same time in development, as shown both by the striped patterns of pair-rule gene expression in the acellular blastoderm and the appearance of all the segments shortly after gastrulation. This type of development is known as **long-germ development**, as the blastoderm corresponds to the whole of the future embryo. Many other insects, such as the flour beetle *Tribolium*, have a **short-germ development**. In short-germ development the blastoderm is short and forms only anterior segments. The posterior segments are formed after completion of the blastoderm stage and gastrulation. Thus, most segments are formed from a cellular blastoderm and posterior segments are added by growth in the posterior region (Fig. 2.42). In spite of early differences between them, the mature germ band stages of both long- and short-germ insect embryos look similar. This, therefore, is a common stage through which all insect embryos develop.

The question clearly arises as to what processes are common to the specification of the body plan in long- and short-germ insects. One clear difference is that whereas the patterning of the body plan in the long-germ band of *Drosophila* takes place before cell boundaries form, much of the body plan in short-germ insects is laid down at a later stage, when the posterior segments are generated during growth. At this stage the embryo is multicellular. So are the same genes involved?

There is very good evidence that the same genes and development processes are involved in the patterning of *Tribolium* and *Drosophila*. For example, the gap gene *Krüppel* is expressed at the posterior end of the *Tribolium* embryo at the blastoderm stage and not in the middle region, as in *Drosophila* (Fig. 2.43). It therefore seems to be specifying the same part of the body in the two insects. Similarly, only two repeats of the pair-rule stripes are present at the blastoderm stage, together with a posterior

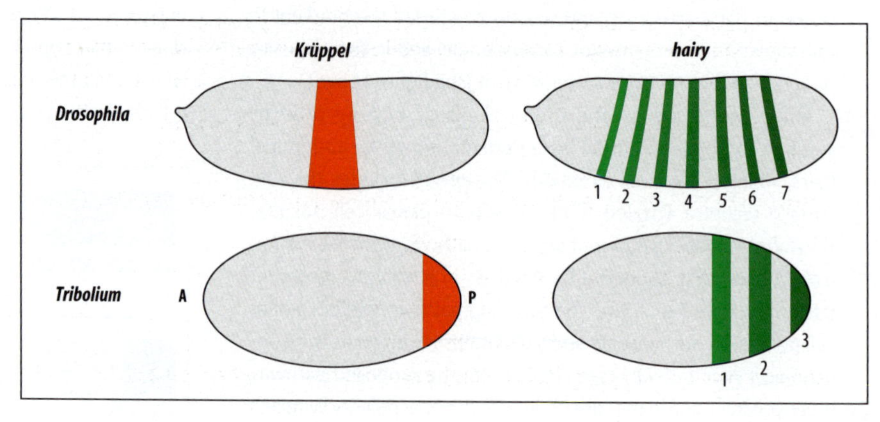

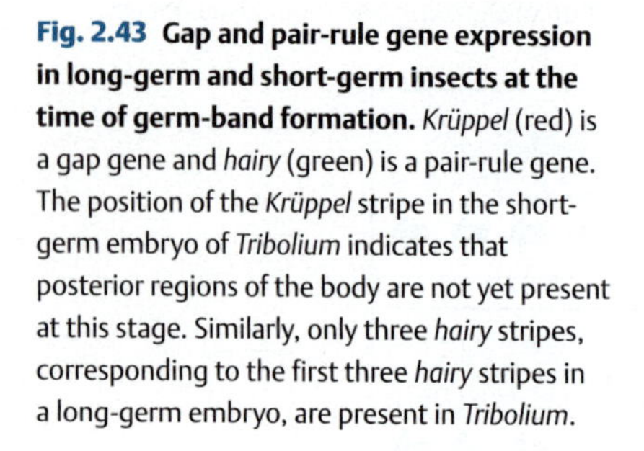
Fig. 2.43 Gap and pair-rule gene expression in long-germ and short-germ insects at the time of germ-band formation. *Krüppel* (red) is a gap gene and *hairy* (green) is a pair-rule gene. The position of the *Krüppel* stripe in the short-germ embryo of *Tribolium* indicates that posterior regions of the body are not yet present at this stage. Similarly, only three *hairy* stripes, corresponding to the first three *hairy* stripes in a long-germ embryo, are present in *Tribolium*.

cap of pair-rule gene expression, in contrast to the seven in *Drosophila*. The genes *wingless* and *engrailed* are also expressed in a similar relation to that in *Drosophila*.

Although not many genes have been studied in detail in other insects, at least one, the gene *engrailed*, is known to be expressed in the posterior region of segments in a variety of insects. The pair-rule gene *even-skipped* (see Section 2.21), while present in the grasshopper (a short-germ insect), may not have a similar role in segmentation. It is, however, involved later in the development of the nervous system in the grasshopper, and is also expressed at the posterior end of the growing germ band.

Looking at some other insects, differences in early development are much more dramatic. In certain parasitic wasps, the egg is small and undergoes cleavage to form a ball of cells, which then falls apart. Each of the resulting small clusters of cells—there can be as many as 400—can develop into a separate embryo. This wasp's development apparently does not depend on maternal information to specify the body axes, and in this respect resembles that of the early mammalian embryo.

Summary

The segmentation genes are involved in patterning the parasegments. One of the first genes to be activated is *engrailed*, which is expressed at the anterior margin of each parasegment, delineating a group of cells which becomes the posterior compartment of a segment. *engrailed* is also a selector gene in that it provides a group of cells with a long-term regional identity. *engrailed* expression shows lineage restriction: cells expressing *engrailed* define the posterior compartment of a segment, and cells never cross from the posterior into the anterior compartment. *engrailed* is turned on by the pair-rule genes, and its expression is maintained by the segmentation genes *wingless* and *hedgehog*, which stabilize the compartment boundary. The compartment boundary is involved in specifying pattern and cell polarity within the segment. Studies in other insects also suggest that a discrete gradient of positional information is set up in each segment, delimited by the boundaries, and this patterns the segment. In contrast to *Drosophila*, some insects have a short-germ pattern of development in which posterior segments are added by growth after the cellular blastoderm stage.

Summary: gene expression defines segment compartments

pair-rule gene expression

⇩

segmentation and selector gene *engrailed* activated in anterior of each parasegment, defining anterior compartment of parasegment and posterior compartment of segment

⇩

engrailed-expressing cells also express segmentation gene *hedgehog*

⇩

cells on other side of compartment boundary express segmentation gene *wingless*

⇩

engrailed expression maintained and compartment boundary stabilized by Wingless and Hedgehog proteins

compartment boundary provides signaling center from which segment is patterned

Specification of segment identity

Each segment has a unique identity, most easily seen in the larva in the characteristic pattern of denticles on the ventral surface (see Fig. 2.35). Since the same segmentation genes are turned on in each segment, what makes the segments different from each other? Their identity is specified by a class of master regulatory genes, known as **homeotic selector genes**, which set the future developmental pathway of each segment. A selector gene controls the activity of other genes and is required throughout development to maintain this pattern of gene expression. Homologs of the *Drosophila* selector genes that control segment identity have subsequently been discovered in virtually all other animals, where they broadly control identity along the antero-posterior axis, as we shall see in Chapter 4.

2.26 Segment identity in *Drosophila* is specified by genes of the Antennapedia and bithorax complexes

The first evidence for the existence of genes that specify segment identity came from unusual and striking mutations in *Drosophila* that produced **homeotic transformations**—the conversion of one segment into another. Eventually the genes that produced these mutations were identified and the intricate way in which they specified segment identity was worked out. The discovery of these genes, now called the **Hox genes**, had an enormous impact on developmental biology. Homologous genes were found in other animals where they acted in a similar way, essentially specifying identity along the antero-posterior axis. Hox genes are now considered as one of the fundamental defining features of multicellular animals. In 1995, the American geneticist Edward Lewis shared the Nobel Prize in Physiology or Medicine for his pioneering work on the *Drosophila* homeotic gene complexes and how they function. In *Drosophila* the Hox genes are organized into two gene clusters which are called the HOM complexes (for homeotic; Fig. 2.44; see also Box 4A, p. 156). The Hox genes all encode transcription factors and take their name from the distinctive **homeobox** DNA motif that encodes part of the DNA-binding site of these proteins.

The two homeotic gene clusters in *Drosophila* are called the bithorax and Antennapedia gene complexes and are named after the mutations that first revealed their existence. Flies with the *bithorax* mutation have part of the haltere (the balancing organ on the third thoracic segment) transformed into part of a wing (Fig. 2.45), whereas flies with the dominant *Antennapedia* mutation have their antennae transformed into legs. Genes identified by such mutations are called

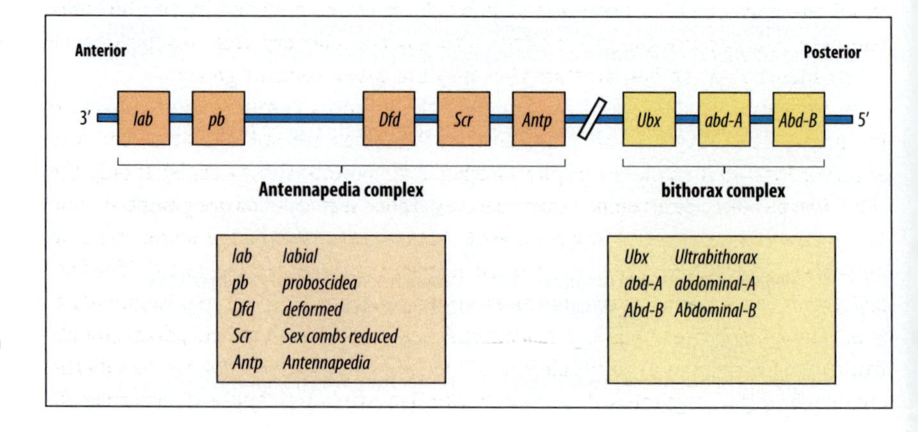

Fig. 2.44 The Antennapedia and bithorax homeotic selector gene complexes. The order of the genes from 3′ to 5′ in each complex reflects both the order of their spatial expression (anterior to posterior) and the timing of expression (3′ earliest).

Fig. 2.45 Homeotic transformation of the wing and haltere by mutations in the bithorax complex. Top panel: in the normal adult, both wing and haltere are divided into an anterior (A) and a posterior (P) compartment. Middle panel: the *bithorax* mutation transforms the anterior compartment of the haltere into an anterior wing region. The mutation *postbithorax* acts similarly on the posterior compartment, converting it into posterior wing (not illustrated). Bottom panel: if both mutations are present together, the effect is additive and the haltere is transformed into a complete wing, producing a four-winged fly.

Photograph courtesy of E. Lewis, from Bender, W., et al.: 1983. Illustration after Lawrence, P.: 1992.

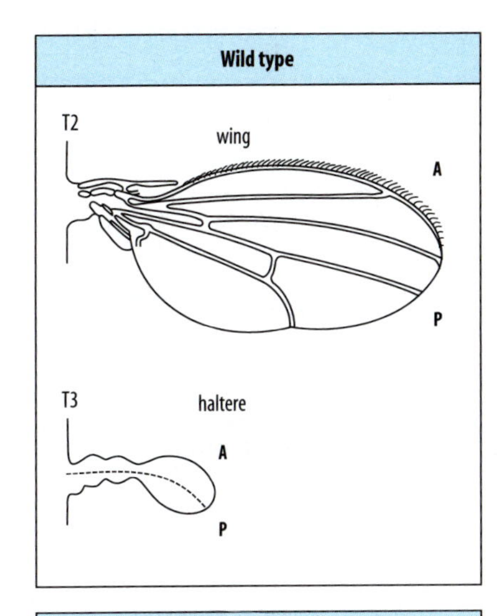

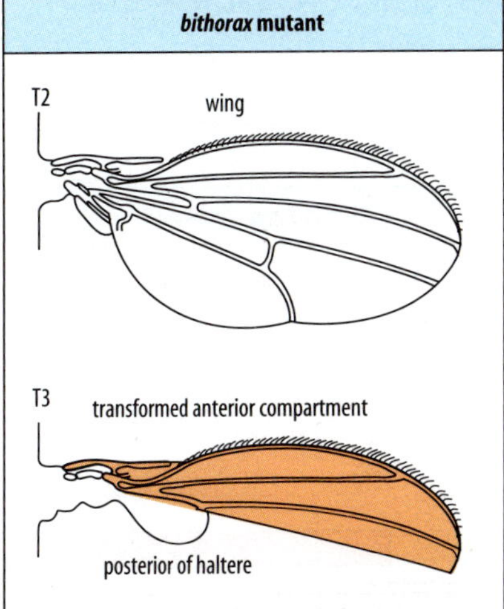

homeotic genes because when mutated they result in **homeosis**—the transformation of a whole segment or structure into another related one, as in the transformation of antenna to leg. These bizarre transformations arise out of the homeotic selector genes' key role as identity specifiers. They control the activity of other genes in the segments, thus determining, for example, that a particular imaginal disc will develop as wing or haltere. The bithorax complex controls the development of parasegments 5–14, while the Antennapedia complex controls the identity of the more anterior parasegments. The action of the bithorax complex is the best understood and will be discussed first.

2.27 Homeotic selector genes of the bithorax complex are responsible for diversification of the posterior segments

The bithorax complex of *Drosophila* comprises three homeobox genes: *Ultrabithorax*, *adbominal-A*, and *Abdominal-B*. These genes are expressed in the parasegments in a combinatorial manner (Fig. 2.46, top panel). *Ultrabithorax* is expressed in all parasegments from 5 to 12, *abdominal-A* is expressed more posteriorly in parasegments 7–13, and *Abdominal-B* more posteriorly still from parasegment 10 onwards. Because the genes are also active to varying extents in different parasegments, their combined activities define the character of each parasegment. *Abdominal-B* also suppresses *Ultrabithorax*, such that *Ultrabithorax* expression is very low by parasegment 14, as *Abdominal-B* expression increases. The pattern of activity of the bithorax complex genes is determined by gap and pair-rule genes.

The role of the bithorax complex was first indicated by classical genetic experiments. In larvae lacking the whole of the bithorax complex (see Fig. 2.46, second panel), every parasegment from 5–13 develops in the same way and resembles parasegment 4. The bithorax complex is therefore essential for the diversification of these segments, whose basic pattern is represented by parasegment 4. This parasegment can be considered as being a type of 'default' state, which is modified in all the parasegments posterior to it by the proteins encoded by the bithorax complex. It is because the genes of the bithorax complex are able to superimpose a new identity on the default state that they are called selector genes.

An indication of the role of each gene of the bithorax complex can be deduced by looking at embryos constructed so that the genes are put one at a time into embryos lacking the whole complex (see Fig. 2.46, bottom three panels). If only the *Ultrabithorax* gene is present, the resulting larva has one parasegment 4, one parasegment 5, and eight parasegments 6. Clearly, *Ultrabithorax* has some effect on all parasegments from 5 backward and can specify parasegments 5 and 6. If *abdominal-A* and *Ultrabithorax* are put into the embryo, then the larva has parasegments 4, 5, 6, 7, and 8, followed by five parasegments 9. So *abdominal-A* affects parasegments from 7 backward, and in combination, *Ultrabithorax* and *abdominal-A* can specify the character of parasegments 7, 8, and 9. Similar principles apply to *Abdominal-B*,

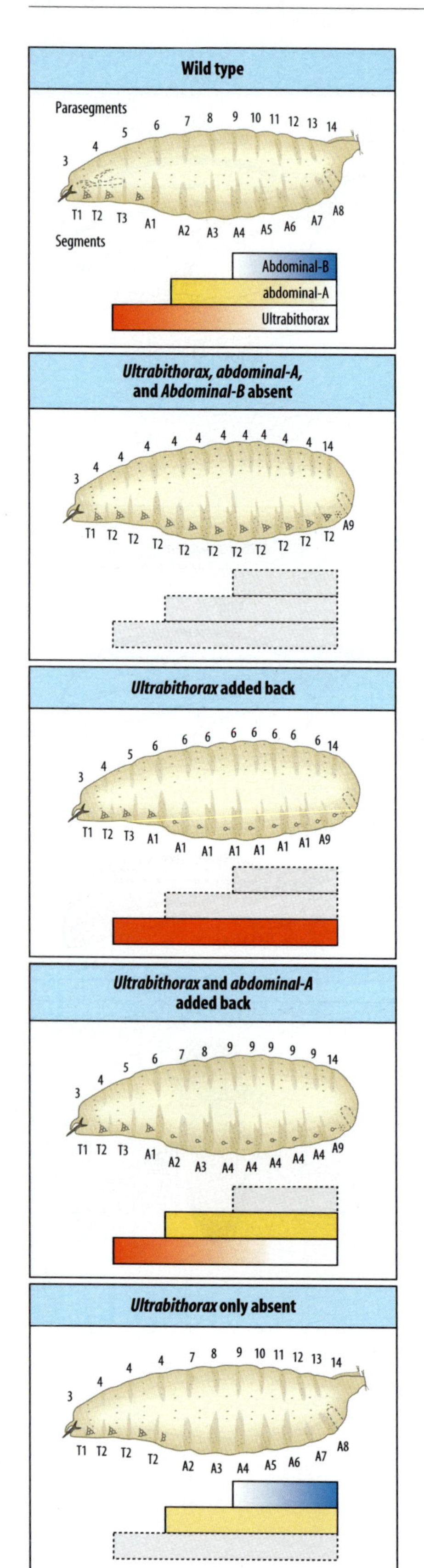

Fig. 2.46 The spatial pattern of expression of genes of the bithorax complex characterizes each parasegment. In the wild-type embryo (top panel) the expression of the genes *Ultrabithorax*, *abdominal-A*, and *Abdominal-B* are required to confer an identity on each parasegment. Mutations in the bithorax complex result in homeotic transformations in the character of the parasegments and the segments derived from them. When the bithorax complex is completely absent (second panel), parasegments 5–13 are converted into nine parasegments 4 (corresponding to segment T2 in the larva), as shown by the denticle and bristle patterns on the cuticle. The three lower panels show the transformations caused by the absence of different combinations of genes. When the *Ultrabithorax* gene alone is absent (bottom panel), parasegments 5 and 6 are converted into 4. In each case the spatial extent of gene expression is detected by *in situ* hybridization (see Box 3D, pp. 112–13). Note that the specification of parasegment 14 is relatively unaffected by the bithorax complex.

whose domain of influence extends from parasegment 10 backward, and which is expressed most strongly in parasegment 14. Differences between individual segments may reflect differences in the spatial and temporal pattern of Hox gene expression.

These results illustrate an important principle, namely that the character of the parasegments is specified by the genes of the bithorax complex acting in a combinatorial manner. Their combinatorial effect can also been seen by taking the genes away, one at a time, from the wild type. In the absence of *Ultrabithorax*, for example, parasegments 5 and 6 become converted to parasegment 4 (see Fig. 2.46, bottom panel). There is a further effect on the cuticle pattern in parasegments 7–14 in that structures characteristic of the thorax are now present in the abdomen, showing that *Ultrabithorax* exerts an effect in all these segments. Such abnormalities may result from expression of 'nonsense' combinations of the bithorax genes. For example, in such a mutant, *abdominal-A* protein is present in parasegments 7–9 without *Ultrabithorax* protein, and this is a combination never found normally.

While the gap and pair-rule proteins control the pattern of Hox gene expression, these proteins disappear after about 4 hours. The continued correct expression of the homeotic genes involves two groups of genes—the Polycomb and Trithorax groups. The proteins of the Polycomb group maintain transcriptional repression of homeotic genes where they are initially off, while the Trithorax group maintain expression in those cells where Hox genes have been turned on. These genes thus maintain expression of the Hox complex.

The downstream targets of Hox proteins are largely unknown, but there may be a very large number of genes whose expression is affected.

2.28 The Antennapedia complex controls specification of anterior regions

The Antennapedia complex comprises five homeobox genes (see Fig. 2.44), which control the behavior of the parasegments anterior to parasegment 5 in a manner similar to the bithorax complex. Since there are no novel principles involved we just mention its role briefly. Several genes within the complex are critically involved in specifying particular parasegments. Mutations in the gene *deformed* affect the ectodermally derived structures of parasegments 0 and 1, those in *Sex combs reduced* affect parasegments 2 and 3, and those in the *Antennapedia* gene affect parasegments 4 and 5, the parasegments that produce the leg imaginal discs. The *Antennapedia* mutation that transforms antennae into legs in the adult fly results from the misexpression of the *Antennapedia* gene in anterior segments in which it is not normally expressed.

2.29 The order of Hox gene expression corresponds to the order of genes along the chromosome

The bithorax and Antennapedia complexes possess some striking features of gene organization. In both, the order of the genes in the complex is the same as the spatial and temporal order in which they are expressed along the antero-posterior axis during development. *Ultrabithorax*, for example, is on the 3′ side of *abdominal-A* on the chromosome, is more anterior in its pattern of expression, and is activated earlier. As we will see, the related Hox gene complexes of vertebrates, whose ancestors diverged from those of arthropods hundreds of millions of years ago, show the same correspondence between gene order and order of expression. This highly conserved temporal and spatial **co-linearity** must be related to the mechanisms that control the expression of these genes.

The complex yet subtle control to which the genes of the bithorax complex are subject is seen in experiments in which production of Ultrabithorax protein is forced in all segments. This targeted gene expression is achieved by linking the Ultrabithorax protein-coding sequence to a heat-shock promoter—a promoter that is activated at 29°C—and introducing the novel DNA construct into the fly genome using a P element (Box 2F, p. 82). When this transgenic embryo is given a heat shock for a few minutes, the extra *Ultrabithorax* gene is transcribed and the protein is made at high levels in all cells. This has no effect on posterior parasegments (in which Ultrabithorax protein is normally present), with the exception of parasegment 5 which, for unknown reasons, is transformed into parasegment 6 (this may reflect a quantitative effect of the Ultrabithorax protein). However, all parasegments anterior to 5 are also transformed into parasegment 6. While this seems a reasonably simple and expected result, consider what happens to parasegment 13.

Transcription of *Ultrabithorax* is normally suppressed in parasegment 13 in wild-type embryos, yet when the protein itself is produced in the parasegment as a result of heat shock, it has no effect. By some mechanism, the Ultrabithorax protein is rendered inactive in this parasegment. This phenomenon is quite common in the specification of the parasegments and is known as phenotypic suppression or posterior prevalence: it means that Hox gene products normally expressed in anterior regions are suppressed by more posterior products.

While the role of the bithorax and Antennapedia complexes in controlling segment identity is well established, we do not yet know very much about how they interact with the genes acting next in the developmental pathway. These will be the genes that actually specify the structures that give the segments their unique identities. What, for example, is the pathway that leads to a thoracic rather than an abdominal segment structure? Does each Hox gene activate just a few or many target genes? Do target genes with different functions act in concert to form a particular structure? Answers to these question will help us understand how changes in a single gene can cause a homeotic transformation like antenna into leg. The role of the Hox genes in the evolution of different body patterns will be considered in Chapter 14.

2.30 The *Drosophila* head region is specified by genes other than the Hox genes

As mentioned earlier in the chapter, the *Drosophila* larval head anterior to the mandibles is formed by the three most anterior parasegments. The segmental structure is also evident in the embryonic brain, which is composed of three segments (neuromeres). The specification of the anterior head and embryonic nervous system is not, however, under the control of either the pair-rule genes or the Hox genes.

Box 2F Targeted gene expression and misexpression screening

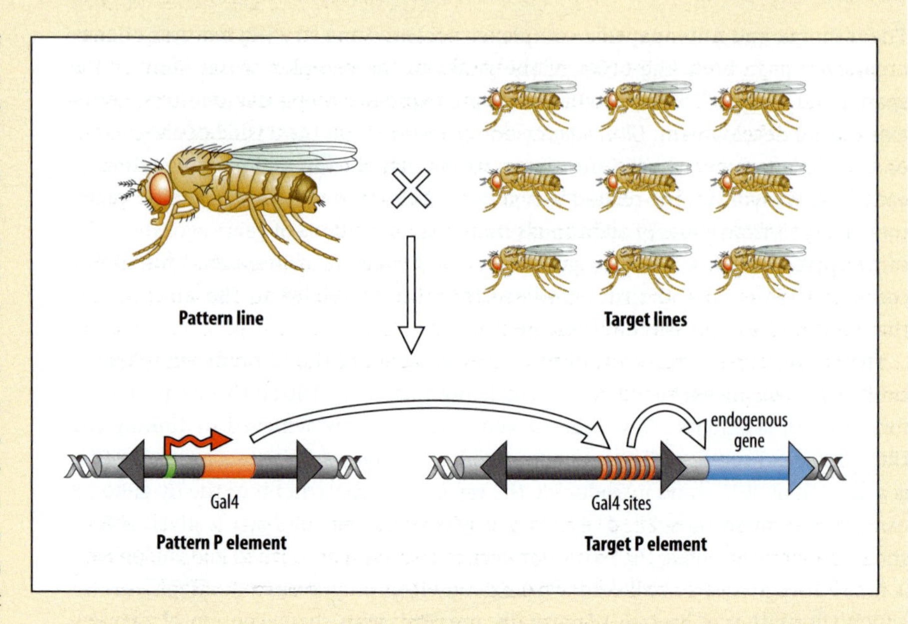

The ability to turn on the expression of a gene in a particular place and time during development is very useful for analyzing its role in development. This is called targeted gene expression and can be achieved in several ways. One approach is to give the selected gene a heat-shock promoter using standard genetic techniques such as P-element-mediated transformation (see Box 2C, p. 59). This enables the gene to be switched on by a sudden rise in the temperature at which the embryos are being kept. By adjusting the temperature, the timing of expression of genes attached to this promoter can be controlled; the effects of expressing a gene at different stages of development can be studied in this way.

Another approach to targeted gene expression uses the transcription factor Gal4 from yeast. This protein can activate transcription of any gene whose promoter has a Gal4-binding site. In *Drosophila*, genes with Gal4-responsive promoters can be created by inserting the Gal4-binding site. To turn on the target gene, Gal4 itself has to be produced in the embryo. In one approach, Gal4 can be produced in a designated region or at a particular time in development by introducing a P element transgene in which the yeast Gal4-coding region is attached to a *Drosophila* regulatory region known to be activated in that situation (see figure). A second, more versatile, approach is based on the so-called **enhancer-trap** technique. The Gal4-coding sequence is attached to a vector that integrates randomly into the *Drosophila* genome. The Gal4 gene will come under the control of the promoter and enhancer region adjacent to its site of integration, and so Gal4 protein will be produced where or when that gene is normally expressed. A variety of *Drosophila* lines with differing patterns of Gal4 expression for different purposes have been produced by this technique.

The selected target gene will be silent in the absence of Gal4. To activate it in a particular tissue, for example, flies that express Gal4 in that tissue are crossed with flies in which the target gene has Gal4-binding sites in its regulatory region. The effect of the novel pattern of target gene expression on development can then be observed. Using the Gal4 system, the pair-rule gene *even-skipped* was expressed in even-numbered parasegments, and this led to changes in the pattern of denticles on the cuticle.

This approach has been adapted to large-scale screens that look systematically for genes whose overexpression or misexpression in a particular tissue causes a mutant phenotype. This is known as misexpression screening. In this case, flies expressing Gal4 in the tissue of interest are crossed with large numbers of different lines of target flies carrying random insertions of the Gal4-binding site, and the progeny screened for a mutant phenotype. This approach is a useful complement to the more conventional genetic screens described in Box 2A (pp. 36–37), which generally detect loss-of-function mutations. If the target flies also carry a mutation in a known gene, misexpression screening can be used to identify genes whose overexpression enhances or suppresses the mutation. This approach can identify genes whose products interact directly or are part of the same pathway.

Instead, it appears to be determined at the blastoderm embryonic stage by the overlapping expression of the genes *orthodenticle*, *empty spiracles*, and *buttonhead*, which all encode gene-regulatory proteins and resemble the gap genes in their effect on phenotype; they are sometimes referred to as the cephalic gap genes. Unlike the gap genes regionalizing the trunk and abdomen, however, they do not regulate each other's activity. Nor is it known whether they work in a combinatorial way like the Hox genes to also confer specific identities on the head segments. *Orthodenticle* and *empty spiracles* have homologs that carry out similar functions in vertebrates, and it is clear that this head-specification system, like the Hox genes, is of ancient evolutionary origin in multicellular animals.

The genes *orthodenticle* and *empty spiracles* are examples of homeobox-containing genes that lie outside the Hox gene clusters. The DNA-binding homeodomain is not confined to the proteins encoded by the Hox gene clusters, and similar domains are found in many other transcription factors involved in development in both *Drosophila* and vertebrates.

Summary

Segment identity is conferred by action of selector or homeotic genes, which direct the development of different parasegments so that each acquires a unique identity. Two clusters of selector genes, known generally as the Hox genes, are involved in specification of segment identity in *Drosophila*: the Antennapedia complex, which controls parasegment identity in the head and first thoracic segment, and the bithorax complex, which acts on the remaining parasegments. Segment identity seems to be determined by the combination of Hox genes active in a particular region. Selector genes need to be switched on continuously throughout development to maintain the required phenotype. The spatial pattern of expression of selector genes along the embryo body is largely determined by the preceding gap gene activity. Mutations in genes of the Antennapedia and bithorax complexes can bring about homeotic transformations, in which one segment or structure is changed into a related one, for example an antenna into a leg. The Antennapedia and bithorax complexes are remarkable in that gene order on the chromosome corresponds to the spatial and temporal order of gene expression along the body.

Summary: homeotic genes and segment identity

gap and pair-rule gene expression expresses parasegments

⇩

selector genes of the HOM complex are expressed along the antero-posterior axis in an order co-linear with the order of genes on the chromosome

⇙ ⇘

Antennapedia complex
lab, pb, Dfd, Scr, Antp

⇩

specify segment identity in head and first thoracic segment

bithorax complex
Ubx, abd-A, Abd-B

⇩

specify segment identity in second and third thoracic segments and abdominal segments

SUMMARY TO CHAPTER 2

The earliest stages of development in *Drosophila* take place when the embryo is a multinucleate syncytium. Maternal gene products deposited in the egg in a particular spatial pattern during oogenesis determine the main body axes and lay down a framework of positional information which activates a cascade of zygotic gene activity that further patterns the body. The first zygotic genes to be activated along the antero-posterior axis are the gap genes, all encoding transcription factors, whose pattern of expression divides the embryo into a number of regions. The transition to a segmented body organization then starts with the activity of the pair-rule genes, whose sites of expression are specified by the gap gene proteins and which divide the axis into 14 parasegments. Along the dorso-ventral axis, zygotic gene expression also defines several regions, including the future mesoderm and future neural tissue. At the time of pair-rule gene expression, the embryo becomes cellularized and is no longer a syncytium. Segmentation genes pattern the segment and compartment boundaries are involved in patterning and polarizing segments. The identity of the segments is determined by two gene complexes which contain homeotic selector genes, known generally as the Hox genes. The spatial pattern of expression of these genes is largely determined by gap gene activity.

GENERAL FURTHER READING

Adams, M.D., *et al.*: **The genome sequence of *Drosophila melanogaster*.** *Science* 2000, **267:** 2185–2195.

Ashburner, M.: Drosophila. *A Laboratory Handbook*. New York: Cold Spring Harbor Laboratory Press, 1989.

Lawrence, P.A.: *The Making of a Fly*. Oxford: Blackwell Scientific Publications, 1992.

Matthews, K.A., Kaufman, T.C., Gelbart, W.M.: **Research resources for *Drosophila*: the expanding universe**. *Nat. Rev. Genet.* **6:** 179–193.

Stolc, V., *et al.*: **A gene expression map for the euchromatic genome of *Drosophilamelanogaster*.** *Science* 2004, **306:** 655–660.

SECTION FURTHER READING

2.2 Cellularization is followed by gastrulation, segmentation, and the formation of the larval nervous system

Stathopoulos, A., Levine, M. **Whole-genome analysis of *Drosophila* gastrulation**. *Curr. Opin. Genet. Dev.* 2004, **14:** 477–484.

2.4 Many developmental mutations have been identified in *Drosophila* through induced mutation and large-scale genetic screening

Nusslein-Volhard. C., Wieschaus, E.: **Mutations affecting segment number and polarity in *Drosophila*.** *Nature* 1980, **287:** 795-801.

Rørth, P.: **A modular misexpression screen in *Drosophila* detecting tissue-specific phenotypes.** *Proc. Natl Acad. Sci. USA* 1996, **93:** 12418–12422.

2.7 Three classes of maternal genes specify the antero-posterior axis

St Johnston, D., Nüsslein-Volhard, C.: **The origin of pattern and polarity in the *Drosophila* embryo.** *Cell* 1992, **68:** 201–219.

2.8 The *bicoid* gene provides an antero-posterior gradient of morphogen

Driever, W., Nüsslein-Volhard, C.: **The bicoid protein determines position in the *Drosophila* embryo in a concentration dependent manner.** *Cell* 1988, **54:** 95–104.

Ephrussi, A., St Johnston, D.: **Seeing is believing: the bicoid morphogen gradient matures.** *Cell* 2004, **116:** 143–152.

Teleman, A.A, Strigini, M., Cohen, S.M.: **Shaping morphogen gradients**. *Cell* 2001, **105:** 559–562.

2.9 The posterior pattern is controlled by the gradients of Nanos and Caudal proteins

Irish, V., Lehmann, R., Akam, M.: **The *Drosophila* posterior-group gene *nanos* functions by repressing *hunchback* activity.** *Nature* 1989, **338:** 646–648.

Murafta, Y., Wharton, R.P.: **Binding of pumilio to maternal *hunchback* mRNA is required for posterior patterning in *Drosophila* embryos.** *Cell* 1995, **80:** 747–756.

Rivera-Pomar, R., Lu, X., Perrimon, N., Taubert, H., Jackle, H.: **Activation of posterior gap gene expression in the *Drosophila* blastoderm**. *Nature* 1995, **376:** 253–256.

Struhl, G.: **Differing strategies for organizing anterior and posterior body pattern in *Drosophila* embryos.** *Nature* 1989, **338:** 741–744.

Summary: main genes involved in specifying pattern in the early Drosophila embryo

	Gene	Maternal/ zygotic	Nature of protein	Transcription factor (T), receptor (R), or signal protein (S)	Function (where known)
Antero-posterior system	*bicoid*	M	Homeodomain	T	Morphogen, provides positional information along AP axis
	hunchback	M/Z	Zinc fingers	T	Morphogen, provides positional information along AP axis
	nanos	M	RNA-binding protein		Helps to establish AP gradient of hunchback protein
	caudal	M	Homeodomain	T	Involved in specifying posterior region
	gurken	M	Secreted protein of TGF-α family	S	Posterior oocyte–follicle cell signaling
	oskar	M			Pole-cell determination
Terminal system	*torso*	M	Receptor tyrosine kinase	R	Terminal specification
	trunk	M		S	Ligand for torso
Gap genes	*hunchback*	Z	Zinc fingers	T	Localize pair-rule gene expression
	Krüppel	Z	Zinc fingers	T	
	knirps	Z	Zinc fingers	T	
	giant	Z	Leucine fingers	T	
	tailless	Z	Zinc fingers	T	
Pair-rule genes	*even-skipped*	Z	Homeodomain	T	Delimits odd-numbered parasegments
	fushi tarazu	Z	Homeodomain	T	Delimits even-numbered parasegments
	hairy	Z	Helix-loop-helix	T	
Segmentation genes	*engrailed*	Z	Homeodomain	T	Defines anterior region of parasegment and posterior region of segment
	hedgehog	Z	Membrane or secreted	S	Components of signaling pathways that pattern segments and stabilize compartment boundaries
	wingless	Z	Secreted	S	
	frizzled	Z	Membrane	R	
	gooseberry	Z	Homeodomain	T	
	patched	Z	Membrane	R	
	smoothened	Z	G-protein-coupled	R	
Selector genes bithorax complex	*Ultrabithorax*	Z	Homeodomain	T	Combinatorial activity confers identity on parasegments 5–13
	abdominal-A	Z	Homeodomain	T	
	Abdominal-B	Z	Homeodomain	T	
Antennapedia complex	*Deformed*	Z	Homeodomain	T	Combinatorial activity confers identity on parasegments anterior to 5
	Sex combs reduced	Z	Homeodomain	T	
	Antennapedia	Z	Homeodomain	T	
	labial	Z	Homeodomain	T	
Maintenance genes	Polycomb group	Z		T	Maintain state of homeotic genes
	Trithorax	Z		T	
Dorso-ventral system					
Maternal genes	*Toll*	M	Membrane	R	Activation results in dorsal protein entering nucleus
	spätzle	M	Extracellular	S	Ligand for Toll protein
	dorsal	M		T	Morphogen, sets dorso-ventral polarity
	tube	M	Adaptor protein		Components of the Toll signaling pathway leading to Dorsal protein entering nuclei
	pelle	M	Protein kinase		
	cactus	M	Cytoplasmic inhibitor		
	gurken	M	Secreted protein of TGF-α family	S	Specifies oocyte axis
	pipe	M	Sulfotransferase	Enzyme	Part of pathway leading to Spätzle processing
Zygotic genes	*twist*	Z	Helix-loop-helix	T	Define mesoderm
	snail	Z	Zinc finger	T	
	rhomboid	Z	Membrane protein	S	Confer regional identity on dorso-ventral axis
	zerknüllt	Z	Homeodomain	T	
	decapentaplegic	Z	Secreted protein of TGF-β family	S	
	tolloid	Z	BMP-2 family	S	
	short gastrulation	Z		S	

AP, antero-posterior; TGF, transforming growth factor.

2.10 The anterior and posterior extremities of the embryo are specified by cell-surface receptor activation

Casanova, J., Struhl, G.: **Localized surface activity of *torso*, a receptor tyrosine kinase, specifies terminal body pattern in *Drosophila*.** *Genes Dev.* 1989, **3:** 2025–2038.

Li, W.X.: **Functions and mechanisms of receptor tyrosine kinase Torso signaling: lessons from *Drosophila* embryonic terminal development.** 2005, **232:** 656–672.

2.11 The dorso-ventral polarity of the embryo is specified by localization of maternal proteins in the egg vitelline envelope

Anderson, K.V.: **Pinning down positional information: dorsal-ventral polarity in the *Drosophila* embryo.** *Cell* 1998, **95:** 439–442.

Sen, J., Goltz, J.S., Stevens, L., Stein, D.: **Spatially restricted expression of *pipe* in the *Drosophila* egg chamber defines embryonic dorsal-ventral polarity.** *Cell* 1998, **95:** 471–481.

2.12 Positional information along the dorso-ventral axis is provided by the Dorsal protein

Belvin, M.P., Anderson, K.V.: **A conserved signaling pathway: the *Drosophila* Toll-dorsal pathway.** *Annu. Rev. Cell Dev. Biol.* 1996, **12:** 393–416.

Imler, J.L., Hoffmann, J.A.: **Toll receptors in *Drosophila*: a family of molecules regulating development and immunity.** *Curr. Top. Microbiol. Immunol.* 2002, **270:** 63–79.

Roth, S., Stein, D., Nüsslein-Volhard, C.: **A gradient of nuclear localization of the dorsal protein determines dorso-ventral pattern in the *Drosophila* embryo.** *Cell* 1989, **59:** 1189–1202.

Steward, R., Govind, R.: **Dorsal-ventral polarity in the *Drosophila* embryo.** *Curr. Opin. Genet. Dev.* 1993, **3:** 556–561.

2.13 The antero-posterior axis of the *Drosophila* egg is specified by signals from the preceding egg chamber and by interactions of the oocyte with follicle cells

González-Reyes, A., St Johnston, D.: **The *Drosophila* AP axis is polarised by the cadherin-mediated positioning of the oocyte.** *Development* 1998, **125:** 3635–3644.

Huynh, J.R., St Johnston, D.: **The origin of asymmetry: early polarization of the *Drosophila* germline cyst and oocyte.** *Curr. Biol.* 2004, **14:** R438–R449.

Riechmann, V., Ephrussi, A.: **Axis formation during *Drosophila* oogenesis.** *Curr. Opin. Genet. Dev.* 2001, **11:** 374–383.

St Johnston, D.: **Moving messages: the intracellular localization of mRNAs.** *Nat. Rev. Mol. Cell Biol.* 2005, **6:** 363–375.

Torres, I.L., Lopez-Schier, H., St Johnston, D.: **A Notch/Delta-dependent relay mechanism establishes anterior-posterior polarity in *Drosophila*.** *Dev. Cell* 2005, **5:** 547–558.

2.14 The dorso-ventral axis of the egg is specified by movement of the oocyte nucleus followed by signaling between oocyte and follicle cells

Jordan, K.C., Clegg, N.J., Blasi, J.A., Morimoto, A.M., Sen, J., Stein, D., McNeill, H., Deng, W.M., Tworoger, M., Ruohola-Baker, H.: **The homeobox gene *mirror* links EGF signalling to embryonic dorso-ventral axis formation through Notch activation.** *Nat. Genet.* 2000, **24:** 429–433.

González-Reyes, A., Elliott, H., St Johnston, D.: **Polarization of both major body axes in *Drosophila* by gurken-torpedo signaling.** *Nature* 1995, **375:** 654–658.

Roth, S., Neuman-Silberberg, F.S., Barcelo, G., Schupbach, T.: **Cornichon and the EGF receptor signaling process are necessary for both anterior-posterior and dorsal-ventral pattern formation in *Drosophila*.** *Cell* 1995, **81:** 967–978.

van Eeden, F., St Johnston, D.: **The polarisation of the anterior-posterior and dorsal-ventral axes during *Drosophila* oogenesis.** *Curr. Opin. Genet. Dev.* 1999, **9:** 396–404.

2.15 The expression of zygotic genes along the dorso-ventral axis is controlled by Dorsal protein

Cowden, J., Levine, M.: **Ventral dominance governs segmental patterns of gene expression across the dorsal-ventral axis of the neurectoderm in the *Drosophila* embryo.** *Dev. Biol.* 2003, **262:** 335–349.

Harland, R.M.: **A twist on embryonic signalling.** *Nature* 2001, **410:** 423–424.

Markstein, M., Zinzen, R., Markstein, P., Yee, K.P, Erives, A., Stathopoulos, A., Levine, M.: **A regulatory code for neurogenic gene expression in the *Drosophila* embryo.** *Development* 2004, **131:** 2387–2394.

Rusch, J., Levine, M.: **Threshold responses to the dorsal regulatory gradient and the subdivision of primary tissue territories in the *Drosophila* embryo.** *Curr. Opin. Genet. Dev.* 1996, **6:** 416–423.

Stathopoulos, A., Levine, M.: **Dorsal gradient networks in the *Drosophila* embryo.** *Dev. Biol.* 2002, **246:** 57–67.

2.16 The Decapentaplegic protein acts as a morphogen to pattern the dorsal region

Ashe, H.L., Levine, M.: **Local inhibition and long-range enhancement of Dpp signal transduction by Sog.** *Nature* 1999, **398:** 427–431.

Mizutani, C.M., Nie, Q., Wan, F.Y., Zhang, Y.T., Vilmos, P., Sousa-Neves, R., Bier, E., Marsh, J.L., Lander, A.D.: **Formation of the BMP activity gradient in the *Drosophila* embryo.** *Dev. Cell* 2005, **8:** 915–924.

Shimmi, O., Umulis, D., Othmer, H., O'Connor, M.B.: **Facilitated transport of a Dpp/Scw heterodimer by Sog/Tsg leads to robust patterning of the *Drosophila* blastoderm embryo.** *Cell* 2005, **120:** 873–886.

Srinivasan, S., Rashka, K.E., Bier, E.: **Creation of a Sog morphogen gradient in the *Drosophila* embryo.** *Dev. Cell* 2002, **2:** 91–101.

Wang, Y-C., Ferguson, E.L.: **Spatial bistability of Dpp-receptor interactions driving *Drosophila* dorsal-ventral patterning.** *Nature* 2005, **434:** 229–234.

Wharton, K.A., Ray, R.P., Gelbart, W.M.: **An activity gradient of decapentaplegic is necessary for the specification of dorsal pattern elements in the *Drosophila* embryo.** *Development* 1993, **117:** 807–822.

2.17 The antero-posterior axis is divided up into broad regions by gap-gene expression

Hülskamp, M., Tautz, D.: **Gap genes and gradients—the logic behind the gaps.** *BioEssays* 1991, **13:** 261–268.

2.18 Bicoid protein provides a positional signal for the anterior expression of hunchback

Brand, A.H., Perrimon, N.: **Targeted gene expression as a means of altering cell fates and generating dominant phenotypes**. *Development* 1993, **118:** 401–415.

Havchmandzadeh, B., Weischaus, E., Leiber, S.: **Establishment of developmental precision and proportions in the early *Drosophila* embryo**. *Nature* 2002, **415:** 798–802.

Simpson-Brose, M., Treisman, J., Desplan, C.: **Synergy between the Hunchback and Bicoid morphogens is required for anterior patterning in *Drosophila***. *Cell* 1994, **78:** 855–865.

Struhl, G., Struhl, K., Macdonald, P.M.: **The gradient morphogen Bicoid is a concentration-dependent transcriptional activator**. *Cell* 1989, **57:** 1259–1273.

2.19 The gradient in Hunchback protein activates and represses other gap genes

Rivera-Pomar, R., Jäckle, H.: **From gradients to stripes in *Drosophila* embryogenesis: filling in the gaps**. *Trends Genet.* 1996, **12:** 478–483.

Struhl, G., Johnston, P., Lawrence, P.A.: **Control of *Drosophila* body pattern by the hunchback morphogen gradient**. *Cell* 1992, **69:** 237–249.

Wu, X., Vakani, R., Small, S.: **Two distinct mechanisms for differential positioning of gene expression borders involving the *Drosophila* gap protein giant**. *Development* 1998, **125:** 3765–3774.

2.21 Gap-gene activity positions stripes of pair-rule gene expression

Small, S., Levine, M.: **The initiation of pair-rule stripes in the *Drosophila* blastoderm**. *Curr. Opin. Genet. Dev.* 1991, **1:** 255–260.

2.22 Expression of the *engrailed* gene delimits a cell-lineage boundary and defines a compartment

Dahmann, C., Basler, K.: **Compartment boundaries: at the edge of development**. *Trends Genet.* 1999, **15:** 320–326.

Gray, S., Cai, H., Barolo, S., Levine, M.: **Transcriptional repression in the *Drosophila* embryo**. *Phil. Trans. R. Soc. Lond.* 1995, **349:** 257–262.

Harrison, D.A., Perrimon, N.: **Simple and efficient generation of marked clones in *Drosophila***. *Curr. Biol.* 1993, **3:** 424–433.

Vincent, J.P., O'Farrell, P.H.: **The state of *engrailed* expression is not clonally transmitted during early *Drosophila* development**. *Cell* 1992, **68:** 923–931.

2.23 Segmentation genes stabilize parasegment boundaries and set up a focus of signaling at the boundary that patterns the segment

Alexandre, C., Lecourtois, M., Vincent, J-P.: **Wingless and hedgehog pattern *Drosophila* denticle belts by regulating the production of short-range signals**. *Development* 1999, **126:** 5689–5698.

Bejsovec, A, Wieschaus, E.: **Segment polarity gene interactions modulate epidermal patterning in *Drosophila* embryos**. *Development* 1993, **119:** 501–517.

Briscoe, J., Therond, P. **Hedgehog signaling: from the *Drosophila* cuticle to anti-cancer drugs**. *Dev. Cell* 2005, **8:** 143–151.

Hooper, J.E., Scott, M.P.: **Communicating with Hedgehogs**. *Nat. Rev. Mol. Cell Biol.* 2005, **6:** 306–317.

Johnson, R.L., Scott, M.P.: **New players and puzzles in the Hedgehog signaling pathway**. *Curr. Opin. Genet. Dev.* 1998, **8:** 450–456.

Larsen, C.W., Hirst, E., Alexandre, C., Vincent, J.-P.: **Segment boundary formation in *Drosophila* embryos**. *Development* 2003, **130:** 5625–5635.

Lawrence, P.A., Casal, J., Struhl, G.: **Hedgehog and engrailed: pattern formation and polarity in the *Drosophila* abdomen**. *Development* 1999, **126:** 2431–2439.

Tolwinski, N.S., Wieschaus, E.: **Rethinking Wnt signaling**. *Trends Genet.* 2004, **20:** 177–181

von Dassow, G., Meir, E., Munro, E.M., Odell, G.M.: **The segment polarity network is a robust developmental module**. *Nature* 2000, **406:** 188–192.

Wehrli, M., Dougan, S.T., Caldwell, K., O'Keefe, L., Schwartz, S., Vaizel-Ohayon, D., Schejter, E., Tomlinson, A., DiNardo, S.: **arrow encodes an LDL-receptor-related protein essential for Wingless signalling**. *Nature* 2000, **407:** 527–530.

2.24 Insect epidermal cells become individually polarized in an antero-posterior direction in the plane of the epithelium

Lawrence, P.A., Casal, J., Struhl, G.: **Cell interactions and planar polarity in the abdominal epidermis of *Drosophila***. *Development* 2004, **131:** 4651–4664.

Ma, D., Yang, C.H., McNeill, H., Simon, MA., Axelrod, J.D.: **Fidelity in planar cell polarity signalling**. *Nature* 2003, **421:** 543–546.

Santo, M., McNeill, H.: **Planar polarity from flies to vertebrates**. *J. Cell Sci.* 2004, **117:** 527–533.

2.25 Some insects use different mechanisms for patterning the body plan

Akam, M., Dawes, R.: **More than one way to slice an egg**. *Curr. Biol.* 1992, **8:** 395–398.

Brown, S.J., Parrish, J.K., Beeman, R.W., Denell, R.E.: **Molecular characterization and embryonic expression of the *even-skipped* ortholog of *Tribolium castaneum***. *Mech. Dev.* 1997, **61:** 165–173.

French, V.: **Segmentation (and *eve*) in very odd insect embryos**. *BioEssays* 1996, **18:** 435–438.

French, V.: **Insect segmentation: genes, stripes and segments in "Hoppers"**. *Curr. Biol.* 2001, **11:** R910–R913.

Sander, K.: **Pattern formation in the insect embryo**. In *Cell Patterning, Ciba Found. Symp. 29*. London: Ciba Foundation, 1975: 241–263.

Tautz, D., Sommer, R.J.: **Evolution of segmentation genes in insects**. *Trends Genet.* 1995, **11:** 23–27.

2.26 Segment identity in *Drosophila* is specified by genes of the Antennapedia and bithorax complexes

Lewis E.B.: **A gene complex controlling segmentation in *Drosophila***. *Nature* 1978, **276:** 565-570.

2.27 Homeotic selector genes of the bithorax complex are responsible for diversification of the posterior segments

Castelli-Gair, J., Akam, M.: **How the Hox gene *Ultrabithorax* specifies two different segments: the significance of spatial and temporal regulation within metameres**. *Development* 1995, **121:** 2973–2982.

Duncan, I.: **How do single homeotic genes control multiple segment identities?** *BioEssays* 1996, **18:** 91–94.

Lawrence, P.A., Morata, G.: **Homeobox genes: their function in *Drosophila* segmentation and pattern formation**. *Cell* 1994, **78:** 181–189.

Liang, Z., Biggin, M.D.: **Eve and ftz regulate a wide array of genes in blastoderm embryos: the selector homeoproteins directly or indirectly regulate most genes in *Drosophila*.** *Development* 1998, **125:** 4471–4482.

Mann, R.S., Morata, G.: **The developmental and molecular biology of genes that subdivide the body of *Drosophila*.** *Annu. Rev. Cell Dev. Biol.* 2000, **16:** 143–271.

Mannervik, M.: **Target genes of homeodomain proteins**. *BioEssays* 1999, **4:** 267–270.

Simon, J.: **Locking in stable states of gene expression: transcriptional control during *Drosophila* development**. *Curr. Opin. Cell Biol.* 1995, **7:** 376–385.

2.29 The order of Hox gene expression corresponds to the order of genes along the chromosome

Morata, G.: **Homeotic genes of *Drosophila***. *Curr. Opin. Genet. Dev.* 1993, **3:** 606–614.

2.30 The *Drosophila* head region is specified by genes other than the Hox genes

Rogers B.T., Kaufman, T.C.: **Structure of the insect head in ontogeny and phylogeny: a view from** *Drosophila*. *Int. Rev. Cytol.* 1997, **174:** 1–84.

Patterning the vertebrate body plan I: axes and germ layers

3

- Vertebrate life cycles and outlines of development
- Setting up the body axes
- The origin and specification of the germ layers

We examine here the similarities and differences in how the body axes and germ layers of the early embryo are specified in the four vertebrate model animals —*Xenopus*, zebrafish, chick, and mouse. A key question in the determination of the axes is the role of maternal factors in the egg as distinct from external influences such as gravity and events at fertilization. Another question is how the internal left-right asymmetry of the vertebrate body, as manifested by the positions of the heart and liver, is specified in the early embryo. By the blastula stage the body plan is sufficiently defined for it to be possible to make fate maps that show how each region of cells will normally develop, although considerable regulation occurs while signals from various key regions of the embryo are acting to specify the three germ layers. This specification is key to early development and may involve gradients of morphogens that specify the positional identity of cells.

We saw in Chapter 2 how early development in insects is largely under the control of maternal factors that interact with each other to broadly specify different regions of the body. This initial blueprint is then elaborated on by the embryo's own genes. We shall now look at how the same task of establishing the outline of the body plan is achieved in early vertebrate development. All vertebrates, despite their many outward differences, have a similar basic body plan. The defining vertebrate structures are the segmented backbone or **vertebral column** surrounding the spinal cord, with the brain at the anterior end enclosed in a bony or cartilaginous skull (Fig. 3.1). These prominent structures mark the antero-posterior axis, the main body axis of vertebrates. The head is at the anterior end of this axis, followed by the trunk with its paired appendages—limbs in terrestrial vertebrates (with the exception of snakes) and fins in fish—and in many vertebrates it terminates in a post-anal tail. In addition, the vertebrate body has a distinct dorso-ventral (back–belly) polarity, with the mouth defining the ventral side and the spinal cord the dorsal side. The antero-posterior and dorso-ventral axes together define the left and right sides of the animal. Vertebrates have a general **bilateral symmetry** around the dorsal midline so that outwardly the right and left sides are mirror images of each other. Some internal organs, such as lungs, kidneys, and gonads, are also present as symmetrically placed paired structures, but single organs such as the heart and liver are arranged asymmetrically with respect to the dorsal midline.

We will look at four vertebrates whose early development has been particularly well studied: amphibians are represented by the frog *Xenopus*, fish by the zebrafish,

Fig. 3.1 The skeleton of a mouse embryo illustrates the vertebrate body plan. The skeletal elements in this embryo have been stained with Alcian blue (which stains cartilage) and Alizarin red (which stains blue). The vertebral column, which develops from blocks of somites, is divided up into cervical (neck), thoracic (chest), lumbar (lower back), and sacral (hip and lower) regions. The paired limbs can also be seen. Scale bar = 1 mm.

Photograph courtesy of M. Maden.

birds by the chicken, and mammals by the mouse. In this chapter, we will first outline the life cycles of the four organisms and briefly describe the anatomical course of their early development, providing the context for the later discussion. The rest of the chapter looks in detail at the first two key events in development: the establishment of the antero-posterior and dorso-ventral axes, and the specification and early patterning of the germ layers, in particular the mesoderm. In Chapter 4 we will continue the discussion of early development in vertebrates by looking at the development of two mesodermal derivatives characteristic of vertebrates—the notochord and somites (see Box 1A, pp. 4–5)—and an ectodermal derivative, the neural tube, which will develop into the nervous system. This takes the embryo up to the stage at which it has become recognizably vertebrate. The development of particular organs and structures, such as the limbs, the head, and the nervous system, will be covered in later chapters.

Vertebrate life cycles and outlines of development

All vertebrate embryos pass through a broadly similar set of developmental stages, which were outlined for *Xenopus* in Chapter 1 (see Box 1A, pp. 4–5). After fertilization, the zygote undergoes cleavage. These are rapid cell divisions by which the embryo becomes divided into a number of smaller cells, initially without any increase in overall mass. This is followed by gastrulation, in which cell movements result in the germ layers—ectoderm, mesoderm and endoderm (see Box 1B, p. 15)—moving into the correct places for further development. At the end of gastrulation, the ectoderm covers the embryo, and the mesoderm and endoderm have moved inside. The endoderm gives rise to the gut, and to its derivatives such as liver and lungs; the mesoderm forms skeletal structures, muscle, connective tissue, kidneys, heart, and blood, as well as some other tissues; and the ectoderm gives rise to the epidermis and the nervous system.

One of the earliest mesodermal structures that can be recognized in vertebrates is the rod-shaped **notochord**, which forms along the antero-posterior axis of the body. This later becomes incorporated into the column of vertebrae that form the spine. The rest of the vertebral column, the skeleton of the trunk, and the muscles of the trunk and limbs develop from blocks of mesoderm called **somites** that form in an antero-posterior sequence on either side of the notochord. The brain and spinal cord are derived from the ectoderm, which forms the neural tube immediately above the notochord. The overall similarity of the body plan in all vertebrates suggests that the developmental processes that establish it are broadly similar in the different animals. This is largely the case, although there are considerable differences in development before the onset of gastrulation.

These differences relate particularly to how and when the axes are set up, and how the germ layers are established, and are mainly related to the different modes of reproduction. Yolk provides all the nutrients for fish, amphibian, and bird embryonic development. Mammalian eggs, by contrast, are small and non-yolky, and the embryo is nourished for the first few days by fluids in the oviduct and uterus. Once implanted in the uterus wall, the embryo develops specialized **extra-embryonic membranes** that both surround and protect the embryo and through which it receives nourishment from the mother via the placenta. Avian embryos also develop extra-embryonic membranes for obtaining nutrients from the yolk, for oxygen and carbon dioxide exchange, and for waste disposal.

Fig. 3.2 shows the differences in form and shape of the early embryos from the four model species. After gastrulation, all vertebrate embryos pass through a

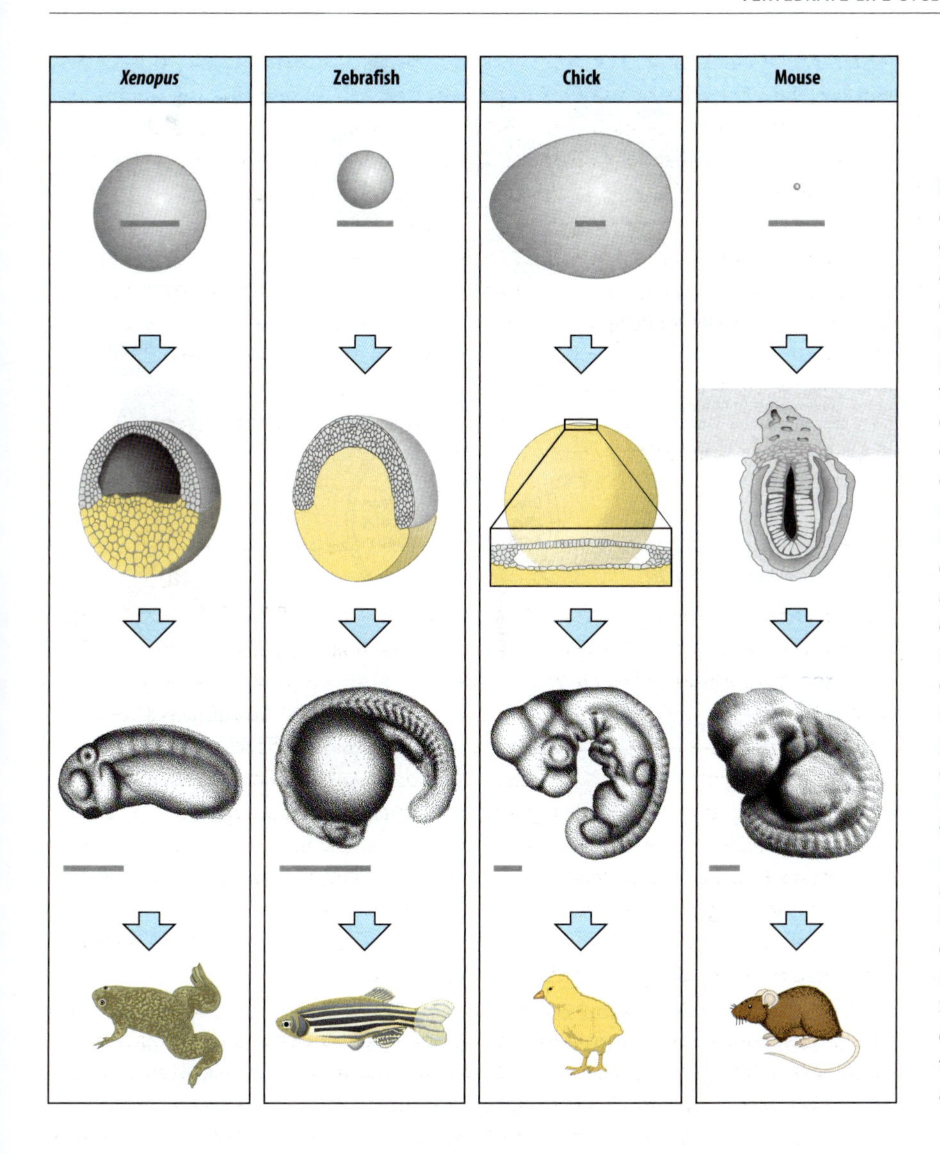

Fig. 3.2 Vertebrate embryos show considerable differences in form before gastrulation but subsequently all go through a stage at which they look similar. The eggs of the frog (*Xenopus*), zebrafish, chicken, and mouse are very different in size (top row). Scale bars in this row all represent 1 mm, except for the chicken egg which represents 10 mm. Their early development (second row) is rather different. In this row, the embryos are shown in cross-section at the stage corresponding roughly to the *Xenopus* blastula (left panel) just before gastrulation commences. The main determinant of tissue arrangement is the amount of yolk (yellow) in the egg. The mouse embryo (far right) at this stage has implanted into the uterine wall and thus has already developed some extra-embryonic tissues required for implantation. The mouse embryo proper is the small cup-shaped structure at the center, seen here in cross-section as a U-shaped layer of epithelium. After gastrulation and formation of the neural tube, vertebrate embryos pass through an embryonic stage at which they all look rather similar (third row), and which is known as the phylotypic stage. The body has developed, and neural tube, somites, notochord, and head structures are present. Scale bars = 1 mm. After this stage their development diverges again. Paired appendages, for example, develop into fins in fish, and wings and legs in the chick (bottom row).

so-called **phylotypic stage**, at which they all more or less resemble each other and show the specific features of embryos of chordates, the phylum to which vertebrates belong. The head is distinct and a neural tube runs along the dorsal midline, under which runs the notochord, flanked on either side by the mesodermal somites. Features special to different groups, such as beaks, wings, and fins, appear later.

To study and analyze development, it is necessary to have a reliable way of identifying and referring to a particular stage of development. Simply measuring the time from fertilization is not satisfactory for most species. Amphibians, for example, develop quite normally over a range of temperatures, but the rate of development changes considerably at different temperatures. Developmental biologists therefore divide the normal embryonic development of each species into a series of numbered stages, which are identified by their main features rather than by time after fertilization. A stage 10 *Xenopus* embryo, for example, refers to an embryo at a very early stage of gastrulation, while the free-swimming tadpole is stage 45. Numbered stages have similarly been characterized for the chick embryo, as here one does not always know the time of fertilization or when the fertilized egg has been placed in the incubator. Even for mouse embryos,

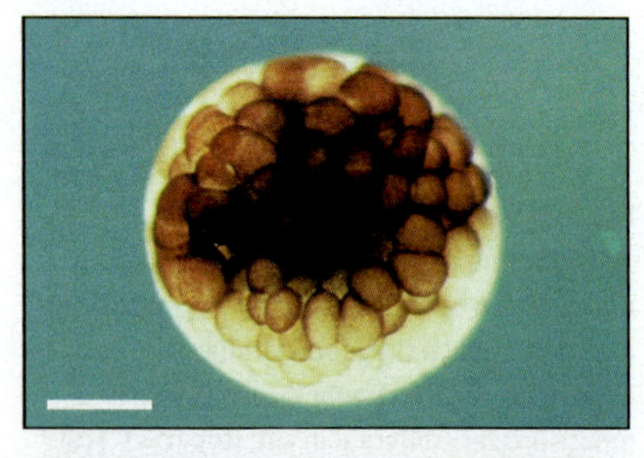

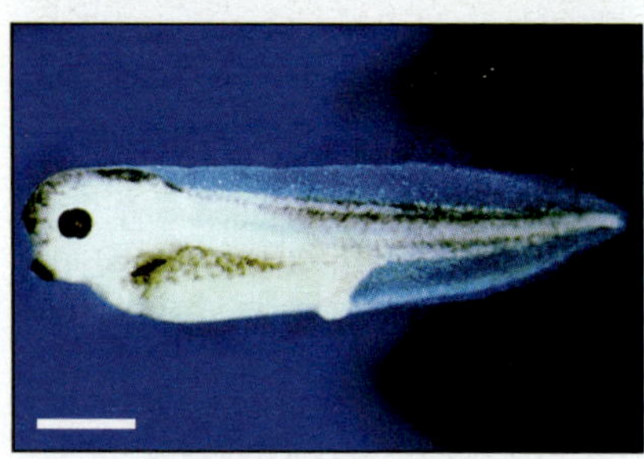

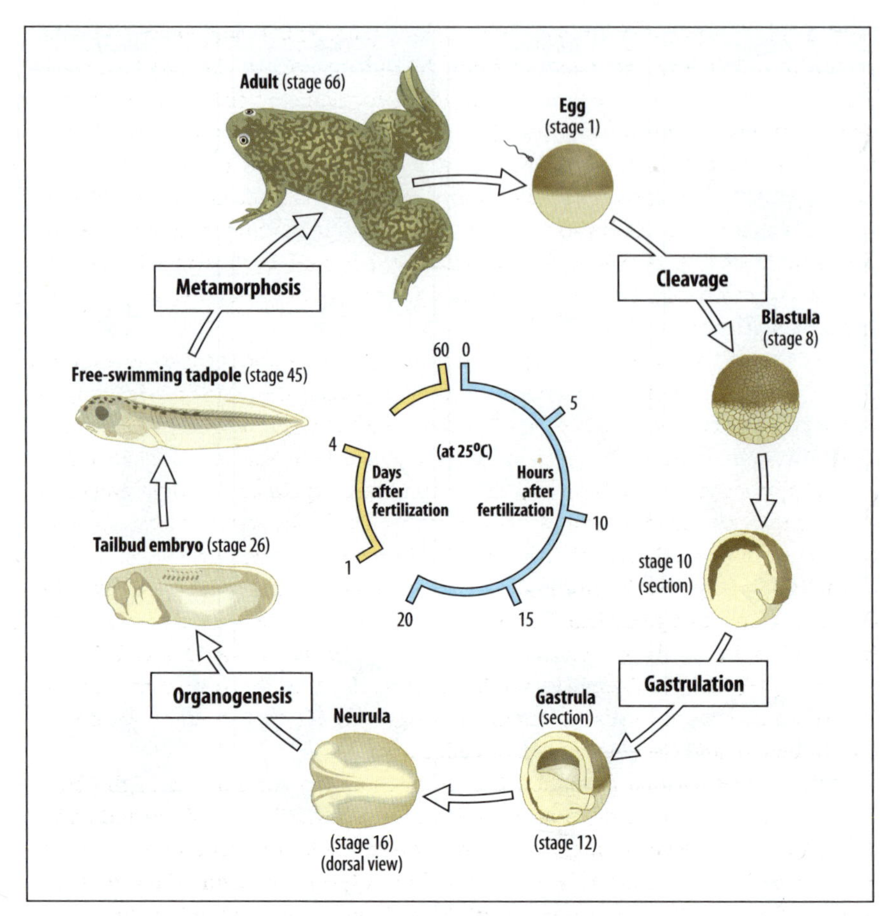

Fig. 3.3 Life cycle of the African claw-toed frog *Xenopus laevis*. The numbered stages refer to standardized stages of *Xenopus* development. More stages can be seen in the larger figure in Box 1A, pp. 4–5. The photographs show: an embryo at the blastula stage (top, scale bar = 0.5 mm); a tadpole at stage 41 (middle, scale bar = 1 mm); and an adult frog (bottom, scale bar = 1 cm).

Photographs courtesy of J. Slack (top, from Alberts, B., et al.: 1994) and J. Smith (middle and bottom).

which develop in a much more constant environment, staging is again by reference to the structure of the embryo; somite number is often used as an indication of developmental stage. For earlier stages of mouse development, before the somites have formed, one has to resort to days post fertilization or *post coitum* (days p.c.); that is, days after mating. This is a reliable measure as the temperature is more constant.

3.1 The frog *Xenopus laevis* is the model amphibian for developmental studies

The amphibian species most commonly used for developmental work is the African claw-toed frog, *Xenopus laevis*, which is able to develop normally in tap water. The diploid species *Xenopus tropicalis* is now also used (see Section 1.6). A great advantage of *Xenopus* is that its fertilized eggs are easy to obtain; females and males injected with the human hormone chorionic gonadotropin and put together overnight will mate and the female will lay hundreds of eggs in the water, which are fertilized by sperm released by the male. Eggs can also be fertilized in a dish by adding sperm to eggs released after hormonal stimulation of the female. The eggs are large (1.2–1.4 mm in diameter) and are thus quite readily manipulated. The embryos of *Xenopus* are extremely hardy and are highly resistant to infection after microsurgery. It is also easy to culture fragments of early *Xenopus* embryos in a simple, chemically defined solution.

The key developmental stages of *Xenopus* are illustrated in Box 1A (pp. 4–5) and its life cycle is summarized in Fig. 3.3. The mature *Xenopus* egg has a distinct polarity, with a dark, pigmented **animal region** and a pale, yolky, and heavier **vegetal region** (Fig. 3.4). The axis running from the animal pole to the vegetal pole is known as the **animal–vegetal axis**. Before fertilization, the egg is enclosed in a protective **vitelline membrane**, which is embedded in a gelatinous coat. Meiosis is not yet complete and, while the first meiotic division has resulted in a small cell—a **polar body**—forming at the animal pole, the second meiotic division is completed only after fertilization, when the second polar body also forms at the animal pole (Box 3A, below). At fertilization, one sperm enters the egg in the animal region. The egg completes meiosis and the egg and sperm nuclei then fuse to form the diploid zygote nucleus.

The first cleavage division of the fertilized egg occurs about 90 minutes after fertilization and is along the plane of the animal–vegetal axis, dividing the embryo into equal left and right halves (Fig. 3.5). Further cleavages follow rapidly at intervals of about 20 minutes. The second cleavage is also along the animal–vegetal axis but at right angles to the first. The third cleavage is equatorial, at right angles to the first two, and divides the embryo into four animal cells and four larger vegetal cells. There is no cell growth between cell divisions at this early stage and so continued cleavage results in the formation of smaller and smaller cells; the cells deriving from cleavage divisions in animal embryos are often called **blastomeres**. Cleavage occurs synchronously and divisions occur in such a way that cells in the yolky vegetal half of the embryo are larger than those in the animal half. Inside this spherical mass of cells a fluid-filled cavity—the **blastocoel**—develops in the animal region, and the embryo is now called a **blastula**.

At the end of blastula formation the *Xenopus* embryo has gone through about 12 cell divisions and is made up of several thousand cells. The mesoderm and endoderm, which will develop into internal structures, are located around the equator in the **marginal zone** and in the vegetal region, respectively, while the ectoderm, which will eventually cover the whole of the embryo, is still confined to the animal region (Fig. 3.6, first panel).

Fig. 3.4 A late-stage *Xenopus* oocyte. The surface of the animal half (top) is pigmented and the paler, vegetal half of the egg is heavy with yolk. Scale bar = 1 mm.
Photograph courtesy of J. Smith.

Box 3A Polar bodies

Polar bodies are small cells formed by meiosis during the development of an oocyte into an egg. In this highly schematic illustration, the segregation of only one pair of chromosomes is shown for simplicity. There are two cell divisions associated with meiosis, and one daughter from each division is almost always very small compared with the other, which becomes the egg—hence the term polar bodies for these smaller cells.

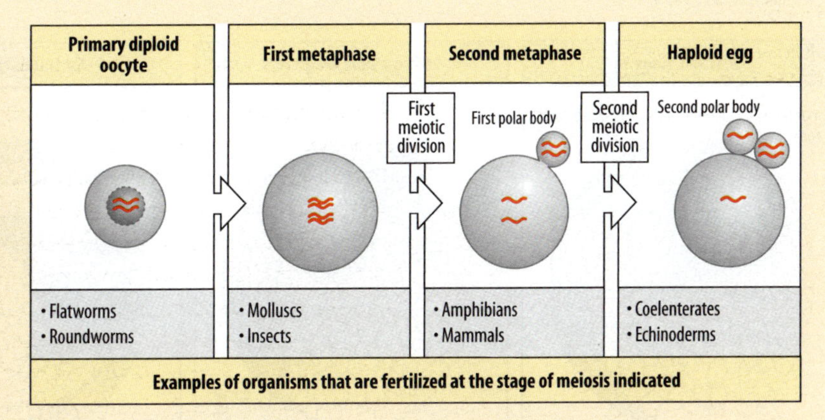

The timing of meiosis in relation to the development of the oocyte varies in different animals, and in some species meiosis is completed and the second polar body formed only after fertilization. In general, polar-body formation is of little importance for later development, but in some animals the site of formation is a useful marker for the embryonic axes.

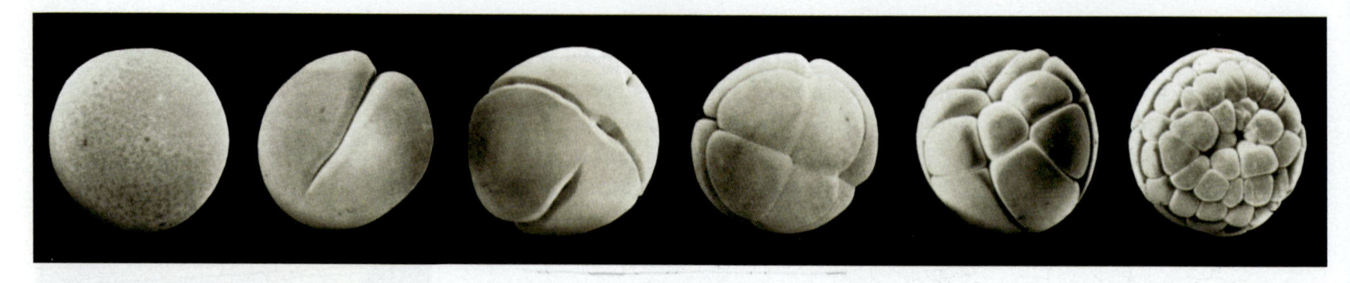

Fig. 3.5 Cleavage of the *Xenopus* embryo. The *Xenopus* embryo undergoes successive cleavages at intervals of about 20 minutes.

Photographs courtesy of R. Kessel, from Kessel, R.G., et al.: 1974.

The next stage is **gastrulation**, which involves extensive cell movements and rearrangement of the tissues of the blastula so that they become located in their proper positions in the body plan. Because it involves changes in form in three dimensions, gastrulation can be quite difficult to visualize. It is initiated by a small slit-like infolding—the **blastopore**—that forms on the surface of the blastula in the marginal zone on the future dorsal side (see Fig. 3.6, second panel). This region is of particular importance in development, as it is the site of the embryonic organizer, known as the Spemann organizer in amphibians, without which dorsal and axial development will not occur (see Fig. 1.4). The future endoderm and mesoderm in the marginal zone move inside the gastrula through the blastopore by rolling under the lip as coherent sheets of cells. This type of inward movement is known as **involution**. Once inside, the tissues converge and extend along the antero-posterior axis beneath the dorsal ectoderm. At the same time the ectoderm spreads downward to cover the whole embryo by a process known as **epiboly**. The involuting layer of dorsal endoderm is closely applied to the mesoderm, and the space between it and the yolky vegetal cells is known as the **archenteron** (see Fig. 3.6, third panel), which is the precursor of the gut cavity. The inward movement of endoderm and mesoderm begins dorsally and then spreads to form a complete circle around the blastopore (see Fig. 3.6, fourth panel). Once gastrulation has started, the embryo is known as a **gastrula**. By the end of gastrulation, the blastopore has closed, the dorsal mesoderm lies beneath the dorsal ectoderm, and the lateral and ventral mesodermal tissues have reached their definitive positions. The gut will form from the cavity that is the archenteron, and by this time the ectoderm has spread to cover the entire embryo. There is still a large amount of yolk

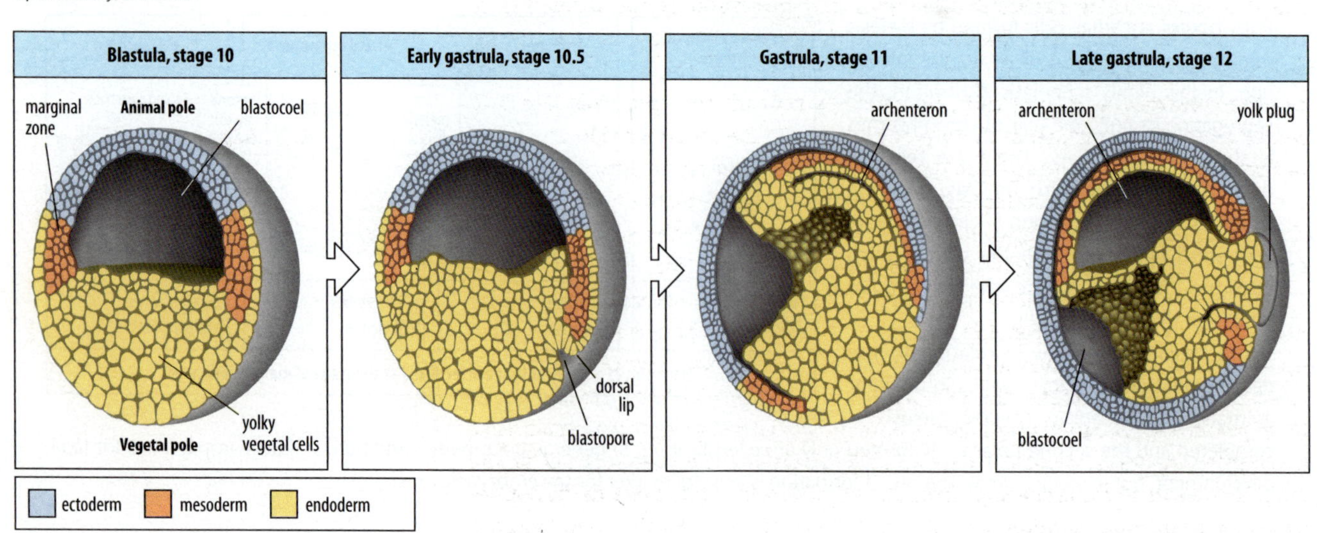

Fig. 3.6 Gastrulation in amphibians. The blastula (first panel) contains several thousand cells and there is a fluid-filled cavity, the blastocoel, beneath the cells at the animal pole. Gastrulation begins (second panel) at the blastopore, which forms on the dorsal side of the embryo. Future mesoderm and endoderm of the marginal zone move inside at this site through the dorsal lip of the blastopore, the mesoderm ending up sandwiched between the endoderm and ectoderm in the animal region (third panel). The tissue movements create a new internal cavity—the archenteron—that will become the gut. Endoderm in the ventral region also moves inside through the ventral lip of the blastopore (fourth panel) and will eventually completely line the archenteron. At the end of gastrulation the blastocoel has considerably reduced in size.

After Balinsky, B.I.: 1975.

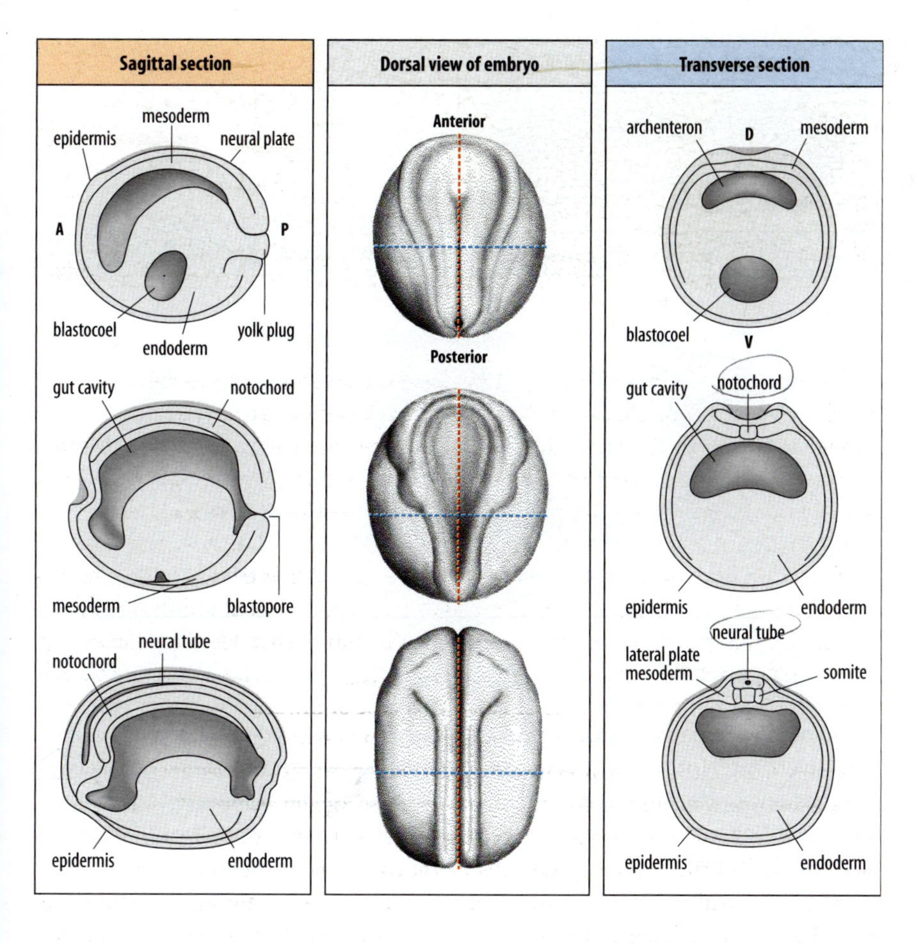

Fig. 3.7 Neurulation in amphibians. Top row: the neural plate develops neural folds just as the notochord begins to form in the midline (see middle row). Middle and bottom rows: the neural folds come together in the midline to form the neural tube, from which the brain and spinal cord will develop. During neurulation, the embryo elongates along the antero-posterior axis. The left panel shows sections through the embryo in the planes indicated by the red dashed lines in the center panel. The center panel shows dorsal surface views of the amphibian embryo. The right panel shows sections through the embryo in the planes indicated by the blue dashed lines in the center panel. This diagram shows neurulation in a urodele amphibian embryo rather than *Xenopus*, as the neural folds in urodele embryos are more clearly defined.

present, which provides nutrients until the larva—the tadpole—starts feeding. During gastrulation, the mesoderm in the dorsal region starts to develop into the notochord and the somites, while the more lateral mesoderm—the **lateral plate mesoderm**—will form mesoderm-derived internal organs such as the kidneys.

Gastrulation is succeeded by **neurulation**, the formation of the **neural tube**, the early embryonic precursor of the central nervous system. The embryo is then called a **neurula**. The earliest visible sign of neurulation is the formation of the **neural folds**, which form on the edges of the **neural plate**, an area of ectoderm overlying the notochord. The folds rise up, fold toward the midline and fuse together to form the neural tube, which sinks beneath the epidermis (Fig. 3.7). **Neural crest cells** detach from the top of the neural tube on either side of the site of fusion and migrate throughout the body to form a variety of structures, as we describe in Chapter 9. The anterior neural tube gives rise to the brain; further back, the neural tube overlying the notochord will develop into the spinal cord. The formation of notochord, somites, and neural tube is considered in Chapter 4. The embryo now begins to look something like a tadpole and we can recognize the main vertebrate features (Fig. 3.8). At the anterior end the brain is already divided up into a number of regions, and the eye and ear have begun to develop. There are also three branchial arches, the most anterior of which will form the lower jaw. More posteriorly, the somites and notochord are well developed. The mouth breaks through at stage 40, about 2.5 days after fertilization. The post-anal tail of the tadpole is formed last. It develops from the **tailbud**, which gives rise to the continuation of notochord, somites, and neural tube in the tail. Further development

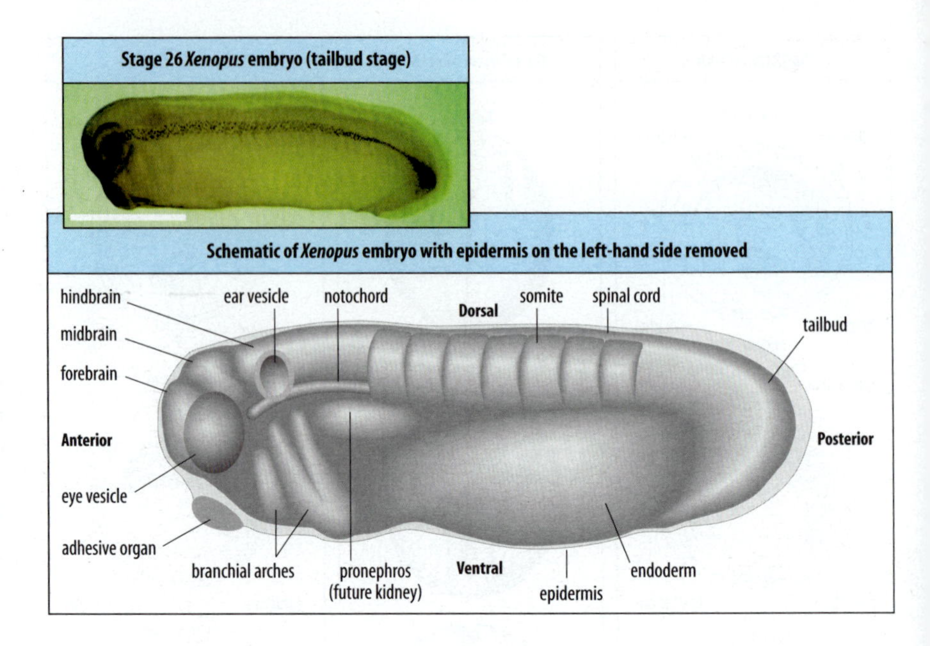

Fig. 3.8 The early tailbud stage (stage 26) of a *Xenopus* embryo. At the anterior end, in the head region, the future eye is prominent and an ear vesicle has formed. The brain is divided into forebrain, midbrain, and hindbrain. Just posterior to the site at which the mouth will form are the branchial arches, the first of which will form the lower jaw. More posteriorly, a succession of somites lies on either side of the notochord (stained brown). The embryonic kidney (pronephros) is beginning to form from lateral mesoderm. Ventral to these structures is the gut (not visible in this picture). The tailbud will give rise to the tail of the tadpole, forming a continuation of somites, neural tube, and notochord. Scale bar = 1 mm.

Photograph courtesy of B. Herrmann.

gives rise to a variety of organs and tissues such as blood and heart, kidneys, lungs, and liver. After organ formation, or **organogenesis**, is completed, the tadpole hatches out of its jelly covering and begins to swim and feed. Later, the tadpole larva will undergo metamorphosis to give rise to the adult frog; the tail regresses and the limbs form.

3.2 The zebrafish embryo develops around a large, undivided yolk

The zebrafish has two great advantages as a vertebrate model for development: its short life cycle of approximately 12 weeks makes genetic analysis relatively easy, and the transparency of the embryo means that the fate of individual cells during development can be observed. The life cycle is shown in Fig. 3.9. The zebrafish egg is about 0.7 mm in diameter, with a clear animal–vegetal axis: the cytoplasm and nucleus at the animal pole sit upon a large mass of yolk. After fertilization, the zygote undergoes cleavage, but cleavage does not extend into the yolk and results in a mound of blastomeres perched above it. The first five cleavages are all vertical, and the first horizontal cleavage gives rise to the 64-cell stage about 2 hours after fertilization (Fig. 3.10).

Further cleavage leads to the **sphere stage**, in which the embryo is now in the form of a **blastoderm** of around 1000 cells lying over the yolk. The hemispherical blastoderm has an outer layer of flattened cells, one cell thick, known as the outer enveloping layer, and a deep layer of more rounded cells (Fig. 3.11). Although different in shape, this stage corresponds to the amphibian blastula. During the early blastoderm stage, blastomeres at the margin of the blastoderm merge and collapse into the yolk cell, forming a continuous layer of multinucleate non-yolky cytoplasm underlying the blastoderm that is called the **yolk syncytial layer**. The blastoderm, together with the yolk syncytial layer, spreads in a vegetal direction by epiboly, as in the *Xenopus* gastrula, to cover the yolk cell. By about 5.5 hours after fertilization it has spread halfway to the vegetal pole. This stage is known as the **shield stage**, after a shield-shaped region overlying the dorsal yolk syncytial layer; this region is analogous to the Spemann organizer of *Xenopus*. Gastrulation then

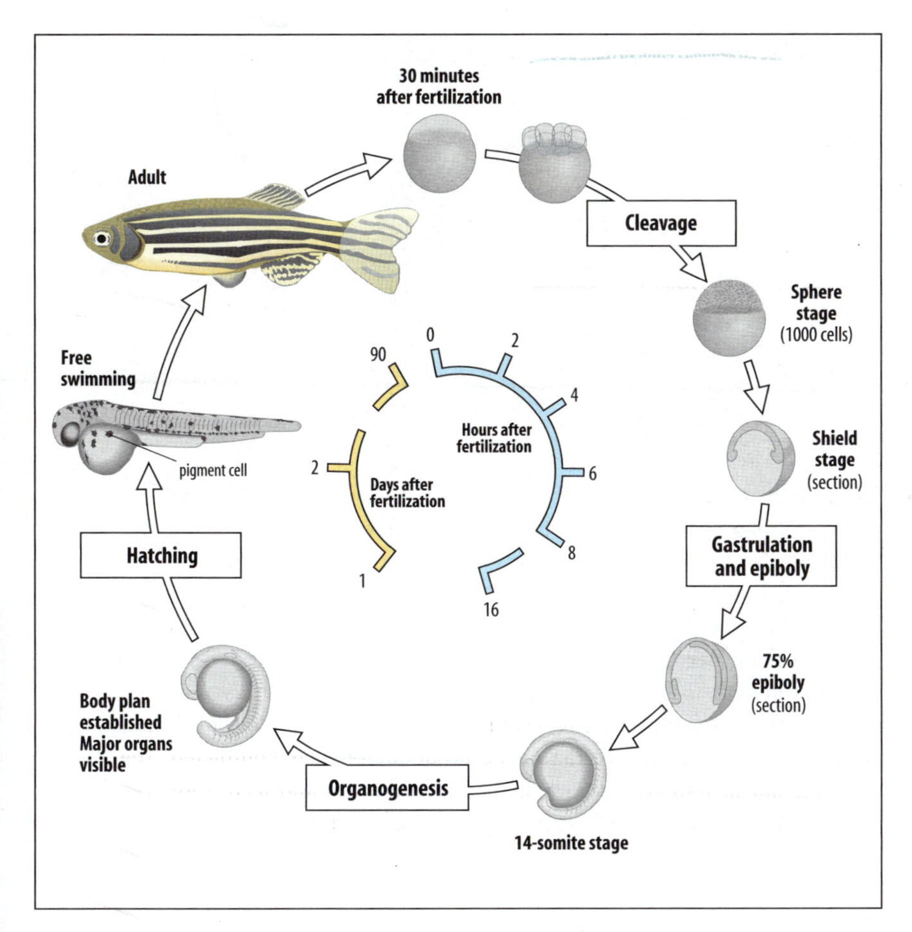

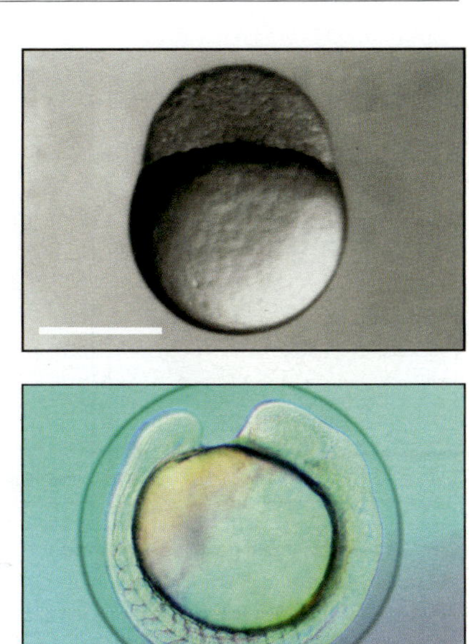

Fig. 3.9 Life cycle of the zebrafish. The zebrafish embryo develops as a cup-shaped blastoderm sitting on top of a large yolk cell. It develops rapidly and by 2 days after fertilization the tiny fish, still attached to the remains of its yolk, hatches out of the egg. The top photograph shows a zebrafish embryo at the sphere stage of development, with the embryo sitting on top of the large yolk cell (scale bar = 0.5 mm). The middle photograph shows an embryo at the 14-somite stage, showing developing organ systems. Its transparency is useful for observing cell behavior (scale bar = 0.5 mm). The bottom photograph shows an adult zebrafish (scale bar = 1 cm).

Photographs courtesy of C. Kimmel (top, from Kimmel, C.B., et al.: 1995), N. Holder (middle), and M. Westerfield (bottom).

begins, with involution of the prospective endodermal and mesodermal cells of the deep layer at the margin of the blastoderm.

Once internalized into the gastrulating embryo, these cells migrate toward the future dorsal side, the tissue converging toward the dorsal midline of the embryo and extending, elongating the embryo in an antero-posterior direction. The future mesoderm and endoderm come to lie beneath the ectoderm. Gastrulation in the zebrafish has features in common with gastrulation in *Xenopus*, but also some differences, one of which is that involution occurs all around the periphery of the

Fig. 3.10 Cleavage of the zebrafish embryo is initially confined to the animal (top) half of the embryo.

Photographs courtesy of R. Kessel, from Kessel, R.G., et al.: 1974.

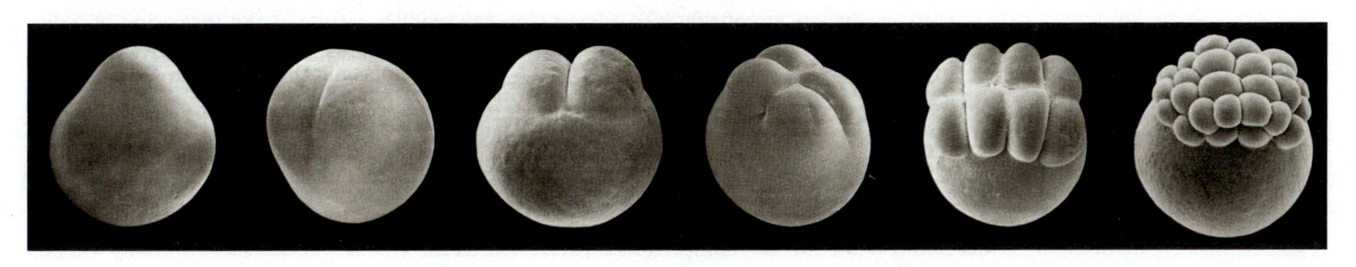

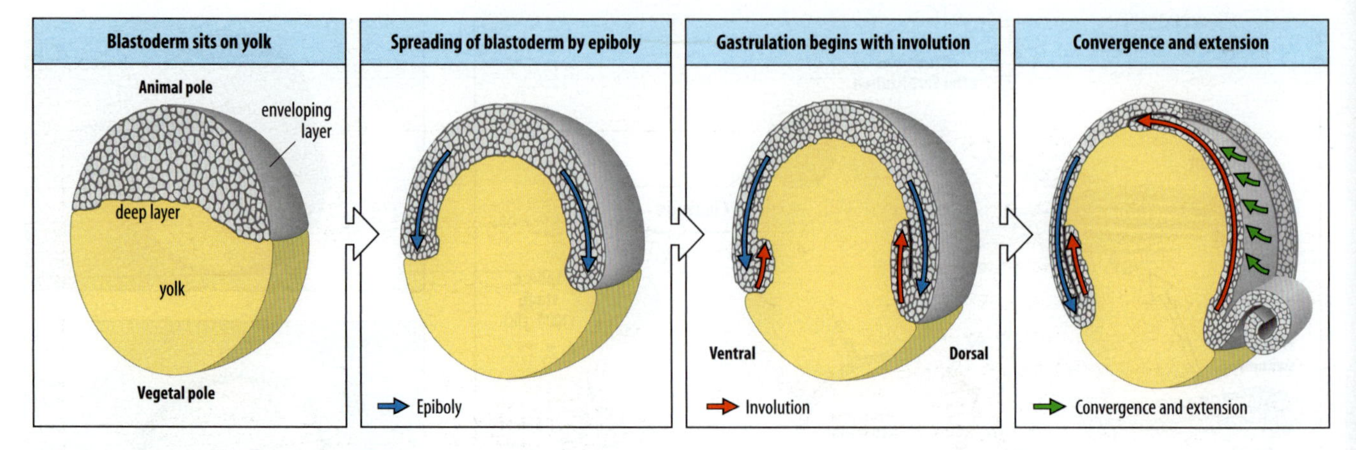

Fig. 3.11 Epiboly and gastrulation in the zebrafish. At the end of the first stage of cleavage the zebrafish embryo is composed of a cluster of blastomeres sitting on top of the yolk. With further cleavage and spreading out of the layers of cells (epiboly), the upper half of the yolk becomes covered by a blastoderm. Gastrulation occurs by involution of cells in a ring around the edge of the blastoderm. The involuting cells converge on the dorsal midline to form the body of the embryo encircling the yolk.

blastoderm at about the same time. We shall examine some of the other differences in Chapter 7. By 9 hours the notochord becomes distinct, and gastrulation is complete by 10 hours. Somite formation, neurulation, and migration of neural crest cells then follow.

Over the next 12 hours the embryo elongates, and the rudiments of the primary organ systems become recognizable. Somites first appear anteriorly at about 10.5 hours, and new ones are formed at intervals of initially 2 hours and then 3 hours; by 18 hours, 18 somites are present. The nervous system develops rapidly. Optic vesicles, which give rise to the eyes, can be distinguished at 12 hours as bulges from the brain, and by 18 hours the body starts to twitch. At 48 hours the embryo hatches, and the young fish begins to swim and feed. One of the advantages of the zebrafish as a developmental model organism is its suitability for large-scale genetic screening because of its small size, short generation time, and transparency (Box 3B, opposite).

3.3 The early chicken embryo develops as a flat disc of cells overlying a massive yolk

Avian embryos are very similar to those of mammals in the morphological complexity of the embryo and the general course of embryonic development, but are easier to obtain and observe. Many observations and manipulations can be carried out simply by opening the egg, but the embryo can also be cultured outside the egg. This is particularly convenient for some experimental microsurgical manipulations and investigation of the effects of chemical compounds. Despite the considerable differences in the very early stages of development between chick and mouse embryos (see Fig. 3.2), gastrulation and later development are very similar in both, and the chick provides a complement to studies of mouse embryology.

The large yolky egg cell is fertilized and begins to undergo cleavage while still in the hen's oviduct. Because of the mass of yolk, cleavage is confined to a small patch of cytoplasm several millimeters in diameter, which contains the nucleus and lies on top of the yolk. Cleavage in the oviduct results in the formation of a disc several cells thick called a **blastodisc** or blastoderm. During the 20-hour passage down the oviduct, the egg becomes surrounded by albumen (egg white), the shell membranes, and the shell (Fig. 3.12). At the time of laying, the blastoderm, which is analogous to the amphibian blastula, is composed of some 60,000 cells. The chick developmental cycle is shown in Fig. 3.13.

The early cleavage furrows extend downward from the surface of the cytoplasm but do not completely separate the cells, whose ventral faces initially remain open to the yolk. The central region of the blastoderm, under which a cavity

Box 3B Large-scale mutagenesis in zebrafish

Zebrafish offer a potentially very valuable vertebrate system for large-scale mutagenesis because large numbers can be handled, and the transparency and large size of the embryos make it relatively easy to identify developmental abnormalities. However, unlike the case of *Drosophila* described earlier in the book (see Box 2A, pp. 36–37), as yet there is no genetic means of eliminating unaffected individuals automatically. This means that all the progeny of a cross have to be examined visually.

A screening program using zebrafish involves breeding for three generations (see figure). Male fish treated with a chemical mutagen are crossed with wild-type females; their F_1 male offspring are crossed again with wild-type females, and the female and male siblings from each of these crosses are themselves crossed. The offspring from each of these pairs are examined separately for homozygous mutant phenotypes. If the F_1 fish carry a mutation, then in 25% of the F_2 matings two heterozygotes will mate and 25% of their offspring will be homozygous for the mutation. Zebrafish can also be made to develop as haploids by fertilizing the egg with sperm heavily irradiated with ultraviolet light. This allows one to detect early-acting recessive mutations without having to breed the fish to obtain homozygous embryos.

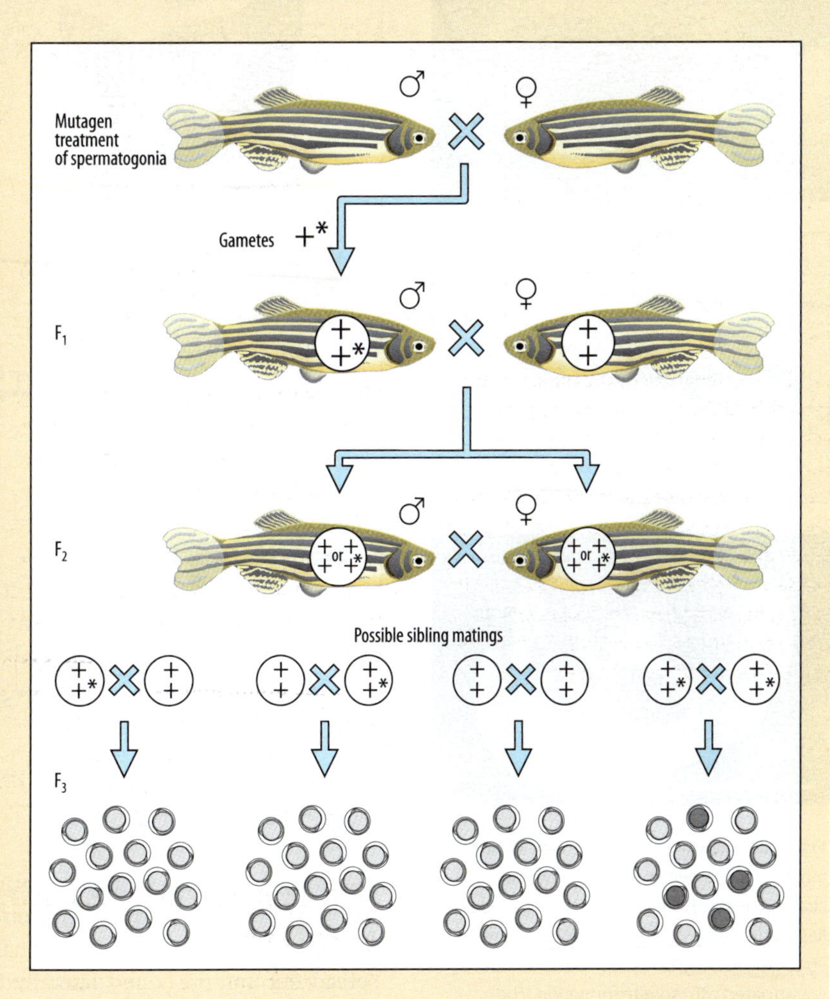

called the **subgerminal space** develops, is translucent and is known as the **area pellucida**, in contrast to the outer region, which is the darker **area opaca** (Fig. 3.14). A layer of cells called the **hypoblast** develops over the yolk to form the floor of the cavity. The hypoblast eventually gives rise to extra-embryonic structures such as those that connect the embryo to its source of nutrients in the yolk. The embryo proper is formed from the remaining blastoderm, known as the **epiblast**.

A crucial region in the development of the chick embryo is the **posterior marginal zone**. This is a slightly thickened region of the epiblast that lies at the junction between the area opaca and area pellucida at the posterior of the embryo, and is the site at which gastrulation begins. A crescent-shaped region of small cells called **Koller's sickle** lies at the front of the posterior marginal zone. Gastrulation in birds and mammals is marked by the development of the **primitive streak**, which is the forerunner of the main body axis of the embryo. In the chicken embryo, the streak is visible as a denser region that starts at Koller's sickle and gradually extends as a narrow strip to just over half way across the area pellucida, forming a furrow in the dorsal face of the epiblast (see Fig. 3.14). It is the first visible indication of the antero-posterior axis of the embryo. Unlike *Xenopus*, cell proliferation

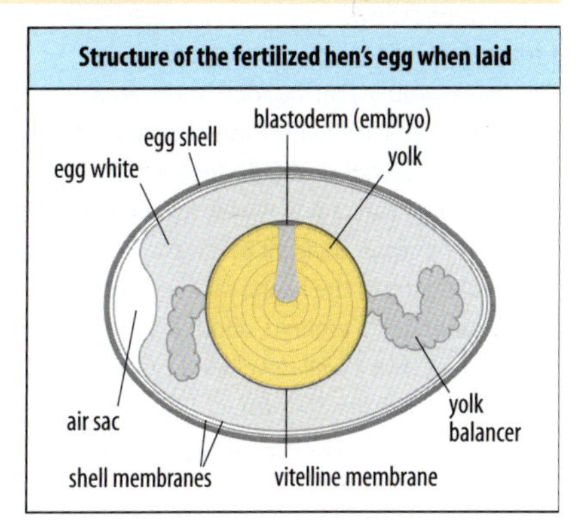

Fig. 3.12 The development of a hen's egg at the time of laying. Cleavage begins after fertilization while the egg is still in the oviduct. The albumen (egg white) and shell are added during the egg's passage down the oviduct. At the time of laying the embryo is a disc-shaped cellular blastoderm lying on top of a massive yolk, which is surrounded by the egg white and shell.

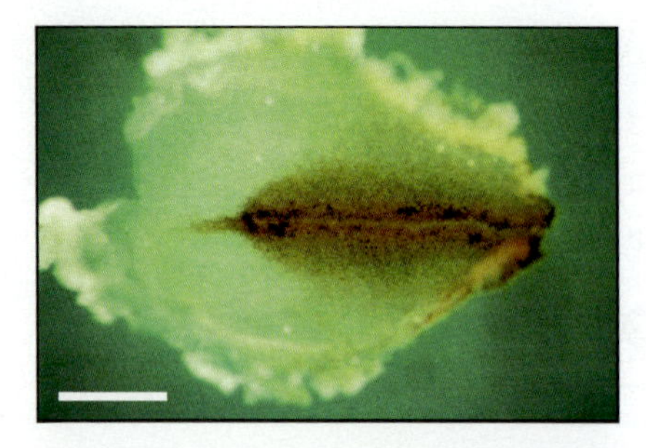

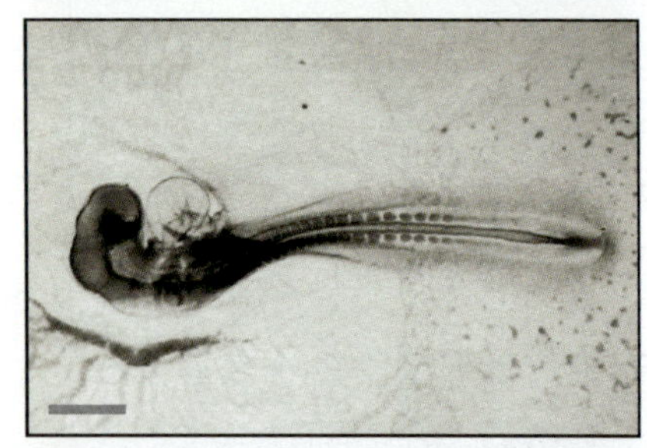

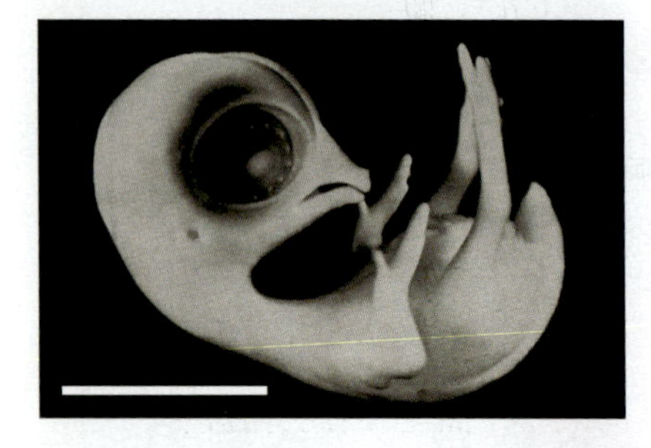

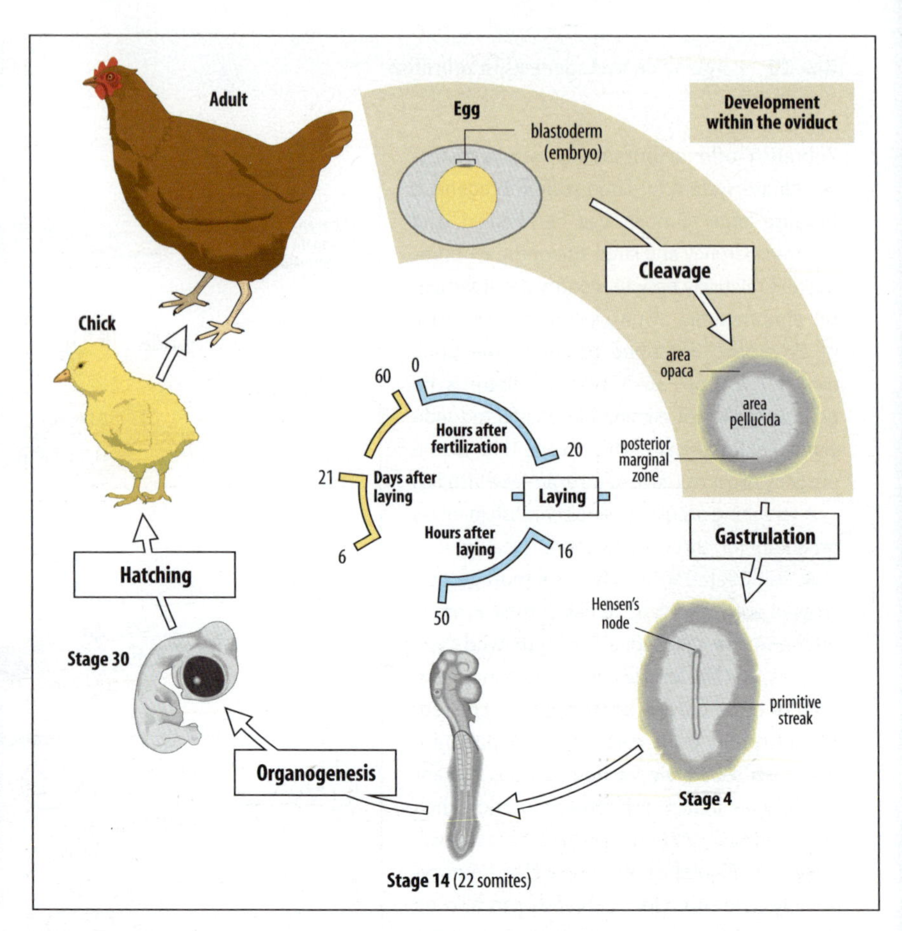

Fig. 3.13 Life cycle of the chicken. The egg is fertilized in the hen and by the time it is laid cleavage is complete and a cellular blastoderm lies on the yolk. After gastrulation, the primitive streak forms. Regression of Hensen's node is associated with somite formation. The photographs show: the primitive streak (stained brown by staining with antibody against Brachyury protein) surrounded by the area pellucida (top, scale bar = 1 mm); a stage 14 embryo (50–53 hours after laying) with 22 somites (the head region is well defined and the transparent organ adjacent to it is the ventricular loop of the heart; middle, scale bar = 1 mm); a stage 35 embryo, about 8.5–9 days after laying, with a well-developed eye and beak (bottom, scale bar = 10 mm).

Top photograph courtesy of B. Herrmann, from Kispert, A., et al.: 1995.

and growth of the embryo continue throughout gastrulation in birds and mammals. Epiblast cells converge on the primitive streak, and as the streak moves forward from the posterior marginal zone, cells in the furrow move inward and spread out anteriorly and laterally beneath the upper layer, forming a layer of loosely connected cells, or **mesenchyme**, in the subgerminal space (Fig. 3.15). The primitive streak is thus similar in some respects to the blastopore region of amphibians, but cells move inwards individually, rather than as a coherent sheet, and this type of inward movement is known as **ingression**. Those cells that move inwards will give rise to mesoderm and endoderm, whereas the cells that remain on the surface of the epiblast give rise to the ectoderm.

The primitive streak starts to form when the hypoblast becomes displaced forward from the posterior marginal zone by a new layer of cells called the secondary hypoblast or **endoblast**, which grows out from the zone. Eventually, this layer in turn becomes displaced by epiblast-derived endoderm cells that have moved in through the streak. The primitive streak in the chick embryo is fully extended by 16 hours after laying. At the anterior end of the streak a condensation of cells known as **Hensen's node** becomes apparent, where cells are also moving inward. Hensen's node is the major organizing center for the early chick embryo, equivalent to the Spemann organizer in amphibians, and is formed of cells derived from the posterior marginal zone and Koller's sickle as well as cells recruited from the epiblast. When the inward movement of cells has ceased, the primitive streak begins to regress, Hensen's node moving toward the posterior end of the embryo (Fig. 3.16). As the node regresses, the notochord is formed immediately

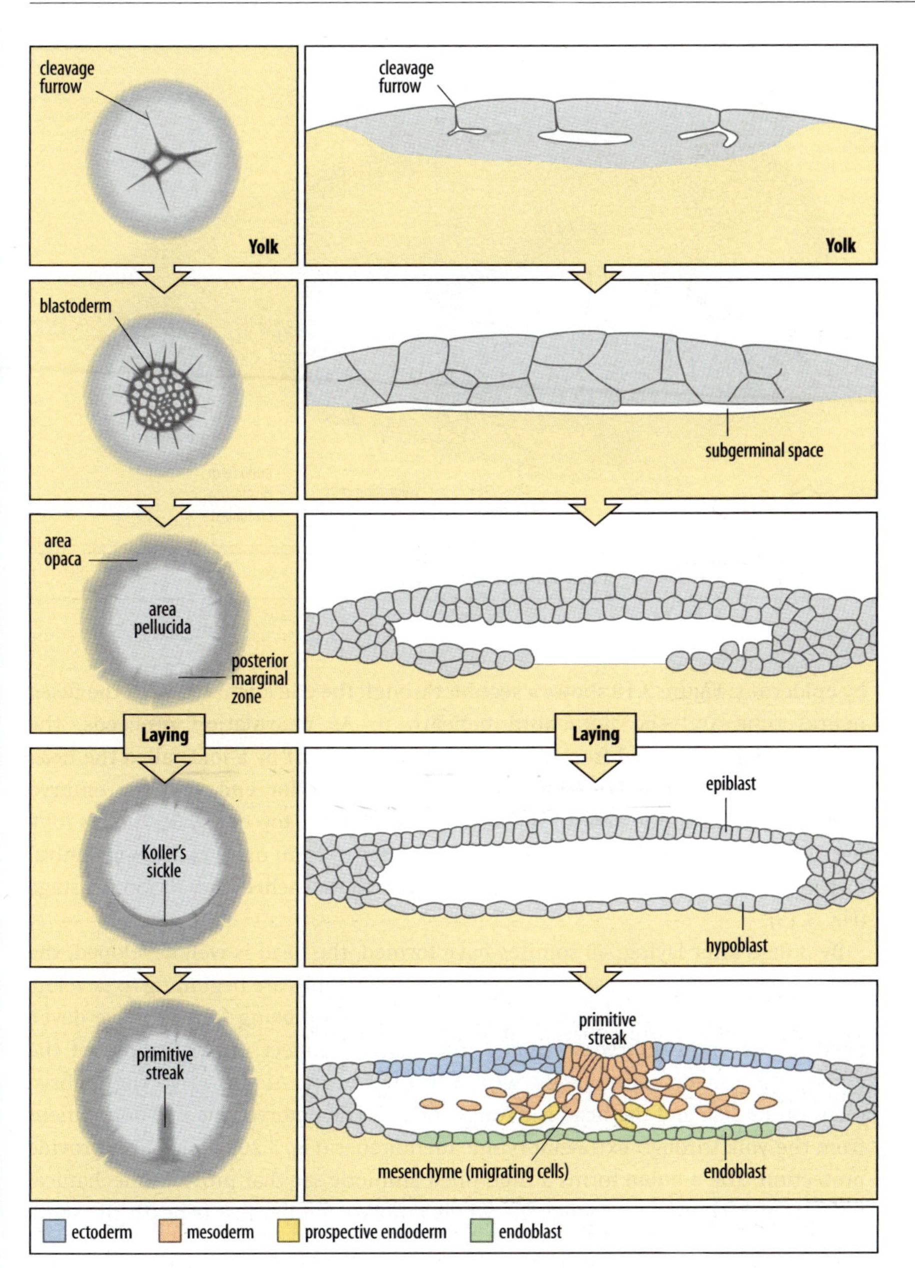

Fig. 3.14 Cleavage and epiblast formation in the chick embryo. By the time the egg is laid, cleavage has divided the small area of egg cytoplasm free from yolk into a disc-shaped cellular blastoderm. The panels on the left show the view of the embryo from above, while the panels on the right show cross-sections through the embryo. The first cleavage furrows extend downward from the surface of the egg cytoplasm and initially do not separate the blastoderm completely from the yolk. In the cellular blastoderm the central area overlying the subgerminal space is called the area pellucida and the marginal region the area opaca. A layer of cells develops immediately overlying the yolk and is known as the hypoblast. This will give rise to extra-embryonic structures, while the upper layers of the blastoderm—the epiblast—give rise to the embryo proper.

anterior to it from internalized mesoderm along the central dorsal midline. At this early stage, the notochord is sometimes called the head process and this developmental stage is known as the head-process stage. The mesoderm immediately on each side of the midline will form the somites, as discussed in Chapter 4. The rest of the mesoderm lateral to the somites is the lateral plate mesoderm and will develop into organs such as heart, kidneys, and the vascular system and blood.

As the notochord forms, the neural tube begins to develop as a pair of folds on either side of the midline of the neural plate ectoderm above the notochord (Fig. 3.17). The folds fuse in the dorsal midline to form the neural tube, initially leaving the anterior and posterior ends open, and neural crest cells detach on either side of the site of fusion. The fused neural tube becomes covered over

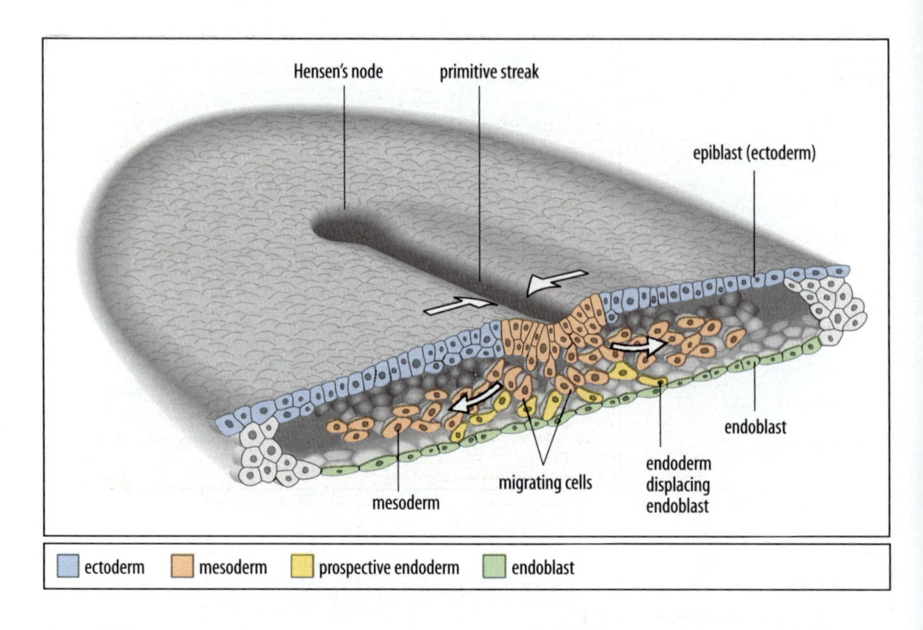

Fig. 3.15 Ingression of mesoderm and endoderm during gastrulation in the chick embryo. Gastrulation begins with the formation of the primitive steak, a region of proliferating and migrating cells, which elongates from the posterior marginal zone. Future mesodermal and endodermal cells migrate through the primitive streak into the interior of the blastoderm. During gastrulation, the primitive streak extends about halfway across the area pellucida (see Fig. 3.14). At its anterior end an aggregation of cells known as Hensen's node forms. As the streak extends, cells of the epiblast move toward the primitive streak (arrows), move through it, and then outward again underneath the surface to give rise internally to the mesoderm and endoderm internally, the latter displacing the hypoblast.
Adapted from Balinsky, B.I., et al.: 1975.

by epidermis; Figure 3.18 shows a section through the chick embryo with the fused neural tube and the notochord beneath it. As neurulation proceeds, the head becomes separated from the surface of the epiblast by a fold called the head fold, the somites start to form starting at the anterior end, and the embryo eventually folds on the ventral side to form the gut (see top of Fig. 3.17). This folding also brings the two heart rudiments together to form one organ lying ventral to the gut. By 2 days after laying, the embryo has reached the 20-somite stage (Fig. 3.19).

By 3 days after laying, 40 somites have formed, the head is well developed, the heart is formed, blood vessels are forming, and the limbs are beginning to develop. Blood vessels and blood islands, where blood cells are being formed, have developed in the extra-embryonic tissues; the vessels connect up with those of the embryo to provide a circulation with a beating heart. At this stage the embryo turns on its side and the head is strongly flexed. The embryo gets its nourishment from the yolk through extra-embryonic membranes (Fig. 3.20), which also provide protection. The **amnion** forms a fluid-filled amniotic sac that provides mechanical protection; a **chorion** surrounds the whole embryo and lies just beneath the shell;

Fig. 3.16 Regression of Hensen's node. After extending about halfway across the blastoderm, the primitive streak begins to regress, with Hensen's node moving in a posterior direction as the head fold and neural plate begin to form anterior to it. As the node moves backward, the notochord develops in the area anterior to it and somites begin to form on either side of the notochord.

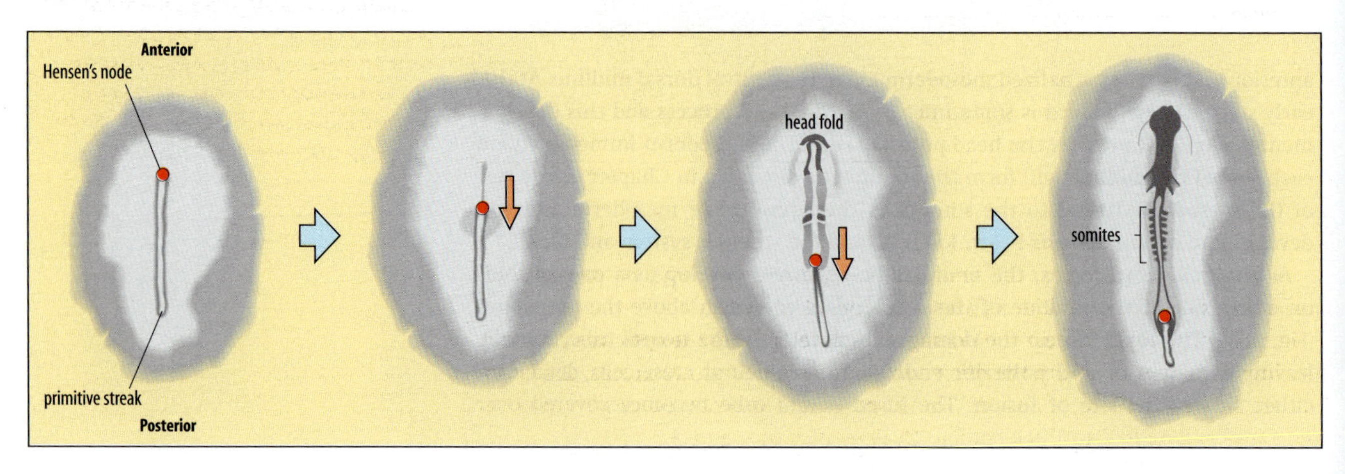

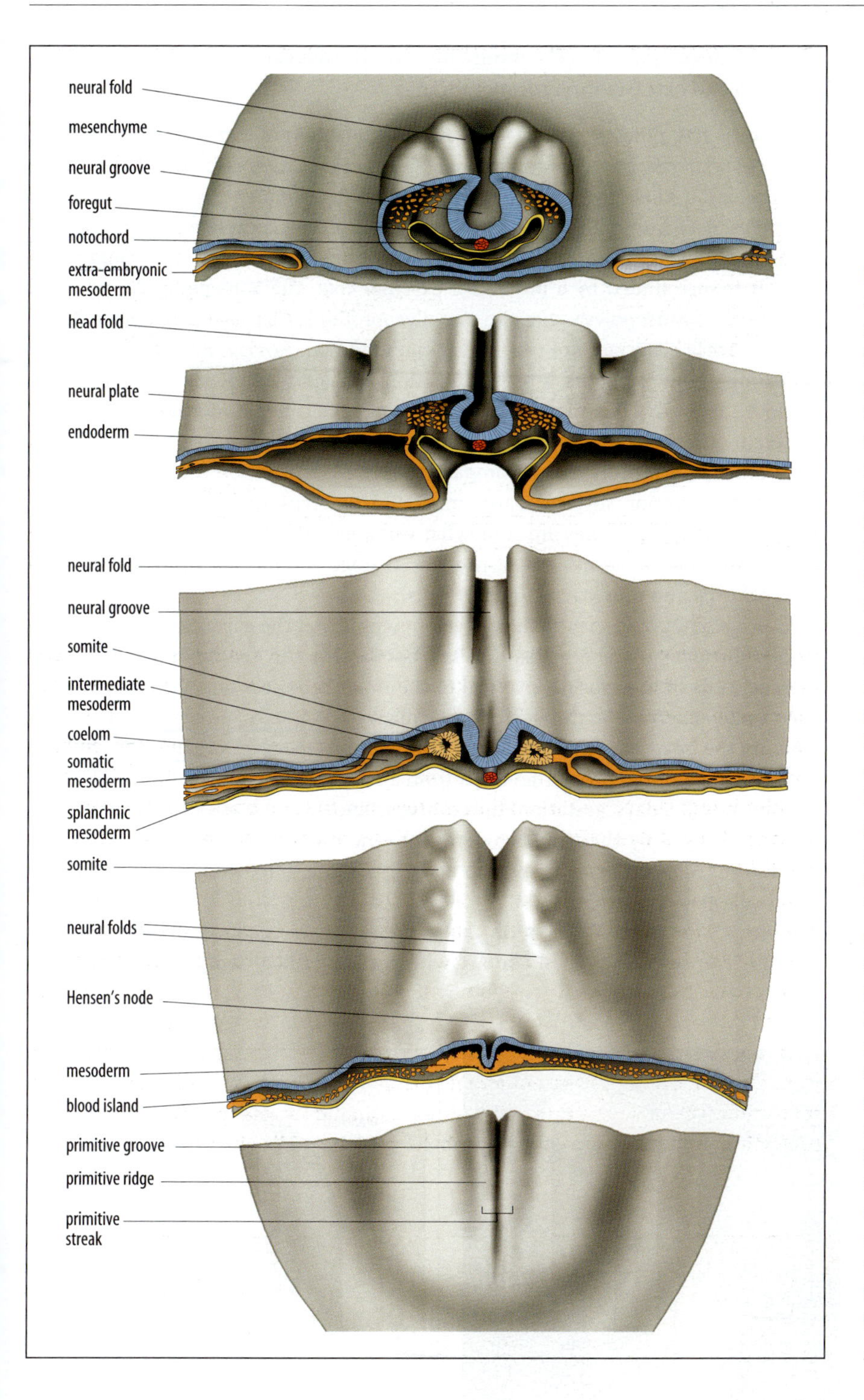

Fig. 3.17 Development of the neural tube and mesoderm in the chick embryo. Once the notochord has formed, neurulation begins, following notochord formation in an anterior to posterior direction. The figure shows a series of sections along the antero-posterior axis of a chick embryo. Neural-tube formation is well advanced at the anterior end (top two sections), where the head fold has already separated the future head from the rest of the blastoderm and the ventral body fold has brought endoderm from both sides of the body together to form the gut. During neurulation, the neural plate changes shape: neural folds rise up on either side and form a tube when they meet in the midline. The mesenchymal mesoderm in this region will give rise to head structures. Further back (middle sections), in the future trunk region of the embryo, notochord and somites have formed and neurulation is starting. At the posterior end, behind Hensen's node (bottom section), notochord formation, somite formation, and neurulation have not yet begun. The mesoderm internalized through the primitive streak starts to form structures appropriate to its position along the antero-posterior and dorso-ventral axes. For example, in the future trunk region, the intermediate mesoderm will form the mesodermal parts of the kidney, and the splanchnic mesoderm will give rise to the heart. The body fold will continue down the length of the embryo, forming the gut and also bringing paired organ rudiments that initially form on each side of the midline (e.g. those of the heart and dorsal aorta) together to form the final organs lying ventral to the gut. Blood islands, from which the first blood cells are produced, form from the ventral-most part of the lateral mesoderm.
After Patten, B.M.: 1971.

an **allantois** receives excretory products and provides the site of oxygen and carbon dioxide exchange; and a **yolk sac** surrounds the yolk.

In the remaining time before hatching, eyes develop from the optic vesicles, and the inner ear develops from the otic vesicles. The embryo grows in size, the internal organs develop, wings, legs, and beak are formed, and down feathers grow on the wings and body. The chick hatches 21 days after the egg is laid.

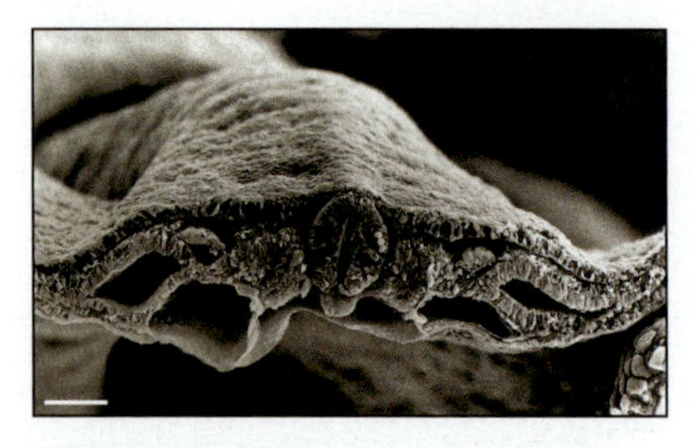

Fig. 3.18 Scanning electron micrograph of chick early somites and neural tube. There are blocks of somites adjacent to the neural tube and the notochord lies beneath it. The lateral plate mesoderm flanks the somites. Scale bar = 0.1 mm.

Photograph courtesy of J. Wilting.

Fig. 3.19 Development of the chick embryo. Left panel: at the 13-somite stage. At the anterior end (top), the head fold has formed. The dark region at the posterior end is Hensen's node. The somites can be seen on either side of the notochord as blocks of white tissue. Between the node and the most recently formed somite is mesoderm that will segment into somites. Center panel: at the 22-somite stage. Right panel: at the 40-somite stage. Development of the head region and the heart are quite well advanced, and the wing and leg buds are present as small protrusions. Brown staining is immunostaining with antibody against the Brachyury protein, a marker of early mesoderm and notochordal mesoderm. Scale bars = 1 mm.

Photographs courtesy of B. Hermann, from Kispert, A., et al.: 1995.

3.4 Early development in the mouse involves the allocation of cells to form the placenta and extra-embryonic membranes

The mouse has a life cycle of 9 weeks, from fertilization to mature adult (Fig. 3.21), which is relatively short for a mammal, and is one of the reasons that the mouse has become a model organism for mammalian development. Fertilization of the egg takes place internally in the oviduct; meiosis is then completed and the second polar body forms. The egg is small, about 100 μm in diameter, and contains no yolk. It is surrounded by a protective external coat, the **zona pellucida**, which is composed of mucopolysaccharides and glycoproteins. Cleavage takes place in the oviduct. Early cleavages are very slow compared with *Xenopus* and chick, the first occurring about 24 hours after fertilization and subsequent cleavages at about 12-hour intervals. They produce a solid ball of cells called a **morula** (Fig. 3.22). At the eight-cell stage the blastomeres increase the area of cell surface in contact with each other in a process called **compaction**. After compaction, the cells are polarized; their exterior surfaces carry microvilli whereas their inner surfaces are smooth. Further cleavages are somewhat variable and are both radial and tangential, so that by the equivalent of the 32-cell stage the morula contains about 10 internal cells and more than 20 outer cells.

A special feature of mammalian development is that the early cleavages give rise to two distinct groups of cells—the **trophectoderm** and the **inner cell mass**. The internal cells of the morula give rise to the inner cell mass and the outer cells to the trophectoderm. The trophectoderm will give rise to extra-embryonic structures such as the placenta, through which the embryo gains nourishment from the mother. The embryo proper develops from a subset of cells in the inner cell mass. At this stage (3.5 days' gestation) the embryo is known as a **blastocyst** (see Fig. 3.22). Fluid is pumped by the trophectoderm into the interior of the blastocyst, which causes the trophectoderm to expand and form a fluid-filled vesicle containing the inner cell mass at one end.

From 3.5 to 4.5 days' gestation the inner cell mass becomes divided into two regions. The surface layer in contact with the fluid-filled cavity of the blastocyst

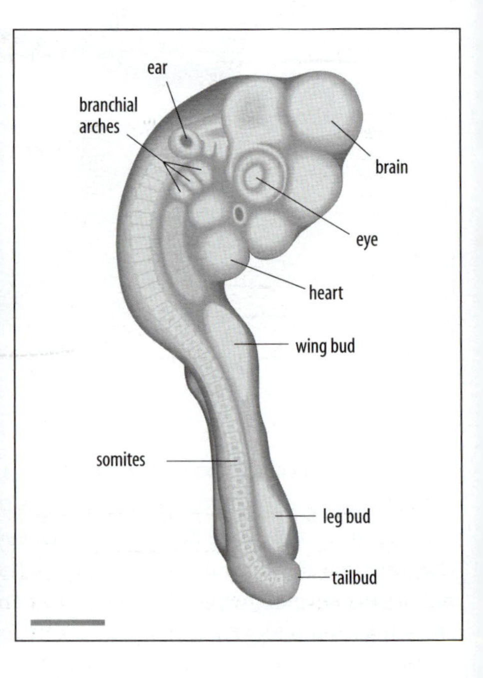

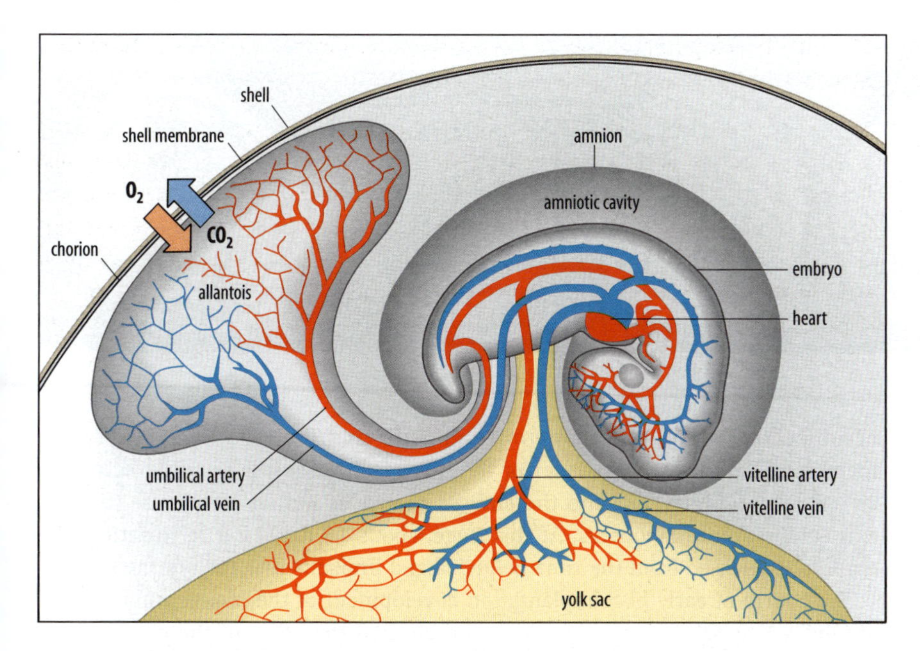

Fig. 3.20 The extra-embryonic structures and circulation of the chick embryo. A chick embryo at the same stage as that shown in the third panel in Fig. 3.19 is depicted *in situ*. The embryo has turned on its side, and its heart is beating. The yolk is surrounded by the yolk sac membrane. The vitelline vein takes nutrients from the yolk sac to the embryo and the blood is returned to the yolk sac via the vitelline artery. The umbilical artery takes waste products to the allantois and the umbilical vein brings oxygen to the embryo. The amnion and fluid-filled amniotic cavity provide a protective chamber for the embryo.

After Patten, B.M.: 1951.

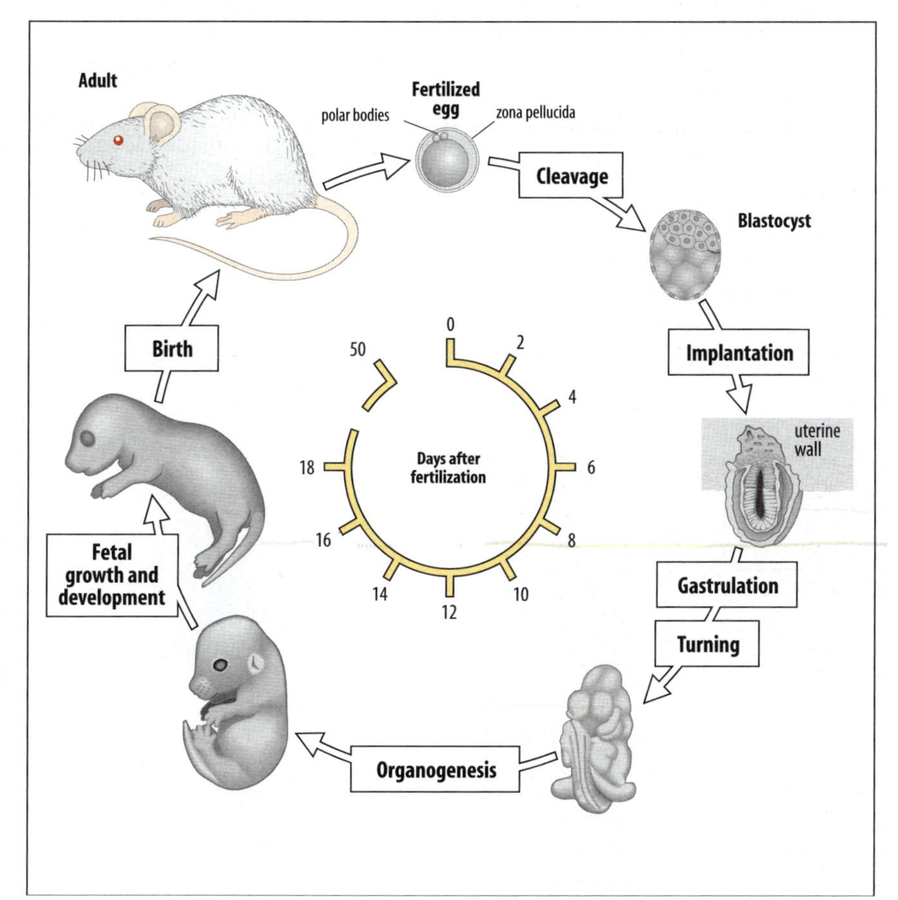

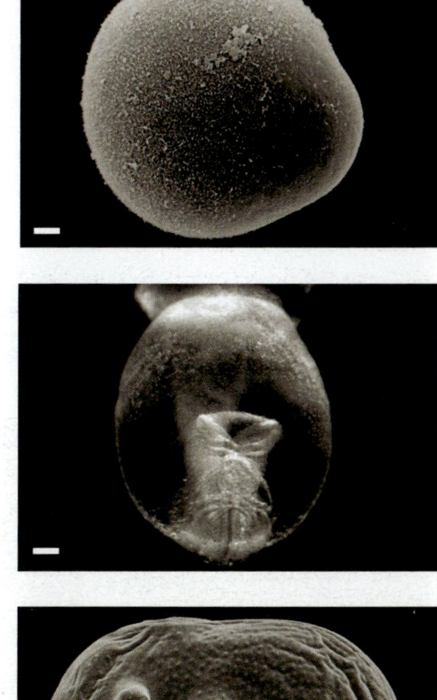

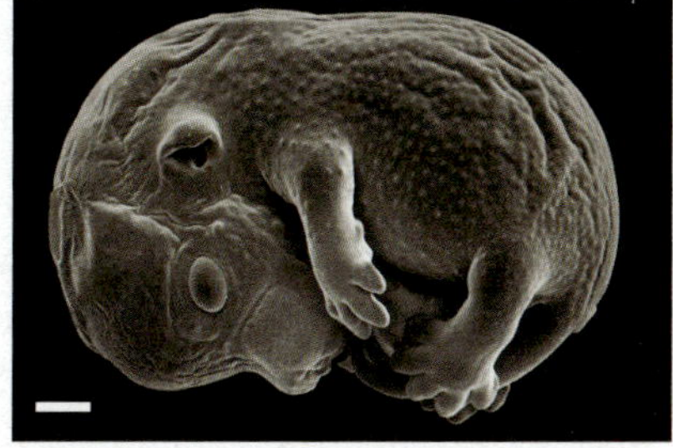

Fig. 3.21 The life cycle of the mouse. The egg is fertilized in the oviduct, where cleavage also takes place before implantation of the blastocyst in the uterine wall at 5 days after fertilization. Gastrulation and organogenesis then take place over a period of about 7 days and the remaining 6 days before birth are largely a time of overall growth. After gastrulation the mouse embryo undergoes a complicated movement known as turning, in which it becomes surrounded by its extra-embryonic membranes (not shown here). The photographs show (from top): a fertilized mouse egg just before the first cleavage (scale bar = 10 μm); the anterior view of a mouse embryo at 8 days after fertilization (scale bar = 0.1 mm); and a mouse embryo at 14 days after fertilization (scale bar = 1 mm).

Photographs courtesy of: T. Bloom (top, from Bloom, T.L.: 1989); N. Brown (middle); and J. Wilting (bottom).

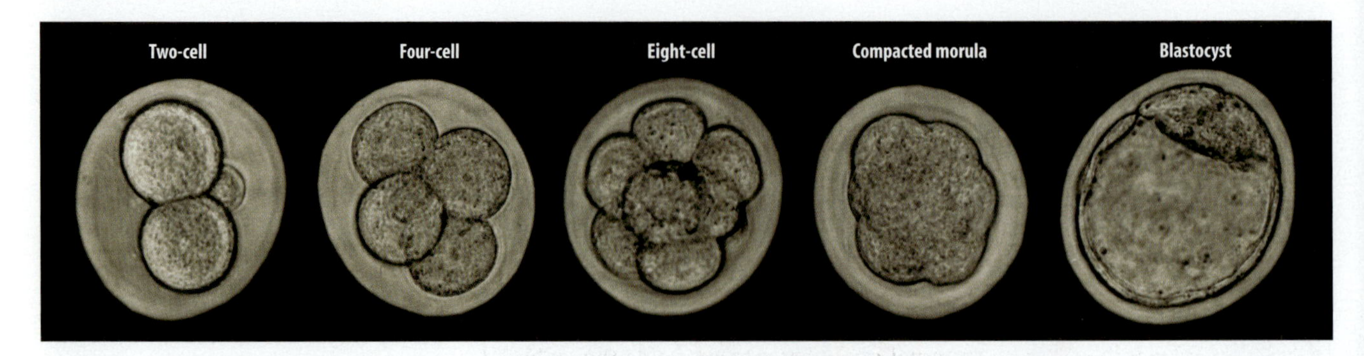

Fig. 3.22 Cleavage in the mouse embryo. The photographs show the cleavage of a fertilized mouse egg from the two-cell stage through to the formation of the blastocyst. After the eight-cell stage, compaction occurs, forming a solid ball of cells called the morula, in which individual cell outlines can no longer be discerned. The internal cells of the morula give rise to the inner cell mass, which can be seen as the compact clump at the top of the blastocyst. It is from this that the embryo proper forms. The outer layer of the hollow blastocyst—the trophectoderm—gives rise to extra-embryonic structures.

Photographs courtesy of T. Fleming.

becomes the **primitive endoderm,** and will contribute to extra-embryonic membranes, while the remainder of the inner cell mass—the **primitive ectoderm** or **epiblast**—will develop into the embryo proper as well as giving rise to some extra-embryonic components. At this stage, about 4.5 days after fertilization, the embryo is released from the zona pellucida and implants into the uterine wall.

The course of early post-implantation development of the mouse embryo from around 4.5 to 8.5 days appears more complicated than that of the chick, partly because of the need to produce a larger variety of extra-embryonic membranes, and partly because the epiblast is distinctly cup-shaped in the early stages. This is a peculiarity of mouse and other rodent embryos. The epiblasts of human and rabbit embryos, for example, are flat, much more resembling that of the chick. Despite the different topology, however, gastrulation and the later development of the mouse embryo is in essence very similar to that of the chick.

The first 2 days of post-implantation development are shown in Fig. 3.23. At implantation, the cells of the mural trophectoderm—the region not in contact with

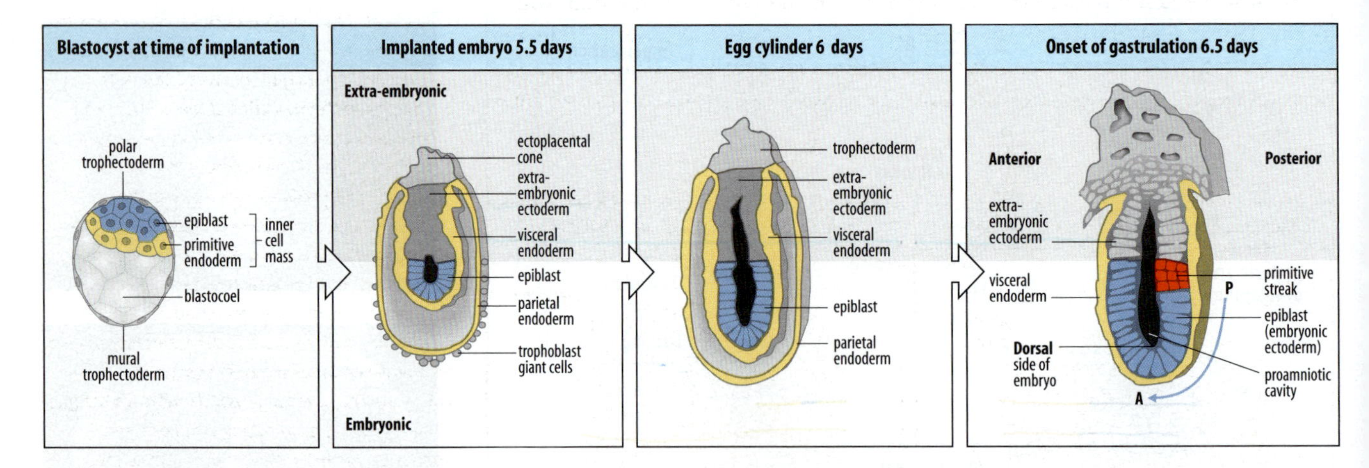

Fig. 3.23 Early post-implantation development of the mouse embryo. First panel: before implantation, the fertilized egg has undergone cleavage to form a hollow blastocyst, in which a small group of cells, the inner cell mass, will give rise to the embryo, while the rest of the blastocyst forms the trophectoderm, which will develop into extra-embryonic structures. At the time of implantation the inner cell mass divides into two regions: the primitive ectoderm or epiblast, which will develop into the embryo proper, and the primitive endoderm, which will contribute to extra-embryonic structures. Second panel: the polar trophectoderm in contact with the epiblast forms extra-embryonic tissues, the ectoplacental cone, and extra-embryonic ectoderm, which contributes to the placenta. The mural trophectoderm gives rise to the trophoblast giant cells. The epiblast elongates and develops an internal cavity (proamniotic cavity), which gives it a cup-shaped form. Third panel: the cylindrical structure containing both the epiblast and the extra-embryonic tissue derived from the polar trophectoderm is known as the egg cylinder. Fourth panel: the beginning of gastrulation is marked by the appearance of the primitive streak at the posterior of the epiblast (P). It starts to extend anteriorly (arrow) towards the bottom of the cylinder. The parietal endoderm and trophoblast giant cells are not shown in this or subsequent figures.

the inner cell mass—replicate their DNA without cell division (endo-reduplication), giving rise to trophoblast giant cells which invade the uterus during implantation. The rest of the trophectoderm grows to form the ectoplacental cone and the **extra-embryonic ectoderm**, which both contribute to the formation of the placenta. Some cells from the primitive endoderm migrate to cover the whole inner surface of the mural trophectoderm. They become the parietal endoderm, which eventually becomes Reichert's membrane, a sticky layer of cells and extracellular matrix that has a barrier function. The remaining primitive endoderm cells form the **visceral endoderm**, which covers the elongating **egg cylinder** containing the epiblast.

→ primitive ectoderm

By 6 days after fertilization, an internal cavity has formed inside the epiblast, which becomes cup-shaped—U-shaped when seen in cross-section (see Fig. 3.23, third panel). The epiblast is now a curved single layer of epithelium, which at this stage contains about 1000 cells. The embryo develops from this layer. The first easily visible sign of the embryonic antero-posterior axis is at about 6.5 days, with the beginning of gastrulation and the appearance of the primitive streak. The streak starts as a localized thickening at the edge of the cup on one side; this is the future posterior end of the embryo. The initial development of the primitive streak in the mouse is similar to that in the chick. Over the next 12–24 hours it elongates until it reaches the bottom of the cup. A condensation of cells—the **node**—becomes distinguishable at the anterior end of the extended streak and corresponds to Hensen's node in the chick embryo. To make primitive streak formation easier to compare with that of the chick, imagine the epiblast cup spread out flat. As in the chick, epiblast cells converge on the primitive streak, and proliferating cells migrate through it to spread out laterally and anteriorly between the ectoderm and the visceral endoderm to form a mesodermal layer (Fig. 3.24). Development from day 7 is shown in Fig. 3.25. Some epiblast-derived cells pass through the mesoderm and enter the visceral endoderm, gradually displacing it to form the definitive embryonic endoderm on the outside of the cup, which is the future ventral side of the embryo. Cells migrating anteriorly from the node form the notochord, while prospective mesodermal cells from the region surrounding it migrate anteriorly to form the somites. Cell proliferation continues

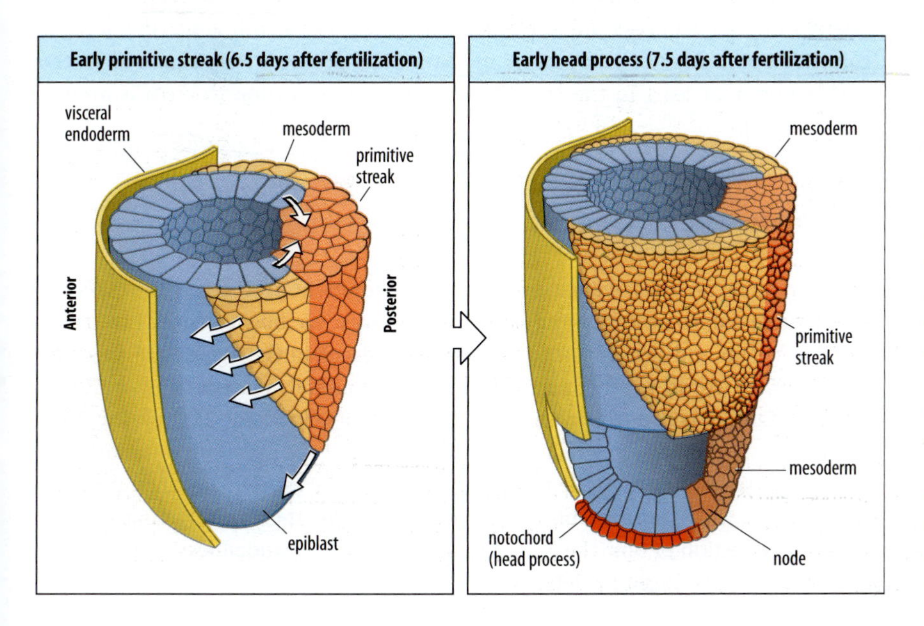

Fig. 3.24 Gastrulation in the mouse embryo. Left panel: as in the chick, gastrulation in the mouse embryo begins when epiblast cells converge on the posterior of the epiblast and move under the surface, forming the denser primitive streak (brown) where the cells are becoming internalized. Once inside, the proliferating cells spread out laterally between the epiblast surface and the visceral endoderm to give a layer of prospective mesoderm (light brown). Some of the internalized cells will eventually replace the visceral endoderm to give definitive endoderm (not shown on these diagrams for simplicity), which will form the gut. Right panel: as gastrulation proceeds, the primitive streak lengthens and reaches the bottom of the cup, with the node at the anterior end. The node gives rise to the notochord, which forms the structure known as the head process. Part of the visceral endoderm and the mesoderm has been cut away in this diagram to show the node and notochord. Note that, given the topology of the mouse embryo at this stage, the germ layers appear inverted (ectoderm on the inner surface of the cup, endoderm on the outer) if compared with the frog gastrula.

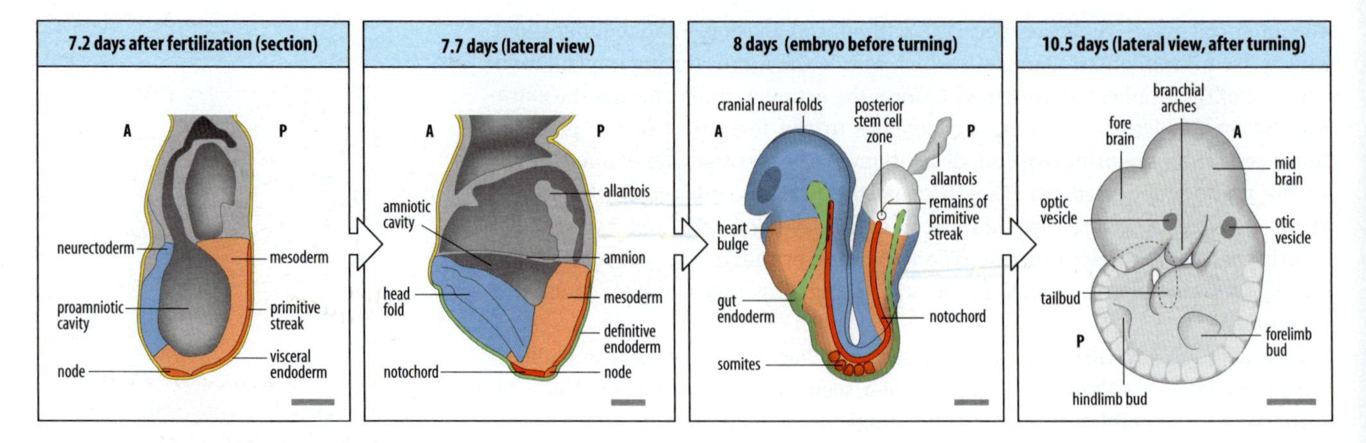

Fig. 3.25 Schematic views of the early development of the mouse embryo to the completion of gastrulation and neurulation. First panel: by 7 days, the primitive streak (brown) has extended to the bottom of the epiblast and the node has formed. The anterior ectoderm (blue) becomes prospective neurectoderm, which will give rise to the brain and spinal cord. Mesoderm is shown as light brown. Second panel: the anterior part of the embryo grows in size and the head fold appears. Definitive endoderm (green) replaces visceral endoderm (yellow) to form an outer layer on the ventral surface of the embryo. The notochord (red) begins to form. Third panel: by 8 days there has been further growth of the embryo anterior to the node, the head is distinct, the neural folds have formed, the foregut and hindgut have closed, and somites are beginning to form on either side of the notochord. The embryo is covered by a layer of ectoderm that will form the epidermis, which is not shown on this diagram. Fourth panel: by 10.5 days, gastrulation and neurulation are complete. The embryo has undergone a complicated turning process around day 9 that brings the dorsal and ventral sides into their final positions. Scale bars for the first three panels = 100 μm; fourth panel, scale bar = 75 μm.

during gastrulation, and the embryo anterior to the node grows rapidly in size; the node eventually forms a center of stem cells in the tailbud that gives rise to the post-anal tail.

As in the chick, somite formation and organogenesis start at the anterior end of the mouse embryo and proceed posteriorly. At around 8.5 days, neural folds have started to form at the anterior end on the dorsal side of the embryo and a head is apparent (see Fig. 3.25). The embryonic endoderm—initially on the ventral surface of the embryo—becomes internalized to form the foregut and hindgut; the ventral surfaces eventually fold together to internalize the gut completely. The heart and liver move into their final positions relative to the gut, and the head becomes distinct. By midway through embryonic development, at 9 days, gastrulation and neurulation are complete: the mouse embryo has a distinct head and the forelimb buds are starting to develop. At 9 days there is a complicated turning process that produces a more recognizable mouse embryo, surrounded by its extra-embryonic membranes (Fig. 3.26). As a result of turning, the initial cup-shaped epiblast has turned inside out so that the dorsal surface is now on the outside; the ventral surface, with the umbilical cord that connects it to the placenta, is facing inwards. (Turning is another developmental quirk peculiar to rodents; human embryos are surrounded by their extra-embryonic membranes from the beginning.) Organogenesis in the mouse proceeds very much as in the chick embryo, at least in the initial stages. From fertilization to birth is around 18 to 21 days.

Setting up the body axes

In the previous sections, we have seen how development proceeds with respect to well defined antero-posterior and dorso-ventral axes. We now consider how these axes are set up and whether they are already present in the egg or are specified later. In other words, to what extent is the embryo already patterned as a result of maternal factors laid down in the egg during its development in the ovary? We begin with amphibians, in which the establishment of the axes is best understood, and the zebrafish, in which the process is thought to be similar, and then compare this strategy with those of birds and mammals. We then briefly consider the intriguing question of how the left-right asymmetry, or 'handedness', of a number of internal organs might be determined.

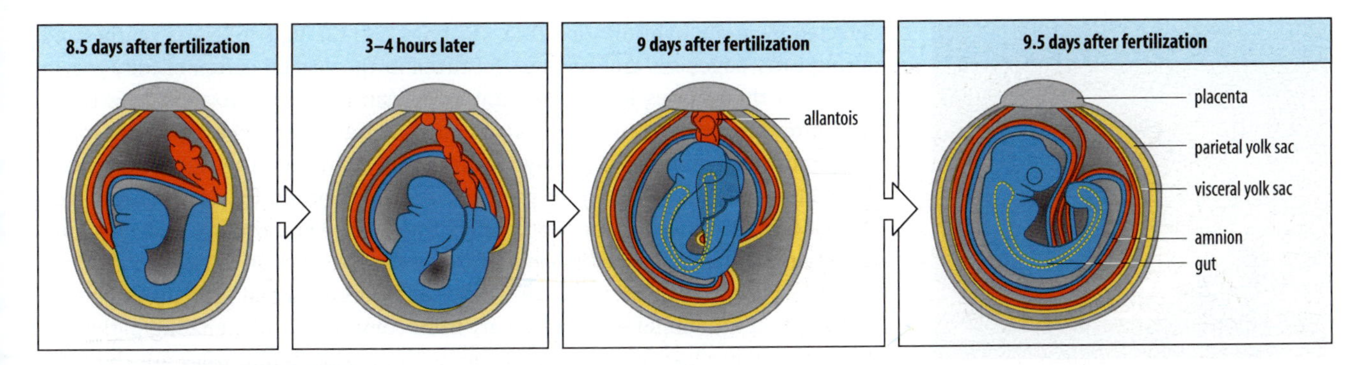

Fig. 3.26 Turning in the mouse embryo. Between 8.5 and 9.5 days, the mouse embryo becomes entirely enclosed in the protective amnion and amniotic fluid. The visceral yolk sac, a major source of nutrition, surrounds the amnion and the allantois connects the embryo to the placenta.

After Kaufman, M.H.: 1992.

3.5 The animal-vegetal axis is maternally determined in *Xenopus* and zebrafish

The very earliest stages in development of both *Xenopus* and the zebrafish are exclusively under the control of maternal factors present in the egg. As we saw in *Drosophila* (Chapter 2), maternal factors are the mRNA and protein products of genes that are expressed in the mother during oogenesis and laid down in the egg while it is being formed. The *Xenopus* egg possesses a distinct polarity even before it is fertilized, and this polarity influences the pattern of cleavage. One end of the egg, the **animal pole,** which sits uppermost when the egg is in water, has a heavily pigmented surface, whereas most of the yolk is located toward the opposite, unpigmented end, the **vegetal pole** (see Fig. 3.4). These differences define the animal-vegetal axis. The pigment itself has no role in development but is a useful marker for the real developmental differences that exist between the animal and vegetal halves of the egg. The animal half contains the nucleus, which is located close to the animal pole. The planes of early cleavages are related to the animal-vegetal axis. The first cleavage is parallel with the axis and often defines a plane corresponding to the midline, dividing the embryo into left and right sides. The second cleavage is in the same plane at right angles to the first and divides the egg into four cells, while the third cleavage is at right angles to the axis and divides the embryo into animal and vegetal halves, each composed of four blastomeres (see Fig. 3.5).

Maternal factors are of key importance in the very early development of *Xenopus* and zebrafish as the embryo's own genes do not generally begin to be transcribed until after a stage called the mid-blastula transition, which will be discussed later in the chapter. The *Xenopus* egg contains large amounts of mRNAs and there are also large amounts of stored proteins; there is, for example, sufficient histone protein for the assembly of more than 10,000 nuclei, quite enough to see the embryo through the first 12 cleavages until its own genes begin to be expressed. There are also a number of localized maternal mRNAs that encode proteins with specifically developmental roles. These are differentially localized along the animal-vegetal axis of the egg before cleavage, with most of the developmentally important maternal products localized in the vegetal hemisphere. The expression of most of the maternal mRNAs in the egg is regulated so that they are only translated after fertilization.

The proteins encoded by several of these mRNAs are candidates for the signals involved in specifying early polarity and setting further development in train. One maternal mRNA encodes the protein Vg-1, which is a member of the transforming growth factor-β (TGF-β) family of signaling proteins (Box 3C, p. 111). Vg-1 mRNA is localized in the vegetal half of the egg (see Fig. 3.27), where its presence can be

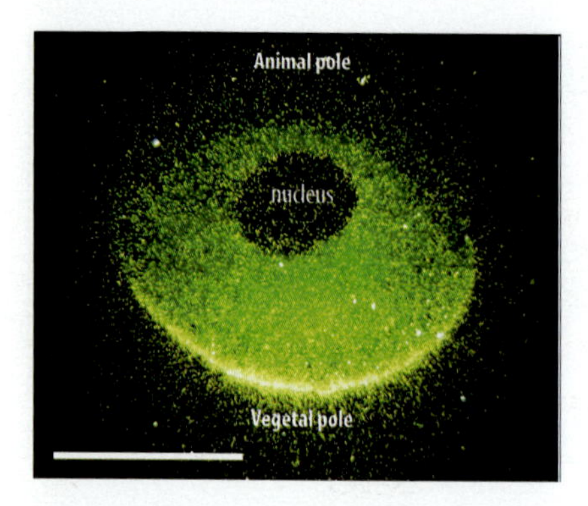

Fig. 3.27 Distribution of mRNA for the growth factor Vg-1 in the amphibian egg. *In situ* hybridization with a radioactive probe for maternal Vg-1 mRNA shows its localization (yellow) in the vegetal region. Scale bar = 1 mm.

Photograph courtesy of D. Melton.

detected by *in situ* hybridization (Box 3D, pp. 112–113). Vg-1 mRNA is synthesized during early oogenesis and becomes localized in the vegetal cortex of fully grown oocytes. It then moves into the vegetal cytoplasm before fertilization. It is translated after fertilization and functions as an early signal for some aspects of mesoderm induction. Another vegetally localized mRNA encodes the signaling protein Xwnt-11, which provides one of the early signals required for dorso-ventral axis specification. Members of the Wnt family of vertebrate proteins are related to, and were named after, the *wingless* gene of *Drosophila* (see Section 2.23), and the *Int-1* gene (now known as *Wnt1*) in mice. They are important signal proteins in pattern formation in all animal embryos. Apart from components of signaling pathways, the other main class of maternal factors with developmental roles are the transcription factors that switch on and regulate zygotic gene expression. Maternal mRNA encoding the T-box transcription factor VegT is localized to the vegetal hemisphere and is translated after fertilization. VegT has a key role in specifying the germ layers and its localization in the vegetal region is crucial to this function.

The animal–vegetal axis of the egg is certainly related to the antero-posterior axis of the tadpole, as the head will form from the animal region, but the axes of the tadpole and the fertilized egg are not directly comparable. Which side of the animal region will form the head, for example, is not determined until the dorsal side of the embryo is specified, which does not happen until after fertilization. The precise position of the embryo's antero-posterior axis thus depends on the specification of the dorso-ventral axis.

In the zebrafish fertilized egg, maternal factors are also distributed along the animal–vegetal axis and key factors for axis formation are present in the vegetal region. Removal of the vegetal-most yolk during early cleavage results in radially symmetric embryos that lack **axial structures**.

3.6 Localized stabilization of the transcriptional regulator β-catenin specifies the future dorsal side and the location of the main embryonic organizer in *Xenopus* and zebrafish

The spherical unfertilized egg of *Xenopus* is radially symmetric about the animal–vegetal axis, and this symmetry is broken only when the egg is fertilized. Sperm entry sets in motion a series of events that defines the dorso-ventral axis of the gastrula (see Section 3.1 for a general outline of *Xenopus* development), with the dorsal side forming more or less opposite the sperm's entry point.

Sperm entry can occur anywhere in the animal hemisphere of the *Xenopus* egg and it causes the outer layer of cytoplasm, the cortex, to loosen from the inner dense cytoplasm so that it is able to move independently. The cortex is a gel-like layer rich in actin filaments and associated material. During the first cell cycle, the cortex rotates such that vegetal cortex moves in a direction away from the site of sperm entry towards one side, which is thus specified as the future dorsal side. This **cortical rotation** involves an array of microtubules that become oriented away from the site of sperm entry on fertilization and which act as tracks for the movement of molecules to their new location.

Cortical rotation appears to relocate maternal factors originally located at the vegetal pole to a site nearer the equator on the side opposite sperm entry, creating a new asymmetry in the fertilized egg. These factors specify their new location as the future dorsal side of the embryo, thus conferring another axis of symmetry, the dorso-ventral axis, on the fertilized egg (Fig. 3.28). For this reason, they are often called **dorsalizing factors**. This is a key event in *Xenopus* development as it sets up the conditions necessary for the specification of the main embryonic organizer,

Box 3C Intercellular signals in development

Proteins that are known to act as signals between cells during development belong to seven main families. Some of these families, such as the fibroblast growth factors (FGFs), were originally identified because they were essential for the survival and proliferation of mammalian cells in tissue culture. The members of the seven families are either secreted or are part of the cell's surface, and they provide intercellular signals in both vertebrates and invertebrates at many stages of development. Other signal molecules, such as insulin, the insulin-like growth factors, and the neurotrophins, are used by embryos for particular tasks, but the eight families listed in the table are the main families involved throughout development.

The signal molecules listed in the table all act by binding to receptors on the surface of a target cell. This produces a signal that is passed, or transduced, across the cell membrane by the receptor to the intracellular biochemical signaling pathways. Each type of signal has a corresponding receptor or set of receptors, and only cells with the appropriate receptors on their surface can respond. The intracellular pathways can be complex and involve numerous different proteins. In the developmental context, the important responses of cells are either a change in gene expression, with specific genes being switched on or off, or a change in the cytoskeleton, leading to changes in cell shape and motility.

Common intercellular signals in vertebrate embryonic development

Family	Receptors	Examples of roles in development
Fibroblast growth factor (FGF)		
Twenty-five FGF genes have been identified in vertebrates, not all of which are present in all vertebrates	FGF receptors (receptor tyrosine kinases)	Roles at all stages in development. Maintenance of mesoderm formation. Induction of spinal cord (Chapter 4). Signal from apical ridge in limb bud (Chapter 9)
Epidermal growth factor (EGF)		
Epidermal growth factor	EGF receptors (receptor tyrosine kinases)	Cell proliferation and differentiation. Limb patterning (Chapter 9)
Transforming growth factor-β (TGF-β)		
A large protein family that includes activin, Vg-1, bone morphogenetic proteins (BMPs), Nodal, and Nodal-related proteins	Type I and Type II receptor subunits (receptor serine/threonine kinases) Receptors act as heterodimers	Roles at all stages in development. Mesoderm induction and patterning
Hedgehog		
Sonic hedgehog and other members	Patched	Positional signaling in limb and neural tube (Chapters 9 and 10). Involved in determination of left–right asymmetry
Wingless (Wnt)		
Twenty-one Wnt genes have been identified in vertebrates, not all of which are present in all vertebrates	Frizzled family of receptors (seven-span transmembrane proteins)	Roles at all stages in development. Dorso-ventral axis specification in *Xenopus*. Limb development (Chapter 9)
Delta and Serrate		
Transmembrane signaling proteins	Notch	Roles at all stages in development. Left–right asymmetry
Ephrins		
Transmembrane signaling proteins	Eph receptors (receptor tyrosine kinases)	Guidance of developing blood vessels (Chapter 9). Guidance of developing neurons in the nervous system (Chapter 10)
Retinoic acid		
Small molecule related to vitamin D	Retinoic acid receptors (RARs) Intracellular proteins related to the steroid receptors. They bind to retinoic acid to form a complex that acts as a transcriptional regulator	Somite formation (Chapter 4). Limb patterning (Chapter 9)

Some signal proteins, such as those of the transforming growth factor-β (TGF-β) family, act as dimers; two molecules form a covalently linked complex that binds to and activates the receptor, bringing two receptor proteins together to form a dimer. In some cases, the active form of the protein signal is a heterodimer, made up of two different members of the same family. Secreted signals can diffuse or be transported short distances through tissue and set up concentration gradients. Some signal molecules, however, remain bound to the cell surface and thus can only interact with receptors on a cell that is in direct contact. One such is Delta protein, which is a membrane-bound protein that interacts with a receptor protein called Notch on an adjacent cell.

For consistency in this book we have hyphenated the abbreviations for all secreted signaling proteins where the name allows—for example Vg-1, BMP-4, and TGF-β.

Box 3D *In situ* detection of gene expression

In order to understand how gene expression is guiding development, it is essential to know exactly where and when particular genes are active. Genes are switched on and off during development and patterns of gene expression are continually changing. Several powerful techniques are available that show where a gene is being expressed, both within whole intact embryos and in sections of embryos.

One set of techniques uses *in situ* hybridization to detect the mRNA that is being transcribed from a gene. If an RNA probe is complementary in sequence to an mRNA being transcribed in the cell, it will base-pair precisely, or hybridize, with the mRNA (see figure, right). A DNA probe of the appropriate sequence can therefore be used to locate its complementary mRNA in a tissue slice or a whole embryo. The probe may be labeled in various ways—with a radioactive isotope, a fluorescent tag, or an enzyme—to enable it to be detected. Radioactively labeled probes are detected by autoradiography, whereas probes labeled with fluorescent dyes are observed by fluorescence microscopy. Enzyme-labeled probes are detected by their reaction with a substrate that produces a colored product (see figure, opposite). The probes can be applied to both tissue sections and whole-mount preparations. Labeling with fluorescent dyes or enzymes has the advantage over radioactive labeling (see figure, below) in that the expression of several different genes can be detected simultaneously.

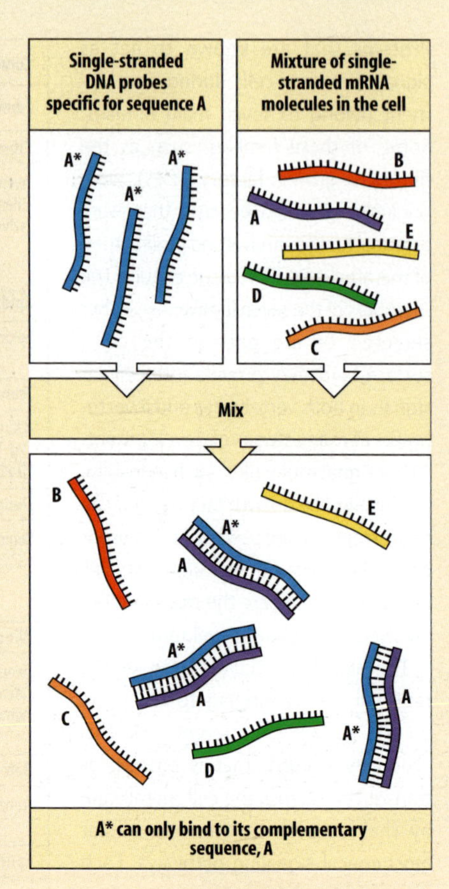

To detect protein expression and localization, the detection agent is an antibody specific for the protein. These are tagged either with fluorescent dyes as shown in Fig. 9.30, or with enzymes that can be detected by reaction with a colored substrate. An example of the histochemical detection of the expression of two different proteins in the same embryo is shown in Fig. 2.29.

The pattern and timing of gene expression can also be followed by inserting a reporter gene into an animal using transgenic techniques. The reporter gene codes for some easily detectable protein, such as green fluorescent protein (GFP), and is placed under the control of an appropriate promoter that will switch on its expression at a time and place desired by the experimenter. GFP and similar small fluorescent proteins, which are isolated from jellyfish and other marine organisms, have proved extremely useful in this regard as they are harmless to living cells and can readily be made visible by illuminating the cells or even whole embryos with light of a suitable wavelength. Fusion proteins in which the GFP coding sequence is attached to one end of a normal protein coding sequence are also used. When the gene is activated, the expression of the normal protein can be followed by means of its GFP tag.

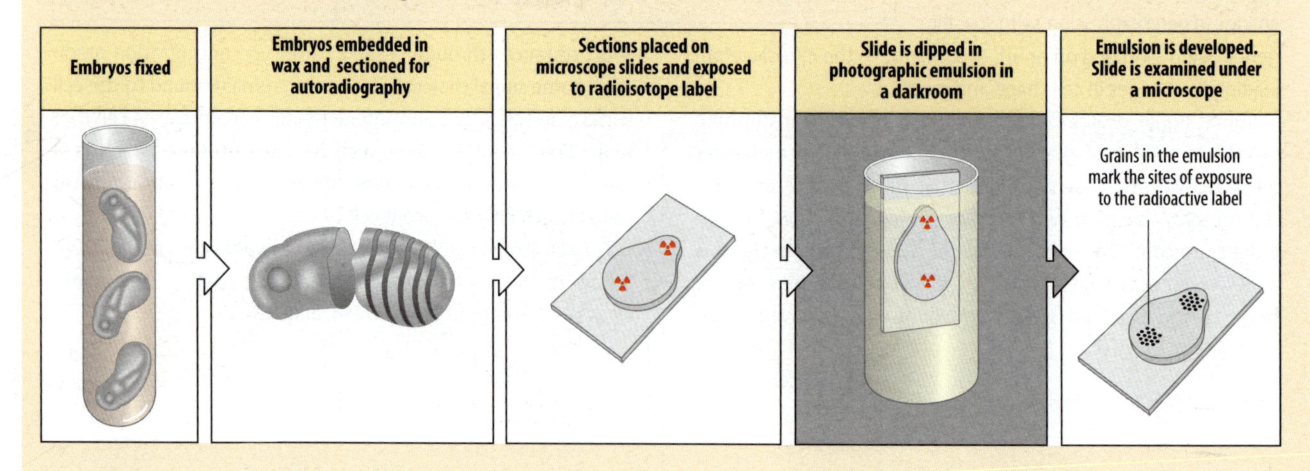

Continued

Box 3D (continued) *In situ* detection of gene expression

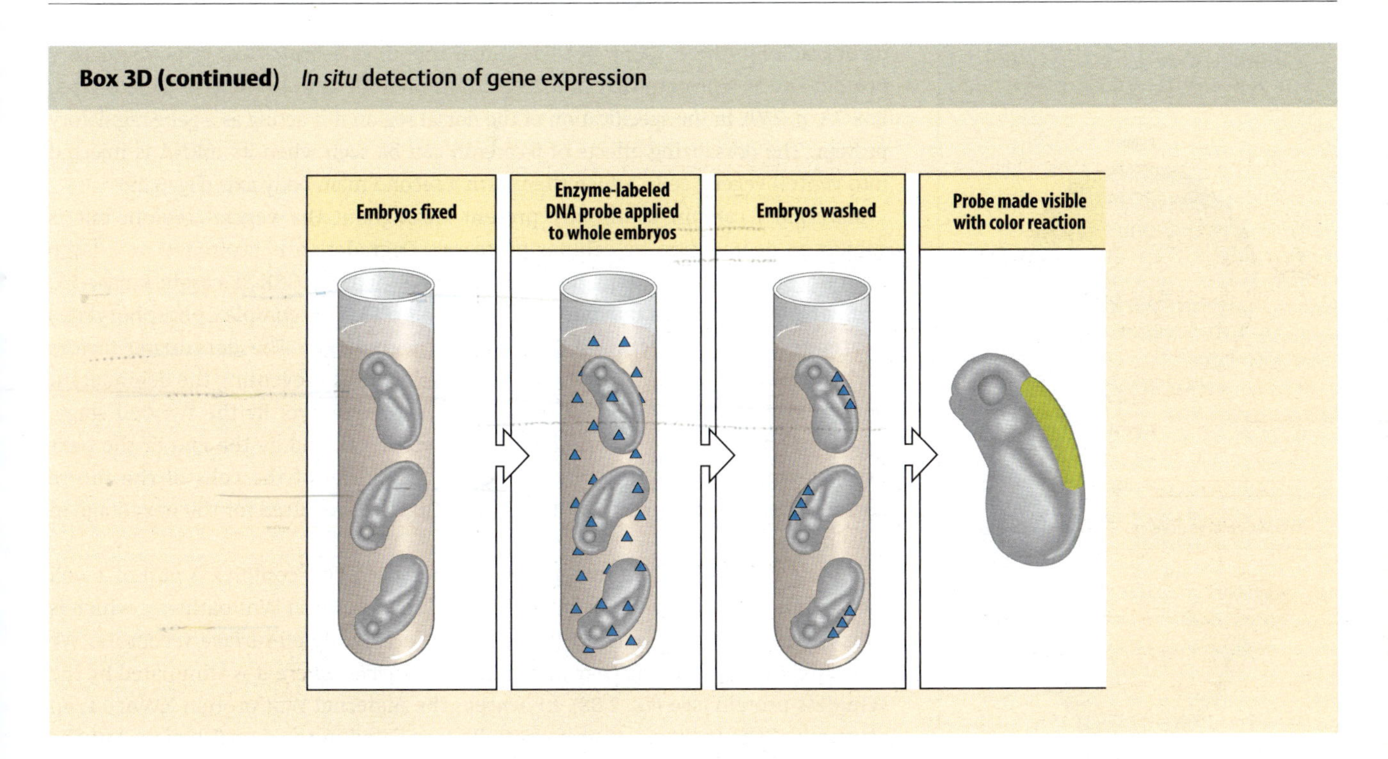

the Spemann organizer, to form in this future dorsal region. As we saw in Chapter 1, the embryonic organizer is a small region of the embryo that is active in the early stages of development and without which the embryo cannot form dorsal or axial structures. Any failure of cortical rotation and translocation of the dorsalizing factors blocks all further anterior and dorsal development, leading to a highly abnormal and non-viable embryo.

Cortical rotation can be prevented by irradiating the ventral side of the egg with ultraviolet (UV) light, which disrupts the microtubule array responsible for the movement. Embryos developing from such treated eggs are ventralized; that is, they are deficient in structures normally formed on the dorsal side and develop excessive amounts of blood-forming mesoderm, a tissue normally only present at the embryo's ventral midline. With increasing doses of radiation, both dorsal and anterior structures are lost, and the ventralized embryo appears as little more than a small distorted cylinder.

The maternal dorsalizing factors act by stimulating a pathway that leads to the local stabilization of the protein **β-catenin** in the future dorsal region by preventing

Fig. 3.28 The future dorsal side of the amphibian embryo develops opposite the site of sperm entry as a result of the redistribution of dorsalizing factors by cortical rotation. After fertilization (first panel), the cortical layer just under the cell membrane rotates, which results in vegetal cortex and associated dorsalizing factors, such as *Wnt-11* mRNA and Dishevelled (Dsh) protein (a component of the Wnt signaling pathway) being relocated at a site approximately opposite to the site of sperm entry (second panel). The Wnt pathway is activated where these factors are present (third panel), which results in an accumulation of β-catenin in nuclei on the future dorsal side of the blastula (not shown). V = ventral; D = dorsal.

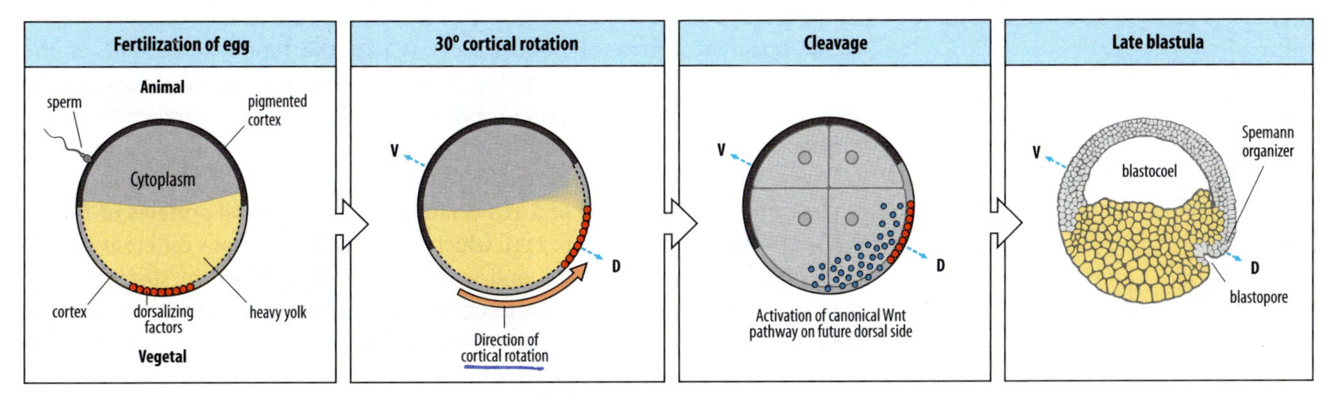

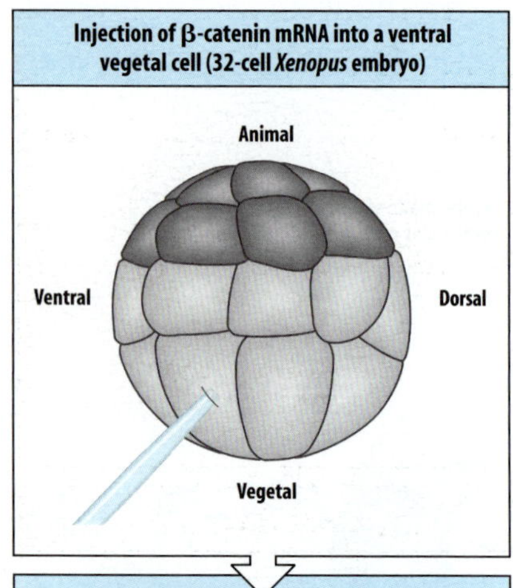

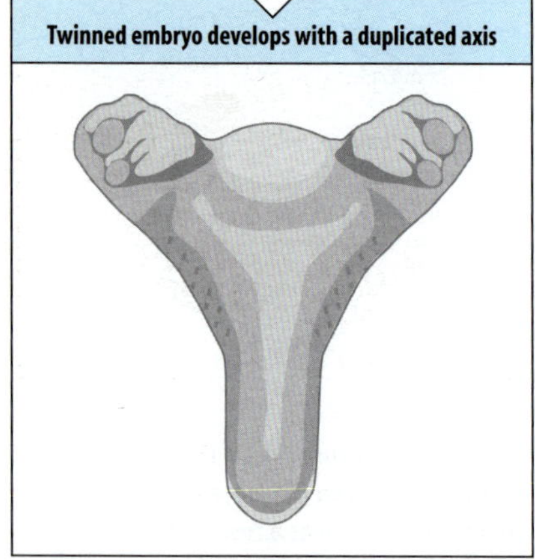

Fig. 3.29 Induction of a new dorsal side by injection of β-catenin mRNA. Injection of mRNA encoding β-catenin into ventral vegetal cells can specify a new Nieuwkoop center (discussed in Section 3.7) at the site of injection, leading to a twinned embryo. Injection of some other vegetally localized maternal mRNAs, such as *Vg-1*, have similar effects.

its degradation there. β-Catenin has a dual role in vertebrates as a gene-regulatory protein and as a protein that links cell-adhesion molecules to the cytoskeleton (see Box 7A, p. 259). In the specification of the dorsal region it is acting as a gene-regulatory protein. The dorsalizing effects of β-catenin can be seen when its mRNA is injected into ventral vegetal cells, which then form a second main body axis (Fig. 3.29).

Maternal β-catenin is initially present throughout the vegetal region, but is subject to degradation by cellular proteases. Degradation is promoted by a large protein complex containing, glycogen synthase kinase-3 (GSK-3), casein kinase-1α, axin, and adenomatous polyposis coli protein (APC). This complex phosphorylates β-catenin, thereby targeting it for protease degradation. The dorsalizing factors act by blocking the activity of the GSK-3 complex, thus preventing the degradation of β-catenin on the prospective dorsal side of the embryo. By the two-cell stage, β-catenin has started to accumulate in the cytoplasm, and by the end of the next two cleavage divisions it is present within the nuclei of the cells of the future dorsal region, where it helps switch on zygotic genes required for the next stage in development.

The stabilization of β-catenin by inhibition of the GSK-3 complex is part of a well known intracellular signaling pathway called the canonical Wnt pathway, which is usually stimulated by secreted Wnt proteins acting at cell-surface receptors. We have already seen this pathway in action in *Drosophila*, where it is stimulated by the Wingless protein (see Fig. 2.38). In *Xenopus* the maternal Wnt protein XWnt-11, an activator of the pathway, is preferentially translated in the dorsal region, but the pathway may also be activated directly by pathway components such as Dishevelled (Dsh), which is moved into the future dorsal region by cortical rotation (see Fig. 3.28), or by inhibitory proteins acting on the GSK-3 complex. The key role played by inhibition of GSK-3 activity in establishing the future dorsal region also explains the long-established demonstration that treatment of embryos with lithium chloride at this stage dorsalizes them, promoting the formation of dorsal and anterior structures at the expense of ventral and posterior structures. Lithium is known to block GSK-3 activity.

There are similarities between the mechanism of dorso-ventral axis specification in *Xenopus* and zebrafish, although in zebrafish the site of sperm entry is not involved in determining the future dorsal region. As in *Xenopus*, specification of a dorsal region in the zebrafish also appears to involve activation of β-catenin by a dorsalizing activity that moves from the vegetal region under some as yet unknown stimulus. Removal of this region results in a radially symmetric embryo that lacks dorsal and anterior structures. During cleavage, the activity in this vegetal region moves towards the animal pole and comes to lie beneath the cellular blastoderm. As in *Xenopus*, this movement appears to depend on microtubules. The maternal factors activate the β-catenin pathway in the dorsal yolk syncytial layer and dorsal margin blastomeres, and β-catenin accumulates in nuclei there.

One apparent difference from *Xenopus* is that the future dorsal side of the zebrafish embryo can be recognized as early as the four- and eight-cell stages by the localization of a maternal mRNA encoding a protein called Nodal-related 1 (Ndr1; previously known as Squint) within two cells on one side of the blastoderm. These two cells give rise to the dorsal region of the embryo (Fig. 3.30). The Ndr1 protein is a member of a subfamily of a TGF-β family called the **Nodal proteins** or **Nodal-related proteins** (see Box 3C, p. 111), which are important in many different aspects of development in vertebrates, and it is likely to be involved in specifying the dorso-ventral axis along with β-catenin. Removal of the cells containing the *Ndr1* mRNA or inhibition of *Ndr1* expression using antisense morpholino oligonucleotides (see Box 5A, p. 189) causes a loss of dorsal structures.

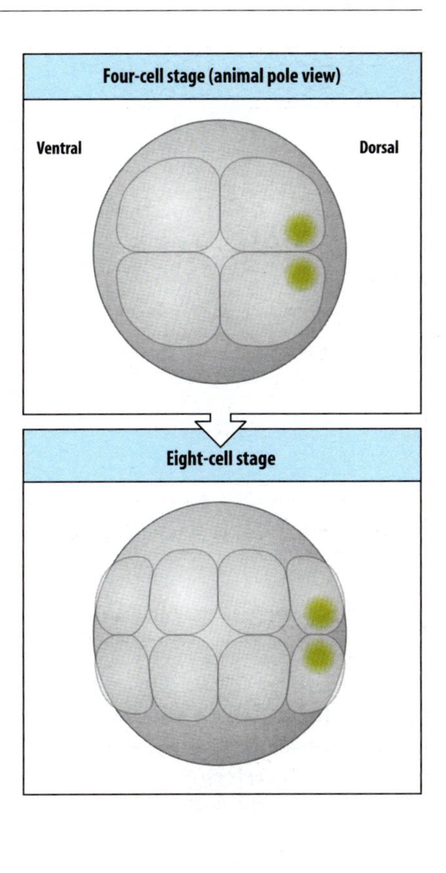

Fig. 3.30 The mRNA for a Nodal-related protein in zebrafish is localized to the future dorsal side of the embryo during early cleavage. *Ndr1* mRNA is initially distributed throughout the zygote. It becomes partly localized within two blastomeres at the four-cell stage. By the eight-cell stage, it is exclusively localized to the two blastomeres that will give rise to the dorsal side of the embryo. At the early gastrula stage, *Ndr1* mRNA is found in the future shield region.

3.7 Signaling centers develop on the dorsal side of *Xenopus* and zebrafish

The combined actions of β-catenin and other maternal factors present in the vegetal region on the future dorsal side lead to the formation of a **signaling center** in this region. A signaling center is a localized region of the embryo that exerts a special influence on surrounding cells and can thus determine how they will develop, in the same way that signaling emanating from the parasegment boundary in *Drosophila* directs the development of the segments (see Section 2.23). The signaling center that develops in the dorsal vegetal region in the *Xenopus* blastula is known as the blastula organizer or **Nieuwkoop center**, after the Dutch embryologist Pieter Nieuwkoop who discovered it. The Nieuwkoop center sets the initial dorso-ventral polarity in the blastula and its main role is to specify the establishment of the Spemann organizer, which arises just above the Nieuwkoop center at the late blastula–early gastrula stage. As we shall see in Chapter 4, signals originating in the Spemann organizer are crucial for further patterning along both the antero-posterior and dorso-ventral axes of the embryo and in inducing the formation of the central nervous system from ectoderm.

The importance of the Nieuwkoop center is shown by experiments that divide the embryo into two at the four-cell stage in such a way that one half contains the site at which the Nieuwkoop center will form while the other does not. The half containing the center will develop most structures of the embryo, although it will lack a gut, which develops from ventral blastomeres. The half without the Nieuwkoop center will develop much more abnormally, with little relevance to the cells' normal fates, producing a distorted, radially symmetric ventralized embryo that lacks all dorsal and anterior structures (Fig. 3.31); it has developed in the same way as an embryo in which cortical rotation has been completely blocked by UV irradiation. This experiment shows that dorsal and ventral regions are established by the four-cell stage.

The influence of the Nieuwkoop center can also be seen dramatically in experiments in which cells from the dorso–vegetal region of the 32-cell *Xenopus* embryo are grafted into the ventral side of another embryo. This gives rise to a twinned

Fig. 3.31 The Nieuwkoop center is essential for normal development. If a *Xenopus* embryo is divided into a dorsal and a ventral half at the four-cell stage, the dorsal half containing the Nieuwkoop center develops as a dorsalized embryo lacking a gut, while the ventral half, which has no Nieuwkoop center, is ventralized and lacks both dorsal and anterior structures.

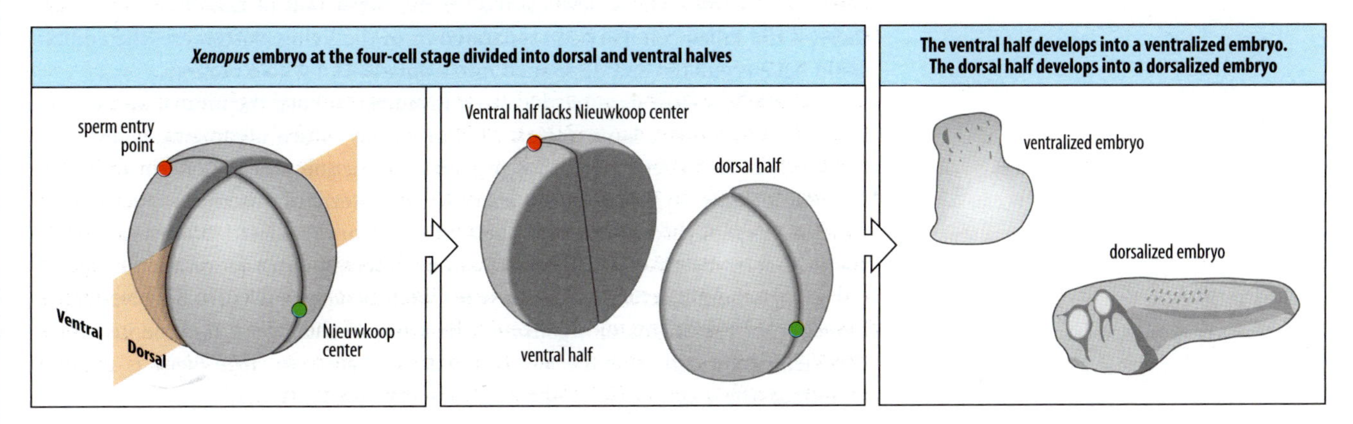

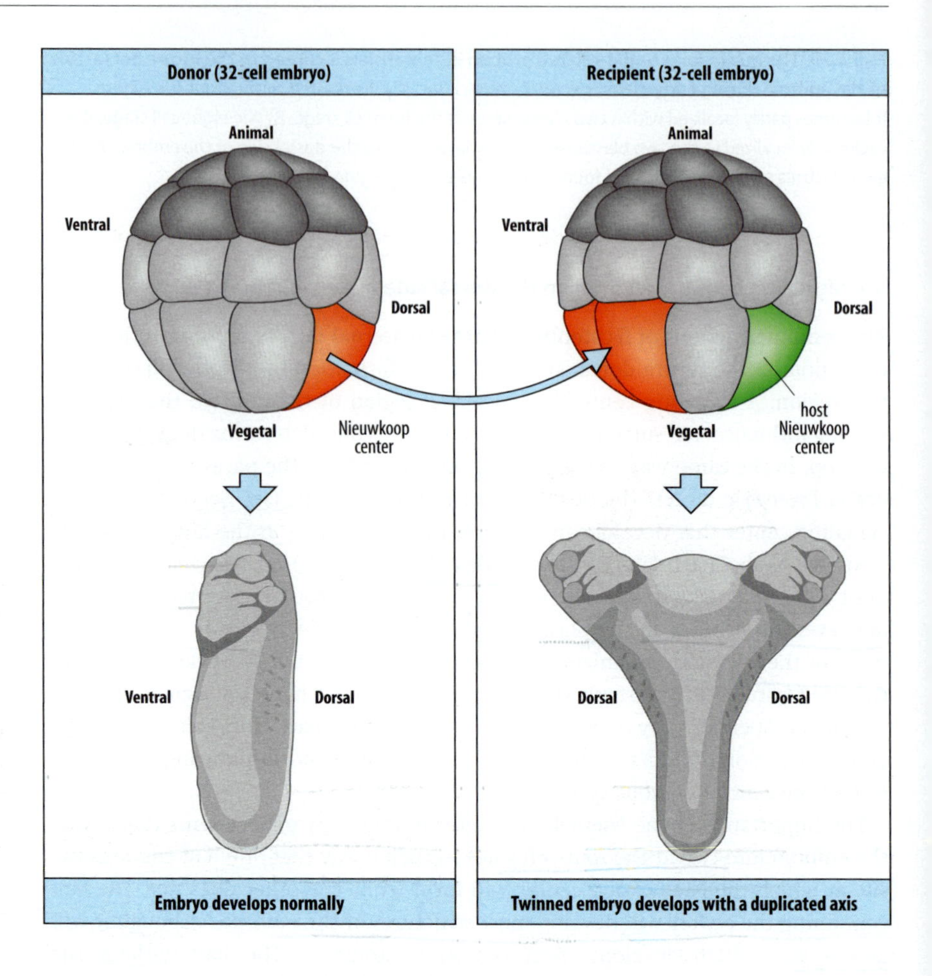

Fig. 3.32 The Nieuwkoop center can specify a new dorsal side. Grafting vegetal cells containing the Nieuwkoop center from the dorsal side to the ventral side of a 32-cell *Xenopus* blastula results in the formation of a second axis and the development of a twinned embryo. The grafted cells signal but do not contribute to the second axis.

embryo with two dorsal sides (Fig. 3.32). Cells from the graft itself contribute to endodermal tissues but not to the mesodermal and neural tissues of the new axis. This shows that the grafted cells are able to induce these fates in the cells of the host embryo. In contrast, grafting ventral cells to the dorsal side has no effect.

We can now understand the result of Roux's classic experiment (see Fig. 1.8) in which he destroyed one cell of a frog embryo at the two-cell stage and got a half-embryo rather than a half-sized whole embryo. The crucial feature of his experiment, it turns out, was that the killed cell remained attached, but the embryo did not 'know' it was dead. The remaining living cell could still develop a functional Nieuwkoop center but it developed as if the other half of the embryo was still there. If the killed cell had been separated from the living blastomere, the embryo could have regulated and developed into a small-sized whole embryo.

Although β-catenin becomes localized in nuclei in both the animal and vegetal regions, specifying a large region as dorsal, the future organizing centers are localized to the vegetal region, as organizing function and mesoderm induction depend not only on β-catenin but also on the maternal transcription factor VegT, which is present throughout the vegetal region from fertilization onwards. By the 16- to 32-cell stage, β-catenin protein can be detected in nuclei on the dorsal side (Fig. 3.33), where it activates zygotic genes such as *siamois*, which is involved in the induction of the Spemann organizer. This arises on the dorsal midline just above the Nieuwkoop center at the late blastula–early gastrula stage. VegT is required to induce the expression of the Nodal-related proteins that induce mesoderm

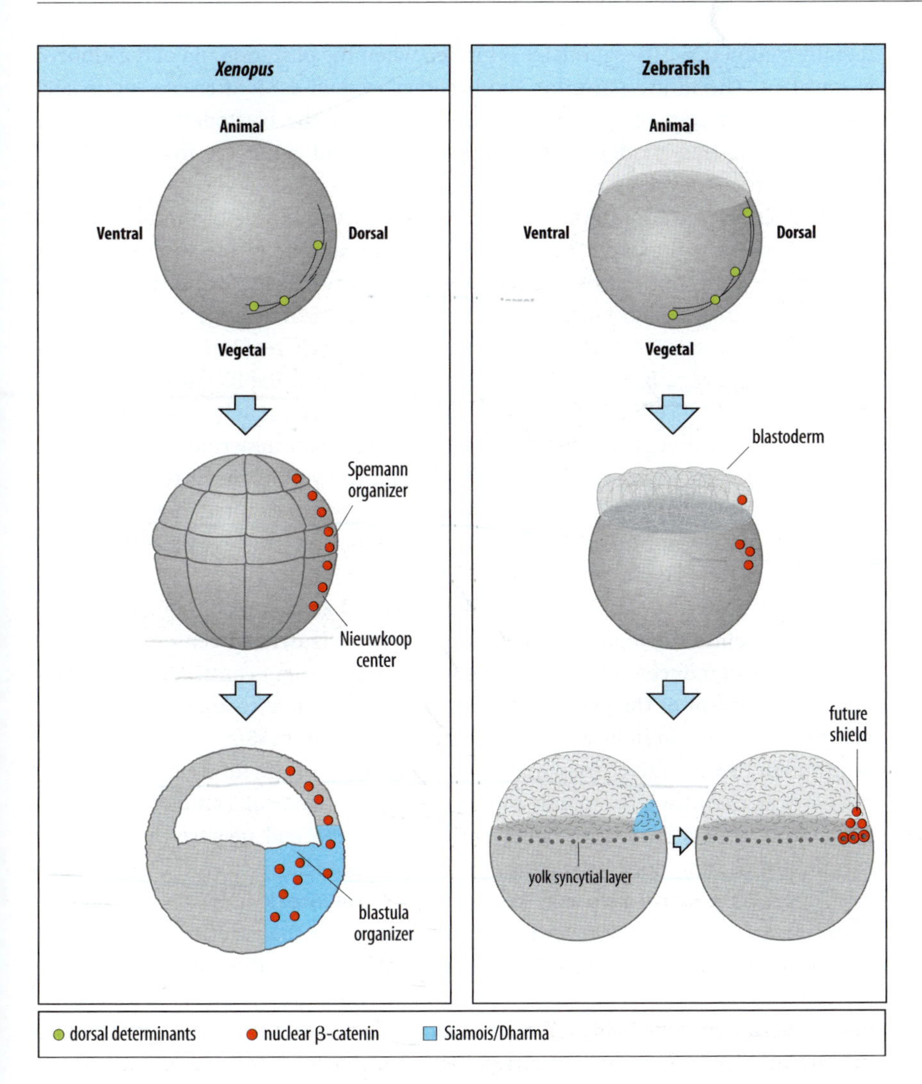

Fig. 3.33 The establishment of the dorsal organizer in *Xenopus* and zebrafish embryos. In *Xenopus*, cortical rotation moves dorsal determinants towards the future dorsal side of the embryo, creating a large region where, starting at the 32-cell stage, β-catenin moves from the cytoplasm into nuclei. The region in which β-catenin is nuclear expresses the transcription factor Siamois and also defines the Nieuwkoop center in the dorsal vegetal region. In zebrafish, dorsal determinants are transported towards the future dorsal side of the embryo and enter the blastoderm. At the mid-blastula transition, β-catenin is translocated into dorsal yolk syncytial layer nuclei where it induces the expression of the transcription factor Dharma in the future shield region.

(see Box 3C, p. 111). Named after the mouse protein Nodal, members of this family of proteins are key factors in mesoderm induction in all vertebrates, and in the frog they are known as the *Xenopus* Nodal-related factors (Xnrs). We will discuss them further when we look at the induction and patterning of the mesoderm in the last part of this chapter.

As noted in Section 3.6, in the zebrafish embryo β-catenin becomes localized in nuclei of the dorsal yolk syncytial layer and the marginal blastomeres overlying it, which will become the shield region. This region contains an organizing center analogous to the Spemann organizer. β-Catenin induces the expression of a transcription factor, called Dharma (or Boz), which is essential for the induction of the organizing function of the shield and the development of head and trunk regions (see Fig. 3.33).

3.8 The antero-posterior and dorso-ventral axes of the chick blastoderm are related to the primitive streak

The chick embryo starts off as a circular blastoderm sitting on the top of the yolk (see Section 3.3). Like the amphibian blastula, the chick blastoderm is initially

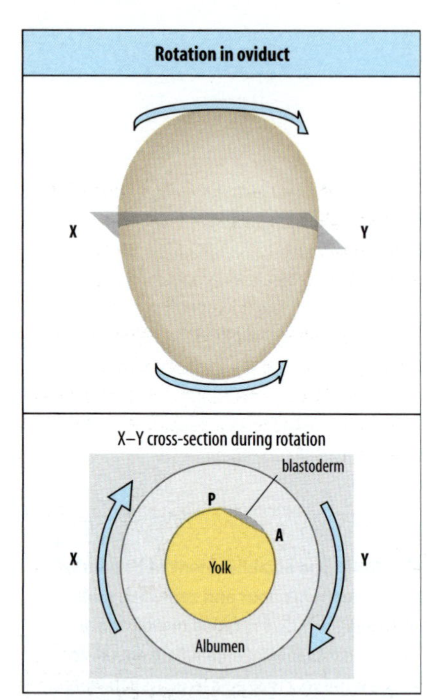

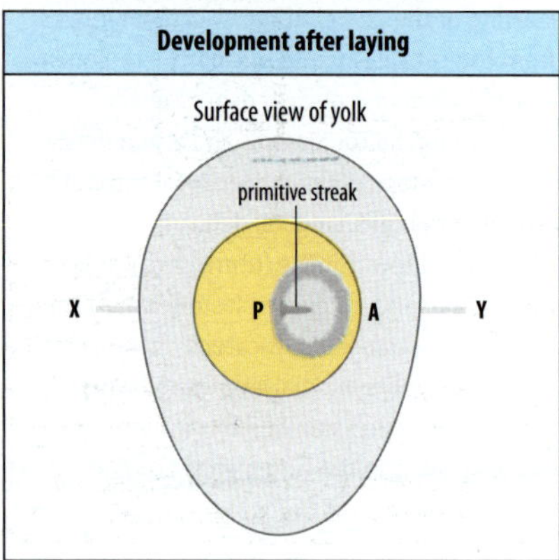

Fig. 3.34 Gravity defines the antero-posterior axis of the chick. Rotation of the egg in the oviduct of the mother results in the blastoderm being tilted in the direction of rotation, though it tends to remain uppermost. The posterior marginal zone (P) develops at that side of the blastoderm which is uppermost and initiates the primitive streak. A = anterior.

radially symmetric. This symmetry is broken when the posterior end of the embryo is specified. The future posterior end becomes evident soon after the egg is laid, when a denser area of cells appears at one side of the blastoderm. As we saw in Section 3.3, this is the posterior marginal zone and is the site from which the primitive streak develops at gastrulation.

The position of the posterior marginal zone is crucial in determining both the ventral end of the dorso-ventral axis and the posterior end of the antero-posterior axis—and it turns out that it is determined by gravity. During its passage through the hen's uterus, which takes about 20 hours, the fertilized egg moves pointed end first and rotates slowly around its long axis, each revolution taking about 6 minutes. Cleavage has already started at this stage, so the blastoderm contains many thousands of cells when the egg is laid. The egg is obliquely tilted in the gravitational field, although the blastoderm tends to remain uppermost. As the shell and albumen (egg white) rotate, the embryo and the yolk under it tend to return to the vertical. Thus, the future blastoderm region becomes tilted in the direction of rotation of the egg. The future posterior marginal zone will develop at the uppermost edge of the blastoderm (Fig. 3.34).

The most posterior part of the posterior marginal zone can be thought of as an organizing center analogous in some ways to the Nieuwkoop center in *Xenopus*, since it too can induce formation of a new body axis without contributing any cells to it. If a fragment of the posterior marginal zone is grafted to another part of the marginal zone, it can induce a new primitive streak (Fig. 3.35).

Although the primitive streak indicates the position of the future head–tail axis, the early streak also corresponds to the future dorso-ventral axis of the embryo, with more ventral structures being derived from the end nearest the posterior marginal zone, and dorsal structures such as the notochord from the node. Once node regression starts to occur (see Fig. 3.16), the streak indicates the antero-posterior axis of the embryo, with the head forming at the anterior end. We shall return to this when we discuss dorso-ventral patterning of the mesoderm later in the chapter.

Various signaling proteins are expressed in the posterior marginal zone before the initiation of the primitive streak. These include Wnt-8c, which is present throughout the marginal zone in a gradient with its high point at the posterior, and Vg-1 (which we have already encountered in *Xenopus*), which is localized to the most posterior region of the marginal zone. Their activity is required to set up the organizer in the posterior marginal zone. Vg-1 and Wnt-8c induce the expression of Nodal. This is expressed in the posterior marginal zone and the streak. Together with fibroblast growth factor (FGF), which is expressed in Koller's sickle, and the protein Chordin, Nodal is required for the formation of the streak and a fully functional Hensen's node (Fig. 3.36). We shall return to these signals when we consider their role in mesoderm induction.

Primitive-streak formation only starts when the hypoblast is displaced from the posterior marginal region by the endoblast (see Section 3.3). The hypoblast actively inhibits streak formation, as shown by experiments in which its complete removal leads to the formation of multiple streaks in random positions. This inhibition is due to a protein, Cerberus, which is produced by the hypoblast and antagonizes the function of Nodal (see Fig. 3.36).

When cells expressing Vg-1 are grafted to another part of the marginal zone in another embryo, they can induce a complete new primitive streak. In general, however, only one axis develops in the grafted embryos—either the host's normal axis or the one induced by the graft. This suggests that the more advanced of the two organizing centers inhibits streak formation elsewhere. It has now been

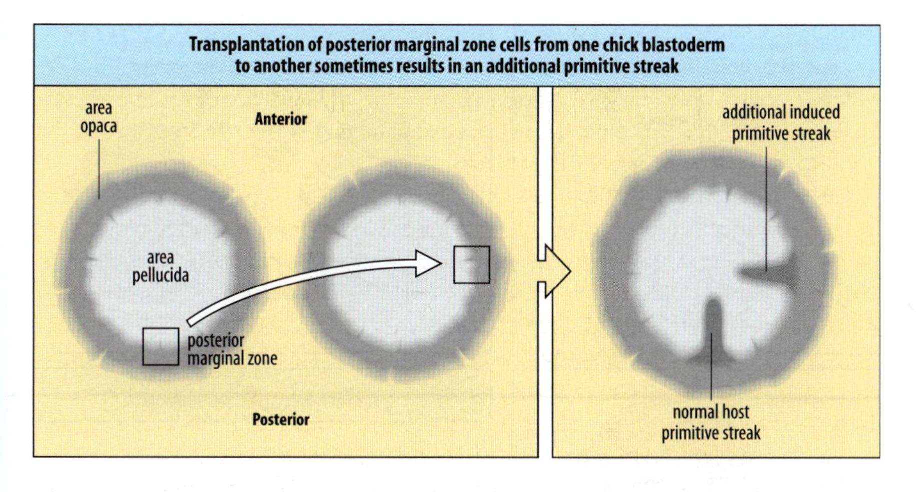

Fig. 3.35 The posterior marginal zone of the chick specifies the posterior end of the antero-posterior axis. Grafting posterior marginal zone cells to another site in the marginal zone can result in the formation of an extra primitive streak, which defines a new antero-posterior axis. This does not always occur. Usually the more advanced of the streaks is the only one to develop because it inhibits the development of the other.

shown that Vg-1 induces an inhibition that can travel across the embryo to prevent another streak forming—this inhibitory signal travels a distance of around 3 mm in about 6 hours. Vg-1 thus appears to be required both for initiating streak formation and for inhibiting the formation of a second streak.

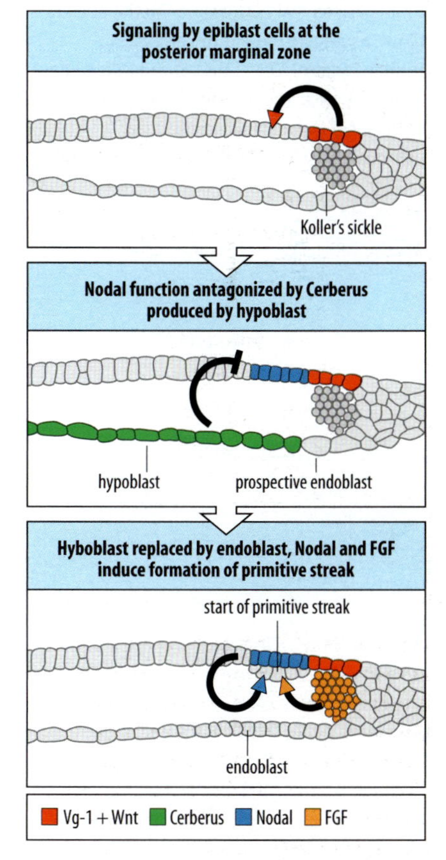

Fig. 3.36 Signals at the posterior marginal zone of the chick epiblast initiate the primitive streak. Epiblast cells in the posterior marginal zone overlying Koller's sickle secrete Vg-1 and Wnts. These signals induce expression of *nodal* in neighboring epiblast cells, but Nodal protein function is blocked by Cerberus, which is produced by the hypoblast. When the hypoblast has been displaced by endoblast, Nodal signaling in the epiblast and FGF signals from Koller's sickle induce the internalization of epiblast cells and formation of the primitive streak.

3.9 The axes of the mouse embryo are not recognizable early in development

The very earliest stages of development of a mammalian egg differ considerably from those of either *Xenopus* or the chick, as the mammalian egg contains no yolk. The extra-embryonic yolk sac membranes and the placenta connecting the embryo to the mother must therefore develop to nourish the embryo. There is no clear sign of polarity in the mouse egg, and any contribution of maternal factors is difficult to determine as zygotic gene activation occurs in the one-cell embryo and is essential for development beyond the two-cell stage. Several claims have been made that the first cleavage of the zygote is related to an axis of the embryo. This is an interesting possibility but not yet clear, as different workers get different results.

The early cleavages do not follow a well ordered pattern. Some cleavages are parallel to the egg's surface, so that a solid ball of cells (the morula) is formed with distinct outer and inner cell populations (see Fig. 3.22). By the 32-cell stage, the mouse embryo has developed into a blastocyst, a hollow sphere of epithelium containing the inner cell mass—some 10–15 cells—attached to the epithelium at one end. The epithelium will form the trophectoderm, which gives rise to extra-embryonic structures associated with implantation and formation of the placenta, a structure unique to mammalian development (see Section 3.4).

The mouse embryo is highly regulative, and the specification of cells as either inner cell mass or trophectoderm depends on their relative positions in the cleaving embryo. Determination of their fate occurs only after the 32-cell stage, and during earlier stages all the cells seem to be equivalent in their ability to give rise to either tissue. The most direct evidence for the effect of position comes from taking individual cells (blastomeres) from disaggregated eight-cell embryos, labeling them, and then combining the labeled blastomeres in different positions with respect to unlabeled blastomeres from another embryo. If the labeled cells are placed on the outside of a group of unlabeled cells they usually give rise to trophectoderm; if they are placed inside, so that they are surrounded by unlabeled cells, they more often give rise to inner cell mass (Fig. 3.37). If a whole embryo is surrounded by blastomeres from another embryo, it can all become part of the

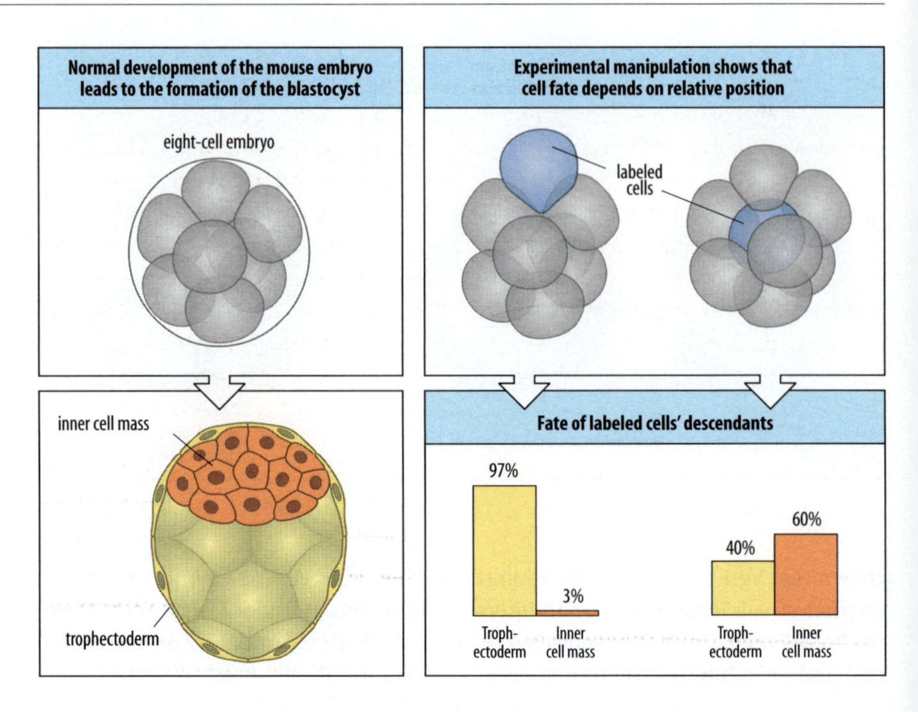

Fig. 3.37 The specification of the inner cell mass of a mouse embryo depends on the relative position of the cells with respect to the inside and outside of the embryo. If labeled blastomeres from a four-cell mouse embryo are separated and combined with unlabeled blastomeres from another embryo, it can be seen that blastomeres on the outside of the aggregate more often give rise to trophectoderm, 97% of the labeled cells ending up in that layer. The reverse is true for the origin of the inner cell mass, which derives predominantly from blastomeres on the inside.

inner cell mass of a giant embryo. Aggregates composed entirely of either 'outside' or 'inside' cells of early embryos can also develop into normal blastocysts, showing that there is no specification of these cells, other than by their position, at this stage.

The segregation of trophectoderm and the pluripotent inner cell mass lineages is the first differentiation event in mammalian embryonic development, and is essential for the formation of the placenta. The transcription factors Cdx2 and Oct3/4 are involved in the differentiation of trophectoderm on the one hand, and keeping the inner cell mass pluripotent on the other. Oct3/4 is associated with pluripotency and is expressed in the morula from the two-cell stage. At the eight-cell stage both transcription factors are expressed in all blastomeres. By the 10–16-cell stage, Cdx2 expression has been lost from a few inner cells, whereas Oct3/4 is still expressed throughout, and by the blastocyst stage Cdx2 expression is confined to the trophectoderm and Oct3/4 to the cells of the inner cell mass. The downregulation of Oct3/4 in prospective trophectoderm cells, possibly by interaction with Cdx2, is essential for their differentiation. The mechanism that restricts the expression of Cdx2 specifically to the outer cells of the morula is unknown. At an earlier stage in development, there is still controversy over whether individual blastomeres from two-cell-stage mouse embryos normally have identical developmental properties and fate, or whether maternal determinants are distributed differently between them.

Before the blastocyst stage the embryo is a symmetrical spheroidal ball of cells. The first asymmetry appears in the blastocyst, where the blastocoel cavity develops asymmetrically, leaving the inner cell mass attached to one part of the trophectoderm (Fig. 3.38). The blastocyst now has one distinct axis running from the site where the inner cell mass is attached—the **embryonic pole**—to the opposite, **abembryonic**, pole, with the blastocoel occupying most of the abembryonic half. This axis is known as the **embryonic–abembryonic axis**. It corresponds to the dorso-ventral axis of the epiblast, but only in a geometrical sense, and is not related to specification of cell fate.

By about 4.5 days after fertilization, the inner cell mass has become differentiated into two tissues: primary or primitive endoderm on its blastocoelic surface (which will form extra-embryonic structures); and the epiblast within, from which the embryo and some extra-embryonic structures will develop. The blastocyst implants in the uterine wall, and the trophectoderm at the embryonic pole proliferates to form the ectoplacental cone, producing extra-embryonic ectoderm that pushes the inner cell mass across the blastocoel. A cavity—the proamniotic cavity—is then formed within the epiblast as its cells proliferate. The mouse epiblast is now a monolayer and is formed into a cup shape, U-shaped in section, with a layer of visceral endoderm on the outside. At this stage, about 5.5 days after fertilization, the egg cylinder has a proximal–distal polarity in relation to the site of implantation—the ectoplacental cone—from which the epiblast is separated by the extra-embryonic ectoderm (see Fig. 3.38).

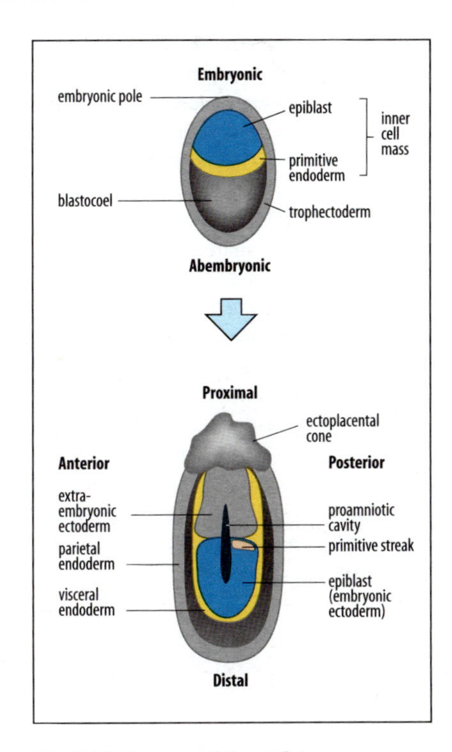

Fig. 3.38 The axes of the early mouse embryo. At the blastocyst stage (top), at about 4 days, the inner cell mass is confined to the embryonic region and this defines an embryonic–abembryonic axis (which relates geometrically, although not in terms of cell fate, to the dorso-ventral axis of the future epiblast). The inner cell mass is oval, and thus also has an axis of bilateral symmetry. At about 5.5 days after fertilization (bottom) the antero-posterior axis in the epiblast becomes visible, with the formation of the primitive streak at the posterior end. The interior of the epiblast cup corresponds to the dorsal side of the future embryo, and the outer side to the future ventral side (see Fig. 3.25).

The true antero-posterior axis of the mouse embryo is thought to be formed at this stage, about 24 hours before gastrulation begins, by a symmetry-breaking event that determines the anterior end of the embryo. A small region of visceral endoderm at the distal-most end of the cup is induced as **anterior visceral endoderm** (**AVE**) by Nodal signaling from the epiblast and it starts to express the homeodomain transcription factors Hex, Lim-1 (Lhx1), and Cerberus-like, which are all markers of anterior tissues. The AVE proliferates and moves forward, forming a strip of cells that displaces the more proximal visceral endoderm on one side of the cup and specifies this side as anterior (Fig. 3.39). The underlying mechanism that controls this asymmetric movement remains unknown. The AVE then sends a signal to the overlying epiblast that specifies this region as anterior ectoderm, which will eventually form the brain and other anterior structures. Thus in mammals, specification of anterior fate precedes the formation of the main embryonic organizing regions—the primitive streak and the node. Indeed, the specification of the anterior region leads directly to the initiation of the primitive streak and its positioning on the opposite side of the cup, which becomes the posterior end of the embryo. The independence of this anterior determination from the development of the rest of the body can be seen when a node is transplanted to another part of the epiblast. The node is able to induce a secondary axis, as in the chick, but this invariably lacks the most anterior regions, including the forebrain.

The extra-embryonic ectoderm plays a crucial role in initially restricting the AVE fate to the most distal region of the visceral endoderm. Nodal is present throughout the epiblast and can in principle induce anterior fate throughout the visceral endoderm. If the extra-embryonic ectoderm is removed, all the visceral endoderm starts to express anterior markers. Anterior fate is restricted by an inhibitory signal, as yet unidentified, from the extra-embryonic ectoderm that prevents the genes specifying anterior fate being expressed in proximal regions. A signal from the extra-embryonic ectoderm also appears to be involved in inducing the movement of the anterior visceral endoderm.

3.10 The bilateral symmetry of the early embryo is broken to produce left–right asymmetry of internal organs

Vertebrates are bilaterally symmetric about the midline of the body for many structures, such as eyes, ears, and limbs. But, while the vertebrate body is outwardly symmetric, most internal organs are in fact asymmetric with respect to the left and right sides. This is known as **left–right asymmetry**. In mice and humans, for example, the heart is on the left side, the right lung has more lobes than the left,

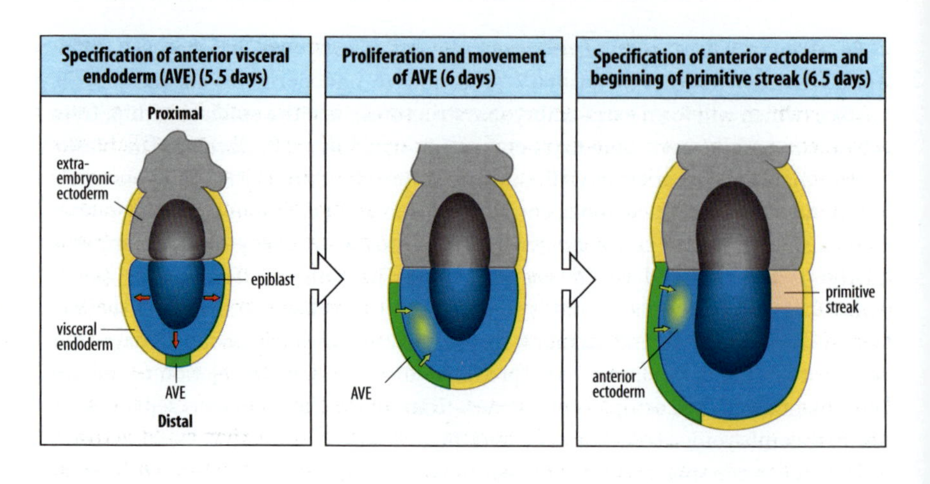

Fig. 3.39 The symmetry-breaking event in the early mouse embryo is the specification of the anterior visceral endoderm. At around 5.5 days after fertilization, before primitive streak formation, visceral endoderm (yellow) at the distal-most end of the cup is specified as anterior visceral endoderm (AVE; green) by signals from the epiblast (red arrows) and begins to proliferate and extend to one side of the epiblast. The anterior fate is restricted to the distal region by inhibitory signals from the extra-embryonic ectoderm to the proximal visceral endoderm (barred black lines). The extended AVE signals to the adjacent epiblast (green arrows) and induces anterior ectoderm (which includes neural ectoderm; pale blue). At 6.5 days the primitive streak (brown) begins to form on the opposite side of the epiblast, thus marking the posterior end of the axis.

the stomach and spleen lie to the left, and the liver has a single left lobe. This handedness of organs is remarkably consistent, but there are rare individuals, about one in 10,000 in humans, who have the condition known as **situs inversus,** a complete mirror-image reversal of handedness. Such people are generally asymptomatic even though all their organs are reversed. A similar condition is produced in mice that carry the *iv* mutation, in which organ asymmetry is randomized (Fig. 3.40). In this context, randomized means that some individuals will have the normal organ asymmetry and some the reverse.

Specification of left and right is fundamentally different from specifying the other axes of the embryo, as left and right have meaning only after the antero-posterior and dorso-ventral axes have been established. If one of these axes were reversed, then so too would be the left–right axis (it is for this reason that handedness is reversed when you look in a mirror: your dorso-ventral axis is reversed, and hence left becomes right and *vice versa*). One suggestion to explain the development of left–right asymmetry is that an initial asymmetry at the molecular level is converted to an asymmetry at the cellular and multicellular level. If that were so, the asymmetric molecules or molecular structure would need to be oriented with respect to both the antero-posterior and dorso-ventral axes. While the mechanism by which left–right symmetry is initially broken is not known, the subsequent cascade of events that leads to organ asymmetry is now rather well understood. Key proteins in establishing leftness are the extracellular signal protein Nodal and the transcription factor Pitx2, which are expressed on the left side of the

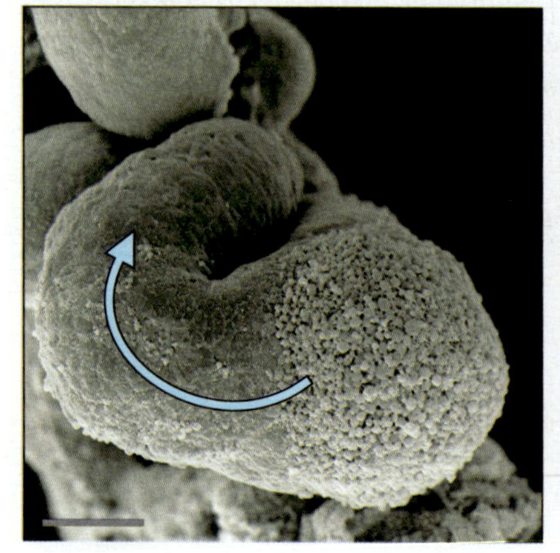
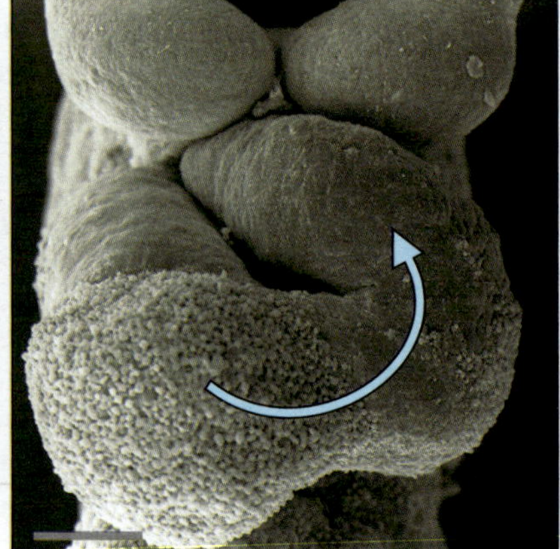

Fig. 3.40 Left–right asymmetry of the mouse heart is under genetic control. Each photograph shows a mouse heart viewed anteriorly after the loops have formed. The normal asymmetry of the heart results in it looping to the left, as indicated by the arrow (left panel). 50% of mice that are homozygous for the mutation in the *iv* gene have hearts that loop to the right (right panel). Scale bar = 0.1 mm.

Photographs courtesy of N. Brown.

node only; if the expression of either of these is made symmetric then organ asymmetry is randomized.

Left-right symmetry is broken at an early stage in chick and *Xenopus* by the asymmetric activity of a proton-potassium pump (H^+/K^+-ATPase). This ATPase can be detected as early as the third cleavage division—the eight-cell stage—in *Xenopus*, and treatment of early embryos with drugs that inhibit the pump cause randomization of asymmetry. In the chick, a mechanism for left-right asymmetry based on the asymmetric activity of this pump in Hensen's node has been proposed (Fig. 3.41). It has been found that the activity of the pump is reduced on the left side of the node, leading to membrane depolarization and release of calcium from cells to the extracellular space on the left-hand side (how the asymmetry of the H^+/K^+-ATPase itself originates is not known). This in turn leads to the expression of the Notch ligands Delta-like and Serrate in a spatial pattern that could activate Notch on the left side of the node only. Notch activity, in conjunction with Sonic hedgehog, an extracellular signal protein homologous with *Drosophila* Hedgehog (see Section 2.23), leads to the left-sided expression of Nodal in the lateral plate mesoderm that gives rise to internal organs such as the heart. The expression of the Nodal antagonist Lefty in the left half of the notochord and neural tube floorplate forms a barrier that prevents these Nodal signals crossing the midline into the right half of the embryo. Nodal in turn activates the expression of the transcription factor Pitx2—a key determinant of leftness. Sonic hedgehog is also expressed on the left side only of Hensen's node, possibly by a separate mechanism. The TGF-β family member activin and its receptor are expressed on the right side, and may repress Sonic hedgehog expression on this side. If the pattern of *nodal* expression is made symmetrical by placing a pellet of cells secreting Sonic hedgehog on the right-hand side, then organ asymmetry is randomized.

Expression of Nodal and Pitx2 on the left side of the node is a highly conserved feature, being found in the mouse as well as the chick, and ectopic expression of either Nodal or Pitx2 on the right side leads to randomization of asymmetry in both animals. Studies in the mouse also implicate the release of calcium from cells in the left side of the node as part of the symmetry-breaking process. In this case

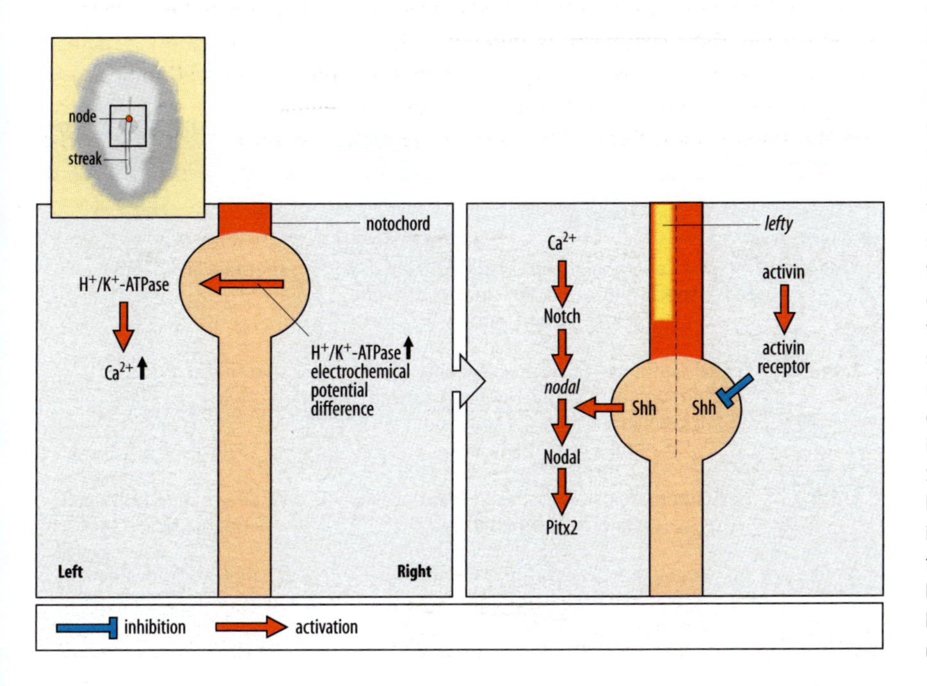

Fig. 3.41 Determination of left-right asymmetry in the chick. The activity of a proton-potassium pump (H^+/K^+-ATPase) in Hensen's node is reduced on the left side of the node, leading to a membrane-potential difference across the node and an increased release of calcium (Ca^{2+}) into the extracellular space on the left side of the node (left panel). This leads to a greater activation of Notch signaling on the left side of the node, which in turn activates expression of the gene *nodal* in cells on the left side of the node. Nodal signaling together with Sonic hedgehog (Shh) then switches on *nodal* expression in the lateral plate mesoderm on the left side, which leads to expression of the transcription factor Pitx2, an important determinant of leftness (right panel). Shh activity is inhibited by activin on the right-hand side. The Nodal antagonist Lefty expressed in the left half of the notochord and the floorplate of the neural tube provides a midline barrier that prevents Nodal crossing to the right-hand side. This is a simplified version of a much more complex set of interactions.

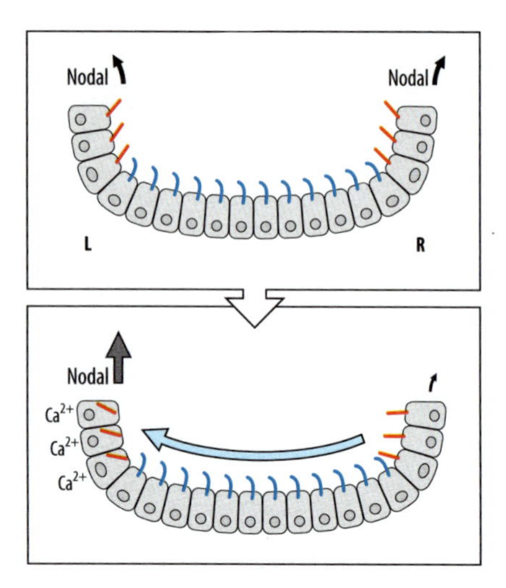

Fig. 3.42 Cilia-directed flow generates left-right asymmetry across the node in the mouse embryo. In the mouse, the initial event that breaks left-right symmetry may be a short period of directed leftward flow of extracellular fluid at the node at around the late head-fold stage. Before this flow, there is a low level of expression of *nodal* mRNA on both sides of the node (top panel). There are two types of cilia in the node: central cilia (blue) that generate the flow and peripheral monocilia (red) that may act as mechanosensors for the direction of flow. Coordinated beating of cilia generates a leftward flow of fluid that results in release of intracellular Ca^{2+} in the sensory cells on the left side of the node. Propagation of the Ca^{2+} signal to nearby cells leads to the upregulation of Nodal expression on the left side.

the asymmetry appears to be due to motile cilia on cells in the mouse node, which produce a leftward flow of the extra-embryonic fluid (Fig. 3.42). This flow activates sensory cilia on the left side of the node, which leads to the asymmetric release of calcium. Reversal of the flow by an imposed fluid movement reverses left-right asymmetry. Mutations that affect the motor protein dynein and block the movement of cilia, as in Kartagener's syndrome in humans and the *iv* mouse mutant, result in randomized asymmetry (see Fig. 3.40). In the mouse *nodal* is initially expressed symmetrically at both edges of the node, and its expression becomes asymmetrical over time, via a positive feedback loop.

Summary

Setting-up the body axes in vertebrates involves maternal factors, external influences, and cell-cell interactions. In the amphibian embryo, maternal factors determine the animal-vegetal axis, which corresponds approximately to the antero-posterior axis, whereas the dorso-ventral axis is specified by the site of sperm entry and the resulting cortical rotation, which leads to the establishment of the Nieuwkoop center. In the zebrafish, maternal factors also specify the dorsal side and the embryonic shield, the site of the embryonic organizer, although the symmetry-breaking event has not yet been identified in this case. In chick embryos, the dorso-ventral axis is specified at cleavage in relation to the yolk while the setting of the antero-posterior axis involves gravity, which determines the side of the blastoderm at which the posterior marginal zone, and thus the primitive streak, will form. It is not clear yet whether specification of the axes in the mouse embryo involves any maternal component. They are definitively established later in the epiblast under the direction of signals from the extra-embryonic tissues. The symmetry-breaking event in the mouse appears to be the specification of the anterior end of the embryo, which in turn determines the position of the primitive streak at the posterior end of the epiblast. The generation of the consistent left-right organ asymmetry found in vertebrates is marked by the expression of the secreted protein Nodal and the transcription factor Pitx2 on the left side only. This asymmetry appears to be generated by a variety of mechanisms, involving a proton-potassium pump asymmetry (in *Xenopus* and chick), the movement of cilia (in mouse and humans), and asymmetrically located calcium signaling, Notch pathway activation, and Nodal signaling.

Summary: vertebrate axis determination

	Dorso-ventral axis	Antero-posterior axis
Xenopus	sperm entry point and cortical rotation lead to specification of dorsal side opposite the point of sperm entry	specified in organizer
Zebrafish	initial symmetry-breaking event unknown. Localization of maternal Ndr1 and nuclear β-catenin specifies dorsal side and position of organizer	specified in organizer
Chick	posterior marginal zone specifies ventral end of dorso-ventral axis	gravity determines position of posterior marginal zone
Mouse	interaction between inner cell mass and trophectoderm	specification and movement of of anterior visceral endoderm

The origin and specification of the germ layers

We have seen in the preceding sections how the main axes are laid down in various vertebrate embryos. We now focus on the earliest patterning of the embryo with respect to these axes: the specification of the three germ layers—endoderm, mesoderm, and ectoderm—and their further diversification.

All the tissues of the body are derived from these three germ layers. The mesoderm becomes subdivided into cells that give rise to notochord, to muscle, to heart and kidney, and to blood-forming tissues, among others. The ectoderm becomes subdivided into cells that give rise to the epidermis and that develop into the nervous system. The endoderm gives rise to the gut and organs such as the lungs, liver, and pancreas. We first look at the **fate maps** (see Section 1.12) of early embryos of different vertebrates, which tell us which tissues the different regions of the embryo give rise to. We then consider how the germ layers are specified and subdivided, with the main focus on *Xenopus*, in which these processes are best understood and where many of the genes and proteins involved have been identified. The fate maps and the genes involved in germ-layer specification are more similar in all our vertebrate models than are the mechanisms involved in early axis specification.

3.11 A fate map of the amphibian blastula is constructed by following the fate of labeled cells

The external appearance of the *Xenopus* blastula at the 32-cell stage gives no indication of how the different regions will develop, but individual cells can be identified with respect to the animal–vegetal axis (defined by pigmentation) and dorso-ventral axes (defined by the sperm entry point). By following the fate of individual cells, or groups of cells, we can make a map on the blastula surface showing the regions that will give rise to, for example, notochord, somites, nervous system, or gut. The fate map shows where the tissues of each germ layer normally come from, but it indicates neither the full potential of each region for development nor to what extent its fate is already specified or determined in the blastula. In other words, we know what each of the early cells will give rise to—the embryo does not. Early vertebrate embryos have considerable capacity for regulation when pieces are removed or transplanted to different parts of the same embryo (see Section 1.12). This implies considerable developmental plasticity at this early stage and also that the actual fate of cells is heavily dependent on the signals they receive from neighboring cells.

One way of making a fate map is to stain various parts of the surface of the early embryo with a lipophilic dye such as diI, and observe where the labeled region ends up. Individual cells can also be labeled by injection of stable high-molecular-weight molecules such as rhodamine-labeled dextran, which cannot pass through cell membranes and so are restricted to the injected cell and its progeny; as rhodamine fluoresces red in UV light, the rhodamine-labeled dextran can be easily detected under a UV microscope. The fluorescent protein green fluorescent protein (GFP) is now widely used for such purposes. Fig. 3.43 shows a *Xenopus* embryo labeled for fate mapping.

The fate map of the late *Xenopus* blastula (Fig. 3.44) shows that the yolky vegetal region, which occupies the lower third of the spherical blastula, gives rise to most of the endoderm. The yolk, which is present in all cells, provides all the nutrition for the developing embryo, and is gradually used up as development proceeds. At the other pole, the animal hemisphere becomes ectoderm, which becomes

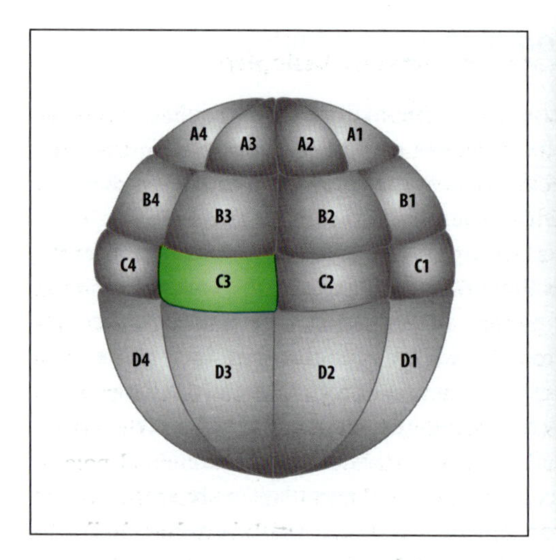

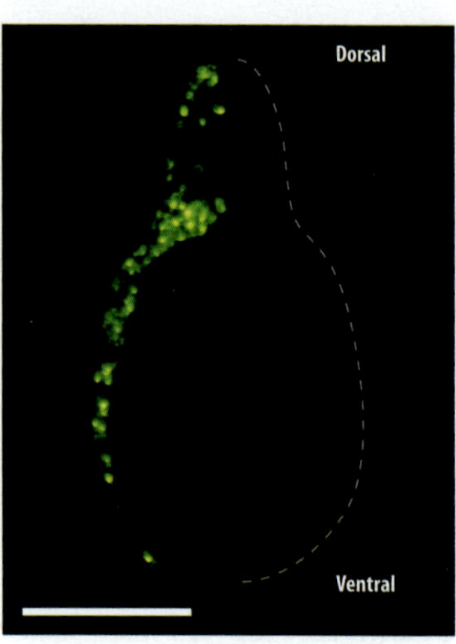

Fig. 3.43 Fate mapping of the early *Xenopus* embryo. Left panel: a single cell in the embryo, C3, is labeled by injection of fluorescein-dextran-amine, which fluoresces green under UV light. Right panel: a cross-section of the embryo, made at the tailbud stage, shows that the labeled cell has given rise to mesoderm cells on one side of the embryo. Scale bar = 0.5 mm.

Photograph courtesy of L. Dale.

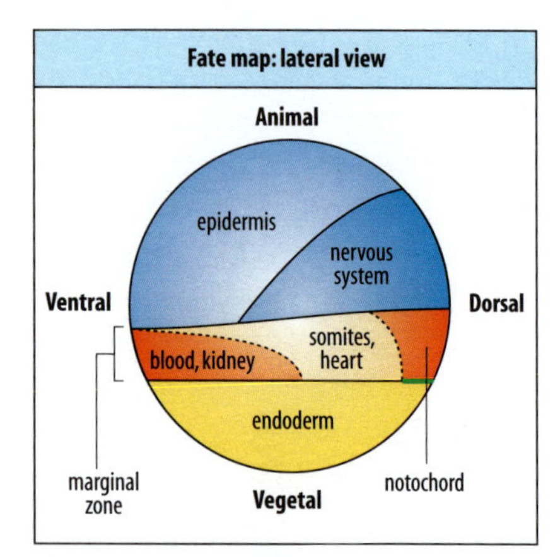

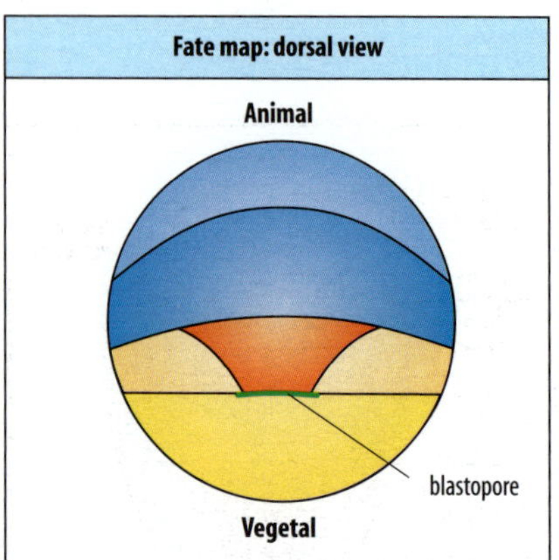

Fig. 3.44 Fate map of a late *Xenopus* blastula. The ectoderm gives rise to the epidermis and nervous system. Along the dorso-ventral axis the mesoderm gives rise to notochord, somites, heart, kidneys, and blood. Note that blood can also form in more dorsal regions. In *Xenopus*, although not in all amphibians, there is also endoderm (not shown here) overlying the mesoderm in the marginal zone.

further diversified into epidermis and the future nervous tissue. The future mesoderm forms from a belt-like region, the marginal zone, around the equator of the blastula. In *Xenopus*, but not in all amphibians, a thin outer layer of presumptive endoderm overlies the presumptive mesoderm in the marginal zone.

The fate map of the blastula makes clear the function of gastrulation. At the blastula stage, the endoderm which gives rise to the gut is on the outside and so must move inside. Similarly, mesodermal tissues that will form internal tissues and organs such as muscle, bone, heart, kidneys, and blood, are on the outside of the embryo and must move inwards. During gastrulation, the marginal zone moves into the interior through the dorsal lip of the blastopore, which lies above the Nieuwkoop center. The fate map of the mesoderm (see Fig. 3.44) shows that it becomes subdivided along the dorso-ventral axis of the blastula. The most dorsal mesoderm gives rise to the notochord, followed, as we move ventrally, by somites (which give rise to muscle tissue), lateral plate (which contains heart and kidney mesoderm), and blood islands (tissue where hematopoiesis first occurs in the embryo). There are also important differences between the future dorsal and ventral sides of the animal hemisphere: the epidermis comes mainly from the ventral side of the animal hemisphere, whereas the nervous system comes from the dorsal side. The epidermis spreads to cover the whole of the embryo after neural-tube formation.

The terms dorsal and ventral in relation to the fate map can be somewhat confusing, because the fate map does not correspond exactly to a neat set of axes at right angles to each other. As a result of cell movements during gastrulation, cells from the dorsal side of the blastula give rise to some ventral parts of the anterior end of the embryo, such as the head, as well as to dorsal structures, and will also form some other ventral structures, such as the heart. The ventral region gives rise to ventral structures in the anterior part of the embryo but will also form some dorsal structures posteriorly. This is why the dorsalized embryos described in Section 3.7 also have overdeveloped anterior structures and lack posterior regions. Indeed, a somewhat different interpretation of the standard fate map for *Xenopus* has been proposed in which the 'dorso-ventral' axis of the blastula is considered to be the future antero-posterior axis of the embryo.

3.12 The fate maps of vertebrates are variations on a basic plan

Fate maps of the early embryos of zebrafish, chick, and mouse have been prepared using techniques essentially similar to those used for *Xenopus*: cells in the early embryo are labeled and their fate followed. In the zebrafish embryo there is extensive cell mixing during the transition from blastula to gastrula, and so it is not possible to construct a reproducible fate map at cleavage stages. In this respect the zebrafish resembles the mouse. The zebrafish late blastula comprises a blastoderm of deep cells and a thin overlying layer (see Fig. 3.11). The overlying layer is largely protective and is eventually lost. At the beginning of gastrulation, the fate of deep-layer cells, from which all the cells of the embryo will come, is correlated with their position in respect to the animal pole. Cells at the margin of the blastoderm give rise to the endoderm, cells slightly further toward the animal pole to mesoderm, while ectodermal cells come from the blastoderm nearest the animal pole (Fig. 3.45). In general terms, the fate map of the zebrafish is rather similar to that of an amphibian, if one imagines the vegetal region of the amphibian blastula being replaced by one large yolk cell.

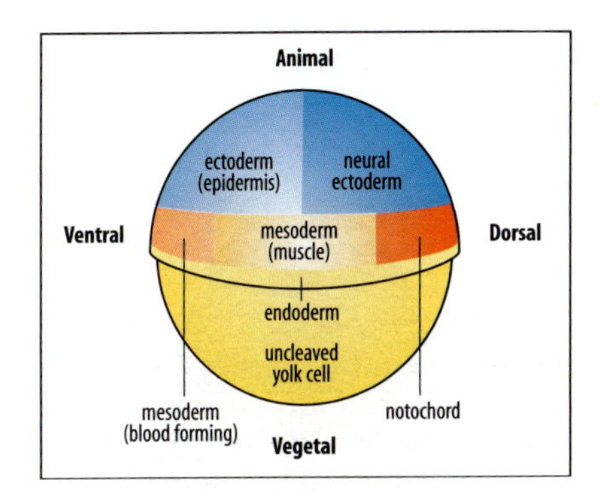

Fig. 3.45 Fate map of zebrafish at the early gastrula stage. The three germ layers come from the blastoderm, which sits on the lower hemisphere composed of an uncleaved yolk cell. The endoderm comes from the margin of the blastoderm and some has already moved inside.

A fate map of the chick embryo cannot be made at the early blastoderm stage that roughly corresponds to the *Xenopus* blastula. This is partly because so much of the chick embryo derives from the posterior marginal zone, which is still a very small region of the total blastoderm at this stage (see Fig. 3.14). Unlike *Xenopus*, there is considerable cell proliferation and growth in the chick embryo during primitive streak formation and gastrulation. There are also extensive cell movements both before and during the emergence of the primitive streak. Once the primitive streak has formed, the picture becomes clearer, and presumptive endoderm, mesoderm, and ectoderm can be mapped (Fig. 3.46).

At the stage shown in Fig. 3.46, the chick blastoderm has become a three-layered structure. Cells have ingressed through the primitive streak into the interior to form mesodermal and endodermal layers. Most of the cells that now form the outer surface of the blastoderm are prospective ectoderm and will form neural tube and epidermis, but there are still regions of the outer blastoderm that will move through the streak and give rise to mesoderm. Hensen's node at the anterior end of the streak is prospective mesoderm; as the node regresses it leaves cells behind that will form the notochord and also contribute to the somites. The mesoderm lying along the antero-posterior midline will give rise to somites and is surrounded by cells that will form the lateral plate mesoderm and structures such as the heart and kidney. In the lowest layer of the embryo, closest to the yolk, the presumptive endoderm is surrounded by cells that will form extra-embryonic structures.

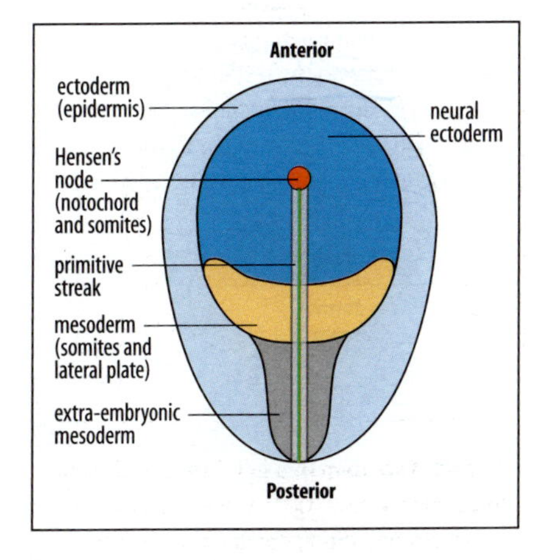

Fig. 3.46 Fate map of a chick embryo when the primitive streak has fully formed. The diagram shows a view of the dorsal surface of the embryo. Almost all the endoderm has already moved through the streak to form a lower layer, so is not shown.

In the case of the mouse, most cells from the inner cell mass of embryos less than 3.5 days old can give rise to many different embryonic tissues, as well as to some extra-embryonic structures, such as the visceral and parietal endoderm, and so at this stage a fate map like that of *Xenopus* cannot be constructed. At 6–7 days' gestation the mouse epiblast becomes transformed into the three germ layers by gastrulation. Gastrulation in the mouse is essentially very similar to gastrulation in the chick, but at this stage the mouse epiblast is folded into a cup, which makes the process of primitive streak and node formation more difficult to follow. A detailed fate map of this stage has been established by tracing the descendants of single cells that have been labeled by injection with a dye. There is, however, extensive cell mixing and cell proliferation in the epiblast. Descendants of a single cell can spread widely and give rise to cells of different germ layers, so that only about 50% of the labeled clones have progeny in only one germ layer.

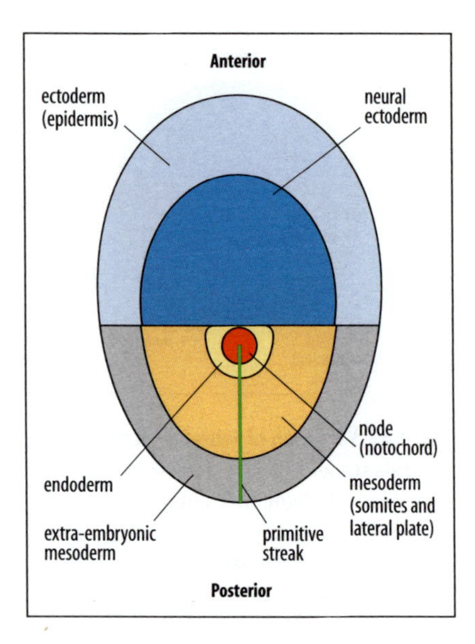

Fig. 3.47 Fate map of a mouse at the late gastrula stage. The embryo is depicted as if the 'cup' has been flattened and is viewed from the dorsal side. At this stage the primitive streak is at its full length.

The fate map obtained is basically similar to that of the chick at the primitive-streak stage (Fig. 3.47). There are some differences; one is that the node gives rise exclusively to the notochord. Cells from the anterior and middle regions of the streak move forward round the node to give rise to the other mesodermal tissues. Somites and gut endoderm derive from the anterior primitive streak, while the middle part of the streak gives rise mainly to lateral-plate mesoderm. The posterior part of the streak gives rise to the tailbud and also provides the extra-embryonic mesoderm that forms the extra-embryonic membranes—the amnion, visceral yolk sac, and allantois.

The fate maps of the different vertebrates are thus similar when one looks at the relationship between the germ layers and the site of the inward movement of cells at gastrulation (Fig. 3.48). The differences are due mainly to the yolkiness of the different eggs, which determines the patterns of cleavage and influences the shape of the early embryo. The similarity in relationship between the germ layers implies that similar mechanisms must be involved in their specification. Fate maps are not specification maps (see Section 1.12), and they do not reflect the full potential for development of the cells of these early embryos.

3.13 Cells of early vertebrate embryos do not yet have their fates determined and regulation is possible

All early vertebrate embryos have considerable powers of regulation (see Section 1.3) when parts of the embryo are removed or rearranged. Experiments show that at the blastula stage, and even later, many cells are not yet determined or specified (see Section 1.12); their potential for development is greater than their position on the fate map suggests.

The state of determination of cells, or of small regions of an embryo, can be studied by transplanting them to a different region of a host embryo and seeing how they develop. If they are already determined, they will develop according to their original position. If they are not yet determined, they will develop in line with their new position. This can be shown experimentally by introducing a single labeled cell from a *Xenopus* blastula into the blastocoel of a later-stage host embryo and following its fate. The transferred cell divides, and during gastrulation its progeny become distributed to different parts of the embryo.

In general, cells in transplants made from early blastulas are not yet determined; their progeny will differentiate according to the signals they receive at their

Fig. 3.48 The fate maps of vertebrate embryos at comparable developmental stages. In spite of all the differences in early development, the fate maps of vertebrate embryos at stages equivalent to a late blastula or early gastrula show strong similarities. All maps are shown in a dorsal view. The future notochord mesoderm occupies a central dorsal position. The neural ectoderm lies adjacent to the notochord, with the rest of the ectoderm anterior to it. The mouse fate map depicts the late gastrula stage. The future epidermal ectoderm of the zebrafish is on its ventral side.

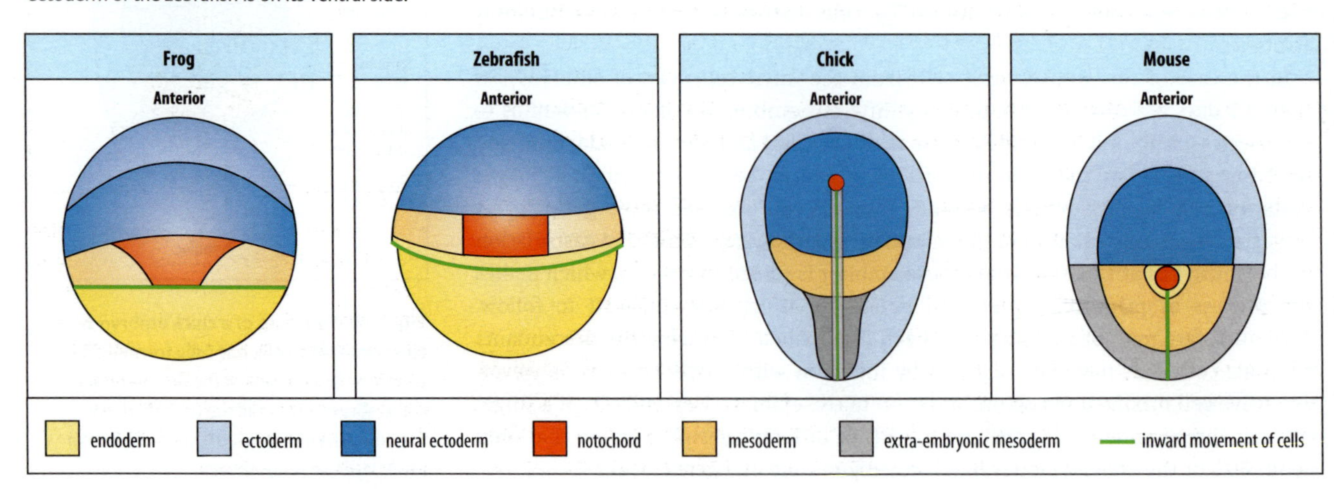

new location. So cells from the vegetal pole, which would normally form endoderm, can contribute to a wide variety of other tissues such as muscle or nervous system, when grafted at an early stage. Similarly, early animal pole cells, whose normal fate is epidermis or nervous tissue, can form endoderm or mesoderm. With time, cells gradually become determined, so that similar cells taken from later-stage blastulas and early gastrulas develop according to their fate at the time of transplantation.

Fragments of a fertilized *Xenopus* egg that are only one-fourth of the normal volume develop into more-or-less normally proportioned but small embryos. There must therefore be a patterning mechanism involving cell interactions that can cope with such differences in size. There are, however, limits to the capacity for regulation. Isolated animal and vegetal halves of an eight-cell *Xenopus* embryo do not develop normally; and while dorsal halves of eight-cell embryos regulate to produce a reasonably normal embryo, early ventral halves do not.

Early mouse embryos can regulate to achieve the correct size. Giant embryos formed by aggregation of several embryos in early cleavage stages can achieve normal size within about 6 days by reducing cell proliferation. The mouse embryo retains considerable capacity to regulate until late in gastrulation. Even at the primitive-streak stage, up to 80% of the cells of the epiblast can be destroyed by treatment with the cytotoxic drug mitomycin C, and the embryo can still recover and develop with relatively minor abnormalities. The early chick embryo also has remarkable powers of regulation, and fragments of the blastoderm can give rise to whole embryos.

Further evidence for regulation in mammals comes from twinning. Twins can result from a separation of the cells at the two-cell stage, but twinning can also occur in humans at a stage as late as 7 days of gestation, when the primitive streak has already started to form. Since early vertebrate embryos show considerable capacity for regulation and many of the cells are not determined, this implies that cell–cell communication must determine cell fate.

To study this question, one can create **chimeric** mice—that is, mice that are mosaics of cells with two different genetic constitutions—by fusing two embryos. Chimeras made from a normal embryo and one that is genetically similar, but homozygous for a mutated version of a single gene, can be used to find out whether the effects of the gene are **cell autonomous** or **non cell autonomous**. If only the mutant cells exhibit the mutant phenotype and are not 'rescued' by the normal cells, the gene is acting cell-autonomously. This means that the product of the gene is acting solely within the cell in which it is made, and is not influencing other cells. The cells of a black mouse, for example, remain black when put into a white mouse, and they do not change the white cells to black; they are thus autonomous with respect to pigmentation (Fig. 3.49). In contrast, a gene is acting non-autonomously when either the mutant cells in the chimera appear to act normally or the normal cells start to show the mutant phenotype. Non-autonomous action is typically due to a gene product that is secreted by one cell and acts on the others.

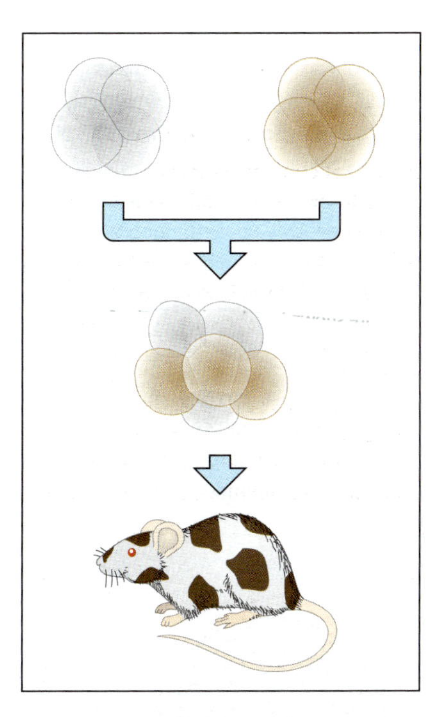

Fig. 3.49 Fusion of mouse embryos gives rise to a chimera. If an eight-cell stage embryo of an unpigmented strain of mouse is fused with a similar embryo of a pigmented strain, the resulting embryo will give rise to a chimeric animal, with a mixture of pigmented and unpigmented cells. The distribution of the different cells in the skin gives this chimera a patchy coat.

The cells in the inner cell mass of the mouse embryo are not yet determined. We have already described how cells of the inner cell mass and the trophectoderm are specified purely by their relative position on the inside or the outside of the embryo (see Fig. 3.37). The cells of the inner cell mass itself remain pluripotent up to 4.5 days after fertilization—they can give rise to many cell types. If 4.5-day cells are injected into the inner cell mass of a 3.5-day blastocyst, they can contribute to all tissues of the embryo, including the germ cells. This provides another way of producing chimeric mice. Cells of the inner cell mass can be cultured to produce so-called **embryonic stem cells** (**ES cells**) and mice with a novel genetic constitution can

Box 3E Producing developmental mutations in mice

When studying the role of a particular gene in development, it is an enormous advantage to be able to study the effects of a mutation in that gene. One way of obtaining an animal with the desired mutation is simply to wait for it to turn up in the population, but in vertebrates the wait may be long indeed. Developmental mutations, in particular, are rarely identified. Mutagenesis screens such as those described in Box 3B (p. 99) for zebrafish are sometimes carried out for mice, but because of the animal's larger size, longer generation time, and the fact that embryos cannot be screened so easily, they are more difficult and costly. In mice, on the other hand, it is possible to routinely produce animals with a particular mutant genetic constitution using transgenic techniques.

Mice that overexpress a given gene in particular cells or at particular times during development can be produced by the injection of the appropriate transgene into the nucleus of a fertilized egg. The transgene will become incorporated into the genome and will be expressed according to the promoter it contains. The effect of misexpressing a gene in a tissue or at a time where it is not usually active can be studied in this way. A simpler and more versatile technique for producing mice with a desired mutation, however, is by introducing **embryonic stem cells** (**ES cells**) carrying the mutation into the blastocysts. ES cells are cultured cells derived from the inner cell mass; they can be maintained in culture indefinitely and grown in large numbers. Inner cell mass cells introduced into the inner cell mass of another embryo will populate all of the mouse's tissues and will contribute to the germ cells. For example, if ES cells from a black-pigmented mouse are introduced into the inner cell mass of an embryo of a white mouse, the mouse that develops from this embryo will be a chimera of 'black' and 'white' cells. In the skin, this mosaicism is visible as patches of black and white hairs (see figure).

ES cells can be genetically manipulated in culture to produce mutant cells in which a particular gene or genes have been inactivated or new genes introduced. This technique is particularly powerful for creating loss-of-function mutations to ascertain the role of particular genes in development (see Box 4B, pp. 158–159). Mutations that lead to the complete absence of the function of the gene are known as gene knock-outs. Some mutations do not lead to a loss of function but to a change or diminution of function.

ES cells (carrying a mutation in a single gene) in culture

ES cells injected into inner cell mass of normal blastocyst

Chimeric animal produces sperm carrying the mutation

Because the animals initially produced by introduction of ES cells are a mixture of mutant and normal cells they may show few, if any, effects of the mutation. If they carry the mutant gene in their germ cells, however, interbreeding can produce a permanent line of non-chimeric animals in which the mutation is present in either the heterozygous or the homozygous state.

ES cells are pluripotent and they can be manipulated in culture to give rise to a wide variety of differentiated cell types, from eggs to muscle and neurons. This capacity is discussed in more detail in Chapter 8.

be generated by the introduction of ES cells carrying particular mutations. If the ES cells contribute to the germ cells, a line of mice that carry the introduced mutation in all their cells can be bred (Box 3E, above).

3.14 In *Xenopus* the endoderm and ectoderm are specified by maternal factors, but the mesoderm is induced from ectoderm by signals from the vegetal region

When explants from different regions of the early blastula are cultured in a simple medium containing the necessary salts for ion balance, tissue from the region nearest the animal pole will form a ball of epidermal cells, while explants from the

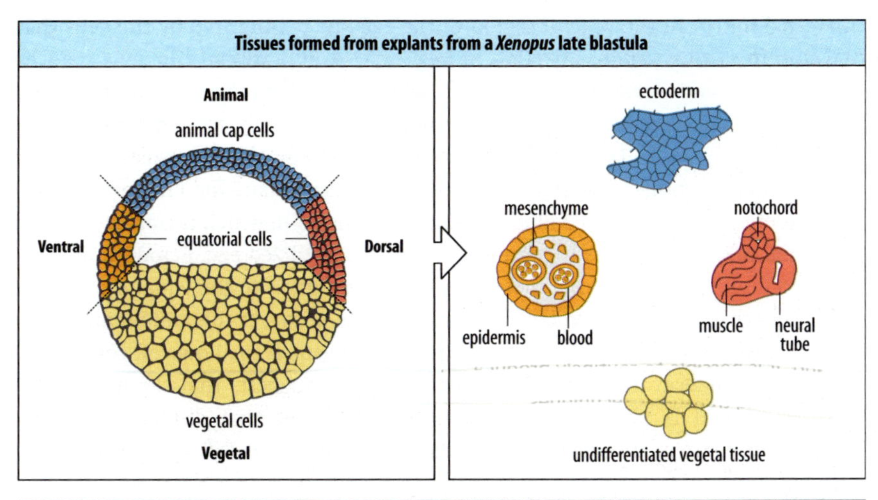

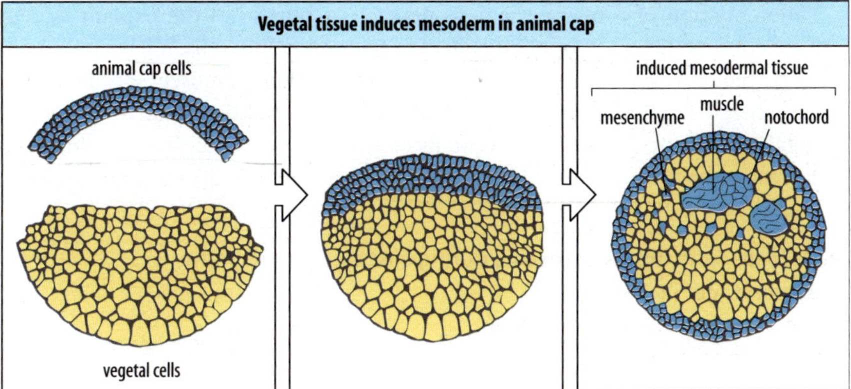

Fig. 3.50 Induction of mesoderm by the vegetal region in the *Xenopus* blastula. Top panels: explants of animal cap cells or vegetal cells on their own from a late blastula form only ectoderm or endoderm, respectively. Explants from the equatorial region, where animal and vegetal regions are adjacent, form both ectodermal (epidermis and neural tube) and mesodermal (mesenchyme, blood cells like erythrocytes, notochord, and muscle) tissues, showing that mesoderm induction has taken place. The reason for the differences in the mesodermal tissues formed by ventral and dorsal explants is explained later. Bottom panels: the standard experiment for studying mesoderm induction is to take explants from animal and vegetal regions that would normally make only ectoderm and endoderm, respectively, and culture them together for 3 days. When such explants from an early blastula are combined and cultured, mesoderm is induced from the animal cap tissue. This mesoderm contains notochord, muscle, blood, and loose mesenchyme.

vegetal region are endodermal in their development (Fig. 3.50). These results are in line with the normal fates of these regions. It is thus generally accepted that in *Xenopus* all the ectoderm and most of the endoderm are specified by maternal factors in the egg. There is no evidence that any signals from other regions of the embryo are necessary for their initial specification. The mesoderm, however, is different. It is formed in a band round the equator of the blastula in response to signals from the prospective endoderm.

The formation of mesoderm can be observed by culturing explants from the blastula. When explants from the equatorial region of the late blastula, where animal and vegetal cells are adjacent, are cultured in isolation, they form ectodermal derivatives and mesoderm (see Fig. 3.50, top panels). When explants of animal and vegetal regions distant from the equator are cultured in contact with each other, mesodermal tissues are also formed (see Fig. 3.50, bottom panels). Confirmation that it is the **animal cap** cells, and not the vegetal cells, that are forming mesoderm was obtained by pre-labeling the animal region of the blastula with a cell-lineage marker and showing that the labeled cells form the mesoderm. Clearly, the vegetal region is producing a signal or signals that can induce mesoderm. Mesoderm can be distinguished by its histology as, after a few days culture, it can contain muscle, notochord, blood, and loose mesenchyme (connective tissue). It can also be identified by the typical proteins that cells of mesodermal origin produce, such as muscle-specific actin.

In contrast, the endoderm is maternally specified in the vegetal region. The transcription factor VegT has a key role in endoderm specification. It is translated from

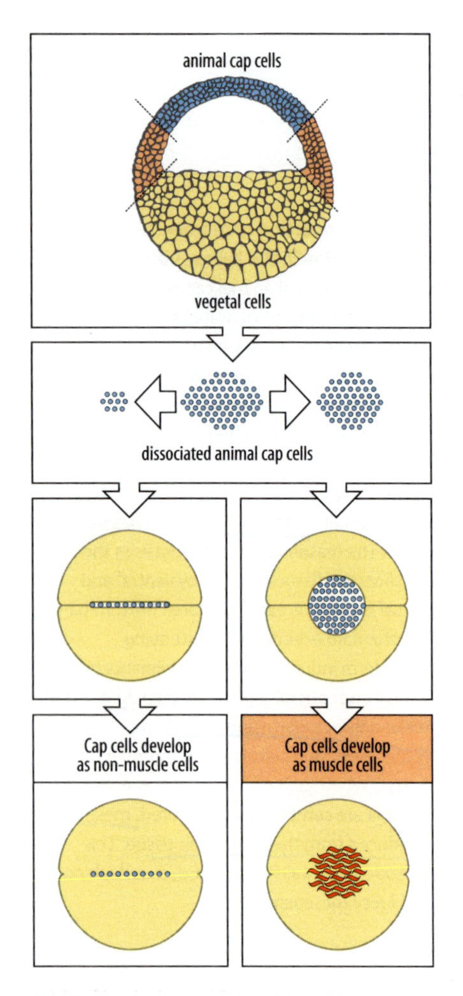

Fig. 3.51 The community effect. One or a small number of animal cap cells in contact with vegetal tissue are not induced to become mesodermal cells and do not begin to express mesodermal markers such as muscle-specific proteins. A sufficiently large number of animal cap cells, about 200, must be present for induction of muscle differentiation to occur.

CONC DEPENDENT

maternal mRNA in the vegetal region of the egg and is inherited by the cells that develop from this region. Injection of VegT mRNA into animal cap cells induces expression of endoderm-specific markers, whereas blocking the translation of VegT in the vegetal region, by injecting oligonucleotides which are antisense to the mRNA (see Box 5A, p. 189), results in a loss of endoderm. VegT is also essential for mesoderm induction, as we shall see later. A strong candidate for a maternal determinant of ectoderm is the E3 ubiquitin-ligase ectodermin, which is translated from maternal mRNA located at the animal pole of the fertilized egg. Ectodermin functions to limit the effect of the mesoderm-inducing signals to the equatorial zone.

In the *Xenopus* blastula explant system described above, if the animal and vegetal explants are separated by a filter with pores too small to allow cell contacts to develop, mesoderm induction still takes place. This suggests that the mesoderm-inducing signal is in the form of secreted molecules that diffuse across the extracellular space, and does not pass directly from cell to cell via cell junctions (see Fig. 1.23).

Differentiation of a mesodermal tissue such as muscle in the explant system appears to depend on a **community effect** in the responding cells. A few animal cap cells placed on vegetal tissue will not be induced to express muscle-specific genes. Even when a small number of individual cells are placed between two groups of vegetal cells, induction does not occur. By contrast, larger aggregates of animal cap cells respond by strongly expressing muscle-specific genes (Fig. 3.51).

3.15 Mesoderm induction occurs during a limited period in the blastula stage

Mesoderm induction is one of the key events in vertebrate development, as the mesoderm itself is a source of essential signals for further development of the embryo. In *Xenopus*, mesoderm induction is largely completed before gastrulation begins. Explant experiments like those described above have shown that in *Xenopus* there is a period of about 7 hours at the blastula stage during which animal cells are competent to respond to a mesoderm-inducing signal; cells lose their competence to respond about 11 hours after fertilization. An exposure of about 2 hours to inducing signal is sufficient for the induction of at least some mesoderm. One might expect that the timing of expression of mesoderm-specific genes, such as those of muscle, would be closely coupled to the time at which the mesoderm is induced—but it is not. Explant experiments have shown that irrespective of when during the competent period the cells are exposed to the 2-hour induction, muscle-specific gene expression always starts at about 5 hours after the end of the competent period, a time that corresponds to mid-gastrula stage in the embryo (Fig. 3.52).

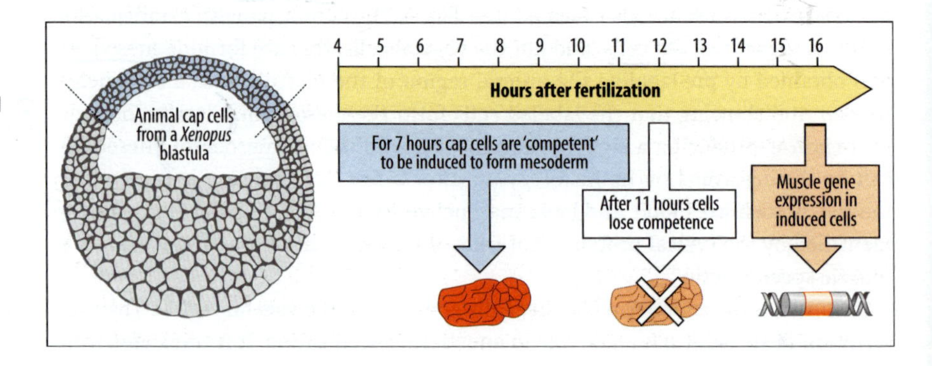

Fig. 3.52 Timing of muscle gene expression is not linked to the time of mesoderm induction. Animal cap cells isolated from an early *Xenopus* blastula are competent to respond to mesoderm-inducing signals only for a period of about 7 hours, between 4 and 11 hours after fertilization. For expression of target genes to occur, exposure to inducer must be for at least 2 hours within this period. Irrespective of when the induction occurs within this competence period, muscle-gene activation occurs at the same time: 16 hours after fertilization.

Muscle gene expression can occur as early as 5 hours after induction, if induction occurs late in the period of competence, or as late as 9 hours after induction, if induction occurs early in the competent period. These results suggest that there is an independent timing mechanism by which the cells monitor the time elapsed since fertilization and then, provided that they have received the mesoderm-inducing signal, they express muscle-specific genes.

The extended period of competence gives the embryo a certain latitude as to when mesoderm induction actually takes place, and means that the timing of the inductive signal does not have to be rigorously linked to a short time period. Given this latitude, an independent timing mechanism for mesodermal gene expression is one way to keep subsequent development coordinated, whatever the actual time of mesoderm induction.

3.16 Zygotic gene expression is turned on at the mid-blastula transition

Another event that appears to be under intrinsic temporal control, although not linked to mesoderm induction, is the beginning of transcription from the embryo's own genes. The *Xenopus* egg contains quite large amounts of maternal mRNAs, which are laid down during oogenesis. After fertilization, the rate of protein synthesis increases 1.5-fold, and a large number of new proteins are synthesized by translation of the maternal mRNA. There is, in fact, very little new mRNA synthesis until 12 cleavages have taken place and the embryo contains 4096 cells. At this stage, transcription from zygotic genes begins, cell cycles become asynchronous, and various other changes occur. This is known as the **mid-blastula transition**, although it does in fact occur in the late blastula, just before gastrulation starts. A few zygotic genes are expressed earlier, most notably two Nodal-related genes, *Xnr-5* and *Xnr-6*, which are expressed as early as the 256-cell stage.

How is the mid-blastula transition triggered? Suppression of cleavage but not DNA synthesis does not alter the timing of transcriptional activation, and so the timing is not linked directly to cell division. Nor are cell–cell interactions involved, as dissociated blastomeres undergo the transition at the same time as intact embryos. The key factor in triggering the mid-blastula transition seems to be the ratio of DNA to cytoplasm—the quantity of DNA present per unit mass of cytoplasm.

Direct evidence for this comes from increasing the amount of DNA artificially by allowing more than one sperm to enter the egg or by injecting extra DNA into the egg. In both cases, transcriptional activation occurs prematurely, suggesting that there may be some fixed amount of a general repressor of transcription present initially in the egg cytoplasm. As the egg cleaves, the amount of cytoplasm does not increase, but the amount of DNA does. The amount of repressor in relation to DNA gets smaller and smaller until there is insufficient to bind to all the available sites on the DNA and the repression is lifted. Timing of the mid-blastula transition thus seems to fit with a timing model of the hourglass egg-timer type (Fig. 3.53). In such a model something has to accumulate, in this case DNA, until a threshold is reached. The threshold is determined by the initial concentration of the cytoplasmic factor, which does not increase.

Zebrafish embryos also go through a mid-blastula transition at which zygotic transcription begins. It occurs at the 512-cell stage, coincidentally with the formation of the syncytial layer at the margin between the blastoderm and yolk (see Section 3.2). Shortly after this, β-catenin activates zygotic genes such as *dharma*, *chordino*, and *nodal-related* in the future dorsal region (see Section 3.6).

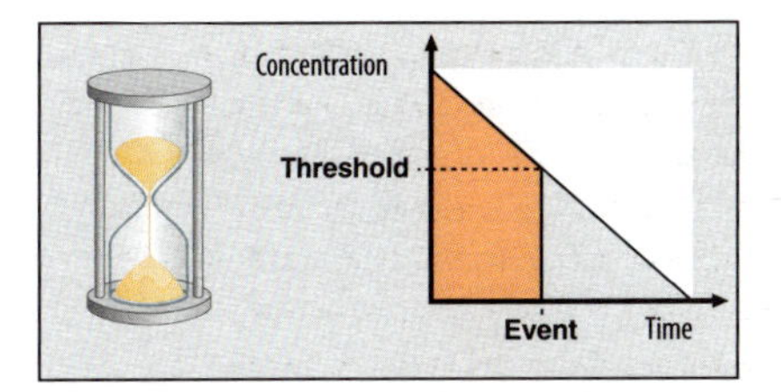

Fig. 3.53 Timing mechanism that could operate in development. A mechanism based on an analogy to an egg-timer could measure time to the mid-blastula transition. The decrease in the active concentration of some molecule, such as a repressor, could occur with time, and the transition could occur when the repressor reaches a critically low threshold concentration. This would be equivalent to all the sand running into the bottom of the egg-timer. In the embryo, a reduction in repressor activity per cell would in fact occur because repressor concentration in the embryo as a whole remains constant, but the number of nuclei increases as a result of cell division. The amount of repressor per nucleus thus gets progressively smaller over time.

3.17 Mesoderm-inducing and patterning signals are produced by the vegetal region, the organizer, and the ventral mesoderm

In the *Xenopus* blastula fate map the mesoderm is divided into a number of regions along the dorso-ventral axis, with the notochord originating in the most dorsal region, then muscle, and the blood-forming tissue most ventrally (significant amounts of blood also form from dorsal regions). If the fate map is compared with the **specification map** of the blastula, however, we can see some differences, particularly in the ectodermal and mesodermal regions (Fig. 3.54). The specification map is constructed by culturing small pieces of the blastula in a simple culture medium and observing what tissues they form. It therefore gives a picture of the actual state of specification of different regions at this stage, as opposed to their normal fate. We see that only a small region on the dorsal side is specified as muscle in the specification map, whereas the fate map at the same blastula stage shows that a great deal of muscle will come from more lateral and ventral regions. Thus, explants from the dorsal marginal zone of a blastula, taken after mesoderm induction has begun but before it is completed, behave much in line with their normal fate; they develop into notochord and muscle, and the explants even mimic gastrulation movements by cellular rearrangements that narrow the explant in one dimension and extend it another. By contrast, ventral and lateral marginal zone explants develop into mesenchyme and blood-forming tissue only. They do not give rise to any muscle, although their normal fate in the embryo is to form considerable amounts.

These results, together with other evidence discussed later, allow us to construct a model of mesoderm induction that involves at least four different signals. Induction by the vegetal region involves at least two sets of signals: one is a general mesoderm inducer, broadly specifying a ventral-type mesoderm, which can be considered the ground or default state; the second signal, from the Nieuwkoop center, acts simultaneously or a little later, and specifies the dorsal-most mesoderm that will contain the Spemann organizer and form notochord. Two further sets of signals pattern the ventral mesoderm along the dorso-ventral axis, subdividing it into prospective muscle, kidney, and blood. The third set of signals comes from the organizer region and modifies the ventralizing action of the fourth set of signals, which comes from the ventral region (Fig. 3.55). This model in no way implies that only four distinct signaling molecules are required, or indeed that all four signals are qualitatively different. It is quite possible that each 'signal' represents the actions of more than one molecule, or that different signals represent the same molecule acting at different concentrations.

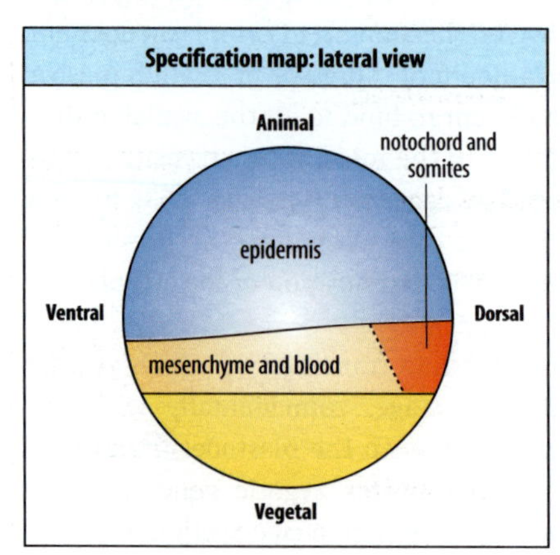

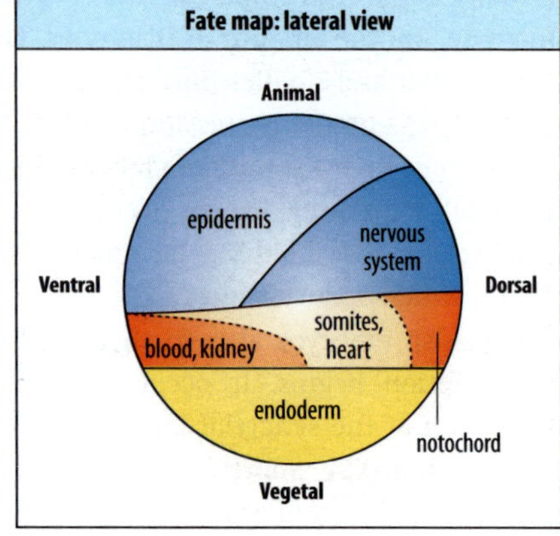

Fig. 3.54 The difference between the fate map and specification map of a *Xenopus* blastula. The fate of a region when isolated and placed in culture is shown on the specification map (left panel), while the normal fate of the blastula regions is shown in the fate map (right panel). There is a clear difference in specification of dorsal and ventral mesoderm. While the notochord's fate and specification maps correspond at this stage, that of the rest of the mesoderm is much more labile, and the specification of most of the somites and other mesodermal tissues has yet to occur. This involves signals from the region of the Spemann organizer, which acts just before and during gastrulation, as well as signals from the ventral region.

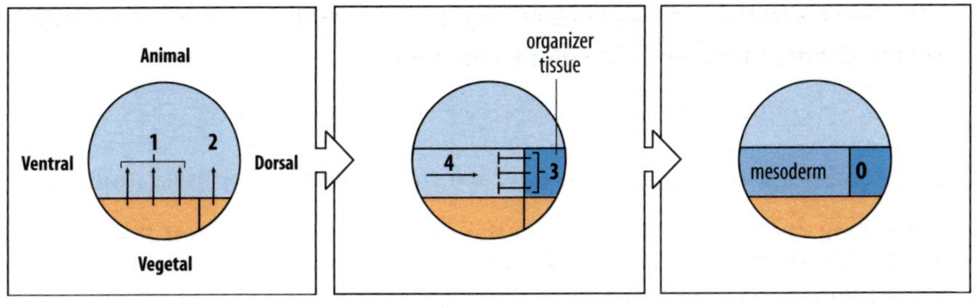

Fig. 3.55 Four signals involved in mesoderm induction. Two signals originate in the vegetal region, the first from the ventral region (1) and the second on the dorsal side from the region of the Nieuwkoop center (2). The first signal specifies ventral mesoderm, and the dorsal signal specifies the Spemann organizer (O) and dorsal mesoderm. The third signal (3) emanates from the organizer and dorsalizes the adjacent mesoderm by inhibiting the ventralizing action of the fourth signal (4), which comes from the ventral region.

Direct evidence for at least two signals from the vegetal region is provided by comparing the inducing effects of dorsal and ventral vegetal regions (Fig. 3.56). Dorsal vegetal tissue containing the Nieuwkoop center induces notochord and muscle from animal cap cells, whereas ventral vegetal tissue induces mainly blood-forming tissue and little muscle. A minimum of two signals from the vegetal region can thus specify the broad differences between dorsal and ventral mesoderm. These signals are, however, insufficient to explain all of the patterning. In normal development, the ventral mesoderm makes a major contribution to the somites, and hence to muscle, yet isolated explants of presumptive ventral mesoderm from an early blastula do not make muscle. Another signal or set of signals is required to pattern the ventral mesoderm further.

This third signal, which has a dorsalizing effect, originates in the Spemann organizer itself. Evidence for this signal comes from combining a fragment of dorsal marginal zone from a late blastula with a fragment of the ventral presumptive mesoderm. The ventral fragment will form substantial amounts of muscle, whereas ventral mesoderm isolated from an early blastula after induction by the vegetal region only will form mainly blood-forming tissue and mesenchyme. A dramatic demonstration of the action of the third signal is to graft the Spemann organizer into the ventral marginal zone of an early gastrula (Fig. 3.57). The graft induces a complete new dorsal axis, resulting in a twinned embryo.

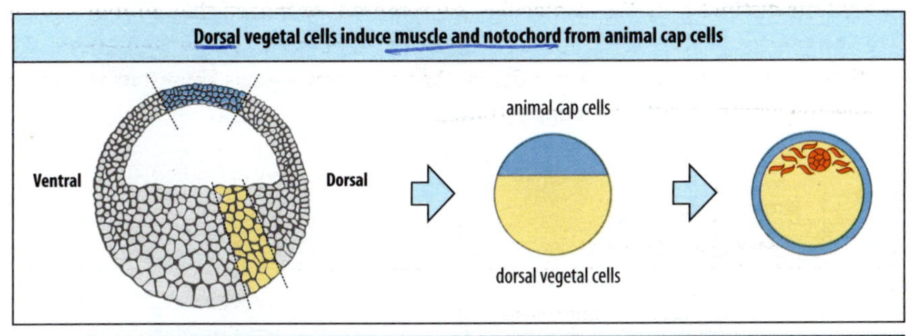

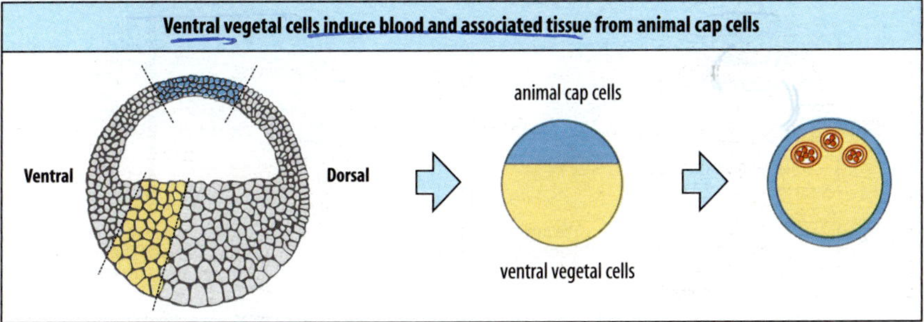

Fig. 3.56 Differences in mesoderm induction by dorsal and ventral vegetal regions. The dorsal vegetal region of the *Xenopus* blastula, which contains the Nieuwkoop center, induces notochord and muscle from animal cap tissues, while ventral vegetal cells induce blood and associated tissues. This is good evidence for different inducing signals coming from the dorsal and ventral vegetal regions.

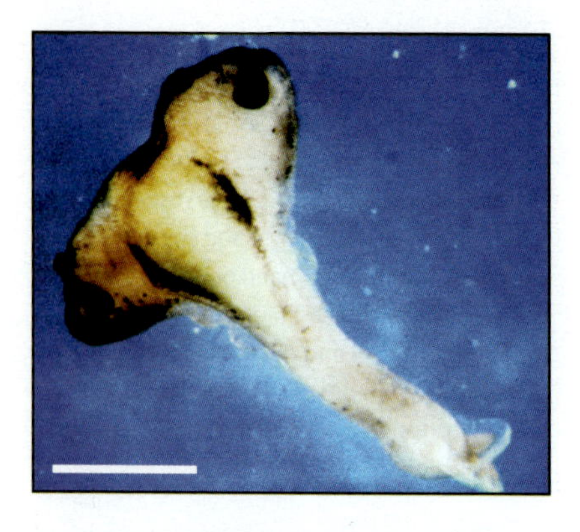

Fig. 3.57 Transplantation of the Spemann organizer can induce a new axis in *Xenopus*. The third set of signals required for mesoderm induction and patterning come from the Spemann organizer region. Their effect can be seen by transplanting the Spemann organizer into the ventral region of another gastrula. The resulting embryo has two distinct heads, one of which was induced by the Spemann organizer. The organizer therefore produces signals that not only pattern the mesoderm dorso-ventrally, but induce neural tissue and anterior structures. Scale bar = 1 mm.

Photograph courtesy of J. Smith.

The fourth signal, which emanates from the ventral region of the embryo, ventralizes the mesoderm and interacts with the dorsalizing signal, which limits its influence.

3.18 Members of the TGF-β family have been identified as mesoderm inducers

The explant experiments discussed in the previous sections suggest that secreted extracellular factors are involved in mesoderm induction. Two main approaches have been used to try and identify these factors in *Xenopus*. One is to apply the candidate factor directly to isolated animal caps in culture. The other is to inject mRNA encoding the suspected inducer into animal cells of the early blastula and then culture the cells. By itself, the ability to induce mesoderm in culture does not prove that a particular protein is a natural inducer in the embryo. Rigorous criteria must be met before such a conclusion can be reached. These include the presence of the protein and its receptor in the right concentration, place, and time in the embryo; the demonstration that the appropriate cells can respond to the factor; and the demonstration that blocking the response prevents induction taking place. On all these criteria, growth factors of the TGF-β family have been identified as the most likely candidates for mesoderm inducers. Members of this family known to be involved in early *Xenopus* development include the *Xenopus* Nodal-related proteins (Xnrs), the protein Derrière, the bone morphogenetic proteins (BMPs), and activin.

The signaling pathways stimulated by TGF-β growth factors are illustrated in Fig. 3.58. Confirmation of the role of TGF-β growth factors in mesoderm induction

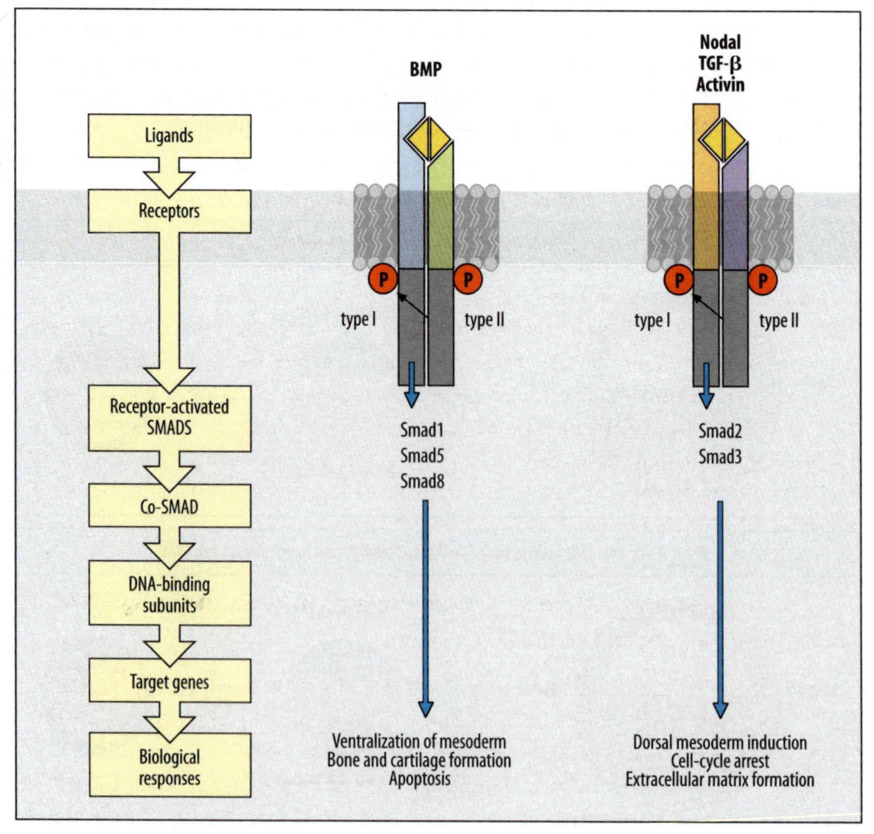

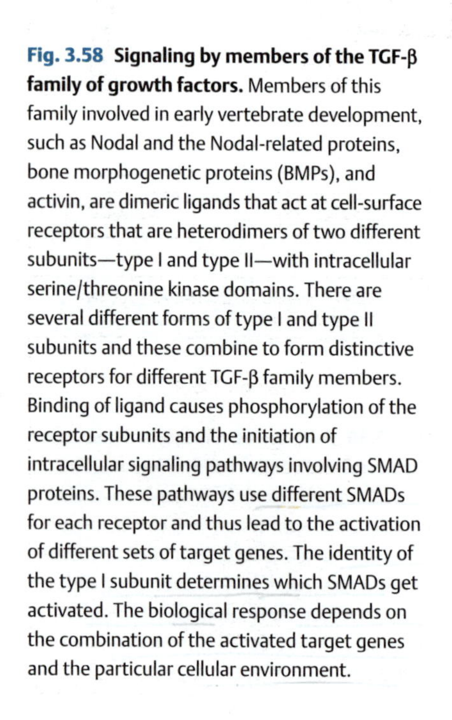

Fig. 3.58 Signaling by members of the TGF-β family of growth factors. Members of this family involved in early vertebrate development, such as Nodal and the Nodal-related proteins, bone morphogenetic proteins (BMPs), and activin, are dimeric ligands that act at cell-surface receptors that are heterodimers of two different subunits—type I and type II—with intracellular serine/threonine kinase domains. There are several different forms of type I and type II subunits and these combine to form distinctive receptors for different TGF-β family members. Binding of ligand causes phosphorylation of the receptor subunits and the initiation of intracellular signaling pathways involving SMAD proteins. These pathways use different SMADs for each receptor and thus lead to the activation of different sets of target genes. The identity of the type I subunit determines which SMADs get activated. The biological response depends on the combination of the activated target genes and the particular cellular environment.

Fig. 3.59 A mutant activin receptor blocks mesoderm induction. Receptors for proteins of the TGF-β family function as dimers (see Fig. 3.58). Receptor function in a cell can be blocked by introducing mRNA encoding a mutant receptor subunit that lacks most of the cytoplasmic domain, and so cannot function. The mutant receptor subunit can bind ligand and forms heterodimers with normal receptor subunits, but the complex cannot signal. The mutant subunit thus acts as a dominant-negative mutation of receptor function. When mRNA encoding the mutant receptor subunit is injected into cells of the two-cell *Xenopus* embryo, subsequent mesoderm formation is blocked. No mesoderm or axial structures are formed except for the cement gland, the most anterior structure of the embryo.

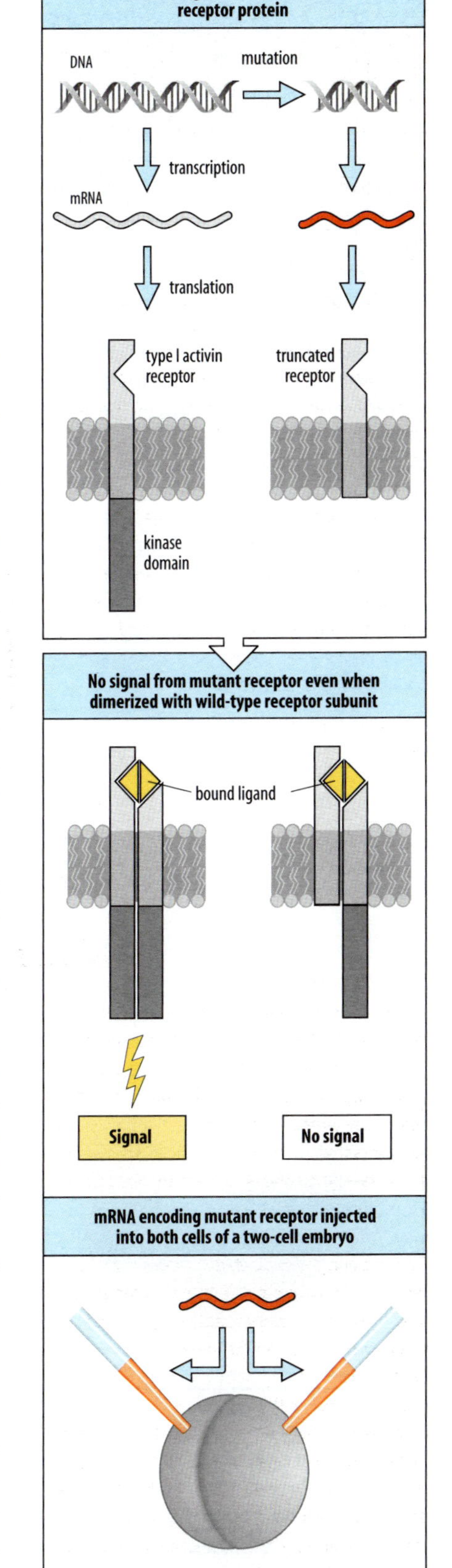

has come from experiments that block activation of the receptors for these factors (Fig. 3.59). The activin type II receptor is a receptor for several TGF-β family growth factors and is expressed and uniformly distributed throughout the early *Xenopus* blastula. When mRNA for a mutant receptor subunit is injected into the early embryo, mesoderm formation is prevented. However, different TGF-β growth factors can act through the same receptor, so these experiments were not able to pinpoint the individual proteins responsible.

Some maternally provided TGF-β family proteins are located vegetally, and among these Vg-1 has recently been shown to have a role in mesoderm induction in *Xenopus*. Another maternal vegetal factor of great importance in mesoderm induction is the transcription factor VegT (see Section 3.5). Directly and indirectly, VegT activates the zygotic expression of the Xnr proteins and Derrière. If VegT is severely depleted, expression of the zygotic *Xnr* genes and *Derrière* is downregulated and almost no mesoderm develops, showing that VegT is crucial for its induction. Injection of mRNA for the Xnrs or Derrière can, however, rescue mesoderm formation in the VegT-depleted embryos, suggesting that these are most likely to be the direct mesoderm inducers. The mRNAs for the genes *Xnr-1, Xnr-4,* and *Xnr-5* rescue head, trunk, and tail mesoderm, indicating that they are involved in the general induction of all mesoderm, whereas Derrière mRNA rescues only trunk and tail, indicating that it cannot induce the anterior-most mesoderm. Nuclear β-catenin also stimulates Xnr gene transcription and thus Xnr levels will be highest in the dorsal side where the VegT signal and the β-catenin signal overlap (Fig. 3.60). This network of events could provide us with the first two signals of the four-signal model. Maternal vegetal factors such as Vg-1 and VegT would provide a signal for general 'ventral' mesoderm induction, which β-catenin converts more locally to induction of dorsal mesoderm. FGF appears to be necessary but not sufficient by itself for general mesoderm induction. The mesoderm-inducing signal itself is one or more of the Xnrs, and the difference between a dorsal and a ventral mesoderm fate is probably due to expression of different Xnrs or differences in Xnr levels along the dorsal axis. Activin is a powerful mesoderm inducer *in vitro*, and is involved in later mesoderm patterning, but its role in primary mesoderm induction *in vivo* remains unclear.

3.19 The dorso-ventral patterning of the mesoderm involves the antagonistic actions of dorsalizing and ventralizing factors

We now consider the other signals that pattern the mesoderm along the dorso-ventral axis once it has been induced. One component of the third signal—the dorsalizing signal—is the protein Noggin, which is a secreted protein unrelated to any of the common growth factor families. The gene *noggin* was first discovered in a screen for factors that could rescue UV-irradiated ventralized *Xenopus* embryos in

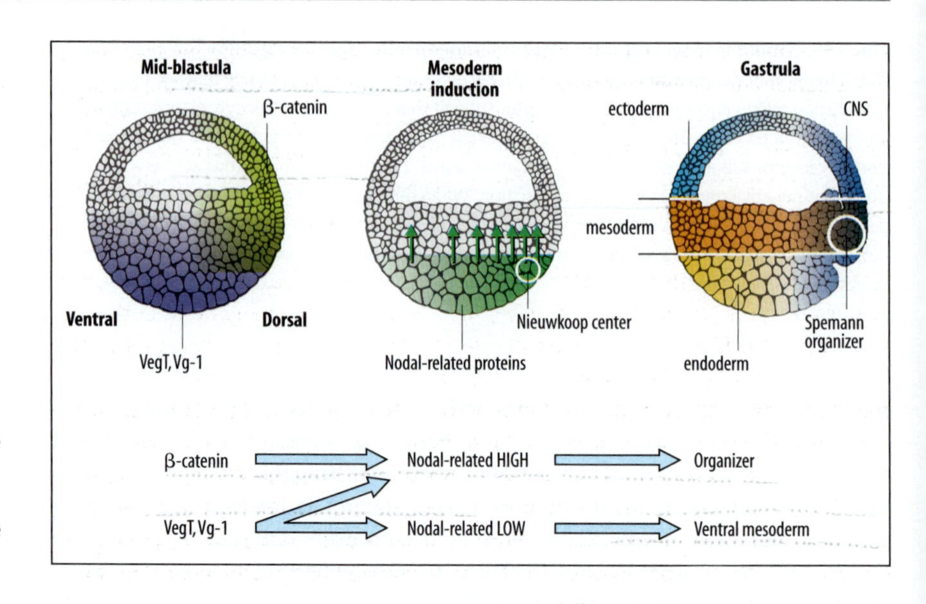

Fig. 3.60 A gradient in Nodal-related proteins may provide the first two signals in mesoderm induction. Maternal VegT in the vegetal region activates the transcription of Nodal-related genes (*Xnrs*). The presence of β-catenin on the dorsal side results in a dorsal-to-ventral gradient in the Xnr proteins. These induce mesoderm and, at high doses, specify the Nieuwkoop center, which in turn induces the Spemann organizer on the dorsal side. This is a simplified version of the actual picture, in which other signals, such as Vg-1 and activin, also have roles. CNS, central nervous system.

which a dorsal region had not been specified by the normal actions of the β-catenin pathway (see Section 3.6). It is strongly expressed at the gastrula stage in the Spemann organizer (Fig. 3.61), which is the source of the third signal. Noggin does not induce mesoderm in animal-pole explants but can dorsalize explants of ventral marginal zone tissue, thus making it a good candidate for a signal that patterns the mesoderm along the dorso-ventral axis. The proteins Chordin, Follistatin, and Frizbee, also secreted by the organizer, are other proposed components of the third signal. Surprisingly, the action of all these signals is not on the cells themselves but on a fourth set of signals.

The fourth set of signals, emanating from the ventral region of the mesoderm, promotes ventralization of the mesoderm. Candidates for these signals are BMP-4 and Xwnt-8. BMP-4 is expressed uniformly throughout the late *Xenopus* blastula, and Xwnt-8 is expressed in the future ventral and lateral mesoderm. As gastrulation proceeds, BMP-4 is no longer expressed in dorsal regions. When the action of BMP-4 is blocked throughout by introducing a **dominant-negative** mutant receptor, the embryo is dorsalized, with ventral cells differentiating as both muscle and notochord. Conversely, overexpression of BMP-4 ventralizes the embryo.

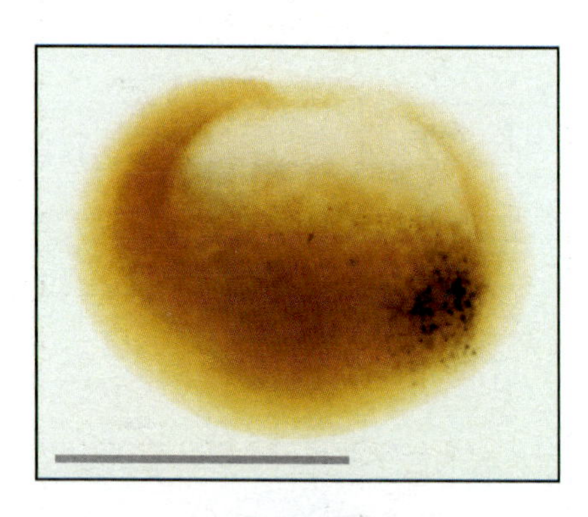

Fig. 3.61 Expression of *noggin* in the *Xenopus* blastula. *noggin* expression is shown as the dark-staining area in the region of the Spemann organizer. Scale bar = 1 mm.

Photograph courtesy of R. Harland, from Smith, W.C., et al.: 1992.

How do the dorsalizing factors, representing signal 3, and the ventralizing factors, representing signal 4, interact? Noggin and Chordin interact with BMP-4 and prevent it from binding to its receptor (Fig. 3.62). In this way, a functional gradient of BMP-4 activity is set up across the dorso-ventral axis with its high point ventrally and little or no activity in the presumptive dorsal mesoderm. Frizbee generates a similar ventral to dorsal gradient of Wnt activity by binding to Wnt proteins and preventing them acting on the presumptive dorsal mesoderm. Another BMP antagonist that may contribute to the third signal is Cerberus, which is produced in the organizer region. Cerberus inhibits the actions of BMPs, Nodal-related proteins and Wnt proteins; it inhibits mesoderm formation and is involved in the induction of anterior structures, and in particular the head. The summary table on p. 144 lists some of the main proteins so far identified as mesoderm inducers and mesoderm-patterning factors in *Xenopus*.

The antagonistic action of Chordin on BMP-4 mirrors that of the *Drosophila* Chordin homolog Sog on the BMP-4 homolog Decapentaplegic in the patterning of the dorso-ventral axis in the fly (see Section 2.16). In the fly, however, the dorso-ventral axis is reversed compared with that of vertebrates, and so Decapentaplegic specifies a dorsal fate. Other components of the mechanism used to form the gradient of Decapentaplegic activity in *Drosophila* are also conserved in vertebrates. Xolloid, for example, is the *Xenopus* equivalent of the fly metalloproteinase Tolloid, and degrades Chordin. Xolloid is thought to act as a clearing agent for Chordin, reducing the extent of its long-range diffusion and helping to maintain a gradient of Chordin dorsalizing activity.

Mutagenesis screens have identified genes in zebrafish similar to those involved in dorso-ventral patterning in *Xenopus*, and they confirm the general outline of germ-layer specification. The zebrafish Ndr1 protein (Squint) is expressed in the dorsal yolk syncytial layer and overlying cells as a result of β-catenin signaling. Together with another Nodal-related protein, Ndr2 (Cyclops), Ndr1 is required for the specification of the marginal blastomeres as **mesendoderm**: prospective endoderm and mesoderm. High levels of Nodal signaling are thought to specify endoderm and lower levels the mesoderm. Double mutants of *Ndr1* and *Ndr2* lack both head and trunk mesoderm, but some mesoderm still develops in the tail region. Zebrafish also have a homolog of the Nodal antagonist Lefty, which is an inhibitor of Ndr1 and Ndr2 and acts over a long range. Like BMP-4 in *Xenopus*, zebrafish BMP-2 (Swirl) and BMP-7 (Snailhouse) are expressed on the ventral side of the embryo and mutants that do not produce these proteins are dorsalized. FGF is involved in the restriction of BMP gene expression to the ventral side. The zebrafish gene *chordino* codes for Chordin, which antagonizes the actions of BMPs. As one might expect, *chordino* mutants show an expansion of ventral and lateral mesoderm.

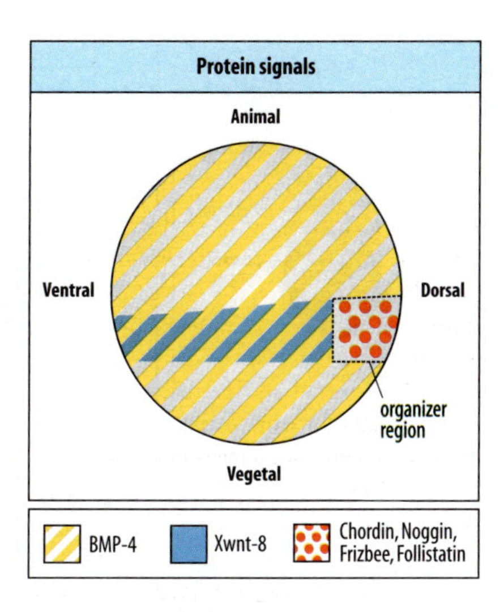

Fig. 3.62 Distribution of protein signals in the *Xenopus* blastula. The signals from the organizer block the action of BMP-4 and Xwnt-8.

3.20 Mesoderm induction and patterning in the chick and mouse occurs during primitive-streak formation

In the chick, most mesoderm induction occurs at the primitive streak. Chick epiblast isolated before streak formation will form some mesoderm containing blood vessels, blood cells, and some muscle, but no dorsal mesodermal structures such as notochord. Treatment of the isolated epiblast with activin, however, results in the additional appearance of notochord and more muscle, suggesting that TGF-β family members such as Nodal act as mesoderm-inducing and/or -patterning signals in chick embryos as they do in *Xenopus*. Other experiments suggest that the full development of the axis, including anterior structures, requires the action of Wnt proteins as well as TGF-β family members. As we saw in Section 3.8, the posterior marginal zone in the chick embryo is crucial for the formation of the primitive streak, and it also supplies signals for mesoderm induction and patterning. Wnt-8c and the TGF-β family member Vg-1 are expressed in the posterior marginal zone and induce the expression of chick Nodal in cells that will move to form the primitive streak and Hensen's node (see Fig. 3.36). Indeed, Vg-1 can induce a whole new axis when cells secreting it are grafted to the margin of an early chick blastoderm before primitive-streak formation, which indicates a role for Vg-1 in mesoderm induction. Chordin is expressed in cells in the streak, and FGF signals from adjacent tissue are also required for streak initiation. Mesodermal patterning also occurs in the streak, with different regions of the streak giving rise to different mesodermal tissues along the dorso-ventral axis (Fig. 3.63). Thus, as well as indicating the direction of the antero-posterior axis, the streak also represents the dorso-ventral axis.

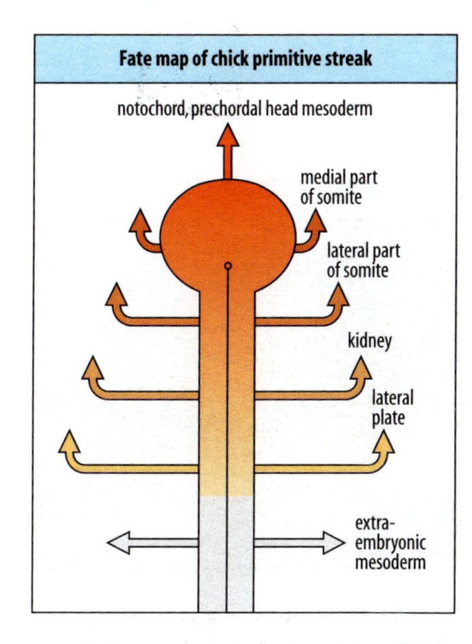

Fig. 3.63 Mesoderm is patterned within the chick primitive streak. Different parts of the primitive streak give rise to mesoderm with different fates. The arrows indicate the direction of movement of the mesodermal cells.

In the mouse, mesoderm induction occurs only in the primitive streak, which is initiated at around 6.5 days, starting from one small region at one side of the

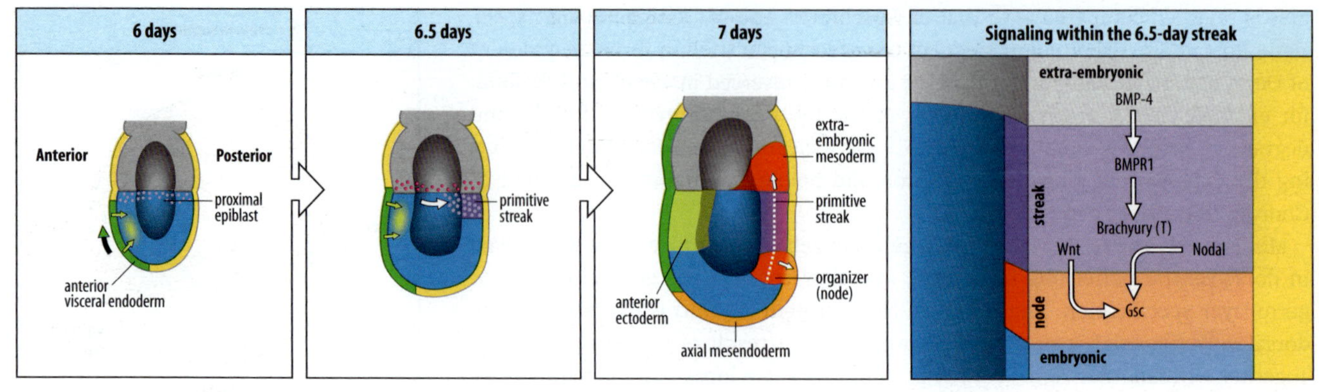

Fig. 3.64 Specification of the primitive streak in the mouse embryo. First panel: by 6 days the anterior visceral endoderm (AVE) has induced anterior character in the underlying epiblast (small arrows) and early markers of the future primitive streak are restricted to the proximal rim of the epiblast. Second panel: BMP-4 (dark blue) is transiently expressed in the adjacent extra-embryonic ectoderm, and the posterior movement (white arrow) of cells expressing primitive-streak markers (purple) results in the primitive streak forming opposite the AVE at 6.5 days. Third panel: by 7 days, extra-embryonic mesoderm (red) is produced from the posterior end of the streak while the node forms at the anterior end, which gives rise to axial mesendoderm, which will form dorsal mesoderm and gut endoderm. Fourth panel: a possible scheme of the molecular interactions within the streak at 6.5 days.

epiblast (see Fig. 3.24). This region is specified as a result of the determination of the anterior end of the embryo by the movement of the AVE (see Section 3.9). As the anterior ectoderm becomes determined, proximal epiblast cells that express genes characteristic of prospective mesoderm start to converge on the posterior proximal region and internalize at the point of convergence to initiate the primitive streak (Fig. 3.64). BMP-4 is expressed in the extra-embryonic ectoderm above the proximal rim of the epiblast and activates the expression of mesodermal markers in adjacent proximal epiblast cells. Nodal is required both to establish the posterior end of the primitive streak and for mesoderm formation, and is expressed in the primitive streak along with Wnts. In mutants lacking Nodal function, mesoderm does not form. It is thought that BMP-4, Nodal and Wnt signaling are inhibited in the AVE and the overlying ectoderm, which allows this ectoderm to develop as anterior tissues. In the mouse, as in the chick, mesoderm patterning also occurs in the primitive streak and the position at which mesodermal cells are induced along the primitive streak determines their eventual fate.

3.21 Gradients in signaling proteins and threshold responses could pattern the mesoderm

The mesoderm-inducing and -patterning signals described in previous sections exert their developmental effects by inducing the expression of specific genes that are required for the further development of the mesoderm into, for example, notochord, muscle, and blood. Expression of the gene *Brachyury*, which encodes a T-box family transcription factor, is one of the earliest markers of mesoderm in all vertebrates, and Brachyury is thought to act as a key transcription factor in mesoderm specification and patterning. In *Xenopus*, it is expressed throughout the presumptive mesoderm in the blastula (Fig. 3.65), later becoming confined to the notochord (the dorsal-most derivative of the mesoderm), and to the tailbud (posterior mesoderm). *Brachyury* is essential for the development of posterior mesoderm. The gene was originally discovered in the mouse as a result of semi-dominant mutations that produced short tails; when homozygous, this mutation is lethal in the embryo as a result of a lack of development of posterior mesoderm (see Fig. 1.13).

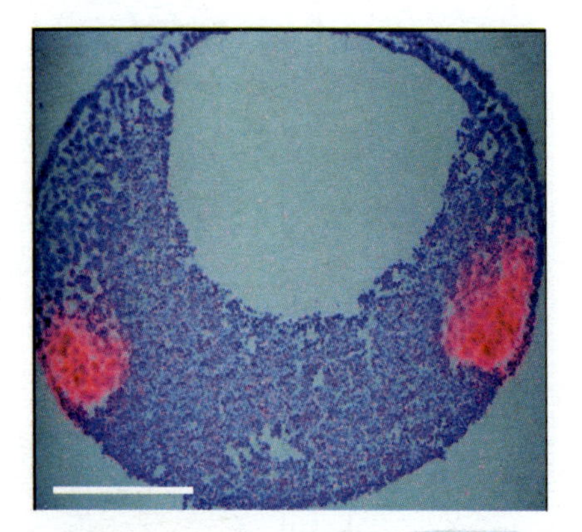

Fig. 3.65 Expression of *Brachyury* in the *Xenopus* blastula. A cross-section through the embryo along the animal–vegetal axis shows that *Brachyury* (red) is expressed in the prospective mesoderm. Scale bar = 0.5 mm.

Photograph courtesy of M. Sargent and L. Essex.

One of the first zygotic genes to be expressed specifically in the *Xenopus* organizer region is *goosecoid*, which encodes a transcription factor somewhat similar to both the Gooseberry and Bicoid proteins of *Drosophila*—hence the name. In line with its presence in the organizer region, microinjection of *goosecoid* mRNA into the ventral region of the blastula mimics to some extent transplantation of the

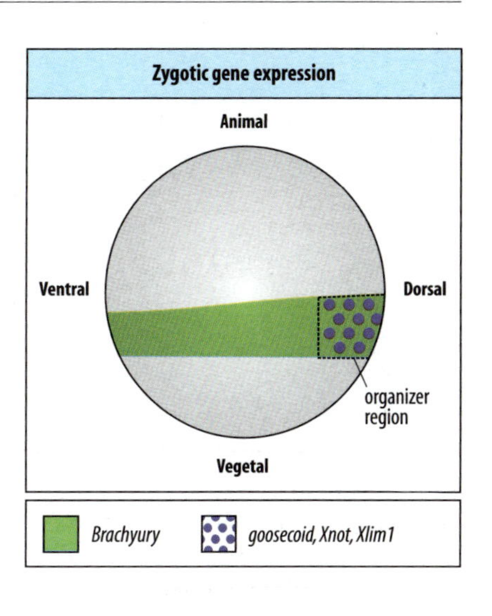

Fig. 3.66 Zygotic gene expression in a late *Xenopus* blastula. The expression domains of a number of zygotic genes that code for transcription factors correspond quite well to demarcations on the specification map. The gene *Brachyury* is expressed in a ring around the embryo corresponding quite closely to the future mesoderm. A number of transcription factors are specifically expressed in the region of the dorsal mesoderm that corresponds to the Spemann organizer and are required for its function. Xnot appears to have a role in the specification of notochord; Goosecoid and Xlim1 are required for head organizer function.

Spemann organizer (see Fig. 3.57), resulting in the formation of a secondary axis. Genes for other transcription factors are specifically expressed in the organizer region (Fig. 3.66 and summary table, p. 144).

It is still not clear how the secreted signaling proteins present in the mesoderm turn on genes like *goosecoid* and *Brachyury* in the right place. For example, *Brachyury* can be experimentally switched on in *Xenopus* ectoderm by mesoderm-inducing factors such as activin, but what induces it normally in the presumptive mesoderm *in vivo* is still unclear; a combination of Vg-1 and Xnrs is thought to be most likely, and its expression is probably maintained by FGF. One model for patterning the mesoderm and other tissues proposes that positional information is provided by a dorso-ventral gradient of a morphogen. Several of the proteins identified as possible mesoderm-patterning agents in *Xenopus*, such as the Xnrs, are indeed expressed in a graded fashion. Just how gradients are set up with the necessary precision is not clear, however. Simple diffusion of the morphogen molecules may play a part, but more complex cellular processes are likely to be involved (see Box 9A, p. 347).

Experiments with activin provide a beautiful example of how a diffusible protein could pattern a tissue by turning on particular genes at specific threshold concentrations. Although activin itself may not be responsible for primary mesoderm patterning in this way *in vivo*, animal cap cells from a *Xenopus* blastula respond to increasing doses of activin by activation of different genes at different threshold concentrations. Increasing activin concentration by as little as 1.5-fold results in a dramatic alteration in the pattern of marker proteins expressed and the tissues that differentiate. For example, this small increase causes a change from homogeneous formation of muscle to formation of notochord. Increasing concentrations of activin can specify several different cell states that correspond to the different regions along the dorso-ventral axis. At the lowest concentrations of activin, only epidermis develops. Then, as the concentration increases, *Brachyury* is expressed, together with muscle-specific genes such as that for actin. With a further increase in activin, *goosecoid* is expressed, and this corresponds to the dorsal-most region of the mesoderm—the Spemann organizer (Fig. 3.67). There is evidence that activin molecules move away from the source and so would be able to set up a gradient. Similar results can be obtained by injecting increasing quantities of activin mRNA.

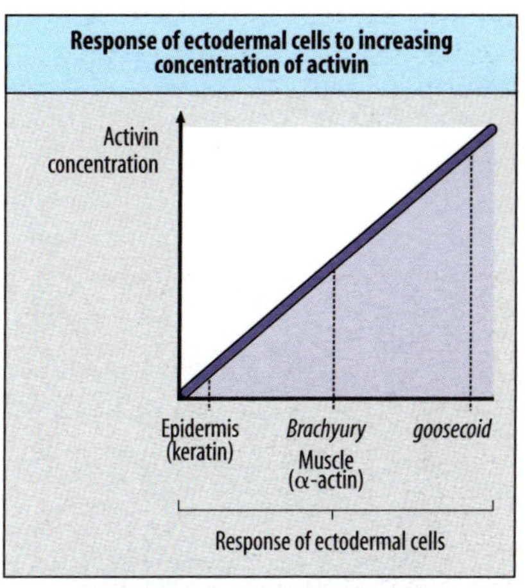

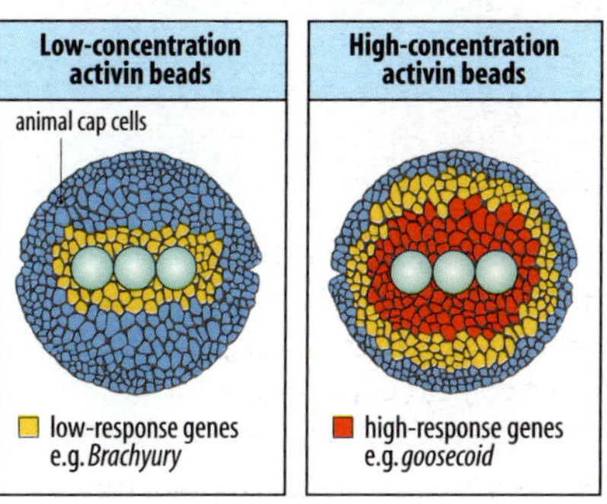

Fig. 3.67 Graded responses of early *Xenopus* tissue to increasing concentrations of activin. When animal cap cells are treated with increasing concentrations of activin, particular genes are activated at specific concentrations, as shown in the top panel. At intermediate concentrations of activin, *Brachyury* is induced, whereas *goosecoid*, which is typical of the organizer region, is only induced at high concentrations. If beads releasing a low concentration of activin are placed in the center of a mass of animal cap cells (lower left panel), expression of low-response genes such as *Brachyury* is induced immediately around the beads. With a high concentration of activin in the beads (lower right panel), *goosecoid* and other high-response genes are now expressed around the beads and the low-response genes farther away.

One can thus see how graded signals could activate transcription factors in particular regions and thus pattern tissues.

How do the cells distinguish between different concentrations of activin? Occupation of just 2% of the activin receptors is required to activate expression of *Brachyury*, while 6% occupation is required for *goosecoid* to be expressed. The link between the signal concentration and gene expression may not be so simple, however. There appear to be additional layers of intracellular regulation; cells expressing *goosecoid* at high activin concentrations also repress *Brachyury* expression, for example, and this involves the action of the Goosecoid protein itself together with other proteins.

Excellent evidence for a secreted morphogen turning on genes at specific threshold concentrations also comes from zebrafish. The putative morphogen in this case is the protein Ndr1, which is involved in patterning the mesoderm (see Section 3.19). Injection of *Ndr1* mRNA into a single cell of an early zebrafish embryo resulted in high-threshold Ndr1 target genes being activated in adjacent cells, whereas low-threshold genes were activated in more distant cells. Graded Nodal signals also pattern the mouse embryo, with the highest levels of Nodal becoming restricted to the posterior region of the epiblast and being required to induce primitive-streak formation and to specify the definitive gut endoderm and anterior mesendoderm—the cells at the anterior end of the primitive streak that move anteriorly along the midline and pattern the overlying neurectoderm (see Fig. 3.64).

We are now in a position to consider the final emergence of the typical vertebrate body plan. Further patterning of the germ layers along both the antero-posterior and dorso-ventral axes occurs during gastrulation, and the antero-posterior patterning is discussed in Chapter 4.

Summary

Once the antero-posterior and dorso-ventral axes are established, one can begin to construct a fate map for the germ layers. There are strong similarities in the fate maps of the frog, zebrafish, chick, and mouse at later stages. Even though there is good evidence for the maternal specification of some regions such as the future endoderm in amphibians, the embryo can still undergo considerable regulation at the blastula stage. This implies that interactions between cells, rather than intrinsic factors, have a central role even in early amphibian development. This strategy is particularly pronounced in the mouse and chick, where it appears to be position that determines cell fate.

In *Xenopus*, the mesoderm and some endoderm are induced from animal cap tissue by the vegetal region, which contains the Nieuwkoop center. Early patterning of the mesoderm can be accounted for by a four-signal model. The first signal is a general mesoderm inducer, specifying a ventral-type mesoderm. The second specifies dorsal mesoderm, including the organizer, while the third signal comes from the ventral side and ventralizes the mesoderm. The fourth signal originates from the organizer and establishes further pattern within the mesoderm by interacting with and inhibiting the third signal.

Protein growth factors of the TGF-β family are excellent candidates for the natural mesoderm-inducing factors as well as for patterning the mesoderm. Other signaling proteins such as Noggin and Chordin inhibit the action of BMP-4, and so are involved in specifying the dorsal and ventral mesoderm. *Brachyury* and *goosecoid* are early mesodermally expressed genes encoding transcription factors and their pattern of expression may be specified by gradients in signaling proteins, the genes being turned on at particular threshold concentrations.

Summary: mesoderm induction in *Xenopus*

VegT in vegetal region

signal 1: general inducing signal
e.g. Nodal-related

↓

induction

↙ ventral mesoderm

↘ signal 2
dorsal mesoderm with Spemann organizer

signal 4: ventralizing signals
e.g. BMP-4, Xwnt-8 ↘

↙ signal 3: dorsalizing signals
e.g. Noggin, Chordin

patterning of mesoderm

SUMMARY TO CHAPTER 3

All vertebrates have the same basic body plan. During early development, the antero-posterior and dorso-ventral axes of this body plan are set up. The mechanism is different in frog, chick, zebrafish, and mouse but can involve localized maternal determinants, external signals, and cell-cell interactions. This early patterning also establishes bilateral asymmetry. It is possible to construct a fate map in the early embryo for the three germ layers—mesoderm, endoderm, and ectoderm. The fate maps of the different vertebrates have strong similarities. At this early stage the embryos are still capable of considerable regulation and this emphasizes the essential role of cell-cell interactions in development. In *Xenopus*, at least four separate signals are involved in mesoderm induction and early patterning. Good candidates for these signals have been identified and include members of the TGF-β family. These signals activate mesoderm-specific genes such as *Brachyury* at particular concentrations and so their gradients could pattern the mesoderm. The summary table on p. 144 lists all genes considered in this chapter in relation to *Xenopus*.

GENERAL FURTHER READING

Bard, J.B.L.: *Embryos. Color Atlas of Development*. London: Wolfe, 1994.

Carlson, B.M.: *Patten's Foundations of Embryology*. New York: McGraw-Hill, 1996.

SECTION FURTHER READING

3.1 The frog *Xenopus laevis* is the model amphibian for developmental studies

Hausen, P., Riebesell, H.: *The Early Development of* Xenopus laevis. Berlin: Springer-Verlag, 1991.

Nieuwkoop, P.D., Faber, J.: *Normal Tables of* Xenopus laevis. Amsterdam: North Holland, 1967.

3.2 The zebrafish embryo develops around a large, undivided yolk

Kimmel, C.B., Ballard, W.W., Kimmel, S.R., Ullmann, B., Schilling, T.F.: **Stages of embryonic development of the zebrafish.** *Dev. Dyn.* 1995, **203:** 253–310.

Westerfield, M. (ed.): *The Zebrafish Book; A Guide for the Laboratory Use of Zebrafish* (Brachydanio rerio). Eugene, Oregon: University of Oregon Press, 1989.

3.3 The early chicken embryo develops as a flat disc of cells overlying a massive yolk

Bellairs, R., Osmond, M.: *An Atlas of Chick Development*. London: Academic Press, 1998.

Hamburger, V., Hamilton, H.L.: **A series of normal stages in the development of a chick.** *J. Morph.* 1951, **88:** 49–92.

Lillie, F.R.: *Development of the Chick: An Introduction to Embryology.* New York: Holt, 1952.

Patten, B.M.: *The Early Embryology of the Chick*. New York: McGraw-Hill, 1971.

Stern, C.D.: **The chick: a great model system becomes even greater.** *Dev. Cell* 2005, **8:** 9–17.

3.4 Early development in the mouse involves allocation of cells to form the placenta and extra-embryonic membranes

Hogan, H., Beddington, R., Costantini, F., Lacy, E.: *Manipulating the Mouse Embryo. A Laboratory Manual* (2nd edn). New York: Cold Spring Harbor Laboratory Press, 1994.

Summary: main genes involved in patterning of axes and germ layers in *Xenopus*

Gene	Maternal/ zygotic	Type of protein	Where expressed	Function of protein
Specification of germ layers and dorso-ventral axis				
VegT	M	Transcription factor	Vegetal region	Endoderm specification; activates expression of mesoderm inducers
Ectodermin	M	Ubiquitin ligase	Animal hemisphere	Ectoderm specification; inhibits mesoderm formation
Vg-1	M	TGF-β family	Vegetal region; RNA enriched dorsally after fertilization	Mesoderm induction
Xwnt-11	M	Wnt family	Vegetal region; RNA translocated to dorsal side after fertilization	Specification of dorsal structures and organizer formation
Dishevelled	M	Wnt pathway signaling protein	Protein associated with vesicles that move to dorsal side of embryo after fertilization	Specification of dorsal structures and organizer formation
β-catenin	M	Acts in Wnt pathway; regulates gene expression	Protein enriched in dorsal nuclei in response to Wnt signaling	Specification of dorsal structures and organizer formation
GSK-3	M	Protein kinase	Protein depleted on dorsal side	Suppression of dorsalizing signals
axin	M	Binds β-catenin	RNA throughout zygote; more protein dorsally than ventrally	Suppression of dorsalizing signals
Derrière	Z	TGF-β family	Vegetal hemisphere and marginal region	Posterior-mesoderm induction
Xnr-1,2,4,5,6	Z	TGF-β family	Vegetal hemisphere and marginal region; higher dorsally	Mesoderm induction
FGF (several types)	Z	Secreted signals	Vegetal hemisphere and marginal region	Posterior-mesoderm development
Mesoderm patterning				
Brachyury	Z	Transcription factor	Throughout prospective mesoderm	Posterior-mesoderm formation
Xwnt-8	Z	Wnt family	Ventral and lateral regions of prospective mesoderm	Mesoderm ventralization
BMP-4	Z	TGF-β family	Throughout late blastula; then excluded from organizer	Mesoderm ventralization
Activin	Z	TGF-β family	Throughout late blastula/ early gastrula	Mesoderm induction and patterning
Organizer function				
siamois	Z	Transcription factor	Nieuwkoop center	Induction of organizer
noggin	Z	Secreted signal	Spemann organizer	Mesoderm dorsalization by antagonizing BMP-4
chordin	Z	Secreted signal	Spemann organizer	Mesoderm dorsalization by antagonizing BMP-4
frizbee	Z	Secreted signal	Spemann organizer	Mesoderm dorsalization by antagonizing Xwnt-8
cerberus	Z	Secreted signal	Spemann organizer	Promotes head development by inhibiting Wnt, Nodal-related, and BMP signaling
goosecoid	Z	Transcription factor	Spemann organizer	Organizer function
Xlim-1	Z	Transcription factor	Spemann organizer	Gastrulation and head formation
Xnot	Z	Transcription factor	Spemann organizer	Notochord specification

Kaufman, M.H.: *The Atlas of Mouse Development*. London: Academic Press, 1992.

3.5 The animal-vegetal axis is maternally determined in *Xenopus* and zebrafish

Birsoy, B., Kofron, M., Schaible, K., Wylie, C., Heasman, J.: **Vg1 is an essential signaling molecule in Xenopus development.** *Development* 2006, **133:** 15–20.

Heasman, J.: **Patterning the early *Xenopus* embryo.** *Development* 2006, **133:** 1205–1217.

Schier, A.F., Talbot, W.S.: **Molecular genetics of axis formation in zebrafish.** *Annu. Rev. Genet.* 2005, **39:** 561–613.

Weaver, C., Kimelman, D.: **Move it or lose it: axis specification in *Xenopus*.** *Development* 2004, **131:** 3491–3499.

3.6 Localized stabilization of the transcriptional regulator β-catenin specifies the future dorsal side and the location of the main embryonic organizer in *Xenopus* and zebrafish

Dosch, R., Wagner, D.S., Mintzer. K.A., Runke, G., Wiemelt, A.P., Mullins, M.C.: **Maternal control of vertebrate development before the midblastula transition: mutants from the zebrafish I.** *Dev. Cell* 2004, **6:** 771–780.

Gerhart, J., Danilchik, M., Doniach, T., Roberts, S., Browning, B., Stewart, R.: **Cortical rotation of the *Xenopus* egg: consequences for the antero-posterior pattern of embryonic dorsal development.** *Development* **(Suppl.)** 1989, 37–51.

Gore, A.V., Maegawa, S., Cheong, A., Gilligan, P.C., Weinberg, E.S., Sampath, K.: **The zebrafish dorsal axis is apparent at the four-cell stage.** *Nature* 2005, **438:** 1030–1035.

He, X., Saint-Jennet, J-P., Woodgett, J.R., Varmus, H.E., Dawid, I.B.: **Glycogen synthase kinase-3 and dorsoventral patterning in *Xenopus* embryos.** *Nature* 1995, **374:** 617–622.

Kodjabachian, L., Dawid, I.B., Toyama, R.: **Gastrulation in zebrafish: what mutants teach us.** *Dev. Biol.* 1999, **126:** 5309–5317.

Pelegri, F.: **Maternal factors in zebrafish development.** *Dev. Dyn.* 2003, **228:** 535–554.

Tao, Q., Yokota, C., Puck, H., Kofron, M., Birsoy, B., Yan, D., Asashima, M., Wylie, C.C., Lin, X., Heasman, J.: **Maternal wnt11 activates the canonical wnt signaling pathway required for axis formation in *Xenopus* embryos.** *Cell* 2005, **120:** 857–871.

Wodarz, A., Nusse, R.: **Mechanisms of Wnt signaling in development.** *Annu. Rev. Cell Dev. Biol.* 1998, **14:** 59–88.

3.7 Signaling centers develop on the dorsal side of *Xenopus* and zebrafish

Kodjabachian, L., Lemaire, P.: **Embryonic induction: is the Nieuwkoop centre a useful concept?** *Curr. Biol.* 1998, **8:** R918–R921.

Smith, J.: **T-box genes: what they do and how they do it.** *Trends Genet.* 1999, **15:** 154–158.

Sokol, S.Y.: **Wnt signaling and dorso-ventral axis specification in vertebrates.** *Curr. Opin. Genet. Dev.* 1999, **9:** 405–410.

3.8 The antero-posterior and dorso-ventral axes of the chick blastoderm are related to the primitive streak

Bertocchini, F., Skromne, I., Wolpert, L., Stern, C.D.: **Determination of embryonic polarity in a regulative system: evidence for endogenous inhibitors acting sequentially during primitive streak formation in the chick embryo.** *Development* 2004, **131:** 3381–3390.

Khaner, O., Eyal-Giladi, H.: **The chick's marginal zone and primitive streak formation. I. Coordinative effect of induction and inhibition.** *Dev. Biol.* 1989, **134:** 206–214.

Kochav, S., Eyal-Giladi, H.: **Bilateral symmetry in chick embryo determination by gravity.** *Science* 1971, **171:** 1027–1029.

Ohta, K., Lupo, G., Kuriyama, S., Keynes, R., Holt, C.E., Harris, W.A., Tanaka, H., Ohnuma, S.: **Tsukushi functions as an organizer inducer by inhibition of BMP activity in cooperation with chordin.** *Dev. Cell* 2004, **7:** 347–358.

Seleiro, E.A.P., Connolly, D.J., Cooke, J.: **Early developmental expression and experimental axis determination by the chicken Vg-1 gene.** *Curr. Biol.* 1996, **11:** 1476–1486.

3.9 The axes of the mouse embryo are not recognizable early in development

Beddington, R.S.P., Robertson, E.J.: **Axis development and early asymmetry in mammals.** *Cell* 1999, **96:** 195–209.

Deb, K., Sivaguru, M., Yul Yong, H., Roberts, M.: **Cdx2 gene expression and trophectoderm lineage specification in mouse embryos.** *Science* 2006, **311:** 992–996.

Hiiragi, T., Solter, D.: **First cleavage plane of the mouse egg is not predetermined but defined by the topology of the two apposing pronuclei.** *Nature* 2004, **430:** 360–364.

Hillman, N., Sherman, M.I., Graham, C.: **The effect of spatial arrangement on cell determination during mouse development.** *J. Embryol. Exp. Morph.* 1972, **28:** 263–278.

Lu, C.C., Brennan, J., Robertson, E.J.: **From fertilization to gastrulation: axis formation in the mouse embryo.** *Curr. Opin. Genet. Dev.* 2001, **11:** 384–392.

Perea-Gomez, A., Camus, A., Moreau, A., Grieve, K., Moneron, G., Dubois, A., Cibert, C., Collignon, J.: **Initiation of gastrulation in the mouse embryo is preceded by an apparent shift in the orientation of the antero-posterior axis.** *Curr. Biol.* 2004, **14:** 197–207.

Rana, A.A., Martinez Barbera, J.P., Rodriguez, T.A., Lynch, D., Hirst, E., Smith, J.C., Beddington, R.S.P.: **Targeted deletion of the novel cytoplasmic dynein mD2LIC disrupts the embryonic organiser, formation of body axes and specification of ventral cell fates.** *Development* 2004, **131:** 4999–5007.

Robertson, E.J., Norris, D.P., Brennan, J., Bikoff, E.K.: **Control of early anterior-posterior patterning in the mouse embryo by TGF-beta signalling.** *Phil. Trans. R. Soc. Lond. B Biol. Sci.* 2003, **358:** 1351–1357.

Rodriguez, T.A., Srinivas, S., Clements, M.P., Smith, J.C., Beddington, R.S.: **Induction and migration of the anterior visceral endoderm is regulated by the extra-embryonic ectoderm.** *Development* 2005, **132:** 2513–2520.

Srinivas, S., Rodriguez, T., Clements, M., Smith, J.C., Beddington, R.S.P.: **Active cell migration drives the unilateral movements of the anterior visceral endoderm.** *Development* 2004, **131:** 1157–1164.

Weber, R.J., Pedersen, R.A., Wianny, F., Evans, M.J., Zernicka-Goetz, M.: **Polarity of the mouse embryo is anticipated before implantation.** *Development* 1999, **126:** 5591–5596.

Zernicka-Goetz, M.: **Developmental cell biology: cleavage pattern and emerging asymmetry of the mouse embryo.** *Nat. Rev. Mol. Cell Biol.* 2005, **6:** 919–928.

3.10 The bilateral symmetry of the early embryo is broken to produce left–right asymmetry of internal organs

Brennan, J., Norris, D.P., Robertson, E.J.: **Nodal activity in the node governs left-right asymmetry.** *Genes Dev.* 2002, **16:** 2339–2344.

Brown, N.A., Wolpert, L.: **The development of handedness in left/right asymmetry.** *Development* 1990, **109:** 1-9.

Capdevila, J., Vogan, K.J., Tabin, C.J., Izpisúa Belmonte, J.C.: **Mechanisms of left-right determination in vertebrates.** *Cell* 2000, **101:** 9-21.

Jost, H.J.: **Diverse initiation in a conserved left/right pathway.** *Curr. Opin. Genet. Dev.* 1999, **9:** 422-426.

Levin, M.: **Left-right asymmetry in embryonic development: a comprehensive review.** *Mech. Dev.* 2005, **122:** 3-25.

Levin M.: **Motor protein control of ion flux is an early step in embryonic left-right asymmetry.** *BioEssays* 2003, **25:** 1002-1010.

McGrath, J., Somlo, S., Makova, S., Tian, X., Brueckner, M.: **Two populations of node monocilia initiate left-right asymmetry in the mouse.** *Cell* 2003, **114:** 61-73.

Raya, A., Izpisua Belmonte, J.C.: **Unveiling the establishment of left-right asymmetry in the chick embryo.** *Mech. Dev.* 2004, **121:** 1043-1054.

Raya, A., Kawakami, Y., Rodriguez-Esteban, C., Ibanes, M., Rasskin-Gutman, D., Rodriguez-Leon, J., Buscher, D., Feijo, J.A., Izpisua Belmonte, J.C.: **Notch activity acts as a sensor for extracellular calcium during vertebrate left-right determination.** *Nature* 2004, **427:** 121-128.

Ryan, A.K., Blumberg, B., Rodriguez-Esteban, C., Yonei-Tamura, S., Tamura, K., Tsukui, T., de la Pena, J., Sabbagh, W., Greenwald, J., Choe, S., Norris, D.P., Robertson, E.J., Evans, R.M., Rosenfeld, M.G., Izpisua Belmonte, J.C.: **Pitx2 determines left-right asymmetry of internal organs in vertebrates.** *Nature* 1998, **394:** 545-551.

Shimeld, S.M.: **Calcium turns sinister in left-right asymmetry.** *Trends Genet.* 2004, **20:** 277-280.

3.11 A fate map of the amphibian blastula is constructed by following the fate of labeled cells

Dale, L., Slack, J.M.W.: **Fate map for the 32 cell stage of *Xenopus laevis*.** *Development* 1987, **99:** 527-551.

Gerhart J.: **Changing the axis changes the perspective.** *Dev. Dyn.* 2002, **225:** 380-383.

Lane, M.C., Smith, W.C.: **The origins of primitive blood in *Xenopus*: implications for axial patterning.** *Development* 1999, **126:** 423-434.

3.12 The fate maps of vertebrates are variations on a basic plan

Beddington, R.S.P., Morgenstern, J., Land, H., Hogan, A.: **An *in situ* transgenic enzyme marker for the midgestation mouse embryo and the visualization of inner cell mass clones during early organogenesis.** *Development* 1989, **106:** 37-46.

Gardner, R.L., Rossant, J.: **Investigation of the fate of 4-5 day post-coitum mouse inner cell mass cells by blastocyst injection.** *J. Embryol. Exp. Morph.* 1979, **52:** 141-152.

Helde, K.A., Wilson, E.T., Cretehos, C.J., Grunwald, D.J.: **Contribution of early cells to the fate map of the zebrafish gastrula.** *Science* 1994, **265:** 517-520.

Kimmel, C.B., Warga, R.M., Schilling, T.F.: **Origin and organization of the zebrafish fate map.** *Development* 1990, **108:** 581-594.

Lawson, K.A., Meneses, J.J., Pedersen, R.A.: **Clonal analysis of epiblast fate during germ layer formation in the mouse embryo.** *Development* 1991, **113:** 891-911.

Smith, J.L., Gesteland, G.M., Schoenwolf, G.C.: **Prospective fate map of the mouse primitive streak at 7.5 days of gestation.** *Dev. Dyn.* 1994, **201:** 279-289.

Stern, C.D.: **The marginal zone and its contribution to the hypoblast and primitive streak of the chick embryo.** *Development* 1990, **109:** 667-682.

Stern, C.D., Canning, D.R.: **Origin of cells giving rise to mesoderm and endoderm in chick embryo.** *Nature* 1990, **343:** 273-275.

3.13 Cells of early vertebrate embryos do not yet have their fates determined and regulation is possible

Lewis, N.E., Rossant, J.: **Mechanism of size regulation in mouse embryo aggregates.** *J. Embryol. Exp. Morph.* 1982, **72:** 169-181.

Snape, A., Wylie, C.C., Smith, J.C., Heasman, J.: **Changes in states of commitment of single animal pole blastomeres of *Xenopus laevis*.** *Dev. Biol.* 1987, **119:** 503-510.

Tam, P.P., Rossant, J.: **Mouse embryonic chimeras: tools for studying mammalian development.** *Development* 2003, **130:** 6155-6163.

Wylie, C.C., Snape, A., Heasman, J., Smith, J.C.: **Vegetal pole cells and commitment to form endoderm in *Xenopus laevis*.** *Dev. Biol.* 1987, **119:** 496-502.

3.14 In *Xenopus* the endoderm and ectoderm are specified by maternal factors, but the mesoderm is induced from ectoderm by signals from the vegetal region

Dale, L.: **Vertebrate development: multiple phases to endoderm formation.** *Curr. Biol.* 1999, **9:** R812-R815.

Dupont, S., Zacchigna, L., Cordenonsi, M., Soligo, S., Adorno, M., Rugge, M., Piccolo, S.: **Germ-layer specification and control of cell growth by Ectodermin, a Smad4 ubiquitin ligase.** *Cell* 2005, **121:** 87-99.

Kimelman, D., Griffin, J.P.G.: **Vertebrate mesoderm induction and patterning.** *Curr. Opin. Genet. Dev.* 2000, **10:** 350-356.

Taverner, N.V., Kofron, M., Shin, Y., Kabitschke, C., Gilchrist, M.J., Wylie, C., Cho, K.W., Heasman, J., Smith, J.C.: **Microarray-based identification of VegT targets in *Xenopus*.** *Mech. Dev.* 2005, **122:** 333-354.

Xanthos, J.B., Kofron, M., Wylie, C., Heasman, J.: **Maternal VegT is the initiator of a molecular network specifying endoderm in *Xenopus laevis*.** *Development* 2001, **128:** 167-180.

3.15 Mesoderm induction occurs during a limited period in the blastula stage

Gurdon, J.B., Lemaire, P., Kato, K.: **Community effects and related phenomena in development.** *Cell* 1993, **75:** 831-834.

3.16 Zygotic gene expression is turned on at the mid-blastula transition

Davidson, E.: *Gene Activity In Early Development*. New York: Academic Press, 1986.

Yasuda, G.K., Schübiger, G.: **Temporal regulation in the early embryo: is MBT too good to be true?** *Trends Genet.* 1992, **8:** 124-127.

3.17 Mesoderm-inducing and patterning signals are produced by the vegetal region, the organizer, and the ventral mesoderm

De Robertis, E.M., Larrain, J., Oelgeschläger, M., Wessely, O.: **The establishment of Spemann's organizer and patterning of the vertebrate embryo.** *Nat. Rev. Genet.* 2000, **1:** 171-181.

Kofron, M., Demel, T., Xanthos, J., Lohr, J., Sun, B., Sive, H., Osada, S-I., Wright, C., Wylie, C., Heasman, J.: **Mesoderm induction in *Xenopus* is a zygotic event regulated by maternal VegT via TGFβ growth factors.** *Development* 1999, **126:** 5759-5770.

Slack, J.M.W.: **Inducing factors in *Xenopus* early embryos.** *Curr. Biol.* 1994, **4:** 116–126.

3.18 Members of the TGF-β family have been identified as mesoderm inducers

Amaya, E., Musci, T.J., Kirschner, M.W.: **Expression of a dominant negative mutant of the FGF receptor disrupts mesoderm formation in *Xenopus* embryos.** *Cell* 1991, **66:** 257–270.

Massagué, J.: **How cells read TGF-β signals.** *Nat. Rev. Mol. Cell Biol.* 2000, **1:** 169–178.

Schier, A.F.: **Nodal signaling in vertebrate development.** *Annu. Rev. Cell. Dev. Biol.* 2003, **19:** 589–621.

Schier, A.F., Talbot, W.S.: **Nodal signaling and the zebrafish organizer.** *Int. J. Dev. Biol.* 2001, **45:** 289–297.

3.19 The dorso-ventral patterning of the mesoderm involves the antagonistic actions of dorsalizing and ventralizing factors

Dale, L.: **Pattern formation: A new twist to BMP signalling.** *Curr. Biol.* 2000, **10:** R671–R673.

Gonzalez, E.M., Fekany-Lee, K., Carmany-Rampey, A, Erter, C, Topczewski, J., Wright, C.V., Solnica-Krezel, L.: **Head and trunk in zebrafish arise via coinhibition of BMP signaling by bozozok and chordino.** *Genes Dev.* 2000, **14:** 3087–3092.

Leyns, L., Bouwmeester, T., Kim, S-H., Piccolo, S., De Robertis, E.M.: **Frzb-1 is a secreted antagonist of Wnt signaling expressed in the Spemann organizer.** *Cell* 1997, **88:** 747–756.

Moon, R.T., Brown, J.D., Yang-Snyder, J.A., Miller, J.R.: **Structurally related receptors and antagonists compete for secreted Wnt ligands.** *Cell* 1997, **88:** 725–728.

Oelgeschläger, M., Larrain, J., Geissert, D., De Robertis, E.M.: **The evolutionarily conserved BMP-binding protein Twisted gastrulation promotes BMP signalling.** *Nature* 2000, **405:** 757–763.

Piccolo, S., Sasai, Y., Lu, B., De Robertis, E.M.: **Dorsoventral patterning in *Xenopus*: inhibition of ventral signals by direct binding of chordin to BMP-4.** *Cell* 1996, **86:** 589–598.

Piepenburg, O., Grimmer, D., Williams, P.H., Smith, J.C.: **Activin redux: specification of mesodermal pattern in *Xenopus* by graded concentrations of endogenous activin B.** *Development* 2004, **131:** 4977–4986.

Schier, A.F.: **Axis formation and patterning in zebrafish.** *Curr. Opin. Genet. Dev.* 2001, **11:** 393–404.

Smith, J.: **Angles on activin's absence.** *Nature* 1995, **374:** 311–312.

Zimmerman, L.B., De Jesús-Escobar, J.M., Harland, R.M.: **The Spemann organizer signal noggin binds and inactivates bone morphogenetic protein 4.** *Cell* 1996, **86:** 599–606.

3.20 Mesoderm induction and patterning in the chick and mouse occurs during primitive-streak formation

Chu, G.C., Dunn, N.R., Anderson, D.C., Oxburgh, L., Robertson, E.J.: **Differential requirements for Smad4 in TGFbeta-dependent patterning of the early mouse embryo.** *Development* 2004, **131:** 3501–3512.

Dunn, N.R., Vincent, S.D., Oxburgh, L., Robertson, E.J., Bikoff, E.K.: **Combinatorial activities of Smad2 and Smad3 regulate mesoderm formation and patterning in the mouse embryo.** *Development* 2004, **131:** 1717–1728.

Lu, C.C., Robertson, E.J.: **Multiple roles for Nodal in the epiblast of the mouse embryo in the establishment of anterior-posterior patterning.** *Dev. Biol.* 2004, **273:** 149–159.

Zakin, L., Reversade, B., Kuroda, H., Lyons, K.M., De Robertis, EM.: **Sirenomelia in Bmp7 and Tsg compound mutant mice: requirement for Bmp signaling in the development of ventral posterior mesoderm.** *Development* 2005, **132:** 2489–2499.

3.21 Gradients in signaling proteins and threshold responses could pattern the mesoderm

Chen, Y., Schier, AF.: **The zebrafish Nodal signal Squint functions as a morphogen.** *Nature* 2001, **411:** 607–609.

Green, J.B.A., New, H.V., Smith, J.C.: **Responses of embryonic *Xenopus* cells to activin and FGF are separated by multiple dose thresholds and correspond to distinct axes of the mesoderm.** *Cell* 1992, **71:** 731–739.

Gurdon, J.B., Standley, H., Dyson, S., Butler, K., Langon, T., Ryan, K., Stennard, F., Shimizu, K., Zorn, A.: **Single cells can sense their position in a morphogen gradient.** *Development* 1999, **126:** 5309–5317.

Jones, C.M., Armes, N., Smith, J.C.: **Signaling by TGF-β family members: short-range effects of Xnr-2 and BMP-4 contrast with the long-range effects of activin.** *Curr. Biol.* 1996, **6:** 1468–1475.

Papin, C., Smith, J.C.: **Gradual refinement of activin-induced thresholds requires protein synthesis.** *Dev. Biol.* 2000, **217:** 166–172.

Schulte-Merker, S., Smith, J.C.: **Mesoderm formation in response to *Brachyury* requires FGF signalling.** *Curr. Biol.* 1995, **5:** 62–67.

Tada, M., Smith, J.C.: ***Xwnt11* is a target of Brachyury: regulation of gastrulation movements via Dishevelled, but not through the canonical Wnt pathway.** *Development* 2000, **127:** 2227–2228.

Vincent, S.D., Dunn, N.R., Hayashi, S., Norris, D.P., Robertson, E.J.: **Cell fate decisions within the mouse organizer are governed by graded Nodal signals.** *Genes Dev.* 2003, **17:** 1646–1662.

Patterning the vertebrate body plan II: the somites and early nervous system

4

- Somite formation and antero-posterior patterning
- The role of the organizer and neural induction

During and after gastrulation, the vertebrate embryo becomes patterned along the antero-posterior and dorso-ventral axes. This patterning is carried out by a combination of signals from various regions of the embryo and the interpretation of positional identity by cells. The expression of genes involved in encoding position along the antero-posterior axis is key to this patterning, as is the mechanism by which their spatial patterns of expression are initially specified. Key morphogenetic events at this stage are the formation of the somites, which give rise to muscles, skeleton, and dermis, by segmentation of blocks of mesoderm along the antero-posterior axis. The nervous system is induced in the dorsal ectoderm by signals from adjacent tissues, particularly the organizer region, and this too raises interesting problems, which include the possible similarities in the four model vertebrates.

In Chapter 3, we saw how the body axes are set up and how the three germ layers are initially specified in vertebrate embryos. Although amphibian, fish, chick, and mouse embryos share some features at these stages, there are many significant differences. As we approach the phylotypic stage—the embryonic stage common to all vertebrates (see Fig. 3.2)—the similarities between vertebrate embryos become greater, and so we can consider the patterning of the vertebrate body plan in a general way.

During gastrulation, the germ layers—mesoderm, endoderm, and ectoderm—move to the positions in which they will develop into the structures of the larval or adult body. The antero-posterior body axis of the vertebrate embryo emerges clearly, with the head at one end and the future tail at the other (Fig. 4.1). In this chapter, we focus mainly on the formation and patterning of the mesoderm that forms the **somites**, the blocks of cells that give rise to the skeleton and muscles of the trunk, and on the patterning of the ectoderm that will develop into the nervous system. The cell movements of gastrulation and the action of the organizer region are crucial to establishing the vertebrate body plan and will be discussed in this chapter in relation to their role in the patterning processes. A detailed discussion of the movements of cells and tissues during gastrulation is, however, deferred to Chapter 7.

At gastrulation, the part of the mesoderm that comes to lie along the dorsal midline of the embryo, under the ectoderm, gives rise to the notochord and somites, and to head mesoderm anterior to the notochord. After gastrulation in *Xenopus* the internalized cells of the dorsal-most mesoderm (the organizer region)

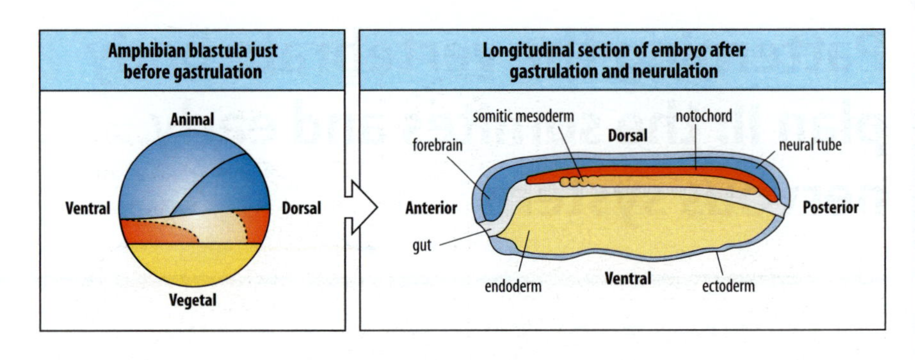

Fig. 4.1 Rearrangement of the presumptive germ layers during gastrulation and neurulation in *Xenopus*. The mesoderm (pink and red on the left), which is in an equatorial band at the blastula stage, moves inside to give rise to the notochord (red on the right), somites (orange), and lateral mesoderm (not shown). The endoderm (yellow) moves inside to line the gut. The neural tube (dark blue) forms and the ectoderm (light blue) covers the whole embryo. The antero-posterior axis emerges, with the head at the anterior end.

have formed a rigid rod-like notochord along the dorsal midline, flanked on each side by mesoderm, which is beginning to form blocks of somites (see Fig. 4.1). In the chick embryo, the notochord forms in the dorsal midline anterior to Hensen's node, from the axial mesoderm left behind as the node and primitive streak regress (Fig. 4.2), and somites are formed from mesoderm on either side of it. Notochord and somite formation proceed similarly in the mouse (see Figs 3.24 and 3.25). Each region, such as an individual somite, now develops largely independently, and can be considered as a developmental module, rather in the same way as the individual segments in *Drosophila*. The vertebrate notochord is a transient structure, and its cells eventually become incorporated into the vertebrae.

As gastrulation proceeds, neurulation begins. The neural tube forms from the ectoderm overlying the notochord and the somites become positioned on either side of it. The internal structure of the *Xenopus* embryo just after the end of neurulation is illustrated in Fig. 4.3. The main structures that can be recognized at this stage are the neural tube, the notochord, the somites, the lateral plate mesoderm, and the endoderm lining the gut. By this stage, different parts of the somites can be distinguished.

Both the mesodermal structures along the antero-posterior axis of the trunk and the ectodermally derived nervous system have a distinct antero-posterior organization. Some of this antero-posterior patterning is controlled by genes called the Hox genes, which are the vertebrate equivalent of the Hox genes responsible for antero-posterior patterning in *Drosophila* (see Chapter 2). In the first part of this chapter we describe somite development and the role of the Hox genes in the antero-posterior patterning of the somitic mesoderm, as shown by their effects on the spinal column, whose vertebrae derive from the somites. In the second part of the chapter we consider the function of the vertebrate embryonic organizer in establishing the antero-posterior organization of the embryo, focusing on its role in the induction of the nervous system. We will also look at the early antero-posterior patterning of the hindbrain, in which Hox genes are again involved.

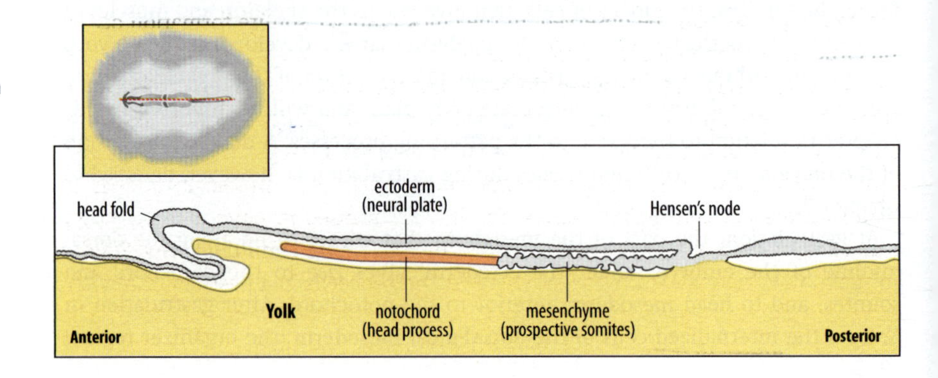

Fig. 4.2 Notochord and head-fold formation in the chick embryo. The diagram shows a sagittal section through the chick embryo (inset, dorsal view) at the stage of head-fold formation as Hensen's node starts to regress. As the node regresses, the notochord (sometimes called the head process at this stage) starts to form anterior to it. The undifferentiated mesenchyme on either side of the notochord will form somites.

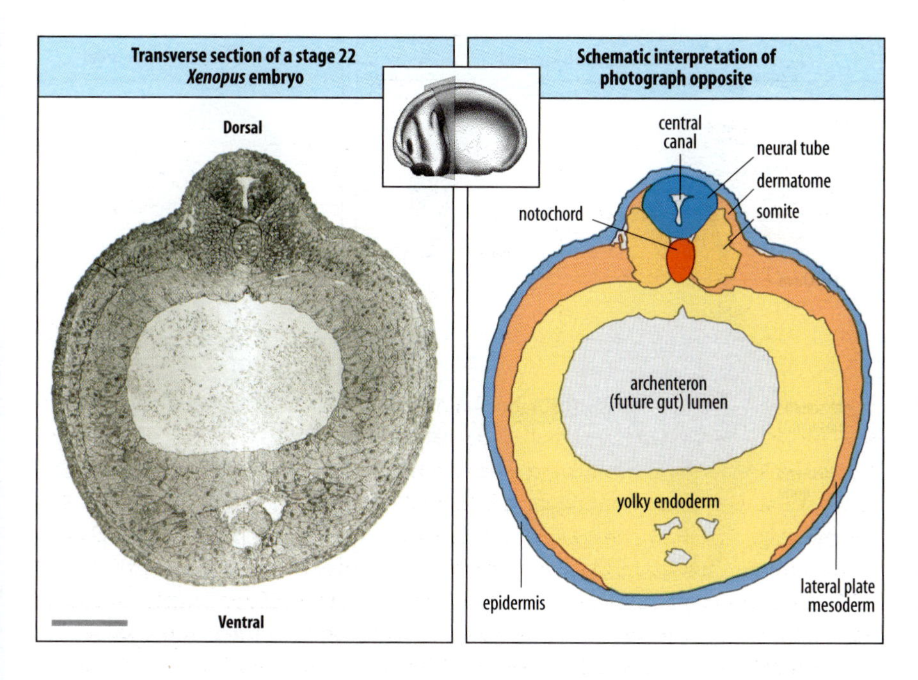

Fig. 4.3 A cross-section through a stage 22 *Xenopus* embryo just after gastrulation and neurulation are completed. The germ layers are now all in place for future development and organogenesis. The most dorsal parts of the somites have already begun to differentiate into the dermatome, which will give rise to the dermis. Scale bar = 0.2 mm.

Photograph from Hausen, P. and Riebesell, M.: 1991.

Somite formation and antero-posterior patterning

The fate maps of the various vertebrates (see Fig. 3.48) show that the notochord develops from the most dorsal region of the mesoderm, and somites from more ventro-lateral mesoderm on either side of the midline, which is known as the **paraxial mesoderm**. The somites give rise to the bone and cartilage of the trunk, the skeletal muscles, and the dermis of the skin on the dorsal side of the body, and their patterning provides much of the body's antero-posterior organization. The vertebrae, for example, have characteristic shapes at different positions along the spine. We will first examine the development of the somites and how their different developmental identities along the antero-posterior axis are specified. We conclude our look at somite formation by discussing how individual somites are patterned and which structures they give rise to.

4.1 Somites are formed in a well defined order along the antero-posterior axis

Somite number can be highly variable between different vertebrates; birds and humans have around 50 whereas snakes have up to several hundred. Our main model organism in the discussion of somite formation will be the chick. Much of the work on somite formation has been done on the chick because of the ease with which the process can be observed. In the chick embryo, somite formation occurs on either side of the notochord in the lateral mesoderm anterior to the regressing Hensen's node (Fig. 4.4). Between the node and the most recently formed somite, there is an unsegmented region—the **pre-somitic mesoderm**—which segments altogether into about 12 somites in the chick embryo. Changes in cell shape and intercellular contacts in the pre-somitic mesoderm result in the formation of distinct blocks of cells—the somites. Somites are formed in pairs, one on either side of the notochord, with each pair of somites forming simultaneously. Somite formation begins at the anterior or head end and proceeds in a posterior direction. A somite forms every 90 minutes in the chick, every 120 minutes in the mouse, every 45 minutes in *Xenopus*, and every 30 minutes in zebrafish.

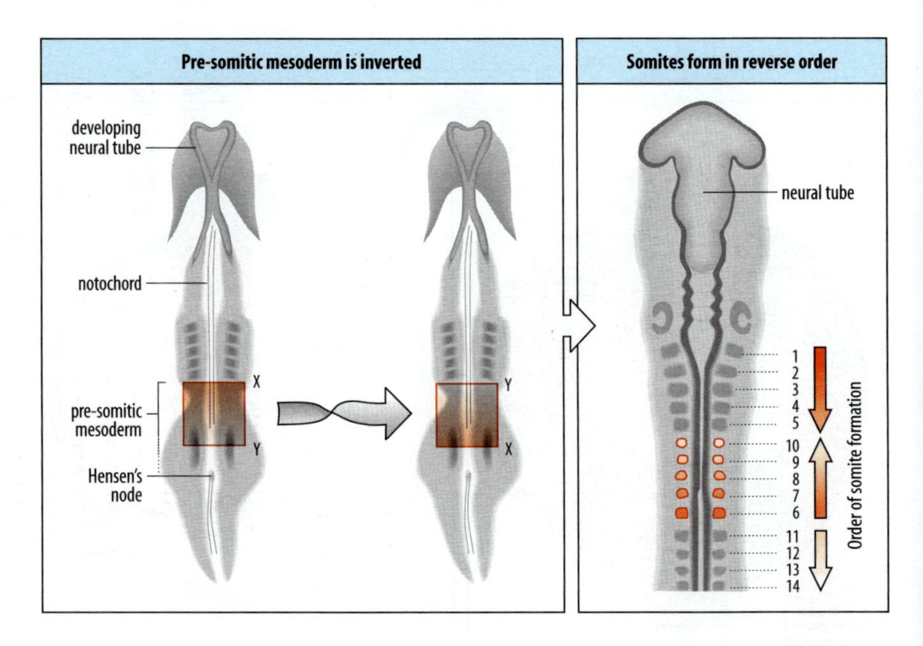

Fig. 4.4 The temporal order of somite formation is specified early in embryonic development. Somite formation in the chick proceeds in an antero-posterior direction. Somites form sequentially in the pre-somitic region between the last-formed somite and Hensen's node, which moves posteriorly. If the antero-posterior axis of the pre-somitic mesoderm is rotated through 180°, as shown by the arrow, the temporal order of somite formation is not altered—somite 6 still develops before somite 10.

The actual sculpting of the somites is quite a complex process and involves tissue separation and cell movements, as well as integration of cells at the anterior and posterior somite borders. Somite formation in *Xenopus*, in particular, displays several unique features. A group of cells of the unsegmented mesoderm, which are oriented perpendicular to the notochord, separates from the rest of the mesoderm by formation of an intersomitic furrow and forms a distinct block—the prospective somite. The block then undergoes a 90° rotation, which involves a series of cellular movements and reorientations that have rarely been investigated, but which appear to depend on a cell's position within the prospective somite. Cells sense their locations within the block and undergo morphological changes and cellular rearrangements according to their positions.

In the chick, the cells that give rise to the somites originate in the epiblast on either side of the anterior primitive streak and move into it at gastrulation to form a population of somitogenic stem cells around Hensen's node. The stem cells divide, and those that remain in the stem-cell region continue to be self-renewing stem cells (see Section 1.17), but those that leave the region as the node regresses form the pre-somitic mesoderm. As new cells are being added to the pre-somitic mesoderm at the posterior end of the chick embryo, somites are forming at the anterior end.

The sequence of somite formation in the unsegmented region is unaffected by transverse cuts in the plate of pre-somitic mesoderm, suggesting that somite formation is an autonomous process and that, at this time, no extracellular signal specifying antero-posterior position or timing is involved. Even if a small piece of the unsegmented mesoderm is rotated through 180°, each somite still forms at the normal time, but with the sequence of formation running in the opposite direction to normal in the inverted tissue (see Fig. 4.4). So, before somite formation begins, a molecular pattern that specifies the time of formation of each somite has already been laid down in the pre-somitic mesoderm, and we shall return to this patterning process later. Given the existence of this pattern, the prospective identity of each somite is related to the temporal order in which their cells entered the pre-somitic mesoderm.

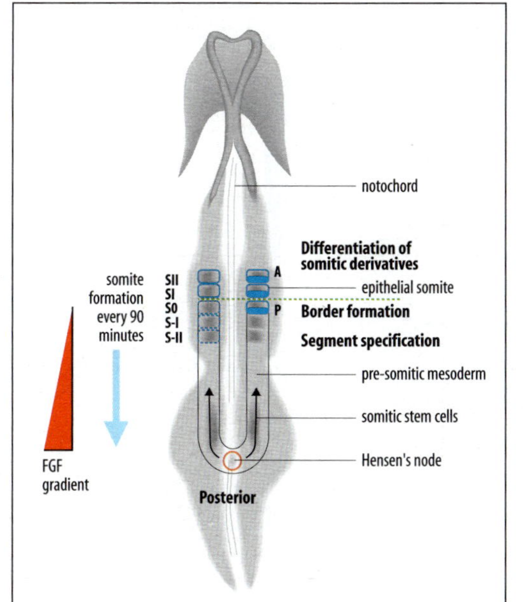

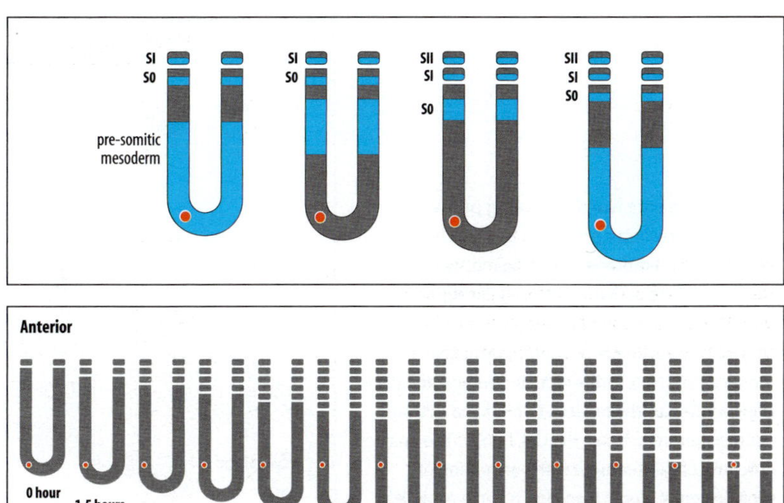

Fig. 4.5 Somite formation in the chick. As shown in the left-hand panel, the somites are generated successively from pre-somitic mesoderm, which is derived from somitic stem cells in the primitive streak. As pre-somitic cells are released into the posterior pre-somitic mesoderm, a new pair of somites buds from the anterior end every 90 minutes. For clarity, in this diagram general somite formation is shown for the left-hand somites while the somites on the right show how each somite becomes internally patterned during its formation. SI, the most recently formed somite; SII, the last but one somite formed; S0, somite in the process of formation, whose boundaries are not yet set; S-I, S-II, blocks of pre-somitic cells that will form somites. The border between each somite is specified by the FGF-8 gradient. At formation, each somite acquires antero-posterior polarity, after which it can respond to the signals that pattern it along the antero-posterior and dorso-ventral axes. The top right-hand panel shows stages in one of the cycles of *c-hairy 1* expression (blue) that sweep from posterior to anterior of the pre-somitic mesoderm every 90 minutes. During each cycle, a given pre-somitic cell (red dot) experiences distinct phases when *c-hairy 1* is expressed and when it is not expressed. The lower right-hand panel shows the progress of a pre-somitic cell (red dot) from the time it enters the pre-somitic mesoderm until it is incorporated into a somite. Somitic cells in the anterior somites will have experienced fewer cycles of *c-hairy 1* expression before they leave the node region for the pre-somitic mesoderm than will cells in posterior somites, and this could define a clock that is both linked to somite segmentation and 'tells' the somite its position along the antero-posterior axis.

Somite formation is largely determined by an internal 'clock' in the pre-somitic mesoderm. This clock is represented by periodic cycles of gene expression, such as that of the gene *c-hairy 1* in the chick embryo, whose expression sweeps from the posterior to the anterior end of the pre-somitic mesoderm with a period of 90 minutes, the time it takes for a pair of somites to form. In a newly formed somite, *c-hairy 1* expression becomes restricted to the posterior end of the somite, where it persists, while a new wave of *c-hairy 1* expression starts at the tail end of the pre-somitic mesoderm (Fig. 4.5). The connection between these oscillations and somite formation is not yet clear, but one of the proteins whose expression cycles is Lunatic fringe, which potentiates activity of the Notch–Delta signaling pathway (Fig. 4.6). This pathway is widely involved in determining cell fate and delimiting boundaries, and is involved in setting the somite boundaries. It is an example of the transmission of signals by direct cell–cell contact, as both Notch and Delta are transmembrane proteins. Mice mutant for the Delta–Notch pathway often do not form somites, and if they do, the somites vary in size and are different on each side of the body. In zebrafish, expression of genes of the *hairy* family also oscillates and models for the oscillation based on feedback inhibition have been proposed.

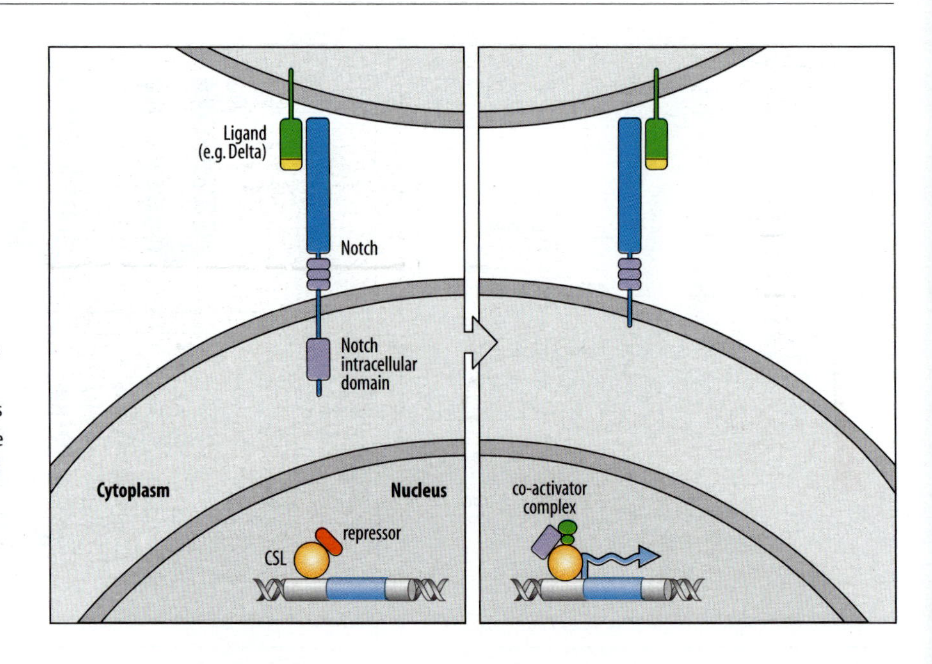

Fig. 4.6 The core Notch signaling pathway. Binding of a receptor protein of the Notch family to its membrane-bound ligand such as Delta or Serrate activates intracellular signaling from the receptor. This is thought to involve the enzymatic cleavage of the cytoplasmic tail of the receptor (Notch intracellular domain) and its translocation to the nucleus to bind and activate a transcription factor of the CSL family. The final outcome of this pathway is the activation of specific genes. Notch signaling is very versatile. In different organisms and in different developmental circumstances, activation of Notch switches different genes on or off.

The timing and position of somite formation is determined by the interaction of the segmentation clock with the growth factor FGF-8. In the chick and the mouse this forms a gradient in both mesoderm and ectoderm with its high point at the node. The gradient also regresses in a posterior direction along with the node. The reason for this dynamic behavior is that *FGF-8* mRNA is only made in cells in and around the node and is progressively degraded in cells that leave the region as the node moves posteriorly. This results in a gradient in intracellular *FGF-8* mRNA, which is translated and secreted to give an extracellular head-to-tail FGF-8 gradient with its eventual high point in the tailbud of the embryo (Fig. 4.7). Somite formation occurs where the level of FGF-8 is at a sufficiently low threshold. As the gradient moves, therefore, it successively defines regions in the pre-somitic mesoderm where somites start to form (see Fig. 4.5). There is also a gradient of retinoic acid, a small, secreted, signaling molecule derived from vitamin A, in the opposite direction, which antagonizes the FGF gradient. Retinoic acid thus keeps the pre-somitic region from continually getting longer. The retinoic acid is synthesized in the somites and diffuses both posteriorly and anteriorly. It is also involved in maintaining the bilateral symmetry of somite development by, in some way not yet

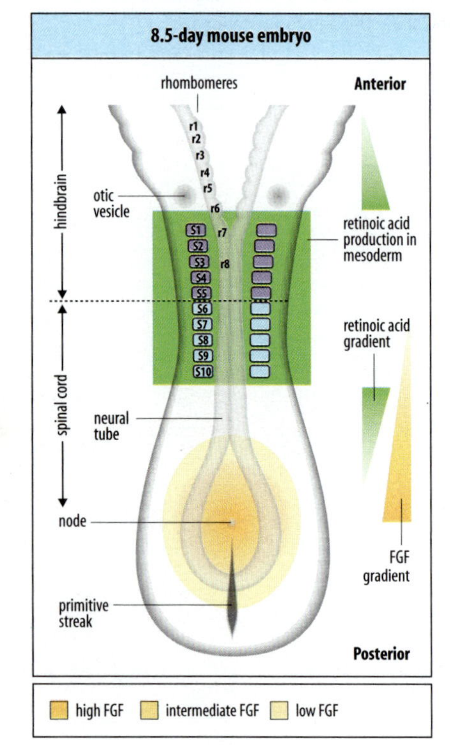

Fig. 4.7 FGF and retinoic acid gradients help to pattern the antero-posterior axis in the mouse embryo. An antero-posterior gradient of FGF develops in the embryo, with its high point at the node. This schematic shows the dorsal view of a 8.5-day mouse embryo with 10 pairs of somites formed and the neural tube partly closed but still open in the region of the future brain (r1, r2, etc., are the rhombomeres of the hindbrain) and at the posterior end of the future spinal cord. The FGF gradient is formed by cells in and around the node synthesizing *FGF* mRNA, which is then gradually degraded when cells leave the region anteriorly to form the pre-somitic mesoderm. This results in a gradient of translated and secreted FGF protein that is continuously moving posteriorly as the embryo elongates. The gradient in the mesoderm fades out at around the position of the last-formed somite, suggesting that somites are formed when the FGF level drops to some low threshold. Retinoic acid synthesized and secreted by the somites forms an opposing gradient that antagonizes FGF. Retinoic acid may also help to pattern the embryo by switching on the expression of specific genes. It also diffuses in an anterior direction from its site of synthesis in the somites and may help to pattern the future hindbrain.

Adapted from Deschamps, J. and Van Nes, J.: 2005.

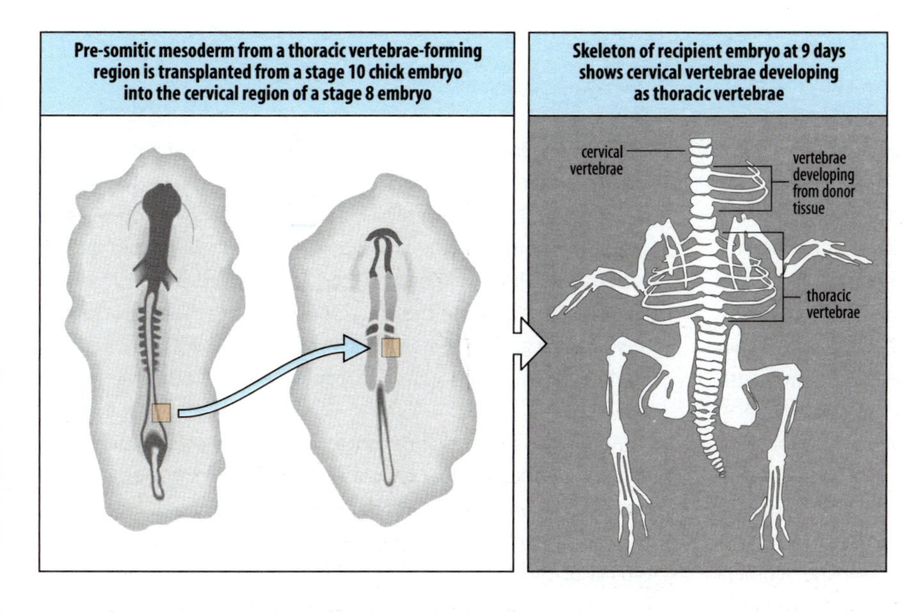

Fig. 4.8 The pre-somitic mesoderm has acquired a positional identity before somite formation. Pre-somitic mesoderm that will give rise to thoracic vertebrae is grafted to an anterior region of a younger embryo that would normally develop into cervical vertebrae. The grafted mesoderm develops according to its original position and forms ribs in the cervical region.

understood, buffering the somites against the signals that are setting up left-right asymmetry in the lateral mesoderm (see Chapter 3).

Somites differentiate into particular axial structures depending on their position along the antero-posterior axis. The anterior-most somites contribute to the skull, those posterior to them will form cervical vertebrae, and more posterior ones will develop as thoracic vertebrae with ribs. Specification by position has occurred before somite formation begins during gastrulation: if unsegmented somitic mesoderm from, for example, the presumptive thoracic region of the chick embryo is grafted to replace the presumptive mesoderm of the neck region, it will still form thoracic vertebrae with ribs (Fig. 4.8). How then is the pre-somitic mesoderm patterned so that somites acquire their identity and form particular vertebrae?

4.2 Identity of somites along the antero-posterior axis is specified by Hox gene expression

The antero-posterior patterning of the mesoderm is most clearly seen in the differences in the vertebrae, each vertebra having well defined anatomical characteristics depending on its location along the axis. The most anterior vertebrae are specialized for attachment and articulation of the skull, while the cervical vertebrae of the neck are followed by the rib-bearing thoracic vertebrae and then those of the lumbar region, which do not bear ribs, and finally, those of the sacral and caudal regions. This antero-posterior patterning occurs early, while the pre-somitic mesoderm is still unsegmented. Patterning of the skeleton along the body axis is based on the mesodermal cells acquiring a positional value that reflects their position along the axis and so determines their subsequent development. Mesodermal cells that will form thoracic vertebrae, for example, have different positional values from those that will form cervical vertebrae.

Patterning along the antero-posterior axis in all vertebrates involves the expression of a set of genes that specify positional identity along the axis. These are the Hox genes, members of the large family of homeobox genes that are involved in many aspects of development (Box 4A, p. 156). The concept of positional identity,

Box 4A The Hox genes

The Hox genes of vertebrates belong to a large group of gene-regulatory proteins that all contain a similar DNA-binding region of around 60 amino acids known as the **homeodomain**, which contains a helix-turn-helix DNA-binding motif that is characteristic of many DNA-binding proteins. The homeodomain is encoded by a DNA motif of around 180 base pairs termed the **homeobox**, a name that came originally from the fact that this gene family was discovered through mutations that produce a homeotic transformation—a mutation in which one structure replaces another. For example, in one homeotic mutation in *Drosophila*, a segment in the fly's body that does not normally bear wings is transformed to resemble the adjacent segment that does bear wings, resulting in a fly with four wings instead of two.

Clusters of homeotic genes involved in specifying segment identity were first discovered in *Drosophila*. There is one Hox gene cluster in *Drosophila* (known as HOM-C), which is organized into two distinct gene complexes, the Antennapedia complex and the bithorax complex. Similar clusters of homeotic genes have been identified in many animals. In vertebrates, the clusters are known as the Hox complexes, and the homeoboxes of the genes are related in sequence to the homeobox of genes of the Antennapedia complex in *Drosophila*. In each Hox cluster the order of the genes from 3′ to 5′ in the DNA is the order in which they are expressed along the antero-posterior axis and specify positional identity. In the mouse, there are four unlinked Hox complexes, designated Hoxa, Hoxb, Hoxc, and Hoxd (originally called Hox1, Hox2, Hox3, and Hox4), located on chromosomes 6, 11, 15, and 2, respectively (see figure). The vertebrate clusters have arisen by duplication of an ancestral cluster, possibly related to the single Hox cluster in the lancelet (amphioxus), a simple chordate. All Hox genes thus resemble each other to some extent; the homology is most marked within the homeobox and less marked in sequences outside it. Genes that have arisen by duplication and divergence within a species are known as **paralogs**, and the corresponding genes in the different clusters (e.g. *Hoxa4, Hoxb4, Hoxc4, Hoxd4*) are usually known as a **paralogous subgroup**. In the mouse there are 13 paralogous groups.

The Hox gene clusters and their role in development are of ancient origin. The mouse and frog genes are similar to each other and to those of *Drosophila*, both in their coding sequences and in their order on the chromosome. In both *Drosophila* and vertebrates, these homeotic genes are involved in specifying regional identity along the antero-posterior axis. The Hox clusters in mice and in *Drosophila* almost certainly arose by gene duplication in some common ancestor of vertebrates and insects.

Most genes that contain a homeobox do not, however, belong to a homeotic complex, nor are they involved in homeotic transformations. Other subfamilies of homeobox genes in vertebrates include the **Pax genes**, which contain a homeobox typical of the *Drosophila* gene *paired*. All these genes encode transcription factors with various functions in development and cell differentiation.

Illustration after Coletta, P., et al.: 1994.

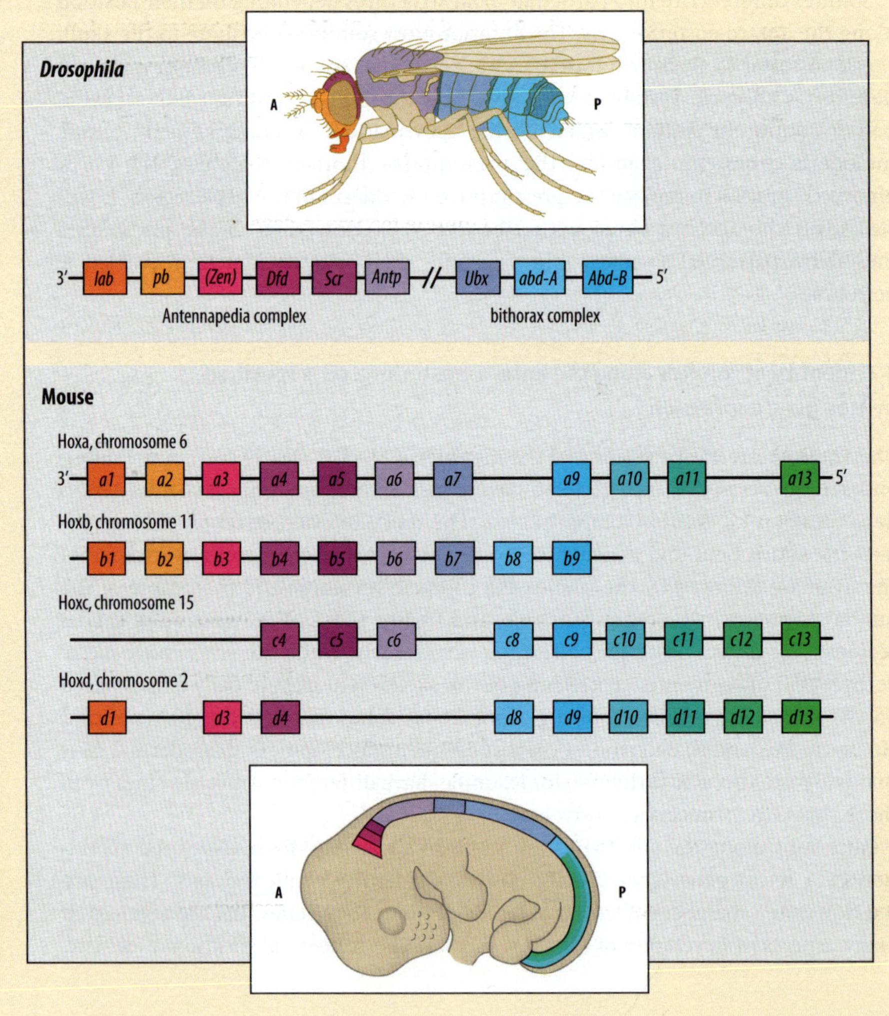

or positional value, has important implications for developmental strategy; it implies that a cell or a group of cells in the embryo acquires a unique state related to its position at a given time, and that this determines its later development (see Section 1.15).

Homeobox genes that specify positional identity along the antero-posterior axis were originally identified in *Drosophila* (see Chapter 2) and it turned out that related genes are involved in patterning the vertebrate axis. As we shall see in the final part of this chapter, patterning along the antero-posterior axis by Hox genes and other homeobox genes is not confined to mesodermal structures; the hindbrain, for example, is also divided into distinct regions. The homeobox genes are the most striking example of a widespread conservation of developmental genes in animals. It is widely believed that there are common mechanisms underlying the development of all animals. This implies that if a gene is identified as having a central role in the development of one animal, it is worth looking to see whether it is present in another animal and whether it has a similar function. This strategy of comparing genes by sequence homology has proved extremely successful in identifying genes involved in development in vertebrates. Numerous genes first identified in *Drosophila*, in which the genetic basis of development is far better understood than in any other animal, have proved to have counterparts involved in development in vertebrates.

All the homeobox genes whose functions are known encode transcription factors. The subset known as the Hox genes are the vertebrate counterparts of a cluster of homeobox genes in *Drosophila* that are involved in specifying the identities of the different segments of the insect body (see Chapter 2). Most vertebrates have four separate clusters of Hox genes that are thought to have arisen by duplications of the genes within a cluster, and of the clusters themselves (see Box 4A, opposite). The zebrafish is unusual in having seven clusters, as a result of further duplication. A particular feature of Hox gene expression in both insects and vertebrates is that the genes in each cluster are expressed in a temporal and spatial order that reflects their order on the chromosome. This is a unique feature in development, as it is the only known case where a spatial pattern of genes on a chromosome corresponds to a spatial pattern in the embryo.

A simple idealized model illustrates the key features by which a Hox gene cluster records positional identity. Consider four genes—I, II, III, and IV—arranged along a chromosome in that order (Fig. 4.9). The genes are expressed in a corresponding order along the antero-posterior axis of a tissue. Thus, gene I is expressed throughout the tissue with its anterior boundary at the anterior end. Gene II has its anterior boundary in a more posterior position and expression continues posteriorly. The same principles apply to the two other genes. This pattern of expression defines four distinct regions, coded for by the expression of different combinations of genes. If the amount of gene product is varied within each expression domain, for example by interactions between the genes, many more regions can be specified.

The role of the Hox genes in vertebrate axial patterning has been best studied in the mouse, because it is possible to either knock out particular Hox genes or to alter their expression (Box 4B, pp. 158–159). As in all vertebrates, the Hox genes start to be expressed in mesoderm cells at an early stage of gastrulation when they begin to move away from the primitive streak and towards the anterior. The 'anterior' genes are expressed first. As the posterior pattern develops later, clearly defined patterns of Hox gene expression are most easily seen in the mesoderm and the neural tube after somite formation and neurulation, respectively (Fig. 4.10). More Hox genes are expressed as gastrulation proceeds. Typically, the pattern of

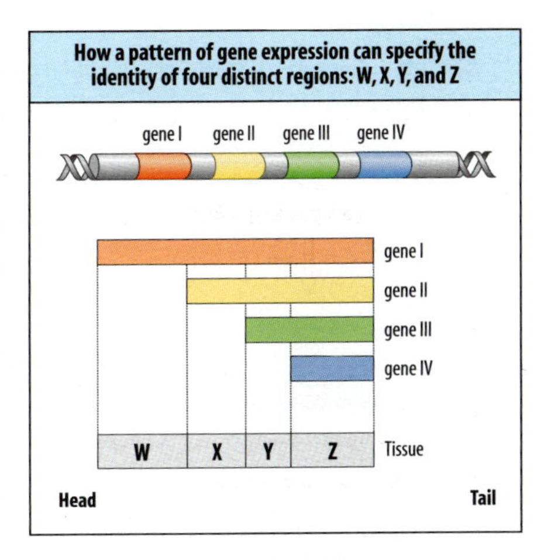

Fig. 4.9 Gene activity can provide positional values. The model shows how the pattern of gene expression along a tissue can specify the distinct regions W, X, Y, and Z. For example, only gene I is expressed in region W but all four genes are expressed in region Z.

Box 4B Gene targeting: insertional mutagenesis and gene knock-out

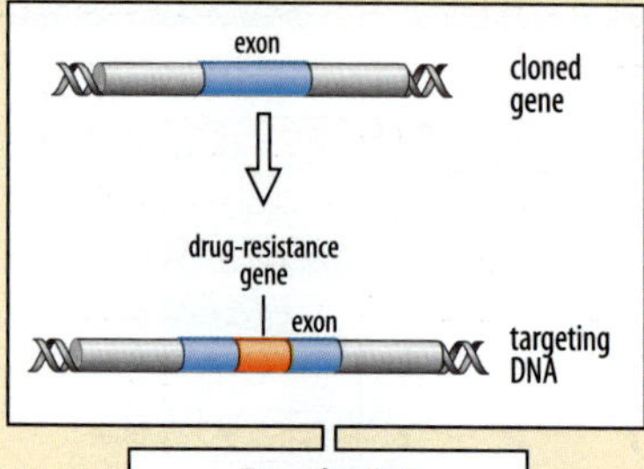

Targeting DNA introduced into ES cells

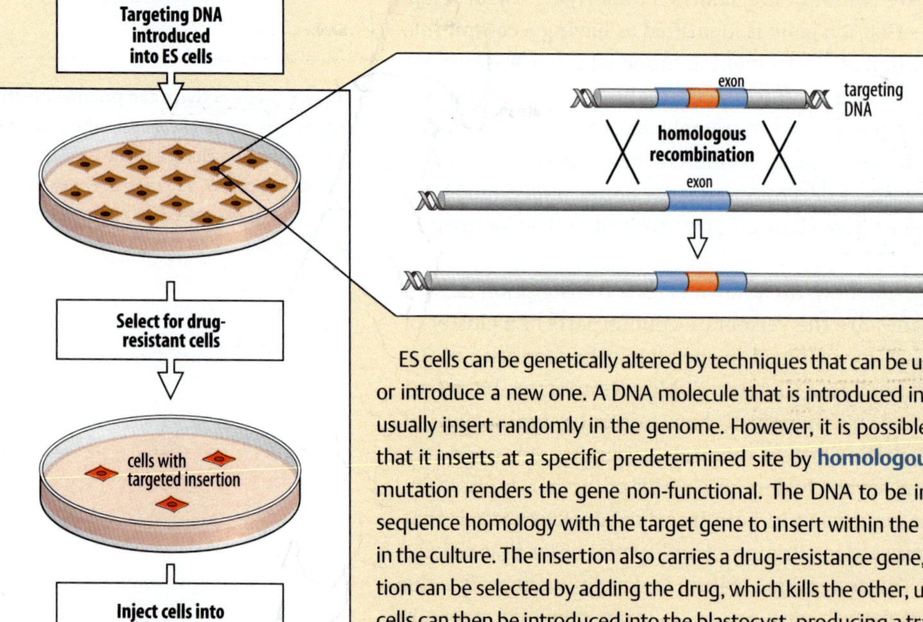

Select for drug-resistant cells

cells with targeted insertion

Inject cells into blastocyst

Chimera
(mutant gene in germ line)

To study the function of a gene controlling development, it is highly desirable to be able to introduce an altered version of the gene into the animal to see what effect it has. Mice into which an additional or altered gene has been introduced are known as transgenic mice (see Box 3E, p. 130). Two main techniques for generating transgenic mice are currently in use. One is to inject DNA containing the required gene directly into the nuclei of fertilized eggs; the other is to alter or add a gene to the genome of embryonic stem cells (ES cells) in culture, and then to inject the genetically altered cells into the blastocyst, where they become part of the inner cell mass.

ES cells can be genetically altered by techniques that can be used to inactivate a particular gene or introduce a new one. A DNA molecule that is introduced into an ES cell by **transfection** will usually insert randomly in the genome. However, it is possible to tailor the DNA in such a way that it inserts at a specific predetermined site by **homologous recombination**; this insertion mutation renders the gene non-functional. The DNA to be introduced must contain enough sequence homology with the target gene to insert within the target gene in at least a few cells in the culture. The insertion also carries a drug-resistance gene, and so cells containing the insertion can be selected by adding the drug, which kills the other, unmodified, cells. The mutated ES cells can then be introduced into the blastocyst, producing a transgenic mouse carrying a mutation in a known gene (see figure, left). The use of homologous recombination to inactivate a gene is known as **gene knock-out** when the animal is homozygous for the inactivated gene. The same technique can be used to insert a novel gene.

The mutated ES cells are then introduced into the cavity of an early blastocyst, which is then returned to the uterus. They become incorporated into the inner cell mass and thus into the embryo, where they can give rise to germ cells and gametes. Once the mutant gene has entered the germ line, strains of mice heterozygous for the altered gene can be intercrossed to produce either viable homozygotes or homozygous lethals, depending on the gene involved, and the effect of completely inactivating and so knocking-out the gene can be examined.

A technique for targeting a gene knock-out to a specific tissue and/or a particular time in development is provided by the *Cre-lox* system. The target gene is first 'loxed' by inserting a *loxP* sequence of 34 base pairs on either side of the gene. These transgenic mice are then crossed with another line of transgenic mice carrying the gene for the recombinase Cre. *loxP* sequences are recognized by Cre, which will excise all the DNA between the two *loxP* sites. In the offspring, if Cre is expressed in all cells, then all cells will excise the 'loxed' target, causing a ubiquitous knock-out of the target gene. However, if the gene for Cre is under the control of a tissue-specific promoter, so that, for example, it is only expressed in heart tissue, the target gene will only be excised in heart tissue (see figure on p. 159). If the *Cre* gene is linked to an inducible control region, it is also possible to induce excision of the target gene at will by exposing the mice to the inducing stimulus.

Continued

Box 4B (continued) Gene targeting: insertional mutagenesis and gene knock-out

A significant number of knock-outs of a single gene result in mice developing without any obvious abnormality, or with fewer and less severe abnormalities than might be expected from the normal pattern of gene activity. A striking example is that of *myoD*, a key gene in muscle differentiation. In *myoD* knock-outs, the mice are anatomically normal, although they do have a reduced survival rate. This could mean that other genes can substitute for some of the functions of *myoD*.

However, it is unlikely that any gene is without any value at all to an animal. It is much more likely that there is an altered phenotype in these apparently normal animals, which is too subtle to be detected under the artificial conditions of life in a laboratory. Redundancy is thus probably apparent rather than real. A further complication is the possibility that, under such circumstances, related genes with similar functions may increase their activity to compensate for the mutated gene.

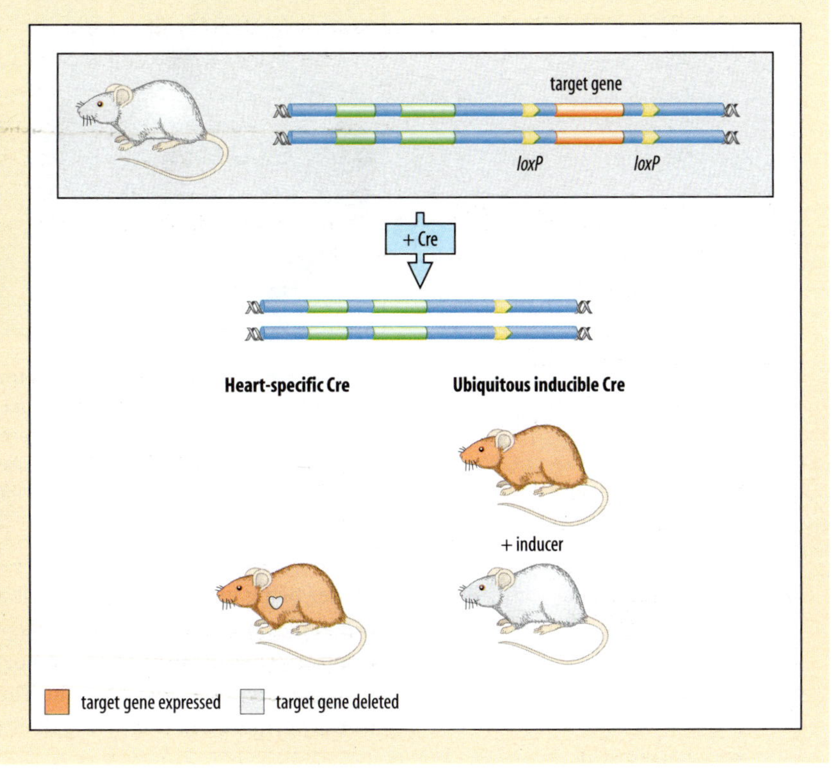

expression of each gene is characterized by a relatively sharp anterior border and, usually, a much less well defined posterior border. Although there is considerable overlap in expression, almost every region in the mesoderm along the antero-posterior axis is characterized by a particular set of expressed Hox genes (Fig. 4.11). For example, the most anterior somites are characterized by expression of genes *Hoxa1* and *Hoxb1*, and no other Hox genes are expressed in this region. By contrast, all the Hox genes are expressed in the most posterior regions. The Hox genes thus provide a code for regional identity. The most anterior expression of Hox genes is in the hindbrain; the more anterior regions of the vertebrate body—the anterior head, forebrain, and midbrain—are characterized by expression of other homeobox genes such as *Emc* and *Otx*, and not by Hox genes. The spatial and temporal order of expression is similar in the mesoderm and the ectodermally derived nervous system, but the boundaries between the regions of gene expression in these two germ layers do not always correspond.

If we focus on just one set of Hox genes, those of the Hoxa complex, we find that the most anterior border of expression in the mesoderm is that of *Hoxa1* in the posterior head mesoderm, while *Hoxa11*, the most posterior gene in the Hoxa complex, has its anterior border of expression in the sacral region (see Fig. 4.11). This exceptional correspondence, or **co-linearity**, between the order of the genes on the chromosome and their order of spatial and temporal expression along the antero-posterior axis, is typical of all the Hox clusters. The genes of each Hox complex are expressed in an orderly sequence, with the gene lying most 3′ in the

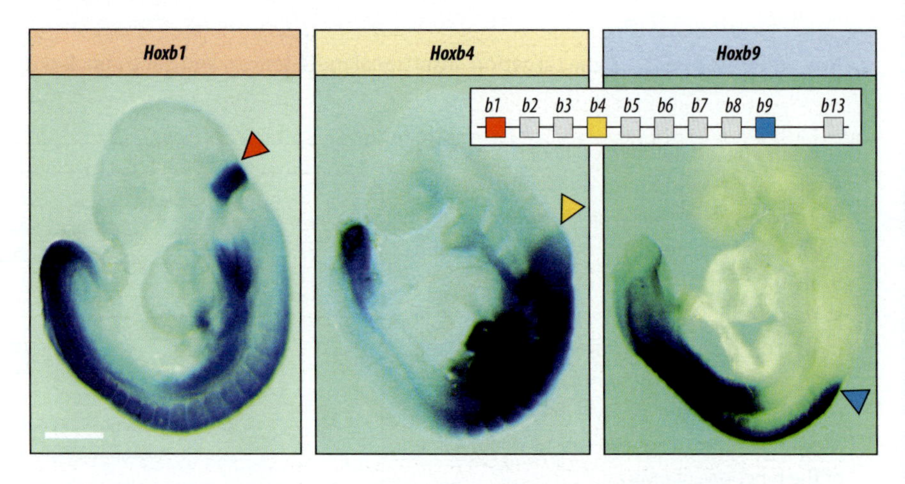

Fig. 4.10 Hox gene expression in the mouse embryo after neurulation. The three panels show lateral views of 9.5 days *post-coitum* embryos immunostained with antibodies specific for the protein products of the *Hoxb1*, *Hoxb4*, and *Hoxb9* genes. The arrowheads indicate the anterior boundary of expression of each gene within the neural tube. The position of the three genes within the Hoxb gene complex is indicated (inset). Scale bar = 0.5 mm.

Photographs courtesy of A. Gould.

cluster being expressed the earliest and in the most anterior position. The correct expression of the Hox genes is dependent on their position in the cluster, and anterior genes must be expressed before more posterior genes.

Support for the idea that the Hox genes are involved in controlling regional identity comes from comparing their patterns of expression in mouse and chick with

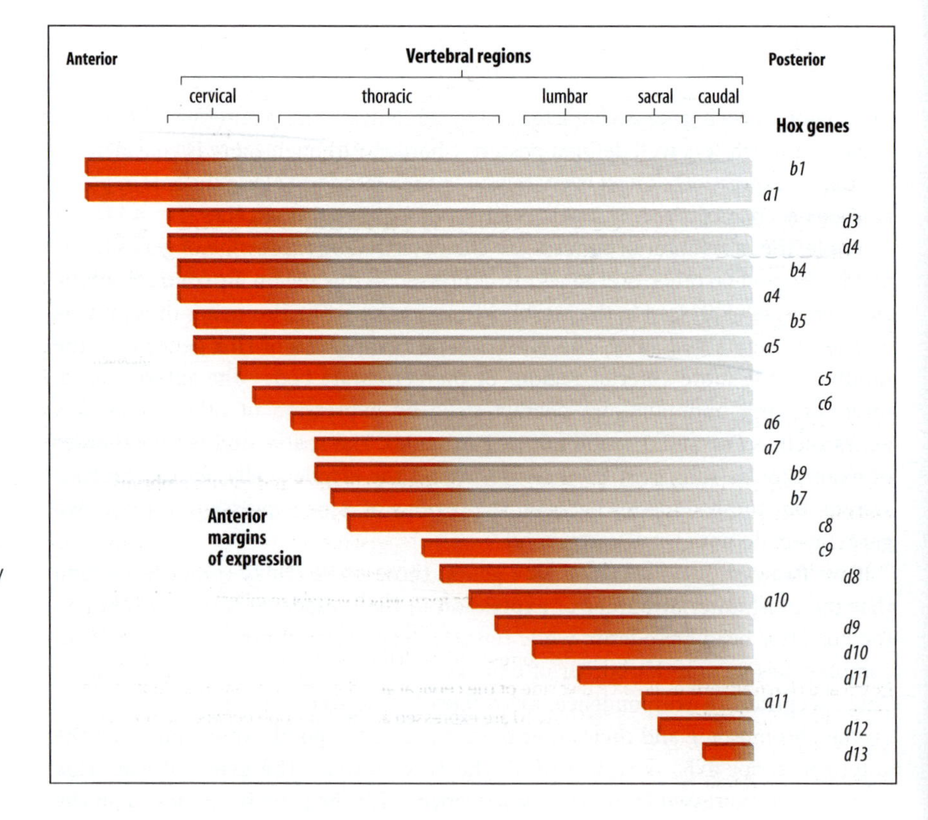

Fig. 4.11 Hox gene expression along the antero-posterior axis of the mouse mesoderm. The anterior border of each gene is shown by the dark red blocks. Expression usually extends backward some distance but the posterior margin of expression may be poorly defined. The pattern of Hox gene expression could specify the identity of the tissues at different positions. For example, the pattern of expression is quite different in anterior and posterior regions of the body axis.

the well defined anatomical regions—cervical, thoracic, sacral and lumbar (Fig. 4.12). Hox gene expression corresponds well with the different regions. For example, even though the number of cervical vertebrae in birds (14) is twice that of mammals, the anterior boundaries of *Hoxc5* and *Hoxc6* gene expression in both chick and mouse lie on either side of the cervical/thoracic boundary. A correspondence between Hox gene expression and region is also similarly conserved among vertebrates at other anatomical boundaries.

It must be emphasized that the summary picture of Hox gene expression given in Fig. 4.11 does not represent a 'snapshot' of expression at a particular time but rather an integrated overall pattern of expression. Some genes are switched on early and are then downregulated, while others are expressed considerably later; the most posterior Hox genes, such as *Hoxd12* and *Hoxd13*, are expressed in the post-anal tail, which develops later. Moreover, this summary picture reflects the general expression of the genes in embryonic regions; not all Hox genes expressed in a region are expressed in all the cells of that region. Nevertheless, the overall pattern suggests that the combination of Hox genes provides positional identity. In the cervical region, for example, each somite, and thus each vertebra, could be specified by a unique pattern of Hox gene expression.

As we saw in Fig. 4.8, grafting experiments show that the character of the somites is already determined in the pre-somitic mesoderm and that somitic tissue transplanted to other levels along the axis retains its original identity. This includes its original pattern of Hox gene expression. By contrast, transplanted lateral plate mesoderm takes on the Hox expression pattern of its new site. Hox genes providing positional specification in somites and lateral plate seem to be separate systems, though similar mechanisms may be involved.

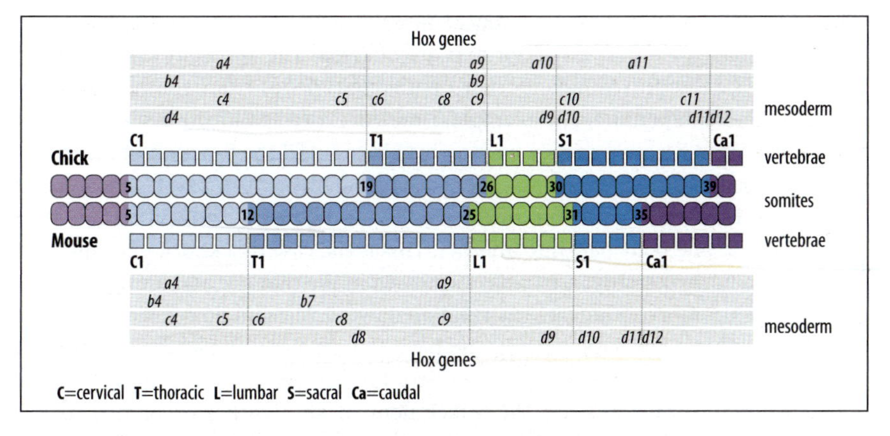

Fig. 4.12 Patterns of Hox gene expression in the mesoderm of chick and mouse embryos, and their relation to regionalization. The posterior margins of expression of Hox genes in the mesoderm vary along the axes. The vertebrae are derived from somites, 40 of which are shown. The vertebrae have characteristic shapes in each of the five regions: cervical (C), thoracic (T), lumbar (L), sacral (S), and caudal (Ca). Which somites form which vertebrae differs in chick and mouse. For example, thoracic vertebrae start at somite 20 in the chick, but at somite 12 in the mouse. The transition from one region to another corresponds with the pattern of Hox gene expression, so *Hoxc5* and *Hoxc6* are expressed on either side of the cervical and thoracic vertebral transition in both chick and mouse. Similarly, *Hoxd9* and *Hoxd10* are expressed at the transition between lumbar and sacral regions.

After Burke, A.C.: 1995.

4.3 Deletion or overexpression of Hox genes causes changes in axial patterning

If the Hox genes do provide positional values that determine a region's subsequent development, then one would expect morphological changes if their pattern of expression is altered. This is indeed the case. In order to see how Hox genes control patterning, either their expression can be prevented by mutation, or they can be expressed ectopically, in abnormal positions. Hox gene expression can be eliminated from the developing mouse embryo by gene knock-out techniques (see Box 4B, pp. 158–159). Experiments along these lines have shown that the absence of a given Hox gene affects patterning in a way that accords with the idea that Hox gene activity provides the cells with positional identity. For example, mice in which the gene *Hoxa3* has been deleted show structural defects in the region of the head and thorax, where this gene is normally strongly expressed, and tissues derived from both ectoderm and mesoderm are affected. But the Hox genes seem to specify positional identity in rather complex ways. There is undoubtedly some apparent **redundancy** between the effects of some of the genes, and when one gene is removed, another may serve in its place. This can make it difficult to interpret the results of Hox gene inactivation. There is also interaction between the individual genes, and this can further complicate results. For example, with the mutated *Hoxa3* gene described above, more posterior axial structures, where the inactivated gene is also normally expressed, show no evident defects.

This observation illustrates a general principle of Hox gene expression, which is that more posteriorly expressed Hox genes tend to inhibit the action of the Hox genes normally expressed anterior to them; this phenomenon is known as **posterior dominance** or **posterior prevalence**. This means that a change in Hox gene expression usually affects the most anterior regions in which the gene is expressed, leaving posterior structures relatively unaffected. The effects of a Hox gene knock-out can also be tissue specific, so that certain tissues in which a Hox gene is normally expressed appear normal, while other tissues at the same position along the antero-posterior axis are affected. The apparent absence of an effect may be due to redundancy, with **paralogous genes** from another complex being able to compensate. For example, *Hoxb1* is expressed in the same region as *Hoxa1* (see Fig. 4.11), and so may be largely able to fulfill the function of an absent *Hoxa1* gene.

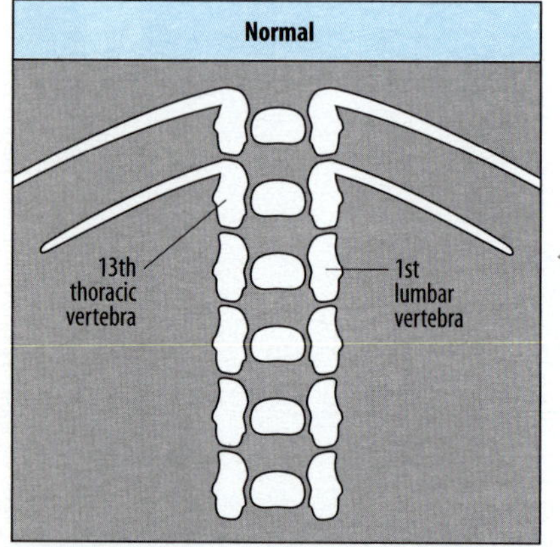

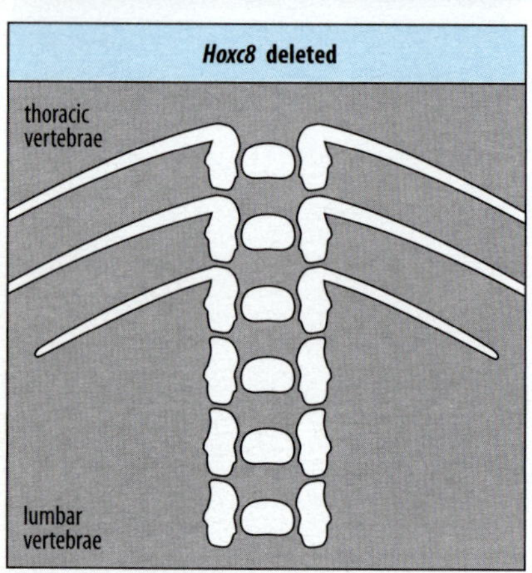

Fig. 4.13 Homeotic transformation of vertebrae due to deletion of *Hoxc8* in the mouse. In loss-of-function homozygous mutants of *Hoxc8*, the first lumbar vertebra is transformed into a rib-bearing thoracic vertebra. The mutation has resulted in the transformation of the lumbar vertebra into a more anterior structure.

Loss of Hox gene function often results in **homeotic transformation**—the conversion of one body part into another. This is the case in a knock-out mutation of *Hoxc8*. In normal embryos, *Hoxc8* is expressed in the thoracic and more posterior regions of the embryo from late gastrulation onward. Mice homozygous for mutant *Hoxc8* die within a few days of birth, and have abnormalities in patterning between the seventh thoracic vertebra and the first lumbar vertebra. The most obvious homeotic transformations are the attachment of an eighth pair of ribs to the sternum and the development of a 14th pair of ribs on the first lumbar vertebra (Fig. 4.13). Thus, the absence of *Hoxc8* modifies the development of some of the cells that would normally express it. Its absence gives them a more anterior positional value, and they develop accordingly. In mice in which *Hoxd11* is mutated, anterior sacral vertebrae are transformed into lumbar vertebrae. Another example of the homeotic transformation of a structure into one normally anterior to it can be seen in knock-out mutations of *Hoxb4*. In normal mice, *Hoxb4* is expressed in the axis (the second cervical vertebra), but not in that giving rise to the atlas (the first cervical vertebra). In *Hoxb4* knock-out mice, the axis is transformed into another atlas.

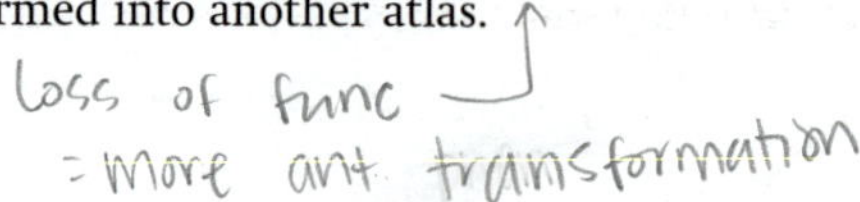

By contrast, abnormal expression of Hox genes in anterior regions that normally do not express them can result in transformations of anterior structures into structures that are normally more posterior. For example, when *Hoxa7*, whose normal anterior border of expression is in the thoracic region, is expressed throughout the whole antero-posterior axis, the basal occipital bone of the skull is transformed into a pro-atlas structure, normally the next most posterior skeletal structure. Overexpression of *Hoxa2* in the first chick branchial arch leads to transformation of the first arch cartilages such as the quadrate and Meckel's cartilage, which is a precursor element of the lower jaw, into second arch cartilages such as those of the tongue skeleton.

In mice, in the absence of all the Hox10 paralogous group genes, there are no lumbar vertebrae and there are ribs on all posterior vertebrae; in the absence of all the Hox11 paralogous group several vertebrae become lumbar. These homeotic transformations do not occur if only some members of the paralogous group are mutated, suggesting apparent redundancy. There are also synergistic interactions between Hox genes of the same paralogous group. Thus, knock-outs of mouse *Hoxa3* do not affect the first cervical vertebra—the atlas—or the basal occipital bone of the skull to which it connects, even though *Hoxa3* is expressed in the mesoderm that gives rise to these bones. However, knock-outs of *Hoxd3* (which is also expressed in this region) cause a homeotic transformation of the atlas into the adjacent basal occipital bone. A double knock-out of *Hoxa3* and *Hoxd3* results in complete deletion of the atlas. The complete absence of this bone in the absence of Hox gene expression suggests that one target of Hox gene action is the cell proliferation required to build such a structure from the somite cells. Unfortunately, very few direct targets of Hox proteins have yet been identified.

In vertebrates, unlike *Drosophila*, we also do not know how the pattern of Hox gene expression is specified. Retinoic acid has been shown experimentally to alter the expression of Hox genes, but whether it is involved in regulating Hox gene expression *in vivo* is not clear. The gradient in retinoic acid that is present from anterior to posterior along the main axis of the mouse embryo (see Fig. 4.7) could be involved in activating Hox genes in normal antero-posterior patterning. Recent evidence suggests that microRNA genes may be embedded in some Hox clusters and these may be involved in post-transcriptional regulation of Hox gene expression (see Box 5B, p. 197, for how microRNAs work).

4.4 Hox gene activation is related to a timing mechanism

In all vertebrates, the Hox genes begin to be expressed at an early stage of gastrulation, when the mesodermal cells begin their gastrulation movements. The anterior-most genes, which correspond to those at the 3′ end of the cluster, are expressed first. If an 'early' Hoxd gene is relocated to the 5′ end of the Hoxd complex, for example, its expression pattern then resembles that of the neighboring *Hoxd13*. This shows that the structure of the Hox complex is crucial in determining the pattern of Hox gene expression.

Unlike the situation in *Drosophila*, where the activation of Hox genes depends on factors unequally distributed along the antero-posterior axis (see Chapter 2), the mechanism of activation in vertebrates is more complex and less well understood. One way the antero-posterior pattern of Hox gene expression might be established in the somitic mesoderm is through linking gene activation to the time spent in the somite stem-cell region (see Fig. 4.5). In the chick and mouse, the whole of the mesoderm alongside the dorsal midline derives from a small

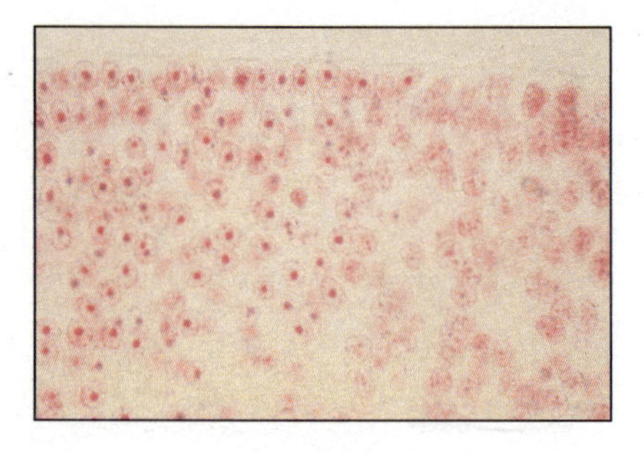

Fig. 4.14 Photograph of quail-chick chimeric tissue. The quail cells are on the left and the chick cells are on the right.

Photograph courtesy of Nicole Le Douarin.

population of stem cells located in the anterior primitive streak and later in the tailbud. It has been shown that genes of the Hoxb cluster follow a strict temporal order of co-linear activation in this stem-cell region, the 3′ genes being activated before the 5′ ones. Expression of genes that are activated in this region is maintained when the cells leave to form the pre-somitic mesoderm, and in this way the temporal pattern of Hox gene activation can be converted into positional information as the cells expressing different Hox genes become distributed along the antero-posterior axis. This model is similar to that proposed for specifying position along the proximo-distal axis of the vertebrate limb bud (see Chapter 9).

We have concentrated here on the expression of Hox genes in the mesoderm, but they are also expressed in a patterned way in the neural tube after its induction, and we shall return to this aspect of antero-posterior regionalization later in the chapter.

4.5 The fate of somite cells is determined by signals from the adjacent tissues

We shall now return to the individual somite and see how it is patterned. This patterning process is quite independent of the global antero-posterior patterning of the whole pre-somitic mesoderm by Hox gene expression. The somites of the vertebrate embryo give rise to major axial structures: the cartilage cells of the embryonic axial skeleton—the vertebrae and ribs; all the skeletal muscles, including those of the limbs; and much of the dermis. The fate maps for particular somites have been made by grafting somites from a quail into a corresponding position in a chick embryo at a similar stage of development and following the fate of the quail cells. These can be distinguished from chick cells by their distinctive nuclei, which can be detected in histological sections (Fig. 4.14). The cells that form the lateral (away from the midline) and medial (nearest the midline) parts of chick somites are of different origins, and are brought together during gastrulation. The medial portion comes from cells in the primitive streak close to Hensen's node, whereas the lateral portion comes from more posterior cells.

Cells located in the dorsal and lateral regions of a newly formed somite make up the **dermomyotome**, which expresses the gene *Pax3*, a homeobox-containing gene of the *paired* family (see Box 4A, p. 156). The dermomyotome is made up of the **myotome**, which gives rise to muscle cells, and the **dermatome**, an epithelial sheet over the myotome which gives rise to the dermis. Cells from the medial region of the somite form mainly axial and back muscles, and express the muscle-specific transcription factor MyoD and related proteins, whereas lateral

Fig. 4.15 The fate map of a somite in the chick embryo. The ventral medial quadrant (blue) gives rise to the sclerotome cells, which migrate to form the cartilage of the vertebrae. The rest of the somite—the dermomyotome—forms the dermatome and myotome, which give rise to the dermis and all the trunk muscles, respectively. The dermomyotome also gives rise to muscle cells that migrate into the limb bud.

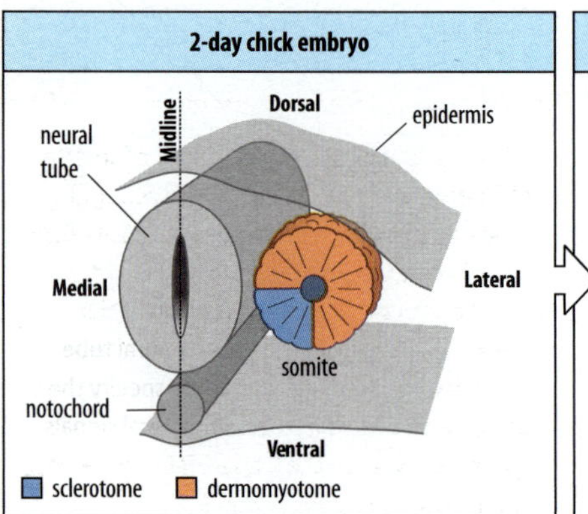

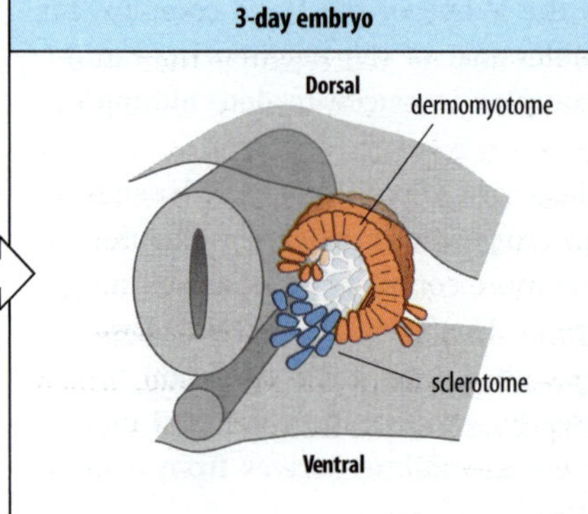

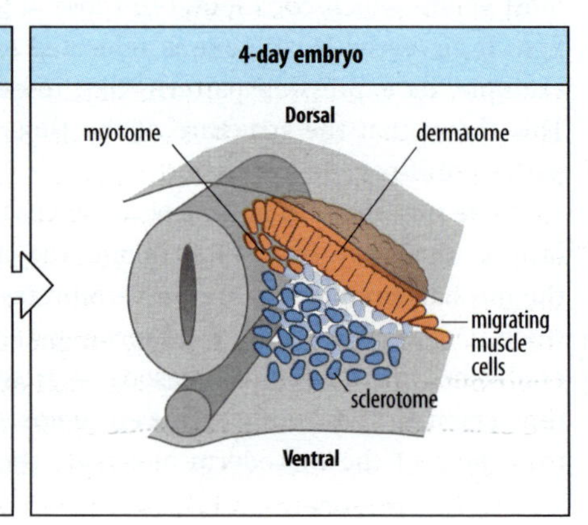

cells migrate to give rise to abdominal and limb muscles. The ventral part of the medial somite contains **sclerotome** cells that express the *Pax1* gene and migrate ventrally to surround the notochord and develop into vertebrae and ribs (Fig. 4.15).

Which cells will form cartilage, muscle, or dermis is not yet determined at the time of somite formation. Specification of these fates requires signals from tissues adjacent to the somite. This is clearly shown by experiments in which the dorso-ventral orientation of newly formed somites is inverted; they still develop normally. In the chick, determination of myotome occurs within hours of somite formation, whereas the future sclerotome is only determined later. Both the neural tube and notochord produce signals that pattern the somite and are required for its future development. If the notochord and neural tube are removed, the cells in the somites undergo apoptosis; neither vertebrae nor axial muscles develop, although limb musculature still does.

The role of the notochord in specifying somitic cells has been shown by experiments in the chick, in which an extra notochord is implanted to one side of the neural tube, adjacent to the somite. This has a dramatic effect on somite differentiation, provided the operation is carried out on unsegmented pre-somitic mesoderm. When the somite develops, there is an almost complete conversion to cartilage precursors (Fig. 4.16), suggesting that the notochord is an inducer of cartilage. The neural tube also has a cartilage-inducing effect on somites, which is mediated by the most ventral region of the tube, the floor plate (see Section 10.7). There is also evidence for a signal from the lateral plate mesoderm, which is involved in specifying the lateral part of the dermomyotome, and for a signal from the overlying ectoderm (Fig. 4.17).

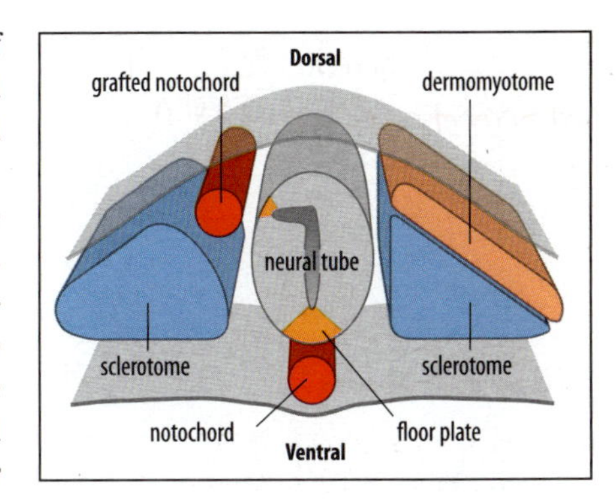

Fig. 4.16 A signal from the notochord induces sclerotome formation. A graft of an additional notochord to the dorsal region of a somite in a 10-somite embryo suppresses the formation of the dermomyotome from the dorsal portion of the somite, and induces the formation of sclerotome, which develops into cartilage. The graft also affects the shape of the neural tube.

Some of the signals that pattern the somite have been identified. In the chick, both the notochord and the floor plate express the gene *Sonic hedgehog*, which encodes a secreted protein that is a key molecule for positional signaling in a number of developmental situations. We met Sonic hedgehog in Chapter 3 as a signal involved in the asymmetry of structures about the midline and we shall meet it again in Chapter 9 in connection with limb development. In somite patterning, the signal generated by *Sonic hedgehog* specifies the ventral region of the somite and is required for sclerotome development. Signals from the dorsal neural tube and from the overlying non-neural ectoderm specify the dorsal region. Secreted signaling proteins of the Wnt family (which we also encountered in Chapter 3) are good candidates for the lateral and dorsal signals. Tendons arise from cells that come from the dorso-lateral domain of the sclerotome and specifically express the transcription factor Scleraxis. This tendon-progenitor region is induced at the boundary of the sclerotome and myotome.

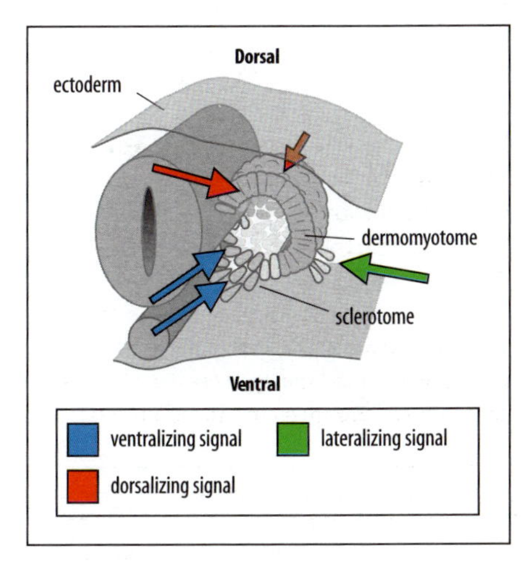

Fig. 4.17 A model for patterning of somite differentiation. The sclerotome is thought to be specified by a diffusible signal, probably the Sonic hedgehog protein, from the notochord and the floor plate of the neural tube (blue arrows). Signals from the dorsal neural tube and ectoderm (red arrows) would specify the dermomyotome, together with lateral signals (green arrow) from the lateral plate mesoderm.

After Johnson, R.L.: 1994.

Regulation of the Pax homeobox genes in the somite by signals from the notochord and neural tube seems to be important in determining cell fate. *Pax3* is initially expressed in all cells that will form somites. Its expression is then modulated by signals from BMP-4 and Wnt proteins so that it becomes confined to muscle precursors. It is then further downregulated in cells that differentiate as the muscles of the back, but remains switched on in the migrating presumptive muscle cells that populate the limbs. Mice that lack a functional *Pax3* gene—*Splotch* mutants—lack limb muscles. In the chick, *Pax1* has been implicated in the formation of the scapula, a key element in the shoulder girdle, part of which is contributed by somites. Unlike the *Pax1*-expressing cells of the vertebrae, which are of sclerotomal origin, the blade of the scapula is formed from dermomyotome cells of chick somites 17–24, whereas the head of the scapula is derived from lateral plate mesoderm. All the scapula-forming cells express *Pax1*.

Summary

Somites are blocks of mesodermal tissue that are formed after gastrulation. They form sequentially in pairs on either side of the notochord, starting at the anterior end of the embryo. The somites give rise to the vertebrae, to the muscles of the trunk and limbs, and to the dermis of the skin. The pre-somitic mesoderm is patterned along its antero-posterior dimension while cells are in the node, and the first manifestation of this pattern is the expression of the Hox genes in the pre-somitic mesoderm before somite formation. The somites are also patterned by signals from the notochord, neural tube, and ectoderm, which induce particular regions of each somite to give rise to muscle, cartilage, or dermis.

The regional character of the mesoderm that gives rise to somites is specified even before the somites form. The positional identity of the somites is specified by the combinatorial expression of genes of the Hox complexes along the antero-posterior axis, from the hindbrain to the posterior end, with the order of expression of these genes along the axis corresponding to their order along the chromosome. Mutation or overexpression of a Hox gene results, in general, in localized defects in the anterior parts of the regions in which the gene is expressed, and can cause homeotic transformations. We can think of Hox genes as providing positional information that specifies the identity of a region and its later development. They act on downstream targets about which we know relatively little.

The role of the organizer and neural induction

We shall now look at the role of the crucially important organizer region both in neural induction and in organizing the antero-posterior axis in vertebrate embryos. The Spemann organizer of amphibians, the shield in zebrafish, Hensen's node in the chick, and the equivalent node region in the mouse all have a similar global organizing function in vertebrate development. They can induce a complete body axis if transplanted to another embryo at an appropriate stage, and so are able to organize and coordinate both dorso-ventral and antero-posterior aspects of the body plan, as well as induce neural tissue from ectoderm. In the mouse, the anterior visceral endoderm is required as well as the node for the induction of structures anterior to the end of the notochord, such as the head and forebrain (see Section 3.9).

During gastrulation, the ectoderm lying along the dorsal midline of the embryo becomes specified as neurectoderm, the **neural plate**. During the subsequent stage of neurulation, this forms the neural tube, which eventually differentiates into the central nervous system—the brain and the spinal cord—and the nerves that innervate the skeletal muscles. The neural tube also throws off neural crest cells, which migrate throughout the body to give rise to the sympathetic and parasympathetic nervous systems and other structures. The nervous system must develop in the correct relationship with other body structures, particularly the mesodermally derived structures that give rise to the skeleto-muscular system. Thus, patterning of the nervous system must be linked to that of the mesoderm, and this is coordinated through the organizer, which is involved in both. In this part of the chapter, we consider the patterning of the neural tube up to shortly after its closure and also consider the specification and formation of the neural crest cells. We will then look at the hindbrain at a later stage, when that region becomes segmented and neural crest cells have migrated.

The function of the organizer has been best studied in amphibians, and we have already described its role in the dorso-ventral patterning of the mesoderm of *Xenopus* (see Section 3.19). We now discuss its profound effects on the antero-posterior axis. To do this we must first return to the late blastula and early gastrula, before the stage of somite formation discussed in the previous part of the chapter.

4.6 The inductive capacity of the organizer changes during gastrulation

In amphibians, the action of the organizer is dramatically demonstrated in what is classically known as **primary embryonic induction**. In the early amphibian gastrula, the Spemann organizer is located in the dorsal lip of the blastopore. If this is grafted to the ventral side of the marginal zone of another gastrula, it can induce a complete second embryo (Fig. 4.18). This second embryo can have a well defined head and trunk region, and even a tail, but is joined to the main embryo along the axis (see Fig. 1.10). A variety of other treatments, such as grafting dorsal vegetal blastomeres containing the Nieuwkoop center to the ventral side, produce a similar result (see Fig. 3.32), but what all these treatments have in common is that, directly or indirectly, they result in the formation of a new Spemann-organizer region. One important question, which is still unresolved, is whether or not the organizer contains functionally separate organizers for the head, trunk, and tail regions. This question arises from classic experiments that showed that a graft of the blastopore dorsal lip (which contains the organizer region) from an early gastrula induces a complete body axis and central nervous system, whereas a dorsal lip from a mid-gastrula induces a trunk and tail but no head, while a dorsal lip from a late gastrula induces only a tail (see Fig. 4.18). This is interpreted to mean that as gastrulation proceeds, the antero-posterior axis becomes specified, and so the cells that make up the organizer at later stages of gastrulation induce only posterior structures. Is this due to a change in the quantity of inducing signals produced by the organizer as gastrulation proceeds, or are different signals involved in specifying different regions of the antero-posterior axis?

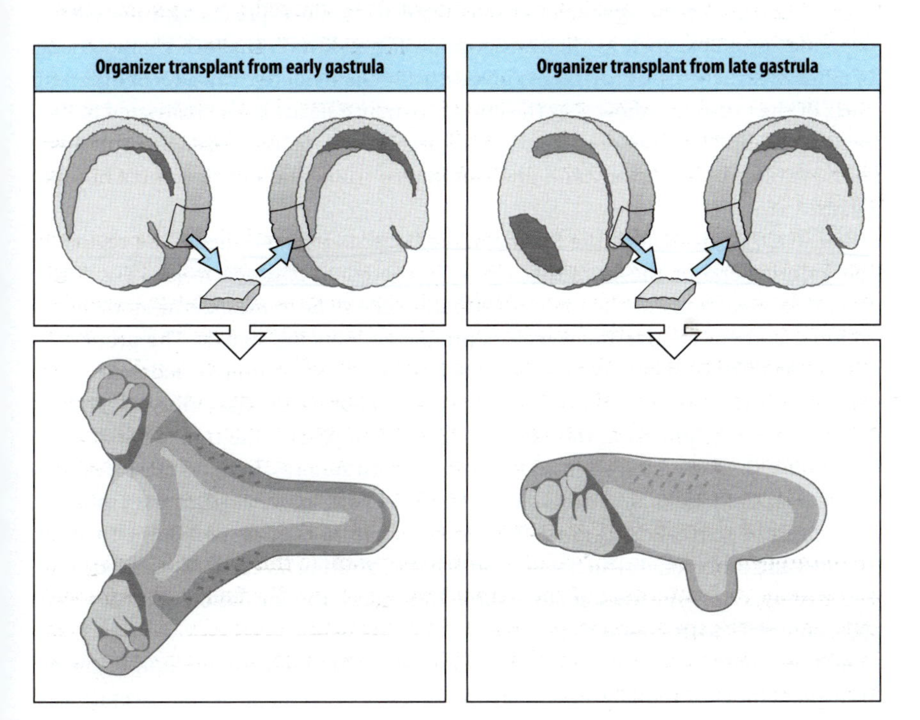

Fig. 4.18 The inductive properties of the organizer change during gastrulation. A graft of the organizer region, from the dorsal lip of the blastopore of an early frog gastrula to the ventral side of another early gastrula, results in the development of an additional anterior axis at the site of the graft (left panels). A graft from the dorsal lip region of a late gastrula to an early gastrula only induces formation of tail structures (right panels).

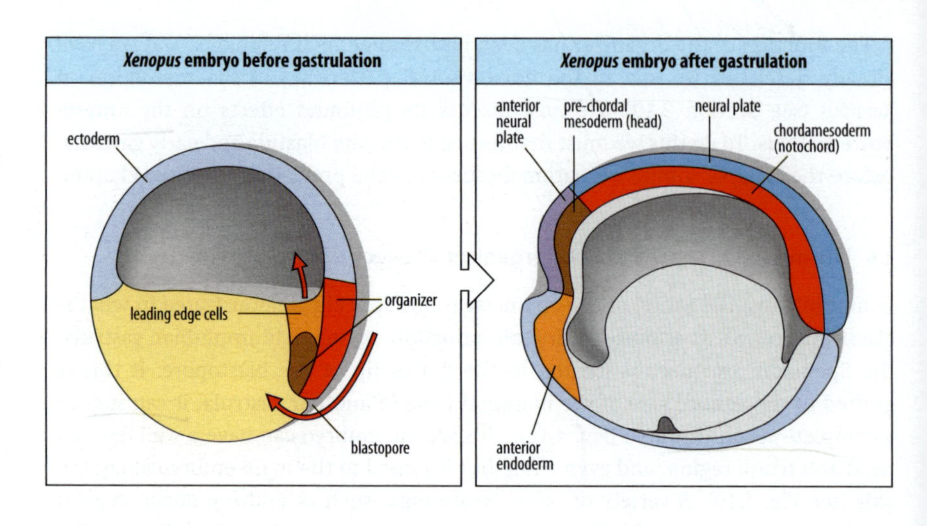

Fig. 4.19 Different parts of the *Xenopus* organizer region give rise to different tissues. In the early gastrula (left panel) the organizer is located in the dorsal lip of the blastopore. The cells of the leading edge (orange) are the first to internalize and give rise to anterior endoderm of the neurula stage (right panel). The deep cells (pale brown) are next to internalize and give rise to the pre-chordal plate, the mesoderm anterior to the notochord, which forms much of the head. The remainder of the organizer gives rise to notochord (red).

Adapted from Kiecker, C. and Niehrs, C.: 2001.

The cells present at the dorsal lip of the blastopore in a *Xenopus* early gastrula give rise during gastrulation to anterior endoderm, prospective head mesoderm (the pre-chordal plate), and the chordamesoderm that forms the notochord (Fig. 4.19). As well as providing cells for these axial structures, the organizer region has patterning and inductive properties: it helps to pattern the adjacent more vegetal mesoderm, as we saw in Chapter 3, and it induces the neural plate in the adjacent dorsal ectoderm. The early gastrula organizer is a complex signaling center with the different parts expressing different genes, having different inductive capacities, and giving rise to different structures. Because of these complex properties, understanding how the organizer region organizes the overall pattern of the antero-posterior axis is not straightforward. As gastrulation proceeds and cells move inwards, the cellular composition of the dorsal lip changes, and at one level this explains the different inductive properties displayed by the organizer over time in experiments such as those described above. For example, in the early gastrula, the vegetal portion of the organizer that is fated to produce pre-chordal mesoderm expresses proteins, such as the transcription factor XOtx2, that are characteristic of anterior structures. Experiments investigating the inductive capacity of different parts of the organizer show that the ability to induce heads is also restricted to this vegetal region. The more dorsal part of the organizer is characterized by expression of the transcription factor Xnot, and can induce trunk and tail structures but not heads.

The avian equivalent to the Spemann organizer is Hensen's node, the region at the anterior end of the primitive streak in the chick blastoderm (see Fig. 3.16). In normal development the node contributes cells to head mesoderm, notochord, somites, and gut endoderm, as well as producing inducing signals. The properties of the avian node have been investigated by transplanting quail nodes to chick embryos. If, for example, a quail node from a head-process-stage embryo is grafted beneath a chick epiblast at the same stage of development, the transplanted node can induce the formation of an additional axis with somites, but no anterior neural tissue (Fig. 4.20). Induction occurs if the graft is placed quite close to the embryo's own primitive streak; the graft then induces the non-axial mesoderm to form somites and other axial structures. As in *Xenopus*, grafting an earlier-stage node into the area opaca of an earlier stage embryo can induce the formation of a new body axis, complete with neural tissue.

In the mouse, the node precursors can induce a similar axis duplication on transplantation to the lateral epiblast of an early embryo, with the exception of

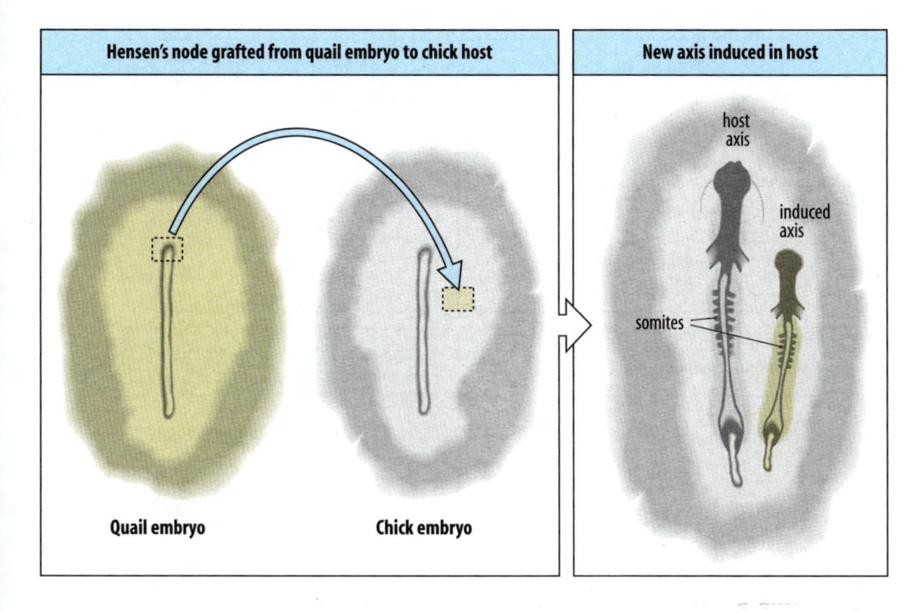

Fig. 4.20 Hensen's node can induce a new axis in avian embryos. When Hensen's node from a quail embryo at the head-process stage is grafted to a position lateral to the primitive streak of a chick embryo at the same stage of development, a new axis forms at the site of transplantation. (At the head-process stage, elongation of the primitive streak is complete, the notochord—the head process—has started to form anterior to the node, but the node has not yet started to regress.) Histological examination shows that although some of the somites of this new axis are formed from the graft itself (quail tissue can easily be distinguished from chick tissue, see Fig. 4.14), others have been induced from host tissue that does not normally form somites. Grafting a node at this stage produces a new axis lacking neural tissue. Grafting at an earlier stage into the area opaca will induce a complete new body axis, complete with neural tissue.

the forebrain, which requires additional signals from the anterior visceral endoderm (see Fig. 3.39).

A number of proteins are specifically expressed in the organizer and are known to be required for its function (Fig. 4.21). Goosecoid, for example, is an early and excellent marker of the organizer that is expressed in the cells that will give rise to foregut, pre-chordal plate, and notochord, and which have internalized by the mid-gastrula stage. Goosecoid can mimic almost all the properties of the organizer, but although it is required for head formation in the normal course of development, *goosecoid* mRNA injected into ventral blastomeres induces a secondary axis lacking a head. This indicates that additional signals are involved in the induction of the head and the central nervous system. Signaling proteins secreted by cells of the organizer are probable candidates. Head formation in *Xenopus* apparently requires the inhibition of both BMP and Wnt, and even Nodal signals, which are being produced in the gastrula at this time: an extra head can be induced by the simultaneous ectopic inhibition of BMP and Wnt signaling in early gastrulas. As we saw in Chapter 3, antagonists of BMPs and Wnts are secreted by the organizer: the proteins Chordin, Noggin, and Follistatin can antagonize BMP, and Dickkopf antagonizes Wnt signaling. The protein Cerberus can inhibit Wnt, Nodal, and BMP signaling. It is important to bear in mind, however, that although many of the same proteins are produced in and around the organizer in different vertebrates, they do not all have precisely the same functions in all our model animals.

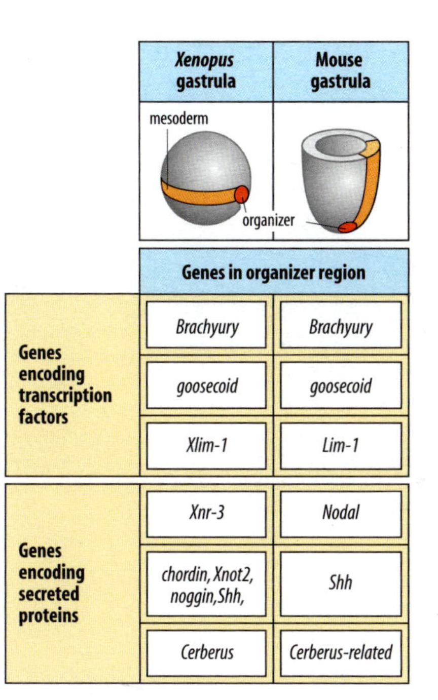

	Xenopus gastrula	Mouse gastrula
Genes in organizer region		
Genes encoding transcription factors	*Brachyury*	*Brachyury*
	goosecoid	*goosecoid*
	Xlim-1	*Lim-1*
Genes encoding secreted proteins	*Xnr-3*	*Nodal*
	chordin, Xnot2, noggin, Shh,	*Shh*
	Cerberus	*Cerberus-related*

Fig. 4.21 Genes expressed in the Spemann organizer region of the *Xenopus* gastrula, and in Hensen's node in the mouse gastrula. There is a similar pattern of gene activity in the two animals, with homologous genes being expressed. The expression of some of these genes, such as *Brachyury*, is not confined to the organizer. *Shh* = *Sonic hedgehog*.

4.7 The neural plate is induced in the ectoderm

The induction of neural tissue from ectoderm was first indicated by the organizer-transplant experiment in frogs illustrated in Fig. 4.18; in the secondary embryo that forms at the site of transplantation, a nervous system develops from the host ectoderm that would normally have formed ventral epidermis. This suggested that neural tissue could be induced from as yet unspecified ectoderm by signals emanating from the organizer mesoderm. The requirement for induction was confirmed by experiments that exchanged prospective neural plate ectoderm for prospective epidermis before gastrulation; the transplanted prospective epidermis developed

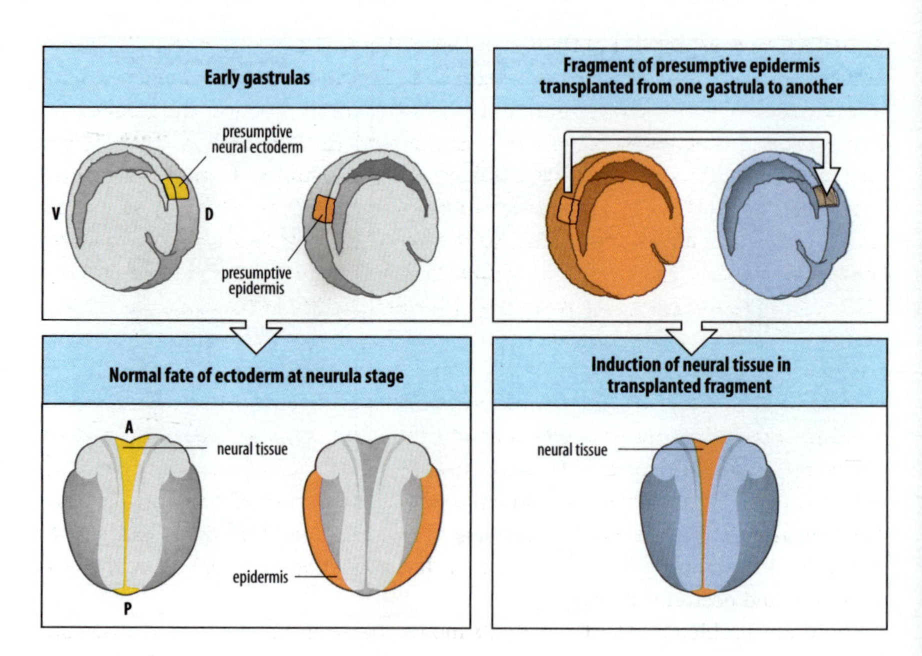

Fig. 4.22 The nervous system of *Xenopus* is induced during gastrulation. The left panels show the normal developmental fate of ectoderm at two different positions in the early gastrula. The right panels show the transplantation of a piece of ventral ectoderm, whose normal fate is to form epidermis, from the ventral side of an early gastrula to the dorsal side of another, where it replaces a piece of dorsal ectoderm whose normal fate is to form neural tissue. In its new location, the transplanted prospective epidermis develops not as epidermis but as neural tissue, and forms part of a normal nervous system. This shows that the ventral tissue has not yet been determined at the time of transplantation, and that neural tissue is induced during gastrulation.

into neural tissue, and the transplanted prospective neural tissue into epidermis (Fig. 4.22). This shows that the formation of the nervous system is dependent on an inductive signal.

An enormous amount of effort was devoted in the 1930s and 1940s to trying to identify the signals involved in neural induction in amphibians. Researchers were encouraged by the finding that a dead organizer region could still induce neural tissue. It seemed to be merely a matter of hard work to isolate the chemicals responsible. Alas, the search was fruitless, for it appeared that an enormous variety of substances were capable of varying degrees of neural induction. As it turned out, this was because newt ectoderm, the main experimental material used, seems to have a high propensity to develop into neural tissue on its own. This is not the case with *Xenopus* ectoderm, although prolonged culture of disaggregated ectodermal cells can result in their differentiation as neural cells after reaggregation. In experiments in *Xenopus*, it was found that the inducing signal could diffuse through a nucleopore filter (which prevents cell contact but allows the passage of quite large molecules, such as proteins), and that contact with organizer mesoderm lasting about 2 hours is required for induction to occur. The molecules responsible for neural induction have still not been definitively identified, although there are now some strong candidates.

Neural tissue can be induced in the chick epiblast—in both the area pellucida and the area opaca—by grafts from the primitive-streak mesoderm. Inductive activity is initially located in the anterior primitive streak and Hensen's node, and later, during regression of the node, becomes confined to the region of pre-somitic mesoderm just anterior to the node. By the four-somite stage inductive activity has disappeared; the competence of the ectoderm to respond disappears at about the same time.

A key point in the study of neural induction was the finding that the disaggregation of *Xenopus* gastrula-stage animal caps removed an inhibitor of neural development; this was subsequently identified as BMP-4. In the late blastula, BMPs are expressed throughout the ectoderm (see Fig. 3.62) but expression is subsequently lost in the neural plate. Neural induction could therefore be due to the production of proteins by the organizer that bound to BMPs and lifted their inhibitory action.

As BMPs induce expression of their own genes, this would also suppress BMP gene expression. BMP inhibitors such as Noggin and Chordin are produced in the organizer (see Section 4.6), and their inhibition of BMP proteins produces the gradient of BMP signaling that helps pattern the early mesoderm (see Section 3.19). These observations led to the so-called default model for neural induction in *Xenopus*. This proposes that the default state of the dorsal ectoderm is to develop as neural tissue, but that this pathway is blocked by the presence of BMPs, which promote the epidermal fate. The role of the organizer is to lift this inhibition by blocking BMP activity; the affected region of ectoderm will then develop as neural ectoderm. Proteins produced by the organizer can act on ectoderm that lies adjacent to it at the beginning of gastrulation (see Fig. 4.19), and as gastrulation proceeds, internalized cells derived from the organizer influence the fate of the ectoderm overlying them.

Elimination of organizer signals individually in amphibians has rather modest effects on neural induction. But when the BMP antagonists Chordin, Noggin, and Follistatin were simultaneously depleted in the organizer of the frog *Xenopus tropicalis* using antisense morpholino oligonucleotides (see Box 5A, p. 189), there was a dramatic failure of neural and other dorsal development and an expansion of ventral and posterior fates.

There are problems with the default model, however, as neural induction in both *Xenopus* and the chick has been shown also to require the growth factor FGF, even when BMP inhibition is lifted by the presence of Noggin and Chordin. In the chick, FGF is secreted by the hypoblast, which as gastrulation proceeds becomes progressively displaced towards the anterior of the embryo by the endoblast (see Section 3.3). Also, some genes that appear to be required for neural induction in chick are not activated by BMPs in ectoderm, but are activated by FGF. One such gene is *churchill*; when its expression is downregulated using an antisense morpholino oligonucleotide (see Box 5A, p. 189) the neural plate does not form. *churchill* can be activated by FGF, and this in turn leads to repression of genes characteristic of mesoderm and activation of the gene for the neural-specific transcription factor *Sox2*.

FGF is not thought to be the sole inducer of the neural tube in vertebrates, however, and inhibition of BMP signaling is likely to have a part to play. FGF activates the mitogen-activated protein (MAP) kinase intracellular signaling pathway (see Fig. 8.4) and there is evidence that MAP kinase can interfere with BMP signaling, causing an inhibitory phosphorylation of Smad-1, which is part of the BMP intracellular signaling pathway (see Fig. 3.58).

Other experiments in chick embryos suggest that BMPs are involved in setting the boundary of the neural plate. BMPs are expressed at the border of the neural plate, and the application of BMP to the border region causes inward displacement and narrowing of the plate, while the application of BMP antagonists causes widening. It seems that, in the chick embryo at least, two separate decisions are required for neural-plate formation: one is between mesendoderm or neural ectoderm fate in epiblast cells that sets the medial edge of the prospective neural plate (the edge that will be adjacent to the streak when gastrulation stops), and involves FGF and *churchill*; the other is made at the lateral edges of the neural plate, where neural and epidermal ectoderm meet, and is likely to involve inhibition of BMP signaling.

In zebrafish, there is a clear difference in the mechanism of induction of anterior neural tissue close to the shield in the dorsal region, and the induction of the posterior neural ectoderm that will give rise to the spinal cord. Whereas the organizer in zebrafish contributes to the induction of anterior neural tissue by inhibiting BMP signaling, the ectoderm that develops into the spinal cord is some distance away from the organizer on the ventral-vegetal side of the embryo,

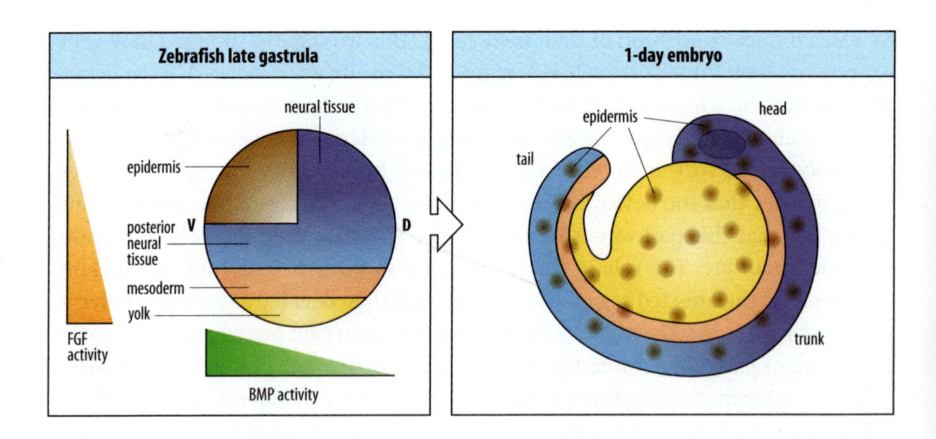

Fig. 4.23 Prospective spinal cord in the zebrafish embryo is distant from the organizer. The ectoderm that will form the spinal cord is situated on the other side of the gastrula from the organizer, in the ventral–vegetal ectoderm, and is too far away to be influenced by signals from the organizer. FGF signals in the ventral–vegetal region induce this ectoderm as neurectoderm and BMPs promote its development as posterior neural tissue.

Adapted from Kudoh, T., et al.: 2004.

where organizer signals do not reach (Fig. 4.23). The initiator of neural development in this ventral–vegetal ectoderm is FGF, and BMP signals, which are high in the ventral region, in this case act to push the neurectoderm towards a posterior fate.

Neural induction is a complex multistep process and the very first stages are likely to occur in the blastula, even before a distinct organizer region becomes detectable. The individual signaling proteins have multiple roles in development, and their roles change over time, which makes it difficult to disentangle their real contributions to any particular process. Their developmental roles can also differ between different vertebrates. An essential similarity in the mechanism of neural induction among vertebrates is likely, however, as Hensen's node from a chick embryo can induce neural gene expression in *Xenopus* ectoderm (Fig. 4.24), which suggests that there has been an evolutionary conservation of inducing signals. Moreover, early nodes induce gene expression characteristic of anterior amphibian neural structures, whereas older nodes induce expression typical of posterior structures. These results are in line with the theory that the vertebrate node specifies different antero-posterior positional values at different times, and they confirm the essential similarity of Hensen's node and the Spemann organizer.

In chick embryos, the spinal cord is not simply an extension of the posterior-most neural plate. The complete spinal cord is generated from a small region of

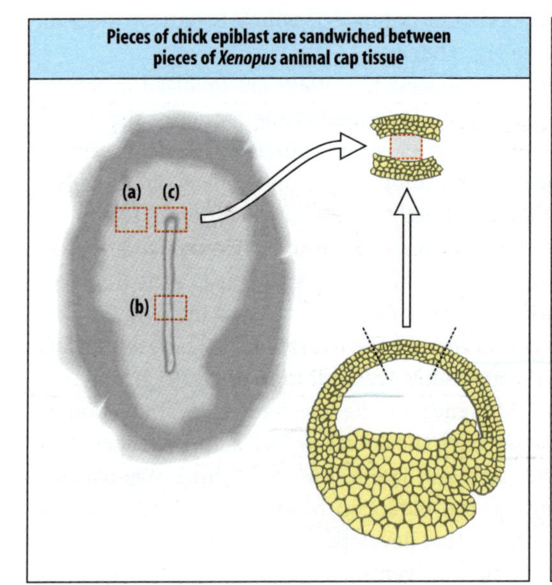

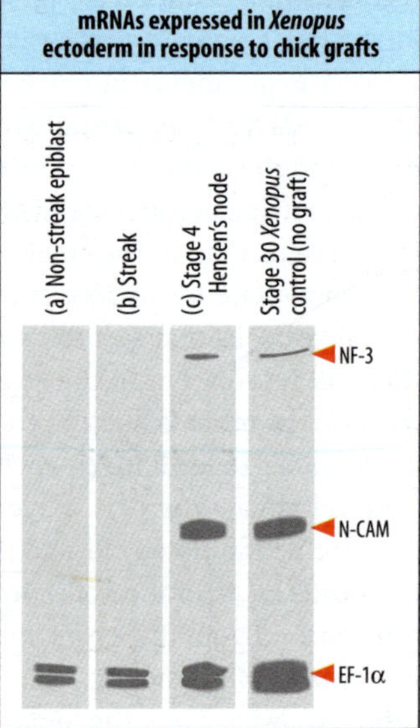

Fig. 4.24 Hensen's node from a chick embryo can induce gene expression characteristic of neural tissue in *Xenopus* ectoderm. Tissues from different parts of the primitive-streak stage of a chick epiblast are placed between two fragments of animal cap tissue (prospective ectoderm) from a *Xenopus* blastula. The induction in the *Xenopus* ectoderm of genes that characterize the nervous system is detected by looking for expression of mRNAs for neural cell adhesion molecule (N-CAM) and neurogenic factor-3 (NF-3), which are expressed specifically in neural tissue in stage 30 *Xenopus* embryos. Only transplants from Hensen's node induce the expression of these neural markers in the *Xenopus* ectoderm. (EF-1α is a common transcription factor expressed in all cells.)

After Kintner, C.R., et al.: 1991.

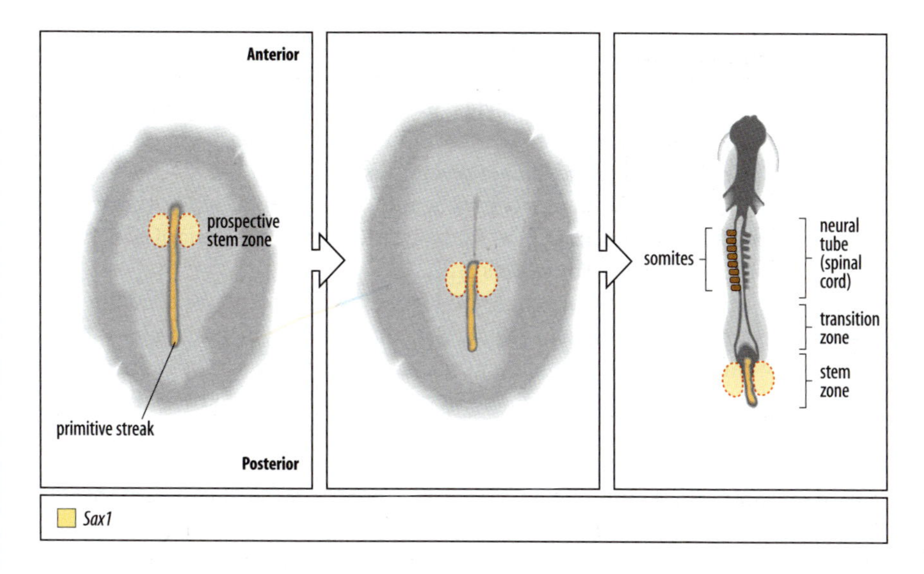

Fig. 4.25 The spinal cord is formed from a zone of stem cells that arises in the posterior neural plate. In the chick embryo, a restricted region of stem cells forms in the posterior neural plate alongside the node and anterior primitive streak just before the onset of somitogenesis (yellow). The region is marked by expression of *Sax1*. This stem-cell region moves posteriorly with the node as it regresses, leaving behind progenitor neural cells that will form the spinal cord. The complete spinal cord is formed from this stem-cell zone.

Adapted from Delfino-Machin, M., et al.: 2005.

proliferating stem cells that develops in the ectoderm at the posterior end of the neural plate on both sides of the node and primitive streak (Fig. 4.25). This zone becomes distinct just before somite formation begins. The stem-cell zone moves posteriorly along with the node, leaving behind progenitor cells that produce the spinal cord.

4.8 The nervous system can be patterned by signals from the mesoderm

Whatever the mechanisms of neural induction eventually turn out to be, it is clear that the neural plate can be patterned by signals from the mesoderm. Pieces of mesoderm taken from different positions along the antero-posterior axis of a newt neurula and placed in the blastocoel of an early newt embryo induce neural structures at the site of transplantation. The structures that are formed correspond more or less to the original position of the transplanted mesoderm: pieces of anterior mesoderm induce a head with a brain, whereas posterior pieces induce a trunk with a spinal cord (Fig. 4.26). Another indication of positional specificity in

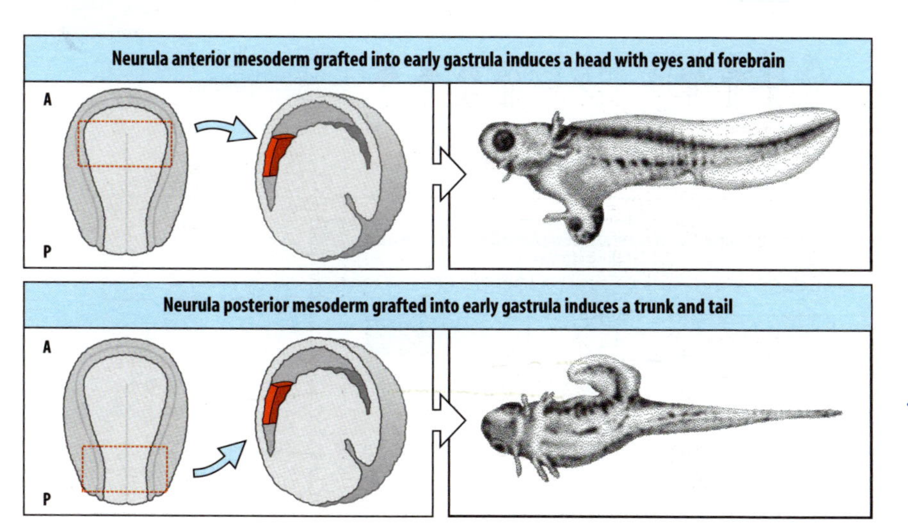

Fig. 4.26 Induction of the nervous system by the mesoderm is region specific. Mesoderm from different positions along the dorsal antero-posterior axis of early newt neurulas induces structures specific to its region of origin when transplanted to ventral regions of early gastrulas. Anterior mesoderm induces a head with a brain (top panels), whereas posterior mesoderm induces a posterior trunk with a spinal cord ending in a tail (bottom panels).

After Mangold, O.: 1933.

induction comes from the observation that pieces of the neural plate themselves induce similar regional neural structures in adjacent ectoderm when transplanted beneath the ectoderm of another gastrula. Indications that gene expression in the mesoderm may be influencing gene expression in the ectoderm come from the observation of coincident expression of several Hox genes in the notochord and in the pre-somitic mesoderm and ectoderm at the same position along the antero-posterior axis: *Xlhbox1* in *Xenopus* and *Hoxb1* in the mouse are examples of genes that have coincident expression of this type.

Hox genes are involved in patterning the hindbrain, as we shall see later, but Hox gene expression cannot be detected in the anterior-most neural tissue of the mouse—the midbrain and forebrain. Instead, homeodomain transcription factors such as Otx and Emc are expressed anterior to the hindbrain and specify pattern in the anterior brain in a manner similar to the Hox genes more posteriorly. The *Drosophila orthodenticle* gene and the mouse *Otx* genes are homologous and provide a good example of the conservation of gene function during evolution. *orthodenticle* is expressed in the posterior region of the future *Drosophila* brain and mutations in the gene result in a greatly reduced brain. In mice, *Otx1* and *Otx2* are expressed in overlapping domains in the developing forebrain and hindbrain, and mutation in *Otx1* leads to brain abnormalities and epilepsy. Mice with a defective *Otx* gene can be partly rescued by replacing it with *orthodenticle*, even though the sequence similarity of the two proteins is confined to the homeodomain region. Human *Otx* can even rescue *orthodenticle* mutants in *Drosophila*.

One model for neural ectoderm patterning in *Xenopus* suggests that qualitatively different inducers are present in the mesoderm at different positions along the antero-posterior axis. Many experiments fit quite well with a simpler two-signal model of neural patterning, however, in which differences are due to quantitative rather than qualitative differences in the inducing signals (Fig. 4.27). In this model, the first, activating, signal is produced by the whole mesoderm and induces the ectoderm to become anterior neural tissue. A second signal transforms part of this tissue so that it acquires a more posterior identity. This latter transforming signal would be graded in the mesoderm with the highest concentration at the posterior end. In *Xenopus*, chick, and zebrafish, Wnts are the posteriorizing factors,

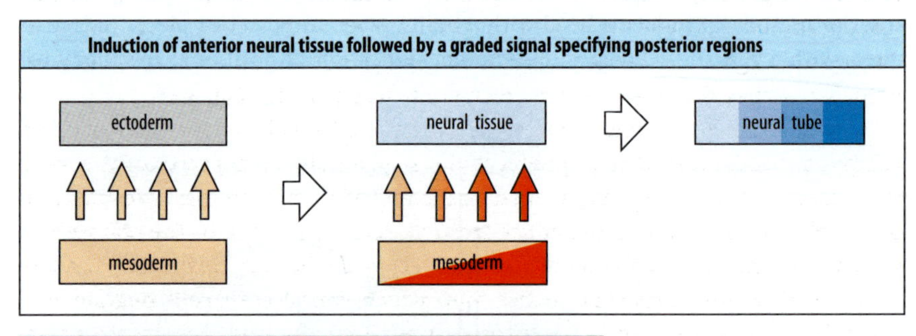

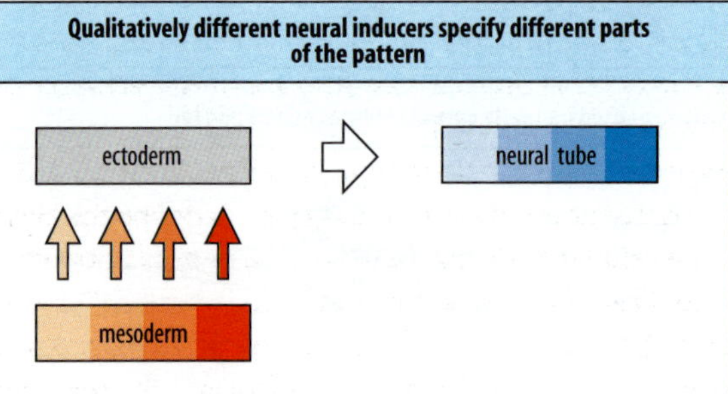

Fig. 4.27 Models of neural patterning by induction. Top panel: in the two-signal model, one signal from the mesoderm first induces anterior tissue throughout the corresponding ectoderm. A second, graded, signal from the mesoderm then specifies more posterior regions. Bottom panel: in an alternative model, qualitatively different inducers are localized in the mesoderm.

After Kelly, O.G., et al.: 1995.

as increased concentrations of these proteins cause cells in the neural plate to adopt a more posterior fate. Another component of the posteriorizing signal is retinoic acid. In the mouse, the anterior visceral endoderm is the source of inhibitory signals that protect anterior tissue from these posteriorizing signals.

4.9 There is an organizer at the midbrain–hindbrain boundary

The developing brain is divided into three main regions—forebrain, midbrain, and hindbrain (Fig. 4.28). The hindbrain has a segmented organization, which will be discussed in the next section, whereas the midbrain and forebrain are unsegmented. An organizing region located at the midbrain–hindbrain boundary, also known as the isthmus, regulates patterning of the midbrain. If this region is grafted into the anterior midbrain it respecifies the tissue as posterior midbrain, and transplanting it into the forebrain converts that tissue into midbrain. FGF-8 is expressed at the isthmus and is a good candidate for signaling posterior midbrain structures; induced expression of FGF-8 in the chick anterior midbrain results in the expression of genes characteristic of the posterior midbrain. Zebrafish that have the mutation *acerebellar*, which is due to loss of FGF-8 function, lack the posterior midbrain regions associated with the organizer. The development of the organizer region is linked to the expression of *Otx2* in the midbrain and *Gbox2* in the hindbrain; the border between the expression of these two genes is important for organizer development. FGF-8 is subsequently expressed on the midbrain side of the border and Wnt on the hindbrain side.

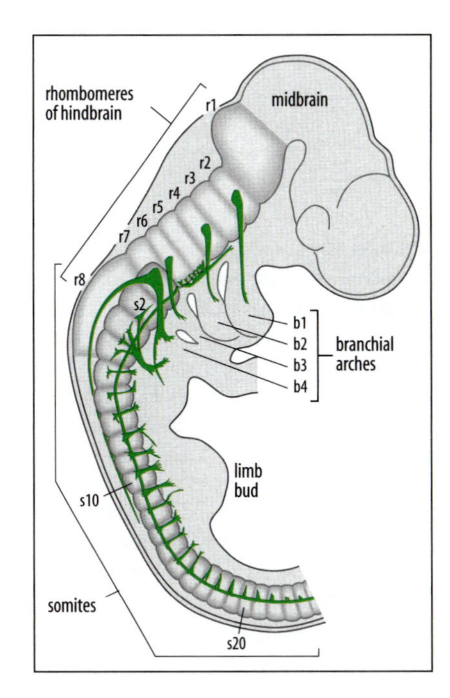

Fig. 4.28 The nervous system in a 3-day chick embryo. The hindbrain is divided into eight rhombomeres (r1 to r8). The positions of the cranial nerves III to XII are shown in green. b1 to b4 are the four branchial arches. b1 gives rise to the jaws. s = somite.

Adapted from Lumsden, A.: 1991.

4.10 The hindbrain is segmented into rhombomeres by boundaries of cell-lineage restriction

Patterning of the posterior region of the head and the hindbrain involves segmentation of the neural tube along the antero-posterior axis. This type of patterning does not occur along the spinal cord, where the pattern of dorsal root ganglia and ventral motor nerves at regular intervals—one pair per somite—is imposed by the somites. In the chick embryo, three segmented systems can be seen in the posterior head region by 3 days of development: the mesoderm on either side of the notochord is subdivided into somites, the hindbrain (the rhombencephalon) is divided into eight **rhombomeres**, and the lateral mesoderm has formed a series of branchial arches that are populated by neural crest cells (see Fig. 4.28).

Development of the hindbrain region of the head involves several interacting components. The neural tube produces the segmentally arranged cranial nerves that innervate the face and neck and neural crest cells, which in turn give rise both to peripheral nerves and most of the facial skeleton. In addition, the otic vesicle gives rise to the ear. The main skeletal elements of the head in this region develop from the first three branchial arches, into which neural crest cells migrate. For example, the first arch gives rise to the jaws, while the second arch develops into the bony parts of the ear. This region of the head is a particularly valuable model for studying patterning along the antero-posterior axis because of the presence of the numerous different structures ordered along it.

Immediately after the neural tube of the chick embryo closes in this region, the future hindbrain becomes constricted at evenly spaced positions to define the eight rhombomeres (see Fig. 4.28). The cellular basis for these constrictions is not understood but may involve differential cell division or changes in cell shape. Whatever the underlying cause, it seems that the boundaries between rhombomeres are barriers of cell-lineage restriction; that is, once the boundaries form, cells and their

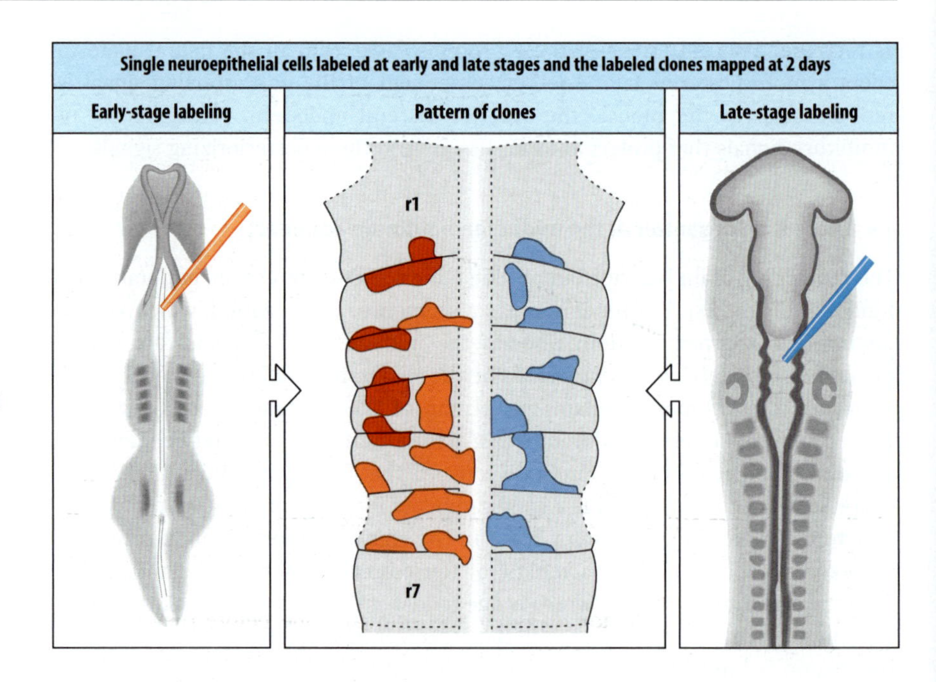

Fig. 4.29 Lineage restriction in rhombomeres of the embryonic chick hindbrain. Single cells are injected with a label (rhodamine-labeled dextran) at an early stage (left panel) or a later stage (right panel) of neurulation, and their descendants are mapped 2 days later. Cells injected before rhombomere boundaries form give rise to some clones that span two rhombomeres (dark red) as well as those that do not cross boundaries (red). Clones marked after rhombomere formation never cross the boundary of the rhombomere that they originate in (blue).

Adapted from Lumsden, A.: 1991.

descendants are confined within a rhombomere and do not cross from one side of a boundary to the other. Marking of individual cells shows that before the constrictions become visible, the descendants of a given labeled cell can populate two adjacent rhombomeres. After the constrictions appear, however, descendants of cells then within a rhombomere never cross the boundaries and are thus confined to a single rhombomere (Fig. 4.29). It seems that the cells of a rhombomere share some adhesive property that prevents them mixing with those of adjacent rhombomeres. This property involves **ephrins**, membrane-bound proteins that interact with **Eph receptors** on adjacent cells and which can generate bidirectional signals (Fig. 4.30). Eph receptors and ephrins are separately expressed in alternating rhombomeres, thus preventing cell mixing at the boundaries. This implies that cells in each rhombomere may be under the control of the same genes, and that the rhombomere is a developmental unit. A rhombomere is thus behaving like a compartment, which is an important feature of insect development, as we saw in Chapter 2, but seems to be rare in vertebrates.

The idea that each rhombomere is a developmental unit is supported by the observation that when an odd-numbered and an even-numbered rhombomere from different positions along the antero-posterior axis are placed next to each other after a boundary between them has been surgically removed, a new boundary forms. No boundaries form when different odd-numbered rhombomeres are placed next to each other, however, suggesting that their cells have similar surface properties. Rhombomeres r3 and r5 express *Eph4*, which is activated by the transcription factor Krox-20, which is expressed in two stripes in the neural plate that become r3 and r5.

The division of the hindbrain into rhombomeres has functional significance in that each rhombomere has a unique identity and this determines how it will develop. As we will see later, their development is under the control of Hox gene expression. We will first consider the behavior of the neural crest cells that originate from the neural tube in the hindbrain and initially migrate over the rhombomeres. They populate the branchial arches, subsequently giving rise to structures such as the lower jaw.

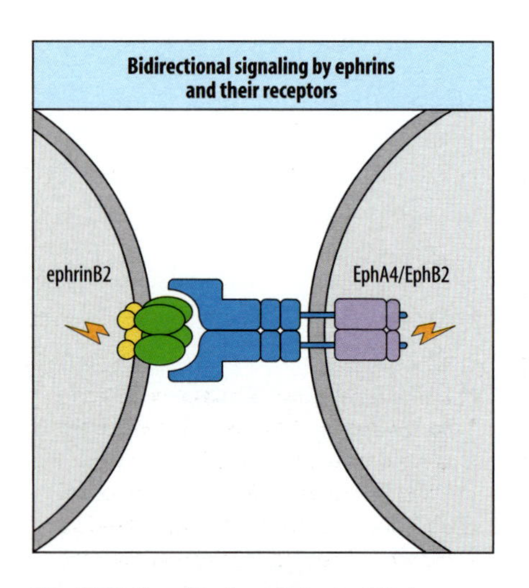

Fig. 4.30 Signaling by ephrins and their receptors. Ephrins binding to their receptors (Eph) can generate a bidirectional signal, in which the ephrin itself also generates a signal in the cell that carries it.

4.11 Neural crest cells arise from the borders of the neural plate

Neural crest cells are induced at the borders of the neural plate, which come together to form the dorsal portion of the neural tube. From there, they migrate to give rise to a wide variety of cell types, as described in Chapter 8. The induction of the neural crest is a multistep process that starts in the early gastrula stage and continues until neural-tube closure. A current model for neural-crest induction is that crest cells form in two bands in the ectoderm along each lateral border of the neural plate, where the level of BMP signaling is just above the level that would block neural-plate formation. In addition, Wnt, FGF, and retinoic acid signals appear to be involved. Induction leads to activation of the transcription factors Sox9 and Sox10, which in turn activate the gene *snail*, an early marker of neural crest.

The cranial neural crest that migrates from the rhombomeres of the dorsal region of the hindbrain has a segmental arrangement, correlating closely with the rhombomere from which the crest cells come. This has been revealed by labeling chick neural crest cells *in vivo* and following their migration pathways. Crest cells from rhombomeres 2, 4, and 6 populate the first, second, and third branchial arches, respectively (Fig. 4.31).

The crest cells have already acquired a positional value before they begin to migrate. When crest cells of rhombomere 4 are replaced by cells from rhombomere 2 taken from another embryo, these cells enter the second branchial arch but develop into structures characteristic of the first arch, to which they would normally have migrated. This can result in the development of an additional lower jaw in the chick embryo. However, neural crest cells have some developmental plasticity and their ultimate differentiation depends on signals from the tissues into which they migrate.

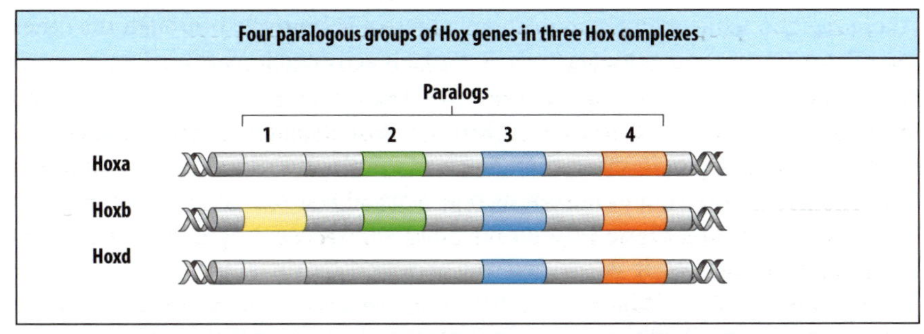

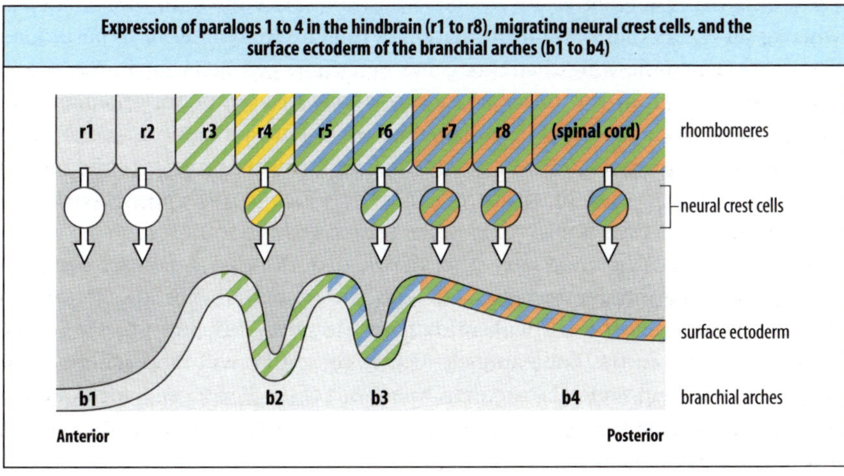

Fig. 4.31 Expression of Hox genes in the branchial region of the head. The expression of genes of three paralogous Hox complexes in the hindbrain (rhombomeres r1 to r8), neural crest, branchial arches (b1 to b4), and surface ectoderm is shown. *Hoxa1* and *Hoxd1* are not expressed at this stage. The arrows indicate the migration of neural crest cells into the branchial arches. Note the absence of neural crest migration from r3 and r5.

After Krumlauf, R.: 1993.

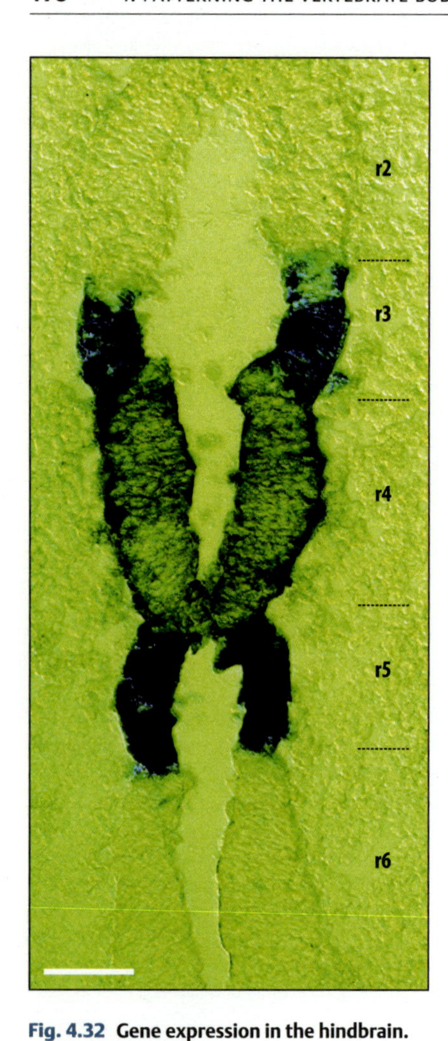

Fig. 4.32 Gene expression in the hindbrain. The photograph shows a coronal section through the hindbrain of a 9.5-day *post-coitum* mouse embryo, which is transgenic for two reporter constructs. The first construct contains the *lacZ* gene under the control of an enhancer from *Hoxb2*, which directs expression in rhombomeres 3 and 5 (revealed as blue staining). The second construct contains an alkaline phosphatase gene under the control of an enhancer from *Hoxb1*, which directs expression in rhombomere 4 (revealed as greenish-brown staining). A similar enhancer directing expression in rhombomere 4 exists for *Hoxb2*. Anterior is uppermost, and the positions of five of the rhombomeres are indicated (r2 to r6). Scale bar = 0.1 mm.

Photograph courtesy of J. Sharpe, from Lumsden, A. and Krumlauf, R.: 1996.

4.12 Hox genes provide positional information in the developing hindbrain

Hox gene expression provides a possible molecular basis for the identities of both the rhombomeres and the neural crest at different positions in the hindbrain. No Hox genes are expressed in the most anterior part of the head but Hox genes are expressed in the mouse embryo hindbrain in a well defined pattern, which closely correlates with the segmental pattern (see Fig. 4.31). It is clear that the three paralogous groups involved have different anterior margins of expression, paralog 1 (i.e. *Hoxb1*, *Hoxc1*, etc.) being most anterior, followed by paralogs 2 and 3. For example, *Hoxb3* has its most anterior region of expression at the border of rhombomeres 4 and 5, while *Hoxb2* has its anterior border at the border of rhombomeres 2 and 3 (Fig. 4.32). In general, the paralogous genes of the different Hox complexes have similar patterns of expression.

The pattern of Hox gene expression in the ectoderm and branchial arches at a particular position along the antero-posterior axis is similar to that in the neural tube and neural crest, and it may be that the crest cells induce their positional values in the overlying ectoderm during their migration.

Transplantation of rhombomeres from an anterior to a more posterior position alters the pattern of Hox gene expression so that it becomes the same as that normally expressed at the new location. The signals responsible for this reprogramming originate from the neural tube itself and not the surrounding tissues. Studies of the control of Hox gene expression at the molecular level have provided some indication as to how their pattern of expression is controlled. For example, although the *Hoxb2* gene is expressed in the three contiguous rhombomeres 3, 4, and 5, its expression in rhombomeres 3 and 5 is controlled quite independently from its expression in rhombomere 4. The regulatory regions of the *Hoxb2* gene carry two separate *cis*-regulatory regions that regulate its expression in these three rhombomeres. Expression in rhombomeres 3 and 5 is controlled through one of these regions, while expression in rhombomere 4 is controlled through the other (see Fig. 4.32). In rhombomeres 3 and 5, *Hoxb2* is activated in part by the transcription factor Krox-20, which is expressed in these rhombomeres but not in rhombomere 4. There are binding sites for Krox-20 in the regulatory DNA element that activates expression of *Hoxb2* in rhombomeres 3 and 5. How the spatially organized expression of transcription factors such as Krox-20 is achieved is not yet known. In the case of *Hoxb4*, whose anterior boundary of expression is at rhombomere 6, there is evidence that it is induced and localized by signaling both within the neural tube and from adjacent somites. There is also a role for retinoic acid in patterning the hindbrain, as its absence leads to loss of hindbrain rhombomeres, whereas an excess causes a transformation of cell fate from anterior to posterior.

An experiment showing that Hox genes determine cell behavior in the rhombomeres comes from the misexpression of *Hoxb1* in an anterior rhombomere. Motor axons from a pair of rhombomeres project to a single branchial arch; rhombomere r2 projects to the first branchial arch while r4 projects to the second arch. *Hoxb1* is expressed in r4 but not in r2. If *Hoxb1* is misexpressed in an r2 rhombomere, this then sends axons to the second arch.

Gene knock-outs in mice have also shown that the Hox genes are involved in patterning the hindbrain region, though the results are not always easy to interpret; knock-out of a particular Hox gene can affect different populations of neural crest cells in the same animal, such as those that will form neurons and those that will form skeletal structures. Knock-out of the *Hoxa2* gene, for example, results in skeletal defects in that region of the head corresponding to the normal domain of expression of the gene, which extends from rhombomere 3 backwards.

Segmentation itself is not affected, but the skeletal elements in the second branchial arch, all of which come from neural crest cells derived from rhombomere 4, are abnormal. The usual elements, such as the stapes of the inner ear, are absent, but instead some of the skeletal elements normally formed by the first branchial arch develop, such as Meckel's cartilage, which is a precursor element in the lower jaw. Thus, suppression of *Hoxa2* causes a partial homeotic transformation of one segment into another. The converse effect is seen when *Hoxa2* is misexpressed in a more anterior position (see also Section 4.4). If *Hoxa2* is misexpressed in all tissues of the first branchial arch, in which no Hox genes are normally expressed, there is a homeotic transformation into second branchial arch. In addition, if *Hox2a* is transfected into prospective anterior neural crest, the cells do not develop in their normal way and the skeletal structures they give rise to are abolished.

These observations, together with those described earlier in this chapter, show that during gastrulation the cells of vertebrates acquire positional values along the antero-posterior axis, and that this positional identity is encoded by the genes of the Hox complexes. Many of the anatomical differences between vertebrates are probably simply due to differences in the subsequent targets of Hox gene actions, which result in the emergence of different but homologous skeletal structures—the mammalian jaw or the bird's beak, for example.

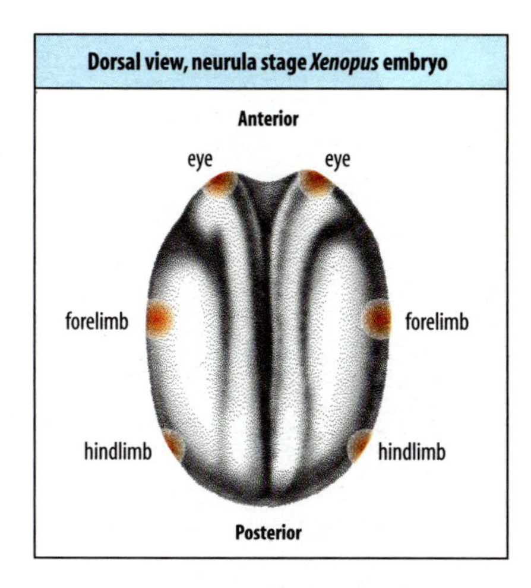

Fig. 4.33 The *Xenopus* embryo has become regionalized by the neurula stage. Various organs such as limbs, heart, and eyes will develop from specific regions (red) of the neurula after gastrulation is complete. Some of these regions, like the limb buds, are already determined at this stage and will not form any other structure. The boundaries of the regions are not sharply defined, however, and within each region or 'field' considerable regulation is still possible.

4.13 The embryo is patterned by the neurula stage into organ-forming regions that can still regulate

At the neurula stage, the body plan has been established and the regions of the embryo that will form limbs, eyes, heart, and other organs have become determined (Fig. 4.33). This contrasts sharply with the blastula stage, at which time no such determination has occurred. The basic vertebrate phylotypic body plan is thus established during gastrulation. But although the positions of various organs are fixed, there is no overt sign yet of differentiation. Numerous grafting experiments have shown that the potential to form a given organ is now confined to specific regions. Each of these regions has, however, considerable capacity for regulation, so that if part of the region is removed a normal structure can still form. For example, the region of the neurula that will form a forelimb will, when transplanted to a different region, still develop into a limb. If part of a future limb region is removed, the remaining part can still regulate to develop a normal limb. The development of limbs and other organs is discussed in Chapter 8.

Summary

Patterning along both the antero-posterior and dorso-ventral axes is closely related to the action of the Spemann organizer and its morphogenesis during gastrulation. When grafted to the ventral side of an early gastrula, the Spemann organizer induces both a new dorso-ventral axis and a new antero-posterior axis, with the development of a second twinned embryo. In the zebrafish the embryonic shield is the organizer, while in chick development, Hensen's node serves a function similar to that of the Spemann organizer, and it too can specify a new antero-posterior axis. In the mouse, the node can specify a new axis apart from the most anterior forebrain, for which induction by the anterior visceral endoderm is also necessary.

The vertebrate nervous system, which forms from the neural plate, is induced both by early signals within the ectoderm and by signals from the mesoderm that comes to lie beneath prospective neural plate ectoderm during gastrulation.

For some proteins that can induce neural tissue in *Xenopus*, such as Noggin, this ability is due to their inhibition of BMP signaling. Patterning of the neural plate can partly be accounted for by a two-signal model: the ectoderm is first specified as anterior neural tissue and then a second set of signals, possibly graded, specifies more posterior structures.

The hindbrain is segmented into rhombomeres, with the cells of each rhombomere respecting their boundaries. Neural crest cells from the hindbrain populate specific mesodermal regions such as the branchial arches in a position-dependent fashion. A Hox gene code provides positional values for the rhombomeres and neural crest cells of the hindbrain region, while other genes specify more anterior regions. By the neurula stage, after gastrulation, the body plan has been established.

Summary: patterning of the vertebrate axial body plan

gastrulation and organizer activity

⇩

the Hox gene complexes are expressed along the antero-posterior axis

⇩

Hox gene expression establishes positional identity for mesoderm, endoderm, and ectoderm

⇙ ⇘

mesoderm develops into notochord, somites, and lateral plate mesoderm

⇩

somites receive signals from notochord, neural tube, and ectoderm

⇩

somite develops into sclerotome and dermomyotome

early signals and mesoderm induce neural plate from ectoderm

⇩

mesoderm signals give regional identity to neural tube

⇩

rhombomeres and neural crest in the hindbrain are characterized by regional patterns of Hox gene expression

SUMMARY TO CHAPTER 4

The germ layers specified during blastula formation become patterned along the antero-posterior and dorso-ventral axes during gastrulation. The primary embryonic organizer is involved in the initial patterning that underlies the regionalization of the antero-posterior axis. Positional identity of cells along the antero-posterior axis is encoded by the combinatorial expression of genes of the four Hox complexes, which provide a code for regional identity. There is both spatial and temporal co-linearity between the order of Hox genes on the chromosomes and the order in which they are expressed along the antero-posterior axis of the embryo from the hindbrain backwards. Inactivation or overexpression of Hox genes can lead both to localized abnormalities and to homeotic transformations of one 'segment' of the axis into another, indicating that these genes are crucial in specifying regional identity. At the end of gastrulation, the basic body plan has been laid down and the nervous system induced. Specific regions of each somite give rise to cartilage, muscle, and dermis, and these regions are specified by signals from the notochord, neural tube, and epidermis. Induction and patterning of the nervous system involves both signals in the early embryo and from the underlying mesoderm. In the hindbrain, Hox gene expression provides positional values for both neural tissue and neural crest cells.

SECTION FURTHER READING

4.1 Somites are formed in a well defined order along the antero-posterior axis

Artavanis-Tsakonas, S., Rand, M.D., Lake, R.J.: **Notch signaling: cell fate control and signal integration in development.** *Science* 1999, **284:** 770–776.

Dale, K.J., Pourquié, O.: **A clockwork somite.** *BioEssays* 2000, **22:** 72–83.

Deschamps, J., van Nes, J.: **Developmental regulation of the Hox genes during axial morphogenesis in the mouse.** *Development* 2005, **132:** 2931–2942.

Dubrulle, J., Pourquié, O.: ***fgf8* mRNA decay establishes a gradient that couples axial elongation to patterning in the vertebrate embryo.** *Nature* 2004, **427:** 419–422.

Dubrulle, J., Pourquié, O: **Coupling segmentation to axis formation.** *Development* 2004, **131:** 5783–5793.

Kawakami, Y., Raya, A., Raya, R.M., Rodriguez-Esteban, C., Izpisua Belmonte, J.C.: **Retinoic acid signalling links left-right asymmetric patterning and bilaterally symmetric somitogenesis in the zebrafish embryo.** *Nature* 2005, **435:** 165–171.

Kieny, M., Mauger, A., Sengel P.: **Early regionalization of somitic mesoderm as studied by the development of the axial skeleton of the chick embryo.** *Dev. Biol.* 1972, **28:** 142–161.

Kulesa, P.M., Fraser, S.E.: **Cell dynamics during somite boundary formation revealed by time-lapse analysis.** *Science* 2002, **298:** 991–995.

Lai, E.: **Notch signaling: control of cell communication and cell fate.** *Development* 2004, **131:** 965–973.

Smith, J.C., Gurdon, J.B.: **Many ways to make a gradient.** *BioEssays* 2004, **26:** 705–706.

Stern, C.D., Charité, J., Deschamps, J., Duboule, D., Durston, A.J., Kmita, M., Nicolas, J.-F., Palmeirim, I., Smith, J.C., Wolpert, L.: **Head-tail patterning of the vertebrate embryo: one, two or many unsolved problems?** *Int. J. Dev. Biol.* 2006, **50:** 3–15.

Vermot, J., Pourquié, O.: **Retinoic acid coordinates somitogenesis and left-right patterning in vertebrate embryos.** *Nature* 2005, **435:** 215–220.

4.2 Identity of somites along the antero-posterior axis is specified by Hox gene expression

Burke, A.C., Nelson, C.E., Morgan, B.A., Tabin, C.: **Hox genes and the evolution of vertebrate axial morphology.** *Development* 1995, **121:** 333–346.

Godsave, S., Dekker, E.J., Holling, T., Pannese, M., Boncinelli, E., Durston, A.: **Expression patterns of Hoxb in the *Xenopus* embryo suggest roles in antero-posterior specification of the hindbrain and in dorso-ventral patterning of the mesoderm.** *Dev. Biol.* 1994, **166:** 465–476.

Kondo, T., Duboule, D.: **Breaking colinearity in the mouse HoxD complex.** *Cell* 1999, **97:** 407–417.

Krumlauf, R.: **Hox genes in vertebrate development.** *Cell* 1994, **78:** 191–201.

Nowicki, J.L., Burke, A.C.: **Hox genes and morphological identity: axial versus lateral patterning in the vertebrate mesoderm.** *Development* 2000, **127:** 4265–4275.

4.3 Deletion or overexpression of Hox genes causes changes in axial patterning

Condie, B.G., Capecchi, M.R.: **Mice with targeted disruptions in the paralogous genes Hoxa3 and Hoxd3 reveal synergistic interactions.** *Nature* 1994, **370:** 304–307.

Conlon, R.A.: **Retinoic acid and pattern formation in vertebrates.** *Trends Genet.* 1995, **11:** 314–319.

Duboule, D.: **Vertebrate Hox genes and proliferation: an alternative pathway to homeosis?** *Curr. Opin. Genet. Dev.* 1995, **5:** 525–528.

Favier, B., Le Meur, M., Chambon, P., Dollé, P.: **Axial skeleton homeosis and forelimb malformations in Hoxd11 mutant mice.** *Proc. Natl Acad. Sci. USA* 1995, **92:** 310–314.

Grammatopoulos, G.A., Bell, E., Toole, L., Lumsden, A., Tucker, A.S.: **Homeotic transformation of branchial arch identity after hoxa2 overexpression.** *Development* 2000, **127:** 5355–5365.

Kessel, M., Gruss, P.: **Homeotic transformations of moving vertebrae and concomitant alteration of the codes induced by retinoic acid.** *Cell* 1991, **67:** 89–104.

Wellik, D.M., Capecchi, M.R.: **Hox10 and Hox11 genes are required to globally pattern the mammalian skeleton.** *Science* 2003, **301:** 363–367.

Rossant, J., Spence, A.: **Chimeras and mosaics in mouse mutant analysis.** *Trends Genet.* 1998, **14:** 358–363.

Ruiz-i-Altaba, A., Jessell, T.: **Retinoic acid modifies mesodermal patterning in early *Xenopus* embryos.** *Genes Dev.* 1991, **5:** 175–187.

Yekta, S., Shih, I.H., Bartel, D.P.: **MicroRNA-directed cleavage of HOXB8 mRNA.** *Science* 2004, **304:** 594–596.

4.4 Hox gene activation is related to a timing mechanism

Duboule, D.: **Vertebrate Hox gene regulation: clustering and/or colinearity?** *Curr. Opin. Genet. Dev.* 1998, **8:** 514–518.

Vasiliauskas, D., Stern, C.D.: **Patterning the embryonic axis: FGF signaling and how vertebrate embryos measure time.** *Cell* 2001, **106:** 133–136.

Wacker, S.A., Janse, H.J., McNulty, C.L., Houtzager, E., Durston, A.J.: **Timed interactions between the Hox expressing non-organiser mesoderm and the Spemann organiser generate positional information during vertebrate gastrulation.** *Dev. Biol.* 2004, **268:** 207–219.

4.5 The fate of somite cells is determined by signals from the adjacent tissues

Brand-Saberi, B., Christ, B.: **Evolution and development of distinct cell lineages derived from somites.** *Curr. Topics Dev. Biol.* 2000, **48:** 1–42.

Brent, A.E., Braun, T., Tabin, C.J.: **Genetic analysis of interactions between the somitic muscle, cartilage and tendon cell lineages during mouse development.** *Development* 2005, **132:** 515–528.

Brent, A.E., Schweitzer, R., Tabin, C.J.: **A somitic compartment of tendon progenitors.** *Cell* 2003, **113:** 235–248.

Brent, A.E., Tabin, C.J.: **Developmental regulation of somite derivatives: muscle, cartilage and tendon.** *Curr. Opin. Genet. Dev.* 2002, **12:** 548–557.

Huang, R., Zhi, Q., Patel, K., Wilting, J., Christ, B.: **Dual origin and segmental organisation of the avian scapula.** *Development* 2000, **127:** 3789–3794.

Olivera-Martinez, I., Coltey, M., Dhouailly, D., Pourquié, O.: **Mediolateral somitic origin of ribs and dermis determined by quail-chick chimeras.** *Development* 2000, **127:** 4611–4617.

Pourquié, O., Fan, C-M., Coltey, M., Hirsinger, E., Watanabe, Y., Bréant, C., Francis-West, P., Brickell, P., Tessier-Lavigne, M., Le Douarin, N.M.: **Lateral and axial signals involved in avian somite patterning: a role for BMP-4.** *Cell* 1996, **84:** 461–471.

Selleck, M., Stern, C.D.: **Fate mapping and cell lineage analysis of Hensen's node in the chick embryo.** *Development* 1991, **112:** 615–626.

4.6 The inductive capacity of the organizer changes during development

Beddington, R.S.P., Robertson, E.H.: **Axis development and early asymmetry in mammals.** *Cell* 1999, **96:** 195–209.

Brickman, J.M., Jones, C.M., Clements, M., Smith, J.C., Beddington, R.S.P.: **Hex is a transcriptional repressor that contributes to anterior identity and suppresses Spemann organiser function.** *Development* 2000, **127:** 2303–2315.

Gad, J.M., Tam, P.P.L.: **The mouse becomes a dachshund.** *Curr. Biol.* 1999, **9:** R783–R786.

Griffin, K., Patient, R., Holder, N.: **Analysis of FGF function in normal and no tail zebrafish embryos reveals separate mechanisms for formation of the trunk and tail.** *Development* 1995, **121:** 2983–2994.

Jones, C.M., Broadbent, J., Thomas, P.Q., Smith, J.C., Beddington, R.S.P.: **An anterior signalling centre in *Xenopus* revealed by the homeobox gene *XHex*.** *Curr. Biol.* 1999, **9:** 946–954.

Joubin, K., Stern, C.D.: **Molecular interactions continuously define the organizer during the cell movements of gastrulation.** *Cell* 1999, **98:** 559–571.

Kiecker, C., Niehrs, C.: **The role of prechordal mesmedoderm in neural patterning.** *Curr. Opin. Neurobiol.* 2001, **11:** 27–33.

Niehrs, C.: **Head in the Wnt: the molecular nature of Spemann's head organizer.** *Trends Genet.* 1999, **15:** 314–315.

Niehrs, C.: **Regionally specific induction by the Spemann-Mangold organizer.** *Nat. Rev. Genet.* 2004, **5:** 425–434.

Piccolo, S., Agius, E., Leyns, L., Bhattacharya, S., Grunz, H., Bouwmeester, T., De Robertis, E.M.: **The head inducer Cerberus is a multifunctional antagonist of Nodal, BMP and Wnt signals.** *Nature* 1999, **397:** 707–710.

Schneider, V.A., Mercola, M.: **Spatially distinct head and heart inducers within the *Xenopus* organizer region.** *Curr. Biol.* 1999, **9:** 800–809.

Yamaguchi, T.P.: **Heads or tails: Wnts and anterior-posterior patterning.** *Curr. Biol.* 2001, **11:** R713–R724.

4.7 The neural plate is induced in the ectoderm

Bachiller, D., Klingensmith, J., Kemp, C., Belo, J.A., Anderson, R.M., May, S.R., MacMahon, J.A., McMahon, A.P., Harland, R.M., Rossant, J., De Robertis, E.M.: **The organizer factors Chordin and Noggin are required for mouse forebrain development.** *Nature* 2000, **403:** 658–661.

Delfino-Machin, M., Lunn, J.S., Breitkreuz, D.N., Akai, J., Storey, K.G.: **Specification and maintenance of the spinal cord stem zone.** *Development* 2005, **132:** 4273–4283.

De Robertis, E.M., Kuroda, H.: **Dorsal-ventral patterning and neural induction in *Xenopus* embryos.** *Annu. Rev. Dev. Biol.* 2004, **20:** 285–308.

Harland, R.: **Neural induction.** *Curr. Opin. Genet. Dev.* 2000, **10:** 357–362.

Kemmati-Brivanlou, A., Melton, D.: **Vertebrate embryonic cells will become nerve cells unless told otherwise.** *Cell* 1997, **88:** 13–17.

Londin, E.R., Niemiec, J., Sirotkin, H.I.: **Chordin, FGF signaling, and mesodermal factors cooperate in zebrafish neural induction.** *Dev. Biol.* 2005, **279:** 1–19.

Sasai, Y., Lu, B., Steinbesser, H., De Robertis, E.M.: **Regulation of neural induction by the Chd and BMP-4 antagonistic patterning signals in *Xenopus*.** *Nature* 1995, **376:** 333–336.

Stern, C.: **Neural induction: old problems, new findings, yet more questions.** *Development* 2005, **131:** 2007–2021.

Streit, A., Berliner, A.J., Papanayotou, C., Sirulnik, A., Stern, C.D.: **Initiation of neural induction by FGF signalling before gastrulation.** *Nature* 2000, **406:** 74–78.

Streit, A., Stern, C.D.: **Neural induction: a bird's eye view.** *Trends Genet.* 1999, **15:** 20–24.

Wilson, P., Kemmati-Brivanlou, A.: **Induction of epidermis and inhibition of neural fate by BMP-4.** *Nature* 1995, **376:** 331–333.

Wilson, S.L., Rydström, A., Trimborn, T., Willert, K., Nusse, R., Jessell, T. M., Edlund, T.: **The status of Wnt signaling regulates neural and epidermal fates in the chick embryo.** *Nature* 2001, **411:** 325–329.

4.8 The nervous system can be patterned by signals from the mesoderm

Ang, S.L., Rossant, J.: **HNF-3b is essential for node and notochord formation in mouse development.** *Cell* 1994, **78:** 561–574.

Aybar, M.J., Mayor, R.: **Early induction of neural crest cells: lessons learned from frog, fish and chick.** *Curr. Opin. Genet. Dev.* 2002, **12:** 452-458.

Brocolli, V., Boncinelli, E., Wurst, W.: **The caudal limit of Otx2 expression positions the isthmaic organizer.** *Nature* 1999, **401:** 164–168.

Doniach, T.: **Basic FGF as an inducer of antero-posterior neural pattern.** *Cell* 1995, **85:** 1067–1070.

Foley, A.C., Skromne, I., Stern, C.D.: **Reconciling different models of forebrain induction and patterning: a dual role for the hypoblast.** *Development* 2000, **127:** 3839–3854.

Kudoh, T., Concha, M.L., Houart, C., Dawid, I.B., Wilson, S.W.: **Combinatorial Fgf and Bmp signalling patterns the gastrula ectoderm into prospective neural and epidermal domains.** *Development* 2004, **131:** 3581–3592.

Pera, E.M., Ikeda, A., Eivers, E., De Robertis, E.M.: **Integration of IGF, FGF, and anti-BMP signals via Smad1 phosphorylation in neural induction.** *Genes Dev.* 2003, **17:** 3023–3028.

Sasai, Y., De Robertis, E.M.: **Ectodermal patterning in vertebrate embryos.** *Dev. Biol.* 1997, **182:** 5–20.

Sharman, A.C., Brand, M.: **Evolution and homology of the nervous system: cross-phylum rescues of *otd*/*Otx* genes.** *Trends Genet.* 1998, **14:** 211–214.

Sheng, G., dos Reis, M., Stern, C.D.: **Churchill, a zinc finger transcriptional activator, regulates the transition between gastrulation and neurulation.** *Cell* 2003, **115:** 603–613.

Stern, C.D.: **Initial patterning of the central nervous system: how many organizers?** *Nat. Rev. Neurosci.* 2001, **2:** 92–98.

Storey, K., Crossley, J.M., De Robertis, E.M., Norris, W.E., Stern, C.D.: **Neural induction and regionalization in the chick embryo.** *Development* 1992, **114:** 729–741.

4.9 There is an organizer at the midbrain–hindbrain boundary

Rhinn, M., Brand, M.: **The midbrain-hindbrain boundary organizer.** *Curr. Opin. Neurobiol.* 2001, **11:** 34–42.

4.10 The hindbrain is segmented into rhombomeres by boundaries of cell-lineage restriction

Klein, R.: **Neural development: bidirectional signals establish boundaries.** *Curr. Biol.* 1999, **9:** R691–R694.

Lewis, J.: **Autoinhibition with transcriptional delay: a simple mechanism for the zebrafish somitogenesis oscillator.** *Curr. Biol.* 2003, **13:** 1398–408.

Lumsden, A.: **Segmentation and compartition in the early avian hindbrain.** *Mech. Dev.* 2004, **121:** 1081–1088.

Xu, Q., Mellitzer, G., Wilkinson, D.G.: **Roles of Eph receptors and ephrins in segmental patterning.** *Phil. Trans. Roy. Soc. B* 2000, **355:** 993–1002.

4.11 Neural crest cells arise from the borders of the neural plate

Huang, X., Saint-Jeannet, J-P.: **Induction of the neural crest and the opportunities of life on the edge.** *Dev. Biol.* 2004, **275:** 1–11.

Keynes, R., Lumsden, A.: **Segmentation and the origin of regional diversity in the vertebrate central nervous system.** *Neuron* 1990, **4:** 1–9.

Le Douarin, N.M., Creuzet, S., Couly, G., Dupin, E.: **Neural crest plasticity and its limits.** *Development* 2004, **131:** 4637–4650.

4.12 Hox genes provide positional information in the hindbrain region

Bell, E., Wingate, R.J., Lumsden, A.: **Homeotic transformation of rhombomere identity after localized Hoxb1 misexpression.** *Science* 1999, **284:** 2168–2171.

Grapin-Botton, A., Bonnin, M-A., McNaughton, L.A., Krumlauf, R., Le Douarin, N.M.: **Plasticity of transposed rhombomeres: Hox gene induction is correlated with phenotypic modifications.** *Development* 1995, **121:** 2707–2721.

Hunt, P., Krumlauf, R.: **Hox codes and positional specification in vertebrate embryonic axes.** *Annu. Rev. Cell Biol.* 1992, **8:** 227–256.

Krumlauf, R.: **Hox genes and pattern formation in the branchial region of the vertebrate head.** *Trends Genet.* 1993, **9:** 106–112.

Nonchev, S., Maconochie, M., Vesque, C., Aparicio, S., Ariza-McNaughton, L., Manzanares, M., Maruthainar, K., Kuroiwa, A., Brenner, S., Charnay, P., Krumlauf, R.: **The conserved role of Krox-20 in directing Hox gene expression during vertebrate hindbrain segmentation.** *Proc. Natl Acad. Sci. USA* 1996, **93:** 9339–9345.

Rijli, F.M., Mark, M., Lakkaraju, S., Dierich, A., Dolle, P., Chambon, P.: **A homeotic transformation is generated in the rostral branchial region of the head by disruption of Hoxa2, which acts as a selector gene.** *Cell* 1993, **75:** 1333–1349.

4.13 The embryo is patterned by the neurula stage into organ-forming regions that can still regulate

De Robertis, E.M., Morita, E.A., Cho, K.W.Y.: **Gradient fields and homeobox genes.** *Development* 1991, **112:** 669–678.

Development of nematodes, sea urchins, ascidians, and slime molds

5

Having looked at early embryonic development in *Drosophila* and vertebrates, we will now examine some aspects of the early development of four organisms that introduce some different developmental mechanisms. The nematode *Caenorhabditis elegans* is a very important model organism for the specification of fate associated with asymmetric cell division, in which much early development is carried out by patterning on a cell-by-cell basis rather than by morphogens. Sea urchins, which represent the echinoderms, are models for highly regulative development, and many key developmental genes have been identified together with their *cis*-control elements. The ascidians are of interest as they are chordates and are thus more closely related to vertebrates than are other model invertebrates such as *Drosophila* and *Caenorhabditis*. The development of the cellular slime mold *Dictyostelium discoideum* is unlike that of any other species we will study here, in that it does not develop from a fertilized egg but from individual cells coming together to form a fruiting body.

This chapter considers aspects of body-plan development in three model invertebrate organisms—nematodes, sea urchins, representing echinoderms, and the chordate ascidians, commonly known as sea squirts—and ends with a short discussion of the cellular slime molds, which represent a very basic developmental system. The evolutionary relationships between the organisms discussed in this chapter are shown in Fig. 5.1. All, with the exception of the cellular slime mold, conform to the general plan of animal development: cleavage leads to a blastula, which undergoes gastrulation with the emergence of a body plan.

There is an old, and now less fashionable, distinction sometimes made between so-called regulative and mosaic development—the former involving mainly cell–cell interactions, while the latter is based on localized cytoplasmic factors and their distribution through asymmetric cell divisions (see Section 1.17). The organisms discussed in earlier chapters, such as the vertebrates, are examples of mainly regulative development, whereas the nematodes and ascidians discussed in this chapter are examples of mosaic-like development, although there is an element of both processes in most organisms.

An important feature of the development of nematodes and ascidians is that cell fate is often specified on a cell-by-cell basis, rather than in groups of cells as in flies and vertebrates, and in general does not rely on positional information established

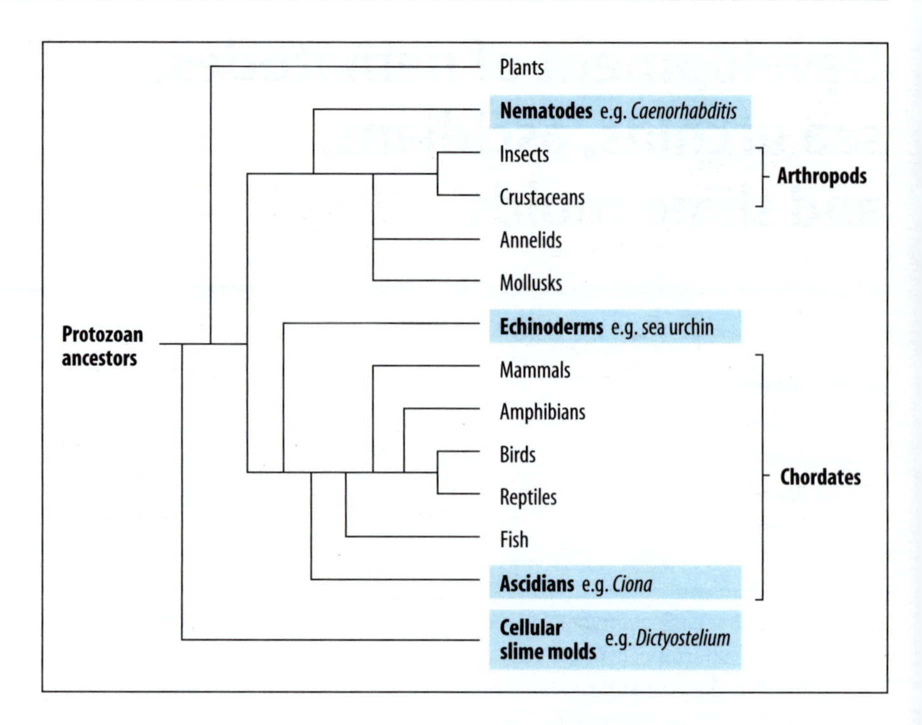

Fig. 5.1 Phylogenetic tree showing relationships between the organisms considered in this book. Those organisms discussed in this chapter are highlighted in blue.

by gradients of morphogens. The early embryos of many invertebrates contain far fewer cells than those of vertebrates or flies, with each cell acquiring a unique identity at an early stage of development. In the nematode, for example, there are only 26 cells when gastrulation starts, compared to thousands in vertebrates. Specification on a cell-by-cell basis often makes use of a developmental mechanism that is less common in early insect and vertebrate development, namely **asymmetric cell division** and the unequal distribution of cytoplasmic factors as a means of determining cell fate (see Section 1.15). Daughter cells resulting from asymmetric cell division often adopt different fates, not as a result of extracellular signals but autonomously, as a result of the unequal distribution of some factor between them. However, asymmetric cell division in the early stages of development does not mean that cell–cell interactions are absent or unimportant in these organisms. Differences in the fates of two daughter cells can also be specified by extracellular factors and cell signaling.

We begin with the nematode *Caenorhabditis elegans*, which has been studied intensively and in which many key developmental genes have been identified. In this animal, specification is largely on a cell-by-cell basis. We then consider sea urchins, which develop much more like vertebrates, in that their embryos rely heavily on intercellular interactions and are highly regulative, and patterning involves groups of cells. Next we consider ascidians, with particular emphasis on the role of the localization of cytoplasmic determinants in their early development. Finally, we look at patterning in the cellular slime mold, which represents a primitive and very different developmental system.

Nematodes

The free-living soil nematode *C. elegans*, whose life cycle is shown in Fig. 5.2, is an important model organism in developmental biology. Its advantages are its

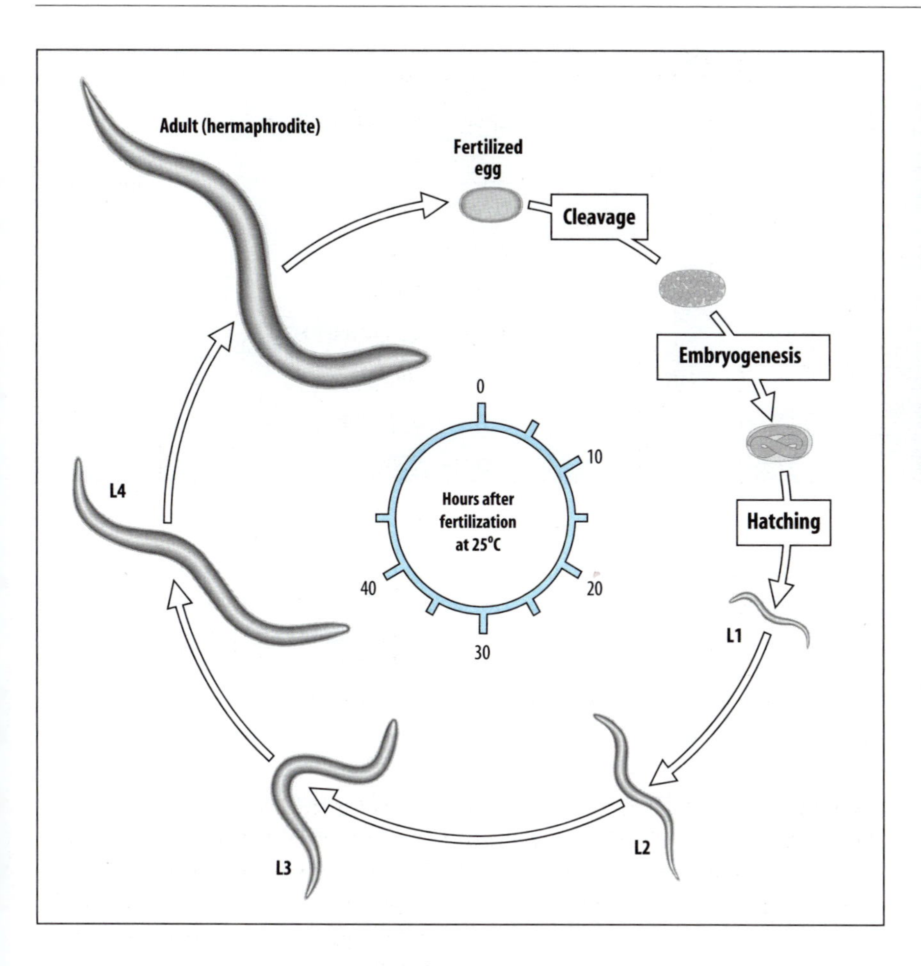

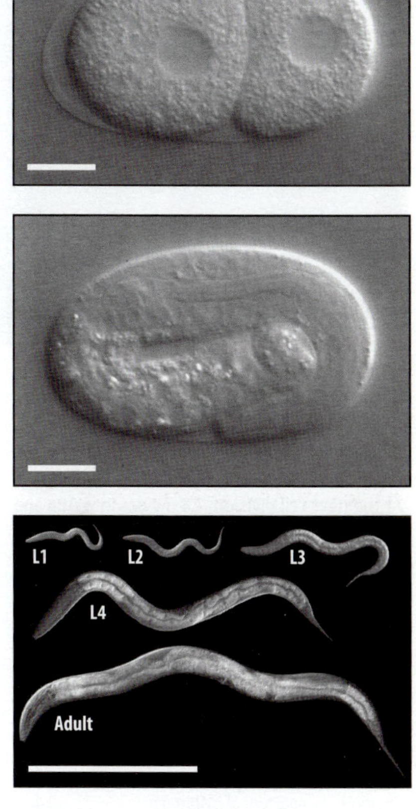

Fig. 5.2 Life cycle of the nematode *Caenorhabditis elegans*. After cleavage and embryogenesis there are four larval stages (L1–L4) before the sexually mature adult develops. Adults of *C. elegans* are usually hermaphrodite, although males can develop. The photographs show: the two-cell stage (top, scale bar = 10 μm); an embryo after gastrulation with the future larva curled up (middle, scale bar = 10 μm); and the four larval stages and adult (bottom, scale bar = 0.5 mm).

Photographs courtesy of J. Ahringer.

suitability for genetic analysis, the small number of cells and their invariant lineage, and the transparency of the embryo, which allows the formation of each cell to be observed. *C. elegans* has a simple anatomy and the adults are about 1 mm long and just 70 μm in diameter. Nematodes can be grown on agar plates in large numbers and early larval stages can be stored frozen and later resuscitated. *C. elegans* adults are mainly **hermaphrodites**; that is, they possess both female and male gonads and reproduce by self-fertilization. Small numbers of males are produced under special conditions. Embryonic development is rapid, the larva hatching after 15 hours at 20°C, although maturation through larval stages to adulthood takes about 50 hours.

The nematode egg is small, only 50 μm in diameter. Polar bodies are formed after fertilization. Before the male and female nuclei fuse, there is what appears to be an abortive cleavage, but after fusion of the nuclei true cleavage begins (Fig. 5.3). The first cleavage is asymmetric and generates an anterior AB cell and a smaller posterior P1 cell. At the second cleavage, AB divides to give ABa anteriorly and ABp posteriorly, while P1 divides to give P2 and EMS. At this stage the main axes can already be identified, since P2 is posterior and ABp dorsal.

The AB cells and the P2 cell will also divide in a well-defined pattern to give rise to the other tissues of the worm. Gastrulation starts at the 28-cell stage, when the descendants of the E cell (produced by division of the EMS cell) that will form the gut move inside. Not all cells that are formed during embryonic development survive; programmed death of specific cells is an integral feature of nematode development.

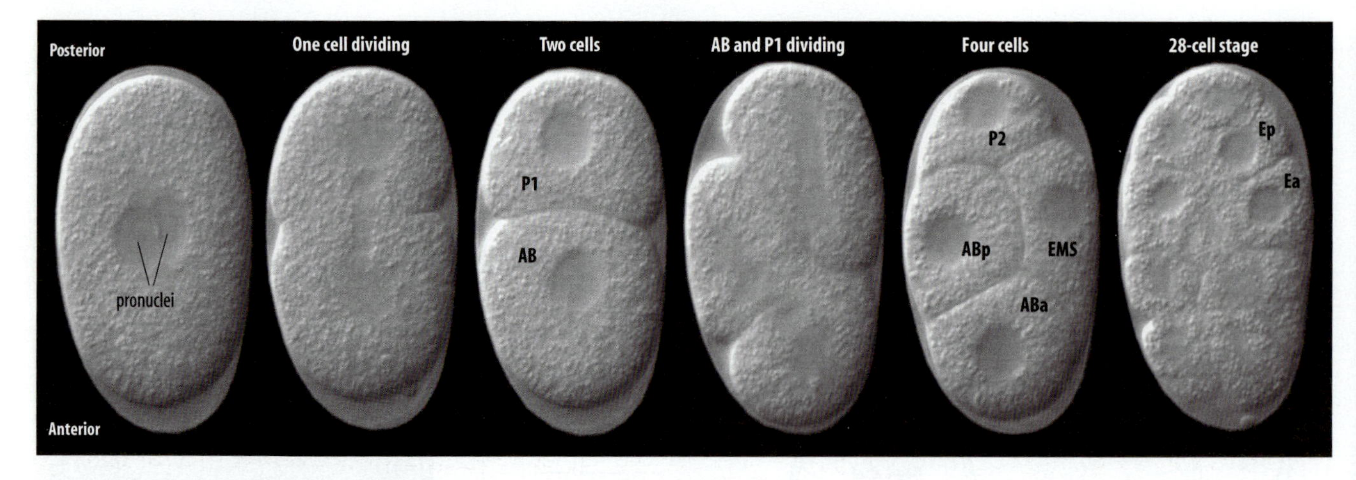

Fig. 5.3 Cleavage of the *C. elegans* embryo. After fertilization, the pronuclei of the sperm and egg fuse. The egg then divides into a large anterior AB cell and a smaller, posterior P1 cell. At the next cell division, AB divides into ABa and ABp, while P1 divides into P2 and EMS. Each of the cells will continue to divide in a well-defined pattern within these groups to give rise to particular cell types and tissues. Gastrulation begins at the 28-cell stage. By this stage, further division of the EMS cell has produced the E cells, which will give rise to the gut, and MS cells (not labeled here) which give rise to a variety of other cell types.

Photographs courtesy of J. Ahringer.

The newly hatched larva (Fig. 5.4), while similar in overall organization to the mature adult, is sexually immature and lacks a gonad and its associated structures, such as the vulva of hermaphrodites, which are required for reproduction. Post-embryonic development takes place during a series of four molts. The additional cells in the adult are derived largely from precursor blast cells (P cells) that are distributed along the body axis. Each of these blast cells founds an invariant lineage involving between one and eight cell divisions. The vulva, for example, is derived from blast cells P5, P6, and P7. One may think of post-embryonic development in the nematode as the addition of adult structures to the basic larval plan.

It is a triumph of direct observation that, with the aid of Nomarski interference microscopy, the complete lineage of every cell in the nematode *C. elegans* has been worked out. The pattern of cell division is invariant—it is the same in every embryo. The larva, when it hatches, is made up of 558 cells, and after four further molts this number has increased to 959, excluding the germ cells, which vary in number. This is not the total number of cells derived from the egg, as 131 cells die during development as a result of programmed cell death or apoptosis, which is discussed in more detail in Section 8.13. As the fate of every cell at each stage is known, a fate map can be accurately drawn at any stage and thus has a precision not found in any vertebrate. As with any fate map, however, even where there is an invariant cell lineage this precision in no way implies that the lineage must determine the fate or that the fate of the cells cannot be altered. As we shall see, cell interactions have a major role in determining cell fate in the nematode.

Fig. 5.4 *C. elegans* larva at the L1 stage (20 hours after fertilization). The vulva will form from the gonad primordium.

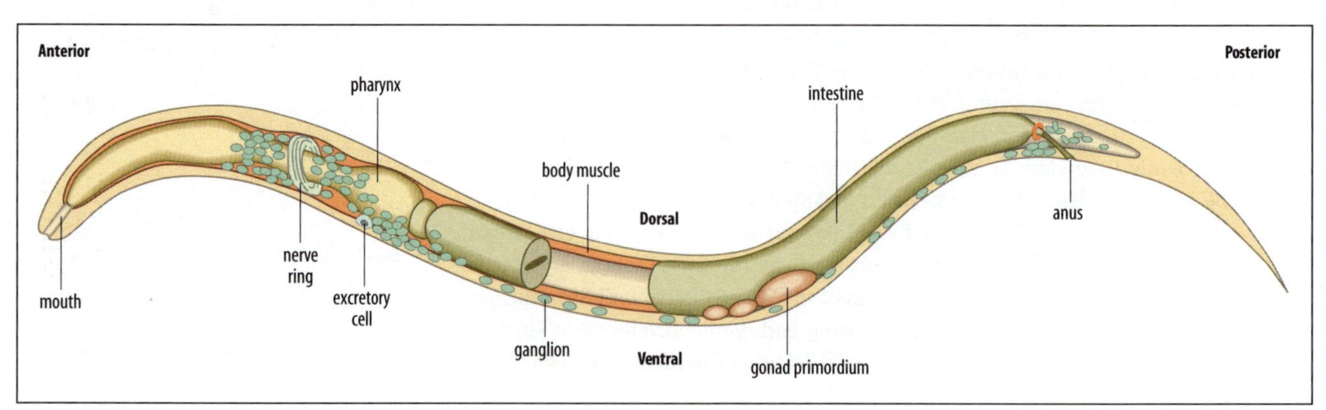

The complete genome of *C. elegans* has been sequenced, and contains nearly 20,000 genes. Around 1700 genes have been identified as affecting development, two-thirds of which were found using the technique of RNA interference (RNAi; Box 5A, below). Many nematode developmental genes are related to the genes that control development in vertebrates and *Drosophila*; they include Hox genes (see Box 4A, p. 156), and genes for signaling molecules of the TGF-β, Wnt, and Notch

Box 5A Gene silencing by RNA interference

Gene knock-outs remove the function of a gene permanently by disrupting its DNA. Another set of techniques for suppressing gene function rely on destroying or inhibiting the mRNA. Because the gene itself is left untouched, and thus suppression can be reversed, this type of loss of function is generally known as **gene silencing**. All methods of gene silencing rely on the introduction into embryos or cultured cells of an RNA with a sequence complementary to that of the specific mRNA being targeted. Depending on the particular technique, the introduced RNA either binds to the mRNA and prevents it from being translated, or targets a nuclease to bind to the mRNA and degrade it.

The first technique of this type to be developed used short synthetic **antisense RNAs**, which are usually chemically modified (for example, morpholino RNAs) to increase their stability within the cell. Morpholino oligonucleotides have been used extensively to block specific gene expression, especially in zebrafish and sea urchins.

A more recent addition to the gene-silencing armoury is the technique of **RNA interference** (**RNAi**), which recruits a natural RNA-degrading mechanism that is apparently ubiquitous in multicellular eukaryotes from plants to mammals. The phenomenon was first uncovered in plants during transgenic experiments, when it was found that introduction of a gene very similar to one of the plant's own genes blocked the expression of both the introduced gene and the endogenous gene. It turned out that suppression was occurring at the level of the mRNAs, which appeared to be rapidly degraded. This phenomenon is now known to be caused by the degradation of double-stranded RNAs by a cellular enzyme called Dicer, which chops them into short lengths of around 21–23 nucleotides. The so-called **short interfering RNA** (**siRNA**) is unwound into single-stranded RNA which becomes incorporated into a nuclease-containing protein complex known as RISC (for RNA-induced silencing complex). The RNA-binding component in RISC is a ribonuclease called Argonaute, also called Slicer in mammalian cells. The siRNA acts a 'guide RNA' to target RISC to any mRNA containing an exactly complementary sequence. The mRNA is then degraded by the nuclease in the RISC (see figure, below). RNAi is particularly effective and easy to carry out in *C. elegans*. Adult worms injected with double-stranded RNA will show suppression of the corresponding gene in the embryos they produce. Worms can also be soaked in the appropriate RNA or fed on bacteria expressing the required double-stranded RNA. The natural role of the cellular RNAi machinery is thought to be defense against viruses, many of which produce double-stranded RNAs.

In plants, fungi, nematodes, *Drosophila*, and most non-mammalian cells, RNAi is carried out by introducing an appropriate double-stranded RNA, which is then processed into siRNAs within the cells. In mammalian cells, however, double-stranded RNAs trigger another defensive response that interferes with the gene-silencing effect, and RNAi is instead achieved by introducing the siRNAs themselves, or by expressing artificial DNA constructs from which siRNAs are transcribed.

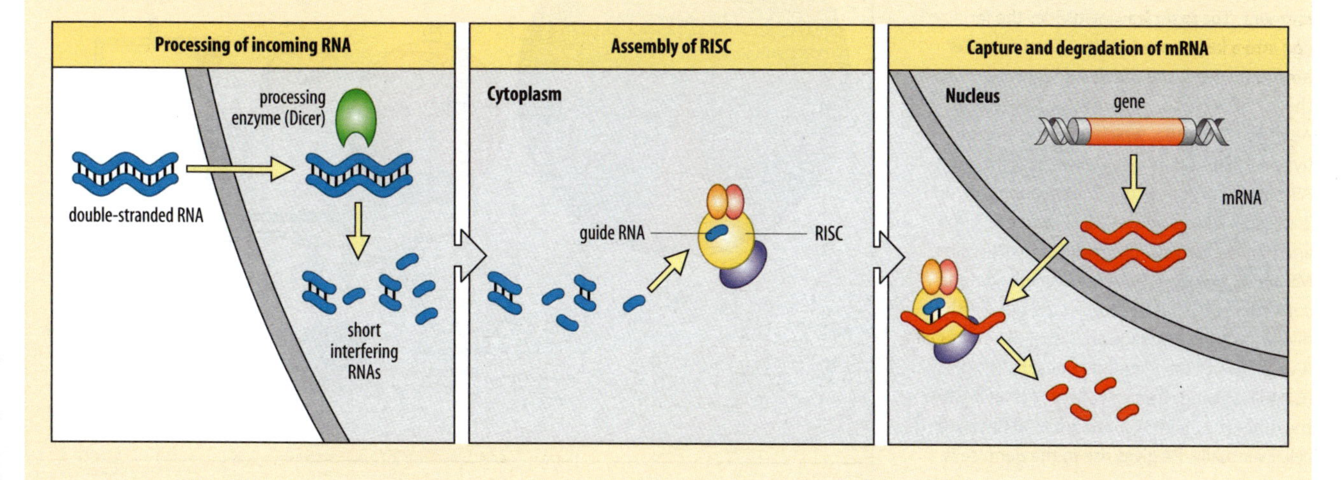

pathways (see Box 3C, p. 111). A notable exception is the Hedgehog signaling pathway (see Section 2.23), which is not present in *Caenorhabditis*. Although zygotic gene expression begins at the four-cell stage, maternal components control almost all of the development up to gastrulation at the 28-cell stage.

5.1 The antero-posterior axis in *C. elegans* is determined by asymmetric cell division

The first cleavage of the nematode egg is unequal, dividing the egg into a large anterior AB cell and a smaller posterior P1 cell. This asymmetry defines the antero-posterior axis and is determined at fertilization. The P1 cell behaves rather like a stem cell; at each further division it produces one P-type cell, and one daughter cell that will embark on another developmental pathway. For the first three divisions, the P-cell daughters give rise to body cells but after the fourth cleavage they only give rise to germ cells (Fig. 5.5). Division of the AB cell gives rise to anterior and posterior AB daughter cells. The anterior daughter, ABa, gives rise to typically ectodermal tissues, such as the epidermis (hypodermis) and nervous system, but also to a portion of the mesoderm of the pharynx. The posterior daughter, ABp, also makes neurons and epidermis as well as some specialized cells. At second cleavage, P1 divides asymmetrically to give P2 and EMS, which will subsequently divide into MS and E. MS gives rise to mesodermal pharynx, while E is the sole precursor of the gut. The C daughter cell, which is derived from the P2 cell at the third cleavage, forms epidermis and muscle; D from the fourth cleavage of the P cell produces only muscle. All the cells undergo further invariant divisions, and at about 100 minutes after fertilization gastrulation begins.

Before fertilization, there is no evidence of any asymmetry in the nematode egg but sperm entry sets up an antero-posterior polarity in the one-celled fertilized egg, which determines the position of the first cleavage division and the future

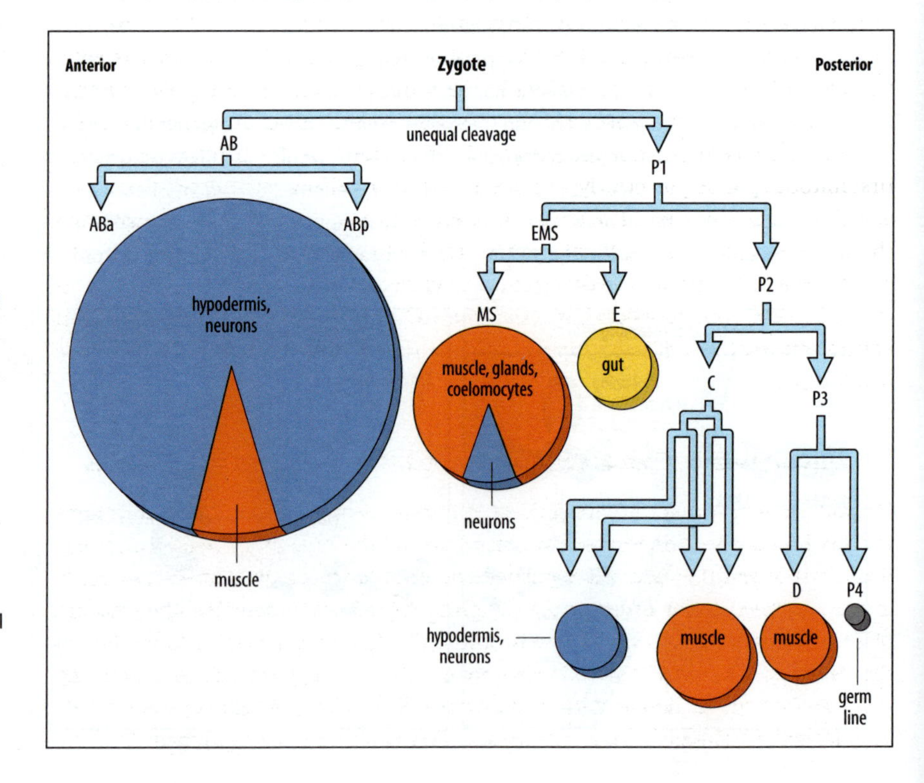

Fig. 5.5 Cell lineage and cell fate in the early *C. elegans* embryo. The cleavage pattern is invariant. The first cleavage divides the fertilized egg into a large anterior AB cell and a smaller posterior P1 cell and the descendants of these cells have constant lineages and fates. AB divides into ABa, which produces neurons, hypodermis, and muscle (of the pharynx), and ABp, which produces neurons, hypodermis, and some specialized cells. P1 divides to give EMS and P2. The EMS cell then divides to give cells MS and E. MS gives rise to muscle, glands, and coelomocytes, and E produces the gut. Further divisions of the P lineage are rather like stem-cell divisions, with one daughter of each division (C and D) giving rise to a variety of tissues while the other (P2 and P3) continues to act as a stem cell. Eventually, P4 gives rise to the germ cells.

embryonic antero-posterior axis. This cleavage is both unequal and asymmetric, forming a large AB cell at the anterior end and the smaller P1 cell at the posterior. The existence of polarity in the fertilized egg becomes evident before the first cleavage. A cap of actin microfilaments forms at the future anterior end, and a set of cytoplasmic granules, the so-called **P granules**, becomes localized at the future posterior end. The P granules are not, however, the determinants of polarization; their redistribution simply reflects it. P granules remain in the P daughters of cell division, eventually becoming localized to the P4 cell, which gives rise to the germline (Fig. 5.6). The P granules are involved in germline cell development.

The initial polarization of the one-celled fertilized egg appears to be due to the centrosome that the sperm nucleus brings into the egg. The egg cortex contains a dense actomyosin network and interaction of the centrosome with this network causes an asymmetric contraction of the network away from the point of sperm entry about 30 minutes after fertilization. This leads to a general flow of cortical components towards the future anterior end of the zygote and cytoplasmic flow towards the future posterior end. Among the proteins localized in the cortex are a diverse group of maternal proteins known as PAR (partitioning) proteins, which are initially uniformly distributed throughout the cortex but rapidly become concentrated in anterior or posterior cortex after the cortical movement. The so-called partitioning-defective mutants that first identified these proteins had defects in the asymmetric divisions that partition the zygote into blastomeres. If the pattern of division is not precisely maintained, future development is affected. PAR proteins similar to those in *C elegans* are involved in establishing and maintaining cell polarity in many different situations, and are found in animals from nematodes to mammals. In *C. elegans*, the correct localization of PAR proteins is required for polarization of the fertilized egg and for the correct positioning of mitotic spindles in the early cleavages.

In the *C. elegans* zygote, cortical contraction and resulting cortical flow leads to the anterior localization of three PAR proteins—PAR-3, PAR-6, and PKC-3—which after localization form a protein complex known as the anterior PAR complex. Their anterior concentration displaces another PAR protein, PAR-2, from the anterior cortex, causing it to accumulate in the posterior along with the PAR protein PAR-1, which is also concentrated there. This asymmetric distribution then results, by mechanisms that are not yet completely understood, in the displacement of the first mitotic spindle posteriorly and thus to the unequal and asymmetric first cleavage. The asymmetrically distributed PAR proteins may be involved in regulating the forces that pull on each pole of the spindle to position them in the cell. The mechanisms that control the first two cleavage divisions are extremely complex, as more than 600 gene products have been shown to be involved by a large-scale RNAi experiment that individually suppressed around 98% of the nematode's genes (see Box 5A, p. 189).

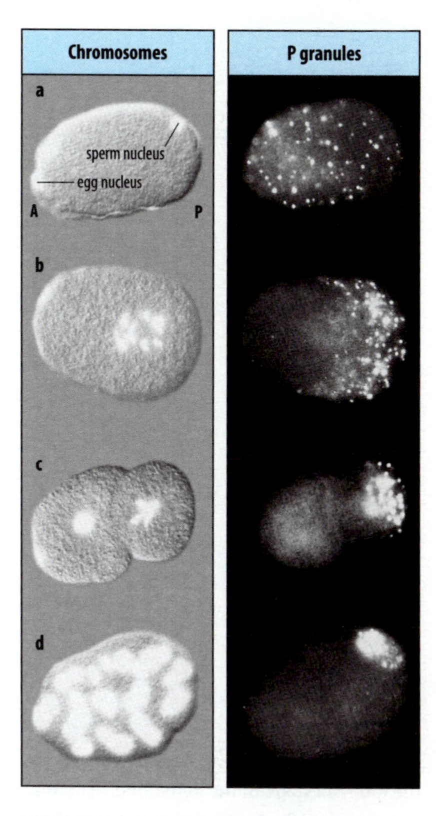

Fig. 5.6 Localization of P granules after fertilization. The movement of P granules is shown during the development of a fertilized egg of *C. elegans* from sperm entry up to the 26-cell stage. In the left panel, the embryo is stained for DNA, to visualize the chromosomes, and in the right panel for P granules. (a) Fertilized egg with egg nucleus at anterior end and sperm nucleus at posterior end. At this stage the P granules are distributed throughout the egg. (b) Fusion of sperm and egg nucleus. The P granules have already moved to the posterior end. (c) Two-cell stage. The P granules are localized in the posterior cell. (d) 26-cell stage. All the P granules are in the P4 cell, which will give rise to the germline only.

Photographs courtesy of W. Wood, from Strome, S., et al.: 1983.

5.2 The dorso-ventral axis in *C. elegans* is determined by cell–cell interactions

Despite the highly determinate cell lineage in the nematode, cell–cell interactions are involved in specifying the dorso-ventral axis. At the time of the second cleavage, if the future anterior ABa cell is pushed and rotated with a glass needle, not only is the antero-posterior order of the daughter AB cells reversed, but the cleavage of P1 is also affected, so that the position of the P1 daughter cell EMS relative to the AB cells is inverted (Fig. 5.7). The manipulated embryo develops completely normally but 'upside down', with its dorso-ventral axis inverted within the eggshell. The normal development that is observed after this manipulation means that the

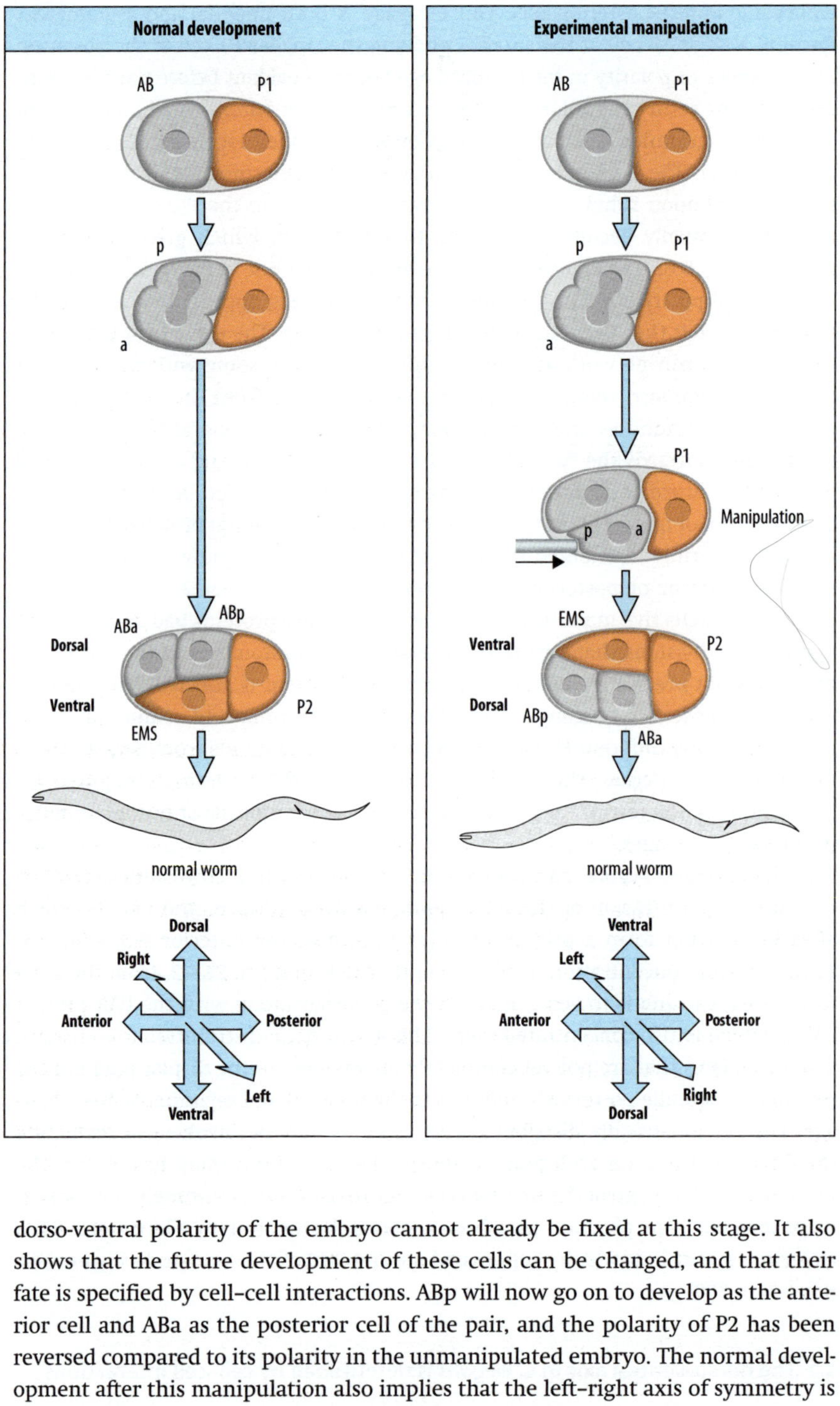

Fig. 5.7 Reversal of dorso-ventral polarity at the four-cell stage of the nematode. In normal embryos, the AB cell rotates at cleavage so that ABa is anterior. If the AB cell is mechanically rotated in the opposite direction at second cleavage, the ABp cell is now anterior. This manipulation also displaces the P1 cell so that when it divides, the position of the EMS daughter cell is reversed with respect to the AB cells. Development is still normal, but the dorso-ventral axis is inverted and so too is left-right asymmetry. a and p show the anterior/posterior orientation of the AB cell prior to and after manipulation.

After Sulston, J.E., et al.: 1983.

Normal development	Experimental manipulation
AB P1	AB P1
p a	p P1 a
	P1 p a Manipulation
Dorsal ABa ABp Ventral EMS P2	EMS Ventral P2 Dorsal ABp ABa
normal worm	normal worm
Dorsal, Right, Anterior, Posterior, Left, Ventral	Ventral, Left, Anterior, Posterior, Right, Dorsal

dorso-ventral polarity of the embryo cannot already be fixed at this stage. It also shows that the future development of these cells can be changed, and that their fate is specified by cell–cell interactions. ABp will now go on to develop as the anterior cell and ABa as the posterior cell of the pair, and the polarity of P2 has been reversed compared to its polarity in the unmanipulated embryo. The normal development after this manipulation also implies that the left–right axis of symmetry is not yet determined; indeed this axis can also be inverted, as we shall now see.

Adult worms have a well defined left–right asymmetry in their internal structures, and this axis is determined after the dorso-ventral axis is established. The pattern of development on the left and right sides of the nematode embryo shows striking differences; indeed, the differences during early development are more marked than in many other animals. Not only do cell lineages on the left and right differ, but some cells even cross from one side to the other. The concept of left and right, as discussed previously (see Section 3.10), only has meaning once the

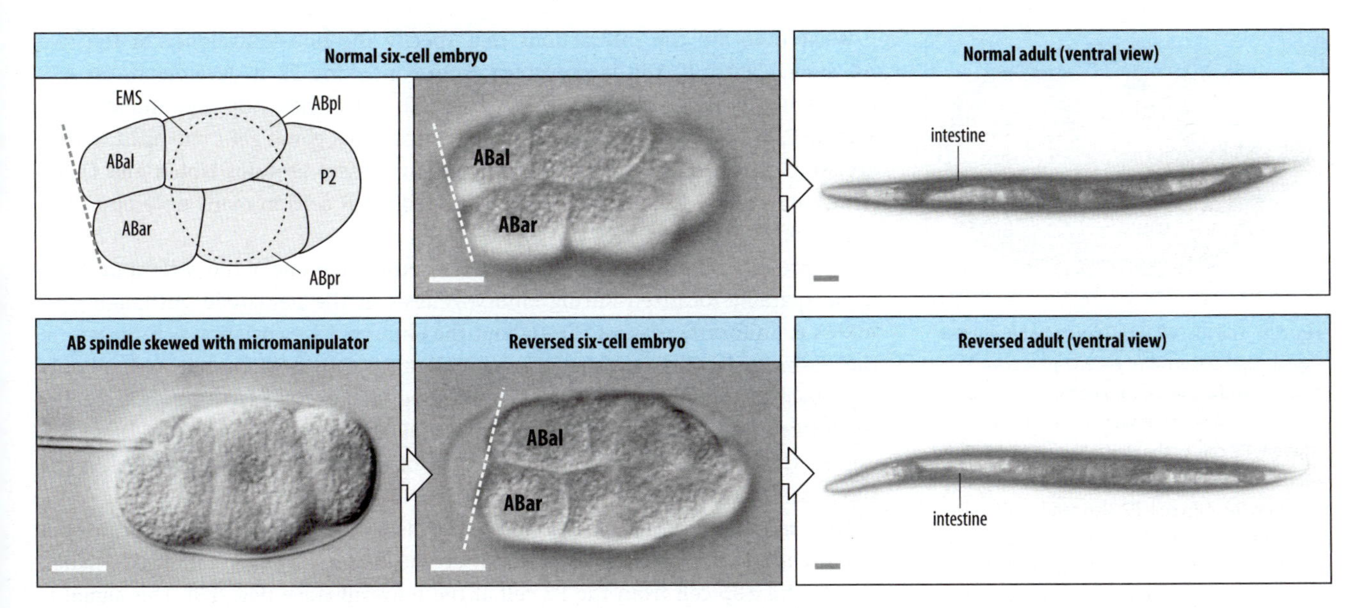

Fig. 5.8 Reversal of handedness in *C. elegans*. At the six-cell stage in a normal embryo, the AB cell on the left side (Abal) is slightly anterior to that on the right (top left panels, scale bar = 10 μm). Manipulation that makes the AB cell on the right side (ABar) more anterior results in an animal with reversed handedness (bottom left panels, scale bars = 10 μm). The right panels show the resulting normal and reversed adults; scale bars = 50 μm.

Photographs courtesy of W. Wood, from Wood, W.B.: 1991.

antero-posterior and dorso-ventral axes are defined. Since the dorso-ventral axis in the nematode can still be reversed at the two-cell to four-cell stage, left and right are also not yet specified.

Specification of left and right occurs at the third cleavage, and handedness can be reversed by experimental manipulation at this stage. Following the division of the AB cell into anterior and posterior blastomeres ABa and ABp, each of these divides at the third cleavage to produce laterally disposed right and left daughter cells. The plane of cleavage is, however, slightly asymmetric so that the left daughter cell lies just a little anterior to its right-hand sister. If the cells are manipulated with a glass rod during this cleavage, their positioning can be reversed so that the right-hand cell lies slightly anterior (Fig. 5.8). This small manipulation is sufficient to reverse the handedness of the animal.

While the molecular mechanisms specifying handedness are not known, mutation of the maternal gene *gpa-16* results in near randomization of spindle orientation and of handedness in the embryo. *gpa-16* encodes a G-protein α subunit and there is evidence that this is involved in the positioning of centrosomes and that it may have a specific role to play at the third cleavage. At a much later stage in nematode development an asymmetry is generated between two bilateral taste receptors, which express different putative chemoreceptor genes; this asymmetric gene expression is controlled by a microRNA, *lsy-6*.

5.3 Both asymmetric divisions and cell–cell interactions specify cell fate in the early nematode embryo

Although cell lineage in the nematode is invariant, experimental evidence such as that described above shows that cell–cell interactions are of crucial importance in specifying cell fate in the early embryo. Otherwise, the reversal of ABa and ABp (which normally have different fates) by micromanipulation just as they are being formed (see Fig. 5.7) would not give rise to a normal worm. ABa and ABp must initially be equivalent, and their fate must be specified by interactions with adjacent cells. Evidence for such an interaction comes from removing P1 at the first cleavage, as pharyngeal cells, a normal product of ABa, are then not made.

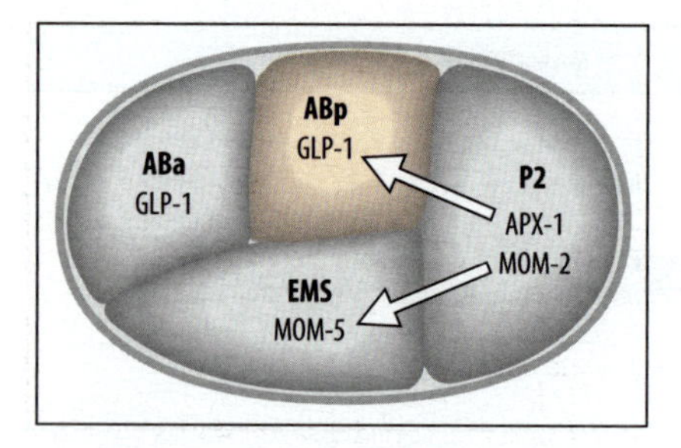

Fig. 5.9 The P2 cell is the source of inductive signals that determine antero-posterior polarity and cell fate. ABp becomes different from ABa as a result of signaling from the adjacent P2 cell by the Delta-like protein APX-1, which interacts with the Notch-like receptor GLP-1 on the ABp cell. P2 also sends a Wnt-like inductive signal, MOM-2, to the EMS cell which interacts with a Frizzled-like receptor MOM-5 on EMS to set up a posterior difference in the cell that determines the different fates of anterior and posterior daughter cells. These interactions depend on cell contact.

What then are the interactions that specify the non-equivalence of the two AB descendants? If ABp is prevented from contacting P2 it develops as an ABa cell. Thus, the P2 blastomere is responsible for specifying ABp. The induction of ABp by P2 involves proteins encoded by the maternal genes *glp-1* and *apx-1*, which are similar, respectively, to the vertebrate and insect proteins Notch and Delta, which are involved in interactions between adjacent cells in many developmental situations.

The protein GLP-1 is a transmembrane receptor and is one of the earliest proteins to be spatially localized during embryogenesis in the nematode. Although *glp-1* mRNA is uniformly present throughout the embryo, its translation is repressed in the posterior P-cell lineage; at the two-cell stage the GLP-1 protein is thus only expressed in the anterior AB cell. This strategy for protein localization is highly reminiscent of the localization of maternal Hunchback protein in *Drosophila* (see Section 2.9); the 3′ untranslated region of the *glp-1* mRNA is involved in the suppression of translation.

After second cleavage the ABa and ABp cells both contain the receptor protein GLP-1. The two cells are then directed to become different by a local inductive signal, sent to the ABp cell from the P2 cell at the four-cell stage (Fig. 5.9). This signal is thought to be the protein APX-1 (produced by the P2 cell) acting as an activating ligand for the receptor GLP-1. As a result of this induction, the descendants of ABa and ABp respond differently to later signals from the adjacent MS cell (a daughter of EMS).

Specification of the EMS cell as a mesendodermal precursor, which gives rise to both muscle cell types (mesodermal) and intestinal cells (endodermal; see Fig. 5.5), also involves both localized maternal determinants and inductive signals from the adjacent P2 cell (see Fig. 5.9). One maternal determinant involved is the transcription factor SKN-1 (the product of the *skin excess* (*skn-1*) gene). *skn-1* mRNA is uniformly distributed at the two-cell stage, but there are much higher levels of SKN-1 protein in the nucleus of P1 than in AB. Maternal-effect mutations in *skn-1* result in the E blastomere derived from EMS making mainly muscle instead of intestinal cells.

Gut formation from the EMS cell also requires inductive signals. If isolated from the influence of its neighbors at the four-cell stage, an EMS cell can develop gut structures, but this property depends on just when it is isolated. If isolated at the beginning of the four-cell stage, EMS cannot develop gut, suggesting that its ability to form gut requires an interaction with other cells at the beginning of this stage. The cell necessary to induce gut formation from EMS is the P2 cell, as removal of P2 at the early four-cell stage results in no gut being formed. Recombining isolated EMS and P2 cells restores gut development, whereas recombining EMS with other cells from the four-cell stage has no effect.

In the embryo, the P2 cell signals to the EMS cell, specifying the ability to make endoderm and also inducing an antero-posterior polarity in EMS that determines which one of the EMS daughter cells will form endoderm and which will form mesoderm. This signal is carried by the Wnt-like protein MOM-2, which interacts with a Frizzled-like receptor (MOM-5) on the surface of the EMS cell (see Fig. 5.9), the point of contact specifying the future posterior cell. The third cleavage divides EMS to give the anterior MS blastomere, which produces muscle cell types, and the posterior E blastomere, which produces the gut (see Fig. 5.5).

We thus begin to see how cell–cell interactions and localized cytoplasmic determinants together specify cell fate in the early nematode embryo. The results of killing individual cells by laser ablation at the 32-cell stage, when gastrulation begins, suggest that many of the cell lineages are now determined, as when a cell is destroyed at this stage there is no regulation and its normal descendants are absent. Cell–cell interactions are, however, required later to effect cells' final differentiation.

Cell differentiation is closely linked to the pattern of cell division. Each blastomere undergoes a unique and nearly invariant series of cleavages that successively divide cells into anterior and posterior daughter cells. Cell fate appears to be specified by whether the final differentiated cell is descended through the anterior (a) or posterior (p) cell at each division. In the lineage generated from the MS blastomere, for example, the cell resulting from the sequence p-a-a-p-p undergoes programmed cell death (Fig. 5.10), whereas that resulting from p-a-a-p-a-a-a gives rise to a particular pharyngeal cell. What is the causal relation between the pattern of division and cell differentiation? One answer appears to lie in the anterior daughter cell having a higher level of the nuclear maternal protein POP-1 compared with its posterior sister. In the case of the division of the EMS blastomere, for example, the anterior cell, which becomes the MS blastomere, has a higher level of POP-1 in the nucleus compared with the posterior cell, which becomes E. In a maternal-effect *pop-1* mutant that does not have any POP-1 protein, MS adopts an E-like fate because the antero-posterior distinction is lost.

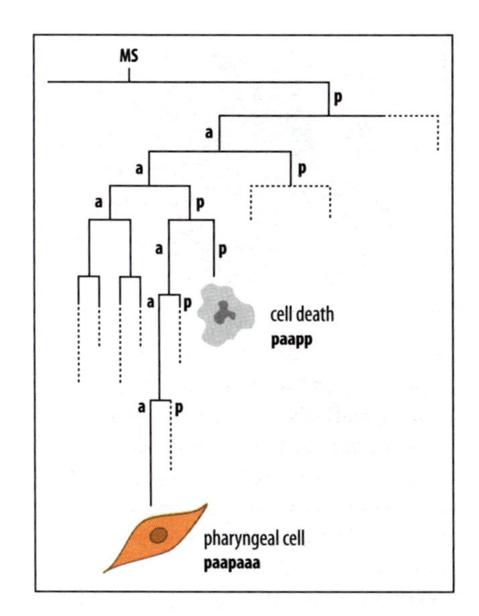

Fig. 5.10 Cell fate is linked to the pattern of cell divisions. The figure illustrates part of the cell lineages generated by the MS blastomere, which give rise to body muscle and the mesodermal cells of the pharynx. Each division produces an anterior (a) and a posterior (p) cell. The lineage paapp, for example, always gives rise to a cell that undergoes programmed cell death and dies by apoptosis. The lineage paapaaa, on the other hand, always results in a particular pharyngeal cell.

The differential distribution of POP-1 in the EMS daughter cells is the result of the Wnt–Frizzled signal from the P2 cell (see Fig. 5.9), and a similar mechanism is thought to specify antero-posterior cell polarity generally. POP-1 is the *C. elegans* version of the vertebrate transcription factor TCF (*Drosophila* Armadillo). As we saw in Figure 2.39, TCF represses gene expression until it interacts with nuclear β-catenin, produced as a result of the Wnt pathway, when the complex becomes a transcriptional activator. In *C. elegans*, a Wnt signal specifies the posterior E cell as endoderm by a similar mechanism (see Fig. 5.9), but with a twist. The system for specifying antero-posterior difference involves signaling through the non-canonical Wnt pathway (see Fig. 7.33), which in this case decreases the amount of POP-1 in the posterior cell nucleus by promoting its export. This appears to enable a β-catenin-like co-activator, which is present at only low levels, to then activate specific gene expression. In the case of the E cell, the genes expressed will specify the cell and its descendants as endoderm. An endodermal fate is normally suppressed in the MS cell because it still contains high levels of the repressor POP-1 in the nucleus.

Signaling involving Wnt proteins themselves seems to be involved in the setting of antero-posterior cell polarities in early cleavage divisions, but the setting of polarity at later divisions may be due to posteriorly localized Frizzled acting on its own, a situation reminiscent of the mechanism by which planar cell polarity may be established in *Drosophila* (see Box 2E, p. 75).

At the 80-cell gastrula stage a fate map can be made for the nematode embryo (Fig. 5.11). At this stage, cells from different lineages that will contribute to the same organ, such as the pharynx or gut, cluster together. Genes that appear to act as organ-identity genes may now be expressed by the cells in the cluster. The *pha-4* gene, for example, seems to be an organ-identity gene for the pharynx. Mutations in *pha-4* result in loss of the pharynx, whereas ectopic expression of normal *pha-4* in all the cells of the embryo leads to all cells expressing pharyngeal cell markers.

5.4 A small cluster of Hox genes specifies cell fate along the antero-posterior axis

Although the body plan of the nematode differs greatly from that of vertebrates or *Drosophila*, and there is no segmental pattern along the antero-posterior axis, homeobox-containing genes are involved in specifying cell fate along this axis in the nematode. The nematode contains a large number of homeobox genes, of which only four are similar to the Antennapedia class of *Drosophila* Hox genes and the Hox genes of vertebrates. These four genes—*lin-39*, *ceh-13*, *mab-5*, and *egl-5*—are arranged in a cluster, the Hox cluster, and in a similar order on the chromosome

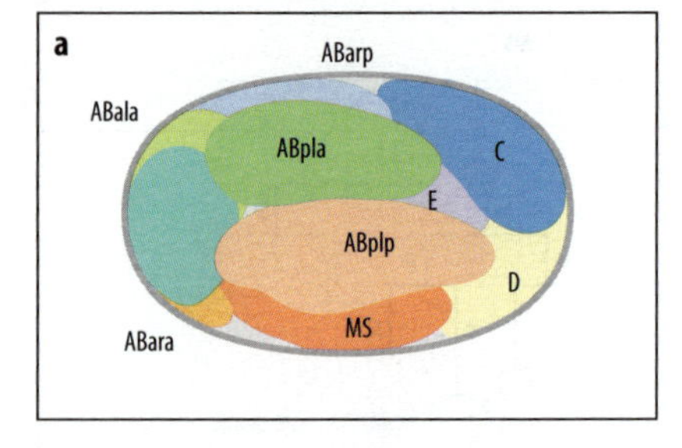

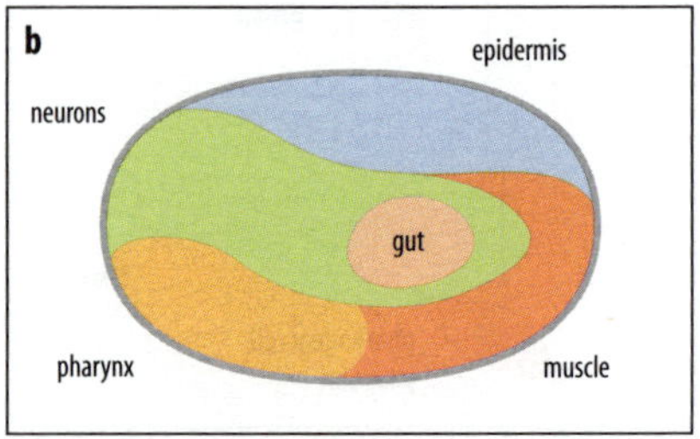

Fig. 5.11 Fate map of the 80-cell gastrula of *C. elegans*. In panel a, the regions of the embryo are color-coded according to their blastomere origin. In panel b they are colored according to the organs or tissues they will ultimately give rise to. At this stage in development, cells from different lineages that contribute to the same tissue or organ have been brought together in the embryo. For clarity, the domains occupied by ABpra and ABprp are not shown in panel a. Anterior is to the left, ventral down.

After Labouesse, M. and Mango, S.E.: 1999.

to their *Drosophila* homologs. A fifth homeobox gene of the nematode cluster, *ceh-23*, is less related to the Antennapedia class. Another two Hox genes with homology to the *Abd-B* gene of *Drosophila* have also been identified outside the Hox cluster, but their functions are not yet known.

lin-39, *mab-5*, and *egl-5* are expressed as zygotic genes in different positions along the antero-posterior axis during development; their spatial order of expression corresponds to their order along the chromosome, as in the other model organisms discussed previously (Fig. 5.12). Although the genes of the Hox cluster are expressed in the embryo, only *ceh-13*, which is required for the anterior organization of the embryo, is essential for embryonic development, and it seems that the other Hox cluster genes carry out their function of regional specification at the post-embryonic larval stage. *egl-5* provides positional identity for posterior structures, while *lin-39* appears to control the fate of mid-body cells and is known to be involved in regulating the development of the vulva in hermaphrodites. *mab-5* controls the development of a slightly more posterior region (see Fig. 5.12). As in other organisms, mutations in Hox genes can cause cells in one region of the body to adopt fates characteristic of other body regions. For example, a mutation in *lin-39* can result in larval mid-body cells expressing fates characteristic of more anterior or posterior body regions.

Despite the fact that nematode Hox gene expression occurs in a regional pattern, the pattern may not be lineage dependent. For example, in the larva, cells expressing the Hox gene *mab-5* all occur in the same region (see Fig. 5.12) but are quite unrelated by lineage. The expression of *mab-5* is due to extracellular positional signals.

5.5 The timing of events in nematode development is under genetic control that involves microRNAs

Embryonic development gives rise to a larva of 558 cells and there are then four larval stages which produce the adult. Because each cell in the developing nematode can be identified by its lineage and position, genes that control the fates of individual cells at specific times in development can also be identified. This enables the genetic control of timing during development to be studied particularly easily in *C. elegans*. The order of developmental processes is of central importance, as well as the time at which they occur. Genes must be expressed both in the right place and at the right time. One well studied example of timing in nematode development is the generation of different patterns of cell division and differentiation in the four larval stages of *C. elegans*, which are easily distinguished in the developing cuticle.

Mutations in a small set of genes in *C. elegans* change the timing of cell divisions in many tissues and cell types. Mutations that alter the timing of developmental events are called **heterochronic**. The first two heterochronic genes to be discovered

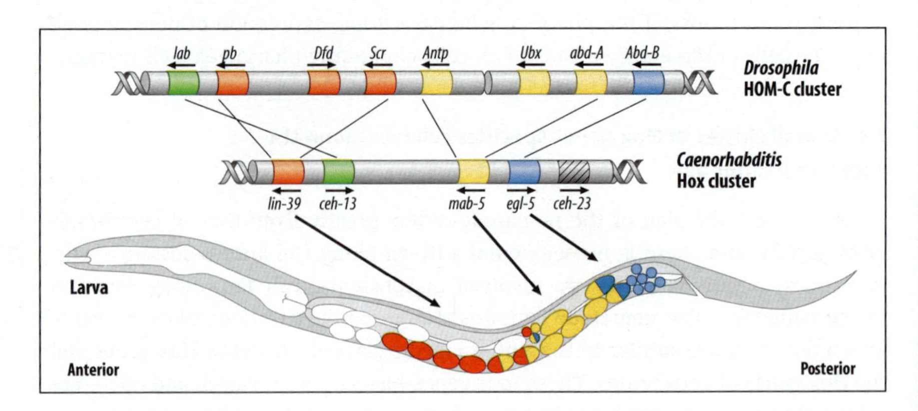

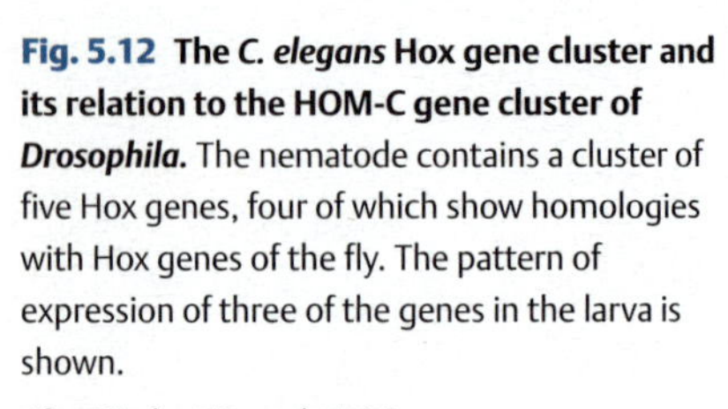

Fig. 5.12 The *C. elegans* Hox gene cluster and its relation to the HOM-C gene cluster of *Drosophila*. The nematode contains a cluster of five Hox genes, four of which show homologies with Hox genes of the fly. The pattern of expression of three of the genes in the larva is shown.

After Bürglin, T.R. et al.: 1993.

in *C. elegans* were *lin-4* and *lin-14*, and we shall use them to illustrate both the phenomenon of heterochronicity and the control of this process by microRNAs (Box 5B, below). Mutations in either *lin-4* or *lin-14* can result in both 'retarded' and 'precocious' development so that, for example, some stage-specific events such as molting and larval cuticle synthesis are repeated at abnormally late stages, leading to retardation of normal events such as adult cuticle synthesis.

Examples of changes in developmental timing that result from mutations in *lin-14* are provided by a cell from the T-cell lineage called T.ap (Fig. 5.13). In wild-type embryos, the T cell generates a lineage that gives rise to epidermal cells, neurons, and their support cells in both the first (L1) and second (L2) larval stages. During the later larval stages, L3 and L4, some of the T-cell descendants divide to give rise to other structures. Gain-of-function mutations in *lin-14* result in retarded development. Post-embryonic development begins normally, but the developmental patterns of the first or second larval stages are repeated. Loss-of-function mutations in *lin-14* result in a precocious phenotype—the pattern of cell divisions seen in

Box 5B Gene silencing by microRNAs

A new category of genes involved in development has been recognized quite recently, which code not for proteins but for specialized short RNAs, called **microRNAs (miRNAs)**. These regulate gene expression by preventing specific mRNA transcripts from being translated. This natural method of gene silencing has similarities with RNA interference by small interfering RNAs (siRNAs) described in Box 5A (p. 189), but appears to be a distinct phenomenon, although there is some overlap in the processing machinery. MicroRNAs were first identified in *C. elegans* as the products of the genes *let-7* and *lin-4*, which control the timing of development (see text). Developmental roles are known for a few other miRNAs in nematodes, and for the miRNA *bantam* in *Drosophila*, which prevents cell death. There is now evidence that as many as 1% of human genes code for miRNAs and regulatory roles for some human miRNAs are being revealed.

The primary miRNA transcript is several hundred nucleotides long and contains an inverted repeat of the mature miRNA sequence. After initial processing in the nucleus, the transcript is folded back on itself to form a double-stranded RNA hairpin, the pre-miRNA, which is exported to the cytoplasm (see figure, below). Here it is cleaved by the enzyme Dicer to produce a short single-stranded miRNA about 22 nucleotides long. As in RNA interference, miRNAs become incorporated as guide RNAs in an RNA-induced silencing protein complex (RISC; see Box 5A, p. 189) and the complex is specifically targeted to an mRNA. Unlike siRNAs, animal miRNAs are typically not perfectly complementary to their target mRNA and have a few mismatched bases. Once bound, the protein complex renders the mRNA inactive and suppresses translation, but the mRNA is not degraded. Many plant miRNAs appear, however, to be perfectly complementary to their target mRNAs, which are degraded by the same mechanism as in RNA interference.

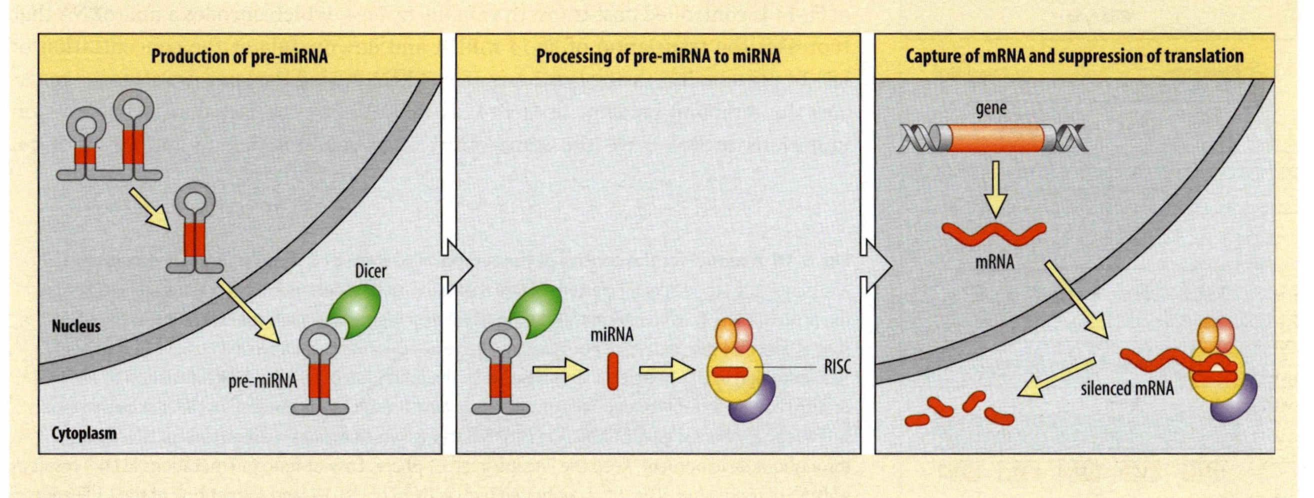

Fig. 5.13 Cell-lineage patterns in wild-type and heterochronic mutants of *C. elegans*. The lineage of the T blast cell (T.ap) continues through four larval stages (left panel). Mutants in the gene *lin-14* show disturbance in the timing of cell division, resulting in changes in the patterns of cell lineage. Loss-of-function mutations result in a precocious lineage pattern, with the pattern of development of early stages being lost (center panel). Gain-of-function mutations result in retarded lineage patterns, with the patterns of the early larval stage being repeated (right panel).

early larval stages is lost and post-embryonic development starts with cell divisions normally seen in the second larval stage.

It has been suggested that the genes that control timing of developmental events may do so by controlling the concentration of some substance, causing it to decrease with time (Fig. 5.14). This temporal gradient could control development in much the same way that a spatial gradient can control patterning. This type of timing mechanism seems to operate in *C. elegans* as the concentration of LIN-14 protein drops tenfold between the first and later larval stages. Differences in the concentration of LIN-14 at different stages of development may specify the fates of cells, with high concentrations specifying early fates and low concentrations later fates. Thus, the decrease of LIN-14 during development could provide the basis of a precisely ordered temporal sequence of cell activities. Dominant gain-of-function mutations of *lin-14* keep LIN-14 protein levels high, and so in these mutants the cells continue to behave as if they are at an earlier larval stage. In contrast, loss-of-function mutations result in abnormally low concentrations of LIN-14, and the larvae therefore behave as if they were at a later larval stage.

Developmental timing in *C. elegans* was one of the first processes discovered in which gene expression was found to be controlled by microRNAs. The expression of *lin-14* is controlled post-transcriptionally by *lin-4*, which encodes a microRNA that represses the translation of *lin-14* mRNA and downregulates the concentration of LIN-14 protein. Increasing synthesis of *lin-4* RNA during the later larval stages generates the temporal gradient in LIN-14, as indicated by the fact that loss-of-function mutations in *lin-4* have the same effect as gain-of-function mutations in *lin-14*.

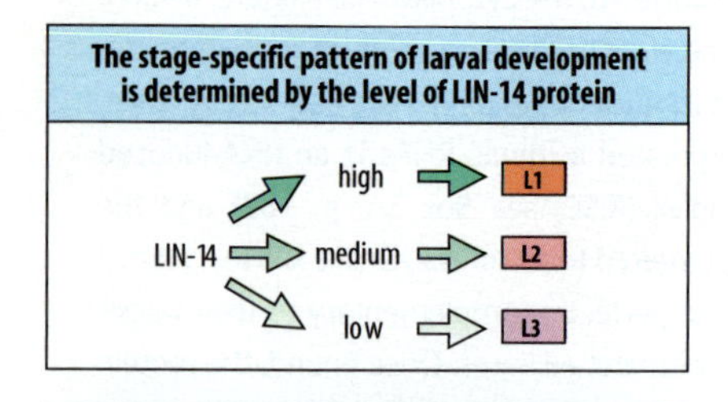

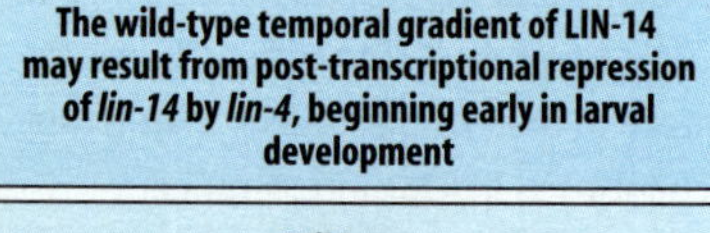

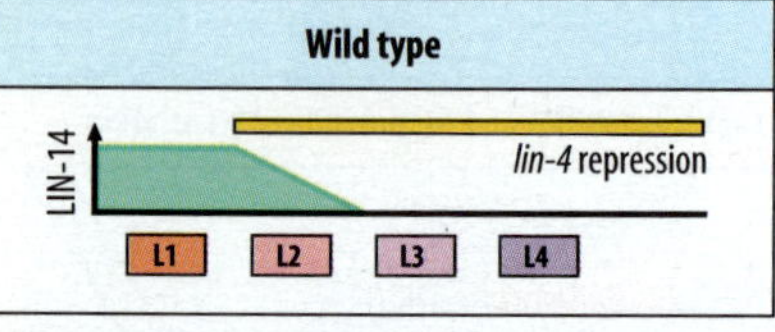

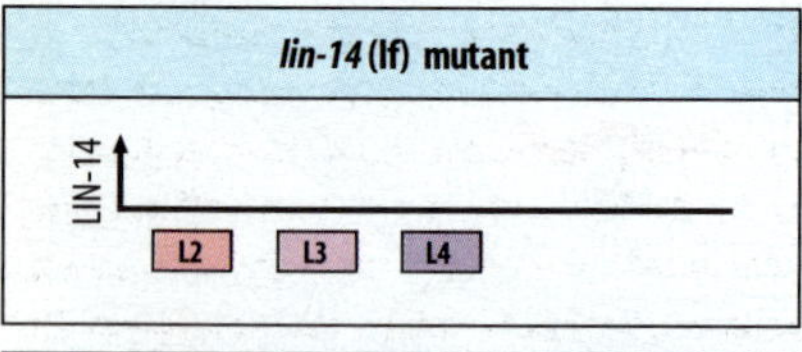

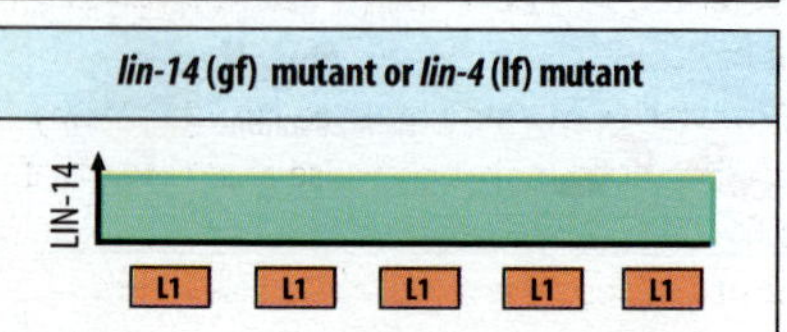

Fig. 5.14 A model for the control of the temporal pattern of *C. elegans* larval development. Top panel: the stage-specific pattern of larval development is determined by a temporal gradient of the protein LIN-14, which decreases during larval development. A high concentration in the early stages specifies the pattern of normal early-stage development as shown in Fig. 5.13 (left panel). The reduction in LIN-14 at later stages is due to the inhibition of *lin-14* mRNA translation by *lin-4* RNA. Bottom panels: loss-of-function (lf) mutations in *lin-14* result in the absence of the first larval stage (L1) lineage, whereas gain-of-function (gf) mutations, which maintain a high level of LIN-14 throughout development, keep the lineage in an L1 phase. Loss-of-function mutations in *lin-4* result in a lifting of repression of *lin-14*, a continued high activity of LIN-14, and a repetition of the L1 lineage.

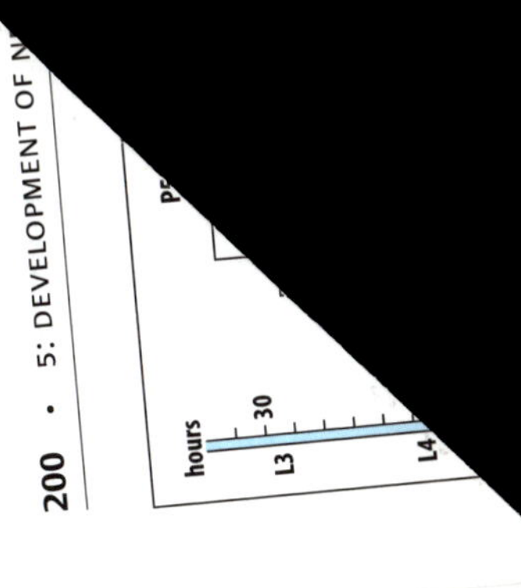

Expression of another gene involved in timing, *lin-41*, is also regulated post-transcriptionally by a microRNA, *let-7*, which represses its translation.

5.6 Vulval development is initiated by the induction of a small number of cells by short-range signals from a single inducing cell

The vulva forms the external genitalia of the adult hermaphrodite worm and connects with the uterus. It is an important model of a developmental structure that is initially specified as a small number of cells—just four, one inducing and three responding. More than 40 genes are involved in its development, but we will deal here with the patterning and induction of the vulva at the level of the individual cells. Unlike the other aspects of *Caenorhabditis* development discussed so far, the vulva is an adult structure that develops in the last larval stage. The mature vulva contains 22 cells, with a number of different cell types. It is derived from ectodermal cells originating from the P blastomere. Six of these cells persist in the larva in a small row aligned antero-posteriorly on the ventral side of the larva, and three of these—P5p, P6p, and P7p—give rise to the vulva, each having a well defined lineage. One of the functions of the Hox gene *lin-39* is to prevent the fusion of these six cells to form hypodermis in the early larval stages, which is the fate of the rest of the ventral ectodermal cells of this lineage. In discussing the development of the vulva, three distinct cell fates are conventionally distinguished—primary (1°), secondary (2°), and tertiary (3°). The primary and secondary fates refer to different cell types in the vulva, whereas the tertiary fate is non-vulval. P6p normally gives rise to the primary lineage, whereas P5p and P7p give rise to secondary lineages. The other three P cells give rise to the tertiary lineage, divide once and then fuse to form parts of the epidermis (Fig. 5.15). Initially, however, all six P cells are equivalent in their ability to develop as vulval cells. One of the questions we address here is how the final three cells become selected as the vulval precursor cells.

The fates of the three vulval precursor cells is specified by an inductive signal from a fourth cell, the gonadal anchor cell, which confers a primary fate on P6p, the cell nearest to it, and a secondary fate on P5p and P7p, the cells lying just beyond it (see Fig. 5.15). In addition, once induced, the primary cell inhibits its immediate neighbors from expressing a primary fate. P cells that do not receive the anchor cell signal adopt a tertiary fate.

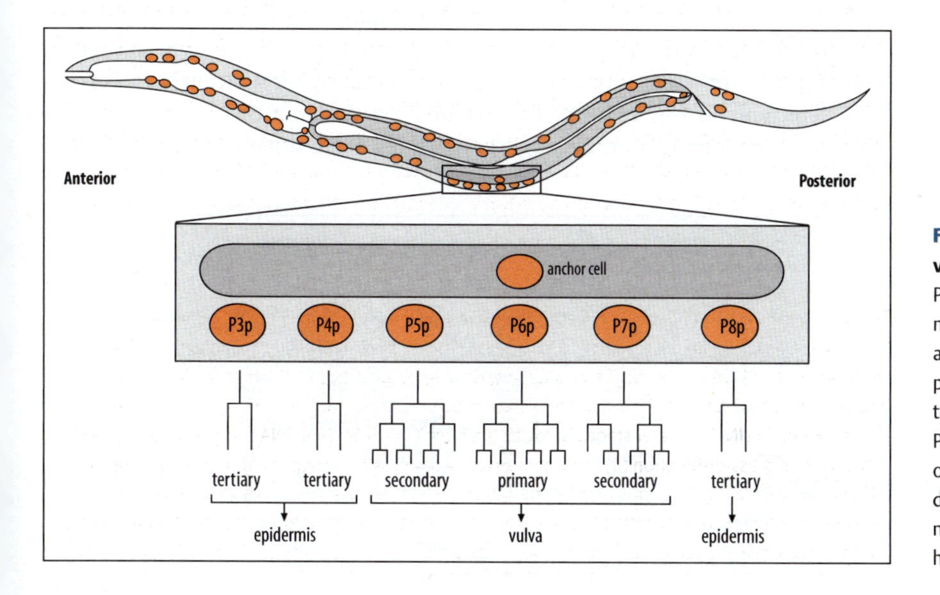

Fig. 5.15 Development of the nematode vulva. The vulva develops from three cells, P5p, P6p, and P7p, in the post-embryonic stages of nematode development. Under the influence of a fourth cell—the anchor cell—P6p undergoes a primary pathway of differentiation that gives rise to eight vulval cells. P6p is flanked by P5p and P7p, which each undergo a secondary pathway of differentiation that gives rise to seven cells of different vulval cell type. Three other P cells nearby adopt a tertiary fate and give rise to hypodermis.

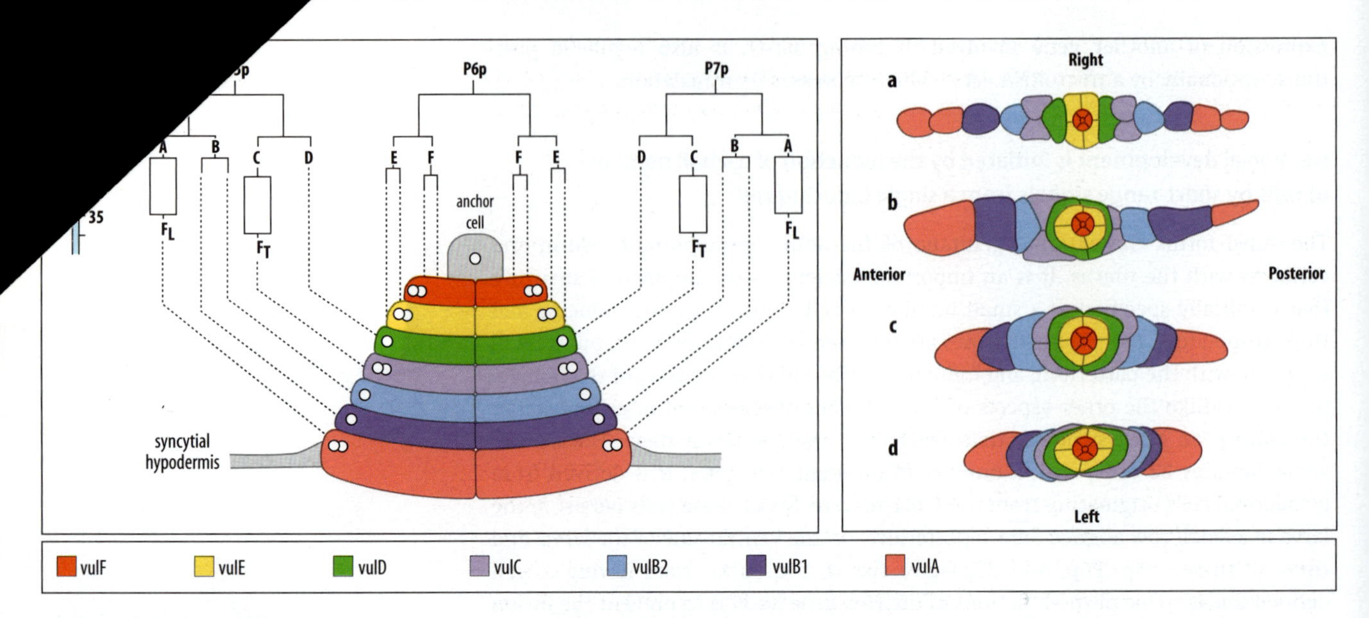

Fig. 5.16 Formation of the vulva by migration and fusion of specified precursors. The left-hand diagram shows a lateral representation of the seven rings of the vulva at around 39 hours and the cell lineages that give rise to them. The daughter cells of the A cell fuse, as do the daughter cells of the C cell. The open circles represent nuclei. At a later stage there will be further fusion of cells in the rings. The blue bar on the left shows the time in hours and larval stages. The right panel shows a ventral view of the developing vulva. The sequence a to d shows morphogenesis of the vulva over a period of 5 hours, from 34 hours. Changes in the shape of the cells and cell fusion give rise to the conical structure seen in the left panel.

After Sharma-Kishore, R., et al.: 1999.

The lineage of the P cells is fixed following induction. If one of the daughter cells of a P cell is destroyed, the other does not change its fate. There is no evidence for cell interactions in the further development of these P-cell lineages, and cell fate is thus probably specified by asymmetric cell divisions. The vulva is formed from the 22 cells derived from the three lineages. Actual morphogenesis of the vulva requires the cells to divide, move and fuse in a precise pattern to form seven toroidal rings (Fig. 5.16).

How are just three P cells specified, and how is the primary fate of the central cell made different from the secondary fate of its neighbors? The six P cells are initially equivalent, in that any of them can give rise to vulval tissue. The key determining signal is that provided by the anchor cell. The vital role of the anchor cell is shown by cell-ablation experiments; when the anchor cell is destroyed with a laser microbeam, the vulva does not develop.

Familiar signaling pathways are involved in vulval induction. The anchor cell's signal is the secreted product of the *lin-3* gene, which is similar to the EGF growth factor of vertebrates. Mutations in *lin-3* result in the same abnormal development that follows removal of the anchor cell: no vulva is formed. The receptor for the inductive signal is a transmembrane tyrosine kinase similar to the EGF receptor (EGFR), encoded by the *let-23* gene (Fig. 5.17). The inductive signal from the anchor cell activates the EGFR intracellular signaling pathway most strongly in P6p but also detectably in P5p and P7p. The maximal activation of this pathway leads to the P6p cell adopting a primary cell fate. It then sends a lateral signal to its two

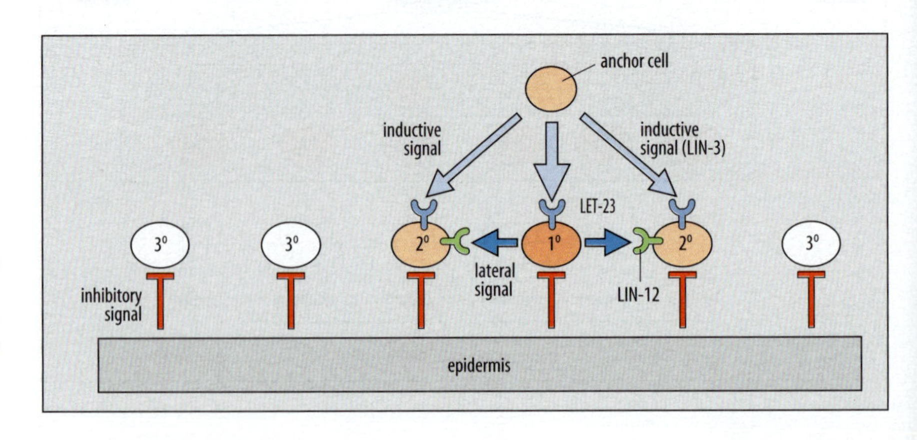

Fig. 5.17 Cell interactions in vulval development. The anchor cell produces a diffusible signal LIN-3, which induces a primary fate in the precursor cell closest to it by binding to the receptor LET-23. It also induces a secondary fate in the two P cells slightly further away, where the concentration of the signal is lower. The cell adopting a primary fate inhibits adjacent cells from adopting the same fate by a mechanism of lateral inhibition involving LIN-12, and also induces a secondary fate in these cells. A constitutive signal from the epidermis inhibits the development of both primary and secondary fates, but is overruled by the initial inductive signal from the anchor cell.

neighbors that both prevents them from adopting a primary fate and induces a secondary fate. The induction of secondary fate and the suppression of primary fate in P5p and P7p involves the Notch signaling pathway, mediated by the transmembrane receptor, LIN-12, a member of the Notch family.

At least one other signal is involved in establishing the vulva: a signal from the hypodermis—the larval nematode's epidermis—which inhibits adoption of a vulval fate in all the six cells of the equivalence group, and which is overruled by the inductive signal.

Summary

Specification of cell fate in the nematode embryo is intimately linked to the pattern of cleavage, and provides an excellent example of the subtle relationships between maternally specified cytoplasmic differences and very local and immediate cell-cell interactions. The antero-posterior axis is specified at the first cleavage by the site of sperm entry and specification of the dorso-ventral and left-right axes involves local cell-cell interactions. Gene products are asymmetrically distributed during the early cleavage stages, but specification of cell fates in the early embryo is crucially dependent on local cell-cell interactions, which are mediated by Wnt and Delta-Notch signaling. The development of the gut, which is derived from a single cell, requires an inductive signal by an adjacent cell. A small cluster of homeobox genes provides positional identity along the antero-posterior axis of the larva. The timing of developmental events in the larva is controlled by changes in the levels of various LIN proteins that decrease over time as a result of post-transcriptional repression by microRNAs. Formation of the adult nematode vulva involves both a well defined cell lineage and inductive cell interactions involving the EGF and Notch signaling pathways.

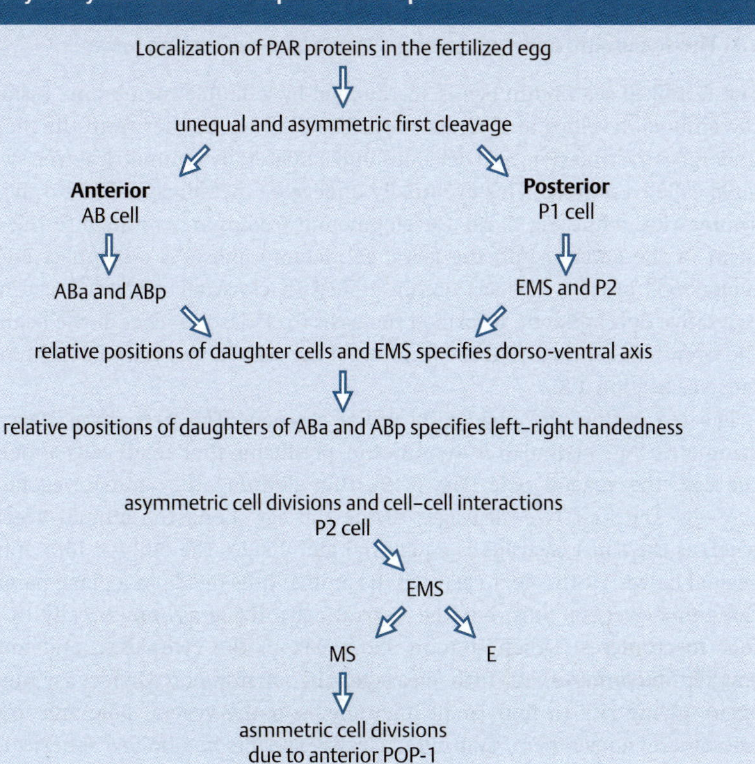

Echinoderms

Echinoderms include the sea urchins and starfish. Because of their transparency and ease of handling, sea urchin embryos have long been used as a model developmental system. Another useful feature is that echinoderms are more closely related to vertebrates than are the other main model invertebrate organisms, *Drosophila* and *Caenorhabditis*. Echinoderms are **deuterostomes**, like vertebrates, and thus have certain basic developmental features in common. Deuterostomes are coelomate animals that have radial cleavage of the egg, and in which the primary invagination of the gut at gastrulation forms the anus, with the mouth developing independently. Arthropods, on the other hand belong to the **protostomes**, in which cleavage of the zygote is not radial and in which gastrulation primarily forms the mouth. Nematodes are acoelomate.

For some years the echinoderms were somewhat neglected as a general developmental model as they are not amenable to investigation by conventional genetics or by the transgenic techniques that have proved so successful in other organisms. With the advent of modern genetic techniques and the ability to identify genes by homology with genes from other organisms, the genetic and molecular basis of sea urchin development can now be studied. A project to sequence the genome of the purple sea urchin *Strongylocentrotus purpuratus* is under way, and the complete gene-regulatory network that governs early development is being worked out.

In the following discussion of early embryonic development in the sea urchin we will focus on an early process that is currently being investigated in detail—the specification of the vegetal region as an organizing center and its subsequent induction of the **endomesoderm**, the cells that will give rise to the sea urchin spicules (skeletal rods), muscle, and gut. The sea urchin is also used as a simple model of gastrulation and to study fertilization, and it is discussed in relation to these topics in Chapters 7 and 11, respectively.

5.7 The sea urchin embryo develops into a free-swimming larva

The fertilized sea urchin egg is surrounded by a double membrane, inside which the embryo develops to blastula stage. The blastula hatches from the membrane, undergoes gastrulation, and develops into a bilaterally symmetrical free-swimming larva called a **pluteus**. This eventually undergoes metamorphosis into the radially symmetrical adult (Fig. 5.18). Developmental studies are confined to the development of the embryo into the larva, as metamorphosis is a complex and poorly understood process. The sea urchin embryo is classically regarded as a model of regulative development. It formed the basis for Driesch's ideas at the beginning of the twentieth century, that the position of the cells in an embryo determined their fate (see Section 1.3).

The sea urchin egg divides by radial cleavage. The first three cleavages are symmetric but the fourth is asymmetric, producing four small cells at one pole of the egg, the vegetal pole (Fig. 5.19), thus defining the animal–vegetal axis of the egg. The first two cleavages divide the egg along the animal–vegetal axis, whereas the third cleavage is equatorial and divides the embryo into animal and vegetal halves. At the next cleavage the animal cells divide in a plane parallel with the animal–vegetal axis, but the vegetal cells divide asymmetrically to produce four **macromeres**, which contain about 95% of the cytoplasm, and four much smaller **micromeres**. At fifth cleavage, the micromeres divide asymmetrically again, giving rise to four small micromeres at the vegetal pole and four larger micromeres above them. Continued cleavage results in a hollow spherical blastula

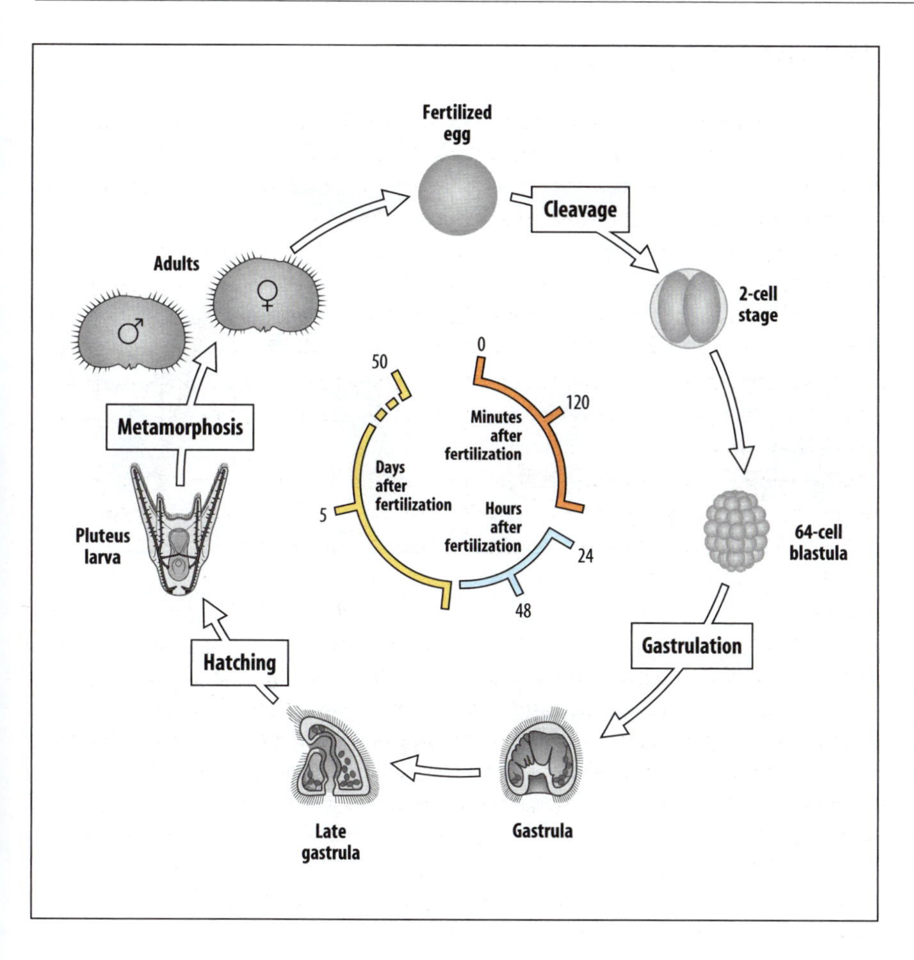

Fig. 5.18 Life cycle of the sea urchin *Strongylocentrotus purpuratus*. Eggs released by the female are fertilized externally by male sperm and develop into a blastula. After cleavage and gastrulation, the embryo hatches as a pluteus larva, which then undergoes metamorphosis into the mature radially symmetrical adult. The photographs show: the blastula (top); the pluteus larva (middle); and the adult (bottom).

Top and middle photographs courtesy of Jim Coffman. Bottom Photograph from Oxford Scientific Films.

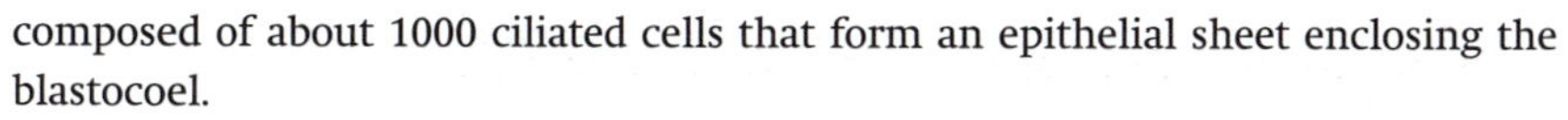

composed of about 1000 ciliated cells that form an epithelial sheet enclosing the blastocoel.

Gastrulation in sea urchin embryos, whose mechanisms we consider in detail in Chapter 8, starts about 10 hours after fertilization, with the mesoderm and endoderm moving inside from the vegetal region. In *S. purpuratus*, the first event is the entry into the blastocoel of about 32 primary mesenchyme (mesoderm) cells at the vegetal pole. They migrate along the inner face of the blastula wall to form a ring in the vegetal region and lay down calcareous skeletal rods. The endoderm, together with the mesodermal secondary mesenchyme, then starts to **invaginate** at the vegetal pole. Invagination involves an inward movement of cells in a sheet, rather like pushing a finger into a balloon. In the embryo, the invagination is driven by local cellular forces, which are discussed in Chapter 7. The invagination eventually stretches right across the blastocoel, where it fuses with a small invagination in the future mouth region on the ventral side. Thus the mouth, gut, and anus are formed. Before the invaginating gut fuses with the mouth, secondary mesenchyme cells at the tip of the invagination migrate out as single cells and give rise to mesoderm, such as muscle and pigment cells. The embryo is then a feeding pluteus larva (see Fig. 5.19).

5.8 The sea urchin egg is polarized along the animal–vegetal axis

Because of its body form, the sea urchin pluteus larva does not have antero-posterior and dorso-ventral axes like insect and vertebrate embryos. Instead the two axes

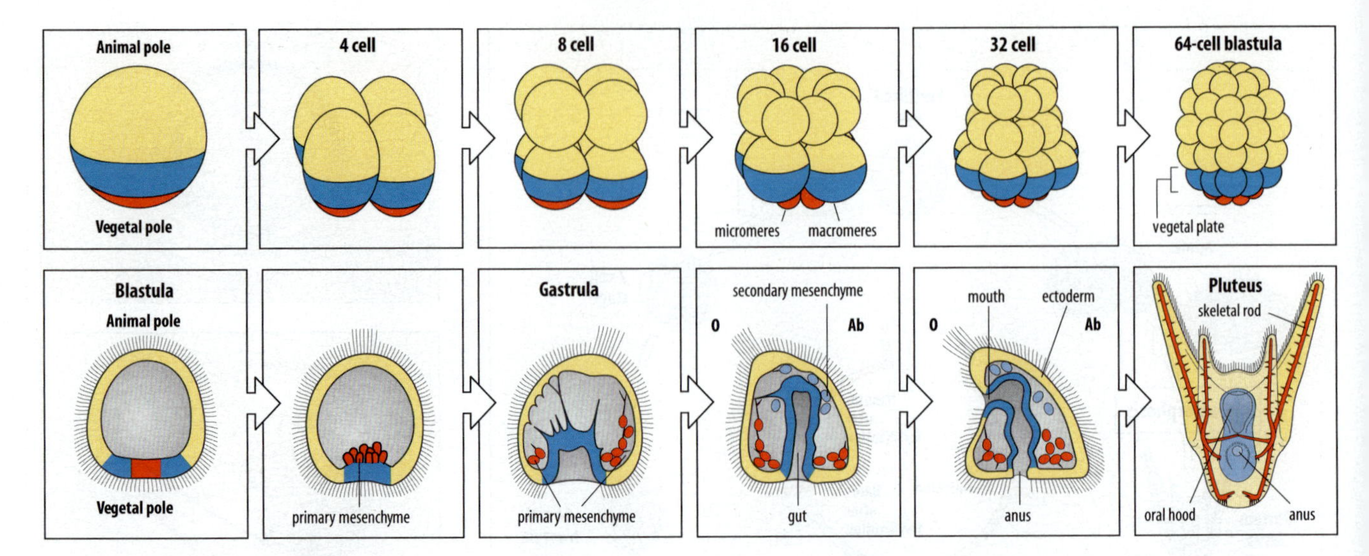

Fig. 5.19 Development of the sea urchin embryo. Top panels: external view of cleavage to the 64-cell stage. The first two cleavages divide the egg along the animal–vegetal axis. The third cleavage divides the embryo into animal and vegetal halves. At the fourth cleavage, which is unequal, four small micromeres (red) are formed at the vegetal pole, only two of which can be seen here. At fifth cleavage, the micromeres divide asymmetrically again, giving rise to four small micromeres and four large micromeres above them, not all of which can be seen here. Further cleavage results in a hollow blastula. Bottom panels: gastrulation and development of the pluteus larva. Sections of the developing embryo through the plane of the animal–vegetal axis are shown, starting from a hollow blastula. Gastrulation begins at the vegetal pole, with the entry of about 40 primary mesenchyme cells into the interior of the blastula. The gut invaginates from this site and fuses with the mouth, which invaginates from the opposite side of the embryo. The mouth defines one end of an oral (O)–aboral (Ab) axis, which will be the main axis of symmetry of the pluteus larva. The secondary mesenchyme comes from the tip of the invagination. During further development, growth of skeletal rods laid down by the primary mesenchyme results in the extension of the four 'arms' of the pluteus larva. The view of the pluteus is from the oral side. The oral hood narrows into the mouth and channels food into it.

that are defined in the embryo are the animal–vegetal axis and an **oral–aboral axis** that is defined by the position of the mouth. An animal–vegetal asymmetry is already present in the unfertilized egg, but the oral–aboral axis is only established after fertilization and the early cleavage stages.

The sea urchin egg has a well defined animal–vegetal polarity that appears to be related to the site of attachment of the egg in the ovary. Polarity is marked in some species by a fine canal at the animal pole, whereas in other species there is a band of pigment granules in the vegetal region. Early development is intimately linked to this egg axis. The first two planes of cleavage are always parallel to the animal–vegetal axis and at the fourth cleavage, which is unequal (see Fig. 5.19), the micromeres are formed at the vegetal pole. The micromeres give rise to the primary mesenchyme, and both the micromeres and the primary mesenchyme may initially be specified by cytoplasmic factors localized at the vegetal pole of the egg.

The animal–vegetal axis is stable and cannot be altered by centrifugation, which redistributes larger organelles such as mitochondria and yolk platelets. Isolated fragments of egg also retain their original polarity. Blastomeres isolated at the two-cell and four-cell stages, each of which contains a complete animal–vegetal axis, give rise to mostly normal but small pluteus larvae (Fig. 5.20, left panel). Similarly, eggs that are fused together with their animal–vegetal axes parallel to each other form giant, but otherwise normal, larvae.

There is, by contrast, a striking difference in the development of animal and vegetal halves if they are isolated at the eight-cell stage. An isolated animal half merely forms a hollow sphere of ciliated ectoderm, whereas a vegetal half develops into a larva that is variable in form but is usually vegetalized; that is, it has a large gut and skeletal rods, and a reduced ectoderm lacking the mouth region (see Fig. 5.20, right panel).

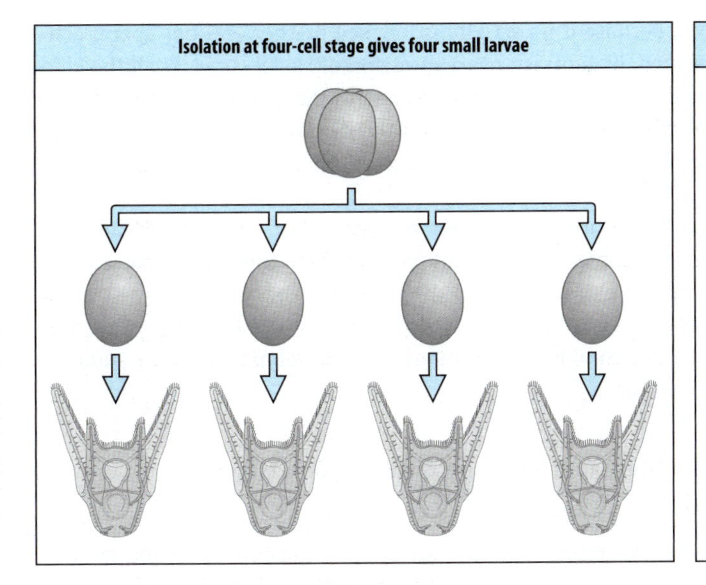

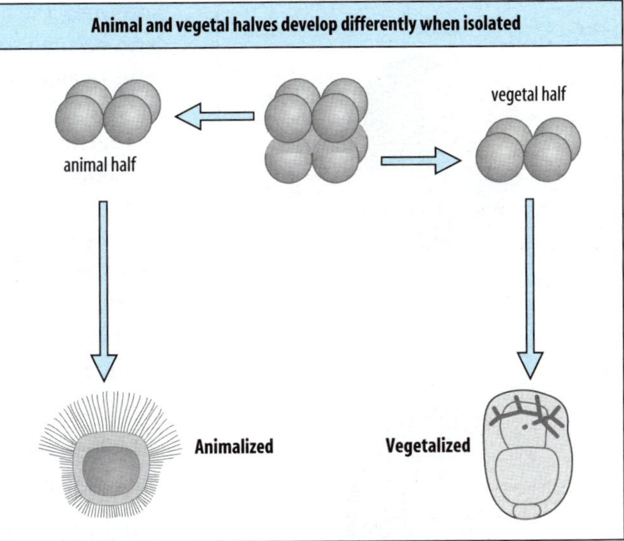

Fig. 5.20 Development of isolated sea urchin blastomeres. Left panel: if isolated at the four-cell stage, each blastomere develops into a small but normal larva. Right panel: an isolated animal half from the eight-cell stage forms a hollow sphere of ciliated ectoderm, whereas an isolated vegetal half usually develops into a highly abnormal embryo with a large gut, and some skeletal structures, but with a reduced ectoderm.

On occasion, however, vegetal halves from the eight-cell stage can form normal pluteus larvae if the third cleavage is slightly displaced toward the animal pole.

All these observations taken together show that there are already maternally determined differences along the animal–vegetal axis that are necessary to specify either animal or vegetal fates. Despite these differences it is clear that the sea urchin embryo has considerable capacity for regulation, implying the occurrence of cell–cell interactions. As we see later, important developmental signals are produced by the micromeres in the vegetal region.

5.9 The oral–aboral axis in sea urchins is related to the plane of the first cleavage

The oral–aboral axis defines the main axis of the pluteus larva and also defines the plane of bilateral symmetry in the larva. There are clear differences in skeletal patterning in the oral and aboral regions and the oral side can be recognized well before invagination forms the mouth, because the skeleton-forming mesenchyme migrates to form two columns of cells on the future oral face. The migration of these cells is considered in Chapter 7.

Unlike the animal–vegetal axis, the oral–aboral axis cannot be identified in the egg and appears to be potentially labile up to as late as the 16-cell stage. In normal development, however, the axis is predicted by the plane of the first cleavage. In the sea urchin *S. purpuratus*, a good correlation was found between the oral–aboral axis and the plane of cleavage: the future axis lies 45° clockwise from the first cleavage plane as viewed from the animal pole (Fig. 5.21). In other species, however, the relationship between the oral–aboral axis and plane of cleavage is different; the axis may coincide with the plane of cleavage or be at right angles to it.

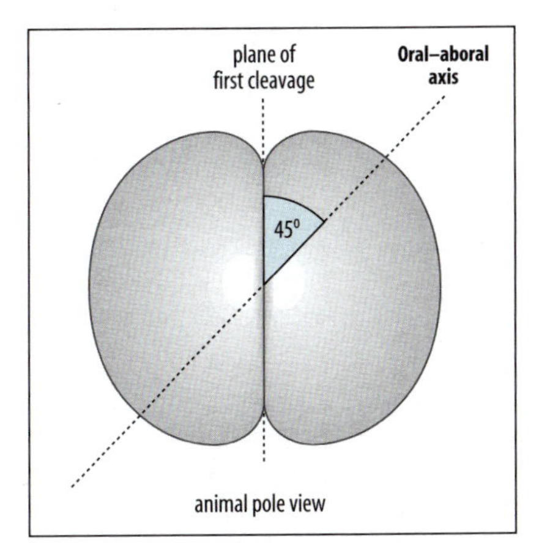

Fig. 5.21 Position of the oral–aboral axis of the sea urchin, *S. purpuratus*, embryo in relation to first cleavage. When viewed from the animal pole, the oral–aboral axis is usually 45° clockwise from the plane of first cleavage. Note that at this stage its polarity is not determined, only the positions of the two ends.

Specification of the oral–aboral axis is not yet well understood. The initial event that breaks the radial symmetry of the fertilized egg is unknown. One clear indication of the establishment of the oral–aboral axis is the expression of the TGF-β family member Nodal in the presumptive oral ectoderm, which occurs around the 60- to 120-cell stage of cleavage. It is one of the earliest zygotic genes to be expressed. This leads to the expression of zygotic genes characteristic of the oral ectoderm, such as *goosecoid* and *Brachyury*, and production of the TGF-β family protein BMP-2/-4 in the oral ectoderm. Nodal signaling appears to be instrumental in organizing and

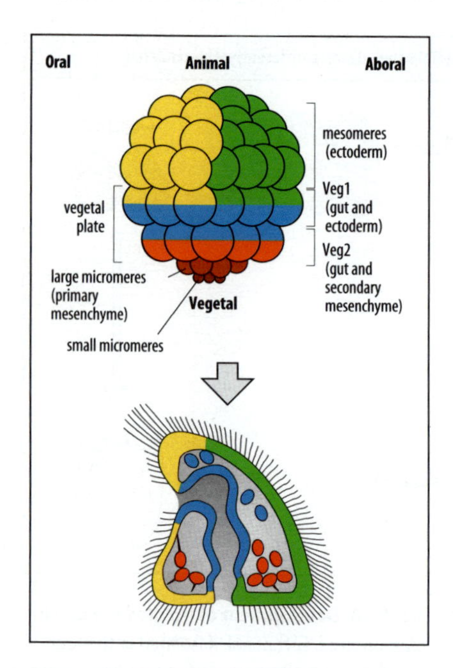

Fig. 5.22 Fate map of the sea urchin embryo. Lineage analysis has shown four main regions at the 60-cell stage. The embryo is divided along the animal–vegetal axis into three bands: the micromeres, which give rise to the primary mesenchyme; the vegetal plate, which gives rise to the endoderm, secondary mesenchyme, and some ectoderm; and the mesomeres, which give rise to ectoderm. The ectoderm is divided into oral and aboral regions.

After Logan, C.Y. and McClay, D.R.: 1997.

patterning the axis, because if its activity is blocked neither oral nor aboral ectoderm become specified. In contrast, overexpression of Nodal converts all the ectoderm to an oral fate. The oral ectoderm gives rise to the cells of the stomodaeum (the mouth), the epithelium that forms the larval outer layer in the oral region, and neurogenic structures in the larva. The aboral ectoderm gives rise to aboral epithelium only.

The Hox cluster in sea urchins contains ten genes but only two of these are expressed during development of the embryo and their function is not known. *SpHox7* is detected at the blastula stage in the ectoderm on the side away from the mouth, and *SpHox11/13b* is widely expressed in the blastula stage. The other Hox genes are expressed in a complex pattern during the development of the adult sea urchin at metamorphosis.

5.10 The sea urchin fate map is finely specified, yet considerable regulation is possible

At the 60-cell stage, five regions of cells can be distinguished along the animal–vegetal axis and a simplified fate map can be constructed (Fig. 5.22). This is composed of three bands of cells: the small and large micromeres at the vegetal pole, which will give rise to mesoderm (the primary mesenchyme that forms the skeleton and some adult structures); the vegetal plate, which comprises the next two tiers of cells (Veg1 and Veg2) and gives rise to endoderm, mesoderm (the secondary mesenchyme that forms muscle and connective tissues), and some ectoderm; and the remainder of the embryo, which gives rise to ectoderm and is divided into future oral and aboral regions. The ectoderm gives rise to the outer epithelium of the embryo and to neurogenic cells.

Using vital dyes as lineage tracers, the pattern of cleavage and of cell fates have both been mapped. The pattern of cleavage has been shown to be invariant. Unlike the nematode, however, this stereotypical cleavage pattern seems irrelevant to normal development. Compressing an early embryo by squashing it with a cover slip, and thus altering the pattern of cleavage, still results in a normal embryo after further development. Thus the stereotypical cleavage pattern does not in most cases predict an invariance of cell fate, and the embryo is highly regulative. Lineage and cell fate are to a large extent separable. The regulative capacity is not uniform, however. The micromeres, for example, have a fixed fate once they are formed and this fate cannot be changed. They give rise only to primary mesenchyme, and no other fate has been observed, even if they are grafted to another site on the embryo. The fate of the vegetal macromeres is not fixed when they are formed. Vegetal macromeres labeled at the 16-cell stage give rise to endoderm, mesoderm, and ectoderm. If cells are labeled at the 60-cell stage, the Veg2 cells derived from these macromeres become either mesoderm (secondary mesenchyme) or endoderm, but never ectoderm, while the Veg1 cells give rise to endoderm and ectoderm but not mesoderm. This shows that endomesoderm specification has occurred but that the boundary between endoderm and ectoderm has not yet been set. This occurs slightly later. Single Veg1 cell progeny labeled after the eighth cleavage give rise to either endoderm or ectoderm, but not both.

When isolated and cultured, some regions of the sea urchin embryo develop more or less in line with their normal fate. Micromeres isolated at the 16-cell stage will form primary mesenchyme cells and may even form skeletal rods. An isolated animal half of presumptive ectoderm forms an animalized ciliated epithelial ball, but there is no indication of mouth formation. Although these experiments suggest that localized cytoplasmic factors are involved in specifying cell fate, this does not tell the full story, as they do not reveal the role of inductive cell–cell interactions in the normal process of development. The ability of a vegetal half to form an almost complete embryo has already been noted. Another dramatic example of regulation is seen when the animal and vegetal blastomeres from half of an eight-cell embryo are combined with the animal blastomeres from another; this unlikely combination can develop into a completely normal embryo. How does this occur? The answer, as we see next, is the ability of the vegetal region to act as an organizer.

5.11 The vegetal region of the sea urchin embryo acts as an organizer

The regulative capacities of the early sea urchin embryo are due in large part to an organizing center that develops in the vegetal region and which, like the amphibian organizer, has the potential ability to induce the formation of an almost complete body axis. This center is initially set up by the activities of maternal factors and becomes located in the micromeres formed at the fourth cleavage division. It produces signals that induce the adjacent vegetal macromeres to adopt an endomesodermal fate and sets in train a relay of events that specify the endomesoderm more finely as mesoderm and endoderm and help set the endoderm–ectoderm boundary. Ablation of the micromeres at the fourth cleavage results in abnormal development, but if they are removed at the sixth cleavage, after 2–3 hours of contact, gastrulation is delayed but normal larvae can develop.

Striking evidence for an organizer came from experiments that combined micromeres isolated from a 16-cell embryo with an isolated animal half from a 32-cell embryo. This combination can give rise to an almost normal larva (Fig. 5.23).

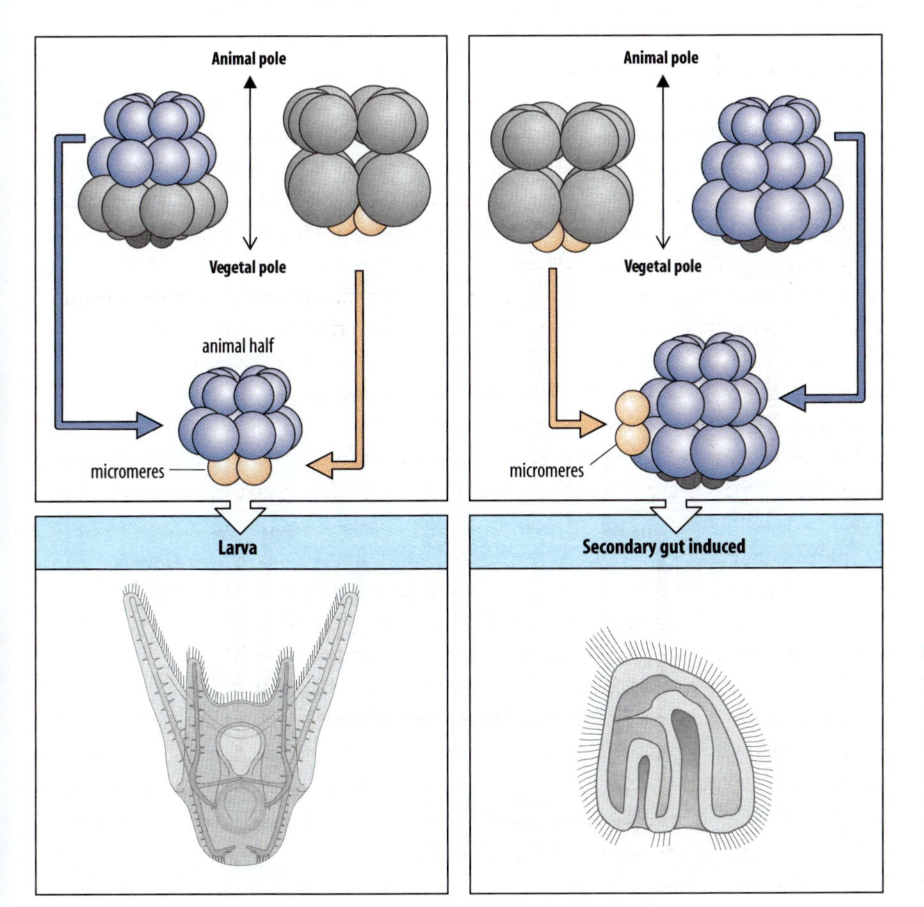

Fig. 5.23 The inductive action of micromeres. Left panels: combining the four micromeres from a 16-cell sea urchin embryo with an animal half from a 32-cell embryo results in a normal larva. An animal half cultured on its own merely forms ectoderm (not shown). Right panels: micromeres implanted into the side of another 32-cell embryo induce the formation of another gut at the implantation site.

The micromeres clearly induce some of the presumptive ectodermal cells in the animal half to form a gut, and the ectoderm becomes correctly patterned. Further evidence for the organizer-like properties of the micromeres comes from grafting them to the side of an intact embryo. Here, they induce endoderm in the presumptive ectoderm, which then invaginates to form a second gut. There is some evidence that the closer the graft is to the vegetal region, the greater the size of the invagination, implying a graded ability along the animal–vegetal axis to respond to a signal from micromeres.

The regulative capacity of the very early embryo can even compensate for loss of the micromeres. If micromeres are removed as soon as they are formed at the fourth cleavage (see Fig. 5.19), regulation occurs and a normal larva will still develop. The most vegetal region assumes the property of micromeres, giving rise to skeleton-forming cells and acquiring organizing properties.

5.12 The sea urchin vegetal region is specified by nuclear accumulation of β-catenin

The sea urchin vegetal region and its organizing properties are specified by a mechanism very similar to the one that specifies the dorsal region in the early *Xenopus* embryo (discussed in Chapter 3). In the sea urchin, the early blastomeres accumulate maternal β-catenin in their nuclei as a result of the activation of part of the canonical Wnt pathway. In the nucleus, β-catenin acts with other transcription factors to activate zygotic gene expression. β-Catenin accumulation does not occur uniformly, however, but is high in the micromere nuclei and almost absent in the most animal region of the future ectoderm. This differential stabilization of β-catenin along the animal–vegetal axis appears to be at least partly controlled by the

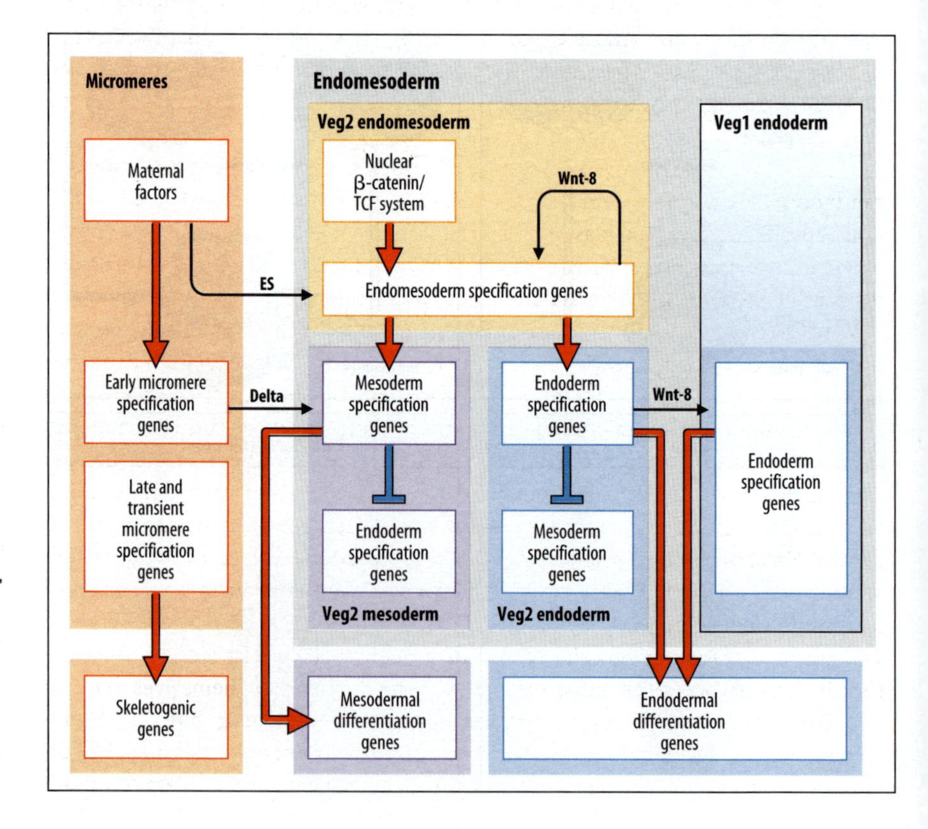

Fig. 5.24 An outline of endomesoderm specification in the sea urchin. The white boxes represent networks of genes that are reponsible for the developmental functions indicated. The background shading represents the embryonic territories involved. Peach, the large micromere precursors that will give rise to skeletal tissue; yellow, the early endomesoderm, which segregates at late cleavage into mesoderm (violet) and endoderm (blue). The red arrows and blue barred lines refer to gene interactions; the thin black arrows represent intercellular signals. ES, early signal.

*Adapted from Oliveri, P. and Davidson, E.H.: Curr. Opin. Genet. Dev. 2004, **14:** 351–360.*

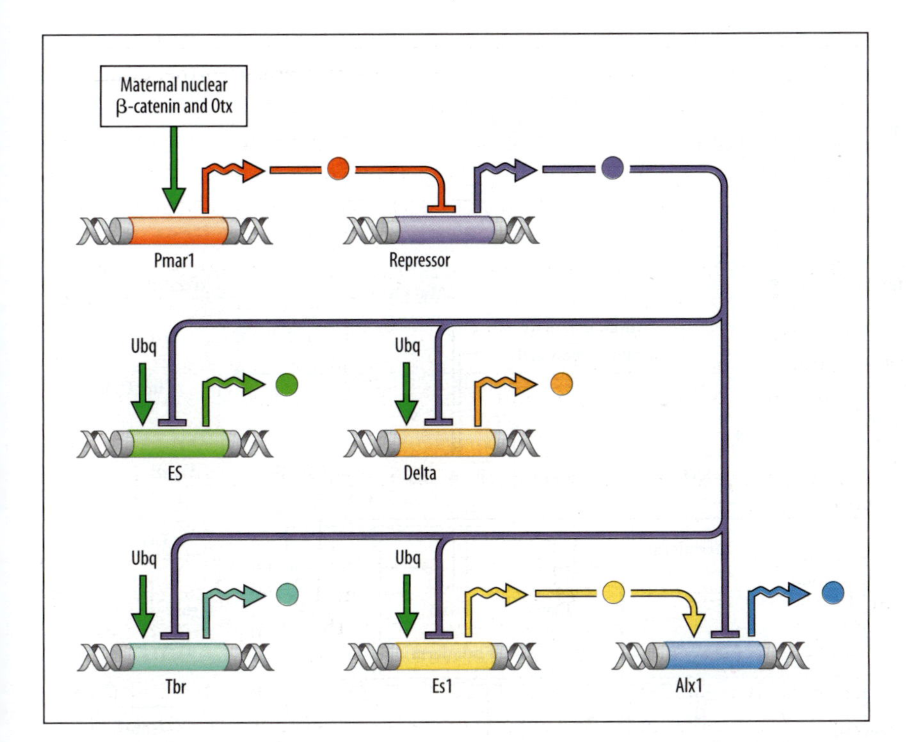

Fig. 5.25 Specification of the sea urchin micromeres and the activation of their organizing function. Two maternal transcription factors, β-catenin and Otx, are localized in micromere nuclei and activate transcription of *pmar1*. This gene encodes a transcriptional repressor that suppresses expression of an inferred 'repressor of micromeres' (Repressor) that is otherwise active throughout the embryo. The lifting of this repression also allows the activation by ubiquitous positive regulators (Ubq) of a set of regulatory genes in the micromeres that are responsible for micromere-organizing properties and for their skeletogenic fate. ES is an as yet unidentified 'early signal' that induces the endomesoderm fate in Veg2 cells between the fourth and sixth cleavages. Delta is the signal that induces secondary mesenchyme fate in Veg2 cells after the seventh cleavage. Tbr, Ets1, and Alx1 are transcriptional regulators.

Adapted from Oliveri, P. and Davidson, E.H.: Curr. Opin. Genet. Dev. 2004, ***14****: 351–360.*

local activation in the vegetal region of maternal Dishevelled protein, which is an intracellular activator of the pathway by which β-catenin is stabilized (see Fig. 2.39).

Nuclear β-catenin promotes a vegetal fate. Treatments that inhibit β-catenin degradation, such as lithium chloride (see Chapter 3), cause vegetalization, resulting in embryos with a reduced ectoderm and an enlarged gut. The main effect of lithium on whole embryos is to cause an expansion of β-catenin accumulation beyond the normal vegetal region, shifting the border between endoderm and ectoderm toward the animal pole. Similarly, overexpression of β-catenin also causes vegetalization and can induce endoderm in animal caps. In contrast, preventing the nuclear accumulation of β-catenin in vegetal blastomeres completely inhibits the development of both endoderm and mesoderm. Micromeres depleted of β-catenin, for example, are unable to induce endoderm when transplanted to animal regions.

Maternal β-catenin and the transcription factor Otx activate the gene *pmar1* in the micromeres at the fourth cleavage. *pmar1* encodes a transcription factor that is produced exclusively in micromeres during cleavage and early blastula stages and is required for the micromeres to express their organizer function and also develop as primary skeletogenic mesenchyme. Soon after they are formed, the micromeres produce a signal, as yet unidentified, that induces the endomesoderm state in the Veg2 cells. A second signal, produced after the seventh cleavage, acts on the Notch receptor in adjacent Veg2 cells to specify them as pigment and secondary mesenchyme cells. At later stages Notch–Delta signaling is also involved in endoderm specification and in delimiting the endoderm–ectoderm boundary. The secreted signaling protein Wnt-8, first expressed by micromeres after the fourth cleavage, acts as an autocrine factor on the micromeres themselves to maintain the stabilization of β-catenin and thus maintain micromere function. Wnt-8 acting later in the Veg2 endoderm is involved in directing the specification of endoderm in adjacent Veg1 cells. In this way, the sea urchin mesoderm and

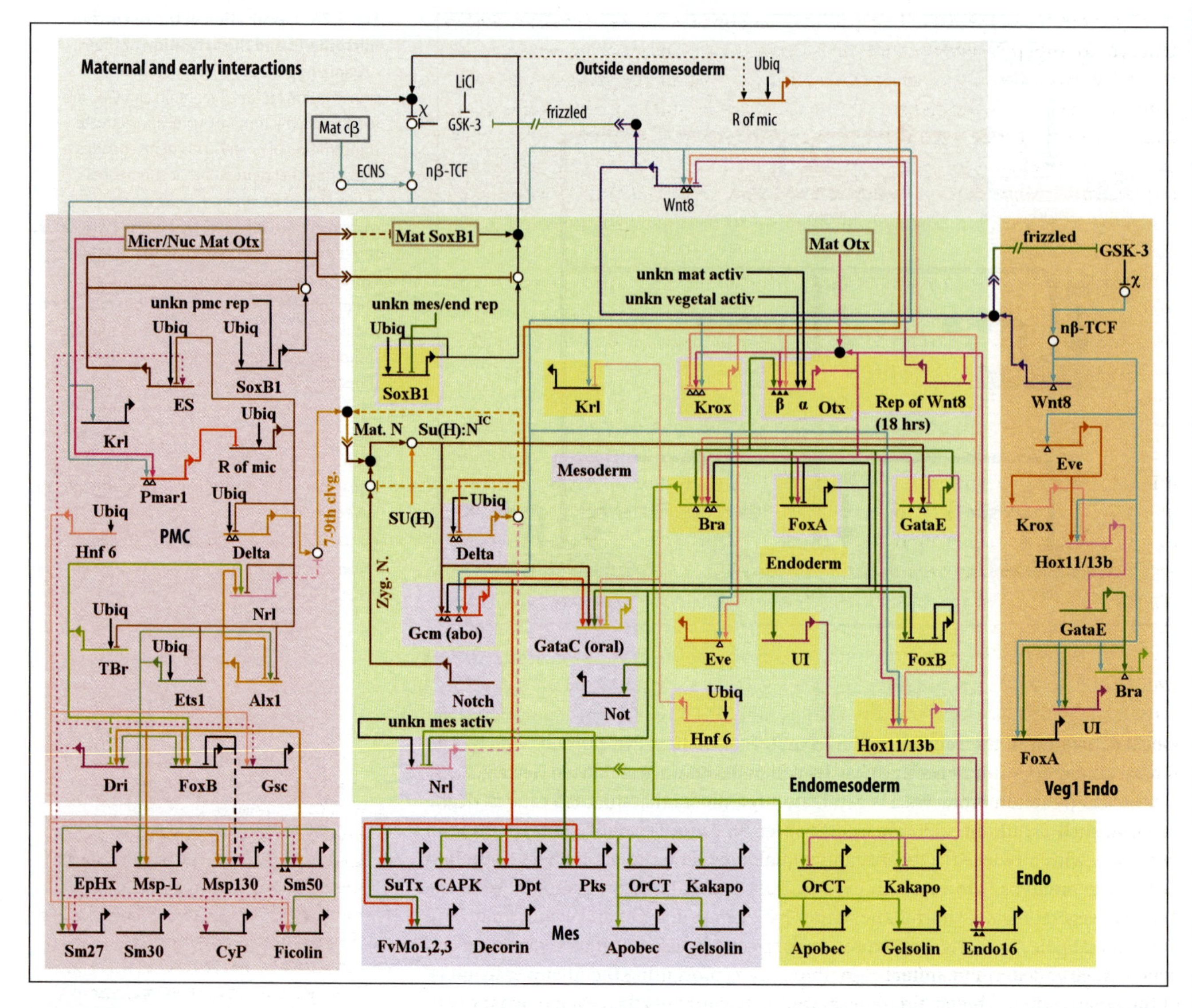

Fig. 5.26 The gene-regulatory network for sea urchin endomesoderm specification. The part of the network shaded in gray at the top corresponds to the nuclear localization of β-catenin under the control of maternal factors. The rest of the network refers to zygotic gene expression. PMC, primary mesenchyme (micromeres); Mes, secondary mesenchyme; Veg1 Endo, Veg1-derived endoderm.

Adapted from Oliveri, P. and Davidson, E.H.: 2004.

endoderm is specified by a series of short-range signals that result in changes in gene expression.

5.13 The genetic control of endomesoderm specification is known in considerable detail

As we have seen in previous sections, specification of the endomesoderm starts with the nuclear accumulation of β-catenin in the ventral region of the blastula. The gene-regulatory network that subsequently governs cell-fate determination (Fig. 5.24) has been worked out in great detail over the past decade. This network comprises a description of the transcription factors involved and the genes they regulate. That part of the network responsible for the specificity of micromeres and primary mesenchyme cells is illustrated in Fig. 5.25. It shows how the maternally supplied transcription factors β-catenin and Otx activate the gene for Pmar1, and how the subsequent interactions lead to a variety of zygotic genes, including those for Delta and Notch, being activated. The main function of Pmarl appears to be to prevent the expression of a repressor that operates throughout the rest of the embryo to suppress endomesoderm formation. The network also shows how control of the micromere specification initially set up by maternal factors is handed over to zygotic genes, thus ensuring the stable maintenance of the endomesodermal state once it is specified. This is just a small part of the larger early gene-regulatory network, which is shown in Fig. 5.26. That is vastly more complicated, but the main elements of Fig. 5.25 can be identified. This figure illustrates the almost frightening complexity of the gene-regulatory networks that are now being revealed.

The complexity of the network is hardly surprising given how complex the regulatory regions of developmental genes can be. As we saw in Chapter 2, the regulatory region of a *Drosophila* pair-rule gene such as *even-skipped* contains many binding sites for transcription factors that can both activate and repress gene activity (see Fig. 2.30). These target sites are grouped into discrete subregions—regulatory modules—which each control expression of the gene at a particular site in the embryo.

Although all the genes controlling sea urchin development have yet to be identified, many whose expression varies with both space and time have been analyzed in detail. Their regulatory regions prove to be complex and modular, as illustrated by the regulatory region of the sea urchin *Endo-16* gene, a gene coding for a secreted glycoprotein of unknown function. *Endo-16* is initially expressed in the vegetal region of the blastula, in the presumptive endoderm (Fig. 5.27). After gastrulation, *Endo-16* expression increases and becomes confined to the middle region of the gut. The regulatory region controlling *Endo-16* expression is about 2200 base pairs long and contains at least 30 target sites, to which 13 different transcription factors can bind. These regulatory sites seem to fall into several subregions, or modules. The function of each module has been determined by attaching it to a reporter gene and injecting the recombinant DNA construct into a sea urchin egg. In such experiments, module A, for example, is found to promote expression of its attached gene in the presumptive endoderm, whereas modules D and C prevent expression in the primary mesenchyme. Midgut expression is under control of module B. Modules D, C, E, and F are activated by the action of lithium and are involved in the vegetalizing action of *Endo-16*.

The modular nature of the *Endo-16* control region is similar to that of the pair-rule genes in *Drosophila*, where different modules are responsible for expression in each of the seven stripes. The sea urchin example illustrates yet again the importance of gene-control regions in integrating and interpreting developmental signals.

Fig. 5.27 Modular organization of the *Endo-16* gene regulatory region. The modular subregions are indicated A to G. Thirteen different transcription factors bind to the regulatory sites in these modules. The spatial domains of the embryo in which each module regulates gene expression are indicated, together with other regulatory modules. Early in development, regulation of gene expression restricts *Endo-16* activity to the vegetal plate and the endoderm of the future gut; later in development it is restricted to the midgut.

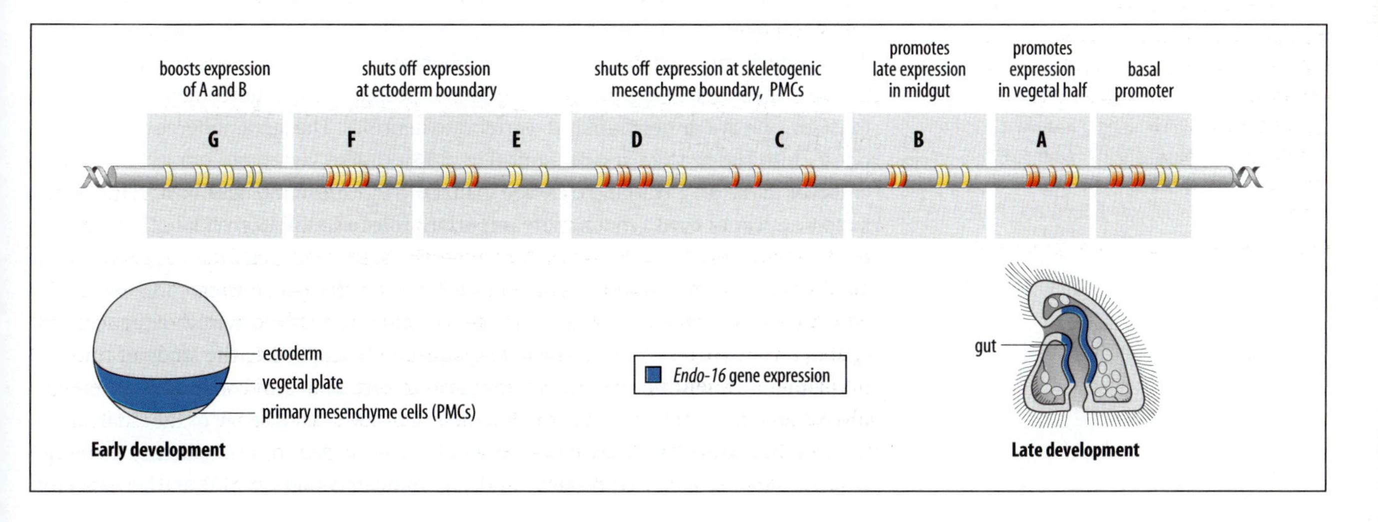

Summary

Sea urchin embryos have a remarkable capacity for regulation even though they also have an invariant cell lineage. The sea urchin egg has a stable animal–vegetal axis before fertilization, which appears to be specified by the localization of maternal cytoplasmic factors in the egg. The oral–aboral axis is determined after fertilization and is related to the plane of first cleavage. After fertilization, maternal factors specify the formation of an organizer region in the most vegetal region of the egg, where the micromeres are formed. This organizer, like the Spemann organizer in amphibians, is able to specify an almost complete body axis. Localized activation of the Wnt pathway and nuclear accumulation of β-catenin specifies the vegetal region and the organizing properties of the micromeres. Signals from the micromeres induce and pattern the endomesoderm along the animal–vegetal axis, while the oral–aboral axis is specified and patterned by the activation of the Nodal signaling pathway in the future oral region. The site and time of expression of developmental genes are controlled by complex gene-regulatory networks.

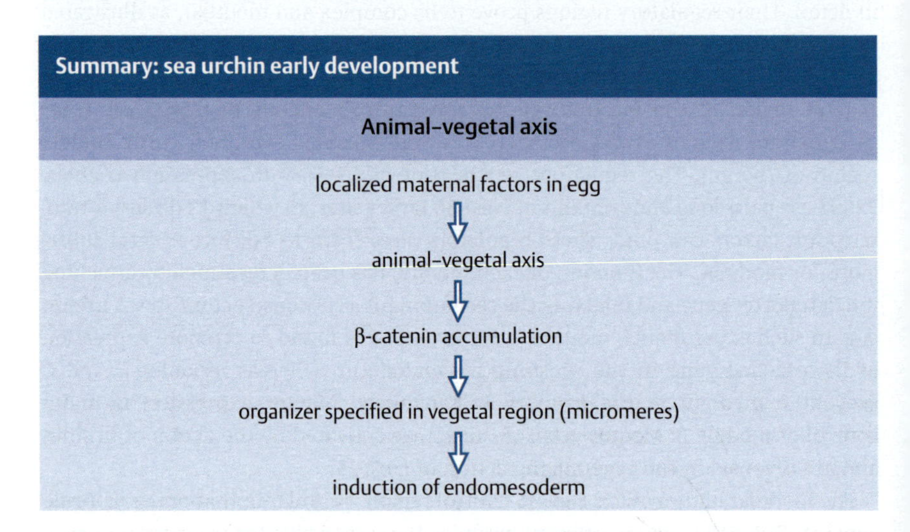

Ascidians

Adult ascidians, also known as tunicates, are sessile marine animals. They are urochordates, which are included in the same phylum—the Chordata—as the vertebrates, because their free-living, tadpole-like larvae possess a notochord, neural tube, and muscles, and are rather similar to vertebrate neurulas. The larva undergoes metamorphosis into the sac-like sessile adult (Fig. 5.28). Unlike vertebrate embryos, ascidian embryos have an invariant cleavage pattern (Fig. 5.29), and localized cytoplasmic factors appear to have a much more important role in specifying cell fate. The genome of the sea squirt *Ciona intestinalis* has recently been sequenced, and sequencing of another species, *Ciona savignyi*, is under way. *C. intestinalis* has an estimated 15,800 genes, which is comparable to the other model invertebrates. The genome sequences will enable the full panoply of genomic techniques to be applied to the study of tunicate development, and similarities and differences to vertebrates and other chordates to be investigated in detail at the genetic and molecular level. Knowledge of the *C. intestinalis* genome has already been exploited in a large-scale screen for *cis*-regulatory elements controlling Hox gene expression, and has indicated that as well as the expected

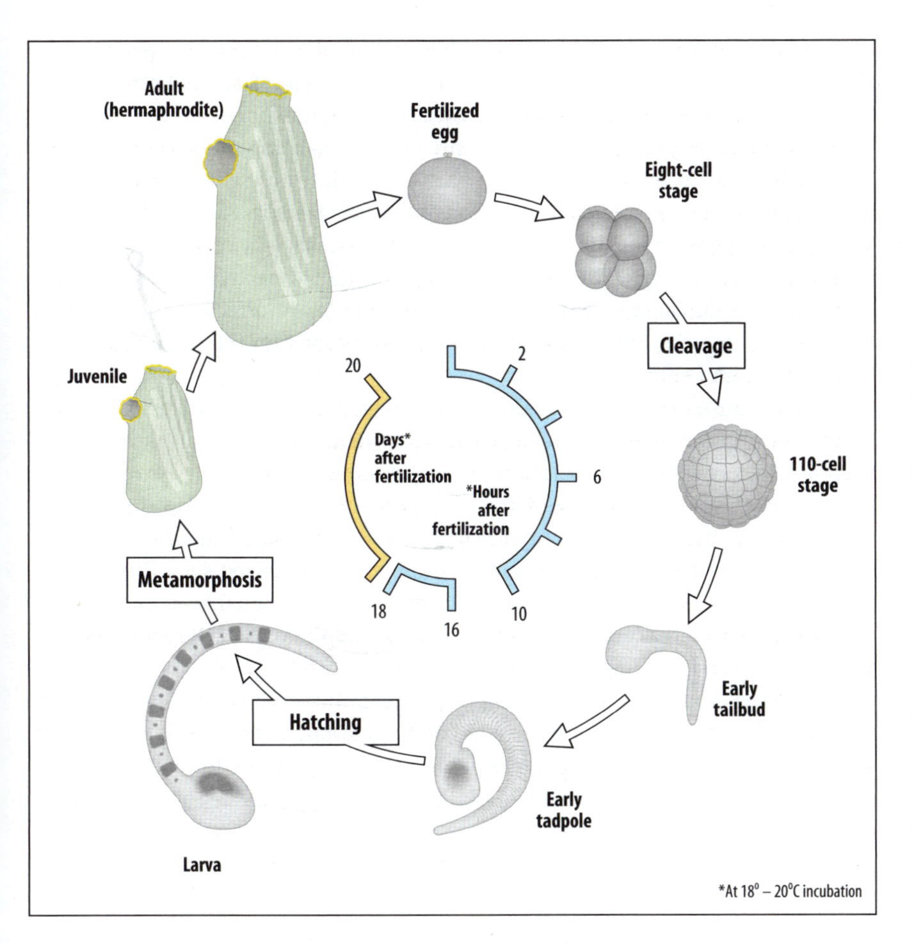

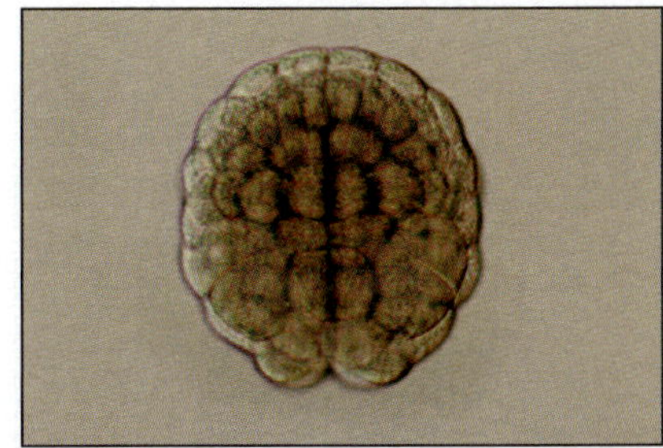

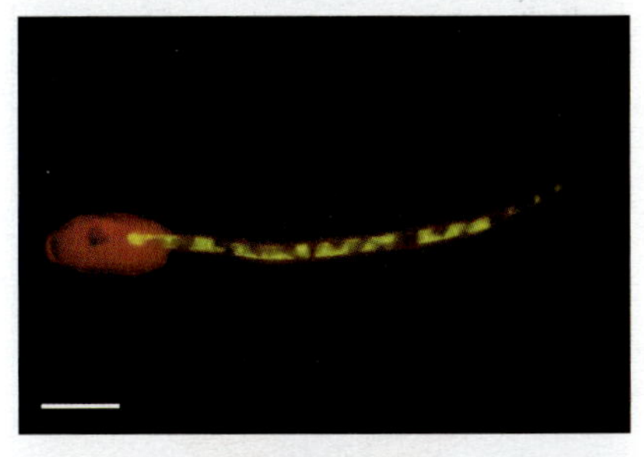

Fig. 5.28 Life cycle diagram of the ascidian *Ciona intestinalis*. *Ciona* is hermaphrodite and eggs are fertilized externally. The fertilized egg takes about 18 hours to hatch into a larva, depending on the temperature of the water. The free-swimming larva undergoes metamorphosis into the sessile juvenile (about 2 cm tall) in around 20 days. Development of the juvenile into a sexually mature adult (which is about 5–8 cm tall) takes a further 2 months. The photographs show: (top) a 110-cell embryo; (middle) a larva (with some notochord cells labeled green; scale bar = 0.1 mm); (bottom) adult *C. intestinalis*.

Top photograph courtesy of Shigeki Fujiwara and Naoki Shimozono; middle photograph courtesy of J. Corbo, from Corbo, J.C., et al.: 1997; bottom photograph courtesy of Andrew Martinez.

antero-posterior expression domains in the central nervous system, ascidians also appear to express Hox genes in nested antero-posterior domains in the epidermis, a feature they share with arthropods (see Chapter 2) but not with vertebrates.

Ascidian embryogenesis has long been regarded as a typical example of mosaic development, with cytoplasmic factors specifying cell fate during cleavage, and cell interactions playing only a relatively minor part. It is now clear, however, that cell interactions are more important in ascidian development than was previously thought.

The developmental axes in an ascidian larva are related to the early pattern of cleavage. Like the amphibian and sea urchin egg, the unfertilized ascidian egg is polarized along an animal–vegetal axis, with the prospective territories of ectoderm, mesoderm, and endoderm lying along this axis from animal to vegetal. Important maternal factors become localized in the vegetal region, among them components of the Dishevelled–β-catenin signaling pathway. As in the sea urchin, localized activation of this pathway stabilizes maternal β-catenin and induces its

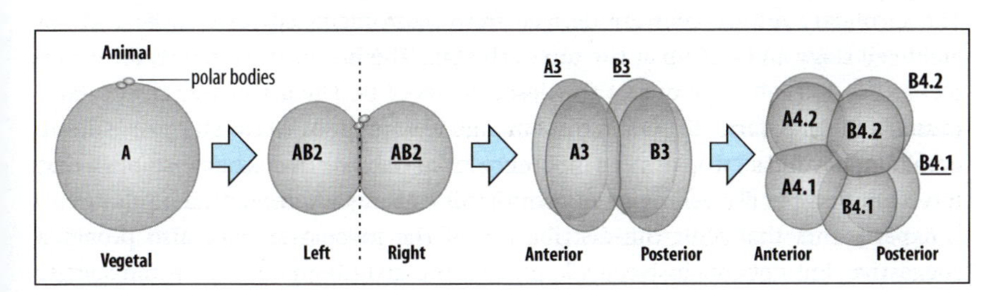

Fig. 5.29 Cleavages in the egg of the ascidian *Halocynthia rosetryi*. The first cleavage divides the embryo into left and right halves. The second cleavage divides the embryo along the antero-posterior axis.

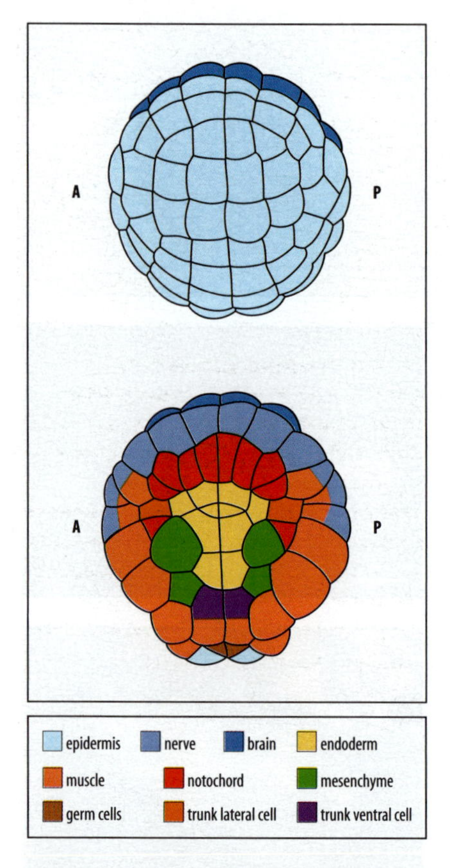

Fig. 5.30 Fate map of the 110-cell ascidian embryo. Top panel: view looking down on the animal pole. Almost all the cells of the animal half become ectoderm. Bottom panel: view from the vegetal pole.

Adapted from Nishida, H.: 2005.

preferential entry into nuclei in vegetal cells, where it is required to specify endoderm in the vegetal region. In contrast, the development of an epidermal fate in animal blastomeres requires suppression of β-catenin function.

The first cleavage passes through the site of polar body formation (see Fig. 5.29), usually dividing the embryo bilaterally, and defines the future antero-posterior axis which runs roughly perpendicular to the animal–vegetal axis. The symmetry of the unfertilized egg is, in fact, broken before the first cleavage by cortical movements that localize certain cytoplasmic determinants to the prospective posterior region. As in the amphibian egg, these cortical movements are directed by the sperm centrosome, and the future posterior pole is established on the same side of the egg as the sperm entry point.

The unfertilized egg itself is highly regulative; if it is divided in a plane parallel to the future antero-posterior axis, each half, when fertilized, will develop into a complete larva. After a few more cleavages, however, the cells become more committed to their prospective fates and there appears to be little capacity for regulation. By the time the embryo reaches the 110-cell stage, at which gastrulation begins, the fate of each blastomere has become restricted such that it gives rise to a single cell type in the larva (Fig. 5.30). When individual blastomeres from this stage are isolated and cultured they also develop into the cell types specified by the fate map.

Despite the prominent role of cytoplasmic determinants in specifying cell fate in ascidians, the development of some types of mesodermal tissue at least depends on inductive signals. Three distinct types of mesodermal tissue are formed in ascidian embryos—the muscles of the larval tail, the mesenchymal tissue that gives rise to internal tissues, and the notochord. Here we shall look at the ways in which the fate of these three mesodermal tissues is determined.

5.14 In ascidians, muscle is specified by localized cytoplasmic factors

The fertilized eggs of the ascidian *Styela* contain a crescent of yellow pigment granules whose fate in the embryo can be followed. The cells that acquire this yellow cytoplasm—the **myoplasm**—during cleavage are those that will give rise to the muscle cells of the larval tail (Fig. 5.31). Before fertilization, the yellow granules are more or less uniformly distributed throughout the egg; following fertilization, there is a dramatic rearrangement of the myoplasm to form the yellow crescent at the equator. Movement is associated with the cytoskeleton, principally cortical actin filaments and a deep network of intermediate filaments. The crescent marks the future posterior end of the embryo, where gastrulation is initiated.

During cleavage, the myoplasm becomes confined to particular cells, and by the eight-cell stage is largely confined to the two dorsal posterior cells, with a small amount in adjacent cells (see Fig. 5.31). In the species *Halocynthia rosetryi*, the cell lineage has been worked out in detail by the injection of a tracer into the early cells. The two posterior B4.1 cells (see Fig. 5.29) that contain myoplasm contribute to the primary muscle cells, 28 of the 42 muscle cells that lie on each side of the tail. The secondary muscle cells are derived from blastomeres adjacent to B4.1 at the eight-cell stage and end up at the top of the tail. The lineage is complex; for example, at the 128-cell stage one of the descendants of B4.1 will still give rise to both muscle and endoderm. So, while there is a good correlation between muscle development and yellow cytoplasm, these observations on their own do not establish that it is something in the yellow cytoplasm that is causing differentiation into muscle.

Experiments that alter the distribution of the myoplasm have also provided suggestive, but not conclusive, evidence that the myoplasm on its own can specify

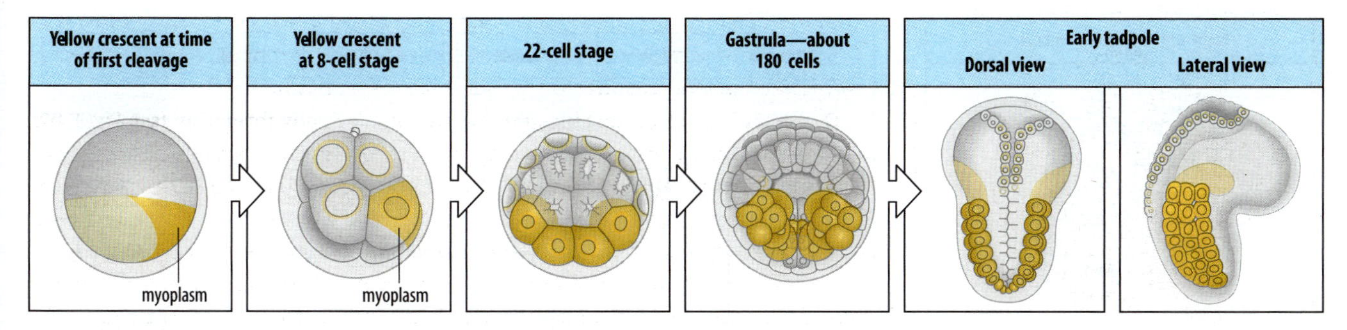

Fig. 5.31 Muscle development and cytoplasmic determinants in the ascidian *Styela*. Following fertilization, the myoplasm, which is colored with yellow granules, moves laterally and toward the equator. This movement forms a yellow crescent at the future posterior end of the embryo. Gastrulation starts at this site. The muscle of the ascidian tadpole's tail comes both from cells that contain the yellow myoplasm and the cells adjacent to them.
After Conklin, E.G.: 1905.

muscle cells. But the most persuasive evidence that a factor in myoplasm can indeed specify muscle cells is provided by experiments with the *macho-1* gene. Maternal *macho-1* mRNA is localized in the myoplasm and its depletion results in loss of the primary muscle cells in the tail. If the posterior-vegetal cytoplasm containing the *macho-1* mRNA is removed from the fertilized egg, the blastomeres that would normally develop as muscle develop as nerve cord. In addition, injection of *macho-1* mRNA into non-muscle cell lineages causes ectopic muscle differentiation.

5.15 Mesenchyme and notochord development in ascidians require signals from the endoderm

Although a muscle fate appears to be determined cell-autonomously, specification of the other two mesodermal cell fates—mesenchyme and notochord—from mesodermal precursors requires signals from the adjacent cells. When blastomeres are isolated at the 32-cell stage, all those blastomeres whose normal fate is mesenchyme will develop into muscle. It appears that a signal from the endoderm, thought to be a fibroblast growth factor (FGF), normally suppresses muscle formation in these blastomeres. In the absence of such signals they develop into muscle, as a result of the presence of the maternal muscle determinants.

The ascidian notochord is similarly specified by the FGF signals from the endoderm. The notochord of the *Halocynthia* larva consists of a single row of 40 cells aligned along the center of the tail. Cells of the A lineage give rise to 32 of the notochord cells, the remaining ones being derived from the B lineage. A-lineage cells that give rise only to notochord can be identified at the 64-cell stage, whereas B-lineage cells that give rise only to notochord cannot be identified until the 110-cell stage (Fig. 5.32). Blastomeres that would normally give rise to notochord do not do so if they are isolated at the 32-cell stage, unless they are induced by being combined with vegetal blastomeres. However, prospective notochord cells of the B lineage isolated at the 110-cell stage do develop into notochord. So, for notochord to develop, induction by vegetal cells is required somewhere between the 32-cell and 110-cell stage. FGF can induce notochord and is probably the inducer *in vivo*. Although the early development of ascidians is very different from that of vertebrates, the presence of a notochord and its induction by vegetal cells present striking parallels between the two groups. Moreover, the same genes are involved in notochord specification in both ascidians and vertebrates.

The gene *Brachyury* is expressed in early mesoderm in vertebrates (see Section 3.22) and then becomes confined to the notochord. Expression of the ascidian homolog of *Brachyury* is first detected at the 64-cell stage in the A-lineage precursors of the notochord, and this stage appears to correspond with the time at which induction is complete. Ectopic expression of the *Brachyury* gene of the ascidian can transform

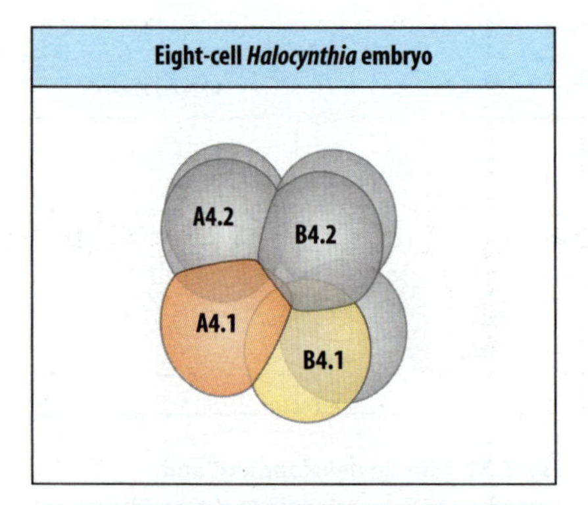

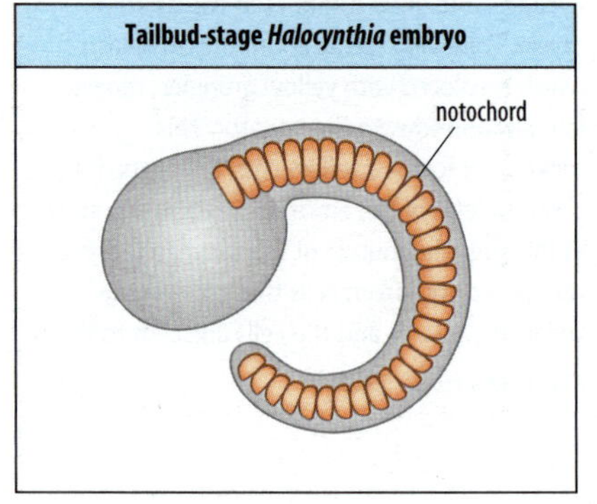

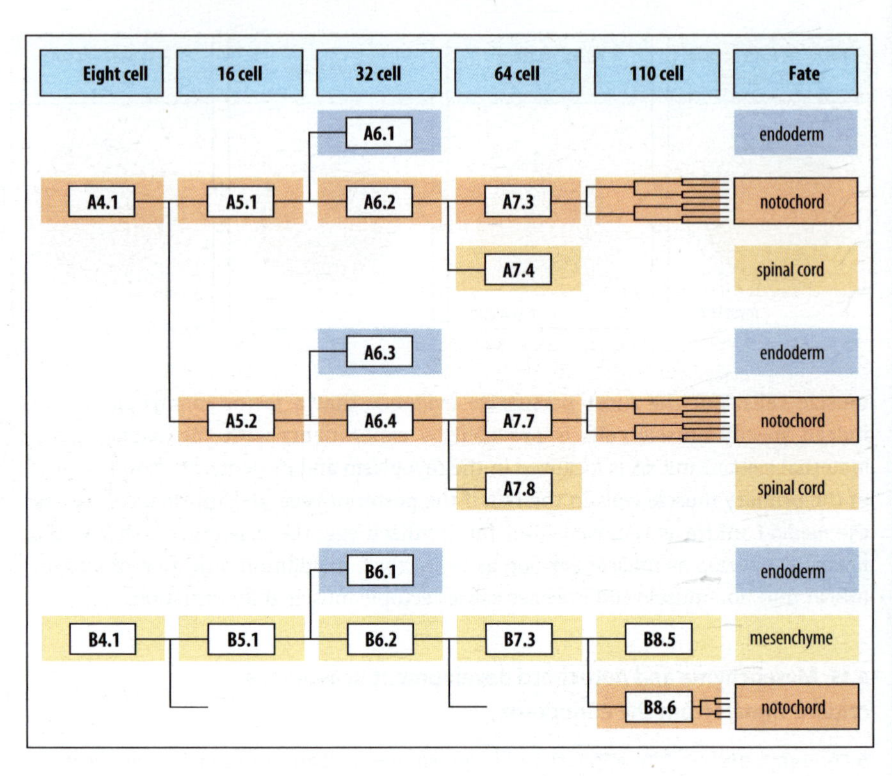

Fig. 5.32 The lineage of the ascidian notochord. All notochord cells are derived from blastomeres A4.1 and B4.1, formed at the eight-cell stage (see Fig. 5.20). Only the lineage on the left side is illustrated, as that on the right side is the same.

After Nakatani, Y., et al.: 1996.

endoderm to notochord. Thus, there appears to be a similar mechanism involved in notochord formation in all chordates.

Summary

In the chordate ascidians, there is evidence that localized cytoplasmic factors are involved in specifying cell fate, particularly of muscle, but that cell interactions are also involved. The notochord develops through a well defined lineage but requires induction. The ascidian homolog of the vertebrate gene *Brachyury*, which is involved in specifying the notochord in vertebrates, is expressed in the presumptive ascidian notochord after induction.

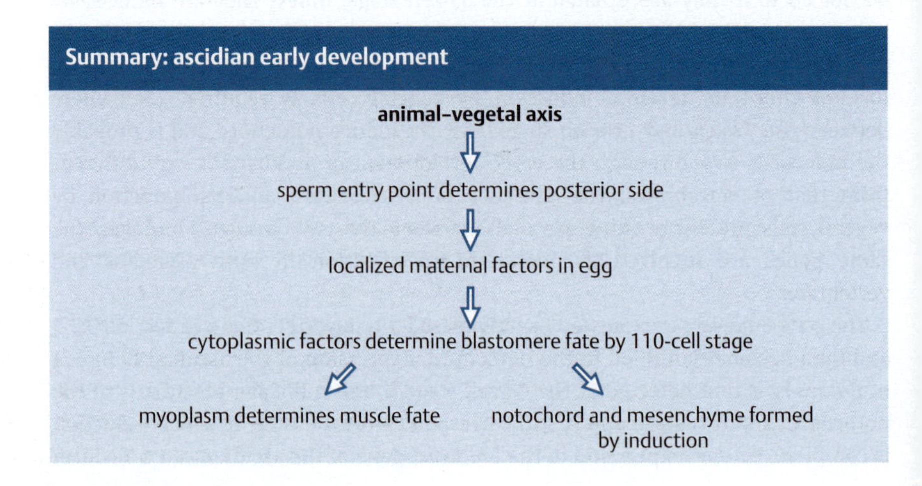

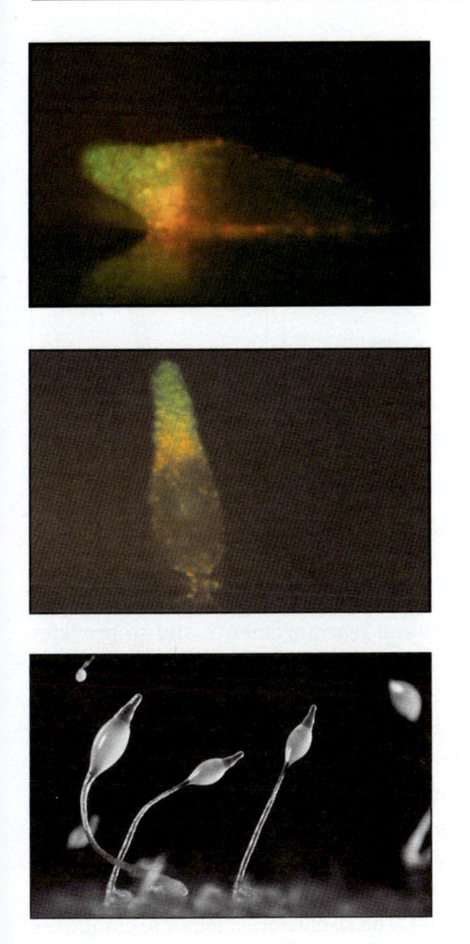

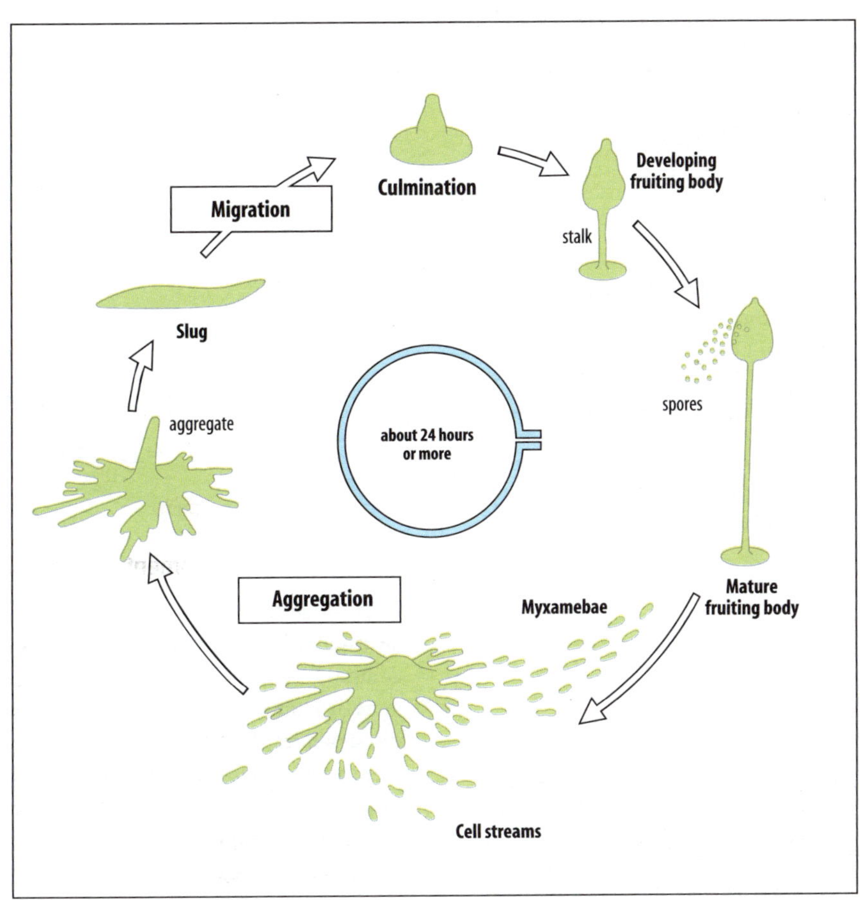

Fig. 5.33 Life cycle of the cellular slime mold *Dictyostelium discoideum*. Unicellular myxamebae aggregate to form a multicellular 'slug' which, after migration, undergoes a process of culmination to form a differentiated, multicellular fruiting body. This is composed of a stalk with spores at the tip. The photographs show: top panel, a living migrating slug in which the prestalk region can be visualized (prestalk A cells, green; prestalk O cells, red) because the prestalk A and O cells are expressing different spectral variants of green fluorescent protein. The prespore region is unstained. Yellow color indicates regions where prestalk A and prestalk O cells physically overlap, and the few cells that co-express the two genes used to distinguish these cell types. Middle panel: a slug that has stopped migrating and is standing on end. Cells in the prestalk region move first to the apex; then they reverse their direction of movement and travel towards the base, as indicated by the green mass of prestalk A cells visible in the prespore region. Bottom panel; mature *Dictyostelium* fruiting bodies.

Photographs in the top and center panels courtesy of P. Dorman, T. Abe, J.G. Williams, and C.J. Weijer. The photograph in the bottom panel courtesy L. Blanton.

Cellular slime molds

The cellular slime molds, such as *Dictyostelium discoideum*, have properties similar to both animals and plants. The cellulose cell walls formed at some stages of their life cycle and the formation of reproductive spores are plant-like, whereas the cell movements involved in their morphogenesis are animal-like. Analysis of protein sequences suggests that slime molds diverged from the eukaryotic lineage after the ancestors of plants but before those of animals (see Fig. 5.1). The slime molds have a developmental strategy with similarities to both animals and plants. The slime mold's simple life cycle (Fig. 5.33) and the ability to grow large numbers of cells easily, together with the possibility of genetic manipulation, make it an attractive system for developmental study.

Slime molds alternate between a unicellular and a multicellular phase. The multicellular phase is not generated through the division of a single large cell to form an embryo, but by the aggregation of hitherto free-living single cells. Cellular slime molds grow and multiply as unicellular social amebae, which feed on bacteria. When their food supply runs out, the cells aggregate to form a migrating multicellular 'slug', typically containing some 100,000 cells. The slug develops into a fruiting body, consisting of a stalk composed of dead vacuolated cells supporting a mass of spores. Each spore develops into a new myxameba, which multiplies by division, and eventually gives rise to a new aggregate and slug. Thus, the simple life cycle involves differentiation into two main cell types—stalk cells and spore

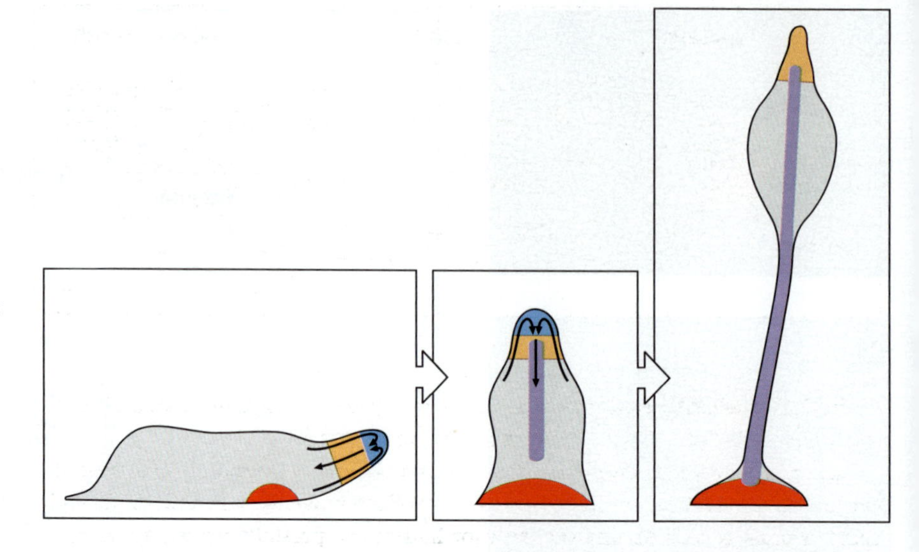

Fig. 5.34 Formation of the fruiting body of *D. discoideum*. In the migrating slug the prestalk cells (blue) are at the front (left panel). At culmination the slug becomes upright and the prestalk cells move up to the apex and down a stalk tube (center panel). The growth of the stalk tube lifts up the prespore region (gray) to the tip. A basal disc (red) forms at the base (right panel).

cells—together with major cell movements and rearrangements, and so provides an interesting example of the diversity of developmental strategies used by multicellular organisms. The formation of the initial aggregate and subsequent morphogenesis is mediated primarily by cyclic AMP acting as a chemoattractant, which is discussed in Chapter 7.

Although many developmental mutants have been isolated from the slime mold *D. discoideum*, until recently it has proved very difficult to isolate the corresponding genes. However, a technique rather similar to P-element insertion in *Drosophila* (see Box 2C, p. 59) is now available, and the inserted transposon can be used to physically detect and isolate the gene. The *D. discoideum* genome comprises around 12,500 genes, not many fewer than the 13,800 estimated for *Drosophila*.

5.16 Patterning of the slime mold slug involves cell sorting and positional signaling

In the migrating slug, prestalk cells are at the anterior end, and prespore cells are at the posterior end. When it is ready to form a fruiting body, the slug rests on its hind end and the prestalk cells form a stalk by migrating down through the prespore region, pushing the prespore region upward (Fig. 5.34).

What specifies the posterior prespore cells and the anterior prestalk cells? This problem is particularly well exemplified by the ability of the slugs to regulate and so maintain the basic pattern. The proportion of stalk to spore cells is more or less constant over a 1000-fold range of slug sizes. The capacity for regulation is also shown by the fact that isolated anterior and posterior regions are capable of forming properly proportioned fruiting bodies. Two classes of mechanism have been proposed to account for this patterning and regulation. The first suggests that there is some mechanism for providing positional information to the cells in the slug so that their relative position is specified. The second suggests that regulation involves the differentiation of cells at random locations, which then sort out to give the normal pattern. It appears that both mechanisms are involved in patterning the slug, with sorting dominant at early stages. The initial ratio of prestalk and prespore cells is based on the stage in the cell cycle when individual cells were starved; cells starved early in the cycle preferentially differentiate into prestalk cells.

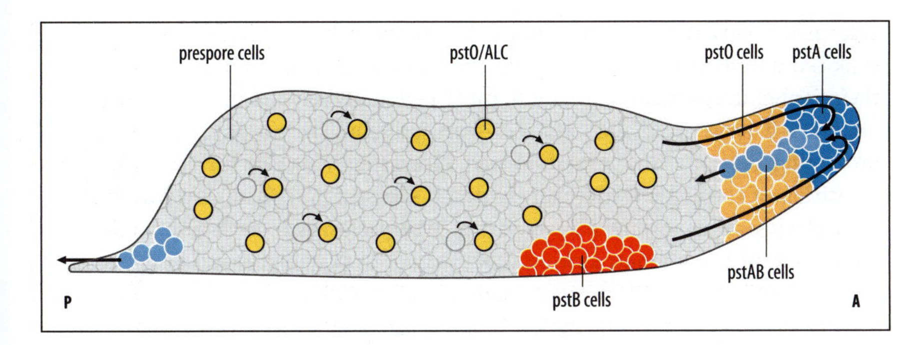

Fig. 5.35 Cell flow and differentiation in the migrating slime mold slug. At the tip are prestalk A (pstA) and prestalk AB (pstAB) cells. Some pstAB cells fall back and are lost at the rear of the slug. The lost pstAB cells are replaced by pstO/ALC cells which are derived from prespore cells. All of the rear of the slug contains prespore cells.

Cyclic AMP and differentiation-inducing factor 1 (DIF-1) control prestalk differentiation. The cells then sort out from the aggregate to form the slug.

Patterning does not simply involve specifying cells as prespore and prestalk. Within the prestalk region there is a diversity of cell types: A, O, B, and AB prestalk cells. Prestalk A cells occupy the anterior half of the prestalk region, while pstO cells occupy the rear half (Fig. 5.35). Prestalk B cells near the slug's prestalk and prespore boundary move downward at the culmination stage (see Fig. 5.34) to form the basal disc. This arrangement of cell types seems to be due to some form of cell sorting, rather than to positional signaling. Differentiation of prestalk A cells occurs at the periphery of the aggregate; only after differentiation do they accumulate at the tip. Also, when cells are removed from the anterior region of a slug and injected into another slug they return to their original position, which suggests that cell sorting is operating. A number of cell adhesion molecules are involved in both migration and cell sorting, and there is a need for migrating cells to overcome any strong cell–cell adhesion.

The migrating slug has a surprisingly dynamic structure, with respect to both cell differentiation and cell movement. At the posterior end of the slug, prestalk AB cells are lost, apparently having drifted backward from the tip region. They are replaced by pstA cells, which in turn are replaced by pstO cells. In the prespore region, posterior to the tip, are another class of cells, pstO/ALC (anterior-like cells), that appear to be very similar to the pstO cells. During slug migration, some of these pstO/ALC cells move to the anterior end and become, at first, pstO cells, and then pstA cells. To maintain the correct proportion of the different types of cells, some prespore cells convert to pstO/ALC. Thus, the correct ratio of prespore to prestalk cells is maintained. The pattern of cell flow during slug migration also foreshadows the behavior of prestalk cells during fruiting body formation, when stalk cells form a stalk tube, which is pushed down through the prespore region.

It seems that the initial specification of prestalk and prespore cells occurs at the time of aggregation. As the cells come together to form a mound, prestalk A cells are confined to an outer ring, suggesting that this specification could result from a positional signal.

5.17 Chemical signals direct cell differentiation in the slime mold

Studies on isolated cells have led to the identification of possible morphogens which can direct stalk and spore cell differentiation: extracellular cyclic AMP is required for prespore cell differentiation, and prespore-specific genes can be induced by adding cyclic AMP. (Cyclic AMP is also involved in cell signaling and chemotaxis in cellular slime molds, as discussed in Chapter 7.) Differentiation-inducing factor 1 (DIF-1) is a chlorinated hexaphenone released by developing cells that induces prestalk and stalk-cell differentiation. Prestalk cells are stimulated to differentiate by DIF-1, but their further differentiation, first into pstAB cells and then into mature stalk cells, is inhibited if both cyclic AMP and DIF-1 are present. The action of cyclic AMP in promoting prespore differentiation and repressing stalk-cell formation involves activation of the enzyme glycogen synthase kinase-3 (GSK-3), an enzyme we have seen playing an important part as part of Wnt pathways in early animal development. Mutant slime molds lacking GSK-3 do not display repression by cyclic AMP, and all cells within the prestalk region undergo precocious differentiation to become pstAB cells. In slime molds, GSK-3 is activated by its specific phosphorylation on tyrosine as a result of signaling via receptors for cyclic AMP on the cell surface. Genetic and biochemical evidence thus combine to indicate that cyclic AMP acts to increase GSK-3 activity and that this prevents premature differentiation as a prestalk AB cell.

During development, DIF-1 is extensively metabolized and thus destroyed by a cytoplasmic enzyme whose activity is induced by DIF-1 itself. This induction by DIF-1 of its own destructive agent seems to provide a negative feedback loop that controls the level of DIF-1, and possibly the number of stalk cells. While this feedback may operate during the aggregation phase, it poses a puzzle with respect to the slug: the apparent concentration of DIF-1 in the prespore zone is 10 times higher than that necessary to inhibit prespore differentiation in isolated cells. Moreover, DIF-1 is graded, with its highest concentration point in the posterior region, whereas the anterior region appears to be a localized sink for DIF-1. One attractive possibility is that, as part of their program of differentiation, prespore cells become insensitive to DIF-1.

Summary

In spite of the cellular slime mold appearing to provide a simple model for pattern formation and differentiation, giving rise to just two main cell types, patterning has turned out to be quite complex. Both positional signaling and cell sorting are involved, and in the migrating slug there is a dynamic balance of cell types as prestalk cells are lost and cells differentiate to replace them. This involves changes in both cell state and cell movement. The identification of cyclic AMP and differentiation-inducing factor (DIF) as factors controlling differentiation, together with the possibility of genetic manipulation, make the cellular slime mold a most promising system for developmental studies.

Summary: slime mold aggregation and differentiation

free-living myxamebae aggregate to form a multicellular 'slug'

⇩ cyclic AMP and DIF-1

specification of prestalk and prespore cells within the slug

⇩

slug migrates with prestalk cells at anterior end

slug stops moving and develops into a fruiting body with a cellular stalk and a head of spores

SUMMARY TO CHAPTER 5

In many invertebrates such as the nematode and the ascidian, cytoplasmic localization and cell lineage, together with local intercellular interactions, seem to be very important in specifying cell fate. However, there is very little evidence in any of these animals for key signaling regions corresponding to the vertebrate organizer, or for morphogen gradients as in *Drosophila*. This may reflect the predominance of mechanisms that specify cell fate on a cell-by-cell basis, rather than by groups of cells as in vertebrates and insects.

Although genetic evidence of the kind available in *Drosophila* is lacking in other invertebrates, except in the case of the nematode *Caenorhabditis elegans*, it is clear that maternal genes are crucial in specifying early patterning. There are numerous examples of the first cleavage in the embryo being related to the establishment of the embryo's axes and evidence that this plane of cleavage is maternally determined. Asymmetric cleavage associated with the unequal distribution of cytoplasmic determinants seems to be very important at early stages. Cytoplasmic movements that are involved in this cytoplasmic localization often occur after fertilization, and in this respect they may resemble the cortical rotation of the amphibian egg.

There can be no doubt that there are cell interactions, even at early stages, in all of the organisms in this chapter; an example is the specification of left–right asymmetry in nematodes, where a small alteration in cell relationships can reverse their handedness. But in most of these embryos the interactions are highly local, only affecting an adjacent cell. This is unlike vertebrates, in which interactions may occur over several cell diameters and involve groups of cells.

By contrast, in sea urchin development, which is highly regulative, cell interactions play a major role, and thus echinoderms resemble the vertebrates in this respect. The micromeres at the vegetal pole have properties similar to those of the vertebrate organizer, and cells are specified in groups rather than on a cell-by-cell basis. Ascidians are members of the chordates, the same phylum as vertebrates, and show a mixture of mosaic development and cell–cell interactions.

Cellular slime molds are organisms of ancient origin, diverging some time after the plant–animal split. Patterning of their multicellular phase involves cell sorting and positional signaling.

FURTHER READING

Nematodes

Sulston, J.E., Scherienberg, E., White, J.G., Thompson, J.N.: **The embryonic cell lineage of the nematode *Caenorhabditis elegans*.** *Dev. Biol.* 1983, **100:** 64–119.

Sulston, J.: **Cell lineage.** In *The Nematode* Caenorhabditis elegans. Edited by Wood, W.B. New York: Cold Spring Harbor Laboratory Press, 1988: 123–156.

Wood, W.B.: **Embryology.** In *The Nematode* Caenorhabditis elegans. Edited by Wood, W.B. New York: Cold Spring Harbor Laboratory Press, 1988: 215–242.

Box 5A Gene silencing by RNA interference

Dykxhoorn, D.M., Novind, C.D., Sharp, P.A.: **Killing the messenger: short RNAs that silence gene expression.** *Nat. Rev. Mol. Cell. Biol.* 2003, **4:** 457–467.

Kamath, R.S., Fraser, A.G., Dong, Y., Poulin, G., Durbin, R., Gotta, M., Kanapin, A., Le Bot, N., Moreno, S., Sohrmann, M., Welchman, D.P., Zipperlen, P., Ahringer, J.: **Systematic functional analysis of the *Caenorhabditis elegans* genome using RNAi.** *Nature* 2003, **421:** 231–237.

Nature Insight: **RNA interference.** *Nature* 2004, **431:** 337–370.

Sonnichsen, B., *et al.*: **Full-genome RNAi profiling of early embryogenesis in *Caenorhabditis elegans*.** *Nature* 2005, **434:** 462–469.

5.1 The antero-posterior axis in *C. elegans* is determined by asymmetric cell division

Cheeks, R.J., Canman, J.C., Gabriel, W.N., Meyer, N., Strome, S., Goldstein, B.: ***C. elegans* PAR proteins function by mobilizing and stabilizing asymmetrically localized protein complexes.** *Curr. Biol.* 2004, **14:** 851–862.

Kemphues, K.: **PARsing embryonic polarity.** *Cell* 2000, **101:** 345–348.

Munro, E., Nance, J., Priess, J.R.: **Cortical flows powered by asymmetrical contraction transport PAR proteins to establish and maintain anterior-posterior polarity in the early *C. elegans* embryo.** *Dev. Cell* 2004, **7:** 413–424.

Wallenfang, M.W., Seydoux, G.: **Polarization of the anterior-posterior axis of *C. elegans* is a microtubule-directed process.** *Nature* 2000, **408:** 89–92.

5.2 The dorso-ventral axis in *C. elegans* is determined by cell–cell interactions

Bergmann, D.C., Lee, M., Robertson, B., Tsou, M.F., Rose, L.S., Wood, W.B.: **Embryonic handedness choice in *C. elegans* involves the Galpha protein GPA-16.** *Development* 2003, **130:** 5731–5740.

Wood, W.B.: **Evidence from reversal of handedness in *C. elegans* embryos for early cell interactions determining cell fates.** *Nature* 1991, **349:** 536–538.

5.3 Both asymmetric divisions and cell–cell interactions specify cell fate in the early nematode embryo

Evans, T.C., Crittenden, S.L., Kodoyianni, V., Kimble, J.: **Translational control of maternal *glp-1* mRNA establishes an asymmetry in the *C. elegans* embryo.** *Cell* 1994, **77:** 183–194.

Kidd, A.R. 3rd, Miskowski, J.A., Siegfried, K.R., Sawa, H., Kimble, J.: **A beta-catenin identified by functional rather than sequence criteria and its role in Wnt/MAPK signaling.** *Cell* 2005, **121:** 761–772.

Labouesse, M., Mango, S.E.: **Patterning the *C. elegans* embryo.** *Trends Genet.* 1999, **15:** 307–313.

Lyczak, R., Gomes, J.E., Bowerman, B.: **Heads or tails: cell polarity and axis formation in the early *Caenorhabditis elegans* embryo.** *Dev. Cell* 2002, **3:** 157–166.

Mello, G.C., Draper, B.W., Priess, J.R.: **The maternal genes *apx-1* and *glp-1* and the establishment of dorsal-ventral polarity in the early *C. elegans* embryo.** *Cell* 1994, **77:** 95–106.

Park, F.D., Priess, J.R.: **Establishment of POP-1 asymmetry in early *C. elegans* embryos.** *Development* 2003 **130:** 3547–3556.

Park, F.D., Tenlen, J.R., Priess, J.R.: ***C. elegans* MOM-5/frizzled functions in MOM-2/Wnt-independent cell polarity and is localized asymmetrically prior to cell division.** *Curr. Biol.* 2004, **14:** 2252–2258.

Strome, S., Wood, W.B.: **Generation of asymmetry and segregation of germ-line granules in early *C. elegans* embryos.** *Cell* 1983, **35:** 15–25.

5.4 A small cluster of Hox genes specifies cell fate along the antero-posterior axis

Bürglin, T.R., Ruvkun, G.: **The *Caenorhabditis* elegans homeobox gene cluster.** *Curr. Opin. Genet. Dev.* 1993, **3:** 615–620.

Clark, S.G., Chisholm, A.D., Horvitz, H.R.: **Control of cell fates in the central body region of *C. elegans* by the homeobox gene *lin-39*.** *Cell* 1993, **74:** 43–55.

Cowing, D., Kenyon, C.: **Correct Hox gene expression established independently of position in *Caenorhabditis elegans*.** *Nature* 1996, **382:** 353–356.

Salser, S.J., Kenyon, C.: **A *C. elegans* Hox gene switches on, off, on and off again to regulate proliferation, differentiation and morphogenesis.** *Development* 1996, **122:** 1651–1661.

Van Auken, K., Weaver, D.C., Edgar, L.G., Wood, W.B.: ***Caenorhabditis elegans* embryonic axial patterning requires two recently discovered posterior-group Hox genes.** *Proc. Natl Acad. Sci. USA* 2000, **97:** 4499–4503.

5.5 The timing of events in nematode development is under genetic control that involves microRNAs

Ambros, V.: **Control of developmental timing in *Caenorhabditis elegans*.** *Curr. Opin. Genet. Dev.* 2000, **10:** 428–433.

Austin, J., Kenyon, C.: **Developmental timekeeping: marking time with antisense.** *Curr. Biol.* 1994, **4:** 366–396.

Grishok, A., Pasquinelli, A.E., Conte, D., Li, N., Parrish, S., Ha, I., Baillie, D.L., Fire, A., Ruvkun, G., Mello, C.C.: **Genes and mechanisms related to RNA interference regulate expression of the small temporal RNAs that control *C. elegans* developmental timing.** *Cell* 2001, **106:** 23–34.

Box 5B Gene silencing by microRNAs

Bartel, D.P.: **MicroRNAS: genomics, biogenesis, mechanism and function.** *Cell* 2004, **116:** 281–297.

He, L., Hannon, G.J.: **MicroRNAs: small RNAs with a big role in gene regulation.** *Nat. Rev. Genet.* 2004, **5:** 522–531.

Murchison, E.P., Hannon, G.J.: **miRNAs on the move: miRNA biogenesis and the RNAi machinery.** *Curr. Opin. Cell Biol.* 2004, **16:** 223–229.

5.6 Vulval development is initiated by the induction of a small number of cells by short-range signals from a single inducing cell

Grunwald, I., Rubin, G.M.: **Making a difference: the role of cell–cell interactions in establishing separate identities for equivalent cells.** *Cell* 1992, **68:** 271–281.

Kenyon, C.: **A perfect vulva every time: gradients and signaling cascades in *C. elegans*.** *Cell* 1995, **82:** 171–174.

Newman, A.P., Sternberg, P.W.: **Coordinated morphogenesis of epithelia during development of the *Caenorhabditis* elegans uterine-vulval connection.** *Proc. Natl Acad. Sci. USA* 1996, **93:** 9329–9333.

Sharma-Kishore, R., White, J.G., Southgate, E., Podbilewicz, B.: **Formation of the vulva in *Caenorhabditis* elegans: a paradigm for organogenesis.** *Development* 1999, **126:** 691–699.

Sundaram, M., Han, M.: **Control and integration of cell signaling pathways during *C. elegans* vulval development.** *BioEssays* 1996, **18:** 473–480.

Yoo, A.S., Bais, C., Greenwald, I.: **Crosstalk between the EGFR and LIN-12/Notch pathways in *C. elegans* vulval development.** *Science* 2004, **303:** 663–636.

Echinoderms

5.7 The sea urchin embryo develops into a free-swimming larva

Horstadius, S.: *Experimental Embryology of Echinoderms*. Oxford: Clarendon Press, 1973.

Wilt, F.H.: **Determination and morphogenesis in the sea urchin embryo.** *Development* 1987, **100:** 559–575.

5.8 The sea urchin egg is polarized along the animal-vegetal axis

Cameron, R.A., Fraser, S.E., Britten, R.J., Davidson, E.H.: **The oral-aboral axis of a sea urchin embryo is specified by first cleavage.** *Development* 1989, **106:** 641–647.

Davidson, E.H., Cameron, R.S., Ransick, A.: **Specification of cell fate in the sea urchin embryo: summary and some proposed mechanisms.** *Development* 1998, **125:** 3269–3290.

Jeffrey, W.R.: **Axis determination in sea urchin embryos: from confusion to evolution.** *Trends Genet.* 1992, **8:** 223–225.

Logan, C.Y., McClay, D.R.: **The allocation of early blastomeres to the ectoderm and endoderm is variable in the sea urchin embryo.** *Development* 1997, **124:** 2213–2223.

5.9 The oral–aboral axis in sea urchins is related to the plane of the first cleavage

Duboc, V., Rottinger, E., Besnardeau, L., Lepage, T.: **Nodal and BMP2/4 signaling organizes the oral-aboral axis of the sea urchin embryo.** *Dev. Cell* 2004, **6:** 397–410.

5.10 The sea urchin fate map is finely specified, yet considerable regulation is possible

Angerer, L.M., Oleksyn, D.W., Logan, C.Y., McClay, D.R., Dale, L., Angerer, R.C.: **A BMP pathway regulates cell fate allocation along the sea urchin animal-vegetal embryonic axis.** *Development* 2000, **127:** 1105–1114.

Angerer, L.M., Angerer, R.C.: **Patterning the sea urchin embryo: gene regulatory networks, signaling pathways, and cellular interactions.** *Curr. Top. Dev. Biol.* 2003, **53:** 159–198.

Sweet, H.C., Hodor, P.G., Ettensohn, C.A.: **The role of micromere signaling in Notch activation and mesoderm specification during sea urchin embryogenesis.** *Development* 1999, **126:** 5255–5265.

Wessel, G.M., Wikramanayake, A.: **How to grow a gut: ontogeny of the endoderm in the sea urchin embryo.** *BioEssays* 1999, **21:** 459–471.

5.11 The vegetal region of the sea urchin embryo acts as an organizer

Ransick, A., Davidson, E.H.: **A complete second gut induced by transplanted micromeres in the sea urchin embryo.** *Science* 1993, **259:** 1134–1138.

5.12 The sea urchin vegetal region is specified by nuclear accumulation of β-catenin

Logan, C.Y., Miller, J.R., Ferkowicz, M.J., McClay, D.R.: **Nuclear β-catenin is required to specify vegetal cell fates in the sea urchin embryo.** *Development* 1999, **126:** 345–357.

Weitzel, H.E., Illies, M.R., Byrum, C.A., Xu, R., Wikramanayake, A.H., Ettensohn, C.A.: **Differential stability of beta-catenin along the animal-vegetal axis of the sea urchin embryo mediated by dishevelled.** *Development* 2004, **131:** 2947–2956.

5.13 The genetic control of endomesoderm specification is known in considerable detail

Davidson, E.H.: **A view from the genome: spatial control of transcription in sea urchin development.** *Curr. Opin. Genet. Dev.* 1999, **9:** 530–541.

Oliveri, P., Davidson, E.H.: **Gene regulatory network controlling embryonic specification in the sea urchin.** *Curr. Opin. Genet. Dev.* 2004, **14:** 351–360.

Kirchnamer, C.V., Yuh, C.V., Davidson, E.H.: **Modular *cis*-regulatory organization of developmentally expressed genes: two genes transcribed territorially in the sea urchin embryo, and additional examples.** *Proc. Natl Acad. Sci. USA* 1996, **93:** 9322–9328.

Ascidians

Keys, D.N., Lee, B.I., Di Gregorio, A., Harafuji, N., Detter, J.C., Wang, M., Kahsai, O., Ahn, S., Zhang, C., Doyle, S.A., Satoh, N., Satou, Y., Saiga, H., Christian, A.T., Rokhsar, D.S., Hawkins, T.L., Levine, M., Richardson, P.M.: **A saturation screen for cis-acting regulatory DNA in the Hox genes of *Ciona intestinalis*.** *Proc. Natl Acad. Sci. USA* 2005, **102:** 679–683.

Nishida, H.: **Specification of embryonic axis and mosaic development in ascidians.** *Dev. Dyn.* 2005, **233:** 1177–1193.

Passamaneck, Y.J., Di Gregorio, A.: ***Ciona intestinalis*: chordate development made simple.** *Dev. Dyn.* 2005, **233:** 1–19.

Satoh, N.: *The Developmental Biology of Ascidians*. Cambridge: Cambridge University Press, 1994.

5.14 In ascidians, muscle may be specified by localized cytoplasmic factors

Bates, W.R.: **Development of myoplasm-enriched ascidian embryos.** *Dev. Biol.* 1988, **129:** 241–252.

di Gregorio, A., Levine, M.: **Ascidian embryogenesis and the origins of the chordate body plan.** *Curr. Opin. Genet. Dev.* 1998, **7:** 457–463.

Kim, G.J., Nishida, H.: **Suppression of muscle fate by cellular interaction as required for mesenchyme formation during ascidian embryogenesis.** *Dev. Biol.* 1999, **214:** 9–22.

Nishida, H.: **Determinative mechanisms in secondary muscle lineages of ascidian embryos: development of muscle-specific features in isolated muscle progenitor cells.** *Development* 1990, **108:** 559–568.

Nishida, H., Sawada, K.: ***macho-1* encodes a localized mRNA in ascidian eggs that specifies muscle fate during embryogenesis.** *Nature* 2001, **409:** 724–728.

5.15 Mesenchyme and notochord development in ascidians require signals from the endoderm

Corbo, J.C., Levine, M., Zeller, R.W.: **Characterization of a notochord-specific enhancer from the *Brachyury* promoter region of the ascidian, *Ciona intestinalis*.** *Development* 1997, **124:** 589–602.

Kim, G.J., Yamada, A., Nishida, H.: **An FGF signal from endoderm and localized factors in the posterior-vegetal egg cytoplasm pattern the mesodermal tissues in the ascidian embryo.** *Development* 2000, **127:** 2853–2862.

Nakatani, Y., Nishida, H.: **Induction of notochord during ascidian embryogenesis.** *Dev. Biol.* 1994, **166:** 289–299.

Nakatani, Y., Yasuo, H., Satoh, N., Nishida, H.: **Basic fibroblast growth factor induces notochord formation and the expression of As-T, a Brachyury homolog, during ascidian embryogenesis.** *Development* 1996, **122:** 2023–2031.

Cellular slime molds

Chisholm, R.L., Firtel, R.A.: **Insights into morphogenesis from a simple developmental system.** *Nat. Rev. Mol. Cell Biol.* 2004, **5:** 531–541.

5.16 Patterning of the slime mold slug involves cell sorting and positional signaling

Abe, T., Early, A., Siegert, F., Weijer, C., Williams, J.: **Patterns of cell movement within the *Dictyostelium* slug revealed by cell type-specific, surface labelling of cells.** *Cell* 1994, **77:** 687–699.

Brown, J.M., Firtel, R.A.: **Regulation of cell-fate determination in *Dictyostelium*.** *Dev. Biol.* 1999, **216:** 426–441.

Brown, J.M., Firtel, R.A.: **Just the right size: cell counting in *Dictyostelium*.** *Trends Genet.* 2000, **16:** 191–193.

Early, A., Abe, T., Williams, J.: **Evidence for positional differentiation of prestalk cells and for a morphogenetic gradient in *Dictyostelium*.** *Cell* 1995, **83:** 91–99.

Jermyn, K., Traynor, D., Williams, J.: **The initiation of basal disc formation in *Dictyostelium* discoideum is an early event in culmination.** *Development* 1996, **122:** 753–760.

Kimmel, A.R., Firtel, R.A.: **Breaking symmetries: regulation of *Dictyostelium* development through chemoattractant and morphogen signal-response.** *Curr. Opin. Genet. Dev.* 2004, **14:** 540–549.

Siu, C.H., Harris, T.J., Wang, J., Wong, E.: **Regulation of cell-cell adhesion during *Dictyostelium* development.** *Semin. Cell. Dev. Biol.* 2004, **15:** 633–641.

5.17 Chemical signals direct cell differentiation in the slime mold

Brisco, F., Firtel, R.A.: **A kinase for cell fate determination?** *Curr. Biol.* 1995, **5:** 228–231.

Gross, J.D.: **Developmental decisions in *Dictyostelium discoideum*.** *Microbiol. Rev.* 1994, **58:** 330–351.

Kay, R.R.: **Differentiation and patterning in *Dictyostelium*.** *Curr. Opin. Genet. Dev.* 1994, **4:** 637–641.

Schilde, C., Araki, T., Williams, H., Harwood, A., Williams, J.G.: **GSK3 is a multifunctional regulator of *Dictyostelium* development.** *Development* 2004, **131:** 4555–4565.

Plant development

6

- Embryonic development
- Meristems
- Flower development and control of flowering

Because of plant cells' rigid cell walls and lack of cell movement within tissues, a plant's architecture is very much the result of patterns of oriented cell divisions. Despite this apparent invariance, however, cell fate in development is largely determined by similar means as in animals—by a combination of positional signals and intercellular communication. As well as communicating by extracellular signals and cell-surface interactions, plant cells are interconnected by cytoplasmic channels, which allow movement of proteins such as transcription factors directly from cell to cell.

The plant kingdom is very large, ranging from the algae, many of which are unicellular, to the multicellular land plants, which exist in a prodigious variety of forms. Plants and animals probably evolved the process of multicellular development independently, their last common ancestor being a unicellular eukaryote some 1.6 billion years ago. Plant development is therefore of interest not simply for its own sake and for its agricultural importance. By looking at the similarities and differences between plant and animal development, a study of plant development can shed light on the way that developmental mechanisms in two groups of multicellular organisms have evolved independently and under different sets of developmental constraints.

Do plants and animals use the same developmental mechanisms? As we shall see in this chapter, the logic behind the spatial layouts of gene expression that pattern a developing flower is similar to that of Hox gene action in patterning the body axis in animals, but the genes involved are completely different. Such similarities between plant and animal development are due to the fact that the basic means of regulating gene expression are the same in both, and thus similar general mechanisms for patterning gene expression in a multicellular tissue are bound to arise. As we shall see, many of the general control mechanisms we have encountered in animal development, such as asymmetric cell division, the response to positional signals, lateral inhibition, and changes in gene expression in response to extracellular signals, are all present in plants. Differences between plants and animals in the way development is controlled arise from some different ways in which plant cells can communicate with each other compared with animal cells, from the existence of rigid cell walls and the lack of large-scale cell movements, to the fact that environment has a much greater impact on plant development than on that of animals.

One general difference between plant and animal development is that most of the development occurs not in the embryo but in the growing plant. Unlike an animal embryo, the mature plant embryo inside a seed is not simply a smaller version of the organism it will become. All the 'adult' structures of the plant—shoots, roots, stalks, leaves, and flowers—are produced after germination from localized groups of undifferentiated cells known as **meristems**. Two meristems are established in the embryo, one at the tip of the root and the other at the tip of the shoot. These persist in the adult plant and almost all the other meristems, such as those in developing leaves and flower shoots, are derived from them. Cells within meristems can divide repeatedly and can potentially give rise to all plant tissues and organs. This means that developmental patterning within meristems to produce organs such as leaves and flowers continues throughout a plant's life.

Plant and animal cells share many common internal features and much basic biochemistry, but there are some crucial differences that have a bearing on plant development. One of the most important is that plant cells are surrounded by a framework of relatively rigid cell walls. There is therefore virtually no cell migration in plants, and major changes in the shape of the developing plant cannot be achieved by the movement and folding of sheets of cells. In plant development, form is largely generated by differences in rates of cell division and by division in different planes, followed by directed enlargement of the cells.

As in animal development, one of the main questions in plant development is how cell fate is determined. Many structures in plants normally develop by stereotyped patterns of cell division, but cell fate is in many cases also known to be determined by factors such as position in the meristem and cell–cell signaling. The cell wall would seem to impose a barrier to the passage of large signaling molecules, such as proteins, although it is very thin in some regions such as the meristems, but all known plant extracellular signaling molecules—such as auxins, gibberellins, cytokinins, and ethylene—are small molecules that easily penetrate cell walls. Plant cells also communicate with each other through fine cytoplasmic channels known as **plasmodesmata**, which link neighboring plant cells through the cell wall, and they may be the channels by which some developmentally important gene-regulatory proteins, and even mRNAs, move directly between cells. The size of the channel can be varied.

Another important difference between plant and animal cells is that a complete, fertile plant can develop from a single differentiated somatic cell and not just from a fertilized egg. This suggests that, unlike the differentiated cells of adult animals, some differentiated cells of the adult plant may retain totipotency and so behave like animal embryonic stem cells (see Chapter 8). Perhaps they do not become fully determined in the sense that adult animal cells do, or perhaps they are able to escape from the determined state, although how this could be achieved is as yet unknown. In any case, this difference between plants and animals illustrates the dangers of the wholesale application to plant development of concepts derived from animal development. Nevertheless, the genetic analysis of development in plants is turning up instances of genetic strategies for developmental patterning rather similar to those of animals.

The small cress-like weed *Arabidopsis thaliana* has become the model plant for genetic and developmental studies, and will provide many of the examples in this chapter. We will begin by describing the main features of its morphology, life cycle, and reproduction.

6.1 The model plant *Arabidopsis thaliana* has a short life cycle and a small diploid genome

The equivalent of *Drosophila* in the study of plant development is the small crucifer *A. thaliana*, commonly called thale cress, which is well suited to genetic and

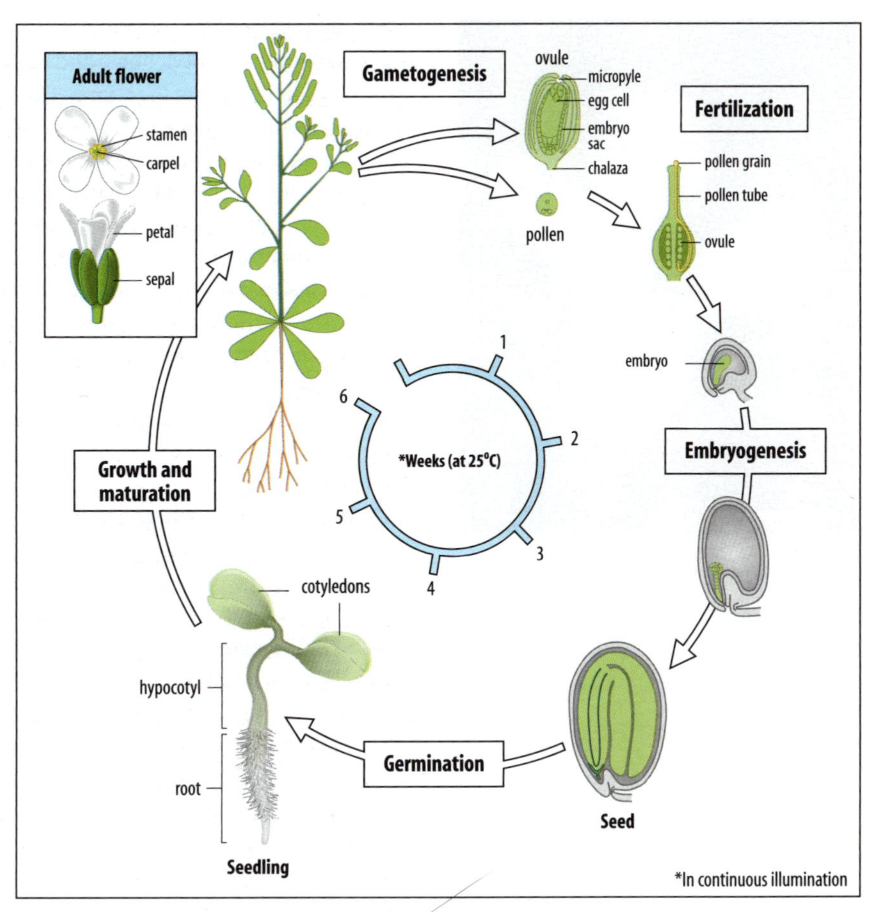

Fig. 6.1 Life cycle of *Arabidopsis*. In flowering plants, egg cells are contained separately in ovules inside the carpels. Fertilization of an egg cell by a male nucleus from a pollen grain takes place inside the ovary. The egg then develops into an embryo contained within the ovule coat, forming a seed. *Arabidopsis* is a dicotyledon, and the mature embryo has two wing-like cotyledons (storage organs) at the apical (shoot) end of the main axis—the hypocotyl—which contains a shoot meristem at one end and a root meristem at the other. Following germination, the seedling develops into a plant with roots, a stem, leaves, and flowers. The photograph shows five mature *Arabidopsis* plants.

developmental studies. It is a diploid (unlike many plants, which are polyploid) and has a relatively small, compact genome which has been sequenced and which contains about 25,000 genes. It is an annual, flowering in the first year of growth, and develops as a small ground-hugging rosette of leaves, from which a branched flowering stem is produced with a flowerhead, or **inflorescence**, at the end of each branch. It develops rapidly, with a total life cycle in laboratory conditions of some 6–8 weeks, and like all flowering plants, mutant strains and lines can easily be stored in large quantities in the form of seeds. The life cycle of *Arabidopsis* is shown in Fig. 6.1.

Each *Arabidopsis* flower (Fig. 6.2) consists of four sepals surrounding four white petals; inside the petals are six stamens, the male sex organs, which produce pollen containing the male gametes. Petals, sepals, and the other floral organs are thought to derive evolutionarily from modified leaves. At the center of the flower are the female sex organs, which consist of an ovary of two carpels, which contain the **ovules**. Each ovule contains an egg cell. Fertilization of an egg cell occurs when a pollen grain deposited on the carpel surface grows a tube that penetrates the carpel and delivers two haploid pollen nuclei to an ovule. One nucleus fertilizes the egg cell while the other fuses with two other nuclei in the ovule. This forms a triploid cell that will develop into a specialized nutritive tissue—the **endosperm**—that surrounds the fertilized egg cells and provides the food source for embryonic development.

Following fertilization, the embryo develops inside the ovule, taking about 2 weeks to form a mature seed, which is shed from the plant. The seed will remain dormant until suitable external conditions trigger germination. The early stages of germination and seedling growth rely on food supplies stored in the **cotyledons**

Fig. 6.2 An individual flower of *Arabidopsis*. Scale bar = 1 mm.

(seed leaves), which are storage organs developed by the embryo. *Arabidopsis* embryos have two cotyledons, and thus belong to the large group of plants known as dicotyledons (dicots for short). The other large group of flowering plants is the monocotyledons (monocots), which have embryos with one cotyledon; adult monocots typically have long narrow leaves and include many important staple crops, such as wheat, rice, and maize. Other agriculturally important dicots include potato, tomato, sugar beet, and most other vegetables.

At germination, the shoot and root elongate and emerge from the seed. Once the shoot emerges above ground it starts to photosynthesize and forms the first true leaves at the shoot apex. About 4 days after germination the seedling is a self-supporting plant. Flower buds are usually visible on the young plant 3–4 weeks after germination and will open within a week. Under ideal conditions, the complete *Arabidopsis* life cycle thus takes about 6–8 weeks (see Fig. 6.1).

Embryonic development

Formation of the embryo, or embryogenesis, occurs inside the ovule, and the end result is a mature dormant embryo enclosed in a seed awaiting germination. During embryogenesis, the shoot–root polarity of the plant body, which is known as the **apical–basal** axis, is established, and the shoot and root meristems are formed. Plants also possess **radial symmetry**, as seen in the concentric arrangement of the different tissues in a plant stem, and this **radial axis** is also set up in the embryo. Development of the *Arabidopsis* embryo involves a rather invariant pattern of cell division (which is not the case in all plant embryos) and this enables structures in the *Arabidopsis* seedling to be traced back to groups of cells in the early embryo to provide a fate map (see Section 6.2).

6.2 Plant embryos develop through several distinct stages

Arabidopsis belongs to the angiosperms, or flowering plants, one of the two major groups of seed-bearing plants; the other is the gymnosperms, or conifers. The typical course of embryonic development in angiosperms is outlined in Box 6A (opposite). Like an animal zygote, the fertilized plant egg cell undergoes repeated cell divisions, cell growth, and differentiation to form a multicellular embryo. The first division of the zygote is at right angles to the long axis, dividing it into an apical cell and a basal cell and establishing an initial polarity that is carried over into the apical–basal polarity of the embryo and into the apical–basal axis of the plant. In many species the first zygotic division is unequal, with the basal cell larger than the apical cell. The basal cell divides to give rise to the **suspensor**, which may be several cells long (see Box 6A). This attaches the embryo to maternal tissue and is a source of nutrients. The apical cell divides vertically to form a two-celled **proembryo** which will give rise to the rest of the embryo. In some species, the basal cell contributes little to the further development of the embryo, but in others, such as *Arabidopsis*, the topmost suspensor cell is recruited into the embryo, where it is known as the **hypophysis**, and contributes to the embryonic root meristem and root cap.

The next two divisions produce an eight-cell **octant-stage** embryo, which develops into a **globular-stage** embryo of around 32 cells (Fig. 6.3). The embryo elongates and the cotyledons start to develop as wing-like structures at one end, while an embryonic root forms at the other. This stage is known as the **heart stage**. Two groups of undifferentiated cells capable of continued division are located at each end of this axis; these are known as the **apical meristems**. The meristem lying

Box 6A Angiosperm embryogenesis

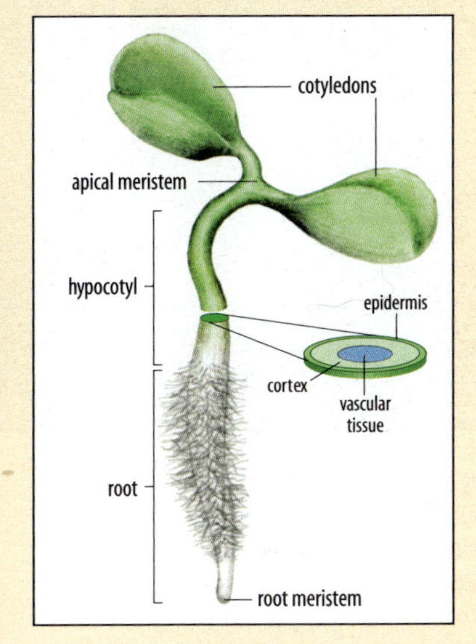

In flowering plants (angiosperms), the egg cell is contained within an ovule inside the ovary in the flower (right inset in the bottom panel). At fertilization, a pollen grain deposited on the surface of the stigma puts out a pollen tube, down which two male gametes migrate into the ovule. One male gamete fertilizes the egg cell while the other combines with another cell inside the ovule to form a specialized nutritive tissue, the endosperm, which surrounds, and provides the food source for, the developing embryo.

The small annual weed *Capsella bursa-pastoris* (shepherd's purse) is a typical dicotyledon. The first, asymmetric, cleavage divides the zygote transversely into an apical and a basal cell (bottom panel). The basal cell then divides several times to form a single row of cells—the suspensor—which in many angiosperm embryos takes no further part in embryonic development, but may have an absorptive function; in *Capsella*, however, it contributes to the root meristem. Most of the embryo is derived from the apical cell. This undergoes a series of stereotyped divisions, in which a precise pattern of cleavages in different planes gives rise to the heart-shaped embryonic stage typical of dicotyledons. This develops into a mature embryo that consists of a cylindrical body with a meristem at either end and two cotyledons.

The early embryo becomes differentiated along the radial axis into three main tissues—the outer epidermis, the prospective vascular tissue, which runs through the center of the main axis and cotyledons, and the ground tissue (prospective cortex) that surrounds it.

The ovule containing the embryo matures into a seed (left inset in the bottom panel), which remains dormant until suitable external conditions trigger germination and growth of the seedling. A typical dicotyledon seedling (top right panel) comprises the shoot apical meristem, two cotyledons, the trunk of the seedling (the hypocotyl), and the root apical meristem. The seedling may be thought of as the phylotypic stage of flowering plants. The seedling body plan is simple. One axis—the apical-basal axis—defines the main polarity of the plant. The shoot forms at the apical pole and the root at the basal pole. The shoot meristem is therefore referred to in the seedling and adult plant simply as the apical meristem. A plant stem also has a radial axis, which is evident in the radial symmetry in the hypocotyl and is continued in the root and shoot. In the center is the vascular tissue, which is surrounded by cortex, and an outer covering of epidermis. At later stages, structures such as leaves and other organs have a dorso-ventral axis running from the upper surface to the lower surface.

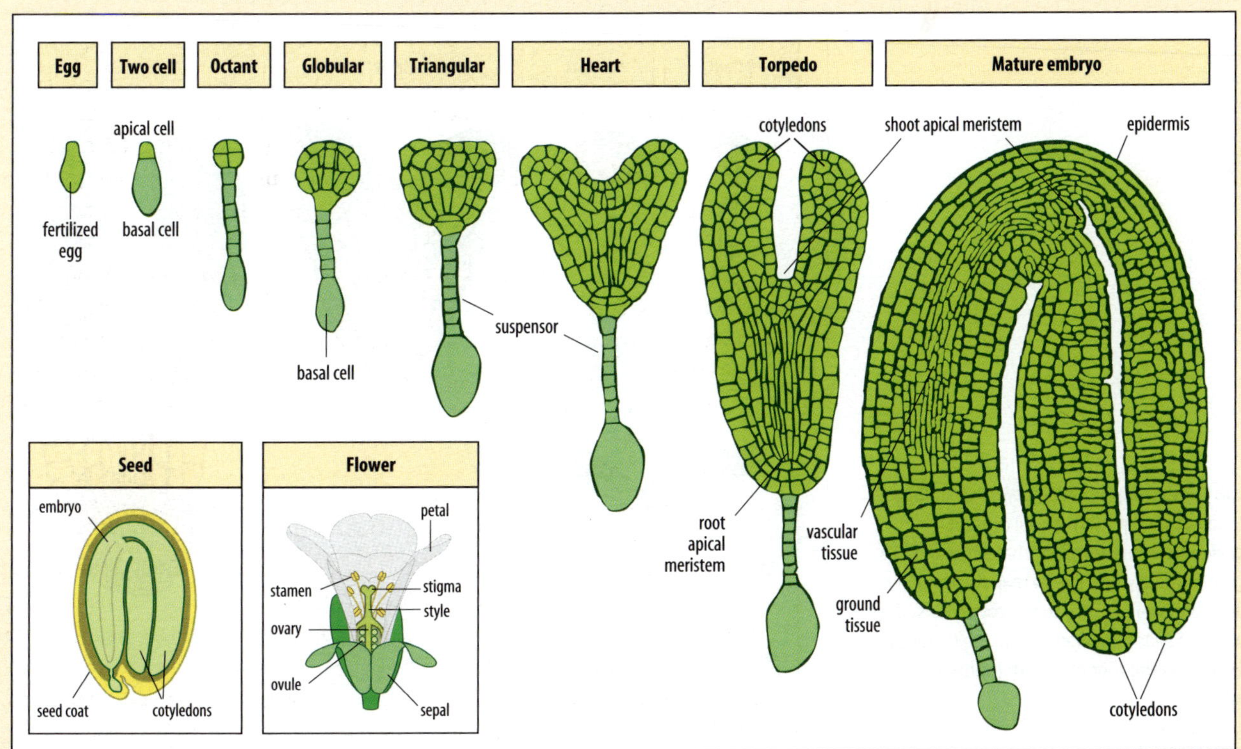

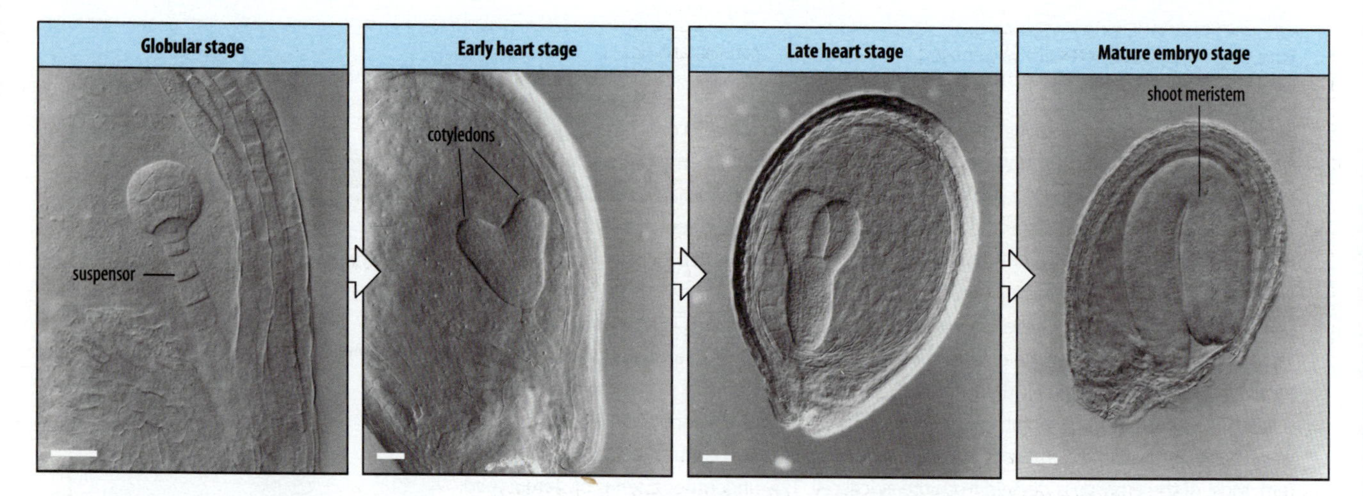

Fig. 6.3 *Arabidopsis* embryonic development. Light micrographs (Nomarski optics) of cleared wild-type seeds of *A. thaliana*. The cotyledons can already be seen at the heart stage. The embryo proper is attached to the seed coat through a filamentous suspensor. Scale bar = 20 μm.

Photographs courtesy of D. Meinke, from Meinke, D.W.: 1994.

between the cotyledons gives rise to the shoot, while the one at the opposite end of the axis, towards the end of the embryonic root, will drive root growth at germination. The region in between the embryonic root and the future shoot will become the seedling stem or **hypocotyl**. Almost all adult plant structures are derived from the apical meristems. The main exception is radial growth in the stem, which is most evident in woody plants, and which is produced by the **cambium**, a ring of secondary meristematic tissue in the stem. After the embryo is mature, the apical meristems remain quiescent until germination.

Seedling structures can be traced back to groups of cells in the early embryo to provide a fate map (Fig. 6.4). In *Arabidopsis*, patterns of cleavage up to the 16-cell stage are highly reproducible, and even at the octant stage it is possible to make

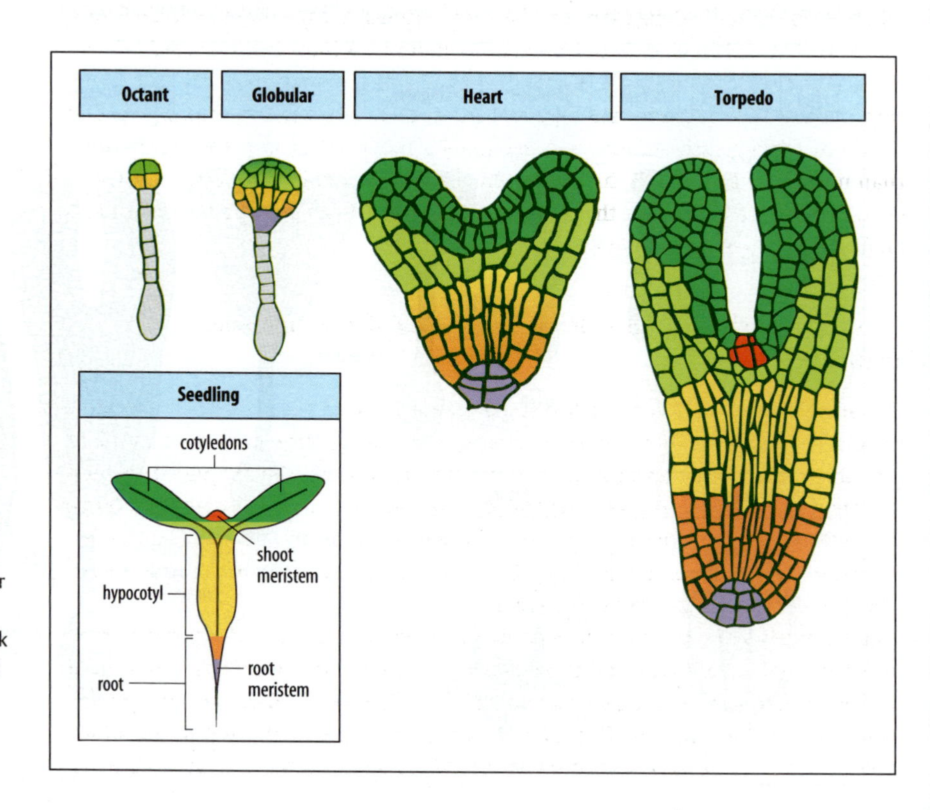

Fig. 6.4 Fate map of the *Arabidopsis* embryo. The stereotyped pattern of cell division in dicotyledon embryos means that at the globular stage it is already possible to map the three regions that will give rise to the cotyledons (dark green) and shoot meristem (red), the hypocotyl (yellow), and the root meristem (purple) in the seedling.

After Scheres, B., et al.: 1994.

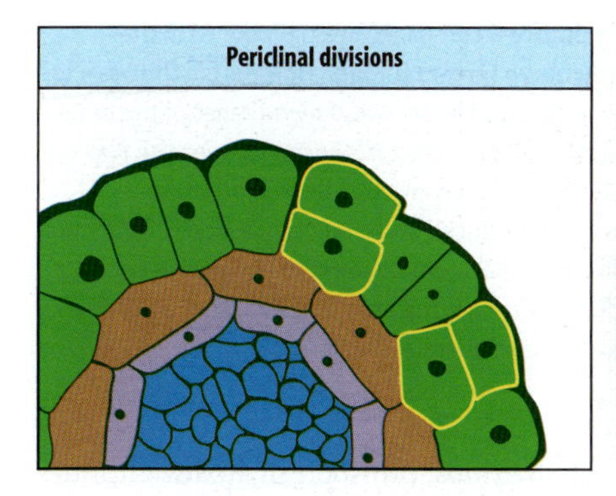

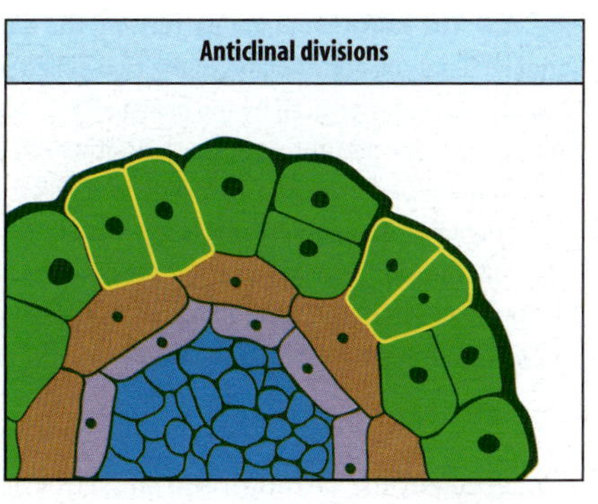

Fig. 6.5 Periclinal and anticlinal divisions. Periclinal cell divisions are parallel to the organ surface, whereas anticlinal divisions are at right angles to the surface.

a fate map for the major regions of the seedling along the apical–basal axis. The upper tier of cells gives rise to the cotyledons and the shoot meristem, the next tier is the origin of the hypocotyl, and the bottom tier together with the region of the suspensor where it joins the embryo will give rise to the root (see Section 6.3). At the heart stage, the fate map is clear.

The radial pattern in the embryo comprises three concentric rings of tissue: the outer **epidermis**, the ground tissue (**cortex** and **endodermis**), and the vascular tissue at the center. This radial axis appears first at the octant stage, when adaxial (central) and abaxial (peripheral) regions become established. Subsequently, **periclinal** divisions, in which the plane of division is parallel to the outer surface, give rise to the different rings of tissue, and **anticlinal** divisions, in which the plane of cell division is at right angles to the outer surface, increase the number of cells in each ring of tissue (Fig. 6.5). At the 16-cell stage the epidermal layer, or dermatogen, is established.

It is not known to what extent the fate of cells is determined at this stage or whether their fate is dependent on the early pattern of asymmetric cell divisions. It seems that cell lineage is not crucial, as mutations have been discovered that show that cell division can be uncoupled from pattern formation. The *fass* mutation in *Arabidopsis* disrupts the regular pattern of cell division by causing cells to divide in random orientation, and produces a much fatter and shorter seedling than normal. But although the *fass* seedling is misshapen, it has roots, shoots, and even eventually flowers in the correct places and the correct pattern of tissues along the radial axis is maintained.

6.3 Gradients of the signal molecule auxin establish the embryonic apical–basal axis

The small organic molecule auxin (indoleacetic acid or IAA) is one of the most important and ubiquitous chemical signals in plant development and plant growth. It causes changes in gene expression by promoting the ubiquitination and degradation of transcriptional regulatory proteins known as Aux/IAA proteins, which leads to the specific activation of auxin-responsive genes. The earliest known function of auxin in *Arabidopsis* is in the very first stage of embryogenesis, where it establishes the apical–basal axis.

Immediately after the first division of the zygote, auxin is actively transported from the basal cell into the apical cell, where it accumulates. It is transported out of the basal cell by the auxin-efflux protein PIN7, which is localized in the apical plasma membrane of the basal cell. The auxin is required to specify the apical cell, which gives rise to all the apical embryonic structures such as shoot apical meristem

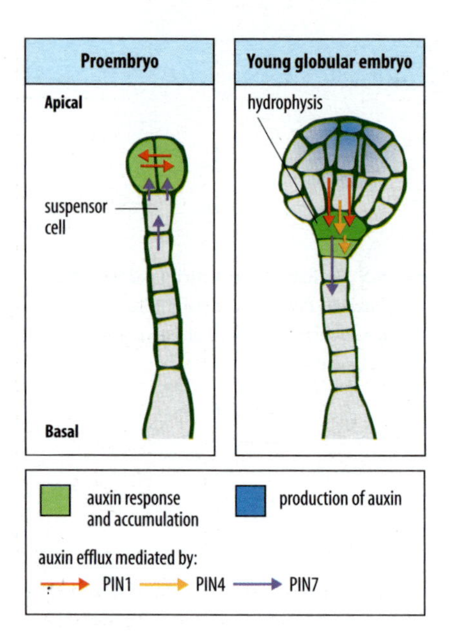

Fig. 6.6 The role of auxin in patterning the early embryo. Left panel: auxin produced in the original basal cell accumulates in the two-celled proembryo (green) through transport in the basal to apical direction mediated by the protein PIN7, which is located in the apical membranes of the basal and suspensor cells (purple arrows). Another PIN protein, PIN1, transports auxin between the two cells of the proembryo (red arrows). Cell division distributes auxin into the developing embryo. Right panel: at the globular stage, free auxin starts to be produced at the apical pole and the direction of auxin transport is reversed. PIN7 becomes localized in basal membranes of the suspensor cells and the proteins PIN1 and PIN4 (orange arrows) transport auxin from the apical region into the most basal cell of the embryo, known as the hypophysis, where it accumulates.

and cotyledons. Through the subsequent cell divisions, transport of auxin continues up through the suspensor cells and into the cell at the base of the developing embryo until the globular-stage embryo of about 32 cells. The apical cells of the embryo then start to produce auxin, and the direction of auxin transport is suddenly reversed. PIN7 proteins in the suspensor cells move to the basal faces of the cells. Other *PIN* genes are activated, and the concerted actions of the PIN proteins cause auxin from the apical region to be transported into the basal region of the globular embryo, from which will develop the hypocotyl, root meristem, and embryonic root (Fig. 6.6).

The importance of auxin in specifying this region is illustrated by the effects of mutations in the cellular machinery for auxin transport or auxin response. The *MONOPTEROS* gene, for example, encodes a protein required for cells to respond to auxin. Embryos mutant for this gene lack the hypocotyl and the root meristem, and abnormal cell divisions are observed in the regions of the octant-stage embryo that would normally give rise to these structures.

Other mutations disrupt patterning along the apical–basal axis but are not linked to auxin. The mutation *topless-1* causes a dramatic switching of cell fate, transforming the whole of the apical region into a second root region and abolishing development of cotyledons and shoot meristem. The normal function of the mutated gene is not known, but appears not to be linked to auxin. The mutation is temperature-sensitive and so can be made to act at different times during development by simply changing the temperature at which the embryos are grown. Such manipulations show that the apical region can be respecified as root between the globular and heart stage, even after it has begun to show molecular signs of developing cotyledons and shoot meristem. This suggests that even though apical cell fate is normally specified by this stage of embryonic development, the decision is not irreversible.

Mutations in the gene *SHOOT MERISTEMLESS* (*STM*) completely block the formation of the shoot apical meristem, but have no effect on the root meristem or other parts of the embryo. *STM* encodes a transcription factor that is also required to maintain cells in the pluripotent state in the adult shoot meristem. The pattern of *STM* expression develops gradually, which is also typical of several other genes that characterize the shoot apical meristem. Expression is first detected in the globular stage in one or two cells and only later in the central region between the two cotyledons (Fig. 6.7).

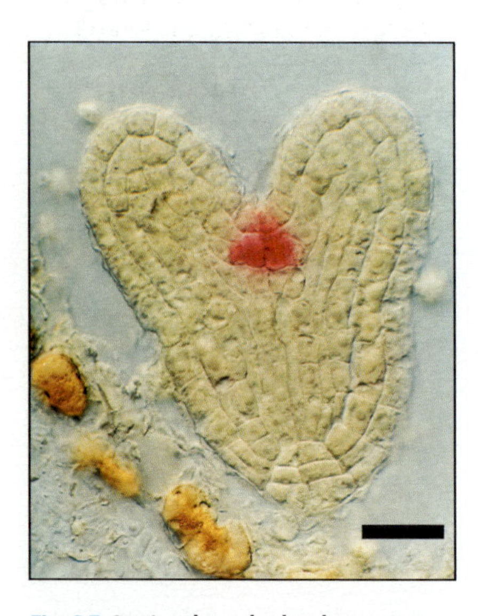

Fig. 6.7 Section through a late heart-stage *Arabidopsis* embryo showing expression of SHOOT MERISTEMLESS (STM). At this stage, the STM RNA (stained red) is expressed in cells located between the cotyledons. Scale bar = 25 μm.

Photograph courtesy of K. Barton. From Long, J.A. and Barton, K.B.: 1998.

6.4 Plant somatic cells can give rise to embryos and seedlings

As gardeners well know, plants have amazing powers of regeneration. A complete new plant can develop from a small piece of stem or root, or even from the cut edge of a leaf. This reflects an important difference between the developmental potential of plant and animal cells. In animals, with few exceptions, cell determination and differentiation are irreversible. By contrast, many somatic plant cells remain totipotent. Somatic cells from roots, leaves, and stems, and even, for some

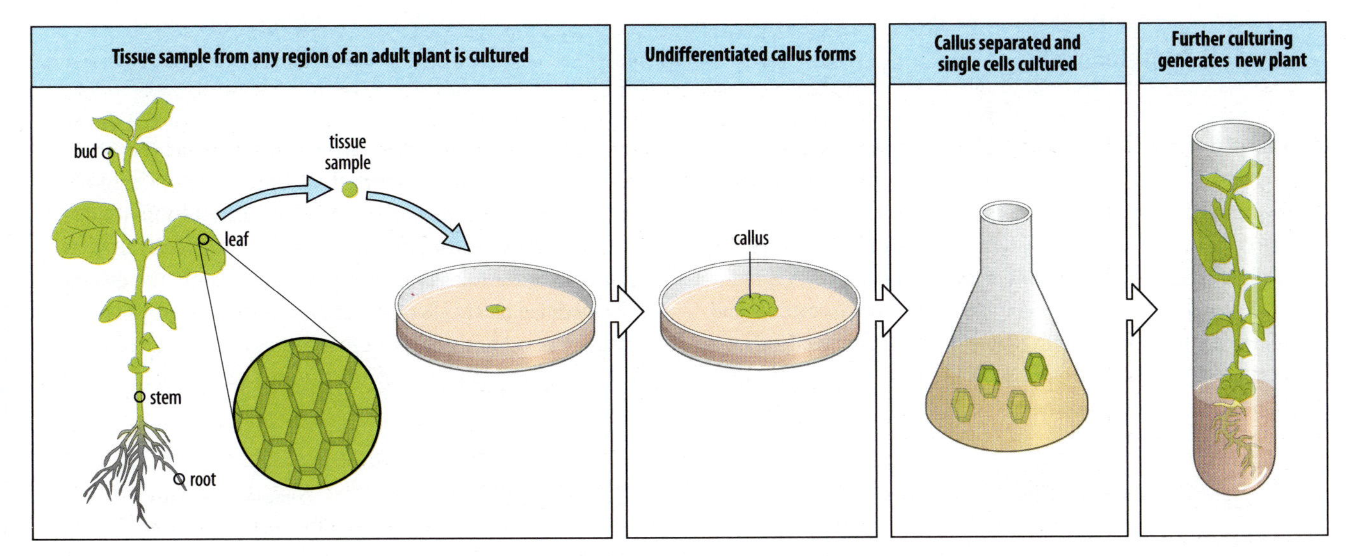

Fig. 6.8 Cultured somatic cells from a mature plant can form an embryo and regenerate a new plant. The illustration shows the generation of a plant from single cells. If a small piece of tissue from a plant stem or leaf is placed on a solid agar medium containing the appropriate nutrients and growth hormones, the cells will start to divide to form a disorganized mass of undifferentiated cells known as a callus. The callus cells are then separated and grown in liquid culture, again containing the appropriate growth hormones. In suspension culture, some of the callus cells divide to form small cell clusters. These cell clusters resemble the globular stage of a dicotyledon embryo, and with further culture on solid medium, develop through the heart-shaped and later stages to regenerate a complete new plant.

species, a single isolated protoplast—a cell from which the cell wall has been removed by enzymatic treatment—can be grown in culture and induced by treatment with the appropriate growth hormones to give rise to a new plant (Fig. 6.8). Careful observation of plant cells proliferating in culture has revealed that some of the dividing cells give rise to cell clusters that pass through a stage strongly resembling normal embryonic development, although the pattern of cell divisions are not the same as those in the embryo. These 'embryoids' can then develop into seedlings. The initial work on plant regeneration in culture was done with carrot cells, but a wide variety of plants can form embryos from single cells in this way. The phenomenon has been studied most in dicotyledons, such as carrot, potato, petunia, and tobacco, which are easy to propagate by these means. The ability of plants to regenerate from somatic cells has also enabled transgenic plants to be easily generated using the plant pathogenic bacterium *Agrobacterium tumefaciens* as the carrier of the gene to be transferred (Box 6B).

The ability of single somatic cells to give rise to whole plants has two important implications for plant development. The first is that maternal determinants may be of little or no importance in plant embryogenesis, as it is unlikely that every somatic cell would still be carrying such determinants. Second, it suggests that many cells in the adult plant body are not fully determined with respect to their fate, but remain totipotent. Of course, this totipotency seems only to be expressed under special conditions, but it is, nevertheless, quite unlike the behavior of animal cells. It is as if such plant cells have no long-term developmental memory, or that such memory is easily reset.

Summary

Early embryonic development in many plants is characterized by asymmetric cell division of the zygote, which specifies apical and basal regions. Embryonic development in flowering plants establishes the shoot and root meristems from which the adult plant develops. Auxin gradients are involved in specifying the apical–basal axis of the *Arabidopsis* embryo. One major difference between plants and animals is that, in culture, a single somatic plant cell can develop through an embryo-like stage and regenerate a complete new plant, indicating that some differentiated plant cells retain totipotency.

Box 6B Transgenic plants

One of the most common ways of generating transgenic plants containing new and modified genes is through infection of plant tissue in culture with the bacterium *Agrobacterium tumefaciens*, the causal agent of crown gall tumors. *Agrobacterium* is a natural genetic engineer. It contains a **plasmid**—the Ti plasmid—that contains the genes required for the transformation and proliferation of infected cells to form a callus. During infection, a portion of this plasmid—the T-DNA (shown in red below)—is transferred into the genome of the plant cell, where it becomes stably integrated. Genes experimentally inserted into the T-DNA will therefore also be transferred into the plant cell chromosomes. Ti plasmids, modified so that they do not cause tumors but still retain the ability to transfer T-DNA, are widely used as vectors for gene transfer in dicotyledonous plants. The genetically modified plant cells of the callus can then be grown into a complete new transgenic plant that carries the introduced gene in all its cells and can transmit it to the next generation.

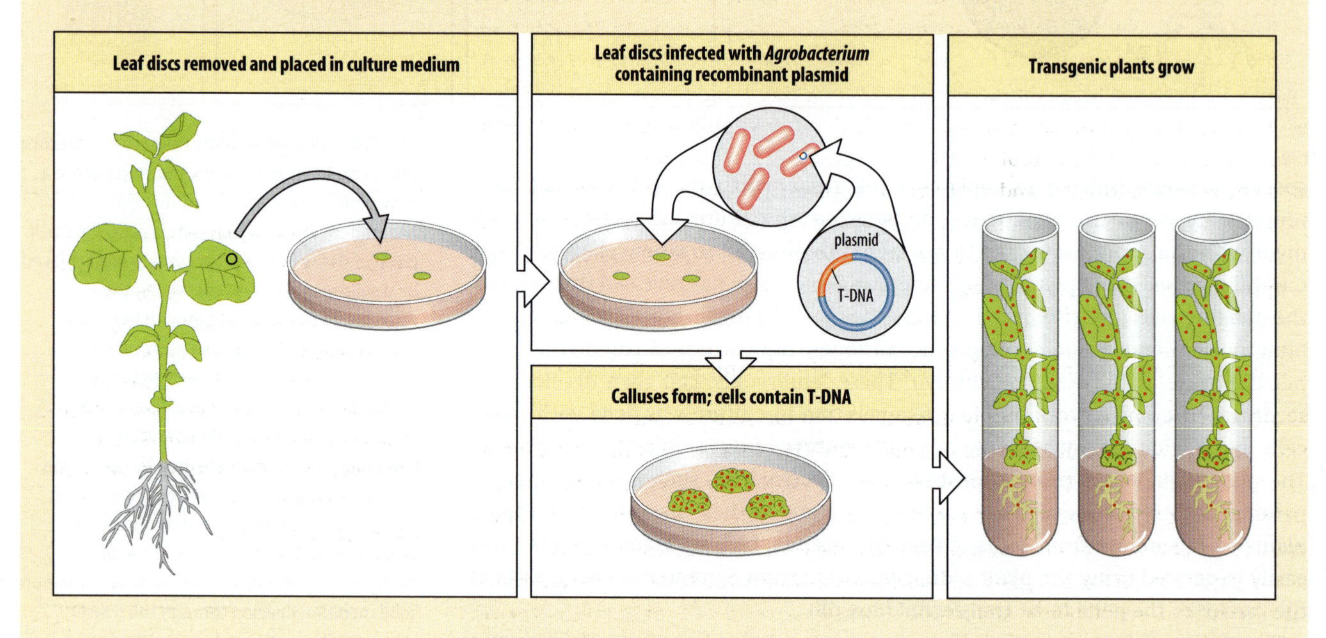

Summary: early development in flowering plants

first asymmetric cell division and auxin signaling in embryo establishes apical–basal axis

⇩

embryonic cell fate is determined by position

⇩

shoot and root meristems of seedling give rise to all adult plant structures

Meristems

In plants, most of the adult structures are derived from just two regions of the embryo, the embryonic shoot and root meristems, which are maintained after germination. The embryonic shoot meristem, for example, becomes the shoot apical meristem of the growing plant, giving rise to all the stems, leaves, and flowers. As the shoot grows, lateral outgrowths from the meristem give rise to leaves and to side shoots. In flowering shoots the vegetative meristem becomes converted into one capable of producing **floral meristems** that make flowers, not leaves. In *Arabidopsis*, for example, the shoot apical meristem changes from a vegetative meristem that makes leaves around it in a spiral pattern to an **inflorescence meristem** that then produces floral meristems, and thus flowers, around it in a spiral pattern, as we shall see later in the chapter. The first stages of future organs are known as **primordia** (singular **primordium**). Each primordium consists of a small number of **founder cells** that produce the new structure by cell division and cell enlargement, accompanied by differentiation.

There is usually a time delay between the initiation of two successive leaves in a shoot apical meristem, and this results in a plant shoot being composed of repeated modules. Each module consists of an **internode** (the cells produced by the meristem between successive leaf initiations), a **node** and its associated leaf, and an axillary bud (Fig. 6.9). The axillary bud itself contains a meristem, known as the **lateral shoot meristem**, which forms at the base of the leaf, and can produce a side shoot when the inhibitory influence of the main shoot tip is removed. Root growth is not so obviously modular, but similar considerations apply, as new lateral meristems initiated behind the root apical meristem give rise to lateral roots.

Shoot apical meristems and root apical meristems operate on the same principles, but there are some significant differences between them. We will use the shoot apical meristem to illustrate the basic principles of meristem structure and properties, and then discuss roots.

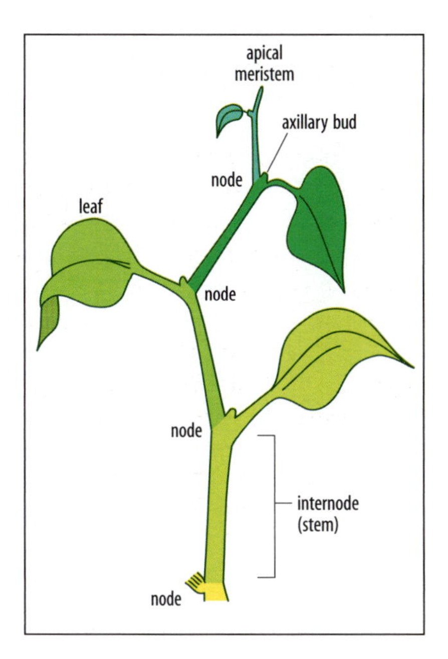

Fig. 6.9 Plant shoots grow in a modular fashion. The shoot apical meristem produces a repeated basic structural module.
The vegetative shoot module typically consists of internode, node, leaf, and axillary bud (from which a side branch may develop). Successive modules are shown here in different shades of green. As the plant grows, the internodes behind the meristem lengthen and the leaves expand.

After Alberts, B., et al.: 2002.

6.5 A meristem contains a small central zone of self-renewing stem cells

Shoot apical meristems are rarely more than 250 μm in diameter in angiosperms and contain a few hundred relatively small, undifferentiated cells that are capable of cell division. Most of the cell divisions in normal plant development occur within meristems, or soon after a cell leaves a meristem, and much of a plant's growth is in fact due to cell elongation and enlargement. Cells leave the periphery of the meristem to form organs such as leaves or flowers, and are replaced from a small central zone of slowly dividing, self-renewing stem cells or **initials** at the tip of the meristem (Fig. 6.10). In *Arabidopsis* this zone comprises around 12 to 20 cells. Initials behave in the same way as animal stem cells. They can divide to give one daughter that remains a stem cell and one that loses the stem-cell property. This daughter cell continues to divide and its descendants are displaced towards the peripheral zone of the meristem, where they become founder cells for a new organ or internode, leave the meristem, and differentiate. A very small number of long-term stem cells at the center may persist for the whole life of the plant.

Meristem stem cells are maintained in the self-renewing state by cells underlying the central zone that form the **organizing center**. As we shall see, it is the microenvironment maintained by the organizing center that gives stem cells their identity.

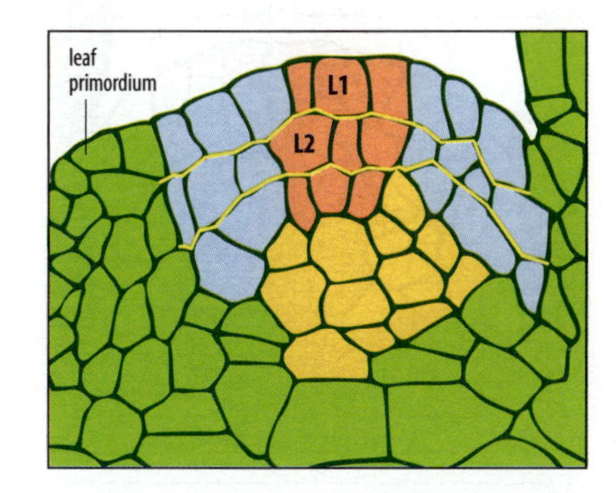

Fig. 6.10 Organization of the *Arabidopsis* shoot meristem. A longitudinal section is shown. The meristem has three main layers, L1, L2, and an inner layer, as indicated by the yellow lines, and is divided functionally into a central zone (CZ), a rib zone (RZ), and a peripheral zone (PZ). The stem cells or initials lie in the central zone, while the peripheral zone contains proliferating cells that will give rise to leaves and side shoots. The rib zone gives rise to the central tissues of the plant stem.

The undetermined state of meristem stem cells is confirmed by the fact that meristems are capable of regulation. If, for example, a seedling shoot meristem is divided into two or four parts by vertical incision, each part becomes reorganized into a complete meristem, which gives rise to a normal shoot. Provided some subpopulation of organizing center cells plus overlying stem cells is present, a normal meristem will regenerate. If a shoot apical meristem is completely removed, no new apical meristem forms, but the incipient meristem at the base of the leaf is now able to develop and form a new side shoot. In the presence of the original meristem, this prospective meristem remains inactive, as active meristems inhibit the development of other meristems nearby—as a result of auxin transport from the shoot apex, among other factors. This regulative behavior is in line with cell–cell interactions being a major determinant of cell fate in the meristem.

6.6 The size of the stem-cell area in the meristem is kept constant by a feedback loop to the organizing center

Numerous genes that control the behavior of the cells in the meristem are known. The gene *SHOOT MERISTEMLESS* (*STM*), which is involved in specifying the shoot meristem in the development of the *Arabidopsis* embryo (see Section 6.3) is, for example, expressed throughout adult shoot meristems but is suppressed as soon as cells become part of an organ primordium. Its role seems to be to maintain meristematic cells in an undifferentiated state, as loss of *STM* function results in all the meristem cells being incorporated into organ primordia. In contrast, mutations in the *Arabidopsis CLAVATA* (*CLV*) genes increase the size of the meristem as a result of an increase in the number of central stem cells. Normally, despite the continual exit of cells from the stem-cell pool, the number of stem cells in a meristem is kept roughly the same throughout a plant's life, by division of the remaining stem cells. The role of the *CLV* genes in regulating stem-cell numbers is understood in some detail, and involves feedback between the stem cells and the organizing center that underlies them.

In *Arabidopsis*, cells of the organizing center express a homeobox transcription factor, WUSCHEL (WUS). This is required to produce a signal (as yet unknown) that gives the overlying cells their stem-cell identity. Mutations in *WUS* result in termination of the shoot meristem and cessation of growth as a result of the loss of stem cells, while its overexpression increases stem-cell numbers. The stem cells express *CLAVATA3* (*CLV3*), which encodes a secreted protein that acts indirectly to repress *WUS*. This feedback loop could control *WUS* activity in the organizing center and suppress *WUS* activation in neighboring cells, thus limiting the extent of *WUS* expression (Fig. 6.11, top panel). In turn, this would regulate the extent of the stem-cell zone above the organizing center. If stem-cell numbers temporarily fall, for example, less CLV3 is produced and *WUS* is expressed more widely, with a consequent increase in stem-cell numbers. More CLV3 is then produced and limits the extent of *WUS* expression. Other CLAVATA proteins are involved in the feedback loop (Fig. 6.11, bottom panel).

Meristems can be induced elsewhere in the plant by the misexpression of genes involved in specifying stem-cell identity, yet another indication that stem-cell identity is conferred by cell–cell interactions and not by an embryonically specified cell lineage.

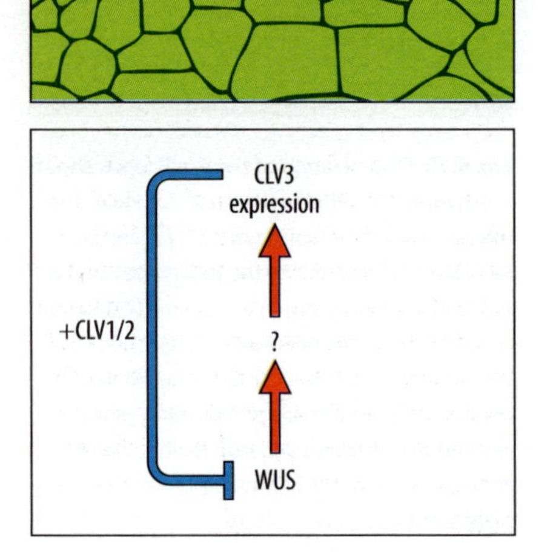

Fig. 6.11 Regulation of the stem-cell population in a shoot meristem. Intercellular signals control the position and size of the stem-cell population in the *Arabidopsis* shoot meristem. Top panel: the organizing center (purple) expresses the transcription factor WUS, which induces the production of an as yet unknown intercellular signal (red arrow) that maintains the overlying stem cells (orange). The stem cells express and secrete the signal protein CLAVATA3 (CLV3; orange dots), which moves laterally and downwards and indirectly represses transcription of the *WUS* gene in the surrounding cells, acting through its receptor proteins CLV1 and CLV2. CLV3 thus limits the extent of the area specified as stem cells. The descendants of the stem cells are continuously displaced into the peripheral zone of the meristem (pale yellow), where they are recruited into leaf primordia. Bottom panel: the negative feedback loop whereby WUS expression is restricted by CLV3. Expression of WUS in the organizing center is responsible for the production of an unknown signal that induces the expression of CLV3. CLV3 in turn signals through CLV1/2 to suppress the expression of *WUS*.

Adapted from Brand, U. et al.: Science 2000, ***289****: 617–619.*

6.7 The fate of cells from different meristem layers can be changed by changing their position

Other evidence that the fate of a meristematic cell is determined by its position in the meristem, and thus the intercellular signals it is exposed to, comes from observing the fates of cells in the different meristem layers. As well as being organized into central and peripheral zones, the apical meristem of a dicotyledon, such as *Arabidopsis*, is composed of three distinct layers of cells (Fig. 6.12). The outermost layer, L1, is just one cell thick. Layer L2, just beneath L1, is also one cell thick. In both L1 and L2, cell divisions are anticlinal—that is, the new wall is in a plane perpendicular to the layer—thus maintaining the two-layer organization. The innermost layer is L3, in which the cells can divide in any plane. L1 and L2 are often known as the tunica, and L3 as the corpus.

To find which tissues each layer can give rise to, the fates of cells in the different layers can be followed by marking one layer with a distinguishable mutation, such as a change in pigmentation or a polyploid nucleus. When a complete layer is genetically different from the others in this way, the organism is known as a

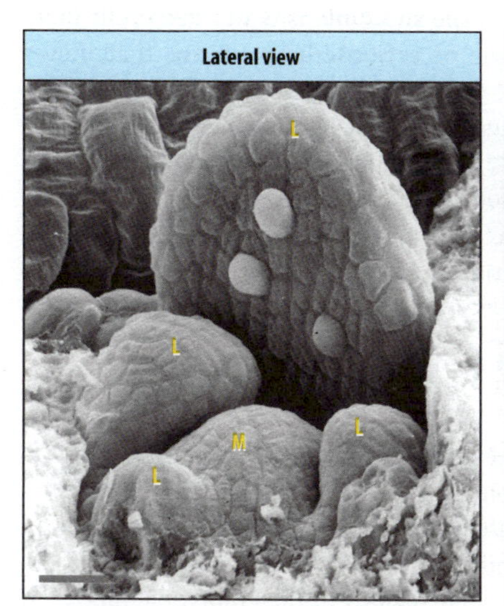

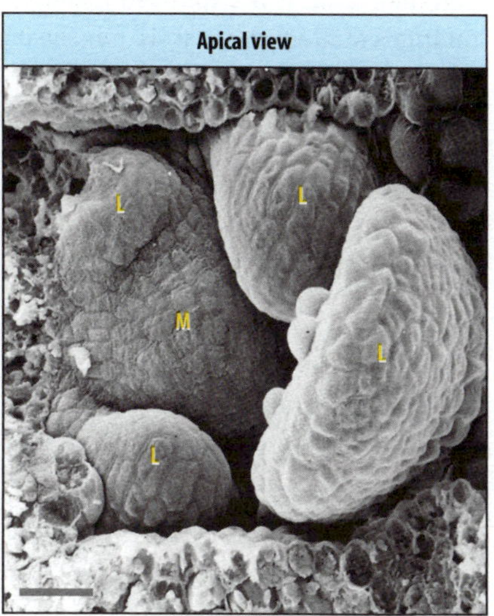

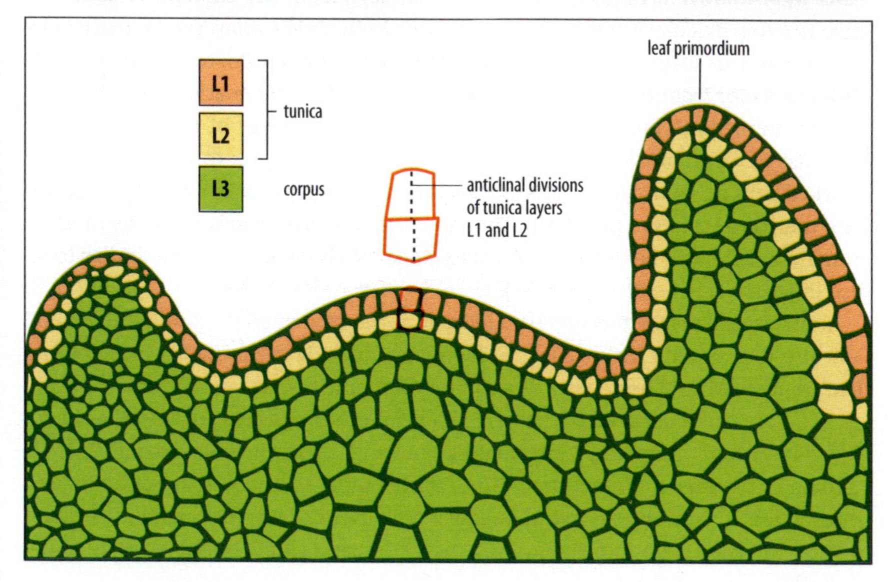

Fig. 6.12 Apical meristem of *Arabidopsis*. Top panels: scanning electron micrographs showing the organization of the meristem at the young vegetative apex of *Arabidopsis*. The plant is a *clavata1* mutant, which has a broadened apex that allows for a clearer visualization of the leaf primordia (L) and the meristem (M). Scale bar = 10 μm. Bottom panel: diagram of a vertical section through the apex of a shoot. The three-layered structure of the meristem is apparent in the most apical region. In layer 1 (L1) and layer 2 (L2), the plane of cell division is anticlinal; that is, at right angles to the surface of the shoot. Cells in layer 3 (L3, the corpus) can divide in any plane. A leaf primordium is shown forming at one side of the meristem.

Photographs courtesy of M. Griffiths.

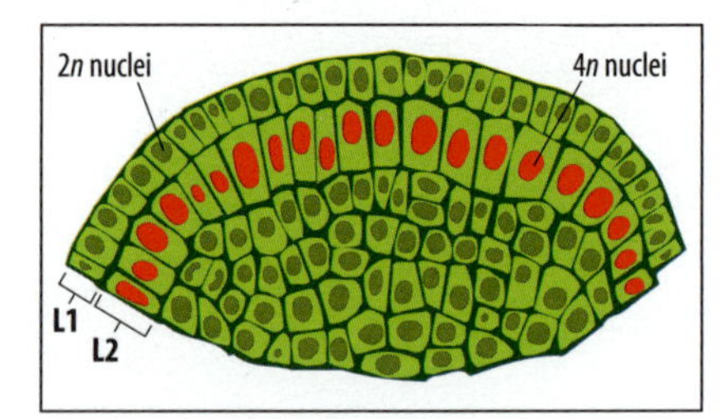

Fig. 6.13 A periclinal meristem chimera composed of cells of two different genotypes. In L1 the cells are diploid whereas the cells of L2 are tetraploid—that is, they have double the normal chromosome number—and are larger and easily recognized.

After Steeves, T.A., et al.: 1989.

periclinal chimera (Fig. 6.13) and the fate of cells from this layer can be traced. As we saw in relation to animal embryos, chimeras are organisms composed of cells of two different genotypes (see Box 3E, p. 130). Plant chimeras can be made by inducing mutation in the apical meristem of a seed or shoot tip by X-irradiation, or by treatment with chemicals such as colchicine that induce polyploidy.

Layer L1 gives rise to the epidermis that covers all structures produced by the shoot, while L2 and L3 both contribute to cortex and vascular structures. Leaves and flowers are produced mostly from L2; L3 contributes mainly to the stem. Although the three layers maintain their identity in the central region of the meristem over long periods of growth, cells in either L1 or L2 occasionally divide periclinally; new cell walls are formed parallel to the surface of the meristem, and thus one of the new cells invades an adjacent layer. This migrant cell now develops according to its new position, showing that cell fate is not necessarily determined by the meristem layer in which the cell originated, and that intercellular signaling is involved in specifying or changing its fate in its new position. The anticlinal pattern of cell division in L2 becomes disrupted when leaf primordia start to form, when the cells divide periclinally as well as anticlinally.

The transcription factor Knotted-1 in maize is homologous to *Arabidopsis* STM and, like STM, it is expressed throughout the shoot meristem to keep cells in an undifferentiated state. It is one example of a transcription factor that moves directly from cell to cell. The *KNOTTED-1* gene is expressed in all layers except L1, but the protein is also found in L1, suggesting that it can move between cells, perhaps via plasmodesmata. In *Knotted-1* gain-of-function mutants, in which the gene is misexpressed in leaves, Knotted-1 protein fused to green fluorescent protein has also been observed to move from the inner layers of the leaf to the epidermis, but not in the opposite direction.

6.8 A fate map for the embryonic shoot meristem can be deduced using clonal analysis

Much of our knowledge about the general properties of meristems outlined in previous sections comes from studies some decades ago that determined how the embryonic shoot apical meristem is related to the development of a plant over its whole lifetime. What was the fate of individual 'embryonic initials', as the stem cells were known at the time? Did particular regions in the embryonic meristem give rise to particular parts of the adult plant? Individual initials can be marked by mutagenesis of seeds by X-irradiation or transposon activation, to give cells with a different color from the rest of the plant, for example. If the marked meristem cell and its immediate progeny populate only part of one layer of the meristem (in contrast to a periclinal chimera), then this area will give rise to visible sectors of marked cells in stems and organs as the plant grows; this type of chimera is known as a **mericlinal chimera** (Fig. 6.14). The fate of individually marked initials in mericlinal chimeras can be determined using clonal analysis in a manner similar to its use in *Drosophila* (see Box 2D, p. 68).

Fig. 6.14 Tobacco plant mericlinal chimera. This plant has grown from an embryonic shoot meristem in which an albino mutation has occurred in a cell of the L2 layer. The affected area occupies about a third of the total circumference of the shoot, suggesting that there are three apical initial cells in the embryonic shoot meristem.

Photograph courtesy of S. Poethig.

In maize, the marked sectors usually start at the base of an internode and extend apically, terminating within a leaf. Some sectors include just a single internode and leaf, representing the progeny of an embryonic initial that is lost from the meristem before the generation of the next leaf primordium. Others, on the other hand, extend through numerous internodes, showing that some embryonic initials remain in the meristem for a long time, contributing to a succession of nodes and internodes. In sunflowers, marked clones have been observed to extend through several internodes up into the flower, showing that a single initial can contribute to both leaves and flowers.

From the analysis of hundreds of mericlinal chimeras, fate maps of the embryonic shoot meristems of several species were constructed, which shed light on the properties of the shoot meristem and how it behaves during normal development. These fate maps are probabilistic because it is not possible to know the location of the marked cell in the embryonic meristem, which is inaccessible inside the seed. The location of the cell that gave rise to a particular sector has to be deduced after looking at a large number of plants and taking into account the number of cells in the embryonic meristem. For example, the frequency with which a marked sector occurs in a given leaf in different plants gives an estimate of the number of cells in the meristem that can give rise to that leaf, while the final size of the clone in relation to the total size of the organ through which it extends gives an estimate of how many cells in the meristem contributed to that organ. If, for example, a marked clone occupies about a quarter of a leaf, then, assuming the clone arises from a single marked initial, around four initials would have contributed to that leaf.

The probabilistic fate map for the maize embryonic shoot apical meristem indicates that the three most apical cells in L1 give rise to the male inflorescence (the tassel and spike; Fig. 6.15). The remainder of the maize meristem can be divided into five tiers of cells that produce internodes and leaves, and which form overlapping concentric domains on the fate map. The outermost domain contributes to the earliest internode–leaf modules, while the inner domains give rise to

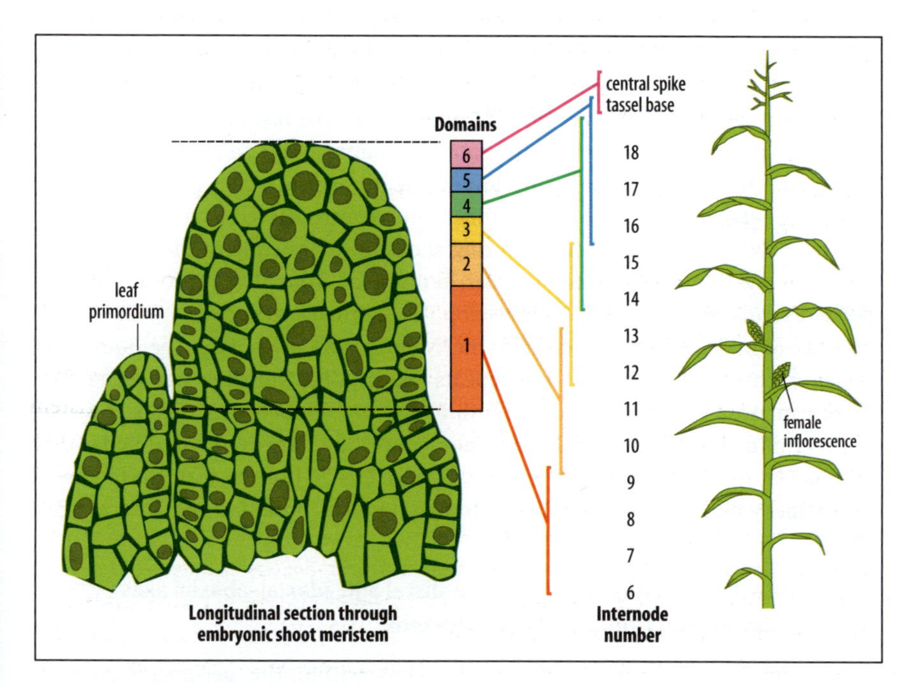

Fig. 6.15 Probabilistic fate map of the shoot apical meristem in the mature maize embryo. In maize, the primordia of the first six leaves are already present in the mature embryo and are excluded from the analysis. At the time the cells were marked, the meristem contained about 335 cells, which will give rise to 12 more leaves, the female inflorescences, and the terminal male inflorescence (the tassel and spike). A longitudinal section through the embryonic apical dome—the shoot meristem—is shown on the left. Clonal analysis shows that it can be divided into six vertically stacked domains, each comprising a set of initials that can give rise to a particular part of the plant. The number of initials in layers L1 and L2 of each domain at the embryonic stage can be estimated from the final extent of the corresponding marked sectors in the mature plant (see text). The fate of each domain, as estimated from the collective results of clonal analysis of many different plants, is shown on the right. Domain 6, comprising the three most apical L1 cells of the meristem, will give rise to the terminal male inflorescence. The fate of cells in the other domains is less circumscribed. Domain 5, for example, which consists of around eight L1 cells surrounding domain 6 and the underlying four or so L2 cells, can contribute to nodes 16–18; domain 4 to nodes 14–18; and domain 3 to nodes 12–15. The female inflorescences, which develop in the leaf axils, are derived from the corresponding domains.

After McDaniel, C.N., et al.: 1988.

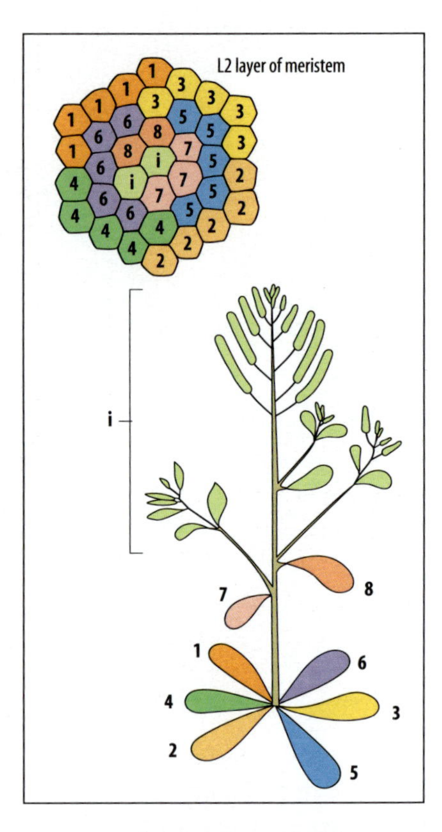

Fig. 6.16 Probabilistic fate map of the embryonic shoot meristem of *Arabidopsis*. The L2 layer of the meristem is depicted as if flattened out and viewed from above. The numbers indicate the leaf, as shown on the plant below, to which each group of meristem cells contributes, and indicate the sequence in which the leaves are formed. The inflorescence shoot (i) is derived from a small number of cells in the center of the layer.

After Irish, V.E.: 1991.

internodes and leaves successively higher up the stem. A fate map has been similarly constructed for the embryonic shoot apical meristem of *Arabidopsis* (Fig. 6.16). Most of the *Arabidopsis* embryonic meristem gives rise to the first six leaves, whereas the remainder of the shoot, including all the flowerheads, is derived from a very small number of embryonic cells at the center of the meristem. Unlike maize, the number of leaves in *Arabidopsis* is not fixed—growth is said to be **indeterminate**. There is no relation between particular cell lineages and particular structures, which indicates that position in the meristem is crucial in determining cell fate. One exception is that germ cells always arise from L2. The L2 layer is a clone, and so there is, in the case of germ cells, a relation between fate and a particular cell lineage.

The conclusions from the clonal analysis experiments are that the initials that contribute to a particular structure are simply those that happen to be in the appropriate region of the meristem at the time; they have not been prespecified in the embryo as, say, flower or leaf.

6.9 Meristem development is dependent on signals from other parts of the plant

To what extent does the behavior of a meristem depend on other parts of the plant? It seems to have some autonomy, because if a meristem is isolated from adjacent tissues by excision it will continue to develop, although often at a much slower rate. Excised shoot apical meristems of a variety of plants can be grown in culture, where they will develop into shoots complete with leaves if the growth hormones auxin and cytokinin are added. The behavior of the meristem *in situ*, however, is influenced in more subtle ways by interactions with the rest of the plant.

As we saw earlier, the apical meristem of maize gives rise to a succession of nodes, and terminates in the male flower. The number of nodes before flowering is usually between 16 and 22. This number is not controlled by the meristem alone, however. Evidence comes from culturing shoot tips consisting of the apical meristem and one or two leaf primordia. Meristems taken from plants that have already formed as many as 10 nodes still develop into normal maize plants with the full number of nodes. The isolated maize meristem has no memory of the number of nodes it has already formed, and repeats the process from the beginning. The meristem itself is therefore not determined in the embryo with respect to the number of nodes it will form. In the plant, control over the number of nodes that are formed must therefore involve signals from the rest of the plant to the meristem, finally directing the meristem to terminate node formation and form a tassel.

6.10 Gene activity patterns the proximo-distal and adaxial-abaxial axes of leaves developing from the shoot meristem

Leaves develop from groups of founder cells within the peripheral zone of the shoot apical meristem. The first indication of leaf initiation in the meristem is usually a swelling of a region to the side of the apex, which forms the **leaf primordium** (Fig. 6.17). This small protrusion is the result of increased localized cell multiplication and altered patterns of cell division. It also reflects changes in polarized cell expansion.

Two new axes that relate to the future leaf are established in a leaf primordium. These are the **proximo-distal axis** (leaf base to leaf tip) and the **adaxial-abaxial axis** (upper surface to lower surface, sometimes called dorsal to ventral). The latter is termed adaxial-abaxial as it is related to the radial axis of the shoot. The upper surface of the leaf derives from cells near the center of this axis (adaxial) while the lower surface derives from more peripheral cells (abaxial). The two leaf surfaces carry out different functions and have different structures, with the top surface being specialized for light capture and photosynthesis. In *Arabidopsis*, flattening of the leaf along the adaxial–abaxial axis occurs after leaf primordia begin developing (see Fig. 6.17), but in monocots like maize the leaf is flattened as it emerges. The establishment of adaxial–abaxial polarity is likely to make use of positional information along the radial axis of the meristem.

Arabidopsis leaf primordia emerge from the shoot meristem with distinct programs of development in the adaxial and abaxial halves. This asymmetry can be seen from the beginning, as the leaf primordium has a crescent shape in cross-section, with a convex outer (abaxial) side and a concave inner (adaxial) surface (Fig. 6.18). Different genes are expressed in the future adaxial and abaxial sides. For example, the *Arabidopsis* gene *FILAMENTOUS FLOWER* (*FIL*) is normally expressed in the abaxial side of the leaf primordium, and specifies an abaxial cell fate. Its ectopic expression throughout the leaf primordium can cause all cells to adopt an abaxial cell fate, and the leaf develops as an arrested cylindrical structure.

The specification of adaxial cell fate in *Arabidopsis* involves the genes *PHABULOSA* (*PHAB*), *PHAVOLUTA* (*PHAV*), and *REVOLUTA* (*REV*). These encode transcription factors and are initially expressed throughout the leaf primordium but become restricted to the adaxial side. Loss-of-function mutations in these genes result in radially symmetrical leaves with only abaxial cell types characteristic of the underside of the leaf. A microRNA (see Box 5B, p. 197) is involved in the restriction of *PHAB*, *PHAV*, and *REV* expression to the adaxial side, targeting and destroying their mRNAs on the abaxial side and thus limiting their activity to the adaxial side.

It has been suggested that, in normal plants, the interaction between adaxial and abaxial initial cells at the boundary between them initiates lateral growth, resulting

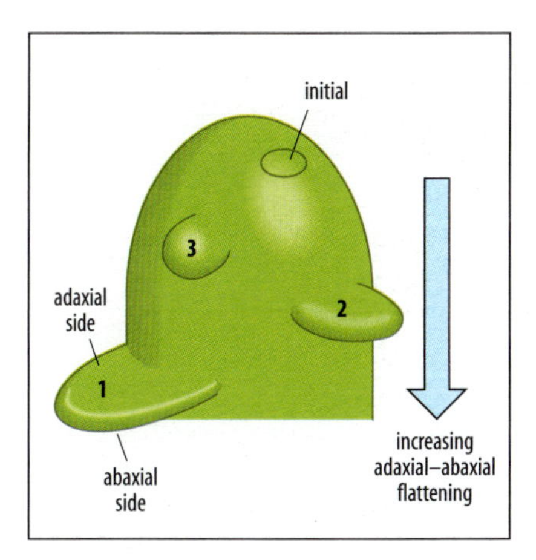

Fig. 6.17 Development of the adaxial–abaxial axis in leaf primordia. In this diagram, the leaf primordia developing from a shoot apical meristem are labeled 1 (oldest) to 3 (youngest). As they mature, the primordia elongate and begin to flatten, acquiring a clear adaxial-abaxial axis, as in 1 and 2. The initial predicts the site of emergence of the next leaf primordium.

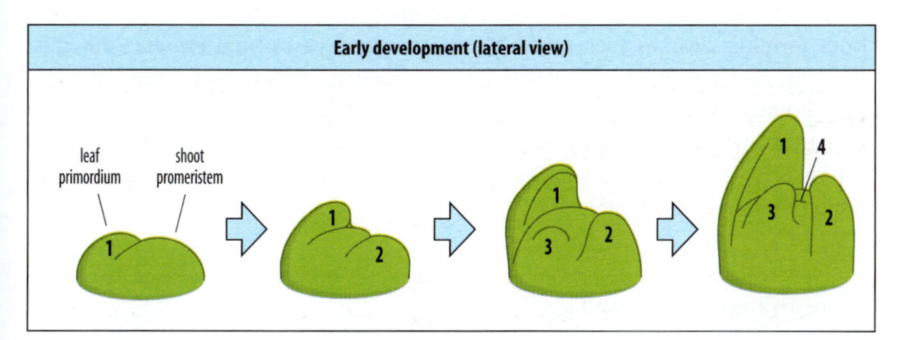

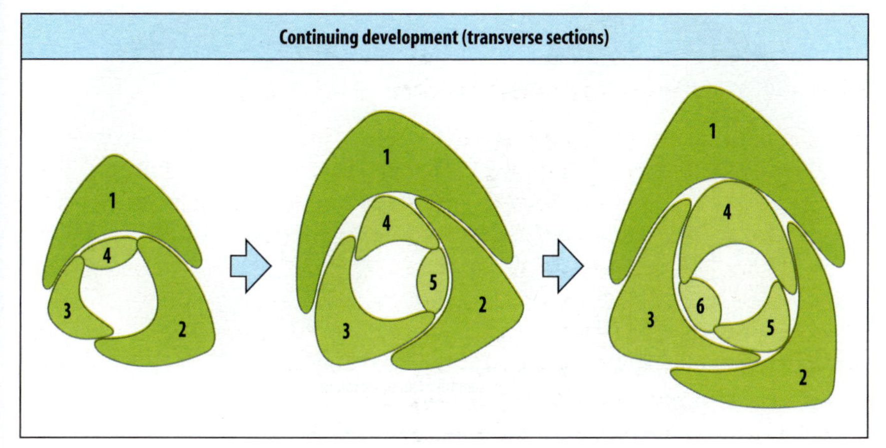

Fig. 6.18 Leaf phyllotaxis. In shoots where single leaves are arranged spirally up the stem, the leaf primordia arise sequentially in a mathematically regular pattern in the meristem. Leaf primordia arise around the sides of the apical dome, just outside the **promeristem** region. A new leaf primordium is formed slightly above and at a fixed radial angle from the previous leaf, often generating a helical arrangement of primordia visible at the apex. Top panel: lateral views of the shoot apex. Bottom panel: view looking down on cross-sections through the apex near the tip, at successive stages from the top panel.

After Poethig, R.S., et al.: 1985 (top panel); and Sachs, T.: 1994 (bottom panel).

in the formation of the leaf blade and the flattening of the leaf. The importance of boundaries in controlling pattern and form has already been seen in the parasegments of *Drosophila* (see Section 2.23) and further examples will be found in Chapter 9.

Development along the leaf proximo-distal axis also appears to be under genetic control. Like other grasses, a maize leaf primordium is composed of prospective leaf-sheath tissue proximal to the stem and prospective leaf-blade tissue distally. Mutations in certain genes result in distal cells taking on more proximal identities, for example, making sheath in place of blade. Similar proximo-distal shifts in pattern occur in *Arabidopsis* as a result of mutation. Positional identity along the proximo-distal axis may reflect the developmental age of the cells, distal cells adopting a different fate from proximal cells because they mature later.

6.11 The regular arrangement of leaves on a stem and trichomes on leaves is generated by competition and lateral inhibition

As the shoot grows, leaves are generated within the meristem at regular intervals and with a particular spacing. Leaves are arranged along a shoot in a variety of ways in different plants, and the particular arrangement, known as **phyllotaxy** or phyllotaxis, is reflected in the arrangement of leaf primordia in the meristem. Leaves can occur singly at each node, in pairs, or in whorls of three or more. A common arrangement is the positioning of single leaves spirally up the stem, which can sometimes form a striking helical pattern in the shoot apex.

In plants in which leaves are borne spirally, a new leaf primordium forms at the center of the first available space outside the central region of the meristem and above the previous primordium (see Fig. 6.18). This pattern suggests a mechanism for leaf arrangement based on lateral inhibition (see Section 1.14), in which each leaf primordium inhibits the formation of a new leaf within a given distance. In this model, inhibitory signals emanating from recently initiated primordia prevent leaves from forming close to each other. There is some experimental evidence for this. In ferns, leaf primordia are widely spaced, allowing experimental microsurgical interference. Destruction of the site of the next primordium to be formed results in a shift toward that site by the future primordium whose position is closest to it (Fig. 6.19). Recent studies indicate, however, that it is competition rather than inhibition that is responsible.

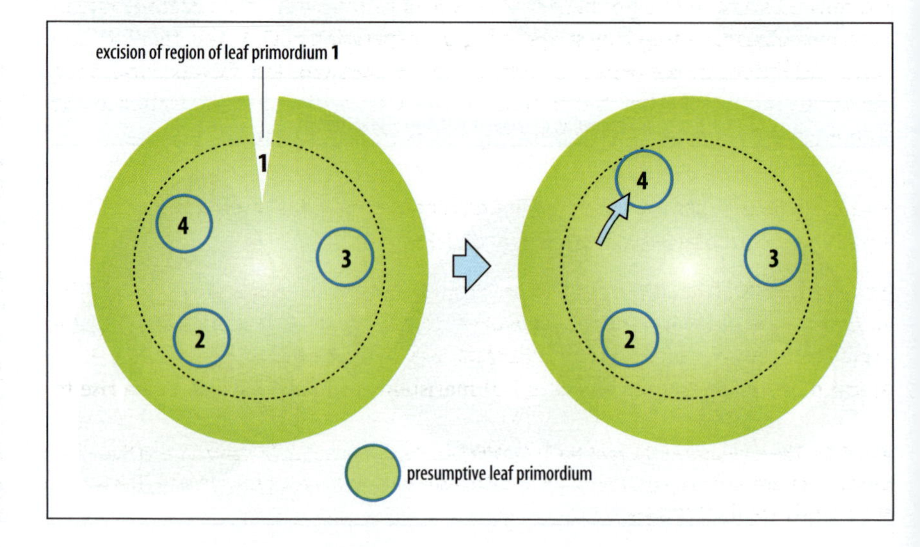

Fig. 6.19 Leaf primordia may be positioned by lateral inhibition or by competition. Leaf primordia on a fern shoot tip form in a regular order in positions 1 to 4. Primordia appear to form as far as possible from existing primordia, so 2 forms almost opposite 1. Normally, 4 will develop between 1 and 2, but if 1 is excised, 4 forms much further away from 2. This result can be interpreted either by the removal of lateral inhibition by primordium 1 or by the removal of the competitive effect of 1 for some primordium-inducing factor such as auxin.

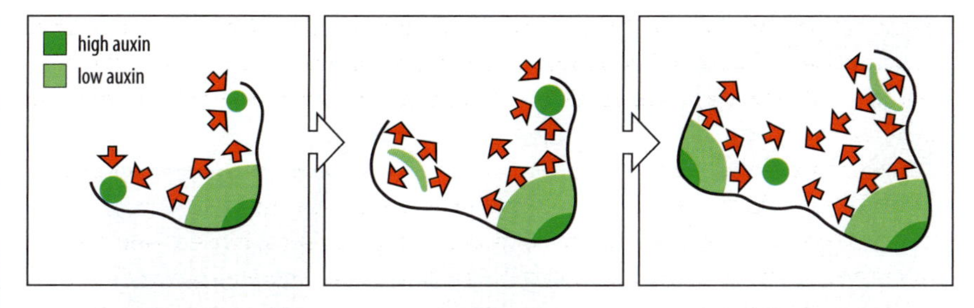

Fig. 6.20 Auxin-dependent mechanism of phyllotaxis in *Arabidopsis*. Auxin is transported through PIN proteins towards areas of high auxin concentration (dark green), at which an organ primordium will form. Cells around the developing primordium become depleted of auxin (light green), which causes the polarity of the PIN proteins to reverse so that auxin then flows away from the primordium. The red arrows indicate the direction of PIN polarity. It is proposed that this pattern of auxin circulation could set up the regular pattern of leaf and flower formation from meristems.

As we have seen in the determination of apical–basal polarity in the embryo, auxin is transported out of cells with the help of proteins such as PIN1 (see Section 6.3). In the shoot, auxin produced in the shoot tip below the meristem is transported upwards into the meristem through the epidermis and the outermost meristem layer. The direction of auxin flow in the shoot apex is controlled by PIN1 and follows a simple rule: the side of the cell on which PIN1 is found is the side nearest the neighbor cell with the highest auxin concentration. Thus auxin transport is always towards a region of higher concentration. A high concentration of auxin is a primordium activator, and initially, auxin is pumped towards a new primordium that is developing at a site of high auxin concentration. This, however, depletes a zone of cells around the primordium of auxin such that cells nearer the center of the meristem now have more auxin than the cells adaxial to the new primordium. This activates a feedback mechanism that causes PIN1 to move to the other side of these cells, and auxin now flows out of the new primordium towards the meristem, creating a new spot of high auxin concentration in the meristem farthest away from any new primordium (Fig. 6.20). This leads to auxin peaks occurring sequentially, at the regular positions later occupied by new leaves.

A clear example of lateral inhibition is involved in the spacing of cells bearing hairs, or **trichomes**, in the leaf epidermis, which are always separated from each other by three or four epidermal cells. The pattern is not related to cell lineage, but arises by lateral inhibition. All cells have the potential to develop as trichomes, but the trichome fate is suppressed in cells immediately surrounding any cell that has begun to develop as a trichome. One network of transcriptional regulators based on the GLABRA (GL) proteins promotes the trichome fate, while another set of transcriptional regulators that includes the protein TRIPTYCHON (TRY) inhibits it. Loss-of-function mutations in the *TRY* gene lead to small clusters of trichomes developing instead of individually spaced ones, which suggests that TRY must be expressed in a trichome-producing cell but acts by inhibiting the trichome fate in adjacent cells, for example by suppressing the expression of GL1. Whether TRY and other inhibitors move between cells, as other transcription factors have been shown to do, or whether the inhibitory effect is mediated by activation of an inhibitory cell–cell signaling pathway by TRY is still not known.

6.12 Root tissues are produced from *Arabidopsis* root apical meristems by a highly stereotyped pattern of cell divisions

The organization of tissues in the *Arabidopsis* root tip is shown in Fig. 6.21. The radial pattern comprises single layers of epidermal, cortical, endodermal, and pericycle cells, with vascular tissue in the center (protophloem and protoxylem). Root apical meristems resemble shoot apical meristems in many ways and give rise to

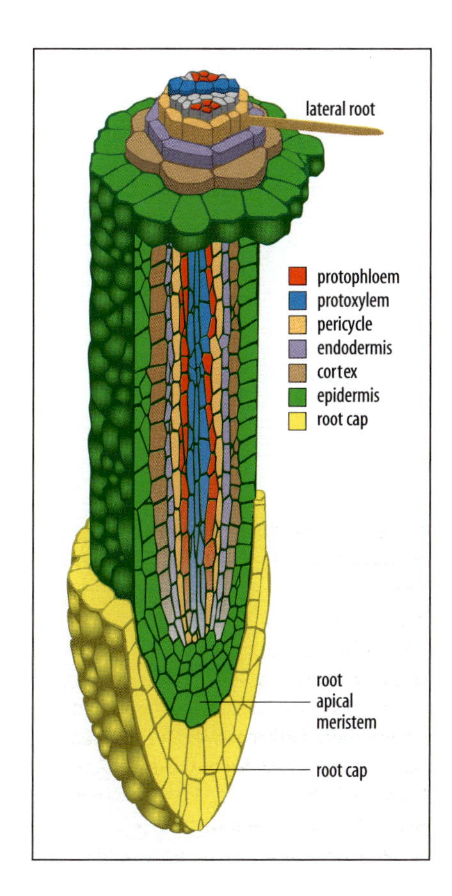

Fig. 6.21 The structure of the root tip in *Arabidopsis*. Roots have a radial organization. In the center of the growing root tip is the future vascular tissue (protoxylem and protophloem). This is surrounded by further tissue layers.

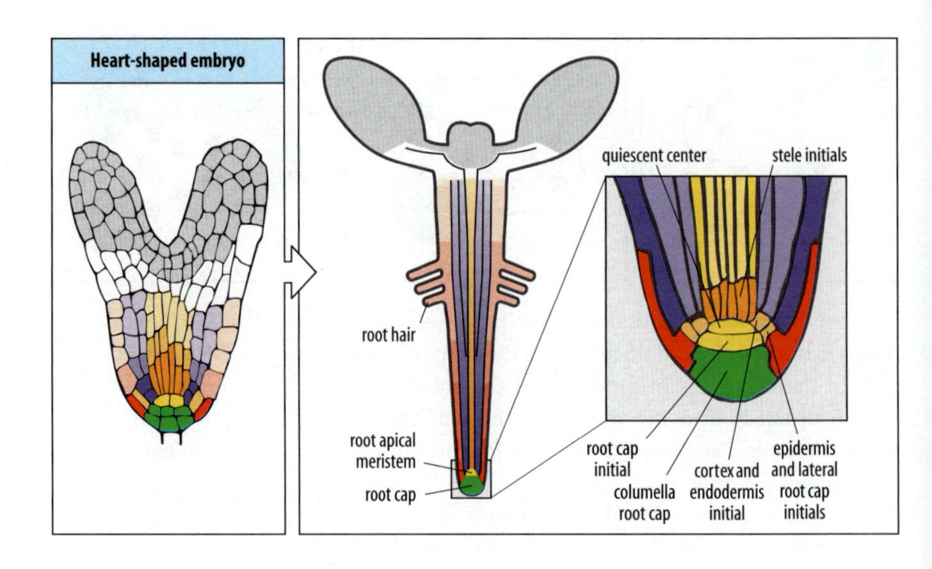

Fig. 6.22 Fate map of root regions in the heart-stage *Arabidopsis* embryo. The root grows by the division of a set of initial cells. The root meristem comes from a small number of cells in the heart-shaped embryo. Each tissue in the root is derived from the division of a particular initial cell. At the center of the root meristem is a quiescent center, which does not divide.

After Scheres, B., et al.: 1994.

the root in a similar manner to shoot generation. But there are some important differences between the root and shoot meristems. The shoot meristem is at the extreme tip of the shoot whereas the root meristem is covered by a root cap (which is itself derived from one of the layers of the meristem); also, there is no obvious segmental arrangement at the root tip resembling the node–internode–leaf module.

The root is set up early (see Section 6.2) and a well organized embryonic root can be identified in the late heart-stage embryo (Fig. 6.22). Clonal analysis has shown that the seedling root meristem can be traced back to a set of embryonic initials that arise from a single tier of cells in the heart-stage embryo. As in the shoot meristem, a root meristem is composed of an organizing center, called the **quiescent center** in roots, in which the cells divide only very rarely, and which is surrounded by stem-cell-like initials that give rise to the root tissue (see Fig. 6.22). The quiescent center is essential for meristem function. When parts of the meristem are removed by microsurgery, it can regenerate, but regeneration is always preceded by the formation of a new quiescent center. Laser destruction of individual quiescent-center cells shows that, as in the shoot meristem, a key function of the quiescent center is to maintain the immediately adjacent initials in the stem-cell state and prevent them from differentiating.

Each initial undergoes a stereotyped pattern of cell divisions to give rise to a number of columns, or **files**, of cells in the growing root (see Fig. 6.21); each file of cells in the root thus has its origin in a single initial. Some initials give rise to both endodermis and cortex, whereas others give rise to both epidermis and the root cap. Before it leaves the meristem, therefore, the undifferentiated progeny of a endodermis/cortex initial, for example, will divide asymmetrically to give one daughter that produces cortex and one that produces endodermis. The gene *SCARECROW* is necessary to confer this asymmetry on the dividing cell, and mutations in this gene give roots with no distinct endodermis or cortex but with a tissue layer with characteristics of both.

The normal pattern of cell divisions is not obligatory, however. As discussed earlier, *fass* mutants, which have disrupted cell divisions, still have relatively normal patterning in the root. In addition, laser ablation of individual meristem cells does not lead to an abnormal root. The remaining initials undergo new patterns of cell division that replace the progeny of the cells that have been destroyed. Such observations show that, as in the shoot meristem, the fate of cells in the developing root meristem depends on their recognition of positional signals and not on their lineage.

As we have seen in Section 6.3, auxin gradients play a major role in patterning the embryo and specifying the root region, and mutations that affect auxin localization lead to root defects. At the globular stage of embryonic development, the auxin-transport protein PIN1 is localized in cells in the future root region and the highest level of auxin is found adjacent to where the quiescent center will develop. The role of auxin in root development continues into the adult plant. Auxin is transported into root-tip cells via the PIN proteins. Cells with raised auxin levels transport auxin better, probably due to an increase in the number of PIN proteins in the membrane as auxin prevents their endocytosis and recycling, and so there is a positive feedback loop that raises the local concentration.

Auxin is also involved in the ability of plants to regenerate from a small piece of stem. In general, roots form from the end of the stem that was originally closest to the root, whereas shoots tend to develop from dormant buds at the end that was nearest to the shoot. This polarized regeneration is related to vascular differentiation and to the polarized transport of auxin. Transport of auxin from its source in the shoot tip toward the root leads to an accumulation of auxin at the 'root' end of the stem cutting, where it induces the formation of roots. One hypothesis suggests that polarity is both induced and expressed by the oriented flow of auxin.

One of the best examples of developmentally important transcription factor movement from one cell to another is found in roots. As noted earlier, expression of the gene *SCARECROW* is required for root cells to adopt an endodermal fate. This expression requires the transcriptional activator SHORT-ROOT (SHR). SHR is, however, not synthesized in the prospective endodermal cells, but in the adjacent cells on the inner side. SHR protein is transported from these cells outwards into the prospective endodermis, and this movement appears to be regulated and not simply due to diffusion.

A further example of cell–cell interactions relates to the two cell types in the root epidermis—the hair-forming trichoblasts and hairless atrichoblasts (Fig. 6.23). These form alternating files of cells on the surface of the developing root and their character appears to depend on positional signals from the underlying cortical cells. Whereas most cell divisions in the future epidermis are horizontal, thus increasing the number of cells per file, occasionally a vertical anticlinal division occurs, pushing one of the daughter cells into an adjacent file. The daughter cell then assumes a fate corresponding to its new position. Like the spacing of trichomes in the leaf, the spacing of trichoblasts within a single file of cells is due to lateral inhibition (see Section 6.11).

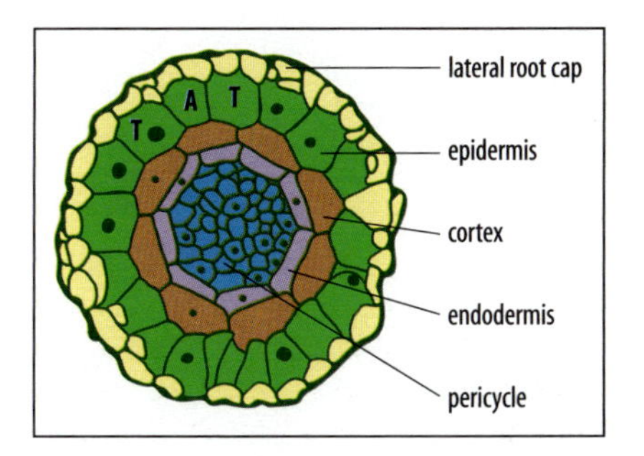

Fig. 6.23 Organization of cell types in the root epidermis. The epidermis is composed of two types of cells: trichoblasts (T), which will form root hairs, and atrichoblasts (A), which will not. Trichoblasts overlie the junction between two cortical cells and atrichoblasts are located over the outer tangential wall of cortical cells.

After Dolan, L., Scheres, B.: 1998.

Summary

Meristems are the growing points of a plant. The apical meristems, found at the tips of shoots and roots, give rise to all the plant organs—roots, stem, leaves, and flowers. They consist of small groups of a few hundred undifferentiated cells that are capable of repeated division. The center of the meristem is occupied by self-renewing stem cells, which replace the cells that are lost from the meristem when organs are formed. The fate of a cell in the shoot meristem depends upon its position in the meristem and interactions with its neighbors, as when a cell is displaced from one layer to another it adopts the fate of its new layer. Meristems can also regulate when parts are removed, in line with cell–cell interactions determining cell fate. Fate maps of embryonic shoot meristems show that they can be divided into domains, each of which normally contributes to the tissues of a particular region of the plant, but the fate of the embryonic initials is not fixed. The shoot meristem gives rise to leaves in species-specific patterns—phyllotaxy—which seem best accounted for in terms of lateral inhibition. Lateral inhibition is also involved in the regular spacing of hairs on root and leaf surfaces. In the root meristem, the cells are organized rather differently from those in the shoot meristem, and there is a much more stereotyped pattern of cell division. A set of initial cells maintains root structure by dividing along different planes.

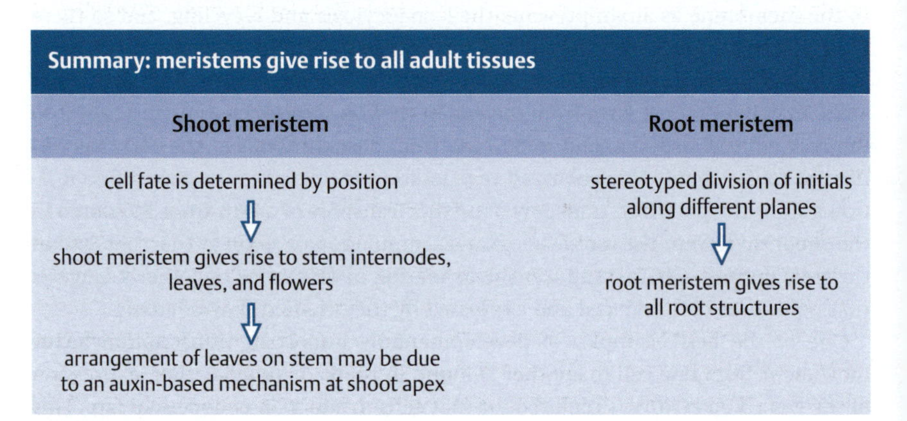

Flower development and control of flowering

Flowers contain the reproductive cells of higher plants and develop from the shoot meristem. In most plants, the transition from a vegetative shoot meristem to a floral meristem that produces a flower is largely or absolutely under environmental control, with day length and temperature being important determining factors. In a plant such as *Arabidopsis*, in which each flowering shoot produces multiple flowers, the vegetative shoot meristem first becomes converted into an inflorescence meristem, which then forms floral meristems, each of which develops completely into a single flower (Fig. 6.24). Floral meristems are thus determinate, unlike the indeterminate shoot apical meristem. Flowers, with their arrangement of floral organs (sepals, petals, stamens, and carpels), are rather complex structures, and it is a major challenge to understand how they arise from the floral meristem.

Fig. 6.24 Scanning electron micrograph of an *Arabidopsis* inflorescence meristem. The central inflorescence meristem (shoot apical meristem, SAM) is surrounded by a series of floral meristems (FM) of varying developmental ages. The inflorescence meristem grows indeterminately, with cell divisions providing new cells for the stem below, and new floral meristems on its flanks. The floral meristems (or floral primordia) arise one at a time in a spiral pattern. The most mature of the developing flowers is on the right (FM1), showing the initiation of sepal primordia surrounding a still-undifferentiated floral meristem. Eventually, such a floral meristem will also form petal, stamen, and carpel primordia.

Photograph from Meyerowitz, E.M., et al.: 1991.

The conversion of a vegetative shoot meristem into one that makes flowers involves the induction of so-called **meristem identity genes**. A key regulator of floral induction in *Arabidopsis* is the meristem identity gene *LEAFY* (*LFY*); a related gene in *Antirrhinum* is *FLORICAULA* (*FLO*). How environmental signals such as day length influence floral induction is discussed later. We will first consider the mechanisms that pattern the flower, in particular those that specify the identity of the floral organs.

6.13 Homeotic genes control organ identity in the flower

The individual parts of a flower each develop from a **floral organ primordium** produced by the floral meristem. Unlike leaf primordia, which are all identical, the

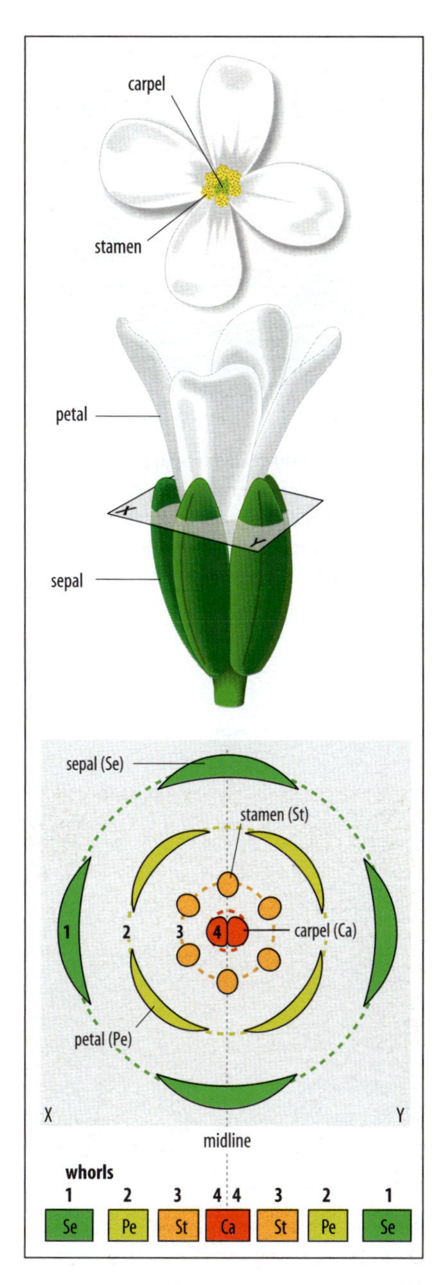

Fig. 6.25 Structure of an *Arabidopsis* flower. *Arabidopsis* flowers are radially symmetrical and have an outer ring of four identical green sepals, enclosing four identical white petals, within which is a ring of six stamens, with two carpels in the center. Bottom: floral diagram of the *Arabidopsis* flower representing a cross-section taken in the plane indicated in the top diagram. This is a conventional representation of the arrangement of the parts of the flower, showing the number of flower parts in each whorl and their arrangement relative to each other.

After Coen, E.S., et al.: 1991.

floral organ primordia must each be given a correct identity and be patterned according to it. An *Arabidopsis* flower has four concentric whorls of structures (Fig. 6.25), which reflect the arrangement of the floral organ primordia in the meristem. The sepals (whorl 1) arise from the outermost ring of meristem tissue, and the petals (whorl 2) from a ring of tissue lying immediately inside it. An inner ring of tissue gives rise to the male reproductive organs—the stamens (whorl 3). The female reproductive organs—the carpels (whorl 4)—develop from the center of the meristem. In a floral meristem of *Arabidopsis*, there are 16 separate primordia, giving rise to a flower with four sepals, four petals, six stamens and a pistil made up of two carpels (see Fig. 6.25).

The primordia arise at specific positions within the meristem, where they develop into their characteristic structures. After the emergence of the primordia in *Antirrhinum*, cell lineages become restricted to particular whorls, rather like the lineage restriction to compartments in *Drosophila* (see Section 2.22). Lineage restriction occurs at the time when the pentagonal symmetry of the flower becomes visible and genes that give the different floral organs their identity are expressed. The lineage compartments within the floral meristem appear to be delineated by narrow bands of non-dividing cells.

Like the homeotic selector genes that specify segment identity in *Drosophila*, mutations in floral identity genes cause homeotic mutations in which one type of flower part is replaced by another. In the *Arabidopsis* mutant *apetala2*, for example, the sepals are replaced by carpels and the petals by stamens; in the *pistillata* mutant, petals are replaced by sepals and stamens by carpels. These mutations identified the floral organ identity genes, and have enabled their mode of action to be determined.

Homeotic floral mutations in *Arabidopsis* fall into three classes, each of which affects the organs of two adjacent whorls (Fig. 6.26). The first class of mutations, of which *apetala2* is an example, affect whorls 1 and 2, giving carpels instead of sepals in whorl 1, and stamens instead of petals in whorl 2. The phenotype of the flower, going from the outside to the center, is therefore carpel, stamen, stamen, carpel.

Fig. 6.26 Homeotic floral mutations in *Arabidopsis*. Left panel: an *apetala2* mutant has whorls of carpels and stamens in place of sepals and petals. Center panel: an *apetala3* mutant has two whorls of sepals and two of carpels. Right panel: *agamous* mutants have a whorl of petals and sepals in place of stamens and carpels. Transformations of whorls are shown inset, and can be compared to the wild-type arrangement, as shown in Fig. 6.25.

Photographs from Meyerowitz, E.M., et al.: 1991 (left panel), and Bowman, J.L., et al.: 1989 (center panel).

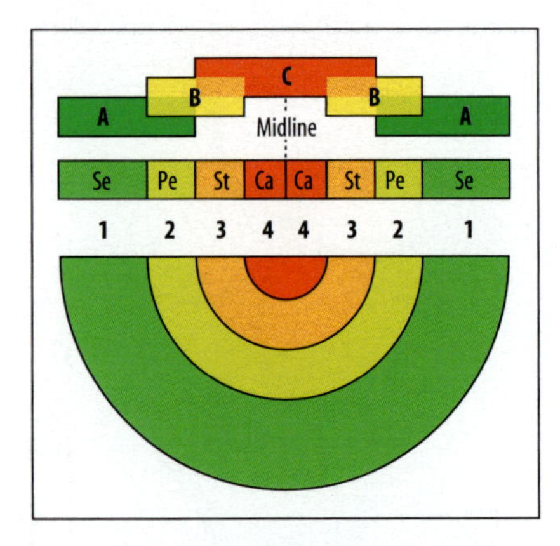

Fig. 6.27 The three overlapping regions of the *Arabidopsis* floral meristem that have been identified by the homeotic floral identity mutations. Region A corresponds to whorls 1 and 2, B to whorls 2 and 3, and C to whorls 3 and 4.

The second class of homeotic floral mutations affects whorls 2 and 3. In this class, *apetala3* and *pistillata* give sepals instead of petals in whorl 2 and carpels instead of stamens in whorl 3, with a phenotype sepal, sepal, carpel, carpel. The third class of mutations affects whorls 3 and 4 and gives petals instead of stamens in whorl 3 and sepals or variable structures in whorl 4. The mutant *agamous*, which belongs to this class, has an extra set of sepals and petals in the center instead of the reproductive organs.

These mutant phenotypes can be accounted for by an elegant model in which overlapping patterns of gene activity specify floral organ identity (Fig. 6.27) in a manner highly reminiscent of the way in which *Drosophila* homeotic genes specify segment identity along the insect's body. In detail, however, there are many differences, and quite different genes are involved. In this instance, plants and animals have, perhaps not surprisingly, independently evolved a similar approach to patterning a multicellular structure, but have recruited different proteins to carry it out.

In essence, the floral meristem is divided by the expression patterns of the homeotic genes into three concentric overlapping regions, A, B, and C, which partition the meristem into four non-overlapping regions corresponding to the four whorls. Each of the A, B, and C regions corresponds to the zone of action of one class of homeotic genes and the particular combinations of A, B, and C functions give each whorl a unique identity and so specify organ identity. Of the genes mentioned in Fig. 6.25, *APETALA1* (*AP1*) and *APETALA2* (*AP2*) are A-function genes, *APETALA3* (*AP3*) and *PISTILLATA* (*P1*) are B-function genes, and *AGAMOUS* (*AG*) is a C-function gene. The expression of *AP3* and *AG* in the developing flower is shown in Fig. 6.28. All the homeotic genes, also known as **floral organ identity genes**, encode transcription factors, and the B- and C-function proteins such as AP3 and AG contain a conserved DNA-binding sequence known as the MADS box. MADS-box genes are present in animals and yeast, but a role in development is known mainly in plants—although a MADS-box transcription factor, MEF2, is involved in muscle differentiation in animals. The original simple model for specifying floral organ identity is presented in more detail in Box 6C (opposite). Since the model was first proposed, more has become known about the activities and functions of the genes identified by the homeotic mutations, more genes controlling flower development have been discovered, and more 'functions' added.

One question that arose when the original floral organ identity model was investigated experimentally was why the ABC genes only showed their homeotic properties in the floral meristem and did not convert leaves into floral organs when artificially overexpressed in vegetative meristems, as might be expected for homeotic genes of this type. The answer came with the discovery of the *SEPALLATA* (*SEP*) genes, which also encode MADS-box proteins. These genes are required for the B and C functions and are only active in floral meristems. The SEP proteins are thought to combine with *B* and *C* gene products to form active gene-regulatory complexes.

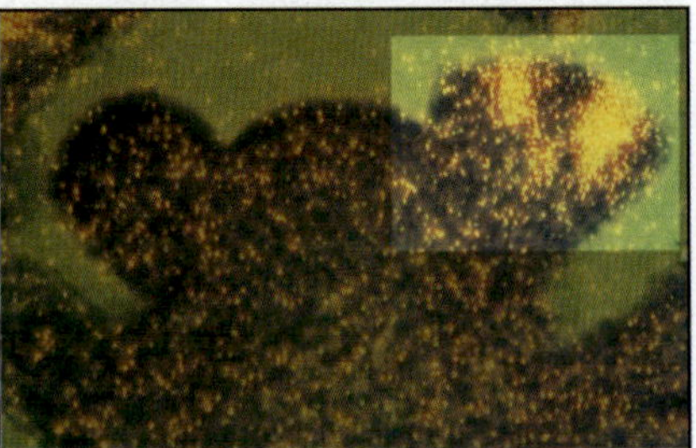

Fig. 6.28 Expression of *APETALA3* and *AGAMOUS* during flower development. *In situ* hybridization shows that *AGAMOUS* is expressed in the central whorls (left panel), whereas *APETALA3* is expressed in the outer whorls that give rise to petals and stamens (right panel).

Box 6C The basic model for the patterning of the *Arabidopsis* flower

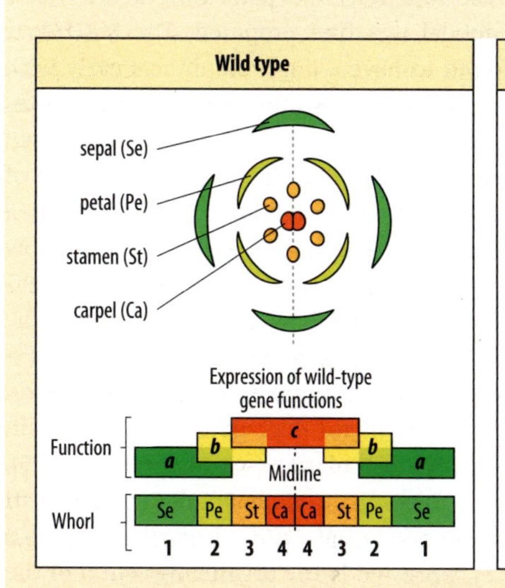

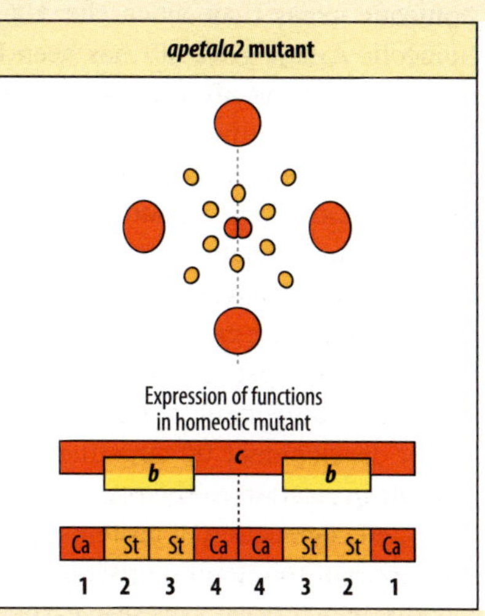

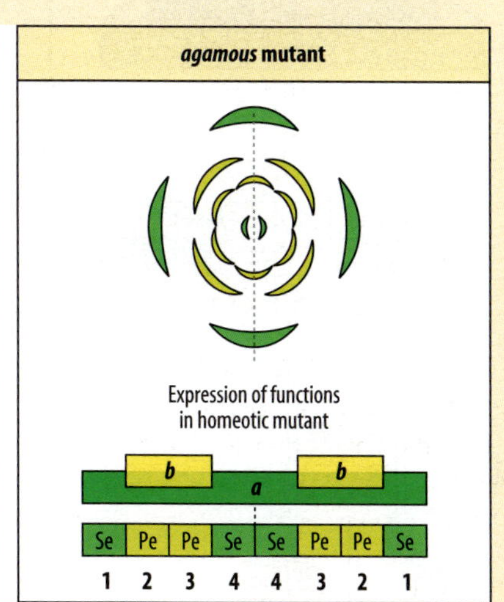

The floral meristem is divided into three overlapping regions, A, B, and C, each region corresponding to a class of homeotic mutations, as shown in Fig. 6.26 (see text). Three regulatory functions—*a*, *b*, and *c*—operate in regions A, B, and C, respectively, as shown in the panels above. In the wild-type flower (top left panel), it is assumed that *a* is expressed in whorls 1 and 2, *b* in 2 and 3, and *c* in whorls 3 and 4. In addition, *a* function inhibits *c* function in whorls 1 and 2 and *c* function inhibits *a* function in whorls 3 and 4—that is, *a* and *c* functions are mutually exclusive. *a* alone specifies sepals, *a* and *b* together specify petals, *b* and *c* stamens, and *c* alone carpels.

The homeotic mutations eliminate the functions of *a*, *b*, or *c*, and alter the regions within the meristem where the various functions are expressed. Mutations in *a*, such as *apetala2* (see center top panel), result in an absence of function *a* and *c* spreads throughout the meristem, resulting in the half-flower pattern of carpel, stamen, stamen, carpel. Mutations in *b*, such as *apetala3* (see Fig. 6.26), result in only *a* functioning in whorls 1 and 2, and *c* in whorls 3 and 4, giving sepal, sepal, carpel, carpel. Mutations in *c* genes (such as *agamous*), result in *a* activity in all whorls, giving the phenotype sepal, petal, petal, sepal (see top right panel).

All the floral homeotic mutants discovered so far in *Arabidopsis* can be quite satisfactorily accounted for by this model (although there are small variations in gene numbers and expression patterns in other species that allow mutant phenotypes not seen in *Arabidopsis*), and particular genes can be assigned to each controlling function. Function *a* corresponds to the activity of genes such as *APETALA2*, *b* to *APETALA3* and *PISTILLATA*, and *c* to *AGAMOUS*. The model also accounts for the phenotype of double mutants, such as *apetala2* with *apetala3*, and *apetala3* with *pistillata*, as shown in the panels on the right.

This system emphasizes the similarity in function between the homeotic genes in animals and those controlling organ identity in flowers, although the genes themselves are completely different. The functional similarity with the Hox complex of *Drosophila* is further illustrated by the role of the *CURLY LEAF* gene of *Arabidopsis*, which is necessary for the stable maintenance of homeotic gene activity. *CURLY LEAF* is related to the *Polycomb* family of genes in *Drosophila* and is similarly required for stable repression of homeotic genes.

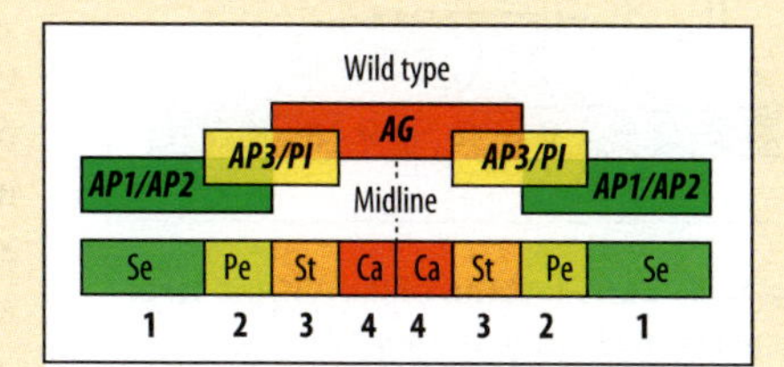

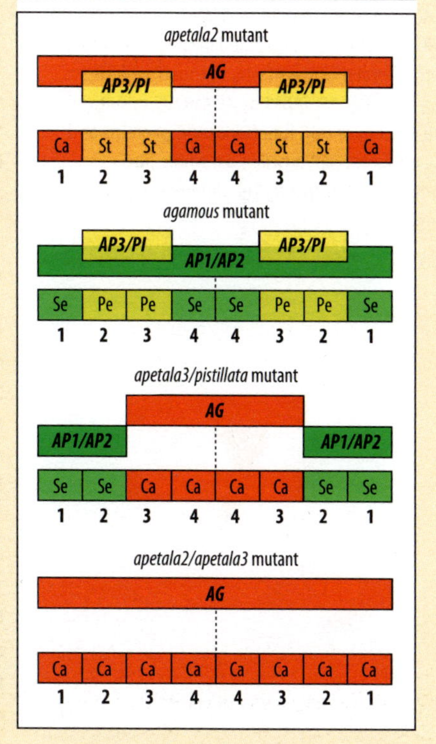

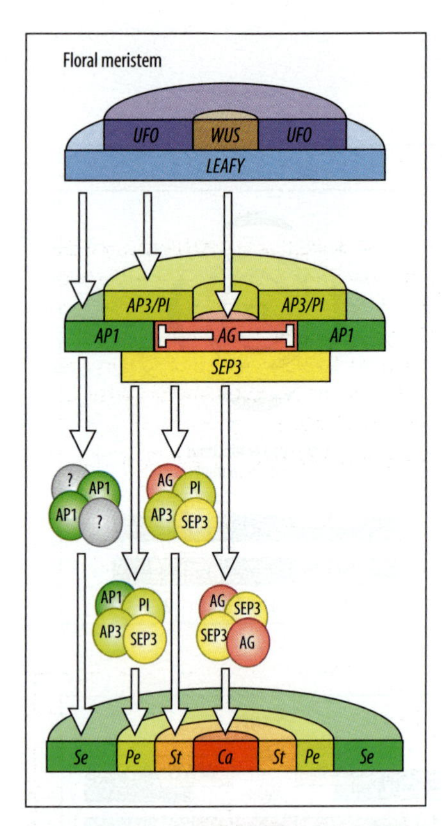

Fig. 6.29 Current status of the ABC model of floral organ identity. The regulatory genes *LEAFY*, *WUSCHEL (WUS)*, and *UNUSUAL FLORAL ORGANS (UFO)* are expressed in specific domains in the floral meristem, which, together with repression of *APETALA1* by *AGAMOUS*, results in the pattern of ABC functions. ABC proteins and the co-factor SEP proteins assemble into complexes that specify the different organ identities.

Adapted from Lohmann, J.U. and Weigel, D.: 2002.

A current view of the mechanism of specifying floral organ identity is shown in Figure 6.29.

There is a better understanding of the functions and patterning of the floral homeotic genes than when the ABC model was first proposed. The MADS-box homeotic A-class gene *AP1* has been found to have a dual role: it acts early with other genes to specify general floral meristem identity and only later contributes to A function. It is induced by the meristem identity gene *LFY*, which is expressed throughout the meristem, and *AP1* is actively inhibited in the central regions of floral meristems by *AGAMOUS*. The expression of the A-function gene *APETALA2* (*AP2*) is translationally repressed by a microRNA, keeping the AP2 protein at a low level. *APETALA3* and *PISTILLATA1* are thought to be activated as a result of a meristem identity gene called *UNUSUAL FLORAL ORGANS* (*UFO*), which is expressed in the meristem in a pattern similar to that of B-function genes (see Fig. 6.29). *UFO* encodes a component of ubiquitin ligase, and is thought to exert its effects on flower development by targeting specific proteins for degradation. As we saw in animals in relation to the control of β-catenin degradation (see Chapters 3 and 5), regulated degradation of proteins can be a powerful developmental mechanism. In the center of the floral meristem, the expression of *AGAMOUS* is partly controlled by *WUS*, which as we have seen earlier, is expressed in the organizing center of the vegetative shoot meristem and continues to be expressed in floral meristems.

Another group of genes that help pattern the floral organ primordia are genes that control cell division. The gene *SUPERMAN* is one example, controlling cell proliferation in stamen and carpel primordia and in ovules. Plants with a mutation in this gene have stamens instead of carpels in the fourth whorl. *SUPERMAN* is expressed in the third whorl, and maintains the boundary between the third and fourth whorls.

Despite the enormous variation in the flowers of different species, the mechanisms underlying flower development seem to be very similar. For example, there are striking similarities between the genes controlling flower development of *Arabidopsis* and *Antirrhinum*, despite the quite different final morphology of the snapdragon flower. In developing *Antirrhinum* flowers, the patterns of activity of the corresponding genes fit well with the cell-lineage restriction to whorls seen in *Arabidopsis*.

6.14 The *Antirrhinum* flower is patterned dorso-ventrally as well as radially

Like *Arabidopsis* flowers, those of *Antirrhinum* consist of four whorls, but unlike *Arabidopsis*, they have five sepals, five petals, four stamens, and two united carpels (Fig. 6.30, left). Floral homeotic mutations similar to those in *Arabidopsis* occur in *Antirrhinum*, and floral organ identity is specified in the same way. Several of the *Antirrhinum* homeotic genes have extensive homology with those of *Arabidopsis*, the MADS box in particular being well conserved.

An extra element of patterning is required in the *Antirrhinum* flower, which has a bilateral symmetry imposed on the basic radial pattern common to all flowers. In whorl 2, the upper two petal lobes have a shape quite distinct from the lower three, giving the flower its characteristic snapdragon appearance. In whorl 3, the uppermost stamen is absent, as its development is aborted early on. The *Antirrhinum* flower therefore has a distinct dorso-ventral axis. Another group of homeotic genes, different from those that govern floral organ identity, appear to act in this dorso-ventral patterning. For example, mutations in the gene *CYCLOIDEA*, which is expressed in the dorsal region, abolish dorso-ventral polarity and produce flowers that are more radially symmetrical (Fig. 6.30, right).

Fig. 6.30 Mutations in *CYCLOIDEA* make the *Antirrhinum* flower symmetrical. In the wild-type flower (left) the petal pattern is different along the dorso-ventral axis. In the mutant (right) the flower is symmetrical. All the petals are like the most ventral one in the wild type and are folded back.

Photograph courtesy of E. Coen, from Coen, E.S., et al.: 1991.

6.15 The internal meristem layer can specify floral meristem patterning

Although all three layers of a floral meristem (Fig. 6.31) are involved in organogenesis, the contribution of cells from each layer to a particular structure may be variable. Cells from one layer can become part of another layer without disrupting normal morphology, suggesting that a cell's position in the meristem is the main determinant of its future behavior. Some insight into positional signaling and patterning in the floral meristem can be obtained by making periclinal chimeras (see Section 6.7) from cells that have different genotypes and that give rise to different types of flower. From such chimeras, one can find out whether the cells develop autonomously according to their own genotype, or whether their behavior is controlled by signals from other cells.

As well as being produced by mutation, chimeras can also be generated by grafting between two plants of different genotypes. A new shoot meristem forms at the junction of the graft, and sometimes contains cells from both genotypes. Such chimeras can be made between wild-type tomato plants and tomato plants carrying the mutation *fasciated*, in which the flower has an increased number of floral organs per whorl. This phenotype is also found in chimeras in which only layer L3 contains *fasciated* cells (Fig. 6.32). The increased number of floral organs is associated with an overall increase in the size of the floral meristem, and in the chimeric plants this cannot be achieved unless the *fasciated* cells of layer L3 induce the wild-type L1 cells to divide more frequently than normal. The mechanism of intercellular signaling between L3 and L1 is not yet known. In *Antirrhinum*, the abnormal expression of *FLORICAULA* in only one meristem layer can result in flower development. These results illustrate the importance of signaling between layers in flower development.

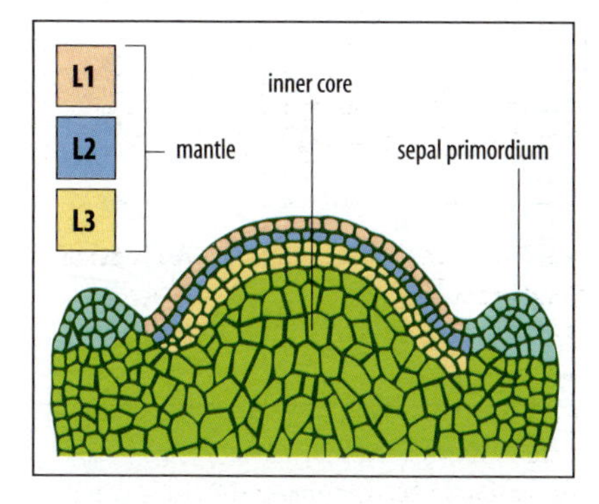

Fig. 6.31 Floral meristem. The meristem is composed of layers L1, L2, and L3. The inner core cells are derived from L3. The sepal primordia are just beginning to develop.

After Drews, G.N., et al.: 1989.

6.16 The transition of a shoot meristem to a floral meristem is under environmental and genetic control

Flowering plants first grow vegetatively, during which time the apical meristem generates leaves. Then, triggered by environmental signals such as increasing

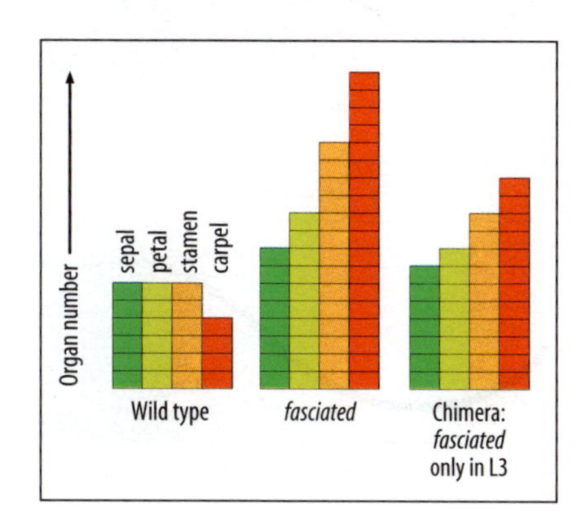

Fig. 6.32 Floral organ number in chimeras of wild-type and *fasciated* tomato plants. In the *fasciated* mutant there are more organs in the flower than in wild-type plants. In chimeras in which only layer L3 of the floral meristem contains *fasciated* mutant cells, the number of organs per flower is still increased, showing that L3 can control cell behavior in the outer layers of the meristem.

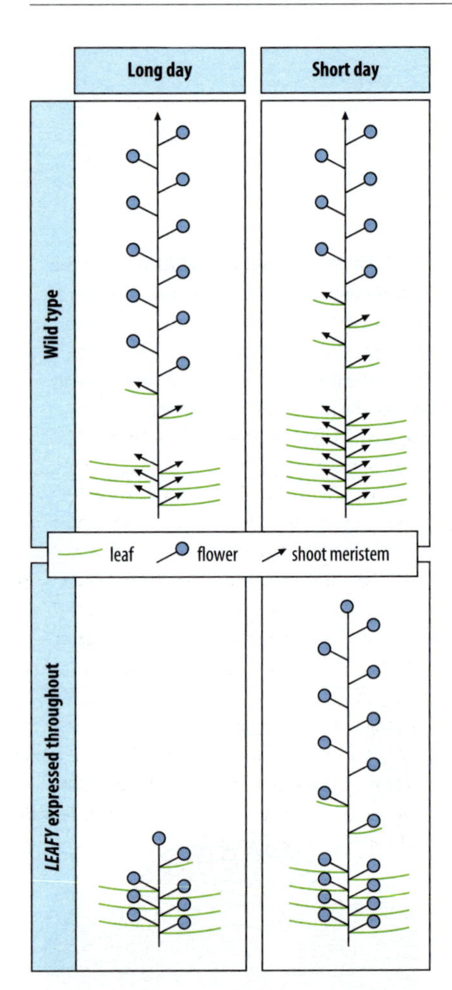

Fig. 6.33 Flowering can be controlled by day length and LEAFY expression. As shown in the top panels, when wild-type *Arabidopsis* is grown under long-day conditions (left), few lateral shoots are formed before the apical shoot meristem begins to form floral meristems. When grown under short-day conditions, flowering is delayed and there are in consequence more lateral shoots. The gene *LEAFY* is normally expressed only in inflorescence and floral meristems, but if it is expressed throughout the plant (bottom panels), all shoot meristems produced are converted to floral meristems in both day lengths.

day length, the plant switches to a reproductive phase and from then on the apical meristem gives rise only to flowers. There are two types of transition from vegetative growth to flowering. In the determinate type, the inflorescence meristem becomes a terminal flower, whereas in the indeterminate type the inflorescence meristem gives rise to a number of floral meristems. *Arabidopsis* is of the indeterminate type (Fig. 6.33). A primary response to floral inductive signals in *Arabidopsis* is the expression of floral meristem identity genes such as *LEAFY* and the dual-function *AP1* (see Section 6.13), which are necessary and sufficient for this transition. *LEAFY* potentially activates *AP1* throughout the meristem while also activating *AGAMOUS* in the center of the flower. *AGAMOUS* then represses the expression of *AP1* in the center, helping to restrict its floral organ identity function to region A (see Fig. 6.27). Mutations in floral meristem identity genes partly transform flowers into shoots. In a *leafy* mutant, which lacks LEAFY function, the flowers are transformed into spirally arranged sepal-like organs along the stem, whereas expression of *LEAFY* throughout a plant is sufficient to confer a floral fate on lateral shoot meristems and they develop as flowers (Fig. 6.33, bottom panels).

In *Arabidopsis*, flowering is promoted by increasing daylength, which predicts the end of winter and the onset of spring and summer (see Fig. 6.33). This behavior is called **photoperiodism**. In some strains, flowering is also accelerated after the plant has been exposed to a long period of cold temperature, a cue that winter has passed. This phenomenon is known as **vernalization**. Grafting experiments have shown that daylength is sensed not by the shoot meristem itself, but by the leaves. When the period of continuous light reaches a certain length, a diffusible flower-inducing signal is produced which is transmitted through the phloem to the shoot meristem. The pathway that triggers flowering involves the plant's **circadian clock**, the internal 24-hour timer that causes many metabolic and physiological processes, including the expression of some genes, to vary throughout the day. One of the genes regulated by the circadian clock is *CONSTANS* (*CO*), which is a key gene in controlling the onset of flowering and provides the link between the plant's day-length-sensing mechanism and production of the flowering signal. The expression of *CO* oscillates on a 24-hour cycle under the control of the circadian clock, and its timing is such that the peak *CO* expression occurs towards the end of the afternoon. This means that in longer days peak expression occurs in the light, whereas in short days, it will already be dark at this time. In the dark, the CO protein is degraded and so the circadian control ensures that CO only accumulates to high enough levels to trigger the flowering pathway when light conditions are favorable (Fig. 6.34).

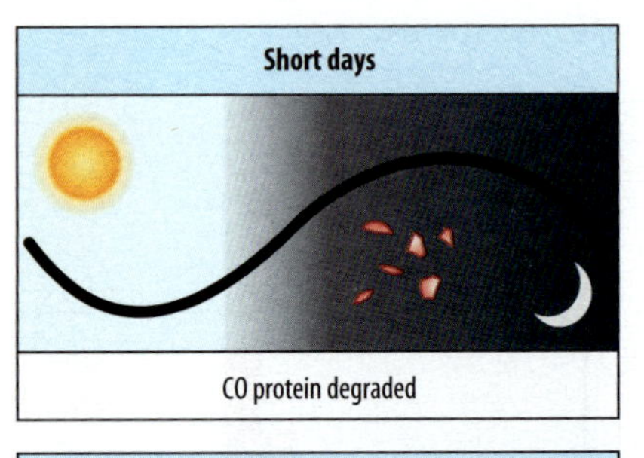

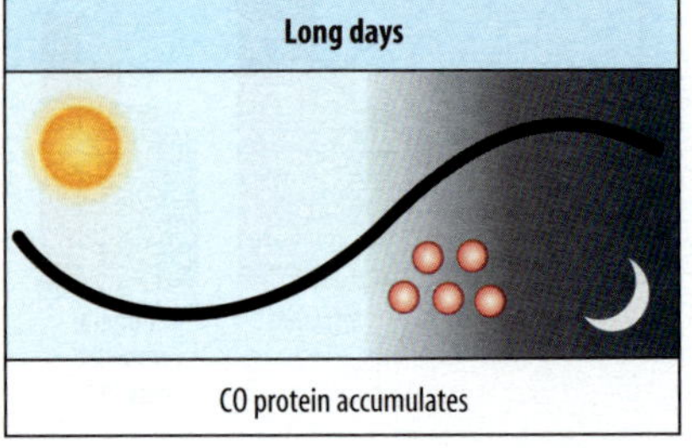

Fig. 6.34 The initiation of flowering is under the dual control of day length and the circadian clock. The transcription factor CONSTANS (CO) is required for production of the flowering signal and is expressed in leaves under the control of the circadian clock. In short days, expression of the *CO* gene peaks in the dark and the protein is rapidly degraded. In long days, peak expression occurs in the light, and the CO protein accumulates.

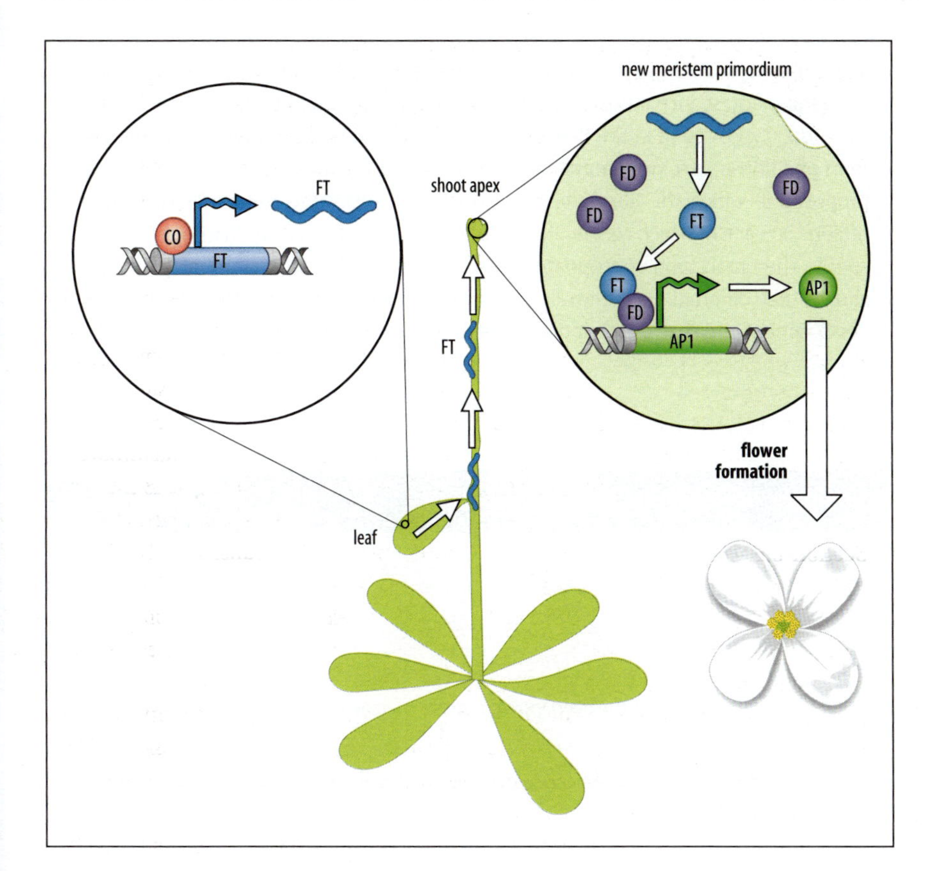

Fig. 6.35 Signals that initiate flowering in *Arabidopsis*. When the day length becomes longer after winter, the transcription factor CO accumulates in the leaf, which in turn activates the transcription of *FT* in leaf phloem cells. The *FT* RNA transcript travels through the phloem to the shoot apex where it is translated into FT protein. This interacts with the transcription factor FD to form a complex, which acts together with the transcription factor LEAFY (whose own expression in upregulated in the shoot apex by FT) to activate key floral meristem identity genes such as *AP1*, which convert the vegetative meristem into one that will produce floral meristems.

CO is a transcription factor that activates a gene known as *FLOWERING LOCUS T* (*FT*), which produces the flowering signal in the form of its RNA transcript. The *FT* RNA is thought to travel from the leaf through the phloem to the shoot apical meristem, where it is translated and acts in a complex with the transcription factor FLOWERING LOCUS D (FD) to turn on the expression of genes such as *AP1* that promote flowering (Fig. 6.35). If FT is activated in a single leaf, this is sufficient to induce flowering. Induction of flowering also requires the downregulation of a set of floral repressor genes. These suppress the transition from a vegetative to the flowering state until the positive signals to flower are received.

Summary

Before flowering, which is triggered by environmental conditions such as day length, the vegetative shoot apical meristem becomes converted into an inflorescence meristem, which either then becomes a flower or produces a series of floral meristems, each of which develops into a single flower. Genes involved in the initiation of flowering and patterning of the flower have been identified in both *Arabidopsis* and *Antirrhinum*. Flowering is induced by day length acting together with the plant's natural circadian rhythms of gene expression to turn on a gene in the leaves that produces a flowering signal that is transported to the shoot meristem. This signal turns on the expression of meristem identity genes that are required for the transformation of the vegetative shoot meristem to an inflorescence meristem and the formation of floral meristems from the inflorescence meristem. Homeotic floral organ identity genes, which specify the organ types found in the flowers have been identified from mutations that transform one flower part into another. On the basis of these mutations, a model has been proposed in which the floral meristem is divided into three concentric overlapping regions, in each of which certain floral identity genes act in a combinatorial manner to specify the organ type appropriate to each whorl. Studies with chimeric plants have shown that different meristem layers communicate with each other during flower development and that transcription factors can move between cells.

Summary: flower development in *Arabidopsis*

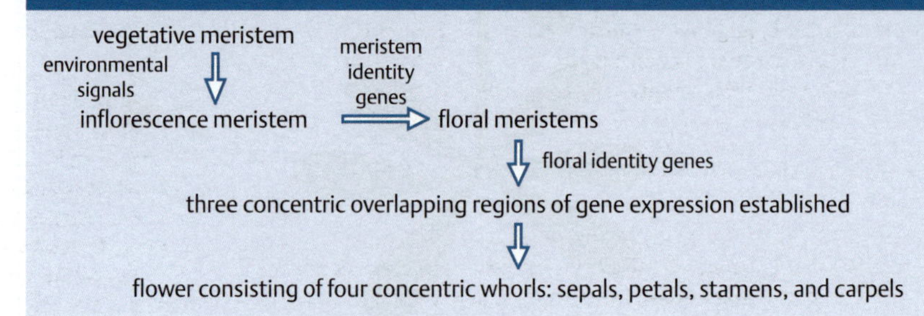

SUMMARY TO CHAPTER 6

A distinctive feature of plant development is the presence of relatively rigid walls and the absence of any cell migration. Another is that a single isolated somatic cell from a plant can regenerate into a complete new plant. Early embryonic development is characterized by asymmetric cell division of the fertilized egg, which specifies the future apical and basal regions. During early development of flowering plants, both asymmetric cell division and cell-cell interactions are involved in patterning the body plan. During this process, the shoot and root meristems are specified and these meristems give rise to all the organs of the plant—stems, leaves, flowers, and roots. The shoot meristem gives rise to leaves in well defined positions, a process involving lateral inhibition. The shoot meristem eventually becomes converted to an inflorescence meristem, which either becomes a floral meristem (in determinate inflorescences) or gives rise to a series of floral meristems, retaining its shoot meristem identity indefinitely (in indeterminate inflorescences). In floral meristems, each of which develops into a flower, homeotic floral organ identity genes act in combination to specify the floral organ types. Increasing day length induces the synthesis of a flowering signal in the leaves that is transported to the shoot meristem where it induces flower formation.

GENERAL FURTHER READING

Meyerowitz, E.M.: ***Arabidopsis*—a useful weed.** *Cell* 1989, **56:** 263–264.

Meyerowitz, E.M.: **Plants compared to animals: the broader comparative view of development.** *Science* 2002, **295:** 1482–1485.

SECTION FURTHER READING

6.2 Plant embryos develop through several distinct stages

Lloyd, C.: **Plant morphogenesis: life on a different plane.** *Curr. Biol.* 1995, **5:** 1085–1087.

Mayer, U., Jürgens, G.: **Pattern formation in plant embryogenesis: a reassessment.** *Semin. Cell Dev. Biol.* 1998, **9:** 187–193.

Meyerowitz, E.M.: **Genetic control of cell division patterns in developing plants.** *Cell* 1997, **88:** 299–308.

Torres-Ruiz, R.A., Jürgens, G.: **Mutations in the *FASS* gene uncouple pattern formation and morphogenesis in *Arabidopsis* development.** *Development* 1994, **120:** 2967–2978.

6.3 Gradients of the signal molecule auxin establish the embryonic apical-basal axis

Friml, J., Vieten, A., Sauer, M., Weijers, D., Schwarz, H., Hamann, T., Offringa, R., Jürgens, G.: **Efflux-dependent auxin gradients establish the apical-basal axis of *Arabidopsis*.** *Nature* 2003, **426:** 147–153.

Jenik, P.D., Barton, M.K.: **Surge and destroy: the role of auxin in plant embryogenesis.** *Development* 2005, **132:** 3577–3585.

Jürgens, G.: **Axis formation in plant embryogenesis: cues and clues.** *Cell* 1995, **81:** 467–470.

Long, J.A., Moan, E.I., Medford, J.I., Barton, M.K.: **A member of the knotted class of homeodomain proteins encoded by the *STM* gene of *Arabidopsis*.** *Nature* 1995, **379:** 66–69.

6.4 Plant somatic cells can give rise to embryos and seedlings

Zimmerman, J.L.: **Somatic embryogenesis: a model for early development in higher plants.** *Plant Cell* 1993, **5:** 1411–1423.

6.5 A meristem contains a small central zone of self-renewing stem cells

Byrne, M.E., Kidner, C.A., Martienssen, R.A.: **Plant stem cells: divergent pathways and common themes in shoots and roots.** *Curr. Opin. Genet. Dev.* 2003, **13:** 551–557.

Großhardt, R., Laux, T.: **Stem cell regulation in the shoot meristem.** *J. Cell Sci.* 2003, **116:**1659–1666.

Ma, H.: **Gene regulation: Better late than never?** *Curr. Biol.* 2000, **10:** R365–R368.

6.6 The size of the stem-cell area in the meristem is kept constant by a feedback loop to the organizing center

Brand, U., Fletcher, J.C., Hobe, M., Meyerowitz, E.M., Simon, R.: **Dependence of stem cell fate in *Arabidopsis* on a feedback loop regulated by *CLV3* activity.** *Science* 2000, **289:** 635–644.

Carles, C.C., Fletcher, J.C.: **Shoot apical meristem maintenance: the art of dynamical balance.** *Trends Plant Sci.* 2003, **8:** 394–401.

Clark, S.E.: **Cell signalling at the shoot meristem.** *Nat. Rev. Mol. Cell Biol.* 2001, **2:** 277–284.

Lenhard, M., Laux, T.: **Stem cell homeostasis in the *Arabidopsis* shoot meristem is regulated by intercellular movement of CLAVATA3 and its sequestration by CLAVATA1.** *Development* 2003, **130:** 3163–3173.

Reddy, G.V., Meyerowitz, E.M.: **Stem-cell homeostasis and growth dynamics can be uncoupled in the *Arabidopsis* shoot apex.** *Science* 2005, **310:** 663–667.

Schoof, H., Lenhard, M., Haecker, A., Mayer, K.F.X., Jürgens, G., Laux, T.: **The stem cell population of *Arabidopsis* shoot meristems is maintained by a regulatory loop between the *CLAVATA* and *WUSCHEL* genes.** *Cell* 2000, **100:** 635–644.

Vernoux, T., Benfey, P.N.: **Signals that regulate stem cell activity during plant development.** *Curr. Opin. Genet. Dev.* 2005, **15:** 388–394.

6.7 Cells from different meristem layers normally give rise to different structures

Castellano, M.M., Sablowski, R.: **Intercellular signalling in the transition from stem cells to organogenesis in meristems.** *Curr. Opin. Plant Biol.* 2005, **8:** 26–31.

Gallagher, K.L., Benfey, P.N.: **Not just another hole in the wall: understanding intercellular protein trafficking.** *Genes Dev.* 2005, **19:** 189–195.

Laux, T., Mayer, K.F.X.: **Cell fate regulation in the shoot meristem.** *Semin. Cell Dev. Biol.* 1998, **9:** 195–200.

Sinha, N.R., Williams, R.E., Hake, S.: **Overexpression of the maize homeobox gene *knotted-1*, causes a switch from determinate to indeterminate cell fates.** *Genes Dev.* 1993, **7:** 787–795.

Turner, I.J., Pumfrey, J.E.: **Cell fate in the shoot apical meristem of *Arabidopsis thaliana*.** *Development* 1992, **115:** 755–764.

Waites, R., Simon, R.: **Signaling cell fate in plant meristems: three clubs on one tousle.** *Cell* 2000, **103:** 835–838.

6.8 A fate map for the embryonic shoot meristem can be deduced using clonal analysis

Irish, V.F.: **Cell lineage in plant development.** *Curr. Opin. Genet. Dev.* 1991, **1:** 169–173.

6.9 Meristem development is dependent on signals from other parts of the plant

Doerner, P.: **Shoot meristems: intercellular signals keep the balance.** *Curr. Biol.* 1999, **9:** R377–R380.

Irish, E.E., Nelson, T.M.: **Development of maize plants from cultured shoot apices.** *Planta* 1988, **175:** 9–12.

Sachs, T.: *Pattern Formation in Plant Tissues.* Cambridge: Cambridge University Press, 1994.

6.10 Gene activity patterns the proximo-distal and adaxial-abaxial axes of leaves developing from the shoot meristem

Bowman, J.L.: **Axial patterning in leaves and other lateral organs.** *Curr. Opin. Genet. Dev.* 2000, **10:** 399–404.

Kidner CA, Martienssen RA.: **Spatially restricted microRNA directs leaf polarity through *ARGONAUTE1*.** *Nature* 2004, **428:** 81–84.

Waites, R., Selvadurai, H.R., Oliver, I.R., Hudson, A.: **The *PHANTASTICA* gene encodes a MYB transcription factor involved in growth and dorsoventrality of lateral organs in *Antirrhinum*.** *Cell* 1998, **93:** 779–789.

6.11 The regular arrangement of leaves on a stem and trichomes on leaves is generated by competition and lateral inhibition

Mitchison, G.J.: **Phyllotaxis and the Fibonacci series.** *Science* 1977, **196:** 270–275.

Reinhardt, D., Pesce, E.R, Stieger, P., Mandel, T., Baltensperger, K., Bennett, M., Traas, J., Friml, J., Kuhlemeier, C.: **Regulation of phyllotaxis by polar auxin transport.** *Nature* 2003, **426:** 255–260.

Reinhardt, D.: **Regulation of phyllotaxis.** *Int. J. Dev. Biol.* 2005, **49:** 539–546.

Scheres, B.: **Non-linear signaling for pattern formation.** *Curr. Opin. Plant Biol.* 2000, **3:** 412–417.

Schiefelbein, J.: **Cell-fate specification in the epidermis: a common patterning mechanism in the root and shoot.** *Curr. Opin. Plant Biol.* 2003, **6:** 74–78.

Smith, L.G., Hake, S.: **The initiation and determination of leaves.** *Plant Cell* 1992, **4:** 1017–1027.

6.12 Root tissues are produced from *Arabidopsis* root apical meristems by a highly stereotyped pattern of cell divisions

Berger, F., Haseloff, J., Schiefelbein, J., Dolan, L.: **Positional information in root epidermis is defined during embryogenesis and acts in domains with strict boundaries.** *Curr. Biol.* 1998, **8:** 421–430.

Costa, S., Dolan, L.: **Development of the root pole and patterning in *Arabidopsis* roots.** *Curr. Opin. Genet. Dev.* 2000, **10:** 405–409.

Dolan, L., Scheres, B.: **Root pattern: shooting in the dark?** *Cell Dev. Biol.* 1998, **9:** 201–206.

Sabatini, S., Beis, D., Wolkenfeldt, H., Murfett, J., Guilfoyle, T., Malamy, J., Benfey, P., Leyser, O., Bechtold, N., Weisbeek, P., Scheres, B.: **An auxin-dependent distal organizer of pattern and polarity in the *Arabidopsis* root.** *Cell* 1999, **99:** 463–472.

Scheres, B., McKhann, H.I., van den Berg, C.: **Roots redefined: anatomical and genetic analysis of root development.** *Plant Physiol.* 1996, **111:** 959–964.

van den Berg, C., Willemsen, V., Hendriks, G., Weisbeek, P., Scheres, B.: **Short-range control of cell differentiation in the *Arabidopsis* root meristem.** *Nature* 1997, **390:** 287–289.

6.13 Homeotic genes control organ identity in the flower

Bowman, J.L., Sakai, H., Jack, T., Weigel, D., Mayer, U., Meyerowitz, E.M.: **SUPERMAN, a regulator of floral homeotic genes in *Arabidopsis*.** *Development* 1992, **114:** 599–615.

Breuil-Broyer, S., Morel, P., de Almeida-Engler, J., Coustham, V., Negrutiu, I., Trehin, C.: **High-resolution boundary analysis during *Arabidopsis thaliana* flower development.** *Plant J.* 2004, **38**:182–192.

Coen, E.S., Meyerowitz, E.M.: **The war of the whorls: genetic interactions controlling flower development.** *Nature* 1991, **353:** 31–37.

Irish, V.F.: **Patterning the flower.** *Dev. Biol.* 1999, **209:** 211–220.

Krizek, B.A., Meyerowitz, E.M.: **The *Arabidopsis* homeotic genes *APETALA3* and *PISTILLATA* are sufficient to provide the B class organ identity function.** *Development* 1996, **122:** 11–22.

Krizek, B.A, Fletcher, J.C.: **Molecular mechanisms of flower development: an armchair guide.** *Nat. Rev. Genet.* 2005, **6:** 688–698.

Lohmann, J.U., Weigel, D.: **Building beauty: the genetic control of floral patterning.** *Dev. Cell* 2002, **2:** 135–142.

Ma, H., dePamphilis, C.: **The ABCs of floral evolution.** *Cell* 2000, **101:** 5–8.

Meyerowitz, E.M., Bowman, J.L., Brockman, L.L., Drews, G.M., Jack, T., Sieburth, L.E., Weigel, D.: **A genetic and molecular model for flower development in *Arabidopsis thaliana*.** *Development Suppl.* 1991, **1:** 157–167.

Meyerowitz, E.M.: **The genetics of flower development.** *Sci. Am.* 1994, **271:** 40–47.

Sakai, H., Medrano, L.J., Meyerowitz, E.M.: **Role of *SUPERMAN* in maintaining *Arabidopsis* floral whorl boundaries.** *Nature* 1994, **378:** 199–203.

Vincent, C.A., Carpenter, R., Coen, E.S.: **Cell lineage patterns and homeotic gene activity during *Antirrhinum* flower development.** *Curr. Biol.* 1995, **5:** 1449–1458.

Wagner, D., Sablowski, R.W.M., Meyerowitz, E.M.: **Transcriptional activation of *APETALA 1*.** *Science* 1999, **285:** 582–584.

6.14 The *Antirrhinum* flower is patterned dorso-ventrally as well as radially

Coen, E.S.: **Floral symmetry.** *EMBO J.* 1996, **15:** 6777–6788.

Luo, D., Carpenter, R., Vincent, C., Copsey, L., Coen, E.: **Origin of floral asymmetry in *Antirrhinum*.** *Nature* 1996, **383:** 794–799.

6.15 The internal meristem layer can specify floral meristem patterning

Szymkowiak, E.J., Sussex, I.M.: **The internal meristem layer (L3) determines floral meristem size and carpel number in tomato periclinal chimeras.** *Plant Cell* 1992, **4:** 1089–1100.

6.16 The transition of a shoot meristem to a floral meristem is under environmental and genetic control

An, H., Roussot, C., Suarez-Lopez, P., Corbesier, L., Vincent, C., Pineiro, M., Hepworth, S., Mouradov, A., Justin, S., Turnbull, C., Coupland, G.: **CONSTANS acts in the phloem to regulate a systemic signal that induces photoperiodic flowering of *Arabidopsis*.** *Development* 2004, **131:** 3615–3626.

Becroft, P.W.: **Intercellular induction of homeotic gene expression in flower development.** *Trends Genet.* 1995, **11:** 253–255.

Blázquez, M.A.: **The right time and place for making flowers.** *Science* 2005, **309:** 1024–1025.

Hake, S.: **Transcription factors on the move.** *Trends Genet.* 2001, **17:** 2–3.

Huang, T., Bohlenius, H., Eriksson, S., Parcy, F., Nilsson, O.: **The mRNA of the *Arabidopsis* gene *FT* moves from leaf to shoot apex and induces flowering.** *Science* 2005, **309:** 1694–1696.

Lohmann, J.U., Hong, R.L., Hobe, M., Busch, M.A., Parcy, F., Simon, R., Weigel, D.: **A molecular link between stem cell regulation and floral patterning in *Arabidopsis*.** *Cell* 2001, **105:** 793–803.

Putterill, J., Laurie, R, Macknight, R.: **It's time to flower: the genetic control of flowering time.** *BioEssays* 2004, **26:** 363–373.

Valverde, F., Mouradov, A., Soppe, W., Ravenscroft, D., Samach, A,. Coupland, G.: **Photoreceptor regulation of CONSTANS protein in photoperiodic flowering.** *Science* 2004, **303:** 1003–1006.

Weigel, D., Nilsson, O.: **A developmental switch sufficient for flower initiation in diverse plants.** *Nature* 1995, **327:** 495–500.

Morphogenesis: change in form in the early embryo

7

- Cell adhesion
- Cleavage and formation of the blastula
- Gastrulation movements
- Neural-tube formation
- Cell migration
- Directed dilation

Change in form in the embryo is brought about by cellular forces that are generated in a variety of ways, including cell division, change in cell shape, rearrangement of cells within tissues, and the migration of individual cells from one part of the embryo to another. In animal embryos, cellular forces responsible for change in form are most frequently generated by contraction of the cell's cytoskeleton, but change in form can also be generated by hydrostatic pressure and by cell division. Resisting the forces are the cells themselves and the adhesive interactions that hold them together in tissues. Gastrulation involves major changes in three-dimensional form and in vertebrates involves the elongation of the embryo. Other examples of morphogenesis discussed here include notochord and neural tube formation. Individual cells migrate in a variety of systems to specific locations, and the mechanism by which they are guided is a key issue.

So far, we have discussed early development mainly from the viewpoint of developmental patterning and the assignment of cell fate. In this chapter, we look at embryonic development from a different perspective. Generation of form in early animal development involves rearrangement of the cell layers and movement of cells from one location to another. All animal embryos, for example, undergo a dramatic change in shape during their early development. This occurs principally during gastrulation, the process that transforms a one-dimensional sheet of cells into the complex three-dimensional animal body. Gastrulation involves extensive rearrangement of cell layers and the directed movement of cells from one location to another.

If pattern formation is likened to painting, morphogenesis is more akin to modeling a formless lump of clay into a recognizable shape. Change in form is largely a problem in cell mechanics: it requires an understanding of the forces involved in bringing about changes in cell shape and cell migration, the cellular machinery that carries them out, and how this is controlled during development.

Two key cellular properties involved in changes in animal embryonic form are cell adhesiveness and cell motility. Animal cells stick to one another, and to the extracellular matrix, through interactions involving cell-surface proteins.

Changes in these proteins can therefore determine both the strength of cell adhesion and its specificity. The second key property, cell motility, encompasses both cell migration and the ability of cells to change their shape within a tissue, for example when a sheet of cells folds—a very common feature in animal embryonic development. The ability of cells to move and change shape is determined by rearrangements in their internal cytoskeletal structures, particularly those that cause contraction. An additional force that operates during morphogenesis, particularly in plants, is hydrostatic pressure, which is generated by osmosis and fluid accumulation. In plants there is no cell movement during growth, and changes in form are generated by oriented cell division and cell expansion, as we saw in Chapter 6.

At the molecular level, changes in embryonic form are the ultimate consequence of the precise spatio-temporal expression of proteins that control cell adhesion, cell motility, oriented cell division, and the generation of hydrostatic pressure. An attractive hypothesis is that pattern-determining genes such as the Hox genes activate other genes that control the expression of such proteins, and there is some evidence for this. It is likely that changes in embryonic form are brought about by an earlier patterning process that determines which cells will express those gene products required to generate and harness the appropriate forces.

The changes in form considered in this chapter are mainly those involved in the development of the animal body plan. First, we look at the mechanisms by which cleavage of the zygote gives rise to the simple shape of the early embryo, of which the spherical blastulas of the mouse, sea urchin, and amphibian are good examples. We then consider the movements that occur during gastrulation and during neurulation—the formation of the neural tube in vertebrates—which also involves folding of cell sheets and rearrangement of cell layers. In vertebrates, migration of cells from the neural crest after neurulation generates a variety of structures in the trunk and head, and we consider how these cells migrate to their correct sites. We also consider the role of chemotaxis and signal propagation in the aggregation of unicellular amebae into the fruiting body in the cellular slime mold *Dictyostelium discoideum*. Finally we look at directed dilation, in which hydrostatic pressure is the force driving changes in shape. Other morphogenetic mechanisms such as cell growth, cell proliferation, and cell death will be considered in relation to the development of particular organs (see Chapter 9) and of the nervous system (see Chapter 10).

We begin by considering how cells adhere to each other, and how differences in adhesiveness and specificity of adhesion are involved in maintaining boundaries between tissues.

Cell adhesion

The late embryo and the adult are composed of a variety of differentiated cell types, which are grouped together in tissues such as skin and cartilage. The integrity of tissues is maintained by adhesive interactions both between cells and between cells and the extracellular matrix; differences in cell adhesiveness also play a part in maintaining the boundaries between different tissues and structures. Cells are stuck together by **cell-adhesion molecules**, which are proteins carried on the cell surface that can bind to other molecules on cell surfaces or in the extracellular matrix (Box 7A, opposite). The particular adhesion molecules expressed by

Box 7A Cell-adhesion molecules and cell junctions

Three classes of adhesion molecules are important in development (see figure). The **cadherins** are transmembrane proteins that, in the presence of calcium ions (Ca^{2+}), adhere to cadherins on the surface of another cell. Calcium-independent cell–cell adhesion involves a different structural class of proteins—members of the large **immunoglobulin superfamily**. The neural cell-adhesion molecule (N-CAM), which was first isolated from neural tissue, is a typical member of this family. Some immunoglobulin superfamily members, such as N-CAM, bind to similar molecules on other cells; others bind to a different class of adhesion molecule, the **integrins**. These are the third class of adhesion molecule involved in development. Integrins also act as receptors for molecules of the extracellular matrix, which they can use to mediate adhesion between a cell and its substratum.

About 30 different types of cadherins have been identified in vertebrates. Cadherins bind to each other through one or more binding sites located within the extracellular amino-terminal 100 amino acids. In general, a cadherin binds only to another cadherin of the same type, but they can also bind to some other molecules. A typical cadherin is E-cadherin (originally known as uvomorulin in mammals), which is involved in the generation of a polarized epithelial sheet in many different situations. Cadherins are the adhesive components in adherens junctions, adhesive cell junctions that are present in many tissues, and in desmosomes, junctions present mainly in epithelia.

As cells approach and touch one another, the cadherins cluster at the site of contact. They interact with the intracellular cytoskeleton through the connection of their cytoplasmic tails with **catenins** and other proteins, and thus can be involved in transmitting signals to the cytoskeleton. This adhesive and signaling role of α-, β-, and γ-catenins is separate from the role of β-catenin as a gene-regulatory protein (see, for example, Section 3.6). The interaction with the cytoskeleton is required for normal strong cell–cell adhesion. In adherens junctions the connection is to actin filaments, whereas in desmosomes it is to intermediate filaments. Nectin, a member of the immunoglobulin superfamily of adhesion molecules, also clusters to adherens junctions in mammalian tissues and connects to the actin cytoskeleton.

Adhesion to the extracellular matrix, which contains proteins such as collagen, fibronectin, laminin, and tenascin, as well as proteoglycans, is mediated by integrins, which bind to these matrix molecules. Each integrin is made up of two different subunits, an α subunit and a β subunit. Twenty-four different integrins are known so far in vertebrates, made up from eight β subunits and 18 α subunits. Many extracellular matrix molecules are recognized by more than one integrin.

Integrins not only bind to other molecules through their extracellular face, but they also associate with the actin filaments or intermediate filaments of the cell's cytoskeleton through complexes of proteins that are in contact with their cytoplasmic region. This association enables integrins to transmit information about the extracellular environment, such as extracellular matrix composition or the type of intercellular contact. Integrins can thus mediate signals from the matrix that affect cell shape, motility, metabolism, and cell differentiation.

Integrins also mediate cell–cell adhesion, binding either to cell adhesion molecules of the immunoglobulin superfamily or via a shared ligand to integrins on another cell surface.

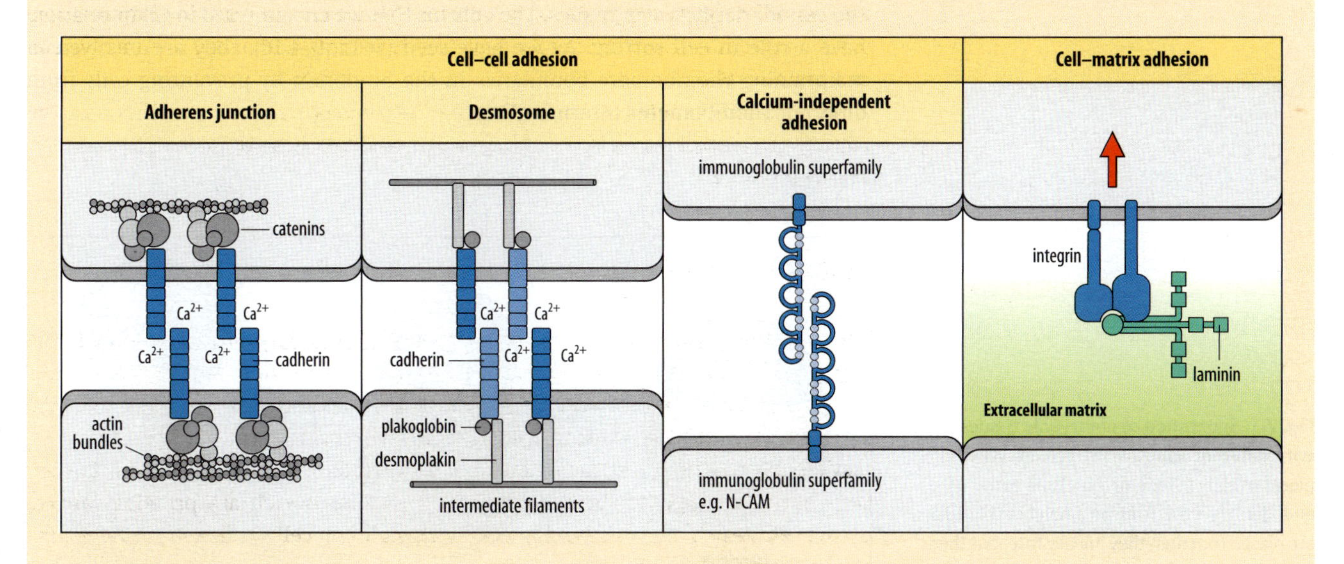

a cell determine which cells it can adhere to, and changes in expression of adhesion molecules are involved in many developmental phenomena, such as neurulation in vertebrates. In tissues adjacent cells are joined together by specialized structures called **adhesive cell junctions** that incorporate cell-adhesion molecules. The junctions that are mainly discussed in this chapter are the **adherens junctions**, in which the adhesion molecule is cadherin and the intracellular linkage is to the actin cytoskeleton (see Box 7A, p. 259).

7.1 Sorting out of dissociated cells demonstrates differences in cell adhesiveness in different tissues

Differences in adhesiveness are illustrated by an experiment in which different tissues are confronted with one another in an artificial setting. Two pieces of early endoderm from an amphibian blastula will fuse to form a smooth sphere, but when a piece of early endoderm and a piece of early ectoderm are combined they initially fuse, but the endodermal and ectodermal cells eventually separate, until only a narrow bridge connects the two types of tissue (Fig. 7.1).

Cellular affinities are further illustrated by disaggregating cells of the presumptive epidermis and presumptive neural plate of an amphibian neurula by treatment with an alkaline solution, then mixing them together and allowing them to reaggregate (Fig. 7.2). The cells of the mixed cell mass exchange neighbors and move in such a way that the epidermal cells are eventually found on the outer face of the aggregate, surrounding a mass of neural cells in the interior. The same types of cell are now in contact with each other. When ectodermal and mesodermal cells are mixed, they similarly sort themselves out to form an aggregate with ectoderm on the outside and mesoderm on the inside. The sorting out of cells of different types is the result of cell movement and differential adhesiveness. Initially, cells move about randomly in the aggregate, exchanging weaker for stronger adhesions. In their final distribution, intercellular binding strengths in the system as a whole are maximized. In general, if the adhesion between unlike cells is weaker than the average of the adhesions between like cells, the cells will segregate type-specifically, with the more cohesive tissue tending to be enveloped by the less cohesive one. These experiments show how differential cell adhesion can stabilize the boundaries between tissues. The ephrins that we encountered in Chapter 4 also have a role in cell sorting. As we have seen (Section 4.10), they are involved in maintaining rhombomere boundaries in the hindbrain by preventing cells from different rhombomeres intermingling.

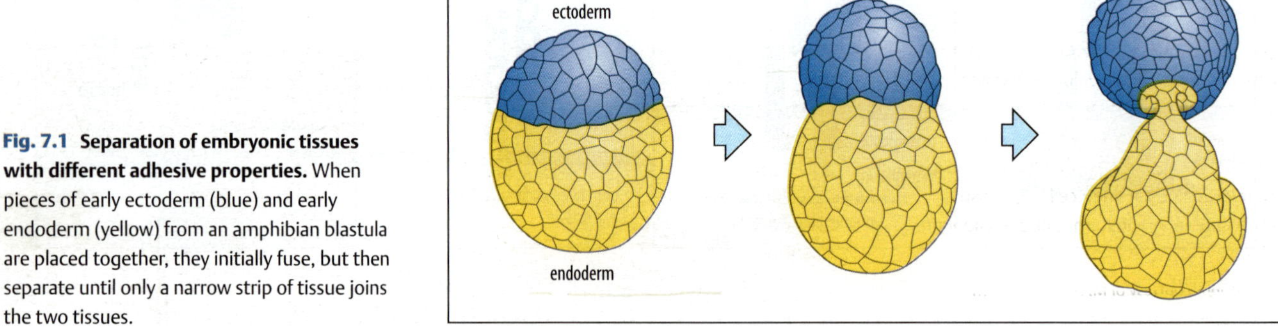

Fig. 7.1 Separation of embryonic tissues with different adhesive properties. When pieces of early ectoderm (blue) and early endoderm (yellow) from an amphibian blastula are placed together, they initially fuse, but then separate until only a narrow strip of tissue joins the two tissues.

7.2 Cadherins can provide adhesive specificity

Differential adhesiveness between cells is the result of the presence of differences in the kinds and numbers of adhesion molecules on cell surfaces. To study the role of these molecules in cells *in vitro*, genes encoding them are introduced and expressed in cells that do not normally produce them, a procedure known as **transfection**. Evidence that the cadherin class of adhesion molecules can provide adhesive specificity comes from studies in which cells with different cadherins on their surface are mixed together.

Cultured fibroblast cells of the L-cell line do not normally adhere strongly to each other, nor do they express cadherins on their surface. L cells transfected with E-cadherin DNA express that cadherin on their surface and adhere to one another, forming a structure resembling a compact epithelium. The adherence is both calcium dependent, indicating that it is due to the cadherin, and specific, as the transfected cells do not adhere to untransfected L cells, which lack surface cadherins. When groups of cells are transfected with different types of cadherin and mixed together in suspension, only those cells expressing the same cadherin adhere strongly to each other: cells expressing E-cadherin adhere strongly to other cells expressing E-cadherin, but only weakly to cells expressing P- or N-cadherin. The amount of cadherin on the cell surface can also have an effect on adhesion. If cells expressing different amounts of the same cadherin are mixed together, those cells expressing more cadherin on their surface form an inner ball, surrounded by the cells expressing less cadherin on their surface (Fig. 7.3). Thus, quantitative differences in cell-adhesion molecules could maintain differential cell adhesion.

Cadherin molecules bind to each other through their extracellular domains, but adhesion is not exclusively controlled by these domains. The cytoplasmic domain associates with the actin filaments of the cytoskeleton by means of a protein complex containing catenins (see Box 7A, p. 259), and failure to make this association results in weak adhesion. In early *Xenopus* blastulas, for example, E-cadherin is expressed in the ectoderm just before gastrulation and N-cadherin appears in the prospective neural plate. If an E-cadherin lacking an extracellular domain is produced in the blastulas from injected mutated mRNA, the mutant cadherin will compete with the embryo's own intact cadherin molecules for the cytoskeletal association sites. The defective cadherin cannot influence cell–cell adhesion as it has no extracellular domain, but it blocks the intact cadherin's access to the cytoskeleton. The result is disruption of the ectoderm during gastrulation, indicating that a cadherin molecule must bind both to its partner on the opposing cell and to the cytoskeleton of its own cell to create a stable adhesion. The initial binding of the extracellular cadherin domains transmits a signal to the cytoskeleton, which then stabilizes the interaction.

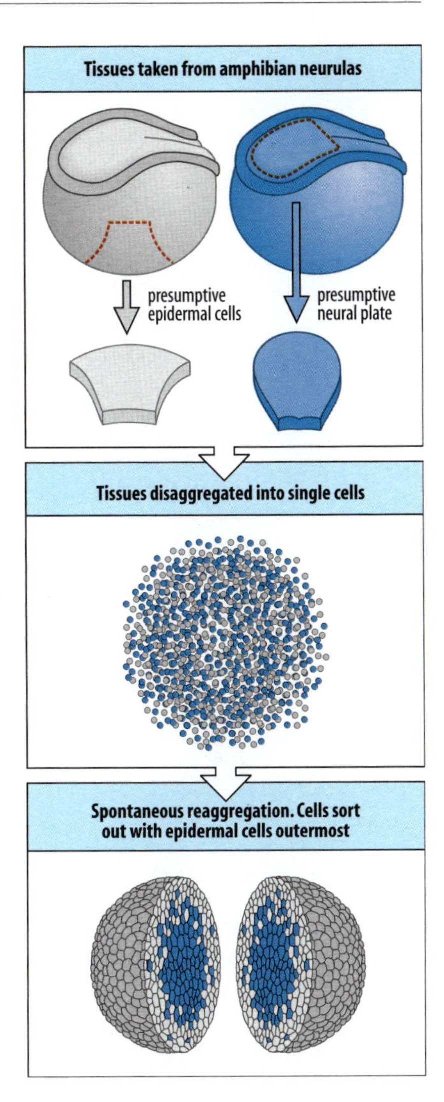

Fig. 7.2 Sorting out of different cell types. Ectoderm from the presumptive neural plate (blue) and from the presumptive epidermis (gray) of early amphibian neurulas are disaggregated into single cells through treatment with an alkaline solution. The cells, when mixed together, sort out with the epidermal cells on the outside.

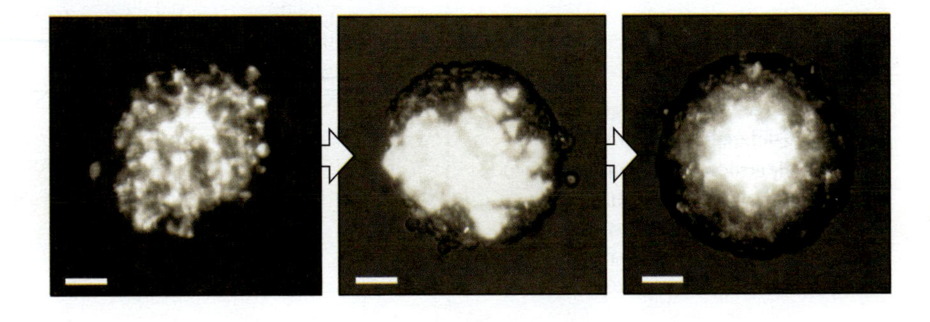

Fig. 7.3 Sorting out of cells with quantitative differences in cell adhesion molecules. Two cell lines with different amounts of P-cadherin on their surface sort out, with the cells containing the most P-cadherin on the inside. The fluorescently stained cells are those expressing the greater amount of P-cadherin. Scale bars = 0.1 mm.

Photographs courtesy of M. Steinberg, from Steinberg, M.S., et al.: 1994.

Summary

The adhesion of cells to each other and to the extracellular matrix maintains the integrity of tissues and the boundaries between them. The associations of cells with each other are determined by the cell-adhesion molecules they express on their surface: cells bearing different adhesion molecules, or different quantities of the same molecule, sort out into separate tissues. Cell–cell adhesion is due mainly to two classes of surface proteins: the cadherins, which bind in a calcium-dependent manner to identical cadherins on another cell surface, and members of the immunoglobulin superfamily, some of which bind to similar molecules on other cells while others bind to different molecules, such as integrins. The binding of this second class of surface proteins is calcium independent. Adhesion to the extracellular matrix is mediated by a third class of adhesion molecule—the integrins. Stable cell adhesion by cadherins involves the external binding domains and interactions with the cytoskeleton, with the cytoplasmic domain of the cadherin molecules binding to a protein complex that contains catenins.

Cleavage and formation of the blastula

In animal eggs, the first step in embryonic development is the division of the fertilized egg by cleavage into a number of smaller cells (blastomeres), leading in many animals to the formation of a hollow sphere of cells—the blastula (see Chapter 3). Cleavage involves short cell cycles, in which cell division and mitosis succeed each other repeatedly without any intervening periods of cell growth. During cleavage therefore, the mass of the embryo does not increase. Early cleavage patterns can vary widely between different groups of animals (Fig. 7.4). The simplest pattern of division is radial, in which successive symmetric cleavages divide the embryo into

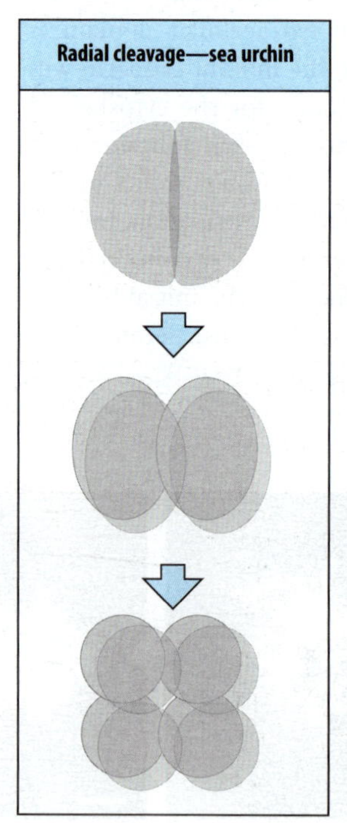

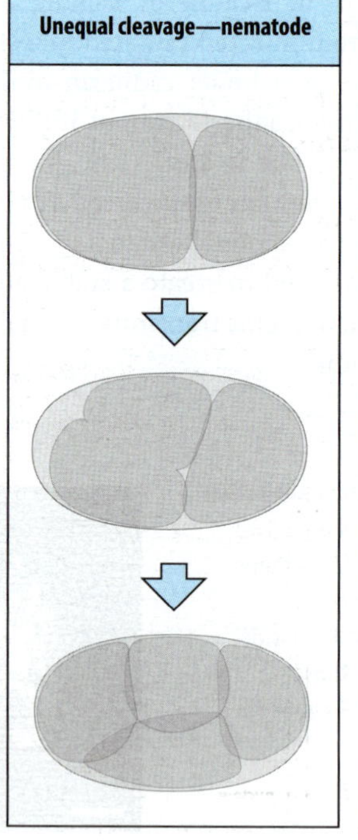

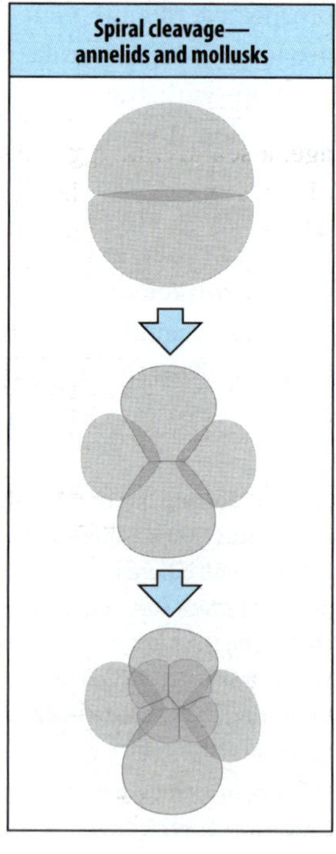

Fig. 7.4 Different patterns of early cleavage are found in different animal groups. Radial cleavages are equal and symmetric (sea urchin). Unequal cleavage (nematode) results in one daughter cell being larger than the other. In spiral cleavage, the mitotic apparatus is oriented at an oblique angle to the long axis of the cell.

equal-sized cells. This pattern is seen in the first three cleavages in the sea urchin. Later cleavages, such as those that form the micromeres, are unequal, one daughter cell being smaller than the other. The first cleavage of the nematode egg produces two cells of unequal size. The spirally cleaving eggs of mollusks and annelids illustrate yet another pattern of cleavage, in which successive divisions of each blastomere, in different planes, produce a spiral arrangement of cells; many of these divisions are unequal.

The amount of yolk in the egg can influence the pattern of cleavage. In yolky eggs undergoing symmetric cleavage, a cleavage furrow develops in the least yolky region and gradually spreads across the egg, but its progress is slowed or even halted by the presence of the yolk. Cleavage may thus be incomplete for some time. This effect is most pronounced in the heavily yolked eggs of birds and zebrafish, where complete divisions are restricted to a region at one end of the egg, and the embryo is formed as a cap of cells sitting on top of the yolk (see Fig. 3.12). Even in moderately yolky eggs, such as those of amphibians, the presence of the yolk can influence cleavage patterns. In frogs, for example, the later cleavages are unequal and asynchronous, resulting in an animal half composed of a mass of small cells and a yolky vegetal region composed of fewer and larger cells.

Two key questions arise relating to early cleavage: how are the positions of the cleavage planes determined; and how can cleavage lead to a hollow blastula (or its equivalent), which has a clear inside–outside polarity?

7.3 The asters of the mitotic apparatus determine the plane of cleavage at cell division

The orientation of the plane of cleavage at cell division is important in a variety of situations. We have already seen its role in relation to unequal cleavages in early invertebrate development, where unequal divisions produce cells with differently specified fates. This is because of the asymmetric distribution of cytoplasmic determinants (see Chapter 5). The plane of cleavage can also be of great importance in later morphogenesis and growth. It determines, for example, whether an epithelial sheet remains as a single cell layer or becomes multilayered.

Experiments show that the plane of cleavage in animal cells is not specified by the mitotic spindle itself, but by the asters at each pole. To disrupt its normal cleavage, a sea urchin egg is deformed by a glass bead. This displaces the mitotic spindle and interrupts the first cleavage furrow, creating a horseshoe-shaped cell in which the two new nuclei are segregated into the separate arms of the horseshoe (Fig. 7.5). At the next division, two cleavage furrows bisect the spindles formed in

Fig. 7.5 The plane of cleavage in animal cells is determined by the asters and not the mitotic spindle. If the mitotic apparatus of a fertilized sea urchin egg is displaced at the first cleavage by a glass bead, the cleavage furrow forms only on the side of the egg to which the mitotic apparatus has been moved. At the next cleavage, furrows bisect each mitotic apparatus, and a furrow also forms between the two adjacent asters, even though there is no spindle between them.

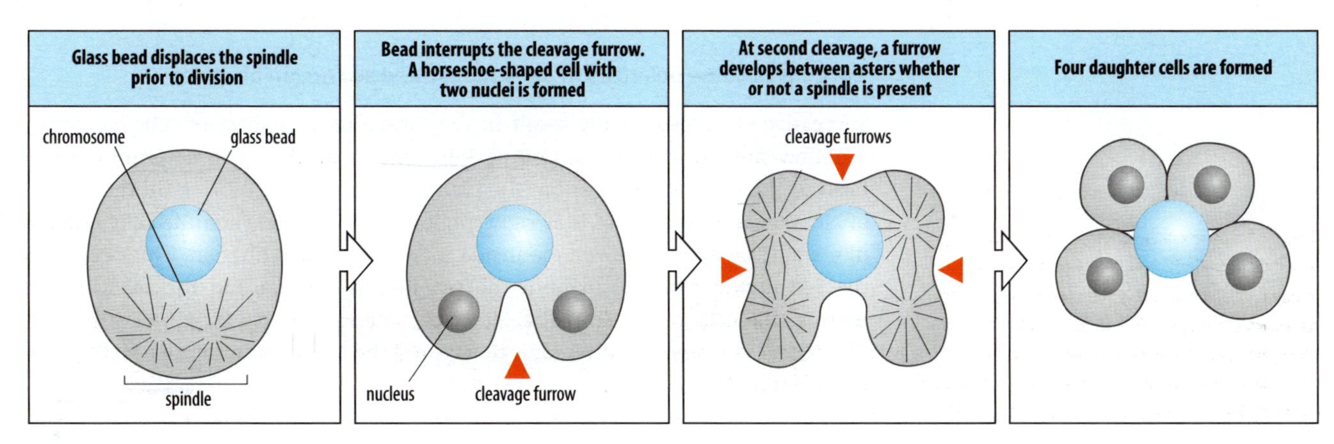

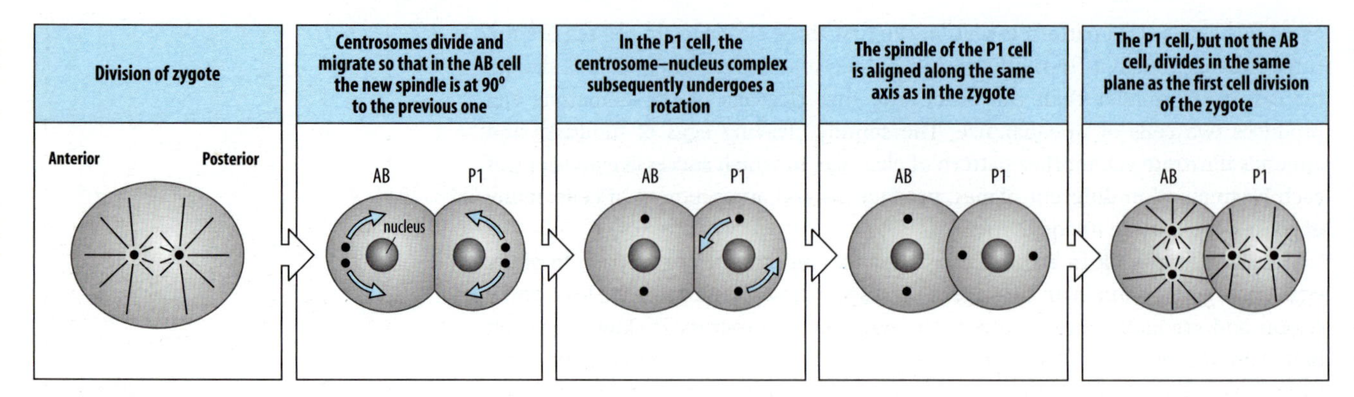

Fig. 7.6 Different cells have different planes of cleavage because of the behavior of their centrosomes. In the nematode, the first cleavage of the zygote divides the cells into an anterior AB cell and a posterior P1 cell. At the next division, the centrosomes of the two cells move in different directions. In the AB cell, the duplicated centrosomes move so that the next cleavage is at right angles to the first cleavage. In the P2 cell, the nucleus and duplicated centrosomes rotate so that the cleavage of P1 is in the same plane as the first cleavage.

After Strome, S.: 1993.

the arms of the horseshoe, but a third furrow also forms between the two adjacent asters not connected by a spindle.

The asters in a dividing cell are composed of microtubules radiating out from a centrosome, which acts as an organizing center for microtubule growth. In most animal cells, the centrosome contains a pair of centrioles, which consist of an array of microtubules. Before mitosis, the centrosome becomes duplicated and the daughter centrosomes move to opposite sides of the nucleus and form asters. In the cells of the early embryo, they usually take up positions that cause the plane of cleavage of the cell to be at right angles to the plane of the previous cleavage. An example of this is cleavage of the AB cell, one of the pair of cells formed by cleavage of the nematode zygote (Fig. 7.6). By contrast, the other member of the pair, the P1 cell, exemplifies the fact that there are many cases where successive cleavages are not at right angles to each other. This is a result of local cytoplasmic factors or cytoskeletal components of the cell causing some centrosome pairs to take up a different orientation with respect to the previous plane of cleavage.

Planes of cleavage can play a very important part in determining the form of the embryo, particularly in early animal and plant development. Higher plant cells lack obvious centrosomes and their spindles lack asters, and instead of a contractile mechanism pinching the cell in two, a new cell wall forms in the plane of division. The plane in which the new cell wall forms is not determined, it seems, by the orientation of the mitotic spindle, but is instead defined before mitosis begins by the appearance of a circumferential band of microtubules and actin filaments.

7.4 Cells become polarized in early mouse and sea urchin blastulas

Successive cleavages usually result in the formation of a sheet of cells enclosing a hollow fluid-filled blastula. Cell packing can be important in determining the form of the early embryo, as can the planes of cleavage (Fig. 7.7). If all the planes of cleavage are radial, that is at right angles to the surface, then the cells remain in a single layer and the volume inside increases at each division. This occurs in the sea urchin blastula. In both the sea urchin blastula and the mammalian morula, the cells of the epithelial sheet become radially polarized, with distinct differences between the outer (apical) face and the inner (basal) surface facing into the blastocoel.

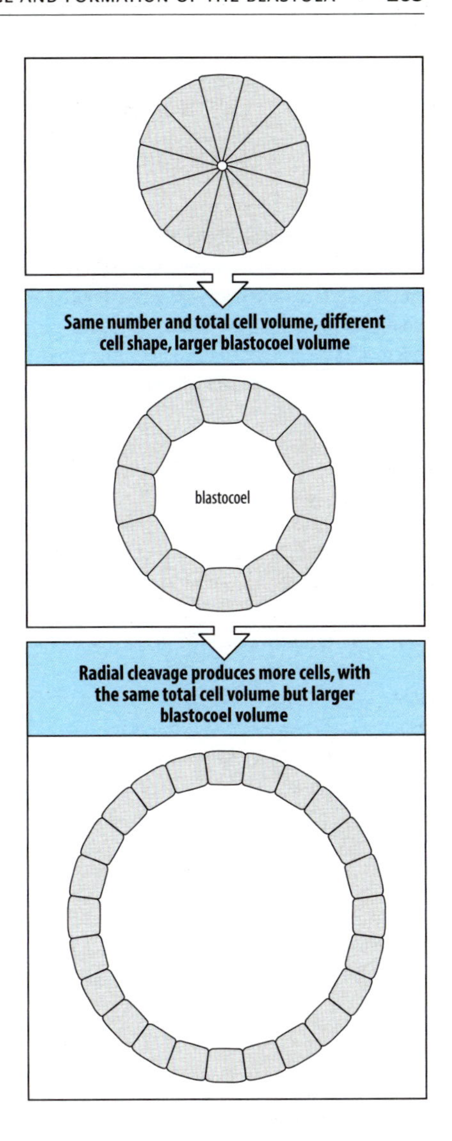

Fig. 7.7 Cell packing can determine the volume of a blastula. Top panel: when adjacent cells in a spherical sheet, such as those of a blastula, make contact with each other over large areas of cell surface, the overall volume of the cell sheet is relatively small as there is little space at its center. Middle panel: with a decrease in the area of cell contact, the size of the internal space (the blastocoel), and thus the overall volume of the blastula, is greatly increased without any corresponding increase in cell numbers or total cell volume. Bottom panel: if the number of cells is increased by radial cleavage, and the packing remains the same, the blastocoel volume will increase further, again without any increase in total cell volume.

In the sea urchin, cleavage eventually produces a hollow spherical blastula composed of a polarized epithelium one cell thick (Fig. 7.8). All planes of cleavage are radial to the surface, thus maintaining a single layer of cells. The surface of the egg, which is covered in microvilli, becomes the outer surface of the blastula and develops an external layer of extracellular matrix, the hyaline layer, to which the cells are attached. Specialized junctions that include desmosomes (see Box 7A, p. 259) develop between adjacent cells and the cell contents become polarized, with the Golgi apparatus oriented toward the apical (outer) surface. On the inner surface of the cell layer, a basal lamina (an organized layer of extracellular matrix) is laid down.

The volume of the hollow interior of the blastula, the blastocoel, increases with each cleavage. While this may partly be due to fluid entering the blastocoel, it is also a result of the cells becoming smaller at each division while the total cell mass remains the same. At each cleavage the number of cells increases, the cells become smaller, and the cell layer thinner. Thus, both the surface area of the blastula and the size of the blastocoel increase (see Fig. 7.7).

In the mouse embryo, the earliest sign of structural differentiation takes place at the eight-cell stage when the embryo undergoes compaction (Fig. 7.9). The blastomeres flatten against each other, thus maximizing cell–cell contacts, and the microvilli, which hitherto were uniformly distributed on cell surfaces, become confined to the apical surface of the cells (Fig. 7.10). Subsequently, some cleavages in the outer cell layer occur tangentially (parallel to the surface), each producing one polarized and one nonpolarized daughter cell. Radial cleavages give two polarized cells. The nonpolarized cells become the inner cell mass, which will give rise to the embryo proper, whereas the outer cells become the trophectoderm and form extra-embryonic structures.

The changes in intercellular contacts that bring about compaction are probably caused by changes in the association between the cell-adhesion molecule E-cadherin (uvomorulin) and the underlying cell cortex. At the two-cell and four-cell stages, E-cadherin is uniformly distributed over blastomere surfaces and contact between cells is not extensive. Only at the eight-cell stage does it become restricted to regions of intercellular contact, where it presumably now acts for the first time as an adhesion molecule. This change in the adhesive properties of E-cadherin may involve an association between the E-cadherin and cytoskeletal elements in the cell cortex (see Box 7A, p. 259). Activation of E-cadherin may involve transduction of

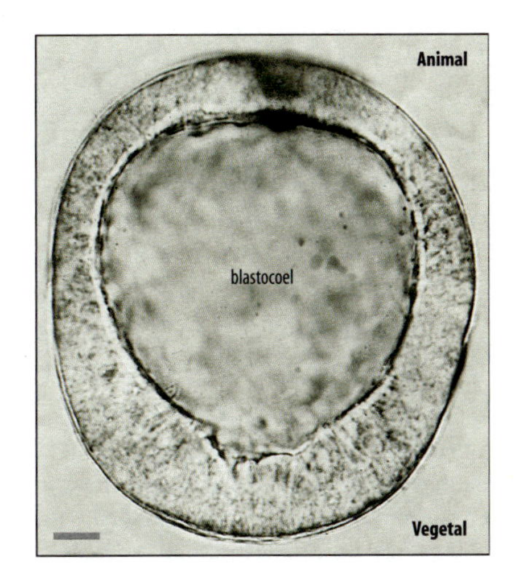

Fig. 7.8 Blastula of a sea urchin embryo. A single layer of cells surrounds the hollow blastocoel. Scale bar = 10 μm.

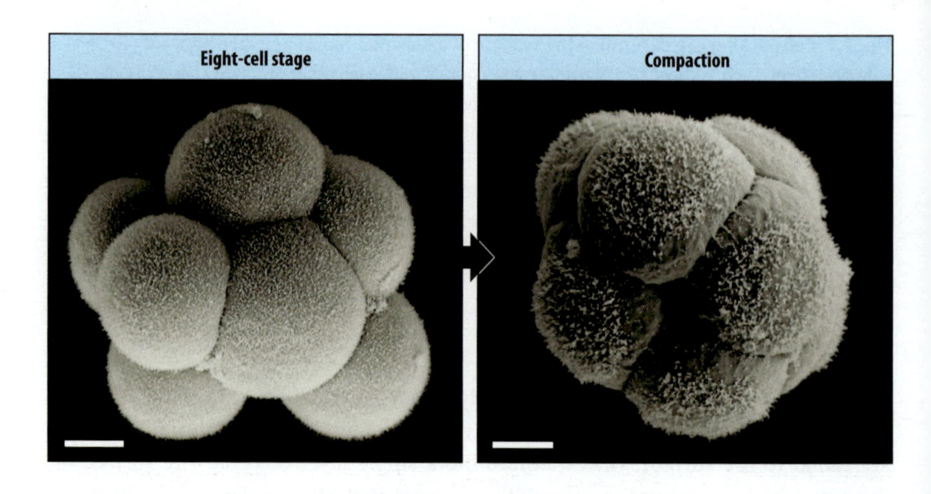

Fig. 7.9 Compaction of the mouse embryo. At the eight-cell stage the cells have relatively smooth surfaces and microvilli are distributed uniformly over the surface. At compaction, microvilli are confined to the outer surface and cells increase their contact with one another. Scale bar = 10 μm.

Photographs courtesy of T. Bloom, from Bloom, T.L.: 1989.

a signal and the action of protein kinase C. If this kinase is activated before the eight-cell stage, compaction occurs prematurely.

A radical remodeling of the cell cortex is associated with compaction and the localization of E-cadherin and the microvilli. The cytoskeletal proteins actin, spectrin, and myosin are cleared from the region of intercellular contact and become concentrated in a band around the apical region of the cell. It has been proposed that both the cell flattening and the redistribution of the cortical elements may be brought about by the contraction of actin filaments, which draws the cortical elements to the apical pole.

7.5 Ion transport is involved in fluid accumulation in the frog blastocoel

Accumulation of fluid in the interior of the blastula exerts an outward pressure on the blastocoel wall, and this hydrostatic pressure is one of the forces involved

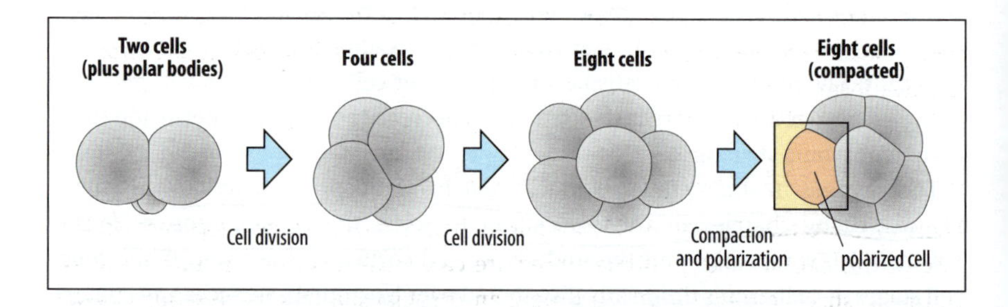

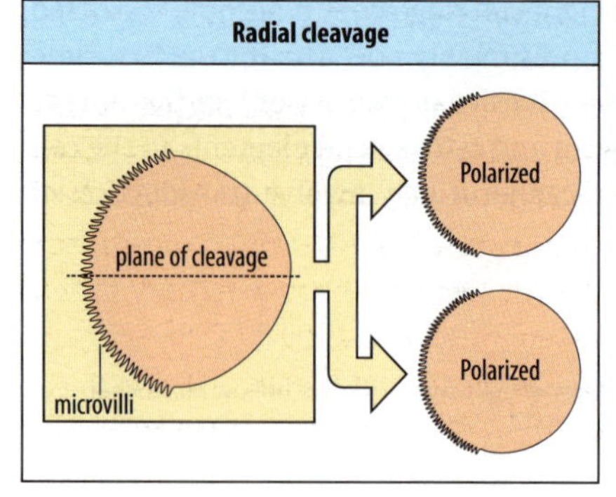

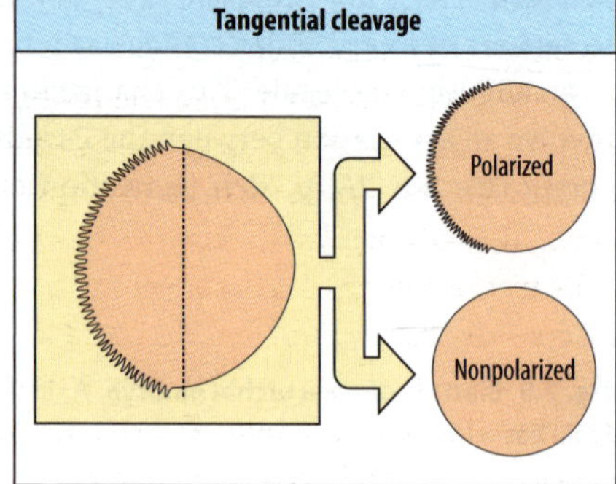

Fig. 7.10 Polarization of cells during cleavage of the mouse embryo. At the eight-cell stage, compaction takes place, with the cells forming extensive contacts (top panel). The cells also become polarized; for example, the microvilli, which were initially uniformly distributed over the cell surface, become confined to the outward-facing cell surface. Future cleavages can thus divide the cell either into two polarized cells (by a radial cleavage, bottom left panel) or into a polarized and a nonpolarized cell (by a tangential cleavage, bottom right panel). The tangential cleavages give rise to the inner cell mass.

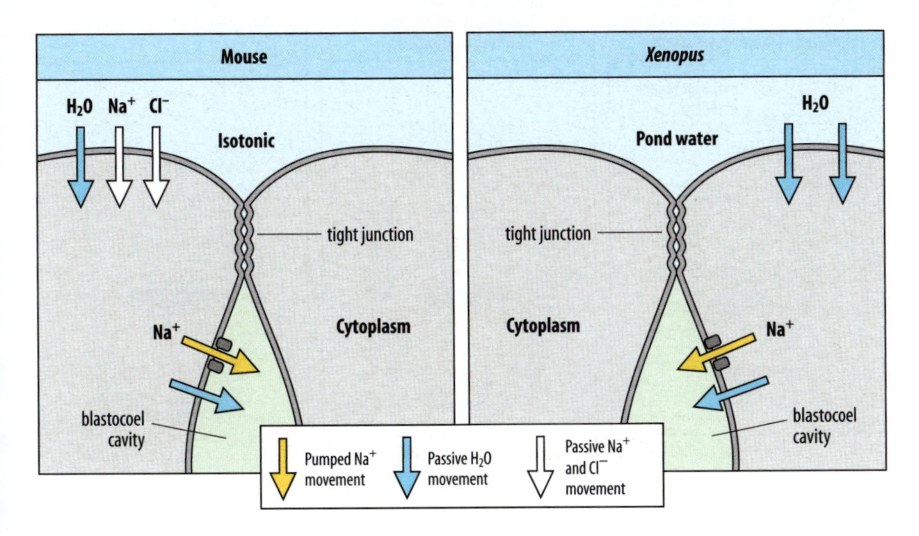

Fig. 7.11 Hydrostatic pressure can inflate the blastula. Left panel: active transport of sodium ions (Na^+) into the blastocoel cavity of the mouse embryo results in water and salts flowing in from the surrounding isotonic solution, creating positive hydrostatic pressure. The blastocoel cavity is sealed off from the outside medium by tight junctions between the epithelial cells at their outer edge. Right panel: the *Xenopus* embryo develops in pond water, so there is no influx of ions and the supply of sodium ions probably comes from the cells themselves, where sodium is stored in an ionically inactive form.

in forming and maintaining the spherical shape of the blastula. In the development of mammalian embryos, there is evidence that the blastocoel fluid is formed by a mechanism that involves active pumping of sodium ions into the blastocoel (Fig. 7.11). Tight junctions between cells of the outer layer appear at the eight-cell stage, and by the 32-cell stage they act as a permeability barrier to the passage of materials across the epithelial cell layer. At the same time, sodium pumps become active in the cell membranes that face the blastocoel. There appears to be a net inward flux of sodium ions: possibly, as in the frog skin, sodium enters passively at the outer face of the cell and is then pumped into the blastocoel across the inner face of the cell. As the ion concentration in the blastocoel fluid increases, water flows into the blastocoel by osmosis and the increase in hydrostatic pressure stretches the surrounding tissue layers. A similar mechanism seems to operate in the *Xenopus* blastula. In this case, as the embryo develops in pond water, which contains very little sodium, the sodium that is pumped into the blastocoel probably comes from the cells themselves, where it is stored in an ionically inactive form.

7.6 Internal cavities can be created by cell death

There are several ways of creating a hollow structure in early development. In the case of the neural tube, a tubular structure can result from the rolling up of an epithelial sheet. Another way of forming an internal space is by creating a cavity in a solid structure. One example of this is the formation of the epithelium of the epiblast in the early mouse embryo.

In the early mouse embryo, the inner cell mass gives rise to the epiblast, which is pushed across the blastocoel (see Section 3.4). Initially the epiblast is a solid mass of cells, but it later turns into an epithelial layer enclosing a fluid-filled cavity. Formation of this cavity results from programmed cell death—apoptosis—of the cells in the center of the epiblast (Fig. 7.12). It is likely that the surrounding cell layers—the visceral endoderm—send a signal to die to all the cells of the epiblast, and that only outer cells in contact with the basement membrane, via integrins, survive. The phenomenon of programmed cell death and its role in development is discussed further in later chapters.

Fig. 7.12 Cavity formation in the epiblast of the mouse embryo. The inner cell mass proliferates and forms a solid mass surrounded by the visceral endoderm. Programmed cell death—apoptosis—in the epiblast results in the formation of a cavity. A signal to die may come from the visceral endoderm; only cells attached to the basement membrane receive a rescue signal and so survive.

After Coucouvanis, E., et al.: 1995.

Summary

In many animals, the fertilized egg undergoes a cleavage stage that divides the egg into a number of small cells (blastomeres) and eventually gives rise to a hollow blastula. Various patterns of cleavage are found in different animal groups. In some animals, such as nematodes and mollusks, the plane of cleavage is of importance in determining the position of particular blastomeres in the embryo, and the distribution of cytoplasmic determinants. The plane of cleavage in animal cells is determined by the orientation of the mitotic spindle, which in turn is due to the action of the asters, whose position is specified by the chromosomes. At the end of the cleavage period, the blastula essentially consists of a polarized epithelium surrounding a fluid-filled blastocoel. Fluid accumulation in the blastocoel may be partly due to active ion transport. Cavity formation in the early mouse epiblast is caused by cell death.

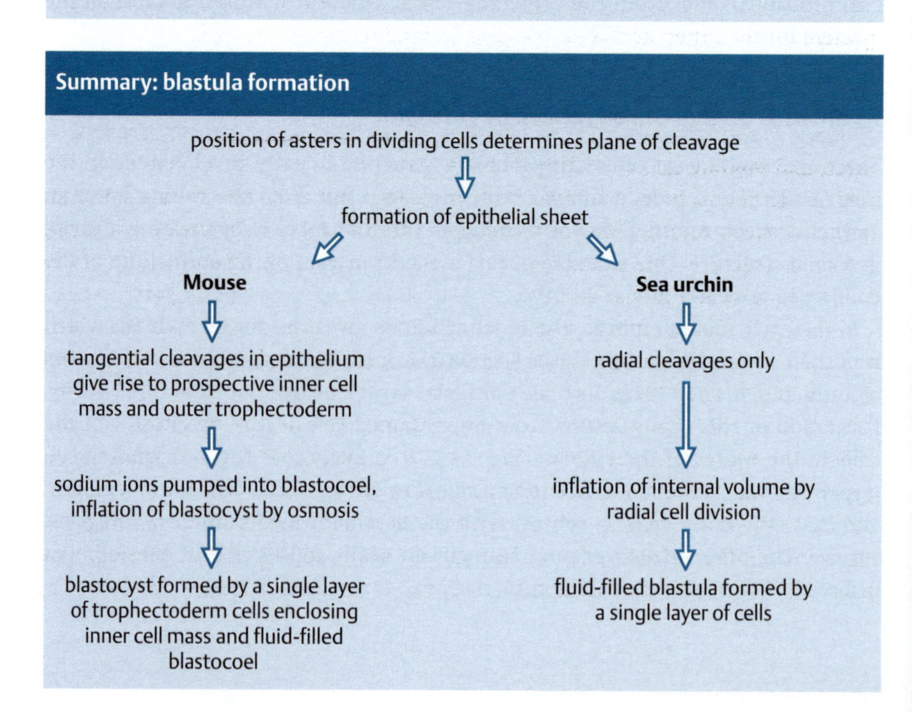

Gastrulation movements

Movements of cells and cell sheets at gastrulation bring most of the tissues of the blastula (or its equivalent stage) into their appropriate position in relation to the body plan. The necessity for gastrulation is clear from fate maps of the blastula in, for example, sea urchins and amphibians, where the presumptive endoderm, mesoderm, and ectoderm can be identified as adjacent regions in an epithelial sheet, and the endoderm and mesoderm must therefore move from the surface to inside the early embryo (see Chapters 3 and 5). After gastrulation, these tissues become completely rearranged in relation to each other: for example, the endoderm develops into an internal gut and is separated from the outer ectoderm by a layer of mesoderm. Gastrulation thus involves dramatic changes in the overall structure of the embryo, converting it into a complex three-dimensional structure. During gastrulation, a program of cell activity involving changes in cell shape and adhesiveness remodels the embryo, so that the endoderm and mesoderm move inside and only ectoderm remains on the outside. The primary force for gastrulation is provided by cell motility (Box 7B, p. 270). In some embryos, all this remodeling occurs with little or no accompanying increase in cell number or total cell mass.

In this section, we first consider gastrulation mechanisms in sea urchins and insects, in which it is a relatively simple process. The mechanisms of gastrulation in chick and mouse are not well understood, and so *Xenopus* serves as our model for the more complex gastrulation process in vertebrates.

7.7 Gastrulation in the sea urchin involves cell migration and invagination

Just before gastrulation begins, the late sea urchin blastula consists of a single layer of cells surrounding a central fluid-filled blastocoel. The future mesoderm occupies the most vegetal region, with the future endoderm adjacent to it (Fig. 7.13). The rest of the embryo gives rise to ectoderm. On their outward-facing surface the cells are attached to an extracellular hyaline layer, which contains the proteins hyalin and echinonectin. Cells are attached to adjacent cells by junctional complexes in the lateral membranes, and form a polarized epithelium. Cadherins and α- and β-catenins are associated with lateral cell–cell contacts and accumulate at adherens junctions from cleavage stages onwards. A basal lamina of extracellular matrix is present on the inner surface of the cells facing into the blastocoel.

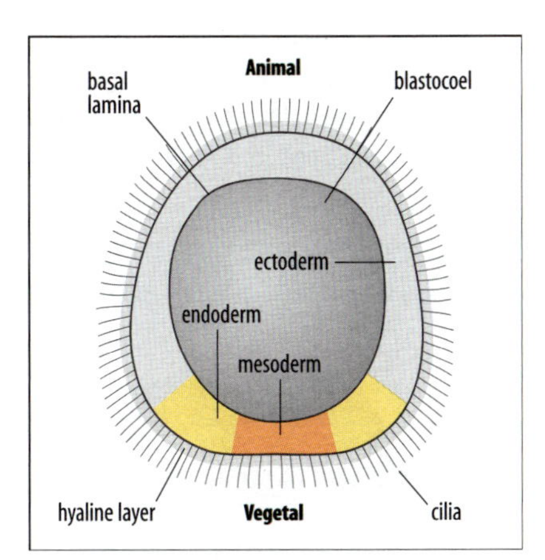

Fig. 7.13 The sea urchin blastula before gastrulation. The prospective endoderm and mesoderm are at the vegetal pole. There is an extracellular hyaline layer and a basal lamina lines the blastocoel.

Gastrulation begins with an epithelial to mesenchymal transition, with the most vegetal mesodermal cells becoming motile and mesenchymal in form. These primary mesenchyme cells become detached from each other and from the hyaline layer, and migrate into the blastocoel as single cells (Fig. 7.14) that have lost both their epithelial polarity and their cuboid shape. The transition to mesenchyme cells and their entry into the blastocoel is foreshadowed by intense pulsatory activity on the inner face of these cells, and on occasion a small transitory infolding or invagination is seen in the surface of the blastula before migration begins properly. Cell internalization requires loss of adhesion and is thus associated with the loss of α- and β-catenins, and the removal of cadherin from the cell-surface by endocytosis.

Having entered the blastocoel, the mesenchyme cells migrate within the blastocoel to become distributed in a characteristic pattern on the inner surface of the blastocoel wall. They first become arranged in a ring around the gut in the vegetal region at the ectoderm–endoderm border. Some then migrate to form two extensions toward the animal pole on the ventral (oral) side (Fig. 7.15). The migration path of individual cells varies considerably from embryo to embryo, but their final

Box 7B Change in cell shape and cell movement

Cells actively undergo major changes in shape during development. The two main changes are associated with cell migration and with the infolding of epithelial sheets. Changes in shape are generated by the cytoskeleton, an intracellular protein framework that controls cell shape and is also involved in cell movement. There are three principal types of protein polymers in the cytoskeleton—**actin filaments** (**microfilaments**), **microtubules**, and **intermediate filaments**—as well as many other proteins that interact with them. Actin filaments and microtubules are dynamic structures, polymerizing and depolymerizing according to the cell's requirements. Intermediate filaments are more stable, forming rope-like structures that transmit mechanical forces, spread mechanical stress, and provide mechanical stability to the cell. Microtubules play an important part in maintaining cell asymmetry and polarity, and provide tracks along which motor proteins convey other molecules and even organelles. Actin filaments can associate with the motor protein myosin to form contractile **actomyosin** complexes, which are primarily responsible for force generation and contractions within cells that lead to a change in shape. Actin filaments are also involved in cell movement. In general, cells generate contractile forces that give rise to intracellular tensions.

Actin filaments are fine threads of protein about 7 nm in diameter and are polymers of the globular protein actin. They are organized into bundles and three-dimensional networks, which in most cells lie predominantly just beneath the plasma membrane, forming the gel-like cell cortex. Numerous actin-binding proteins are associated with actin filaments, and are involved in bundling them together, forming networks, and aiding the polymerization and depolymerization of the actin subunits. Actin filaments can form rapidly by polymerization of actin subunits and can be equally rapidly depolymerized. This provides the cell with a highly versatile system for assembling actin filaments in a variety of different ways and in different locations, as required. The fungal drug cytochalasin D prevents actin polymerization and is a useful agent for investigating the role of actin networks. Actin filaments can also assemble with myosin into contractile structures, which act as miniature muscles. Localized contraction of an actin and myosin network in the cell cortex, for example, constricts animal cells in two at cell division (see figure, left panel) and a similar network can also mediate apical constriction of the cell and its longitudinal elongation (see figure, right panel).

Many embryonic cells are capable of movement over a solid substratum. They move by extending a thin sheet-like layer of cytoplasm known as a lamellipodium (see figure, right panel), or long fine cytoplasmic processes called filopodia. Both these temporary structures are pushed outwards from the cell by the assembly of actin filaments. Contraction of the actomyosin network at the rear of the cell then draws the cell forward. To do this, the contractile system must be able to exert a force on the substratum and this occurs at focal contacts, points at which the advancing filopodia or lamellipodium are anchored to the surface over which the cell is moving. *In vivo* this will often be a layer of extracellular matrix. At focal contacts, integrins (see Box 7A, p. 259) both adhere to extracellular matrix molecules through their extracellular domains, and provide an anchor point for actin filaments through their cytoplasmic domains. It is likely that integrins can mediate signal transduction across the plasma membrane at focal contacts, thus providing the cell with a means of sensing the environment and controlling cell movement.

The small GTP-binding proteins known as the Rho-family GTPases, which include Rho, Rac, and Cdc42, have a key role in regulating the actin cytoskeleton. Rac is required for extension of lamellipodia, for example, and Cdc42 is necessary to maintain cell polarity. Rho is involved in the planar-cell polarity signaling pathway from the cell-surface protein Frizzled (see Fig. 7.33). Signals from the environment relayed to the cytoskeleton via the Rho-family proteins are thus able to control cell movement.

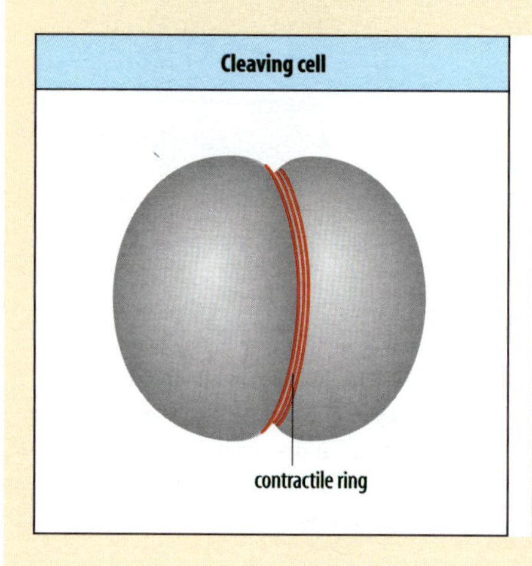

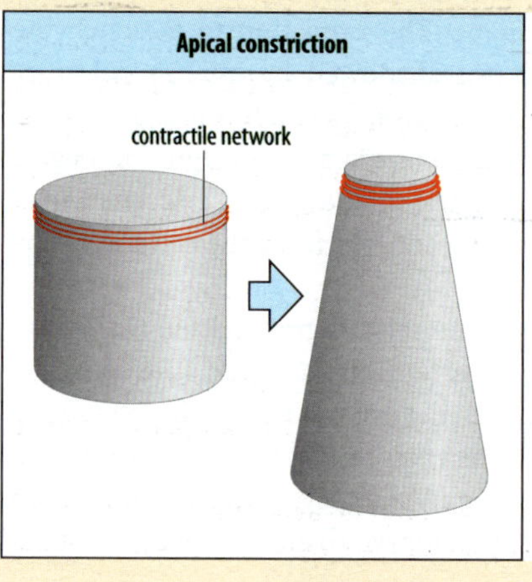

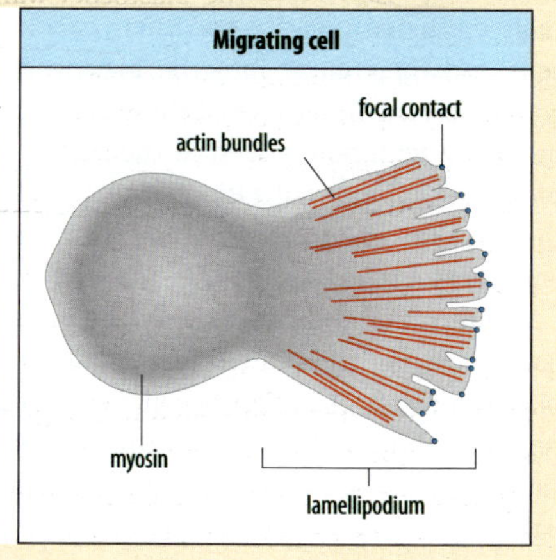

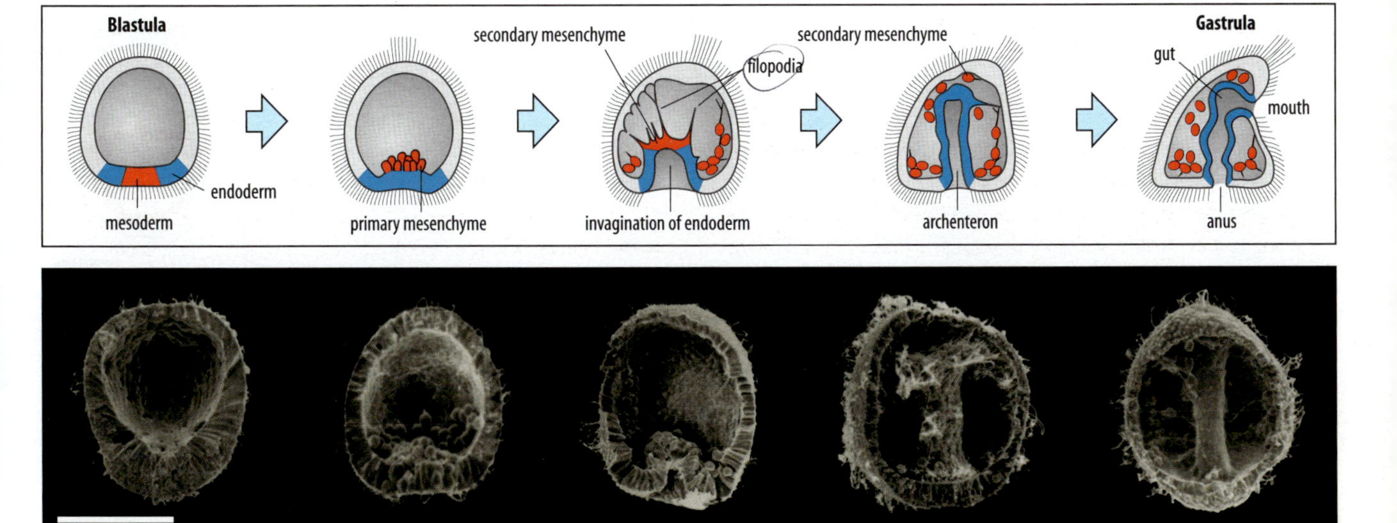

Fig. 7.14 Sea urchin gastrulation. Cells of the vegetal mesoderm undergo the transition to primary mesenchyme cells and enter the blastocoel at the vegetal pole. This is followed by an invagination of the endoderm, which extends inside the blastocoel toward the animal pole, forming a clear archenteron. Filopodia extending from the secondary mesenchyme cells at the tip of the invaginating endoderm contact the blastocoel wall and draw the invagination to the site of the future mouth, with which it fuses, forming the gut. Scale bar = 50 µm.

Photographs courtesy of J. Morrill.

pattern of distribution is fairly constant. The primary mesenchyme cells later lay down the skeletal rods of the sea urchin endoskeleton by secretion and, as these develop, the distribution of the cells changes.

The primary mesenchyme cells move over the inner surface of the blastocoel wall by means of fine filopodia, which can be up to 40 µm long and may extend in several directions. At any one time each cell has on average six filopodia, most of which are branched. When filopodia make contact with, and adhere to, the blastocoel wall, they contract, drawing the cell body toward the point of contact. As each cell extends several filopodia (Fig. 7.16), some or all of which may contract on contact with the wall, there seems to be competition between the filopodia, the cell being drawn toward that region of the wall where the filopodia make the most stable contact. The movement of the primary mesenchyme cells therefore resembles a random search for the most stable attachment. As they migrate, their filopodia fuse, forming cable-like extensions.

Analysis of video films of migrating cells suggests that the most stable contacts are made in the regions where the cells finally accumulate, namely the vegetal ring and the two ventro-lateral clusters. Thus, the pattern of contact stability of the inner surface of the blastocoel wall determines the pattern of migration of the cells, but the molecular basis for this adhesion is not known, although glycosaminoglycans and proteoglycans are involved. The surfaces of the cells over which the mesenchyme cells move are covered with a basal lamina, which has been implicated in influencing the contacts between filopodia and the blastocoel wall. For example, there is a strong association between filopodia and a matrix component.

Primary mesenchyme cells introduced by injection at the animal pole move in a directed manner to their normal positions in the vegetal region. This suggests that guidance cues, possibly graded, are distributed over the blastocoel wall. Even cells that have already migrated will migrate again to form a similar pattern when introduced into a younger embryo.

The entry of the primary mesenchyme is followed by the invagination and extension of the endoderm to form the embryonic gut (the archenteron). The endoderm

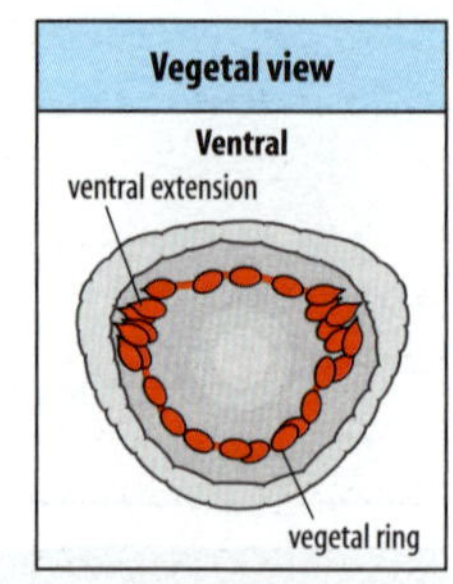

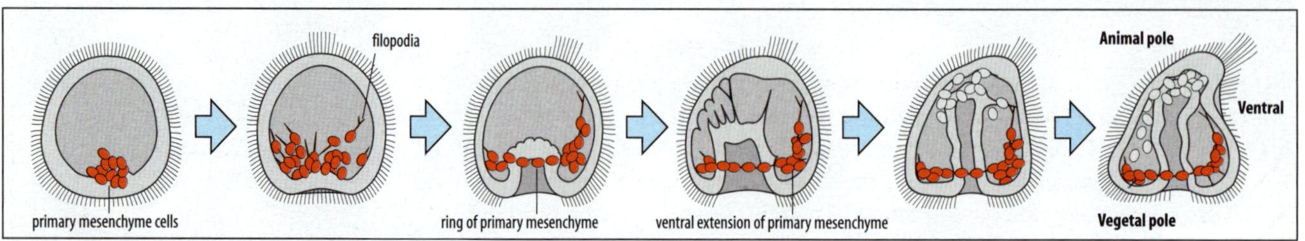

Fig. 7.15 Migration of primary mesenchyme in early sea urchin development. The primary mesenchyme cells enter the blastocoel at the vegetal pole and migrate over the blastocoel wall by filopodial extension and contraction. Within a few hours they take up a well-defined ring-like pattern in the vegetal region with extensions along the ventral side.

invaginates as a continuous sheet of cells (see Fig. 7.14). Formation of the gut occurs in two phases. During the initial phase, the endoderm invaginates to form a short, squat cylinder extending up to halfway across the blastocoel. There is then a short pause, after which extension continues. In this second phase, the cells at the tip of the invaginating gut, which will later detach as the secondary mesenchyme, form long filopodia, which make contact with the blastocoel wall. Filopodial extension and contraction pull the elongating gut across the blastocoel until it eventually comes in contact with and fuses with the mouth region, which forms a small invagination on the ventral side of the embryo (see Fig. 7.14). During this process the number of invaginating cells doubles, and this is in part due to cells around the site of invagination contributing to the hindgut.

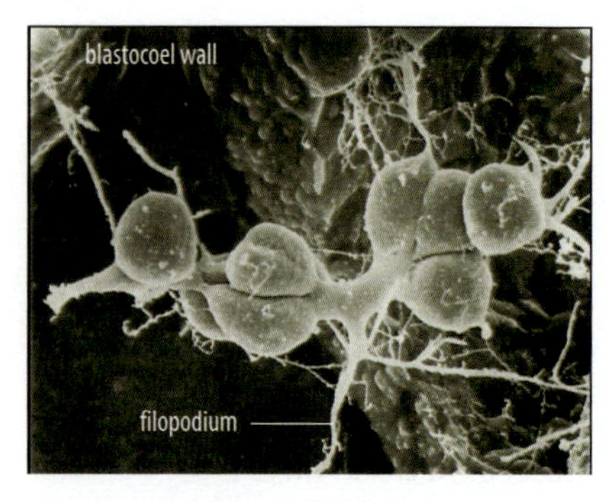

Fig. 7.16 Filopodia of sea urchin mesenchyme cells. The scanning electron micrograph shows a group of primary mesenchyme cells, some of which have fused together, moving over the blastocoel wall by means of their numerous filopodia, which can extend and contract.

Photograph courtesy of J. Morrill, from Morrill, J.B., et al.: 1985.

How is invagination of the endoderm initiated? The simplest explanation is that a change in shape of the endodermal cells causes, and initially maintains, the change in curvature of the cell sheet (Fig. 7.17). At the site of invagination, the initially cuboid cells adopt a more elongated, wedge-shaped form, which is narrower at the outer (apical) face. This change in cell shape is the result of constriction of the cell at its apex, probably by contraction of cytoskeletal elements. The change in cell shape is initially sufficient to pull the outer surface of the cell sheet inward and to maintain invagination, as shown by a computer simulation in which apical constriction, modeled to spread over the vegetal pole, results in an invagination (Fig. 7.18).

Other mechanisms have been proposed to account for the primary invagination. These include secretion of chondroitin sulfate, leading to the swelling of the apical lamina and thus buckling of the vegetal plate, and contraction of the cells along the apico-basal axis, which would cause the cells to round up and so buckle the plate inwards.

The primary invagination only takes the gut about a third of the way to its final destination. The second phase of gastrulation involves two different mechanisms: contraction of filopodia in contact with the future mouth region, as described above, and a filopodial-independent process. Treatments that interfere with filopodial attachment to the blastocoel wall result in failure of the gut to elongate completely, but it still reaches about two-thirds of its complete length. Filopodial-independent extension is due to active rearrangement of cells within the endodermal sheet. If a sector of cells in the vegetal mesoderm is labeled with a fluorescent dye before gastrulation, this sector is seen to become a long, narrow strip as the gut extends (Fig. 7.19). We will look at this type of active cell rearrangement—known as **convergent extension**—in more detail later in this chapter, in relation to similar phenomena in amphibian gastrulation. In the sea urchin, this cell rearrangement requires the interaction of cells with the basal lamina. If antibodies against a basal lamina component are injected into the blastocoel, the second phase of gastrulation is blocked.

What guides the tip of the gut to the future mouth region? As the long filopodia at the tip of the gut initially explore the blastocoel wall, they make more stable contacts at the animal pole and then in the region where the future mouth will form. Filopodia making contact there remain attached for 20–50 times longer than when attached to other sites on the blastocoel wall. Similar differences in adhesiveness enable mesoderm cells to migrate to their correct positions. Gastrulation in the sea urchin clearly shows how changes in cell shape, changes in cell adhesiveness, and cell migration all work together to cause a major change in embryonic form.

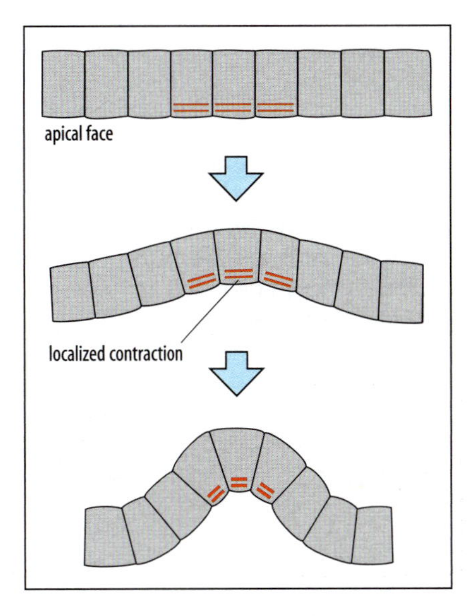

Fig. 7.17 Change in cell shape in a small number of cells can cause invagination of endodermal cells. Bundles of filaments composed of actin and other motor proteins contract at the outer edge of the cells, making them wedge-shaped. As long as the cells remain mechanically linked to adjacent cells in the sheet, this local change in cell shape draws the sheet of cells inward at that point.

7.8 Mesoderm invagination in *Drosophila* is due to changes in cell shape that are controlled by genes that pattern the dorso-ventral axis

At the time gastrulation begins, the *Drosophila* embryo consists of a blastoderm of about 6000 cells, which forms a superficial single-celled layer (see Section 2.2). Gastrulation begins with the invagination of a longitudinal strip of future mesodermal cells (8–10 cells wide) on the ventral side of the embryo, to form a ventral furrow and then a tube inside the body. This breaks up into individual cells, which spread out to form a single layer of mesoderm on the interior face of the ectoderm (Fig. 7.20). The gut develops slightly later, by invaginations of prospective endoderm near the anterior and posterior ends of the embryo (see Fig. 2.3), which we shall not consider here.

Invagination of the mesoderm is rapid: formation of the mesodermal tube takes about 30 minutes and cell spreading about an hour. The invagination initially forms a furrow, which in cross-section looks remarkably similar to the primary invagination of the endoderm in sea urchin gastrulation. Invagination appears to occur in two phases: during the first phase, the central strip of cells develop flattened and smaller apical surfaces, possibly by apical contraction, and their nuclei move away from the periphery. This results in a furrow which folds into the interior of the embryo to form a tube, the central cells forming the tube proper while the peripheral cells form a 'stem'; during the second phase, the tube

Fig. 7.18 Computer simulation of the role of apical constriction in invagination. Computer simulation of the spreading of apical constriction over a region of a cell sheet shows how this can lead to an invagination.

Illustration after Odell, G.M., et al.: 1981.

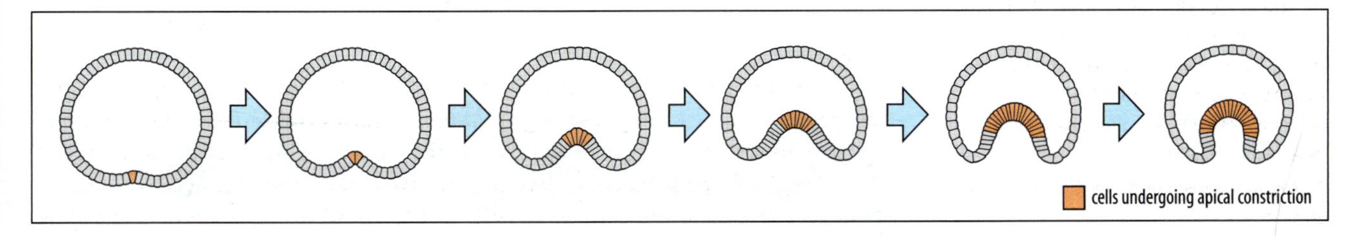

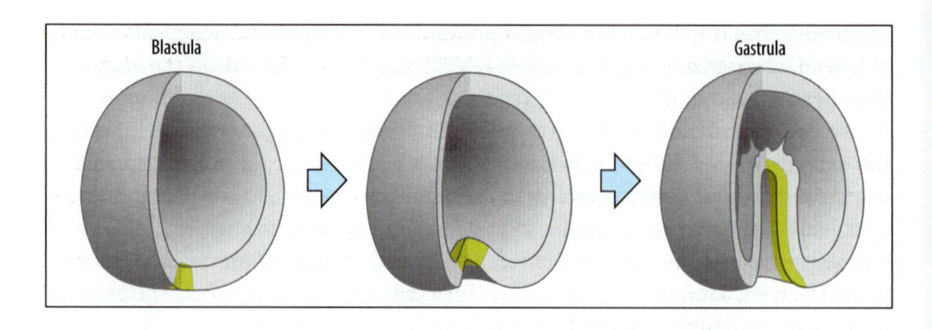

Fig. 7.19 Gut extension during sea urchin gastrulation. Labeling of cells in the vegetal region of the blastula shows that gut extension involves cell rearrangement within the endoderm, causing the labeled cells to become distributed in a long narrow strip.

dissociates into individual cells which proliferate and spread out laterally. The entry of cells into cell division is delayed during invagination and this delay is essential for the cell-shape changes to occur.

Mutant *Drosophila* embryos that have been dorsalized or ventralized (see Section 2.12) show that this behavior of the mesodermal cells is autonomous, as in the sea urchin (see Section 7.7), and is not affected by adjacent tissues. In dorsalized embryos, which have no mesoderm, no cells undergo nuclear migration or apical contraction. In ventralized embryos, in which most of the cells are mesoderm, these changes occur throughout the whole of the dorso-ventral axis.

Gastrulation in *Drosophila* provides us with the possibility of linking gene action with changes in cell shape during morphogenesis. Mesoderm invagination is affected by mutations in the genes *twist* and *snail,* which are expressed in the prospective mesoderm before gastrulation, and which encode transcription factors (see Section 2.15). In mutants that lack *twist* function, a small transient furrow is formed, and in *snail* mutants the prospective mesodermal cells flatten, but no other change occurs. Double mutants of these genes show no cell-shape changes and no invagination. The transcription factors encoded by these genes may therefore be controlling, either directly or indirectly, the expression of cell components such as cytoskeletal proteins, which are required for the shape changes to occur. This has recently been shown for Twist, which induces the expression of proteins in ventral mesoderm cells that activate signaling pathways that lead directly to changes in cell shape.

Twist induces the production of two proteins, Folded gastrulation (Fog) and T48, which activate signaling pathways that converge on RhoGEF2, a guanine-exchange factor. This can, by activating the small GTP-binding protein Rho, induce contractile shape changes by stimulating myosin II, which forms contractile assemblies with actin filaments (see Box 7B, p. 270). Contraction is localized to the apical face of the

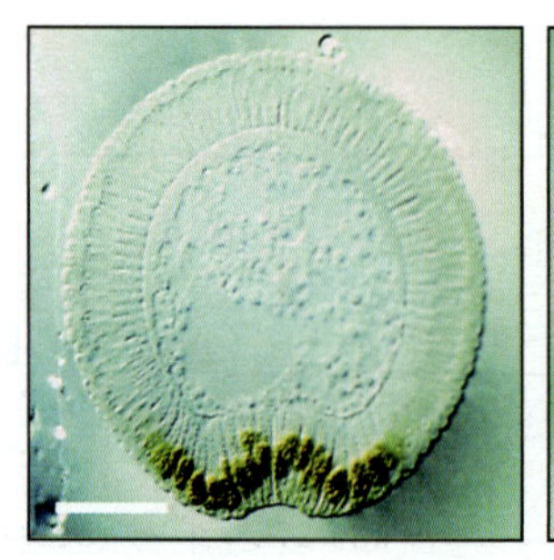
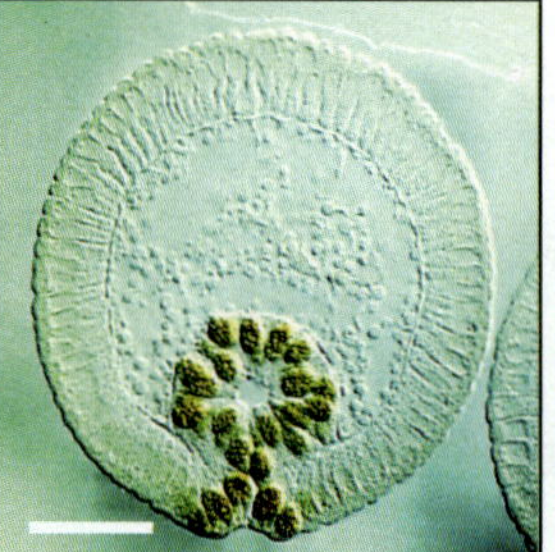
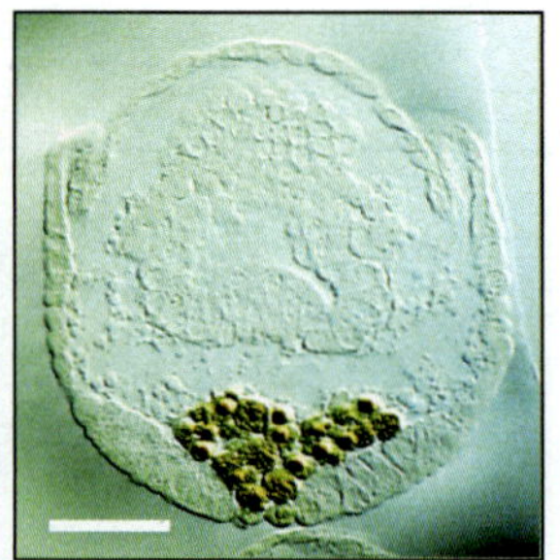

Fig. 7.20 Drosophila gastrulation. Mesodermal cells (stained), in a longitudinal strip on the ventral side, change shape and cause an invagination, which results in a ventral furrow (left panel). Further apical contraction of the cells results in the mesoderm forming a tube on the inside of the embryo (center panel). The mesodermal cells then start to migrate individually to different sites (right panel). Scale bars = 50 μm.

Photographs courtesy of M. Leptin, from Leptin, M. et al.: 1992.

cell by the location there of another G protein, called Concertina, which also interacts with the pathway and activates RhoGEF2, resulting in apical constriction and mesoderm invagination.

Twist and Snail also control all subsequent steps of mesoderm development, including expression of receptors for FGF. When the central part of the mesoderm is fully internalized, its cells make contact with the ectoderm. Adherens junctions that had been established in the blastoderm are now lost, and the mesoderm dissociates into single cells which spread out to cover the ectoderm as a single-cell layer. In gastrulation, signaling from the ectoderm via the FGF receptor Heartless on mesodermal cells is needed for the invaginating mesodermal cells to be able to spread out and cover the inside of the ectoderm. Failure to activate the FGF signaling pathway results in the mesodermal cells remaining near the site of invagination.

Snail and Twist may also be involved in mesodermal cell spreading by their effects on cadherin expression. Mesodermal cell spreading may be caused or facilitated by the mesodermal cells switching from expression of ectodermal E-cadherin to N-cadherin, so that they no longer adhere to ectodermal cells. The change from E- to N-cadherin is under the control of Snail and Twist. The actions of Snail suppress the production of E-cadherin in the mesoderm, while expression of Twist induces the production of N-cadherin.

7.9 Germ-band extension in *Drosophila* involves myosin-dependent intercalation

Another dramatic change in shape that occurs in the *Drosophila* embryo is the extension of the germ band (see Fig. 2.4), which leads to an almost doubling in the length of the epithelial layer that forms the thorax and abdomen of the embryo. This extension is driven not by cell division or changes in cell shape but by a rearrangement of the cells of the ventral part of the epithelium, which converge towards the ventral midline. The rearrangement involves intercalation of adjacent cells so that they converge toward the midline of the tissue, causing extension of the tissue in the antero-posterior direction. In germ-band convergent extension, this remodeling occurs over the tissue as a whole and is due to regulated changes in the cadherin-containing adherens junctions that normally hold the cells tightly together.

At the beginning of germ-band extension the epidermal cells are packed in a regular hexagonal pattern, with their boundaries either parallel to the dorso-ventral axis or at 60° to it, as shown in Fig. 7.21. When extension starts, the adherens

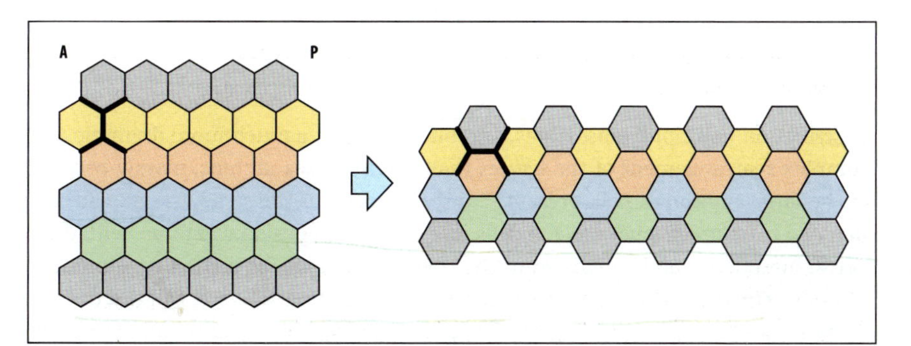

Fig. 7.21 Junction remodeling enables cells to undergo intercalation in *Drosophila* germ-band extension. The disassembly and reassembly of intercellular contacts in the germ-band epithelium drives the tissue to narrow across the dorso-ventral axis and elongate along the antero-posterior axis. The successive changes in intercellular contacts are outlined in black.

Adapted from Bertet, C. et al.: 2004.

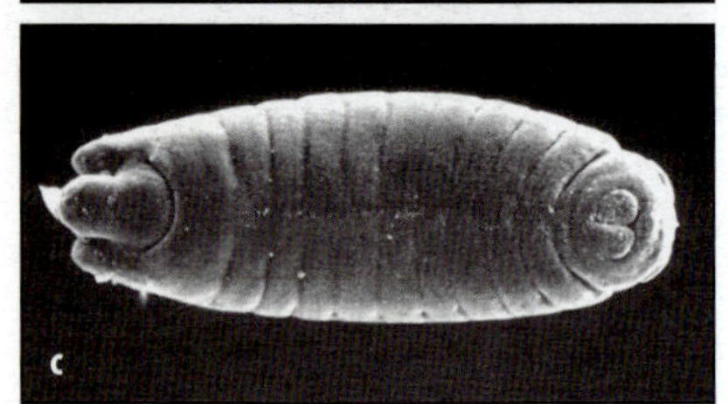

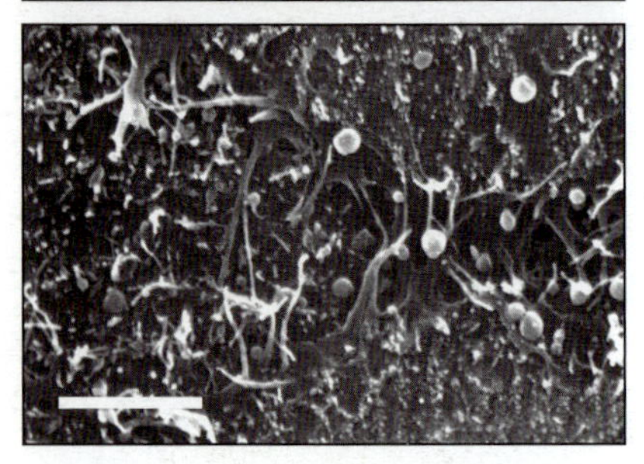

Fig. 7.22 Dorsal closure in *Drosophila*. The closure of the epidermis along the dorsal midline over a period of about 2 hours is shown in the micrographs in panels a–c. Starting at both ends of the opening, the edges of the epidermis come together over the amnioserosa (seen as the central oval area in a and b) and fuse to form the midline seam, matching the segment pattern on either side. The bottom panel shows filopodia acting as a zipper.

Photographs courtesy of A. Jacinto, from Jacinto, A., et al.: 2002.

junctions on the faces parallel to the dorso-ventral axis shrink and disappear and the cells become diamond-shaped. New junctions parallel to the antero-posterior axis appear, making the boundaries hexagonal again and intercalating the cells along the dorso-ventral axis—which makes the tissue extend along the antero-posterior axis. The ability to undergo this process reflects the intrinsic polarization of the cells in the plane of the membrane.

The mechanism of intercalation involves the regulated localization and activity of myosin, which co-localizes with the β-catenin–E-cadherin–actin complexes at the junctions (see Box 7A, p. 259) and is enriched in shrinking junctions. The proposed mechanism is that regulated myosin contraction prevents E-cadherin in the junctions holding the cells together, enabling new contacts to be made that result in cell intercalation.

7.10 Dorsal closure in *Drosophila* and ventral closure in *C. elegans* are brought about by the action of filopodia

At about 11 hours after the beginning of *Drosophila* embryogenesis, when gastrulation is complete and the germ band has undergone retraction (see Fig. 2.4), the amnioserosa on the dorsal surface of the embryo is not yet covered by the epidermis. The epidermis now moves to close this gap, zipping-up the opening from both ends (Fig. 7.22). About 2 hours after this starts, the two epithelial fronts have fused to form a seam along the dorsal midline, with the segment pattern exactly matched. The movement of the epidermis is due to the action of filopodia and lamellipodia, which extend and retract from the cells along the edges of the epithelial sheet. Filopodia are active along the whole epithelial front, extending up to 10 μm from the edge, and their behavior is similar to that seen in sea urchin gastrulation (see Section 7.7). Laser ablation of filopodial cells, especially at the site of 'zippering', prevents dorsal closure. This suggests that the main force for closure is exerted at this site by the filopodia drawing the two edges together. The small GTPase Rac plays a key role in organizing actin filaments at the cell's leading edge, and is also involved through the Jun N-terminal kinase (JNK) intracellular signaling pathway in controlling changes in cell shape and fusion.

The *C. elegans* embryo undergoes a process similar to dorsal closure, but on the ventral surface. At the end of gastrulation, the epidermis only covers the dorsal region and the ventral region is bare. The epidermis then spreads around the embryo until its edges finally meet along the ventral midline. Time-lapse studies show that the initial ventral migration is led by four cells, two on each side, which extend filopodia towards the ventral midline. Blocking filopodial activity by laser ablation or by cytochalasin D, which blocks the formation of actin filaments, prevents ventral closure, showing that the filopodia provide the driving force. Closure probably occurs by a zipper mechanism similar to that in the fly.

7.11 Vertebrate gastrulation involves several different types of tissue movement

Gastrulation in amphibians, fishes, and birds involves a much more dramatic and complex rearrangement of the tissues than occurs in sea urchins, mainly because of the large amount of yolk present. But the outcome is the same: the transformation of a two-dimensional sheet of cells into a three-dimensional embryo, with ectoderm, mesoderm, and endoderm in the correct positions for further development of body structure. The main movements of gastrulation that we will discuss are **involution**, which is the rolling-in of coherent sheets of endoderm and mesoderm

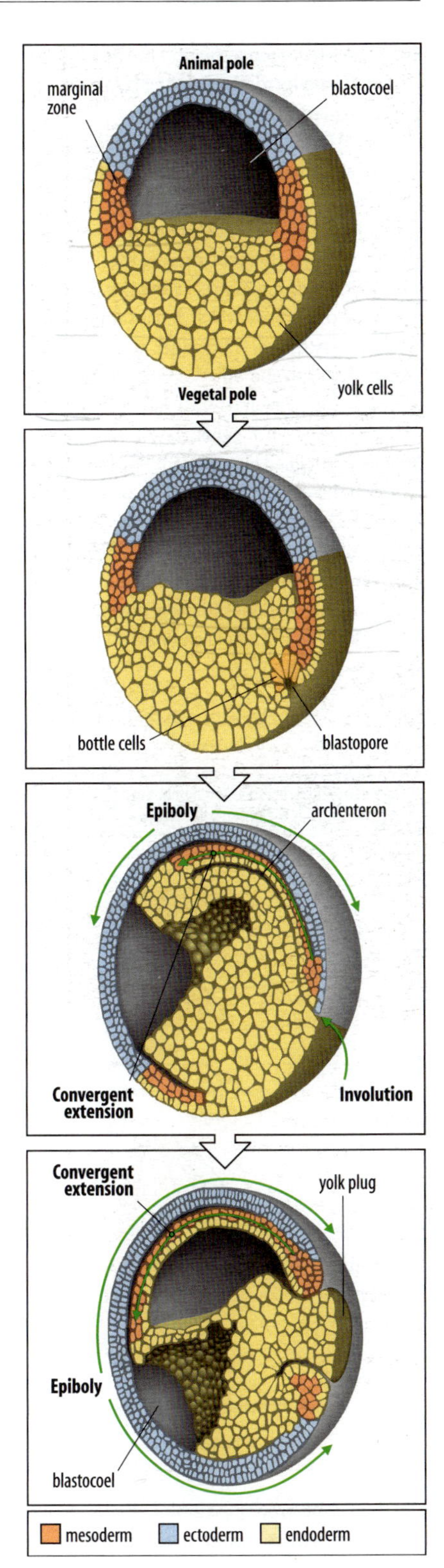

Fig. 7.23 Tissue movements during gastrulation of *Xenopus*. In the late blastula the future mesoderm (red) is in the marginal zone, overlaid by presumptive endoderm. Gastrulation is initiated by the formation of bottle cells in the blastopore region, which is followed by the involution of mesoderm over the dorsal lip of the blastopore. Marginal zone endoderm and mesoderm move inside over the dorsal lip of the blastopore (events in this region are shown in more detail in Fig. 7.24). The marginal zone endoderm, which was on the surface of the blastula, now lies ventral to the mesoderm and forms the roof of the archenteron. At the same time, the ectoderm of the animal cap spreads downward. The mesoderm converges and extends along the antero-posterior axis. The region of involution spreads ventrally to include more endoderm, and forms a circle around a plug of yolky vegetal cells. The ectoderm spreads by epiboly.

After Balinsky, B.I.: 1975.

at the blastopore, as occurs in *Xenopus* (Fig. 7.23), convergent extension, which elongates the body axis, and **epiboly**, which is the spreading of the ectoderm as the endoderm and mesoderm move inside. Gastrulation in mammals and birds occurs in the primitive streak (see Chapter 3) and involves the convergence of epiblast cells on the midline, the separation or **delamination** of cells individually from the epiblast and their internalization, followed by migration internally and convergent extension. We will begin with gastrulation in *Xenopus*.

In the late *Xenopus* blastula, the presumptive endoderm extends from the most vegetal region to cover the presumptive mesoderm. During gastrulation, the presumptive endoderm moves inside through the blastopore to line the gut, while all the cells in the equatorial band of mesoderm move inside to form a layer of mesoderm underlying the ectoderm, extending antero-posteriorly along the dorsal midline of the embryo (Fig. 7.24).

Gastrulation starts at a site on the dorsal side of the blastula, toward the vegetal pole. The first visible sign is the formation of bottle-shaped cells (bottle cells, see Fig. 7.24, second panel) by some of the presumptive mesodermal cells. As in the sea urchin and *Drosophila*, this change in cell shape causes the formation of a small groove in the blastula surface—the blastopore—and defines the dorsal lip of the blastopore, which corresponds to the Spemann organizer. The layer of mesoderm and endoderm starts to roll in around the blastopore, and into the interior of the blastula. This rolling under of a sheet of cells against its own inner surface is referred to as involution (see Fig. 7.24). As gastrulation proceeds, the region of involution spreads laterally and vegetally so that involution involves the vegetal endoderm and eventually forms a circle around a plug of yolky cells (see Fig. 7.23). With time, the blastopore contracts, forcing the yolky vegetal cells into the interior where they form the floor of the gut.

The first mesodermal cells to involute migrate as individual cells over the blastocoel roof to give rise to very anterior mesodermal structures in the head region. Behind them is mesoderm that enters, together with the overlying endoderm, as a single multilayered sheet. For this cell sheet, going through the narrow blastopore is rather like going through a funnel, and the cells become rearranged by convergent extension, which is a major feature of gastrulation. The mesoderm is initially in the form of an equatorial ring, but during gastrulation it converges and extends along the antero-posterior axis; hence the term convergent extension (Fig. 7.25). Thus, cells that initially are on opposite sides of the embryo come to lie next to each other (Fig. 7.26). The migrating mesoderm cells are polarized towards the animal pole and their migration depends on their interaction with a fibronectin-rich extracellular matrix on the blastocoel roof. Migration is directed by platelet-derived growth factor (PDGF).

In *Xenopus*, convergent extension occurs in both the mesoderm and endoderm as they involute, as well as in the future neural tissue overlying the mesoderm, which

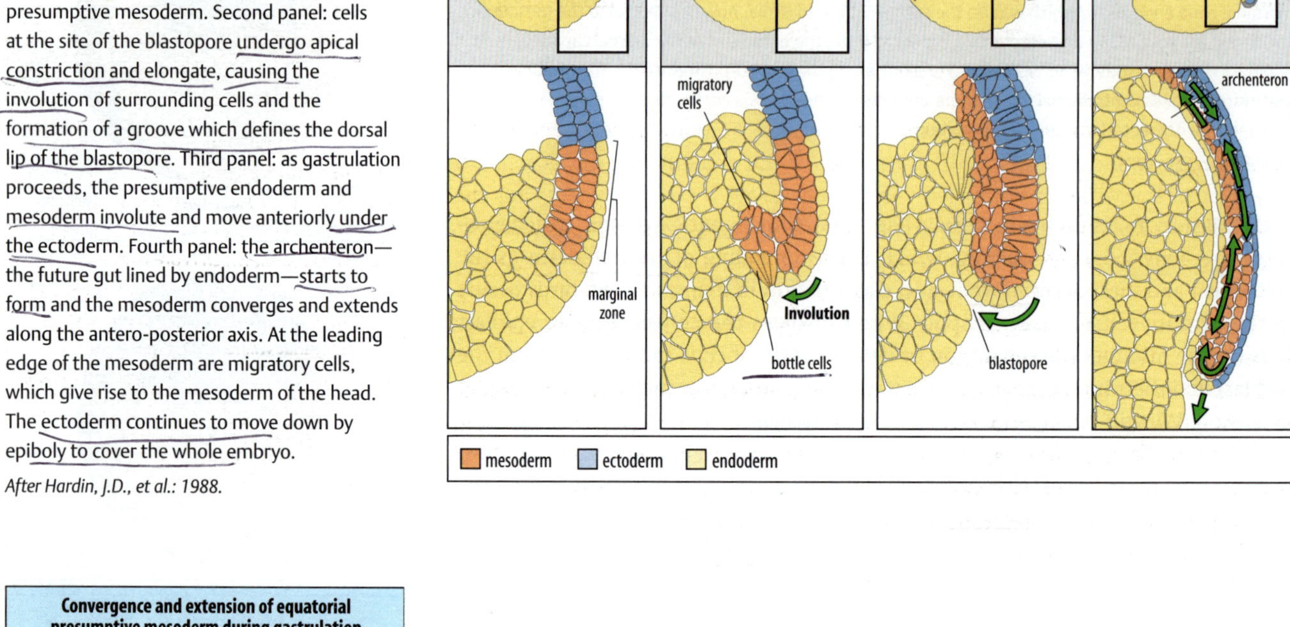

Fig. 7.24 Tissue movement in the dorsal region during formation of the blastopore and gastrulation in *Xenopus*. First panel: late blastula before gastrulation. In the marginal zone, presumptive endoderm overlies the presumptive mesoderm. Second panel: cells at the site of the blastopore undergo apical constriction and elongate, causing the involution of surrounding cells and the formation of a groove which defines the dorsal lip of the blastopore. Third panel: as gastrulation proceeds, the presumptive endoderm and mesoderm involute and move anteriorly under the ectoderm. Fourth panel: the archenteron—the future gut lined by endoderm—starts to form and the mesoderm converges and extends along the antero-posterior axis. At the leading edge of the mesoderm are migratory cells, which give rise to the mesoderm of the head. The ectoderm continues to move down by epiboly to cover the whole embryo.

After Hardin, J.D., et al.: 1988.

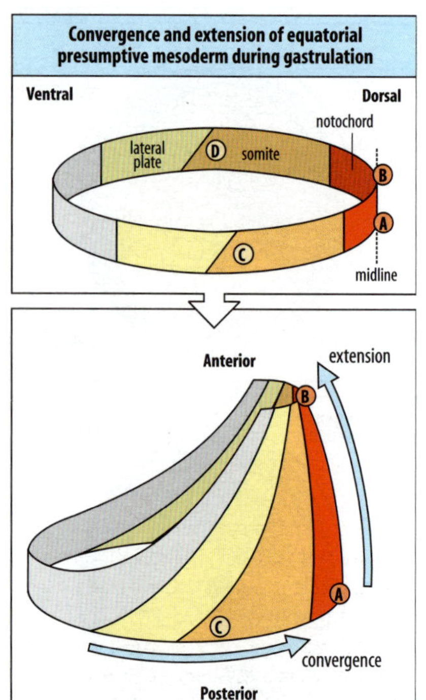

Fig. 7.25 Convergent extension of the mesoderm. Initially the mesoderm is in an equatorial ring, but during gastrulation it converges and extends along the antero-posterior axis. A–D are reference points, used to illustrate the extent of movement during convergent extension. In the bottom panel, D is hidden opposite C.

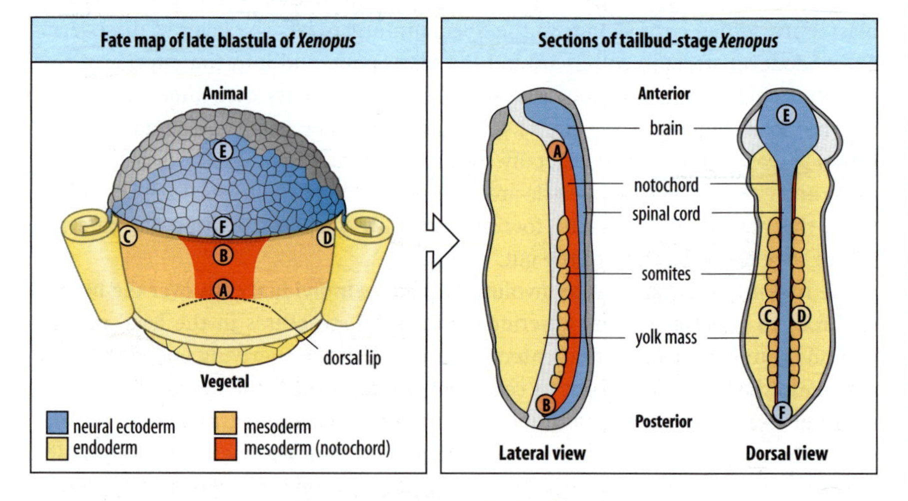

Fig. 7.26 The rearrangement of mesoderm and endoderm during gastrulation in *Xenopus*. The presumptive mesoderm is present in a ring around the blastula underlying the endoderm, which is shown peeled back. During gastrulation, both these tissues move inside the embryo through the blastopore and completely change their shape, converging and extending along the antero-posterior axis, so that points A and B move apart. Thus, the two points C and D, originally on opposite sides of the blastula, come to lie next to each other. In the neural ectoderm there is also convergent extension, as shown by the movement of points E and F. Note that no cell division or cell growth has occurred in these stages, and all these changes have occurred by rearrangement of cells within the tissues.

gives rise to the spinal cord. One can appreciate the dramatic nature of convergent extension by viewing the early gastrula from the blastopore. The future mesoderm can be identified by the expression of the gene *Brachyury*, and can be seen as a narrow ring around the blastopore (Fig. 7.27). As this ring of tissue expressing *Brachyury* enters the gastrula it converges into a narrow band of tissue along the dorsal midline of the embryo, extending in the antero-posterior direction, and eventually only the prospective notochordal mesoderm along the midline itself continues to express *Brachyury*.

As gastrulation proceeds, the mesoderm comes to lie immediately beneath the ectoderm, with the endoderm lining the roof of the archenteron, the future gut (see Fig. 7.24). While the mesoderm is in contact with the blastocoel roof the tissues do not fuse; this is due to cell repulsion mediated by cadherins. As the mesoderm and endoderm move inside, the ectoderm of the animal cap region increases in surface area by a rearrangement and stretching of the cells within it to form a thinner sheet, which spreads down over the vegetal region by epiboly and will eventually cover the whole embryo.

Gastrulation in the zebrafish has both similarities and differences to that of *Xenopus*. It starts when the blastoderm at the animal pole starts to spread out over the yolk cell and down towards the vegetal pole (Fig. 7.28). By the time that the blastoderm margin has reached about half-way down the yolk cell, deep layer cells have accumulated around the blastoderm edge to form the embryonic germ ring, and have formed two layers, prospective ectoderm overlying mesendoderm, which is

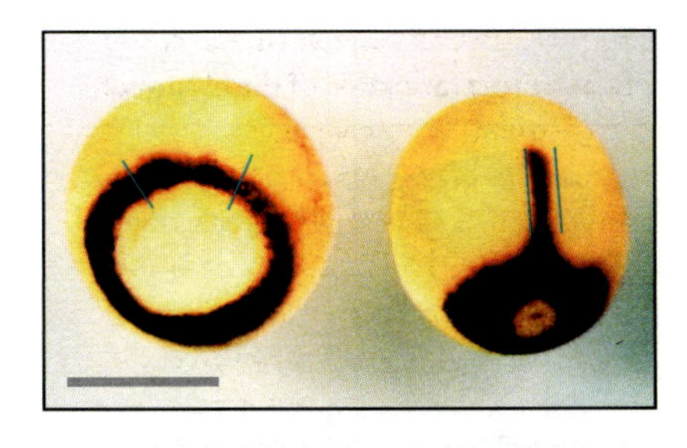

Fig. 7.27 Expression of *Brachyury* during *Xenopus* gastrulation illustrates convergent extension. Left: before gastrulation, the expression of *Brachyury* (dark stain) marks the future mesoderm, which is present as an equatorial ring when viewed from the vegetal pole. Right: as gastrulation proceeds, the mesoderm that will form the notochord (delimited by blue lines) converges and extends along the midline. Scale bar = 1 mm.

Photograph courtesy of J. Smith, from Smith, J.C., et al.: 1995.

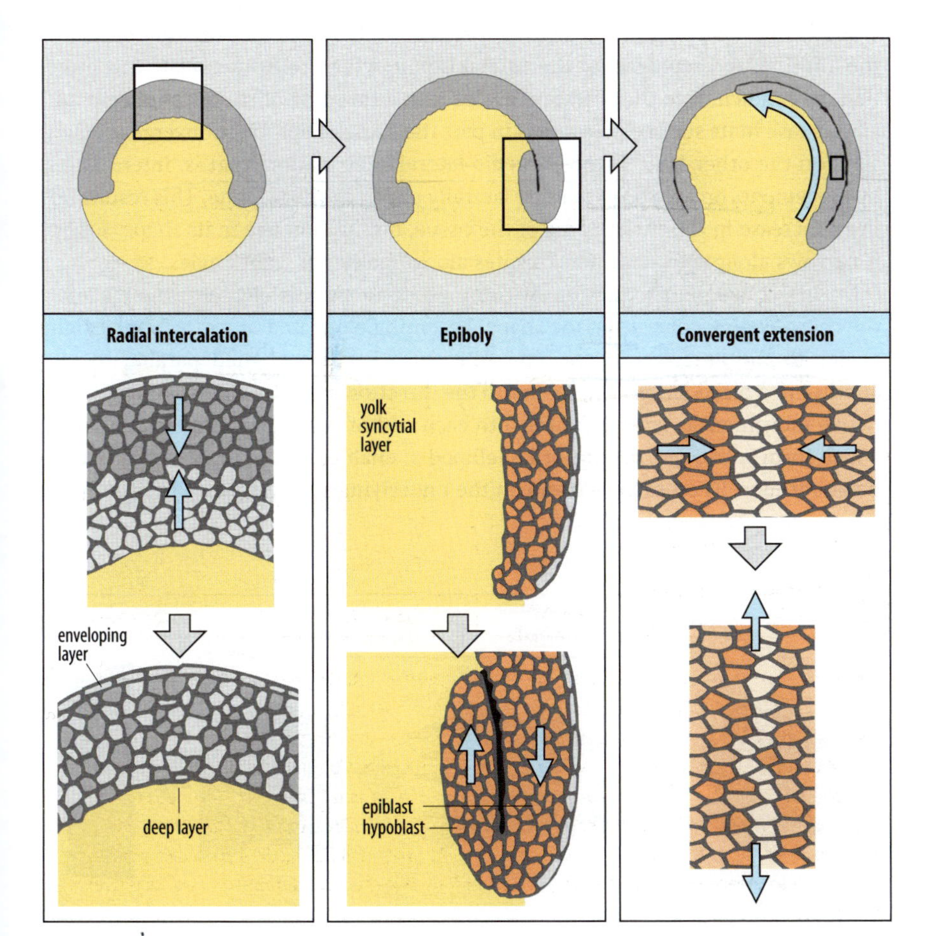

Fig. 7.28 Zebrafish gastrulation movements. Left panel: the first sign of gastrulation in the zebrafish embryo is spreading of the blastoderm over the surface of the yolk cell. This epiboly is caused by a thinning of the deep layer by a radial intercalation process. Center panel: when the blastoderm has reached about half the way down the yolk cell, separate layers of prospective mesendoderm (brown) and ectoderm (white) have formed and the mesendoderm starts to internalize. Ectoderm does not internalize and continues to move over the yolk cell. It will eventually cover the whole embryo. Right panel: convergent extension of the mesendoderm towards the dorsal midline results in elongation of the embryo.

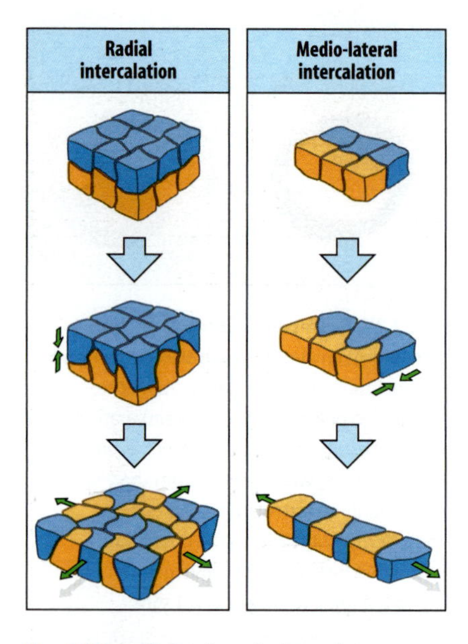

Fig. 7.29 Radial and medio-lateral intercalation. In radial intercalation, the cells intercalate in a direction perpendicular to the surface, producing an increase in surface area of the cell sheet. In medio-lateral intercalation, the sheet of cells narrows and elongates.

prospective mesoderm and endoderm. The mesendodermal cells move internally at the edge of the germ ring and then move upwards under the ectodermal layer towards the future anterior end of the embryo, forming a layer called the hypoblast. The ectoderm, now sometimes called the epiblast, continues to undergo epiboly (spreading) and is therefore moving in the opposite direction to the mesendoderm. How the contact between these layers is maintained is not clear, but E-cadherin is expressed in both layers and may be involved. The next step is convergent extension of the mesendodermal and ectodermal layers towards the dorsal midline and along the future antero-posterior axis, which elongates the embryo.

In the chick and mouse, gastrulation is indicated by formation of the primitive streak, as described in more detail in Chapter 3. As the streak moves forward, epiblast cells separate from the epithelial sheet and move into the interior as individual cells. Cells that move into the streak will give rise to mesoderm and endoderm. The mechanisms of streak formation, movement of epiblast cells, and delamination are not as clearly understood as are the involution movements in *Xenopus*. One view is that convergence at the posterior end plays a key role in providing the force for streak extension, and that contraction and delamination of mesodermal cells in the streak draws lateral cells to the streak.

7.12 Convergent extension and epiboly are due to different types of cell intercalation

Before considering in detail how convergent extension brings about elongation of the body axis in vertebrates, it is helpful to compare the two main types of intercalation that operate during gastrulation (Fig. 7.29). **Radial intercalation** occurs in the multilayered ectoderm of the animal cap, in which cells intercalate in a direction perpendicular to the surface. This leads to a thinning of the sheet of cells and an increase in its surface area and is in part the cause of epiboly. Convergent extension, on the other hand involves **medio-lateral intercalation**; that is, intercalation by movements on an axis extending laterally out from the midline. This results not in an increase in the surface area of the tissue, but in a change in its shape, so that it narrows along one axis and elongates along another at right angles.

During convergent extension of the mesoderm by medio-lateral intercalation the cells take on a characteristic shape, becoming elongated in a direction at right angles to the antero-posterior axis. They also become aligned parallel to one another in a direction perpendicular to the direction of tissue extension (Fig. 7.30). Active movement is largely confined to each end of these elongated bipolar cells, which form active protrusions or lamellipodia, enabling them to exert traction on neighboring cells on each side and on the underlying substratum and to shuffle in

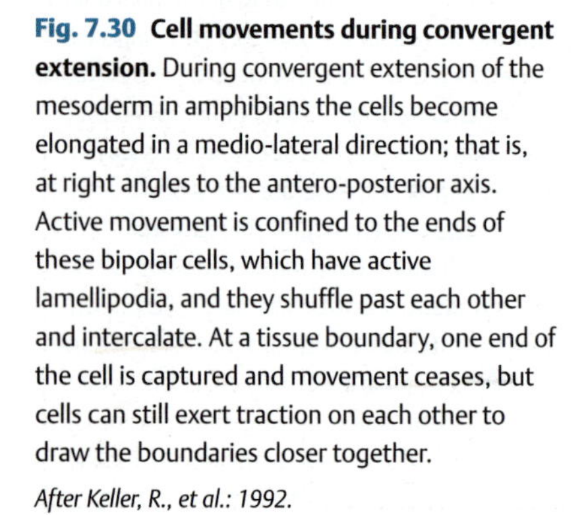

Fig. 7.30 Cell movements during convergent extension. During convergent extension of the mesoderm in amphibians the cells become elongated in a medio-lateral direction; that is, at right angles to the antero-posterior axis. Active movement is confined to the ends of these bipolar cells, which have active lamellipodia, and they shuffle past each other and intercalate. At a tissue boundary, one end of the cell is captured and movement ceases, but cells can still exert traction on each other to draw the boundaries closer together.

After Keller, R., et al.: 1992.

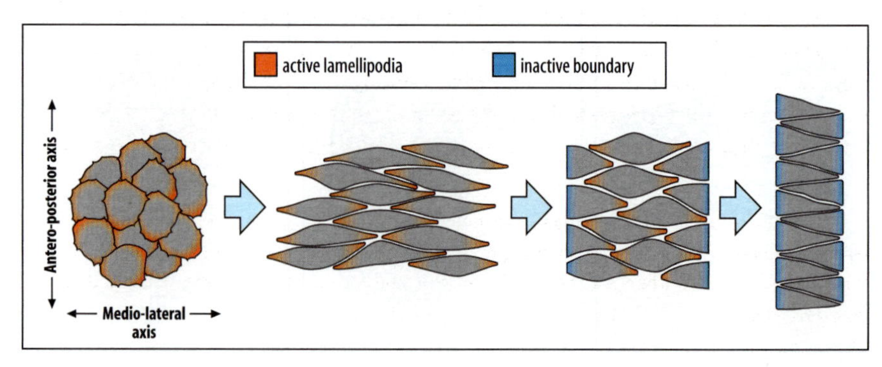

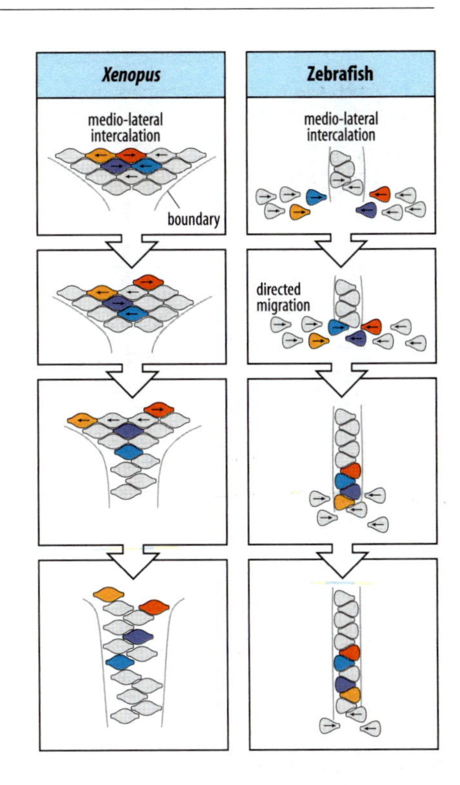

Fig. 7.31 Convergent extension in *Xenopus* and zebrafish. The left panels show convergent extension of mesoderm in *Xenopus* during the formation of the notochord, which occurs by medio-lateral intercalation of cells in a coherent tissue sheet. The right panels show formation of the notochord by convergent extension in the zebrafish, which occurs by directed movement of loosely packed mesenchymal mesodermal cells from the lateral regions towards the midline, insertion into the future notochord and medio-lateral intercalation within the notochord boundary.

between each other, always along the medio-lateral axis, with some cells moving medially and some laterally. There are some small protrusions on the anterior and posterior faces of the cells but these seem to hold apposed surfaces together rather than being involved in cell movement.

The boundary of a tissue undergoing convergent extension is maintained by the behavior of cells at the boundary. When an active cell tip is at the boundary of the tissue, its movement ceases and the cell becomes monopolar (see Fig. 7.30), with only the tip facing into the tissue still active. It is as if the cell tip that has reached the boundary has become fixed there. Precisely what defines a boundary is not clear, but boundaries between mesoderm, ectoderm, and endoderm seem to be specified before gastrulation commences. As gastrulation proceeds, a further boundary appears within the mesoderm, demarcating the future notochord from the future somites.

In amphibian and zebrafish gastrulation, convergent extension is along the antero-posterior axis and cells become polarized in the medio-lateral direction. The contractile forces generated at the tips of the elongated cells both move the cells to the midline and extend the tissue. One can visualize the process of convergent extension in terms of lines of cells pulling on one another in a direction at right angles to the direction of extension. Because the cells at the boundary are trapped at one end, this tensile force causes the tissue to narrow and hence extend anteriorly (Fig. 7.31). In fishes, convergent extension involves an initial directed migration of individual mesodermal cells from lateral regions towards the midline, a movement also known as dorsal convergence, followed by medio-lateral intercalation as cells become incorporated into the future notochord (see Fig. 7.31).

The presumptive notochord mesoderm is among the first of the tissues to internalize during gastrulation, and in amphibians it is the first of the dorsal structures to differentiate. Initially it can be distinguished from the adjacent somitic mesoderm by the packing of its cells and by the slight gap that forms the boundary between the two tissues, possibly reflecting differences in cell adhesiveness. The notochord elongates considerably and develops into a stiff rod composed of a stack of thin flat cells shaped like pizza slices. The two main mechanisms involved in its morphogenesis are medio-lateral intercalation leading to convergent extension, and directed dilation.

After initial elongation by convergent extension, the *Xenopus* notochord undergoes a further dramatic narrowing in width accompanied by an increase in height (Fig. 7.32). At the same time, the cells become elongated at right angles to the main antero-posterior axis, foreshadowing the later pizza-slice arrangement of cells within the notochord (see Fig. 7.32, bottom). The cells intercalate between their

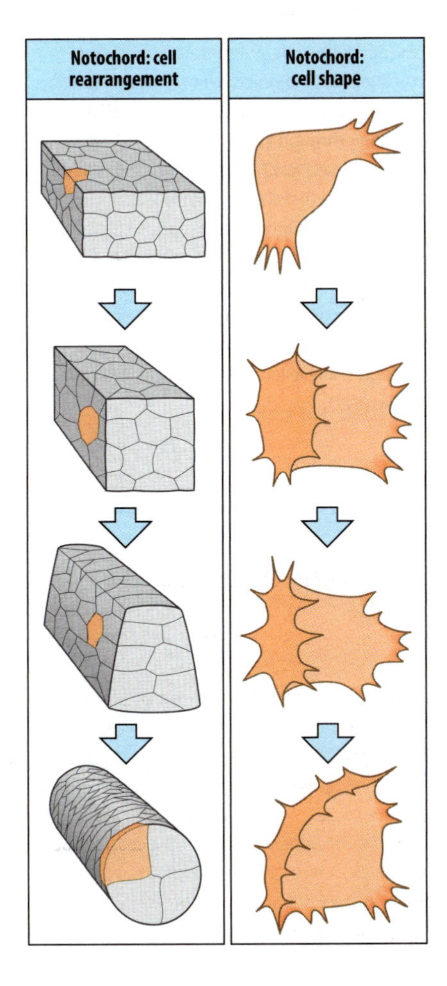

Fig. 7.32 Changes in cell arrangement and cell shape during notochord development in *Xenopus*. During convergent extension of the notochord, its height increases. At the same time, the cells become elongated. The bottom of the figure shows the eventual arrangement of cells, and their pizza-slice shape.

After Keller, R., et al.: 1989.

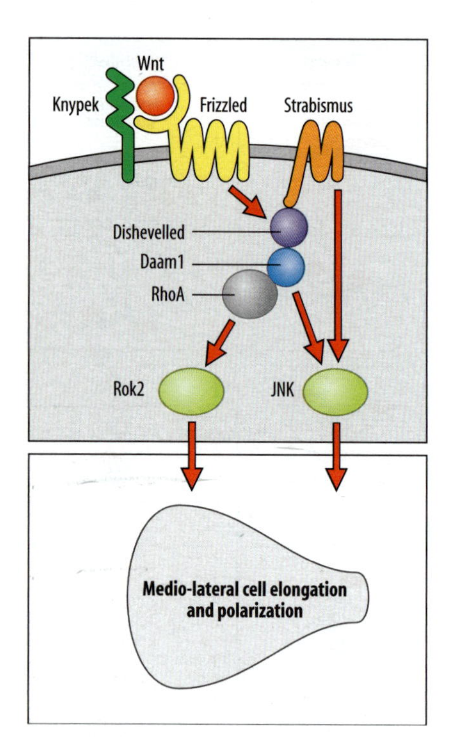

Fig. 7.33 The planar-cell polarity signaling pathway. Planar-cell polarity in converging and extending mesoderm is determined by the so-called non-canonical Wnt signaling pathway. The pathway shown here is from zebrafish. Signaling through Wnt-11 activates RhoA and the Rho kinase Rok2, leading to changes in the cell's actin cytoskeleton (compare with the canonical Wnt pathway that leads to β-catenin activation in Fig. 2.39). In addition to the Wnt receptor Frizzled and the associated cell-surface proteoglycan Knypek, the cell-surface protein Strabismus is also thought to signal through Dishevelled to activate Jun N-terminal kinase (JNK).

neighbors again, thus causing convergent extension. The later stage in notochord elongation, involving directed dilation, is discussed later in this chapter.

The molecular mechanisms controlling convergent extension in vertebrates are related to those controlling planar cell polarity in *Drosophila*, and involve signaling by Wnts through Frizzled via the non-canonical Wnt pathway (Fig. 7.33). Unlike the Wnt pathway that specifies the *Xenopus* dorsal region, this pathway does not end in stabilization of β-catenin but diverges after Dishevelled to activate the small GTPase RhoA and the protein kinase Rok2. Another cell-surface protein, Strabismus, also interacts with Dishevelled to stimulate the activation of the Jun N-terminal kinase (JNK). Both these pathways have effects on the cell's cytoskeleton. As well as polarizing the cells, the planar-cell polarity signaling pathway also appears to be required for the deposition of an extracellular fibronectin matrix over which the cells move in the correct direction. In *Xenopus*, for example, the mesoderm moves anteriorly over the blastocoel roof. When the pathway is disrupted, orderly fibronectin deposition does not occur and the cells fail to become polarized.

For medio-lateral intercalation to function properly, the intercalating cells must be oriented at right angles to the antero-posterior axis. The mechanism of this orientation is not well understood. Various factors appear to ensure that cells become oriented in a medio-lateral direction. The cells of the neural plate are oriented in response to a signal from midline structures such as the underlying notochord. Patterning along the antero-posterior axis also helps to polarize the intercalation process. When 'anterior' and 'posterior' explants of the chordamesoderm—the prospective notochord—of a *Xenopus* early gastrula are disaggregated and then recombined, the cells recognize each other by their current antero-posterior positional identity and sort out so as to end up in the correct position. Antero-posterior patterning in the chordamesoderm, which is situated at and just anterior to the upper lip of the blastopore (see Fig. 7.26), is reflected in two opposing gradients of *Brachyury* and *Chordin* expression, the former highest at the blastopore lip and the latter highest in the anterior chordamesoderm. In further experiments, it was found that only combined explants of anterior and posterior tissue underwent convergent extension by intercalation; combinations of all anterior or all posterior cells did not. This suggests that the differential patterning along the mesoderm is required to set up polarity.

Summary

At the end of cleavage the animal embryo is essentially a closed sheet of cells, which is often in the form of a sphere enclosing a fluid-filled interior. Gastrulation converts this sheet into a solid three-dimensional embryonic animal body. During gastrulation, cells move into the interior of the embryo, and the regions of endoderm and mesoderm, which were originally adjacent in the cell sheet, take up their appropriate positions in the embryo. Gastrulation results from a well-defined spatio-temporal pattern of changes in cell shape, cell movements, and changes in cell adhesiveness, and the main forces underlying the morphological changes are generated by localized cellular contractions. In the sea urchin, gastrulation occurs in two phases. In the first, changes in cell shape and adhesion result in migration of mesoderm cells into the interior, and invagination of the endodermal part of the cell sheet to form the gut. In the second phase, extension of the gut to reach the mouth region on the opposite side of the embryo occurs by cell rearrangement within the endoderm, which causes the tissue to narrow and lengthen (convergent extension), and by traction of filopodia extending from the gut tip against the blastocoel wall. Invagination of the mesoderm in

Drosophila occurs by a similar mechanism to that of the invagination of endoderm in the sea urchin, while germ-band extension is due to myosin-driven intercalation at cell junctions. Dorsal closure in *Drosophila* is driven by filopodial extension and contraction. Gastrulation in vertebrates involves more complex movements of cells and cell sheets and also results in elongation of the embryo along the antero-posterior axis. It involves three processes: involution, in which a double-layered sheet of endoderm and mesoderm rolls into the interior over the lip of the blastopore; convergent extension of endoderm and mesoderm in the antero-posterior direction to form the roof of the gut, and the notochord and somitic mesoderm, respectively; and epiboly—the spreading of the ectoderm from the animal cap region to cover the whole outer surface of the embryo. Both convergent extension and epiboly are due to cell intercalation, the cells becoming rearranged with respect to their neighbors. Medio-lateral rearrangements in the plane of the sheet involve planar cell polarity, with the cells all becoming polarized in the same direction within the sheet.

Summary: gastrulation in *Xenopus* and sea urchin

Xenopus and zebrafish	Sea urchin
mesoderm and endoderm move inside by involution or internalization	formation of blastopore: mesodermal cells migrate into the interior; endoderm moves inside by invagination of the epithelial sheet
⇩	⇩
convergent extension of mesoderm along antero-posterior axis	convergent extension of endoderm and traction via filopodia completes gut extension
⇩	
extension of ectoderm over whole surface by epiboly	

Neural-tube formation

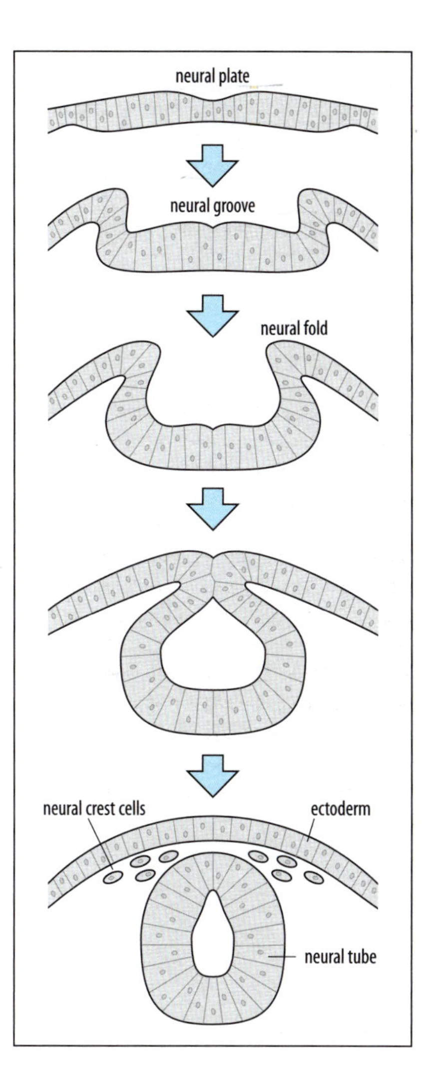

Fig. 7.34 Neural-tube formation results from formation and fusion of neural folds. Bending of the neural plate inwards at both sides creates the neural groove, with the neural folds at the edges. The neural folds come together and fuse to form a tube of epithelium—the neural tube. This detaches from the rest of the ectoderm, which becomes the epidermis. Neural crest cells detach from the dorsal neural tube and migrate away from it.

Neurulation in vertebrates results in the formation of the neural tube, a tube of epithelium derived from the dorsal ectoderm, which develops into the brain and spinal cord (see Section 4.8). The region that gives rise to the neural tube initially appears as a thickened plate of tissue—the neural plate—following induction by the mesoderm.

The vertebrate neural tube is formed by two different mechanisms in different regions of the body. The anterior neural tube forms the brain and the central nervous system in the trunk, and is essentially formed by the folding of the neural plate into a tube. At neurulation, the edges of the neural plate become raised above the surface, forming two parallel neural folds with a depression—the neural groove—between them. The neural folds eventually come together along the dorsal midline of the embryo and fuse at their edges to form the neural tube, which then separates from the adjacent ectoderm. This surface layer of ectoderm becomes epidermis (Fig. 7.34). The posterior neural tube in the lumbar and tail region, in contrast, develops from a solid rod of cells which develops an interior cavity or lumen. There are, however, variations in different vertebrates; in zebrafish and other fishes, for example, the whole of the neural plate first forms a solid rod, which later becomes hollow.

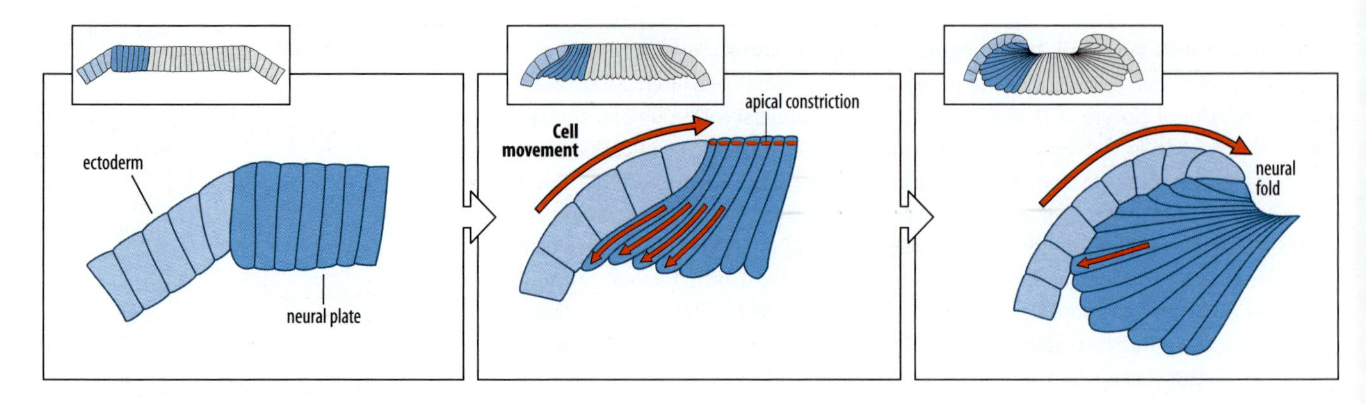

Fig. 7.35 Change in cell shape and cell crawling may drive neural folding. Cells at the edge of the neural plate change their shape and appear to crawl along the under surface of the adjacent ectoderm. This may be partly responsible for causing the neural folds to develop.

7.13 Neural-tube formation is driven by changes in cell shape and cell migration

A very early sign of neurulation in *Xenopus* is the formation of the neural groove along the midline. As with gastrulation, changes in cell shape are associated with the curvature of the neural plate and the formation of the folds, but the mechanism is not well understood. During gastrulation, the cells of the neural plate become longer and narrower than the cells of the adjacent ectoderm (Fig. 7.35). Neurulation commences with the edges of the plate beginning to roll upwards. The cells at the edge of the plate, where it is bending, become constricted at the apical surface, making them wedge-shaped. Apical constriction is induced by the activity of the actin-binding protein Shroom. This change in cell shape could, in principle, draw the edges of the neural plate up into folds. A local change in curvature might also be brought about by the cells at the edge of the plate crawling down the inner surface of the adjacent ectoderm.

The neural folds fuse in the midline, but the prospective dorsal neural tube cells are still some way from the midline and the future lumen of the neural tube is filled with cells. The formation of the dorsal region of the tube and its lumen is due to a complex pattern of movements involving the migration of the neural crest cells to the dorsal midline and radial intercalation of ventral neural cells.

Neurulation does not occur along the whole of the neural plate at the same time, but starts in the region of the midbrain, and proceeds anteriorly and posteriorly (Fig. 7.36). Changes in the shape of neural furrow cells in chick embryos are associated with the folding of the neural tube, but whether they are a cause or a result of folding is not yet clear. The detailed cellular mechanism underlying the shape changes is not yet established and in the chick there is some evidence that tissues lateral to the neural plate could exert forces that cause neural fold formation. Cells in the midline of the chick neural furrow, the so-called hinge point, are wedge-shaped. Later, wedge-shaped cells are seen at additional hinge points on the sides of the furrow, where the furrow curves round further to form a tube (see Fig. 7.36). When the folds meet they zip up into close apposition so that the lumen almost disappears; it only opens up again after the folds fuse.

Although changes in cell shape in general often involve changes in the actin cytoskeleton, it has been established that bending of the neural plate in the mouse is not dependent on actin microfilaments throughout. Treatment of embryos with cytochalasin B, which prevents actin-based contraction, blocks neural fold formation in the head region but not in the spinal cord. By a mechanism that is still unknown, cells at the hinge joints continue to change shape and adopt a wedge-shaped form.

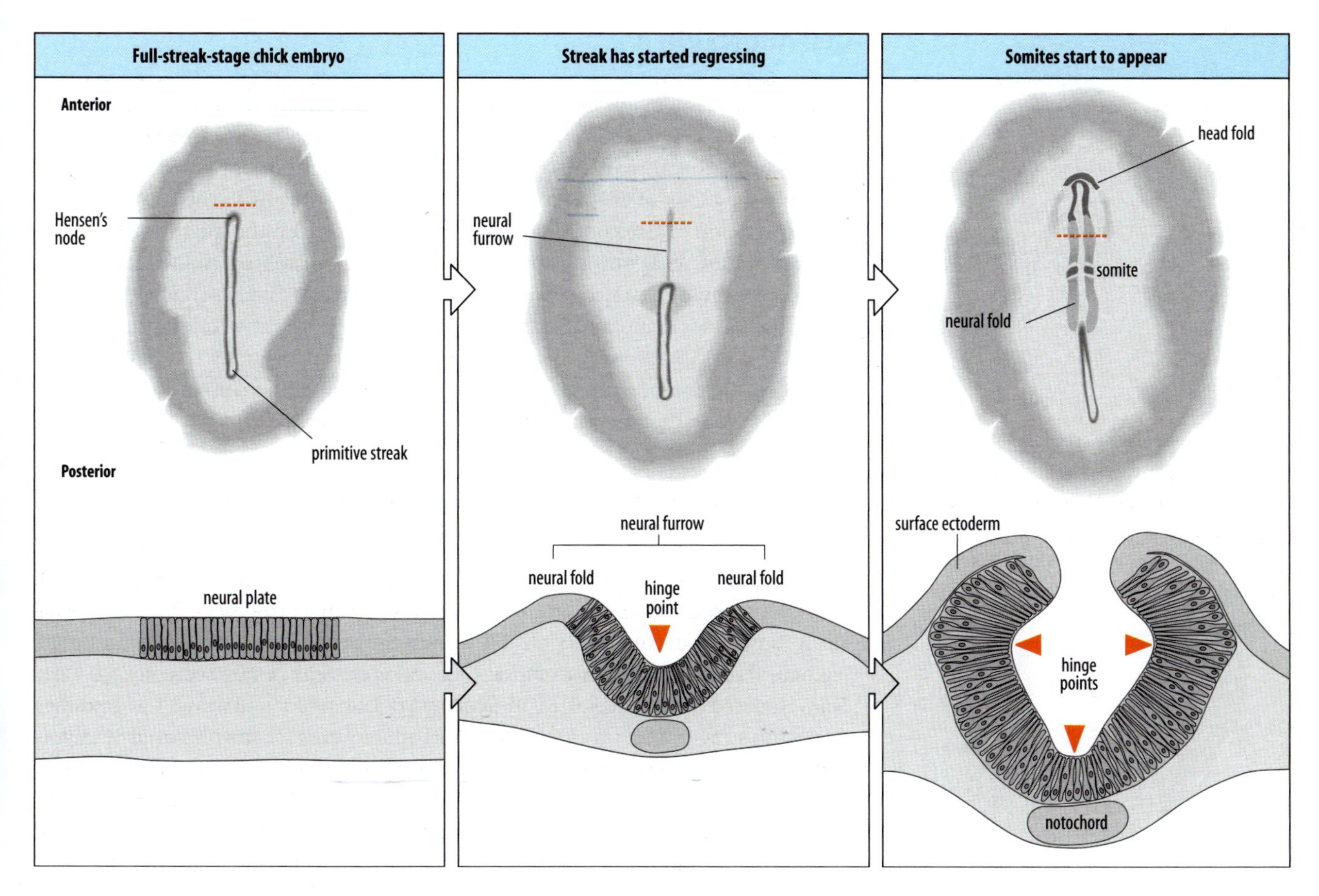

Fig. 7.36 Change in cell shape in the neural plate during chick neurulation. Top: surface views of the chick epiblast. Bottom: cross-sections taken through the epiblast at the sites indicated by the dashed lines in the top diagrams. Cells anterior to Hensen's node become elongated to form the neural plate. Cells in the center of the neural plate become wedge-shaped, defining the hinge point at which the neural plate bends. Additional hinge points with wedge-shaped cells also form on the sides of the furrow.

After Schoenwolf, G.C., et al.: 1990.

The neural tube, which is initially part of the ectoderm, separates from the presumptive epidermis after its formation. This involves changes in cell adhesion. The cells of the neural plate, like the rest of the ectoderm in the chick, initially express the adhesion molecule L-CAM on their surface. However, as the neural folds develop, the neural plate ectoderm begins to express both N-cadherin and N-CAM, whereas the adjacent ectoderm expresses E-cadherin only. These changes in adhesiveness may enable the neural tube to separate from the surrounding ectoderm and allow it to sink beneath the surface, the rest of the ectoderm reforming over it in a continuous layer. The changes in adhesiveness could themselves provide the mechanism for neural-tube formation, in a manner similar to that described in Section 7.1 for the sorting out of cells.

Summary

In vertebrates, neurulation results in neural-tube formation following induction by the mesoderm. Neurulation in mammals, birds, and amphibians results from both the folding of the neural plate into a tube, and the formation of a lumen in a solid rod of cells. Separation of the neural tube from the adjacent ectoderm or rod formation requires changes in cell adhesiveness. The formation of the neural folds and their coming together in the midline is apparently driven by changes in cell shape within the tube itself, as well as by forces generated by adjacent tissues.

Cell migration

Cell migration is a major feature of animal morphogenesis, with cells moving over relatively long distances from one site to another. We have already considered the migrations of the primary mesenchyme cells in the interior of the sea urchin blastula, and now look at the migration of the neural crest cells in the chick embryo. Their migration is also controlled by interactions between the migrating cells and the substratum. As a contrast, we also look at the aggregation of individual amebae of the slime mold *Dictyostelium* into a multicellular fruiting body, which provides an example of how chemotaxis and signal propagation can control migration pattern. We leave until Chapter 10 the migration of immature neurons that is so fundamental to the morphogenesis of the nervous system. Other important examples of cell migration that are dealt with in later chapters include muscle-cell migration in vertebrate limbs (see Chapter 9), and germ-cell migration (see Chapter 11).

7.14 Neural crest migration is controlled by environmental cues and adhesive differences

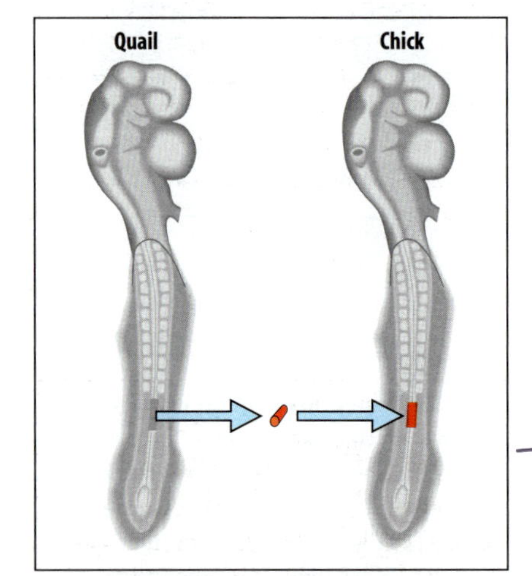

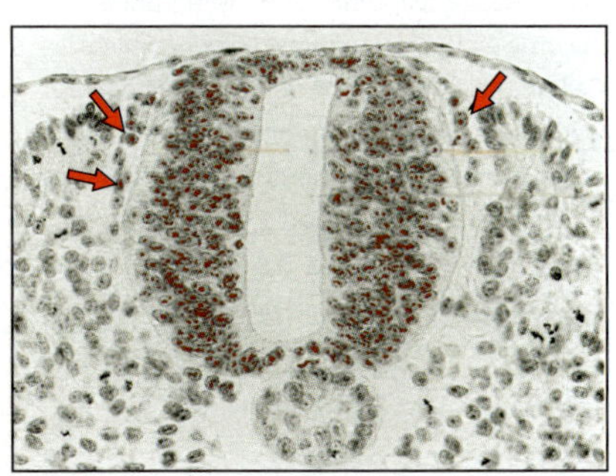

Fig. 7.37 Following cell-migration pathways by grafting a piece of quail neural tube to a chick host. A piece of neural tube from a quail embryo is grafted to a similar position in a chick host. The photograph shows migration of the quail neural crest cells (red arrows). Their migration can be tracked as quail cells have a nuclear marker that distinguishes them from chick cells.

Photograph courtesy of N. Le Douarin.

Neural crest cells of vertebrates have their origin at the edges of the neural plate or tube and first become recognizable during neurulation. At the time of closure of the neural tube in vertebrate embryos, neural crest cells can be seen on each side. They undergo an epithelial to mesenchymal transition and leave the midline, migrating away from it on either side. BMPs are required for the initiation of migration. The epithelial to mesenchymal transition in vertebrates involves the genes *slug* (in chick) and *snail* (in mouse), which control the process by which nonmotile epithelial cells become migrating cells. Inhibition of *slug* expression in the chick inhibits migration.

Neural crest cells give rise to a wide variety of different cell types, which include cartilage in the head, pigment cells in the dermis, the medullary cells of the adrenal gland, glial Schwann cells, and the neurons of both the peripheral and the autonomic nervous systems. The differentiation of neural crest cells is discussed in Chapter 8. Here we focus on the migration of the crest cells in the trunk region of the chick embryo. We have already discussed the migration of neural crest cells from the hindbrain region of the neural tube to form the branchial arches (see Section 4.11).

Various strategies have been employed to follow the migration of neural crest cells. For example, because quail cells have a nuclear marker that distinguishes them from chick cells, grafting a neural tube from a quail embryo into a chick embryo allows the subsequent migration pathways of the quail neural crest cells in the chick embryo to be followed (Fig. 7.37). It is also possible to identify migrating chick neural crest cells by tagging them with labeled monoclonal antibodies, or by labeling them with the dye DiI.

Multiple subtypes of cadherin are dynamically expressed during these processes. In the chick embryo, for example, the ectoderm initially expresses L-CAM; during neural plate invagination, however, L-CAM expression is replaced by that of N-cadherin. At the same time, cadherin-6B begins to be expressed in the invaginating neural plate, most strongly at the neural crest-generating area. When neural crest cells leave the neural tube, these cadherins become undetectable, and cadherin-7 appears instead. Cadherin-7 expression persists during migration of the crest cells. From these observations, it is proposed that changes in cadherin expression during neural crest development may have a role in the segregation of neural crest cells from the neural tube. If N-cadherin or cadherin-7 is constitutively expressed in cells of the dorsal neural tube by injection of retroviral vectors

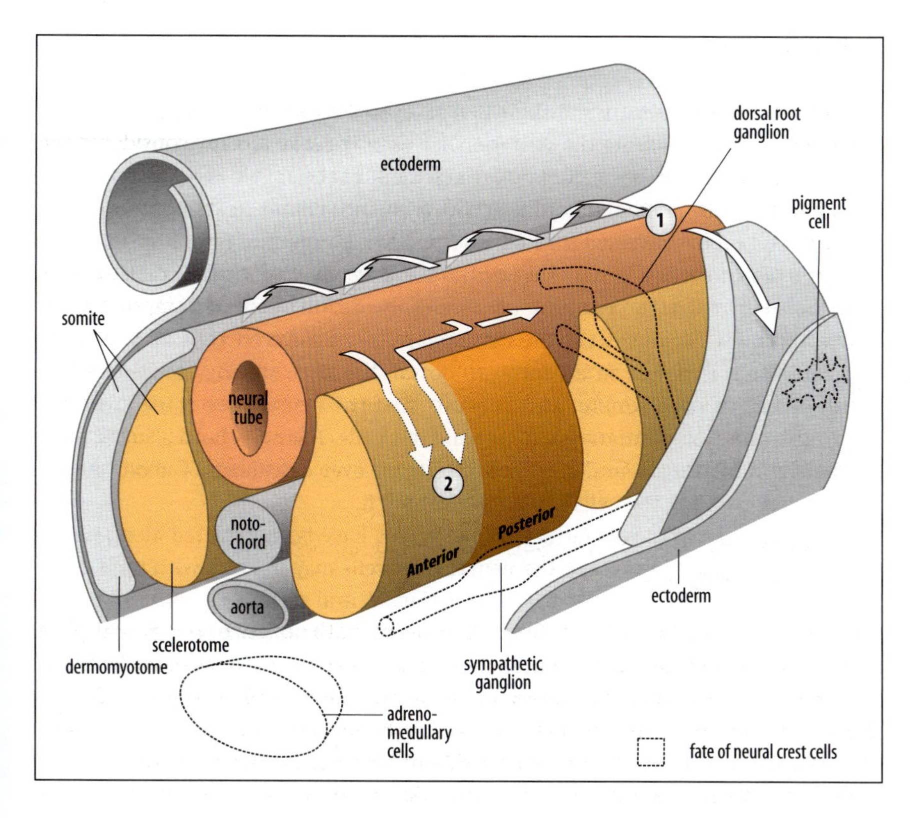

Fig. 7.38 Neural crest migration in the trunk of the chick embryo. One group of cells (1) migrates under the ectoderm to give rise to pigment cells (shown in outline). The other group of cells (2) migrates over the neural tube and then through the anterior half of the somite; the cells do not migrate through the posterior half of the somite. Those that migrate along this pathway give rise to dorsal root ganglia, sympathetic ganglia, and cells of the adrenal cortex, and their future sites are also shown in outline. Those cells opposite the posterior regions of a somite migrate in both directions along the neural tube until they come to the anterior region of a somite. This results in a segmental pattern of migration and is responsible for the segmental arrangement of ganglia.

carrying the cadherin genes, the cells do not turn off N-cadherin expression or they express cadherin-7 prematurely. Under these experimental conditions, the movement of crest cells from the neural tube is dramatically suppressed. Migration also requires loss of adhesion of the cells to the neural tube, and both N-cadherin and E-cadherin are lost from neural crest cells at about the time of migration. Migration of trunk neural crest in the chick embryo is controlled by BMP signaling. A threshold level of BMP is required for migration, and the level is modulated by the activity of the BMP antagonist Noggin in the neural tube.

There are two main migratory pathways for neural crest cells in the trunk of the chick embryo (Fig. 7.38). One goes dorso-laterally under the ectoderm and over the somites; cells that migrate this way mainly give rise to pigment cells, which populate the skin and feathers. The other pathway is more ventral, primarily giving rise to sympathetic and sensory ganglion cells. Some crest cells move into the somites to form dorsal root ganglia; others migrate through the somites to form sympathetic ganglia and adrenal medulla, but appear to avoid the region around the notochord. Trunk neural crest selectively migrates through the anterior (rostral) half of the somite and not through the posterior (caudal) half. Within each somite, neural crest cells are found only in the anterior half, even when they originate in neural crest adjacent to the posterior half of the somite. This behavior is unlike that of the neural crest cells taking the dorsal pathway, which migrate over the whole dorso-lateral surface of the somite. The anterior migration pathway results in the distinct segmental arrangement of spinal ganglia in vertebrates, with one pair of ganglia corresponding to one pair of somites—one segment—in the embryo (Fig. 7.39). The segmental pattern of migration is due to the adhesive properties of the somites. If the somites are rotated through 180° so that their antero-posterior axis is reversed, the crest cells still migrate through the original anterior halves only.

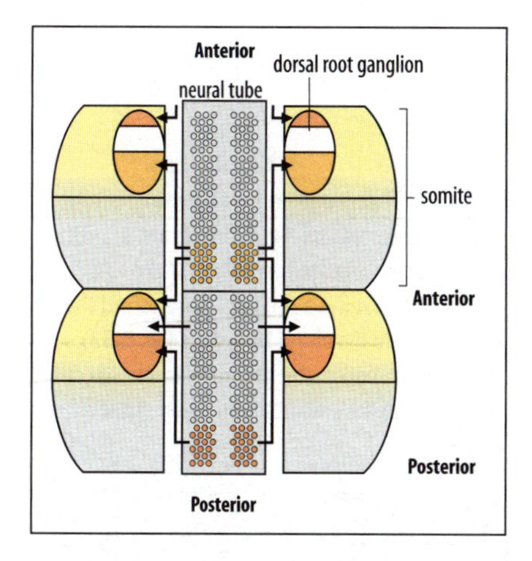

Fig. 7.39 Segmental arrangement of dorsal root ganglia is due to the migration of crest cells through the anterior half of the somite only. Neural crest cells cannot migrate through the posterior (gray) half of a somite but can migrate in either direction along the neural tube and through the anterior half of somites. The dorsal root ganglion in a given segment is thus made up of crest cells from the posterior region of the segment anterior to it, neural crest cells immediately adjacent to it (white), and crest cells from the posterior of its own segment.

Ephrin (ephrinB1 in the chick) is expressed in the posterior halves of the somites, while crest cells have receptors (Eph receptors) for these ligands. Repulsive interactions between the ephrins and their Eph receptors could contribute to the expulsion of the crest cells from the posterior halves of the somites. This would provide a molecular basis for the segmental arrangement of the spinal ganglia.

The neural tube and notochord also both influence neural crest migration. If the early neural tube is inverted through 180°, before neural crest migration starts, so that the dorsal surface is now the ventral surface, one might think that the cells that normally migrate ventrally, now being nearer to their destination, would move ventrally. But this is not the case, and many of the crest cells move upward through the sclerotome in a ventral to dorsal direction, staying confined to the anterior half of each somite. This suggests that the neural tube somehow influences the direction of migration of neural crest cells. The notochord also exerts an influence, inhibiting neural crest cell migration over a distance of about 50 μm, and thus preventing the cells from approaching it.

Many different extracellular matrix molecules have been detected along neural crest migratory pathways, and the neural crest cells may interact with their molecules by means of their cell-surface integrins. Neural crest cells cultured *in vitro* adhere to, and migrate efficiently on, fibronectin, laminin, and various collagens. Blocking the adhesion of the neural crest cells to fibronectin or laminin by blocking the integrin β_1 subunit *in vivo* causes severe deficiencies in the head region but not in the trunk, suggesting that the crest cells in these two regions adhere by different mechanisms, probably involving other integrins. It is striking how neural crest cells in culture will preferentially migrate along a track of fibronectin, although the role of this molecule in guiding the cells in the embryo is still unclear.

7.15 Slime mold aggregation involves chemotaxis and signal propagation

The myxamebae of the cellular slime mold *D. discoideum* feed on bacteria. When a local food source is exhausted, they enter the multicellular stage of the slime mold life cycle (see Fig. 5.33). The first phase of this stage is aggregation, which when observed in time-lapse videos presents a dramatic spectacle, the cells streaming toward a center of aggregation like small rivers of cells converging into a lake. As they approach the focus of aggregation, the cells in the streams adhere to each other at their anterior and posterior ends through a membrane glycoprotein that is expressed at this stage. Eventually the cells collect into a compact multicellular mound (Fig. 7.40), which develops into a stalked fruiting body (see Fig. 5.33).

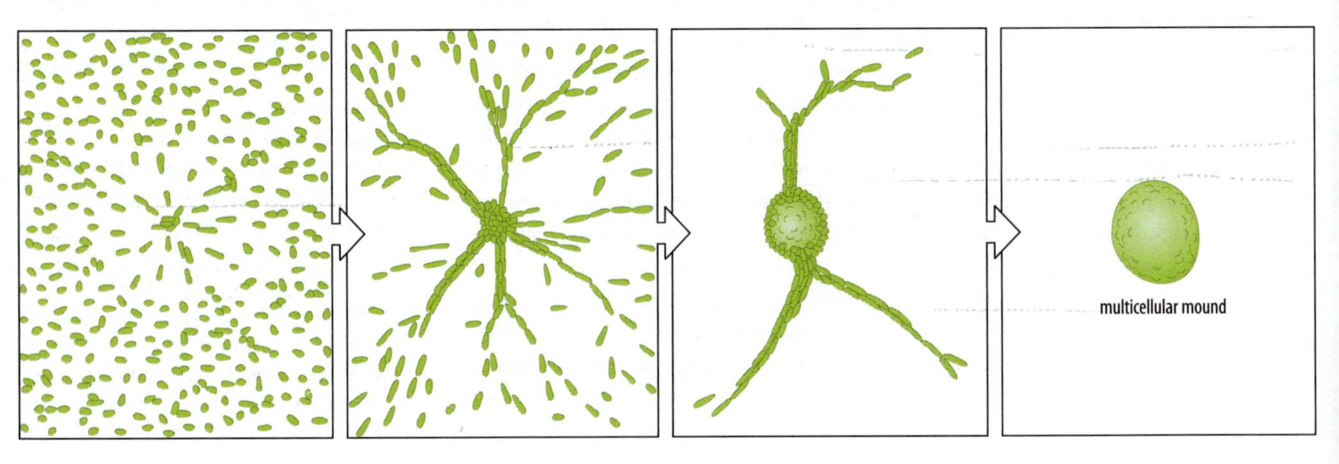

Fig. 7.40 Aggregation in the cellular slime mold *Dictyostelium*. Aggregating amebae of *D. discoideum* stream towards a focal point, eventually forming a multicellular mound that develops into a fruiting body (see Fig. 5.33). In the cell streams, the normally ameboid cells become bipolar and adhere to each other at either end.

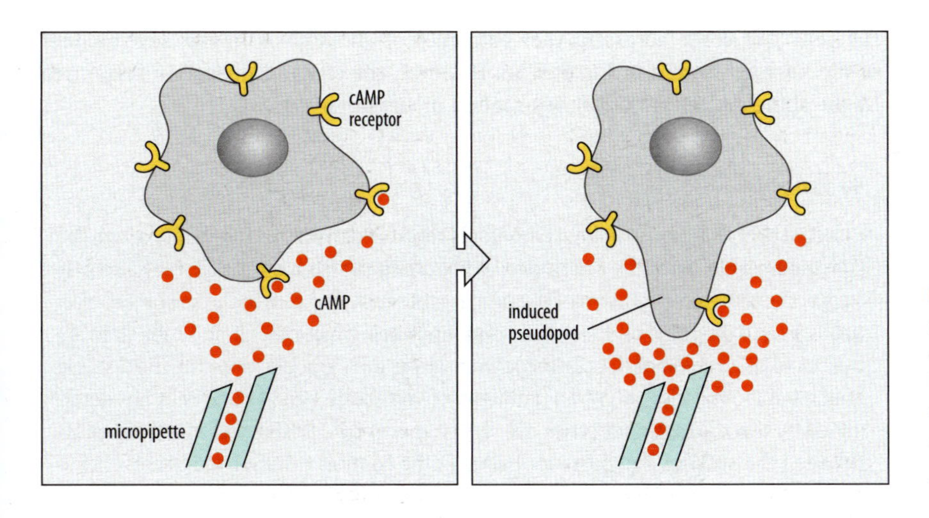

Fig. 7.41 Chemotactic response of slime mold cells. Amebae will move up a gradient of cyclic AMP (cAMP). A localized source of cAMP binds to membrane receptors on the side of the cell facing the source, resulting in the ameba extending a pseudopod toward the source.
After Alberts, B., et al.: 1989.

The cells move intermittently rather than continuously, in pulses of inward movement. The mechanism of aggregation involves both chemotaxis of individual cells, and the propagation of the chemotactic signal from one ameba to another. In *Dictyostelium*, cyclic AMP (cAMP) is the chemoattractant. Amebae respond to an increasing concentration of cAMP, and will thus move up a gradient of cAMP by extending a pseudopod in the direction of the source (Fig. 7.41). Chemotaxis up such a gradient only operates over short distances, much less than 1 mm, because it is difficult to establish a reliable diffusion gradient of a chemoattractant over larger distances. Yet amebae can aggregate from distances up to 5 mm from the center. This is achieved by the propagation of the chemotactic signal in a manner rather similar to the conduction of a nerve impulse. During aggregation the cells move with a periodicity of about 6 minutes. They first become highly polarized as cAMP binds to its receptors, and then move up the gradient for about 1 minute. They then move randomly for 5 minutes. An extracellular phosphodiesterase then clears the medium of cAMP, allowing the amebae to respond to the next wave of the chemoattractant.

Dictyostelium cells can respond chemotactically to a stable gradient in which there is only a 2% difference in cAMP across the cell. When an ameba receives a pulse of cAMP, the cAMP binds to membrane receptors that stimulate the cell not only to respond chemotactically by moving towards the source, but also to propagate the signal by producing a pulse of cAMP itself. By this relay mechanism, a wave of cAMP is propagated across the field of aggregating amebae (Fig. 7.42). The pulsatory nature of the signal is shown by introducing a micropipette containing cAMP into a field of amebae ready to aggregate: it can act as an aggregation center only if it provides cAMP pulses of the correct frequency. An important feature of the propagation of the signal is that the cells are refractory to cAMP for a short time immediately after they have given out a pulse of cAMP. This ensures that the pulse only propagates outward. The presence of the enzyme phosphodiesterase, which breaks down cAMP, prevents its concentration rising to a level where it saturates the system, which would prevent propagation.

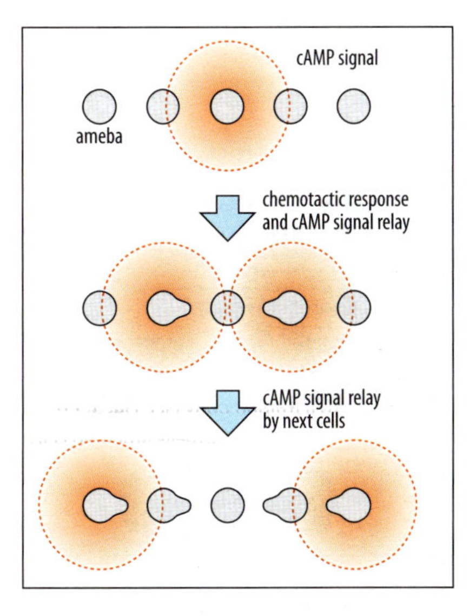

Fig. 7.42 Propagation of a cAMP signal and chemotaxis by slime mold amebae. A cell at the center of the aggregate gives off a pulse of cAMP. This induces pseudopod formation and chemotactic movement in the surrounding cells in the direction of the cAMP source. It also causes the responding cells to produce a pulse of cAMP themselves, and the signal is thus relayed outward. After releasing a pulse of cAMP, a cell becomes refractory to cAMP for a short time. This ensures the signal is propagated outward only.

Aggregation of slime mold cells is the best-understood example of chemotaxis in a developing system. The chemotactic response involves local activation of actin polymerization at the leading edge, causing protrusion of the cell in this direction, accompanied by inhibition of such activity over the rest of the cell. Myosin assembly is inhibited at the front, but myosin and actin cause contraction at the rear of the cells. We come across further examples of chemotaxis in the development of the nervous system in Chapter 10. However, the signal propagation relay used by the slime mold has not yet been found in any other system.

Summary

Directed cell migration in animal embryos is controlled mainly by interactions with the substratum over which the cells move, although signals from other cells may also play a part. The directed migration of the primary mesenchyme cells in the sea urchin blastula is due to filopodia on these cells exploring their environment, and then drawing the cell to that region of the blastocoel wall where the filopodia make the most stable attachment. Similarly, migration pathways of vertebrate neural crest cells are determined by the adhesive properties over which they move. Differences in adhesiveness between the anterior and posterior halves of the somites result in neural crest being prevented from migrating over or through posterior halves. Thus, presumptive dorsal ganglia cells collect adjacent to anterior halves, giving them a segmental arrangement.

Aggregation in the cellular slime mold *Dictyostelium* involves both chemotaxis and signal propagation. The individual amebae extend processes in the direction of an increasing concentration of cAMP, and move toward the source. In addition, the cells respond to a pulse of cAMP by giving out a pulse of cAMP themselves, and this results in propagation of the signal, enabling cells as far away as 5 mm to be attracted to the aggregation center.

Summary: directed cell migration of vertebrate neural crest

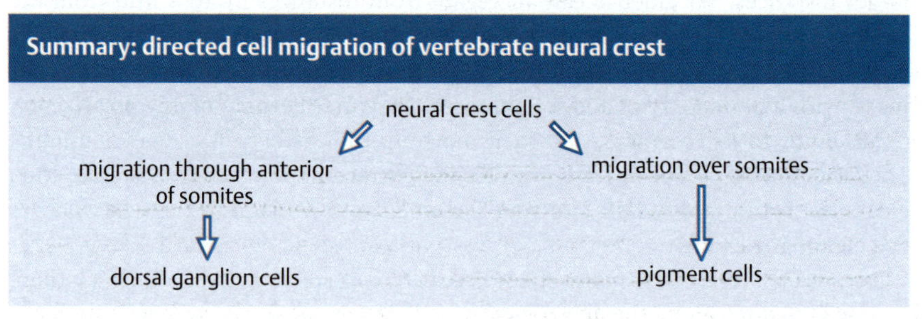

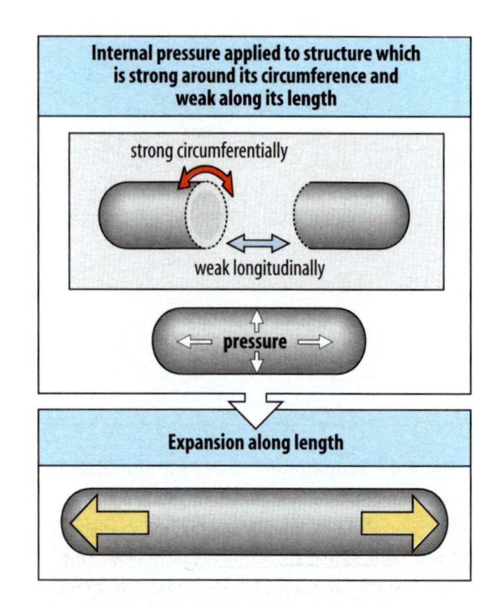

Fig. 7.43 Directed dilation. Hydrostatic pressure inside a constraining sheath or membrane can lead to elongation of the structure. If the circumferential resistance is much greater than the longitudinal resistance, as it is in the notochord sheath, the rod of cells inside the sheath lengthens.

Directed dilation

Hydrostatic pressure can provide the force for morphogenesis in a variety of situations. An increase in hydrostatic pressure inside a spherical sheet of cells causes the sphere to increase in volume. We have already seen how hydrostatic pressure is involved in blastula formation in both the mouse and amphibian. Here, we consider examples of directed dilation where the increase in pressure causes an asymmetric change in shape. If the circumferential resistance to pressure in a tube is much greater than the resistance to longitudinal extension, then an increase in the internal pressure causes an increase in length (Fig. 7.43).

7.16 Later extension and stiffening of the notochord occurs by directed dilation

After the *Xenopus* notochord has formed as described in Section 7.12, its volume increases threefold, and there is considerable further lengthening as it straightens and becomes stiffer. At this stage, the notochord has become surrounded by a sheath of extracellular material, which restricts circumferential expansion but does allow expansion in the antero-posterior direction. The cells within the notochord develop fluid-filled vacuoles and expand in volume as a result. They thus exert hydrostatic pressure on the notochord sheath, resulting in **directed dilation**: circumferential expansion of the notochord is prevented by the resistance of the sheath, and this ensures that the increase in volume (dilation) is directed along the notochord's long axis.

The vacuoles in the notochord cells are filled with glycosaminoglycans which, because of their high carbohydrate content, tend to attract water into the vacuoles by osmosis. It is this that produces the hydrostatic pressure that causes the increase in cell volume, and the consequent stiffening and straightening of the notochord. Changes in the structure of the sheath during the period of notochord elongation fit well with the proposed hydrostatic mechanism. The sheath contains both glycosaminoglycans, which have little tensile strength, and the fibrous protein collagen, which has a high tensile strength; during notochord dilation the density of collagen fibers increases, providing resistance to circumferential expansion. The crucial role of the sheath in dilation and elongation is shown by the fact that if it is digested away, the notochord buckles and folds and the notochord cells, instead of being flat, become rounded.

7.17 Circumferential contraction of hypodermal cells elongates the nematode embryo

During the early development of the nematode there is little change in body shape from the spherical form of the fertilized egg, even during gastrulation. After gastrulation, about 5 hours after fertilization, the embryo begins to elongate rapidly along its antero-posterior axis. Elongation takes about 2 hours, during which time the nematode embryo decreases in circumference about threefold and undergoes a fourfold increase in length.

This elongation is brought about by a change in shape of the hypodermal (epidermal) cells that make up the outermost layer of the embryo; their destruction by laser ablation prevents elongation. During embryo elongation, these cells change shape so that, instead of being elongated in the circumferential direction, they become elongated along the antero-posterior axis (Fig. 7.44). Throughout this elongation the hypodermal cells remain attached to each other by desmosome cell junctions. The desmosomes are also linked within the cells by actin-containing

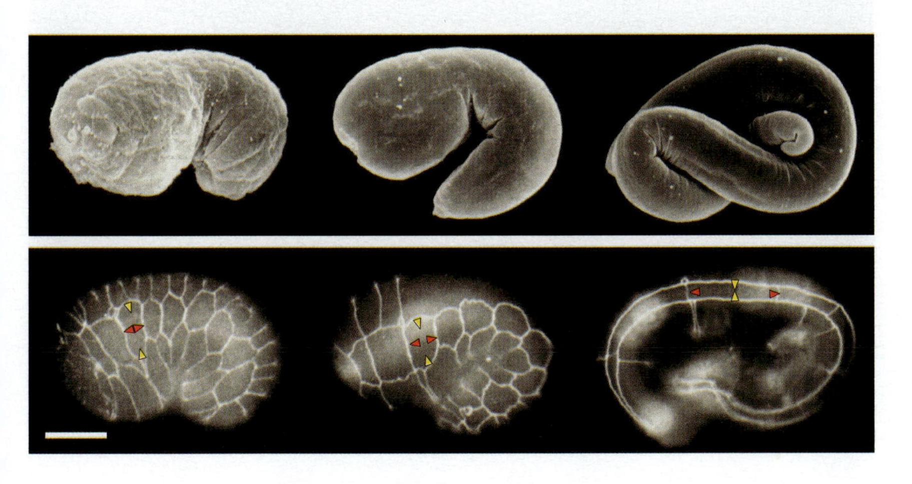

Fig. 7.44 Increase in nematode body length by directed dilation. The change in body shape over 2 hours is illustrated in the top panel. The increase in length is due to circumferential contraction of the hypodermal cells, as shown in the bottom panel. The change in shape of a single cell can be seen in the cell marked with arrows. Scale bar = 10 μm.

Photographs courtesy of J. Priess, from Priess, J.R., et al.: 1986.

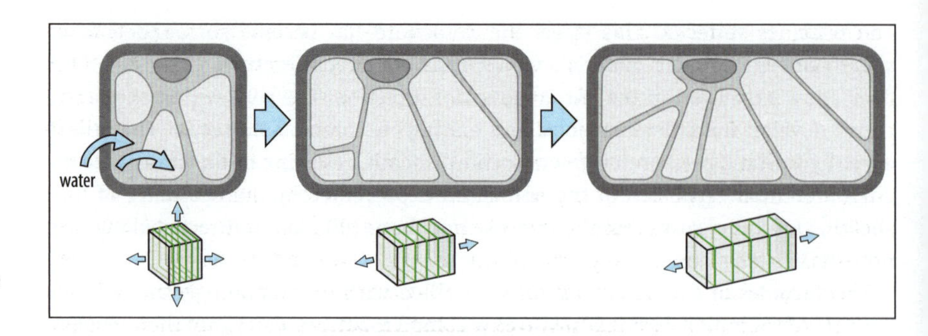

Fig. 7.45 Enlargement of a plant cell. Plant cells expand as water enters the cell vacuoles and thus causes an increase in intracellular hydrostatic pressure. The cell elongates in a direction perpendicular to the orientation of the cellulose fibrils in its cell wall.

fibers that run circumferentially, and these fibers appear to shorten as the cells elongate. The disruption of actin filaments by cytochalasin D treatment blocks elongation, and so it is very likely that their contraction brings about the change in cell shape. Circumferential contraction of the hypodermal cells causes an increase in hydrostatic pressure within the embryo, forcing an extension in an antero-posterior direction. Circumferentially oriented microtubules may also have a mechanical role in constraining the expansion, in the same way as the sheath of the *Xenopus* notochord described above. Increase in nematode body length is thus another example of directed dilation.

7.18 The direction of cell enlargement can determine the form of a plant leaf

Cell enlargement is a major process in plant growth and morphogenesis, providing up to a 50-fold increase in the volume of a tissue. The driving force for expansion is the hydrostatic pressure—turgor pressure—exerted on the cell wall as the protoplast swells as a result of the entry of water into cell vacuoles by osmosis (Fig. 7.45). Plant-cell expansion involves synthesis and deposition of new cell-wall material, and is

Fig. 7.46 The shape of the leaves of *Arabidopsis* is affected by mutations affecting cell elongation. The *angustifolia* mutation results in cells with narrow leaves, while *rotundifolia* mutations cause short, fat leaves to develop.

Photograph courtesy of H. Tsukaya, from Tsuge, T. et al.: 1996.

an example of directed dilation. It is accompanied by an increase in both protein and RNA synthesis. The direction of cell growth is determined by the orientation of the cellulose fibrils in the cell wall. Enlargement occurs primarily in a direction at right angles to the fibrils, where the wall is weakest. The orientation of cellulose fibrils in the cell wall is thought to be determined by the microtubules of the cell's cytoskeleton, which are responsible for positioning the enzyme assemblies that synthesize cellulose at the cell wall. Plant growth hormones, such as ethylene and gibberellins, alter the orientation in which the fibrils are laid down and so can alter the direction of expansion. Auxin aids expansion by loosening the structure of the cell wall.

The development of a leaf involves a complex pattern of cell division and cell elongation, with cell elongation playing a central part in the expansion of the leaf blade. Two mutations that affect the shape of the blade by affecting the direction of cell elongation have been identified. Leaves of the *Arabidopsis* mutant *angustifolia* are similar in length to the wild type but are much thinner (Fig. 7.46). In contrast, the *rotundifolia* mutations reduce the length of the leaf relative to its width. Neither of these mutations affects the number of cells in the leaf. Examination of the cells in the developing leaf shows that these mutations are affecting the direction of elongation of the enlarging cells.

Summary

Directed dilation results from an increase in hydrostatic pressure and unequal peripheral resistance to this pressure. Extension of the notochord is brought about by directed dilation, in which the notochord increases in volume while its circumferential expansion is constrained by the notochord sheath, forcing it to elongate. Similarly, the nematode embryo elongates after gastrulation due to a circumferential contraction in the outer hypodermal cells that generates pressure on the internal cells, forcing the embryo to extend in an antero-posterior direction. In plants, the direction of cell enlargement determines the shape of leaves. In plant cell enlargement, the direction of elongation is yet another example of directed dilation.

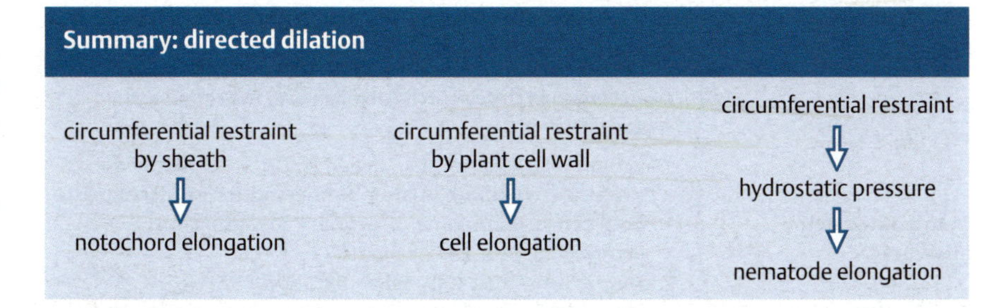

SUMMARY TO CHAPTER 7

Cells are held together by specific adhesion molecules. Changes in the shape of the embryo and cell migration are due to changes in cell adhesion and to forces generated by the cell. Formation of the blastula results from cell division and cell polarization, and in some cases from the movement of water into the blastocoel. Gastrulation involves major movements of cell sheets so that the future endoderm and mesoderm move inside the embryo to their appropriate positions in relation to the main body plan. Invagination of the mesoderm in both sea urchins and *Drosophila* involves similar changes in cell shape. Convergent extension has a key role in both sea urchin and amphibian gastrulation, and is the result of particular patterns of cell intercalation. Epiboly, the spreading of a multilayered sheet, is due to intercalation. Convergent extension, together with directed dilation, is also involved in notochord formation. Directed dilation causes the elongation of the nematode embryo and the direction of enlargement of plant cells. Changes in cell shape and cell adhesion are responsible for neural-tube formation. Directed cell migration of sea urchin mesenchyme and vertebrate neural crest cells is controlled by the adhesiveness of the substratum over which the cells move. Migration and aggregation of slime mold cells is by propagation of a chemotactic signal.

GENERAL FURTHER READING

Alberts, B. *et al.*: *Molecular Biology of the Cell* (4th edn). New York: Garland Science, 2002.

Stern, C. *et al.*: *Gastrulation: From Cells to Embryos*. New York: Cold Spring Harbor Laboratory Press; 2004.

SECTION FURTHER READING

7.1 Sorting out of dissociated cells demonstrates differences in cell adhesiveness in different tissues

Townes, P., Holtfreter, J.: **Directed movements and selected adhesions of embryonic amphibian cells.** *J. Exp. Zool.* 1955, **128:** 53–120.

Box 7A Cell-adhesion molecules and cell junctions

Cunningham, B.A.: **Cell adhesion molecules as morphoregulators.** *Curr. Opin. Cell Biol.* 1995, **7:** 628–633.

Hynes, R.O: **Integrins: bidirectional allosteric signaling mechanisms.** *Cell* 2002, **110:** 673–687.

Tepass, U., Truong, K., Godt, D., Ikura, M., Peifer, M.: **Cadherins in embryonic and neural morphogenesis.** *Nat. Rev. Mol. Cell Biol.* 2000, **1:** 91–100.

Thiery, J.P.: **Cell adhesion in development: a complex signaling network.** *Curr. Opin. Genet. Dev.* 2003, **13:** 365–371.

7.2 Cadherins can provide adhesive specificity

Duguay, D., Foty, R.A., Steinberg, M.S.: **Cadherin-mediated cell adhesion and tissue segregation: qualitative and quantitative determinants.** *Dev. Biol.* 2003, **253:** 309–323.

Levine, E., Lee, C.H., Kintner, C., Gumbiner, B.M.: **Selective disruption of E-cadherin function in early *Xenopus* embryos by a dominant negative mutant.** *Development* 1994, **120:** 901–909.

Takeichi, M., Nakagawa, S., Aono, S., Usui, T., Uemura, T.: **Patterning of cell assemblies regulated by adhesion receptors of the cadherin superfamily.** *Proc. Roy. Soc. Lond. B* 2000, **355:** 885–896.

7.3 The asters of the mitotic apparatus determine the plane of cleavage at cell division

Glotzer, M.: **Cleavage furrow positioning.** *J. Cell Biol.* 2004, **164:** 347–351.

Staiger, C., Doonan, J.: **Cell division in plants.** *Curr. Opin. Cell Biol.* 1993, **5:** 226–231.

7.4 Cells become polarized in early mouse and sea urchin blastulas

Fleming, T.P., Johnson, M.H.: **From egg to epithelium.** *Annu. Rev. Cell Biol.* 1988, **4:** 459–485.

Sobel, J.S.: **Membrane-cytoskeletal interactions in the early mouse embryo.** *Semin. Cell Biol.* 1990, **1:** 341–347.

Sutherland, A.E., Speed, T.P., Calarco, P.G.: **Inner cell allocation in the mouse morula: the role of oriented division during fourth cleavage.** *Dev. Biol.* 1990, **137:** 13–25.

Winkel, G.K., Ferguson, J.E., Takeichi, M., Nuccitelli, R.: **Activation of protein kinase C triggers premature compaction in the four-cell stage mouse embryo.** *Dev. Biol.* 1990, **130:** 1–15.

7.5 Ion transport is involved in fluid accumulation in the frog blastocoel

Warner, A.E.: **Physiological approaches to early development.** *Recent Adv. Physiol.* 1984, **10:** 87–123.

7.6 Internal cavities can be created by cell death

Coucouvanis, E., Martin, G.R.: **Signals for death and survival: a two-step mechanism for cavitation in the vertebrate embryo.** *Cell* 1995, **83:** 279–287.

7.7 Gastrulation in the sea urchin involves cell migration and invagination

Davidson, L.A., Koehl, M.A.R., Keller, R., Oster, G.F.: **How do sea urchins invaginate? Using biomechanics to distinguish between mechanisms of primary invagination.** *Development* 1995, **121:** 2005–2017.

Davidson, L.A., Oster, G.F., Keller, R.E., Koehl, M.A.R.: **Measurements of mechanical properties of the blastula wall reveal which hypothesized mechanisms of primary invagination are physically plausible in the sea urchin *Stronglyocentrotus* purpuratus.** *Dev. Biol.* 1998, **204:** 235–250.

Ettensohn, C.A.: **Cell movements in the sea urchin embryo.** *Curr. Opin. Genet. Dev.* 1999, **9:** 461–465.

Gustafson, T., Wolpert, L.: **Studies on the cellular basis of morphogenesis in the sea urchin embryo. Directed movements of primary mesenchyme cells in normal and vegetalized larvae.** *Exp. Cell Res.* 1999, **253:** 288–295.

Hardin, J., McClay, D.R.: **Target recognition by the archenteron during sea urchin gastrulation.** *Dev. Biol.* 1990, **142:** 86–102.

Hodor, P.G., Illies, M.R., Broadley, S., Ettensohn, C.A.: **Cell-substrate interactions during sea urchin gastrulation: migrating primary mesenchyme cells interact with and align extracellular matrix fibers that contain ECM3, a molecule with NG2-like and multiple calcium-binding domains.** *Dev. Biol.* 2000, **222:** 181–194.

Malinda, K.A., Fisher, G.W., Ettensohn, C.A.: **Four-dimensional microscopic analysis of the filopodial behavior of primary mesenchyme cells during gastrulation in the sea urchin embryo.** *Dev. Biol.* 1995, **172:** 552–566.

Miller, J.R., McClay, D.R.: **Changes in the pattern of adherens junction-associated β-catenin accompany morphogenesis in the sea urchin embryo.** *Dev. Biol.* 1997, **192:** 310–322.

Odell, G.M., Oster, G., Alberch, P., Burnside, B.: **The mechanical basis of morphogenesis. I. Epithelial folding and invagination.** *Dev. Biol.* 1981, **85:** 446–462.

Peterson, R.E., McClay, D.R.: **Primary mesenchyme cell patterning during the early stages following ingression.** *Dev. Biol.* 2003, **254:** 68–78.

Raftopoulou, M., Hall, A.: **Cell migration: Rho GTPases lead the way.** *Dev. Biol.* 2004, **265:** 23–32.

Ridley, A.J., Schwartz, M.A., Burridge, K., Firtel, R.A., Ginsberg, M.H., Borisy, G., Parsons, J.T., Horwitz, A.R.: **Cell migration: integrating signals from front and back.** *Science* 2003, **302:** 1704–1709.

7.8 Mesoderm invagination in *Drosophila* is due to changes in cell shape that are controlled by genes that pattern the dorso-ventral axis

Dawes-Hoang, R.E., Parmar, K.M., Christiansen, A.E., Phelps, C.B., Brand, A.H., Wieschaus, E.F.: **Folded gastrulation, cell shape change and the control of myosin localization.** *Development* 2005, **132:** 4165–4178.

Leptin, M., Affolter, M.: ***Drosophila* gastrulation: identification of a missing link.** *Curr. Biol.* 2004, **14:** R480–R482.

Wilson, R., Leptin, M.: **FGF-receptor dependent morphogenesis of the *Drosophila* mesoderm.** *Proc. Roy. Soc. Lond. B* 2000, **355:** 891–895.

7.9 Germ-band extension in *Drosophila* involves myosin-dependent intercalation

Bertet, C., Sulak, L., Lecuit, T.: **Myosin-dependent junction remodelling controls planar cell intercalation and axis elongation.** *Nature* 2004, **429:** 667–671.

7.10 Dorsal closure in *Drosophila* and ventral closure in *C. elegans* are brought about by the action of filopodia

Chin-Sang, I.D., Chisholm, A.D.: **Form of the worm: genetics of epidermal morphogenesis in *C. elegans*.** *Trends Genet.* 2000, **16:** 544–551.

Jacinto, A., Woolner, S., Martin, P.: **Dynamic analysis of dorsal closure in *Drosophila*: from genetics to cell biology.** *Dev. Cell* 2002, **3:** 9–19.

Woolner, S., Jacinto, A., Martin, P.: **The small GTPase Rac plays multiple roles in epithelial sheet fusion — dynamic studies of *Drosophila* dorsal closure.** *Dev. Biol.* 2005, **282:** 163–173.

7.11 Vertebrate gastrulation involves several different types of tissue movement

Goto, T., Davidson, L., Asashima, M., Keller, R.: **Planar cell polarity genes regulate polarized extracellular matrix deposition during frog gastrulation.** *Curr. Biol.* 2005, **15:** 787–793.

Keller, R.: **Cell migration during gastrulation.** *Curr. Opin. Cell Biol.* 2005, **17:** 533–541.

Montero, J.A., Heisenberg, C.P.: **Gastrulation dynamics: cells move into focus.** *Trends Cell Biol.* 2004, **14:** 620–627.

Montero, J.A., Carvalho, L., Wilsch-Brauninger, M., Kilian, B., Mustafa, C., Heisenberg, C.P.: **Shield formation at the onset of zebrafish gastrulation.** *Development* 2005, **132:** 1187–1198.

Shih, J., Keller, R.: **Gastrulation in *Xenopus laevis*: involution—a current view.** *Dev. Biol.* 1994, **5:** 85–90.

Wacker, S., Grimm, K., Joos, T., Winklbauer, R.: **Development and control of tissue separation at gastrulation in *Xenopus*.** *Dev. Biol.* 2000, **224:** 428–439.

Yang, X., Dormann, D., Munsterberg, A.E., Weijer, C.J.: **Cell movement patterns during gastrulation in the chick are controlled by positive and negative chemotaxis mediated by FGF4 and FGF8.** *Dev. Cell* 2002, **3:** 425–437.

7.12 Convergent extension and epiboly are due to different types of cell intercalation

Adams, D.S., Keller, R., Koehl, M.A.: **The mechanics of notochord elongation, straightening, and stiffening in the embryo of *Xenopus laevis*.** *Development* 1990, **100:** 115–130.

Keller, R., Cooper, M.S., D'Anilchik, M., Tibbetts, P., Wilson, P.A.: **Cell intercalation during notochord development in *Xenopus laevis*.** *J. Exp. Zool.* 1989, **251:** 134–154.

Keller, R., Shih, J., Sater, A.: **The cellular basis of the convergence and extension of the *Xenopus* neural plate.** *Dev. Dyn.* 1992, **193:** 199–217.

Keller, R., Davidson, L., Edlund, A., Elul, T., Ezin, M., Shook, D., Skoglund, P.: **Mechanisms of convergence and extension by cell intercalation.** *Proc Roy. Soc. Lond. B* 2000, **355:** 897–922.

Ninomiya, H., Elinson, R.P., Winklbauer, R.: **Antero-posterior tissue polarity links mesoderm convergent extension to axial patterning.** *Nature* 2004, **430:** 364–367.

Wallingford, J.B., Fraser, S.E., Harland, R.M.: **Convergent extension: the molecular control of polarized cell movement during embryonic development.** *Dev. Cell* 2002, **2:** 695–706.

7.13 Neural-tube formation is driven by changes in cell shape and cell migration

Davidson, L.A., Keller, R.E.: **Neural tube closure in *Xenopus laevis* involves medial migration, directed protrusive activity, cell intercalation and convergent extension.** *Development* 1999, **126:** 4547–4556.

Haigo, S.L., Hildebrand, J.D., Harland, R.M., Wallingford, J.B.: **Shroom induces apical constriction and is required for hingepoint formation during neural tube closure.** *Curr. Biol.* 2003, **13:** 2125–2137.

Lowery, L.A., Sive, H.: **Strategies of vertebrate neurulation and re-evaluation of teleost neural tube formation.** *Development* 2005, **132:** 2057–2067.

Schoenwolf, G.C., Smith, J.L.: **Mechanisms of neurulation: traditional viewpoint and recent advances.** *Development* 1990, **109:** 243–270.

Ybot-Gonzalez, P., Copp, A.J.: **Bending of the neural plate during mouse spinal neurulation is independent of actin microfilaments.** *Dev. Dyn.* 1999, **215:** 273–283.

7.14 Neural crest migration is controlled by environmental cues and adhesive differences

Bronner-Fraser, M.: **Mechanisms of neural crest migration.** *BioEssays* 1993, **15:** 221–230.

Delannet, M., Martin, F., Bussy, B., Chersh, D.A., Reichardt, L.F., Duband, J.L.: **Specific roles of the $\alpha_V\beta_1$, $\alpha_V\beta_3$ and $\alpha_V\beta_5$ integrins in avian neural crest cell adhesion and migration on vitronectin.** *Development* 1994, **120:** 2687–2702.

Erickson, C.A., Perris, R.: **The role of cell-cell and cell-matrix interactions in the morphogenesis of the neural crest.** *Dev. Biol.* 1993, **159:** 60–74.

Holder, N., Klein, R.: **Eph receptors and ephrins: effectors of morphogenesis.** *Development* 1999, **126:** 2033–2044.

Nagawa, S., Takeichi, M.: **Neural crest emigration from the neural tube depends on regulated cadherin expression.** *Development* 1998, **125:** 2963–2971.

Nieto, M.A., Sargent, M.G., Wilkinson, D.G., Cooke, J.: **Control of cell behaviour during vertebrate development by *Slug*, a zinc finger gene.** *Science* 1994, **264:** 835–839.

Poliakoff, A., Cotrina, M., Wilkinson, D.G.: **Diverse roles of Eph receptors and ephrins in the regulation of cell migration and tissue assembly.** *Dev. Cell* 2004, **7:** 465–480.

Tucker, R.P.: **Neural crest cells: a model for invasive behavior.** *Int. J. Biochem. Cell Biol.* 2004, **36:** 173–177.

Xu, Q., Mellitzer, G., Wilkinson, D.G.: **Roles of Eph receptors and ephrins in segmental patterning.** *Proc. Roy. Soc. Lond. B* 2000, **353:** 993–1002.

7.15 Slime mold aggregation involves chemotaxis and signal propagation

Chisholm, R.L., Firtel, R.A.: **Insights into morphogenesis from a simple developmental system.** *Nat. Rev. Mol. Cell Biol.* 2004, **5:** 531–541.

Firtel, R.A., Chung, C.Y.: **The molecular genetics of chemotaxis: sensing and responding to chemoattractant gradients.** *BioEssays* 2000, **22:** 603–615.

Gerisch, G.: **Cyclic AMP and other signals controlling cell development and differentiation in *Dictyostelium*.** *Annu. Rev. Biochem.* 1987, **56:** 853–879.

Sager, B.M.: **Propagation of traveling waves in excitable media.** *Genes Dev.* 1996, **10:** 2237–2250.

Siu, C.H.: **Cell-cell adhesion molecules in *Dictyostelium*.** *BioEssays* 1990, **12:** 357–362.

7.16 Later extension and stiffening of the notochord occurs by directed dilation

Adams, D.S., Keller, R., Koehl, M.A.: **The mechanics of notochord elongation, straightening and stiffening in the embryo of *Xenopus* laevis.** *Development* 1990, **110:** 115–130.

7.17 Circumferential contraction of hypodermal cells elongates the nematode embryo

Priess, J.R., Hirsh, D.I.: ***Caenorhabditis* elegans morphogenesis: the role of the cytoskeleton in elongation of the embryo.** *Dev. Biol.* 1986, **117:** 156–173.

7.18 The direction of cell enlargement can determine the form of a plant leaf

Jackson, D.: **Designing leaves. Plant morphogenesis.** *Curr. Biol.* 1996, **6:** 917–919.

Tsuge, T., Tsukaya, H., Uchimiya, H.: **Two independent and polarized processes of cell elongation regulate leaf blade expansion in *Arabidopsis thaliana* (L.) Heynh.** *Development* 1996, **122:** 1589–1600.

8

Cell differentiation and stem cells

- The control of gene expression
- Models of cell differentiation
- The plasticity of gene expression

The differentiation of unspecialized cells into many different cell types occurs first in the developing embryo and continues after birth and throughout adulthood. The character of specialized cells such as nerve or muscle is the result of a particular pattern of gene activity, which determines which proteins are synthesized. Thus, how this particular pattern develops is a central question in cell differentiation, and is discussed in this chapter in relation to some well studied model systems, including muscle, blood, and neural crest cell differentiation. We shall see that gene expression is under a complex set of controls that include the actions of transcription factors, regulated RNA splicing, chemical modification of DNA, and post-translational modifications to chromatin proteins, and that external signals play a key role in differentiation by triggering intracellular signaling pathways that affect gene expression. Another key question in development is the capacity of differentiated cells to change their state. As we shall see, although one cell type rarely changes into another under normal circumstances, such changes are possible. A notable example is the ability of a nucleus from a differentiated cell to become reprogrammed when transferred into an enucleated egg, where in some species it has proved able to direct embryonic development. Multipotent embryonic and adult stem cells also provide insights into cell plasticity that may help regenerative medicine.

Embryonic cells start out looking identical to each other but eventually become different, acquiring distinct identities and specialized functions. As we have seen in previous chapters, many developmental processes, such as the early specification of the germ layers, involve transient changes in cell form, patterns of gene activity, and the proteins synthesized. **Cell differentiation**, in contrast, involves the gradual emergence of cell types that have a clear-cut identity in the adult, such as muscle cells, nerve cells, skin cells, and fat cells (Fig. 8.1). There are more than 200 clearly recognizable differentiated cell types in mammals. Initially, embryonic cells fated to become different cell types differ primarily from each other only in their pattern of gene activity, and thus in the proteins present. Differentiation occurs over successive cell generations, the cells gradually acquiring new features while their potential fates become more and more restricted.

Differentiated cells serve specialized functions and have achieved a terminal and usually stable state, in contrast to many of the transitory differences characteristic of earlier stages in development. The specified precursors of cartilage and muscle cells have no obvious structural differences from each other and so look the same

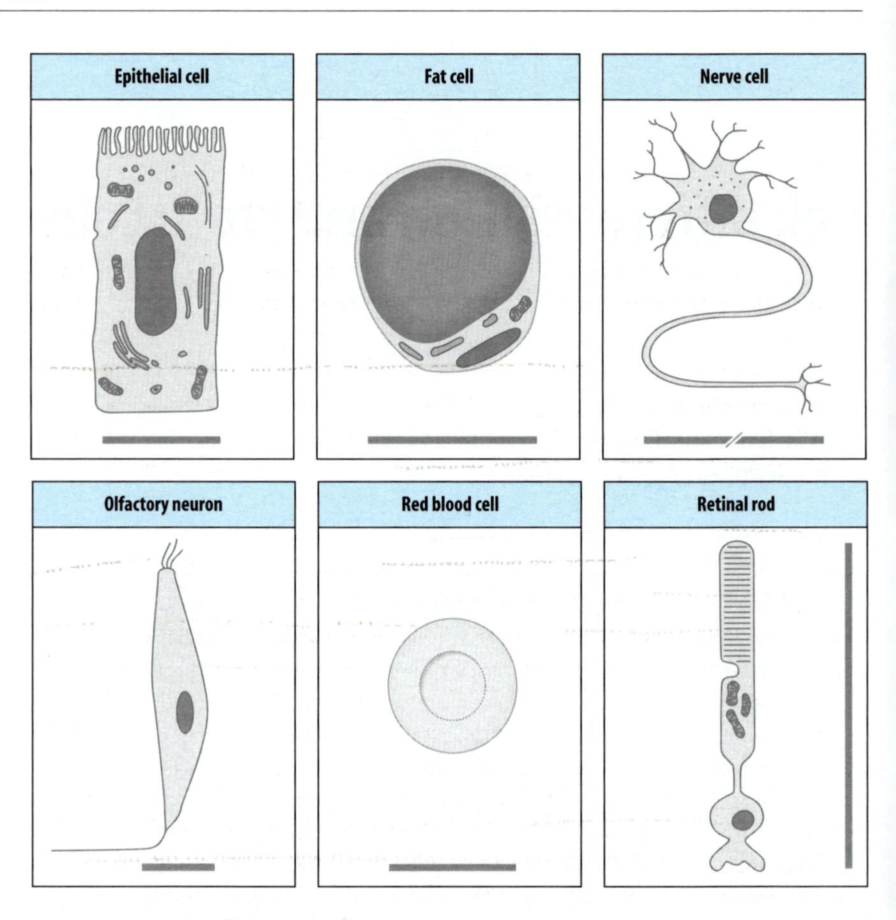

Fig. 8.1 Differentiated cell types. Mammalian cell types come in a variety of shapes and sizes. Scale bars: epithelial cell, 15 μm; fat cell, 100 μm; nerve cell, 100 μm–1m; olfactory neuron, 8 μm; red blood cell, 8 μm; retinal rod, 20 μm.

and might be described as undifferentiated; nevertheless, they will differentiate as cartilage and muscle, respectively, when cultured under appropriate conditions. Similarly, at an early stage in their differentiation, the precursors of white blood cells are structurally indistinguishable from those of red blood cells, but are distinct in the proteins they express. The differences between the precursor cells at these early stages reflect differences in gene activity and thus the proteins they contain, which control their future development.

As with earlier processes in development, the central feature of cell differentiation is a change in gene expression. This eventually leads to the production of so-called luxury or cell-specific proteins that characterize a fully differentiated cell: hemoglobin in red blood cells, keratin in skin epidermal cells, and muscle-specific actins and myosins. Naturally, the genes expressed in a differentiated cell include not only those encoding the luxury or cell-specific proteins, but also those for a wide range of 'housekeeping' proteins, such as the glycolytic enzymes involved in energy metabolism. It is very important to realize how many different genes are active in any given cell in the embryo at any one time—it is of the order of several thousand, only a small number of which may be involved in specifying cell fate or differentiation. The expression levels of large numbers of genes in a tissue can now be determined simultaneously by the technique of DNA microarray (Box 8A, opposite), which can help to detect the genes that are expressed differently in different tissues or at different stages in development, and which may therefore be involved in development and differentiation.

A differentiated cell is characterized by the proteins it contains. The presence of different proteins can result in considerable structural changes; mature mammalian red blood cells lose their nucleus and become biconcave discs stuffed

Box 8A DNA microarrays for studying gene expression

DNA microarrays or **DNA chips** enable the RNA products of thousands of genes to be monitored simultaneously, and the main use of microarrays in developmental biology is to monitor the changes in gene expression that occur within a tissue or embryo at different stages of development or after experimental manipulation. Microarrays come in various formats, but the most commonly used are slides studded with a regular array of clusters, or 'spots', of DNA fragments of known sequence, each cluster representing a DNA sequence found in a specific protein-coding gene. These fragments are known as the probes, and a microarray can contain many thousands of different probes. Probes may either be oligonucleotides specially synthesized to contain the required sequences, or cDNAs that have been made by reverse transcription of mRNA and thus represent the genes being expressed by a given tissue.

To determine which genes are being expressed, the total mRNA from the tissue of interest is extracted, then either amplified as RNA or converted into cDNA and amplified by PCR. It is then tagged with a fluorescent dye and hybridized to the microarray. Those cDNAs that hybridize to their complementary probes make a pattern of fluorescent signals which is scanned and the image converted into digital form. This is analyzed to determine which probes have been bound and thus what genes were being expressed in the cell sample.

To compare gene expression in two tissues or two conditions (such as two stages in development), the two cDNA samples are separately labeled with dyes that fluoresce at different wavelengths; the samples are then mixed and applied to the microarray (see figure). In essence, cDNAs present at higher concentration in one sample compared with the other will win out in the competition to bind to their corresponding probes and will be detected as signals at the respective wavelengths (in the readout one sample is usually represented as green and the other as red). cDNAs present at a higher level in one sample than the other will show up as a shade of the appropriate color, with the signal grading towards yellow for genes that are expressed at much the same levels in both samples (see figure). This raw image is processed to identify the genes that are being differentially expressed in the two samples, and the analyzed microarray data can then be presented in various formats that show up the gene-expression patterns in the data.

Another technique for determining which genes are being differentially expressed in different conditions is suppressive subtractive hybridization. This has been used to determine how gene expression changes after an embryo has had its development altered by manipulation or chemical treatment, for example. The total mRNA from a normal embryo and from the experimental one are isolated and cDNA is made from it. The subtraction step removes those cDNAs that are present in both samples, thus identifying differentially expressed, and possibly developmentally important, genes.

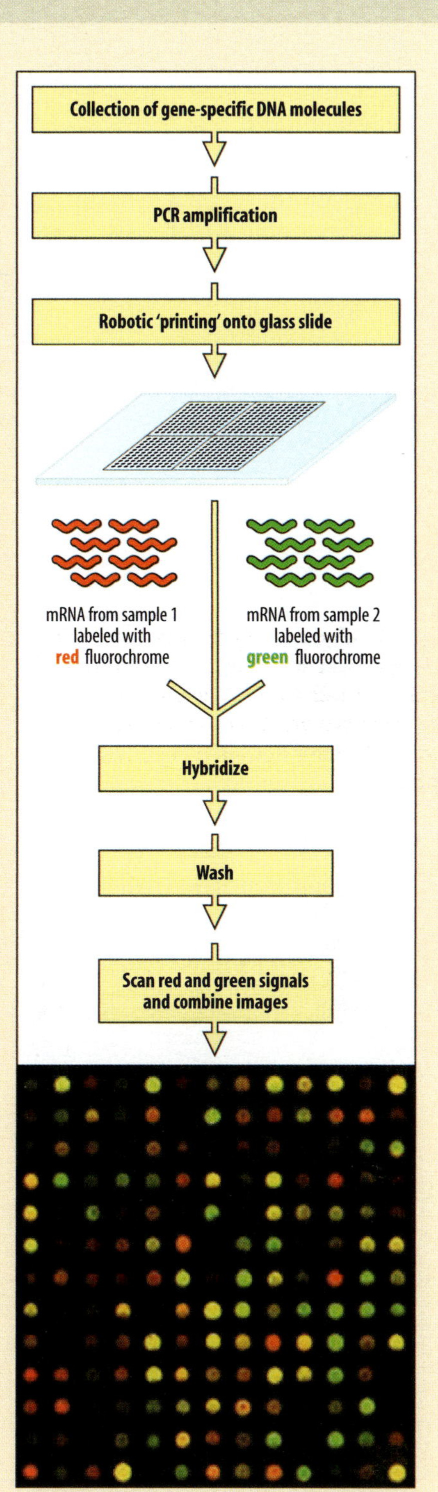

with hemoglobin, whereas neutrophils, a type of white blood cell, develop a multilobed nucleus and a cytoplasm filled with secretory granules. Yet both have a similar early lineage. Expression of a single protein can change a cell's differentiated state. If the gene *myoD* is introduced into fibroblasts, for example, they will develop into muscle cells, as *myoD* encodes a master transcriptional regulator of muscle differentiation.

In the earliest stages of differentiation, differences between cells are not easily detected and probably consist of subtle alterations caused by a change in activity of a few, or perhaps many, genes. At this early stage, cells become **determined**, or committed, with respect to their developmental potential (see Section 1.10): the mesoderm of somites, for example, can give rise in normal development to muscle, cartilage, dermis, and vascular tissue, but not to other cell types. Cells that are determined with respect to their eventual fate will form only the appropriate cell types when grafted to a different site in the embryo; they retain their identity. Once determined for a particular fate, cells pass on that determined state to all their progeny.

Cell differentiation is known to be controlled by a wide variety of external signals, ranging from cell-surface proteins to secreted polypeptide cytokines and molecules of the extracellular matrix. Examples of all of these signals will be encountered in this chapter, but it is important to remember that, while the external signals that stimulate differentiation are often referred to as being instructive, they are in general selective, in the sense that at any stage of development, the number of developmental options open to a cell is very limited; there are rarely more than a few options available to a cell at any given time. These options are set by the cell's internal state, which in turn reflects its developmental history (see Section 1.10). External signals cannot, for example, convert an endodermal cell into a muscle or nerve cell, although they can cause cells that are already determined as either muscle or cartilage to develop into one or the other. Thus the same signals, such as Wnts, FGFs, and members of the TGF-β family, can be used again and again in different circumstances. In contrast, some self-renewing **stem cells** (see Section 1.17) can give rise to a wide range of cell types under the influence of different external signals.

There can be a conflict between cell division and cell differentiation, with a cessation of cell proliferation being necessary for full cell differentiation to occur. Cell proliferation is most evident before the terminal stage of differentiation, and most terminally differentiated cells divide only rarely. Some cells, such as skeletal muscle and nerve cells, do not divide at all after they have become fully differentiated. In cells that do divide after differentiation, the differentiated state, like the determined state, is passed on through all subsequent cell divisions. As the pattern of gene activity is the key feature in cell differentiation, this raises the question of how a particular pattern of gene activity is first established, and then how it is passed on to daughter cells.

We start this chapter with a consideration of the mechanisms by which a pattern of gene activity can be established, maintained, and inherited at cell division. We then turn to a major question, namely the molecular basis of the specificity of cell differentiation, with muscle cells, blood cells, and neural crest cells providing our main model systems. Finally, we look at the reversibility and plasticity of the differentiated state, and particularly the special properties of stem cells and the reprogramming of differentiated nuclei in oocytes that makes it possible to **clone** some animals—that is, to make an animal that is genetically identical with the animal from which the nucleus came. This chapter concentrates on differentiation in animal cells. As we have seen in Chapter 6, plant cells do not have a permanently determined state, and a single somatic cell can give rise to a whole plant.

The control of gene expression

Every nucleus in the body of a multicellular organism is derived from the single zygotic nucleus in the fertilized egg. But patterns of gene activity in differentiated cells vary enormously from one cell type to another. The egg itself has a pattern of gene activity that is different from that of the embryonic cells to which it gives rise. This raises the question of how the particular pattern of gene activity in a differentiated cell is specified and how it is inherited. Control of gene activity is a fundamental feature of development as it determines when and where proteins are made.

To understand the molecular basis of cell differentiation, we first need to know how a gene can be expressed in a cell-specific manner. Why does a certain gene get switched on in one cell and not in another? We focus here on the regulation of **transcription**, which is the first (and generally the most significant) step in the expression of a gene. Control of protein synthesis can, however, also occur after transcription: at the stage of **RNA splicing** where the functional mRNA is produced, or at the **translation** of mRNA to produce protein. We have seen an example of translational regulation in early *Drosophila* development, where the Nanos protein controls the production of Hunchback protein by preventing translation of maternal *hunchback* RNA (see Section 2.9), and in *C. elegans*, where a microRNA prevents the translation of *lin-14* RNA to give LIN-14 protein (see Section 5.5).

8.1 Control of transcription involves both general and tissue-specific transcriptional regulators

The transcription of a gene is initiated when RNA polymerase binds tightly to the start site of transcription in the **promoter** region of the gene (see Section 1.10). The polymerase, with its associated proteins, unwinds a short region of the DNA helix and starts synthesizing RNA, using one of the DNA strands as the template. The promoter region, and the other sites in the DNA at which expression of the gene can be regulated, are known as its ***cis*-regulatory control regions**, in contrast to the **coding regions** that encode the amino-acid sequence of the protein. We have already seen how complex *cis*-regulatory control regions can be in examples from *Drosophila* (see Fig. 2.31) and sea urchin (see Fig. 5.27). The *cis*-regulatory regions contain sites at which **gene-regulatory proteins** or **transcription factors** bind to activate or repress the initiation of transcription. Most of the key genes in development are initially in an inactive state and require activating transcription factors, or **activators**, to turn them on.

The importance of control regions in tissue-specific gene expression can be clearly demonstrated by experiments in which the control region of one tissue-specific gene is replaced by the control region of another. For example, in mice, the enzyme elastase is only synthesized in the pancreas and growth hormone is only synthesized in the pituitary gland. The control region of an isolated mouse elastase gene can be joined to the protein-coding region of an isolated human growth hormone gene, and the resulting DNA injected into the nucleus of a fertilized mouse egg, where it becomes integrated into the genome (see Box 3E, p. 130). In the transgenic mouse embryos that develop from this egg, human growth hormone can be detected in the pancreas, showing that the human growth hormone gene is now expressed under the control of the mouse elastase control region (Fig. 8.2). This experiment shows the importance of the control elements in determining where a gene is expressed.

In eukaryotic cells, binding of RNA polymerase onto the correct region of DNA to start transcription requires the cooperation of a set of so-called **general**

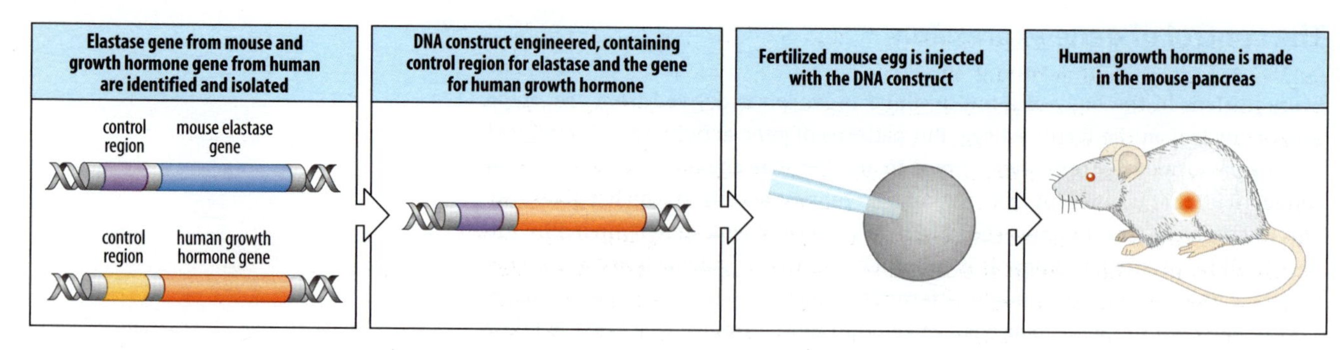

Fig. 8.2 Tissue-specific gene expression is controlled by regulatory regions. The control region of the mouse elastase gene is joined to a sequence of DNA that codes for human growth hormone. This DNA construct is injected into the nucleus of a fertilized mouse egg where it becomes integrated into the genome. When the mouse develops, human growth hormone is made in the pancreas under the control of the mouse elastase promoter. Normally, growth hormone is made only in the pituitary, and elastase only in the pancreas.

transcription factors, which form an initiation complex with the polymerase at the promoter site (Fig. 8.3). One can think of this complex as a transcribing machine. In eukaryotes, most protein-coding genes are transcribed by RNA polymerase II, which forms a multicomponent transcriptional initiation complex with a set of general transcription factors. This complex binds to the promoter at a short DNA sequence known as the TATA box, located near the start point of transcription. By itself, however, this complex is not usually sufficient to turn on transcription in highly regulated genes such as those involved in development; this requires the additional binding of gene-specific regulatory proteins elsewhere in the *cis*-regulatory regions.

For any given gene, the specificity of its activation is due to particular combinations of gene-regulatory proteins binding to individual sites in the control regions. There are at least 1000 different transcription factors in the genomes of *Drosophila* and *C. elegans*, and as many as 3000 in the human genome. On average, around five different transcription factors may act together at any particular control region. The combinatorial action of gene-regulatory proteins is a key principle in the control of gene expression and is fundamental to the exquisite control and complexity of the gene expression that drives development.

Some regulatory sites are within the promoter region, adjacent to the TATA box, and are present in similar positions in most protein-coding genes. Tissue-specific or developmental-stage expression are generally controlled by sites outside the immediate promoter region. These sites vary greatly in type and position from gene to gene, and may be thousands of base pairs away from the start point of transcription. These distant sites are thought to be able to control gene activity because the

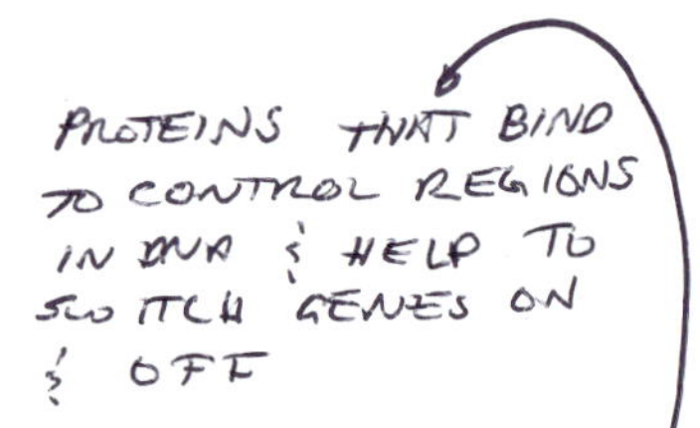

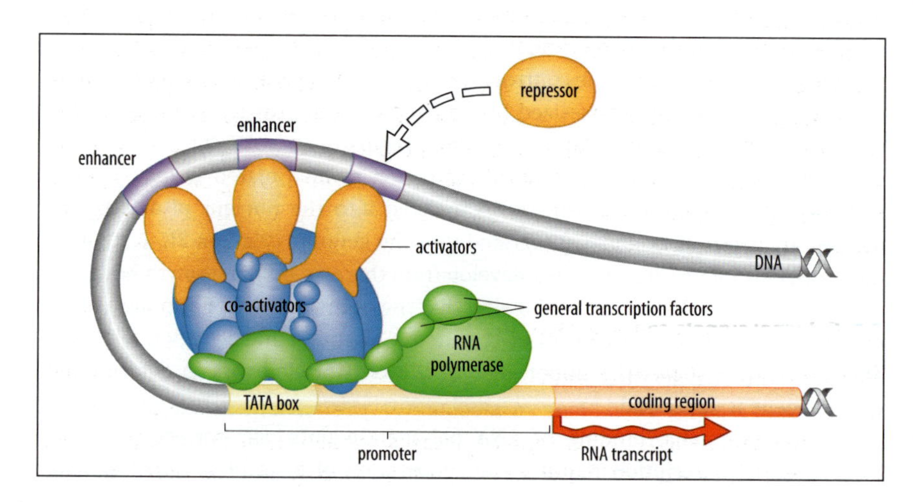

Fig. 8.3 Gene expression is regulated by the coordinated action of gene-regulatory proteins that bind to control regions in DNA. The transcription machinery, composed of RNA polymerase and the general transcription factors that bind to the promoter region, is common to many cells. Other regulatory regions, which have to be bound by gene-regulatory proteins (activators and repressors) before transcription can occur, are located adjacent to the promoter and usually also further upstream, and sometimes downstream, of the gene. The upstream regulatory regions, such as enhancers, may be many kilobases away from the start point of transcription.

After Tijan, R.: 1995.

DNA can form loops, thus bringing the sites into close proximity to the promoter region. Proteins bound at distant sites can thus make contact with the proteins bound at the promoter region, forming a fully active transcription initiation complex (see Fig. 8.3).

An important class of regulatory proteins are those known as co-activators, which link the transcriptional machinery to the activator and repressor proteins that bind to specific sites such as enhancers. One such linking protein is CBP, which binds to several common transcription factors such as CREB (cyclic AMP-response element-binding protein) and NFκB and is an essential co-activator of a variety of genes. CBP is a histone acetyltransferase, an enzyme that acetylates histones with resulting changes in the structure and transcriptional properties of **chromatin**—the complex of DNA, histones, and other proteins of which chromosomes are made. We shall return to the importance of chromatin state and histone modifications in regulating gene activity in Section 8.3.

Transcriptional regulators fall into two main groups: those that are required for the transcription of a wide range of genes, and which are found in a number of cell types; and those that are required for a particular gene or set of genes whose expression is tissue restricted, and which are only found in one or a very few cell types. We shall encounter both types of transcription factor in the following sections, when we look at the expression of globin genes in red blood cell precursors and the regulation of muscle genes in muscle precursor cells. In general, it can be assumed that activation of each gene involves a unique combination of transcription factors. It should also be noted that while differentiation may involve a relatively small number of genes, the number is not easy to establish, as each cell has several thousand active genes, probably most related to housekeeping functions. The expression of any given gene is thus initiated and maintained by a particular combination of gene-regulatory proteins. We have seen an example of this combinatorial control in the early development of *Drosophila*, where the temporal and regional activation of pair-rule genes is brought about by specific combinations of gap-gene proteins binding to the regulatory regions of the pair-rule genes (see Fig. 2.31). An important feature in the control of transcription is interaction between gene-regulatory proteins themselves or between these proteins and other proteins or small molecules.

Small differences in the regulatory region of a gene can have significant effects on the effect of a particular transcription factor. For example, the transcription factor Pit1 is needed to activate the genes for three hormones—growth hormone, prolactin, and thyrotropin—each of which is made by a different type of cell in the pituitary gland. But how does Pit1 turn on the right gene in each cell type without activating the other two? Pit1, like other transcription factors, has to bind to a regulatory sequence on its target genes and just a small sequence variation between the regulatory elements of the prolactin and growth hormone genes causes Pit1 to bind very differently to the two. In prolactin-producing cells, Pit1 is an activator of the prolactin gene, whereas when Pit1 binds to the regulatory region of the growth hormone gene in the same cell, it represses it. Just two additional bases in the growth hormone regulatory region seem to underlie this very different response.

8.2 External signals can activate genes

Most of the molecules that act as signals between cells during development are peptides or proteins. These bind to receptors in the cell membrane and do not themselves enter the cell. The signal is relayed to the cell nucleus by a process known as **signal transduction**, of which we have already seen several examples,

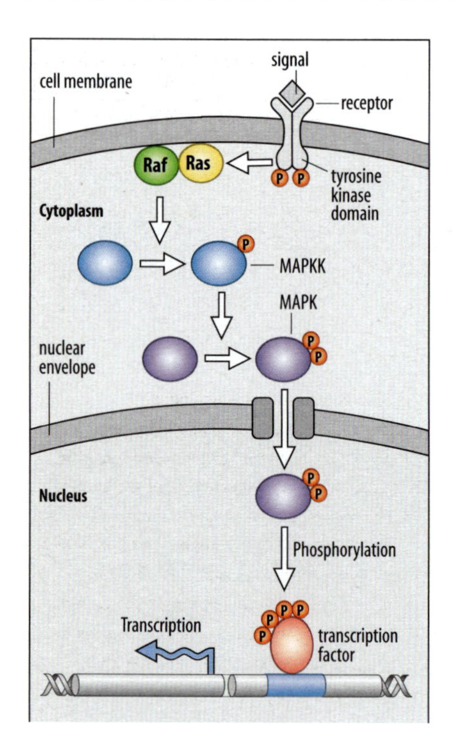

Fig. 8.4 Simplified outline of the Ras–Raf–MAPK intracellular signaling pathway by which signals received at the cell membrane can alter gene expression. Binding of the signal molecule to the cell-surface receptor sets in train an intracellular cascade of protein phosphorylations. The cytoplasmic portion of the receptor comprises a tyrosine protein kinase, which is stimulated to phosphorylate itself when the signal molecule binds. The phosphorylated receptor then stimulates the activation of the small GTPase Ras (details not shown). Its activation leads to the phosphorylation of the serine/threonine protein kinase Raf. This then phosphorylates another protein kinase, MAPK kinase (MAPKK), which phosphorylates mitogen-activated protein kinase (MAPK), which moves into the nucleus, where it can phosphorylate a transcription factor and so alter gene expression.

including the Wnt, TGF-β, and Hedgehog pathways. Signal transduction can be a very complex process, but the outline of one common type of signaling pathway from a type of receptor known as a receptor tyrosine kinase is shown in Fig. 8.4. This pathway is commonly used in response to mitogens, signals that stimulate cell division, and is, for example, one of the pathways leading from the receptors for growth factors such as FGF and EGF. It is often known as the Ras–MAPK pathway because of the key involvement of the small G protein Ras and a core cascade of serine/threonine protein kinases that ends with the activation of a mitogen-activated protein kinase (MAPK). Binding of the external signal to the receptor results in the cytoplasmic domain of the receptor becoming phosphorylated. This leads to the activation of Ras at the plasma membrane, which results in the binding and activation of the first kinase in the cascade, which in the mammalian pathway is called Raf. This phosphorylates and activates the next protein kinase, the MAPK kinase (MAPKK), which phosphorylates and activates MAPK. MAPK may then phosphorylate other kinases and can enter the nucleus and phosphorylate transcription factors, thus activating gene expression (see Fig. 8.4). There are several different MAPKs in mammalian cells that are used in different pathways and have different target transcription factors.

A rather different type of signal transduction pathway is that in which a signal at the cell surface leads to translocation of transcription factors that are stored as inactive cytoplasmic complexes into the nucleus. The activation of the receptor protein Toll in *Drosophila*, leading to Dorsal protein entering the nucleus, is an example of this type of pathway (see Section 2.11).

Unlike protein hormones and growth factors, steroid hormones are lipid soluble; they cross the plasma membrane unaided and enter the cell and stimulate a simpler pathway of signal transduction than do protein or peptide signals. Once inside the cell, steroids bind to receptor proteins and the complex of steroid and receptor is then able to act as a transcriptional regulator, binding directly to control sites in the DNA to activate (or in some cases repress) transcription. In many cases, the control sites, known generally as steroid response elements, act as enhancers. Other lipid-soluble signal molecules that act through similar receptors and this type of pathway are thyroid hormone and retinoic acid.

As we shall see in Chapter 11, steroid hormones produced in the testis are responsible for the secondary sexual characteristics that make male mammals different from females. In insects, the steroid hormone ecdysone is responsible for metamorphosis (see Chapter 12) and induces differentiation in a wide variety of cells. In these cases, the hormone is able to turn a whole set of different genes on and off by acting at shared response elements. An example of the control of tissue-specific expression of a single gene is provided by the steroid hormone estrogen, which causes the chick oviduct to produce the protein ovalbumin, a major component of egg white. The continued presence of estrogen is required for transcription of the ovalbumin gene; when estrogen is withdrawn, ovalbumin mRNA and protein disappear.

The action of estrogen in the chick provides an excellent example of the mechanisms underlying the tissue specificity of the response to this hormone. For example, the circulating estrogen activates the ovalbumin gene in the cells of the oviduct, but the ovalbumin gene in liver cells is unaffected. This tissue-specific response in the oviduct cannot easily be accounted for by additional proteins binding to the regulatory regions. Rather, the explanation for the differential response in these two tissues is thought to be some preceding heritable change in the structure of the chromatin. This may, for example, allow transcription factors such as the steroid-hormone–receptor complexes to have access to the ovalbumin gene in one cell type but not in the other.

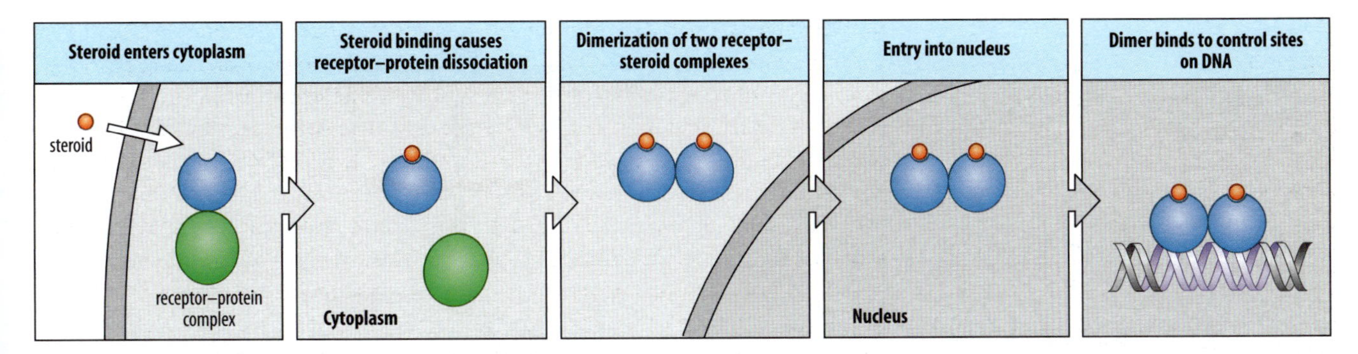

Fig. 8.5 Steroid hormones control transcription by binding to an intracellular receptor to form a transcriptional regulator. In the case of the glucocorticoid receptor shown here, the receptor is present in the cytoplasm as a complex with another protein. Steroid binding causes the receptor to dissociate from this protein and form a dimer with another steroid-bound receptor. This dimer enters the nucleus, where it binds to control sites on DNA and activates transcription.

Some steroid hormones, such as glucocorticoids, bind to a cytoplasmic receptor protein that is then translocated into the nucleus. For example, binding of a steroid to the glucocorticoid receptor in the cytoplasm results in the dissociation of the receptor from a cytoplasmic protein that keeps it inactive in the absence of steroids (Fig. 8.5). Two steroid–receptor complexes then join to form dimers, which move into the nucleus, bind to DNA, and activate specific genes. Other steroids have receptors that are already bound to their target DNA sequences in the nucleus even when the hormone is absent, but hormone binding is necessary to enable them to activate transcription.

8.3 Maintenance and inheritance of patterns of gene activity may depend on chemical and structural modifications to chromatin as well as on regulatory proteins

A central feature of development in general, and cell differentiation in particular, is that some genes are maintained in an active state, while others are repressed and inactive. Many differentiated cells, including fibroblasts, liver cells, and myoblasts (cells that give rise to muscle cells), continue dividing after differentiation, and the particular pattern of gene activity characterizing the differentiated state is reliably transmitted through many cell divisions. It is important to note that genes encoding gene-regulatory proteins can affect the activity of several, often many, other genes (Fig. 8.6).

Developmental genes generally have complex control regions, containing binding sites for a variety of activating or inhibitory transcription factors. Whether or not a gene is activated depends on the precise combination and levels of the regulatory factors present. A mechanism for maintaining a differentiated pattern of gene activity could require the continued presence of these specific gene-regulatory proteins. The continued activity of a gene, in both the differentiated cell itself and in its progeny, would require the continual presence of the appropriate positive regulatory proteins. Continued inactivity of a gene would require the continual presence of repressor proteins.

One way of keeping a gene active is for the gene product itself to act as the positive regulatory protein (Fig. 8.7). All that is then required for the pattern of gene activity to be maintained is the initial event that first activates the gene; once switched on it remains active. This type of positive feedback control of gene expression occurs in muscle-cell differentiation, where the product of the *myoD* gene acts as an activator of *myoD* expression.

In *Drosophila*, the selector genes remain active throughout development and maintain the developmental pathway of a region (see Chapter 2). These genes

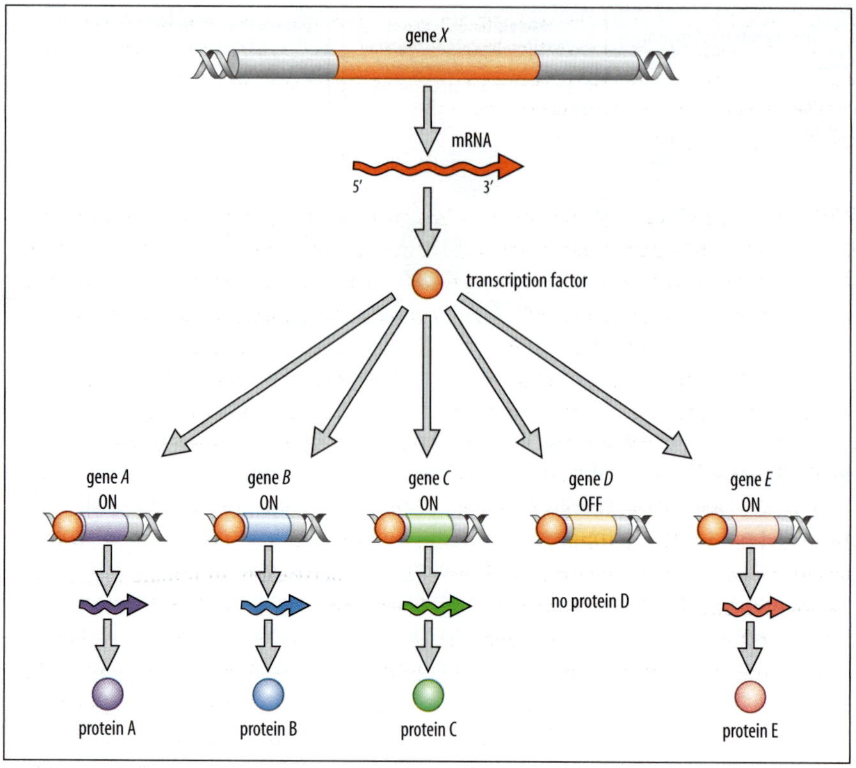

Fig. 8.6 A gene can regulate the activity of other genes when it codes for a transcription factor. Transcription factors can activate or repress the activity of genes by binding to their control regions. In the illustration, the activation of gene *X*, encoding a transcription factor, eventually leads to the production of four new proteins (A, B, C, and E), and the repression of protein D production.

After Alberts, B., et al.: 2002.

encode transcription factors, many of which are homeodomain transcription factors, that act by establishing and maintaining a pattern of gene activity. Such a pattern can be inherited through large numbers of cell divisions if the cytoplasm contains sufficient quantities of the necessary regulatory proteins.

A different mechanism for maintaining and inheriting a pattern of active and inactive genes relies on chemical and structural changes in the chromatin itself.

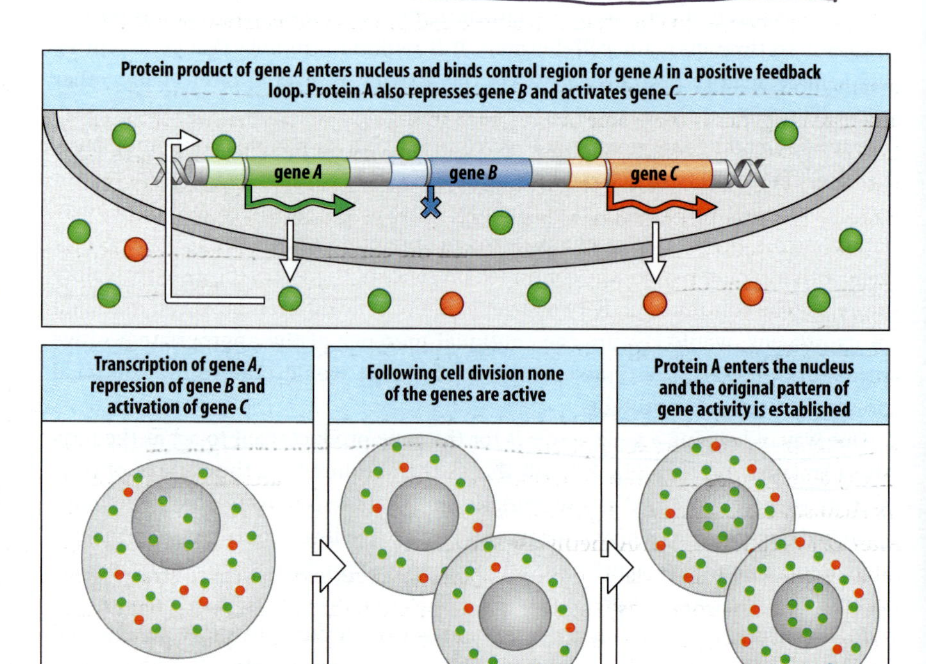

Fig. 8.7 Continued expression of gene regulatory proteins could maintain a pattern of differentiated gene activity. Top panel: transcription factor A is produced by gene *A*, and acts as a positive gene-regulatory protein for its own control region. Once activated therefore, gene *A* remains switched on and the cell always contains A. Transcription factor A also acts on the control regions of genes *B* and *C* to repress and activate them respectively, setting up a cell-specific pattern of gene expression. Bottom panels: after cell division, the cytoplasm of both daughter cells contains sufficient amounts of protein A to reactivate gene *A*, and thus maintain the pattern of expression of genes *B* and *C*.

Fig. 8.8 Inheritance of an inactivated X chromosome. In early mammalian female embryos one of the two X chromosomes, either the paternal X (X_p) or the maternal X (X_m), is randomly inactivated. In the figure, X_p is inactivated and this inactivation is maintained through many cell divisions. The inactivated chromosome becomes highly condensed.

After Alberts, B., et al.: 2002.

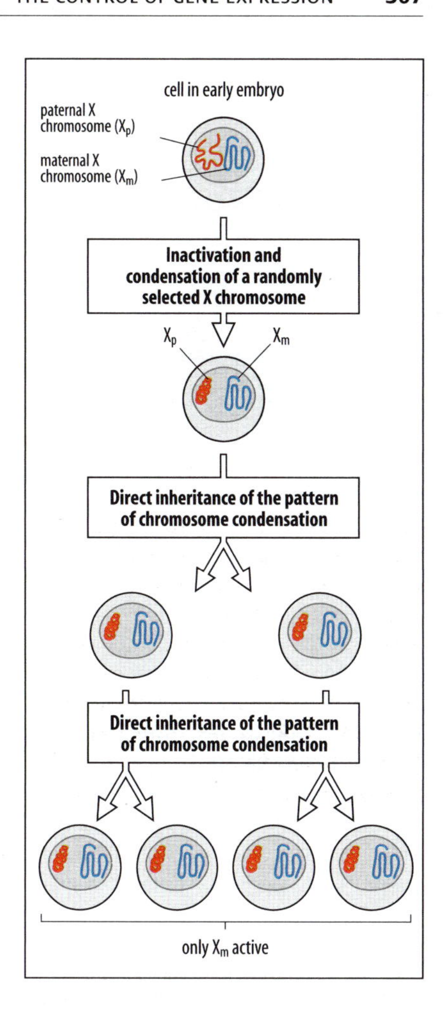

Whether a stretch of chromosome can be transcribed depends on whether or not it is packaged into a chromatin structure that is accessible to transcription factors, RNA polymerase and other necessary proteins. The compaction of chromatin seen at each mitosis, for example, when the condensed chromosomes become visible under the light microscope, renders the chromosomes transcriptionally inactive. Localized structural alterations of this sort that persist throughout the cell cycle could provide a means of shutting down, or silencing, genes during development. Chromatin that is packaged into a structure that cannot be transcribed is called **heterochromatin**.

Evidence that changes in the packing state of chromatin can be inherited through many cell divisions and can keep genes inactive over a long period is provided by the phenomenon of X-chromosome inactivation. In female mammals, one of the two X chromosomes is inactivated and becomes highly condensed heterochromatin early in embryogenesis. The inactive X chromosome is replicated at each division but remains transcriptionally inactive during the rest of the cell cycle. This pattern is preserved through many cell divisions throughout the individual's lifetime (Fig. 8.8). The chromatin in the inactivated X chromosome can be seen to be in a different physical state from that in the active one. During the interphase of the cell cycle—the period between successive mitoses—all the other chromosomes, including the active X chromosome, become extended threads and are no longer visible under the light microscope; the inactive X chromosome, however, remains highly condensed and is visible in human cells as a Barr body (Fig. 8.9). Inactivation is not irreversible, however, as the inactive X chromosome is reactivated during the formation of egg cells, which contain just one X chromosome.

Localized changes in chromatin structure and protein composition in interphase chromosomes are thought to provide a general mechanism for long-term gene inactivation. A reflection of the differences in chromatin structure between active and inactive genes is that some active genes show a heightened sensitivity to digestion with the enzyme DNase I at sites known as hypersensitive sites. This is just one indication that in actively transcribed genes, the chromatin is packaged into a more open structure that allows enzymes and other proteins, including the transcription apparatus, access to the DNA. When the chromatin is packed into heterochromatin, on the other hand, the DNA is inaccessible to DNase I and is protected from digestion.

Underlying these changes in the transcriptional properties of chromatin are chemical modifications to the DNA itself and to the associated histone proteins. In vertebrates, methylation of cytosine at certain sites in the DNA is correlated with the absence of transcription in those regions. Moreover, the pattern of methylation can be faithfully inherited when the DNA replicates, thus providing a mechanism for passing on a pattern of gene activity to daughter cells (Fig. 8.10). After DNA replication, DNA methylase recognizes the presence of a methyl group on a cytosine and methylates the corresponding group on the other strand. The inactive X chromosome has a pattern of methylation that differs from that of the active X, and so methylation is likely to play an important part in its inactivation. DNA methylation is the mechanism used in mammals for the differential imprinting of maternal and paternal genes in the egg, which is discussed in Chapter 11.

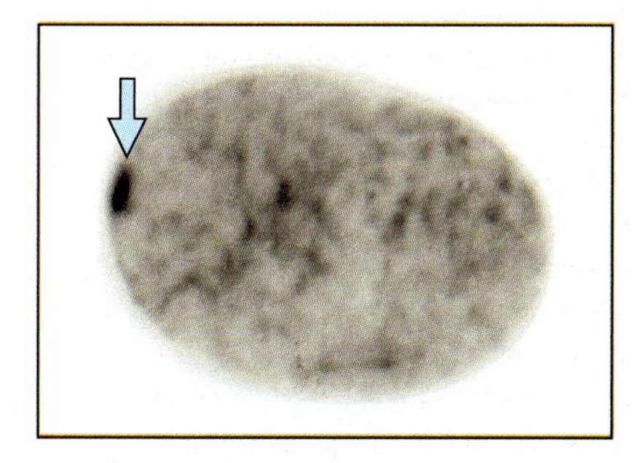

Fig. 8.9 Inactivated X chromosome (the Barr body). The photograph shows the Barr body (arrow) in the interphase nucleus of a female human buccal cell.

Photograph courtesy of J. Delhanty.

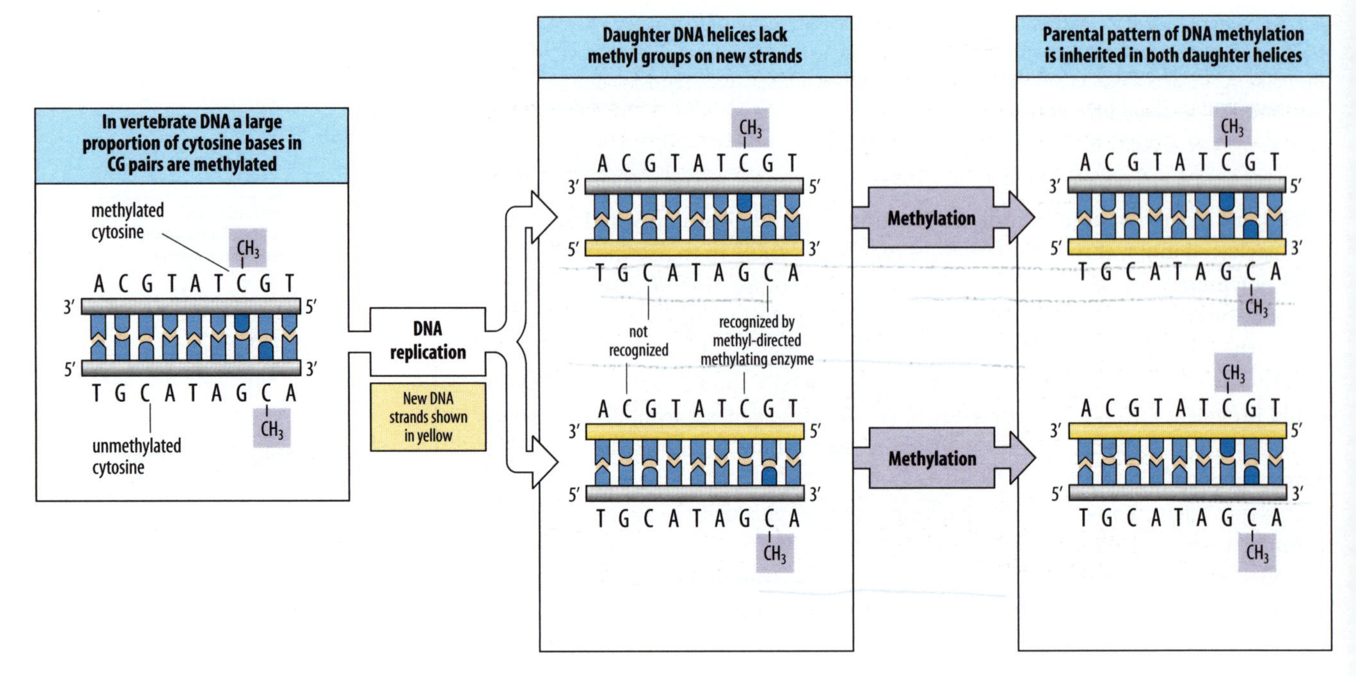

Fig. 8.10 Inheritance of the pattern of methylation on DNA. Many cytosine bases occurring in cytosine–guanine (CG) pairs in vertebrate DNA are methylated (a CH_3 group is added). When the DNA is replicated, the new complementary strand lacks methylation but the pattern can be reinstated on the new strand by methylase enzymes that recognize the methylated cytosine of the CG pair on the old DNA strand and methylate the cytosine of the corresponding CG pair on the new strand.

After Alberts, B., et al.: 2002.

This is the process by which certain genes are made inactive in the maternal genome but not in the paternal genome.

The post-translational modification of histone proteins is also closely linked to regulation of gene activity. The histones in chromatin can undergo modifications such as methylation, acetylation, or phosphorylation on specific amino-acid residues, and this can change the properties of the chromatin. Histone acetylation, for example, tends to be associated with regions of chromatin that are being transcribed, whereas histone methylation is more characteristic of transcriptionally silent heterochromatin. Histones are acetylated by enzymes known as histone acetyltransferases (HATs), which are usually part of large multiprotein complexes known generally as **chromatin-remodeling complexes** because of the effect of their activity on chromatin structure and activity. HATs are targeted to particular genes by interaction with sequence-specific gene-activating proteins. Acetyl groups are removed from histones by enzymes called histone deacetylases, and this can play a part in gene repression. Epigenetic changes such as DNA methylation, histone acetylation, and histone methylation are all thought to exert their effect on chromatin structure and gene expression through the proteins they recruit to the altered sites.

Summary

Transcriptional control is a key feature of cell differentiation. Tissue-specific expression of a eukaryotic gene depends on regulatory sites located in the *cis*-regulatory regions flanking the gene. These comprise both the promoter sequences adjacent to the start site of transcription, which bind RNA polymerase, and more distant sites that control the tissue-specific or developmental-stage—specific expression of the gene. The combination of different regulatory proteins bound to sites in the *cis*-regulatory regions determines whether a gene is active or not, and small differences in regulatory regions between genes can have significant effects on their pattern of expression.

In some cases, cell-specific expression is due to the correct combination of regulatory proteins only being present in that cell type; in other cases, gene expression may be prevented by the gene being packed away in a state in which it is inaccessible to transcription factors and RNA polymerase. Gene-regulatory proteins interact not only with DNA but also with each other, forming multicomponent transcription complexes that are responsible for initiating transcription by RNA polymerase. Extracellular signals such as protein growth factors and hormones do not themselves enter the cell but can specifically influence gene expression by acting at cell-surface receptors to stimulate intracellular signal transduction pathways. In contrast, signal molecules that can diffuse through cell membranes, such as steroid hormones, form a complex with an intracellular receptor protein. The hormone–receptor complex then binds to the DNA at specific control elements and acts as a transcription factor to influence gene expression.

Once established, the pattern of gene activity in a differentiated cell can be maintained over long periods of time and is transmitted to the cell's progeny. Mechanisms that maintain the pattern include the continued action and production of gene regulatory proteins, and long-term localized changes in the structure and properties of chromatin that are due to chemical modifications to DNA and to histones.

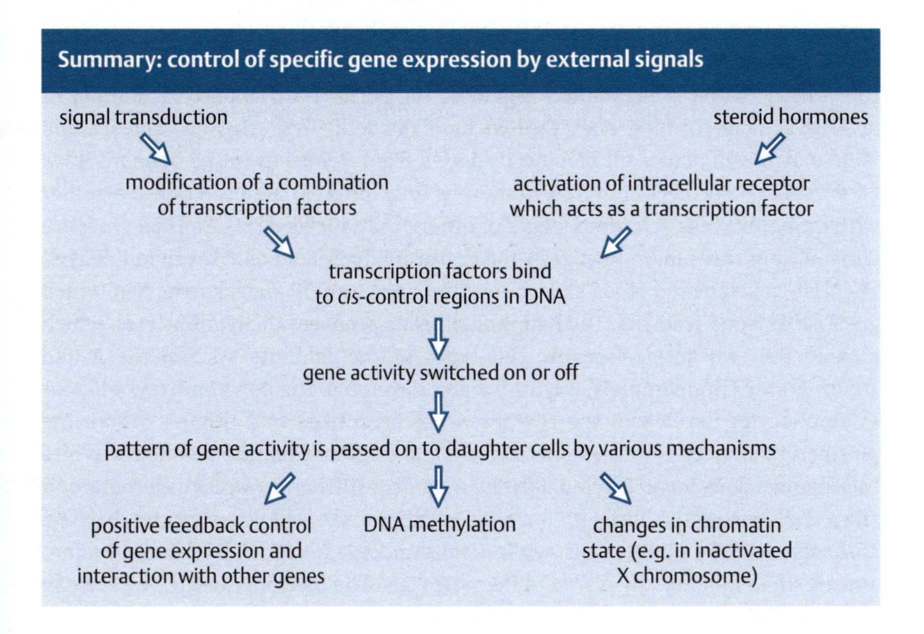

Models of cell differentiation

Cellular differentiation is by no means confined to the embryo. A number of adult tissues, such as blood, skin, and the gut epithelial lining, are continually being renewed from undifferentiated stem cells, and are well studied models of cell differentiation. We shall also consider skeletal muscle, which although mainly differentiating in the embryo, can be renewed if necessary from stem cells in the adult. Embryonic neural crest cells, which we discussed in Chapter 7 in regard to their migration, give rise to a great variety of different structures, and we shall look here at some aspects of their differentiation. Finally, we consider apoptosis, or programmed cell death, which is the fate of very large numbers of developing cells, both in embryonic development and in the adult.

8.4 All blood cells are derived from multipotent stem cells

Hematopoiesis, as blood formation, is a particularly well studied example of differentiation, partly because cells at all stages are relatively accessible and can be grown in culture, and also because of its medical importance. In hematopoiesis, we consider the differentiation of a **multipotent** stem cell—a stem cell that can give rise to many different types of differentiated cell. We first look at the general process of hematopoiesis and at the external signals that influence it, and then consider the control of gene expression in one cell type—the red blood cell.

All the blood cells in the adult mammal originate from a population of multipotent stem cells located in the bone marrow. These stem cells are self-renewing and are the precursors of progenitor cells that become committed to one of the hematopoietic lineages at a later stage. Thus, hematopoiesis is, in effect, a complete developmental system in miniature, in which a single cell—the multipotent stem cell—gives rise to numerous different cell types. Some, such as red blood cells (erythrocytes), are short-lived and must continually be replaced throughout adult life. As a measure of complexity of the process, hematopoietic stem cells from fetal liver express at least 200 transcription factors, a similar number of membrane-associated proteins, and almost 150 signaling molecules.

The stem cells for hematopoiesis in mammals are derived from mesoderm in the yolk-sac blood islands and a region of the aorta. They colonize a number of definitive blood-forming sites that include the fetal liver, thymus, spleen, and bone marrow. In adults, all blood cells derive from stem cells in the bone marrow. Mammalian blood cells comprise numerous fully differentiated cell types, together with immature cells at various stages of differentiation (Fig. 8.11). The cell types are derived from three main lineages—the erythroid, lymphoid, and myeloid lineages. The first yields the red blood cells, or erythrocytes, and the megakaryocytes, which give rise to blood platelets. The lymphoid lineage produces the lymphocytes, which include the two antigen-specific cell types of the immune system, the B and T lymphocytes. In mammals, B lymphocytes develop in the bone marrow, whereas T lymphocytes develop in the thymus, from precursors that originate from the pluripotent stem cells in the bone marrow and migrate from the bloodstream into the thymus. Both B and T lymphocytes undergo a further terminal differentiation after they encounter antigen. Terminally differentiated B lymphocytes become antibody-secreting plasma cells, while T cells undergo terminal differentiation into at least three functionally distinct effector T cells. The myeloid lineage gives rise to the rest of the white blood cells, or leukocytes, which all derive from the bone marrow in adults. These are the eosinophils, neutrophils, and basophils (collectively known as granulocytes or polymorphonuclear leukocytes), mast cells, and monocytes. Monocytes differentiate into macrophages after they have left the bone marrow and entered tissues. Mast cells also reside in tissues.

Within the bone marrow, the different types of blood cells and their precursors are intimately mixed up with connective-tissue cells—the bone marrow stromal cells. The ability of stem cells both to self-renew and to differentiate into different cell types is tightly regulated by the cells and molecules in their immediate environment—the 'niche' that they inhabit. In the case of hematopoiesis, the **stem-cell niche** is provided by the bone marrow stromal cells. Secreted signal proteins of the Wnt family (see Box 3C, p. 111) together with other signals from the stromal cells seem necessary to maintain stem-cell proliferation. The existence of the multipotent

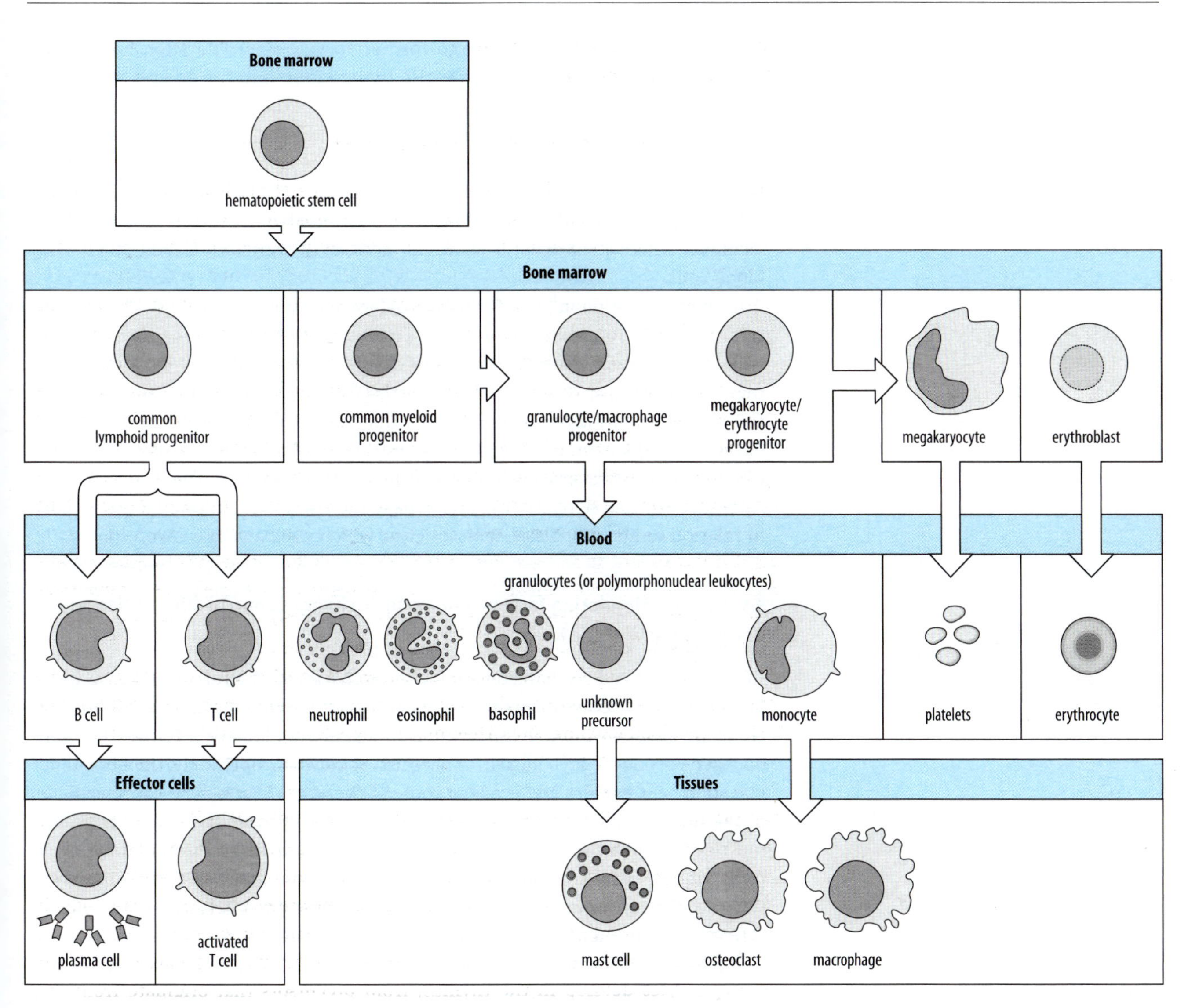

Fig. 8.11 The derivation of blood cells from pluripotent stem cells. The pluripotent stem cells, which are self-renewing, give rise to all of the blood cells, as well as to tissue macrophages, mast cells, and osteoclasts in bone. The stem cells are thought to give rise to uncommitted progenitors, which then split into more committed myeloid, erythroid, and lymphoid precursors. These then give rise to the different blood cell lineages (not all final differentiated cell types are shown). Hematopoietic growth factors influence the proliferation and differentiation of the different lineages.

After Janeway, C.A., et al.: 2005.

stem cell can be inferred from the ability of bone marrow to reconstitute a complete blood and immune system when transplanted into individuals whose own bone marrow has been destroyed. The key experiment is the transfusion of a suspension of bone marrow cells into a mouse that has received a potentially lethal dose of X-irradiation. The mouse would normally die as a result of lack of blood cells, which, like other proliferating cells, are particularly sensitive to radiation, but transfusion of bone marrow cells reconstitutes the hematopoietic system and allows the mouse to recover. Multipotent stem cells can also be grown in culture, where they also proliferate extensively and differentiate into blood cells.

The multipotent stem cell generates progenitor cells that become irreversibly committed to one or other of the lineages leading to the different blood cell types. There is early commitment to either the myeloid or lymphoid lineage, for example, followed by later commitment to the individual lineages leading to the differentiated cell types. The hematopoietic system can thus be considered as a hierarchical system with the multipotent hematopoietic stem cell at the top (see Fig. 8.11). This generates uncommitted precursors that then become committed to the main lineages, after which they undergo further rounds of proliferation and further commitment, finally undergoing overt differentiation into the different cell types.

All this activity occurs in the microenvironment of the bone marrow stroma, and is regulated by external signals that are provided by the stromal cells in the form of hematopoietic growth factors and other cytokines. In this way, the numbers of the different types of blood cells can be regulated according to the individual's physiological need. Loss of blood results in an increase in red cell production, for example, whereas infection leads to an increase in lymphocytes and other white blood cells.

A central question in considering the behavior of stem cells is how a single stem cell can divide to produce two daughter cells, one of which remains a stem cell while the other gives rise to a lineage of differentiating cells. One possibility is that there is an intrinsic difference between the two daughter cells because the stem-cell division is an asymmetric division that results in the two cells acquiring a different complement of proteins. A clear example of this mechanism is neuroblast development in *Drosophila*, which is considered in Chapter 10. The second possibility is that external signals in the niche make them different. In general, particularly in relation to hematopoiesis, it is not clear which mechanism is involved.

8.5 Colony-stimulating factors and intrinsic changes control differentiation of the hematopoietic lineages

Hematopoiesis can be characterized by the activities of a hierarchy of transcription factors, whose overlapping patterns of expression specify the various cell lineages. Some are only expressed in immature cells and are not lineage specific—the proto-oncogene product c-Myb falls into this class. Somewhat suprisingly, the hematopoietic stem cells express low levels of some of the genes that will be expressed later in the myeloid and erythroid lineages. Thus, gene repression as well as activation must occur when the cell becomes committed to a particular lineage. Some of the transcription-factor interactions that occur in hematopoietic stem-cell differentiation into the erythroid and myeloid lineages are illustrated in Fig. 8.12, and include activation, inhibition, and more complex interactions.

In blood cell precursors, the transcription factor GATA1, for example, is only found in erythrocyte precursors and the precursors of some myeloid lineages, and is necessary for the differentiation of these lineages. Its effect can be concentration-dependent: eosinophils form when low levels of GATA1 are present, whereas red cells and megakaryocytes develop at higher concentrations. Another key transcription factor is PU.1, which inhibits the activity of GATA1, and vice versa. As with all cell types, it is the combination of transcription factors, and not any particular one, that is responsible for the specific pattern of gene expression that leads to differentiation into a given cell type.

How then is the activity of all of these factors controlled? Signals delivered by extracellular protein growth factors and differentiation factors have a key role in this. Studies on blood-cell differentiation in culture have identified at least 20 extracellular proteins, known generally as colony-stimulating factors or hematopoietic growth factors, that can affect cell proliferation and cell differentiation at various points in hematopoiesis (Fig. 8.13). Both stimulators and inhibitors have been identified. Not all the factors controlling blood cell production are made by blood cells or stromal cells. The protein erythropoietin, for example, which induces the differentiation of committed red blood cell precursors, is mainly produced by the kidney, in response to physiological signals that indicate red blood cell depletion.

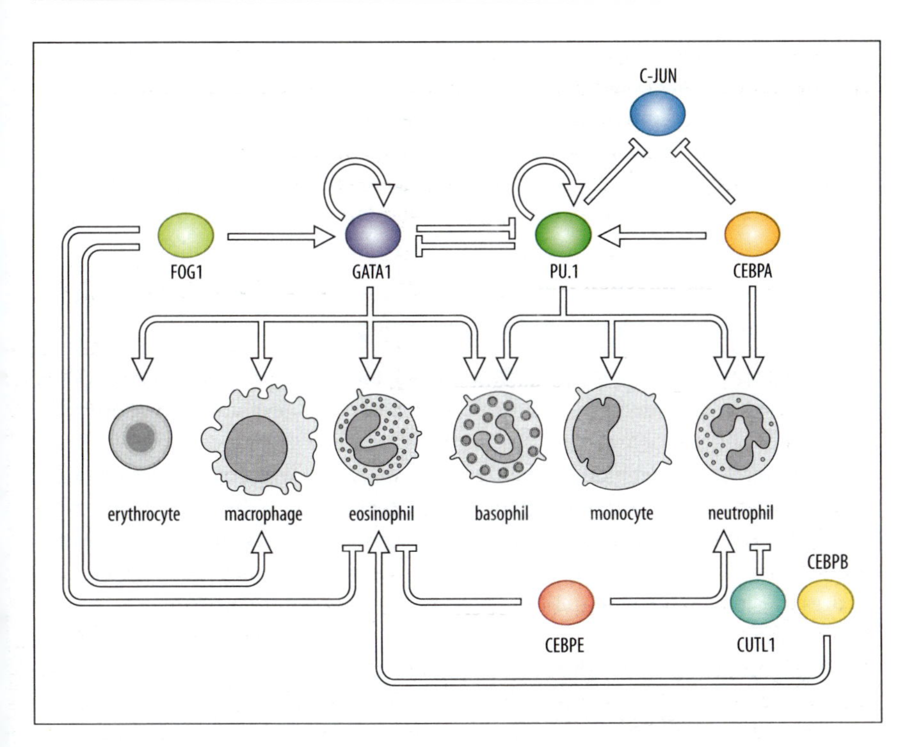

Fig. 8.12 Some transcription factors of importance in blood cell differentiation. Some of the gene-regulatory interactions involved in the differentiation of the erythroid and myeloid lineages of blood cells. Macrophages, eosinophils, basophils, monocytes, and neutrophils belong to the myeloid lineage. Arrows indicate that the transcription factor stimulates differentiation of that lineage, whereas barred lines indicate inhibition. For example, the differentiation of neutrophils requires activation of the transcription factors CEBPE, PU.1, and CEBPA, and inhibition of CUTL1.

Despite the well established role of these factors in blood-cell proliferation and differentiation, there is evidence that the commitment of a cell to one or other pathway in the myeloid lineage might be a chance event, with the growth factors simply promoting the survival of particular cell lineages. Evidence for this comes from experiments where the two daughters of a single early progenitor cell (which can give rise to a variety of blood-cell types) are cultured separately but under the same conditions. Both daughter cells usually give rise to the same combination of cell types, but in about 20% of cases, they give rise to dissimilar combinations.

Among the plethora of growth factors, it is difficult to distinguish those that might be exerting specific effects on differentiation from those that may be required for the survival and proliferation of a lineage or group of lineages. The contrasting functions of three growth factors—granulocyte-macrophage colony-stimulating factor (GM-CSF), macrophage colony-stimulating factor (M-CSF), and granulocyte colony-stimulating factor (G-CSF)—are well established. GM-CSF is generally required for the development of most myeloid cells from the earliest progenitors that can be identified. In combination with G-CSF, however, it tends to stimulate only granulocyte (principally neutrophil) formation from the common granulocyte-macrophage progenitor. This is in contrast to M-CSF, which, in combination

Some hematopoietic growth factors and their target cells	
Type of growth factor	Responding hematopoietic cells
Erythropoietin (EPO)	Erythroid progenitors
Granulocyte colony-stimulating factor (G-CSF)	Granulocytes, neutrophils
Granulocyte-macrophage colony-stimulating factor (GM-CSF)	Granulocytes, macrophages
Interleukin-3	Multipotent precursor cells
Stem cell factor (SCF)	Stem cells
Macrophage colony-stimulating factor (M-CSF)	Macrophages, granulocytes

Fig. 8.13 Hematopoietic factors and their target cells.

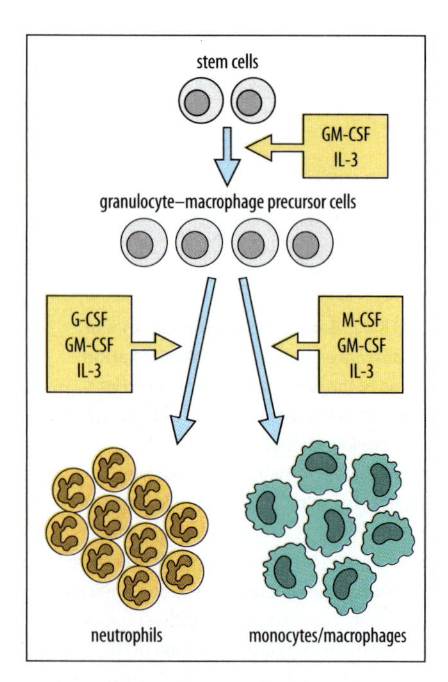

Fig. 8.14 Colony-stimulating factors can direct the differentiation of neutrophils and macrophages. Neutrophils and macrophages are derived from a common granulocyte-macrophage precursor cell. The choice of differentiation pathway can be determined by, for example, the growth factors G-CSF and M-CSF. The growth factors GM-CSF and IL-3 are also required, in combination, to promote the survival and proliferation of cells of the myeloid lineage generally.

After Metcalf, D.: 1991.

with GM-CSF, tends to stimulate differentiation of monocytes (macrophages) from the same progenitors (Fig. 8.14).

These growth and differentiation factors have no strict specificity of activity, in the sense of each factor exerting a specific effect on one type of target cell only; rather, they act in different combinations on target cells to produce different results, the outcome also depending on the developmental history of their target.

8.6 Developmentally regulated globin gene expression is controlled by regulatory sequences far distant from the coding regions

We now focus on one type of differentiated blood cell, the red blood cell or erythrocyte, to consider the transcription factors that are active in these cells during their differentiation, and that control expression of their cell-specific genes. The main feature of red blood cell differentiation is the synthesis of large amounts of the oxygen-carrying protein hemoglobin, which involves the coordinated regulation of two different sets of globin genes. The program of erythropoiesis is primarily driven by the DNA-binding transcription factors GATA2 and GATA1, which bind to sites in the globin genes and recruit other gene regulatory proteins.

Vertebrate hemoglobin is a tetramer of two identical α-type and two identical β-type globin chains. The α-globin and β-globin genes belong to different multigene families. Each family consists of a cluster of genes and the two families are located on different chromosomes. In mammals, different members of each family are expressed at various stages of development so that distinct hemoglobins are produced during embryonic, fetal, and adult life. Hemoglobin switching is a mammalian adaptation to differing requirements for oxygen transport at different stages of life; human fetal hemoglobin, for example, has a higher affinity for oxygen than adult hemoglobin, and thus is able to efficiently take up the oxygen that has been released by maternal hemoglobin in the placenta. In turn, the oxygen affinity of adult hemoglobin is sufficient for it to pick up oxygen in the relatively oxygen-rich environment of the lungs. Here, we look at the control of expression of the β-globin genes, as examples of genes that are not only uniquely cell specific, but are also developmentally regulated.

The human β-globin gene cluster contains five genes—ε, γ_G, γ_A, δ, and β (Fig. 8.15). These are expressed at different times during development: ε is expressed

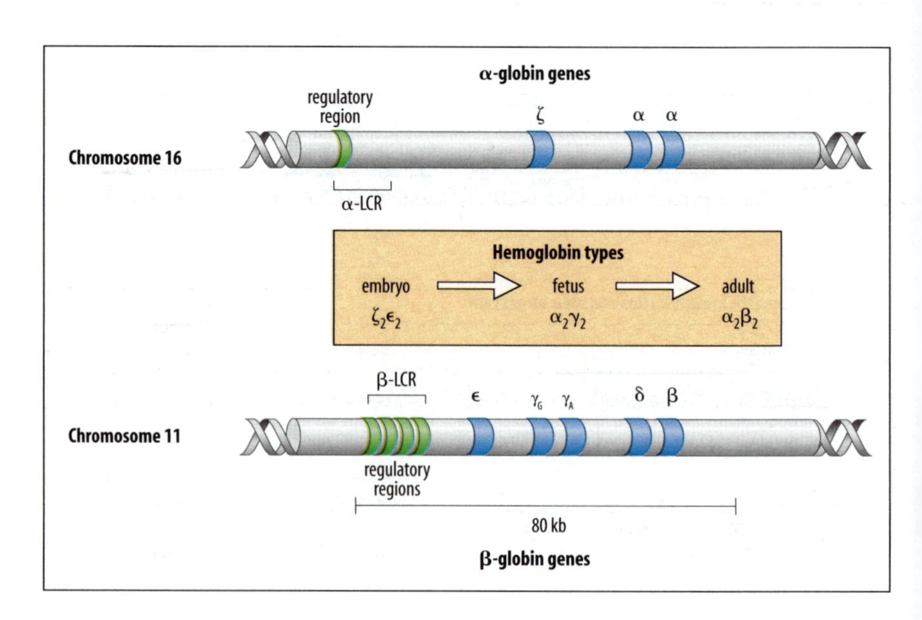

Fig. 8.15 The human globin genes. Human hemoglobin is made up of two identical α-type globin subunits and two identical β-type globin subunits, which are encoded by α-family and β-family genes on chromosomes 16 and 11, respectively. The composition of human hemoglobin changes during development. In the embryo it is made up of ζ (α-type) and ε (β-type) subunits; in the fetal liver, of α and γ subunits; and in the adult most of the hemoglobin is made up of α and β subunits, but a small proportion is made up of α and δ subunits. The regulatory regions α-LCR and β-LCR are locus control regions involved in switching on the hemoglobin genes at different stages.

in the early embryo in the embryonic yolk sac; the two γ genes, which encode proteins differing by only one amino acid, are expressed in the fetal liver; and the δ and β genes are expressed in erythroid precursors in the adult bone marrow. The protein products of all these genes combine with globins encoded by the α-globin complex to form physiologically different hemoglobins at each of these three stages of development.

Mutations in the globin genes are the cause of several relatively common inherited blood diseases. One of these, sickle-cell anemia, is caused by a point mutation in the gene coding for β-globin. In people homozygous for the mutation, the abnormal hemoglobin molecules aggregate into fibers, forcing the cells into the characteristic sickle shape. These cells cannot pass easily through fine blood vessels and tend to block them, causing many of the symptoms of the disease. They also have a much shorter lifetime than normal blood cells. Together these effects cause anemia—a lack of functional red blood cells. Sickle-cell anemia is one of the very few genetic diseases where the link between a mutation and the subsequent developmental effects on health is fully understood.

The control regions that regulate expression of the β-globin gene cluster are complex and extensive. Each gene has a promoter and control sites immediately upstream (to the 5′ side) of the transcription starting point, and there is also an enhancer downstream (to the 3′ side) of the β-globin gene, which is the last gene in the cluster (Fig. 8.16). But these local control sequences, which contain binding sites for transcription factors specific for erythroid cells, as well as for other more widespread transcriptional activators, are not sufficient to provide properly regulated expression of the β-globin genes.

The successive switching on and off of different globin genes during development is an intriguing feature of globin gene expression. This developmentally regulated expression of the β-globin gene cluster depends on a region a considerable distance upstream from the ε gene. This is the locus control region (LCR; see Fig. 8.16), which stretches over some 10,000 base pairs at around 5000 base pairs from the 5′ end of the ε gene. The LCR confers high levels of expression on any β-family gene linked to it, and is also involved in directing the developmentally correct sequence of expression of the whole β-globin gene cluster, even though the β-globin gene itself, for example, is around 50,000 base pairs away from the extreme 5′ end of the LCR. A similar control region has been found upstream of the α-globin gene cluster.

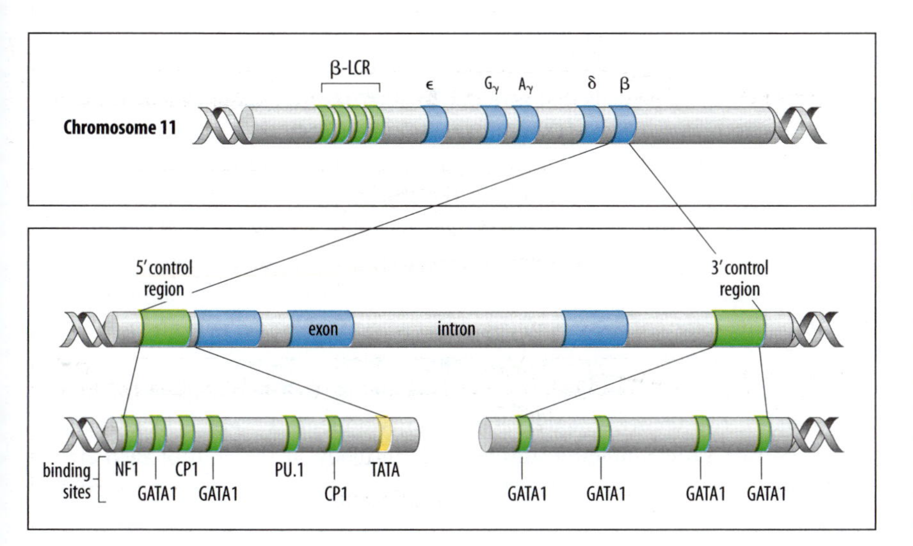

Fig. 8.16 The control regions of the β-globin gene. The β-globin gene is part of a complex of other β-type globin genes and is only expressed in the adult. Binding sites for the relatively erythroid-specific transcription factor GATA1, as well as for other non-tissue-specific transcription factors, such as NF1 and CP1, are located in the control regions. Upstream of the whole β-globin cluster, in the locus control region (LCR), are additional control regions required for high-level expression and full developmental regulation of the β-globin genes.

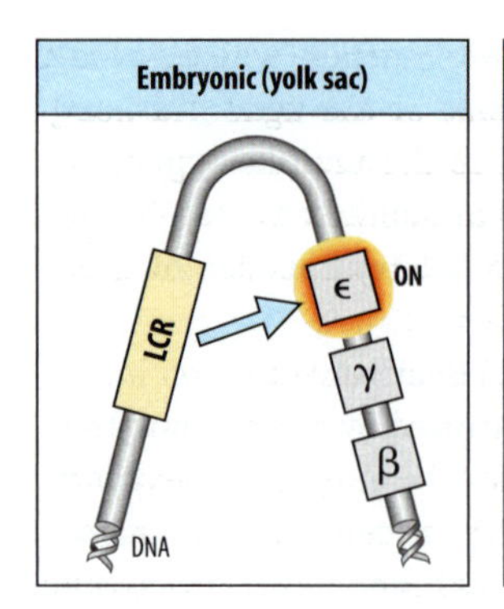

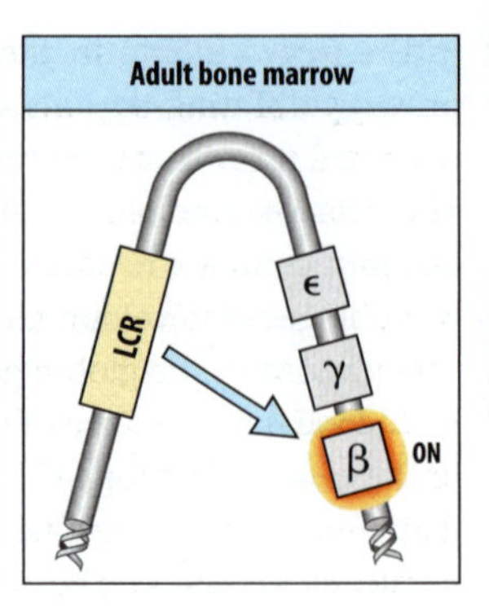

Fig. 8.17 A possible mechanism for the successive activation of β-family globin genes by the LCR during development. The LCR is thought to make contact with the promoter of each gene in succession, at different stages of development, thus controlling their temporal expression.

After Crossley, M., et al.: 1993.

An attractive model for the control of globin-gene switching envisages an interaction of LCR-bound proteins with proteins bound to the promoters of successive globin genes (Fig. 8.17). The DNA between the LCR region and the globin genes is thought to loop in such a way that proteins binding to the LCR can physically interact with proteins bound to the globin gene promoters. Thus, in erythrocyte precursors in the embryonic yolk sac the LCR would interact with the ε promoter, in fetal liver it would interact with the two γ promoters, and in the adult bone marrow with the β-gene promoter. There is some evidence that, for the β-globin locus at least, the LCR acts by promoting the transition from initiation of transcription to transcriptional elongation by RNA polymerase II.

8.7 Differentiation of cells that make antibodies involves irreversible DNA rearrangement

Cell differentiation does not usually involve any permanent alteration to the cell's DNA, but is caused chiefly by reversible associations of gene-regulatory proteins with DNA and reversible enzymatic modifications of DNA and histones that produce changes in the pattern of gene expression. There are, however, rare cases in which the DNA in the differentiated cell becomes irreversibly altered. The only known examples of this in vertebrates are the B and T lymphocytes of the vertebrate immune system, which, like the erythrocytes discussed previously, are derived from hematopoietic stem cells (see Fig. 8.11). In the case of lymphocytes, the changes in gene expression that give these cells their ability to recognize and respond to foreign antigens are accompanied by irreversible DNA rearrangements that involve loss of DNA. We look here at what happens during the differentiation of B lymphocytes—the cells that give rise to antibody-producing plasma cells.

It has been estimated that the human immune system can produce up to 10^{15} different antibody molecules, which vary in their ability to bind different antigens. To encode each antibody separately in the genome is obviously impossible, as it would require a number of genes far greater than any genome could possibly contain. This amazing diversity arises instead from a unique combinatorial mechanism, in which complete antibody genes are assembled during cell differentiation by combining different gene segments from a relatively small collection (no more than a few hundred) encoded in the genome.

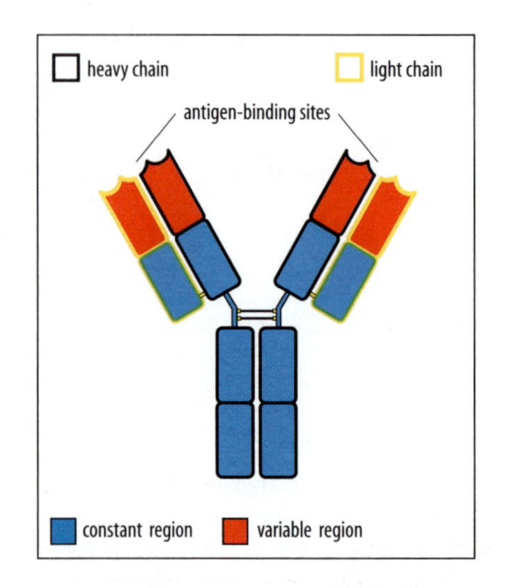

Fig. 8.18 Schematic drawing of an antibody molecule. The molecule is made up of two light chains and two heavy chains. Most of the molecule has a constant structure but the antigen-binding sites are located in the variable region.

After Janeway, C.A., et al.: 2005.

An antibody, or immunoglobulin, molecule is Y-shaped (Fig. 8.18) and is composed of two identical 'light' chains and two identical, larger 'heavy' chains. There are two antigen-binding sites on each antibody, at the ends of each arm of the Y, which are formed by parts of the heavy and light chains. The antigen-binding site determines the antigen specificity of the antibody and varies in structure from one type of antibody molecule to the next. The regions of the light and heavy chains that form the antigen-binding site are known as the variable regions, as they vary in amino-acid sequence from one antibody to another. The remainder of the antibody molecule is more constant between different antibodies. It is the variable regions of the genes for light and heavy chains that are assembled by irreversible DNA rearrangements during B-lymphocyte development.

Each differentiated B cell expresses just one type of heavy chain and one type of light chain, and produces an immunoglobulin of a single antigen specificity. Each heavy-chain variable region is encoded by a DNA sequence made up from three different gene segments (called V, D, and J), whereas the light-chain variable region is encoded by just two gene segments (V and J). During the development of a B lymphocyte, several DNA recombination events occur which assemble complete heavy- and light-chain genes from these segments. Because there are multiple copies of the different types of gene segment, many different combinations can be assembled, and thus many different antibody-binding sites generated. The principle of gene assembly can be illustrated for the light-chain gene. During B-lymphocyte differentiation, one V gene segment is moved by a DNA recombination event to lie next to one of the J gene segments. The rearranged variable region, together with the DNA sequence encoding the constant region, is then transcribed into one long RNA, which is spliced to make mRNA encoding the complete light chain (Fig. 8.19). Millions of B lymphocytes are produced by the immune system each day; by joining up different gene segments in different B cells, considerable antibody diversity can be achieved.

Fig. 8.19 DNA rearrangements generate a functional immunoglobulin light-chain gene. At the immunoglobulin light-chain locus, germline DNA carries a set of V gene segments and a set of J gene segments. During B-lymphocyte development, one V gene segment and one J gene segment are joined at random by somatic recombination to give a DNA sequence coding for a unique light-chain variable region. This is transcribed into RNA, together with the constant (C) region sequence. RNA splicing removes the RNA between the end of the J segment and the beginning of the C region to give a light-chain mRNA, which is then translated into protein. In the antibody molecule (illustrated bottom right), the V, J, and C regions of the light-chain protein are indicated in red, yellow, and blue, respectively.

After Janeway, C.A., et al.: 2005.

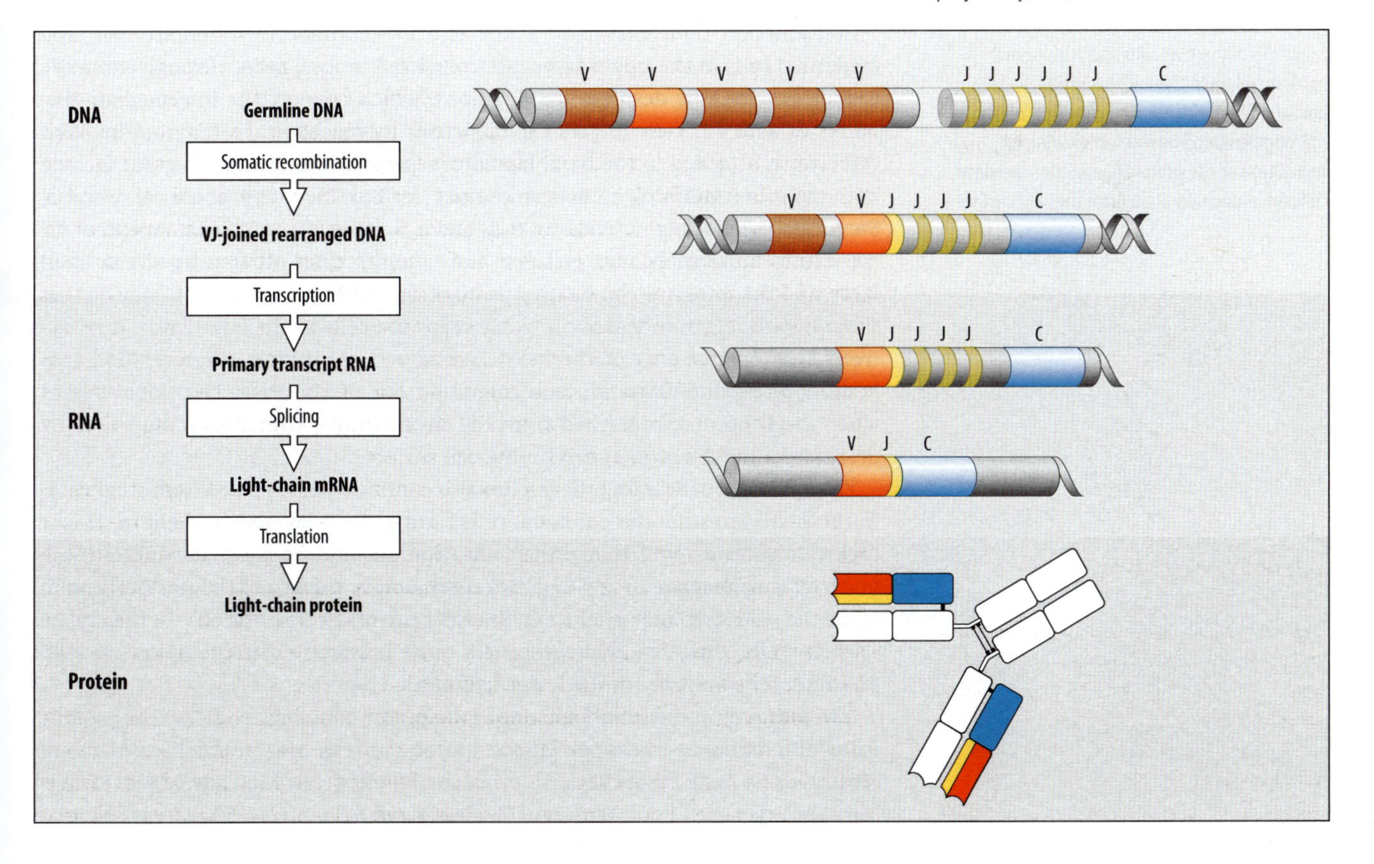

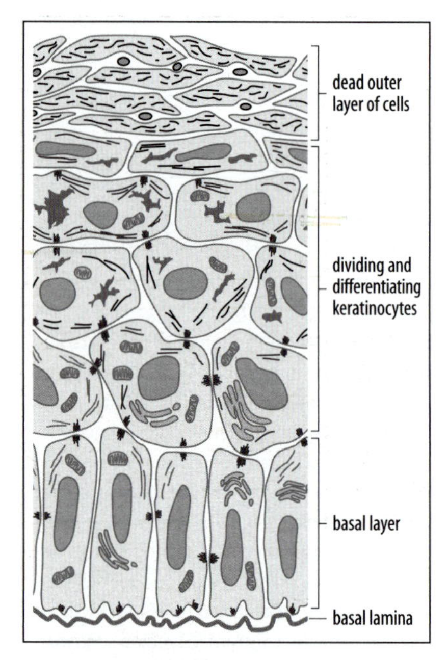

Fig. 8.20 Differentiation of keratinocytes in human epidermis. Descendants of stem cells, which will become keratinocytes, are detached from the basal lamina, divide several times, and leave the basal layer before beginning to differentiate. Differentiation of keratinocytes involves the production of large amounts of the intermediate filament protein keratin. In the intermediate layers, the cells are still large and metabolically active, whereas in the outer epidermal layers, the cells lose their nuclei, become filled with keratin filaments, and their membranes become insoluble owing to deposition of the protein involucrin. The dead cells are eventually shed from the skin surface.

8.8 The epithelia of adult mammalian skin and gut are continually replaced by derivatives of stem cells

Mammalian skin is composed of three layers—the **dermis**, which mainly contains fibroblasts, the protective outer **epidermis**, which mainly contains keratin-filled cells called **keratinocytes**, and the **basal lamina** or basement membrane, composed of extracellular matrix, which separates the epithelial epidermis from the dermis. The epidermis, which we consider here, comprises a multilayered epithelium of keratinocytes (Fig. 8.20) and associated hair follicles and sebaceous and sweat glands. Because of the protective function of the epidermis, cells are continuously lost from its outer surface and must be replaced, and it is maintained throughout life by stem cells that reside in the basal layer of the epidermis in contact with the basal lamina, and at the base of the follicles.

The stem cells form a special population in the basal layer that multiplies relatively slowly and retains its proliferative potential throughout the life of the organism. These stem cells divide asymmetrically, with one daughter remaining a stem cell and the other daughter becoming committed to differentiation. After becoming committed, this cell divides a few times more, and its progeny differentiate further after leaving the basal layer. One terminal differentiation pathway culminates in the production of the outermost, cornified epidermal layers that provide a protective covering for the skin. The terminally differentiated cells of the sebaceous gland are lipid-filled sebocytes that eventually burst, releasing their contents onto the surface of the skin. Differentiation of the hair follicle is more complex, comprising eight different cell lineages. In keratinocyte differentiation, the cells move through the skin as they mature, until by the time they reach the outermost layer they are completely filled with keratin and their membranes are strengthened with the protein involucrin. The dead cells eventually flake off the surface (see Fig. 8.20).

Cell-adhesion molecules play a key role in keratinocyte differentiation. The epidermal cells in the upper layers are connected to each other through extensive cadherin-containing desmosomal junctions, which connect the intermediate filaments of adjacent cells to give the epidermis its mechanical strength. The basal cell layer is attached to the basal lamina by specialized cell junctions that include both integrin-containing hemi-desmosomes (see Box 7A, p. 259) and focal contacts. Stem cells express higher levels of $\alpha_2\beta_1$ and $\alpha_3\beta_1$ integrins, which are receptors for the extracellular molecules collagen and laminin, than other cells of the basal layer, and the integrins can be used as markers to detect stem cells. It may be that these high levels of integrins are what keeps the cells in the basal layer. It is also likely that components of the basal lamina actively inhibit differentiation into keratinocytes; differentiation and migration out of the basal layer only begins when a cell becomes detached from the basal lamina, which is accompanied by decreased amounts of integrins on the cell surface.

The epithelial cells lining the gut are also continuously replaced from stem cells. In the small intestine the epithelial cells form a single-layered epithelium. This is highly folded into villi that project into the gut and crypts that penetrate the underlying connective tissue. Cells are continuously shed from the tips of the villi, while the stem cells that produce their replacements lie near the base of the crypts. The new cells generated by the stem cells move upward toward the tips of the villi. En route, they multiply in the lower half of the crypt (Fig. 8.21).

A single crypt in the small intestine of the mouse contains about 250 cells of four main differentiated cell types. About 150 of the cells are proliferative, dividing about twice a day so that the crypt produces 300 new cells each day; about 12 cells leave the crypt each hour. Wnt signaling leading to β-catenin/TCF activity is essential for stem-cell proliferation and the subsequent lineage differentiation.

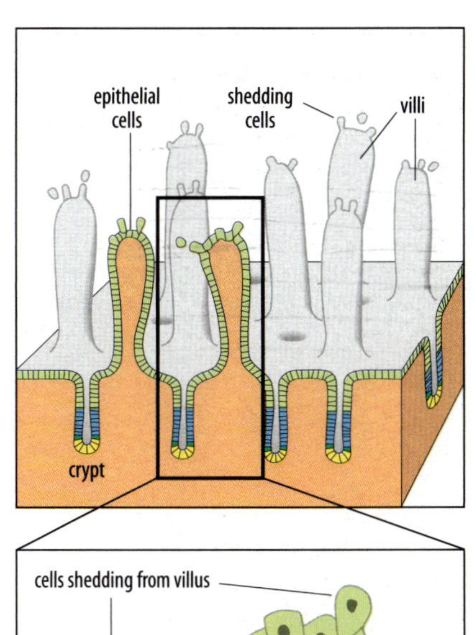

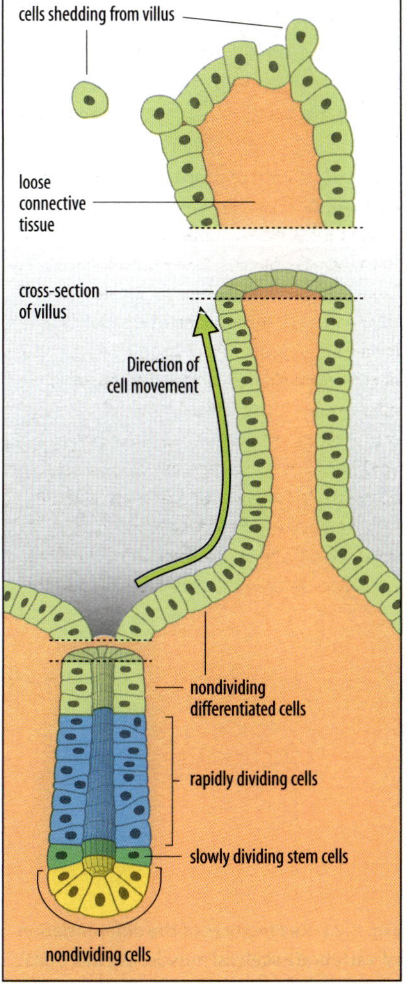

Fig. 8.21 Epithelial cells lining the mammalian gut are continually replaced. The upper panel shows the villi—epithelial extensions covering the wall of the gut. The lower panel is a detail of a crypt and villus. The bases of the crypts contain stem cells. They give rise to cells that proliferate and move upward. These differentiate into epithelium in the projecting villus, and are shed from the tips. The total time of transit is about 4 days.

After Alberts, B., et al.: 2002.

Another epithelial structure, the mammary gland, has recently been shown to be generated from stem cells. A complete mouse mammary gland, capable of producing milk, has been grown from a single stem cell.

8.9 A family of genes can activate muscle-specific transcription

In Chapter 4, we considered the signals involved in specifying vertebrate muscle cells in the somite. Here we focus on the differentiation of cells that are already specified to become skeletal muscle; differentiation of cardiac muscle involves different transcription factors. Differentiation of vertebrate skeletal striated muscle can be studied in cell culture and provides a valuable model system for cell differentiation. Skeletal muscle cells derive from the myotome of the somites. **Myoblasts**, cells that are committed to forming muscle, can be isolated from chick or mouse embryos and mouse cells can be cloned to produce cell cultures derived from a single myoblast. These cells will proliferate until growth factors are removed from the culture medium, whereupon proliferation ceases and overt differentiation into muscle cells begins. The cells then begin to synthesize muscle-specific proteins such as actin, myosin II, and tropomyosin, which are part of the contractile apparatus, and muscle-specific enzymes such as creatine phosphate kinase. Myoblasts also undergo structural change during differentiation. They first become bipolar in shape as a result of reorganization of the microtubule cytoskeleton, and then fuse to form multinucleate **myotubes** (Fig. 8.22). Fusion *in vivo* requires the presence of integrin β_1 on the cell membrane. Within about 20 hours of withdrawal of growth factors, typical striated **muscle fibers** can be seen.

The gene *myoD* is a member of a family of basic helix-loop-helix transcription-factor genes that is expressed only in muscle precursors and muscle cells, and can be considered as key controlling genes for muscle differentiation. Activation of these genes leads to the switching on of muscle-specific genes and differentiation. *myoD* will even induce differentiation into muscle if transfected into fibroblasts, which do not normally express either the MyoD protein or the muscle-specific structural proteins and enzymes. As well as *myoD*, the family includes three other genes—*mrf4*, *myf5*, and *myogenin*—that can all induce muscle differentiation in fibroblasts and other non-muscle cells. *mrf4*, *myf5*, and *myoD* are the first genes to be switched on in muscle precursors in mammals. *myoD* is also activated by the transcription factor Pax3, which becomes restricted to muscle-cell precursors (see Section 4.5). They then turn on *myogenin* and hence muscle differentiation. Both *myoD* and *myf5* are expressed in proliferating, undifferentiated myogenic cells, whereas *myogenin* is only expressed during muscle differentiation (Fig. 8.23). Once *myoD* is turned on it maintains its own expression by means of a positive feedback loop.

Despite the apparently powerful effects of MyoD and Myf5, gene knock-out experiments in transgenic mice (see Box 3E, p. 130) show that mice lacking MyoD make apparently normal skeletal muscle if *myf5* is active, suggesting that there is redundancy in the system. Apparent redundancy is not uncommon in development, and has probably evolved to serve as a back-up mechanism to ensure

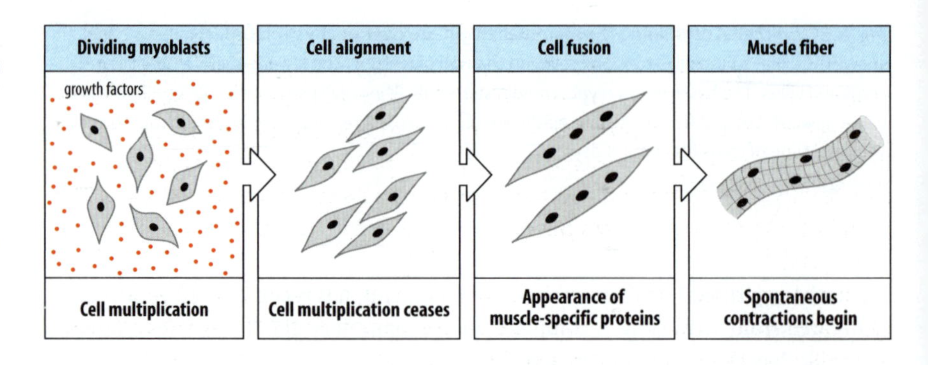

Fig. 8.22 Differentiation of striated muscle in culture. Myoblasts are cells that are committed to becoming muscle but which have yet to show signs of differentiation. In the presence of growth factors they continue to multiply but do not differentiate. When growth factors are removed the myoblasts stop dividing; they align and fuse into multinucleate myotubes, which develop into myofibers that contract spontaneously.

the continuation of development in the face of fluctuations in gene expression (see Section 1.19). In such mice the level of *myf5* expression is raised, implying that the expression of *myoD* normally inhibits *myf5* expression, and that Myf5 protein can compensate for the absence of MyoD. Mice lacking Myf5 also make skeletal muscle although they have other defects; the most obvious abnormality is a shortening of the ribs. Mice lacking both Myf5 and MyoD, however, not only lack all skeletal muscle but also the myogenic precursor cells. This shows that these proteins are acting as myogenic determination factors. Indeed, in embryos in which mutant *myf5* and *myoD* genes have been activated but which do not make functional Myf5 or MyoD protein, prospective muscle cells mislocate, and may be integrated into other differentiation pathways such as cartilage and bone. Mice in which the *myogenin* gene has been knocked out, on the other hand, lack most skeletal muscle but precursor myoblasts are present. Myogenin is thus considered to be a myogenic differentiation factor rather than a protein essential for muscle determination.

The MyoD family of transcription factors activates transcription of muscle-specific genes by binding to a nucleotide sequence called the E-box, which is present in the enhancer and/or promoter regions of these genes. MyoD forms a heterodimer with products of the E2.2 gene such as E_{12}, and this binds to the E-box. As the activity of MyoD and its relatives is essential for muscle-cell differentiation, we must now look at how the activity of MyoD could be regulated.

8.10 The differentiation of muscle cells involves withdrawal from the cell cycle, but is reversible

The proliferation and differentiation of myoblasts are mutually exclusive phenomena. Skeletal myoblasts that are multiplying in culture do not differentiate.

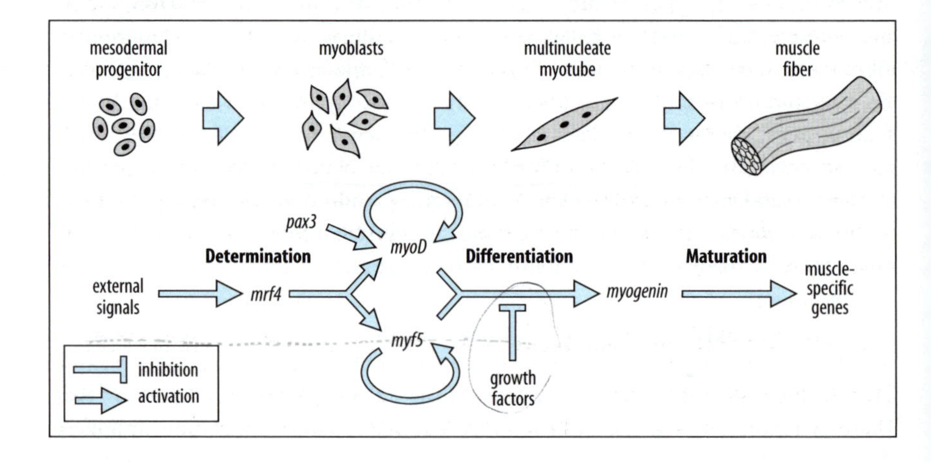

Fig. 8.23 Key features of the differentiation of vertebrate skeletal muscle. External signals initiate muscle differentiation by activating the genes *myoD* and *myf5*. One of these genes is expressed preferentially (depending on the species) and their activity is mutually inhibitory and self-sustaining. The proteins encoded by these genes activate further genes, such as *myogenin*, which in turn activate expression of muscle-specific genes.

Only when proliferation ceases does differentiation begin. In the presence of protein growth factors, myoblasts that are expressing both MyoD and Myf5 proteins continue to proliferate and do not differentiate into muscle. This means that the presence of MyoD and Myf5 is not in itself sufficient for muscle differentiation and that an additional signal or signals are required. In culture, this stimulus to differentiate can be supplied by the removal of growth factors from the medium. The myoblasts then withdraw from the cell cycle, cell fusion occurs, and differentiation takes place.

There is an intimate relationship between the proteins that control progression through the cell cycle and the muscle determination and differentiation factors. MyoD and Myf5 are phosphorylated by cyclin-dependent protein kinases, which become activated at key points in the cell cycle. Phosphorylation affects the activity of MyoD and Myf5 by making them more likely to be degraded. Thus, in actively proliferating cells, the levels of the myogenic factors will be kept low. A number of other proteins also interact with myogenic factors or with their active partners such as E_{12}, and inhibit their activity. One is the transcription factor Id, which is present at high concentrations in proliferating cells, and which probably helps to prevent premature activation of muscle-specific genes.

The myogenic factors themselves actively interfere with the cell cycle to cause cell-cycle arrest. Myogenin and MyoD activate transcription of members of the p21 family of proteins, which block cell-cycle progression, causing the cell to withdraw from the cell cycle and to differentiate. Another key protein in preventing progression through the cell cycle is the retinoblastoma protein pRB. This is phosphorylated in proliferating cells by the cyclin-dependent kinase 4 (Cdk4) and dephosphorylation of pRB is essential both for withdrawal from the cell cycle and for muscle differentiation. MyoD can interact strongly and specifically with Cdk4 to inhibit pRB phosphorylation, which would also help to lead to cell-cycle arrest.

The opposition between cell proliferation and terminal differentiation makes developmental sense. In those tissues where fully differentiated cells do not divide further, it is essential that sufficient cells are produced to make a functional structure, such as a muscle, before differentiation begins. The reliance on external signals to induce differentiation, or to permit a cell already poised for differentiation to start differentiating, is a way of ensuring that differentiation only occurs in the appropriate conditions.

But despite the apparently terminal nature of muscle-cell differentiation, muscle cells provide an example of a differentiated cell that can undergo **dedifferentiation**—the loss of differentiated characteristics—re-enter the cell cycle and even give rise to other cell types. The mouse gene *mox1* encodes a homeodomain-containing transcriptional repressor, which is normally expressed in undifferentiated tissue and whose ectopic expression can block muscle-cell differentiation. When mouse muscle myotubes in culture were transfected with *mox1*, a significant number of the myotubes cleaved to form both smaller multinucleate myotubes and, more significantly, mononuclear cells that could undergo cell division. When these dividing mononuclear cells were cultured in the appropriate conditions, some expressed markers for other cell types such as cartilage or fat cells. A similar reprogramming of muscle cells occurs in the regeneration of amphibian limbs, as we shall see in Chapter 13.

8.11 Skeletal muscle and neural cells can be renewed from stem cells in adults

Once formed, skeletal muscle cells can enlarge by cell growth but do not divide. They can, however, be replaced. In adult mammalian muscle there are cells called

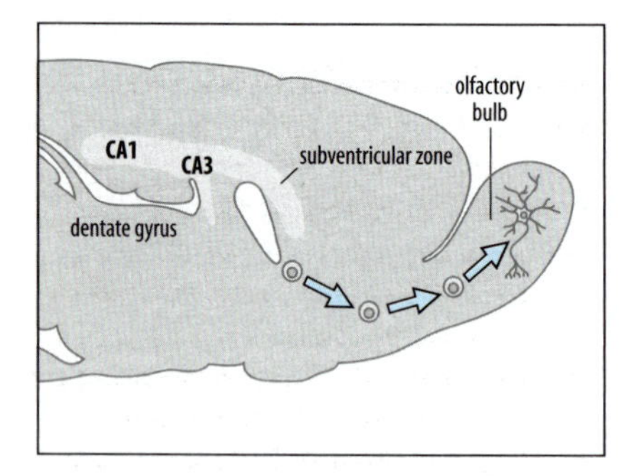

Fig. 8.24 Sites of neurogenesis in the adult mammalian brain. Neurogenesis occurs mainly in the subgranular layer of the dentate gyrus in the hippocampus and the anterior part of the subventricular zone. New neuronal cells generated in the dentate gyrus migrate to the granular layer, while new neuronal cells generated in the subventricular zone migrate to the olfactory bulb and differentiate there into interneurons.

satellite cells that can proliferate and differentiate into new muscle cells if the muscle is damaged. These satellite cells are the muscle stem cells. They lie between the basal lamina and the cell membranes of mature muscle fibers and can be recognized by their expression of characteristic marker proteins such as the cell-surface protein CD34 and the transcription factor Pax7. When satellite cells are activated, for example by injury to the muscle, they proliferate and develop into myoblasts. It is not clear, however, whether one daughter cell from the initial division remains a stem cell or whether a new satellite cell develops later from one of the proliferating cells.

We will see in Chapter 10 how neurons and other nervous system cells develop from embryonic neural stem cells. Once fully differentiated, however, neurons in the mammalian brain do not divide, and for many years it was thought that no new brain neurons at all were produced in the adult mammals. Relatively recently, however, neurogenesis has been demonstrated as a normal occurrence in the adult mammalian brain, and neural stem cells have been identified in adult mammals that can generate neurons, astrocytes, and oligodendrocytes in culture. Neurogenesis in adults occurs in just two regions: the subventricular zone of the lateral ventricle and the subgranular zone of the hippocampus. New neurons generated in the subventricular zone migrate to give rise to interneurons in the olfactory bulb, whereas the stem cells in the hippocampus give rise to new neurons of the dentate granule layer of the hippocampus (Fig. 8.24). Adjacent glial cells in these regions provide a stem-cell niche, producing signals that allow the neural stem cells to operate. Wnt signaling through the β-catenin pathway in neural stem cells has been shown to play an important role in regulating adult neurogenesis in the hippocampus, but the nature of the other signals is unknown.

8.12 Embryonic neural crest cells differentiate into a great variety of different cell types

We now turn to the embryonic neural crest, which gives rise to a remarkable number of cell types. As described in Chapter 4, neural crest cells arise from the dorsal neural tube and are recognizable as discrete cells when they emerge from the neural epithelium. The specification of the neural crest from uncommitted precursor cells at the margin of the neural plate is the result of an inductive interaction between the neural plate and the adjacent presumptive ectoderm, and involves Wnt and BMP signals. Once they have detached from the neural tube, neural crest cells migrate to many sites, as discussed in Chapter 7. Here we shall discuss their potential for differentiation. Neural crest cells give rise to much of the skeleton of the head, the neurons and glia of the peripheral nervous system (which includes the sensory and autonomic nervous systems), endocrine cells, such as the chromaffin cells of the adrenal medulla, and melanocytes, which are pigmented cells found in the skin and other tissues (Fig. 8.25). The neural crest thus gives rise not only to the usual ectodermal derivatives, such as nervous tissue, but to mesodermal types of tissues in the form of the facial bones, cartilage, and dermis. The neural crest with potential to form mesodermal as well as ectodermal derivatives is called the **mesectoderm**. This tremendous potential was originally discovered by removing the neural crest from amphibian embryos and noting which structures failed to develop.

The differentiation of neural crest cells has been studied in detail using quail neural crest transplanted into chick embryos. Quail cells contain a distinctive nucleus that distinguishes them from chick cells. It is therefore possible to track the transplanted quail cells and see which tissues they contribute to and what cell types they differentiate into (see Fig. 7.37). To construct a fate map of the neural crest,

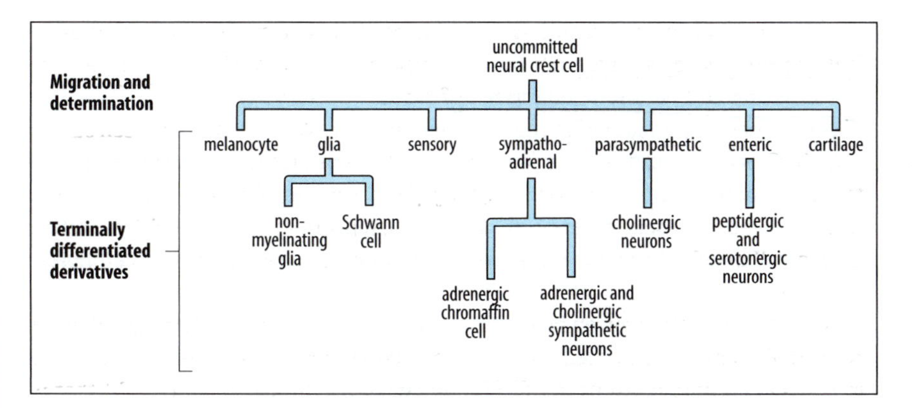

Fig. 8.25 Derivatives of the neural crest. Neural crest cells give rise to a wide variety of cell types that include melanocytes, cartilage, glia, and a variety of neurons distinguished by their functional specializations and the neurotransmitters they produce. Cholinergic neurons use acetylcholine as their neurotransmitter, adrenergic neurons principally use noradrenaline (norepinephrine), and peptidergic and serotonergic neurons produce peptide neurotransmitters and serotonin, respectively.

small regions of quail neural tube are grafted to equivalent positions in chick embryos of the same age, prior to neural crest migration. The fate map shows a general correspondence between the position of crest cells along the antero-posterior axis and the axial position of the cells and tissues they will give rise to (Fig. 8.26). For example, the neural crest cells that give rise to tissues in the facial and pharyngeal regions are derived from crest anterior to somite 5, whereas ganglion cells of the autonomic sympathetic system come from crest posterior to somite 5.

To what extent is the fate of neural crest cells fixed before migration? Some crest cells are unquestionably multipotent. Single neural crest cells injected with a tracer shortly after they have left the neural tube can be seen to give rise to a number of different cell types, both neuronal and non-neuronal. Also, by changing the position of neural crest before cells start to migrate, it has been shown that the developmental potential of these cells is greater than their normal fate would suggest. Multipotent stem cells have been isolated from neural crest that can give rise to neurons, glia, and smooth muscle. Similar multipotent cells have even been isolated from the peripheral nerves of mammalian embryos days after neural crest migration. Almost all of the cell types to which the neural crest can give rise will differentiate in tissue culture, and the developmental potential of single neural crest cells has been studied in this way. Most of the clones that develop from neural crest cells taken at the time of migration contain more than one cell type, showing that the cell was multipotent at the time it was isolated. As neural crest cells migrate, however, their potential decreases progressively: the size of cultured

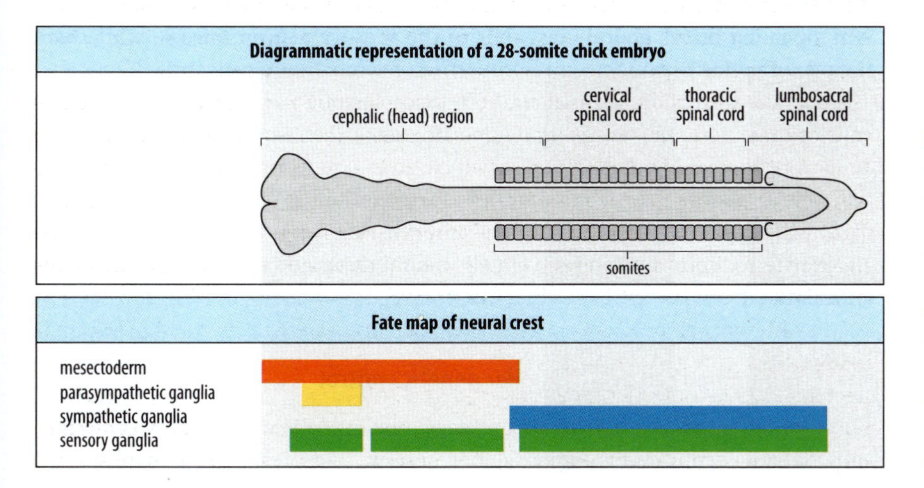

Fig. 8.26 Fate and developmental potential of neural crest cells. The fate map of the neural crest (bottom panel) shows that there is a general correspondence between the position of the neural crest cells along the antero-posterior axis and the position of the structure to which these cells give rise (top panel). For example, the crest that gives rise to the mesectoderm of the head comes from the anterior region. By contrast, the developmental potential, with the exception of the presumptive mesectoderm, is very much greater than the presumptive fate.

clones gets smaller and the diversity of cell types they contain is less. Thus, shortly after the beginning of migration, the neural crest is a mixed population of pluripotent cells and cells whose potential is restricted.

Numerous factors that affect the differentiation and proliferation of neural crest cells have been identified by studies of cultured cells and mutations in mice. For example, melanocyte survival and differentiation is stimulated by Wnts and the cytokines endothelin and stem cell factor (SCF), gliogenesis by glial growth factor (also called neuregulin), and BMPs are required for the development of autonomic neurons. Activation of the Wnt β-catenin pathway in the neural tube promotes the formation of sensory neurons by neural crest at the expense of virtually all other derivatives, and brain-derived neurotrophic factor (BDNF) produced by the neural tube has been implicated in the survival of sensory neurons.

Melanocytes are derived from neural crest cells that follow a dorsal migration route under the ectoderm. A variety of signals have been shown to induce differentiation of melanocytes from neural crest cells, including Wnt signals and SCF. Mice homozygous for either the *white spotting* (*W*) or *Steel* mutations (genes originally identified by unusual coat color) lack melanocytes in the skin and other tissues. (They are also deficient in hematopoietic stem cells, and their germ-cell development is affected.) Although they apparently have the same gross effect, the genes code for two quite different proteins. *white spotting* codes for the cell-surface protein Kit, which is the receptor for SCF. It is expressed on developing melanoblasts, which are the undifferentiated precursors of skin melanocytes. *Steel*, in contrast, codes for SCF, which is produced by the fibroblasts and keratinocytes that surround the melanoblast. Mutations in either gene block melanocyte differentiation, showing that the signal generated by the interaction of SCF and Kit is essential for differentiation (Fig. 8.27). The peptide endothelin stimulates the differentiation of neural crest cells into melanocytes and glia, and can also cause cultured quail pigment cells to dedifferentiate and give rise to both glial cells and melanocytes.

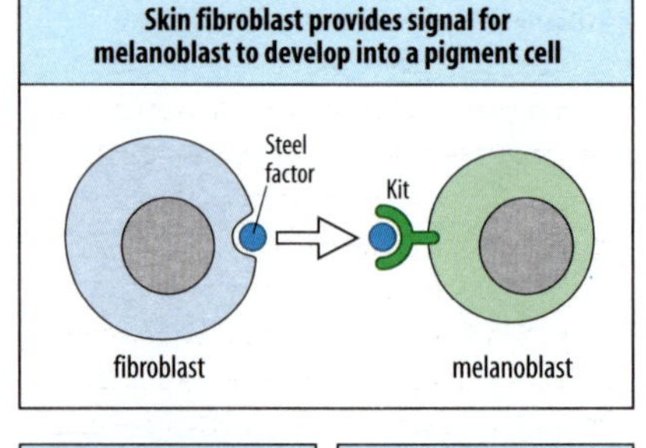

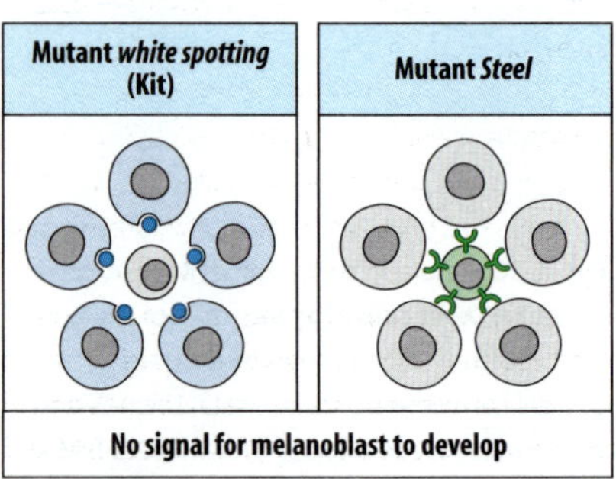

Fig. 8.27 The receptor Kit and its ligand the Steel factor are involved in melanoblast differentiation. When either the gene encoding Kit (the gene *white spotting, or W*) or the *Steel* gene are mutant, melanoblasts fail to differentiate into melanocytes.

Neural crest cells give rise to both the sensory dorsal root ganglia and the autonomic ganglia of the peripheral nervous system. These ganglia contain both neurons and glia, the neurons differentiating first. So what determines whether a neural crest cell gives rise to a neuron or to glia? There is evidence that activation of Notch in neural crest stem cells inhibits neuronal formation and promotes differentiation into glia. There is also evidence that Notch ligands such as Delta are expressed on neuroblasts, the precursors of neurons. Thus, once a neuroblast has embarked on the neurogenic differentiation pathway under the influence of neurogenic signals such as BMP-2, it may provide a feedback signal that prevents the surrounding cells from developing further as neuroblasts and promotes formation of glia instead. Wnt signaling biases neural crest cells to the sensory neuron lineage, while BMP signals influence them to acquire properties of sympathetic neurons.

Hox gene expression in the neural crest (see Chapter 4) is yet another determinant of crest potential. For example, no Hox genes are expressed in the anterior domain of the cephalic neural crest, which gives rise to the crest cells that form the bones of the face, whereas anterior Hox genes are expressed in the posterior part of the cephalic neural crest. If the anterior Hox genes *Hoxa2*, *Hoxa3*, or *Hoxb4* are transfected into the anterior cephalic neural crest, no facial skeletal structures differentiate.

8.13 Programmed cell death is under genetic control

Selective cell death is a normal part of development. While not strictly cell differentiation, it is convenient to consider it here in these terms. It is involved, for

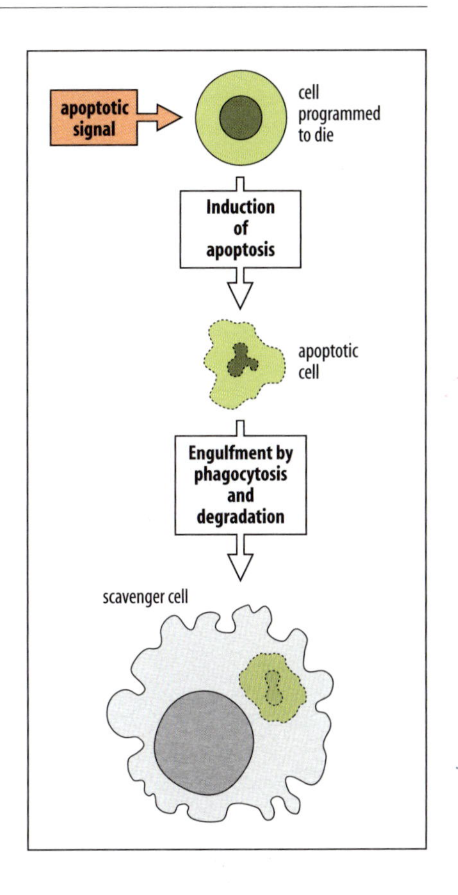

Fig. 8.28 Programmed cell death. During development many cells undergo programmed cell death or apoptosis. Apoptosis may be triggered by an external signal (as shown here) or cells may be programmed to die unless they receive a specific survival signal.

example, in the morphogenesis of the vertebrate limb, where interdigital cell death is essential for separating the digits (see Chapter 9), and the development of the vertebrate nervous system involves the death of large numbers of neurons. Programmed cell death is particularly important in the development of the nematode (see Chapter 5): 959 somatic cells come from the egg and 131 die during development. In all these cases, the dying cell undergoes a type of cell suicide known as **apoptosis**, which requires both RNA and protein synthesis and is quite distinct from the death that occurs as a result of pathological damage. In apoptosis, the cytosolic calcium concentration rises and leads to the activation of an endonuclease, which fragments the chromatin. The cell contents remain bounded by a cell membrane throughout, even though the cell breaks up into fragments. The dying cell is finally phagocytosed by scavenger cells (Fig. 8.28). These features distinguish apoptosis from cell death due to damage, where the whole cell tends to swell and eventually burst open (lyse); this type of cell death is known as **necrosis**.

Studies of apoptosis in the nematode have shown that although many different types of cells die, their death is initiated through a common mechanism centered on the actions of CED-3, which is a type of protease called a **caspase**. CED-3 triggers the cellular changes that lead to apoptotic cell death. The nematode caspase-activation pathway is shown in Figure 8.29 (left panel). CED-4 in this pathway is an adaptor protein that activates CED-3. In nematodes with mutations that inactivate either

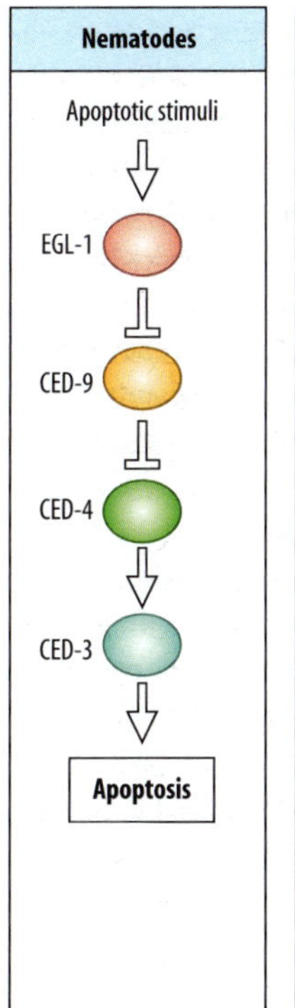

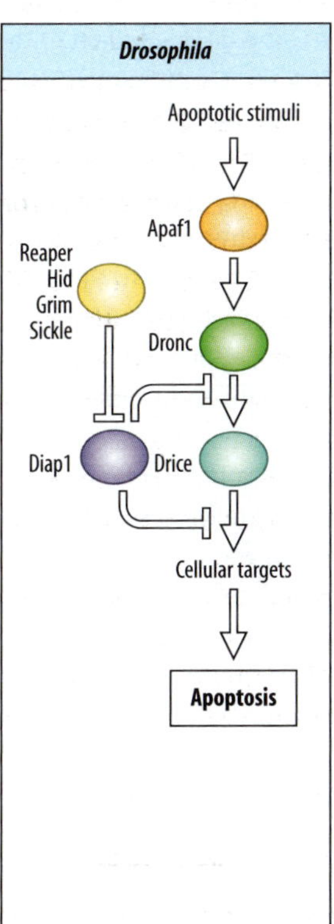

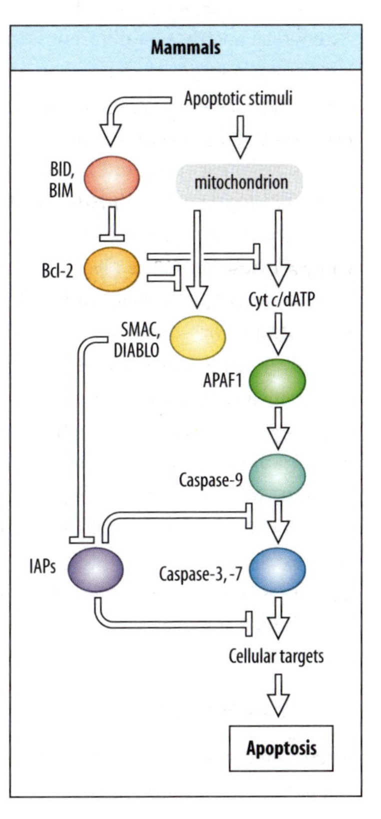

Fig. 8.29 Apoptotic pathways in nematodes, *Drosophila*, and mammals. Left panel: in nematodes, programmed cell death during development requires the activation of a pathway that ends in the activation of the caspase CED-3, a type of protease, by the protein CED-4. CED-3 triggers the cellular changes that lead to cell death. The actions of CED-3 and CED-4 are inhibited by CED-9. If CED-9 is active, the cell will not undergo apoptosis. Inhibition by CED-9 is lifted by the actions of EGL-1, which is expressed in response to pro-apoptotic signals. Middle panel: a similar pathway is present in *Drosophila*, with the addition of a second caspase (Drice). Diap1 is a negative regulator of the caspases that is not present in nematodes. Its inhibition is lifted by the actions of proteins such as Reaper and Sickle. Bottom panel: the apoptotic pathway in mammals is even more complex, but uses the same main components. Apoptotic signals are transduced through the EGL-1 homologs BIM and BID to cause the release of pro-apoptotic factors from mitochondria. One of these is cytochrome *c*, which triggers the caspase-activation pathway. In the absence of apoptotic signals, cell death is inhibited by BCL-2, which is homologous to CED-9. In all three diagrams, the proteins are color-coded to show homology.

ced-3 and *ced-4*, none of the 131 cells that normally die do so, but appear instead to differentiate in the same way as their sister cells. Such worms appear to survive well and die at the usual age of several weeks.

The nematode's cell-death program is under the control of another protein, CED-9, which acts as a brake on the process by preventing the activation of CED-3 by CED-4. If *ced-9* is inactivated by mutation, many cells that normally do not die will die. And if the *ced-9* gene is made abnormally active by mutation, no cell deaths occur. Apoptosis occurs when the actions of the protein EGL-1 remove inhibition by CED-9. EGL-1 is produced in response to pro-apoptotic signals. A similar pathway is present in *Drosophila* and mammals (Fig. 8.29, middle and right panels). The mammalian homologs of *ced-9* are the Bcl-2 family of genes (Fig. 8.29, right panel). The sequence homologies between the gene *Bcl-2* and *ced-9* are so strong that *Bcl-2* introduced into nematodes will function in place of *ced-9*. Mammals also possess caspases homologous to CED-3, which initiate the proteolytic cascade that leads to cell death.

Far from programmed cell death being a rare phenomenon, in animals the cells in all tissues are intrinsically programmed to undergo cell death, and are only prevented from dying by positive control signals from neighboring cells. Cell death also plays a key role both in controlling growth and the prevention of cancer (see Chapter 12).

Summary

Several adult tissues, such as the blood and the skin, are continuously being replaced by the differentiation of new cells from stem cells. Hematopoiesis illustrates how a single progenitor cell, the multipotent stem cell, can be both self-renewing and generate a variety of different lineages. The differentiation of the various lineages of blood cells seems to depend on external signals, although there is some evidence for an intrinsic tendency to diversification. External factors are, however, essential for the survival and proliferation of the specific cell types. The epithelial tissues of adult mammalian skin and gut are also replaced from self-renewing stem cells. The derivation of blood cells from the multipotent stem cell involves a successive restriction of developmental potential as cells become committed to the different lineages. Similar processes occur in the diversification of the embryonic neural crest. Before they leave the crest, some crest cells still have a broad developmental potential, and environmental signals both direct the pathways of neural crest differentiation and promote the survival of particular cell types.

Cellular differentiation does not usually involve irreversible changes to the DNA. The only exception to this in vertebrates occurs in lymphocyte differentiation, during the assembly of the genes for immunoglobulins and T-cell antigen receptors, which involves irreversible DNA rearrangements. In general, cell differentiation is controlled by complex combinations of transcription factors, whose expression and activity are influenced by external signals. Some transcription factors are common to many cell types, but others have a very restricted pattern of expression. Muscle differentiation, for example, can be brought about by the expression of transcription factors specific to the muscle lineage, such as MyoD, which initiate the differentiation program and bind to the control regions of muscle-specific genes. Programmed cell death is a common fate for cells, especially during development, and there is evidence that positively acting signals are generally required to prevent programmed cell death and allow cells to survive.

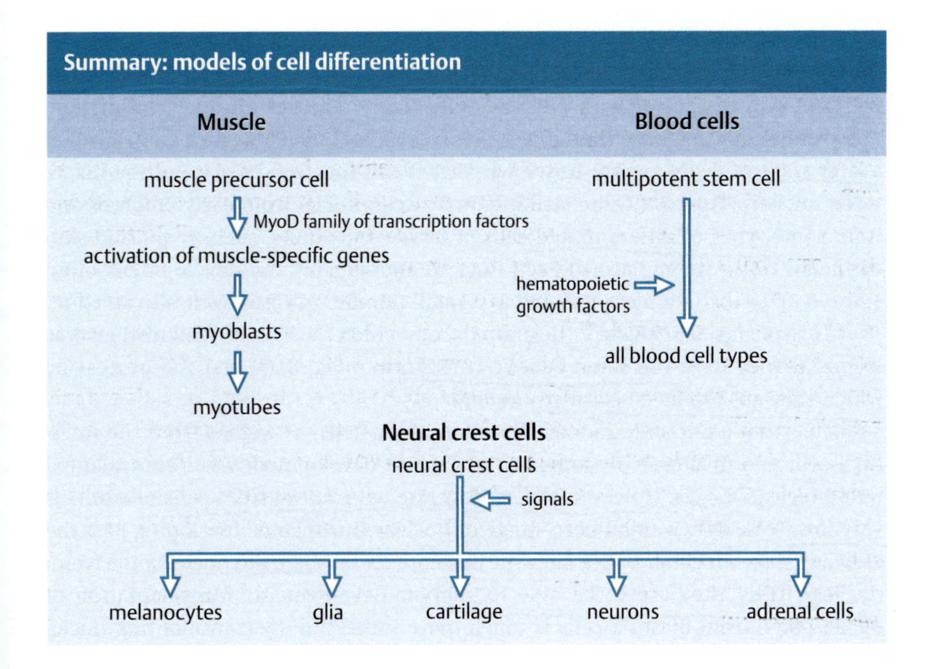

The plasticity of gene expression

In earlier parts of this chapter we have looked at examples of genes being turned on and off during development and have considered some of the mechanisms involved. These chiefly involve the binding of different combinations of gene-regulatory proteins to the control regions of genes. How reversible are these changes, and thus how reversible is cell differentiation? We have also seen several examples of stem cells that both self-renew and give rise to a wide variety of cell types. This is of great importance, as it offers the possibility of using stem cells to replace cells that have been damaged or lost in human disease. Thus the issue is one of how genes can be controlled in stem cells to give the desired cell type, and raises the question of just how multipotent stem cells are. Could, for example, blood stem cells give rise to neurons under appropriate conditions? How plastic is the pattern of gene activity? We start by examining the extent to which the pattern of gene activity in differentiated cells can revert to that found in the fertilized egg. One way of finding out whether it can be reversed in practice is to place the nucleus of a differentiated cell in a different cytoplasmic environment, one that contains a different set of potential gene-regulatory proteins.

8.14 Nuclei of differentiated cells can support development

The most dramatic experiments addressing the reversibility of differentiation have investigated the ability of diploid nuclei from cells at different stages of development to replace the nucleus of an egg and to support normal development. If they can do this, it would indicate that no irreversible changes have occurred to the genome during differentiation. It would also show that a particular pattern of nuclear gene activity is determined by whatever transcription factors and other regulatory proteins are being synthesized in the cytoplasm of the cell. Such experiments were initially carried out using the eggs of amphibians, which are particularly robust to experimental manipulation.

In the unfertilized eggs of *Xenopus*, the nucleus lies directly below the surface at the animal pole. A dose of ultraviolet radiation directed at the animal pole destroys the DNA within the nucleus, and thus effectively removes all nuclear function. These enucleated eggs are then injected with a diploid nucleus taken from a cell at a later stage of development to see whether it can function in place of the inactivated nucleus. The results are striking: in *Xenopus*, nuclei from early embryos and from some types of differentiated cells of larvae and adults, such as gut and skin epithelial cells, can replace the egg nucleus and support the development of an embryo up to the tadpole stage, and in a small number of cases even into an adult. These organisms are **clones** of the animal from which the somatic cell nucleus was taken, as they have the same genetic constitution (Fig. 8.30) and the process by which they are produced is called **cloning**.

Nuclei from adult skin, kidney, heart, and lung cells, as well as from the intestinal cells and myotome of tadpoles, can support development when transplanted into enucleated eggs. However, the success rate with nuclei from cells of adults is very low, with only a small percentage of nuclear transplants developing past the cleavage stage. In general, the later the developmental stage the nuclei come from, the less likely they are to be able to support development. Transplantation of nuclei taken from blastula cells is much more successful; by transplanting nuclei from the same blastula into several enucleated eggs, a clone of genetically identical frogs can be obtained (Fig. 8.31). An even greater success rate is achieved if the process is repeated with nuclei from the blastula stage of the cloned embryo. These results show that, at least in those differentiated cell types tested, the genes required for development are not irreversibly altered. More importantly, when exposed to the egg's cytoplasmic factors, they behave as the genes in a fertilized egg's zygotic nucleus would. So in this sense at least, many of the nuclei of the embryo and the adult are equivalent, and their behavior is determined entirely by factors present in the cell.

What about organisms other than *Xenopus*? Similar results have been obtained in insects, where nuclei from the blastoderm stage can, when returned to the egg, participate in the formation of a wide range of tissue types. In ascidians, the genetic equivalence of nuclei at different developmental stages can also be demonstrated.

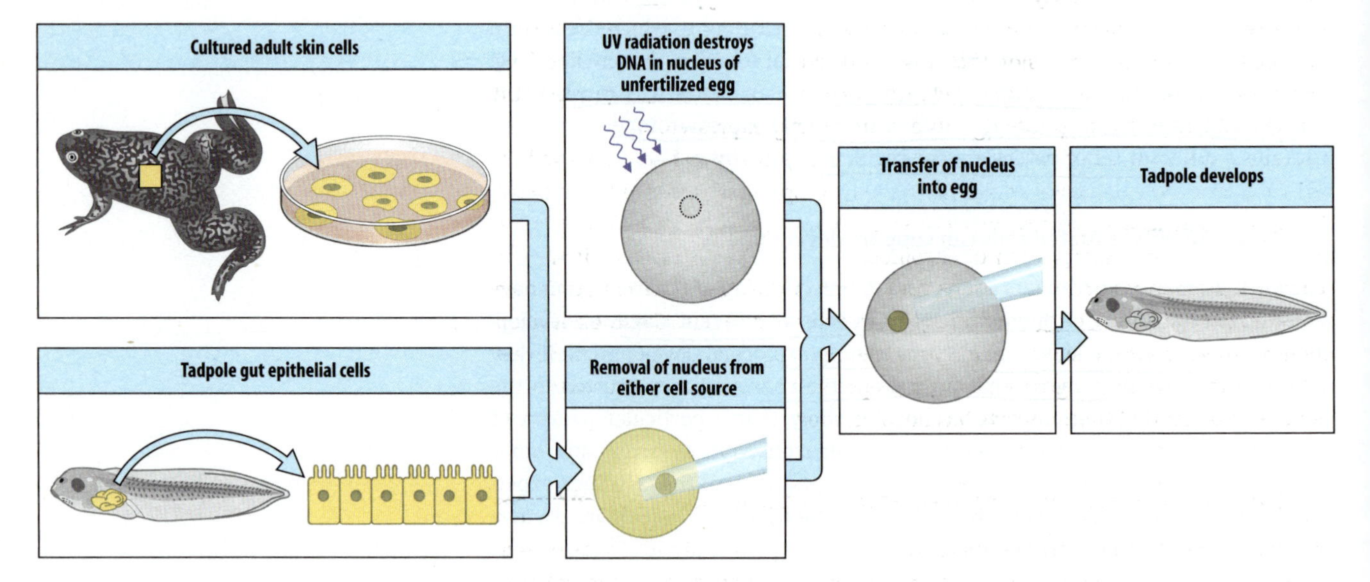

Fig. 8.30 Nuclear transplantation. The haploid nucleus of an unfertilized *Xenopus* egg is rendered non-functional by treatment with ultraviolet radiation. A diploid nucleus taken from the epithelial cells lining the tadpole gut or from cultured adult skin cells is transferred into the enucleated egg, and can support the development of an embryo until at least the tadpole stage.

In plants, single somatic cells can give rise to fertile adult plants, demonstrating complete reversibility of the differentiated state at the cellular level (see Section 6.4). In some mammals, successful development has been achieved following removal of the egg nucleus with a micropipette and injection of a nucleus from a somatic cell, or by fusion of the somatic cell with the egg cell. In cattle and sheep, nuclei from embryonic cell cultures can successfully support development when transplanted into an oocyte.

A lamb, the famous Dolly, was the first mammal to be cloned using a nucleus from an adult somatic cell, in this case a cell line derived from the udder. Mice have been similarly cloned using nuclei from the somatic cumulus cells that surround the recently ovulated egg, and several generations of these mice have been produced by repeating the procedure. Transplantation of nuclei from differentiated neurons from the brain, however, only supported development to the late blastocyst stage, but fertile mice have been produced by transferring the nuclei of post-mitotic sensory neurons into enucleated oocytes. The blastocyst stage of these cloned embryos were then used to generate **embryonic stem cells** (**ES cells**) from the cells of the inner cell mass (see Box 3E, p. 130) and these were injected into host blastocysts made from tetraploid zygotes. All of the embryonic lineage in the embryos could subsequently be seen to come from the injected cells, and the embryos gave rise to normal mice (the takeover of the whole embryonic lineage by the introduced embryonic cells occurs because the tetraploid cells are debilitated and cannot compete).

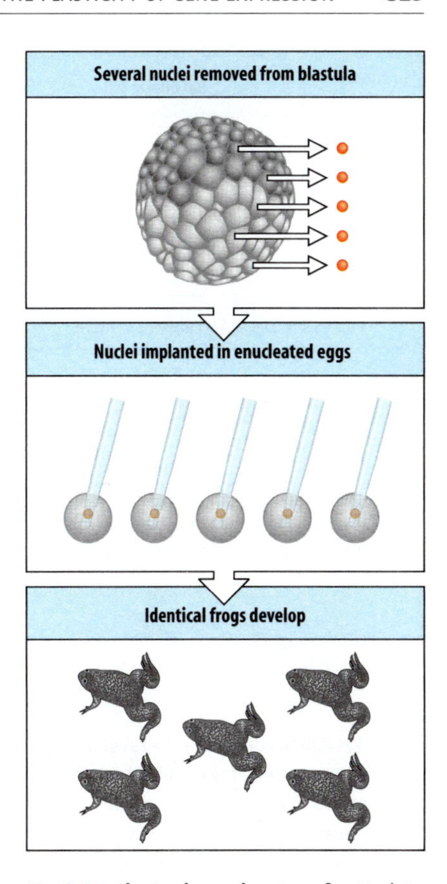

Fig. 8.31 Cloning by nuclear transfer. Nuclei from cells of the same blastula are transferred into enucleated, unfertilized *Xenopus* eggs. The frogs that develop are clones as they all have the same genetic constitution.

The success rate of cloning from somatic cell nuclei is extremely low, and the reasons for this are not yet well understood. Most cloned animals derived from nuclear transplantation die before birth, and those that survive are usually abnormal in some way. The most likely cause of failure is incomplete reprogramming of the donor nucleus to remove all the epigenetic modifications, such as DNA methylation and histone modification, that are involved in determining and maintaining the differentiated cell state, as discussed earlier, and thus to restore the DNA to a state resembling that of a newly fertilized oocyte. A related cause of abnormality may be that the reprogrammed genes have not gone through the normal imprinting process that occurs during germ-cell development, where different genes are silenced in the male and female parents, and which will be discussed in Chapter 11. There are also abnormalities in many of the adults that do develop from cloned embryos. These include early death, limb deformities and hypertension in cattle, and immune impairment in mice, and all these defects are thought to be due to abnormalities of gene expression that arise from the cloning process. Studies have shown that some 5% of the genes in cloned mice are not correctly expressed and that almost half of the imprinted genes are incorrectly expressed.

Cloning from an adult somatic cell nucleus has not yet been achieved in primates. There is evidence that this failure may be specifically due to the removal during the enucleation process of factors required for mitotic spindle assembly, and thus the correct segregation of chromosomes at cell division. In primates, it seems, these factors are in the egg nucleus that is removed, rather than being scattered throughout the egg as in many other types of animals. Cloning a human being by these means, even if it could be achieved, is banned in most countries on both practical and ethical grounds. Despite reports in the media that humans have been cloned, none of these reports has been verified.

8.15 Patterns of gene activity in differentiated cells can be changed by cell fusion

Nuclear transplantation into eggs, particularly frog eggs, is facilitated by their size and the large amount of cytoplasm. With other cell types, particularly differentiated

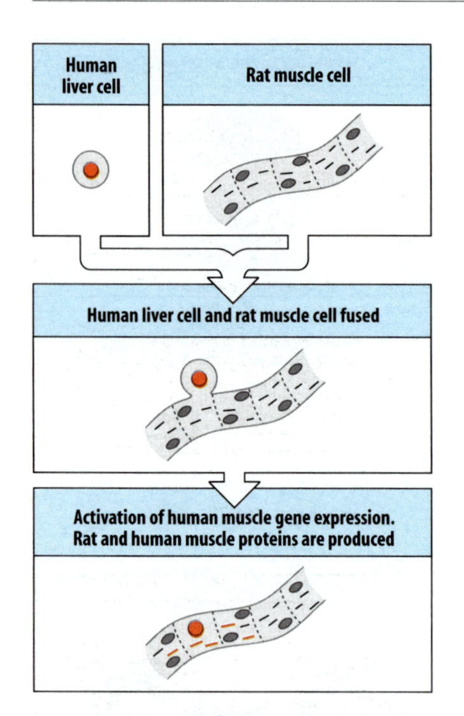

Fig. 8.32 Cell fusion shows the reversibility of gene inactivation during differentiation. A human liver cell is fused with a mouse muscle cell. The exposure of the human nucleus (red) to the mouse muscle cytoplasm results in the activation of muscle-specific genes and repression of liver-specific genes in the human nucleus. Human muscle proteins are made together with mouse muscle proteins. This shows that the inactivation of muscle-specific genes in human liver cells is not irreversible.

cells, injection of nuclei into foreign cytoplasm is not possible. It is possible, however, to expose the nucleus of one cell type to the cytoplasm of another type by fusing the two cells together. In this process, which can be brought about by treatment with certain chemicals or viruses, the plasma membranes fuse and nuclei from different cells come to share a common cytoplasm.

A striking example of the reversibility of gene activity after cell fusion is provided by the fusion of chick red blood cells with human cancer cells in culture. Unlike mammalian red blood cells, mature chick red blood cells have a nucleus. Gene activity in the nucleus is, however, completely turned off and it thus produces no mRNA. When a chick red blood cell is fused with a cell from a human cancer cell line, gene expression in the red blood cell nucleus is reactivated, and chick-specific proteins are expressed. This shows that human cells contain cytoplasmic factors capable of initiating transcription in chick nuclei.

Fusion of differentiated cells with striated muscle cells from a different species provides further evidence for reversibility of gene expression in differentiated cells (Fig. 8.32). Multinucleate striated muscle cells are good partners for cell-fusion studies as they are large, and muscle-specific proteins can easily be identified. Differentiated human cells representative of each of the three germ layers can be fused with mouse multinucleate muscle cells, so that the human cell nuclei are exposed to the mouse muscle cytoplasm. This results in muscle-specific gene expression being switched on in the human nuclei. For example, human liver cell nuclei in mouse muscle cytoplasm no longer express liver-specific genes; instead, their muscle-specific genes are activated and human muscle proteins are made.

These results clearly show that patterns of gene expression in differentiated cells can be changed, and that gene expression can be controlled by factors present in the cytoplasm. Some of these factors at least are transcription factors, as we will see below, and this leads one to consider that the differentiated state may be at least partly maintained by the continuous action of transcription factors on their target genes.

8.16 The differentiated state of a cell can change by transdifferentiation

A fully differentiated cell is generally stable, which is essential if it is to serve a particular function in the mature animal. If it has retained the ability to divide, the cell passes its differentiated state on to all its descendants. In some long-lived cells such as neurons, which do not divide once they are differentiated, the differentiated state must remain stable for many years. Plant cells maintain their differentiated state while in the plant, but do not maintain it when placed in cell culture (see Chapter 6).

Under certain conditions, however, differentiated animal cells are not stable, and this provides yet another demonstration of the potential reversibility of patterns of gene activity. We have discussed the dedifferentiation of muscle cells (see Section 8.10) and, as we shall see in Chapter 13, cells of one type can change into a different, but related, type in regenerating tissues. The classic example of such a change is regeneration of the lens in the adult newt eye, which occurs by recruitment of cells from the dorsal region of the pigmented epithelium of the iris. These first dedifferentiate and then re-differentiate to form a new lens. Another example is the dedifferentiation of skeletal muscle cells in regenerating newt limbs and the ability of these cells to give rise to cartilage. The change of one differentiated cell type into another is known as **transdifferentiation**, and a number of cell types have been found to undergo such a change in tissue culture, particularly when the culture conditions are altered by the addition of chemical agents.

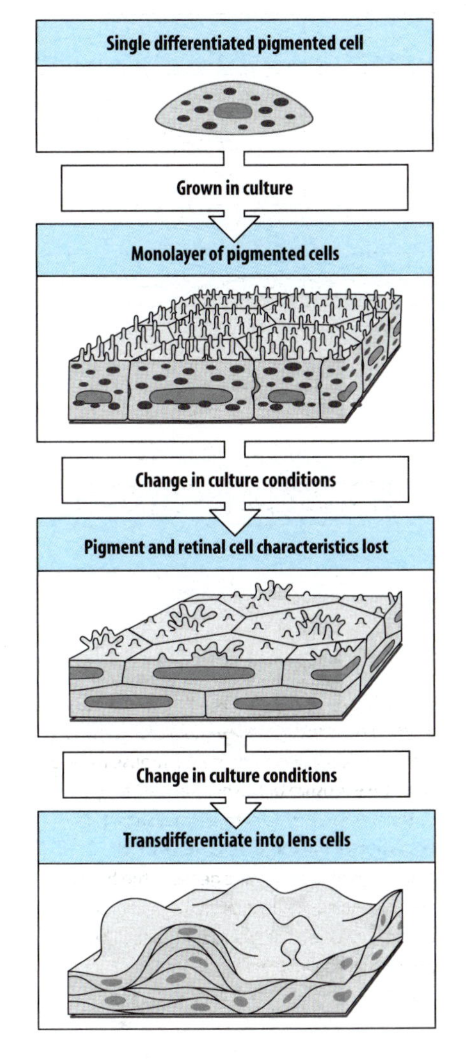

Fig. 8.33 Transdifferentiation of retinal pigment cells. A single pigmented epithelial cell from the embryonic chick retina can be grown in culture to produce a monolayer of pigmented cells. On further culture in the presence of hyaluronidase, serum, and phenylthiourea, they lose their pigment and retinal cell characteristics. If cultured at a high density with ascorbic acid, they differentiate as lens cells and produce the lens-specific protein crystallin.

After Okada, T.S.: 1992.

One well-established case of transdifferentiation occurs when pigmented epithelial cells of the embryonic chick retina are exposed to certain culture conditions. The pigment disappears and the cells start to take on the structural characteristics of lens cells, and to produce the lens-specific protein crystallin (Fig. 8.33). Another example is the transdifferentiation of liver to pancreas. The liver and the pancreas both derive from endoderm, and arise from adjacent regions of endoderm in the embryo. Pdx1 is a transcription factor that is essential for the development of the pancreas and is not expressed in the liver. When Pdx1 is expressed in transgenic *Xenopus* tadpoles under the control of a liver-specific promoter, however, part or all of the liver is converted to pancreatic tissue containing both exocrine and endocrine pancreatic cells, and proteins characteristic of differentiated liver cells disappear from the converted regions.

Other cells that can transdifferentiate are the chromaffin cells of the adrenal gland (Fig. 8.34). These small cells are derived from the neural crest and secrete adrenaline (epinephrine) into the blood. In culture, the chromaffin phenotype can be maintained by the addition of glucocorticoids, but when the steroids are removed and nerve growth factor (NGF) is added to the culture medium, the chromaffin cells transdifferentiate into sympathetic neurons. These neurons are larger than chromaffin cells, with dendritic and axonal processes, and they secrete noradrenaline (norepinephrine) instead of adrenaline. This transformation shows that even terminally differentiated cells can change into another cell type when exposed to the appropriate environmental signals.

It should be noted that in all the cases described above, the transdifferentiation is to a developmentally related cell type. Pigment cells and lens cells are both derived from the ectoderm and are involved in eye development, liver and pancreatic cells are both of endodermal origin, and chromaffin cells and sympathetic neurons both differentiate from neural crest.

A particularly interesting example of transdifferentiation, in which one cell type transdifferentiates into two different types in succession, is provided by jellyfish striated muscle. When a small patch of striated muscle and its associated

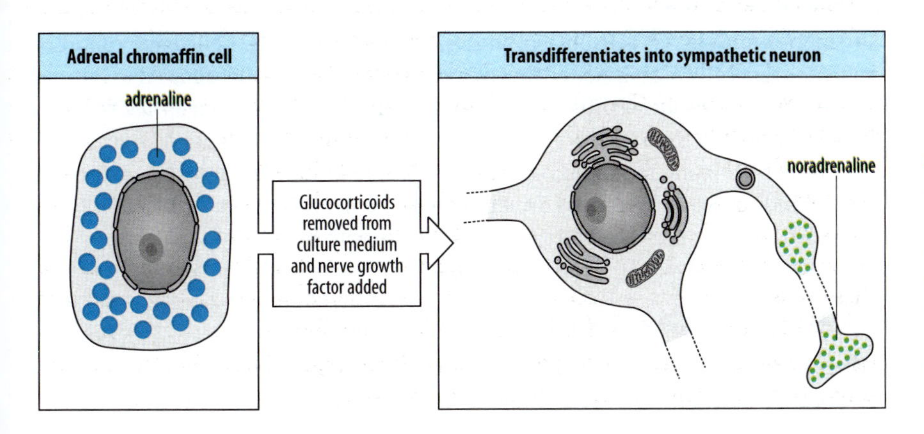

Fig. 8.34 Transdifferentiation of chromaffin cells. Chromaffin cells of the adrenal medulla secrete adrenaline (epinephrine). Their phenotype is maintained in culture in the presence of glucocorticoids. If the steroids are removed and nerve growth factor added, the cells transdifferentiate into sympathetic neurons which secrete noradrenaline (norepinephrine).

After Doupe, A.J., et al.: 1985.

extracellular matrix (mesogloea) is cultured, the striated muscle state is maintained. However, if the cultured tissue is treated with enzymes that degrade the extracellular matrix, the cells form an aggregate, and within 1–2 days some have trans-differentiated into smooth muscle cells, which have a different cellular morphology. This is followed by the appearance of a second cell type, nerve cells. This example indicates a role for the extracellular matrix in maintaining the differentiated state of the striated muscle.

8.17 Embryonic stem cells can proliferate and differentiate into many cell types in culture

The stem cells that are responsible for tissue renewal in adults are limited in the range of cell types they give rise to, and require a specific stem-cell niche to maintain their stem-cell state, such as the bone marrow stroma that supports hematopoietic stem cells (see Section 8.6). In contrast, early embryonic cells can differentiate into numerous cell types and do not require a special niche.

The primary example of almost completely totipotent stem cells in mammals are the ES cells derived from the inner cell mass (see Section 3.13). Mouse ES cells have been studied intensively. They can be maintained in culture for long periods, apparently indefinitely, but if injected into a blastocyst that is then returned to the uterus, they can contribute to all types of cell in the embryo that develops (see Box 3E, p. 130).

To maintain their pluripotent state in culture, ES cells must both express the transcription factor Oct4, which is restricted to pluripotent stem cells, and be exposed to the cytokine leukemia inhibitory factor (LIF). If grown in suspension, ES cells spontaneously form aggregates called embryoid bodies, which can differentiate into many cell types, including heart muscle, blood islands, neurons, pigmented cells, epithelia, fat cells, macrophages, and even germ cells. When cultured in the absence of serum, which contains a variety of growth and differentiation factors, many of which are still unknown, ES cells differentiate into neural cells. Their normal proliferative and stem-cell properties are restored, however, if BMPs are added to the culture medium.

By manipulating the culture conditions, particularly in respect of the growth factors present, ES cells can be made to differentiate into a particular cell type. One example is the development of stem cells for the glia, the supporting cells of the nervous system. Aggregates of ES cells successively treated with particular combinations of FGF-2, EGF, and platelet-derived growth factor continue to proliferate as long as the growth factors are present. But when the factors are withdrawn the cells differentiate into either of two glial cell types—astrocytes or oligodendrocytes. Other combinations of growth factors lead to the differentiation of ES cells into vascular smooth muscle, fat cells, macrophages, or red blood cells. Cell shape can also regulate stem-cell commitment. Human mesenchymal stem cells allowed to flatten in culture differentiate as osteocytes, whereas if they remain rounded they become adipocytes. In regard to the potential use of neural stem cells, it is encouraging that ES cells can be made to develop specifically into neural progenitors by treatment with retinoic acid, and when transplanted into the neural tube of a mouse embryo these progenitors developed into neurons appropriate to their new location.

8.18 Stem cells could be a key to regenerative medicine

The goal of regenerative medicine is to restore the structure and function of damaged or diseased tissues. As stem cells can proliferate and differentiate into a wide variety of cell types, they are strong candidates for use in cell-replacement therapy, the restoration of tissue function by the introduction of new healthy cells. This type of therapy might eventually offer an alternative to conventional organ transplantation from a donor, with its attendant problems of rejection and shortage of organs, and might also be able to restore function to tissues such as brain and nerves. Both ES cells and adult stem cells are being studied with cell-replacement therapy in mind.

ES cells have the advantage that they can differentiate into a wide range of cell types and could, in theory, be used to repair any tissue. Stem cells from an unrelated donor embryo will still cause immune rejection reactions, however, and because of their proliferative capacity and totipotency, the risk of embryonal tumors—tumors derived from embryonic cells—is high. Cultured ES cells introduced into another embryo will develop normally, but when ES cells are introduced under the skin of a genetically identical adult mouse they give rise to a tumor known as a **teratocarcinoma** (Fig. 8.35). These unusual tumors contain a mixture of differentiated cells. There is also the ethical problem that to make human ES cells, a blastocyst has to be destroyed; there are those who believe that, even at this very early stage, this is tantamount to destroying a human being.

The advantage of adult stem cells is that it might be possible to take these from the patient's own undamaged tissues and reimplant them where needed, thus avoiding the problems of tissue rejection. This is already done in the case of skin grafts, where the patient's own skin can be used. But the range of cell types and tissues

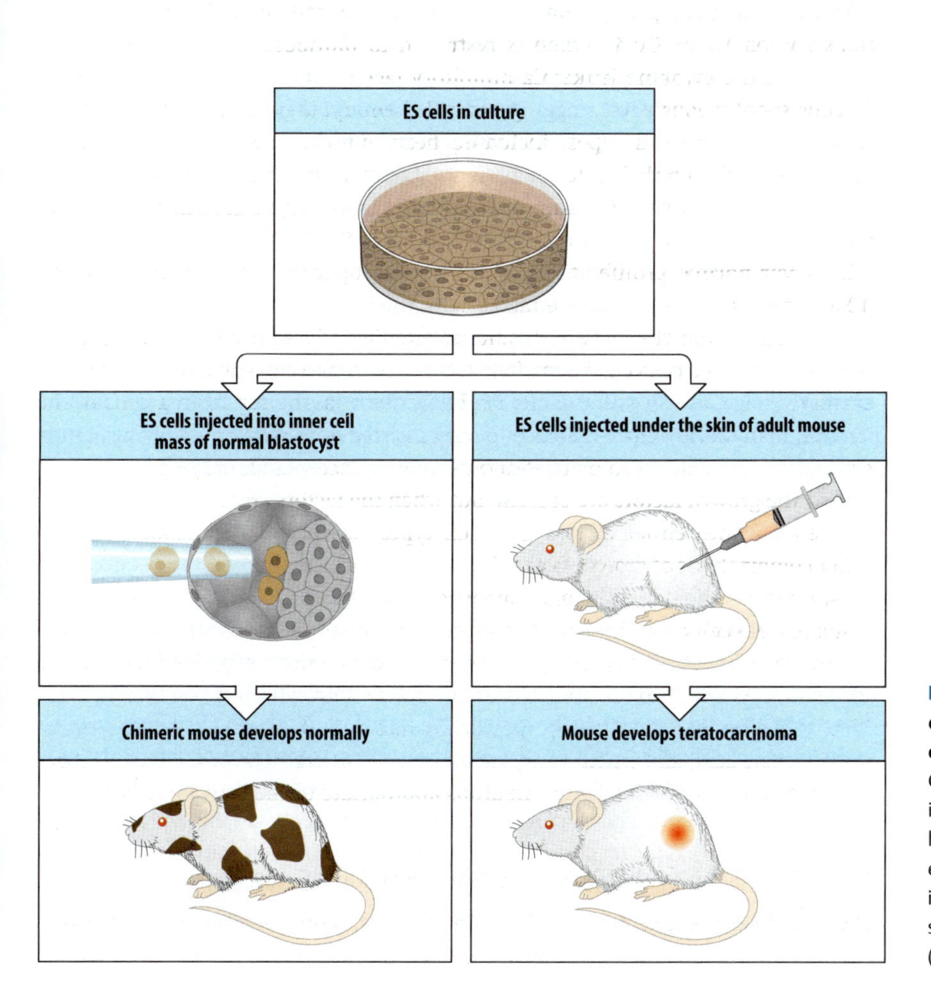

Fig. 8.35 Embryonic stem cells (ES cells) can develop normally or into a tumor, depending on the environmental signals they receive. Cultured ES cells originally obtained from an inner cell mass of a mouse contribute to a healthy chimeric mouse when injected into an early embryo (bottom left panels), but if injected under the skin of an adult mouse, the same cells will develop into a teratocarcinoma (bottom right panels).

that could be repaired is limited. There have been reports over the past few years that mouse adult hematopoietic stem cells will differentiate into cell types quite different from their normal range when transplanted into another type of tissue. If confirmed, this would have been an exciting breakthrough, as blood stem cells are the most easily available of all stem cells. But further work has shown that it is unlikely that these results are due to stem-cell differentiation, and there is in fact little evidence for the developmental plasticity of adult hematopoietic stem cells. In mice in which adult hematopoietic stem cells had apparently repaired liver damage caused by a gene defect in the liver cells, it was found that the stem cells had not given rise to new liver cells themselves but had fused with existing liver cells and provided them with good copies of the necessary genes. Other cases where it initially appeared that hematopoietic stem cells had been able to give rise to cells outside their normal repertoire, such as neural cells, have also been proved to be due to fusion—for example, the fusion of the blood stem cells with Purkinje cells in the brain.

Adult stem cells also cannot be the answer for all tissues. In some tissues that are prime medical targets, the cells of interest are not normally renewed from stem cells but by division of the cells themselves. This is the case for the insulin-producing β cells of the pancreas, whose destruction is the cause of one type of diabetes. But generating specific cell types by manipulating ES cells in culture might one day provide cells that could repair even damaged heart muscle, which is not self-renewing in mammals.

To generate ES cells of the patient's own tissue type, it would be a great advantage to be able to make use of **therapeutic cloning** by somatic cell nuclear transfer (see Section 8.14). In this technique a nucleus from a somatic cell from the patient would be introduced into an enucleated human oocyte that would then be allowed to develop to the blastocyst stage. The blastocyst could then be used as a source of ES cells that could be maintained in culture, and which would not cause immune rejection when implanted into the patient. Making human blastocysts by somatic cell nuclear transfer has, however, proved almost impossible so far, and no ES cells have yet been derived by this procedure.

There are claimed to be ethical issues associated with the use of stem cells and therapeutic cloning. To make human ES cells, the blastocyst must be destroyed and there are those who believe that this is a destruction of a human life. There is good evidence that the blastocyst does not necessarily represent an individual at this very early stage, as it is still capable of giving rise to twins at a later stage. And in practice, many early embryos are lost during assisted reproduction involving *in vitro* fertilization, a widely accepted medical intervention. Acceptance of IVF and rejection of the use of ES cells could be seen as a contradiction.

Summary

Transplantation of nuclei from differentiated animal cells into fertilized eggs, and cell-fusion studies, show that the pattern of gene expression in the nucleus of a differentiated cell can often be reversed, implying that it is determined by factors supplied by the cytoplasm and that no genetic material has been lost. Although the differentiated state of an animal cell *in vivo* is usually extremely stable, some cases of differentiation are reversible. Transdifferentiation of one differentiated cell type into another has been shown to occur during regeneration and in cultured cells. Both amphibians and some mammals can be cloned by nuclear transplantation from an adult somatic cell into an enucleated egg. Stem cells are self-renewing and differentiate into a wide variety of cell types. They are being investigated with a view to being used in cell-replacement therapy.

SUMMARY TO CHAPTER 8

Cell differentiation leads to distinguishable cell types whose specialized characters and properties are determined by their pattern of gene activity and thus the proteins they produce. The pattern of gene activity in a differentiated cell can be maintained over long periods of time and can be transmitted to the cell's progeny. Maintenance and inheritance of a pattern of gene activity probably involves a combination of several mechanisms, including the continued action of gene-regulatory proteins, changes in the packing state of chromatin, and chemical modifications to DNA.

Differentiation reflects the establishment of a particular pattern of gene expression rather than the loss of genetic material. Lymphocytes are an exception as their DNA has undergone irreversible changes during differentiation. The differentiated state of an animal cell *in vivo* is usually extremely stable, but in some cases transdifferentiation of one differentiated cell type into another can occur, for example, during tissue regeneration. The cloning of amphibians and some mammals by somatic cell nuclear transfer into enucleated eggs also shows that the nuclei of differentiated cells can be reprogrammed and restored to a state in which they are able to direct embryonic development. Stem cells that can proliferate and differentiate into a variety of cell types could be a key to regenerative medicine.

Apart from the use of universal mechanisms of gene regulation, there may be little similarity between the detailed mechanisms of differentiation in different cell types. The understanding of any given differentiation pathway lies in the detail of the external signals, internal signal transduction pathways, gene-regulatory proteins, and gene products at work in each particular case. Induction of cell differentiation by signaling molecules such as growth factors requires the execution of a complex program of intracellular events. Different stimuli, or even the same stimulus at different developmental stages, can activate the same internal signaling pathway, yet activate different genes in different cells, because of the cells' own developmental histories. Signal transduction at the cell membrane sets in motion a cascade of events within the cell, which results, for example, in the phosphorylation and activation of transcription factors and the consequent switching on and off of gene expression. Programmed cell death is a common fate of cells during development, and there is evidence that positively acting signals are generally required to prevent it and allow cells to survive.

SECTION FURTHER READING

8.1 Control of transcription involves both general and tissue-specific transcriptional regulators

Cheung, W.L., Briggs, S.D., Allis, C.D.: **Acetylation and chromosomal functions.** *Curr. Opin. Cell Biol.* 2000, **12:** 328–333.

Levine, M., Tjian, R.: Transcription regulation and animal diversity. *Nature* 2003, **424:** 147–151.

Mannervik, M., Nibur, Y., Zhang, H., Levine, M.: **Transcriptional coregulators in development.** *Science*, 1999, **284:** 606–609.

Scully, K.M., Jacobsen, E.M., Jepsen, K., Lunyak, V., Viadiu, H., Carriere, C., Rose, D.W., Hooshmand, F., Aggarwal, A.K., Rosenfeld, M.G.: **Allosteric effects of Pit-1 DNA sites on long-term repression in cell type specificiation.** *Science* 2000, **290:** 1127–1131.

Struhl, K.: **Fundamentally different logic of gene regulation in eukaryotes and prokaryotes.** *Cell* 1999, **98:** 1–4.

Box 8A DNA microarrays

Butte, A.: **The use and analysis of microarray data.** *Nat. Rev. Microarray Collection* 2004 [www.nature.com/reviews/focus/microarrays]

8.2 External signals can activate genes

Karin, M., Hunter, T.: **Transcriptional control by protein phosphorylation: signal transmission from the cell surface to the nucleus.** *Curr. Biol.* 1995, **5:** 747–757.

8.3 Maintenance and inheritance of patterns of gene activity may depend on chemical and structural modifications to chromatin as well as on regulatory proteins

de Laat, W., Grosveld, F.: **Spatial organization of gene expression: the active chromatin hub.** *Chromosome Res.* 2003, **11:** 447–459.

8.4 All blood cells are derived from multipotent stem cells

Dieterlen-Lievre, F.: **Hematopoiesis: Progenitors and their genetic program.** *Curr. Biol.* 1998, **8:** 727–730.

Ema, H., Nakauchi, H.: **Self-renewal and lineage restriction of hematopoietic stem cells.** *Curr. Opin. Genet. Dev.* 2003, **13:** 508–512.

Phillips, R.L., Ernst, R.E., Brunk, B., Ivanova, N., Mahan, M.A., Deanehan, J.K., Moore, K.A., Overton, G.C., Lemischka, I.R.: **The genetic program of hematopoietic stem cells.** *Science* 2000, **288:** 1635–1640.

8.5 Colony-stimulating factors and intrinsic changes control differentiation of the hematopoietic lineages

Anguita, E., Hughes, J., Heyworth, C., Blobel, G.A., Wood, W.G., Higgs, D.R.: **Globin gene activation during haemopoiesis is driven by protein complexes nucleated by GATA-1 and GATA-2.** *EMBO J.* 2004, **23:** 2841–2852.

D'Andrea, A.D.: **Hematopoietic growth factors and the regulation of differentiative decisions.** *Curr. Opin. Cell Biol.* 1994, **6:** 804–808.

Kluger, Y., Lian, Z., Zhang, X., Newburger, P.E., Weissman, S.M.: **A panorama of lineage-specific transcription in hematopoiesis.** *BioEssays* 2004, **26:** 1276–1287.

Metcalf, D.: **Control of granulocytes and macrophages: molecular, cellular, and clinical aspects.** *Science* 1991, **254:** 529–533.

Orkin, S.H.: **Diversification of haematopoietic stem cells to specific lineages.** *Nat. Rev. Genet.* 2000, **1:** 57–64.

8.6 Developmentally regulated globin gene expression is controlled by regulatory sequences far distant from the coding regions

Engel, J.D., Tanimoto, K.: **Looping, linking, and chromatin activity: new insights into β-globin locus regulation.** *Cell* 2000, **100:** 499–502.

Grosveld, F.: **Activation by locus control regions?** *Curr. Opin. Genet. Dev.* 1999, **9:** 152–157.

Tanimoto, K., Liu, Q., Bungert, J., Engel, J.D.: **Effects of altered gene order or orientation of the locus control region on human beta-globin gene expression in mice.** *Nature* 1999, **398:** 344–348.

Tolhuis, B., Palstra, R.J., Splinter, E, Grosveld, F., de Laat, W.: **Looping and interaction between hypersensitive sites in the active beta-globin locus.** *Mol. Cell* 2002, **10:** 1453–1465.

8.7 Differentiation of cells that make antibodies involves irreversible DNA rearrangement

Janeway, C.A., Travers, P., Shlomchik, M., Walport, M.: *Immunobiology: The Immune System in Health and Disease.* 6th edn. London/New York: Garland Publishing, 2005.

8.8 The epithelia of adult mammalian skin and gut are continually replaced by derivatives of stem cells

Gordon, J.I., Hermiston, M.L.: **Differentiation and self-renewal in the mouse gastrointestinal epithelium.** *Curr. Opin. Cell Biol.* 1994, **6:** 795–803.

Lechler, T., Fuchs, E.: **Asymmetric cell divisions promote stratification and differentiation of mammalian skin.** *Nature* 2005, **437:** 275–280.

Niemann, C., Watt, F.M.: **Designer skin: lineage commitment in postnatal epidermis.** *Trends Cell Biol.* 2002, **12:** 185–192.

Shackleton, M., Vaillant, F., Simpson, K.J., Stingl, J., Smyth, G.K., Asselin-Labat, M-L., Wu, L., Lindeman, G.J., Visvader, J.E.: **Generation of a functional mammary gland from a single stem cell.** *Nature* 2006, **439:** 84–88.

Slack, J.M.W.: **Stem cells in epithelial tissues.** *Science* 2000, **287:** 1431–1433.

Watt, F.M., Jones, P.H.: **Expression and function of the keratinocyte integrins.** *Development (Suppl.)* 1993: 185–192.

8.9 A family of genes can activate muscle-specific transcription

Arnold, H.H., Winter, B.: **Muscle differentiation: more complexity to the network of myogenic regulators.** *Curr. Opin. Gen. Dev.* 1998, **8:** 539–544.

Kassar-Duchossoy L, Gayraud-Morel B, Gomes D, Rocancourt D, Buckingham M, Shinin V, Tajbakhsh S.: **Mrf4 determines skeletal muscle identity in Myf5:Myod double-mutant mice.** *Nature* 2004, **431:** 466–471.

Naya, F.J., Olson, E.: **MEF2: a transcriptional target for signaling pathways controlling skeletal muscle growth and differentiation.** *Curr. Opin. Cell Biol.* 1999, **11:** 683–688.

Tapscott, S.J.: **The circuitry of a master switch: Myod and the regulation of skeletal muscle transcription.** *Development* 2005, **132:** 2685–2695.

8.10 The differentiation of muscle cells involves withdrawal from the cell cycle, but is reversible

Novitch, B.G., Mulligan, G.J., Jacks, T., Lassar, A.B.: **Skeletal muscle cells lacking the retinoblastoma protein display defects in muscle gene expression and accumulate in S and G_2 phases of the cell cycle.** *J. Cell Biol.* 1996, **135:** 441–456.

Odelberg, S.J., Kolhoff, A., Keating, M.: **Dedifferentiation of mammalian myotubes induced by *msx1*.** *Cell* 2000, **103:** 1099–1109.

8.11 Skeletal muscle and neural cells can be renewed from stem cells in adults

Blau, H.M., Brazelton, T.R., Weimann, J.M.: **The evolving concept of a stem cell: entity or function?** *Cell* 2001, **105:** 829–841.

Dhawan, J., Rando, T.A.: **Stem cells in postnatal myogenesis: molecular mechanisms of satellite cell quiescence, activation and replenishment.** *Trends Cell Biol.* 2005, **15:** 663–673.

Lie, D-C., Colamarino, S.A., Song, H-J., Desire, L., Mira, H., Consiglio, A., Lein, E.S., Jessberger, S., Lansford, H., Dearier, A.R., Gage, F.H.: **Wnt signalling regulates adult hippocampal neurogenesis.** *Nature* 2005, **437:** 1370–1375.

Morrison, S.J.: **Stem cell potential: Can anything make anything?** *Curr. Biol.* 2001, **11:** R7–R9.

Taupin, P.: **Adult neurogenesis in the mammalian central nervous system: functionality and potential clinical interest.** *Med. Sci. Monit.* 2005, **11:** 247–252.

Watt, F.M., Hogan, B.L.M.: **Out of Eden: stem cells and their niches.** *Science* 2000, **287:** 1430–1431.

Weissman, I.L.: **Stem cells: units of development, units of regeneration, and units of evolution.** *Cell* 2000, **100:** 157–168.

8.12 Embryonic neural crest cells differentiate into a great variety of different cell types

Anderson, D.J.: **Genes, lineages and the neural crest.** *Proc. R. Soc. Lond. B* 2000, **355:** 953–964.

Bronner-Fraser, M.: **Making sense of the sensory lineage.** *Science* 2004, **303:** 966–968.

Dorsky, R.I., Moon, R.T., Raible, D.W.: **Environmental signals and cell fate specification in premigratory neural crest.** *BioEssays* 2000, **22:** 708–716.

Garcia-Castro, M., Bronner-Fraser, M.: **Induction and differentiation of the neural crest.** *Curr. Opin. Cell. Biol.* 1999, **11:** 695–698.

Le Douarin N.M., Dupin E.: **Multipotentiality of the neural crest.** *Curr. Opin. Genet. Dev.* 2003, **13:** 529–536.

Le Douarin, N.M., Creuzet, S., Couly, G., Dupin, E.: **Neural crest cell plasticity and its limits.** *Development* 2004, **131:** 4637-4650.

Lee, H.Y., Kleber, M., Hari, L., Brault, V., Suter, U., Taketo, M.M., Kemler, R., Sommer, L.: **Instructive role of Wnt/beta-catenin in sensory fate specification in neural crest stem cells.** *Science* 2004, **303:** 1020–1023.

Morrison, S.J., White, P.M., Zock, C., Anderson, D.J.: **Prospective identification, isolation by flow cytometry, and *in vivo* self-renewal of multipotent mammalian neural crest stem cells.** *Cell* 1999, **96:** 737–749.

Morrison, S.J., Perez, S.E., Qiao, Z., Verdi, J.M., Hicks, C., Weinmaster, G., Anderson, D.J.: **Transient Notch activation initiates an irreversible switch from neurogenesis to gliogenesis by neural crest stem cells.** *Cell* 2000, **101:** 499–510.

8.13 Programmed cell death is under genetic control

Ellis, R.E., Yuan, J.Y., Horvitz, H.R.: **Mechanisms and functions of cell death.** *Annu. Rev. Cell. Biol.* 1991, **7:** 663–698.

Jacobson, M.D., Weil, M., Raff, M.C.: **Programmed cell death in animal development.** *Cell* 1997, **88:** 347–354.

Raff, M.: **Cell suicide for beginners.** *Nature* 1998, **396:** 119–122.

Riedl, S.J., Shi, Y.: **Molecular mechanisms of caspase regulation during apoptosis.** *Nat. Rev. Mol. Cell Biol.* 2004, **5:** 897–907.

Sancho, E., Batlle, E., Clevers, H.: **Live and let die in the intestinal epithelium.** *Curr. Opin. Cell Biol.* 2003, **15:** 763–770.

Vaux, D.L., Korsmeyer, S.J.: **Cell death in development.** *Cell* 1999, **96:** 245–254.

8.14 Nuclei of differentiated cells can support development of the egg

Eggan, K., Baldwin, K., Tackett, M., Osborne, J., Gogos, J., Chess, A., Axel, R., Jaenisch, R.: **Mice cloned from olfactory sensory neurons.** *Nature* 2004, **428:** 44–49.

Gurdon, J.B.: **Nuclear transplantation in eggs and oocytes.** *J. Cell Sci. Suppl.* 1986, **4:** 287–318.

Gurdon, J.B., Colman, A.: **The future of cloning.** *Nature* 1999, **402:** 743–746.

Gurdon JB, Byrne JA, Simonsson S.: **Nuclear reprogramming and stem cell creation.** *Proc. Natl Acad. Sci. USA* 2003, **100 Suppl 1:** 11819–11822.

Humpherys, D., Eggan, K., Akutsu, H., Friedman, A., Hochedlinger, K., Yanagimachi, R., Lander, E.S., Golub, T.R., Jaenisch, R.: **Abnormal gene expression in cloned mice derived from embryonic stem cell and cumulus cell nuclei.** *Proc. Natl Acad. Sci. USA* 2002, **99:** 12889–12894.

Onishi, A., Iwamoto, M., Akita, T., Mikawa, S., Takeda, K., Awata, T., Hanada, H., Perry, A.C.F.: **Pig cloning by microinjection of fetal fibroblast nuclei.** *Science* 2000, **289:** 1188–1190.

Rhind, S.M., Taylor, J.E., De Sousa, P.A., King, T.J., McGarry, M., Wilmut, I.: **Human cloning: can it be made safe?** *Nat. Rev. Genet.* 2003, **4:** 855–864.

Simerly, C., Dominko, T., Navara, C., Payne, C., Capuano, S., Gosman, G., Chong, K.Y., Takahashi, D., Chace, C., Compton, D., Hewitson, L., Schatten, G.: **Molecular correlates of primate nuclear transfer failures.** *Science* 2003, **300:** 297.

Solter, D.: **Mammalian cloning: advances and limitations.** *Nat. Rev. Genet.* 2000, **1:** 199–207.

8.15 Patterns of gene activity in differentiated cells can be changed by cell fusion

Blau, H.M.: **How fixed is the differentiated state? Lessons from heterokaryons.** *Trends Genet.* 1989, **5:** 268–272.

Blau, H.M., Baltimore, D.: **Differentiation requires continuous regulation.** *J. Cell Biol.* 1991, **112:** 781–783.

Blau, H.M., Blakely, B.T.: **Plasticity of cell fates: insights from heterokaryons.** *Cell Dev. Biol.* 1999, **10:** 267–272.

8.16 The differentiated state of a cell can change by transdifferentiation

Eguchi, G., Kodama, R.: **Transdifferentiation.** *Curr. Opin. Cell Biol.* 1993, **5:** 1023–1028.

Horb, M.E., Shen, C.N., Tosh, D., Slack, J.M.: **Experimental conversion of liver to pancreas.** *Curr. Biol.* 2003, **13:** 105–115.

8.17 Embryonic stem cells can proliferate and differentiate into many cell types in culture

Byrne, J.A., Simonsson, S., Western, P.S., Gurdon, J.B.: **Nuclei of adult mammalian somatic cells are directly reprogrammed to *oct-4* stem cell gene expression by amphibian oocytes.** *Curr. Biol.* 2003, **13:** 1206–1213.

Ho, A.D.: **Kinetics and symmetry of divisions of hematopoietic stem cells.** *Exp. Hematol.* 2005, **33:** 1–8.

Loebel, D.A., Watson, C.M., De Young, R.A., Tam, P.P.: **Lineage choice and differentiation in mouse embryos and embryonic stem cells.** *Dev. Biol.* 2003, **264:** 1–14.

McBeath, R., Pirone, D.M., Nelson, C.M., Bhadriraju, K., Chen, C.S.: **Cell shape, cytoskeletal tension, and RhoA regulate stem cell lineage commitment.** *Dev. Cell* 2004, **6:** 483–495.

Molofsky, A.V., Pardal, R., Morrison, S.J.: **Diverse mechanisms regulate stem cell self-renewal.** *Curr. Opin. Cell Biol.* 2004, **16:** 700–707.

Plachta, N., Bibel, M., Tucker, K.L., Barde, Y.A.: **Developmental potential of defined neural progenitors derived from mouse embryonic stem cells.** *Development* 2004, **131:** 5449–5456.

West, J.A., Daley, G.Q.: ***In vitro* gametogenesis from embryonic stem cells.** *Curr. Opin. Cell Biol.* 2004, **16:** 688–692.

Ying, Q.L., Nichols, J., Chambers, I., Smith, A.: **BMP induction of Id proteins suppresses differentiation and sustains embryonic stem cell self-renewal in collaboration with STAT3.** *Cell* 2003, **115:** 281–292.

8.18 Stem cells could be a key to regenerative medicine

Jaenisch, R.: **Human cloning—the science and ethics of nuclear transplantation.** *N. Engl. J. Med.* 2004, **351:** 2787–2792.

McClaren, A.: **Ethical and social considerations of stem cell research.** *Nature* 2001, **414:** 129–131.

Pera, M.F., Trounson, A.O.: **Human embryonic stem cells: prospects for development.** *Development* 2004, **131:** 5515–5525.

Pomerantz, J., Blau, H.M.: **Nuclear reprogramming: a key to stem cell function in regenerative medicine.** *Nat. Cell Biol.* 2004, **6:** 810–816.

Wagers, A.J., Weissman, I.L.: **Plasticity of adult stem cells.** *Cell* 2004, **116:** 639–648.

Wurmer, A.E., Palmer, T.D., Gage, F.H.: **Cellular interactions in the stem cell niche.** *Science* 2004, **304:** 1253–1255.

Organogenesis

9

- The vertebrate limb
- Insect wings, legs, and eyes
- Internal organs: blood vessels, lungs, kidneys, heart, and teeth

Once the basic animal body plan has been laid down, the development of organs as varied as insect wings and vertebrate limbs begins. One question is whether quite new developmental mechanisms and principles are involved, or does organogenesis involve the same basic mechanisms as those used in early development? Organ development involves very large numbers of genes and, because of this complexity, general principles can be quite difficult to distinguish. Nevertheless, many of the mechanisms used in organogenesis, such as positional information, are similar to those of earlier development, and certain signals are used again and again.

So far, we have concentrated almost entirely on the aspects of development involved in laying down the basic body plan in various organisms and on early morphogenesis and cell differentiation. We now turn to the development of specific organs and structures—**organogenesis**—which is a crucial phase of development that leads to the embryo at last becoming a fully functioning organism, capable of independent survival.

The development of certain organs has been studied in great detail and they provide excellent models for looking at developmental processes such as pattern formation, the specification of positional information, induction, change in form, and cellular differentiation. In this chapter, we first consider the development of model organ systems in a variety of organisms—the chick limb, insect appendages such as legs and wings in *Drosophila*, and the vertebrate heart. We then look at examples of tubular morphogenesis and epithelial patterning in the development of the lungs, kidney, blood vessels and, finally, teeth. The development of other major organs such as the gut, liver, and pancreas involves no other different principles, and we will not consider them here.

The cellular mechanisms involved in organogenesis are essentially similar to those encountered in earlier stages of development; they are merely employed in different spatial and temporal patterns. Many of the genes and signaling molecules involved will be familiar from earlier chapters. We shall, however, see that the mechanisms involved are much more complex and that while one can identify some general principles, such as the use of positional information, there is much additional detail for which there are no unifying principles—it just works that way in that particular organ.

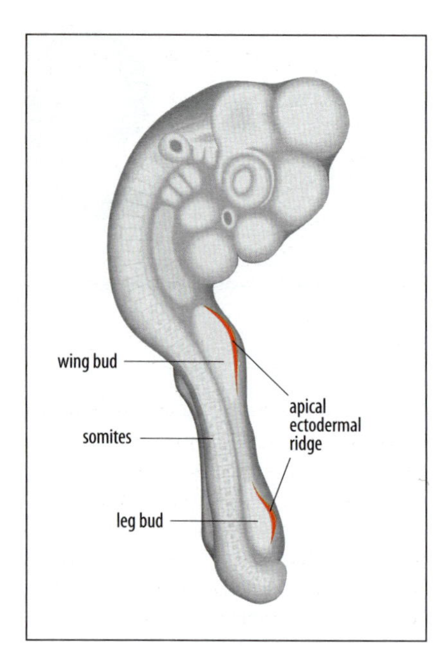

Fig. 9.1 The limb buds of the chick embryo. Limb buds appear on the flanks of the embryo on the third day of incubation (only limbs on the right side are shown here). They are composed of mesoderm, with an outer covering of ectoderm. Along the tip of each runs a thickened ridge of ectoderm, the apical ectodermal ridge.

The vertebrate limb

The vertebrate embryonic limb is a particularly good system in which to study pattern formation. The basic pattern is initially quite simple, and in the chick embryo at least, the limbs are easily accessible for surgical manipulation. The limb is a good model for studying cellular interactions within a structure containing a large number of cells, and for elucidating the role of intercellular signaling in development. In chick embryos, the limbs begin to develop on the third day after the egg is laid, when the structures of the main body axis are already well established. The limbs develop from small protrusions—the **limb buds**—which arise from the body wall of the embryo (Fig. 9.1). By 10 days, the main features of the limbs are well developed (Fig. 9.2). These are the cartilaginous elements (which are later replaced by bone), the muscles and tendons, and the structures in the surface epithelium such as feather buds. The limb has three developmental axes: the **proximo-distal axis** runs from the base of the limb to the tip; the antero-posterior axis runs parallel with the body axis (in the human hand it goes from the thumb to the little finger, and in the chick wing from digit 2 to digit 4); the dorso-ventral axis is the third axis—in the human hand it runs from the back of the hand to the palm.

9.1 The vertebrate limb develops from a limb bud

The early limb bud has two major components—a core of loose mesenchymal cells derived from the lateral plate mesoderm, and an outer layer of ectodermal epithelial cells (Fig. 9.3). The skeletal elements and connective tissues of the limb develop from the mesenchymal core, but the myogenic cells that will give rise to the muscles have a separate lineage and migrate into the bud from the somites (see Section 4.5). At the very tip of the limb bud is a thickening in the ectoderm—the **apical ectodermal ridge** or **apical ridge**, which runs along the dorso-ventral boundary (Fig. 9.4). Directly beneath this lies a region called the **progress zone**, which is composed of rapidly proliferating undifferentiated mesenchymal cells. It is only when cells leave this zone that they begin to differentiate. As the bud grows, the cells start to differentiate and cartilaginous structures begin to appear in the mesenchyme. The proximal part of the limb, that is, the part nearest to the body, is the first to differentiate, and differentiation proceeds distally as the

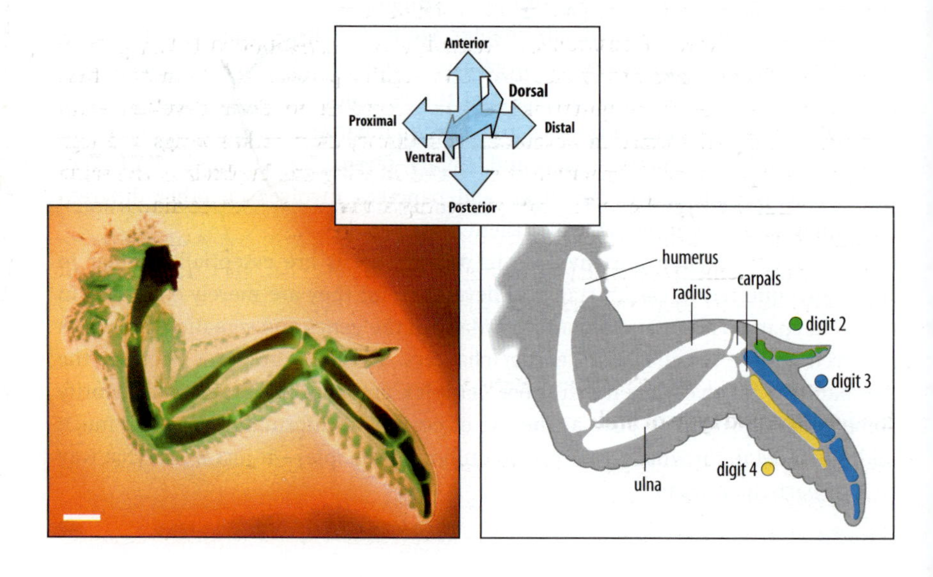

Fig. 9.2 The embryonic chick wing. The photograph shows a stained whole mount of the wing of a chick 10 days after the egg has been laid. By this time, the main cartilaginous elements (e.g. humerus, radius, and ulna) have been laid down. They later become ossified to form bone. The muscles and tendons are also well developed at this stage but cannot be seen in this type of preparation, but the feather buds can be seen. The three developmental axes of the limb are proximo-distal, antero-posterior, and dorso-ventral, as shown in the top panel. Note that the chick wing has only three digits, which have been called 2, 3, and 4. Scale bar = 1 mm.

limb extends. Among the structures found in the developing limb, the patterning of the cartilage has been the best studied, as this can be stained and seen easily in whole mounts of the embryonic limb. The disposition of muscles and tendons is more intricate, and while this can be studied in whole mounts, histological examination of serial sections through the limb may be needed.

The first clear sign of cartilage differentiation is the increased packing of groups of cells, a process known as condensation. The cartilage elements are laid down in a proximo-distal sequence in the chick wing—the humerus, radius, and ulna, the wrist elements (carpals), and then three easily distinguishable digits, 2, 3, and 4 (see Fig. 9.2). Figure 9.5 compares this sequence in the development of a chick wing with the similar sequence in the development of a mouse forelimb.

The chick limb bud at 3 days is about 1 mm wide by 1 mm long, but by 10 days it has grown around tenfold, mostly in length. The basic pattern has been laid down well before then (see Fig. 9.5). But even at 10 days, the limb is still small compared with the newborn limb. The apical ridge disappears as soon as all the basic elements of the limb are in place, and growth occupies most of the subsequent development of the limb, both before and after birth. During the later growth phase, the cartilaginous elements are largely replaced by bone. Nerves only enter the limb after the cartilage has been laid down, and we shall discuss this in detail in Chapter 10. The problem of limb patterning is to understand how the basic pattern of cartilage, muscle, and tendons is formed in the right places and how they make the right connections with each other.

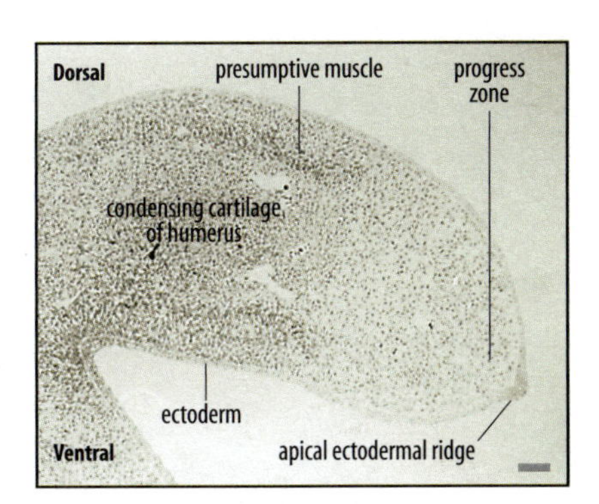

Fig. 9.3 Cross-section through a developing limb bud. The thickened apical ectodermal ridge is at the tip. Beneath the ridge is the progress zone where undifferentiated cells proliferate. Proximal to this zone, mesenchyme cells condense and differentiate into cartilage. Presumptive muscle cells migrate into the limb from the adjacent somites. Scale bar = 0.1 mm.

9.2 Patterning of the limb involves positional information

Some aspects of vertebrate limb development fit very well with a model of pattern formation based on positional information. The developing chick limb behaves as if its cells' future development is determined by their position with respect to the main limb axes while they are in the progress zone (Fig. 9.6). The fate of cells along the proximo-distal axis may involve a timing mechanism, their fate being determined by how long they remain in the progress zone, but a different suggestion is that the pattern is already specified in the early bud, although the mechanism is unknown. Patterning along the antero-posterior axis is specified by a signal, or signals, emanating from a polarizing region at the posterior margin of the limb bud. The dorso-ventral axis is specified by a signal, or signals, from the ectoderm. We shall consider all these regions in more detail later.

Central to the idea of positional information is the distinction between positional specification and interpretation. Cells acquire positional information first and then interpret these positional values according to their developmental history. It is the difference in developmental history that makes wings and legs different, for positional information is signaled in wing and leg buds in the same way. An attractive hypothesis for limb patterning is that a single three-dimensional positional field controls the development of the cells that give rise to all the mesodermal limb elements—cartilage, muscle, and tendons. The specification of all three axes is, as we shall see, linked by molecular signals.

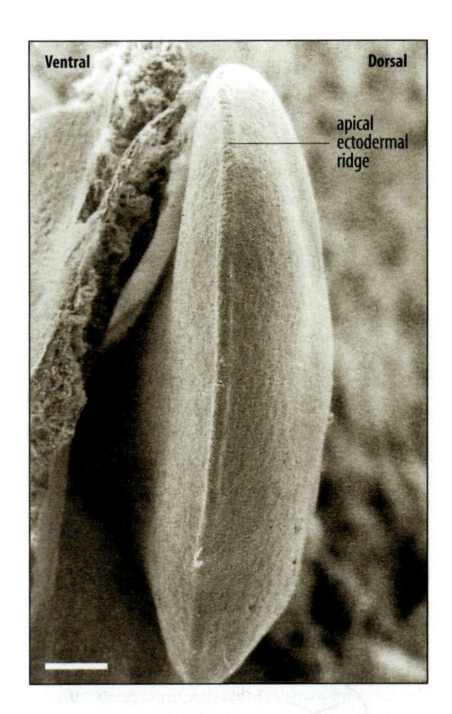

Fig. 9.4 Scanning electron micrograph of a chick limb bud at 4.5 days after laying, showing the apical ectodermal ridge. Scale bar = 0.1 mm.

9.3 Genes expressed in the lateral plate mesoderm are involved in specifying the position and type of limb

The forelimbs and hindlimbs of vertebrates arise at precise positions along the antero-posterior axis of the body. Transplantation experiments have shown that the lateral plate mesoderm that gives rise to the limb-bud mesenchyme becomes

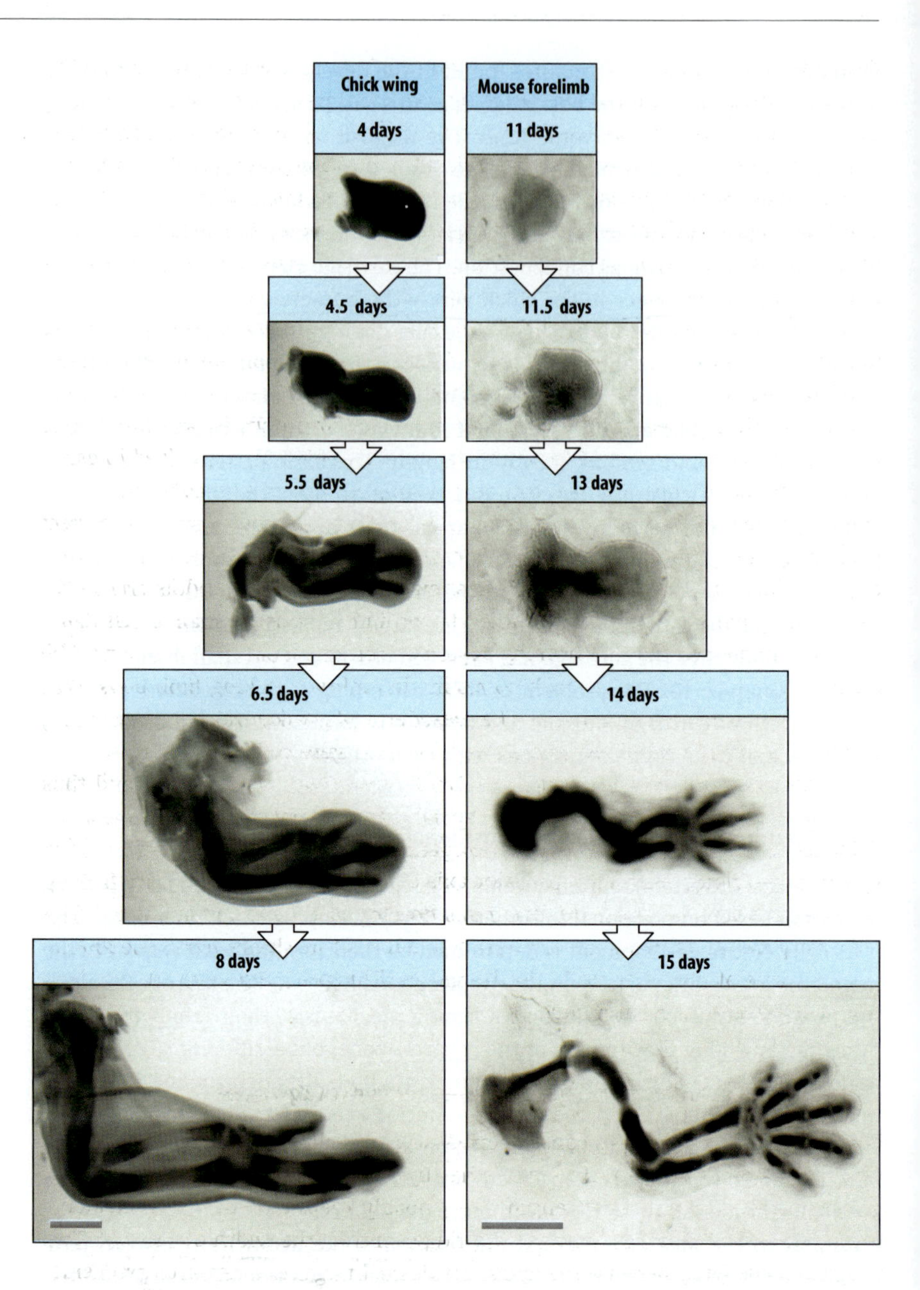

Fig. 9.5 The development of the chick wing and mouse forelimb are similar. The skeletal elements are laid down, as cartilage, in a proximo-distal sequence as the limb bud grows outward. The cartilage of the humerus is laid down first, followed by the radius and ulna, wrist elements, and digits. Scale bar = 1 mm.

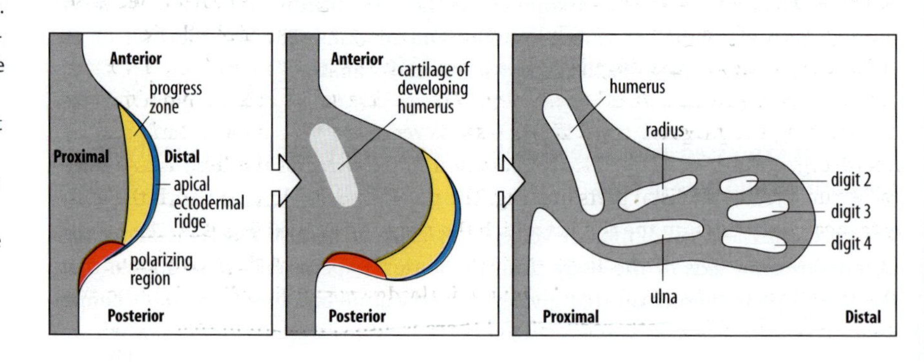

Fig. 9.6 Cells acquire positional values along the antero-posterior and proximo-distal axes. In the early bud there are two signaling regions—the polarizing region at the posterior margin, and the apical ectodermal ridge. The polarizing region specifies position along the antero-posterior axis. There are two main current models for how positional values along the proximo-distal axes may be specified. One proposes that the pattern is specified in the early limb bud; the other proposes that positional values are acquired according to the time the cells spend in the progress zone. When cells leave the progress zone, they differentiate. Thus, cartilage begins to differentiate first in the most proximal region, and the cartilaginous elements are laid down in a proximo-distal sequence.

determined to form limbs in these exact positions long before limb buds are visible. As the lateral plate mesoderm becomes regionalized by Hox gene expression along this body axis, it is possible that Hox gene coding could determine prospective forelimb and hindlimb regions, thus specifying the position and type of a limb bud. This is analogous to the way in which the different thoracic segments and their appendages are specified in insect development (see Chapter 2). The homeodomain transcription factor Pitx1 is expressed in the hindlimb region and plays a key role in determining the differences between hindlimb and forelimb.

Proteins of the fibroblast growth factor (FGF) family appear to be key signaling molecules in initiating both forelimb and hindlimb development in vertebrates. The T-box transcription factors Tbx5 and Tbx4, which are related to the mesodermal marker Brachyury (see Section 3.22), are expressed in forelimbs and hindlimbs, respectively, and are required for limb initiation via FGF. FGF is involved in establishing and maintaining the two main organizing regions of a limb. These are the apical ectodermal ridge, which is essential for limb outgrowth and for correct patterning along the proximo-distal axis of the limb, and a mesodermal polarizing region, which is situated on the posterior side of the limb bud and is crucial for determining pattern along the antero-posterior limb axis, as we shall discuss later. FGF-10 is expressed in limb-forming regions, and knock-out of FGF-10 and FGF receptor genes in the mouse embryo results in embryos lacking limb buds. Wnt proteins produced and secreted by the mesoderm play a key role in determining where FGFs are expressed and maintained. The Hox gene coding in the mesoderm could thus determine where limbs develop by specifying where Wnts, and thus FGFs, are expressed, and thus where an apical ridge and polarizing region develop. Evidence for the combined actions of Hox genes in determining the position of the forelimb bud along the antero-posterior axis comes from knock-out mice lacking *Hoxb5* expression, in which the forelimbs develop at a more anterior level. The different patterns of Hox gene expression at different positions along the antero-posterior axis of the body could also be involved in specifying whether the limb will be a forelimb or a hindlimb.

FGF-10

9.4 The apical ectodermal ridge is required for limb outgrowth

The apical ectodermal ridge consists of closely packed columnar epithelial cells, which are directly connected to each other by gap junctions. Their tight packing gives the ridge a mechanical strength that probably keeps the limb flattened along the dorso-ventral axis; the length of the ridge controls the width of the bud. The progress zone, lying beneath the apical ectodermal ridge, is a region of proliferating undifferentiated mesenchymal cells that produces the initial outgrowth of the limb bud. It may also be the region where the mesenchymal cells acquire the positional information that determines what structures they will develop into, but this is controversial. Perhaps surprisingly, the establishment of the early bud does not reflect a local increase in cell division in the limb-bud region but rather a decrease from a previously high rate of cell proliferation along the rest of the flank.

The apical ectodermal ridge is essential for both the outgrowth of the limb and for its correct proximo-distal patterning. This is due to its role in inducing and maintaining the progress zone. If the ridge is removed from a chick limb bud by microsurgery, there is a significant reduction in growth and the limb is anatomically truncated, with distal parts missing. The proximo-distal level at which the limb is truncated depends on the time at which the ridge is removed (Fig. 9.7). The earlier the ridge is removed, the greater the effect; removal at a later stage only results in loss of the ends of the digits. Following apical-ridge removal, cell proliferation in the progress zone is greatly reduced and there is also cell death in that region.

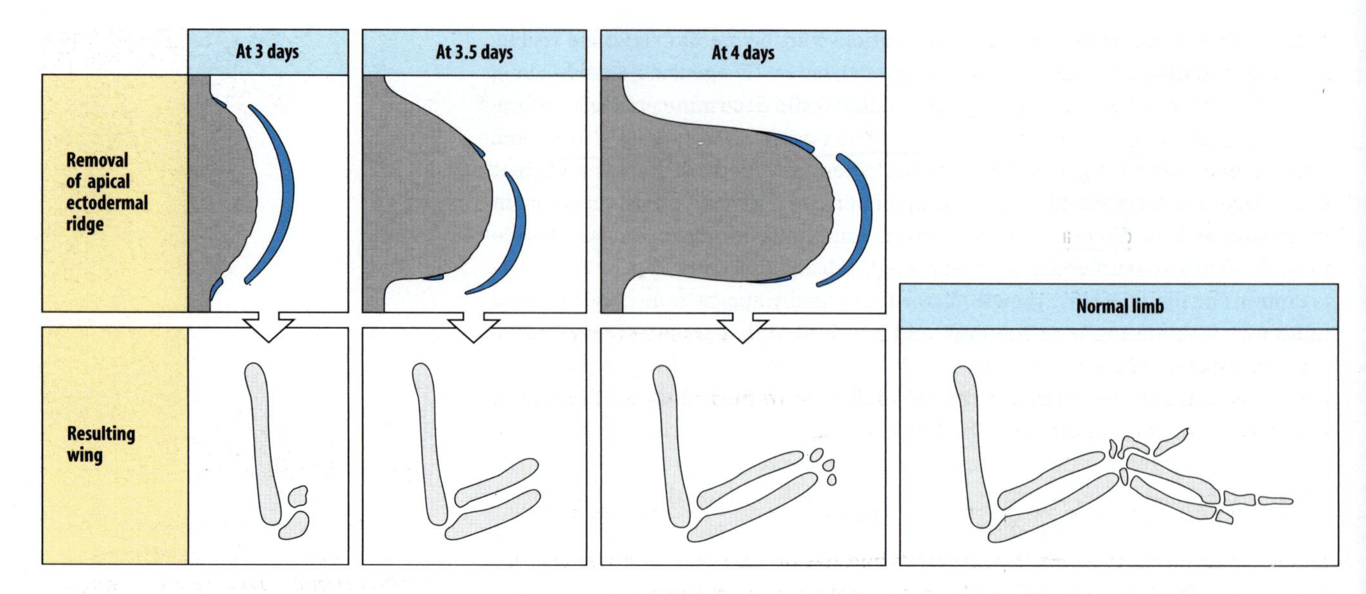

Fig. 9.7 The apical ectodermal ridge is required for proximo-distal development. Limbs develop in a proximo-distal sequence. Removal of the ridge from a developing limb bud leads to truncation of the limb; the later the ridge is removed, the more complete the resulting limb.

Signaling by the apical ectodermal ridge to the underlying mesenchyme has been shown by grafting an isolated ridge to the dorsal surface of an early limb bud. The result is an ectopic outgrowth, which may even develop cartilaginous elements and digits. In the chick, a key signal from the ridge is provided by FGFs. FGF-8 is expressed throughout the ridge and FGF-4 in the posterior region. FGF-8 can act as a functional substitute for an apical ridge. If the ridge is removed and beads that release the growth factor are grafted into the tip of the limb in its place, more-or-less normal outgrowth of the limb continues. If sufficient FGF-8 is provided to the outgrowing cells, a fairly normal limb develops. Similar results are obtained with FGF-4 (Fig. 9.8).

The apical ridge is thus one of the crucial organizing regions of the vertebrate limb. The other is the mesodermal region at the posterior margin of the limb bud, known as the **polarizing region** or the **zone of polarizing activity (ZPA)**, which is involved in patterning the antero-posterior axis (Fig. 9.9). The early limb bud has considerable powers of regulation and pieces of it can generally be removed, rotated, or put in a different position without perturbing the final pattern. But this does not apply to either the apical ridge, as we have just seen, or the polarizing region.

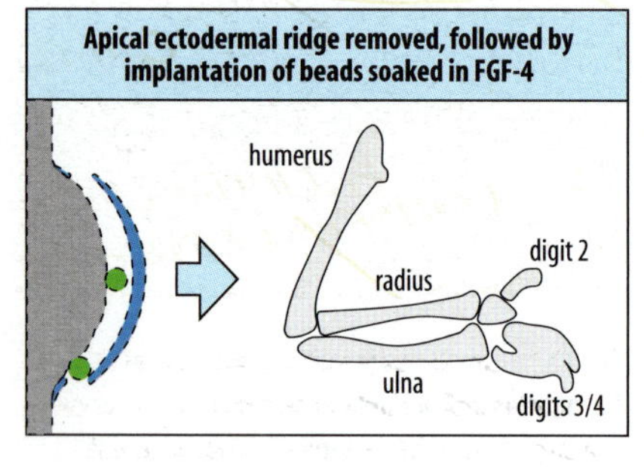

Fig. 9.8 The growth factor FGF-4 can substitute for the apical ectodermal ridge. After the ridge is removed, grafting beads that release FGF-4 into the limb bud results in almost normal limb development.

The apical ridge, polarizing region, and progress zone are mutually dependent on each other for their maintenance and function. The apical ridge is itself maintained by signals from the progress zone and the polarizing region. One of the crucial signaling molecules in this interaction is Sonic hedgehog, which is expressed in the polarizing region (see Fig. 9.9). There is a positive feedback loop between the polarizing region and the apical ectodermal ridge: Sonic hedgehog signaling modulates the level of FGF-4 in the posterior ridge, while FGF-4 maintains Sonic hedgehog expression. Signaling by bone morphogenetic proteins (BMPs) has a complex role in the limb bud tip. A minimum level of BMP is required to maintain the progress zone, but if the level rises too high the apical ridge regresses. Extra BMP in the form of BMP-soaked beads placed under the ridge causes cell death, whereas if BMP activity is blocked, the ridge expands anteriorly and its lifetime is extended. A balance is maintained by a complex feedback signaling loop, in which Sonic hedgehog induces expression of the gene for the protein Gremlin, which blocks BMP activity and thus the signals from the ridge.

Shh FGF-4 FGF-8

The interaction between FGF and Sonic hedgehog can be shown when ectopic limbs are induced in the flank. Localized application of FGF-4 to the flank of a chick embryo, between the wing and leg buds, induces the production of FGF-8 in the ectoderm, initiating formation of a new apical ridge, followed by the ectopic expression of Sonic hedgehog. The protein then feeds back to induce expression of the embryo's own gene for FGF-4, resulting in the maintenance of the ridge and the outgrowth of an additional limb bud at this site. Wing buds develop from application of FGF to the anterior flank, whereas leg buds develop from application to the posterior flank (Fig. 9.10). The specification of the ectopic limb as either fore- or hindlimb is probably controlled by the combinations of Hox genes expressed along the antero-posterior axis.

We now consider how the developing limb is patterned along its three axes: antero-posterior, proximo-distal, and dorso-ventral.

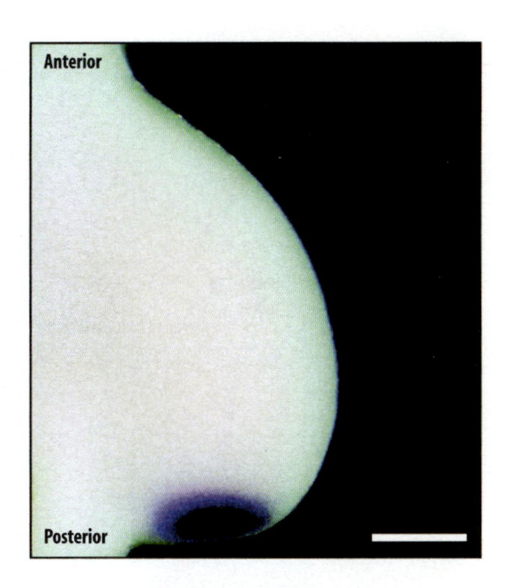

Fig. 9.9 ***Sonic hedgehog*** **is expressed in the polarizing region of a chick limb bud.** *Sonic hedgehog* is expressed at the posterior margin of the limb bud in the polarizing region and provides a positional signal along the antero-posterior axis. Scale bar = 0.1 mm.

Photograph courtesy of C. Tabin.

9.5 The polarizing region specifies position along the limb's antero-posterior axis

The polarizing region of a vertebrate limb bud has organizing properties that are almost as striking as those of the Spemann organizer in amphibians. When the polarizing region from an early wing bud is grafted to the anterior margin of another early chick wing bud, a wing with a mirror-image pattern develops: instead of the normal pattern of digits—2 3 4—the pattern 4 3 2 2 3 4 develops (Fig. 9.11). The pattern of muscles and tendons in the limb shows similar mirror-image changes.

The additional digits come from the host limb bud and not from the graft, showing that the grafted polarizing region has altered the developmental fate of the host cells in the anterior region of the limb bud. The limb bud widens in response to the polarizing graft, which enables the additional digits to be accommodated; widening of a limb bud is always associated with an increase in the extent of the apical ectodermal ridge and the rate of cell division is higher than in the anterior region of a normal bud. A polarizing region graft can also specify another ulna anteriorly.

One way that the polarizing region could specify position along the antero-posterior axis is by producing a morphogen that forms a posterior to anterior gradient (see Box 9A, p. 347). The concentration of morphogen could specify the position of cells along the antero-posterior axis with respect to the polarizing region located at the posterior margin of the limb. Cells could then interpret their positional values by developing specific structures at particular threshold concentrations of morphogen. Digit 4, for example, would develop at a high concentration, digit 3 at a lower one, and digit 2 at an even lower one. According to this model, a graft of an additional polarizing region to the anterior margin would set up a mirror-image gradient of morphogen, which would result in the 4 3 2 2 3 4 pattern of digits that is observed (see Fig. 9.11, center panel).

If the action of the polarizing region in specifying the character of a digit is due to the level of the signal, then when the signal is weakened, the pattern of digits should be altered in a predictable manner. Grafting small numbers of polarizing region cells to a limb bud results only in the development of an additional digit 2 (see Fig. 9.11, right panel). An analogous result can be obtained by leaving the polarizing region graft in place for a short time and then removing it. When left for 15 hours, an additional digit 2 develops, whereas the graft has to be in place for up to 24 hours for a digit 3 to develop.

But just because a morphogen gradient model is consistent with many experimental results, this is not sufficient to prove it. In the mouse limb, for example, cells in the polarizing region end up in digits, and it is suggested that the time

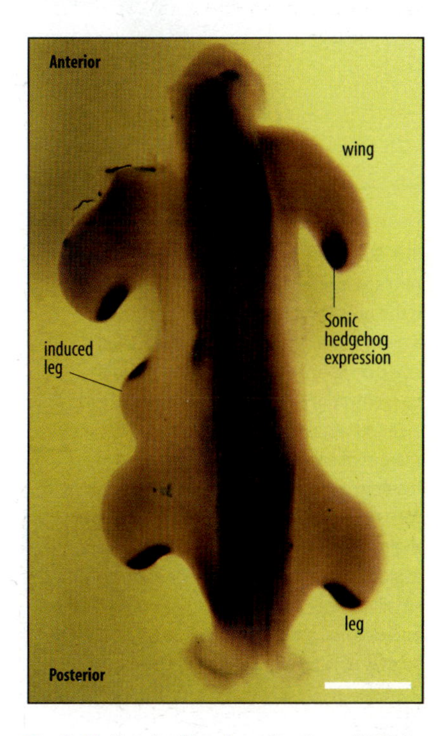

Fig. 9.10 Result of local application of FGF-4 to the flank of a chick embryo. FGF-4 is applied to a site close to the hindlimb bud, where it induces outgrowth of an additional hindward limb. The dark regions indicate expression of *Sonic hedgehog*. Scale bar = 1 mm.

Photograph courtesy of J-C. Izpisúa-Belmonte, from Cohn, M.J., et al.: 1995.

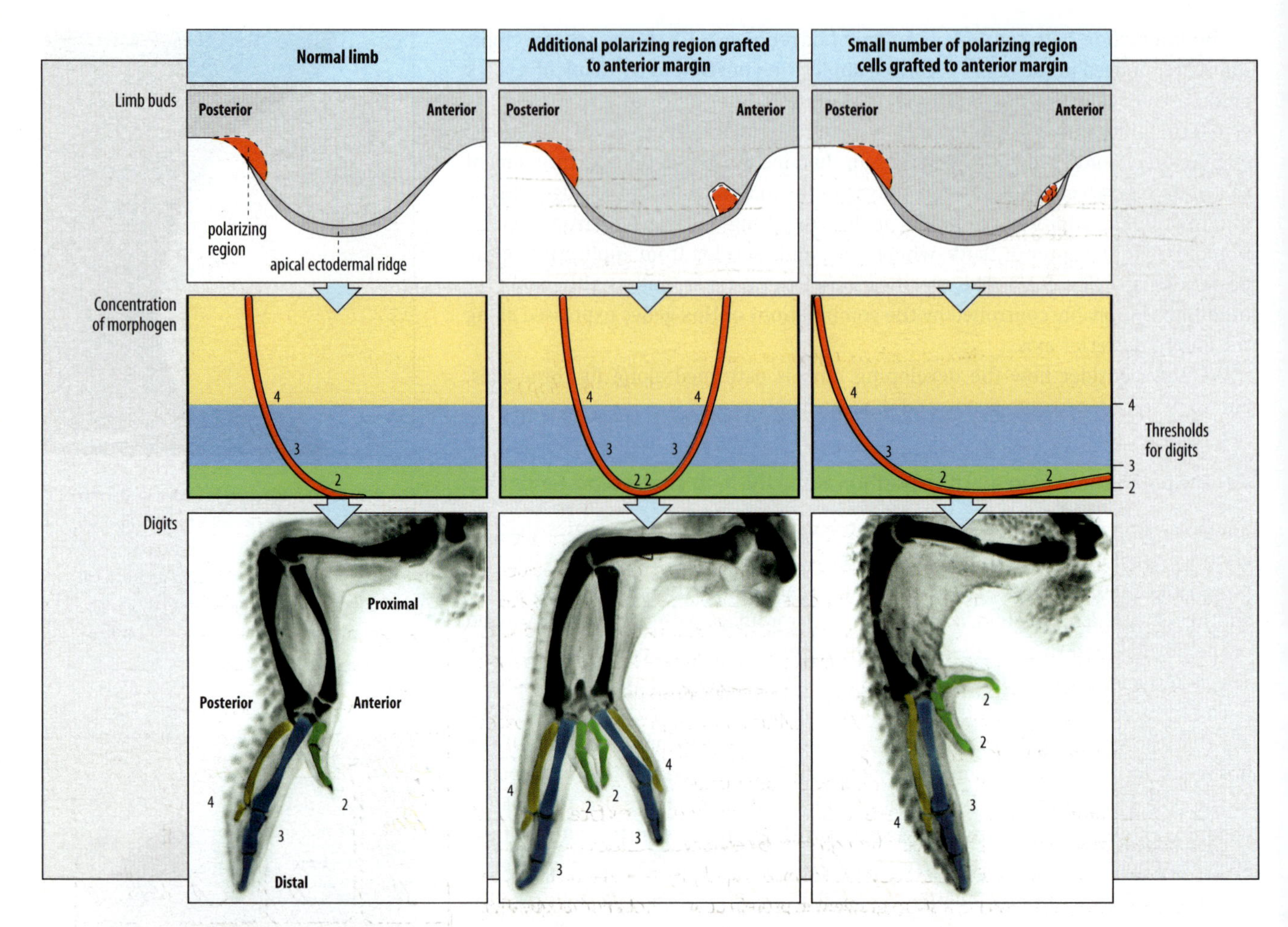

Fig. 9.11 The polarizing region can specify pattern along the antero-posterior axis. If the polarizing region is the source of a graded morphogen, the different digits could be specified at different threshold concentrations of signal (left panels). Grafting an additional polarizing region to the anterior margin of a limb bud (center panels) would result in a mirror-image gradient of signal, and thus the observed mirror-image duplication of digits. The signal from a grafted polarizing region can be attenuated by grafting only a small number of polarizing region cells to the anterior margin of the limb bud (right panels), so that only an extra digit 2 develops.

these cells spend in the polarizing region determines which digit they will become. There is also evidence that mechanisms other than a morphogen gradient cause a set of cartilaginous elements to begin developing, which are then modified by a signal from the polarizing region, as discussed later.

A persuasive case can, however, be made for a gradient of the Sonic hedgehog protein as the key component of the natural polarizing signal. The *Sonic hedgehog* gene is expressed in the polarizing region (see Fig. 9.9) and Sonic hedgehog protein is involved in numerous patterning processes elsewhere in vertebrate development, for example in the establishment of internal left–right asymmetry (see Section 3.10), and in the neural tube, as we shall see in Chapter 10. The *Drosophila* homolog of Sonic hedgehog, the Hedgehog protein, is a key signaling molecule in the patterning of the segments in the *Drosophila* embryo (see Section 2.23), as well as of *Drosophila* legs and wings, as we shall see later in this chapter. Chick fibroblast cells transfected with a retrovirus containing the *Sonic hedgehog* gene acquire the properties of a polarizing region; they cause the development of a mirror-image limb when grafted to the anterior margin of a limb bud, as will beads soaked in Sonic hedgehog protein. The bead must be left in place for 16 to 24 hours for extra digits to be specified, and the pattern of digits depends on the concentration of Sonic hedgehog protein in the bead. A higher concentration is required for a digit 4 to develop than for a digit 2, for example.

Box 9A Positional information and morphogen gradients

Pattern formation in development can be specified by positional information, which is in turn specified by a gradient in some property. The basic idea, related to the French Flag problem (see Fig. 1.24), is that a specialized group of cells at one boundary of the area to be patterned secrete a molecule whose concentration decreases with distance from the source, thus forming a gradient of information. Any cell along the gradient then 'reads' the concentration at that particular point, and interprets this to respond in a manner appropriate to its position by switching on a particular pattern of gene expression, for example. Molecules that can induce such changes in cell fate are known as morphogens. Gradients of positional information are used for various tasks during development: regionalization of the antero-posterior and dorso-ventral axes and patterning of segments and imaginal discs in insects; mesoderm patterning in vertebrates; vertebrate limb patterning; and patterning along the dorso-ventral axis of the neural tube in the chick, among others.

It is still unclear how gradients of positional information are specified, particularly the relative roles of morphogen diffusion and cell–cell interactions, although there is evidence for gradients of morphogen molecules in many cases. There are several questions about positional gradients still to be solved. How is positional information specified over distances of up to 20 cell diameters? Is it the extracellular or intracellular concentration of a morphogen that specifies positional identity? Does endocytosis play a key role?

Morphogens may sometimes travel by simple diffusion through extracellular spaces or, as described for the Decapentaplegic protein in *Drosophila* dorso-ventral patterning (see Section 2.16), a gradient can be formed by the interaction of the morphogen with various other proteins, both secreted extracellular proteins and cell-surface receptors. Morphogens may also be transported through cells by endocytic vesicles (see figure). In other cases, receptor-mediated endocytosis followed rapidly by degradation has been proposed as a way of producing a sharp gradient in protein activity. Yet another question is the role of the responding cells' neighbors in helping to form the gradient.

Cells are thought to sense most morphogens through cell-surface receptors, and it seems that morphogens can act at very low concentrations. It is most likely that a cell measures a particular morphogen concentration according to how many receptors for the morphogen are occupied, and thus by the strength of the signal that is transmitted from the receptors.

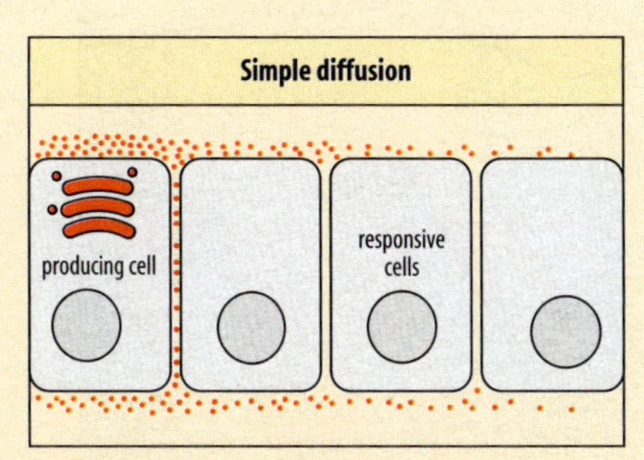

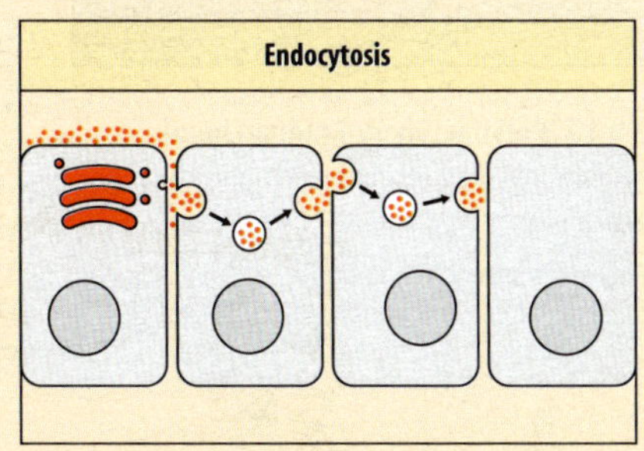

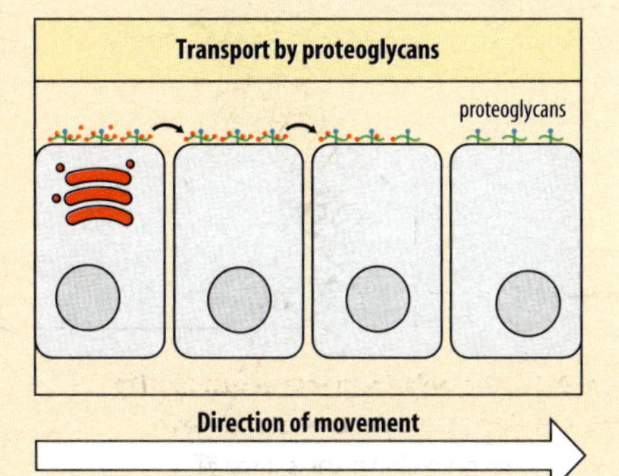

Sonic hedgehog is involved in specifying both digit number and digit identity. The intracellular signaling pathway leading from Sonic hedgehog stimulates the conversion of a repressor form of the transcription factor Gli3 into a transcriptional activator in the same way that the repressor Ci is converted to an activator in the Hedgehog signaling pathway in *Drosophila* (see Fig. 2.37). In the absence of Sonic hedgehog signaling, the repressor form of Gli3 prevents digit formation, but if both Sonic hedgehog and Gli3 are inactivated, then a limb with many digits but no antero-posterior pattern develops. This implies that there is some mechanism that is capable of setting up a set of digit-like elements. It could, for example, be based on a reaction–diffusion type of mechanism, as discussed later. There is also evidence that Sonic hedgehog might exert its effects on specifying digit identity indirectly through its effects on the distribution of bone morphogenetic proteins such as BMP-2, which is expressed in the posterior of the limb. There is evidence

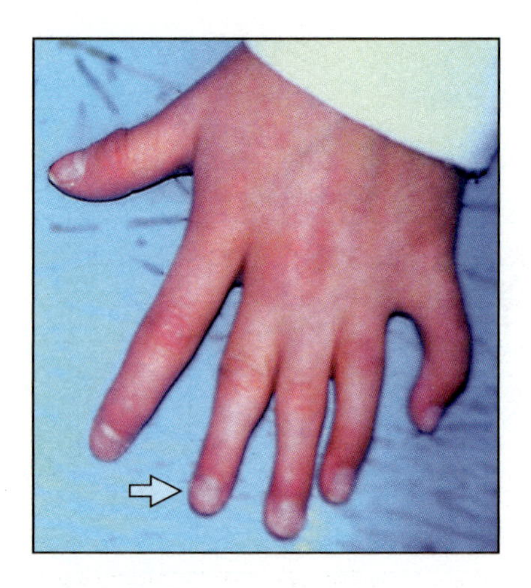

Fig. 9.12 Polydactyly in a human hand. The additional digit (arrow) resembles the adjacent digit.

Photograph courtesy of R. Winter.

that the level of BMPs in the interdigital regions plays a key role in regulating digit identity; if BMP signaling is inhibited in an interdigital region, the adjoining digit takes on the identity of the one anterior to it.

Expression of *Sonic hedgehog* may be specified by early expression of Hox genes and requires the prior expression of the transcription factor dHand. This is produced in the early chick limb bud and becomes restricted to the posterior margin, thus restricting *Sonic hedgehog* expression to this region also. Overexpression of dHand results in *Sonic hedgehog* expression anteriorly as well as posteriorly, and leads to duplications along the antero-posterior axis. Sonic hedgehog may also be involved in determining the size of the polarizing region. Increasing Sonic hedgehog levels in the region by local application leads to reduction in *Sonic hedgehog* gene expression and an increase in cell death. This could be the basis of a negative feedback autoregulatory mechanism to keep signaling more or less constant.

Further evidence for the key role of *Sonic hedgehog* comes from a mouse mutation that results in additional anterior expression of *Sonic hedgehog* and **polydactyly**—the presence of additional digits. In this case, the extra digits are formed anteriorly, a condition known as preaxial polydactyly. One such mutation maps to a long-range global control region that regulates *Sonic hedgehog* expression even though it is located 1 Mb away from the gene. A similarly distant region also regulates limb-specific human Sonic hedgehog expression, and mutations in this region seem to be the most common cause of preaxial polydactyly in humans (Fig. 9.12).

Mice in which the *Sonic hedgehog* gene has been knocked out develop proximal limb structures but lack distal ones, showing that the gene is not required for the actual initiation of limb development. The loss of distal structures probably reflects a requirement for Sonic hedgehog signaling in order to maintain the apical ectodermal ridge, and thus the progress zone, for a sufficient time for these structures to develop (see Section 9.3). Loss of distal limb structures is also seen in the human condition acheiropodia, in which structures below the elbow or knee are missing. This condition is caused by a recessive mutation in the Sonic hedgehog global control region that abolishes *Sonic hedgehog* expression.

T-box transcription factors are also crucial to determining digit identity. In the hindlimb of the chick, expression of Tbx2 and Tbx3 in the most posterior interdigital region specifies the identity of digit 4, while expression of Tbx3 alone in the next interdigital region specifies the identity of digit 3. Misexpression of these transcription factors can change one digit into the other. Tbx2 and Tbx3 may be exerting their effects by regulating the level of interdigital BMP expression, with feedback control loops between Tbx gene expression, BMP signaling, and Hox gene expression being involved. Just how the positional values that determine digit identity are specified still remains unclear.

The signaling molecule retinoic acid, a small lipid-soluble molecule derived from vitamin A, is also present at a higher concentration in posterior than anterior regions and, when applied locally to the anterior margin of a chick wing bud, causes formation of extra digits in a mirror-image pattern similar to that obtained by grafting a polarizing region. Retinoic acid induces a new polarizing region by inducing *Sonic hedgehog* expression, so is unlikely to act as a positional signal itself. It is possible, however, that retinoic acid is required for limb-bud initiation as inhibition of retinoic acid synthesis blocks limb-bud outgrowth. It is also responsible for patterning the proximal region of the limb. Grafting a chick node to the anterior margin of the bud can also induce extra digits in a manner similar to a polarizing region, and this is probably due to the presence of retinoic acid in the node.

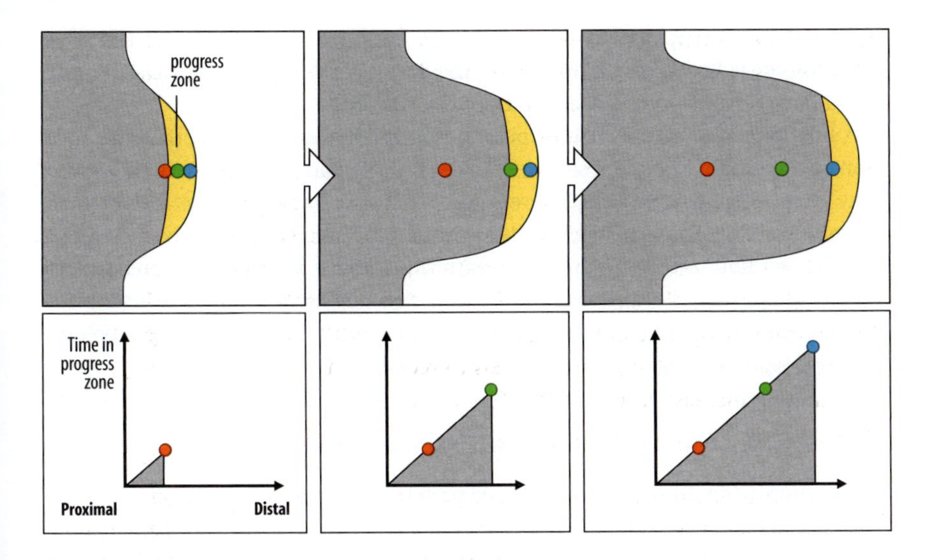

Fig. 9.13 A cell's proximo-distal positional value may depend on the time it spends in the progress zone. Cells are continually leaving the progress zone. If the cells could measure how long they spend in the progress zone, this could specify their position along the proximo-distal axis could be specified. Cells that leave the zone early (red) form proximal structures whereas cells that leave it last (blue) form the tips of the digits.

9.6 Position along the proximo-distal axis may be specified by a timing mechanism

In contrast to the specification of pattern along the limb's antero-posterior axis, patterning along the proximo-distal axis is less well understood, but may be specified by the time cells spend in the progress zone. As the limb bud grows, cells are continually leaving the progress zone. In the forelimb, for example, cells leaving first develop into the humerus and those leaving last form the tips of the digits. If cells could measure the time they spend in the progress zone, for example by counting cell divisions or by increased synthesis or degradation of some molecule, this would give them a positional value along the proximo-distal axis (Fig. 9.13) and the mechanism could be similar to that involved in specifying position along the body axis as the node regresses (see Section 4.4). A timing mechanism of this sort is consistent with the experimental observation that removal of the apical ridge results in a distally truncated limb (see Fig. 9.7). Removal means that cells in the progress zone no longer proliferate, so more distal structures cannot form.

This model has, however, been criticized, as ridge removal results in significant cell death and this, rather than loss of the progress zone, may be the cause of the truncations. It has also been proposed, on the basis of the lineages of labeled cells, that all the elements in the limb are specified in the early bud. If this model were correct then, since there are seven cartilaginous regions along the limb—humerus, radius, ulna, carpal elements in the wrist, and three phalanges—and since the early bud is 300 μm long, each element would be specified as a set of just four cells along the axis, which is most improbable.

Evidence for a mechanism based on time comes from killing cells in the progress zone or blocking their proliferation at an early stage by, for example, X-irradiation. The result is that proximal structures are absent but distal ones are present, and can be almost normal. Because many cells in the irradiated progress zone do not divide, the number of cells that leave the progress zone during each unit of time is much smaller than normal. Proximal structures are thus very small, or even absent. As the bud grows out, however, the progress zone becomes repopulated by the surviving cells, so that distal structures such as digits are formed at almost normal size. It is possible by this mechanism to account for the effect of thalidomide, which when taken by pregnant women resulted in their children lacking proximal limb structures; sometimes just a hand projected from the shoulder. Thalidomide is known to interfere with the development of blood vessels, and this could have resulted in cell death in the progress zone.

A further complication with respect to the proximo-distal axis is that specification of the proximal region, such as shoulder girdle and humerus in the forelimb, may be specified by a process different from that specifying more distal structures. In mice lacking Sonic hedgehog these proximal regions are relatively normal. Instead, retinoic acid is involved in patterning these proximal regions and its targets are the genes of the *meis* family, which encode homeodomain transcription factors that are involved in specifying proximal identity. These genes are repressed in distal regions by epidermal growth factor (EGF).

9.7 The dorso-ventral axis is controlled by the ectoderm

In the chick wing, there is a well defined pattern along the dorso-ventral axis; large feathers are only present on the dorsal surface and the muscles and tendons have a complex dorso-ventral organization. Dorsal and ventral compartments are already specified in the ectoderm of the presumptive limb bud regions in early chick embryos, with the apical ridge developing at the boundary between them. Signals from the somites are involved in specifying the dorsal ectoderm.

In the chick, the establishment of the dorso-ventral ectodermal boundary appears to involve signaling from Notch, modified by the actions of the protein Radical fringe, a glycosyltransferase that is expressed in the presumptive dorsal limb bud ectoderm before formation of the ridge. The chick apical ridge develops at the boundary between cells that express Radical fringe and those that do not. Radical fringe does not appear to be needed for apical ridge formation in mice, however, as the ridge still develops in mice lacking Radical fringe.

The development of pattern along the dorso-ventral axis has been studied by recombining ectoderm taken from left limb buds with mesoderm from right limb buds, so that the dorso-ventral axis of the ectoderm is reversed with respect to the underlying mesoderm. The right and left wing buds are removed from a chick embryo and the ectoderm separated from the mesenchyme. The wing buds are then recombined with the dorso-ventral axis of the ectoderm opposite to that of the mesoderm. The reconstituted limb buds are then grafted to the flank of a host embryo, with just the dorso-ventral axis of the ectoderm inverted with respect to the enclosed mesenchyme, and allowed to develop. In general, the proximal regions of the limbs that develop have the dorso-ventral polarity of a normal limb: that is, their dorso-ventral pattern corresponds to the source of the mesoderm. Distal regions, however, in particular the 'hand' region, have a reversed dorso-ventral axis, with the pattern of muscles and tendons reversed and corresponding to that of the dorso-ventral axis of the ectoderm. The ectoderm can therefore specify dorso-ventral patterning in the limb.

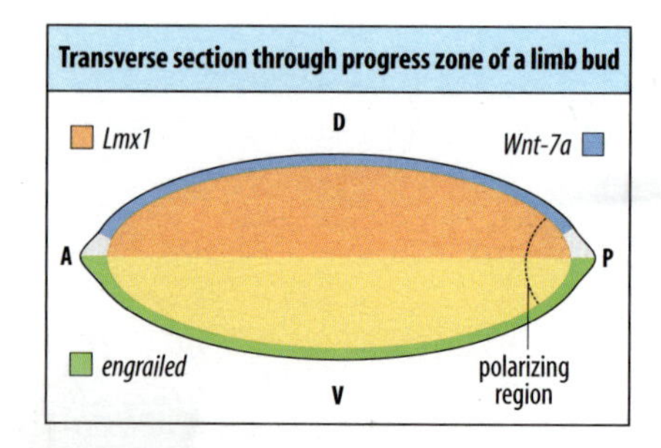

Fig. 9.14 The ectoderm controls dorso-ventral pattern in the developing limb. The gene *Wnt-7a* is expressed in the dorsal ectoderm and *engrailed* (the vertebrate version of the *Drosophila* gene *engrailed*) is expressed in the ventral ectoderm. The gene *Lmx1* is induced by *Wnt-7a* in the dorsal mesoderm and is involved in specifying dorsal structures.

A dorso-ventral pattern is easily seen in the digits of mouse limbs, where ventral surfaces—the palm or paw pads—normally have no fur, whereas dorsal surfaces do. Genes controlling the dorso-ventral axis in vertebrate limbs have been identified from mutations in mice. For example, mutations that inactivate the gene *Wnt-7a* result in limbs in which many of the dorsal tissues adopt ventral fates to give a double ventral limb, the two halves being mirror images. The *Wnt-7a* gene is expressed in the dorsal ectoderm (Fig. 9.14) and this suggests that the ventral pattern may be the ground state and is modified dorsally by the dorsal ectoderm, with Wnt-7a protein playing a key role in patterning the dorsal mesoderm. Expression of the gene *engrailed*, which encodes a transcription factor (see Section 2.2), characterizes ventral ectoderm. Mutations that destroy *engrailed* function result in *Wnt-7a* being expressed ventrally as well, giving a double dorsal limb.

One function of Wnt-7a is to induce expression of the LIM homeobox transcription factor Lmx1 in the underlying mesenchyme (see Fig. 9.14). Mammalian Lmx1 specifies a dorsal pattern in the mesoderm. Ectopic expression of *Lmx1* in the ventral mesoderm of the mouse limb results in the cells adopting a dorsal fate, resulting in a mirror-image dorsal limb. Inhibition of *Lmx1* expression in the chick limb results in the absence of dorsal structures, but the cells do not adopt a ventral fate. Thus other factors are involved in specifying a ventral fate in the chick. In many *Wnt-7a* mutant mice, posterior digits are lacking, suggesting that *Wnt-7a* is also required for normal antero-posterior patterning. Similar results are observed in chick embryos when dorsal ectoderm is removed. This has led to the suggestion that development along all three limb axes is integrated by interactions between the signals Wnt-7a, FGF-4, and Sonic hedgehog.

9.8 Different interpretations of the same positional signals give different limbs

The positional signals controlling limb patterning are the same in chick wing and leg, but they are interpreted differently. A polarizing region from a wing bud, for example, will specify additional mirror-image digits if grafted into the anterior margin of a leg bud. However, these digits develop as toes, not wing digits, as the signal is being interpreted by leg cells. In a similar manner, signals are conserved between different vertebrates: for example, a mouse apical ectodermal ridge grafted in place of a chick apical ectodermal ridge can provide the appropriate signal for early chick limb development. Signals are, however, interpreted according to the origin of the responding cells: when grafted into the anterior margin of a chick wing bud a polarizing region from a mouse limb bud specifies additional wing structures, not mouse structures. Thus, the signals from the different polarizing regions are the same, as are the signals from the different apical ridges; the difference in the structures formed by the limbs is a consequence of how the signals are interpreted and depends on the genetic constitution and developmental history of the responding cells in the limb bud.

A further demonstration of the interchangeability of positional signals comes from grafting limb bud tissue to different positions along the proximo-distal axis of the bud. If tissue that would normally give rise to the thigh is grafted from the proximal part of an early chick leg bud to the tip of an early wing bud, it develops into toes with claws (Fig. 9.15). The tissue has acquired a more distal positional value after transplantation, but interprets it according to its own developmental program, which is to make leg structures. The positional signals must thus be interpreted by limb-specific proteins, such as the transcription factors Tbx2 and Tbx3, which as noted earlier, are involved in the specification of digits. In humans, mutations in the related T-box gene *Tbx5* are associated with the Holt-Oram syndrome, which is most commonly characterized by abnormalities of the thumbs and heart defects. Another limb-specific protein is the transcription factor Pitx1 (which is related to the Otx homeodomain proteins). It is expressed in the leg bud only and its expression in the wing results in the development of leg-like distal structures.

9.9 Homeobox genes also provide positional values for limb patterning

Hox genes specify position along the antero-posterior axis of the vertebrate body (see Section 4.2) and there is now considerable evidence that they also record positional values within the limbs. At least 23 different Hox genes are expressed during

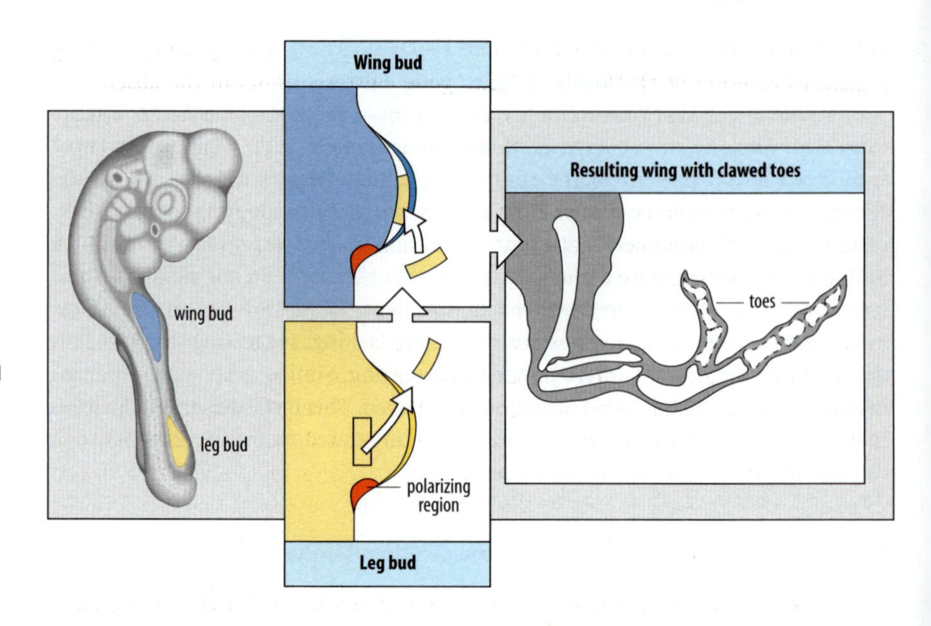

Fig. 9.15 Chick proximal leg bud cells grafted to a distal position in a wing bud acquire distal positional values. Proximal tissue from a leg bud that would normally develop into a thigh is grafted to the tip of a wing bud. In the progress zone, it acquires more distal positional values and interprets these as leg structures, forming clawed toes at the tip of the wing.

chick limb development. Attention has been largely focused on the genes of the Hoxa and Hoxd gene clusters, which are related to the *Drosophila Abdominal-B* gene (see Box 4A, p. 156). These sets of genes are expressed in both forelimbs and hindlimbs, whereas the expression of Hoxb and Hoxc gene clusters is restricted to the forelimb and the hindlimb respectively.

Once a limb bud has started to form, Hox genes expressed within the bud appear to record positional information along the limb axes. The expression of Hox genes during limb development is dynamic, however; the pattern of expression of a particular gene can undergo significant changes as the limb bud grows. Here, we confine our attention to the expression of the Hoxa and Hoxd genes at just one stage of wing development (Fig. 9.16). At this stage, expression of *Hoxa9* through *Hoxa13* is sequentially initiated along the proximo-distal axis, resulting in a nested pattern of expression of different Hoxa genes along this axis. That is, *Hoxa9* expression covers the complete region where the Hoxa genes are expressed, and successive genes occupy more and more distal regions. Hoxd genes form a similar, but skewed, nested pattern at the posterior margin at the distal region of the bud (Fig. 9.16, lower panel).

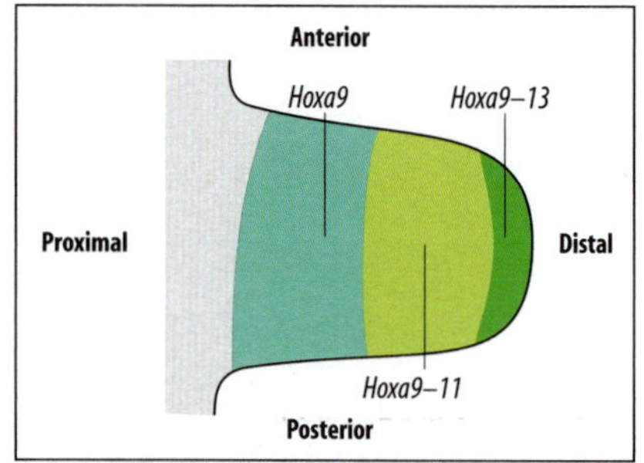

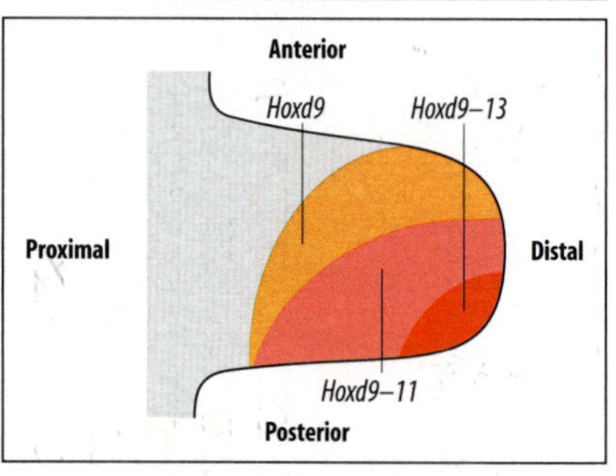

Fig. 9.16 Pattern of Hox gene expression in the chick wing bud. The Hoxa genes (top panel) are expressed in a nested pattern along the proximo-distal axis, *Hoxa13* being expressed most distally, while the Hoxd genes (bottom panel) are expressed in a similarly nested pattern along the antero-posterior axis, *Hoxd13* being expressed most posteriorly.

At the stage of limb development depicted in Fig. 9.16, the proximo-distal sequence of expression of Hoxa genes corresponds to the three main proximo-distal regions of the limb. *Hoxa9* is expressed in the upper limb, where the humerus (or femur in the hindlimb) forms. *Hoxa9–Hoxa11* are expressed in the lower limb, where the radius and ulna (or tibia and fibula) develop, and *Hoxa9–Hoxa13* are expressed in the wrist and digits. There are some differences in the pattern of expression of the paralogous Hox groups 9 to 13 in the forelimb and hindlimb. Hox9 paralogous genes are not expressed in the hindlimb, but there is only relatively small variation in the pattern of expression of the other groups.

If the Hox genes are involved in recording positional information, then experimental manipulations that lead to changes in the pattern of the limb skeletal elements should cause a corresponding change in Hox expression domains. If a polarizing region is grafted to the anterior margin of a wing bud, there is indeed a change to a mirror-image pattern of Hoxd expression (Fig. 9.17). This occurs within 24 hours of grafting, which is about the time required for the polarizing region to exert its effect. Hox genes may also specify the polarizing region, as deletion of all Hoxa and Hoxd gene clusters results in the absence of Sonic hedgehog expression and severe truncations of the limbs (see Section 9.5).

Does alteration of Hox gene expression in the limb lead to homeotic changes, similar to those seen in the vertebrae (see Section 4.4)? The results of gene knock-out experiments in mice designed to answer this question can be quite complicated, but for the proximo-distal axis there is a quite clear pattern of control of the different regions by the Hox genes (Fig. 9.18). This is based on triple mutants of the genes in one paralogous group. For example, paralogous group Hox10 controls proximal elements, whereas Hox11 patterns the zeugopod (the radius and ulna in the forelimb, tibia and fibula in the hindlimb). Inactivation of *Hoxd13* disrupts digit development, where it is normally expressed at a high level. This expression pattern of *Hoxd13* is due to a control region located 100 kb 5′ to the Hoxd cluster, an example of long-range control similar to the control of β-globin expression in red blood cell differentiation (see Section 8.6). Strong activation of *Hoxd13* compared to the other Hoxd genes may be because it is the first gene that this distant enhancer comes in contact with.

It seems that the Hox genes also have an important influence later on the growth of the cartilaginous elements once they are formed. Overexpression of *Hoxa13*, which is normally expressed in the distal region of the limb bud, results in limbs in which the radius and the ulna are reduced in size. This suggests that they are transformed into small wrist-like elements, probably by a change in the control of cell multiplication. Overexpression of *Hoxd13* results in the shortening of the long bones of the leg because of the effects of *Hoxd13* on the rate of cell proliferation in the growing cartilaginous elements. These results show that expression of the Hox genes can control the size of the cartilaginous elements in the limbs at both early and later stages. Mutations in Hox genes are known to underlie human conditions in which limbs are abnormal, such as fusion of digits and some types of polydactyly, but these phenotypes are difficult to explain given our current knowledge. Just how the Hox genes determine patterning along the proximo-distal axis remains unclear and requires identification of their downstream targets. One known example is the regulation of the expression of BMPs in the distal limb by *Hoxa13*; loss of *Hoxa13* function causes digit defects.

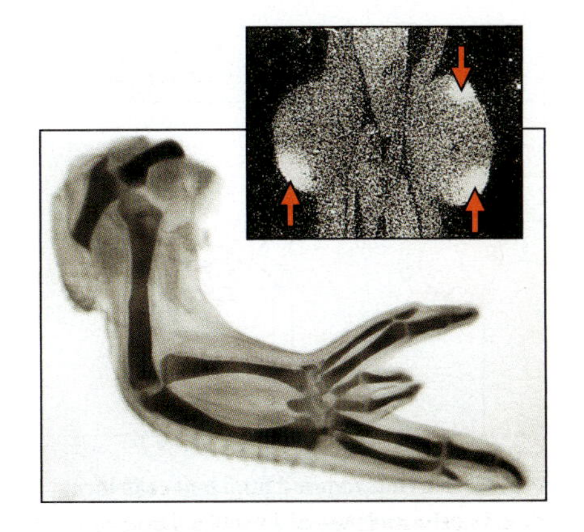

Fig. 9.17 Change in expression of Hoxd genes following a polarizing region graft. Grafting a polarizing region to the anterior margin of a limb bud results in mirror-image expression of the *Hoxd13* genes and a mirror-image duplication of the digits (main panel). The inset shows the expression of *Hoxd13* (arrows) at the posterior margin in a normal limb bud (inset, left). In the operated limb (inset, right) *Hoxd13* is expressed at both the anterior and posterior margins.

9.10 Self-organization may be involved in the development of the limb bud

The development of a series of cartilaginous elements along the antero-posterior axis may involve mechanisms other than signaling by the polarizing region. Evidence for this comes, for example, from observing the development of reaggregated limb buds after the mesenchymal cells of early chick limb buds have first been disaggregated and thoroughly mixed to disperse the polarizing region. They are then reaggregated, placed in an ectodermal jacket, and grafted to a site

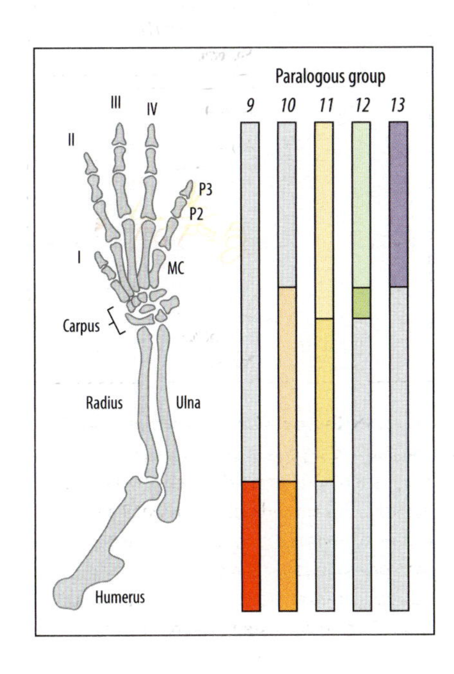

Fig. 9.18 Regions of Hox gene expression in the mouse forelimb. Colored regions denote regions of Hox gene influence on patterning as determined from gene knock-out experiments. Hox9 (red) and Hox10 (orange) paralogs function together to pattern the upper forelimb (humerus), Hox10 paralogs also affect the middle forelimb (radius and ulna), as represented by the lighter orange shading. Hox11 paralogs (yellow) are mostly involved in patterning this middle region, with some influence on pattern in the wrist and 'hand' region (lighter yellow shading). Hox12 paralogs (green) predominantly pattern the wrist, while Hox13 paralogs (purple) predominantly pattern the hand.

Adapted from Wellik, D.M. and Capecchi, M.R.: 2003.

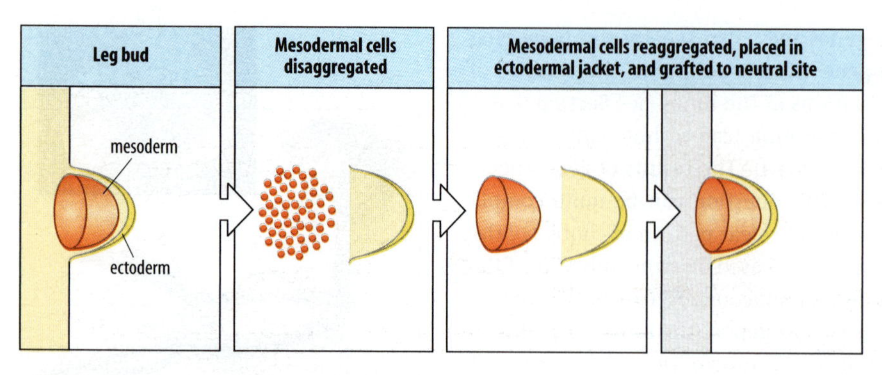

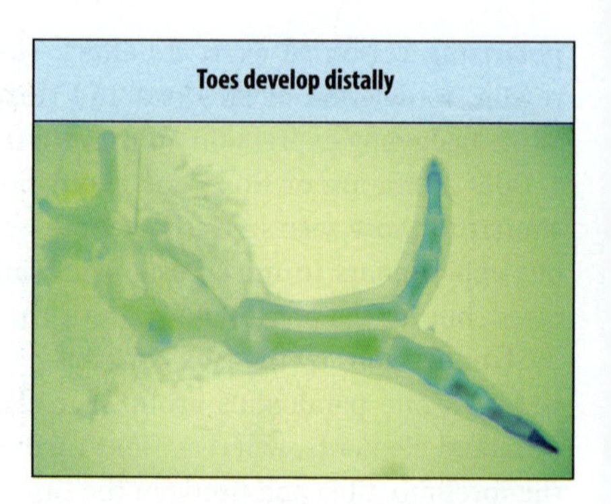

Fig. 9.19 Reaggregated limb bud cells form digits in the absence of a localized polarizing region. Mesodermal cells from a chick leg bud are separated, mixed to disperse all the cells, including the polarizing region, and then reaggregated, placed in an ectodermal jacket, and grafted into a neutral site. Well-formed toes develop distally.

where they can acquire a blood supply, such as the dorsal surface of an older limb. Limb-like structures develop from these grafted buds even though they have developed in the absence of a polarizing region. Several long cartilaginous elements may form in the more proximal regions of these abnormal limbs, although none of the proximal elements can be easily identified with normal structures. More distally, however, reaggregated hindlimb buds develop identifiable toes (Fig. 9.19). The fact that well-formed cartilaginous elements can develop at all in the absence of a discrete polarizing region shows that the bud has a considerable capacity for self-organization.

There may therefore be a mechanism in the limb bud that generates a basic pattern—a **prepattern**—of equivalent cartilaginous elements. These elements could then be given their identities and further refined by response to positional information involving signals such as Sonic hedgehog and the activation of Hox genes and their downstream targets (see Sections 9.5 and 9.9). The mechanism for generating the prepattern could be based on a **reaction–diffusion mechanism** (Box 9B, opposite). In the wing, for example, a reaction–diffusion or related mechanism could result in a single peak in some morphogen forming in the proximal region of the limb, which would specify a prepattern for the humerus. More distally, alterations in the reaction–diffusion conditions, due to changes in proximo-distal positional information, could give rise to three peaks of the morphogen, giving the cartilaginous elements of the three wing digits. These prepatterns could then be modified by signals specifying antero-posterior and dorso-ventral positional information.

In this reaction–diffusion model, polydactyly in humans could simply result from a chance widening of the limb bud. If a reaction–diffusion mechanism generates a periodic pattern of cartilage elements across the limb as the digits are forming, merely widening the limb bud by some small developmental accident would enable a further digit to develop.

9.11 Limb muscle is patterned by the connective tissue

Limb muscle cells have a different origin from that of the limb mesenchymal cells. If quail somites are grafted into a chick embryo at a site opposite to where the wing bud will develop, the wing that subsequently forms will have muscle cells of quail origin, but all its other cells will be of chick origin. This demonstrates that limb muscle cells have a different lineage to that of limb connective tissues (cartilage and tendons). Cells that give rise to muscle migrate into the limb bud from the somites at a very early stage (see Section 4.5). After migration, the future muscle cells multiply and initially form a dorsal and a ventral block of presumptive

Box 9B Reaction–diffusion mechanisms

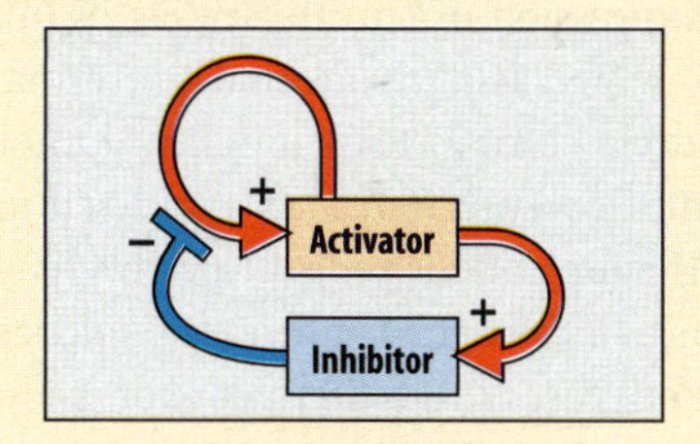

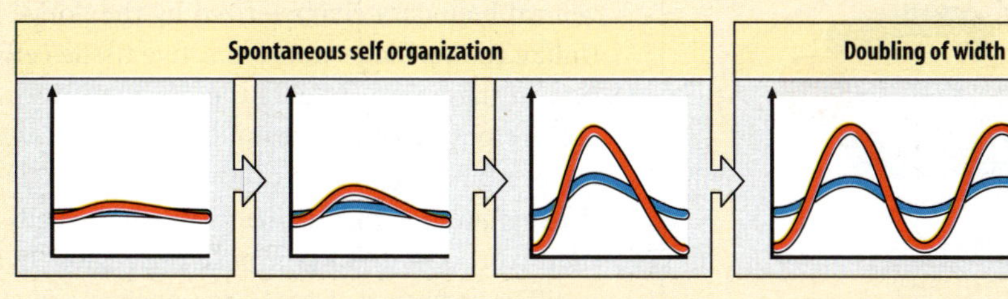

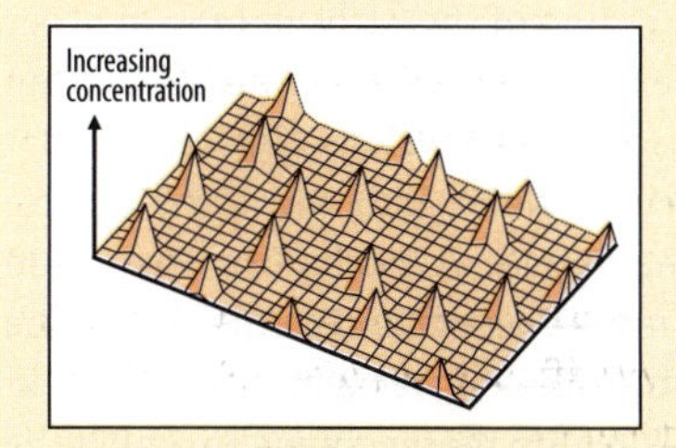

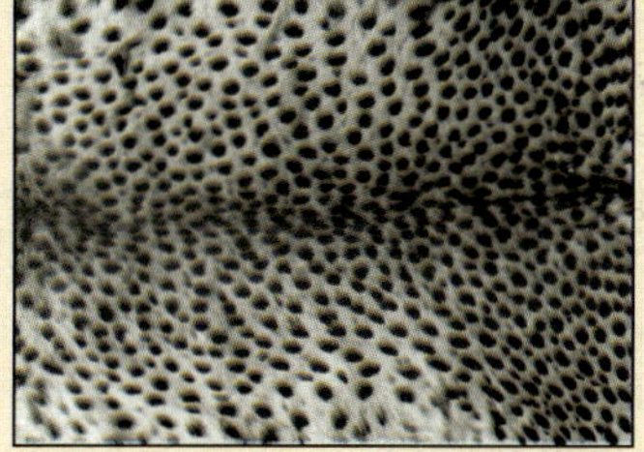

There are some self-organizing chemical systems that spontaneously generate spatial patterns of concentration of some of their molecular components. The initial distribution of the molecules is uniform, but over time the system forms wave-like patterns (see above). The essential feature of such a self-organizing system is the presence of two or more types of diffusible molecules that interact with one another. For that reason, such a system is known as a **reaction–diffusion system**. For example, if the system contains an activator molecule that stimulates both its own synthesis and that of an inhibitor molecule, which in turn inhibits synthesis of the activator, a type of lateral inhibition will occur so that synthesis of activator is confined to one region (see left).

Under appropriate conditions, which are determined by the reaction rates and diffusion constants of the components, a closed system of a certain size can spontaneously develop a spatial pattern of activator with a single concentration peak. If the size of the system is increased, two peaks will eventually develop, and so on. Such a mechanism could thus generate periodic patterns such as the arrangement of digits or the sepals or petals of flowers. When the chemicals can diffuse in two dimensions, the system can give rise to a number of peaks that are somewhat irregularly spaced (see left).

Such a system may underlie some of the patterns of pigmentation that are common throughout the animal kingdom, such as the patterns of spots and stripes seen in the zebra and cheetah (see two panels, left). How these patterns are generated is not yet known, but one possibility is a reaction–diffusion mechanism. Assuming that pigment is synthesized in response to some activator, and that synthesis only occurs at high activator concentrations, some animal color patterns can be mimicked by a reaction–diffusion system in computer simulations.

A characteristic feature of reaction–diffusion patterns is that new intermediate peaks appear as the system grows in size. The angelfish *Pomocanthus semicirculatus* provides a remarkable example of striping that could be generated by a reaction–diffusion mechanism (see below). Juvenile *P. semicirculatus*, less than 2 cm long, have only three dorso-ventral stripes. As the fish grow, the intervals between the stripes increase until the fish is around 4 cm long. New stripes then appear between the original stripes and the stripe intervals revert to those present at the 2 cm stage. As the fish grows larger, the process is repeated. This type of dynamic patterning is what would be expected of a reaction–diffusion mechanism. Computer simulations of reaction–diffusion mechanisms can also generate the patterns seen on a wide variety of mollusk shells. Nevertheless, there is as yet no direct evidence for a reaction–diffusion system patterning any developing organism. Top figure after Meinhardt, H., *et al.*: 1974.

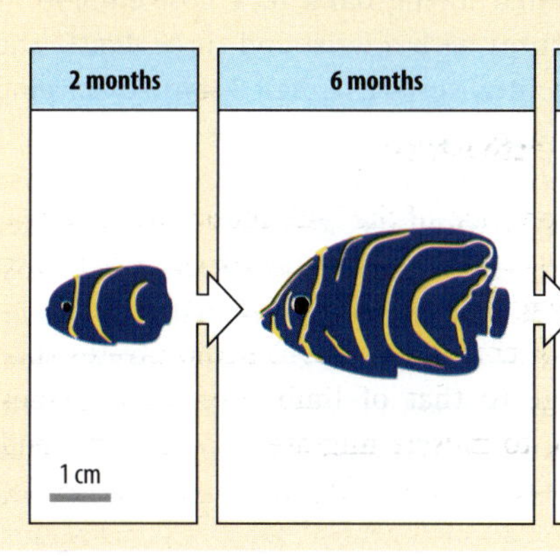

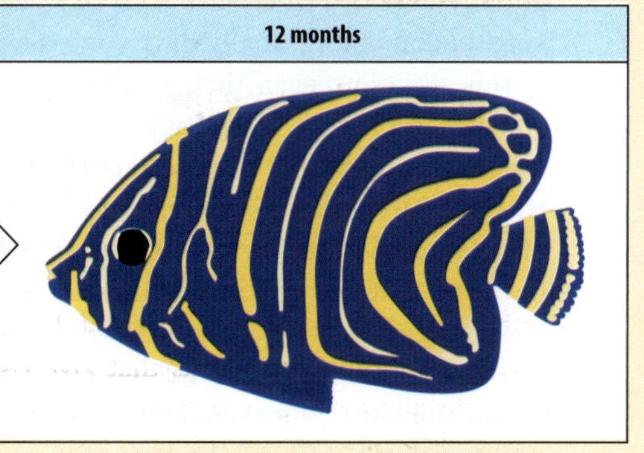

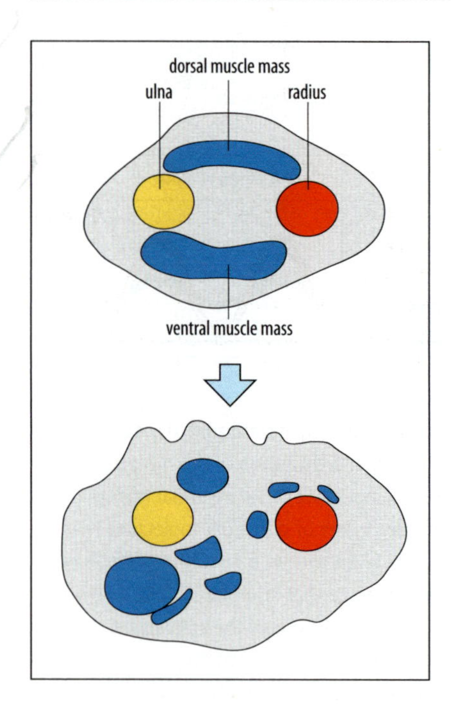

Fig. 9.20 Development of muscle in the chick limb. A cross-section through the chick limb in the region of the radius and ulna shortly after cartilage formation shows presumptive muscle cells present as two blocks—the dorsal muscle mass and the ventral muscle mass. These blocks then undergo a series of divisions to give rise to individual muscles.

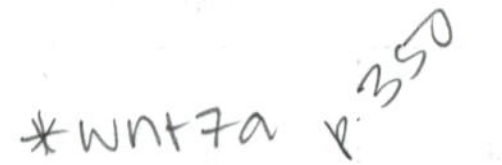

muscle (Fig. 9.20). These blocks undergo a series of divisions to give the final muscle masses. The initial migration of the future muscle cells is restricted to whether they entered in the dorsal or ventral region; they do not cross the dorso-ventral boundary characterized by the dorsal expression of Lmx1b (see Fig. 9.14). Unlike the cartilage and connective tissue cells, which acquire positional information in the progress zone, the presumptive muscle cells, at least initially, do not acquire positional values, and are all equivalent. However, *Hoxa11* is expressed in the presumptive muscle cells when they enter the limb and is probably induced by the surrounding mesenchyme. When the cells are in the dorsal and ventral masses they express *Hoxa13* in the posterior region. In both cases this pattern of expression is different from that of the mesenchyme in the same regions.

Evidence for the early equivalence of muscle cells comes from experiments where somites are grafted from the future neck region of the early embryo in order to replace the normal wing somites. The muscle cells that develop thus come from neck somites, but a normal pattern of limb muscles still develops. This shows that the muscle pattern is determined by the connective tissue into which the muscle cells migrate, rather than by the muscle cells themselves. A mechanism that could pattern the muscle is based on the prospective muscle-associated connective tissue having surface or adhesive properties that the muscle cells recognize, resulting in the migration of muscle cells to these regions. The limb mesenchyme to which the myoblasts migrate is characterized by the transcription factor TCF4, and ectopically expressed TCF4 can attract myoblasts to that site. The pattern of muscle could thus be determined by the pattern of muscle-associated connective tissue, which is specified by mechanisms similar to those that produce the pattern of cartilage. If the pattern of adhesiveness in the connective tissue changed with time, the presumptive muscle cells would migrate to the new sites, and this could account for the splitting of the muscle masses.

Presumptive muscle cells proliferate in the limb and express Pax3. When *Pax3* is downregulated, cell proliferation stops and the cells differentiate (see Section 8.9). The overlying ectoderm is required to prevent premature differentiation, and BMP-4 plays a key role in this. The development of tendons is marked by the expression of the transcription factor Scleraxis (see Section 4.5).

9.12 The initial development of cartilage, muscles, and tendons is autonomous

The pattern of cartilage, tendon, and muscle in the limb may be specified by the same signals, as a polarizing region graft causes the development of a mirror-image pattern of all these elements. Each of the elements develops in its final position and there is little interaction between them. Thus, for example, if just the tip of an early chick wing bud is removed and grafted to the flank of a host embryo, it initially develops into normal distal structures with a wrist and three digits. The long tendon that normally runs along the ventral surface of digit 3 starts to develop in this situation, even though both its proximal end and the muscle to which it attaches are absent. The tendon does not continue to develop, however, because it does not make the necessary connection to a muscle, and so is not put under tension. The mechanism whereby the correct connections between tendons, muscles, and cartilage are established has still to be determined. It is clear, however, that there is little or no specificity involved in making such connections; if the tip of a developing limb is inverted dorso-ventrally, dorsal and ventral tendons may join up with inappropriate muscles and tendons. They simply make connections with those muscles and tendons nearest to their free ends—they are promiscuous.

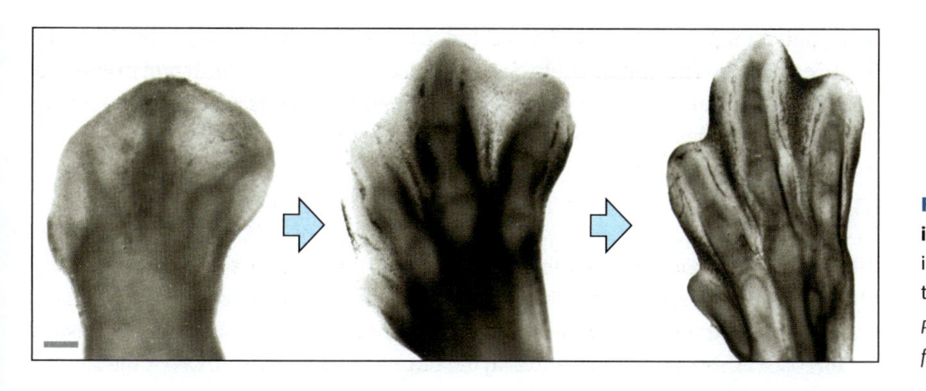

Fig. 9.21 Cell death during leg development in the chick. Programmed cell death in the interdigital region results in separation of the toes. Scale bar = 1 μm.

Photographs courtesy of V. Garcia-Martinez, from Garcia-Martinez, V., et al.: 1993.

9.13 Joint formation involves secreted signals and mechanical stimuli

The first event in joint formation is the formation of an 'interzone' at the site of the prospective joint, in which the cartilage-producing cells become fibroblast-like and cease producing the cartilage matrix, thus separating one cartilage element from another. At this time, the BMP-related protein GDF-5 is expressed in the presumptive joint region together with Wnt-14. Ectopic misexpression of Wnt-14 can induce the early steps in joint formation, and mice lacking GDF-5 have some of their joints missing. A little later, the musculature begins to function and muscle contraction occurs. At this point high levels of hyaluronan, a component of the proteoglycans that lubricate joint surfaces, are made by the cells of the interzone; the hyaluronan is secreted in vesicles that coalesce to form the joint cavity, which separates the articulating surfaces of the joint from each other. Synthesis of hyaluronan by the cells lining these surfaces depends on the mechanical stimulation that results from muscle contraction, and in the absence of this activity, the joints fuse. Cells isolated from the interzone region and grown in culture respond to mechanical stimulation by increased secretion of hyaluronan when the substratum to which they are attached is stretched.

9.14 Separation of the digits is the result of programmed cell death

Programmed cell death by apoptosis plays a key role in molding the form of chick and mammalian limbs, especially the digits. The region where the digits form is initially plate-like, as the limb is flattened along the dorso-ventral axis. The cartilaginous elements of the digits develop from the mesenchyme at the correct positions within this plate. Separation of the digits then depends on the death of the cells between these cartilaginous elements (Fig. 9.21). There is evidence for the involvement of BMPs in cell death: if the functioning of BMP receptors is blocked in the developing chick leg, cell death does not occur and the digits are webbed.

This cell death is a normal programmed part of patterning and cell differentiation (see Chapter 8). The fact that ducks and other waterfowl have webbed feet whereas chickens do not is simply the result of there being less cell death between the digits of waterfowl. If chick-limb mesoderm is replaced with duck-limb mesoderm, cell death between the digits is reduced and the chick limb develops 'webbed' feet (Fig. 9.22). It is the mesoderm that determines the patterns of cell death both within the mesoderm and the overlying ectoderm. In amphibians, digit separation is not due to cell death, but results from the digits growing more than the interdigital region.

Programmed cell death also occurs in other regions of the developing limb, such as the anterior margin of the limb bud, between the radius and ulna, and in the

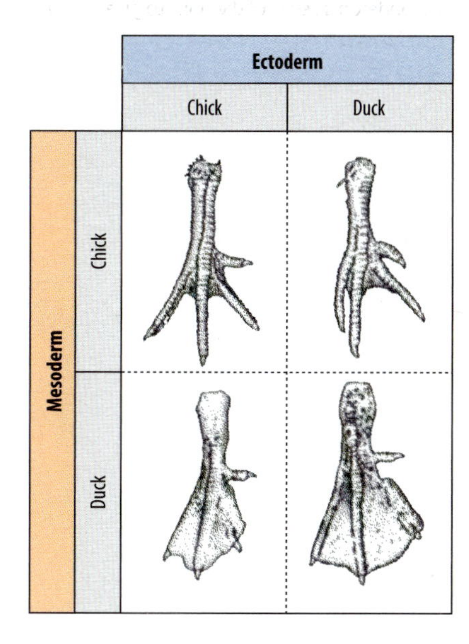

Fig. 9.22 The mesoderm determines the pattern of cell death. The webbed feet of ducks and other water birds form because there is less cell death between the digits than in birds without webbed feet. When the mesoderm and ectoderm of embryonic chick and duck limb buds are exchanged, webbing develops only when duck mesoderm is present.

wing polarizing region at a later stage. Indeed, it was the investigation of cell death in this region by transplanting it to the anterior margin of the limb bud that led to the discovery that it acts as a polarizing region. This suggests that cell death could control the number of cells expressing Sonic hedgehog.

Summary

Positioning and patterning of the vertebrate limb is largely carried out by intercellular interactions that provide the cells with positional information. The position of the limb buds along the antero-posterior axis of the body is probably related to Hox gene expression. There are two key signaling regions within the limb bud. One is the apical ectodermal ridge, which specifies the progress zone in the underlying mesenchyme where cells acquire their positional identity; the second is the polarizing region at the posterior margin of the limb, which specifies pattern along the antero-posterior axis. The signals from the apical ridge are fibroblast growth factors (FGFs) and are essential for limb outgrowth. Sonic hedgehog protein is expressed in the polarizing region and probably provides a graded positional signal that is pivotal for determining digit number and pattern. The dorso-ventral axis is specified by the ectoderm. Hox genes are expressed in a well-defined spatio-temporal pattern within the limb bud and may provide the molecular basis of positional identity. The limb muscle cells are not generated within the limb but migrate in from the somites and are patterned by the limb connective tissue. Patterning of the cartilaginous elements may involve a self-organizing mechanism. In birds and mammals, separation of the digits is achieved by programmed cell death.

Summary: vertebrate limb development

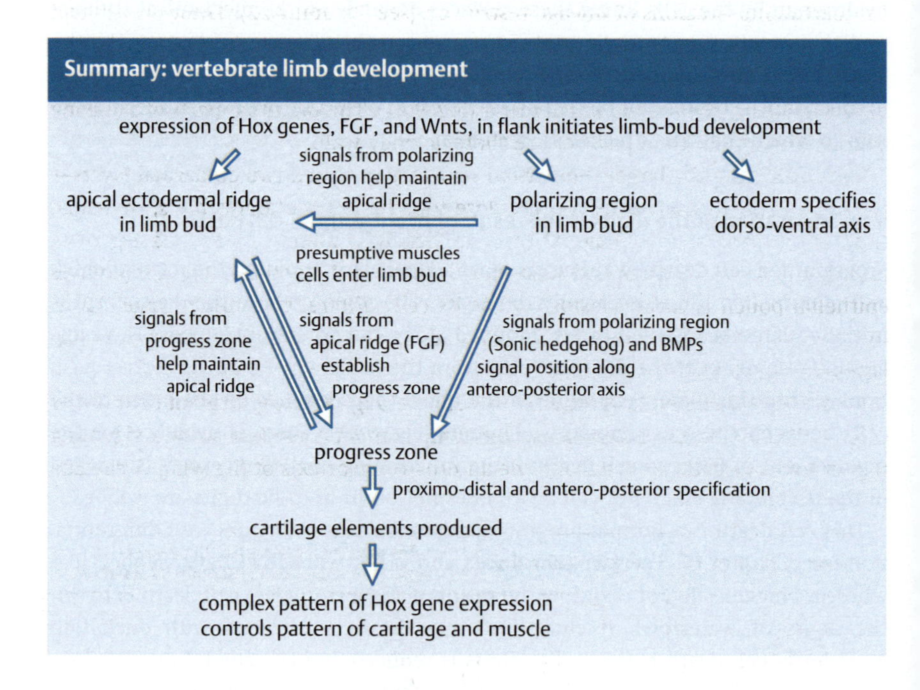

Insect wings, legs, and eyes

The adult organs and appendages of *Drosophila*, such as the eyes, wings, legs, and antennae, develop from imaginal discs, which are excellent systems for analyzing pattern formation. The discs invaginate from the embryonic ectoderm as simple pouches of epithelium during embryonic development and remain as such until metamorphosis (see Fig. 2.6). Although all imaginal discs superficially look rather similar, they develop according to the segment in which they are located. In the wing and leg discs, the specification as wing or leg and the basic patterning occur within the embryonic epithelium, at the time when the segments are being patterned and given their identity. The wing and leg discs are initially specified in the embryo as clusters of 20–40 cells. During larval development, the discs grow about 1000-fold. The adult *Drosophila* compound eyes develop from imaginal discs at the anterior end of the embryo. The eye is made up of some 800 units, each containing about 20 cells, and is an excellent model system for looking at the role of cell–cell interactions in development.

Although insect wings and legs differ so greatly in appearance, they are partially homologous structures, and the strategy of their patterning is very similar. Moreover, the mechanisms by which they are patterned, and even the genes involved, show some amazing similarities to the patterning of the vertebrate limb. We will start by discussing the patterning of the wing disc and its development into the adult wing and then compare the patterning of the leg disc. We will finish this part of the chapter by looking at some aspects of the development of the *Drosophila* eye.

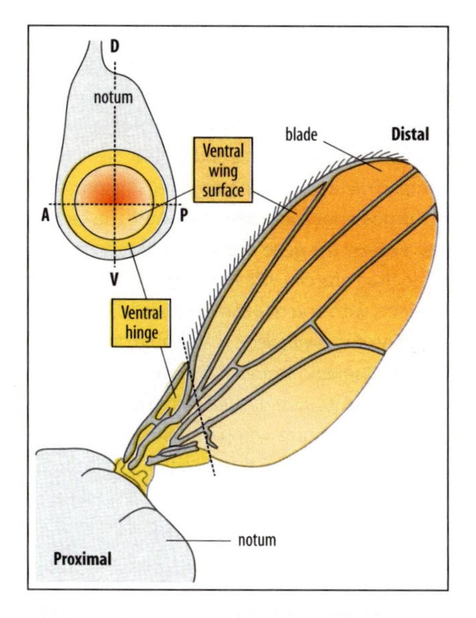

Fig. 9.23 Fate map of the wing imaginal disc of *Drosophila*. The wing disc before metamorphosis is an ovoid epithelial sheet bisected by the boundary between the anterior (A) and the posterior (P) compartments. There is also a compartment boundary between the future ventral (V) and dorsal (D) surfaces of the wing. At metamorphosis, the ventral surface will fold under the dorsal surface (see Fig. 9.24). The disc also contains part of the dorsal region of thoracic segment 2 in the region of wing attachment.

After French, V., et al.: 1994.

9.15 Positional signals from the antero-posterior and dorso-ventral compartment boundaries pattern the wing imaginal disc

Imaginal discs are essentially epithelial sheets, which in the case of wing and leg discs will give rise to epidermis in the adult. The epidermis of body segments, and the wings and legs that derive from them, is divided into anterior and posterior compartments—regions of lineage restriction (see Section 2.22). These correspond to the compartment boundaries in the segments in which they develop. In the wing disc, a second compartment boundary between the dorsal and ventral regions develops during the second larval instar (Fig. 9.23). The proximal parts of the wing and leg discs give rise to parts of the thoracic body wall.

The adult wing is a largely epidermal structure in which two epidermal layers—the dorsal and ventral surfaces—are close together. At metamorphosis, by which time patterning of the imaginal discs is largely complete, wing and leg discs undergo a series of profound anatomical changes. Essentially, the invaginated epithelial pouch is turned inside out, as its cells differentiate and change shape. In the case of the wing, one surface then folds under the other to form the double-layered wing structure (Fig. 9.24).

In the wing disc, signaling regions are set up along the compartment boundaries (Fig. 9.25). Cells at the antero-posterior compartment boundary form a signaling region that specifies pattern along the antero-posterior axis of the wing. A cascade

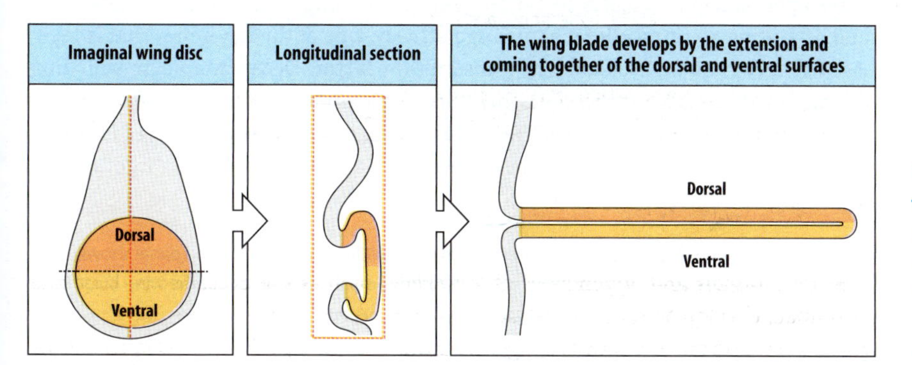

Fig. 9.24 Schematic representation of the development of the wing blade from the imaginal disc. Initially, the dorsal and ventral surfaces are in the same plane. At metamorphosis, the sheet folds and extends, so the dorsal and ventral surfaces come into contact with each other.

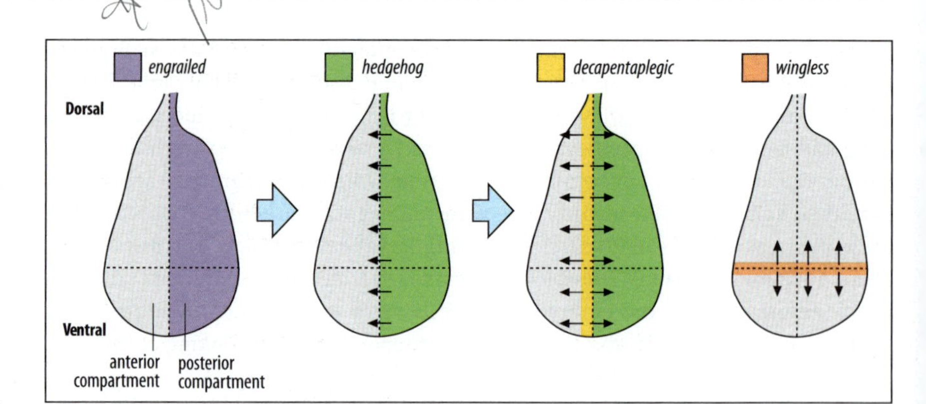

Fig. 9.25 Establishment of signaling regions in the wing disc at the compartment boundaries. The disc is divided into anterior and posterior compartments and, later in development, into dorsal and ventral compartments. The gene *engrailed* is expressed in the posterior compartment, where the cells also express the gene *hedgehog* and secrete the Hedgehog protein. Where Hedgehog protein interacts with anterior compartment cells, the gene *decapentaplegic* is activated, and Decapentaplegic protein is secreted into both compartments, as indicated by the black arrows. At the dorso-ventral compartment boundary, the *wingless* gene is activated and the signaling molecule is the protein Wingless.

of events sets up this signaling center. It begins with the expression of the *engrailed* gene in the posterior compartment of the disc (see Fig. 9.25), which reflects the pattern of gene expression in the embryonic parasegment from which the disc derives. We shall return to the question of what gives wing and leg discs their segmental identity later in this chapter.

The cells that express *engrailed* also express the gene *hedgehog* (see Section 2.23). The maintenance of compartment boundaries in the wing depends partly on communication between compartments. At the compartment boundary, the secreted Hedgehog protein acts as a short-range morphogen over about 10 cell diameters and induces adjacent cells in the anterior compartment to express the *decapentaplegic* gene (see Fig. 9.25). As noted in Box 9A (p. 347), Hedgehog protein may not move by simple diffusion or endocytosis. The Hedgehog signal acts on the anterior cells through a receptor complex involving the proteins Patched and Smoothened (see Fig. 2.37).

The Decapentaplegic protein, which is a member of the TGF-β family of signaling proteins (see Box 3C, p. 111), is secreted at the compartment boundary (see Fig. 9.25) and is probably the positional signal for patterning both the anterior and posterior compartments along the antero-posterior axis. There is evidence that Decapentaplegic forms concentration gradients in the wing that provide a long-range signal controlling the localized expression of the genes for the transcription factors Spalt (now also known as Spalt-related) and Omb in bands overlapping the expression of Decapentaplegic (Fig. 9.26). They provide a further level of patterning in the wing. Decapentaplegic protein acts as a morphogen, specifying the expression of *spalt* and *omb* when particular threshold concentrations are reached (Fig. 9.27); low levels induce *omb* whereas higher levels are required to induce *spalt*. Omb is related to Tbx2 and Tbx3, which are involved in vertebrate limb patterning (see Section 9.5).

Decapentaplegic regulates its target genes by binding to its receptor Thick veins and activating an intracellular signaling pathway. One of the key genes that is regulated is *brinker*, which encodes a transcription factor. Decapentaplegic signaling downregulates *brinker* expression, thus forming a gradient of intranuclear Brinker protein that is the inverse of the Decapentaplegic gradient. Repression of *brinker* is involved in the ability of Decapentaplegic to activate its other target genes;

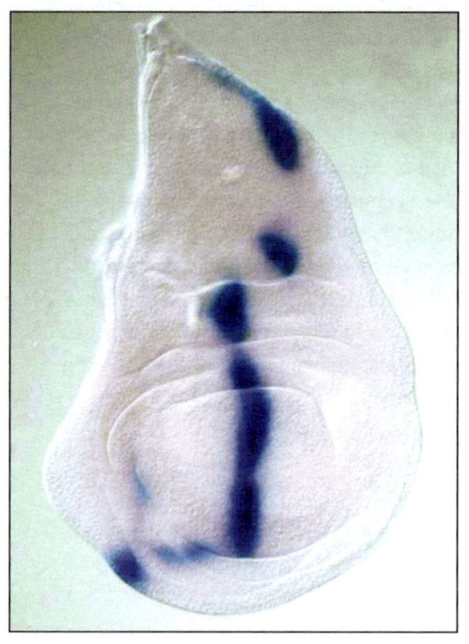

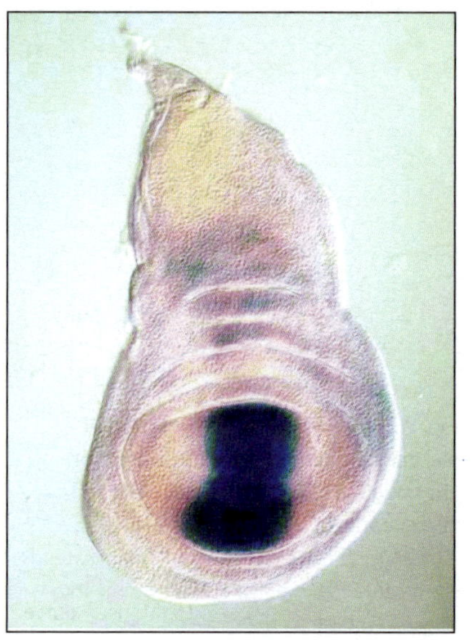

Fig. 9.26 *decapentaplegic* and *spalt* expression in *Drosophila* wing disc. The top panel shows the expression of *decapentaplegic* at the antero-posterior compartment border of the wing blade. The bottom panel shows the expression of *spalt* in the wing blade.

Photographs courtesy of K. Basler, from Nellen, D., et al.: 1996.

these genes will be expressed in patterns that reflect their sensitivity to activation by Decapentaplegic and repression by Brinker. In addition, *brinker* reduces cell growth, while Decapentaplegic promotes it.

The evidence that Decapentaplegic is a morphogen in the wing disc comes from several types of experiments. In the first, clones of cells unable to respond to a Decapentaplegic signal do not express *spalt* or *omb*. In the second, ectopic expression of *decapentaplegic* leads to the localized activation of *spalt* and *omb* around the *hedgehog* clone. Formation of the long-range Decapentaplegic gradient in the wing disc is a complex process and is not yet fully understood. It appears to involve receptor-mediated endocytosis, as blocking the action of dynamin, an intracellular protein required for the internalization of endocytic vesicles, blocks gradient formation. The glypican Dally is also involved in both shaping the Decapentaplegic gradient and influencing its signaling via the receptor Thick veins.

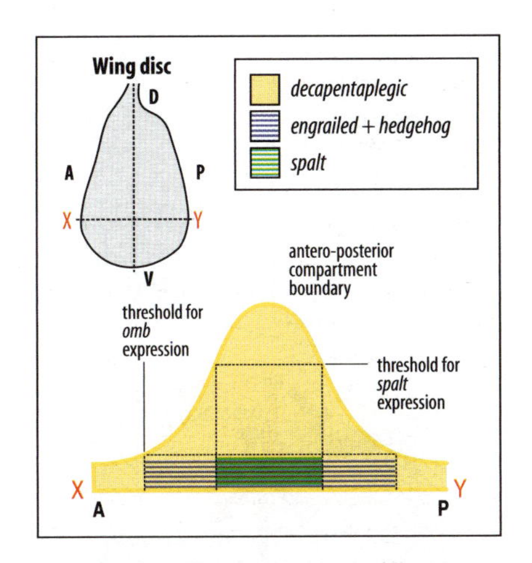

Fig. 9.27 A model for patterning the antero-posterior axis of the wing disc. The decapentaplegic protein is produced at the compartment boundary. The schematic cross-section of the wing disc (X–Y) along the antero-posterior axis shows how the presumed gradient in decapentaplegic in both anterior and posterior compartments activates the genes *spalt* and *omb* at their threshold concentrations. The gradient in *hedgehog* expression at the antero-posterior boundary is not shown.

The pattern of veining on the adult wing blade (Fig. 9.28) is one of the ultimate outcomes of this antero-posterior patterning activity. It is proposed that the gradients in Hedgehog and Decapentaplegic in the wing disc specify the pattern along the antero-posterior axis of the wing. The role of Hedgehog and Decapentaplegic in wing patterning is further shown by ectopically expressing *hedgehog* in genetically marked cell clones generated at random in wing discs. When such clones form in the posterior compartment they have little effect (as Hedgehog is normally expressed throughout this compartment in the disc) and development is more or less normal. With *hedgehog* clones in the anterior compartment, however, a mirror-symmetric repeated pattern is produced along the antero-posterior axis in that compartment as can be seen in the adult wing (see Fig. 9.28). Ectopic expression of *hedgehog* has set up new sites of *decapentaplegic* expression in the anterior compartment which results in new gradients of Hedgehog and Decapentaplegic protein being formed.

The differentiation of a wing vein at a later stage of development requires signaling through the EGF receptor, and one of the targets indirectly regulated by Hedgehog and Decapentaplegic is the gene *rhomboid*. This encodes an intramembrane serine protease that is part of the EGF receptor signaling pathway and is expressed specifically in prospective vein cells. Another of their targets is the gene *blistered*, which encodes a transcription factor that endows cells with an intervein fate. There is no single or simple process by which the morphogens specify the vein pattern: rather, the position of each vein is likely to be specified by a unique combination of factors, including the level of Decapentaplegic and Hedgehog signaling, which compartment the cells are in, and the induction or repression of particular

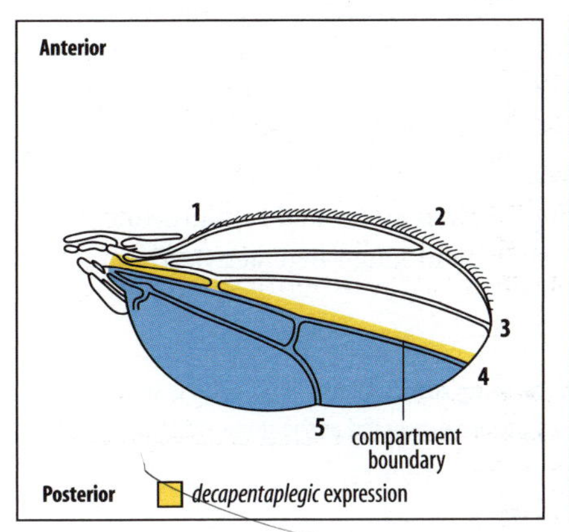

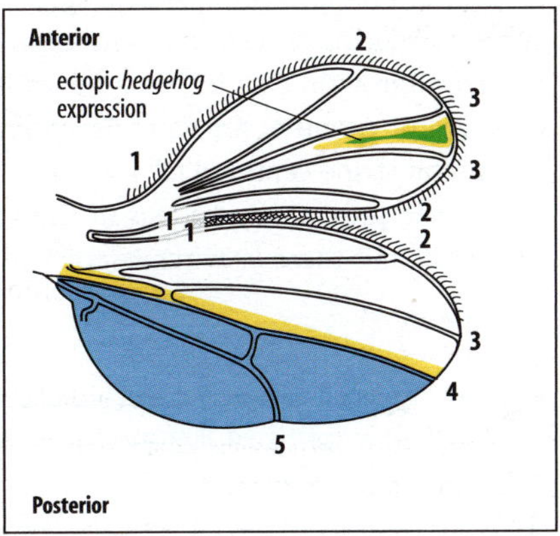

Fig. 9.28 Alteration of wing patterning due to ectopic expression of *hedgehog* and *decapentaplegic*. When *hedgehog* is expressed in a clone of cells in the anterior compartment a new source of Decapentaplegic protein is set up. Left panel: in the normal wing, Decapentaplegic protein is made at the compartment boundary, where Hedgehog is expressed. Right panel: the ectopic expression of *hedgehog* in the anterior compartment results in a new source of Decapentaplegic protein and a new wing pattern develops in relation to this new source.

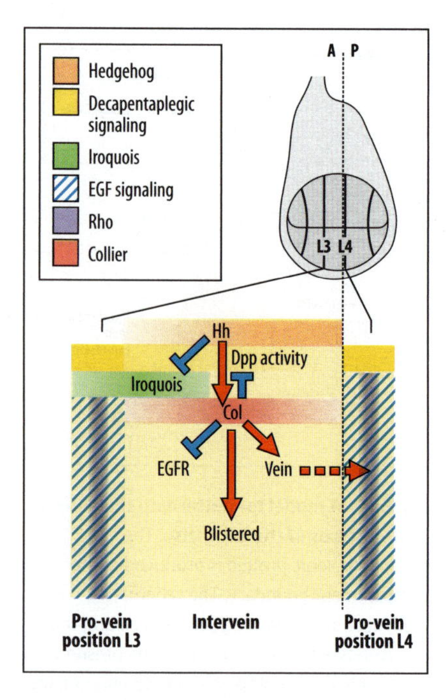

Fig. 9.29 Wing veins are individually specified by specific combinations of transcription factors and signaling proteins. This simplified schematic representation indicates how different combinations of Hedgehog (Hh) and Decapentaplegic (Dpp) signaling could indirectly delimit the future positions of veins L3 and L4 in the *Drosophila* wing imaginal disc by turning on transcription factors (such as Collier, Blistered, and Iroquois) in different antero-posterior regions (see text for details). Eventual differentiation as veins requires Rhomboid (Rho) expression and EGF signaling and. For simplicity, the differences in Hh, Dpp, Iroquois, and Collier activity across the region are shown as bars at the top of the figure. In reality, this pattern extends throughout the whole area.

transcription factors (Fig. 9.29). For example, wing pro-vein L4 is specified at the antero-posterior compartment boundary by the protein Vein. This is a ligand for the EGF receptor and is expressed in cells adjacent to the boundary as an indirect result of the actions of Hedgehog via its induction of the transcription factor Collier. Hedgehog also indirectly defines the position of pro-vein L3 in the anterior compartment via Collier, which together with Blistered, specifies the L3–L4 intervein region. L3 develops immediately anterior to this region from cells that express the transcription factor Iroquois under the indirect control of both Hedgehog and Decapentaplegic. Formation of the more lateral veins L2 (anterior) and L5 (posterior) depends on Decapentaplegic signaling and the expression of Spalt. It seems, therefore, that Decapentaplegic and Hedgehog do not directly pattern the veins, but set up a series of cell–cell interactions that specify positional information across the wing disc, perhaps even to the level of single cells. It is striking how precise the pattern of veins is if one compares the wings on either side of the fly.

The wing disc is also subdivided into dorsal and ventral compartments, which correspond to the dorsal and ventral surfaces of the adult wing (see Fig. 9.23). These compartments develop after the wing disc is formed—during the second larval instar. The dorsal and ventral compartments were originally identified by cell-lineage studies, but they are also defined by expression of the homeotic selector gene *apterous*, which is confined to the dorsal compartment, and which defines the dorsal state. The Apterous protein is a structural relative of Lmx1 from the mouse, which specifies a dorsal pattern in the mesoderm of the mouse limb bud (see Section 9.7). Like the antero-posterior compartment boundary, the dorso-ventral boundary acts as an organizing center, and Wingless is the signaling molecule at this boundary. Wingless is one member of the large Wnt family of secreted signal proteins, and we have already seen it in action in embryonic development (see Section 2.23). Reduction of *wingless* function results in loss of wing structures, and the gene is named after this adult phenotype. Apterous induces the Notch ligand Serrate in dorsal cells, and restricts the expression of the other Notch ligand, Delta, to ventral cells. Notch signaling at the compartment boundary induces expression of Wingless. Wingless in turn induces expression of the Notch ligands Delta and Serrate at the boundary, thus maintaining its own expression. At the early stages, Apterous also induces expression of the glycosyltransferase Fringe in dorsal cells. Fringe promotes Notch signaling in response to Delta, but inhibits it in response to Serrate, which helps to localize Notch signaling, and thus Wingless expression, to the compartment boundary (see Fig. 9.25). Apterous is also thought to induce the expression of cell-surface proteins that, in conjunction with Fringe-modulated Notch signaling, maintain the boundary of lineage restriction once Apterous itself is no longer expressed. Wingless serves a role analogous to Decapentaplegic in the antero-posterior patterning system, and activates expression of various genes at specific thresholds, which further pattern the wing and are responsible for its outgrowth. One of these genes is *vestigial*, which encodes a transcriptional co-activator essential for wing development and outgrowth and which is induced in a broad stripe centered on the dorso-ventral boundary (Fig. 9.30).

How the proximo-distal axis of the wing blade is patterned remains unclear in *Drosophila*, but there is evidence for both short- and long-range activation of genes

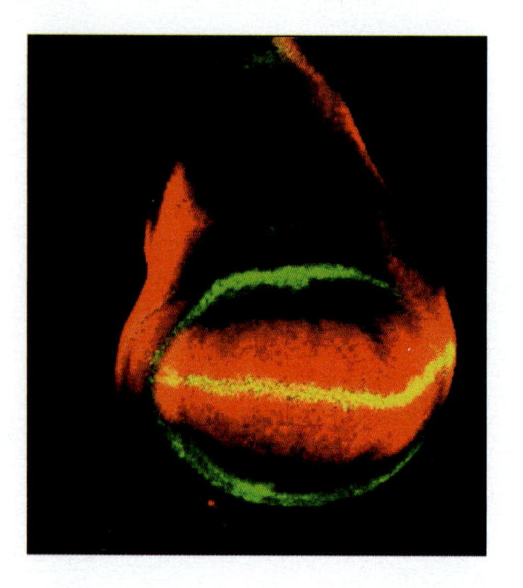

Fig. 9.30 Expression of *wingless* and *vestigial* in the wing disc. *wingless* expression is shown in green, while *vestigial* is red. Both *wingless* and *vestigial* are expressed at the dorso-ventral boundary as indicated by the yellow stripe.

Photograph courtesy of K. Basler, from Zecca, M., et al.: 1996.

in the wing blade by Wingless. We shall see later in the chapter how the colorful pattern on butterfly wings is a manifestation of proximo-distal patterning.

While both Decapentaplegic and Wingless may act as morphogens whose graded concentrations provide cells with positional information, the formation of these gradients is not due to simple diffusion. The gradient of activity can also be modified in other ways, as both Decapentaplegic and Wingless can regulate the expression of their receptors. Decapentaplegic, for example, represses the expression of one of its receptors, Thick veins, so that fewer receptors are present where Decapentaplegic levels are high. Wing disc cells also have thin actin-based extensions called cytonemes that project to the antero-posterior signaling boundary. These projections may also be involved in long-range cell–cell communication.

9.16 *Drosophila* wing epidermal cells show planar cell polarity

As well as the general asymmetry of wing structure, which is due to patterning along the various axes, the individual cells in the adult *Drosophila* wing epidermis become polarized in a proximo-distal direction; this is reflected by the pattern of hairs (trichomes) on the wing surface. Each of the approximately 30,000 cells in the wing epidermis produces a single hair, which points distally (see Fig. 9.28, which shows only those hairs that would be visible on the edge of the wing). This is an example of planar cell polarity. As we saw in Chapter 2 in relation to the polarity of hairs on the fly abdomen (see Box 2E, p. 75), signaling via the cell-surface receptor protein Frizzled plays a key role in setting polarity, and it is also required here for the correct orientation of the wing hairs. If clones of cells in the wing are made to lack *frizzled* activity, they are surrounded by wrongly polarized cells, showing that signaling is occurring between adjacent cells and that cell polarity is not established cell-autonomously. Frizzled is localized to the distal edge of the wing cells where it co-localizes with the signaling pathway component Dishevelled. The initial polarization of the epithelial sheet may be provided by the cell adhesion proteins Dachsous and Fat, which are members of the proto-cadherin family. Dachsous is graded with high levels proximally and inhibits Fat activity, thus creating opposing gradients.

9.17 The leg disc is patterned in a similar manner to the wing disc, except for the proximo-distal axis

Insect legs are essentially jointed tubes of epidermis and thus have a quite different structure from those of vertebrates. The epidermal cells secrete the hard outer cuticle that forms the exoskeleton. Internally, there are muscles, nerves, and connective tissues. The easiest way of relating the leg imaginal disc to the adult leg is to think of it as a collapsed cone. Looking down on the disc, one can imagine it as a series of concentric rings, each of which will form a proximo-distal segment of leg. A change in the shape of the epithelial cells is responsible for the outward extension of the leg at metamorphosis. The process is rather like turning a sock inside out, with the result that the center of the disc ends up as the distal end, or tip, of the leg (Fig. 9.31). The outermost ring gives rise to the base of the leg, which is attached to the body, and those nearer the center give rise to the more distal structures (Fig. 9.32).

The leg disc contains some 30 cells at its initial formation but grows to over 10,000 cells by the third instar. The first steps in the patterning of the leg disc along its antero-posterior axis are the same as in the wing. The *engrailed* gene is expressed in the posterior compartment and induces expression of *hedgehog*. The Hedgehog

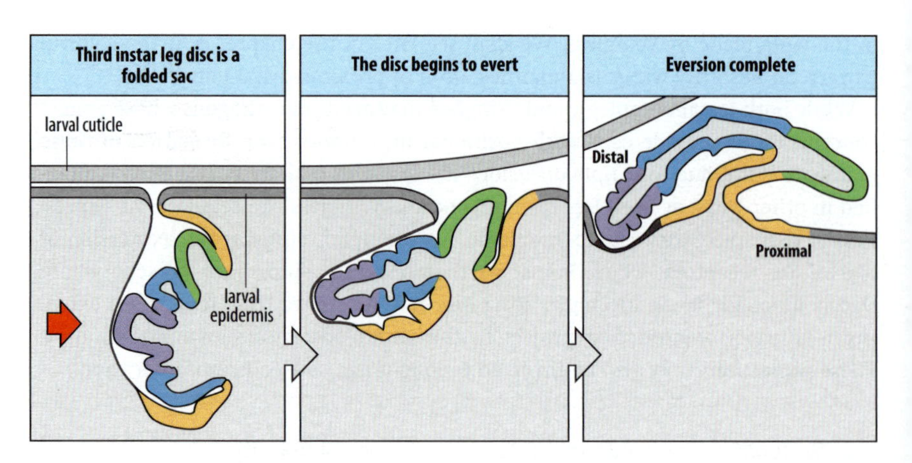

Fig. 9.31 ***Drosophila* leg disc extension at metamorphosis.** The disc epithelium, which is an extension of the epithelium of the body wall, is initially folded internally. At metamorphosis it extends outward, rather like turning a sock inside out. The red arrow in the first panel is the viewpoint that produces the concentric rings shown in Fig. 9.32.

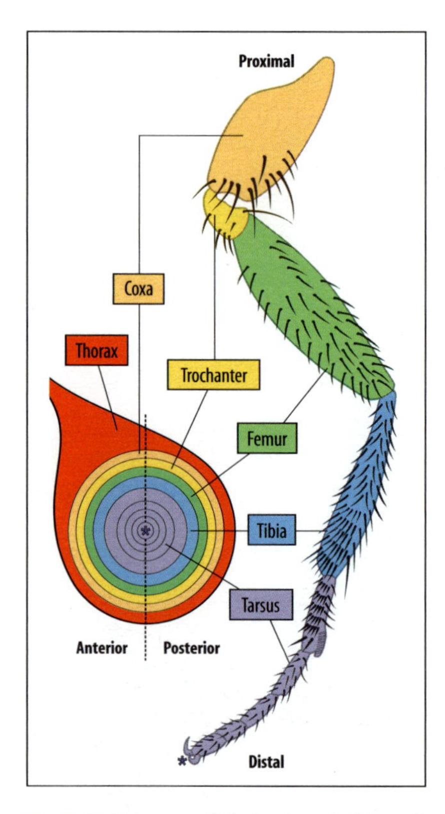

Fig. 9.32 **Fate map of the leg imaginal disc of *Drosophila*.** The disc is a roughly circular epithelial sheet, which becomes transformed into a tubular leg at metamorphosis. The center of the disc becomes the distal tip of the leg and the circumference gives rise to the base of the leg—this defines the proximo-distal axis. The tarsus is divided into five tarsal segments. The presumptive regions of the adult leg, such as the tibia, are thus arranged as a series of circles with the future tip at the center. A compartment boundary divides the disc into anterior and posterior regions.

After Bryant, P.J.: 1993.

protein induces a signaling region at the antero-posterior compartment border. In the dorsal region of the leg disc, the expression of *decapentaplegic* is induced, as it is in the wing. In the ventral region, however, Hedgehog induces expression of *wingless* instead of *decapentaplegic* along the border, and Wingless protein acts as the positional signal (Fig. 9.33). The patterning of the proximo-distal axis is far better understood in the leg disc than in the wing. It also involves interactions between Wingless and Decapentaplegic. The distal end of the proximo-distal axis is the point at which Wingless and Decapentaplegic expression meet, and is marked by expression of the homeodomain transcription factor Distal-less.

Patterning along the proximo-distal axis takes place sequentially. Distal-less is expressed in the future distal region, while another homeodomain transcription factor, Homothorax, is expressed in the peripheral region surrounding Distal-less, which is the future proximal region (Fig. 9.34). Their actions lead to the expression of the gene *dachshund*, which also encodes a transcription factor, in a ring between *Distal-less* and *homothorax*, which eventually leads to some overlap of the expression of *dachshund* with *Distal-less* and *homothorax*. Each of these genes is required for the formation of particular leg regions, but the expression domains do not correspond precisely with leg segments. Expression of *dachshund*, for example, corresponds to femur, tibia, and proximal tarsus. There is no evidence for a proximo-distal compartment boundary; cells expressing *homothorax* and *Distal-less* are, however, prevented from mixing at the interface between their two territories. There are further complex interactions that appear to involve a gradient of EGF receptor activity from distal to proximal. At the third instar larval stage, the EGF receptor ligand Vein and the signaling pathway component Rhomboid are expressed at the central point of the disc. Genes coding for transcription factors Bric-a-brac and Bar are expressed at different levels in the tarsal segments and may determine their identity. The level of activity of these genes may be determined by the gradient of EGF receptor activity. Notch signaling is required for joint formation. Delta and Serrate, ligands for Notch, are expressed as a ring in each leg segment and their activation of Notch results in specification of the joint-forming cells.

9.18 Butterfly wing markings are organized by additional positional fields

The variety of color markings on butterfly wings is remarkable: more than 17,000 species can be distinguished. Many of these patterns are variations on a basic 'ground plan' consisting of bands and concentric eyespots (Fig. 9.35). The wings are

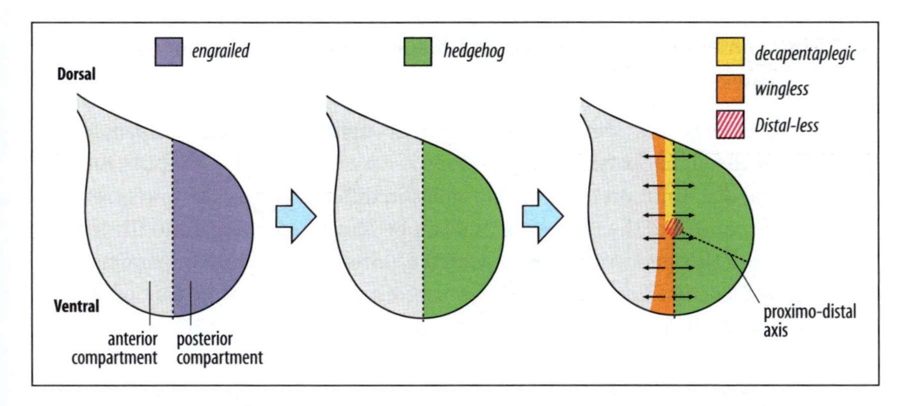

Fig. 9.33 Establishment of signaling centers in the antero-posterior compartment of the leg disc, and the specification of the distal tip. The gene *engrailed* is expressed in the posterior compartment and induces expression of *hedgehog*. Where Hedgehog protein meets and signals to anterior compartment cells, the gene *decapentaplegic* is expressed in dorsal regions and the gene *wingless* is expressed in ventral regions. Both of these genes encode proteins that are secreted. Expression of the gene *Distal-less*, which specifies the proximo-distal axis, is activated where the Wingless and Decapentaplegic proteins meet.

covered with overlapping cuticular scales, which are colored by pigment synthesized and deposited by the epidermal cells. How are these patterns specified? Butterfly wings develop from imaginal discs in the caterpillar in a similar way to *Drosophila* wings. Surgical manipulation has shown that the eyespot is specified at a late stage in the development of the wing disc, and that the pattern is dependent on a signal emanating from the center of the spot. A number of the genes that control wing development in *Drosophila*, such as *apterous*, are expressed in the butterfly in a spatial and temporal pattern similar to that in the fly. Thus, as in *Drosophila*, the shape and structure of the butterfly wing is patterned by a field of positional information. The patterning of the pigmentation, however, involves the establishment of additional fields of positional information.

The expression pattern of *Distal-less* in butterfly wing discs suggests that the mechanism used to delineate the color pattern on the butterfly wing is similar to the mechanism that specifies positional information along the proximo-distal axis of the insect leg. In the butterfly wing, *Distal-less* is expressed in the center of the eyespot, whereas in the *Drosophila* leg disc it is expressed in the central region corresponding to the future tip of the limb. Thus, it is possible that eyespot development and distal leg patterning involve similar mechanisms, although in the wing Notch expression precedes that of Distal-less. The eyespot may be thought of as a proximo-distal pattern superimposed on the two-dimensional wing surface. The center of the eyespot may represent the distal-most positional value, with the surrounding rings representing progressively more proximal positions, as in the leg. The site of the eyespot may be specified with reference to the primary wing pattern (that is, the anterior, posterior, dorsal, and ventral compartments); a secondary coordinate system centered on the eyespot would then be established, with expression of the *Distal-less* gene defining the central focus.

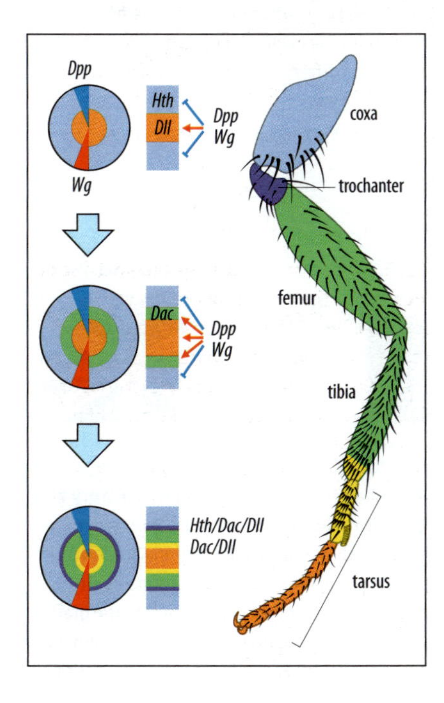

Fig. 9.34 Regional subdivision of the *Drosophila* leg along the proximo-distal axis. The pattern of gene expression in the leg is shown on the right. The stages of gene expression are shown in the two columns on the left, and are viewed as if looking down on the disc. Because of the way in which the leg extends from the disc, the center corresponds to the future tip of the leg, and the more proximal regions to successive rings around it. *decapentaplegic* (*dpp*) and *wingless* (*wg*) are initially expressed in a graded manner along the antero-posterior compartment boundary and together induce *Distal-less* (*Dll*) in the center and repress *homothorax* (*hth*), which is expressed in the outer region. They then induce *dachshund* (*dac*) in a ring between *Dll* and *hth*. Further signaling leads to these domains overlapping. While *hth* expression corresponds to proximal regions and *Dll* to distal regions, there is no simple relation between the other genes and the leg segments.

After Milan, M., Cohen, S.M.: 2000.

Fig. 9.35 Butterfly wing pattern. Ventral view of a female African butterfly *Bicyclus anynana*, showing wing color pattern and prominent eyespots. Scale bar = 5 mm.

Photograph courtesy of V. French and P. Brakefield.

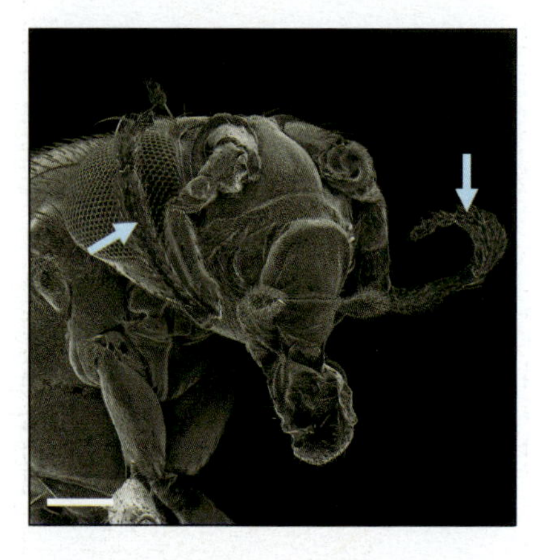

Fig. 9.36 Scanning electron micrograph of *Drosophila* carrying the *Antennapedia* mutation. Flies with this mutation have the antennae converted into legs (arrows). Scale bar = 0.1 mm.

Photograph by D. Scharfe, from Science Photo Library.

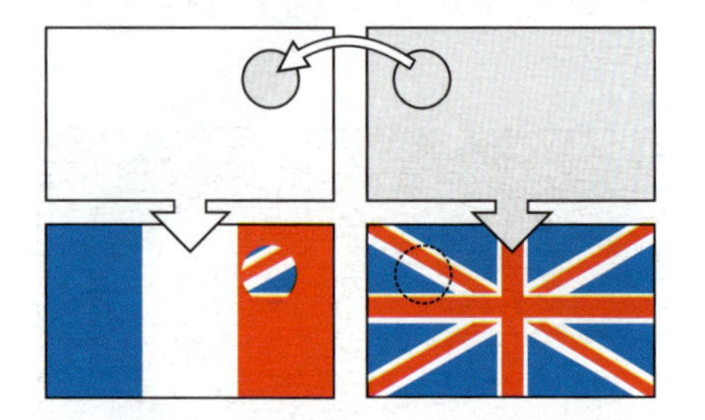

Fig. 9.37 Cells interpret their position according to their developmental history and genetic make-up. For example, if there are two flags which use the same positional information to produce a different pattern, then a graft from one to the other would result in the graft developing according to its intrinsic pattern, but with the pattern appropriate to its new position. This is what happens in imaginal discs.

9.19 The segmental identity of imaginal discs is determined by the homeotic selector genes

Patterning of legs and wings involves similar signals, such as Decapentaplegic protein, yet the actual pattern that develops is very different. This implies that the wing and leg discs interpret positional signals in different ways. This interpretation is under the control of the Hox genes and can be illustrated in regard to the leg and antenna. If the Hox gene *Antennapedia*, which is normally expressed in parasegments 4 and 5 (see Section 2.26) and specifies the discs for the second pair of legs, is expressed in the head region, the antennae develop as legs (Fig. 9.36). No homeotic genes are normally expressed in the antennal disc. Using the technique of mitotic recombination (see Box 2D, p. 68) it is possible to generate a clone of *Antennapedia*-expressing cells in a normal antennal disc. These cells develop as leg cells, but exactly which type of leg cell depends on their position along the proximo-distal axis; if, for example, they are at the tip, they form a claw. It is as if the positional values of the cells in the antenna and leg are similar, and the difference between the two structures lies in the interpretation of these values, which is governed by the expression or non-expression of the *Antennapedia* gene. This can be illustrated with respect to the French flag and the Union Jack (Fig. 9.37)—cells develop according to both their position and their developmental history, which determines the genes that they are expressing at any given time. This principle applies also to wing and haltere imaginal discs. Thus, we find a similarity in developmental strategy between insects and vertebrates; both use the same positional information in appendages like legs and wings, and interpret it differently. Still to be resolved is the question of how the expression of a single transcription factor like Antennapedia can transform an antenna into a leg. This requires an understanding of its downstream targets. So far, there is evidence that Antennapedia acts as a repressor of antennal identity in the leg by, for example, preventing the coexpression of *homothorax* and *Distal-less* in the femur region, which occurs in the antenna but not in the leg disc, where these two genes are expressed in adjacent and non-overlapping domains (see Fig. 9.34). In evolutionary terms, the antennal state may be the ground state for limb development.

The character of a disc and how positional information is interpreted is thus determined by the pattern of Hox gene expression (or the absence of Hox genes in the case of the antennal disc). Insect legs develop only on the three thoracic segments and not on the abdominal segments and, in *Drosophila*, wings develop only on the second thoracic segment. These adult structures are segment-specific because the particular type of imaginal disc that gives rise to them is only formed by certain parasegments. There are no appendages on abdominal segments in *Drosophila* because the genes required for formation of leg and wing discs are suppressed in the abdomen. The type of disc formed in a particular thoracic segment is typically specified by the action of one of the Hox genes expressed in the segment. For example, expression of the genes *Antennapedia* and *Ultrabithorax* specifies the second and third pair of legs, respectively.

The leg imaginal discs arise from small clusters of ectodermal cells in parasegments 3–6, which contribute to the future thoracic segments of the embryo (Fig. 9.38). In the embryo, each disc initially contains around 25 cells and is formed during growth of the blastoderm. The discs arise at the parasegment boundaries, with anterior and posterior compartments of adjacent parasegments contributing to each disc. In the future second thoracic segment, the leg disc splits off a second disc early in its development, which becomes the wing disc. Similarly, an additional disc, which develops into the haltere, a balancing organ, is formed in the future third thoracic segment.

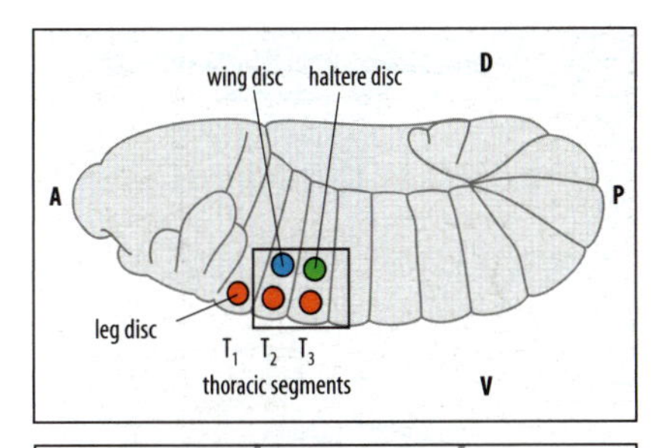

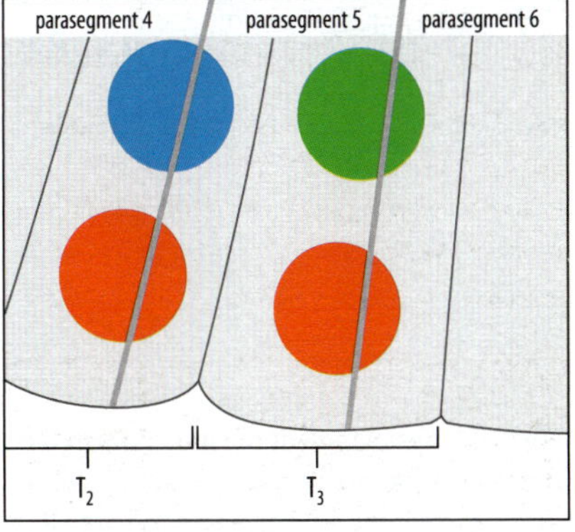

Fig. 9.38 The position in the late *Drosophila* embryo of the imaginal discs that give rise to the adult thoracic appendages. The imaginal discs for the legs, wings, and halteres lie across the parasegment boundaries in the thoracic segments as shown for the wing and haltere discs in the lower panel.

Because imaginal discs are formed across parasegment boundaries, mutations in Hox genes in *Drosophila* adults can cause compartment-specific homeotic transformations of halteres into wings. In normal *Drosophila* adults, the wing is on the second thoracic segment and the haltere on the third thoracic segment; they arise from imaginal discs originating at the boundaries of parasegments 4/5 and 5/6, respectively (see Fig. 9.38). In the normal embryo, the *Ultrabithorax* gene, one of the genes of the bithorax complex (see Box 4A, p. 156), is expressed in parasegments 5 and 6 and is involved in specifying their identity. The *bithorax* mutation (*bx*), which causes the *Ultrabithorax* gene to be misexpressed, can transform the anterior compartment of the third thoracic segment, and thus of the haltere, into the corresponding anterior compartment of the second segment: the anterior half of the haltere thus becomes a wing (see Fig. 2.45). The *postbithorax* mutation (*pbx*), which affects a regulatory region of the *Ultrabithorax* gene, transforms the posterior compartment of the haltere into a wing (Fig. 9.39). If both mutations are present in the same fly, the effect is additive and the result is a fly that has four wings but cannot fly. Another mutation, *Haltere mimic*, causes a homeotic transformation in the opposite direction: the wing is transformed into a haltere.

As with the antenna and leg, it is possible to generate a mosaic haltere with a small clone of cells containing an *Ultrabithorax* mutation (such as *bithorax*) in the disc; the cells in the clone make wing structures, which correspond exactly to those that would form in a similar position in a wing. It is as if the positional values in haltere and wing discs are identical and all that has been altered in the mutant is how this positional information is interpreted (see Fig. 9.33). In fact, other imaginal discs seem to have similar positional fields.

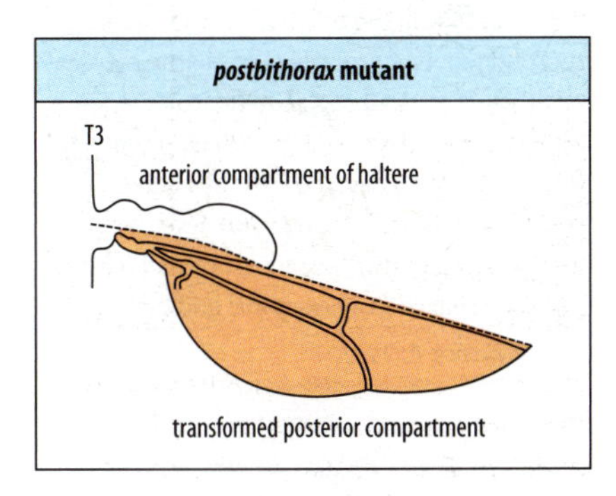

Fig. 9.39 Effects of the *postbithorax* mutation in *Drosophila*. The mutation *postbithorax* acts on the posterior compartment of a haltere, converting it into the posterior half of a wing.

9.20 Patterning of the *Drosophila* eye involves cell–cell interactions

The *Drosophila* compound eye is composed of about 800 identical photoreceptor organs called **ommatidia** (singular **ommatidium**) arranged in a hexagonal array of crystalline regularity (Fig. 9.40). Each ommatidium is made up of eight photoreceptor neurons (R1–R8), together with four overlying cone cells (which secrete the lens) and eight additional pigment cells. The genetic analysis of ommatidium development has provided one of the best model systems for studying the patterning of a small group of cells. An important early finding from lineage analysis was that the pattern of each ommatidium is specified by cell–cell interactions and is not based on cell lineages.

The eye develops from the single-layered epithelial sheet of the eye imaginal disc, located at the anterior end of the larva. Specification and patterning of the cells of the ommatidia begins in the middle of the third larval instar. Patterning starts at the posterior of the eye disc and progresses anteriorly, taking about 2 days, during which time the disc grows eight times larger. One of the earliest events in

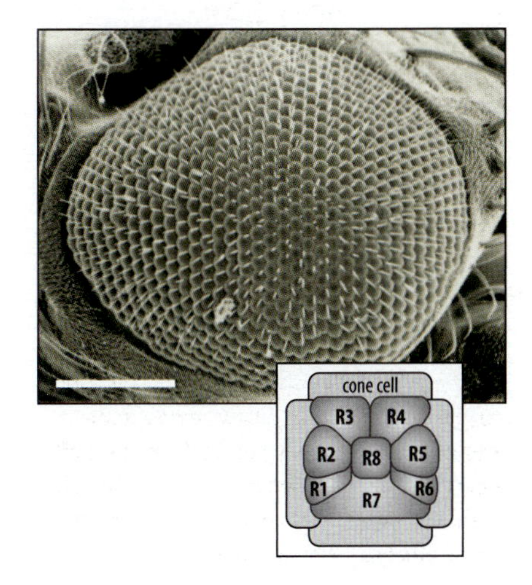

Fig. 9.40 Scanning electron micrograph of the compound eye of *Drosophila*. Each unit is an ommatidium. At the third instar, the ommatidial cluster (inset) is made up of eight photoreceptor neurons (R1–R8) and four cone cells. Scale bar = 50 μm.

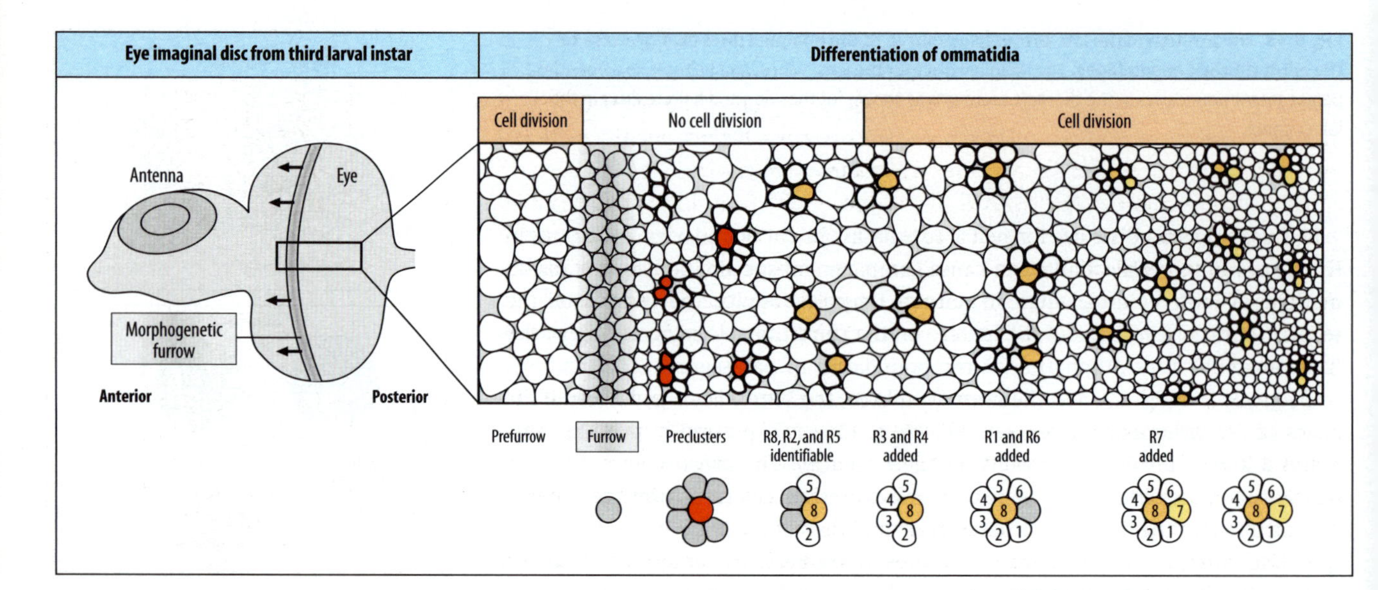

Fig. 9.41 Development of ommatidia in the *Drosophila* compound eye. The compound eye develops from the eye imaginal disc, which is part of a larger disc that also gives rise to an antenna. During the third larval instar, the morphogenetic furrow develops in the eye disc and moves across it in a posterior to anterior direction. Ommatidia develop behind the furrow, the photoreceptor neurons being specified in the order shown, R8 developing first and R7 last. The individual ommatidia are regularly spaced in a hexagonal grid pattern.

After Lawrence, P.: 1992.

eye differentiation is the formation of a groove, the **morphogenetic furrow**, which sweeps across the disc from posterior to anterior in response to a wave of signals that initiate development of ommatidia from the eye disc cells. As the furrow moves across the epithelium from posterior to anterior, clusters of cells that will give rise to the ommatidia appear behind it, spaced in a hexagonal array (Fig. 9.41). The furrow moves slowly across the disc, at a rate of 2 hours per row of ommatidial clusters. As the morphogenetic furrow moves forward, the cells behind it start to differentiate to form regularly spaced ommatidia. The ommatidia are arranged in rows, with each row half an ommatidium out of register with the previous one. This gives the characteristic hexagonal packing arrangement (see Fig. 9.41). The first cells to differentiate are the R8 photoreceptor neurons. These appear as regularly spaced cells in each row, separated from each other by about eight cells. This separation sets the spacing pattern for the ommatidia.

The passage of the morphogenetic furrow is essential for differentiation of the ommatidia, as mutations that block its progress also block the differentiation of new rows of ommatidia, resulting in a fly with abnormally small eyes. Although there are no anterior and posterior compartments in the eye discs, the cells just behind the furrow can be regarded as resembling posterior cells, similar to the situation in the wing-disc posterior compartment (see Section 9.15). They secrete Hedgehog protein, which triggers the expression of Decapentaplegic, which in turn causes the cells to become competent to form neural tissue. The gene Wingless also plays a part in eye disc patterning. It is expressed at the lateral edges of the eye disc, and prevents the furrow starting in these regions. We thus see that, even though imaginal discs give rise to very diverse structures, the key signals involved in patterning the leg, wing, and eye are similar, although they have different roles in each case.

Once specified, each R8 cell initiates a cascade of signals that recruits a surrounding cluster of 20 cells to form an ommatidium (see Fig. 9.41). R2 and R5 differentiate on either side of it, to form two functionally identical neurons. R3 and R4, which are a slightly different type of photoreceptor, form next. All these cells become arranged in a semi-circle with R8 at the center. R1 and R6 differentiate next and almost complete the circle, which is finally closed by the differentiation of R7 adjacent to R8. The clusters rotate 90°, so that R7 comes to be closest to the equator of the disc and R3 furthest away; rotation is in the opposite direction in the dorsal and ventral halves. This polarization of the ommatidium is yet another example of planar cell polarity and involves the activation of the Frizzled protein in R3/R4. The dorsal and ventral regions of the eye have distinct polarity, with the ommatidia in each half of the eye oriented towards the equator.

The determination of the equator of the eye results from the specification of the dorsal region of the eye disc by members of the *Iroquois* gene complex, which is followed by specification of the equator itself, involving the actions of Notch, its ligands Serrate and Delta and the secreted protein Fringe in a similar way to the specification of the dorso-ventral compartment boundary in the wing (see Section 9.15).

The regular spacing of the ommatidia within the eye involves a lateral inhibition mechanism that spaces the R8 cells. All cells in the eye disc initially have the capacity to differentiate as R8 cells, and as the morphogenetic furrow passes they start to do so. But some inevitably gain a lead and are thus able to inhibit the differentiation of another R8 cell over a range of around three cell diameters. Cells that will give rise to R8 express the gene *atonal*. Inhibitors of *atonal* that space the R8 cells are the secreted Scabrous protein and Notch.

In the eye, cell fate is specified and determined cell by cell, not in groups of cells. Two proteins crucial in the patterning of an individual ommatidium are the EGF receptor and its ligand Spitz. One model for patterning the ommatidium is based on both EGF receptor activation and the age of the cells (Fig. 9.42). Spitz is produced by the three earliest-specified and most centrally located cells—R8, R2, and R5. It activates the EGF receptor on neighboring cells, and this recruits R3, R4, R1, R6, and R7 to a photoreceptor fate. The responding cells also secrete the protein Argos. This diffuses away and inhibits more distant cells from being activated by Spitz, so that no more cells in the prospective ommatidial cluster develop as photoreceptors. The actual character of each photoreceptor may be determined by the age of the cell, with cells passing through a series of 'states', each representing a potential fate. Other signals are also involved, however. The difference between R3 and R4, for example, involves Notch signaling; the cell with the high level of Notch activity is inhibited from becoming R3, and becomes R4. After the photoreceptors have differentiated, the four lens-producing cone cells develop, and finally the surrounding ring of accessory cells.

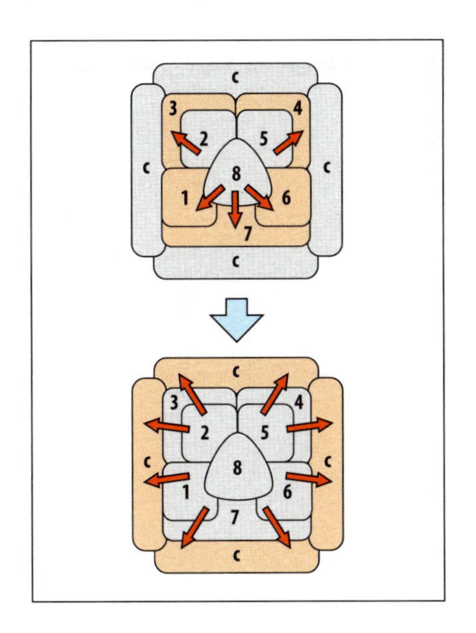

Fig. 9.42 Sequential recruitment of photoreceptors and cone cells during ommatidial development. The three earliest-specified and most centrally located photoreceptor cells—R8, R2, and R5—produce the protein Spitz. It activates the *Drosophila* EGF receptor DER on neighboring cells, which recruits R3, R4, R1, R6, and R7 to a photoreceptor fate (top). These then also produce Spitz, which interacts with DER on the prospective cone cells (c) to recruit them to the ommatidium. Once cells are determined they secrete the protein Argos. This diffuses away and inhibits more distant cells from being activated by Spitz, so that no more cells in the prospective ommatidial cluster develop as photoreceptors.

Adapted from Freeman, 1997.

The specification of R7 as a photoreceptor cell is one of the best understood cases of the specification of cell fate on an individual cell basis. It requires expression of the gene *sevenless* in the prospective R7 and *bride-of-sevenless* (*boss*) in R8. When either gene is inactivated, the phenotype is the same: R7 does not develop and an extra cone cell is formed. The Sevenless protein is a transmembrane receptor tyrosine kinase and Boss is its ligand. Sevenless protein is produced not only by R7 but also by other cells in the ommatidium, including lens cells. Thus, expression of Sevenless is a necessary, but not sufficient, condition for R7 specification. Using genetic mosaics it can be shown that for R7 to develop, only R8 need express Boss protein and that this is the signal by which R8 induces R7. Boss is also an integral membrane protein; it is present on the apical surface of the R8 cell, where it makes contact with R7. A second signal for R7 is provided by the R1/R6 pair, which must activate Notch in the R7 cell.

9.21 Activation of the gene *eyeless* can initiate eye development

The *eyeless* gene is involved in initiating eye development throughout the animal kingdom, even though the resulting eyes are very different in structure. In *Drosophila*, mutations in *eyeless* result in the reduction or complete absence of the fly's compound eye. This gene is expressed in the region of the eye disc anterior to the morphogenetic furrow. Induction of expression of the *eyeless* gene in other imaginal discs results in the development of ectopic eyes, and these have been induced in this way in wings, legs, antennae, and halteres. The fine structure of these ectopic eyes is remarkably normal, and distinct ommatidia are present. Therefore, *eyeless* is clearly a key gene for eye development and its expression switches on the requisite developmental pathway. It is estimated that some 2000 genes are eventually activated as a result of *eyeless* activity, and all of them are required for eye morphogenesis. The action of *eyeless* may be similar to that of the Hox genes, as it appears to change the interpretation of positional information in discs in which it is ectopically expressed.

The *eyeless* gene is an example of the remarkable conservation of developmental genes throughout the animal kingdom. Its homolog in vertebrates is called *Pax6*. Like *Drosophila* without *eyeless* function, mice with defective *Pax6* function have smaller eyes than normal, or no eyes at all. People heterozygous for mutations in the human *Pax6* gene have a variety of eye malformations collectively known as aniridia, because of the partial or complete absence of the iris; the patients also have cognitive defects. Ectopic expression of *Pax6* in the fly can substitute for *eyeless*, resulting in the development of an ectopic eye. In *Xenopus*, *Pax6* can induce ectopic eyes when its mRNA is injected into a blastomere at the 16-cell stage.

Summary

The legs and wings of *Drosophila* develop from epithelial sheets—imaginal discs—that are set aside in the embryo. Hox genes acting in the parasegments specify which sort of appendage will form and can direct the interpretation of positional information. The leg and wing imaginal discs are divided at an early stage into anterior and posterior compartments. The boundary between the compartments is a pattern-organizing center and a source of signals that pattern the disc. In the wing disc, expression of *decapentaplegic* is activated by the Hedgehog protein at the antero-posterior compartment boundary and the Decapentaplegic and Hedgehog proteins act as antero-posterior patterning signals. The wing disc also has dorsal and ventral compartments, with Wingless probably acting as the patterning signal along the dorso-ventral boundary. In the leg disc, the antero-posterior compartment boundary is established in a very similar way, except that Hedgehog activates *wingless* instead of *decapentaplegic* in the ventral regions of the leg and the Wingless protein acts as the patterning signal in these regions. There is no evidence of dorsal and ventral compartments in the leg disc. The proximo-distal axis of the leg is specified by the interaction between Decapentaplegic and Wingless proteins, which activates genes such as *dachshund* and *homothorax* in concentric domains corresponding to regions along the leg. The colorful eyespots on butterfly wings may be patterned by a mechanism similar to that used to organize the proximo-distal axis of the insect leg. The *Drosophila* compound eye contains around 800 individual ommatidia, arranged in a regular hexagonal pattern, and develops from an imaginal disc. The regular spacing of the ommatidia is achieved by lateral inhibition of the photoreceptor neuron R8. The patterning of the eight photoreceptor neurons in each ommatidium is due to local cell–cell interactions, in which the photoreceptor cells are specified and differentiate in a strict order, initiated by R8. *eyeless* is a key gene in eye development and it can induce ectopic eyes in other imaginal discs. It is related to *Pax6* in vertebrates.

Summary: pattern formation in leg and wing imaginal discs of *Drosophila*

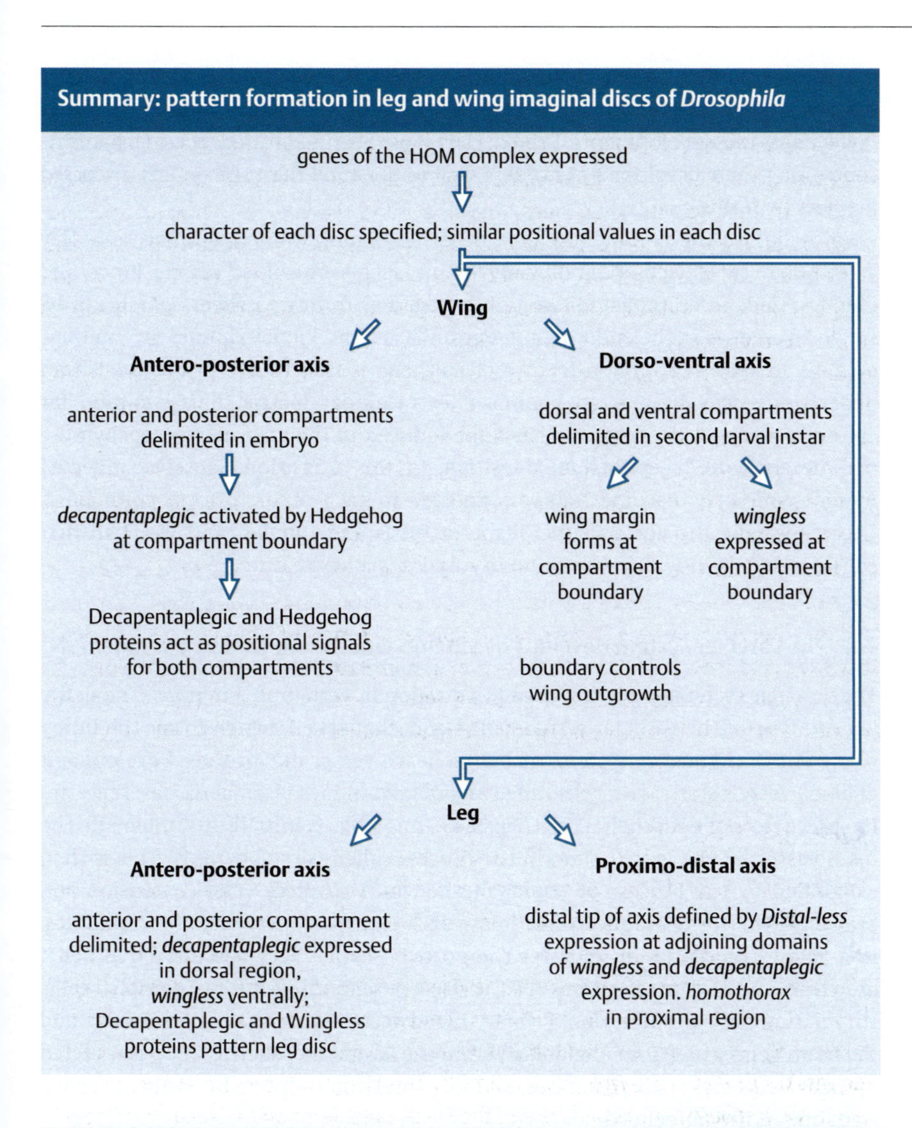

Summary: development of the ommatidia of the *Drosophila* eye

eyeless gene expression required for eye development. Morphogenetic furrow moves across eye disc, associated with hedgehog and decapentaplegic signals

the eight photoreceptors of the future ommatidia begin to develop behind the furrow, R8 developing first and R7 last. Ommatidia spaced by lateral inhibition

photoreceptors specified by DER activation, other signals, and time; R7 specification depends on a signal from R8

Internal organs: blood vessels, lungs, kidneys, heart, and teeth

All the structures discussed so far are external, and this has made them relatively easy to study. We now consider some, mainly vertebrate, internal organs. Their development illustrates principles underlying developmental phenomena we have not yet encountered, such as branching morphogenesis and the development of loosely organized mesenchyme into epithelium. Although numerous genes involved in the development of the organs have been identified, the mechanisms controlling their development are less well understood than the organs discussed previously in this chapter.

Epithelia are the commonest type of tissue organization in animals and play a key role in the development of many organs, such as the blood vessels, lungs, and kidneys. Cells in epithelia adhere tightly together to form a sheet, which can be single layered, as in the endothelium of capillaries and kidney tubules, or multilayered, as in skin. A common feature of epithelia is that they are separated from underlying tissue by a basal lamina of extracellular matrix. It is common for epithelia involved in organogenesis to be induced in the adjacent mesenchyme—the **mesenchyme-to-epithelium transition**. In this transition, loosely connected mesenchyme cells become tightly connected to each other to form epithelium. These epithelia can form in the shape of tubes and tubules, and we shall first consider organs in which formation of tubules is a key feature.

9.22 The vascular system develops by vasculogenesis followed by angiogenesis

The vascular system is the first organ to develop in vertebrate embryos. The defining cell type of the vascular system is the endothelial cell, which forms the lining of the entire circulatory system, including heart, veins, and arteries. Development of blood vessels starts with cells called **angioblasts** in mesodermal tissues; these are the precursors of endothelial cells (Fig. 9.43). Angioblasts initially assemble into the main vessels of the vasculature, in the process called **vasculogenesis**. This is then elaborated by the process of **angiogenesis**, which involves vessel extension and branching to form venules, arterioles, and extensive networks of capillaries. Angioblast differentiation requires the growth factor VEGF (vascular endothelial growth factor) and its receptors. VEGF is also a potent mitogen for endothelial cells, stimulating their proliferation. Primary blood vessels such as the dorsal aorta and the main veins arise from angioblasts from the lateral mesoderm. VEGF is secreted by axial structures such as somites, and its expression is driven by Sonic hedgehog signaling in the notochord.

New blood capillaries are formed by sprouting from preexisting blood vessels (see Fig. 9.43). The capillary grows by degradation of the extracellular matrix and proliferation of cells at the tip of the sprout. Cells at the tip extend filopodia-like processes that guide and extend the sprout. During their development, blood vessels navigate along specific paths towards their targets, with the filopodia at the leading edge responding to both attractant and repellant cues on other cells and in the extracellular matrix. Sprouting is prevented by chemorepellant cues in the extracellular environment such as those provided by proteins of the netrin and semaphorin families, which act on the receptors Unc5 and plexins, respectively, at the tips of growing blood vessels to block filopodial activity. Netrins and semaphorins are also involved in guidance of axons, as we shall see in Chapter 11, and there is a striking similarity in the signals and mechanisms that guide developing blood vessels and neurons. Blood-vessel morphogenesis also requires modulation of adhesive interactions between endothelial cells and the extracellular matrix and between the endothelial cells themselves. Changes in integrin-mediated cell adhesion are particularly important in this context.

VEGF stimulates endothelial cells to degrade the surrounding basement membrane matrix and to migrate and proliferate. In the skin of mice, embryonic

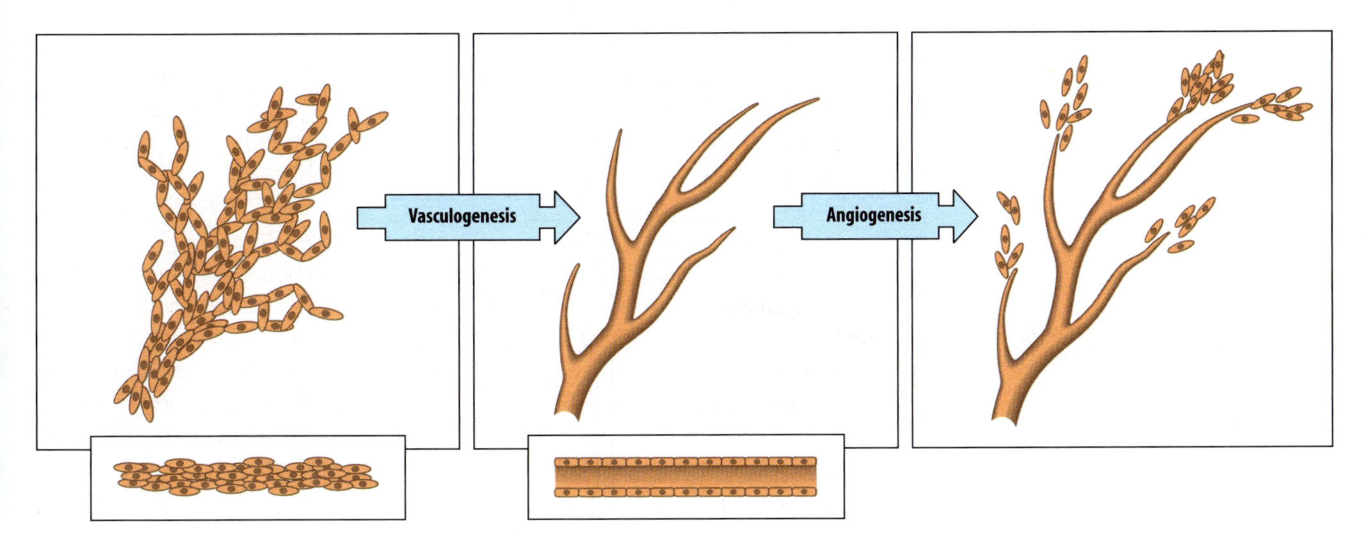

Fig. 9.43 Vasculogenesis and angiogenesis. Angioblasts assemble into simple tubes during vasculogenesis. These are then elongated during angiogenesis, which involves vessel extension, branching, and growth.

nerves form a template that directs the growth of arteries. The nerves secrete VEGF, which may both attract blood vessels and specify them as arteries. Expression of the *Vegf* gene is induced by lack of oxygen or hypoxia, and thus an active organ using up oxygen promotes its own vascularization. Many solid tumors produce VEGF and other growth factors that stimulate angiogenesis, and isolated endothelial cells in culture can be stimulated with tumor-conditioned medium to form capillary-like networks. The monoclonal antibody bevacizumab (Avastin), an anti-angiogenic agent that targets VEGF, has been approved for clinical use in combination chemotherapy for colorectal cancer. The Tie family of receptors and their ligands, the angiopoietins, are also required for angiogenesis. If endothelial cells are cultured on beads and then embedded in a fibrin gel, treatment with angiopoietin-1 will induce the formation of capillary sprouts from the cells on the beads.

The first tubular structures are formed by endothelial cells and the vessels are then covered by pericytes and smooth muscle cells. Arteries and veins are defined by the direction of blood flow as well as by structural and functional differences. There is evidence from lineage tracing that angioblasts are specified as arterial or venous before they form blood vessels; their identities are still undetermined, however, and they can switch identity. Arterial and venous capillaries join each other in capillary beds, which form the interchange sites between the arterial and venous blood systems. Ephrin B2 is expressed in arterial blood vessels while its putative receptor EphB4 is expressed in venous vessels, and interaction between them may be required for capillary bed formation. Lymphatic vessels are also developmentally part of the vascular system and originate by budding from veins. The expression of the homeobox gene Prox1 marks an early stage of commitment to the lymphatic lineage.

Angiogenesis is not just a property of the embryo. It can occur throughout life and, properly regulated, is the means of repairing damaged blood vessels. Excessive or abnormal angiogenesis is, however, the hallmark of many human diseases. These include cancer, obesity, psoriasis, atherosclerosis, and arthritis. Human disorders relating to the nervous system that involve abnormalities in the vasculature include Alzheimer's disease, where blockage of the vasculature is a major cause of disease symptoms, and motor neuron disease.

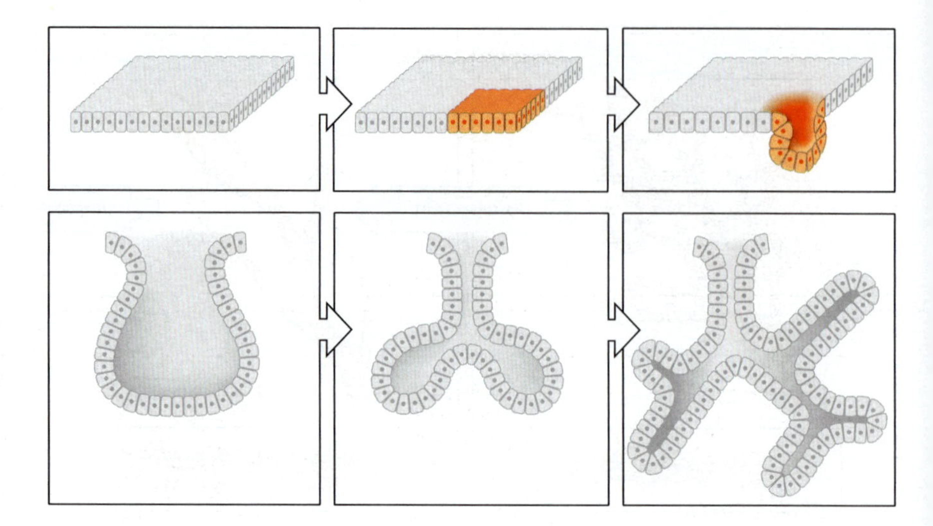

Fig. 9.44 The formation of tubular structures. Top row: a small section of an epithelial sheet becomes deformed and forms a pouch. Bottom row: the process is repeated to form successive branches. As illustrated here, branching in the formation of the mammalian lung is accompanied by cell proliferation.

9.23 The tracheae of *Drosophila* and the lungs of vertebrates branch using similar mechanisms

The mammalian lung and the **tracheal system** of *Drosophila* are both involved in the delivery of oxygen from the air to the tissues, and both are elaborately branched. These essentially tubular structures both begin as an epithelial monolayer from which a simple sac or tube grows out and then branches repeatedly. Tube formation and branching occurs in a sequence of steps. First the initial site of tube formation is specified, often as a placode—a thickened region of epithelium. This region bulges outwards or **evaginates** to form a tube, and this is then followed by outgrowth. Evagination and outgrowth are repeated at the end of the tube to initiate formation of two branches, and then repeated at the end of these tubes and so on (Fig. 9.44). Evagination is likely to be caused by the constriction of both the apical and the basal faces of the cells.

Air enters the tracheal system of the *Drosophila* larva through the openings in the body wall called spiracles, and oxygen is delivered to the tissues by some 10,000 or so fine tubules which develop from 20 ectodermal sacs or placodes, of about 80 cells each, during embryogenesis. In *Drosophila*, the positions in the epithelial sheet at which tracheae develop are controlled by specific genes. Each sac gives rise to six primary tracheal branches. These branches form when the epithelium deforms locally in particular directions in response to extracellular cues. Cells at the tips of the branch develop filopodia and move towards the source of the cue. Remarkably, tracheal extension and branching pattern is due to a combination of cell intercalation, change in cell shape, and directed cell movement, and no cell proliferation is involved. Secondary branches form from the ends of the primary branches and will also branch further.

More than 50 genes required for tracheal branching have been identified in *Drosophila*. *Trachealess* is turned on in the sac epithelium itself, while six domains of *branchless* expression are found in the cells surrounding the sac and determine where the primary branches will form. *Branchless* codes for an FGF-like protein that signals primary branch formation and interacts with an FGF receptor, Breathless, on the tracheal cells (Fig. 9.45). As the primary branch moves towards the source of Branchless, the gene is turned off. Later, *branchless*-expressing cells again stimulate branch formation from the tips of the primary branches. Branchless induces expression of a new set of genes, which include a gene called *sprouty*. The Sprouty

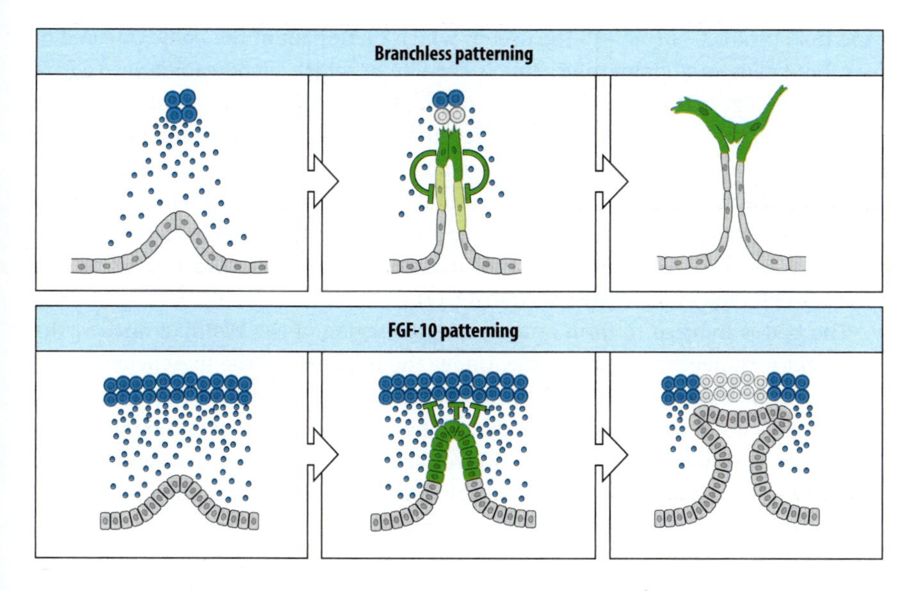

Fig. 9.45 Models of branching patterns in the tracheae of *Drosophila* and the mouse lung. Top panels: in *Drosophila*, localized secretion of Branchless protein (blue) attracts the tracheal cells to move towards it, creating a primary branch (left panel). At the top of the branch, the gene *sprouty* is induced, and the Sprouty protein (green) inhibits branching further away from the source of Branchless (center panel). Branchless also induces secondary branches (right panel). Bottom panel: in the mouse, FGF-10 (blue) induces outgrowth of a branch of a lung tracheole towards the FGF-10 source. At the tip of the branch, secretion of an inhibitor (green) of *FGF-10* expression (possibly Sonic hedgehog) splits the region of attraction, forming two secondary branches.

After Metzger, R.J. and Krasnow, M.A.: 1999.

protein blocks Branchless signaling to sites further away from the source, and so limits branching to the tip of the tubule.

The trachea and lungs of the mouse develop from the foregut when it is still a simple tube of endoderm. The trachea is unbranched, and is formed by the division of the foregut by a longitudinal septum into two tubes—the trachea and the esophagus. The development of mouse lungs starts with left and right epithelial lung buds, which derive from the endoderm of the foregut. These move into the surrounding mesenchyme, and give rise to millions of subdivisions by successive branching due to evagination and outgrowth. In contrast to the *Drosophila* tracheal system, outgrowth is by cell proliferation (see Fig. 9.45). Early lung development has a reproducible and asymmetric pattern, even in culture. The left primary lung bud sprouts several secondary buds along its lateral face in a precise spatial and temporal sequence. The pattern of branching of the right bud is quite different; for example, the first branch arises from the dorsal surface. Transplantation studies show that the mesenchyme plays a key role in controlling branching and bud formation. Mesenchyme from distal lung buds can induce ectopic buds when grafted next to tracheal endoderm. By contrast, tracheal mesenchyme inhibits outgrowth of lung buds.

FGF-10 is the most likely candidate for regulating lung budding since the buds grow towards areas in which FGF-10 is being expressed. Sonic hedgehog is proposed to act as a negative feedback signal that shuts off *FGF-10* gene expression in the mesenchyme as the branch approaches it (see Fig. 9.45). Two new buds then sprout towards each of the two lateral domains that continue to secrete FGF-10. Embryos lacking FGF-10 develop neither lung buds nor limbs. However, a bead loaded with FGF-10 cannot induce a bud from tracheal endoderm although it will stimulate outgrowth when placed near distal lung buds. A second regulator of lung development is BMP-4, which is expressed in the same region as FGF-10 and controls cell proliferation and differentiation of cells along the branches.

9.24 The development of kidney tubules involves reciprocal induction by the ureteric bud and surrounding mesenchyme

The mammalian kidney is composed mainly of a set of convoluted tubules that receive fluid and salts from the blood, balance its composition, and then convey the final product—urine—to the ureter, which carries out of the body. One end of a tubule ends in a glomerulus; this is the site at which blood capillaries release fluid and salts that are then taken up by the tubule. The other end of the tubule connects to the ureter. The kidney is formed from two regions of mesoderm, the ureteric bud and the metanephric mesenchyme (another name for the mammalian kidney is the metanephros). The ureteric bud branches off the Wolffian duct, which develops from a primitive form of kidney, the pronephros, that functions as the adult kidney in amphibians and fish. In mammals, the Wolffian ducts develop into the vas deferens in males (see Section 11.11).

The bud is induced to form at the posterior region of the Wolffian duct by the adjacent metanephric mesenchyme. Under the influence of the mesenchyme, the ureteric bud branches to form the proximal portions of the urine-carrying tubules, which are known as the collecting ducts. The ureteric bud induces the mesenchymal cells to condense around it and form epithelial structures that develop into renal tubules or nephrons. Each tubule elongates to form a glomerulus at one end, where filtration of the blood will occur, while the other end fuses to the ureteric collecting duct connecting to the ureter (Fig. 9.46). The intermediate portion of the tubule is convoluted and its epithelial cells become specialized for the resorption of ions from urine. The mammalian kidney is a good developmental system to study as it can develop in organ culture; explanted metanephric mesenchyme will form many glomeruli and tubules over a period of about 6 days, even though a blood supply is absent.

The development of the ureteric bud and the mesenchyme depends on mutual inductive interactions, neither being able to develop in the absence of the other. The mesenchyme induces the ureteric bud to grow out by secreting glial-cell-derived neurotrophic factor (GDNF). The site of expression of GDNF is critical for positioning ureteric bud development and restricting it to a single site, and positioning of GDNF expression is under the control of the proteins Slit2 and its receptor Robo2. As we shall see in Chapter 10, these proteins are also involved in the guidance of axons in nervous-system development. GDNF in the mesenchyme is the main controller of ureteric bud branching, and its receptor, Ret, is present in the bud. Knock-out of the gene for either GDNF or Ret results in the absence of ureteric bud outgrowth. The cellular basis for branching is similar to that just described for lung development.

As the bud branches it induces the overlying mesenchyme at the tips to condense and to form epithelium that then forms the nephrons. Among the growth factors secreted by the bud and involved in this induction are FGF-2, leukemia inhibitory factor (LIF), and TGF-β2, A key protein that must be expressed in the mesenchyme for tubule formation to occur is the zinc-finger transcription factor WT1. Loss-of-function mutations in the *WT1* gene are associated with Wilm's tumor, a cancer of the kidney in children, and the gene is therefore known as the Wilm's tumor suppressor gene.

A very early response to induction is the expression of the transcription factor Pax2 in the mesenchyme, which is essential for subsequent tubule development, probably through its stimulation of the expression of GDNF. The expression of *WT1* then increases, the loose mass of mesenchyme cells condenses, and the cells start to express a matrix glycoprotein, syndecan, on their surface. This condensing stage is followed by the formation of distinct cellular aggregates, in which the mesenchymal cells become polarized and acquire an epithelial character and secrete the signal protein Wnt-4. Each aggregate then forms an S-shaped tube, which elongates and differentiates to form the functional unit of a renal tubule and glomerulus. During this transition, the composition of the extracellular matrix secreted by the

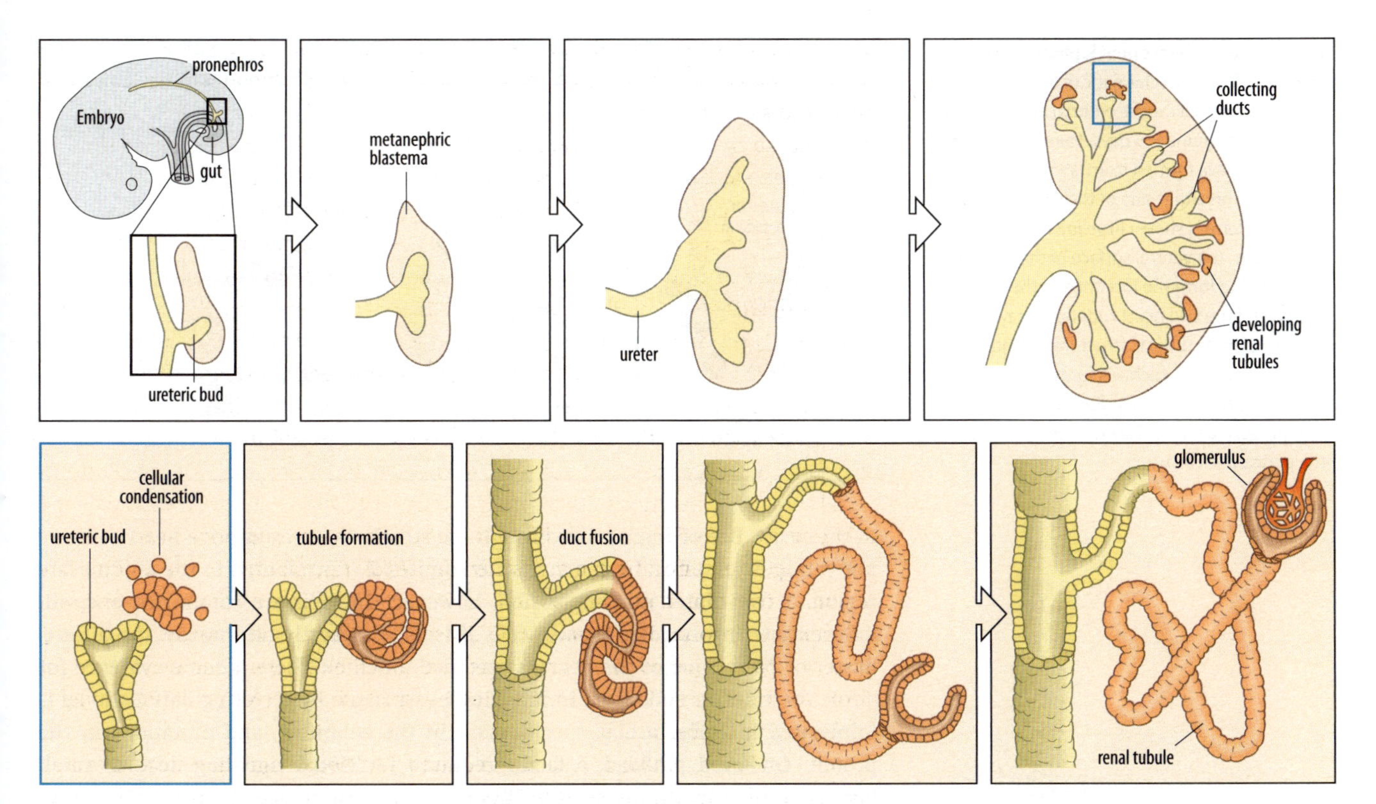

Fig. 9.46 Kidney development from mesenchyme. The kidney develops from a loose mass of mesenchyme, the metanephric blastema, which is induced to form tubules by the ureteric bud. The ureteric bud is itself induced by the mesenchyme to grow and branch to form the collecting ducts of the kidney that connect to the ureter. The mesenchyme cells form cellular condensations that become epithelial tubules, which open into the collecting system formed from the ureteric bud. Each tubule develops a glomerulus, through which waste products are filtered out of the blood.

cells changes: mesenchymal collagen I is replaced by basal lamina proteins, such as collagen IV and laminin, that are typically secreted by epithelial cells. The adhesion molecules (see Box 7A, p. 259) expressed by the cells also change; for example, the N-CAM expressed by the mesenchymal cells is replaced by E-cadherin in the epithelial cells. Integrins are also involved in the epithelial–mesenchyme interactions.

The number of nephrons per kidney varies widely, from around 230,000 to 1,800,000. Nephron number is determined by the number of bud branches, and how branching is terminated is not well understood—TGF-β may be involved. Many different proteins appear to be involved in determining nephron number.

9.25 The development of the vertebrate heart involves specification of a mesodermal tube that is patterned along its long axis

The heart is one of the first structures to form during organogenesis. It is of mesodermal origin and is first established as a single tube consisting of two epithelial layers—the inner **endocardium**, which is an endothelial sheet, and the outer **myocardium**, which is contractile. During development, this tube becomes divided longitudinally into two chambers, the atrial and ventricular chambers. A two-chambered heart is the basic adult form in fish, but in higher vertebrates, such as birds and mammals, looping and further partitioning give rise to the four-chambered heart. In humans about 1 in 100 live-born infants have some congenital heart malformation, while *in utero* heart malformation leading to death of the embryo is between 5 and 10%.

In the chick and mouse, the lateral plate mesoderm is the major source of heart mesoderm. In the chick, ingression of heart precursor cells through the primitive streak occurs throughout most of the period of streak formation. The heart precursors ingress in a roughly similar antero-posterior order to their eventual position

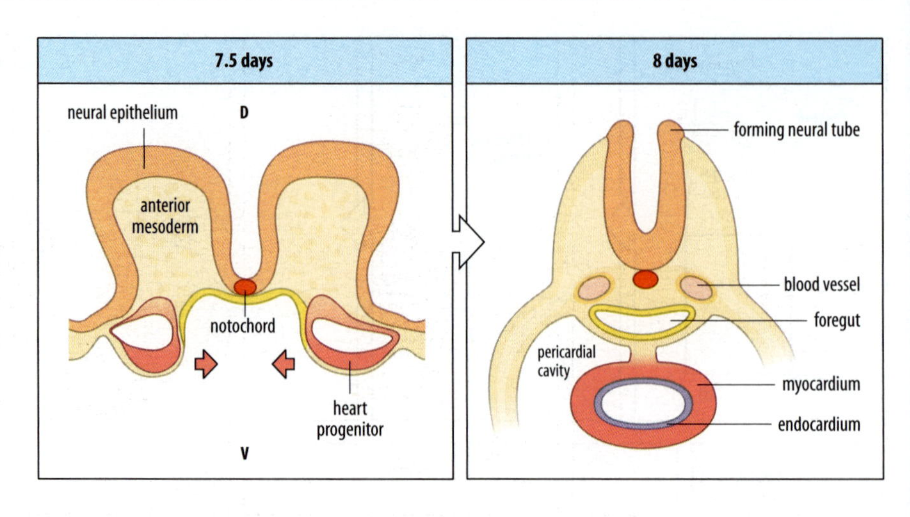

Fig. 9.47 Early heart development in the mouse. Left panel: schematic of a transverse section through a mouse embryo at the level of the heart at around 7.5 days. Paired heart rudiments have formed from lateral plate mesoderm on either side of the ventral midline. Gut endoderm is shown in yellow. Right panel: by 8.5 days, the ventral surface of the embryo has folded to form the foregut and the body cavity, and the two heart rudiments have fused to form a single tubular structure composed of the myocardium, which will form heart muscle, and the epithelial endocardium lining the internal surface. The heart is surrounded by the pericardial cavity.

in the heart. Before ingression through the streak, the presumptive heart cells are not yet determined and can give rise to somites if transplanted to the appropriate region. A few hours after ingression, however, the cells are committed and will differentiate into cardiac muscle cells if isolated. Shortly afterwards, the antero-posterior patterning of the heart is specified. In chick, there is some evidence for a role for anterior endoderm in inducing heart tissue by secreting BMP-2. Nodal is implicated in early cardiac specification in the zebrafish, and mutations in the protein One-eyed pinhead, a factor required for Nodal signaling, lead to small hearts, while mutations in the T-box transcription factor Spadetail lead to complete lack of a heart. Many factors, including Wnts and FGF, seem to be involved in cardiac induction. In *Xenopus*, heart mesoderm is specified by signals from the organizer region (see Fig. 3.54).

In all vertebrates, the precursor heart cells come to lie in two patches of lateral plate mesoderm on either side of the midline. They then undergo a complex morphogenetic process, move towards the midline and fuse to form the heart tube (Fig. 9.47). A number of mutations in the zebrafish disrupt this process, leading to two laterally positioned hearts—a condition known as cardia bifida. One of the mutated genes is called *miles apart*, and codes for a receptor that binds lysosphingolipids; sphingosine 1-phosphate is the likely ligand in this case. *Miles apart* is not expressed in the migrating heart cells themselves, but in cells on either side of the midline, and thus may be involved in directing the migration of the presumptive heart cells. Lineage studies of individual cells in the early heart region showed that they gave rise to two lineages that aggregate from a common precursor. One lineage gives rise to the primitive left ventricle and outflow tract, all other regions are colonized by both lineages. There is no relationship between lineage and antero-posterior patterning in the heart.

There is some evidence that the heart develops as a modular organ in which each anatomical region is controlled by a distinct transcriptional program. The regions are the segments along the heart tube—atria, left ventricle, right ventricle, and ventricular outflow tract in the mammalian heart (Fig. 9.48). Precursor cells of these regions appear to have separate lineages that develop according to their position along the antero-posterior axis and which have specific patterns of gene expression. The homeodomain transcription factor Nkx2.5, for example, is required for heart development and is one of the first markers of early heart cells; mutations in the *Nkx2.5* gene in mice and humans result in heart abnormalities. The expression of *Nkx2.5* in the developing heart is complex and somewhat modular. Seven different activating regions and three repressor regions have been

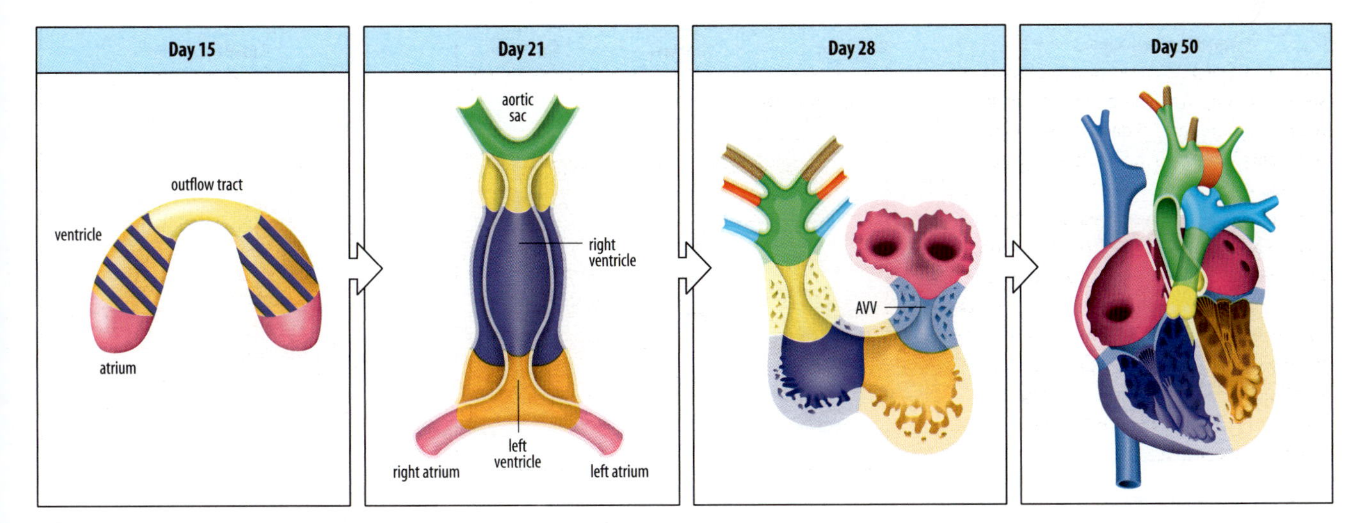

Fig. 9.48 Schematic of human heart development. By day 15 of human embryonic dvelopment, cardiogenic precursors have formed a crescent, as shown in the first panel, in which the main regions of the heart are already specified. The two arms of the crescent fuse along the midline to give a linear heart tube, which is patterned along the antero-posterior axis with the regions and chambers of the mature heart (second panel). After looping (third panel) these regions are disposed approximately in their eventual positions. Later development results in further patterning (fourth panel), and the formation of valves between, for example, the atria and ventricles. AVV, atrioventricular valve region.

After Srivastava, D., Olson, E.N.: 2000.

identified in the control region of *Nkx2.5*. Activating regions include those that lead to its expression in the entire heart tube, or just in the right ventricle and outflow tract. There is evidence that retinoic acid acts as a signal for antero-posterior positional activity in heart development, its source being the mesoderm posterior and lateral to the node.

The later development of the heart involves asymmetric looping of the heart tube. This is not well understood but is related to the left–right asymmetry of the antero-posterior axis (see Section 3.10). There is evidence that the extracellular matrix molecule plectin is expressed earlier on the left side due to the expression of Pitx2 on that side. Then there is the formation of the separate chambers—four in the case of the mammalian heart. In addition, cells of neural crest origin contribute to the outflow tracts; they are essential, for example, to formation of the pulmonary artery and aorta from the single embryonic outflow tract, the truncus arteriosus. Defects in these regions account for 30% of congenital heart defects in humans, some of which are the result of developmental perturbations in neural crest cell specification and migration.

There are remarkable similarities—although we should no longer be surprised—between the genes involved in heart development in *Drosophila* and in vertebrates. The homeobox gene *tinman* is required for heart formation in *Drosophila*, and is a homolog of vertebrate *Nkx2.5*. When vertebrate *tinman*-like genes are expressed in *Drosophila* they can substitute for *Drosophila tinman* and rescue some of the abnormalities caused by lack of normal *tinman* functions. In *Drosophila*, *decapentaplegic* maintains *tinman* expression in the dorsal mesoderm; in vertebrates, BMPs have been shown to induce *Nkx2.5* and are also able to induce cardiac cell differentiation. Nkx2.5 or Tinman act together with GATA transcription factors to turn on cardiac-specific gene expression, although GATA transcription factors are not expressed exclusively in cardiac precursors.

9.26 A homeobox gene code specifies tooth identity

Teeth develop from a well defined series of interactions between the oral epithelium and the mesenchyme of the jaws, which is of neural crest origin. Teeth develop from areas of thickened oral epithelium called the tooth placodes. The epithelium invaginates and the surrounding mesenchyme condenses, forming a tooth primordium or tooth germ (Fig. 9.49). In each primordium, a specialized

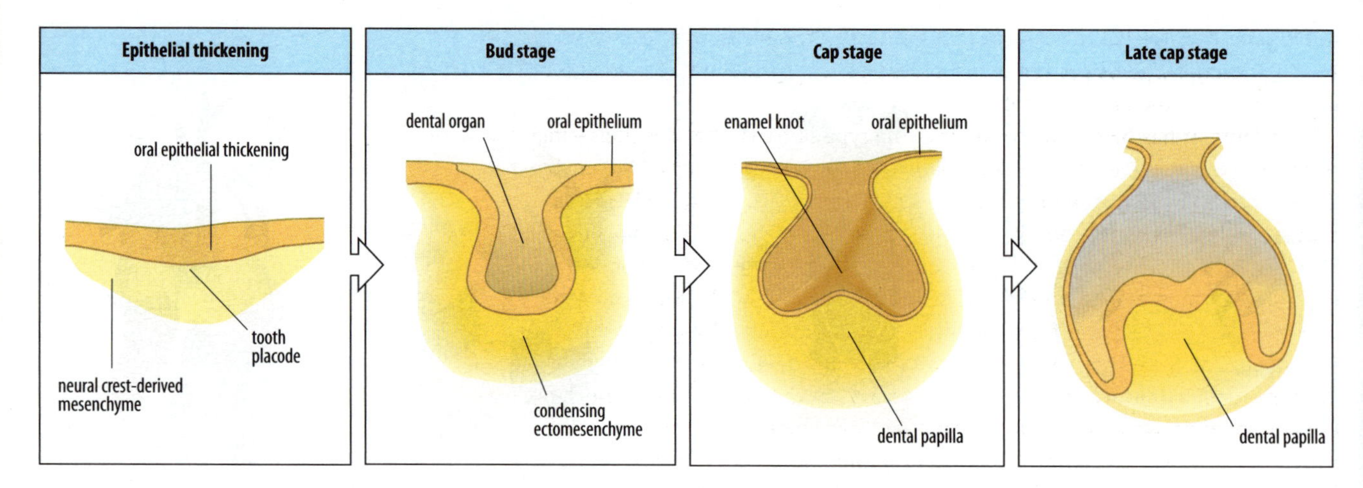

Fig. 9.49 Schematic diagram of the stages of tooth development.

group of cells—the enamel knot—acts as a signaling center for later development, such as cusp formation. The mesenchyme forms the dental papilla, which gives rise to the pulp and dentine, while the enamel is secreted by the epithelial cells. Tooth primordia that give rise to different types of teeth, such as incisors and molars, are specified at precise positions, and it is this positioning that we will discuss here.

Early tooth development is controlled by reciprocal interactions, by way of secreted signals, between the oral epithelium and the underlying mesenchyme. The sites of tooth formation are determined initially by signals produced by the epithelium. BMP-4 is expressed in the epithelium at all sites of tooth formation, and induces, or maintains, expression of the homeodomain transcription factor Msx1 in the underlying mesenchyme. FGF signaling is also involved in a similar pattern. Sonic hedgehog is expressed in the epithelium at sites of tooth formation throughout tooth development, whereas Wnt-7b is expressed elsewhere in the oral epithelium except at sites of tooth formation. The reciprocal expression of these latter two signaling molecules establishes the boundaries between the oral epithelium and the dental epithelium that invaginates to form the tooth. As tooth-bud

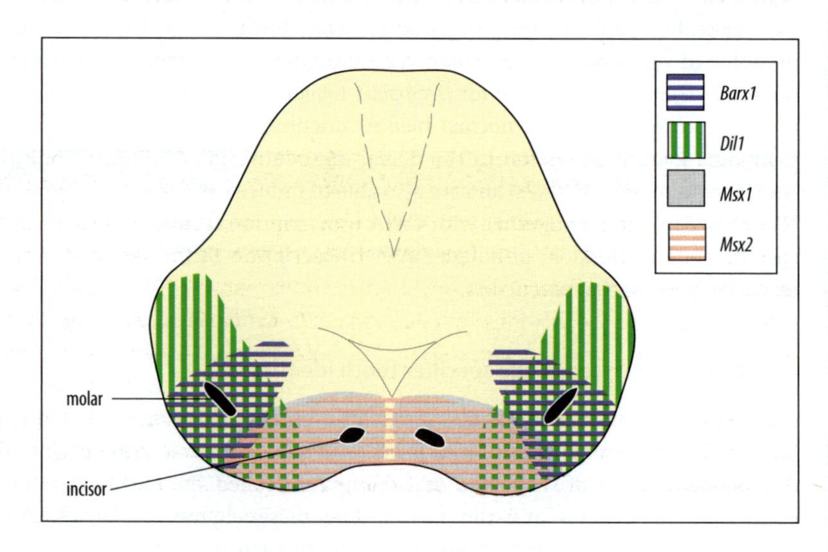

Fig. 9.50 The expression domains of four homeobox genes in the mesoderm of the mandible before initiation of tooth primordia. The positions at which the incisors and molars will develop are indicated in black.

After Ferguson, C.A., et al.: 2000.

development proceeds, the direction of signaling changes so that signals secreted by the mesenchyme direct the development of the epithelium, for example the development of the enamel knot.

In rodents, it has been established that the type of tooth that develops—incisor or molar—is determined by the expression of homeobox genes in the mesenchyme. These homeobox genes, which include *Barx1*, *Dlx1*, *Dlx2*, *Msx1*, and *Msx2* among others, provide a spatial code in the facial mesoderm (Fig. 9.50), analogous to the Hox gene code along the antero-posterior body axis Thus, *Dlx2* and *Barx1* are expressed in mesenchymal cells that will form molars, while *Msx1* and *Msx2* are expressed in cells that will form incisors. In mouse mutants lacking both *Dlx2* and *Dlx1* (a similar gene that can substitute for *Dlx2* in development), molars do not develop, but ectopic cartilage structures form at the molar sites. The homeodomain transcription factor Islet1, which we shall meet again in neural development, is expressed exclusively in developing incisors. It has a positive regulatory interaction with BMP-4, which induces Msx1 in the incisor region.

Summary

Development of internal organs such as the vascular system, lungs, kidneys, heart, and teeth are intimately involved with epithelial patterning and morphogenesis, and a mesenchyme-to-epithelial transition is common. The vascular system develops from endothelial cells that initially form the precursors to the large blood vessels of the vascular system. These then grow and branch in the process of angiogenesis. Branching of the tracheae of *Drosophila* and the vertebrate lung involve similar principles and are controlled by the surrounding mesenchyme. The development of the mammalian kidney involves both mesenchyme-to-epithelium transitions and reciprocal induction. The ureteric bud is induced by the prospective kidney mesenchyme to grow and branch, while the ureteric bud induces the cells in the mesenchyme to condense and form tubules. The heart initially forms as a linear tube made up of an inner endocardium and an outer myocardium. Its later development may be modular and is controlled by genes similar to those that regulate heart formation in *Drosophila*. Teeth form by an invagination of the oral epithelium and the underlying mesenchyme and their positioning and development is controlled by reciprocal signaling between epithelium and mesenchyme. A homeobox gene code in the mesenchyme specifies tooth identity.

SUMMARY TO CHAPTER 9

The development of organs involves similar processes, and in some cases the same genes, as those used earlier in development. The pattern of the vertebrate limb is specified along three axes, at right angles to each other. Signaling molecules provide positional information, and interpretation of this information involves Hox genes. The development of *Drosophila* wings and legs from imaginal discs depends on positional signals generated at compartment boundaries, and some of these signal proteins are similar to those used in vertebrate limb patterning. The development of internal vertebrate organs such as the vascular system, lungs, and teeth involves epithelial morphogenesis and, in some cases, branching of tubular structures.

SECTION FURTHER READING

9.1 The vertebrate limb develops from a limb bud

Tickle, C.: **Patterning systems—from one end of the limb to the other.** *Dev. Cell* 2003, **4:** 449–458.

9.2 Patterning of the limb involves positional information

Cohn, M.J., Tickle, C.: **Limbs: a model for pattern formation within the vertebrate body plan.** *Trends Genet.* 1996, **12:** 253–257.

9.3 Genes expressed in the lateral plate mesoderm are involved in specifying the position and type of limb

Charité, J., De Graaff, W., Shen, S., Deschamps, J.: **Ectopic expression of *Hoxb-8* causes duplication of the ZPA in the forelimb and homeotic transformation of axial structures.** *Cell* 1994, **78:** 589–601.

Kawakami, Y., Capdevila, J., Buscher, D., Itoh, T., Rodriguez Esteban, C., Izpisua Belmonte, J.C.: **WNT signals control FGF-dependent limb initiation and AER induction in the chick embryo.** *Cell* 2001, **104:** 891–900.

Minguillon, C., Buono, J.D., Logan, M.P.: ***Tbx5* and *Tbx4* are not sufficient to determine limb-specific morphologies but have common roles in initiating limb outgrowth.** *Dev. Cell* 2004, **8:** 75–84.

Tickle, C., Münsterberg, A.: **Vertebrate limb development—the early stages in chick and mouse.** *Curr. Opin. Genet. Dev.* 2001, **11:** 476–481.

9.4 The apical ectodermal ridge is required for limb outgrowth

Dudley, A.T., Tabin, C.J.: **Constructive antagonism in limb development.** *Curr. Opin. Genet. Dev.* 2000, **10:** 387–392.

Lallemand, Y., Nicola, M.A., Ramos, A., Bach, A., Clement, C.S., Robert, B.: **Analysis of Msx1:Msx2 double mutants reveals multiple roles for Msx genes in limb development.** *Development* 2005, **132:** 3003–3014.

Niswander, L., Tickle, C., Vogel, A., Booth, I., Martin, G.R.: **FGF-4 replaces the apical ectodermal ridge and directs outgrowth and patterning of the limb.** *Cell* 1993, **75:** 579–587.

Pizette, S., Niswander, L.: **BMPs negatively regulate structure and function of the limb apical ectodermal ridge.** *Development* 1999, **126:** 883–894.

9.5 The polarizing region specifies position along the limb's antero-posterior axis

Ahn, S., Joyner, A.L.: ***In vivo* analysis of quiescent adult neural stem cells responding to Sonic hedgehog.** *Nature* 2005, **437:** 894–897.

Dahn, R.D., Fallon, J.F.: **Interdigital regulation of digit identity and homeotic transformation by modulated BMP signaling.** *Science* 2000, **289:** 438–441.

Drossopoulou, G., Lewis, K.E., Sanz-Ezquerro, J.J., Nikbakht, N., McMahon, A.P., Hofmann, C., Tickle, C.: **A model for anteroposterior patterning of the vertebrate limb based on sequential long- and short-range Shh signalling and Bmp signalling.** *Development* 2000, **127:** 1337–1348.

Harfe, B.D., Scherz, P.J., Nissim, S., Tian, H., McMahon, A.P., Tabin, C.J.: **Evidence for an expansion-based temporal Shh gradient in specifying vertebrate digit identities.** *Cell* 2004, **118:** 517–528.

Hill, R.E., Heaney, S.J., Lettice, L.A.: **Sonic hedgehog: restricted expression and limb dysmorphologies.** *J. Anat.* 2003, **202:** 13–20.

Riddle, R.D., Johnson, R.L., Laufer, E., Tabin, C.: ***Sonic hedgehog* mediates polarizing activity of the ZPA.** *Cell* 1993, **75:** 1401–1416.

Suzuki, T., Takeuchi, J., Koshiba-Takeuchi, K., Ogura, T.: ***Tbx* genes specify posterior digit identity through Shh and BMP signaling.** *Dev. Cell* 2004, **6:** 43–53.

Tickle, C.: **Retinoic acid and chick limb development.** *Development (Suppl.)* 1991, **1:** 113–121.

Tickle, C.: **Making digit patterns in the vertebrate limb.** *Nat. Rev. Mol. Cell Biol.* 2006, **7:** 1–9.

Yashiro, K., Zhao, X., Uehara, M., Yamashita, K., Nishijima, M., Nishino, J., Saijoh, Y., Sakai, Y., Hamada, H.: **Regulation of retinoic acid distribution is required for proximodistal patterning and outgrowth of the developing mouse limb.** *Dev. Cell* 2004, **6:** 411–422.

Zuniga, A.: **Globalisation reaches gene regulation: the case for vertebrate limb development.** *Curr. Opin. Genet. Dev.* 2005, **15:** 403–409.

9.6 Position along the proximo-distal axis may be specified by a timing mechanism

Dudley, A.T., Ros, M.A., Tabin, C.J.: **A re-examination of proximo-distal patterning during vertebrate limb development.** *Nature* 2002, **418:** 539–544.

Wolpert, L., Tickle, C., Sampford, M.: **The effect of cell killing by X-irradiation on pattern formation in the chick limb.** *J. Embryol. Exp. Morph.* 1979, **50:** 175–193.

Wolpert, L.: **Limb patterning: reports of model's death exaggerated.** *Curr. Biol.* 2002, **12:** R628–R638.

9.7 The dorso-ventral axis is controlled by the ectoderm

Altabef, M., Clarke, J.D.W., Tickle, C.: **Dorso-ventral ectodermal compartments and origin of apical ectodermal ridge in developing chick limb.** *Development* 1997, **124:** 4547–4556.

Geduspan, J.S., MacCabe, J.A.: **The ectodermal control of mesodermal patterns of differentiation in the developing chick wing.** *Dev. Biol.* 1987, **124:** 398–408.

Riddle, R.D., Ensini, M., Nelson, C., Tsuchida, T., Jessell, T.M., Tabin, C.: **Induction of the LIM homeobox gene *Lmx1* by *Wnt-7a* establishes dorsoventral pattern in the vertebrate limb.** *Cell* 1995, **83:** 631–640.

9.8 Different interpretations of the same positional signals give different limbs

Krabbenhoft, K.M., Fallon, J.F.: **The formation of leg or wing specific structures by leg bud cells grafted to the wing bud is influenced by proximity to the apical ridge.** *Dev. Biol.* 1989, **131:** 373–382.

Logan, M.: **Finger or toe: the molecular basis of limb identity.** *Development* 2003, **130:** 6401–6410.

9.9 Homeobox genes may also provide positional values for limb patterning

Goff, D.J., Tabin, C.J.: **Analysis of *Hoxd-13* and *Hoxd-11* misexpression in chick limb buds reveals that Hox genes affect both bone condensation and growth.** *Development* 1997, **124:** 627–636.

Goodman, F.R.: **Limb malformations and the human HOX genes.** *Am. J. Med. Genet.* 2002, **112:** 256–265

Kmita, M., Fraudeau, N., Herault, Y., Duboule, D.: **Serial deletions and duplications suggest a mechanism for the collinearity of Hoxd genes in limbs.** *Nature* 2002, **420:** 145–150.

Kmita, M., Tarchini, B., Zàkàny, J., Logan, M., Tabin, C.J., Duboule, D.: **Early developmental arrest of mammalian limbs lacking *HoxA/HoxD* gene function.** *Nature* 2005, **435:** 1113–1116.

Knosp, W.M., Scott, V., Bachinger, H.P., Stadler, H.S.: **HOXA13 regulates the expression of bone morphogenetic proteins 2 and 7 to control distal limb morphogenesis.** *Development* 2004, **131:** 4581–4592.

Nelson, C.E., Morgan, B.A., Burke, A.C., Laufer, E., DiMambro, E., Muytaugh, L.C., Gonzales, E., Tessarollo, L., Parada, L.F., Tabin, C.: **Analysis of Hox gene expression in the chick limb bud.** *Development* 1996, **122:** 1449–1466.

Vogt, T.F., Duboule, D.: **Antagonists go out on a limb.** *Cell* 1999, **99:** 563–566.

Wellik, D.M., Capecchi, M.R.: **Hox10 and Hox11 genes are required to globally pattern the mammalian skeleton.** *Science* 2003, **301:** 363–367.

Zakany, J., Kmita, M., Duboule, D.: **A dual role for Hox genes in limb anterior-posterior asymmetry.** *Science* 2004, **304:** 1669–1672.

9.10 Self-organization may be involved in the development of the limb bud

Hardy, A., Richardson, M.K., Francis-West, P.N., Rodriguez, C., Izpisúa-Belmonte, J.C., Duprez, D., Wolpert, L.: **Gene expression, polarising activity and skeletal patterning in reaggregated hind limb mesenchyme.** *Development* 1995, **121:** 4329–4337.

Box 9B Reaction–diffusion mechanisms

Kondo, S., Asai, R.: **A reaction-diffusion wave on the skin of the marine angelfish *Pomocanthus*.** *Nature* 1995, **376:** 765–768.

Meinhardt, H., Gierer, A.: **Pattern formation by local self-activation and lateral inhibition.** *BioEssays* 2000, **22:** 753–760.

Murray, J.D.: **How the leopard gets its spots.** *Sci. Am.* 1988, **258:** 80–87.

9.11 Limb muscle is patterned by the connective tissue

Amthor, H., Christ, B., Patel, K.: **A molecular mechanism enabling continuous embryonic muscle growth—a balance between proliferation and differentiation.** *Development* 1999, **126:** 1041–1053.

Hashimoto, K., Yokouchi,Y., Yamamoto, M., Kuroiwa, A.: **Distinct signaling molecules control *Hoxa-11* and *Hoxa-13* expression in the muscle precursor and mesenchyme of the chick limb bud.** *Development* 1999, **126:** 2771–2783.

Knight, B., Laukaitis, C., Akhtar, N., Hotchin, N.A., Edlung, M., Horwitz, A.R.: **Visualizing muscle cell migration *in situ*.** *Curr. Biol.* 2000, **10:** 576–595.

Robson, L.G., Kara, T., Crawley, A., Tickle, C.: **Tissue and cellular patterning of the musculature in chick wings.** *Development* 1994, **120:** 1265–1276.

Schnorrer, F., Dickson, B.J.: **Muscle building; mechanisms of myotube guidance and attachment site selection.** *Dev. Cell* 2004, **7:** 9–20.

Schweiger, H., Johnson, R.L., Brand-Sabin, B.: **Characterization of migration behaviour of myogenic precursor cells in the limb bud with respect to Lmx1b expression.** *Anat. Embryol.* 2004, **208:** 7–18.

9.12 The initial development of cartilage, muscles, and tendons is autonomous

Kardon, G.: **Muscle and tendon morphogenesis in the avian hind limb.** *Development* 1998, **125:** 4019–4032.

Ros, M.A., Rivero, F.B., Hinchliffe, J.R., Hurle, J.M.: **Immunohistological and ultrastructural study of the developing tendons of the avian foot.** *Anat. Embryol.* 1995, **192:** 483–496.

9.13 Joint formation involves secreted signals and mechanical stimuli

Spitz, F., Duboule, D.: **The art of making a joint.** *Science* 2001, **291:** 1713–1714.

9.14 Separation of the digits is the result of programmed cell death

Garcia-Martinez, V., Macias, D., Gañan, Y., Garcia-Lobo, J.M., Francia, M.V., Fernandez-Teran, M.A., Hurle, J.M.: **Internucleosomal DNA fragmentation and programmed cell death (apoptosis) in the interdigital tissue of the embryonic chick leg bud.** *J. Cell. Sci.* 1993, **106:** 201–208.

Zou, H., Niswander, L.: **Requirement for BMP signaling in interdigital apoptosis and scale formation.** *Science* 1996, **272:** 738–741.

Zuzarte-Luis, V., Hurel, J.M.: **Programmed cell death in the embryonic vertebrate limb.** *Semin. Cell Dev. Biol.* 2005, **16:** 261–269.

9.15 Positional signals from the antero-posterior and dorso-ventral compartment boundaries pattern the wing imaginal disc

Baeg, G.H., Selva, E.M., Goodman, R.M., Dasgupta, R., Perrimon, N.: **The Wingless morphogen gradient is established by the cooperative action of Frizzled and heparan sulfate proteoglycan receptors.** *Dev. Biol.* 2004, **276:** 89–100.

Blair, S.S.: **Notch signaling: Fringe really is a glycosyltransferase.** *Curr. Biol.* 2000, **10:** R608–R612.

Crozatier, M., Glise, B., Vincent, A.: **Patterns in evolution: veins of the *Drosophila* wing.** *Trends Genet.* 2004, **20:** 498–505.

Entchev, E.V., Schwabedissen, A., González-Gaitán, M.: **Gradient formation of the TGF-β homolog Dpp.** *Cell* 2000, **103:** 981–991.

Fujise, M., Takeo, S., Kamimura, K., Matsuo, T., Aigaki, T., Izumi, S., Nakato, H.: **Dally regulates Dpp morphogen gradient formation in the *Drosophila* wing.** *Development* 2003, **130:** 1515–1522.

Han, C., Belenkaya, T.Y., Wang, B., Lin, X.: ***Drosophila* glypicans control the cell-to-cell movement of Hedgehog by a dynamin-independent process.** *Development* 2004, **131:** 601–611.

Kruse, K., Pantazis, P., Bollenbach, T., Julicher, F., González-Gáitan, M.: **Dpp gradient formation by dynamin-dependent endocytosis: receptor trafficking and the diffusion model.** *Development* 2004, **131:** 4843–4856.

Lawrence, P.: **Morphogens: how big is the picture?** *Nat. Cell Biol* 2001, **3:** E151–E154.

Lawrence, P.A., Struhl G.: **Morphogens, compartments, and pattern: lessons from *Drosophila*?** *Cell* 1996, **85:** 951–961.

Matakatsu, H., Blair, S.S.: **Interactions between Fat and Daschous and the regulation of planar cell polarity in the *Drosophila* wing.** *Development* 2004, **131:** 3785–3794.

Milán, M., Cohen, S.M.: **A re-evaluation of the contributions of Apterous and Notch to the dorsoventral lineage restriction boundary in the *Drosophila* wing.** *Development* 2003, **130:** 553–562.

Morata, G.: **How *Drosophila* appendages develop.** *Nature Rev. Mol. Cell Biol.* 2001, **2:** 89–97.

Muller, B., Hartmann, B., Pyrowolakis, G., Affolter, M., Basler, K.: **Conversion of an extracellular Dpp/BMP morphogen**

gradient into an inverse transcriptional gradient. *Cell* 2003, **113:** 221–233.

Rauskolb, C., Correia, T., Irvine, K.D.: **Fringe-dependent separation of dorsal and ventral cells in the *Drosophila* wing.** *Nature* 1999, **401:** 476–480.

Strutt, D.I.: **Asymmetric localization of Frizzled and the establishment of cell polarity in the *Drosophila* wing.** *Mol. Cell* 2001, **7:** 367–375.

Tabata, T.: **Genetics of morphogen gradients.** *Nature Rev. Genet.* 2001, **2:** 620–630.

Teleman, A.A., Strigini, M., Cohen, S.M.: **Shaping morphogen gradients.** *Cell* 2001, **105:** 559–562.

Zecca, M., Basler, K., Struhl, G.: **Sequential organizing activities of engrailed, hedgehog and decapentaplegic in the *Drosophila* wing.** *Development* 1995, **121:** 2265–2278.

9.16 *Drosophila* wing epidermal cells show planar cell polarity

Bastock, R., Strutt, H., Strutt, D.: **Strabismus is asymmetrically localised and binds to Prickle and Dishevelled during *Drosophila* planar polarity patterning.** *Development* 2003, **130:** 3007-3014.

Cho, E., Irvine, K.D.: **Action of fat, four-jointed, dachsous and dachs in distal-to-proximal wing signaling.** *Development* 2004, **131:** 4489–4500.

Ma, D., Yang, C-H., McNeill, H., Simon, M.A., Axelrod, J.D.: **Fidelity in planar cell polarity signaling.** *Nature* 2003, **421:** 543–547.

9.17 The leg disc is patterned in a similar manner to the wing disc, except for the proximo-distal axis

Emerald, B.S., Cohen, S.M.: **Spatial and temporal regulation of the homeotic selector gene *Antennapedia* is required for the establishment of leg identity in *Drosophila*.** *Dev. Biol.* 2004, **267:** 462–472.

Kojima, T.: **The mechanism of *Drosophila* leg development along the proximodistal axis.** *Dev. Growth Differ.* 2004, **46:** 115–129.

9.18 Butterfly wing markings are organized by additional positional fields

Brakefield, P.M., French, V.: **Butterfly wings: the evolution of development of colour patterns.** *BioEssays* 1999, **21:** 391–401.

French, V., Brakefield, P.M.: **Pattern formation: a focus on Notch in butterfly spots.** *Curr. Biol.* 2004, **14:** R663–R665.

9.19 The segmental identity of imaginal discs is determined by the homeotic selector genes

Carroll, S.B.: **Homeotic genes and the evolution of arthropods and chordates.** *Nature* 1995, **376:** 479–485.

Morata, G.: **How *Drosophila* appendages develop.** *Nature Rev. Mol. Cell Biol.* 2001, **2:** 89–97.

Si Dong, P.D., Chu, J., Panganiban, G.: **Coexpression of the homeobox genes *Distal-less* and *homothorax* determines *Drosophila* antennal identity.** *Development* 2000, **127:** 209–216.

9.20 Patterning of the *Drosophila* eye involves cell-cell interactions

Baonza, A., Casci, T., Freeman, M.: **A primary role for the epidermal growth factor receptor in ommatidial spacing in the *Drosophila* eye.** *Curr. Biol.* 2001, **11:** 396–404.

Bonini, N.M., Choi, K.W.: **Early decisions in *Drosophila* eye morphogenesis.** *Curr. Opin. Genet. Dev.* 1995, **5:** 507- 515.

Chou, W.H., Huber, A., Bentrop, J., Schulz, S., Schwab, K., Chadwell, L.V., Paulsen, R., Britt, S.G.: **Patterning of the R7 and R8 photoreceptor cells of *Drosophila*: evidence for induced and default cell-fate specification.** *Development* 1999, **126:** 607-616.

Frankfort, B.J., Mardon, G.: **R8 development in the *Drosophila* eye: a paradigm for neural selection and differentiation.** *Development* 2002, **129:** 1295–1306.

Freeman, M.: **Cell determination strategies in the *Drosophila* eye.** *Development* 1997, **124:** 261–270.

Strutt, H., Strutt, D.: **Polarity determination in the *Drosophila* eye.** *Curr. Opin. Genet. Dev.* 1999, **9:** 442–446.

Tomlinson, A., Struhl, G.: **Delta/Notch and Boss/Sevenless signals act combinatorially to specify the *Drosophila* R7 photoreceptor.** *Mol. Cell* 2001, **7:** 487–495.

9.21 Activation of the gene *eyeless* can initiate eye development

Halder, G., Callaerts, P., Gehring, W.J.: **Induction of ectopic eyes by targeted expression of the eyeless gene in *Drosophila*.** *Science* 1995, **267:** 1788–1792.

Gehring, W.J., Kazuho, I.: **Mastering eye morphogenesis and eye evolution.** *Trends Genet.* 1999, **15:** 371–377.

9.22 The vascular system develops by vasculogenesis followed by angiogenesis

Carmeliet, P.: **Angiogenesis in health and disease.** *Nat. Med.* 2003, **9:** 653–660.

Carmeliet, P., Tessier-Lavigne, M.: **Common mechanisms of nerve and blood vessel wiring.** *Nature* 2005, **436:** 193–200.

Coultas, L., Chawengsaksophak, K., Rossant, J.: **Endothelial cells and VEGF in vascular development.** *Nature* 2005, **438:** 937–945.

Harvey, N.L., Oliver, G: **Choose your fate: artery, vein or lymphatic vessel?** *Curr. Opin. Genet. Dev.* 2004, **14:** 499–505.

Hogan, B.L.M., Kolodziej, P.A.: **Molecular mechanisms of tubulogenesis.** *Nat. Rev. Genet.* 2002, **3:** 513–523.

Jain, R.K.: **Molecular recognition of vessel maturation.** *Nat. Med.* 2003, **9:** 685–692.

Lu, X., Le Noble, F., Yuan, L., Jiang, Q., De Lafarge, B., Sugiyama, D., Breant, C., Claes, F., De Smet, F., Thomas, J.L., Autiero, M., Carmeliet, P., Tessier-Lavigne, M., Eichmann, A.: **The netrin receptor UNC5B mediates guidance events controlling morphogenesis of the vascular system.** *Nature* 2004, **432:** 179–186.

Rafii, S., Lyden, D: **Therapeutic stem and progenitor cell transplantation for organ vascularization and regeneration.** *Nat. Med.* 2003, **9:** 702–712.

Reese, D.E., Hall, C.E., Mikawa, T.: **Negative regulation of midline vascular development by the notochord.** *Dev. Cell* 2004, **6:** 699–708.

Roman, B.L., Weinstein, B.M.: **Building the vertebrate vasculature: research is going swimmingly.** *BioEssays* 2000, **22:** 882–893.

Rossant, J., Hirashima, M.: **Vascular development and patterning: making the right choices.** *Curr. Opin. Genet. Dev.* 2003, **13:** 408–412.

Wang, H.U., Chen, Z-F., Anderson, D.J.: **Molecular distinction and angiogenic interaction between embryonic arteries and veins revealed by ephrin-B2 and its receptor Eph-B4.** *Cell* 1998, **93:** 741–753.

9.23 The tracheal system of *Drosophila* and the lungs of vertebrates branch using similar mechanisms

Affolter, M., Bellusci, S., Itoh, N., Shilo, B., Thiery, J.P., Werb, Z.: **Tube or not tube: remodeling epithelial tissues by branching morphogenesis.** *Dev. Cell* 2003, **4:** 11–18.

Hogan, B.L.M., Kolodziej, P.A.: **Molecular mechanisms of tubulogenesis.** *Nat. Rev. Genet.* 2002, **3:** 513–523.

Metzger, R.J., Krasnow, M.A.: **Genetic control of branching morphogenesis.** *Science* 1999, **284:** 1635–1639.

Ribeiro, C., Neumann, M., Affolter, M.: **Genetic control of cell intercalation during tracheal morphogenesis in *Drosophila*.** *Cell* 2004, **14:** 2197–2207.

9.24 The development of kidney tubules involves reciprocal induction by the ureteric bud and surrounding mesenchyme

Barasch, J., Yang, J., Ware, C.B., Taga, T., Yoshida, K., Erdjument-Bromage, H., Tempst, P., Parravicini, E., Malach, S., Aranoff, T., Oliver, J.A.: **Mesenchymal to epithelial conversion in rat metanephros is induced by LIF.** *Cell* 1999, **99:** 377–386.

Gao, X., Chen, X., Taglienti, M., Rumballe, B., Little, M.H., Kreidberg, J.A.: **Angioblast-mesenchyme induction of early kidney development is mediated by Wt1 and Vegfa.** *Development* 2005, **132:** 5437–5449.

Grieshammer, U., Le, Ma, Plump, A.S., Wang, F., Tessier-Lavigne, M., Martin, G.R.: **SLIT2-mediated ROBO2 signaling restricts kidney induction to a single site.** *Dev. Cell* 2004, **6:** 709–717.

Kuure, S., Vuolteenaho, R., Vainio, S.: **Kidney morphogenesis: cellular and molecular regulation.** *Mech. Dev.* 2000, **94:** 47–56.

Schedl, A., Hastie, N.D.: **Cross-talk in kidney development.** *Curr. Opin. Genet. Dev.* 2000, **10:** 543–549.

Shah, M.M., Sampogna, R.V., Sakurai, H., Bush, K.T., Nigam, S.K.: **Branching morphogenesis and kidney disease.** *Development* 2004, **131:** 1449–1462.

Shakya, R., Watanabe, T., Costantini, F.: **The role of GDNF/Ret signaling in ureteric bud cell fate and branching morphogenesis.** *Dev. Cell* 2005, **8:** 65–74.

Vainio, S., Muller, U.: **Inductive tissue interactions, cell signaling, and the control of kidney organogenesis.** *Cell* 1997, **90:** 975–978.

9.25 The development of the vertebrate heart involves specification of a mesodermal tube that is patterned along its long axis

Brand, T.: **Heart development: molecular insights into cardiac specification and early morphogenesis.** *Dev. Biol.* 2003, **288:** 1–19.

Carmeliet, P.: **Angiogenesis in health and disease.** *Nat. Med.* 2003, **9:** 653–660.

Driever, W.: **Bringing two hearts together.** *Nature* 2000, **406:** 141–142.

Harvey, R.P.: **Patterning the vertebrate heart.** *Nat. Rev. Genet.* 2002, **3:** 544–556.

Kelly, R.G., Buckingham, M.E.: **The anterior heart-forming field: voyage to the arterial pole of the heart.** *Trends Genet.* 2002, **18:** 210–216.

Linask, K.K., Yu, X., Chen, Y., Han, M.D.: **Directionality of heart looping: effects of Pitx2c misexpression on flectin asymmetry and midline structures.** *Dev. Biol.* 2002, **246:** 407–417.

Meilhac, S.M., Esner, M., Kelly, R.G., Nicolas, J.F., Buckingham, M.E.: **The clonal origin of myocardial cells in different regions of the embryonic mouse heart.** *Dev. Cell* 2004, **6:** 685–698.

Moorman, A.F., Christoffels, V.M.: **Cardiac chamber formation: development, genes and evolution.** *Physiol Rev.* 2003, **83:** 1223–1267.

Rosenthal, N., Xavier-Neto, J.: **From the bottom of the heart: anteroposterior decisions in cardiac muscle differentiation.** *Curr. Opin. Cell Biol.* 2000, **12:** 742–746.

Srivastava, D., Olson, E.N.: **A genetic blueprint for cardiac development.** *Nature* 2000, **407:** 221–226.

Stainier, D.Y.R.: **Zebrafish genetics and vertebrate heart formation.** *Nature Rev. Genet.* 2001, **2:** 39–48.

9.26 A homeobox gene code specifies tooth identity

Ferguson, C.A., Hardcastle, Z., Sharpe, P.T.: **Development and patterning of the dentition.** *Linn. Soc. Symp.* 2000, **20:** 188–201.

Mitsiadis, T.A., Angeli, I., James, C., Lendahl, U., Sharpe, P.T.: **Role of Islet1 in the patterning of murine dentition.** *Development* 2003, **130:** 4451–4460.

Sarkar, L., Cobourne, M., Naylor, S., Smalley, M., Dale, T., Sharpe, P.T.: **Wnt/Shh interactions regulate ectodermal boundary formation during mammalian tooth development.** *Proc. Natl Acad. Sci. USA* 2000, **97:** 4520–4524.

Tucker, A., Sharpe, P.: **The cutting-edge of mammalian development: how the embryo makes teeth.** *Nat. Rev. Genet.* 2004, **5:** 499–508.

Development of the nervous system

10

- Specification of cell identity in the nervous system
- Neuronal migration
- Synapse formation and refinement

The nervous system is the most complex of all the organ systems in the animal embryo. In mammals, for example, billions of nerve cells, or neurons, develop a highly organized pattern of connections, creating the neuronal network that makes up the functioning brain and the rest of the nervous system. There are many hundreds of different types of neurons, differing in identity and connectivity, even though they may look quite similar. Understanding these developmental processes is not simple, as they involve neuronal cell differentiation, morphogenesis, and migration. Neurons send out long processes from the cell body and these must be guided to find their targets. The picture is further complicated by the activity of the nerve cells, which can refine the pattern of connections they make.

All nervous systems, vertebrate and invertebrate, provide a system of communication through a network of electrically excitable cells—the nerve cells or **neurons**—of varying sizes, shapes, and functions (Fig. 10.1). The nervous system also contains non-neuronal cells known collectively as **glia**, which have a variety of supporting roles. The Schwann cells in the peripheral nervous system are glial cells, as are the oligodendrocytes and astrocytes of the central nervous system. Neurons connect with each other and with other target cells, such as muscle, at specialized junctions known as **synapses**. A neuron receives input from other neurons through its highly branched **dendrites**, and if the signals are strong enough to activate the neuron it generates a new electrical signal—a nerve impulse, or action potential—at the **cell body**. This electrical signal is then conducted along the **axon** to the **axon terminal**, or nerve ending, which makes a synapse with the dendrites or cell body of another neuron or with the surface of a muscle cell. The dendrites and axon terminals of individual neurons can be extensively branched, and a single neuron in the central nervous system can receive as many as 100,000 different inputs. At a synapse, the electrical signal is converted into a chemical signal, in the form of a chemical neurotransmitter, which is released from the nerve ending and acts on receptors in the membrane of the opposing target cell to generate or suppress a new electrical signal. The nervous system can only function properly if the neurons are correctly connected to one another, and thus a central question surrounding nervous-system development is how the connections between neurons develop with the appropriate specificity.

Yet for all its complexity, the nervous system is the product of the same kind of cellular and developmental processes as those involved in the development of

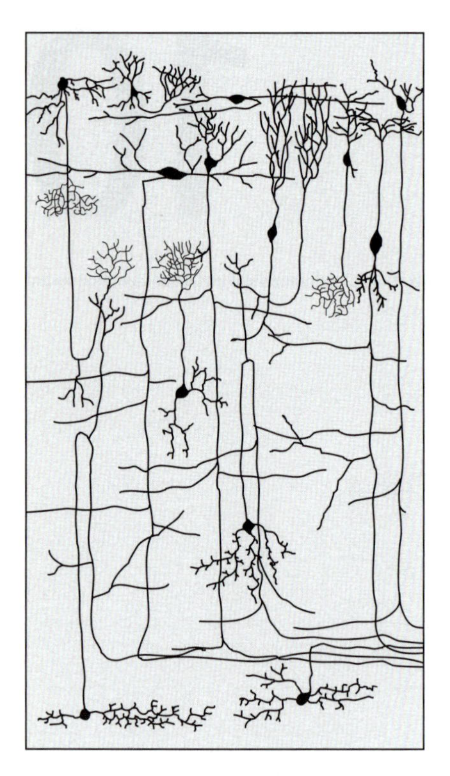

Fig. 10.1 An assemblage of the various neurons found in the optic lobe of the avian brain. Axons and dendrites extend from the cell bodies, which are usually small and rounded, and contain the nucleus. In the brain there are many more nervous connections that are not shown here.

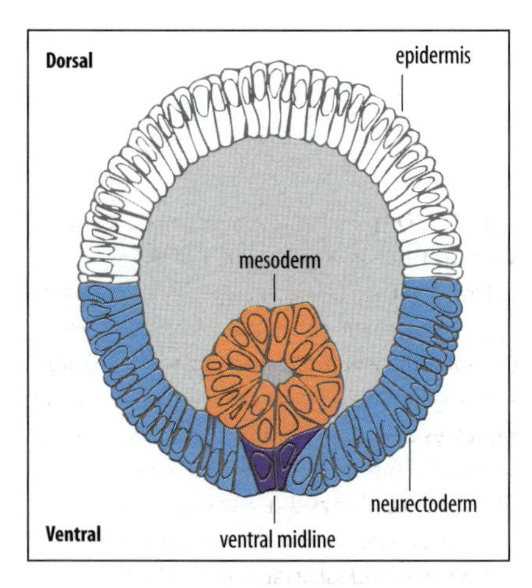

Fig. 10.2 Cross-section of a *Drosophila* early gastrula showing the location of the neurectoderm (pale blue). The mesoderm (red), which originally lies along the ventral midline, has already been internalized.

other organs. The overall process of nervous-system development can be divided up into four major stages: the specification of neural cell (neuron or glial cell) identity; the migration of neurons and the outgrowth of axons to their targets; the formation of synapses with target cells, which can be other neurons, muscle or gland cells; and the refinement of synaptic connections through the elimination of axon branches and cell death.

We have already considered some aspects of the development of the vertebrate nervous system, including neural induction (Chapter 4), the formation of the neural tube (Chapter 7), the segmental arrangement of sensory dorsal root ganglia (Chapter 7), and the differentiation of neurons from neural crest cells (Chapter 8). We will re-examine some of these key steps when we consider the specification of neural cells. The general focus of this chapter is on the mechanisms that control the identity of cells in the nervous system and that pattern their synaptic connections.

Specification of cell identity in the nervous system

We start by considering how neural cells acquire their individual identities, focusing primarily on the specification of neurons. As in other developing systems, nerve cell specification is governed both by external signals and by intrinsic differences generated through asymmetric cell divisions. We first consider neural specification in *Drosophila*, focusing on the specification of the neuronal precursors, which has revealed many key developmental processes in neurogenesis, and then turn to vertebrates, where similar mechanisms are involved, this time focusing on motor neurons.

10.1 Neurons in *Drosophila* arise from proneural clusters

In *Drosophila*, the presumptive central nervous system is specified at an early stage of embryonic development as two longitudinal stripes of cells, just dorsal to the ventral mesoderm (see Chapter 2). This region is called the neurogenic zone, or **neurectoderm** (Fig. 10.2), and comprises ectodermal cells with the potential to form neural cells and epidermis. The neurectoderm is subdivided along the antero-posterior and dorso-ventral axes into a precise orthogonal pattern of **proneural clusters**. Within each cluster, cell–cell interactions direct one cell into a neural precursor or **neuroblast** fate. The rest become epidermal cells. The neuroblasts enlarge, and delaminate into the interior of the embryo to form an invariant pattern of about 30 neuroblasts per hemisegment. A **hemisegment** is the lateral half of a segment, one on each side of the midline, and is the developmental unit for the nervous system. Each neuroblast that forms within a hemisegment expresses a unique set of genes, and divides in a specific lineage to produce neurons. Most glial cells in the embryonic central nervous system develop from ectodermal cells that express the gene *glial cells missing* (*gcm*) and become glioblasts, of which there are two per hemisegment.

Neurectodermal cells are given the potential to become neuronal precursors by the expression of genes known as **proneural genes**. Of particular importance in *Drosophila* are the genes of the *achaete–scute* complex. These encode DNA-binding basic helix-loop-helix transcription factors that form homodimers and heterodimers with each other, and initiate neural specification by regulating the transcription of target genes.

Neuroblast formation in *Drosophila* occurs in five temporally and spatially distinct waves. Each neuroblast acquires a unique fate based on the particular hemisegment it is part of, its position within the hemisegment, and its time of formation. Neuroblasts that form in identical positions in different hemisegments at the same time express the same set of genes and follow the same fate. Segmentation genes such as *wingless* divide the hemisegment into regions transversely, while genes acting along the dorso-ventral axis divide the hemisegment into three longitudinal columns of prospective neuroblasts along each side of the midline (Fig. 10.3). The identity of the columns is specified by the expression of the homeodomain transcription factor genes *msh*, *ind*, and *vnd*, in dorsal to ventral order in the ventral neurectoderm. The developing vertebrate neural plate also has three longitudinal domains (see Fig. 10.3) which express genes homologous to those in the fly—*Msx*, *Gsh*, and *Nkx2*, respectively. This is a striking example of the evolutionary conservation of a mechanism for regional specification.

Shortly after the formation of a proneural cluster in the *Drosophila* embryonic neurectoderm, one cell in the cluster starts to express Achaete protein at a higher level than the other cells. This will become the neuroblast. Neuroblasts leave the surface layer during germ-band extension and eventually give rise to neurons that will form the ventral nerve cord of the larva. Not all of the cells in the neurectoderm become neural cells. Initially, all the cells within a proneural cluster are capable of becoming neuroblasts, but one cell, through an apparently random event, then produces a signal that prevents its immediate neighbors from developing further along the neural pathway (Fig. 10.4, top row). This lateral inhibition eventually leads to just one cell developing as a neuroblast, with the rest of the cluster

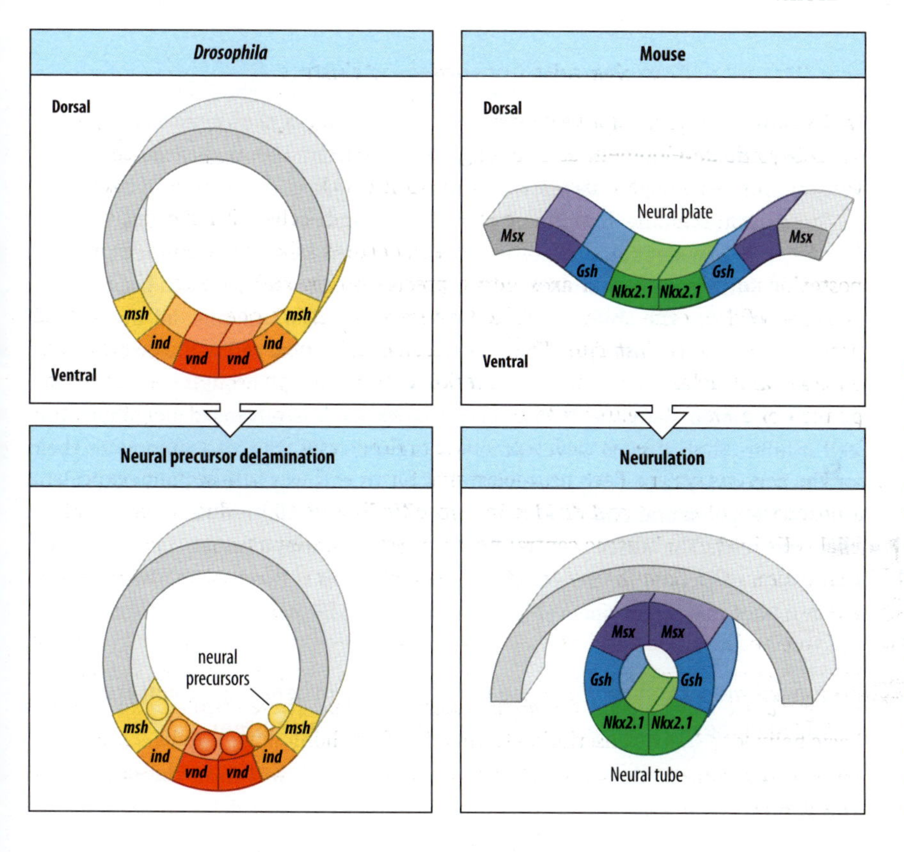

Fig. 10.3 Organization of invertebrate and vertebrate neuroepithelium into three columns of neural precursors. The developing *Drosophila* embryonic central nervous system (left) and the vertebrate neural plate (right) are both organized into three columns. Each column expresses a specific homeobox gene, and similar genes are expressed in the same medial-lateral order in *Drosophila* and in vertebrates, with *vnd*/*Nkx2.1* expressed most medially (nearest the midline), *ind*/*Gsh* in an intermediate position, and *msh*/*Msx* expressed most laterally. The expression in the neural tube is inverted compared with the neural plate as a result of the tube's formation by invagination from the plate (see Chapter 7).

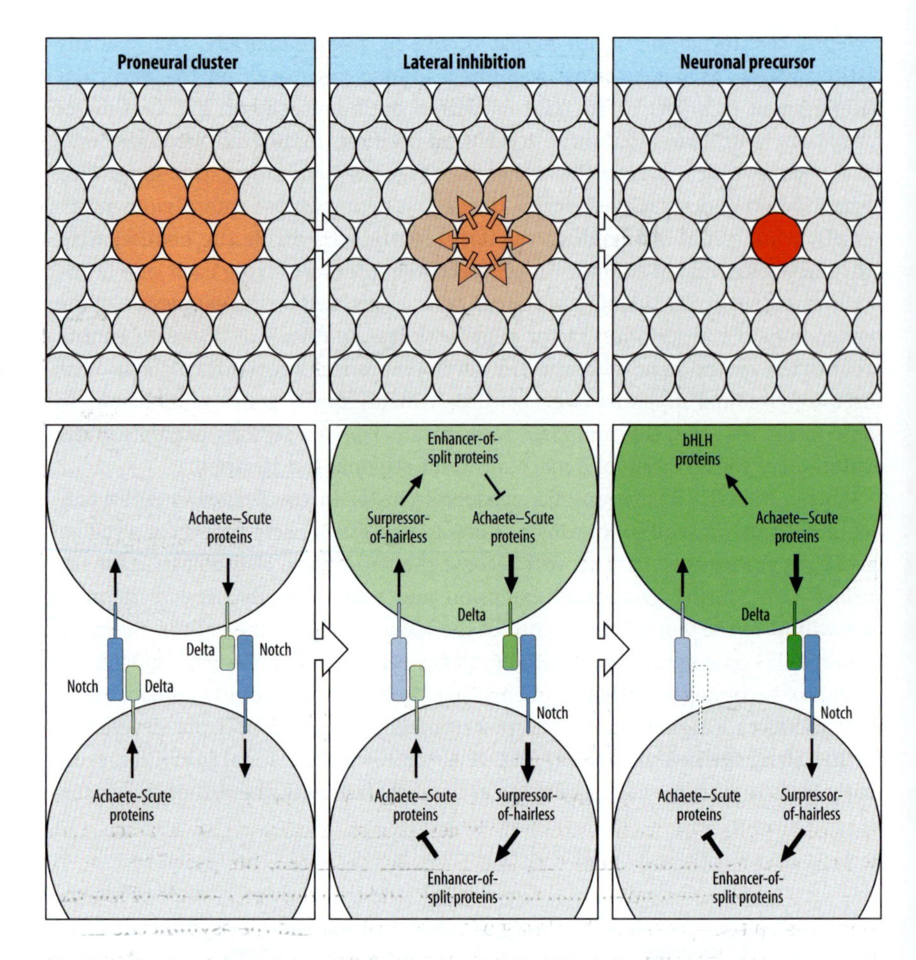

Fig. 10.4 The role of Notch signaling in lateral inhibition in neural development in *Drosophila*. As shown in the top row, the proneural cluster gives rise to a single neuronal precursor cell, the neuroblast or sensory organ precursor, by means of lateral inhibition. The rest of the cells in the cluster become epidermal or support cells. As shown in the bottom panels, Notch signaling between cells of the proneural cluster is initially similar and keeps the cells in the proneural state. An imbalance develops when one cell begins to express higher levels of Delta, the ligand for Notch, and thus activates Notch to a higher level in neighboring cells. This initial imbalance is rapidly amplified by a feedback pathway involving two transcription factors, Suppressor-of-hairless and Enhancer-of-split. Their activities repress the production of Achaete and Scute proteins and Delta protein in the affected cells. This prevents them from proceeding along the pathway of neuronal development and from delivering inhibitory Delta signals to the prospective neuroblast. bHLH, basic helix-loop-helix.

After Kandel, E.R., et al.: 2000.

becoming epidermal cells. In the grasshopper embryo, if a neuroblast is killed just as it begins to develop, a neighboring cell develops into a neuroblast instead. Once specified, the neuroblasts migrate individually from the superficial cell layer inward and go through a stereotyped series of cell divisions to produce neurons.

The transmembrane receptor protein Notch and its cell-surface ligand Delta (see Fig. 4.6) play key roles in lateral inhibition in the neurectoderm. Activation of Notch by Delta leads to inhibition of the proneural genes in the Notch-bearing cell, which thus loses the ability to give rise to a neuroblast. Initially, all the cells in a proneural cluster can express both Notch and Delta. One cell will, however, start to express Delta sooner or more strongly than the others. Through the interaction of Delta protein with the Notch receptors on its less advanced neighbors this cell inhibits their further development as neural cells, and also suppresses their production of Delta (see Fig. 10.4, bottom row). In this way, the initially equivalent cells of the proneural cluster are resolved by lateral inhibition into one neuronal precursor cell. The other cells go on to develop into epidermal cells and cease expression of Delta. If either Notch or Delta function is inactivated, the neurectoderm makes many more neuroblasts and fewer epidermal cells than normal.

10.2 Asymmetric cell divisions and timed changes in gene expression are involved in the development of the *Drosophila* nervous system

After specification, the neuroblasts and glioblasts that will form the neurons of the ventral nerve cord of the fly's central nervous system delaminate from the

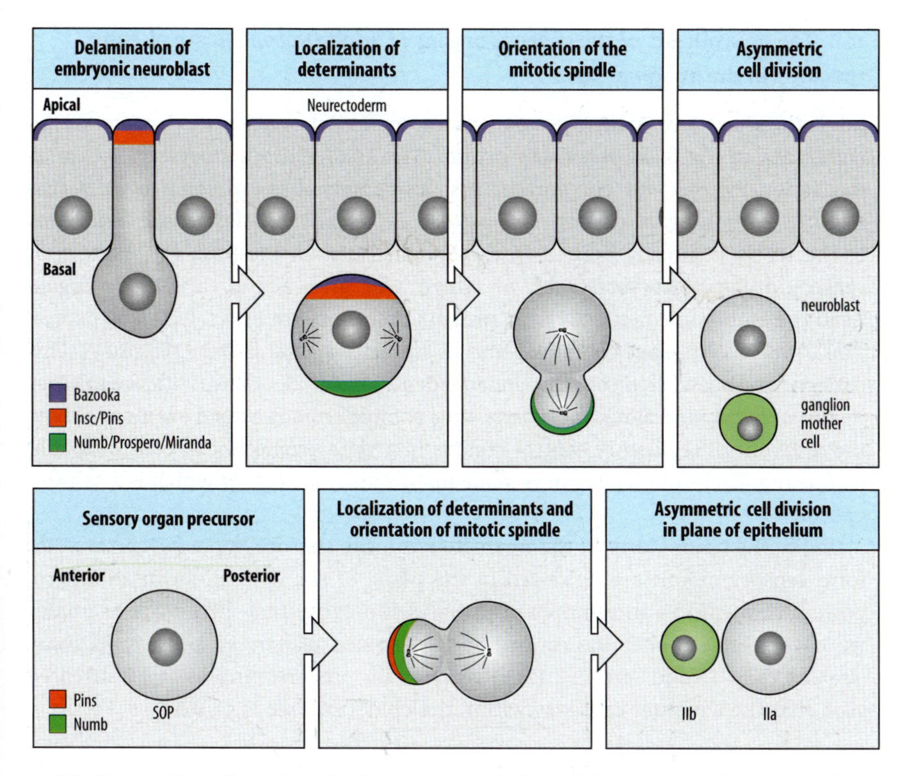

Fig. 10.5 The formation of neuronal cells from neuroblasts in *Drosophila* by asymmetric cell division. Top: generation of ganglion mother cells of the central nervous system by the asymmetric division of neuroblasts in the *Drosophila* embryo. Once specified, neuroblasts move out of the neurectodermal epithelium and then behave as neural stem cells. Each neuroblast divides asymmetrically to give an apical cell, which remains a neuroblast stem cell, and a smaller basal cell—the ganglion mother cell. The orientation of cell division and cell fate are specified by localized protein determinants. Numb protein is localized to one end of the neuroblast before division and the daughter that receives it becomes the ganglion mother cell that will give rise to neurons. Prospero and Miranda are required for the proper localization of Numb, and Inscuteable (Insc) and Pins are required for the correct orientation of cell division. Bottom: the sensory organ precursor (SOP) that gives rise to the cells of a sensory bristle in the adult *Drosophila* epidermis contains Numb protein on one side only. Other proteins, such as Pins, are also located on this side and are involved in orienting the plane of cell division. The SOP undergoes an asymmetric division in the plane of the epithelium and the Numb protein is passed on to just one of the daughter cells (IIb). Numb protein inhibits Notch function. This cell will give rise to a sensory neuron and a sheath cell. The cell that does not receive Numb protein (IIa) displays Notch activity and gives rise to a hair cell and a socket cell, which are non-neuronal cells.

epithelium to lie adjacent to the inner, or basal, face of the neurectoderm, and then behave as stem cells. Each neuroblast divides asymmetrically to give an apical cell, which remains a neural stem cell, and a smaller basal cell, the **ganglion mother cell**, which will differentiate into neurons (Fig. 10.5). A complex cascade of interactions sets up the apical–basal polarity of the neuroblast and the asymmetric distribution of cell determinants that specify the different fates of the two daughter cells. At the top of this cascade is the protein Bazooka, which is located in the apical face of the prospective neuroblast. Bazooka is required for the asymmetric distribution of two other proteins, Numb and Prospero, which are localized in the basal cortex of the neuroblast as it enters mitosis, and after division are localized mainly in the ganglion mother cell. The correct orientation of the mitotic spindle requires the presence of the protein Inscuteable (Insc) (see Fig. 10.5). Numb may act to help specify ganglion mother cell fate by antagonizing Notch activity, thus making the two daughter cells different in their response to a Notch signal (this is a later action of Notch than its action in specifying neural fate described above).

The neuroblast keeps its position at the inner surface of the ectoderm and with each further division pushes the older cells deeper in. A ganglion mother cell divides once again to give two neurons, which do not divide further. The layers of neurons thus formed have different fates and are defined by the sequential expression of four transcription factors—Hunchback (Hb), Krüppel (Kr), Pdm1, and Castor (Cas)—in the originating neuroblasts. Hb is expressed earliest and the ganglion mother cells produced from the Hb-expressing neuroblasts at this time form the deepest neuronal layer, while Cas is expressed last and the Cas-expressing ganglion mother cells form the most superficial layer. The precise timing of the transition from expression of one transcription factor to the next in the neuroblasts is thus crucial for the correct formation of the ventral nerve cord. The timing of the transition from Hunchback to Krüppel expression is linked in some way to neuroblast division, while the later transitions involve an intrinsic timing mechanism in the neuroblast that is independent of cell division.

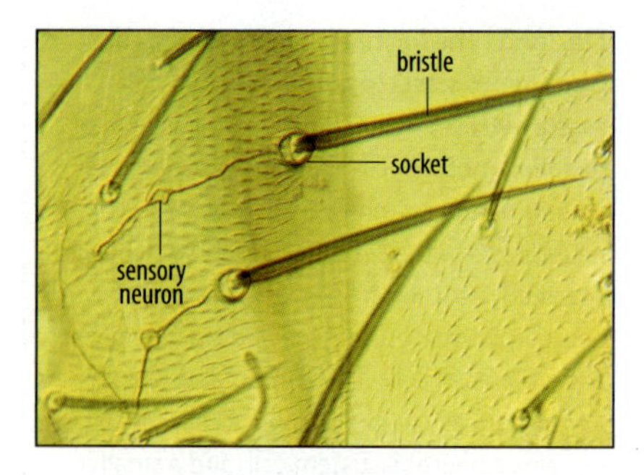

Fig. 10.6 Sensory bristles in the epidermis of adult *Drosophila*. The bristles respond to external stimuli and stimulate the sensory neurons.

Photograph courtesy of Y. Jan.

10.3 The neuroblasts of the sensory organs of adult *Drosophila* are already specified in the imaginal discs

Adult flies have two main types of sensory organ: external sensory organs in the epidermis, in the form of sensory bristles (Fig. 10.6) and other organs that act as mechanoreceptors and chemoreceptors, and internal chordotonal organs that monitor stretching of tissues. The pattern of sensory bristles in the epidermis of the wings is already present in the imaginal disc in the form of a pattern of **sensory organ precursors**, each of which will give rise to a sensory organ. As in the embryonic neurectoderm, proneural clusters are first selected by expression of the *achaete–scute* complex genes in a precise spatial pattern (Fig. 10.7). This pattern is constructed by the independent action of site-specific regulatory modules distributed along the *achaete–scute* complex and is a good example of how the different *cis*-regulatory control regions in a gene promoter can control spatial patterning. Each sensory organ is made up of four cells, one of which is a sensory neuron.

Unlike the neuroblasts in the neurectoderm, the sensory organ precursors that form sensory organs are polarized in the plane of the epithelium in an antero-posterior direction—an example of planar cell polarity that, like other examples we have seen, is under the control of the Frizzled planar cell polarity pathway (see Box 2E, p. 75 and Section 7.12). Division of the precursor in this case is oriented such that the daughter cells stay within the epidermis (see Fig. 10.5).

10.4 The vertebrate nervous system is derived from the neural plate

All the cells of the vertebrate nervous system derive from the neural plate, a region of columnar epithelium induced from the ectoderm on the dorsal surface of the embryo during gastrulation (see Chapter 4), and from sensory placodes in the head region that contribute to the cranial nerves. Toward the end of gastrulation, the neural plate begins to fold and form the neural tube (Fig. 10.8). Neural crest cells migrate away from the dorsal part of the tube and give rise to both the sensory neurons of the peripheral nervous system and to the autonomic nervous system (see Chapter 8). The cells that remain within the neural tube give rise to the brain and spinal cord, which together comprise the central nervous system.

The neural tube becomes regionalized along the antero-posterior axis, as described in Chapter 4, and this regionalization is particularly clear in the hindbrain region,

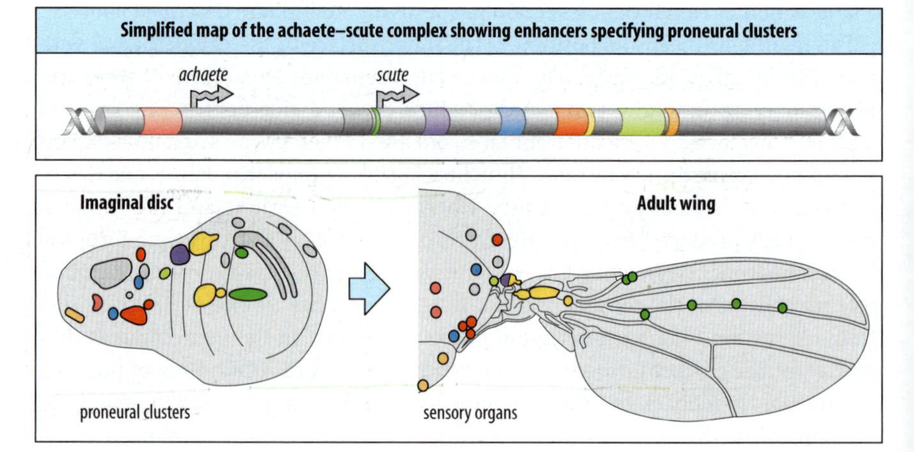

Fig. 10.7 Specification of sensory bristles in the *Drosophila* wing. The location of proneural clusters in *Drosophila* wing imaginal discs is largely under the control of the *achaete–scute* gene complex, the expression of which is activated at specific sites through different enhancers. Top panel: the position of the various enhancers that determine the site-specific expression of the *achaete* and *scute* genes. These enhancers are located both upstream and downstream of the *achaete* and *scute* genes. Bottom panel: location of the proneural clusters in the imaginal disc (left) and of the corresponding sensory bristles in the adult wing and adjacent thoracic segments (right), with colors corresponding to the enhancer responsible for the site-specific expression of the genes.

After Campuzano, S., et al.: 1992.

where the Hox genes are expressed in a well defined pattern. The expression of these genes gives the cells of the neural tube their positional value along the antero-posterior axis of the hindbrain. The neural tube also becomes patterned along the dorso-ventral axis.

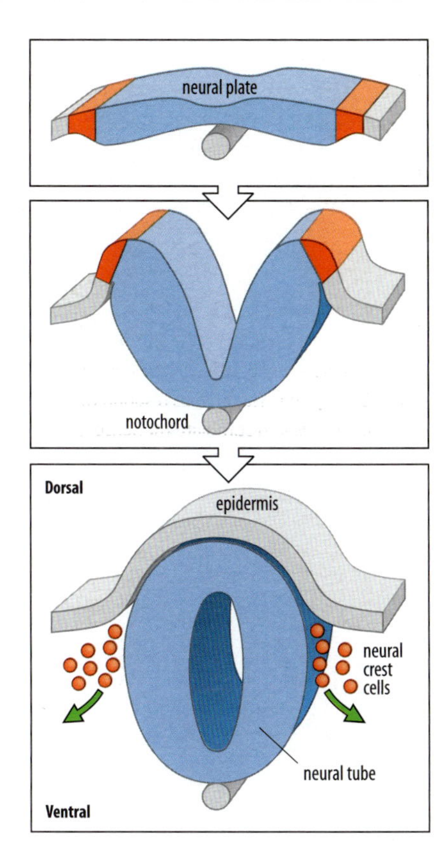

Fig. 10.8 The vertebrate nervous system is derived from the neural plate. The neural tube forms by folding of the neural plate. It sinks beneath the surface and is overlain by the epidermis. The neural tube gives rise to both the brain and the spinal cord. Neural crest cells originating from the line of fusion of the neural folds migrate away from the neural tube to form sensory nerves and the autonomic nervous system (among other structures).

10.5 Specification of vertebrate neuronal precursors involves lateral inhibition

In frog embryos, neurons are not generated simultaneously throughout the neural plate but are initially confined to three longitudinal stripes on each side of the midline. In *Xenopus*, the medial stripe—the one nearest the midline—which becomes the ventral region of the neural tube, gives rise to motor neurons. The more lateral stripes give rise to interneurons and sensory neurons. The *neurogenin* gene is expressed in the cells of the three longitudinal stripes and its activity is required for neuronal differentiation. *Neurogenin* codes for a basic helix-loop-helix transcription factor that is distantly related to the Achaete and Scute proteins of *Drosophila*.

The initial selection of cells to form the stripes involves lateral inhibition, in which the proteins Delta and Notch again play a key role, as in *Drosophila*. Binding of Delta to the Notch protein provides a signal that inhibits neuronal differentiation by inhibiting the synthesis of Neurogenin. Initially, all cells in the neural plate are able to make Neurogenin, Delta, and Notch, and neuronal differentiation is prevented by mutual inhibition. When one cell starts to express Delta more strongly than its neighbors, the stronger signal delivered to the adjacent cells both prevents neuronal differentiation in these cells, by shutting down *neurogenin* expression, and represses the synthesis of Delta, stopping the cells from delivering reciprocal inhibition. The actions of Neurogenin in the one Delta-expressing cell lead to activation of the gene *neuroD*, which encodes a transcription factor required for neural differentiation (Fig. 10.9). Confirmation of this proposed mechanism comes from the results of overexpression of Delta or expression of a mutant Notch that signals continuously. These both result in a reduction in the number of neurons formed. Inhibition of Delta, on the other hand, leads to an increase in the number of neurons, as does overexpression of Neurogenin. An important feature of this mechanism is that a pool of undifferentiated neural precursors is maintained in the form of the cells in which *neurogenin* is inhibited, thus ensuring that neurogenesis can occur over long periods of time.

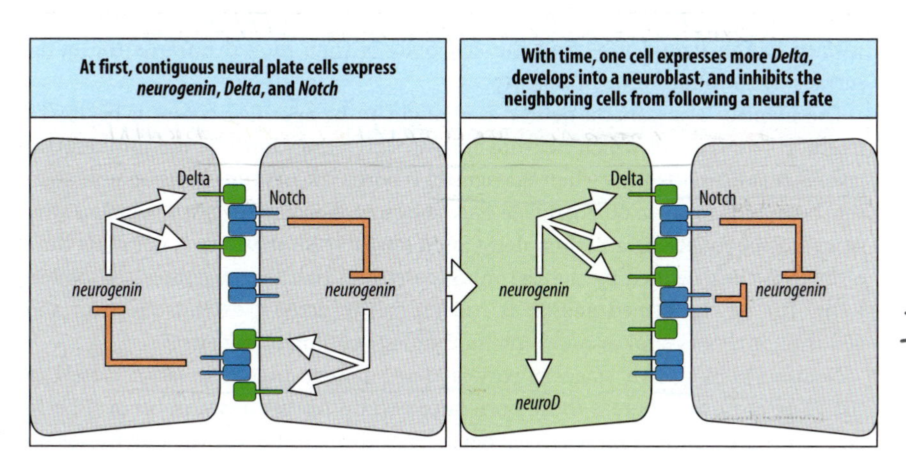

Fig. 10.9 Lateral inhibition specifies single cells as neuronal precursors in the vertebrate nervous system. The *neurogenin* gene is initially expressed in stripes of contiguous cells in the neural plate, and these cells also express the Delta and Notch proteins. The Delta–Notch interaction between adjacent cells mutually inhibits expression of *neurogenin*. When, by chance, one cell expresses more Delta than its neighbors, it inhibits the expression of Delta protein in the neighboring cells and so develops as a neuron, expressing *neurogenin* and *neuroD*.

10.6 Neurons are formed in the proliferative zone of the neural tube and migrate outwards

The neural tube generates a large number of different neuronal and glial cell types. In both brain and spinal cord, all neurons and glia arise from a proliferative layer of epithelial cells that lines the lumen of the neural tube and is called the **ventricular zone**, as later it lines the lateral ventricles of the brain. The neural tube is initially made up of a single layer of epithelium; a proliferative zone composed of rounded-up dividing cells develops at its inner surface. Once formed, neurons migrate outward from this zone and build up the neural tissue in concentric layers. The hollow neural tube thus eventually develops into the hollow spinal cord, and into the brain with its fluid-filled ventricles that represent the remains of the neural tube lumen.

The cells in the ventricular zone are the neural progenitor cells, which give rise to both neurons and glia. In mammals, neurons are typically produced first and glial cells later. The initial specification of neural progenitor cell daughters as neurons involves transcription factors of the Ngn family (Ngn2 in the spinal cord, Ngn1 in the future cortex). Other transcription factors can maintain a cell in the progenitor state after cell division by preventing the expression of genes characteristic of post-mitotic neurons. This action is counteracted by the actions of the Ngn factors which promote a neuronal fate.

An important question is the relationship between the daughter cells of a neural progenitor cell and the progenitor cell itself. Progenitor cells are multipotent, and there must be a mechanism for both generating neurons and maintaining the progenitor cell state. It seems that two mechanisms are involved. In one, the progenitor cell gives rise to two similar cells, both progenitor cells or both neurons; in the other the division is asymmetric, and one daughter remains a progenitor stem cell while the other develops into a neuron.

We shall consider the process that links neuronal production with neuronal migration with reference to the mammalian cerebral cortex. The cortex is organized into six layers (I–VI), numbered from the cortical surface inward. They each contain neurons with distinctive shapes and connections. For example, large pyramidal cells are concentrated in layer V, and smaller stellate neurons predominate in layer IV. All of these neurons have their origin in the ventricular zone and migrate out to their final positions along radial glial cells—greatly elongated cells that extend across the developing neural tube (Fig. 10.10). Another mode of migration, termed tangential migration, has been observed that is independent of glial fibers. The leading process of the neuron is initially extended to attach to the outer surface of the neural tube and the cell body is then moved towards the outer surface by the shortening of this process.

The identity of a cortical neuron is thought to be specified before it begins to migrate. The layer to which neurons migrate after their birth in the proliferative layer is related to the time when the neuron is born. The neurons of the mammalian central nervous system do not divide once they have become specified, which is why newly formed neurons are often designated **post-mitotic neurons**. A neuron's time of birth is thus defined by the last mitotic division that its progenitor cell underwent. The newly formed neuron is still immature; later it extends an axon and dendrite processes, and assumes the morphology of a mature neuron.

Neurons born at early stages of cortical development migrate to layers closest to their site of birth, whereas those born later end up further away, in more superficial layers (Fig. 10.11). The younger neurons must migrate past the older ones on their way to their correct position, giving rise to layers and columns of cells. The exception is the first-formed layer, which remains as the outer layer. Thus, there is

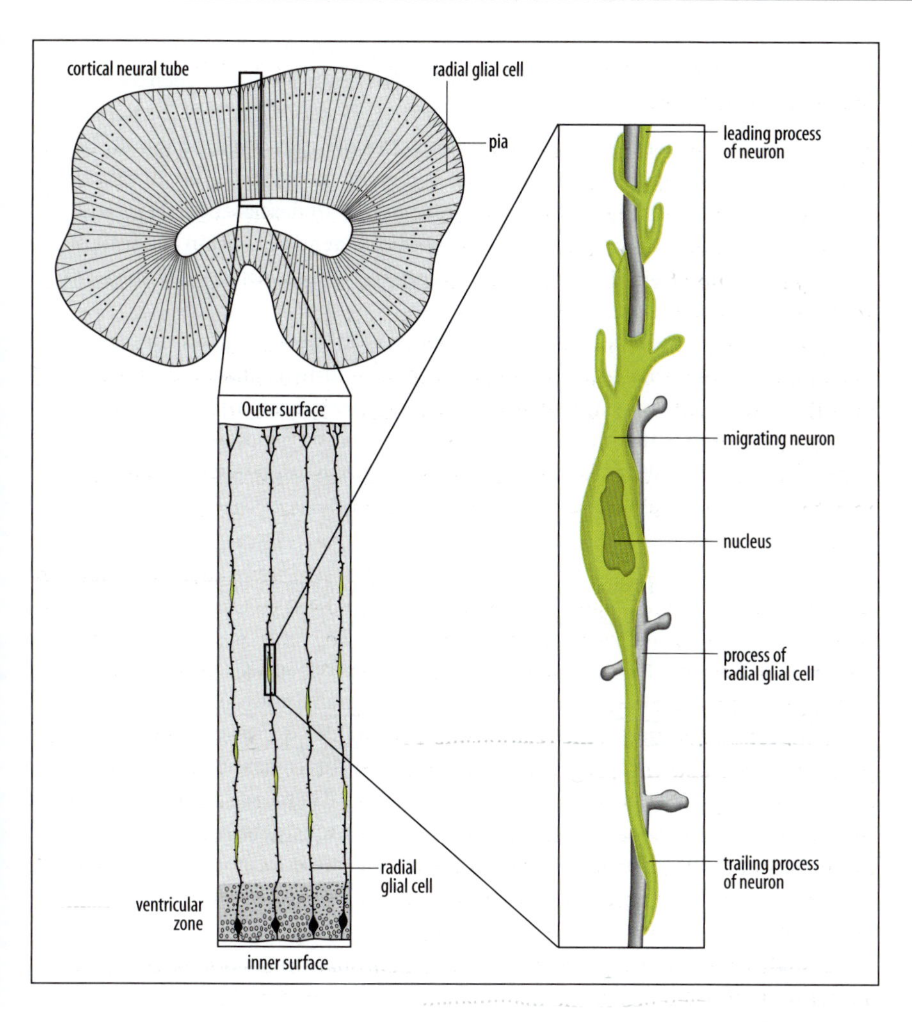

Fig. 10.10 Cortical neurons migrate along radial glial cells. Neurons generated in the ventricular zone migrate to their final locations in the cortex along radial glial cells that extend right across the wall of the neural tube.

After Rakic, P.: 1972.

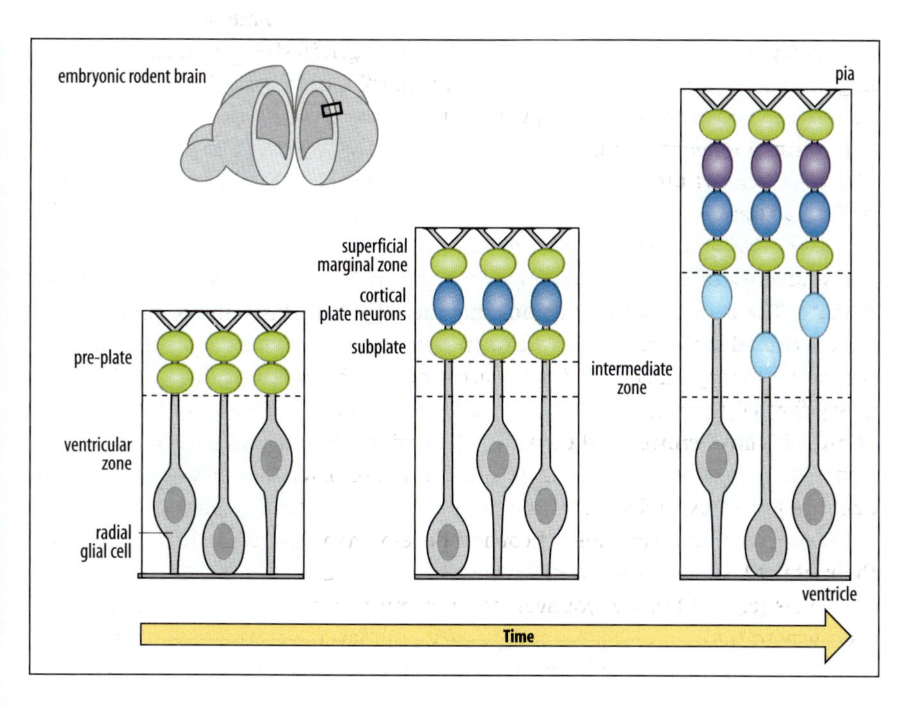

Fig. 10.11 The generation of neuronal layers in the mammalian cortex. Left: development of the cortex begins with the migration of the first wave of post-mitotic neurons (green) towards the outer surface of the neural tube to form the pre-plate above the ventricular zone. Middle: the second wave of newly formed neurons (blue) migrates past the first layer of the pre-plate to form the first layer of so-called cortical plate neurons. The upper layer of the pre-plate neurons becomes the superficial marginal zone and will become cortical layer I (cortex layers are numbered from the outside inwards). Right: a third wave of neurons (purple) migrates to form a second layer of cortical plate neurons above the first. Subsequent waves of neurons will each form a more superficial layer of cortical plate. The intermediate zone contains migrating neurons (light blue).

Adapted from Honda, T., et al.: 2003.

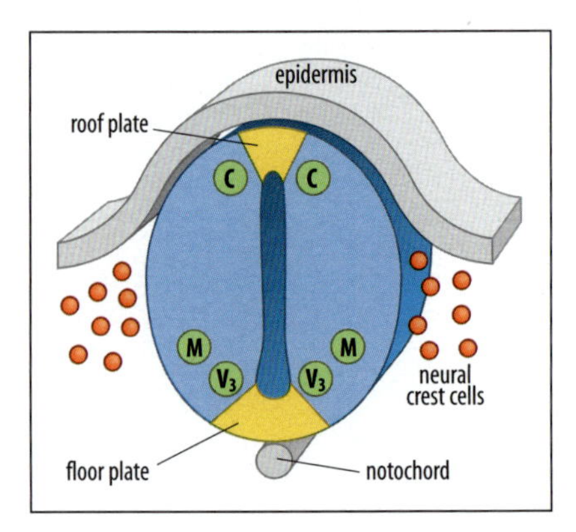

Fig. 10.12 Dorso-ventral organization in embryonic spinal neural tube. In neural tube that develops into spinal cord, a floor plate of non-neuronal cells develops along the ventral midline, and a roof plate of non-neuronal cells along the dorsal midline. Commissural neurons (C) differentiate in the dorsal region, near the roof plate. Motor neurons (M) and V3 interneurons differentiate close to the floor plate. (see Fig. 10.15).

an inside to outside sequence of neuronal differentiation in the neural tube that gives rise to the cerebral cortex and other layered brain structures such as the cerebellum. Mutations in the gene *reelin*, which is expressed in the first-formed, outermost layer of the cortex, disrupt migration: the neurons seem unable to bypass their predecessors and proper cortical layers do not form. The migration of neurons in the brain is in fact more complicated than described here; in addition to radial migration, brain neurons also migrate tangentially within the wall of the tube. Radial migration of neurons also occurs in the spinal cord and gives rise to a layered structure.

There is now evidence for the generation of new neurons in the adult mammalian brain. One region where neurons are formed is the wall of the lateral ventricle; these neurons contribute to the olfactory system. Another region of neuron formation is the hippocampus, where a special type of neuron, the granule cell, is generated. Neuronal stem cells from adult brain can be made to differentiate into neurons and glial cells in culture (see Chapter 8).

10.7 The pattern of differentiation of cells along the dorso-ventral axis of the spinal cord depends on ventral and dorsal signals

There is a distinct dorso-ventral pattern in the developing spinal cord. Future motor neurons are located ventrally, whereas commissural neurons, which send axons across the cord, differentiate primarily in the dorsal region (Fig. 10.12). In addition to the neuronal cell types, there is, in the ventral midline, a group of non-neuronal cells, which forms the **floor plate**. There are also non-neuronal **roof plate** cells in the dorsal midline. Sensory neurons develop from the neural crest cells that arise laterally and dorsally and migrate into the dorsal root ganglia (Fig. 10.13). Each cell type is present symmetrically on either side of the midline.

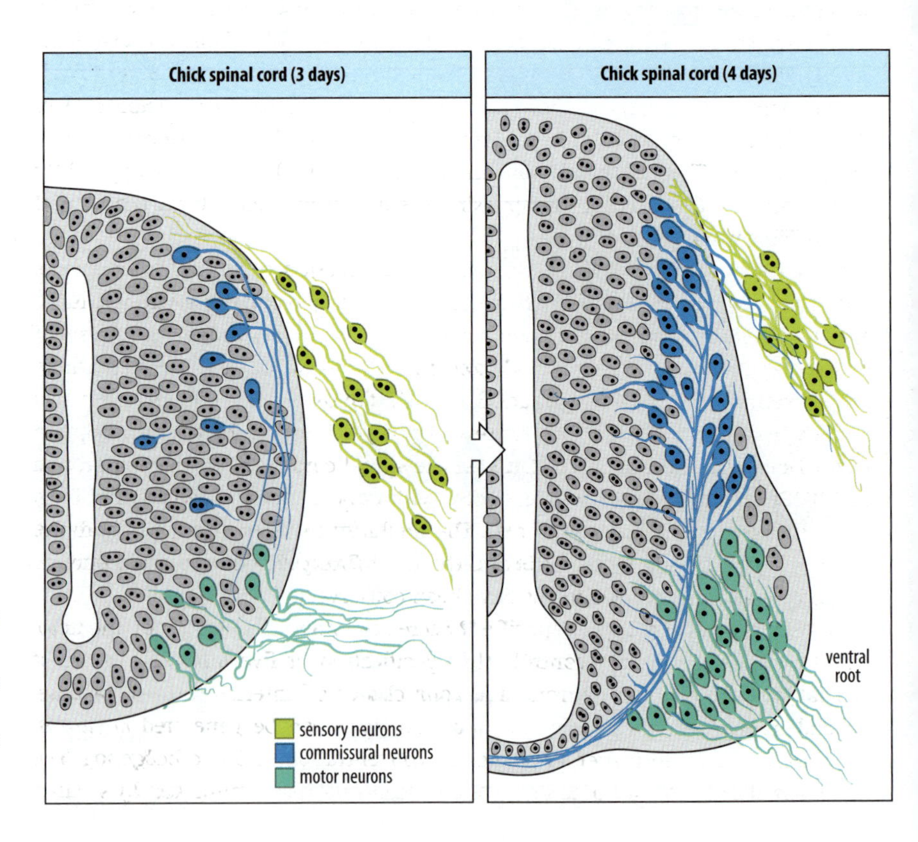

Fig. 10.13 Formation of sensory and motor neurons in the chick spinal cord. Three days after an egg has been laid, motor neurons are beginning to form within the chick embryo neural tube. A day later, the motor neurons send at axons to form the ventral root. The sensory neurons that migrate from the dorsal part of the spinal cord are derived from neural crest cells and form the segmental dorsal root ganglia (see Section 7.14). The commissural neurons send axons towards and across the ventral midline.

The floor plate can be induced by the notochord, but the relationship between notochord and floor plate does not fit snugly into standard models of induction. One reason is that the presumed inducer—the notochord—produces the same signal as the tissue induced—the floor plate. This signal is Sonic hedgehog, which we have already encountered as an important signal in limb development (see Chapter 9). Another reason is that precursor cells at the anterior end of the primitive streak contribute to both notochord and floor plate. A substantial proportion of floor plate cells are, however, specified quite late from cells in the neural tube. The patterning activity of the notochord can be shown by grafting a segment of notochord to a site lateral or dorsal to the neural tube; a second floor plate is induced in neural tube in direct contact with the transplanted notochord, and additional motor neurons are generated. Molecular markers of dorsal cell differentiation are suppressed in the region of the graft.

The proliferative zone on the inner surface of the neural tube gives rise to all the motor neurons, to interneurons and to the oligodendrocytes that will form myelin sheaths around them. The conversion of neuro-epithelial cells in the ventral neural tube to motor neuron progenitor cells is driven by Sonic hedgehog from the floor plate and retinoic acid secreted by the adjacent mesoderm. Control of neural progenitor cell proliferation by mitogenic Wnt signals is also important in controlling the number of neurons produced. In line with the presumed role of Sonic hedgehog and retinoic acid in neural differentiation, they can direct the development of embryonic stem cells into motor neurons and interneurons.

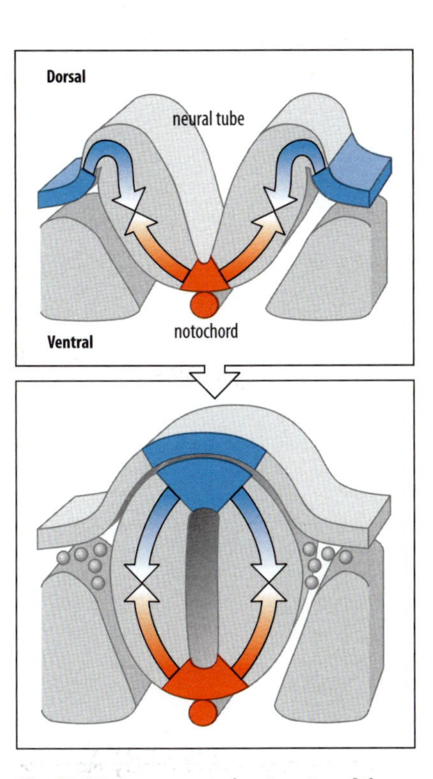

Fig. 10.14 Dorso-ventral patterning of the spinal cord involves both dorsal and ventral signals. As the neural tube forms (top panel), dorsal and ventral signaling occurs. The ventral signal from the floor plate is the Sonic hedgehog protein (red), whereas the dorsal signal (blue) includes BMP-4, which is induced during neural tube formation by the adjacent ectoderm. Ventral and dorsal signals oppose each other's action.

The dorsal epidermal ectoderm, and later the roof plate, patterns the dorsal half of the neural tube. In addition to the signal provided by Sonic hedgehog from the floor plate and notochord in the ventral midline, there are also dorsal signals. Using genetic knock-out and cell fate mapping, it has been shown that selective removal of the roof plate from the neural tube of the mouse embryo results in the loss of the most dorsal interneurons. BMPs from the ectoderm specify the roof plate, and the signals from the roof plate include BMPs which pattern the dorsal region of the neural tube. BMP-4 and BMP-7 are expressed in the epidermal ectoderm and, in experiments in chick embryos, have been shown to mimic the effect of the dorsalization signal. These BMP signals apparently ensure dorsal differentiation by opposing the action of signals from the ventral region (Fig. 10.14).

The induction of dorsal fate results in the expression of several BMPs in the dorsal region of the neural tube. The pattern of expression of BMPs is strongly influenced by ventral signals. This system bears a strong similarity to the mechanism of patterning of the dorso-ventral body axis in early *Drosophila* development, where the expression of *decapentaplegic*, another TGF-β family gene, is confined to dorsal regions because its expression is repressed in ventral regions by the transcription factor Dorsal (see Section 2.12). Thus, in the spinal cord, Sonic hedgehog protein and BMPs may provide positional signals with opposing actions emanating from the two ends of the dorso-ventral axis. The similarity to the patterning of both the vertebrate somite (see Section 4.5) and the early *Drosophila* embryo again emphasizes how patterning mechanisms have been conserved.

How are neuronal subtypes specified? Secretion of Sonic hedgehog by the notochord and the floor plate controls the specification of five different classes of ventral neurons—motor neurons and four classes of interneurons—in a dose-dependent manner; these different neuronal types can be generated *in vitro* in response to a two- to threefold change in concentration of Sonic hedgehog. The incremental changes in Sonic hedgehog concentration are mimicked by smaller

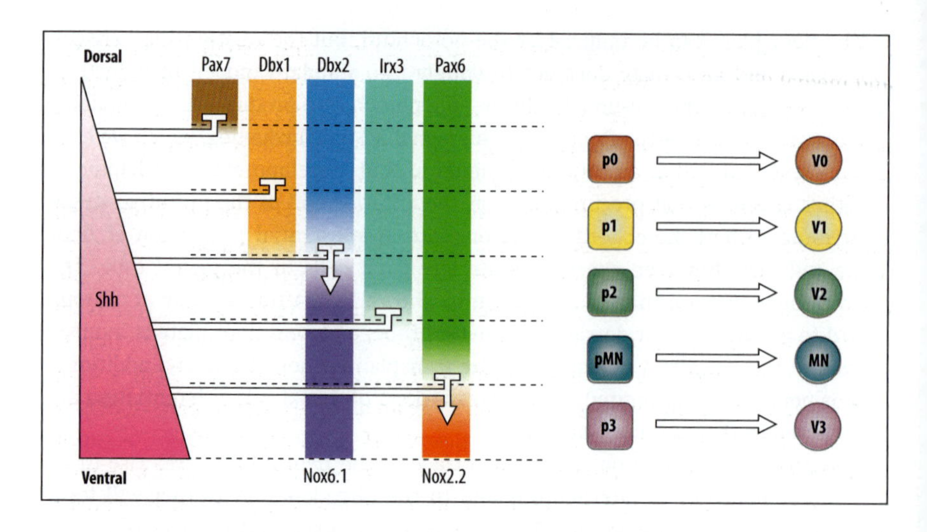

Fig. 10.15 Specification of neuronal subtypes in the ventral neural tube. Different neural subtypes (shown as different colors) are specified along the dorso-ventral axis of the ventral neural tube in response to a gradient of Sonic hedgehog (Shh) with its source in the floor plate. Shh protein mediates the repression of the so-called Class I homeodomain protein genes (*Pax7*, *Pax6*, *Dbx1*, *Dbx2*, and *Irx3*), whereas Class II homeodomain protein genes (*Nkx2.2* and *Nkx6.1*), are activated. Interactions between Class I and Class II genes pattern this expression further. Five neuronal subtypes are generated from the five domains. MN, motor neurons.

Adapted from Jessell, T.M.: 2000.

changes in the activity of the transcription factor Gli3 (see Section 9.5), which provides the key intracellular message in a graded form. A number of homeobox genes are regulated in ventral progenitor cells in response to Sonic hedgehog signaling, and can be divided into two classes. Class I genes are generally repressed by Sonic hedgehog, and include *Pax7*, *Pax6*, *Dbx1*, *Dbx2*, and *Irx3*, whereas Class II genes, such as *Nkx2.2* and *Nkx6.1*, are activated (Fig. 10.15). Class I genes have ventral limits of expression whereas Class II genes have dorsal limits. Interactions between the regions expressing Class I and Class II proteins establish sharp boundaries between progenitor domains that give rise to the different types of neurons (see Fig. 10.15).

Motor neurons developing in the ventral region of the spinal cord can be classified on the basis of the position of their cell bodies along the cord dorso-ventral axis, and the muscles that their axons innervate. In the chick embryo, motor neurons in the ventral region are subdivided into three main longitudinal columns on each side of the midline—a median column nearest the midline, and

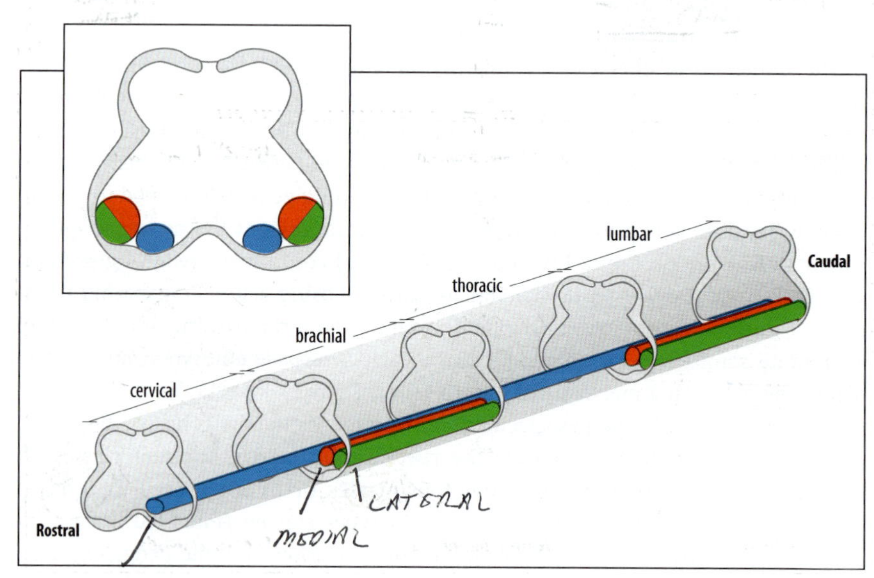

Fig. 10.16 The organization of the motor columns in the developing chick spinal cord. The organization of motor columns along the chick embryo spinal cord at stages 28–29. The median motor column is in blue, the medial division of the lateral column in red and the lateral division of the lateral motor column in green. The lateral motor columns are only present in the brachial and lumbar regions, in register with the position of limb formation.

a lateral motor column (LMC), which is further divided into two divisions, lateral and medial (Fig. 10.16). Motor neurons in the median motor column send axons to axial and body wall muscles, whereas neurons of the LMC project to ventral and dorsal limb muscles. The LMC is only present in the **brachial** and lumbar regions, where the forelimbs and hindlimbs, respectively, develop.

Neuronal fate along the antero-posterior axis of the spinal cord is determined primarily by signals from the node and later from the adjacent paraxial mesoderm. In quail-to-chick grafts, grafting of quail brachial paraxial mesoderm to the thoracic level in chick embryos at the time of neural tube closure results in presumptive thoracic neurons becoming specified as brachial neurons. The signals from the node and mesoderm—particularly FGF, which is graded from a high posterior level to a low anterior level—generate an antero-posterior pattern of Hox gene expression in the spinal cord. For example, *Hoxc6* is expressed in the motor neurons adjacent to the forelimb, while *Hoxc9* is expressed in the thoracic region. Motor neuron specification in anterior regions of the spinal cord is also dependent on retinoic acid, and blocking the retinoic acid receptor in the brachial region converts many of the neurons to thoracic subtypes.

Each motor neuron subtype expresses a different combination of LIM-homeodomain-transcription factors, which provide the neurons with positional identity within the spinal cord and enable their axons to select a particular pathway to innervate target muscles (Fig. 10.17). Interactions between the LIM proteins Lim1 and Isl1 give the neurons their medio-lateral positional identity in the LMC. Neurons expressing Lim1 settle in the lateral part of the LMC, while those expressing Isl1

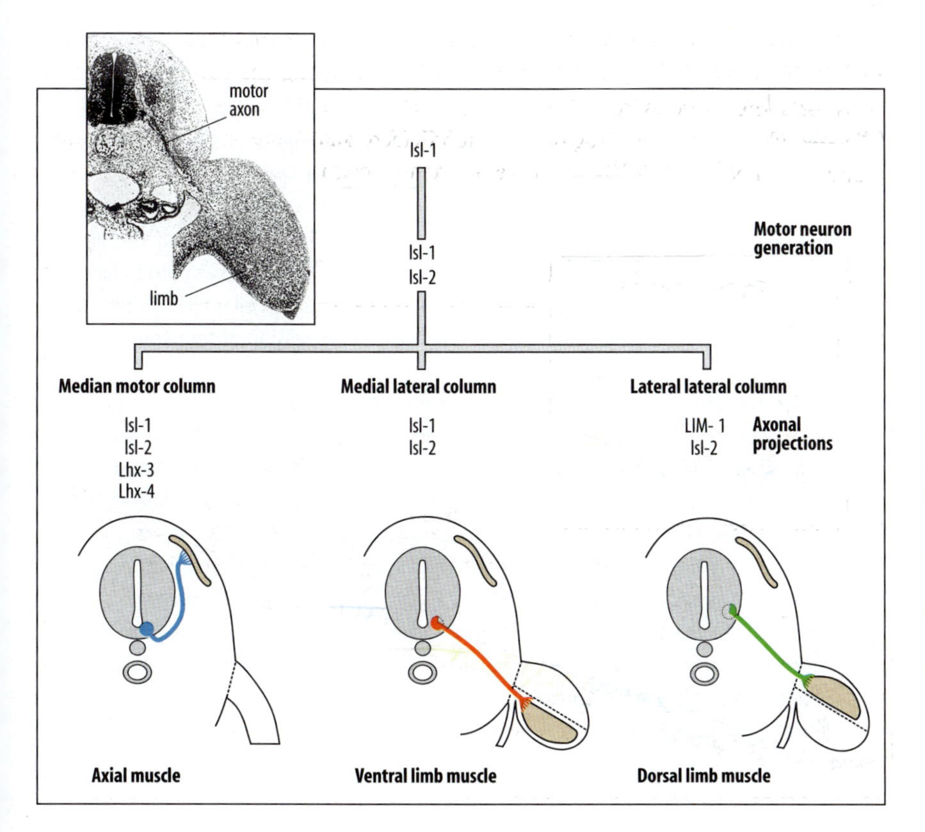

Fig. 10.17 Different motor neurons in the chick spinal cord express different LIM homeodomain proteins. The photograph (inset) shows a transverse section through a chick embryo at the level of the developing wing. Motor neurons can be seen entering the limb. Initially, all developing motor neurons express the LIM proteins Isl1 and Isl2. At the time of axon extension, motor neurons that innervate different muscle regions express different combinations of Isl1, Isl2, Lim1, Lhx3, and Lhx4.

Micrograph courtesy of K. Tosney.

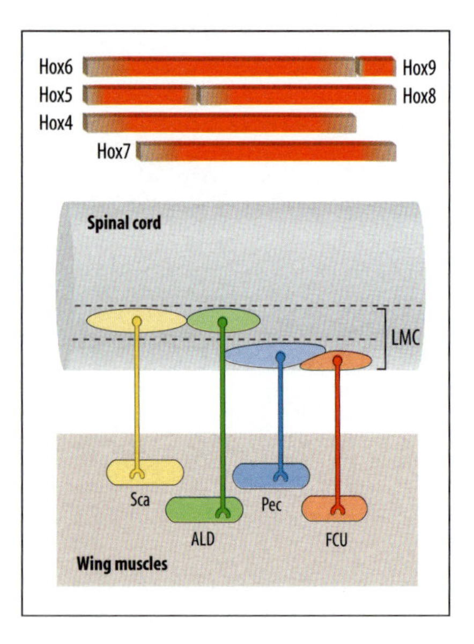

Fig. 10.18 Pools of motor neurons in the developing spinal cord make connections with specific limb muscles. The lateral columns of motor neurons in the brachial region of the developing chick spinal cord are organized into a number of groups, or pools, that each make connections to a specific muscle group in the developing wing. The identity of each pool of neurons is specified by interactions between the Hox genes expressed in the spinal cord in this region. LLC, lateral lateral column; LMC, lateral medial column (see Fig. 10.16). Muscles: Sca, scapulohumeralis posterior; Pec, pectoralis; ALD, anterior latissimus dorsi; FCU, flexor carpi ulnaris.

Adapted from Dasen et al.: 2005.

settle in the medial division of the LMC. Lim1 is also required for the axons of the lateral neurons to select a dorsal trajectory in the limb. The expression of Lmx1b in dorsal limb mesenchyme is also involved in controlling axonal outgrowth. If either of these proteins is absent, the motor axons select dorsal and ventral pathways randomly. Medial LMC neurons express Isl1 and send axons ventrally into the limb. The actual migration of the motor neuron axons is controlled partly by repulsive interactions between the EphA family of tyrosine kinase receptors on the neurons and their ephrin ligands in the limb mesenchyme, which is considered in the next part of the chapter. The distribution and level of both EphA and ephrins is defined by the LIM family proteins.

As well as being organized into the lateral column, and into divisions within the LMC with ventral and dorsal trajectories into the limb, motor neurons serving the limbs are further subdivided into so-called pools along the dorso-ventral and antero-posterior axes. There can be over 50 muscle groups in a typical avian or mammalian limb and thus there is a corresponding number of pools of motor neurons, with the neurons in each pool directing their axons to a particular target muscle.

The identity of the motor neurons in each pool is specified by the combinations of Hox genes expressed in different regions along the axis (see Chapters 4 and 9). For example, in the chick, the posterior limit of the region containing brachial motor neurons maps to the posterior boundary of the expression of Hox6 genes while the anterior boundary is defined by the anterior limit of expression of *Hoxc6*. Interactions between Hox 8, Hox7, Hox6, Hox5, and Hox4 genes specify the identities of pools of motor neurons that innervate specific muscles in the chick wing (Fig. 10.18).

Summary

In *Drosophila*, prospective neural tissue is specified early in development as a band of neurectoderm along the dorso-ventral axis. Within the neurectoderm, expression of proneural genes, such as those of the *achaete-scute* complex, provides clusters of neurectoderm cells with the potential to form neural cells. Only one cell of the cluster finally gives rise to a neuroblast because of lateral inhibition, involving the Notch receptor protein and its ligand Delta. Neuroblasts in the ventral nerve cord behave as stem cells, dividing asymmetrically to give rise to a daughter cell which gives rise to neurons. Layers are formed which are defined by the expression of four transcription factors.

The vertebrate nervous system is derived from the neural plate, which is specified during gastrulation. It contains the cells that form the brain and spinal cord, as well as the neural crest cells that contribute to the peripheral nervous system. Neurons are specified within the neural plate by a mechanism of lateral inhibition, similar to that in *Drosophila*. Neurons are generated in the proliferative zone on the inner surface of the neural tube and then migrate to different locations in the cortex. Patterning of neuronal cell types along the dorso-ventral axis of the spinal cord involves signals from both the ventral and dorsal regions. Secretion of Sonic hedgehog by the notochord and floor plate provides a graded signal that leads to specification of different neuronal types in the ventral region of the spinal cord. Neuronal identity along the antero-posterior axis is specified by the combinatorial expression of Hox genes.

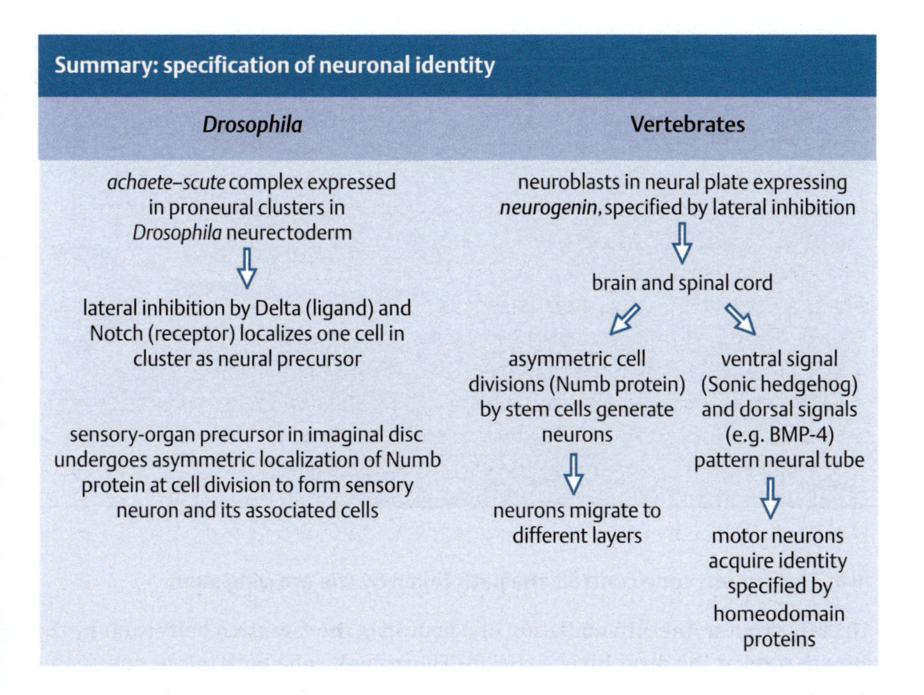

Neuronal migration

The working of the nervous system depends on discrete neuronal circuits, in which neurons make numerous connections with each other (Fig. 10.19). In the rest of this chapter, we will look at how these connections are set up. The functioning network of neurons is established by migration of immature neurons and by the guided outgrowth of axons toward their target cells. What guides the migrating neurons and controls the growth of axons and the contacts they make with other cells?

We have already discussed one example of extensive cell migration in relation to neural development—the migration of neural crest cells to give rise to the segmental arrangement of dorsal root ganglia (see Section 7.14). In that case, the migration is controlled by signals from the adjacent somites. Here, we deal with axon guidance in the central and peripheral nervous systems. We will first look at mechanisms for guiding axons and evidence for specificity of guidance and connections.

Fig. 10.19 Neurons make precise connections with their targets. Neurons (green) and their target cells usually develop in different locations. Connections between them are established by axonal outgrowth, guided by the movement of the axon tip (growth cone). Often, the initial set of relatively non-specific synaptic connections is then refined to produce a more precise pattern of connectivity.

After Alberts, B., et al.: 2002.

Birth of neurons | Outgrowth of axons and dendrites | Synaptic connections made | Refinement of synaptic connections

growth cone

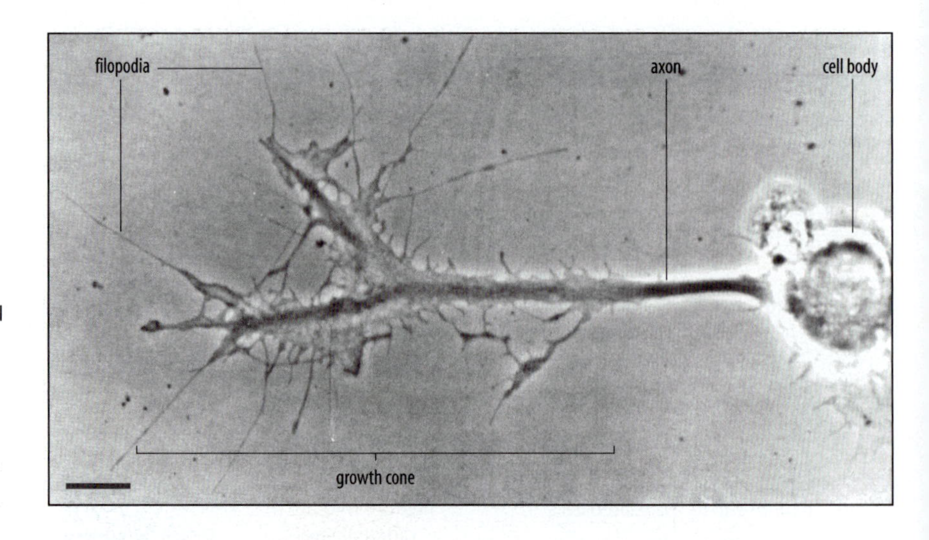

Fig. 10.20 A developing axon and growth cone. The axon growing out from a neuron's cell body ends in a motile structure called the growth cone. Many filopodia are continually extended and retracted from the growth cone to explore the surrounding environment. Scale bar = 10 μm.

Photograph courtesy of P. Gordon-Weeks.

10.8 The growth cone controls the path taken by the growing axon

An early event in the differentiation of a neuron is the extension of its axon by the **growth cone** at the axon tip (Fig. 10.20). The growth cone both moves and senses its environment. As in other cells capable of migration, for example the primary mesenchyme of the sea urchin embryo (see Section 7.7), the growth cone can continually extend and retract filopodia, which help to pull the axon tip forward over the underlying substratum. Between the filopodia, the edge of the growth cone forms thin ruffles—lamellipodia—similar to those on a moving fibroblast (see Box 7B, pp. 269–270). Indeed, in its ultrastructure and mechanism of movement, the growth cone closely resembles the leading edge of a fibroblast crawling over a surface. Unlike a moving fibroblast, however, the extending axon also grows in size, with an accompanying increase in the total surface area of the neuron's plasma membrane. The additional membrane is provided by intracellular vesicles, which fuse with the plasma membrane.

The activity of the growth cone guides axon outgrowth, and is influenced by the contacts the filopodia make with other cells and with the extracellular matrix. In general, the growth cone moves in the direction in which its filopodia make the most stable contacts. In addition, diffusible substances can bind to receptors on the growth cone surface, and so influence its direction of migration. Some of these cues promote axon extension, whereas others inhibit it by causing the collapse of the growth cone. The extension and retraction of filopodia involves the assembly and disassembly of the actin cytoskeleton. Members of a family of intracellular signaling proteins, the Ras-related GTPases, are involved in the reorganization of the actin cytoskeleton. Activation of one of these, Rho, causes growth cones to stop extending, whereas Rac and Cdc42, other members of the family, are involved in growth-cone extension. But just how growth cones transduce extracellular signals so as to extend or collapse filopodia is not fully understood.

Axon growth cones are guided by two main types of cue—attractive and repulsive. In addition, cues can act either at long or short range, thus giving four ways in which the growth cone can be guided (Fig. 10.21). Long-range attraction involves diffusible **chemoattractant** molecules released from the target cells and is, in principle, similar to the chemotaxis of motile cells such as those of slime molds (see Section 7.15). In contrast, **chemorepellants** are molecules that repel migrating cells or axons. Both chemoattractant and chemorepellant proteins have been identified in the developing nervous system. Short-range guidance is mediated

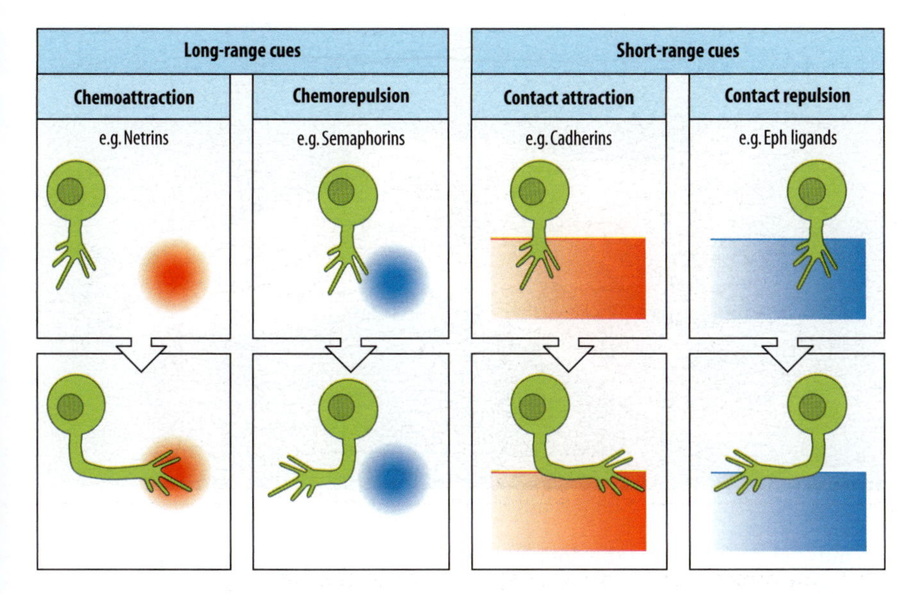

Fig. 10.21 Axon-guidance mechanisms. Four types of mechanism can contribute to controlling the direction in which a growth cone moves. Attractant molecules can either be present as a diffusible gradient of chemoattractant, or be bound to the substratum, leading to contact-dependent attraction. Similarly, molecules that repel axon growth can be either bound or diffusible.

by contact-dependent mechanisms involving molecules bound to other cells or to the extracellular matrix; again, such interactions can be either attractive or repulsive. Cadherins, for example, can act as short-range attractors of neurons, while ephrins can mediate repulsion on contact. However, signals often function as both attractants and repellants, depending on the cellular context.

A number of different extracellular molecules that guide axons have been identified. The semaphorins are among the most prominent of the conserved families of axon-guidance molecules. There are seven different subfamilies, three of which are secreted whereas the others are associated with the cell membrane. There are two classes of neuronal receptors for semaphorins—the plexins and neuropilins. Axon repulsion is a major theme of semaphorin action. Some, however, can both attract and repel growth cones; which of these they do will depend on the nature of the neuron and on the semaphorin receptors present in the neuron membrane. Even different regions of the same neuron can respond in opposite ways to the same signal—the apical dendrites of pyramidal neurons in the cortex grow towards a source of semaphorin 3A, whereas their axons are repelled. This difference in response appears to be due to the presence of guanylate cyclase, a component of the pathway that transduces the semaphorin signal, in the dendrites but not in the axon. Other classes of neuronal guidance proteins, such as the secreted netrins, are bifunctional, in that they attract some neurons and repel others. The attractant function of netrins is mediated by the DCC ('deleted in colorectal cancer') family of receptors on growth cones. The Slit proteins are secreted glycoproteins with a repellant function. They repel a variety of axon classes by acting on receptors of the Robo family on neurons, but can also stimulate elongation and branching in sensory axons.

10.9 Motor neurons from the spinal cord make muscle-specific connections

An animal's muscular system enables it to carry out an enormous variety of movements, which all depend on motor neurons having made the correct connections with muscles. The development of the chick limb provides a good model for studying how these connections are set up. Motor neurons at different dorso-ventral positions within the spinal cord innervate distinct target muscles within the limb. Those near the midline project to ventral muscle, whereas lateral neurons project

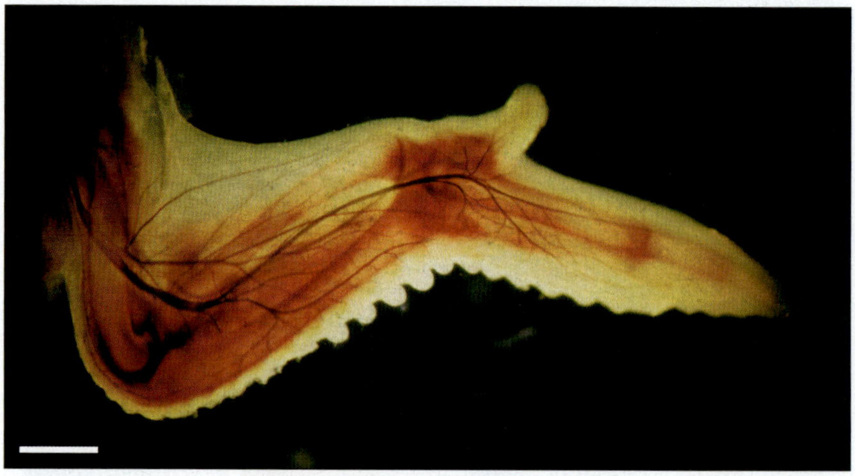

Fig. 10.22 Innervation of the embryonic chick wing. The nerves are stained brown and include both sensory and motor neurons. Scale bar = 1 mm.

Photograph courtesy of J. Lewis.

to muscles derived from the dorsal muscle mass. Each neuronal subtype expresses a combination of LIM homeodomain proteins (see Fig. 10.17), which may provide the neurons with positional identity and enable their axons to select a particular pathway (see Section 10.7).

The pattern of innervation of the limb muscles of the chick is rather precise (Fig. 10.22). In the chick embryo, the motor neuron axons that enter a developing limb appear to make their connections with particular muscles from an early stage. When the motor axons growing out from the spinal cord first reach the base of the limb, they are all mixed up in a single bundle. At the base of the developing limb, however, the axons separate out and form new nerves containing only those axons that will make connections with particular muscles.

Evidence that local cues in the limb are responsible for this sorting-out comes from inverting, along the antero-posterior axis, a section of the spinal cord whose motor neurons will innervate the hindlimb. After inversion, motor axons from lumbo-sacral segments 1–3, which normally enter anterior parts of the limb, initially grow into posterior limb regions, but then take novel paths to innervate the correct muscles. So, even when the axon bundles enter in reverse order, the correct relationship between motor neurons and muscles is achieved. Motor axons can thus clearly find their appropriate muscles when their displacement relative to the limb is small. But with a large displacement, this is not the case. A complete dorso-ventral inversion of the limb bud results in the axons failing to find their appropriate targets. In such cases, the axons follow paths normally taken by other neurons. The mesoderm of the limb thus provides local cues for the motor axons and the axons themselves have an identity that allows them to choose the correct pathway.

Motor neurons in the medial division of the LMC (see Fig. 10.17) innervate ventral muscle masses in the limb, while those in lateral divisions innervate dorsal limb muscles. EphA4 signaling in the lateral set of motor axons is a major determinant of the dorsal route they take into the limb. EphA4 is thought to mediate a repulsive interaction with its ligand ephrin A, which is restricted to the ventral limb mesenchyme. Migration of medial motor axons into the ventral mesenchyme appears to be regulated by another set of receptors that similarly respond to repulsive signals from the dorsal limb mesenchyme. The expression of EphA4 is determined by the LIM homeodomain proteins (see Section 10.7), with Lim1 expression in the lateral motor neurons leading to raised levels of EphA4 whereas Isl1 expression in the medial axons leads to a reduction in EphA4. The concentration

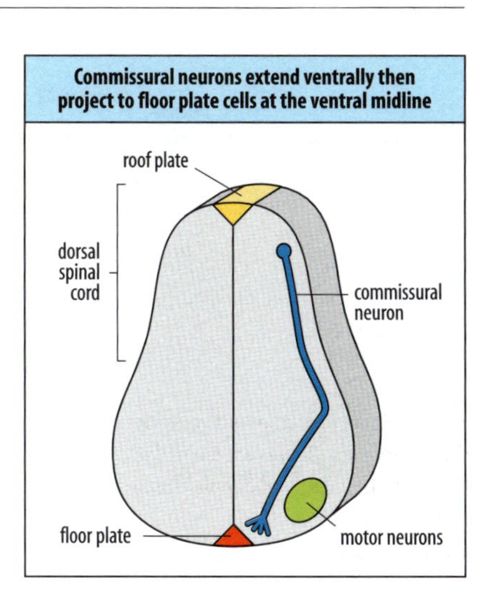

Fig. 10.23 Chemotaxis can guide commissural axons in the spinal cord. In the chick, spinal cord commissural neurons extend ventrally, and then toward the floor plate cells.

After Tessier-Lavigne, M., et al.: 1991.

of ephrin A in the dorsal limb mesenchyme is apparently kept at a low level by the expression there of the LIM protein Lmxb1.

10.10 Axons crossing the midline are both attracted and repelled

Early in the development of the central nervous system, neurons have to make a decision with respect to the midline—should they cross it or not? If they do cross it, will the axons cross it again? These choices are very clear in the insect and vertebrate spinal cords. Whereas the development of the nervous system is largely symmetrical about the midline of the body, and proceeds independently on either side, one half of the body needs to know what the other half is doing. So, in *Drosophila*, although some neurons remain entirely within the lateral half of the body in which they were formed, many axons cross the ventral midline of the embryo, and those that do cross it do not cross back. In vertebrates, axons of commissural neurons, which link the two sides of the cord, similarly cross the ventral midline of the developing brain and spinal cord and then move away. Mechanisms that allow crossing, but not recrossing, of the midline are involved in both cases.

In the vertebrate spinal cord, commissural neurons develop in the dorsal region and send their axons ventrally along the lateral margin of the cord (Fig. 10.23). About halfway, the axons make an abrupt turn and project to the floor plate in the ventral midline, bypassing the motor neurons. Netrin-1 is a key attractant produced in the floor plate and in the midline, and knock-out of the mouse *netrin-1* gene, or the gene for its receptor DCC, results in abnormal commissural axon pathways (Fig. 10.24). Other proteins involved in guiding neuronal migration include semaphorins and ephrins.

Axons approaching the midline provide a good example of the long-range attraction and repulsion of growth cones. Axons not destined to cross the midline are repelled by the protein Slit, present at the midline. Slit binds to receptors called Robo1 and Robo2 on the axon and exerts a chemorepellant effect.

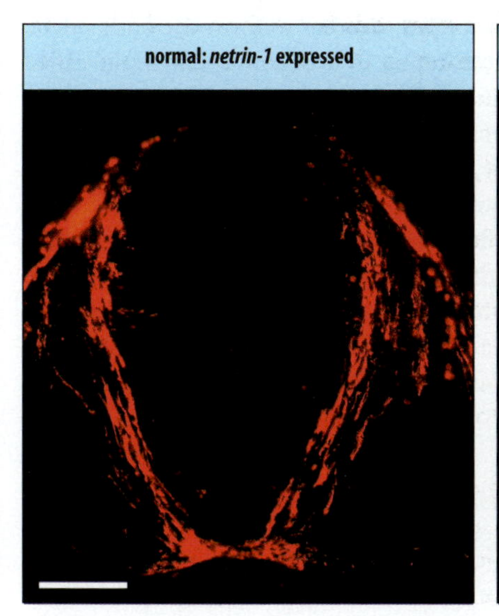

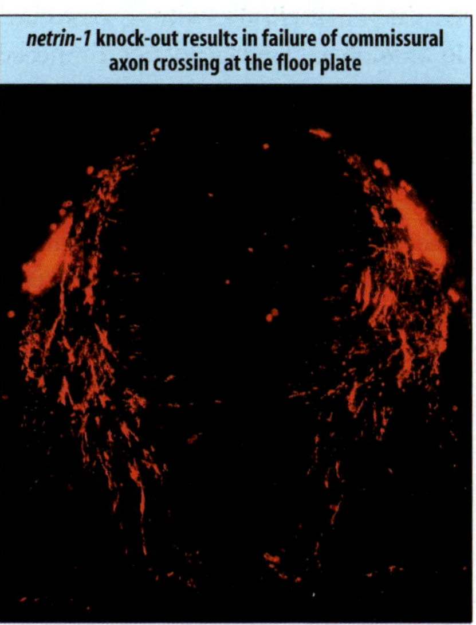

Fig. 10.24 Effect of *netrin-1* gene knock-out in mice. In mice lacking *netrin-1*, the commissural axons do not migrate toward the floor plate. Scale bar = 0.1 mm.

Photographs courtesy of M. Tessier-Lavigne, from Serafini, T., et al.: 1996.

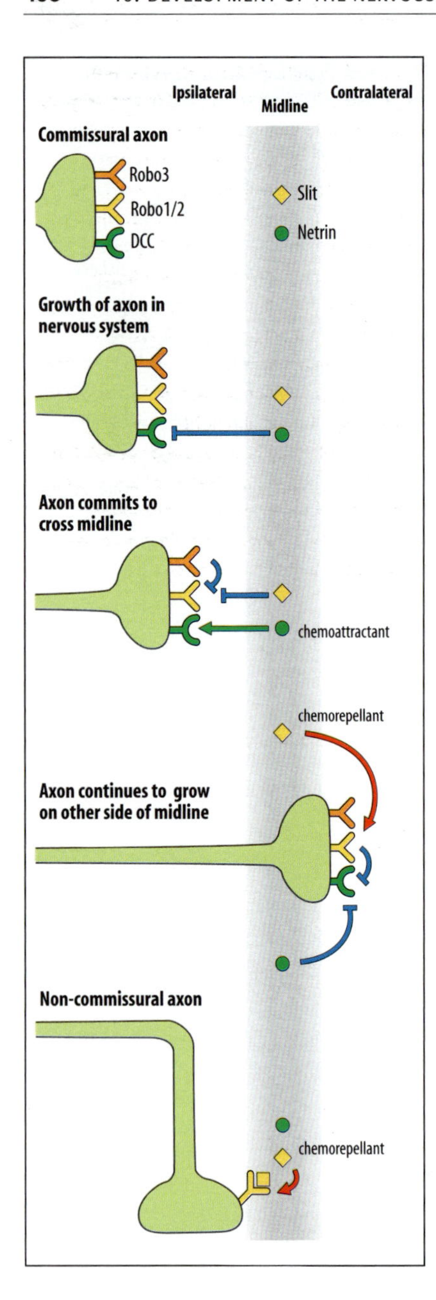

Fig. 10.25 Commissural neurons are attracted to the floor plate at the ventral midline by the chemoattractant Netrin-1, which signals attraction via the netrin-1 receptor DCC. Once commissural axons reach and cross the midline, they remain on the contralateral side through a mechanism that involves a second set of floor-plate guidance factors, the Slit proteins. Slit proteins, acting via Robo receptors, function to repel commissural axons once they have crossed the midline. But since commissural axons express Robo1 and Robo2 as they approach the floor plate, the repellant activities of Slit proteins must be inactivated prior to the crossing of commissural axons. The inactivation process is achieved by the expression of a Robo-receptor-related protein termed Rig-1 or Robo3. Expression of Robo3 prevents Robo1 and Robo2 from transducing Slit signals. But once commissural axons have crossed the midline, Robo3 expression ceases, permitting Slit to activate Robo1 and Robo2 and repel commissural axons.

For commissural axons approaching the midline, the chemorepellant effect of Slit is overcome by the Robo receptor-related protein Robo3 expressed on the neuron, which interferes with Slit signaling through Robo1 and Robo2 (Fig. 10.25). Once the axons cross the midline, Robo3 ceases to be expressed and the growth cone is repelled from the midline again by Slit proteins. After commissural neurons have crossed the midline in the mouse, they make a sharp anterior turn toward the brain. This is thought to be due to a gradient of Wnt proteins in the floor plate with the high point anterior.

In the fly, a protein called Commissureless expressed in the growth cones of commissural axons approaching the midline reduces the number of Robo1 receptors on the axonal surface, rendering the axons insensitive to Slit and enabling them to cross. Commissureless thus plays the same role as Robo3 in vertebrates.

10.11 Neurons from the retina make ordered connections on the tectum to form a retino-tectal map

The vertebrate visual system has been the subject of intense investigation for many years, and the highly organized projection of neurons from the eye to the brain is one of the best models we have to show how topographic neural projections are made. A characteristic feature of the vertebrate brain is the presence of topographic maps. That is, the neurons from one region of the nervous system project in an ordered manner to one region of the brain, so that nearest-neighbor relations are maintained. The most studied topographical projection is that of the optic nerve from the retina into the brain.

The retina develops light-sensitive cells that indirectly activate neurons—the retinal ganglion cells—whose axons are bundled together to form the optic nerve. In amphibians and birds this connects the retina to a region of the brain called the **optic tectum**. In mammals, the main target for retinal neurons is the **lateral geniculate nuclei** in the forebrain, from which other neurons convey the signal to the visual cortex, but there is also a pathway from the retina to another region of the brain, the **superior colliculus**, which is equivalent to the retino-tectal pathway. In all cases, retinal neurons map in a highly ordered manner onto their target structures.

In amphibians, the optic nerve from the right eye makes connections with the left optic tectum, while the nerve from the left eye makes connections to the right side of the brain. The optic nerve from each eye is made up of thousands of axons, which connect with the optic tectum in a highly ordered manner: there is a

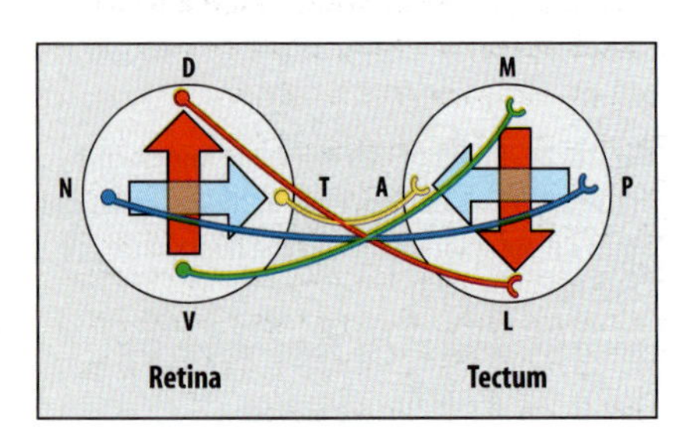

Fig. 10.26 The retina maps onto the tectum. Dorsal (D) retinal neurons connect to the lateral (L) side of the tectum and ventral (V) retinal neurons connect to the medial (M) tectum. Similarly, temporal (T) (or posterior) retinal neurons connect to anterior (A) tectum and nasal (N) (or anterior) retinal neurons connect to the posterior (P) tectum.

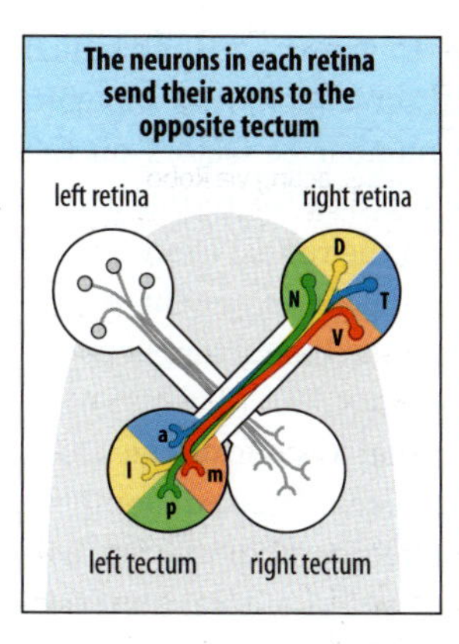

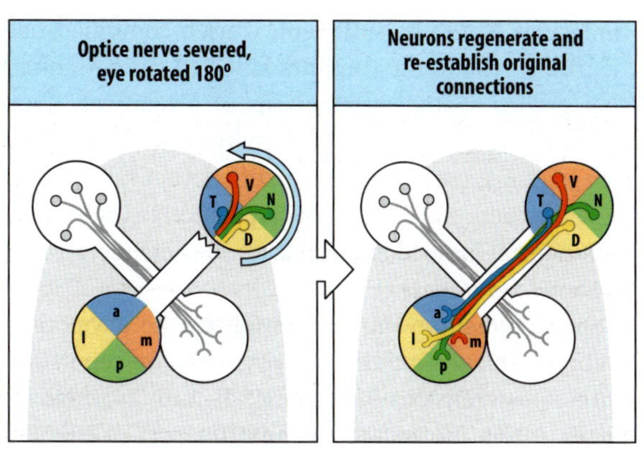

Fig. 10.27 Retino-tectal connections in amphibians are re-established in the original arrangement after severance of the optic nerve and rotation of the eye. Left panel: neurons in the optic nerve from the left eye mainly connect to right optic tectum and those from the right eye to the left optic tectum. There is a point-to-point correspondence between neurons from different regions in the retina (nasal (N), temporal (T), dorsal (D), ventral (V)) and their connections in the tectum (posterior (p), anterior (a), lateral (l), medial (m), respectively). Center and right panels: if one optic nerve of a frog is cut and the eye rotated dorso-ventrally through 180°, the severed ends of the axons degenerate. When the neurons regenerate, they make connections with their original sites of contact in the tectum.

point-to-point correspondence between a position on the retina and one on the tectum. Neurons in the dorsal region of the retina project to the ventral region of the tectum and regions in the anterior (nasal) region of the retina project to the posterior region of the tectum (Fig. 10.26). The development of the projection from the retina to the tectum is initially only reasonably precise but, in mammals at least, is fine-tuned later by nerve impulses from the retina.

Remarkably, in some lower vertebrates, such as fish and amphibians, the pattern of connections can be re-established with precision when the optic nerve is cut. The ends of the axons distal to the cut die, new growth cones form, and axon outgrowth reforms connections to the tectum. In frogs, even if the eye is inverted through 180°, the axons still find their way back to their original sites of contact (Fig. 10.27). However, the animals subsequently behave as if their visual world has been turned upside down: if a visual stimulus, such as a fly, is presented above the inverted eye, the frog moves its head downward instead of upward (Fig. 10.28), and can never learn to correct this error.

From such experiments it was suggested that each retinal neuron carries a chemical label that enables it to connect reliably with an appropriately chemically labeled cell in the tectum. This is known as the **chemoaffinity hypothesis** of connectivity. The chemical labels probably do not provide unique lock-and-key interactions although the genome very probably has the capacity to do so: 50 unique

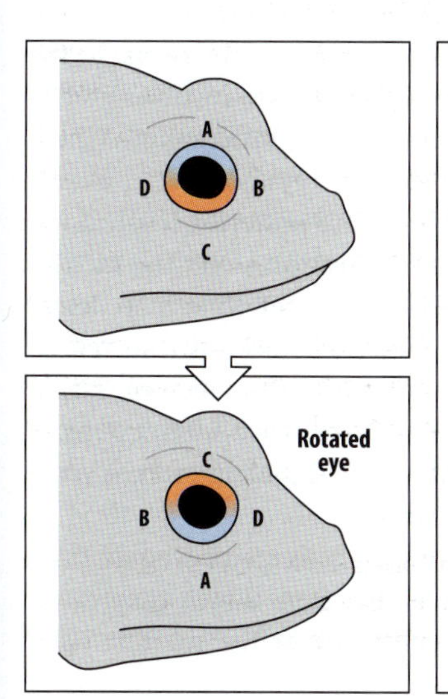

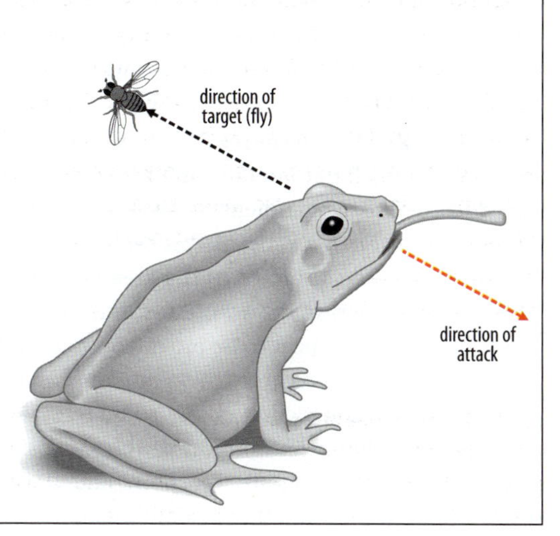

Fig. 10.28 The effect of eye rotation on the visual behavior of a frog. After severance of the optic nerve and rotation of the eye, the retina re-establishes its original connections with the optic tectum (see Fig. 10.27). However, because the eye has been rotated, the image falling on the tectum is upside down compared with normal. When the frog sees a fly above its head, it thinks the fly is below it, and moves its head downward to try to catch it.

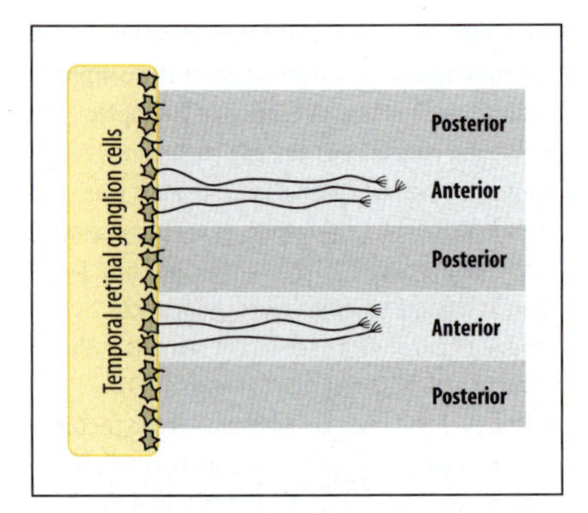

Fig. 10.29 Choice of targets by retinal axons. If pieces of temporal retina are placed next to a 'carpet' of alternating stripes (90 μm wide) of anterior and posterior tectal membranes, the temporal retinal cells only extend axons onto anterior tectal membranes.

molecules would be sufficient, if used combinatorially, to giving each individual neuron in the brain a unique label. In the retino-tectal projection, it is thought, that graded spatial distributions of a relatively small number of factors on the tectum provide positional information, which can be detected by the retinal axons. The spatially graded expression of another set of factors on the retinal axons would provide them with their own positional information. The development of the retino-tectal projection could thus, in principle, result from the interaction between these two gradients.

Such gradients were first found in the developing visual system of the chick embryo. An axon-guiding activity based on repulsion has been detected along the antero-posterior axis of the chick tectum. Normally, the temporal (posterior) half of the retina projects to the anterior part of the tectum and the nasal (anterior) retina projects to the posterior tectum. When offered a choice between growing on posterior or anterior tectal cells, temporal axons from an explanted chick retina show a preference for anterior tectal cells (Fig. 10.29). This choice is mediated by repulsion of the axons—as shown by the collapse of their growth cones—by a factor located on the surface of posterior tectal cells.

Ephrins and their receptors provide the gradients that mediate specificity as they are expressed in reciprocal gradients in retina and tectum. There are complementary gradients on both the neurons from the retina and their targets in the brain. In the retino-collicular map of the mouse, members of the EphA family of receptors together with their ephrin A ligands are the basis of patterning the map in the antero-posterior dimension. EphAs are expressed in a gradient, low in the nasal region, high in the temporal region, of the mouse retina, while ephrin A ligands are graded in the superior colliculus from high posterior to low anterior. Thus retinal neurons with the most EphA receptors connect to collicular targets with the fewest ephrin As. Retinal axons bearing low levels of the receptor move into areas containing high levels of the ligands, whereas retinal axons bearing high levels of receptor stop before they reach areas with high levels of the ligand (Fig. 10.30). One explanation for this is that binding of ligand to the receptor sends a repulsive signal, and the axon ceases to migrate when this signal reaches a threshold value. The strength of the signal received by the migrating axon will be proportional to the product of the concentrations of receptor and ligand. Thus, the threshold will be reached when there is a high level of both receptor and ligand, but not when there is a low level of receptor, even when ligand level is high. In this model, the axons are competing with each other to connect to the target tissue.

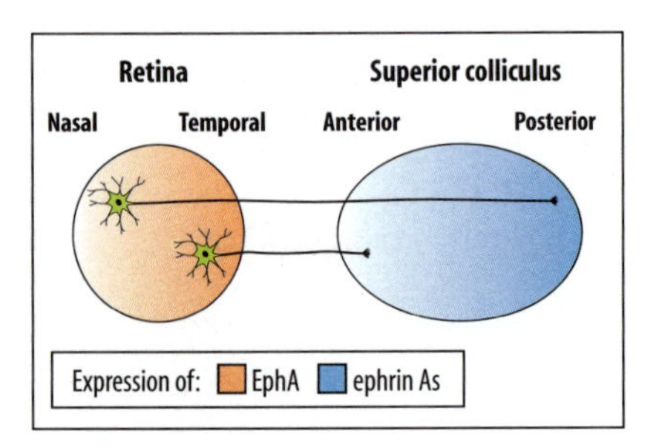

Fig. 10.30 Complementary expression of the ephrins and their receptors in the mouse retino-collicular projection. The receptor EphA is a receptor tyrosine kinase and ephrin As are its ligands. Temporal retinal neurons, which express high levels of EphA on their surface, are repulsed by the posterior tectum, where high levels of ephrins A2 are also present, but can make connections with the anterior tectum. Nasal neurons from the retina make the best contacts in the posterior tectum, as they express low levels of EphA and thus are only repulsed by the high levels of ephrins in this area.

Retino-collicular mapping from the dorso-ventral dimension of the retina to the medial-lateral dimension of the tectum or superior colliculus also involves graded expression of Ephs and ephrins, in this case EphB–ephrin B interactions. These mediate adhesive and attractive interactions rather than repulsion, and enable axon terminals that have reached their appropriate antero-posterior level in the tectum to branch out laterally and arrive at their final position along the medial-lateral axis. But this gradient is not sufficient to account for the final map, based as it is on attraction only: if this were the only factor at work, all the retinal neurons would be attracted to the area of highest ephrin B in the tectum, rather than mapping across the whole tectum. The necessary repulsive interactions in these dimensions appear to be provided by gradients of the signaling protein Wnt3 in the tectum and a receptor, Ryk, on the retinal axons.

In species with binocular vision, some retinal ganglion cells are prevented from crossing the midline and project to the same side as that from which they came. These axons express EphB1 receptors and sense ephrin B2 at the midline optic chiasm, which prevents them from crossing.

Summary

Growth cones at the tip of the extending axon guide it to its destination. Filopodial activity at the growth cone is influenced by environmental factors, such as contact with the substratum and with other cells, and can also be guided by chemotaxis. Guidance involves both attraction and repulsion. In the development of motor neurons that innervate vertebrate limb muscles, the growth cones guide the axons so as to make the correct muscle-specific connections, even when their normal site of entry into the limb is disturbed.Transcription factors give motor neurons an identity that determines their pathway. Attraction and repulsion control axons crossing the midline in both *Drosophila* and vertebrates. Gradients in diffusible molecules are probably responsible for the directional growth of commissural axons in the spinal cord. Neuronal guidance of axons from the retina to make the correct connections with the optic tectum involves gradients in cell-surface molecules both on the tectal neurons and on the retinal axons that can promote or repel growth cone approach, and there is competition between neurons for sites.

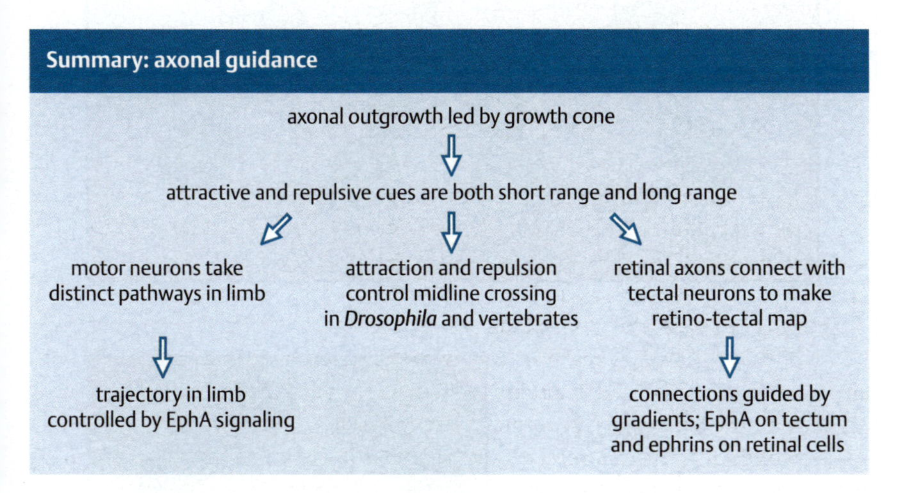

Synapse formation and refinement

When axons reach their targets they make specialized connections—**synapses**—which are essential for signaling between neurons and their target cells. Formation of synapses in the correct pattern is a basic requirement of any developing nervous system. The connections may be made with other nerve cells, with muscles, and also with certain glandular tissues. Here, we focus mainly on the development and stabilization of synapses at the junctions between motor neurons and muscle cells in vertebrates, which are called **neuromuscular junctions**. There are very many neuronal cell types—hundreds if not thousands—and how they match up to make synapses reliably is a central problem. The molecular mechanisms are only beginning to be known. Studies on the vertebrate neuromuscular junction have guided much of the understanding of synapse formation.

Setting up the organization of a complex nervous system in vertebrates involves refining an initially rather imprecise organization by extensive programmed cell death (see Section 8.13). The establishment of a connection between a neuron and its target appears to be essential not only for the functioning of the nervous system but also for the very survival of many neurons. Neuronal death is very common in

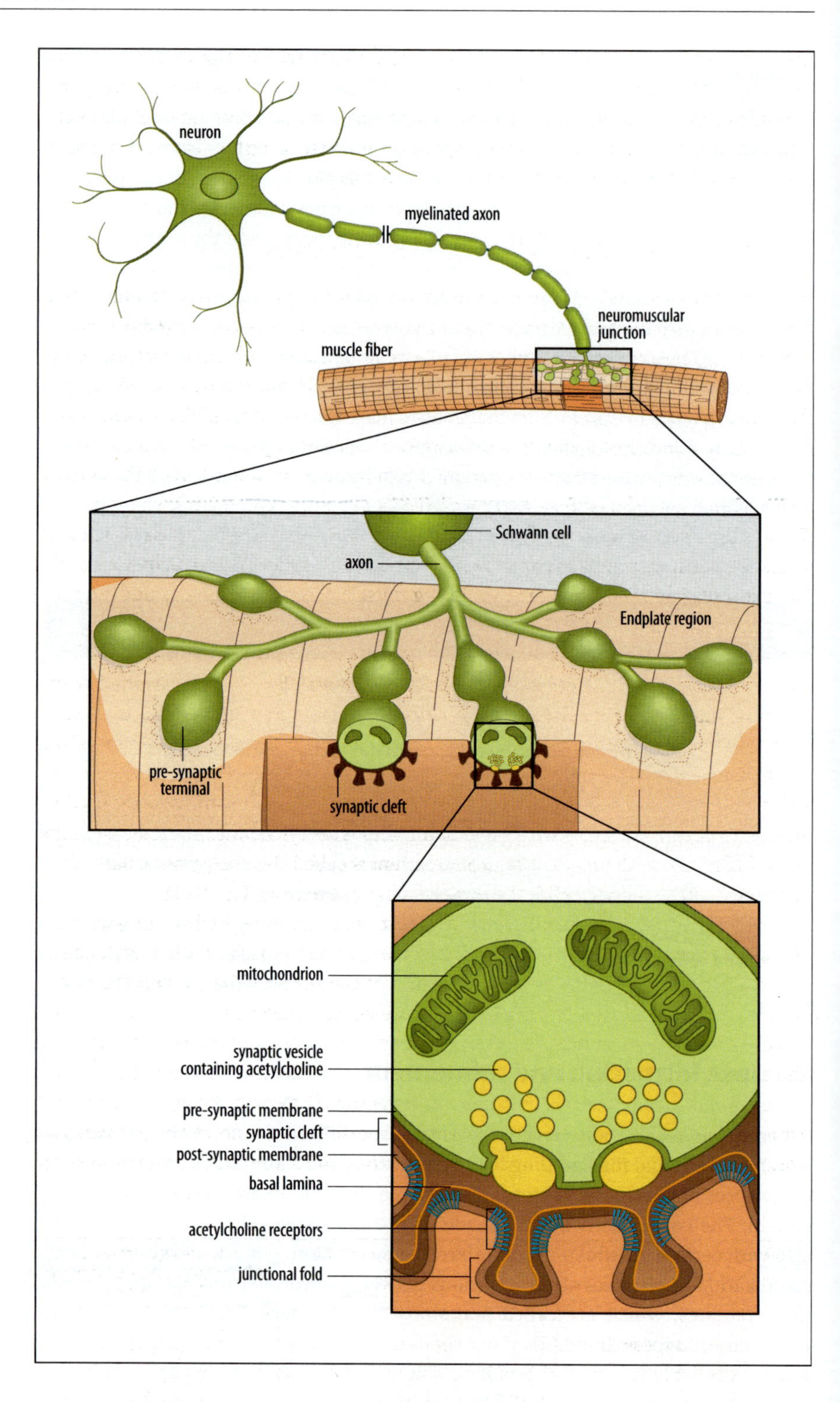

Fig. 10.31 Structure of the vertebrate neuromuscular junction. The motor neuron axon, which is covered in a myelin sheath produced by the glial Schwann cell, innervates a muscle fiber. At the neuromuscular junction, the axon branches in the endplate region, and makes synaptic connections with the muscle membrane. Communication between nerve and muscle is by release of the neurotransmitter acetylcholine from synaptic vesicles into the synaptic cleft. Acetylcholine diffuses across the cleft and binds to acetylcholine receptors on the muscle cell membrane.

After Kandel, E.R., et al.: 1991.

the developing vertebrate nervous system; too many neurons are produced initially, and only those that make appropriate connections survive. Survival depends on the neuron receiving neurotrophic factors, such as nerve growth factor, which are produced by the target tissue and for which neurons compete.

A special feature of nervous system development is that fine-tuning of synaptic connections depends on the interaction of the organism with its environment and the consequent neuronal activity. This is particularly true of the vertebrate visual system, where sensory input from the retina in a period immediately after birth modifies synaptic connections so that the animal can perceive fine detail. Again, this refinement seems to involve competition for neurotrophic factors. We return to this topic after first considering synapse formation.

10.12 Synapse formation involves reciprocal interactions

We will start our look at synapse formation with the neuromuscular junction as this is one of the most intensively studied and best understood types of synapse. The mature neuromuscular junction is a complex structure involving extensive modification of the nerve ending and the muscle cell membrane (Fig. 10.31). Just before the junction, the axon terminal branches into a network-like arrangement. Each branch ends in a swelling, which is in contact with a special endplate region on the muscle fiber. The axon's plasma membrane is separated from the muscle cell's plasma membrane by a narrow cleft (the **synaptic cleft**) filled with extracellular material secreted by both the neuron and the muscle cell. A basal lamina of extracellular material forms around the muscle cell surface. The structure comprising axon plasma membrane, the opposing muscle cell plasma membrane, and the cleft between them, is a synapse.

Electrical signals cannot pass across the synaptic cleft, and for the neuron to signal to the muscle, the electrical impulse propagated down the axon is converted at the terminal into a chemical signal. This is a chemical neurotransmitter that is released into the synaptic cleft from **synaptic vesicles** in the axon terminal. Molecules of the neurotransmitter diffuse across the cleft and interact with receptors on the muscle cell membrane, causing the muscle fiber to contract. The neurotransmitter used by motor neurons connecting with skeletal muscles is acetylcholine. Because the signal travels from nerve to muscle, the axon terminal is called the **pre-synaptic** part of the junction and the muscle cell is the **post-synaptic** partner (see Fig. 10.31).

Development of a neuromuscular junction is progressive. Before the arrival of the axon there is already some pre-patterning on the muscle, with acetylcholine receptors concentrated in the central region of the muscle fiber. When the motor neuron terminal arrives, it releases the proteoglycan Agrin into the basal lamina, which promotes the clustering of acetylcholine receptors and the specialization of the muscle cell surface at the site (Fig. 10.32). Activation of the muscle specific kinase MuSK in the muscle cell induces receptor clustering through the protein Rapsyn. Signals from the muscle in turn induce differentiation of the pre-synaptic zone on the axon terminal and align this zone with the post-synaptic area on

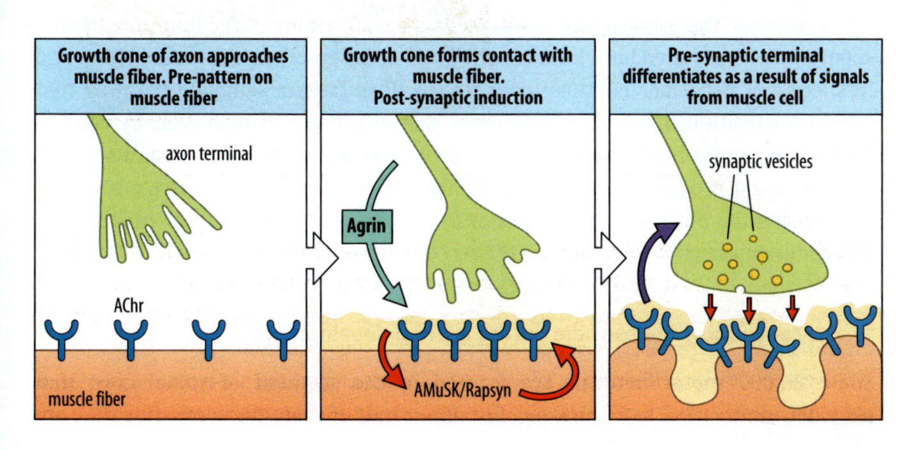

Fig. 10.32 Development of the neuromuscular junction. Before contact by an axon, acetylcholine receptors (AChr) are gathered in a central region of the muscle fiber. When a motor axon growth cone contacts a muscle fiber it releases the proteoglycan Agrin (blue arrow) into the extracellular material. This promotes acetylcholine receptor clustering in the muscle cell membrane and specialization of the post-synaptic muscle cell surface. A signal (as yet unknown) from the muscle cell (purple arrow) then helps to induce the differentiation of the pre-synaptic terminal, with the formation of synaptic vesicles. Synaptic signaling (small red arrows) can then begin.

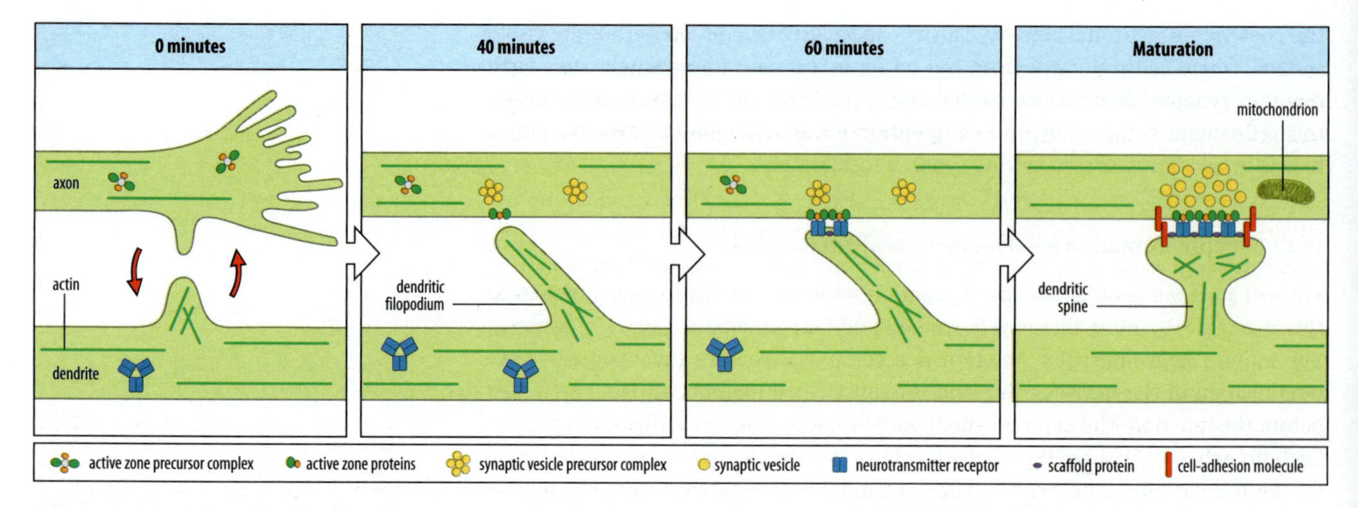

Fig. 10.33 Interneuronal synapse formation. A dendrite sends out filopodia that contact the axon growth cone. Reciprocal signaling (red arrows) initiates synapse formation aided by cell-adhesion molecules. In the axon, proteins that compose the pre-synaptic active zone gather at the site of contact along with synaptic vesicle precursors. In the dendrite, neurotransmitter receptors are synthesized and inserted into the post-synaptic membrane along with post-synaptic scaffold proteins (red). The post-synaptic connection develops into typical dendritic spines.

the muscle. These signals are not yet clearly identified, but the glycoprotein laminin β2 in the basal lamina is required for correct alignment and full pre-synaptic differentiation.

The neuromuscular junction is a highly specialized type of synapse, but synapse formation within the vast network of neurons that form the central nervous system follows similar principles. Synapses are made between the axon terminal of one neuron and a dendrite, cell body, or more rarely the axon, of another neuron. An individual motor neuron in the spinal cord, for example, will receive input from thousands of other neurons on its dendrites and cell body. As with neuromuscular junctions, interneuronal synapse formation is likely to involve reciprocal signaling between axon and dendrite. Synapses between neurons can form quite rapidly, on a time scale of about an hour, and observations suggest that filopodia on the dendrite initiate synapse formation by reaching out to the axon (Fig. 10.33). Signals from the axon, including release of neurotransmitter, may initially guide the dendrite filopodia. Many other factors have been shown to promote synapse formation, including cell-adhesion molecules and Wnt signaling.

Cell-adhesion molecules implicated in synapse formation include the neuroligins, which are post-synaptic membrane proteins that bind to proteins called β-neurexins on the pre-synaptic axon surface and induce differentiation in the pre-synaptic terminal. Neuroligins promote the function of excitatory synapses rather than inhibitory synapses. Cadherins, protocadherins, and members of the immunoglobulin superfamily (see Box 7A, p. 259) have also been found localized to synapses.

In *Drosophila*, the cell-adhesion molecule Dscam, a member of the immunoglobulin superfamily, is required both for axon guidance in the embryonic nervous system and in specifying particular neuronal connections. It initiates a signaling pathway that leads to activation of Rho-family GTPases that affect axonal migration by causing changes in the actin cytoskeleton. The Dscam gene is exceptional in that alternative splicing could produce an estimated 38,000 distinct forms of cell-surface receptor, thus raising the possibility that different forms of the protein might be produced in different neurons and contribute to the recognition of self from non-self.

10.13 Many motor neurons die during normal development

Some 20,000 motor neurons are formed in the segment of spinal cord that provides innervation to a chick leg, but about half of them die soon after they are formed (Fig. 10.34). Cell death occurs after the axons have grown out from the cell bodies and entered the limb, at about the time that the axon terminals are reaching their potential targets—the skeletal muscles of the limb. The role of the target muscles in preventing cell death is suggested by two experiments. If the leg bud is removed, the number of surviving motor neurons decreases sharply. When an additional limb bud is grafted at the same level as the leg, providing additional targets for the axons, the number of surviving motor neurons increases.

Survival of a motor neuron depends on its establishing contacts with a muscle cell. Once a contact is established, the neuron can activate the muscle, and this is followed by the death of a proportion of the other motor neurons that are approaching the muscle cell. Muscle activation by neurons can be blocked by the drug curare, which prevents neuromuscular signal transmission, and this block results in a large increase in the number of motor neurons that survive.

Even after neuromuscular connections have been made, some are subsequently eliminated. At early stages of development, single muscle fibers are innervated by axon terminals from several different motor neurons. With time, most of these connections are eliminated, until each muscle fiber is innervated by the axon terminals from just one motor neuron (Fig. 10.35). This is due to competition between the synapses, with the most powerful input to the target cell destabilizing the less powerful inputs to the same target. This elimination process and the maturation of a stable neuromuscular junction take up to 3 weeks in the rat.

The well-established matching of the number of motor neurons to the number of appropriate targets by these mechanisms suggests that a general mechanism for the development of nervous system connectivity in all parts of the vertebrate nervous system is that excess neurons are generated and only those that make the required connections are selected for survival. This mechanism is well suited to regulating cell numbers by matching the size of the neuronal population to its targets. We now look in more detail at the neurotrophic factors that promote neuronal survival, and then consider the role of neural activity in the elimination of neuromuscular synapses.

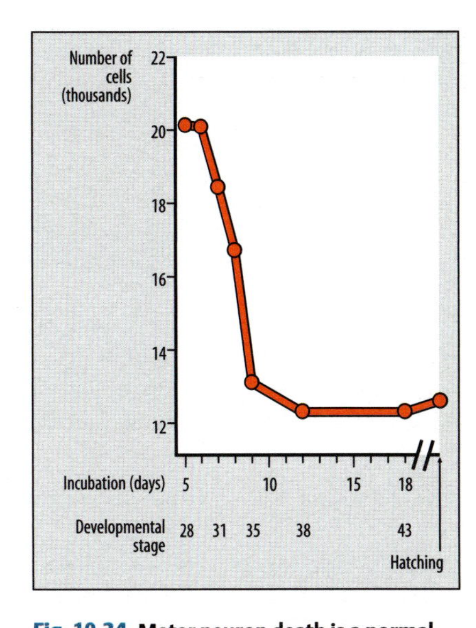

Fig. 10.34 Motor neuron death is a normal part of development in the chick spinal cord. The number of motor neurons innervating a limb decreases by about half before hatching, as a result of programmed cell death during development. Most of the neurons die over a period of 4 days.

10.14 Neuronal cell death and survival involve both intrinsic and extrinsic factors

Neuronal death during development takes place as a result of the activation of the apoptotic pathway (see Section 8.13). Apoptosis can be triggered or prevented either by an intrinsic developmental program or by external factors. In the development of the nematode, for example, particular neurons and other cells are intrinsically programmed to die (see Chapter 5), and the specificity of cell death can be shown to be under genetic control. Loss of transcription factors involved in apoptosis, for example, leads to the survival of two pharyngeal neurosecretory neurons that would normally die.

As we saw in Section 8.13, members of the Bcl-2 protein family control the apoptotic pathway in vertebrates—some acting to promote apoptosis and some to prevent it. A key family member promoting neuronal apoptosis during development is *Bax*; mice mutant for *Bax* have increased numbers of motor neurons in the face. A protein called Survivin is known to inhibit apoptosis during early neurogenesis. Mutant mice in which Survivin is deleted from embryonic day 10.5 onwards are born with much smaller brains than normal and numerous foci of apoptosis throughout the brain and spinal cord and die soon after birth.

In many developmental situations it seems that cells are programmed to die unless they receive specific survival signals. Extracellular factors play a key role in

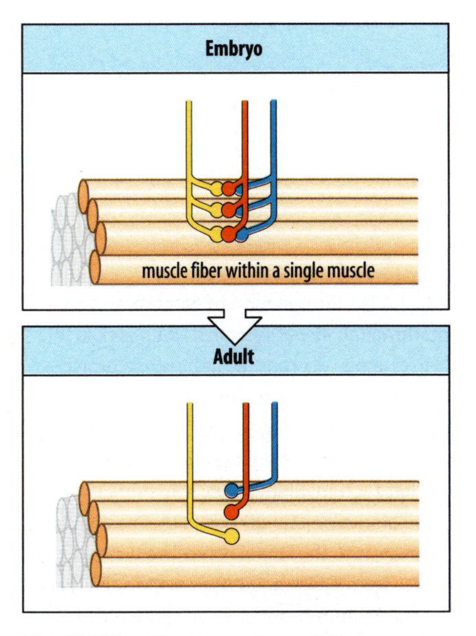

Fig. 10.35 Refinement of muscle innervation by neural activity. Initially, several motor neurons innervate the same muscle fiber. Elimination of synapses means that each fiber is eventually innervated by only one neuron.

After Goodman, C.S. and Shatz, C.J.: 1993.

neuronal survival in vertebrates. The first one identified was nerve growth factor (NGF), whose discovery was due to the serendipitous observation that a mouse tumor implanted into a chick embryo evoked extensive growth of nerve fibers towards the tumor. This suggested that the tumor was producing a factor that promoted axonal outgrowth. The factor was eventually identified as a protein, NGF, using axon outgrowth in culture as an assay. NGF is necessary for the survival of a number of types of neurons, particularly those of the sensory and sympathetic nervous systems.

NGF is a member of a family of proteins known as the **neurotrophins**. As well as NGF, they include brain-derived neurotrophic factor (BDNF), neurotrophin-3 (NT-3), neurotrophin-4/5 (NT-4/5), and glial cell-line-derived neurotrophic factor (GDNF). The receptors for the neurotrophins are receptor tyrosine kinases known as Trk proteins; the receptor for NGF, for example, is called TrkA. Different types of neuron require different neurotrophins for their survival, and the requirement for certain neurotrophins also changes during development. GDNF, for example, prevents the death of facial motor neurons, and is also expressed in developing limb buds. There is increasing evidence that individual neurotrophins act on a variety of neural cell types.

10.15 The map from eye to brain is refined by neural activity

The development and function of the nervous system depends not only on synapse formation but also on the regulated disassembly of previously functional synaptic connections. We have already considered how axons from the retina make connections with the tectum so that a retino-tectal map is established. This map is initially rather coarse-grained, in that axons from neighboring cells in the retina make contacts over a large area of the tectum. This area is much larger early on than at later stages of development, when the retino-tectal map is more finely tuned. Fine-tuning of the map results, as in muscle, from the withdrawal of axon terminals from most of the initial contacts, and requires neural activity. This requirement is seen particularly clearly in the development of visual connections in mammals. If a mammal, including a human, is deprived of vision during a critical period, then it will suffer from poor visual acuity.

In the mammalian visual system axons from the retina first connect to the lateral geniculate nucleus, onto which they map in an ordered manner (Fig. 10.36).

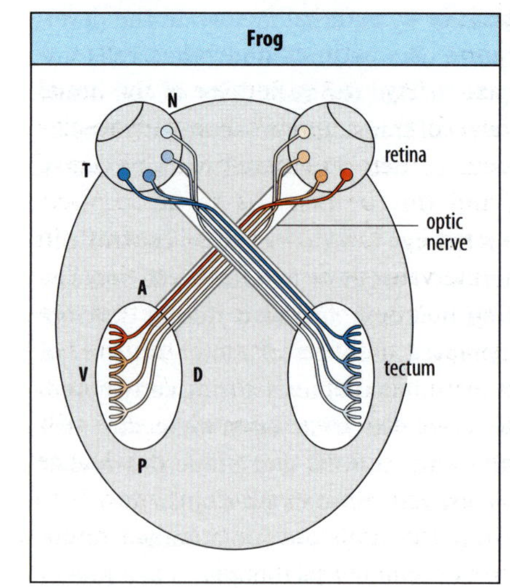

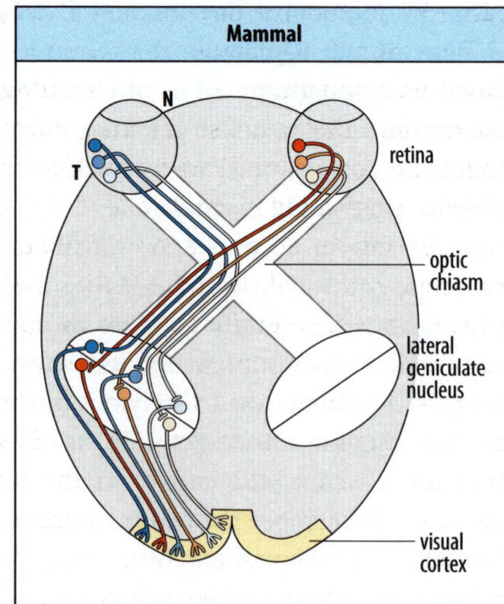

Fig. 10.36 Comparison of amphibian and mammalian visual systems. Left panel: in the frog tadpole, neurons from the retina project directly to the optic tectum, neurons from the left retina connecting to the right optic tectum and those from the right retina connecting to the left tectum. Right panel: in mammals with good binocular (stereo) vision, the main visual pathway is to the lateral geniculate nucleus (LGN), from which neurons project to the visual cortex. Neurons from one half of the retina project to the LGN on the same side as the eye, and those from the other half cross over to the opposite LGN at the optic chiasm. For the sake of clarity, the neurons from only one half of the retina are shown. N, nasal; T, temporal.

After Goodman, C.S. and Shatz, C.J.: 1993.

Input from one half of each eye goes to the opposite side of the brain, whereas input from the other half of each eye goes to the same side of the brain. Neurons from the lateral geniculate nucleus then send axons to the visual cortex. When there is a visual stimulus, the inputs from the retinal axons activate neurons in the lateral geniculate nucleus, which then activate neurons in the corresponding region of the visual cortex. There is thus input from both eyes at the same location in the cortex. The adult visual cortex consists of six cell layers, but we need only focus here on layer 4, which is where many axons from the lateral geniculate nucleus make connections.

In the lateral geniculate nucleus, as in the visual cortex, the neurons are arranged in layers. Each layer receives input from retinal axons from either the right or the left eye, but not both. Thus, from the beginning, inputs from left and right eyes are separated. However, layers in the lateral geniculate nucleus that receive left or right inputs both make connections with layer 4 of the visual cortex. At birth, these inputs overlap and are mixed, but with time the inputs from left and right eyes become separated into blocks of cortical cells about 0.5 mm wide, which are known as **ocular dominance columns** (Fig. 10.37). Adjacent columns respond to the same stimulus in the visual field; one column responding to signals from the left eye, and the next to signals from the right eye. This arrangement is essential for good binocular vision. The columns can be detected and mapped by making electrophysiological recordings. They can also be directly observed by injecting a tracer, such as radioactive proline, into one eye. The tracer is taken up by retinal neurons, transported by the optic nerve to the lateral geniculate, and from there to the visual cortex, where its pattern can be detected by autoradiography. This reveals a striking array of stripes representing the input from one eye (see Fig. 10.37).

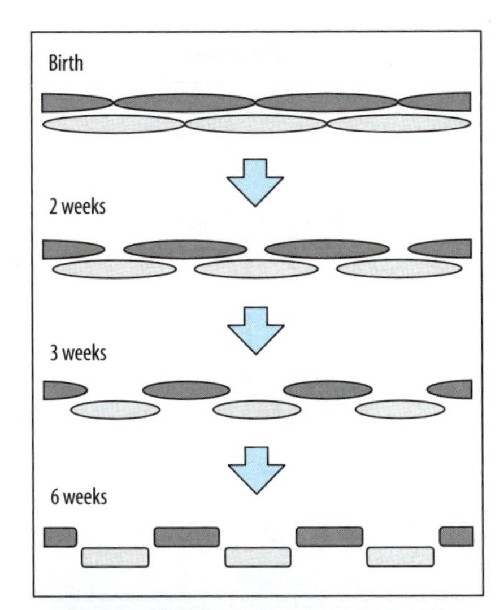

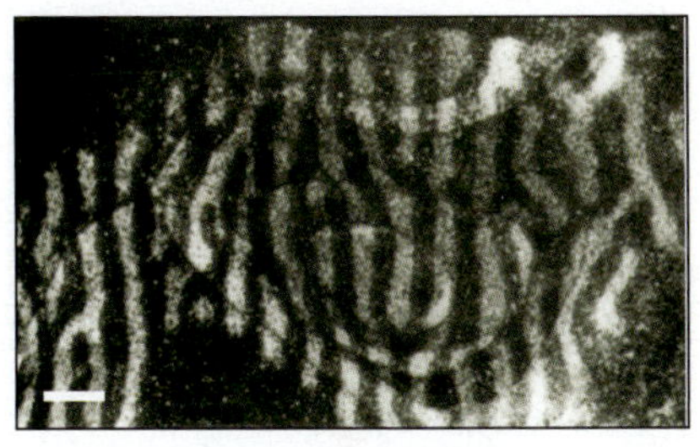

Fig. 10.37 Visualization of ocular dominance columns in the cortex. A radioactive tracer is injected into one eye, from where it is transported to the cortex through the neurons. Tracer injected at birth is broadly distributed in the cortex. Tracer injected at later times becomes confined to alternating columns of cortical cells (brighter stripes), representing the ocular dominance columns for that eye, as seen in the photograph. Scale = 1 mm.

After Kandel, E.R., et al.: 1995.

Neural activity and visual input are essential for the development and maintenance of the ocular dominance columns. While sensory input is important, spontaneous activity plays a key role. Initial formation of the stripes in non-human primates occurs before visual experience and thus involves spontaneously generated waves of action potentials in the mammalian retina. These waves may act through the release of neurotrophins that can remodel synaptic connections. If neural activity is blocked during development by the injection of tetrodotoxin, ocular dominance columns do not develop, and the inputs from the two eyes into the visual cortex remain mixed. If the input from one eye is blocked, the territory in the visual cortex occupied by the other eye's input expands at the expense of the blocked eye. The release of the inhibitory neurotransmitter γ-aminobutyric acid (GABA) is required to initiate the critical period.

The favored explanation for the formation of ocular dominance columns is based on competition between incoming axons (Fig. 10.38). Because of the initial connections that establish the map, individual cortical neurons can initially receive input from both eyes. Within a particular region, there will therefore be overlap of stimuli originating from the two eyes, and this overlap has to be resolved. Neighboring cells carrying input from the same eye tend to fire simultaneously in response to a visual stimulus; if they both innervate the same target cell, they can thus cooperate to excite it. As in muscle, stimulation of electrical activity in the target cell tends to strengthen the active synapses and suppress those that are not active at the time—cells that fire together, wire together. As there is competition between neurons for targets, this could generate discrete regions of cortical cells that respond only to one eye or the other, and so form the ocular dominance columns. Such a mechanism explains why experimental exposure of animals to continuous strobe lighting after birth, which causes simultaneous firing of neurons in both eyes, prevents the formation of ocular dominance columns.

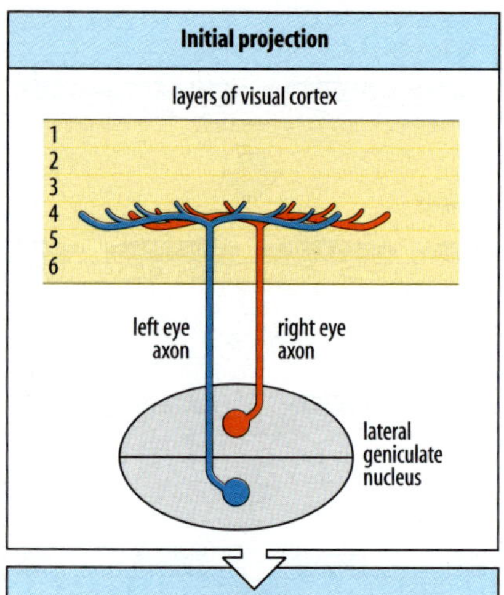

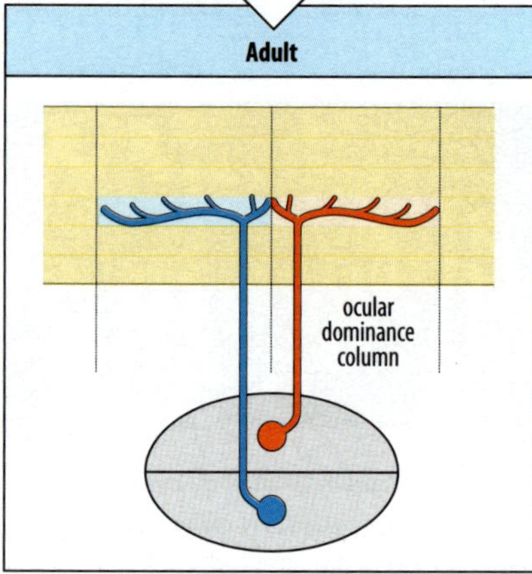

Fig. 10.38 Development of ocular dominance columns. Initially, neurons from the lateral geniculate nucleus, representing projections from both eyes and stimulated by the same visual stimulus, project to the same region of the visual cortex. (Only projection to layer 4 is shown here.) With visual stimulation, the neuronal connections separate out into columns, each representing innervation from only one eye. If stimulation by vision is blocked, ocular dominance columns do not form.

After Goodman, C.S. and Shatz, C.J.: 1993.

A possible mechanism for refining connections in response to neuronal activity involves local release of neurotrophins. A certain level of activation, or activation by two axons simultaneously, may induce the release of neurotrophins from the target cells, and only those axons that have been recently active may be able to respond to them. Neural activity is also important in the development of the visual nervous system in vertebrates other than mammals. When neural activity in developing chick or amphibian embryos is blocked with drugs that prevent neuronal firing, no fine-grained retino-tectal map develops.

Summary

Neurons communicate with each other and with target cells such as muscles by means of specialized junctions called synapses. The formation of a neuromuscular junction involves changes in both pre-synaptic (neuronal) and post-synaptic (muscle cell) membranes once contact is made and depends on reciprocal signaling between muscle and nerve cell. Reciprocal interactions also occur in the formation of synapses between axons and dendrites, and cell-adhesion molecules are likely to be involved. Initially, most mammalian muscle fibers are innervated by two or more motor axons, but nervous activity results in competition between synapses so that a single fiber is eventually innervated by only one motor neuron. Many neurons produced in the developing nervous system die. About half of the motor neurons that initially innervate the vertebrate limb undergo cell death; those that survive do so because they make functional connections with muscles. Many neurons depend on neurotrophins, such as nerve growth factor, for their survival, with different classes of neurons requiring different neurotrophins. Brain function is based not only on the assembly of synapses but also on their regulated disassembly. Neural activity has a major role in refining the connections between the eye and the brain. In mammals, input from the left and right eyes is required for the development of ocular dominance columns in the visual cortex. These are adjacent columns of cells responding to the same stimulus from left and right eyes, respectively. Formation of the columns, which are essential for binocular vision, is a result of competition for cortical targets by axons carrying visual input from different eyes.

Summary: synapse formation and refinement

Neuromuscular junction	Neuron–neuron synapse
axon of motor neuron releases Agrin at junction with muscle	dendrite extends filopodia which contact axon terminal; cell-adhesion molecules form initial contact between dendrite and axon
⇩	⇩
acetylcholine receptors cluster at junction and local synthesis of receptors occurs	pre-synaptic active zone assembles; neurotransmitter receptors cluster on pre-synaptic membrane
⇩	
electrical activity of muscle reduces receptor synthesis elsewhere	
neurotrophins are required for neuronal survival—about 50% of limb motor neurons die	retino-tectal map refined by neuronal activity, leading to ocular dominance columns

SUMMARY TO CHAPTER 10

The processes involved in the development of the nervous system, with its numerous connections, are similar to those found in other developmental systems. Presumptive neural tissue is specified early in development—during gastrulation in vertebrates, and when the dorso-ventral axis is patterned in *Drosophila*. Within the neurectodermal tissue, the specification of cells that give rise to neural cells involves lateral inhibition. The further development of neurons from neuronal precursors involves both asymmetric cell divisions and cell–cell signaling. The patterning of different types of neurons within the vertebrate spinal cord is due to both ventral and dorsal signals. As they develop, neurons extend axons and dendrites. The axons are guided to their destination by growth cones at their tips. Guidance is due to the growth cone's response to attractive and repulsive signals, which may be diffusible or bound to the substratum. Gradients in such molecules can guide the axons to their destination, as in the retino-tectal system of vertebrates. The functioning of the nervous system depends on the establishment of specific synapses between axons and their targets. Specificity appears to be achieved by an initial overproduction of neurons that compete for targets, with many neurons dying during development. Refinement of synaptic connections involves further competition. Neural activity plays a major role in refining connections, such as those between the eye and the brain.

GENERAL FURTHER READING

Kandel, E.R., Schwartz, J.H., Jessell, T.H.: *Principles of Neural Science* (4th edn). New York: McGraw-Hill, 2000.

Kerszberg, M.: **Genes, neurons and codes: remarks on biological communication.** *BioEssays* 2003, **25:** 699–708.

SECTION FURTHER READING

10.1 Neurons in *Drosophila* arise from proneural clusters

Cornell, R.A., Ohlen, T.V.: **Vnd/nkx, ind/gsh, and msh/msx: conserved regulators of dorsoventral neural patterning?** *Curr. Opin. Neurobiol.* 2000, **10:** 63–71.

Gibert, J.M., Simpson, P.: **Evolution of *cis*-regulation of the proneural genes.** *Int. J. Dev. Biol.* 2003, **47:** 643–651.

Gómez-Skarmeta, J.L., Rodriguez, I., Martinez, C., Culí, J., Ferrés-Marco, D., Beamonte, D., Modolell, J.: **Cis-regulation of *achaete* and *scute*: shared enhancer-like elements drive their coexpression in proneural clusters of imaginal discs.** *Genes Dev.* 1995, **9:** 1809–1882.

Jones, B.W., Fetter, R.D., Tear, G., Goodman, C.S.: ***glial cells missing*: a genetic switch that controls glial versus neuronal fate.** *Cell* 1995, **82:** 1013–1023.

Skeath, J.B.: **At the nexus between pattern formation and cell-type specification: the generation of individual neuroblast fates in the *Drosophila* embryonic central nervous system.** *BioEssays* 1999, **21:** 922–931.

Udolph, G., Lüer, K., Bossing, T., Technau, G.M.: **Commitment of CNS progenitor along the dorso-ventral axes of *Drosophila* neurectoderm.** *Science* 1995, **269:** 1278–1281.

Weiss, J.B., Von Ohlen, T., Mellerick, D.M., Dressler, G., Doe, C.Q., Scott, M.P.: **Dorsoventral patterning in the *Drosophila* central nervous system: the intermediate neuroblasts defective homeobox gene specifies intermediate column identity.** *Genes Dev.* 1998, **12:** 3591–3602.

10.2 Asymmetric cell divisions and timed changes in gene expression are involved in the development of the *Drosophila* nervous system

Brody, T., Odenwald, W.: **Regulation of temporal identities during *Drosophila* neuroblast lineage development.** *Curr. Opin. Cell Biol.* 2005, **17:** 672–675.

Chan, Y.-M., Jan, Y.N.: **Conservation of neurogenic genes and mechanisms.** *Curr. Opin. Neurobiol.* 1999, **9:** 582–588.

Grosskortenhaus, R., Pearson, B.J., Marusich, A., Doe, C.Q.: **Regulation of temporal identity transitions in *Drosophila* neuroblasts.** *Dev. Cell* 2005, **8:** 193–202.

Karcavich, R.E.: **Generating neuronal diversity in the Drosophila central nervous system: a view from the ganglion mother cells.** *Dev. Dyn.* 2005, **232:** 609–616.

10.3 The neuroblasts of the sensory organs of adult *Drosophila* are already specified in the imaginal discs

Guo, M., Jan, L.Y., Jan, Y.N.: **Control of daughter cell fates during asymmetric division: interaction of Numb and Notch.** *Neuron* 1996, **17:** 27–41.

Jan, Y.-N., Jan, L.Y.: **Polarity in cell division: what frames thy fearful asymmetry?** *Cell* 2000, **100:** 599–602.

Knoblich, J.A.: **Asymmetric cell division during animal development.** *Nat. Rev. Mol. Cell. Biol.* 2001, **2:** 11–20.

10.5 Specification of vertebrate neuronal precursors involves lateral inhibition

Chitins, A., Henrique, D., Lewis, J., Ish-Horowitcz, D., Kintner, C.: **Primary neurogenesis in *Xenopus* embryos regulated by a homologue of the *Drosophila* neurogenic gene *Delta*.** *Nature* 1995, **375:** 761–766.

Ma, Q., Kintner, C., Anderson, D.J.: **Identification of *neurogenin*, a vertebrate neuronal determination gene.** *Cell* 1996, **87:** 43–52.

10.6 Neurons are formed in the proliferative zone of the neural tube and migrate outwards

D'Arcangelo, G., Curran, T.: ***Reeler*: new tales on an old mutant mouse.** *BioEssays* 1998, **20:** 235–244.

Gage, F.H.: **Mammalian neural stem cells.** *Science* 2000, **287:** 1433–1438.

Lee, S.K., Lee, B., Ruiz, E.C., Pfaff, S.L.: **Olig2 and Ngn2 function in opposition to modulate gene expression in motor neuron progenitor cells.** *Genes Dev.* 2005, **19:** 282–294.

Lyuksyutova, A.I., Lu, C.C., Milanesio, N., King, L.A., Guo, N., Wang, Y., Nathans, J., Tessier-Lavigne, M., Zou, Y.: **Anterior-posterior guidance of commissural axons by Wnt-frizzled signaling.** *Science* 2003, **302:** 1984–1988.

Maricich, S.M., Gilmore, E.C., Herrup, K.: **The role of tangential migration in the establishment of the mammalian cortex.** *Neuron* 2001, **31:** 175–178.

Sauvageot CM, Stiles CD.: **Molecular mechanisms controlling cortical gliogenesis.** *Curr. Opin. Neurobiol.* 2002, **12:** 244-249.

Wichterle, H., Lieberam, I., Porter, J.A., Jessell, T.M.: **Directed differentiation of embryonic stem cells into motor neurons.** *Cell* 2002, **110:** 385–397.

10.7 The pattern of differentiation of cells along the dorso-ventral axis of the spinal cord depends on ventral and dorsal signals

Briscoe, J., Pierani, A., Jessell, T.M., Ericson, J.: **A homeodomain protein code specifies progenitor cell identity and neuronal fate in the ventral neural tube.** *Cell* 2000, **101:** 435–445.

Ensini M., Tsuchida T.N., Belting, H.G., Jessell, T.M.: **The control of rostrocaudal pattern in the developing spinal cord: specification of motor neuron subtype identity is initiated by signals from paraxial mesoderm.** *Development* 1998, **125:** 969–982.

Jacob, J., Hacker, A., Guthrie, S.: **Mechanisms and molecules in motor neuron specification and axon pathfinding.** *BioEssays* 2001, **23:** 582–595.

Jessell, T.M.: **Neuronal specification in the spinal cord: inductive signals and transcriptional controls.** *Nat. Rev. Genet.* 2000, **1:** 20–29.

Kania, A., Johnson, R.L., Jessell, T.M.: **Coordinate roles for LIM homeobox genes in directing the dorsoventral trajectory of motor axons in the vertebrate limb.** *Cell* 2000, **102:** 161–173.

Kania, A., Jessell, T.M.: **Topographic motor projections in the limb imposed by LIM homeodomain protein regulation of ephrin-A:EphA interactions.** *Neuron* 2003, **38:** 581–596.

Megason, S.G., McMahon, A.P.: **A mitogen gradient of dorsal midline Wnts organizes growth in the CNS.** *Development* 2002, **129:** 2087–2098.

Sockanathan, S., Perlmann, T., Jessell, T.M.: **Retinoid receptor signaling in postmitotic motor neurons regulates rostrocaudal positional identity and axonal projection pattern.** *Neuron* 2003, **40:** 97–111.

Stamataki, D., Ulloa, F., Tsoni, S.V., Mynett, A., Briscoe, J.: **A gradient of Gli activity mediates graded Sonic Hedgehog signaling in the neural tube.** *Genes Dev.* 2005, **19:** 626–641.

Tanabe, Y., Jessell, T.M.: **Diversity and pattern in the developing spinal cord.** *Science* 1996, **274:** 1115–1123.

Wilson, L., Maden, M.: **The mechanisms of dorsoventral patterning in the vertebrate neural tube.** *Dev. Biol.* 2005, **282:** 1–13.

10.8 The growth cone controls the path taken by the growing axon

Carmeliet, P., Tessier-Lavigne, M.: **Common mechanisms of nerve and blood vessel wiring.** *Nature* 2005, **436:** 193–200.

Tear, G.: **Neuronal guidance: a genetic perspective.** *Trends Genet.* 1999, **15:** 113–118.

Tessier-Lavigne, M., Goodman, C.S.: **The molecular biology of axon guidance.** *Science* 1996, **274:** 1123–1133.

10.9 Motor neurons from the spinal cord make muscle-specific connections

Dasen, J.S., Tice, B.C., Brenner-Morton, S., Jessell, T.M.: **A Hox regulatory network establishes motor neuron pool identity and target-muscle connectivity.** *Cell* 2005, **123:** 477–491.

Lance-Jones, C., Landmesser, L.: **Pathway selection by embryonic chick motoneurons in an experimentally altered environment.** *Proc. R. Soc. Lond. B* 1981, **214:** 19–52.

Tosney, K.W., Hotary, K.B., Lance-Jones, C.: **Specifying the target identity of motoneurons.** *BioEssays* 1995, **17:** 379–382.

10.10 Axons crossing the midline are both attracted and repelled

Giger, R.J., Kolodkin, A.L.: **Silencing the siren: guidance cue hierarchies at the CNS midline.** *Cell* 2001, **105:** 1–4.

Long, H., Sabatier, C., Ma, L., Plump, A., Yuan, W., Ornitz, D.M., Tamada, A., Murakami, F., Goodman, C.S., Tessier-Lavigne, M.: **Conserved roles for Slit and Robo proteins in midline commissural axon guidance.** *Neuron* 2004, **42:** 213–223.

Simpson, J.H., Bland, K.S., Fetter, R.D., Goodman, C.S.: **Short-range and long-range guidance by Slit and its Robo receptors: a combinatorial code of Robo receptors controls lateral position.** *Cell* 2000, **103:** 1019–1032.

Woods, C.G.: **Crossing the midline.** *Science* 2004, **304:** 1455–1456.

10.11 Neurons from the retina make ordered connections on the tectum to form a retino-tectal map

Feldheim, D.A., Kim, Y-I., Bergemann, A.D., Frisen, J., Barbacid, M., Flanagan, J.G.: **Genetic analysis of ephrin A2 and ephrin A5 shows their requirement in multiple aspects of retinocollicular mapping.** *Neuron* 2000, **25:** 563–574.

Hansen, M.J., Dallal, G.E., Flanagan, J.G.: **Retinoic axon response to ephrin-As shows a graded concentration-dependent transition from growth promotion to inhibition.** *Neuron* 2004, **42:** 707–730.

Klein, R.: **Eph/ephrin signaling in morphogenesis, neural development and plasticity.** *Curr. Opin. Cell Biol.* 2004, **16:** 580–589.

Löschinger, J., Weth, F., Bonhoeffer, F.: **Reading of concentration gradients by axonal growth cones.** *Phil. Trans. R. Soc. Lond. B* 2000, **355:** 971–982.

McLaughlin, T., Hindges, R., O'Leary, D.D.: **Regulation of axial patterning of the retina and its topographic mapping in the brain.** *Curr. Opin. Neurobiol.* 2003, **13:** 57–69.

Reber, M., Bursold, P., Lemke, G.: **A relative signalling model for the formation of a topographic neural map.** *Nature* 2004, **431:** 847–853.

Schmitt, A.M., Shi, J., Wolf, A.M., Lu, C-C., King, L.A., Zou, Y.: **Wnt-Ryk signalling mediates medial-lateral retinotectal topographic mapping.** *Nature* 2006, **439:** 31–37.

Wilkinson, D.G.: **Topographic mapping: organising by repulsion and competition?** *Curr. Biol.* 2000, **10:** R447–R451.

10.12 Synapse formation involves reciprocal interactions

Buffelli, M., Burgess, R.W., Feng, G., Lobe, C.G., Lichtman, J.W., Sanes, J.R.: **Genetic evidence that relative synaptic efficacy biases the outcome of synaptic competition.** *Nature* 2003, **424:** 430–434.

Chess, A.: **Monoallelic expression of protocadherin genes.** *Nat. Genet.* 2005, **37:** 120–121.

Goda, Y., Davis, G.W.: **Mechanisms of synapse assembly and disassembly.** *Neuron* 2003, **40:** 243–264.

Jan, Y.-N., Jan, L.Y.: **The control of dendritic development.** *Neuron* 2003, **40:** 229–242.

Hua, J.Y., Smith, S.J.: **Neural activity and the dynamics of central nervous system development.** *Nat. Neurosci.* 2004, **7:** 327–332.

Levinson, J.N., El-Husseini, A.: **Building excitatory and inhibitory synapses: balancing neuroligin partnerships.** *Neuron* 2005, **48:** 171–174.

Li, Z., Sheng, M.: **Some assembly required: the development of neuronal synapses.** *Nat. Rev. Mol. Cell Biol.* 2003, **4:** 833–841.

Lin, W., Burgess, R.W., Dominguez, B., Pfaff, S.L., Sanes, J.R., Lee, K.-F.: **Distinct roles of nerve and muscle in postsynaptic differentiation of the neuromuscular synapse.** *Nature* 2001, **410:** 1057–1064.

Schmucker, D., Clemens, J.C., Shu, H., Worby, C.A., Xiao, J., Muda, M., Dixon, J.E., Zipursky, S.L.: ***Drosophila* Dscam is an axon guidance receptor exhibiting extraordinary molecular diversity.** *Cell* 2000, **101:** 671–684.

Wallace, B.G.: **Signaling mechanisms mediating synapse formation.** *BioEssays* 1996, **18:** 777–780.

10.13 Many motor neurons die during normal development

Oppenheim, R.W.: **Cell death during development of the nervous system.** *Annu. Rev. Neurosci.* 1991, **14:** 453–501.

10.14 Neuronal cell death and survival involve both intrinsic and extrinsic factors

Birling, M.C., Price, J.: **Influence of growth factors on neuronal differentiation.** *Curr. Opin. Cell Biol.* 1995, **7:** 878–884.

Burden, S.J.: **Wnts as retrograde signals for axon and growth cone differentiation.** *Cell* 2000, **100:** 495–497.

Davies, A.M.: **Neurotrophic factors. Switching neurotrophin dependence.** *Curr. Biol.* 1994, **4:** 273–276.

Henderson, C.E.: **Programmed cell death in the developing nervous system.** *Neuron* 1996, **17:** 579–585.

Jiang, Y., de Bruin, A., Caldas, H., Fangusaro, J., Hayes, J., Conway, E.M., Robinson, M.L., Altura, R.A.: **Essential role for surviving in early brain development.** *J. Neurosci.* 2005, **25:** 6962–6970.

Pettmann, B., Henderson, C.E.: **Neuronal cell death.** *Neuron* 1998, **20:** 633–647.

Serafini, T.: **Finding a partner in a crowd: neuronal diversity and synaptogenesis.** *Cell* 1999, **98:** 133–136.

Šestan, N., Artavanis-Tsakonas, S., Rakic, P.: **Contact-dependent inhibition of cortical neurite growth mediated by Notch signaling.** *Science* 1999, **286:** 741–746.

10.15 The map from eye to brain is refined by neural activity

Katz, L.C., Shatz, C.J.: **Synaptic activity and the construction of cortical circuits.** *Science* 1996, **274:** 1133–1138.

Germ cells, fertilization, and sex

11

- The development of germ cells
- Fertilization
- Determination of the sexual phenotype

Animal embryos develop from a single cell, the fertilized egg, which is the product of the fusion of egg and sperm. The germline cells—the germ cells—that give rise to eggs and sperm in the adult are specified in the early embryo, and in this chapter we shall consider their specification and how they differentiate into eggs and sperm. An important property of germ cells is that they remain pluripotent, as discussed in Chapter 8, but we shall see here that mammalian eggs and sperm have certain genes differentially silenced by a process known as genomic imprinting, and we will look at the implications of this for development. Embryonic development is initiated by the process of fertilization and we shall discuss the mechanisms in sea urchins and mammals that ensure that only one sperm enters the egg. In the animals we cover in this chapter sex is determined by the chromosomal constitution, and we shall see that mechanisms of sex determination and of compensating for the unequal complement of sex chromosomes between males and females differ considerably in different species.

In sexually reproducing organisms, there is a fundamental distinction between the **germline** cells, or **germ cells**, and the somatic or body cells (see Section 1.2). The former give rise to the **gametes**—eggs and sperm in animals—whereas somatic cells make no genetic contribution to the next generation. So far in this book, we have looked at the somatic development of our model organisms. Not surprisingly, a great deal of the biology of animals and plants is devoted to reproduction and sex, and in this chapter, we look at germ-cell formation, fertilization, and sex determination, mainly in the mouse, *Drosophila*, and *Caenorhabditis*. As germ cells are the cells that will give rise to the next generation, their development is crucial, and in animals, germ cells are usually specified and set aside early in embryonic development. Plants, although reproducing sexually, differ from most animals in that their germ cells are not specified early in embryonic development, but during the development of the flowers. It is worth noting that some simple animals, such as the coelenterate *Hydra*, can reproduce asexually, by budding, and that even in some vertebrates, such as turtles, the eggs can develop without being fertilized.

We will begin this chapter by considering how the germ cells are specified in the early animal embryo and how they differentiate into eggs and sperm. Then, we will look at the **fertilization** and activation of the egg by the sperm—the vital step that initiates development. Finally, we turn to **sex determination**—why males and females are different from each other. In all the animals we look at here, sex is determined genetically, by the number and type of specialized chromosomes known as the **sex chromosomes**. Male and female embryos initially look the same, with sexual differences only emerging as a result of the activity of sex-determining genes located on the sex chromosomes.

The development of germ cells

In all but the simplest animals, the cells of the germline are the only cells that can give rise to a new organism. So, unlike somatic cells, which eventually all die, germ cells in a sense outlive the bodies that produced them. They are, therefore, very special cells. The outcome of germ-cell development is either a male gamete (the sperm in animals), or a female gamete (the egg). The egg is a particularly remarkable cell, as all the cells in an organism are derived from it. In species whose embryos receive no nutrition from the mother after fertilization, the egg must also provide everything necessary for development, as the sperm contributes virtually nothing to the organism other than its chromosomes and a centrosome.

Animal germ cells typically differ from somatic cells by dividing less often during early embryogenesis. Later, they are the only cells to undergo the type of nuclear division called meiosis, by which gametes with half the number of chromosomes of the germ-cell precursor are produced. In many animals, primordial germ cells are formed in locations that seem to protect them from the inductive signals that specify the fate of somatic cells. And in *C. elegans*, for example, there is evidence for mechanisms that generally repress transcription in germ cells compared to somatic cells. In some animals, once the primordial germ cells are specified, they migrate into **gonads**—the ovary and testis—somatic structures that usually develop some distance away from the site of germ-cell origin. Once within the gonads, the germ cells differentiate as either male or female gametes.

Germ cells are specified very early in some animals, although not in mammals, by cytoplasmic determinants present in the egg. We therefore start our discussion of germ-cell development by looking at the specification of primordial germ cells by special cytoplasm—the **germplasm**—which is already present in the egg.

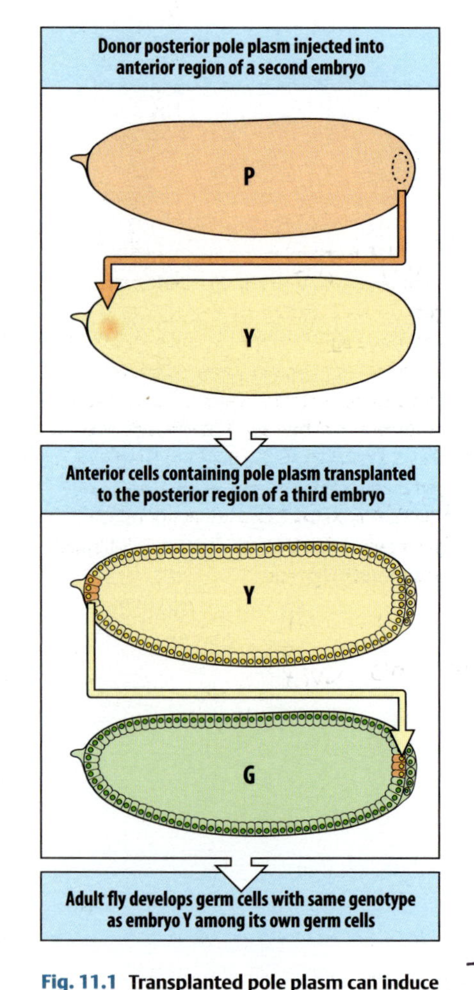

Fig. 11.1 Transplanted pole plasm can induce germ-cell formation in Drosophila. Pole plasm from a fertilized egg of genotype P (pink) is transferred to the anterior end of an early cleavage-stage embryo of genotype Y (yellow). After cellularization, cells containing pole plasm induced at the anterior end of embryo Y are transferred to the posterior end (a site from which germ cells can migrate into the gonad) of another embryo, of genotype G (green). The adult fly that develops from embryo G contains germ cells of genotype Y as well as those of G.

11.1 Germ-cell fate can be specified by a distinct germplasm in the egg

In flies, nematodes, and frogs, molecules localized in specialized cytoplasm in the egg are involved in specifying the germ cells. The clearest example of this is in *Drosophila*, where primordial germ cells known as **pole cells** become distinct at the posterior pole of the egg about 90 minutes after fertilization, more than an hour before the cellularization of the rest of the embryo. The cytoplasm at the posterior pole is called the **pole plasm** and is distinguished by large organelles, the polar granules, which contain both proteins and RNAs. That there is something special about this posterior cytoplasm is demonstrated by two key experiments. First, if the posterior end of the egg is irradiated with ultraviolet light, which destroys the pole plasm activity, no germ cells develop, although the somatic cells in that region do develop. Second, if pole plasm from an egg is transferred to the anterior pole of another embryo, the nuclei that become surrounded by the pole plasm are specified as germ cells (pole cells; Fig. 11.1). If these pole cells are then transplanted into the posterior pole region of a third embryo, they develop as functional germ cells. We shall discuss the localization of the pole plasm in the next section.

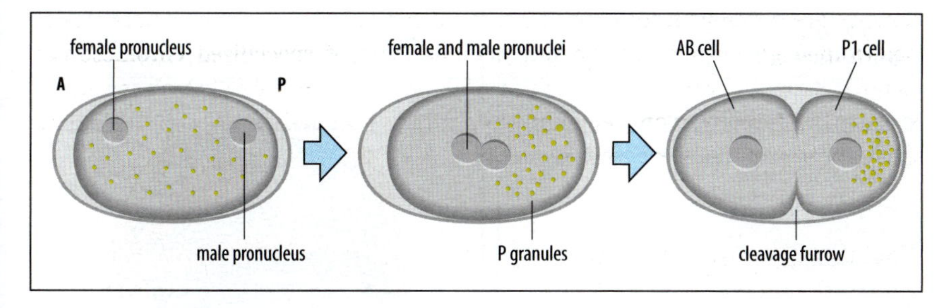

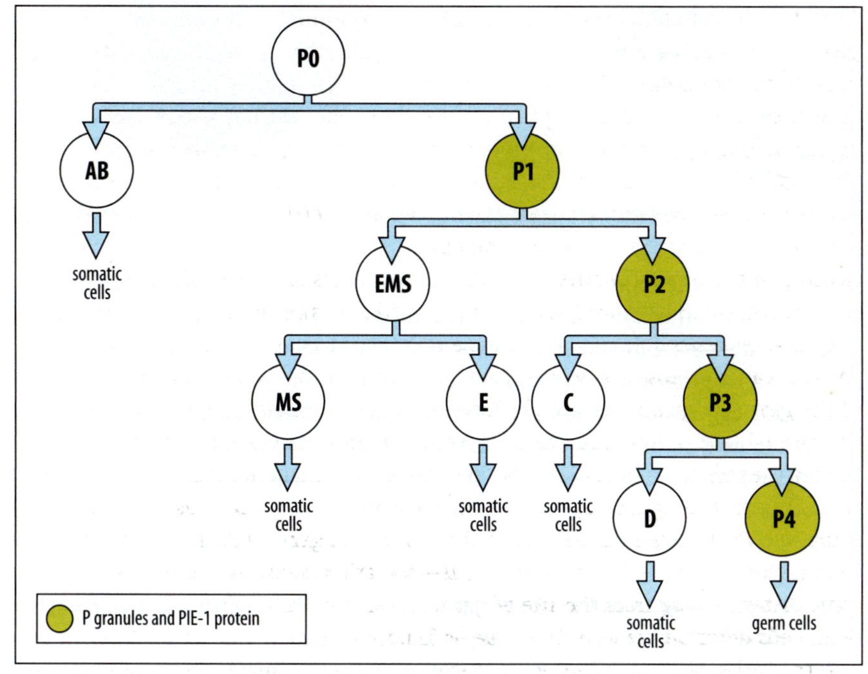

Fig. 11.2 P granules and PIE-1 protein become asymmetrically distributed to germline cells during cleavage of the nematode egg. Before fertilization, P granules are distributed throughout the egg. After fertilization, P granules become localized at the posterior end of the egg. At the first cleavage, they are only included in the P1 cell (top panel), and thus become confined to the P cell lineage. The PIE-1 protein is only present in P cells. All germ cells are derived from P4, which is formed at the fourth cleavage.

In the nematode, a germ-cell lineage is set up at the end of the fourth cleavage division, with all the germ cells being derived from the P4 blastomere (see Section 5.1). The P4 cell is derived from three stem-cell like divisions of the P1 cell. At each of these divisions, one daughter produces somatic cells whereas the other divides again to produce a somatic cell progenitor and a P cell. The egg contains P granules in its cytoplasm which become asymmetrically distributed after fertilization, and are subsequently confined to the P-cell lineage (Fig. 11.2). The association of germ-cell formation with the P granules suggested that they might have a role in germ-cell specification, and at least one P-granule component, the product of the *pgl-1* gene, has been shown to be necessary for germ-cell development. PGL-1 may specify germ cells by regulating some aspect of mRNA metabolism. The gene *pie-1* is involved in maintaining the stem-cell property of the P blastomeres. It encodes a nuclear protein that is expressed maternally and is not a component of P granules; the PIE-1 protein is only present in the germline blastomeres, and represses new transcription of zygotic genes in these blastomeres until it disappears at around the 100-cell stage. This general repression may protect the germ cells from the actions of transcription factors that promote development into somatic cells.

There is also evidence for germ plasm in *Xenopus* eggs. After fertilization, distinct yolk-free patches of cytoplasm aggregate at the yolky vegetal pole. When the

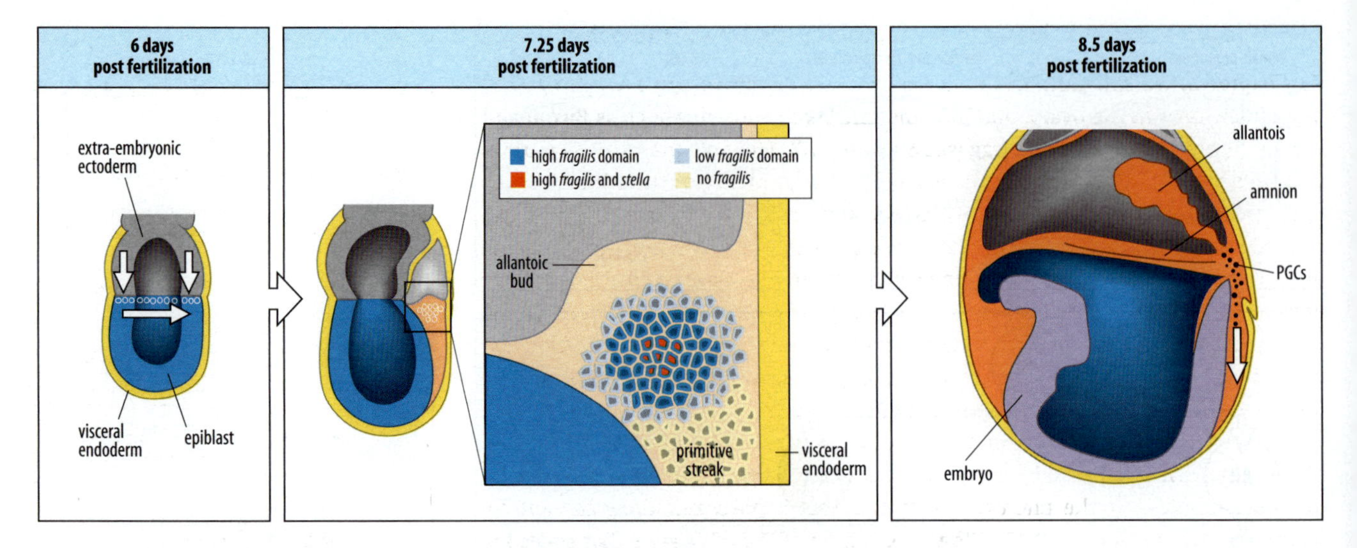

Fig. 11.3 Germ-cell formation in the mouse. The precursors (white) of primordial germ cells (PGCs) and extra-embryonic mesoderm are induced in the proximal epiblast by signals from the extra-embryonic ectoderm that include BMP-4. During gastrulation, these cells move to the posterior end of the embryo above the primitive streak. Here they form a cluster in which the central cells become specified as primordial germ cells and the peripheral cells as extra-embryonic mesoderm. After their formation, the primordial germ cells migrate to the gonads.

blastomeres at the vegetal pole cleave, this cytoplasm is distributed asymmetrically so that it is retained only in the most vegetal daughter cells, from which the germ cells are derived. Ultraviolet irradiation of the vegetal cytoplasm abolishes the formation of germ cells, and transplantation of fresh vegetal cytoplasm into an irradiated egg restores germ-cell formation. At gastrulation, the germ plasm is located in cells in the floor of the blastocoel cavity, among the cells that give rise to the endoderm. Cells containing the germ plasm are, however, not yet determined as germ cells and can contribute to all three germ layers if transplanted to other sites. At the end of gastrulation, the primordial germ cells are determined, and migrate out of the presumptive endoderm and into the **genital ridge**, which develops from mesoderm lining the abdominal cavity and will form the gonads.

In the zebrafish, germ-cell specification is related to the localization of the mRNA for the gene *vasa*. The mRNA is synthesized during oogenesis and becomes localized to the cleavage furrows during the first embryonic cleavages. By the 32-cell stage it has been drawn into clumps that segregate into four cells that become the primordial germ cells.

There is no evidence for germ plasm being involved in germ-cell formation in the mouse or other mammals. Indeed, it seems that although germ plasm is present in several of our developmental model organisms, it is the less prevalent mode of specification in animals generally. Germ-cell specification in the mouse involves cell–cell interactions, as cultured embryonic stem cells can, when injected into the inner cell mass, give rise to both germ cells and somatic cells (see Box 3E, p. 130). Primordial germ cells can be identified in the mouse midway through gastrulation in the extra-embryonic mesoderm posterior to the primitive streak (Fig. 11.3). The prospective germ cells and the extra-embryonic mesoderm have their origin in common precursor cells that are induced in the proximal epiblast by BMP-4 and other signals from the extra-embryonic ectoderm. These cells migrate to a position at the posterior end of the primitive streak. The primordial germ cells then express high levels of the transmembrane protein Fragilis, which is thought to be involved in their aggregation and separation from other cells, and they subsequently express the proteins Stella and Oct-4, which are involved in maintaining pluripotency. They also express high levels of the enzyme alkaline phosphatase which provides a convenient means of detecting them histochemically. The germ cells migrate anteriorly and enter the hindgut, from where they finally migrate to the genital ridge.

11.2 Pole plasm becomes localized at the posterior end of the *Drosophila* egg

In Chapter 2, we saw how the main axes of the *Drosophila* egg are specified by the follicle cells in the ovary, and how the mRNAs for proteins such as Bicoid and Nanos become localized in the egg (see Section 2.13). The pole plasm also becomes localized at the posterior end of the egg under the influence of the follicle cells.

Several maternal genes are involved in pole plasm formation in *Drosophila*. Mutations in any of at least eight genes result in the affected homozygous individual being 'grandchildless'. Its offspring lack a proper pole plasm, and although they may develop normally in other ways, they lack germ cells and are therefore sterile. One of these eight genes is *oskar*, which plays a central role in the organization and assembly of the pole plasm (Fig. 11.4); of the genes involved in pole plasm formation, *oskar* is the only one to have its mRNA localized at the posterior pole. The signal for localization is contained in the 3′ untranslated region of the mRNA. Staufen protein is required for the localization of *oskar* mRNA and may act by linking the mRNA to the microtubule system that is polarized along the antero-posterior axis (see Section 2.13). When the region of the *oskar* gene that codes for the 3′ localization signal is replaced by the *bicoid* 3′ localization signal, flies made transgenic with this DNA construct have *oskar* mRNA localized at the anterior end of the egg (see Fig. 11.4).

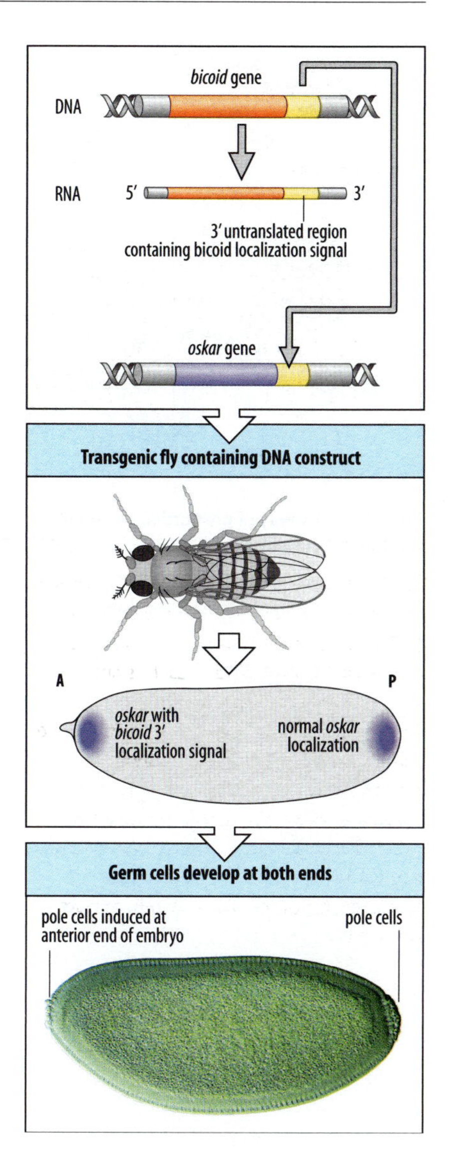

Fig. 11.4 The gene *oskar* is involved in specifying the germ plasm in *Drosophila*. In normal eggs, *oskar* mRNA is localized at the posterior end of the embryo, whereas *bicoid* mRNA is at the anterior end. The localization signals for both *bicoid* and *oskar* mRNA are in their 3′ untranslated regions. By manipulating the *Drosophila* DNA, the localization signal of *oskar* can be replaced by that of *bicoid* (top panel). A transgenic fly is made containing the modified DNA. In its egg, *oskar* becomes localized at the anterior end (middle panel). The egg therefore has *oskar* mRNA at both ends, and germ cells develop at both ends of the embryo, as shown in the photograph (bottom panel). Thus, *oskar* alone is sufficient to initiate the specification of germ cells.

Photograph courtesy of R. Lehmann.

11.3 Germ cells migrate from their site of origin to the gonad

In many animals, germ cells develop at some distance from the gonads, and only later migrate to them, where they differentiate into eggs or sperm. The reason for this separation of site of origin from final destination is not known, but it may be a mechanism for excluding germ cells from the general developmental upheaval involved in laying down the body plan, or a mechanism for selecting the healthiest of the germ cells, namely those that survive migration. Their pathway of migration is controlled by the environment, and in *Xenopus*, for example, germ cells transplanted to the wrong place in the blastula do not end up in the gonad. In zebrafish, the primordial germ cells are specified in four separate positions relative to the embryonic axis.

The vertebrate gonad develops the mesoderm lining the abdominal cavity, which is known as the genital ridge. The primordial germ cells migrate there from distant sites. In *Xenopus*, the primordial germ cells originate in the endoderm (which forms the gut) and migrate to the future gonad along a cell sheet that joins the gut to the genital ridge. Only a small number of cells start this journey, dividing about three times before arrival, so that about 30 germ cells colonize the gonad. The number of primordial germ cells that arrive at the genital ridge of the mouse, by a very similar pathway (Fig. 11.5), is about 8000, starting from a population of around 45 at the posterior end of the primitive streak (see Section 11.1). In chick embryos, the pattern of migration is different: the germ cells originate at the head end of the embryo, and most arrive at their destination via the blood vessels, leaving the bloodstream at the hindgut and then migrating along the epithelial sheet.

In the mouse gastrula, germ cells first become readily detectable by staining for alkaline phosphatase posterior to the primitive streak (see Fig. 11.3). They become incorporated into the hindgut and then migrate into the adjacent connective tissue. Two genes involved in controlling proliferation of migrating germ cells are *White spotting* (*W*) and *Steel*, which we have already met in relation to melanocyte differentiation (see Section 8.12). Mutations that inactivate either of these genes cause a decrease in germ-cell numbers. *White spotting* codes for the cell-surface

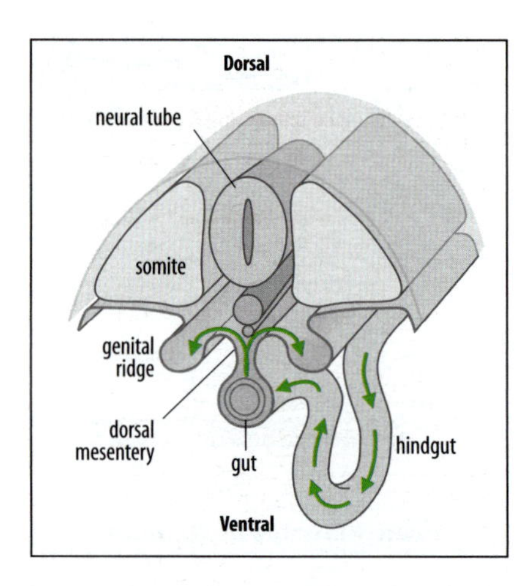

Fig. 11.5 Pathway of primordial germ-cell migration in the mouse embryo. In the final stage of migration, the cells move from the gut tube into the genital ridge, via the dorsal mesentery.

After Wylie, C.C., et al.: 1993.

receptor Kit, which is expressed in the migrating germ cells. Its ligand, the Steel protein, is expressed in the cells along which the germ cells migrate. The requirement for these two genes indicates that the migrating cells are continuously receiving signals from the surrounding tissue. A similar function in germ-cell guidance is played in zebrafish by the chemokine receptor CXCR4 and its ligand SDF-1. Once they arrive at the gonad, the germ cells begin to differentiate into sperm or eggs, as described in the next section.

In *Drosophila*, prospective germ cells are formed next to the posterior midgut, and during gastrulation they are carried along the dorsal side of the embryo with the midgut primordium. They then migrate across the epithelium to the adjacent mesoderm to join with gonadal precursor cells. Various enzymes are involved in guiding their migration. Lipidphosphate phosphatases are expressed in the ventral region of the posterior midgut and are involved in repelling germ cells from this region, while the enzyme 3-hydroxy-3-methylglutaryl coenzyme A reductase is involved in producing chemoattractant cues that guide the germ cells to the gonadal cells.

11.4 Germ-cell differentiation involves a reduction in chromosome number

Germ cells are haploid—that is, they contain just one copy of each chromosome in contrast to the two copies present in the diploid somatic cells. At fertilization, therefore, the diploid chromosome number is reinstated. Primordial germ cells are diploid, and reduction from the diploid to the haploid state during germ-cell development occurs during the specialized nuclear division known as **meiosis** (Fig. 11.6). Meiosis comprises two cell divisions, in which the chromosomes are replicated before the first division, but not before the second, so that their number is reduced by half.

Development of eggs and sperm follow different courses, even though they both involve meiosis. The development of the egg is known as **oogenesis**. The main stages of mammalian oogenesis are shown in Fig. 11.7 (left panel). In mammals, germ cells undergo a small number of proliferative mitotic cell divisions as they migrate to the gonad. In the case of developing eggs, the diploid oogonia continue to divide mitotically for a short time in the ovary. After entry into meiosis, the primary oocytes become arrested in the prophase of the first meiotic division. They never proliferate again; thus, the number of oocytes at this embryonic stage is generally considered to be the total number of eggs the female mammal ever has; recent studies in mice, however, suggest that oocytes and follicles are produced in the postnatal ovary. In humans, most of these oocytes degenerate before puberty, leaving about 40,000 out of an original 6 million. In mammals and many other vertebrates, oocyte development may be held in suspension after birth for months (mice) or years (humans). The female becomes sexually mature at puberty, when the oocytes undergo further development, growing in size up to 1000-fold. In mammals, meiosis resumes at the time of ovulation, proceeds as far as the metaphase of the second meiotic division, where it becomes arrested again, and is completed after fertilization.

The strategy for **spermatogenesis**—the production of sperm—is different. Diploid germ cells that give rise to sperm do not enter meiosis in the embryo, but become arrested at an early stage of the mitotic cell cycle in the embryonic testis. They resume mitotic proliferation after birth. Later, in the sexually mature animal, spermatogonial stem cells give rise to differentiating spermatocytes, which undergo meiosis, each forming four haploid spermatids that mature into sperm (see Fig. 11.7, right panel). Thus, unlike the fixed number of oocytes in female mammals, sperm continue to be produced throughout life.

In *Drosophila*, there is a continuous production of both eggs and sperm from a population of stem cells. Oogenesis begins with the division of a stem cell

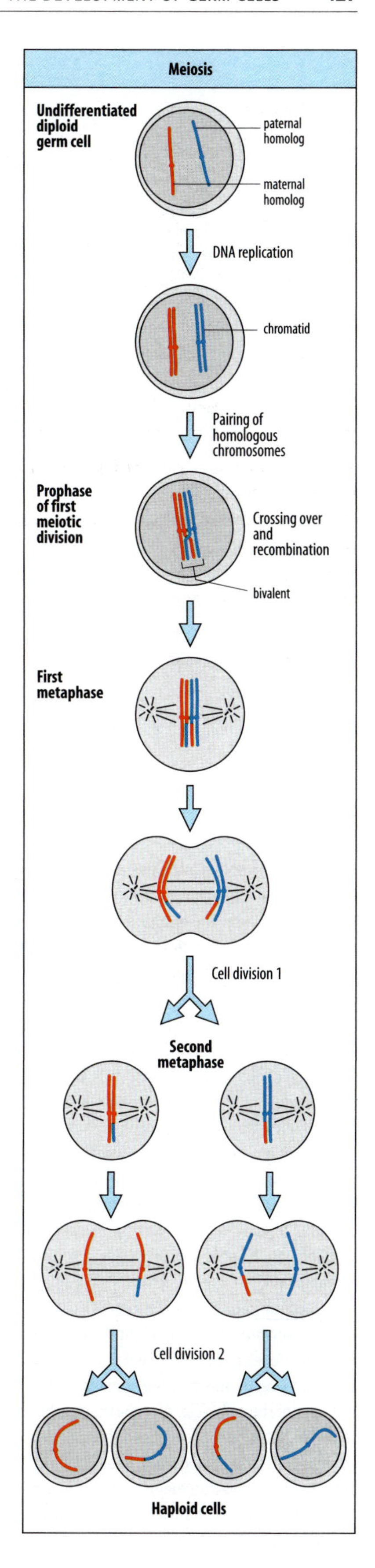

Fig. 11.6 Meiosis produces haploid cells. Meiosis reduces the number of chromosomes from the diploid to the haploid number. Only one pair of homologous chromosomes is shown here for simplicity. Before the first meiotic division the DNA replicates, so that each chromosome entering meiosis is composed of two identical chromatids. The paired homologous chromosomes (known as a bivalent) undergo crossing over and recombination, and align on the meiotic spindle at the metaphase of the first meiotic division. The homologous chromosomes separate, and each is segregated into a different daughter cell at the first cell division. There is no DNA replication before the second meiotic division. The daughter chromatids of each chromosome separate and segregate at the second cell division. The chromosome number of the resulting daughter cells is thus halved.

(see Fig. 2.15), and so there is no intrinsic limitation to the number of eggs that a female fly can produce. *Drosophila* oogenesis and the localization of cytoplasmic determinants in the egg have already been discussed in Section 2.13.

11.5 Oocyte development can involve gene amplification and contributions from other cells

Eggs vary enormously in size among different animals, but they are always larger than the somatic cells. A typical mammalian egg is about 0.1 mm in diameter, a frog's egg about 1 mm, and a hen's egg about 5 cm (see Fig. 3.1). To achieve these large sizes—and even the mammalian egg has a mass 100-fold greater than that of a typical mammalian somatic cell—a variety of mechanisms have evolved, some organisms using several of them together. One strategy is to increase the overall number of gene copies in the developing oocyte, as this proportionately increases the amount of mRNA that can be transcribed, and thus the amount of protein that can be synthesized. This strategy is generally adopted by vertebrate oocytes, which are arrested in the prophase of the first meiotic division and thus have double the normal diploid number of genes, as do all mitotic cells at this stage. Transcription and oocyte growth continue while the oocyte is in the arrested stage. Another strategy, adopted by insects and amphibians, is to produce many extra copies of those genes whose products are needed in large quantities in the egg. Thus, the ribosomal RNA genes of amphibians are amplified during oocyte development from hundreds to millions. In insects, the genes that encode the proteins of the egg membrane, the chorion, become amplified in the surrounding follicle cells.

A different strategy is for the oocyte to rely on the synthetic activities of other cells. In insects, the nurse cells adjacent to the oocyte, which are its sister cells, make and deliver many mRNAs and proteins to the oocyte (see Fig. 2.15). Yolk proteins in birds and amphibians are made by liver cells, and are carried by the blood to the ovary, where the proteins enter the oocyte by endocytosis and become packaged in yolk platelets. From an early stage, the oocyte is polarized, and the yolk platelets accumulate at the vegetal pole. Developmentally important mRNAs are produced by the mother and delivered into the egg, where they become localized to the vegetal cortex by a microtubule-based mechanism. In *C. elegans*, yolk proteins are made in intestinal cells, and in *Drosophila* in the fat body, and then transported to the oocytes.

11.6 Some genes controlling embryonic growth are imprinted

Like all other cells in an animal, mature gametes have completed a program of differentiation. Yet at the end of that process they must have a genome that, after fertilization, is capable of controlling the development of the embryo. Of all the

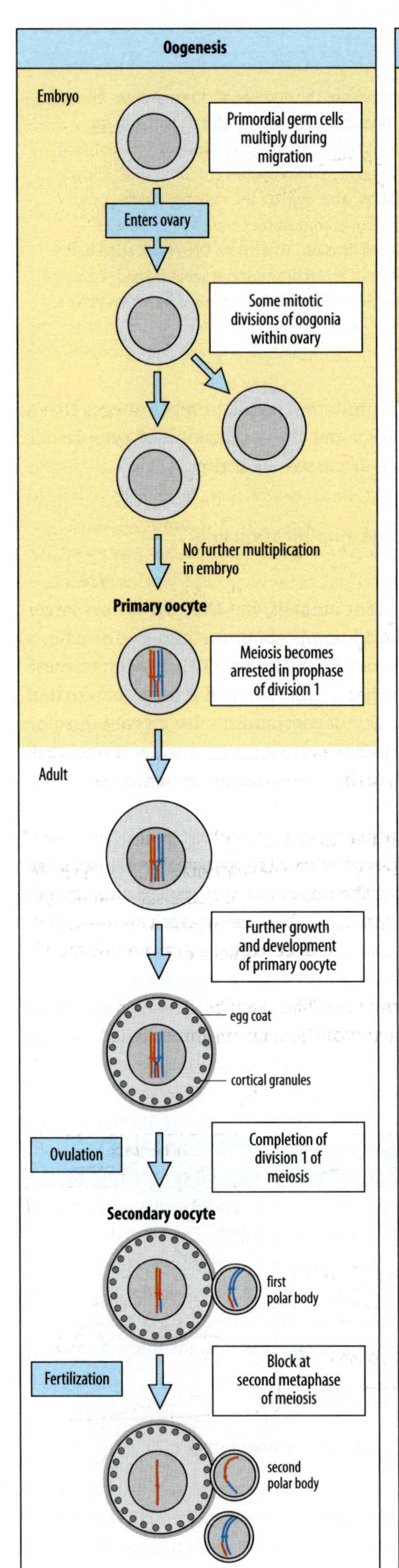

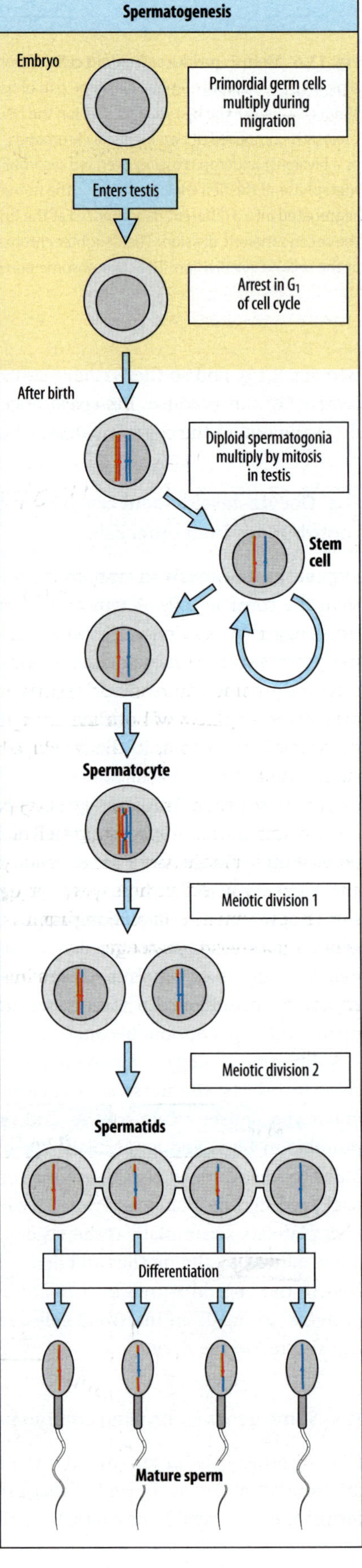

Fig. 11.7 Oogenesis and spermatogenesis in mammals. Left panel: after the germ cells that form oocytes enter the embryonic ovary, they divide mitotically a few times and then enter the prophase of the first meiotic division. No further cell multiplication occurs. Further development occurs in the sexually mature adult female. This includes a 100-fold increase in mass, the formation of external cell coats, and the development of a layer of cortical granules located under the oocyte plasma membrane. In each cycle, a group of follicles starts to grow, oocyte growth and maturation follows; a few eggs are ovulated but most degenerate. Eggs continue to mature in the ovary under hormonal influences, but become blocked in the second metaphase of meiosis, which is only completed after fertilization. Polar bodies are formed at meiosis (see Box 3A, p. 93). Right panel: germ cells that develop into sperm enter the embryonic testis and become arrested at the G_1 stage of the cell cycle. After birth, they begin to divide mitotically again, forming a population of stem cells (spermatogonia). These give off cells that then undergo meiosis and differentiate into sperm. Sperm can therefore be produced indefinitely.

cells in the body, they are the only ones with genomes that are passed on to future generations. Their genomes must revert to a state from which all the cells of the organism can be derived, and so there must be no permanent alterations in the genetic constitution. As we saw in Chapter 8, in the cloning procedure the egg cytoplasm can 'reprogram' a diploid nucleus from a somatic cell so that it is able to give rise to a new organism. Cloning in mammals has, however, turned out to be a much more difficult than in some other animals, and this is most likely to be due to the phenomenon of **imprinting**, the fact that certain genes in eggs and sperm are specifically programmed to be switched off during development. Evidence for imprinting comes from the different contributions of the maternal and paternal genomes to the development of the embryo.

Mouse eggs can be manipulated by nuclear transplantation to have either two paternal genomes or two maternal genomes, and can be reimplanted into a mouse for further development. The embryos that result are known as **androgenetic** and **gynogenetic** embryos, respectively. Although both kinds of embryo have a diploid number of chromosomes, their development is abnormal. The embryos with two paternal genomes have well-developed extra-embryonic tissues, but the embryo itself is abnormal, and does not proceed beyond a stage at which several somites are present. By contrast, the embryos with diploid maternal genomes have relatively well-developed embryos, but the extra-embryonic tissues—placenta and yolk sac—are poorly developed (Fig. 11.8). These results clearly show that both maternal and paternal genomes are necessary for normal mammalian development: the two parental genomes function differently in development, and are required for the normal development of both the embryo and the placenta. This is the reason that mammals cannot be naturally produced parthenogenetically, by activation of an unfertilized egg.

Such observations showed that the paternal and maternal genomes are modified, or imprinted, during germ-cell differentiation. The paternal and maternal genomes contain the same set of genes, but the imprinting process turns on or off certain genes in either the sperm or egg, so that they are or are not expressed during development. Imprinting implies that the affected genes carry a 'memory' of being in a sperm or an egg.

Imprinting in mammals is a reversible process. The reversibility requirement is important, because in the next generation any of the chromosomes could end up

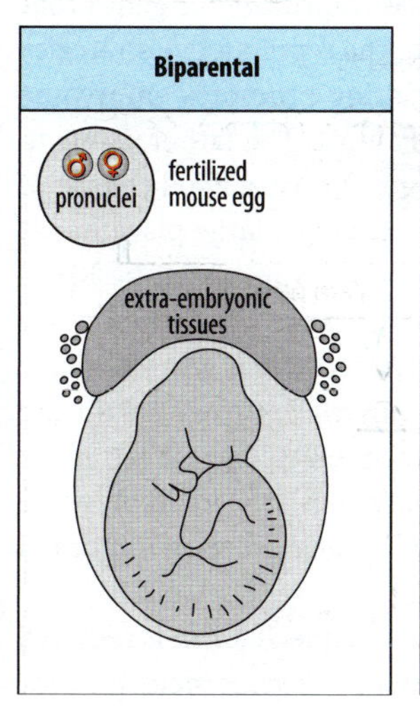

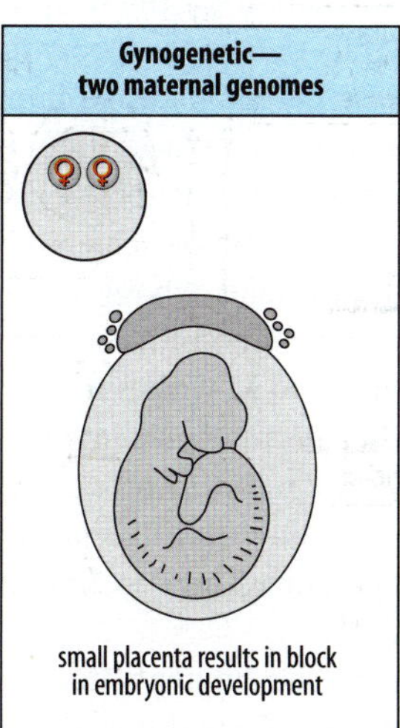

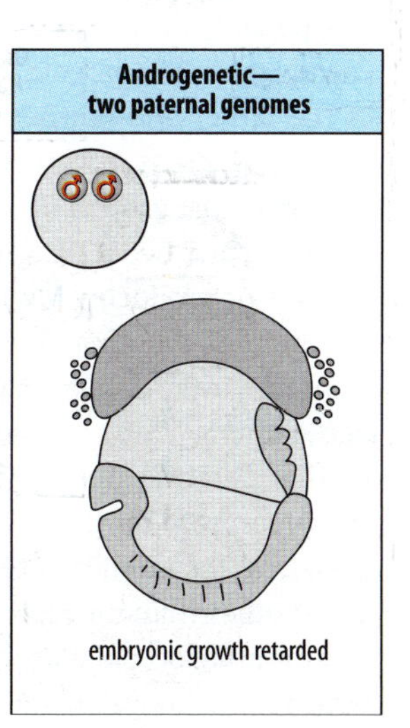

Fig. 11.8 Paternal and maternal genomes are both required for normal mouse development. A normal biparental embryo has contributions from both the paternal and maternal nuclei in the zygote after fertilization (left panel). Using nuclear transplantation, an egg can be constructed with two paternal or two maternal nuclei from an inbred strain. Embryos that develop from an egg with two maternal genomes—gynogenetic embryos (center panel)—have underdeveloped extra-embryonic structures. This results in development being blocked, although the embryo itself is relatively normal and well developed. Embryos that develop from eggs with two paternal genomes—androgenetic embryos (right panel)—have normal extra-embryonic structures, but the embryo itself only develops to a stage where a few somites have formed.

in male or female germ cells. Inherited imprinting is probably erased during early germ-cell development, and imprinting is later established afresh during germ-cell differentiation. When mammals are cloned using donor nuclei from differentiated somatic cells, these nuclei will not have gone through the normal reprogramming and imprinting process that occurs during germ-cell formation and this could account for the large number of failures and abnormalities in the cloned embryos.

The imprinted genes affect not only early development but also the later growth of the embryo. Further evidence that imprinted genes are involved in the growth of the embryo comes from studies on chimeras made between normal embryos and androgenetic or gynogenetic embryos. When inner cell mass cells from gynogenetic embryos are injected into normal embryos, growth is retarded by as much as 50%. But when androgenetic inner cell mass cells are introduced into normal embryos, the chimera's growth is increased by up to 50%. The imprinted genes on the male genome thus significantly increase the growth of the embryo.

At least 70 imprinted genes have been identified in mice, and a significant number are for non-coding RNA. Some are involved in growth control. The insulin-like growth factor IGF-2 is required for embryonic growth; its gene, *Igf2*, is imprinted in the maternal genome—that is, it is turned off, so that only the paternal gene is active. Direct evidence for the *Igf2* gene being imprinted comes from the observation that when sperm carrying a mutated, and thus defective, *Igf2* gene fertilize a normal egg, small offspring result. This is because only very low levels of IGF-2 are produced from the imprinted maternal gene and this is not enough to make up for the loss of IGF-2 expression from the paternal genome. In contrast, if the non-functional mutant gene is carried by the egg's genome, development is normal, with the required IGF-2 activity being provided from the normal paternal gene.

Closely linked to the *Igf2* gene on mouse chromosome 7 is the gene *H19* (of unknown function), which is imprinted in the opposite direction. It is expressed from the maternal chromosome but not from the paternal one (Fig. 11.9). By manipulating the chromosomes of an egg it was possible to make a diploid egg with two maternal genomes but with one genome having the normal female imprinting of *Igf2* and *H19* while in the other *Igf2* was not imprinted, as in the male germ cell (*H19* was absent in this genome). This egg could develop normally, unlike an embryo with two female genomes, which fails to develop properly, and this normal development must involve major changes in gene activity.

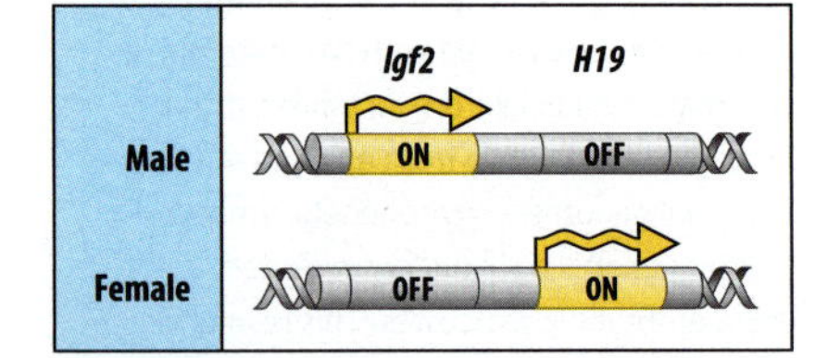

Fig. 11.9 Imprinting of genes controlling embryonic growth. In mouse embryos, the paternal gene for insulin-like growth factor 2 (*Igf2*) is on, but the gene on the maternal chromosome is off. In contrast, the closely linked gene *H19* is switched on in the maternal genome but silenced in the paternal genome.

A possible evolutionary explanation for the reciprocal imprinting of genes that control growth is parental-conflict theory: the theory that the reproductive strategies of the father and mother are different. Paternal imprinting promotes embryonic growth, whereas maternal imprinting reduces it, for example. The father wants to have maximal growth for his own offspring so that his genes have a good chance of surviving and being carried on. This can be achieved by having a large placenta, as a result of producing growth hormone, whose production is stimulated by IGF-2. The mother, who may mate with different males, benefits more by spreading her resources over all her offspring, and so wishes to prevent too much growth in any one embryo. Thus, a gene that promotes embryonic growth is turned off in the mother. Paternal genes expressed in the offspring could be selected to extract more resources from mothers, because an offspring's paternal genes are less likely to be present in the mother's other children. There are, however, many effects of imprinted genes other than on growth.

Imprinting occurs during germ-cell differentiation, and so a mechanism is required both for maintaining the imprinted condition throughout development and for wiping it out during the next cycle of germ-cell development. One mechanism for maintaining imprinting is DNA methylation (see Fig. 8.10). Evidence that DNA methylation is required for imprinting comes from mice in which DNA methylation is aberrant. In these mice, the *Igf2* gene is no longer imprinted and is expressed from the maternal chromosome.

A number of developmental disorders in humans are associated with specific imprinted genes. One is Prader-Willi syndrome, which is linked to a loss of expression of the paternal copy of a gene on chromosome 15, usually as a result of a deletion of a small region of the chromosome containing that gene. Infants fail to thrive and later can become extremely obese; they also show mental retardation and mental disturbances such as obsessional-compulsive behavior. Angelman syndrome results from the loss of the same region of the maternal chromosome 15. The effect in this case is severe motor and mental retardation. Beckwith-Wiedemann syndrome is due to a generalized disruption of imprinting on a region of chromosome 11. There is excessive fetal overgrowth and a disposition to cancer.

Summary

In many animals, germ cells are specified by localized cytoplasmic determinants in the egg. Localization of cytoplasmic determinants is controlled by cells surrounding the oocyte. Germ cells in mammals are specified by intercellular interactions at a later embryonic stage. Once the germ cells are determined, they migrate from their site of origin to the gonads, where further development and differentiation takes place. In the gonads, the diploid germ cells undergo meiosis, eventually producing haploid eggs and sperm. The number of oocytes in a female mammal is fixed before birth, whereas sperm production in male mammals is continuous throughout adult life. Eggs are always larger than somatic cells, and some are very large indeed. In order to achieve their increased size, specialized cells surrounding the developing oocyte may provide some of the constituents, such as yolk; in addition, some genes producing materials required in large amounts may be amplified in the oocyte.

Both maternal and paternal genomes are necessary for normal mammalian development. Embryos with diploid maternal or paternal genomes develop abnormally. Certain genes in eggs and sperm are imprinted, so that the activity of the same gene is different depending on whether it is of maternal or paternal origin. Several imprinted genes are involved in growth control of the embryo. Improper imprinting can lead to developmental abnormalities in humans.

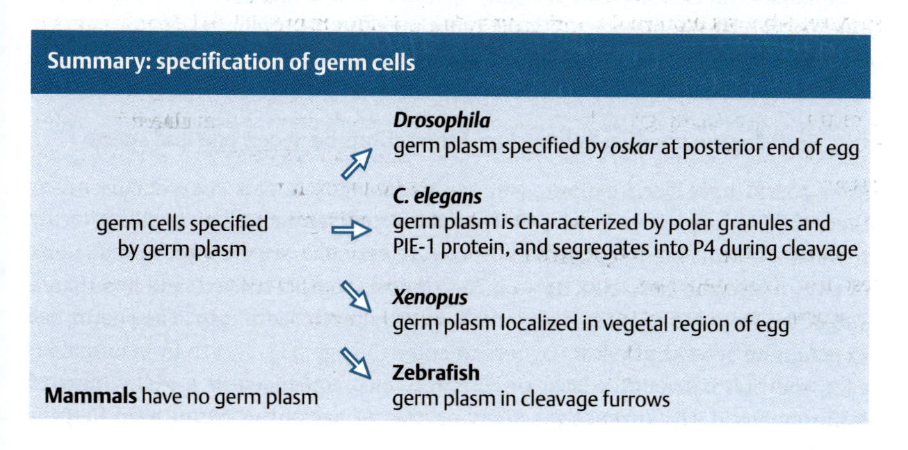

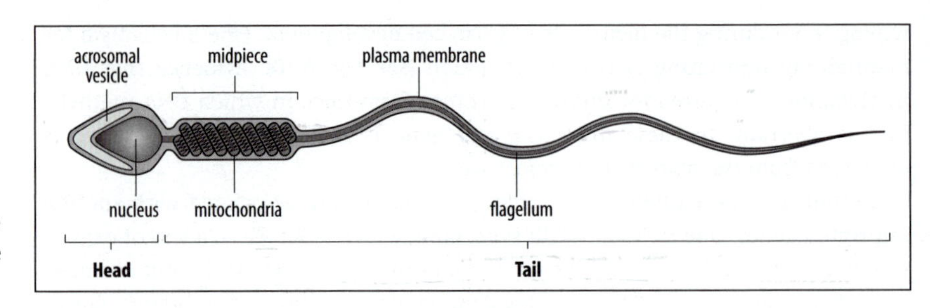

Fig. 11.10 A human sperm. The acrosomal vesicle at the anterior end of the sperm contains enzymes that are used to digest the protective coats around the egg. The plasma membrane on the head of the sperm contains various specialized proteins that bind to the egg coats and facilitate entry. The sperm moves by its single flagellum, which is powered by mitochondria. The overall length from head to tail is about 60 μm.

Fertilization

Fertilization—the fusion of egg and sperm—is the trigger that initiates development. The membranes of the egg and sperm fuse, the sperm nucleus enters the egg cytoplasm, becoming the sperm **pronucleus**. In mammals and many other animals, fertilization triggers the completion of meiosis in the egg and the set of maternal chromosomes retained in the egg becomes the egg pronucleus. The sperm and egg pronuclei form the zygotic nucleus. At fertilization, the egg is activated to start dividing and embark on its developmental program. Fertilization can be either external, as in frogs, or internal, as in *Drosophila*, mammals, and birds. Of all the sperm released by a male animal, only one fertilizes each egg. In many animals, including mammals, sperm penetration activates a blocking mechanism in the egg that prevents any further sperm entering the egg—the so-called block to **polyspermy**. This is necessary because if more than one sperm nucleus enters the egg there will be additional sets of chromosomes and centrosomes, resulting in abnormal development. There are a variety of mechanisms to ensure that only one sperm nucleus contributes to the zygote. In some animals, such as birds, many sperm penetrate the egg but all but one are destroyed in the cytoplasm.

Both eggs and sperm are structurally specialized for fertilization. The specializations of the egg are directed to preventing fertilization by more than one sperm, whereas those of the sperm are directed to facilitating penetration of the egg. Eggs are usually surrounded by several protective layers, and the eggs of many organisms have a layer of cortical granules just beneath the plasma membrane. The mammalian egg is bounded by a plasma membrane with cortical granules beneath it, and around it the zona pellucida, which serves to make the egg impenetrable to more than one sperm. All sperm are motile cells, typically designed for activating the egg, and at the same time delivering their nucleus into the egg cytoplasm. They essentially consist of a nucleus, mitochondria to provide an energy source, and a flagellum for movement. The anterior end is highly specialized, aiding penetration (Fig. 11.10). The sperm of the nematode *C. elegans* and some other invertebrates is unusual as it more closely resembles a normal cell and moves by ameboid motion.

11.7 Fertilization involves cell-surface interactions between egg and sperm

After sperm have been deposited in the mammalian female reproductive tract, they undergo a process known as **capacitation** which facilitates fertilization by removing certain inhibitory factors. There are very few mature eggs—usually one or two in humans and about 10 in mice—waiting to be fertilized, and less than a 100 of the millions of sperms deposited actually reach these eggs. The sperm has to penetrate several physical barriers to enter the egg (Fig. 11.11). In mammalian eggs, the first barrier is a layer of cumulus cells, embedded in a sticky mass of hyaluronic acid. Hyaluronidase activity on the surface of the sperm head helps it

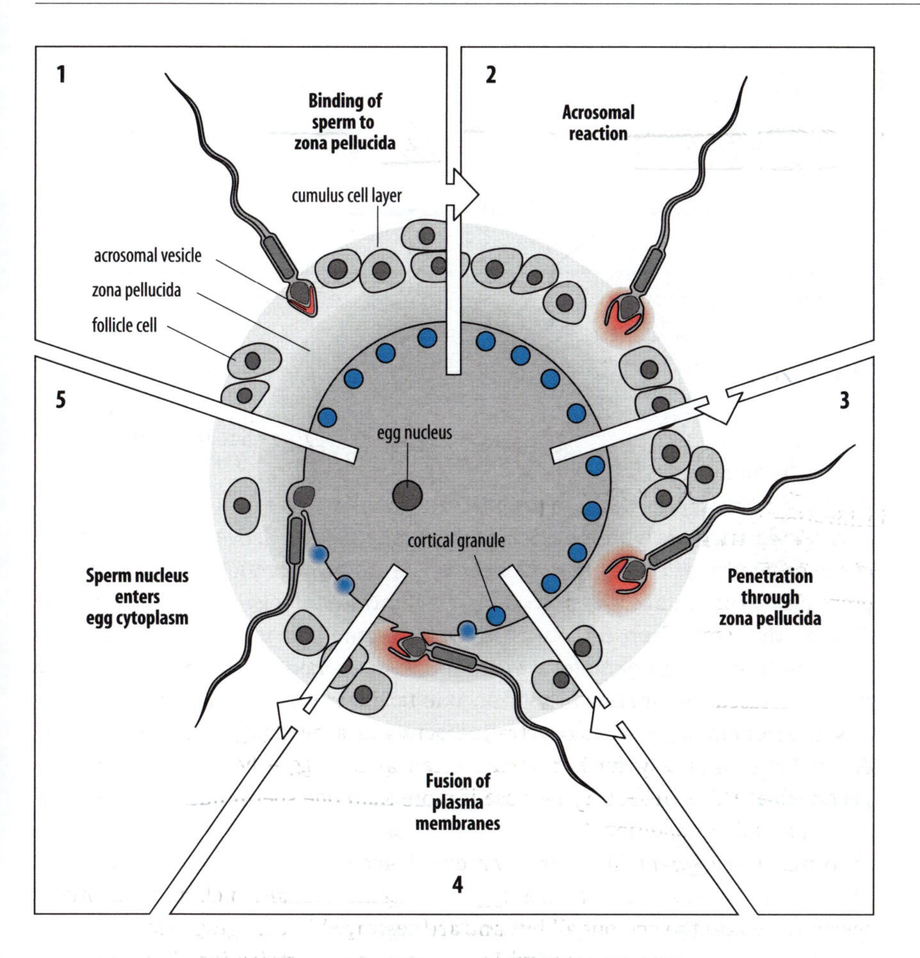

Fig. 11.11 Fertilization of a mammalian egg. After penetrating the follicle-derived cumulus cell layer, the sperm binds to the zona pellucida (1). This triggers the acrosomal reaction (2), in which enzymes are released from the acrosomal vesicle and break down the zona pellucida. This enables the sperm to penetrate the zona pellucida (3), and bind to the egg plasma membrane. The plasma membrane of the sperm head fuses with the egg plasma membrane (4). This activates the egg, causing a release of cortical granules and the sperm nucleus then enters the egg (5).

After Alberts, B., et al.: 1989.

to penetrate this layer. The sperm next encounters the **zona pellucida**, a layer of glycoproteins surrounding the egg. This also acts as a physical barrier, but sperm are helped to penetrate it by the **acrosomal reaction**—the release of the contents of the acrosomal vesicle located in the sperm head. The zona pellucida contains three glycoproteins, including ZP3, which is a receptor for species-specific binding of the sperm. On the sperm head plasma membrane is a protein SED1, which binds to the zona pellucida. When the sperm binds to ZP3, the contents of the acrosome are released by exocytosis. The enzymes released include β-N-acetylglucosaminidase, which breaks down the oligosaccharide side chains on the zona pellucida glycoproteins, and a protease called acrosin. These enzymes allow the sperm to approach the egg plasma membrane.

The acrosomal reaction also exposes proteins on the sperm surface that can bind to the egg membrane and are involved in the fusion of sperm and egg membranes. A potential component is the protein fertilin, which may bind to an integrin-like receptor on the egg plasma membrane. A key egg receptor for the sperm is the protein CD9. Interaction of a sperm with CD9 could initiate sperm and egg fusion. In the sperm of many invertebrates, such as the sea urchin, the acrosomal reaction results in the extension of a rod-like acrosomal process. This forms by the polymerization of actin and facilitates contact with the egg membrane.

It is possible to fertilize human eggs in culture, and then transfer a very early embryo to the mother. This technique of *in vitro* fertilization (IVF) has been of great help to couples who have, for a variety of reasons, difficulty in conceiving. It is even possible to fertilize a human egg by injecting a single intact sperm directly into the egg.

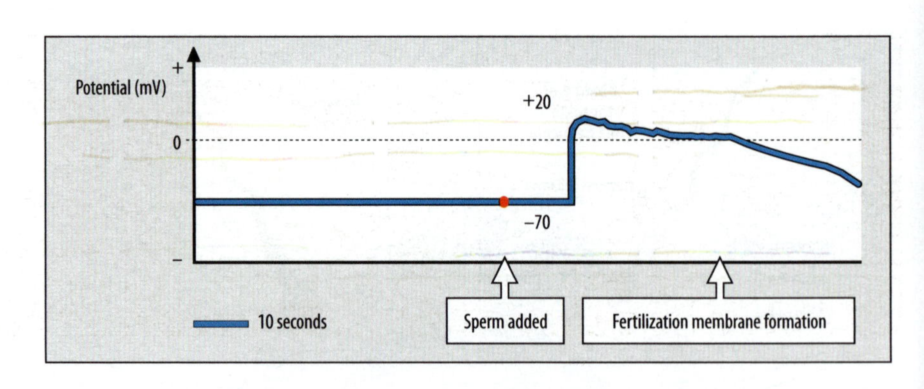

Fig. 11.12 Depolarization of the sea urchin egg plasma membrane at fertilization. The resting membrane potential of the unfertilized sea urchin egg is –70 mV. At fertilization, it changes rapidly to +20 mV, and then slowly returns to the original value. This depolarization may provide a fast block to polyspermy.

11.8 Changes in the egg membrane at fertilization block polyspermy

Although many sperm attach to the coats surrounding the egg, it is important that only one sperm fuses with the egg plasma membrane and delivers its nucleus into the egg cytoplasm. There are thus mechanisms to prevent more than one sperm entering. The main block to polyspermy in many animals is brought into play as soon as the first sperm fuses with the plasma membrane. This stimulates the release of the cortical granules which, in mammals, contain enzymes that block further sperm from binding to the zona pellucida. In sea urchins, which we shall consider in more detail, the enzymes contribute to the formation of an impenetrable fertilization membrane around the fertilized egg.

In the sea urchin egg, a rapid block to polyspermy is triggered by a transient depolarization of the egg plasma membrane, caused by sperm–egg fusion. The electrical membrane potential across the plasma membrane goes from –70 mV to +20 mV within a few seconds of sperm entry (Fig. 11.12). The membrane potential slowly returns to its original level while the fertilization membrane, which is made from the cortical granules and vitelline membrane, is formed. If depolarization is prevented, polyspermy occurs, but how depolarization blocks polyspermy is not yet clear. (In mouse fertilization there is no change in membrane potential.) The slower cortical reaction in the sea urchin is better understood. Sperm entry into the sea urchin egg results in the initiation of a wave of calcium release and leads to the cortical granules, which lie just beneath the plasma membrane in the mature egg, releasing their contents to the outside of the plasma membrane by exocytosis. This results in the vitelline membrane lifting off the plasma membrane. The cortical granule contents contribute to the vitelline membrane to form the fertilization membrane, and also provide a hyaline layer between it and the egg's plasma membrane (Fig. 11.13). Both these structures prevent sperm from binding to the egg plasma membrane.

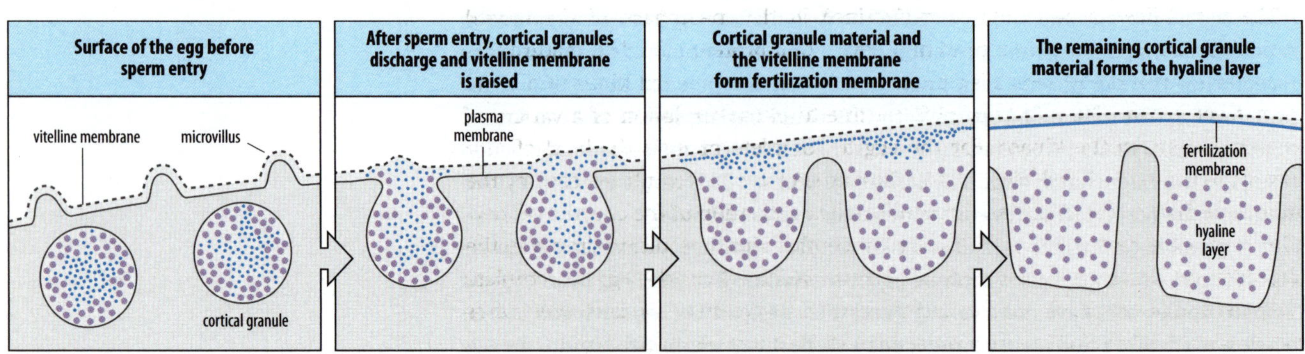

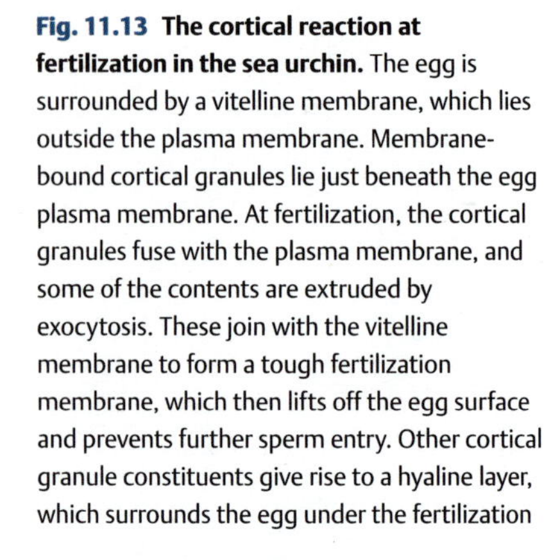

Fig. 11.13 The cortical reaction at fertilization in the sea urchin. The egg is surrounded by a vitelline membrane, which lies outside the plasma membrane. Membrane-bound cortical granules lie just beneath the egg plasma membrane. At fertilization, the cortical granules fuse with the plasma membrane, and some of the contents are extruded by exocytosis. These join with the vitelline membrane to form a tough fertilization membrane, which then lifts off the egg surface and prevents further sperm entry. Other cortical granule constituents give rise to a hyaline layer, which surrounds the egg under the fertilization

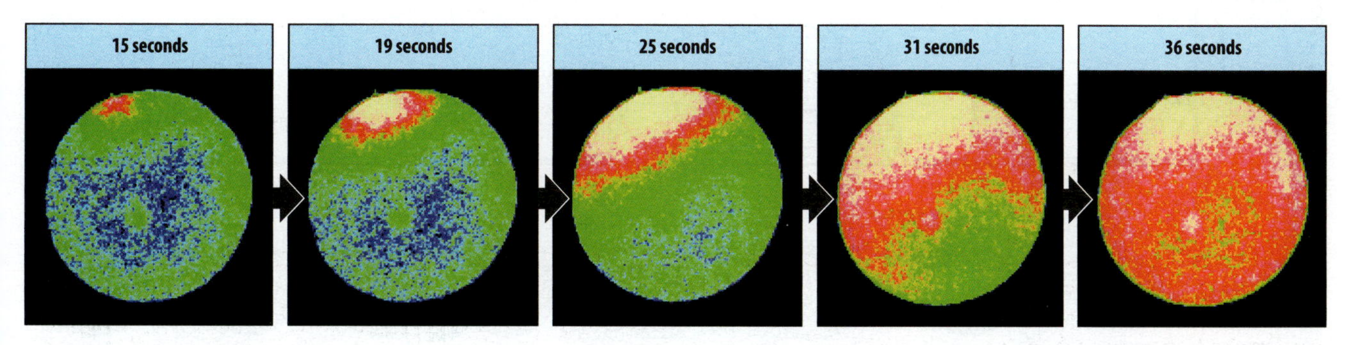

Fig. 11.14 Calcium wave at fertilization. A series of images showing an intracellular calcium wave at fertilization in a sea urchin egg. The fertilizing sperm has fused just to the left of the top of the egg, and triggered the wave. Calcium ion concentration is monitored with a calcium-sensitive fluorescent dye, using confocal fluorescence microscopy. Calcium concentration is shown in false color: red is the highest concentration, then yellow, green, and blue. Times shown are seconds after sperm entry.

Photographs courtesy of M. Whitaker.

11.9 A calcium wave initiated at fertilization results in egg activation

The activation of the egg at fertilization initiates a series of events that result in the start of development. For example, in the sea urchin egg there is a several-fold increase in protein synthesis, and there are often changes in egg structures, such as the cortical rotation that occurs in amphibian eggs (see Section 3.6). But the main events are that the egg, which has been blocked at a stage in meiosis, now completes meiosis, whereupon the egg and sperm nuclei fuse to form the diploid zygotic genome, and the fertilized egg enters mitosis. In mice and humans, the pronuclear membranes disappear before the pronuclei come together.

Fertilization and egg activation in mammals and sea urchins are associated with an explosive release of free Ca^{2+} ions within the egg, producing a wave of calcium that travels across it (Fig. 11.14). The calcium release is triggered by sperm entry, and is both necessary and sufficient to initiate the onset of development. The wave starts at the point of sperm entry and crosses the egg at a speed of 5–10 μm per second. In all mammals, oscillations in calcium concentration occur for several hours after fertilization. The mechanism for the Ca^{2+} release at fertilization is not known, but it is possible that the sperm introduces a specific protein factor that initiates Ca^{2+} release after granule fusion.

The sharp increase in free Ca^{2+} is crucial for egg activation. Eggs in a variety of animals are activated if the Ca^{2+} concentration in the egg cytosol is artificially increased, for example, by direct injection of Ca^{2+}. Conversely, preventing calcium increase by injecting agents that bind it, such as the calcium chelator EGTA, blocks activation. It has been known for many years that *Xenopus* eggs can be activated simply by prodding them with a glass needle; this is due to a local influx in calcium at the site of insertion that triggers a calcium wave. Calcium initiates development of the fertilized egg by its action on proteins that control the cell cycle.

The unfertilized *Xenopus* egg is maintained in the metaphase of the second meiotic division by the presence of high levels of a protein complex, maturation-promoting factor (MPF), which is a complex of a cyclin-dependent kinase (Cdk) and its partner cyclin. The effects of MPF are due to phosphorylation of a variety of protein targets by the kinase. For the egg to complete meiosis the level of MPF activity must be reduced (Fig. 11.15). Similar Cdk–cyclin complexes control the mitotic cell cycle. The calcium wave results in the activation of the enzyme calmodulin-dependent protein kinase II. The activity of this kinase results indirectly in the degradation of the cyclin component of MPF, which allows the egg to complete meiosis. The pronuclei then fuse, and the zygote moves on to the next stage of its development, which is the entry into the mitotic cell cycles of cleavage.

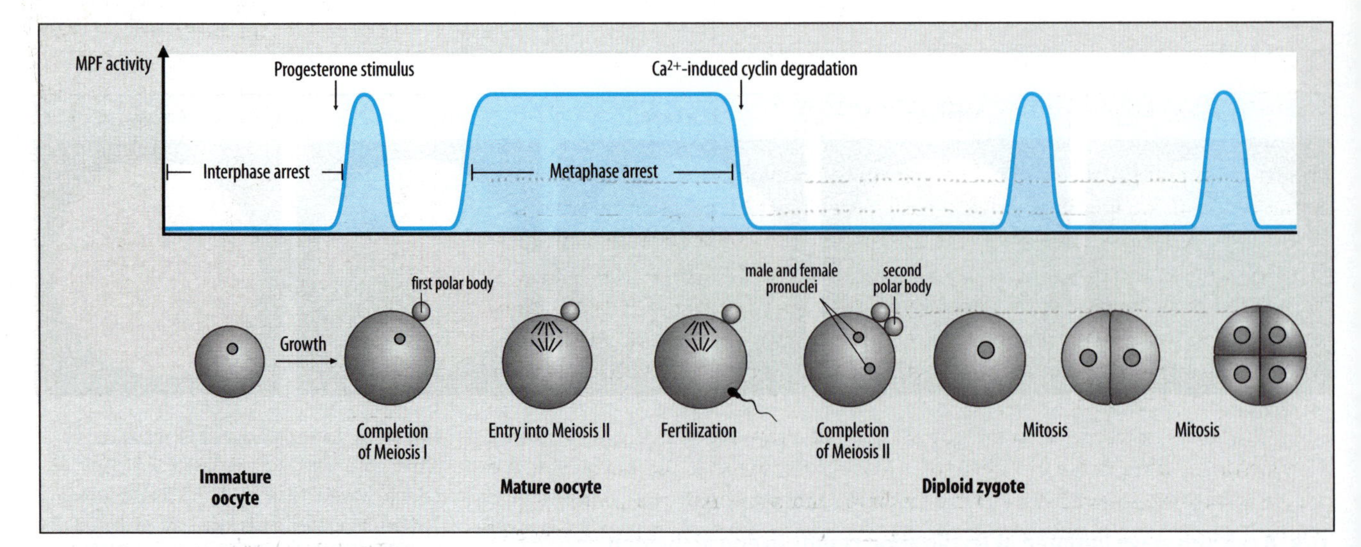

Fig. 11.15 Profile of maturation-promoting factor (MPF) activity in early *Xenopus* development. The immature *Xenopus* oocyte cell cycle is arrested. On receipt of a hormonal progesterone stimulus it enters, and completes, the first meiotic division, with the formation of the first polar body. It enters the second meiotic division, but becomes arrested again in metaphase. The egg is laid at this point. At fertilization, the calcium wave leads to completion of meiosis, and the second polar body is formed. The zygote starts to cleave rapidly by mitotic divisions. Maturation-promoting factor (MPF) rises sharply just before each division of the meiotic and mitotic cell cycles, remains high during mitosis, and then decreases abruptly and remains low between successive mitoses.

Summary

The fusion of sperm and egg at fertilization stimulates the egg to start dividing and developing. Both sperm and egg have specialized structures relating to fertilization. The initial binding of sperm to the mammalian egg is mediated by zona pellucida molecules, and leads to the release of the contents of the sperm acrosome, which facilitates the penetration of the sperm through the layers surrounding the egg and allows it to reach the egg plasma membrane. A block to polyspermy allows only one sperm to fuse with the egg and deliver its nucleus into the egg cytoplasm. In sea urchins, the first block is rapid and partial, and the second results from the release of egg cortical granule contents to the exterior to form an impenetrable fertilization membrane. A key role in egg activation after fertilization is played by the release of free calcium ions into the cytosol, which spread in a wave from the site of sperm fusion. In mammals, and most other vertebrates, fertilization triggers the completion of the second meiotic division; the sperm and egg haploid pronuclei give rise to the zygote nucleus, and the egg divides.

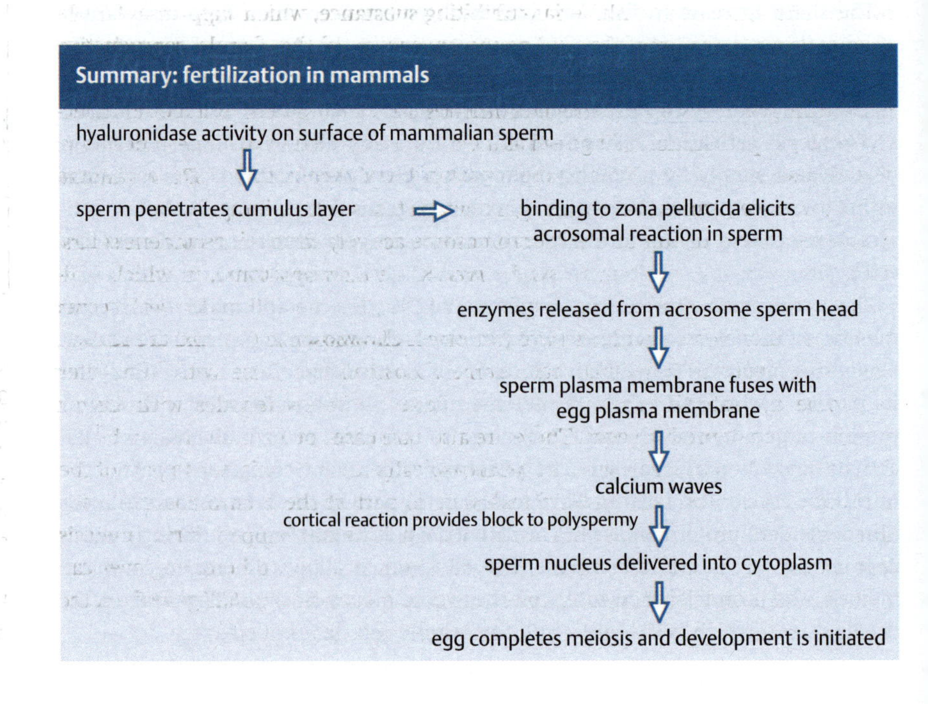

Determination of the sexual phenotype

In organisms that produce two phenotypically different sexes, sexual development is the result of the modification of a basic developmental program in order that one of the sexes can develop. The early embryo is similar in both males and females, with sexual differences only developing at later stages. In the organisms considered here, somatic sexual phenotype is genetically fixed at fertilization by the chromosomal content of the gametes (the reproductive cells) that fuse to form the fertilized egg. In mammals, for example, sex is determined by the Y chromosome; males are XY, females are XX.

Even among vertebrates, however, sex is not always determined by which chromosomes are present; in alligators, it is determined by the environmental temperature during incubation of the embryo, and some fish can switch sex as adults in response to environmental conditions. In insects, there is a wide range of different sex-determining mechanisms. Intriguing though these are, we focus here on those organisms in which the genetic and molecular basis of sex determination is best understood—mammals, *Drosophila*, and the nematode—in which sex is determined by chromosomal content, although by quite different mechanisms.

We first consider the determination of the somatic sexual phenotype—the development of the individual as either male or female. We then deal with the determination of the sexual phenotype of the germ cells—whether they become eggs or sperm—and finally consider how the embryo compensates for the difference in chromosomal composition between males and females.

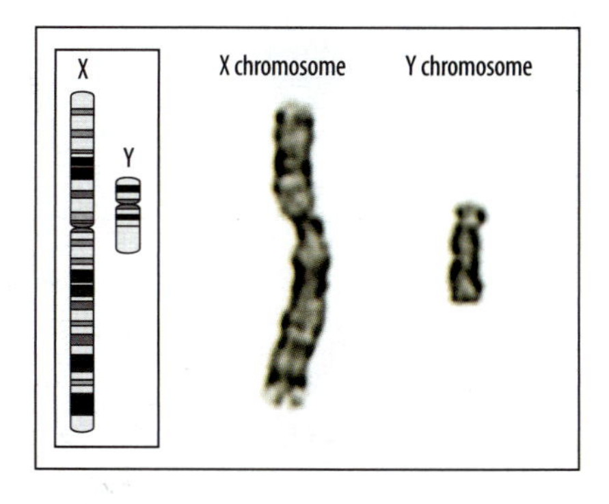

Fig. 11.16 The sex chromosomes in humans. If two X chromosomes (XX) are present, a female develops, whereas the presence of a Y chromosome (XY) leads to development of a male. The inset shows a diagrammatic representation of the banding in the chromosomes, which represent regions of increased chromatin condensation.

11.10 The primary sex-determining gene in mammals is on the Y chromosome

The genetic sex of a mammal is established at the moment of conception, when the sperm introduces into the egg either an X or a Y chromosome (Fig. 11.16). Eggs contain one X chromosome; if the sperm introduces another X the embryo will be female, if a Y it will be male. The presence of a Y chromosome results in the somatic cells of the embryo's gonads developing into testes rather than into ovaries. The testes secrete Müllerian-inhibiting substance, which suppresses female development by causing the embryonic precursor of the female reproductive organs to regress. It also induces cells to become Leydig cells, which secrete the male hormone testosterone, which stimulates the development of male reproductive organs. Specification of a gonad as a testis is controlled by a single gene on the Y chromosome, the **sex-determining region of the Y chromosome** (*SRY* in humans and *Sry* in mice), which was formerly known as testis-determining factor.

Evidence that a region on the Y chromosome actively determines maleness first came from two unusual human syndromes: Klinefelter syndrome, in which individuals have two X chromosomes and one Y (XXY), but are still males; and Turner syndrome, in which individuals have just one X chromosome (XO) and are female. Both these types of individual have some abnormalities; those with Klinefelter syndrome are infertile males with small testes, whereas females with Turner syndrome do not produce eggs. There are also rare cases of XY individuals who are female, and XX individuals who are phenotypically male. This is due to part of the Y chromosome being lost (in XY females) or to part of the Y chromosome being transferred to the X chromosome (in XX males). This can happen during meiosis in the male germ cells; the X and Y chromosomes pair up, and crossing over can occur between them. Very rarely, this crossing over transfers the *SRY* gene from the Y chromosome onto the X (Fig. 11.17), thus leading to sex reversal.

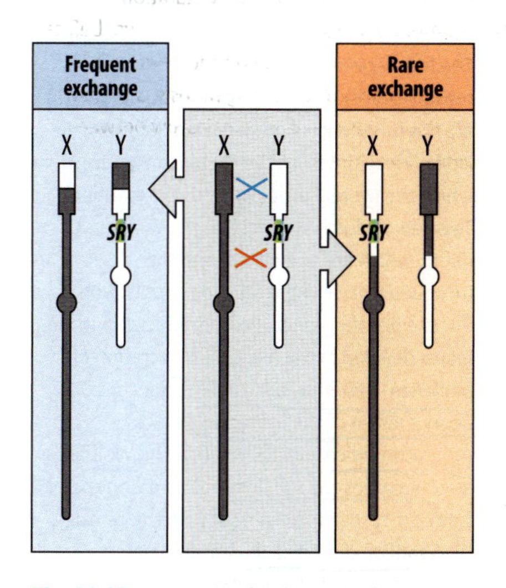

Fig. 11.17 Sex reversal in humans due to chromosomal exchange. At meiosis in male germ cells, the X and Y chromosomes pair up (center panel) and there is crossing over of the distal region (blue cross), which does not affect sexual development (left panel). On rare occasions, crossing over involves a larger segment that includes the *SRY* gene (red cross), so that the X chromosome now carries this male-determining gene (right panel).

After Goodfellow, P.N., et al.: 1993.

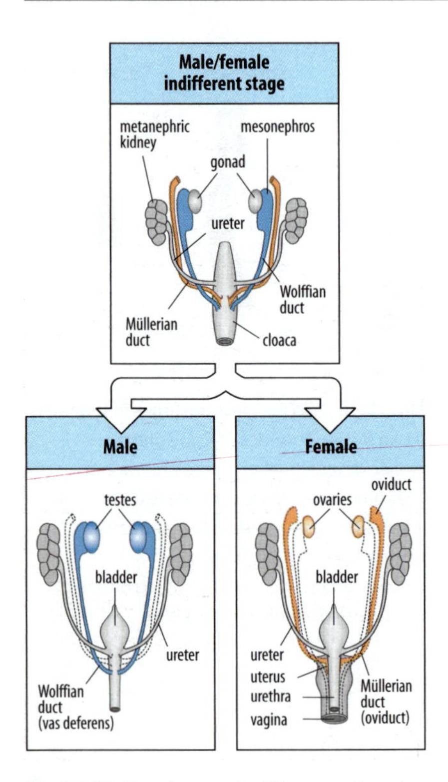

Fig. 11.18 Development of the gonads and related structures in mammals. Top panel: early in development, there is no difference between males and females in the structures that give rise to the gonads and related organs. The future gonads lie adjacent to the mesonephros, which are embryonic kidneys that are not functional in adult mammals. (The true kidney develops from the metanephros, from which the ureter carries urine to the bladder.) Two sets of ducts are present; the Wolffian ducts, which are associated with the mesonephros, and the Müllerian ducts. Both ducts enter the cloaca. Bottom left panel: after testes develop in the male, their secretion of Müllerian-inhibiting substance results in degeneration of the Müllerian duct by programmed cell death, whereas the Wolffian duct becomes the vas deferens, carrying sperm from the testis. Bottom right panel: in females, the Wolffian duct disappears, also by programmed cell death, and the Müllerian duct becomes the oviduct. The uterus forms at the end of the Müllerian ducts.

After Higgins, S.J., et al.: 1989.

The sex-determining region alone is sufficient to specify maleness, as shown by an experiment in which the mouse equivalent of the *SRY* gene (*Sry*) was introduced into the eggs of XX mice. These transgenic embryos developed as males, even though they lacked all the other genes on the Y chromosome. The presence of the *Sry* gene, which encodes a transcription factor, resulted in these XX embryos developing testes instead of ovaries. The gene is expressed in the developing gonad just before its differentiation as a testis and activates another transcription factor Sox9. However, these transgenic mice were not completely normal males, as other genes on the Y chromosome are necessary for the development of the sperm. These XX males were therefore infertile.

11.11 Mammalian sexual phenotype is regulated by gonadal hormones

All mammals, whatever their genetic sex, start off as embryos along a sexually neutral developmental pathway. The presence of a Y chromosome causes testes to develop, and the hormones they produce switch the development of all somatic tissues to a characteristic male pathway. In the absence of a Y chromosome, the development of somatic tissues is along the female pathway. Thus, while the sex of the gonads—the testis and ovary—is genetically determined, all the other cells in the mammalian body are neutral, irrespective of their chromosomal sex. It does not matter if they are XX or XY, as any future sex-specific development they undergo is controlled by hormones. The primary role of the testis in directing male development was originally demonstrated by removing the prospective gonadal tissue from early rabbit embryos. All the embryos developed as females, irrespective of their chromosomal constitution. To develop as a male, therefore, a testis has to be present. The testis exerts its effect on sexual differentiation of somatic tissues primarily by secreting the hormone testosterone.

The gonads in mammals develop in close association with the **mesonephros**; this is an embryonic kidney that contributes to both the male and female reproductive organs. Associated with the mesonephros on each side of the body are the **Wolffian ducts**, which run down the body to the cloaca, an undifferentiated opening. Another pair of ducts, the **Müllerian ducts**, run parallel to the Wolffian ducts and also open into the cloaca. In early mammalian development, before gonadal differentiation, both sets of ducts are present (Fig. 11.18). In females, in the absence of the testes, the Müllerian ducts develop into the **oviducts** (Fallopian tubes), which transport eggs from the ovaries to the uterus, while the Wolffian ducts degenerate. In males, the expression of *Sry* results in the differentiation of Sertoli cells, which are the somatic cells of the testis and are essential for testis formation and for spermatogenesis, as they retain the germ cells that migrate into the gonad. Sertoli cell development involves upregulation of the gene *Sox-9* and the secretion of **Müllerian-inhibiting substance**, which induces regression of the Müllerian duct, largely by apoptosis. The interstitial cell lineage in the testis then differentiates into Leydig cells, which produce testosterone. This causes the Wolffian duct to become the vas deferens, the duct that carries sperm to the penis. There is evidence that the extracellular signaling molecule Wnt-4 represses testosterone production in the undifferentiated gonad, and that one of the functions of the *Sry* gene is to cause the downregulation of *Wnt-4* expression, which occurs when Sertoli cells differentiate. FGF-9 is required for Sertoli cell differentiation and male mice lacking it develop as females.

The main secondary sexual characters that distinguish males and females are the reduced size of mammary glands in males, and the development of a penis and a scrotum in males instead of the clitoris and labia of females (Fig. 11.19). At early

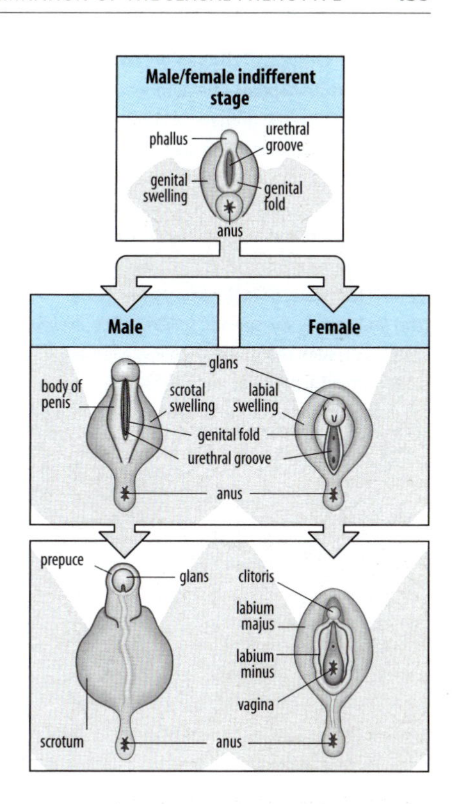

Fig. 11.19 Development of the genitalia in humans. At an early embryonic stage, the genitalia are the same in males and females (top panel). After testis formation in males, the phallus and the genital fold give rise to the penis, whereas in females they give rise to the clitoris and the labia minus. The genital swelling forms the scrotum in males and the labia majus in females.

stages of embryonic development the genital region of males and females is indistinguishable. Differences only arise after gonad development, as a result of the action of gonadal hormones. For example, in humans, the phallus gives rise to the clitoris in females and the end of the penis in males.

The role of gonadal hormones in sexual development is illustrated by rare cases of abnormal sexual development. Certain XY males develop as phenotypic females in external appearance, even though they have testes and secrete testosterone. They have a mutation that renders them insensitive to testosterone because they lack the testosterone receptor, which is present throughout the body. Conversely, genetic females with a completely normal XX constitution can develop as phenotypic males in external appearance if they are exposed to male hormones during their embryonic development.

Sex-specific behavior is also affected by the hormonal environment as a result of the effects of hormones on the brain. For example, male rats castrated after birth develop the sexual behavioral characteristics of genetic females.

11.12 The primary sex-determining signal in *Drosophila* is the number of X chromosomes, and is cell autonomous

The external sexual differences between *Drosophila* males and females are mainly in the genital structures, although there are also some differences in bristle patterns and pigmentation, and male flies have a sex comb on the first pair of legs. In flies, sex determination of the somatic cells is cell autonomous—that is, it is specified on a cell-by-cell basis—and there is no process resembling the control of somatic sexual differentiation by hormones. The pathway of somatic sexual development is the result of a series of gene interactions that are initiated by the primary sex signal, and which act on a binary genetic switch. The end result of this cascade is the expression of just a few effector genes, whose activity controls the subsequent male or female differentiation of the somatic cells.

Like mammals, fruit flies have two unequally sized sex chromosomes, X and Y, and males are XY and females XX. But these similarities are misleading. In flies, sex is not determined by the presence of a Y chromosome, but by the number of X chromosomes. Thus, XXY flies are female and X flies are male. The chromosomal composition of each somatic cell determines its sexual development. This is beautifully illustrated by the creation of genetic mosaics in which the left side of the animal is XX and the right side X: the two halves develop as female and male, respectively (Fig. 11.20).

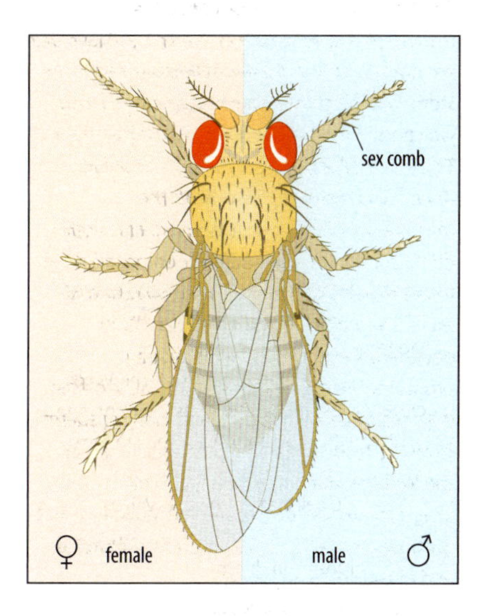

Fig. 11.20 A *Drosophila* female/male genetic mosaic. The left side of the fly is composed of XX cells and develops as a female, whereas the right side is composed of X cells and develops as a male. The male fly has smaller wings, a special structure, the sex comb, on its first pair of legs, and different genitalia at the end of the abdomen (not shown).

In flies, the presence of two X chromosomes results in the production of the protein Sex-lethal, coded for by the gene *Sex-lethal*, which is located on the X chromosome. This leads to female development through the activation of a series of genes that determine the sexual state and then produce the sexual phenotype (Fig. 11.21). The end of the sex-determination pathway is the *transformer* gene, which determines how the mRNA of the *doublesex* gene is spliced; *doublesex* encodes a transcription factor whose activity ultimately produces most aspects of somatic sex. The *doublesex* gene is active in both males and females, but different protein

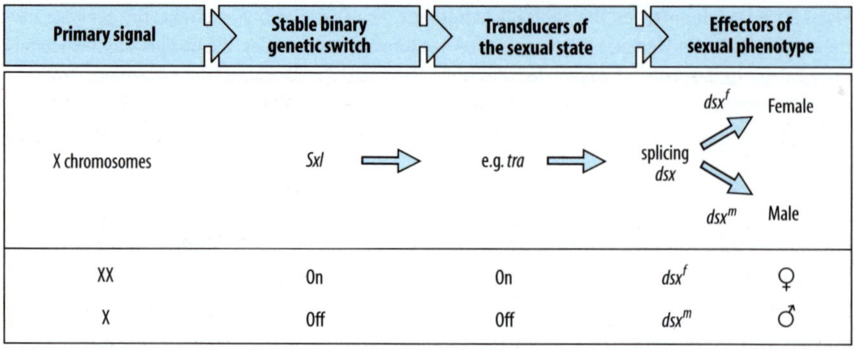

Primary signal	Stable binary genetic switch	Transducers of the sexual state	Effectors of sexual phenotype	
X chromosomes	*Sxl* ⇨	e.g. *tra* ⇨	splicing *dsx* → *dsx*^{f} / *dsx*^{m}	Female / Male
XX	On	On	*dsx*^{f}	♀
X	Off	Off	*dsx*^{m}	♂

Fig. 11.21 Outline of the sex-determination pathway in *Drosophila*. The number of X chromosomes is the primary sex-determining signal, and in females the presence of two X chromosomes activates the gene *Sex-lethal* (*Sxl*). This produces Sex-lethal protein, whereas no Sex-lethal protein is made in males, who have only one X chromosome. The activity of *Sex-lethal* is transduced via the *transformer* gene (*tra*) and causes sex-specific splicing of *doublesex* RNA (*dsx*^{f}), such that the cells follow a female developmental pathway. In the absence of Sex-lethal protein, the splicing of *doublesex* RNA to give *dsx*^{m} RNA leads to male development.

products are produced in the two sexes as a result of sex-specific RNA splicing. Males and females thus express similar but distinct Doublesex proteins, which act in somatic cells to induce expression of sex-specific genes, as well as to repress characteristics of the opposite sex. Expression of the *transformer* gene is controlled by *Sex-lethal*. In the presence of Sex-lethal protein, *transformer* RNA is productively spliced and this, together with the Transformer-2 protein, leads to the female form of the Doublesex protein being made, resulting in female differentiation. Production of the male form of the protein is the default pathway.

Sex-lethal is only turned on in females with two X chromosomes. In the absence of early *Sex-lethal* expression, male development occurs. Once *Sex-lethal* is activated in females, it remains activated through an autoregulatory mechanism. Early expression of *Sex-lethal* in females occurs through activation of a promoter P_e (e stands for establishment) at about the time of syncytial blastoderm formation. At this stage, Sex-lethal protein is synthesized and accumulates in the blastoderm of female embryos. At the cellular blastoderm stage, another promoter for *Sex-lethal*, P_m (m stands for maintenance), becomes active in both males and females, and P_e is shut off, but the sex is already determined. The *Sex-lethal* RNA transcribed from P_m in contrast to that from P_e can only be spliced into an mRNA for the Sex-lethal protein if some Sex-lethal protein is already present. Only females contain any Sex-lethal protein, and so only in females is the mRNA productively spliced and more Sex-lethal protein synthesized (Fig. 11.22). This autoregulatory loop at the post-transcriptional level results in Sex-lethal protein being synthesized throughout female development.

How does the number of X chromosomes control these key sex-determining genes? The mechanism in *Drosophila* involves interactions between the products of so-called numerator genes on the X chromosome and of genes on the autosomes, as well as maternally specified factors. In females, the twofold higher level of numerator proteins activates the *Sex-lethal* gene by binding to sites in the P_e promoter, and so overcomes the repression of *Sex-lethal* exerted by the products of maternal genes on the autosomes.

The pathway outlined in Fig. 11.21 is an oversimplification, as *doublesex* does not control all aspects of somatic sexual differentiation in *Drosophila*. There is an additional branch of the pathway downstream of *transformer*, which controls sexually dimorphic aspects of the nervous system and sexual behavior. This branch includes the genes *fruitless*, whose activity has been shown to be necessary for male sexual behavior. The *transformer* gene may act directly on *fruitless*.

In other dipteran insects the same general strategy for sex determination is used, but there are marked differences at the molecular level. Only *doublesex* has been found in dipterans distantly related to *Drosophila*; *Sex-lethal* has been found in other dipterans but is not involved in sex determination there.

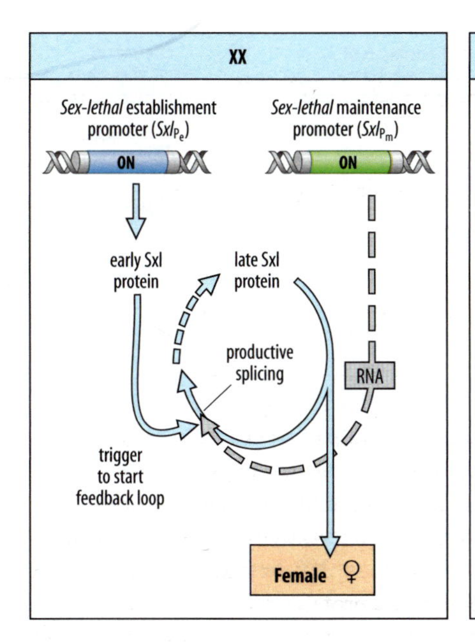

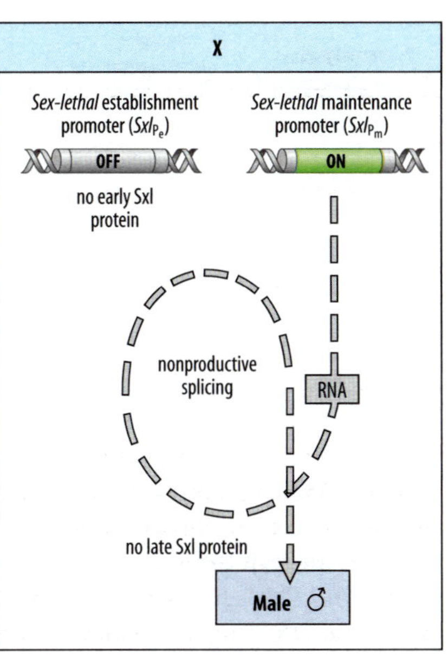

Fig. 11.22 Production of Sex-lethal protein in *Drosophila* sex determination. When two X chromosomes are present, the early establishment promoter (P_e) of the *Sex-lethal* (*Sxl*) gene is activated at the syncytial blastoderm stage in future females, but not in males. This results in the production of Sxl protein. Later, at the blastoderm stage, the maintenance promoter (P_m) of *Sxl* becomes active in both females and males, and P_e is turned off. The *Sxl* RNA is only correctly spliced if Sxl protein is already present, which is only in females. A positive feedback loop for Sxl protein production is thus established in females. The continued presence of Sxl protein initiates a cascade of gene activity leading to female development. If no Sxl protein is present, male development ensues.

After Cline, T.W.: 1993.

11.13 Somatic sexual development in *Caenorhabditis* is determined by the number of X chromosomes

In the nematode *C. elegans*, the two sexes are self-fertilizing hermaphrodite (essentially a modified female) and male (Fig. 11.23), although in other nematodes they are male and female. Hermaphrodites produce a limited amount of sperm early in development, with the remainder of the germ cells developing into oocytes. Sex in *C. elegans* (and other nematodes) is determined by the number of X chromosomes: the hermaphrodite (XX) has two X chromosomes, whereas the presence of just one X chromosome leads to development as a male (XO). The primary sex signal in *C. elegans* acts on the gene *XO lethal* (*xol-1*). With two X chromosomes, *xol-1* expression is low, resulting in the development of a hermaphrodite. *xol-1* is repressed by the SEX-1 protein, which is encoded on the X chromosome and is a key element in 'counting' the number of X chromosomes present.

A cascade of gene activity converts the level of *xol-1* expression into the somatic sexual phenotype (Fig. 11.24). Genes involved include those for nuclear proteins, such as SDC-2, and for a secreted protein, Hermaphrodite-1 (HER-1). At the end of

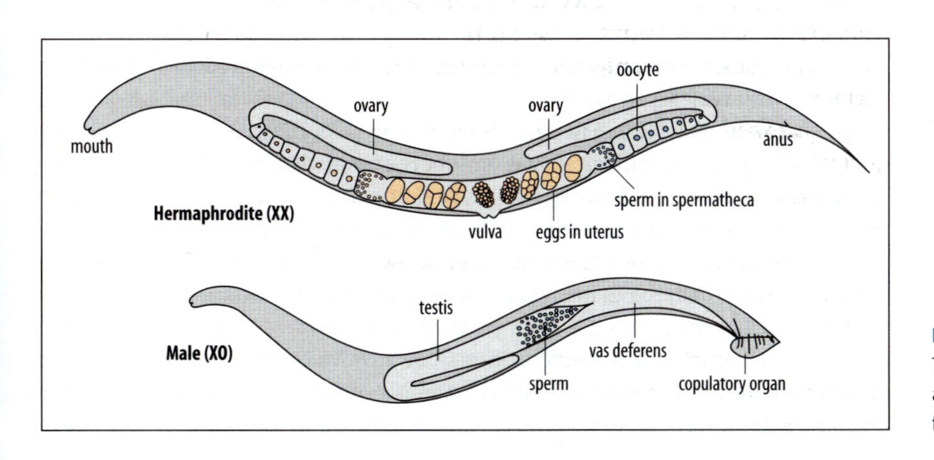

Fig. 11.23 Hermaphrodite and male *C. elegans*. The hermaphrodite has a 'two-armed' gonad and makes both eggs and sperm. The eggs are fertilized internally. The male makes sperm only.

Fig. 11.24 Outline of the somatic sex determination pathway in *C. elegans*. The primary signal for sex determination is set by the number of X chromosomes. When two X chromosomes are present, the expression of the gene *XO lethal* (*xol-1*) is low, leading to hermaphrodite development, whereas *xol-1* is expressed at a high level in males. There is a cascade of gene expression starting from *xol-1* that leads to the gene *transformer-1* (*tra-1*), which codes for a transcription factor. If *tra-1* is active, development as a hermaphrodite occurs, but if it is expressed at low levels, males develop. The product of the *hermaphrodite-1* (*her-1*) gene is a secreted protein, which probably binds to a receptor encoded by *transformer-2* (*tra-2*), inhibiting its function.

Primary signal	Binary genetic switch		Transducers of the sexual state			Effectors of sexual phenotype
X chromosomes	*xol-1* (inhibition) ⊣	*sdc-1* *sdc-2* *sdc-3* ⊣	*her-1* ⊣	*tra-2* *tra-3* ⊣	*fem-1* *fem-2* *fem-3* ⊣	*tra-1* → Hermaphrodite / Male
XX	Low	High	Low	High	Low	High ⚥
X	High	Low	High	Low	High	Low ♂

the cascade is the gene *transformer-1* (*tra-1*), which encodes a transcription factor. Expression of TRA-1 protein is both necessary and sufficient to direct all aspects of hermaphrodite (XX) somatic cell development. A gain-of-function mutation in *tra-1* leads to hermaphrodite development in an XO animal, irrespective of the state of any of the regulatory genes that normally control its activity. Mutations that inactivate *tra-1* lead to complete masculinization of XX hermaphrodites. Unlike *Drosophila*, sex determination in *C. elegans* requires cell–cell interactions, as the process involves secreted proteins. The *mab-3* gene, which is related to *Drosophila doublesex*, acts downstream of *tra-1* to promote male-specific development in the peripheral nervous system and the gut. The gene *egl-1* (see Section 8.12) is turned on in males and early in male development this results in the death of certain neurons associated in the hermaphrodite with egg-laying.

Before considering how the sex of the germ cells is determined and how the imbalance of X-linked genes between the sexes is dealt with in animals, we need to digress briefly to touch upon sex determination in flowering plants.

11.14 Most flowering plants are hermaphrodites, but some produce unisexual flowers

As we have seen, the flowers of angiosperms share a common organization in which the four types of floral organs are arranged in concentric whorls (see Chapter 6). The two inner whorls are the sexual organs—the stamens and carpels. Sepals and petals in the outer whorls are not sexual organs, but may serve to attract pollinators. Stamens produce pollen, which contains the male gametes corresponding to the sperm of animals. The female reproductive structures of flowers are the carpels, which are either free, or are fused to form a compound ovary. Carpels are the site of ovule formation, and each ovule produces an egg cell. Most flowering plants are hermaphrodites, bearing flowers with functional male and female sexual organs. However, not all flowering plants are of this type.

In about 10% of flowering plants, flowers of just one sex are produced. Flowers of different sexes may occur on the same plant, or be confined to different plants. The development of male or female flowers usually involves the selective resorption of either the stamens or pistil after they have been specified and have started to grow.

In maize, male and female flowers develop at particular sites on the shoot. The tassel at the tip of the main stem only bears flowers with stamens; the 'ears' at the ends of the lateral branches bear female flowers containing pistils. Sex determination becomes visible when the flower is still small, with the stamen primordia being larger in males and the pistil longer in females. The smaller organs eventually degenerate.

The plant hormone gibberellic acid may be involved in sex determination, as differences in gibberellin concentration are associated with the different sexual organs. In the maize tassel, gibberellin concentration is 100-fold lower than in the developing ears. If the concentration of gibberellic acid is increased in the tassel, pistils can develop.

11.15 Germ-cell sex determination can depend both on cell signals and genetic constitution

The determination of the sex of animal germ cells—that is, whether they will develop into eggs or sperm—is strongly influenced by the signals they receive when they enter the gonad. In the mouse, the primordial diploid germ cells continue to proliferate for a few days after entering the genital ridge, which is where the gonads form. Their future development is determined largely by the sex of the gonad in which they reside, and not by their own chromosomal constitution. Initially, they undergo mitotic division in both males and females. In females, diploid germ cells enter prophase of the first meiotic division in the embryo; they then arrest at this stage until the mouse becomes a sexually mature female. In male embryos, diploid germ cells divide mitotically for some time in the genital ridge, but then also stop dividing, becoming arrested in the G_1 phase of the cell cycle (Fig. 11.25). They start dividing again after birth and enter meiosis some 7 to 8 days after birth.

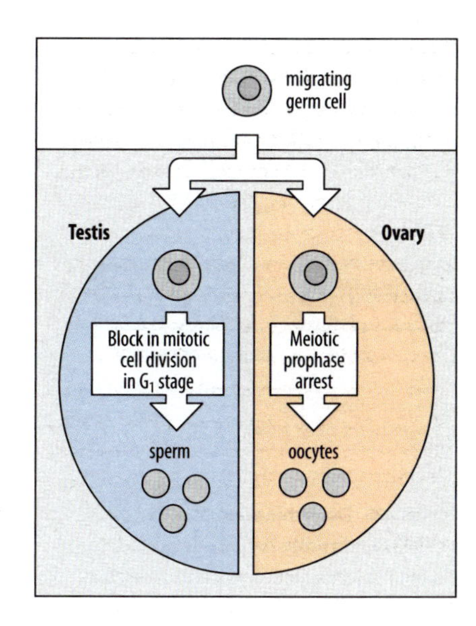

Fig. 11.25 Environmental signals specify germ-cell sex in mammals. Migrating germ cells, whether XX or XY, enter meiotic prophase and start developing as oocytes unless they enter a testis. In the testis, the germ cells receive an inhibitory signal that blocks mitotic division and prevents them entering meiotic prophase.

All mouse germ cells that enter meiosis before birth develop as eggs, whereas those not entering meiosis until after birth develop as sperm. Germ cells, whether XX or XY, that fail to enter the genital ridge and instead end up in adjacent tissues such as the embryonic adrenal gland or mesonephros, enter meiosis and begin developing as oocytes in both male and female embryos; thus, the default germ-cell sex appears to be female. In XX/XY chimeras, XX germ cells that are surrounded by testis cells develop along the spermatogenesis pathway. However, the later development of germ cells that develop in these inappropriate sites is abnormal. There is little reproductive future for XY germ cells in the ovaries and none for XX germ cells in the testes.

In *Drosophila*, the difference in the behavior of XY and XX germ cells depends initially on the number of X chromosomes, as in somatic cells; the *Sex-lethal* gene again plays an important role, although most other elements in the sex-determination pathway may differ from those in somatic cells. Both chromosomal constitution and cell interactions are involved in the development of germ-cell sexual phenotype. Transplantation of genetically marked pole cells (see Section 11.1) into a *Drosophila* embryo of the opposite sex shows that male XY germ cells in a female XX embryo become integrated into the ovary and begin to develop as sperm; that is, their behavior is autonomous with respect to their genetic constitution. By contrast, XX germ cells in a testis attempt to develop as sperm, showing a role for environmental signals. In neither case, however, are functional sperm produced.

The hermaphrodite of *C. elegans* provides a particularly interesting example of germ-cell differentiation, as both sperm and eggs develop within the same gonad. Unlike the somatic cells, which have a fixed lineage and number (see Section 5.1), the number of germ cells in an adult nematode is indeterminate, with about 1000 germ cells in each 'arm' of the gonad. At hatching of the first-stage larva, there are just two founder germ cells, which proliferate to produce the germ cells. The germ cells are flanked on each side by cells called distal tip cells, and their proliferation is controlled by a signal from the distal tip cells. This signal is the protein LAG-2, which is homologous to the Notch ligand Delta. The receptor for LAG-2 on the germ cells is GLP-1, which is similar both to nematode LIN-12, which is involved in vulva formation (see Section 5.6) and to Notch. We have already met GLP-1 acting to determine cell fate in the early embryo (see Section 5.3).

In *C. elegans*, entry of germ cells into meiosis from the third larval stage onward is controlled by the distal tip signal. In the presence of this signal, the cells

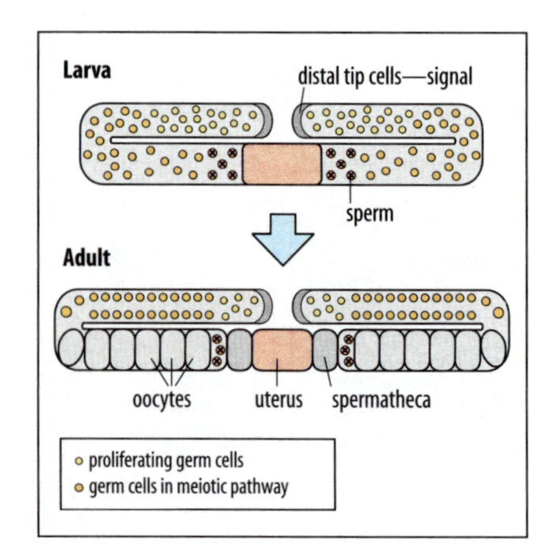

Fig. 11.26 Determination of germ-cell sex in the hermaphrodite nematode gonad. Top: during the larval stage, germ cells in a zone close to the distal tips of the gonad multiply; when they leave this zone in the larval stage they enter meiosis and develop into sperm. Bottom: in the adult, cells that leave the proliferative zone develop into oocytes. The eggs are fertilized as they pass into the uterus.

After Clifford, R., et al.: 1994.

proliferate, but as they move away from it, they enter meiosis and develop as sperm (Fig. 11.26, top). In the hermaphrodite gonad, all the cells that are initially outside the range of the distal tip signal develop as sperm, but cells that later leave the proliferative zone and enter meiosis develop as oocytes (see Fig. 11.26, bottom). The eggs are fertilized by stored sperm as they pass into the uterus. The male gonad has similar proliferative meiotic regions, but all the germ cells develop as sperm.

Sex determination of the nematode germ cells is somewhat similar to that of the somatic cells, in that the chromosomal complement is the primary sex-determining factor and many of the same genes are involved in the subsequent cascades of gene expression. The terminal regulator genes required for spermatogenesis are called *fem* and *fog*. In hermaphrodites, there must be a mechanism for activating the *fem* genes in some of the XX germ cells, so allowing them to develop as sperm.

11.16 Various strategies are used for dosage compensation of X-linked genes

In all the animals we have considered in this chapter there is an imbalance of X-linked genes between the sexes. One sex has two X chromosomes, whereas the other has one. This imbalance has to be corrected to ensure that the level of expression of genes carried on the X chromosome is the same in both sexes. The mechanism by which the imbalance in X-linked genes is dealt with is known as **dosage compensation**. Failure to correct the imbalance leads to abnormalities and arrested development. Different animals deal with the problem of dosage compensation in different ways (Fig. 11.27).

Mammals, such as mice and humans, achieve dosage compensation by inactivating one of the X chromosomes in females after the blastocyst has implanted in the uterine wall. Once X inactivation is initiated in female embryos, it is maintained in all somatic cells throughout life. For example, female mice heterozygous for a coat pigment gene carried on the X chromosome have patches of color on their coat, produced by clones of epidermal cells that express the X chromosome with the pigment gene. The rest of the epidermis is composed of cells in which that

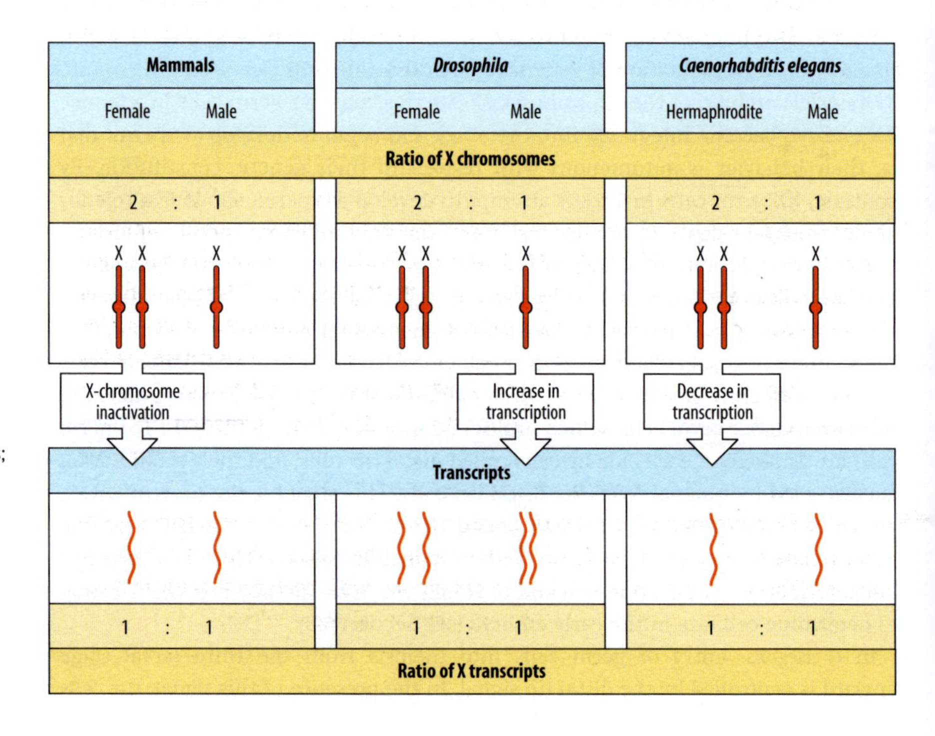

Fig. 11.27 Mechanisms of dosage compensation. In mammals, *Drosophila*, and *C. elegans* there are two X chromosomes in one sex and only one in the other. Mammals inactivate one of the X chromosomes in females; in *Drosophila* males there is an increase in transcription from the single X chromosome; and in *C. elegans* there is a decrease in transcription from the X chromosomes in hermaphrodites. The result of these different dosage compensation mechanisms is that the level of X chromosome transcripts is approximately the same in males and females.

X chromosome has been inactivated. In some tissues, the inactive X chromosome can be identified in the nucleus as a Barr body (see Section 8.3) and X inactivation provides an important model for the inheritance of gene expression. Whatever the sex chromosome constitution, for example, XY, XX, XXY, or XXXY, there is only one active X chromosome per somatic cell, with all the other X chromosomes inactivated. During early cleavage stages, the paternal X is inactivated, and this persists in the trophectoderm. In the inner cell mass, however, the inactivation of the paternal X is reversed and then random X inactivation takes place; how just one X per diploid nucleus is randomly inactivated is not fully understood.

X inactivation is dependent on a small region of the X chromosome, the inactivating center. It contains a proposed switch gene, *Xist*, which produces a non-coding RNA. Inactivation seems to be caused by the chromosome being coated by *Xist* RNA, which is expressed by the inactive X but not by an active X chromosome. If the *Xist* gene is introduced into another chromosome, that chromosome becomes silenced. This inactivation is correlated with methylation of the DNA.

Dosage compensation in *Drosophila* works in a different way to that in the mouse; instead of repression of the 'extra' X activity in females, transcription of the X chromosome in males is increased nearly twofold. A set of male-specific genes, the MSL complex, controls most dosage compensation, and these are repressed in females by Sex-lethal protein, thereby preventing excessive transcription of the X chromosome. The increased activity in males is regulated by the primary sex-determining signal, which results in the dosage compensation mechanism operating when the *Sex-lethal* gene is 'off'. In females, where *Sex-lethal* is 'on', it turns off the dosage compensation mechanism. This regulatory mechanism again involves non-coding RNA.

In *C. elegans*, dosage compensation is achieved by reducing the level of X chromosome expression in XX animals to that of the single X chromosome in XO males. The number of X chromosomes is communicated by a set of X-linked genes that can repress the master gene *xol-1*. A key event in initiating nematode dosage compensation is the expression of the protein SDC-2, which occurs only in hermaphrodites. It forms a complex specifically with the X chromosome and triggers assembly of the dosage compensation complex, which also includes the protein DPY-27. The complex binds to a specific region on the X chromosome and reduces transcription. SDC-2 also represses *her-1* on the sex-determination pathway.

Summary

The development of early embryos of both sexes is very similar. A primary sex-determining signal sets off development toward one or the other sex, and in mammals, *Drosophila*, and *C. elegans*, this signal is determined by the chromosomal complement of the fertilized egg. In mammals, the *Sry* gene on the Y chromosome is responsible for the embryonic gonad developing into a testis and producing hormones that determine male sexual characteristics. The sexual phenotype of the somatic cells is determined by the gonadal hormones. In *C. elegans* and *Drosophila*, the primary sex-determining signal is the number of X chromosomes. In *Drosophila*, the gene *Sex-lethal* is turned on in females but not in males, in response to this signal. In both cases, this results in further gene activity in which sex-specific RNA splicing is involved. In *C. elegans*, the gene *XO lethal* is turned off in hermaphrodites and on in males, eventually leading to sex-specific expression of the gene *transformer-1*, which determines the sexual phenotype. Somatic sexual differentiation in *Drosophila* is cell autonomous and is controlled by the number of X chromosomes; in *C. elegans*, cell–cell interactions are also involved. Most flowering plants are hermaphrodites, having flowers with both male and female organs.

In mammals, signals from the gonads determine whether the germ cells develop into oocytes or sperm. Male germ cells in *Drosophila* develop along the sperm pathway even in an ovary, but female germ cells develop along the sperm pathway when placed in a testis. Most *C. elegans* adults are hermaphrodites, and produce both sperm and eggs from the same gonad.

Various strategies of dosage compensation are used to correct the imbalance of X chromosomes between males and females. In female mammals, one of the X chromosomes is inactivated; in *Drosophila* males the activity of the single X chromosome is upregulated; and in *C. elegans* the activity of the X chromosomes in XX hermaphrodites is downregulated to match that from the single X chromosome in males.

Summary: determination of sexual phenotype

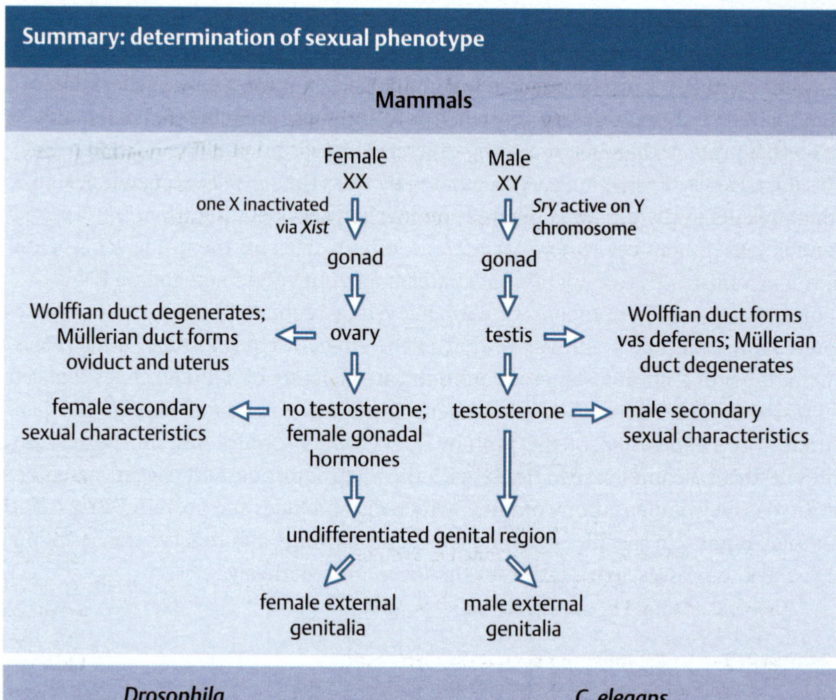

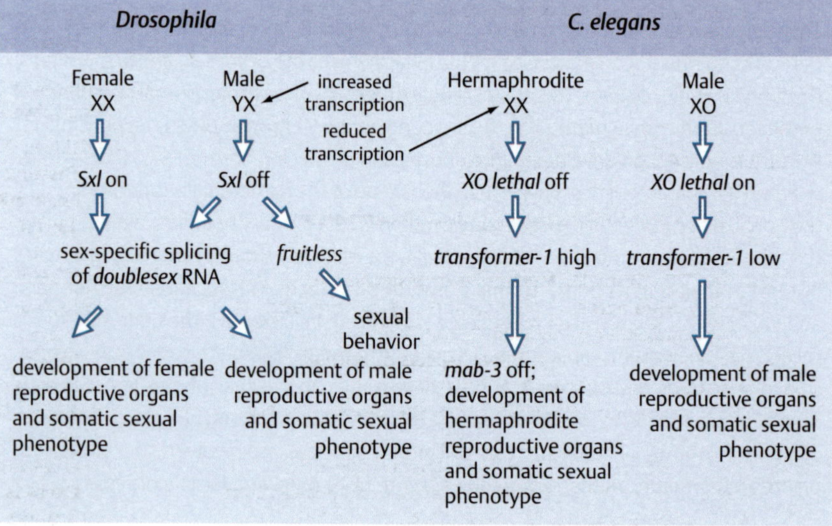

SUMMARY TO CHAPTER 11

In many animals, the future germ cells are specified by localized cytoplasmic determinants in the egg. Germ cells in mammals are unusual, as they are specified entirely by cell–cell interactions, and the same is true of flowering plants, in which the germ cells are only specified at a late stage, as the flowers develop. In animals, the development of germ cells into sperm or egg depends both on chromosomal constitution and interactions with the cells of the gonad. At fertilization, fusion of sperm and egg initiates development, and there are mechanisms to ensure that only one sperm enters an egg. Both maternal and paternal genomes are required for normal mammalian development, as some genes are imprinted; for such genes, whether they are expressed or not during development depends on whether they are derived from the sperm or the egg. In many animals, the chromosomal constitution of the embryo determines which sex will develop. In mammals, the Y chromosome is male-determining; it specifies the development of a testis, and the hormones produced by the testis cause the development of male sexual characteristics. In the absence of a Y chromosome, the embryo develops as a female. In *Drosophila* and *C. elegans*, sexual development is initially determined by the number of X chromosomes, which sets in train a cascade of gene activity. In nematodes, the somatic sexual phenotype is determined by cell–cell interactions; in the fly, somatic cell sexual differentiation is cell autonomous. Animals use a variety of strategies of dosage compensation to correct the imbalance in the number of X chromosomes in males and females.

GENERAL FURTHER READING

Chadwick, D., Goode, J.: *The Genetics and Biology of Sex Determination 2002.* Novartis Foundation Symposium 244. New York: John Wiley, 2002.

Crews, D.: **Animal sexuality.** *Sci. Am.* 1994, **270:** 109–114.

Zarkower, D.: **Establishing sexual dimorphism: conservation amidst diversity?** *Nat. Rev. Genet.* 2001, **2:** 175–185.

SECTION FURTHER READING

11.1 Germ-cell fate can be specified by a distinct germ plasm in the egg

Extavour, C.G., Akam, M.: **Mechanisms of germ cell specification across the metazoans: epigenesis and preformation.** *Development* 2003, **130:** 5869–5884.

Matova, N., Cooley, L.: **Comparative aspects of animal oogenesis.** *Dev. Biol.* 2001, **231:** 291–320.

McLaren, A.: **Primordial germ cells in the mouse.** *Dev. Biol.* 2003, **262:** 1–15.

Mello, C.C., Schubert, C., Draper, B., Zhang, W., Lobel, R., Priess, J.R.: **The PIE-1 protein and germline specification in *C. elegans* embryos.** *Nature* 1996, **382**: 710–712.

Ray, E.: **Primordial germ-cell development: the zebrafish perspective.** *Nat. Rev. Genet.* 2003, **4:** 690–700.

Saitou, M., Barton, S.C., Surani, M.A.: **A molecular programme for the specification of germ cell fate in mice.** *Nature* 2002, **418:** 293–300.

Saitou, M., Payer, B., Lange, U.C., Erhardt, S., Barton, S.C., Surani, M.A.: **Specification of germ cell fate in mice.** *Philos. Trans. R. Soc. Lond. B Biol. Sci.* 2003, **358:** 1363–1370.

Williamson, A., Lehmann, R.: **Germ cell development in *Drosophila*.** *Annu. Rev. Cell Dev. Biol.* 1996, **12:** 365–391.

11.2 Pole plasm becomes localized at the posterior end of the *Drosophila* egg

Micklem, D.R., Adams, J., Grunert, S., St. Johnston, D.: **Distinct roles of two conserved Staufen domains in *oskar* mRNA localisation and translation.** *EMBO J.* 2000, **19:** 1366–1377.

11.3 Germ cells migrate from their site of origin to the gonad

Doitsidou, M., Reichman-Fried, M., Stebler, J., Koprunner, M., Dorries, J., Meyer, D., Esguerra, C.V., Leung, T., Raz, E.: **Guidance of primordial germ cell migration by the chemokine SDF-1.** *Cell* 2002, **111:** 647–659.

Knaut, H., Werz, C., Geisler, R., Nusslein-Volhard, C; Tubingen 2000 Screen Consortium: **A zebrafish homologue of the chemokine receptor Cxcr4 is a germ-cell guidance receptor.** *Nature* 2003, **421:** 279–282.

Santos, A.C., Lehmann, R.: **Germ cell specification and migration in *Drosophila* and beyond.** *Curr. Biol.* 2004, **14:** R578–R589.

Santos, A.C., Lehmann, R.: **Isoprenoids control germ cell migration downstream of HMGCoA reductase.** *Dev. Cell* 2004, **6:** 283–293.

Wylie, C.: **Germ cells.** *Cell* 1999, **96:** 165–174.

11.4 Germ-cell differentiation involves a reduction in chromosome number

De Rooij, D.G., Grootegoed, J.A.: **Spermatogonial stem cells.** *Curr. Opin. Cell Biol.* 1998, **10:** 694–701.

11.5 Oocyte development can involve gene amplification and contributions from other cells

Browder, L.W.: *Oogenesis.* New York: Plenum Press, 1985.

Choo, S., Heinrich, B., Betley, J.N., Chen, A., Deshler, J.O.: **Evidence for common machinery utilized by the early and late RNA localization pathways in *Xenopus* oocytes.** *Dev. Biol.* 2004, **278:** 103–117.

de Rooij, D.G., Grootegoed, J.A.: **Spermatogonial stem cells.** *Curr. Opin. Cell Biol.* 1998, **10:** 694–701.

Spradling, A.: **Developmental genetics of oogenesis.** In Drosophila *Development.* Edited by Bate, M., Martinez-Arias, A. New York: Cold Spring Harbor Laboratory Press, 1993: 1–69.

11.6 Some genes controlling embryonic growth are imprinted

Kono, T., Obata, Y., Wu, Q., Niwa, K., Ono, Y., Yamamoto, Y., Park, E.S., Seo, J.S., Ogawam H.: **Birth of parthenogenetic mice that can develop to adulthood.** *Nature* 2004, **428:** 860–864.

Morison, I.M., Ramsay, J.P., Spencer, H.G.: **A census of mammalian imprinting.** *Trends Genet.* 2005, **21:** 457–465.

Reik, W., Walter, J.: **Genomic imprinting: parental influence on the genome.** *Nat. Rev. Genet.* 2001, **2:** 21–32.

11.7 Fertilization involves cell-surface interactions between egg and sperm

Jungnickel, M.K., Sutton, K.A., Florman, H.M.: **In the beginning: lessons from fertilization in mice and worms.** *Cell* 2003, **114:** 401–404.

Singson, A.: **Every sperm is sacred: fertilization in *Caenorhabditis* elegans.** *Dev. Biol.* 2001, **230:** 101–109.

11.8 Changes in the egg membrane at fertilization block polyspermy

Ohelndieck, K., Lennarz, W.J.: **Role of the sea urchin egg receptor for sperm in gamete interactions.** *Trends Biochem. Sci.* 1995, **20:** 29–33.

Tian, J., Gong, H., Thomsen, G.H., Lennarz, W.J.: ***Xenopus laevis* sperm–egg adhesion is regulated by modifications in the sperm receptor and the egg vitelline envelope.** *Dev. Biol.* 1997, **187:** 143–153.

11.9 A calcium wave initiated at fertilization results in egg activation

Fissore, R.A., *et al.*: **Sperm induced calcium oscillations. Isolation of the Ca^{2+}-releasing component(s) of mammalian sperm extracts: the search continues.** *Mol. Hum. Reprod.* 1999, **5:** 189–192.

Swann, K., Parrington, J.: **Mechanism of Ca^{2+} release at fertilization in mammals.** *J. Exp. Zool.* 1999, **285:** 267–275.

Whitaker, M.: **Calcium at fertilization and in early development.** *Physiol. Rev.* 2006, **86:** 25–88.

11.10 The primary sex-determining gene in mammals is on the Y chromosome

Capel, B. **The battle of the sexes.** *Mech. Dev.* 2000, **92:** 89–103.

Koopman, P.: **The genetics and biology of vertebrate sex determination.** *Cell* 2001, **105:** 843–847.

Schafer, A.J., Goodfellow, P.N.: **Sex determination in humans.** *BioEssays* 1996, **18:** 955–963.

11.11 Mammalian sexual phenotype is regulated by gonadal hormones

Swain, A., Lovell-Badge, R.: **Mammalian sex determination: a molecular drama.** *Genes Dev.* 1999, **13:** 755–767.

Vainio, S., Heikkila, M., Kispert, A., Chin, N., McMahon, AP.: **Female development in mammals is regulated by Wnt-4 signalling.** *Nature* 1999, **397:** 405–409.

11.12 The primary sex-determining signal in *Drosophila* is the number of X chromosomes, and is cell autonomous

Brennan, J., Capel, B.: **One tissue, two fates: molecular genetic events that underlie testis versus ovary development.** *Nat. Rev. Genet.* 2004, **5:** 509–520.

Cline, T.W.: **The *Drosophila* sex determination signal: how do flies count to two?** *Trends Genet.* 1993, **9:** 385–390.

Hodgkin, J.: **Sex determination compared in *Drosophila* and *Caenorhabditis*.** *Nature* 1990, **344:** 721–728.

MacLaughlin, D.T., Donahoe, M.D.: **Sex determination and differentiation.** *New Engl. J. Med.* 2004, **350:** 367–378.

11.13 Somatic sexual development in *Caenorhabditis* is determined by the number of X chromosomes

Cline, T.W., Meyer, B.J.: **Vive la difference: males vs. females in flies vs. worms.** *Annu. Rev. Genet.* 1996, **30:** 637–702.

Raymond, C.S., Shamu, C.E., Shen, M.M., Seifert, K.J., Hirsch, B., Hodgkin, J., Zarkower, D.: **Evidence for evolutionary conservation of sex-determining genes.** *Nature* 1998, **391:** 691–695.

11.14 Most flowering plants are hermaphrodite, but some produce unisexual flowers

Irisa, E.N.: **Regulation of sex determination in maize.** *BioEssays* 1996, **18:** 363–369.

11.15 Germ-cell sex determination can depend both on cell signals and genetic constitution

McLaren, A.: **Signaling for germ cells.** *Genes Dev.* 1999, **13:** 373–376.

Seydoux, G., Strome, S.: **Launching the germline in *Caenorhabditis* elegans: regulation of gene expression in early germ cells.** *Development* 1999, **126:** 3275–3283.

11.16 Various strategies are used for dosage compensation of X-linked genes

Akhtar, A.: **Dosage compensation: an intertwined world of RNA and chromatin remodeling.** *Curr. Opin. Genet. Dev.* 2003, **13:** 161–169.

Avner, P., Heard, E.: **X-chromosomes inactivation: counting, choice and initiation.** *Nat. Rev. Genet.* 2001, **2:** 59–67.

Csankovszki, G., McDonel, P., Meyer, B.J.: **Recruitment and spreading of the *C. elegans* dosage compensation complex along X chromosomes.** *Science* 2004, **303:** 1182–1185.

Heard, E.: **Recent advances in X-chromosome activation.** *Curr. Opin. Cell Biol.* 2004, **16:** 247–255.

Latham, K.E.: **X chromosome imprinting and inactivation in preimplantation mammalian embryos.** *Trends Genet.* 2005, **21:** 120–127.

Meller, V.H.: **Dosage compensation: making 1X equal 2X.** *Trends Cell Biol.* 2000, **10:** 54–59.

Meyer, B.J.: **Sex in the worm: counting and compensating X-chromosome dose.** *Trends Genet.* 2000, **16:** 247–253.

Growth and post-embryonic development

12

- Growth
- Molting and metamorphosis
- Aging and senescence

Patterning of the embryo occurs on a small scale followed by growth. The control of growth, and therefore size, is a key problem, involving at its core the control of cell proliferation. To what extent is cell proliferation programmed during embryonic development and what role do systemic factors such as hormones play? Another aspect of post-embryonic development in many invertebrates is metamorphosis, in which the form of the animal is completely changed. Finally, there is aging, which is controlled but not programmed.

Development does not stop once the embryonic phase is complete. Most, but by no means all, of the growth in animals and plants occurs in the post-embryonic period, when the basic form and pattern of the organism has already been established. The basic patterning is done on a small scale, over dimensions of less than a millimeter. In many animals, the embryonic phase is immediately succeeded by a free-living larval or immature adult stage. In others, such as mammals, considerable growth occurs during a late embryonic or fetal period, while the embryo is still dependent on maternal resources. Growth then continues after birth. Growth is a central aspect of all developing systems, determining the final size and shape of the organism and its parts. Animals with a larval stage not only grow in size, but may undergo **metamorphosis**, in which the larva is transformed into the adult form. Metamorphosis often involves a radical change in form and the development of new organs.

We first consider the roles of intrinsic growth programs and of factors such as growth hormones in controlling both embryonic and post-embryonic growth. This is followed by a discussion of metamorphosis in insects and amphibians. Finally, we look at what might be considered an abnormal aspect of post-embryonic development—aging.

Growth

Growth is defined as an increase in the mass or overall size of a tissue or organism; this increase may result from cell proliferation, cell enlargement without division, or by accretion of extracellular material, such as bone matrix or even water (Fig. 12.1). In early embryonic development, there is little growth during cleavage and blastula formation, and cells get smaller at each cleavage division.

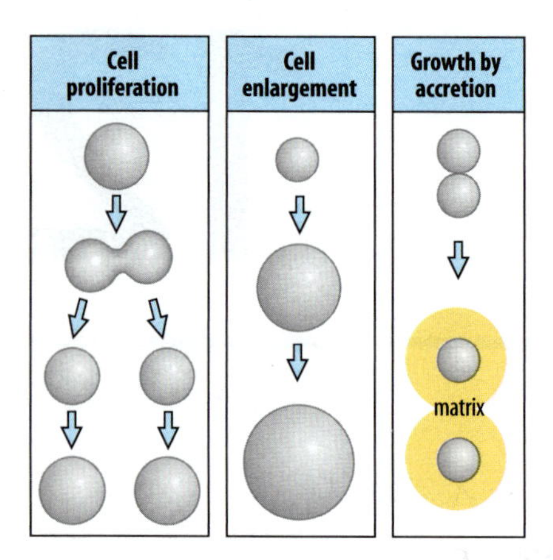

Fig. 12.1 The three main strategies for growth in vertebrates. The most common mechanism is cell proliferation—cell growth followed by division. A second strategy is cell enlargement, in which cells increase their size without dividing. The third strategy is to increase size by accretionary growth, such as secretion of a matrix.

In animals, the basic body pattern is laid down when the embryo is still small. The growth program—that is, how much an organism or an individual organ grows and how it responds to factors like hormones—may also be specified at an early stage in development. Overall growth of the organism mainly occurs in the period after the basic pattern of the embryo has been established; there are, however, many examples where earlier organogenesis involves localized growth, as in the vertebrate limb bud (see Chapter 10), and the developing nervous system (see Chapter 11). Different rates of growth in different parts of the body, or at different times, during early development profoundly affect the shape of organs and the organism.

Unlike the situation in animals, where the embryo is essentially a miniature version of the free-living larva or adult, plant embryos bear little resemblance to the mature plant. Most of the adult plant structures are generated after germination by the shoot or root meristems, which have a capacity for continual growth (see Chapter 6). In woody plants, the cambial layer in the trunk, branches, and roots also retains proliferative capacity. It can give rise to the main tissues of the plant axis, enabling trees to increase in girth year after year.

12.1 Tissues can grow by cell proliferation, cell enlargement, or accretion

Although growth often occurs through an increase in the number of cells, this is only one of three main means of growth (see Fig. 12.1). There is also a significant amount of cell death in many growing tissues, and the overall growth rate is determined by the rates of cell death and cell proliferation.

A second strategy is growth by cell enlargement—that is, by individual cells increasing their mass and getting bigger; this is the case in *Drosophila* larval growth, for example. Most of the growth in size of plants is due to cell enlargement. Once differentiated, skeletal and heart muscle cells and neurons never divide again, although they do increase in size. Neurons grow by the extension and growth of axons and dendrites, whereas muscle growth involves an increase in mass, as well as the fusion of satellite cells to pre-existing muscle fibers to provide new nuclei. Cell enlargement is also a major feature of plant growth, as we shall see. And differences in size between closely related species of *Drosophila* are partly the result of the larger species having larger cells. Some growth is through a combination of cell proliferation and cell enlargement. For example, lens cells are produced by cell division from a proliferative zone for an extended period, while their differentiation involves considerable cell enlargement.

The third growth strategy, accretionary growth, involves an increase in the volume of the extracellular space, which is achieved by secretion of large quantities of extracellular matrix by cells. This occurs in both cartilage and bone, where most of the tissue mass is extracellular.

Certain vertebrate tissues, including the blood (see Section 8.4) and epithelia (see Section 8.8), are continually renewed throughout an animal's lifetime by cell division and differentiation from a stem cell population. The end products of this type of proliferative system, such as mature red blood cells and keratinocytes, are themselves incapable of division and eventually die.

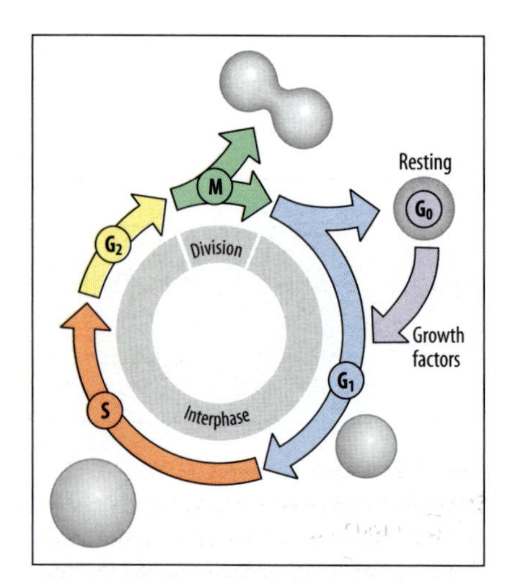

Fig. 12.2 The eukaryotic cell cycle. After mitosis (M), the daughter cells can either enter a resting phase (G_0), in which they effectively withdraw from the cell cycle, or proceed through G_1 to the phase of DNA synthesis (S). This is followed by G_2, and then by mitosis. Cell growth occurs throughout G_1, S, and G_2. The decision to enter G_0 or to proceed through G_1 may be controlled by both intracellular status and extracellular signals such as growth factors. Cells such as neurons and skeletal muscle cells, which do not divide after differentiation, are permanently in G_0.

12.2 Cell proliferation can be controlled by an intrinsic program

When a eukaryotic cell duplicates itself it goes through a fixed sequence of events called the **cell cycle** (Fig. 12.2). The cell grows in size, the DNA is replicated, and these replicated chromosomes then undergo mitosis, and become segregated into two daughter nuclei. Only then can the cell divide to form two daughter cells, which can go through the whole sequence again. During cleavage of the fertilized egg there is no cell growth, and cell size decreases with each division, but in other proliferating cells the cytoplasmic mass must double in preparation for cell division.

The standard eukaryotic mitotic cell cycle is divided into well marked phases. At the M phase, mitosis and cell cleavage give rise to two new cells. The rest of the cell cycle, between one M phase and the next, is called interphase. Replication of DNA occurs during a defined period in interphase, the S phase. Preceding S phase is a period known as G_1 (the G stands for gap), and after it another interval known as G_2, after which the cells enter mitosis (see Fig. 12.2). G_1, S phase, and G_2 collectively make up interphase, the part of the cell cycle during which cells synthesize proteins and grow, as well as replicating their DNA. Particular phases of the cell cycle are absent in some cells: during cleavage of the fertilized egg, G_1 and G_2 are virtually absent; in meiosis (see Section 11.4) there is no DNA replication at the second division; and in *Drosophila* salivary glands there is no M phase, as the DNA replicates repeatedly without mitosis or cell division, leading to the formation of giant polytene chromosomes.

Growth factors and other signaling proteins play a key role in controlling cell growth and proliferation. Studies of cell cultures show that growth factors are essential for cells to multiply, with the particular growth factor or factors required depending on the cell type. When somatic cells are not proliferating they are usually in a state known as G_0, into which they withdraw after mitosis (see Fig. 12.2). Growth factors enable the cell to proceed out of G_0 and progress through the cell cycle. Numerous growth factors that can control cell proliferation have been discovered, but in general their precise roles in normal development are not yet known. Some exceptions include erythropoietin, which promotes proliferation of red blood cell precursors (see Section 8.5). The role of hormones is considered later.

The timing of events in the cell cycle is controlled by a set of 'central' timing mechanisms. One set of proteins known as **cyclins** controls the passage through key transition points in the cell cycle. Cyclin concentrations oscillate during the cell cycle, and these oscillations correlate with transitions from one phase of the cycle to the next. Cyclins act by forming complexes with, and helping to activate, protein kinases known as **cyclin-dependent kinases** (**CDKs**). These phosphorylate proteins that trigger the events of each phase, such as DNA replication in S phase or mitosis in M phase.

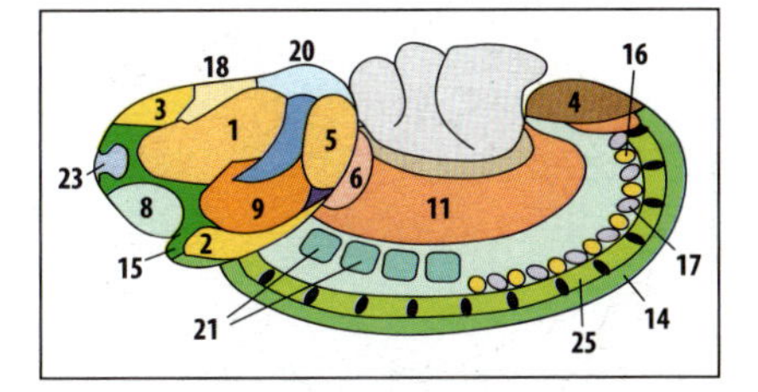

Fig. 12.3 Domains of mitosis in the *Drosophila* blastoderm. Areas composed of cells that divide at the same time are indicated by the various colors, and the numbers indicate the order in which these domains undergo mitosis at the 14th cell cycle, when zygotic String protein is first expressed. The schematic illustrates a lateral view of the embryo corresponding to a stage after mesoderm has been internalized and segmentation has begun (segments are indicated by the black marks along the lower surface). Anterior is to the left and dorsal is up. The gray region on the dorsal surface is the amnioserosa. The internalized mesoderm and some other domains are not visible in this view.

After Edgar, B.A., et al.: 1994.

In general, the mechanisms controlling the pattern of cell division in the embryo are not well understood. One example of intrinisic control comes from *Drosophila*, where the early cell cycles are under the control of genes that pattern the embryo; these exert their effect by their influence on the cyclins. The first cell cycles in the *Drosophila* embryo are represented by rapid and synchronous nuclear divisions, without any accompanying cell divisions. This creates a syncytial blastoderm (see Fig. 2.2). There are virtually no G phases, just alternations of DNA synthesis (S phase) and mitosis. But at cycle 14, there is a major transition to a different type of cell cycle, a transition similar to the mid-blastula transition in frogs (see Section 3.16). At cycle 14 and subsequently, the cell cycles have a well defined G_2 phase, and the blastoderm becomes cellularized. After the 17th or 18th cycle, cells in the epidermis and mesoderm stop dividing, and differentiate. This cessation of proliferation is caused by the exhaustion of maternal cyclin E originally laid down in the egg, which is required for progression through the cell cycle.

At the 14th cell cycle, distinct spatial domains with different cell-cycle times can be seen in the *Drosophila* blastoderm (Fig. 12.3). This patterning of cell cycles is produced by a change in the synthesis and distribution of a protein phosphatase called String, which exerts control on the cell cycle by dephosphorylating and activating a cyclin-dependent kinase. In the fertilized egg, the String protein is of maternal origin and is uniformly distributed. It therefore produces a synchronized pattern of nuclear division throughout the embryo. After cycle 13, the maternal String protein disappears and zygotic String protein becomes the controlling factor.

Zygotic *string* gene transcription occurs in a complex spatial and temporal pattern. Only cells in which the *string* gene is expressed enter mitosis. This results in variation in the rate of cell division in different parts of the blastoderm, which ensures that the correct number of cells are generated in different tissues. The pattern of zygotic *string* gene expression is controlled by transcription factors encoded by the early patterning genes, such as the gap and pair-rule genes and those patterning the dorso-ventral axis. One exception to the rule that expression of *string* leads to cell proliferation is the presumptive mesoderm, which is the first domain in which *string* is expressed but the tenth to divide. The failure to divide is due to the expression of the gene *tribble* in this region, as Tribble protein degrades String. The delay is necessary to allow ventral furrow formation and thus mesoderm invagination (see Section 7.8), as cell division inhibits ventral furrow formation. Prospective mesodermal cells proliferate after they have been internalized. Thus, the early cell cycles in *Drosophila* development provide a good example of how genes can control the pattern of cell divisions. In contrast to this intrinsic program in the *Drosophila* embryo, the cell proliferation and growth that occurs in the imaginal discs that give rise to adult structures is modulated by cell–cell interactions, that is, by extracellular signals, and how this is linked to pattern formation is far from clear.

In early mammalian embryos, cell proliferation times vary with developmental time and place. The first two cleavage cycles in the mouse last about 24 hours, and subsequent cycles take about 10 hours each. After implantation, the cells of the epiblast proliferate rapidly; cells in front of the primitive streak have a cycle time of just 3 hours, but the signals involved in this proliferation have not yet been identified.

12.3 Organ size can be controlled by external signals and intrinsic growth programs

A striking discovery has been the recognition that cells must receive signals, such as growth factors, not only for them to divide, but also simply to survive. In the absence of all growth factors, cells commit suicide by apoptosis, as a result of activation of an internal cell death program (see Section 8.13). There is a significant amount of cell death in all growing tissues, so that overall growth rate depends on the rates of both cell death and cell proliferation.

Both intrinsic growth programs and systemic circulating factors can determine organ size, and their relative importance in the development of different organs varies a great deal. Some organs control their own size. For example, if several fetal thymus glands are transplanted into a developing mouse embryo, each grows to its normal size, showing that the thymus follows an intrinsic growth program. If the same experiment is done with spleens, however, each spleen grows much less than normal so that the final total mass of the spleens is equivalent to one normal spleen. For the spleen, therefore, systemic factors produced by the spleen itself are a major controlling factor in its growth. For muscle, the protein myostatin, which is produced by the myoblast, may be a systemic negative regulator of growth, as mutation in the *myostatin* gene in mice leads to a significant increase in muscle mass; the number of muscle fibers and their size are both increased. There is now some evidence for a similar negative signal controlling cell number in neural tissue. In butterflies, there is evidence for interactions between imaginal discs; the removal of hindwing discs results in larger forewings and forelegs.

Even when they are capable of division, the cells of many adult tissues do not divide, or divide infrequently. Cells can, however, be induced to divide by injury or other stimuli, such as a reduction in the mass of the tissue. Liver cells divide relatively infrequently but if, for example, two thirds of a rats's liver is removed, the cells in the remaining third proliferate and restore the liver to its normal size within a few weeks. This restorative capacity indicates the presence of factors in the circulation that control cell proliferation. Removal of a kidney leads to an increase in size of the remaining kidney, but in this case the growth is mainly the result of cell enlargement rather than cell proliferation. And in Chapter 13 we shall see how intercalary growth is stimulated to replace missing positions of amputated limbs when cells with disparate positional values are put next to each other.

Patterning of the embryo occurs while the organs are still very small. For example, human limbs have their basic pattern established when they are less than 1 cm long. Yet, over the years, the limb grows to be at least one hundred times longer. How is this growth controlled? It appears that each of the cartilaginous elements in the limb has its own individual growth program. In the chick wing, the initial size of the cartilaginous elements in the long bones—the humerus and the ulna—are similar to the elements in the wrist (Fig. 12.4). Yet, with growth, the humerus and ulna increase many times in length compared with the wrist bones. These growth programs are specified when the elements are initially patterned and involve both cell multiplication and matrix secretion. Each skeletal element follows its own growth program even when grafted to neutral sites, provided that a good blood supply is established.

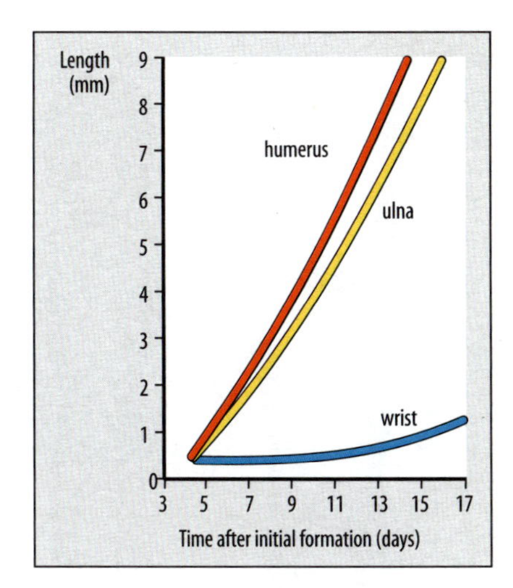

Fig. 12.4 Comparative growth of the cartilaginous elements in the embryonic chick wing. When first laid down, the cartilaginous elements of the humerus, ulna, and wrist are the same size, but the humerus and ulna then grow much more than the wrist element.

A classic illustration of an intrinsic growth program comes from grafting limb buds between large and small species of salamanders of the genus *Ambystoma*. A limb bud from the larger species grafted to the smaller species initially grows slowly, but eventually ends up its normal size, which is much larger than any of the limbs of the host (Fig. 12.5). Whatever the circulatory factors, such as hormones, that influence growth, the intrinsic response of different tissues is therefore crucial. Such intrinsically patterned differential growth can affect the overall form of an organism considerably, as can be seen in Figure 12.6, even though in many cases circulating hormones are also required for growth.

Growth of the different parts of the body is not uniform, and different organs grow at different rates. At 9 weeks of development, the head of a human embryo is more than a third of the length of the whole embryo, whereas at birth it is only about a quarter. After birth, the rest of the body grows much more than the head, which is only about an eighth of the body length in the adult (see Fig. 12.6).

Fig. 12.5 The size of limbs is genetically programmed in salamanders. An embryonic limb bud from a large species of salamander, *Ambystoma tigrinum*, grafted to the embryo of a smaller species, *Ambystoma punctatum*, grows much larger than the host limbs—to the size it would have grown in *Ambystoma tigrinum*.

Photograph from Harrison, R.G.: 1969.

12.4 Organ size may be determined by absolute dimension rather than cell number

How do cells in an organ know when to stop growing, particularly when they are not relating their size to that of the whole body? Evidence that animals can monitor the linear dimensions of their organs comes from organisms with either less or more than the diploid number of chromosomes—that is, haploid and polyploid organisms. For a given cell type, cell size is usually proportional to ploidy, so haploid cells are half the volume of diploid cells, and tetraploid cells twice the volume. Animals such as salamanders with unusual ploidy grow to a normal size, and tetraploid salamanders have only half the number of cells. Tetraploid mouse embryos compensate for larger cells by having fewer cells, but usually die before birth.

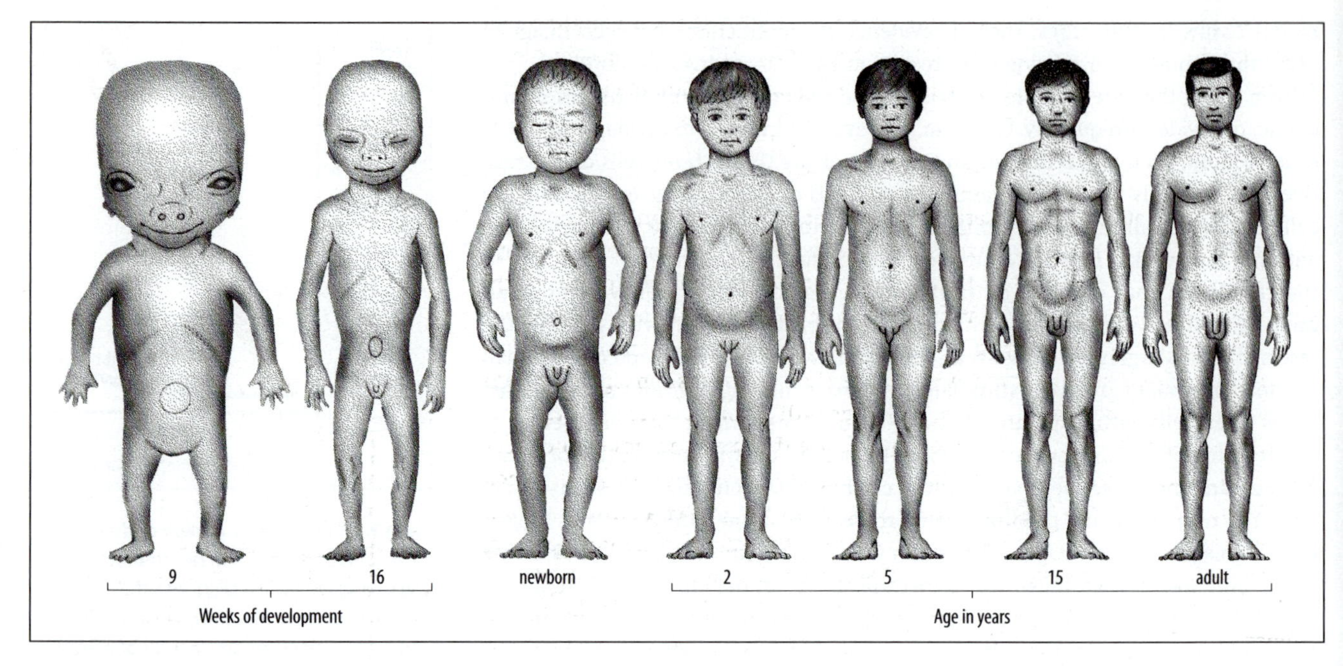

Fig. 12.6 Different parts of the human body grow at different rates. At 9 weeks of development the head is relatively large but, with time, other parts of the body grow much more than the head.

After Gray, H.: 1995.

In *Drosophila*, the final size of mosaics of haploid and diploid cells is normal, with the haploid region containing smaller but more numerous cells. Polyploid plants are in fact larger than normal, but in mosaics of diploid and polyploid cells it is possible to find the sort of size compensation seen in animals.

The growth of the *Drosophila* wing provides an excellent model for this mechanism of growth control. The wing disc (see Section 9.15) initially contains about 40 cells, and this grows in the larva to about 50,000 cells before metamorphosis. A striking feature is that mosaics of slow-growing *Minute* cells (see Box 2D, p. 68) and normal cells result in competition between these cell types, so that the wing is made almost entirely of normal cells. The clonal descendants of a single cell can contribute from a tenth to as much as a half of the wing. That cell proliferation does not equal growth is clearly demonstrated by blocking cell division in either an anterior or posterior compartment, in which case the cells become larger but wing size is normal.

We have already seen how Decapentaplegic and Wingless proteins play a key role in patterning the limb, and if the production of either protein is defective, growth is stunted. However, these proteins should not necessarily be thought of as growth factors. Rather, as one model proposes, local growth may depend on reading the steepness of the concentration gradients of these morphogens. As long as the gradient is sufficiently steep, the cells will grow and divide. This would account for growth not depending on cell size or cell number.

Wing disc cells can apparently sense each other's growth rates and this can result in competitive interactions between cells with different growth rates, leading to elimination of the slower-growing cells. One instance in which this competition is induced experimentally is when slow-growing *Minute* mutant cells are eliminated and die in the presence of faster-growing clones of wild-type cells, so that the overall size of the wing remains the same (see Box 2D, p. 68). It is likely that some mechanism of cell competition and apoptosis occurs during normal wing growth. The *myc* gene is an important regulator of growth, and if clones of wing cells are made that locally overexpress the *Drosophila* version of Myc, they grow faster than the surrounding wild-type cells. The resulting cell competition leads to apoptosis in nearby cells through the induction of the pro-apoptotic gene *hid*, and the wing is almost normal in size. Production of the Hid protein can be suppressed by a microRNA called *bantam*, resulting in promotion of cell growth.

12.5 Growth can be dependent on growth hormones

Size is affected not only by the rate of growth but by how long growth continues. Hormones control the ultimate size of insects through their influence on the duration of the larval feeding period; they do not control growth itself. In insects, the insulin signaling pathway plays an important role. Insulin is produced in the brain and can control organ growth, and can drive storage of nutrients in fat body cells. In *Drosophila*, size is strongly influenced by the effect of the insulin signaling pathway on both the duration and rate of larval growth. Growth occurs in the larval instars and the adult size is determined by the size the larva reaches before metamorphosis, which occurs as a result of release of ecdysone by the prothoracic gland. It may be that this release occurs when the gland reaches a critical size and so determines body size. There is also evidence that during the larval phases ecdysone suppresses growth by antagonizing insulin activity.

Growth in mammals, on the other hand, is directly influenced by hormones. Human growth during the embryonic, fetal, and post-natal periods provides a good model for mammalian growth. The human embryo increases in length from 150 mm at implantation to about 50 cm over the 9 months of gestation. During the first 8 weeks after conception, the embryonic body does not increase greatly in size, but the basic human form is laid down in miniature. The greatest rate of growth occurs at about 4 months, when the embryo grows as much as 10 cm per month. Growth after birth follows a well defined pattern (Fig. 12.7, left panel). During the first year after birth, growth occurs at a rate of about 2 cm per month. The growth rate then declines steadily until the start of a characteristic adolescent growth spurt at puberty at about 11 years in girls and 13 years in boys (see Fig. 12.7, right panel). In pygmies, this adolescent growth spurt does not occur, hence their characteristic short stature.

The maternal environment plays an important role in controlling fetal growth. This is well illustrated by crossing a large shire horse with a much smaller Shetland pony. When the mother is a shire mare, the newborn foal is similar in size to a normal shire foal, but when the mother is a Shetland, the newborn is much smaller. However, with growth after birth, the offspring of both these crosses become similar in size, and achieve a final size intermediate between shires and Shetlands.

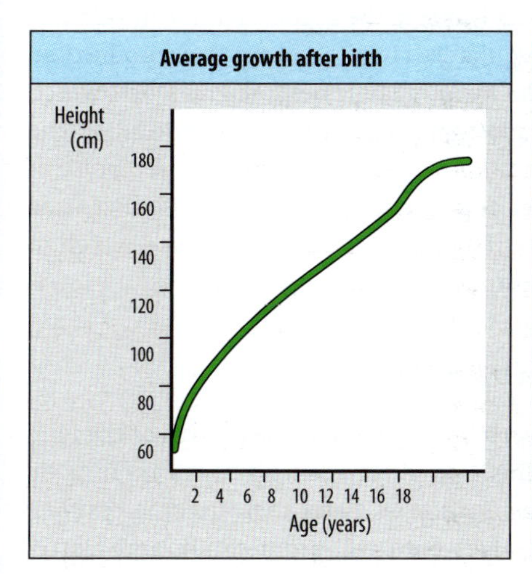

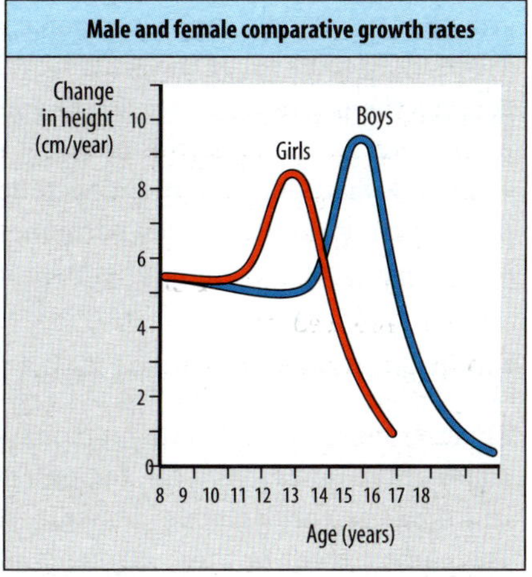

Fig. 12.7 Normal human growth. Left panel: an average growth curve for a human male after birth. Right panel: comparative growth rates of boys and girls. There is a growth spurt at puberty in both sexes, which occurs earlier in girls.

The amount of nourishment an embryo receives can have profound effects in later life. In early intrauterine life, undernutrition tends to produce small but normally proportioned animals. In contrast, undernutrition during the post-natal growth period leads to selective organ damage. For example, rats that are undernourished immediately after weaning have normal skeletal growth, but the liver and kidneys do not grow normally and are permanently small. Epidemiological studies in humans have shown that small size at birth is associated with increased death rates from cardiovascular diseases and non-insulin-dependent diabetes. The mechanisms underlying these long-term effects are not understood, but the studies emphasize the importance of adequate fetal growth for adult life.

Embryonic growth is dependent on growth factors. We have seen that fibroblast growth factors (FGFs) control cell proliferation in the progress zone of the developing chick limb, and are required for proliferation and outgrowth of the bud (see Section 9.3). Evidence for the role of other growth factors in embryonic growth comes from the technique of gene knock-out (see Box 4B, pp. 158–159). The **insulin-like growth factors 1 and 2** (**IGF-1** and **IGF-2**) are single-chain protein growth factors that closely resemble insulin and each other in their amino-acid sequence. They have a key role not only in post-natal mammalian growth, but also in growth during embryonic development. Newborn mice lacking a functional *Igf2* gene develop relatively normally, but weigh only 60% of the normal newborn body weight. Mice in which the *Igf1* gene has been inactivated are also growth retarded, and IGF-1 also has an important role in post-embryonic growth. Both factors and their receptors are present as early as the eight-cell stage of mouse development. *Igf2* is one of the genes that are imprinted in mammals; it is inactivated in maternal germ cells (see Section 11.6).

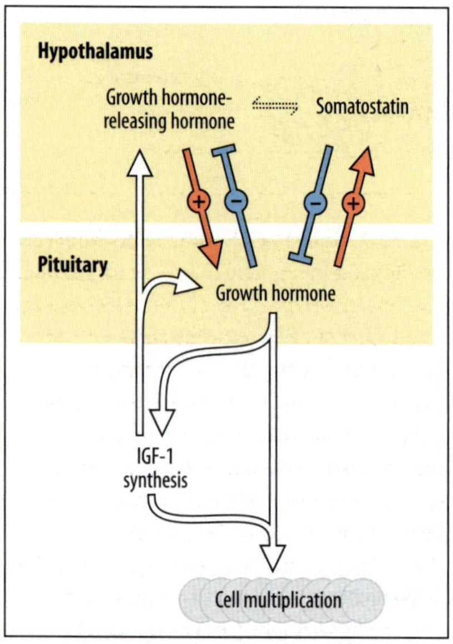

Growth hormone is a protein hormone that is essential for the post-embryonic growth of humans and other mammals; it is also synthesized in the embryo and can affect the growth of fetal tissues. Within the first year of birth, the pituitary gland begins to secrete growth hormone. A child with insufficient growth hormone grows less than normal, but if growth hormone is given regularly, normal growth is restored. In this case, there is a catch-up phenomenon, with a rapid initial response that tends to restore the growth curve to its original trajectory.

Production of growth hormone in the pituitary is under the control of two hormones produced in the hypothalamus: growth hormone-releasing hormone, which promotes growth hormone synthesis and secretion, and somatostatin, which inhibits its production and release. Growth hormone produces many of its effects by inducing the synthesis of IGF-1 (Fig. 12.8) and, to a lesser extent, IGF-2. Post-natal growth, as well as embryonic growth, is largely due to the actions of these insulin-like growth factors, and complex hormonal regulatory circuits control their production.

Puberty is initiated by the activity of the hypothalamic neuronal network, which governs the intermittent release of gonadotropin-releasing hormone (GuRH). The mechanism determining this timing is not known. One of the results of a pulse of GuRH is a sharp increase in the secretion of gonadotropins; these cause increased production of the steroid sex hormones—the estrogens and androgens—which in turn cause increased pulses of growth hormone production.

Fig. 12.8 Growth hormone production is under the control of the hypothalamic hormones. Growth hormone is made in the pituitary gland and is secreted. Growth hormone-releasing hormone from the hypothalamus promotes growth hormone synthesis, while somatostatin inhibits it. Growth hormone controls its own release by negative feedback signals to the hypothalamus. Growth hormone causes the synthesis of the insulin-like growth factor IGF-1 and this promotes the production of growth hormone.

12.6 Growth of the long bones occurs in the growth plates

An important aspect of post-embryonic vertebrate growth is the growth of the long bones of the limbs (humerus, femur, radius, and ulna). The long bones are initially laid down as cartilaginous elements (see Section 9.1) and then become ossified. The early growth of these elements involves both cell proliferation and matrix

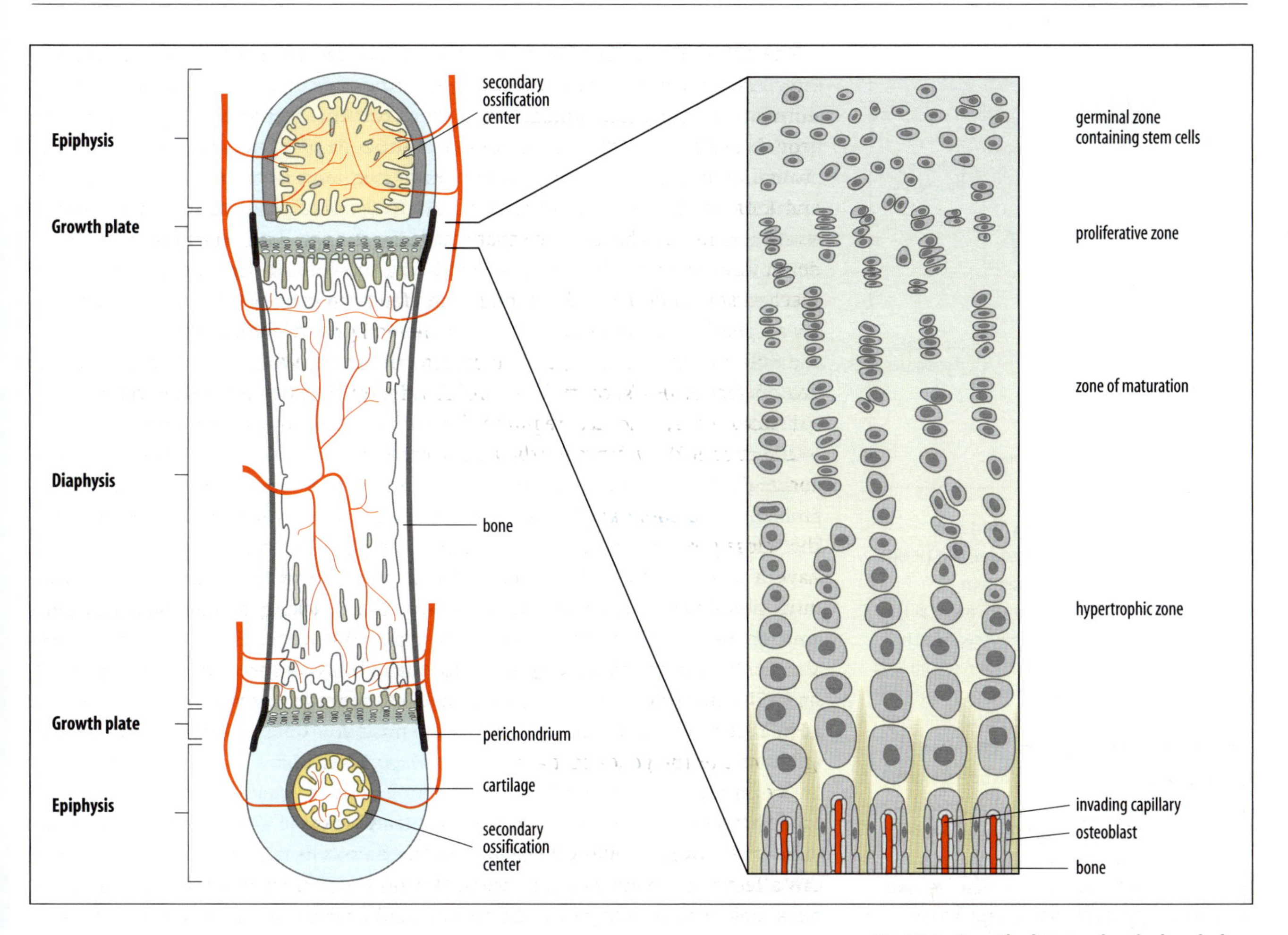

Fig. 12.9 Growth plates and endochondral ossification in the long bone of a vertebrate. The long bones of vertebrate limbs increase in length by growth from cartilaginous growth plates. The growth plates are cartilaginous regions that lie between the epiphysis of the future joint and the central region of the bone, the diaphysis. In the figure, bone has already replaced cartilage in the diaphysis, and more bone is being added at the growth plates. Within the growth plates, cartilage cells multiply in the proliferative zone, then mature and undergo hypertrophy (cell enlargement). They are then replaced by bone, which is laid down by specialized cells called osteoblasts. These derive both from perichondrial cells and from cells that invade the bone along with the blood vessels. Secondary sites of ossification are located within the epiphyses.

After Wallis, G.A.: 1993.

secretion in a well-defined pattern. The end of a long bone is known as the epiphysis and the central region as the diaphysis. In both fetal and post-natal growth, the cartilage is replaced by bone in a process known as **endochondral ossification**, in which ossification starts in the centers of the long bones and spreads outward (Fig. 12.9). Secondary ossification centers then develop at each end of the bone. The adult long bones thus have a bony shaft with cartilage confined to the articulating surfaces at each end, and to two internal regions near each end—the **growth plates**—in which growth occurs. In the growth plates, the cartilage cells, or **chondrocytes**, are usually arranged in columns, and various zones can be identified. Just next to the bony epiphysis is a narrow germinal zone, which contains stem cells. Next is a proliferative zone of cell division, followed by a zone of maturation, and a hypertrophic zone, in which the cartilage cells increase in size. Finally, there is a zone in which the cartilage cells die and are replaced by bone laid down by cells called **osteoblasts**, which differentiate from cells that form the perichondrium surrounding the cartilage. This stage involves Wnt signaling. There is a strong similarity to the development of skin, where basal stem cells give rise to dividing cells, which differentiate into keratinocytes and finally die (see Section 8.8).

The proliferation of chondrocytes at the ends of the long bones, and later in the growth plate, is controlled so that at a given distance from the end of the bone they stop dividing and enlarge to form the scaffolding for bone as appropriate. In mice, the proliferation of chondrocytes is controlled by the secreted signaling proteins parathyroid-hormone-related protein (PHRP) and Indian hedgehog. Indian hedgehog

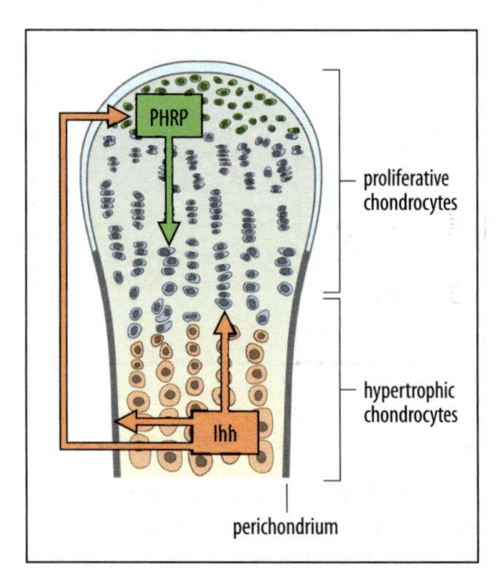

Fig. 12.10 Indian hedgehog (Ihh) and parathyroid-hormone-related protein (PHRP) form feedback loops that maintain chondrocyte proliferation and growth at the ends of developing bones. PHRP secreted by the chondrocytes at the ends of developing long bones acts on proliferating chondrocytes (blue) to maintain proliferation and thus prevents the expression of Indian hedgehog. Chondrocytes farther away from the end of the bone (orange) escape the influence of PHRP and express Ihh. This acts on adjacent chondrocytes to increase the rate of proliferation and also acts on perichondrial cells to form osteoblasts. In some way that is not yet understood, production of Ihh also stimulates PHRP synthesis by the chondrocytes at the end of the bone, thus forming a positive feedback loop that maintains PHRP production.

Adapted from Kronenberg, H.M.: 2003.

belongs to the large family of Hedgehog signaling proteins. PHRP is secreted by the chondrocytes and perichondrial cells at the ends of the prospective bones and stimulates chondrocytes to proliferate, which prevents them from expressing Indian hedgehog. Once chondrocytes move out of the zone of influence of PHRP they stop proliferating, start to express Indian hedgehog, and become hypertrophic (Fig. 12.10). Indian hedgehog diffuses back into the pool of proliferating chondrocytes, where it increases their rate of proliferation, and it also, by some mechanism as yet unknown, stimulates production of PHRP by the cells at the ends of the bone. It also acts on the adjacent perichondrial cells to form bone-producing osteoblasts. In the absence of Indian hedgehog in mice there is an accelerated rate of hypertrophy and differentiation of chondrocytes to osteoblasts, resulting in short, stubby limbs. Cell proliferation is also suppressed, and Indian hedgehog production increased, by FGFs that are produced in the perichondrium.

Growth hormone affects bone growth by acting on the growth plates. The cells in the germinal zone have receptors for growth hormone, and growth hormone is probably directly responsible for stimulating these stem cells to proliferate. Further growth, however, is probably mediated by IGF-1, whose production in the growth plate is stimulated by growth hormone. Thyroid hormones are also necessary for optimal bone growth; they act both by increasing the secretion of growth hormone and IGF-1, and by stimulating hypertrophy of the cartilage cells. FGF is also important for bone growth; the genetic defect that gives rise to achondroplasia (short-limbed dwarfism) is a dominant mutation in the FGF receptor-3, whose normal function is to limit, rather than promote, bone formation. When it is mutated, it limits growth abnormally in response to FGF.

The rate of increase in the length of a long bone is equal to the rate of new cell production per column multiplied by the mean height of an enlarged cell. The rate of new cell production depends both on the time cells take to complete a cycle in the proliferative zone, and the size of this zone. Different bones grow at different rates, and this can reflect the size of the proliferative zone, the rate of proliferation, and the degree of cell enlargement in the growth plate.

In view of the complexity of the growth plate, it is remarkable that human bones in limbs on opposite sides of the body can grow for some 15 years independently of each other, and yet eventually match to an accuracy of about 0.2%. This is achieved by having many columns of cells in each plate, so that growth variations are averaged out. Growth of a bone ceases when the growth plate ossifies, and this occurs at different times for different bones. Ossification of growth plates occurs in a strict order in different bones and can therefore be used to provide a measure of physiological age. The timing of growth cessation in the growth plate appears to be instrinsic to the plate itself rather than to hormonal influence. Cell senescence, and thus growth cessation, may be due to the chondrocyte stem cells having only a finite proliferative potential.

12.7 Growth of vertebrate striated muscle is dependent on tension

The number of striated (skeletal) muscle fibers in vertebrates is determined during embryonic development. Once differentiated, striated muscle cells lose the ability to divide. Post-embryonic growth of muscle tissue results from an increase in individual fiber size, both in length and girth. The number of myofibrils within the enlarged muscle fiber can increase more than 10-fold. Additional nuclei for the much-enlarged cell are provided by the fusion of satellite cells with the fiber. Satellite cells, which are undifferentiated cells lying adjacent to the differentiated muscle, also act as a reserve population of stem cells that can replace damaged muscle (see Section 8.11).

The increase in the length of a muscle fiber is associated with an increase in the number of sarcomeres—the functional contractile units—it contains. For example, in the soleus muscle of the mouse leg, as the muscle increases in length, the number of sarcomeres increases from 700 to 2300 at 3 weeks after birth. This increase in number seems to depend on the growth of the long bones putting tension on the muscle through its tendons. If the soleus muscle is immobilized by placing the leg in a plaster cast at birth, sarcomere number increases slowly over the next 8 weeks, but then increases rapidly when the cast is removed (Fig. 12.11). One can thus see how bone and muscle growth are mechanically coordinated.

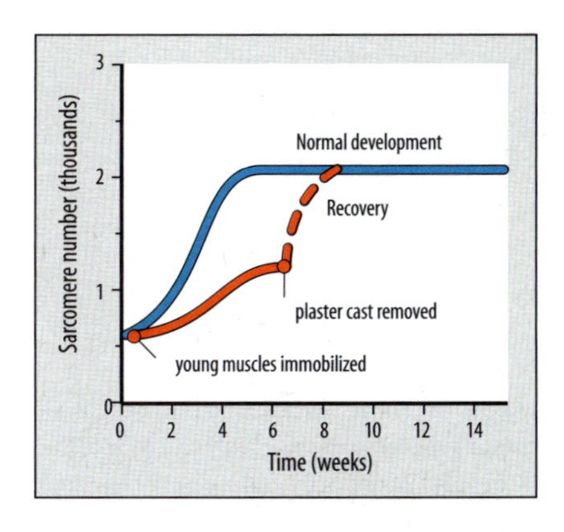

Fig. 12.11 The growth in length of the muscles attached to the long bones of the mouse leg depends on the tension provided by long bone growth. The length of a muscle is related to the number of sarcomeres—the basic contractile unit. If the limb is immobilized by a plaster cast in a position in which there is no tension on the muscle, there is little increase in muscle length until the cast is removed. There is then a rapid increase in length.

12.8 Cancer can result from mutations in genes that control cell multiplication and differentiation

Cancer can be regarded as a major perturbation of normal cellular behavior that results from certain mutations in somatic cells. Creating and maintaining tissue organization requires strict controls on cell division, differentiation, and growth. In cancer, cells escape from these normal controls, and proceed along a path of uncontrolled growth and migration that can kill the organism. There is usually a progression from a benign localized growth to malignancy in which the cells **metastasize**—migrate to many parts of the body where they continue to grow. Most cancers derive from a single abnormal cell that has acquired a number of mutations. Progression of a mutant cell to becoming a tumor-producing cell, a process known as tumor progression, is an evolutionary process, involving both further mutations and selection of those cells best able to proliferate.

The cells most likely to give rise to cancer are those that are undergoing continual division. Because they replicate their DNA frequently, they are more likely than other cells to accumulate mutations that arise from errors in DNA replication. In almost all cancers, the cancer cells are found to have a mutation in one or more genes. Not all mutations give rise to a cancer, however, and particular genes in which mutation can contribute to cancer formation have been identified in humans and other mammals. These genes are known as **proto-oncogenes**; when such a gene undergoes mutation it becomes an **oncogene**. In some cases, the presence of a single oncogene is capable of making a cell cancerous. At least 70 proto-oncogenes have been identified in mammals.

There is another group of genes in which mutations can also lead to cancer. These are the **tumor suppressor genes**, in which inactivation or deletion of both copies of the gene is required for a cell to become cancerous. The classic example of a tumor caused by the loss of such a gene is the childhood tumor retinoblastoma, which is a tumor of retinal cells. Although retinoblastoma is normally very rare, there are families in which an inherited predisposition to it is determined by a single gene, and they have been the means of identifying the gene responsible for this cancer. The inherited defect in some of these families turns out to be a deletion of a particular region on one of the two copies of chromosome 13. This on its own does not cause the cells to be cancerous. However, if any retinal cell also acquires a deletion of the same region on the other copy of chromosome 13, a retinal tumor develops. The gene in this region that is responsible for susceptibility to retinoblastoma is known as the *retinoblastoma* (*RB*) gene. Both copies of the *RB* gene must be lost or inactivated for a cell to become cancerous (Fig. 12.12), and hence *RB* is regarded as a tumor suppressor gene. The *RB* gene encodes a protein, pRB, that is involved in the regulation of the cell cycle.

The tumor suppressor gene *p53* plays a key role in many cancers; about half of all human tumors contain a mutated form of *p53*. This gene is not required for

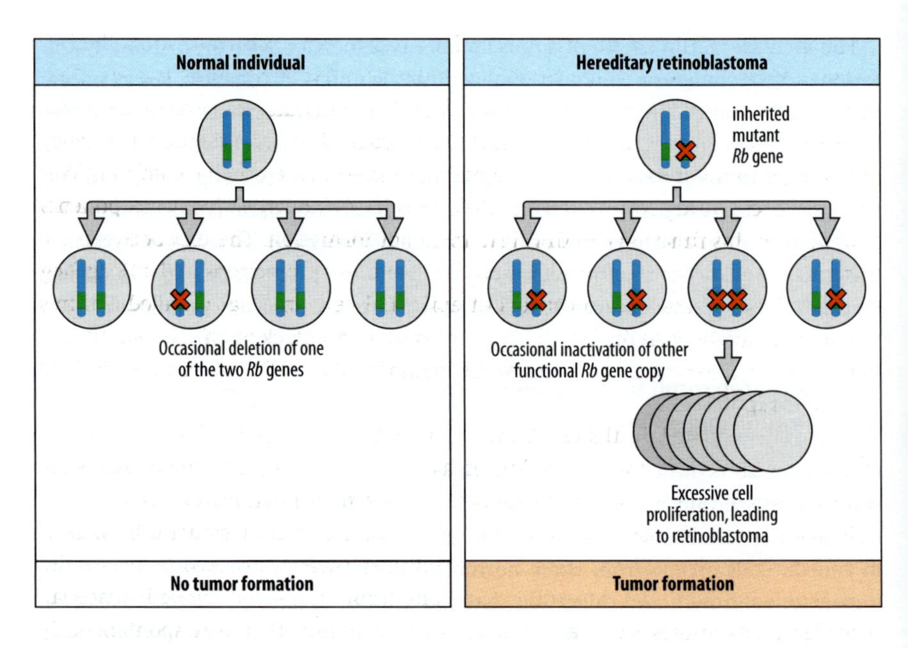

Fig. 12.12 The *retinoblastoma* (RB) gene is a tumor suppressor gene. If only one copy of the *RB* gene is lost or inactivated, no tumor develops (left panel). In individuals already carrying an inherited mutant *RB* gene, if the other copy of the gene is lost or inactivated in a cell, that cell will generate a retinal tumor (right panel). Such individuals are thus at a much greater risk of developing retinoblastoma, and usually do so at a young age.

development per se, but when cells are exposed to agents that damage DNA, then *p53* is activated and the p53 protein prevents the cell from replicating the damaged DNA and thus giving rise to mutant cells. Instead, p53 can cause the cell to die by apoptosis. The mutant forms of *p53* found in cancers do not carry out this function and make cells more likely to accumulate mutations.

A major feature of cancer is the failure of the cells to differentiate properly. The majority of cancers—over 85%—occur in epithelia. This is not surprising when it is recalled that many epithelia (such as the epidermis and the lining of the gut) are constantly being renewed by division and differentiation of stem cells (see Section 8.8). In normal epithelia, cells generated by stem cells continue to divide for a little time until they undergo differentiation, when they stop dividing. By contrast, cancerous epithelial cells continue to divide, although not necessarily more rapidly, and usually fail to differentiate. Another feature of cancer cells, unlike developing cells, is that when they divide, they are genetically unstable; the gain or loss of chromosomes is common in solid tumors.

This failure of cancer cells to differentiate is also clearly seen in certain leukemias—cancers of white blood cells. All blood cells are continually renewed from a pluripotent stem cell in the bone marrow, by a process in which steps in differentiation are interspersed with phases of cell proliferation. The pathway eventually culminates in terminal cell differentiation and a complete cessation of cell division (see Section 8.4). Several types of leukemia are caused by cells continuing to proliferate instead of differentiating. These cells become stuck at a particular immature stage in their normal development, a stage that can be identified by the molecules expressed on their cell surfaces.

A number of developmental genes that we have considered elsewhere are also involved in cancer. One view is that many cancers arise from tissue stem cells and that the Hedgehog and Wnt signaling pathways may promote cancer growth by promoting stem-cell renewal. For example, the first known mammalian member of the Wnt family was an oncogene (the mouse Int-1 protein, which is involved in brain development), and abnormal expression of Wnts can block cell differentiation. The Wnt pathway is crucial to the proliferation of intestinal cells and overactivation of this pathway is known to be involved in colorectal cancer, most commonly through mutations in the gene for APC (adenomatous polyposis coli). Normally, APC helps to keep the pathway inhibited in the absence of Wnt signaling (see Fig. 2.39) and thus acts as a tumor suppressor. Mutations that inactivate APC lead to the constitutive activation of the pathway and thus to unregulated cell proliferation. Hedgehog signaling is thought to control Wnt signaling in intestinal epithelia, confining Wnt activity to stem and progenitor cells. Mutations in Hedgehog pathway components that disrupt this function can thus lead to tumor formation. The link between the Hedgehog pathway and cancer was initially recognized by mutations in the Hedgehog receptor Patched that led to cancers of epithelial cells in what is called Gorlin's syndrome. Mutations in the Notch pathway can lead to a block in differentiation, and can therefore also result in a cancer, while members of the TGF-β family are involved in tumor suppression.

Most cancer-related deaths result from tumors that have spread from their site of origin to other tissues, the process known as **metastasis**. Central to metastasis is the ability of tumor cells to undergo an epithelial-to-mesenchyme transition, as the tumor cells must lose their tissue organization and migrate as mesenchymal cells. Loss of E-cadherin-mediated cell–cell adhesion present in epithelia is involved in metastasis.

Rarely, cancers can develop without any alteration in the cell's genetic material. The clearest examples are **teratocarcinomas**, solid tumors that arise spontaneously from germ cells. Teratocarcinomas are unusual tumors, in that they can contain a bizarre mixture of differentiated cell types. Spontaneous teratocarcinomas usually occur in the ovary or testis, and are derived from the germ cells. In the mouse ovary, accidental activation of an unfertilized egg results in its development *in situ* to the stage of epiblast formation; this epiblast then gives rise to a tumor. Similarly, if the epiblast of an early mouse embryo is transplanted to any site in the body of an adult mouse in which it receives a good blood supply, it gives rise to a teratocarcinoma.

There is good reason to believe that teratocarcinomas are not caused by genetic alterations. A mouse inner cell mass placed in culture gives rise to embryonic stem cells (ES cells), which can grow indefinitely in culture. These cells maintain their embryonic character, and when put back into the inner cell mass of another embryo contribute normally to many different tissues, including the germ line, to produce a chimeric mouse. However, when the same ES cells are placed under the skin of an adult mouse, they develop into a teratocarcinoma (see Fig. 8.35). Transgenic mice containing tissues derived from ES cells do not have an increased probability of forming tumors, yet those same cells consistently form tumors when grafted into adult mice. This indicates that the teratocarcinoma must be the result of the inner mass cells receiving the wrong developmental signals, and not of a genetic change. This raises a warning in relation to the therapeutic use of ES cells as discussed in Chapter 8.

12.9 Hormones control many features of plant growth

Plant growth is achieved by cell division in meristems and organ primordia, followed by irreversible cell enlargement, which achieves most of the increase in size (see Section 7.18). Unlike the protein growth hormones of animals, plant hormones are typically small-molecular-weight organic molecules. Auxin (indole-3-acetic acid) is one of the main regulators of plant growth, and is implicated in a large number of developmental processes, including embryonic development (see Section 6.2), growth towards the light, tissue polarity, vascular tissue differentiation, and apical dominance, which is the phenomenon of the suppression of the growth of lateral buds immediately below the apical bud. Apical dominance is caused by a diffusible inhibitor of bud outgrowth produced by the shoot apex, as shown by putting an

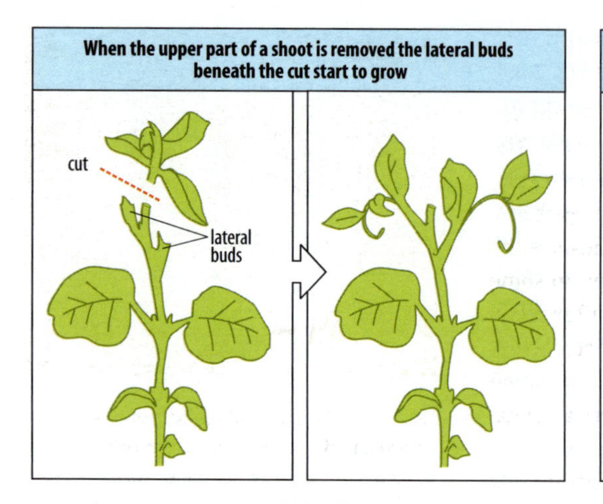

Fig. 12.13 Apical dominance in plants. Left panels: the growth of lateral buds is inhibited by the apical meristem above them. If the upper part of the stem is removed, lateral buds start growing. Right panels: an experiment to show that apical dominance is due to inhibition of lateral bud outgrowth by a substance secreted by the apical region. This substance is the plant hormone auxin.

excised apex in contact with an agar block, which can then, on its own, inhibit lateral growth. The inhibitor is auxin. It is produced by the apical bud, transported down the stem, and suppresses the outgrowth of buds that fall within its sphere of influence. If the apical bud is removed, apical dominance is also removed, and lateral buds start to grow (Fig. 12.13). Application of auxin to the cut tip replaces the suppressive effect of the apical bud.

Another family of plant hormones, the gibberellins, regulate stem elongation and have some similar effects to auxin. Cytokinins, which can stimulate cell proliferation in culture, are derivatives of adenine. The chemical nature of plant hormones is very different from those of animals; nevertheless, they are thought to act through specific hormone-binding receptors and to stimulate intracellular signal transduction.

Plant growth is affected by a variety of environmental factors, such as temperature, humidity, and light. A seedling grown in the dark takes on a characteristic etiolated form in which chloroplasts do not develop, there is extensive elongation of internodes, and the leaves do not expand. The effect of light on plant growth (photomorphogenesis) is mediated through a family of intracellular receptor proteins called phytochromes, which respond to red light and regulate many aspects of plant development and growth.

Summary

Growth in both animals and plants mainly occurs after the basic body plan has been laid down and the organs are still very small. Organ size can be controlled both by external signals and by intrinsic growth programs. Organ size can be determined by absolute dimensions rather than cell number or size. In animals, growth can occur by cell multiplication, cell enlargement, and secretion of large amounts of extracellular matrix. A plant's growth in size is both by cell enlargement and cell division. In mammals, insulin-like growth factors (IGFs) are required for normal embryonic growth, and they also mediate the effects of growth hormone after birth. Human post-natal growth is largely controlled by growth hormone, which is made in the pituitary gland. Growth in the long bones occurs in response to growth hormone stimulation of the cartilaginous growth plates at either end of the bone. Cancer is the result of the loss of growth control and differentiation. Plant growth is due to cell division in meristems followed by cell enlargement, and is dependent on auxins, gibberellins, and other growth hormones.

Molting and metamorphosis

Many animals do not develop directly from an embryo into an 'adult' form, but into a larva from which the adult eventually develops by metamorphosis. The changes that occur at metamorphosis can be rapid and dramatic, the classic examples being the metamorphosis of a caterpillar into an adult butterfly, a maggot into a fly, and a tadpole into a frog. Another striking example of metamorphosis is the transformation of the pluteus stage larva of the sea urchin into the adult. In some cases it is hard to see any resemblance between the animal before and after metamorphosis. The adult fly does not resemble the larva at all, because adult structures develop from the imaginal discs and so are completely absent from the larval stages (see Chapters 2 and 9). In frogs, the most obvious external changes at metamorphosis are the regression of the tadpole's tail and the development of limbs, although many other structural changes occur. In some insects, the entire body plan is transformed, with most larval tissues undergoing cell death as the adult tissues develop from the imaginal discs and histoblasts. In arthropods and nematodes, increase in size in larval and pre-adult stages requires shedding of the external cuticle, which is rigid, a process known as **molting**.

A number of features distinguish early embryogenesis from molting, metamorphosis, and other aspects of post-embryonic development. Whereas the signal molecules in early development act over a short range and are typically protein growth factors, many signals in post-embryonic development are produced by specialized endocrine cells, and include both protein and non-protein hormones. The synthesis of these hormones is orchestrated by the central nervous system in response to environmental cues, and there are complex interactions between the endocrine glands and their secretions.

Fig. 12.14 Growth and molting of the caterpillar of the tobacco hawkmoth (*Manduca sexta*). The caterpillar, known as the tobacco hornworm, goes through a series of molts. The tiny hatchling (1)—indicated by the arrow—molts to become a caterpillar (2), and then undergoes three further molts (3, 4, and 5). The increase in size between molts is about twofold. The caterpillars are sitting on a lump of caterpillar food. Scale bar = 1 cm.

Photograph courtesy of S.E. Reynolds.

12.10 Arthropods have to molt in order to grow

Arthropods have a rigid outer skeleton, the cuticle, which is secreted by the epidermis. This makes it impossible for the animals to increase in size gradually. Instead, increase in body size takes place in steps, associated with the loss of the old outer skeleton and the deposition of a new larger one. This process is known as **ecdysis** or molting. The stages between molts are known as instars. *Drosophila* larvae have three instars and molts. The increase in overall size between molts can be striking, as illustrated in Figure 12.14 for the tobacco hornworm.

At the start of a molt, the epidermis separates from the cuticle in a process known as apolysis, and a fluid (molting fluid) is secreted into the space between the two (Fig. 12.15). The epidermis then increases in area by cell multiplication or cell enlargement, and becomes folded. It begins to secrete a new cuticle, and the old cuticle is partly digested away, eventually splits, and is shed.

Fig. 12.15 Molting and growth of the epidermis in arthropods. The cuticle is secreted by the epidermis. At the start of molting, the cuticle separates from the epidermis—apolysis—and a fluid is secreted between them. The epidermis grows, becomes folded, and begins to secrete a new cuticle. Enzymes weaken the old cuticle, which is shed.

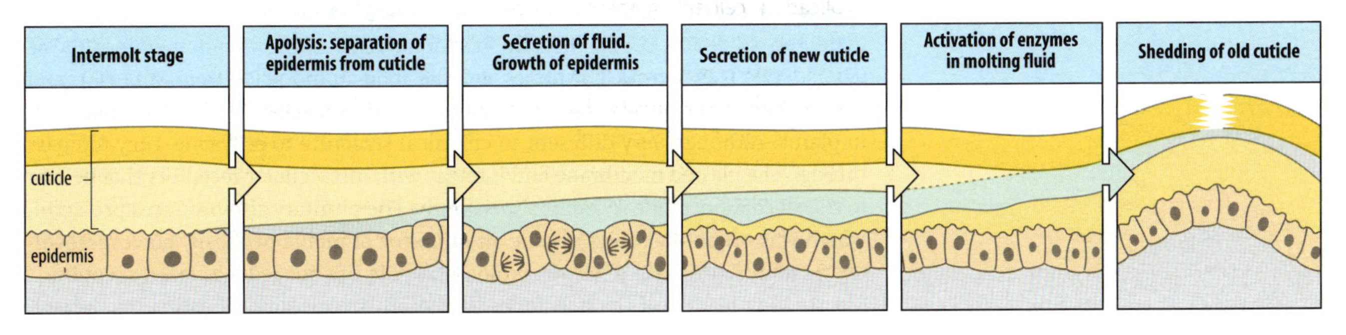

Molting is under hormonal control. Stretch receptors that monitor body size are activated as the animal grows, and this results in the brain secreting prothoracicotropic hormone. This activates the prothoracic gland to release the steroid hormone ecdysone, which is the hormone that causes molting. A similar hormonal circuit controls metamorphosis, and is described more fully in the next section.

12.11 Metamorphosis is under environmental and hormonal control

When an insect larva has reached a particular stage it does not grow and molt any further, but undergoes a more radical metamorphosis into the adult form. Metamorphosis occurs in many animal groups other than arthropods, including amphibians. In both insects and amphibians, environmental cues, such as nutrition, temperature, and light, as well as the animal's internal developmental program, control metamorphosis through their effects on neurosecretory cells in the brain. There are two groups of hormone-producing cells, one of which promotes metamorphosis, and the other inhibits it. Metamorphosis occurs when the inhibition, which is predominant in the larval stage, is overcome in response to environmental cues. The signals produced by the two sets of endocrine cells control the development of all the cells involved in metamorphosis. Larval tissues such as gut, salivary glands, and certain muscles undergo programmed cell death. The imaginal discs now develop into rudimentary adult appendages like wings, legs, and antennae. The nervous system is also remodeled.

In insects, temperature and light cues stimulate neurosecretory cells in the larva's central nervous system to release signals that act on a neurosecretory release site behind the brain, which then secretes prothoracicotropic hormone. This acts on the prothoracic gland to stimulate the production of the steroid hormone ecdysone. It is ecdysone that promotes metamorphosis, as demonstrated by its ability to induce premature metamorphosis in fly larvae. However, the action of ecdysone can be counteracted by another hormone—juvenile hormone—produced by the corpus allatum, an endocrine gland located just behind the brain. As its name implies, juvenile hormone maintains the larval state. In butterflies, a pulse of ecdysone in the final instar larva triggers the beginning of pupa formation, and another pulse some days later initiates the later stages of metamorphosis (Fig. 12.16).

Ecdysone crosses the plasma membrane, where it interacts with intracellular ecdysone receptors that belong to the steroid hormone receptor superfamily. These receptors are gene-regulatory proteins, which are activated by binding their hormone ligand (see Section 8.2). The hormone-receptor complex binds to the regulatory regions of a number of different genes, inducing a new pattern of gene activity characteristic of metamorphosis.

In amphibians, in response to nutritional status and environmental cues, such as temperature and light, the neurosecretory cells of the hypothalamus release corticotropin-releasing hormone, which acts on the pituitary gland, causing it to release thyroid-stimulating hormone. This in turn acts on the thyroid gland to stimulate the secretion of the thyroid hormones that bring about metamorphosis (Fig. 12.17). The thyroid hormones are the iodo-amino acids thyroxine (T_4), and tri-iodothyronine (T_3). They are signaling molecules of ancient origin, occurring even in plants. Although very different in chemical structure to ecdysone, they too pass through the plasma membrane and interact with intracellular receptors that belong to the steroid hormone receptor superfamily. The pituitary also produces prolactin, which was originally thought to be an inhibitor of metamorphosis; however, overexpression of prolactin does not prolong tadpole life but reduces tail resorption.

A striking feature of the hormones that stimulate metamorphosis is that as well as affecting a wide variety of tissues, they affect different tissues in different ways,

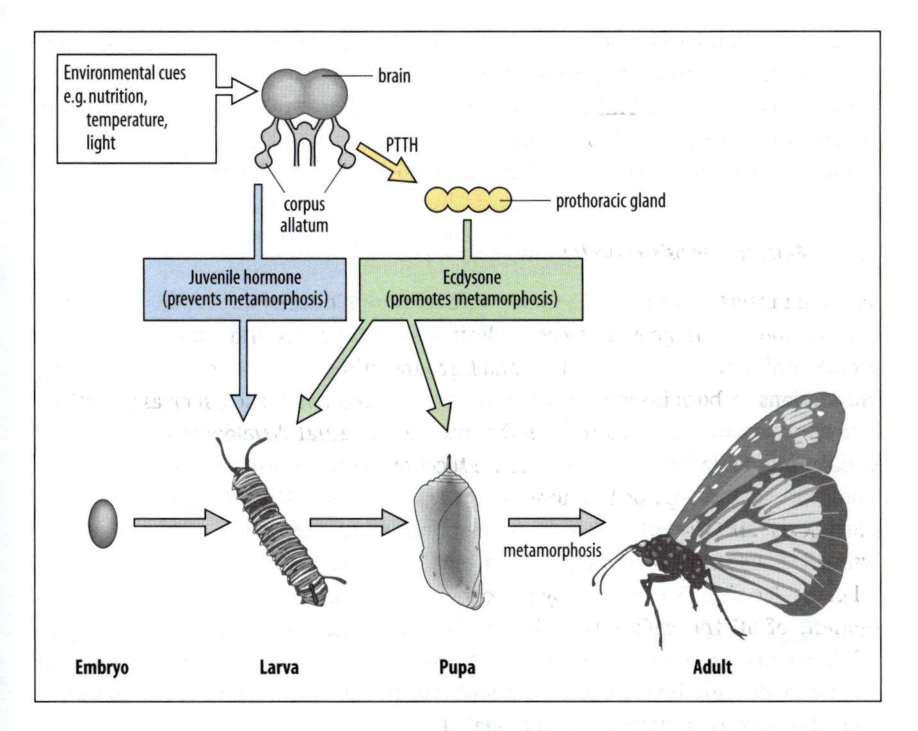

Fig. 12.16 Insect metamorphosis. The corpus allatum of a butterfly larva secretes juvenile hormone, which inhibits metamorphosis. In response to environmental changes, such as an increase in light and temperature, the corpus allatum of the final instar larva begins to secrete prothoracicotropic hormone (PTTH). This acts on the prothoracic gland to stimulate the secretion of ecdysone, the hormone that overcomes the inhibition by juvenile hormone and causes metamorphosis.

After Tata, J.R.: 1998.

their effects varying from the subtle to the gross. In the tadpole limb, for example, thyroid hormones promote development and growth, whereas they cause cell death and degeneration in the tail. Fast muscles are the first to go, and the notochord later collapses. Metamorphosis also leads to changes in the responsiveness of cells to other signals; for example, in *Xenopus*, estrogen can only induce the synthesis of vitellogenin, a protein required for the yolk of the egg, after metamorphosis.

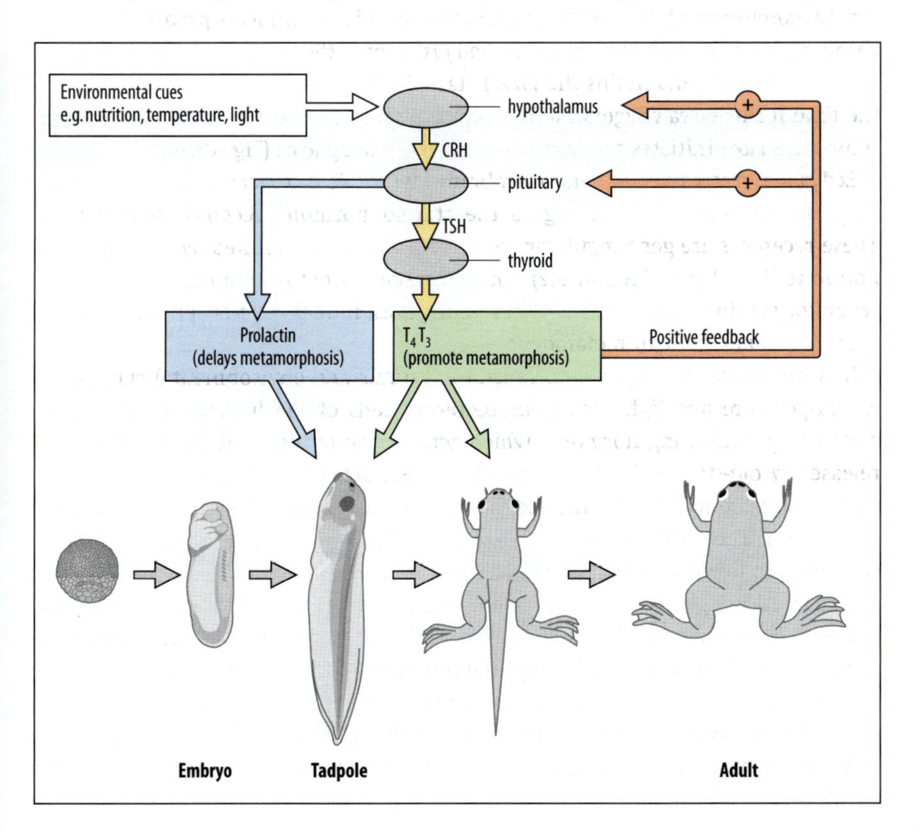

Fig. 12.17 Amphibian metamorphosis. Changes in the environment, such as an increase in nutritional levels, cause the secretion of corticotropin-releasing hormone (CRH) from the hypothalamus, which acts on the pituitary to release thyroid-stimulating hormone (TSH). This in turn acts on the thyroid glands to stimulate secretion of the thyroid hormones thyroxine (T_4) and tri-iodothyronine (T_3), which cause metamorphosis. The thyroid hormones also act on the hypothalamus and pituitary to maintain synthesis of CRH and TSH.

After Tata, J.R.: 1998.

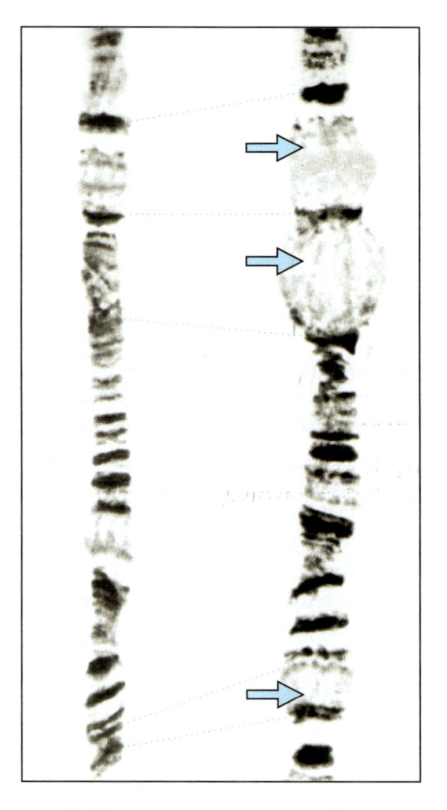

Fig. 12.18 Gene activity seen as puffs on the polytene chromosomes of Drosophila. A region of a chromosome is shown from a young third instar (left), and from an older larva (right), after ecdysone has induced puffs at three loci (arrowed).

Photograph courtesy of M. Ashburner.

Each tissue has its own response to the hormones causing metamorphosis, and some of these effects can be reproduced in culture. When excised *Xenopus* tadpole tails are exposed to thyroid hormones in culture, for example, they cause cell death and complete tissue regression.

There are alterations in the expression of many genes—several hundred at least—during *Drosophila* metamorphosis. In *Drosophila*, because of the special characteristics of polytene chromosomes, it is possible to see some changes in gene activity that occur during metamorphosis. Cells in some tissues of the larva grow and repeatedly pass through S phase without undergoing mitosis and cell division. The cells become very large and can have several thousand times the normal complement of DNA. In salivary gland cells, many copies of each chromosome are packed side by side to form giant polytene chromosomes. When a gene is active, the chromosome at that site expands into a large localized 'puff', which is easily visible (Fig. 12.18). The puff represents the unfolding of chromatin and the associated transcriptional activity. When the gene is no longer active the puff disappears. During the last days of larval life, a large number of puffs are formed in a precise sequence, a pattern that is under the direct influence of ecdysone.

Summary

Arthropod larvae grow by undergoing a series of molts in which the rigid cuticle is shed. Metamorphosis during the post-embryonic period can result in a dramatic change in the form of an organism. In insects, it is hard to see any resemblance between the animal before and after metamorphosis, whereas in amphibians, the change is somewhat less dramatic. Environmental and hormonal factors control metamorphosis. In both insects and amphibians there are two sets of hormonal signals, one promoting, the other delaying, metamorphosis. Thyroid hormones cause metamorphosis in amphibians, and ecdysone does the same in insects. In *Drosophila*, gene activity during metamorphosis can be monitored by localized puffing on the giant polytene

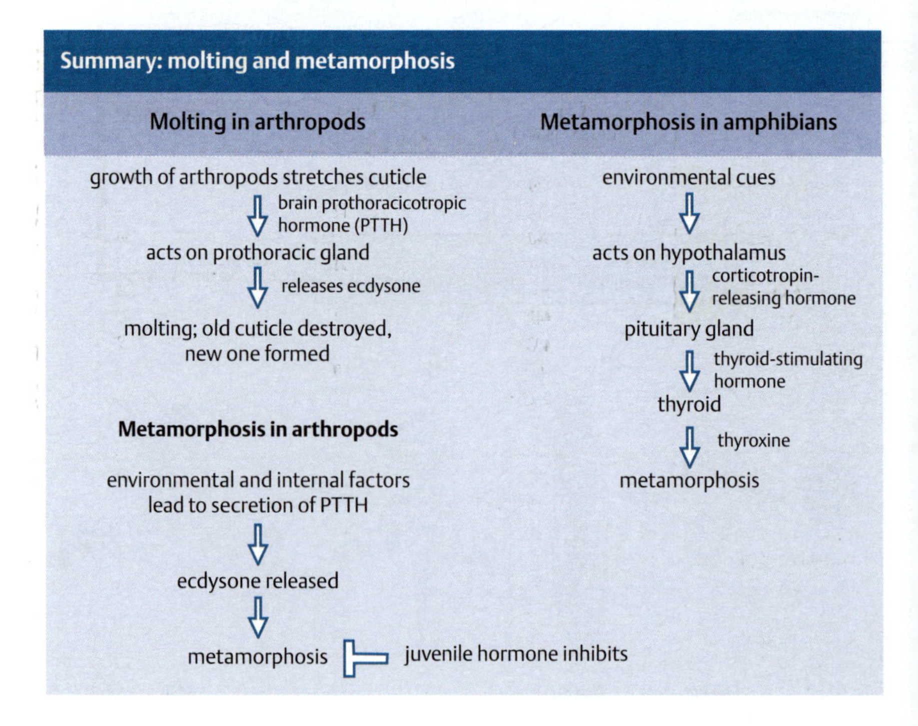

Aging and senescence

Organisms are not immortal, even if they escape disease or accidents. With aging—the passage of time—comes an increasing impairment of physiological functions, which reduces the body's ability to deal with a variety of stresses, and an increased susceptibility to disease. This age-related decline in function is known as **senescence**. As we shall see in Chapter 13, there are some simple animals, such as the cnidarian *Hydra*, which do not show senescence. The phenomenon of senescence raises many unanswered questions as to its underlying mechanisms, but we can at least consider some general questions, such as whether senescence is part of an organism's post-embryonic developmental program or whether it is simply the result of wear and tear.

Although individuals may vary in the time at which particular aspects of aging appear, the overall effect is summed up as an increased probability of dying in most animals, including humans, with increased age. This life pattern, which is illustrated in relation to *Drosophila* (Fig. 12.19), is typical of many animals, but there is little evidence that aging contributes to mortality in the wild; more than 90% of wild mice die during their first year. There are, however, exceptions, such as the Pacific salmon, in which death does not come after a process of gradual aging, but is linked to a certain stage in the life cycle, in this case to spawning.

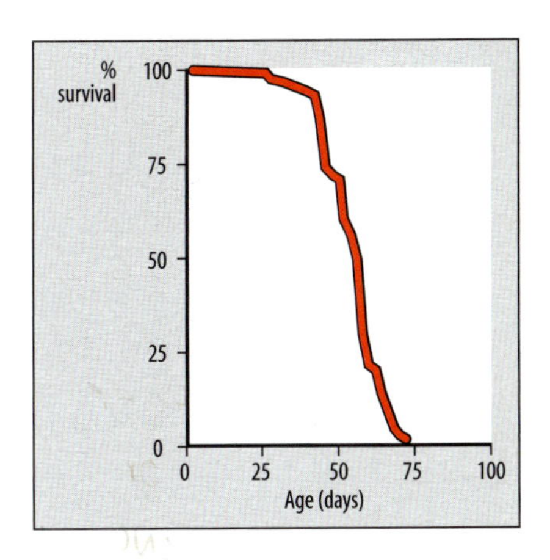

Fig. 12.19 Aging in *Drosophila*. The probability of dying increases rapidly at older ages.

One view of senescence is that it is the outcome of an accumulation of damage that eventually outstrips the ability of the body to repair itself, and so leads to the loss of essential functions. For example, some old elephants die of starvation because their teeth have worn out. The nematode *C. elegans* lives for an average or around 20 days and the major cellular change that occurs with age is the progressive deterioration of muscle. This deterioration has a random effect on the lifespan, which varies from 10 to 30 days. Nevertheless, there is clear evidence that senescence is under genetic control, as different animals age at vastly different rates, as shown by their different lifespans (Fig. 12.20). An elephant, for example, is born after 21 months' embryonic development, and at that point shows few, if any, signs

Longevity and time to attain reproductive maturity at puberty for various mammals

	Maximum lifespan (months)	Length of gestation (months)	Age at puberty (months)
Human	1440	9	144
Finback whale	960	12	–
Indian elephant	840	21	156
Horse	744	11	12
Chimpanzee	534	8	120
Brown bear	442	7	72
Dog	408	2	7
Cattle	360	9	6
Rhesus monkey	348	5.5	36
Cat	336	2	15
Pig	324	4	4
Squirrel monkey	252	5	36
Sheep	240	5	7
Gray squirrel	180	1.5	12
European rabbit	156	1	12
Guinea-pig	90	2	2
House rat	56	0.7	2
Golden hamster	48	0.5	2
Mouse	42	0.7	1.5

Fig. 12.20 Table showing lifespan, length of gestation, and age at puberty for various mammals.

of aging, whereas a 21-month-old mouse is already well into middle age and beginning to show signs of senescence. This genetic control of aging can be understood in terms of the disposable soma theory, which puts it into the context of evolution. The disposable soma theory proposes that natural selection tunes the life history of the organism so that sufficient resources are invested in maintaining the repair mechanisms that prevent aging at least until the organism has reproduced and cared for its young. Thus mice, which start reproducing when just a few months old, need to maintain their repair mechanisms for much less time than do elephants, which only start to reproduce when around 13 years old. In most species of animals in the wild, few individuals live long enough to show obvious signs of senescence and senescence need only be delayed until reproduction is complete.

12.12 Genes can alter the timing of senescence

The maximum recorded life spans of animals show dramatic differences in length (see Fig. 12.20). Humans can live as long as 120 years, some owls 68 years, cats 28 years, *Xenopus* 15 years, mice 3.5 years, and the nematode about 25 days. Mutations in genes that affect lifespan have been identified in *C. elegans*, *Drosophila*, mice, and humans, and may give clues to the mechanisms involved; both the ability to resist damage to DNA and the effects of oxygen radicals are important.

A *C. elegans* worm that hatches as a first instar larva in an uncrowded environment with ample food grows to adulthood and can survive for 25 days. In crowded conditions and when food is short, however, the animal enters a quiescent third-instar larval state known as the **dauer** state, where it neither eats nor grows until food becomes available again. When conditions become favorable the dauer larva molts and becomes a fourth instar larva. The dauer state can last for 60 days and has no effect on the post-dauer life span: it is therefore considered to be a state in which the larva does not age. An insulin/IGF-1 signaling system plays an important role both in controlling entry into the dauer state in response to stress and in the overall control of fertility, life span, and metabolism in *C. elegans*. Mutations that cause a strong reduction in the expression of *daf-2*, which encodes a receptor in this pathway, arrest development in the dauer state. The normal role of DAF-2 is to antagonize the activity of DAF-16, a transcription factor whose activity lengthens life span and increases resistance to some types of stress. A partial loss of DAF-2 function results in longer adult life span after the dauer state but with reduced fertility and embryonic viability, while a further reduction in DAF-2 at the larval stage by RNA interference increases life span even more without shortening the dauer state. Removal of the reproductive system of these animals resulted in their having a mean life span of 125 days and remaining quite healthy—in humans, this would equate to a life span of 500 years. But when *daf-2* mutants were cultured alongside wild-type worms, the mutants became extinct in just a few generations, partly due to their reduced fertility. Microarray analysis of the effects of DAF-16 show that it activates stress-response and antimicrobial genes. In laboratory conditions it seems that aging animals, in which DAF-16 is inhibited, are killed by the bacteria on which they feed.

A similar system regulates aging in *Drosophila* and mutations in the insulin/IGF-1 pathway almost double the life span. The flies, rather like the dauer larvae, enter into a state of reproductive diapause. These effects on life span may be due to resistance to oxidative stress. Aging in mice can be retarded by reduction in pituitary activity, and there is also evidence that female mice with a mutation in the IGF-1 receptor (which is the receptor activated by IGF-1 and IGF-2) can live 33% longer than usual.

There is evidence that oxidative damage accelerates aging and that reactive oxygen radicals are key players. Reduction in food intake increases life span in other animals; rats on a minimal diet live about 40% longer than rats allowed to eat as much as they like, which is thought to be partly due to a reduced exposure to free radicals. These are formed during the oxidative breakdown of food. They are highly reactive, and can damage both DNA and proteins. A long-lived rodent species generates less reactive oxygen than the laboratory mouse. Oxidative stress also exerts its effects by damaging mitochondria.

Humans that are homozygous for the recessive gene defect known as Werner's syndrome show striking effects of premature aging. There is growth retardation at puberty, and by their early twenties those affected by this syndrome have gray hair and suffer from a variety of illnesses, such as heart disease, that are typical of old age. Most die before the age of 50. Fibroblasts taken from Werner's syndrome patients undergo fewer cell divisions in culture before becoming senescent and dying than do fibroblasts from unaffected people of the same age. The gene affected in Werner's syndrome has been isolated, and is thought to encode a protein involved in unwinding DNA. Such unwinding is required for DNA replication, DNA repair, and gene expression. The inability to carry out DNA repair properly in Werner's syndrome patients could subject the genetic material to a much higher level of damage than normal. The link between Werner's syndrome and DNA thus fits with the possibility that aging is linked to the accumulation of damage in DNA, but it may also be related to cell senescence.

12.13 Cultured mammalian cells undergo cell senescence

One might think that when cells are isolated from an animal, placed in culture, and provided with adequate medium and growth factors, they would continue to proliferate almost indefinitely. But this is not the case. For example, mammalian fibroblasts—connective tissue cells—will only go through a limited number of cell doublings in culture; the cells then stop dividing, however long they are cultured (Fig. 12.21). For normal fibroblasts, the number of cell doublings depends

Fig. 12.21 Vertebrate fibroblasts can only go through a limited number of divisions in culture. Fibroblasts placed in culture are subcultured until they stop growing (top panels). The number of cell doublings in culture before they stop dividing is related to their maximum age, as indicated by the figures in brackets on the graph (bottom panel).

Cells divide until they completely cover the dish and continue to divide when placed in fresh culture medium

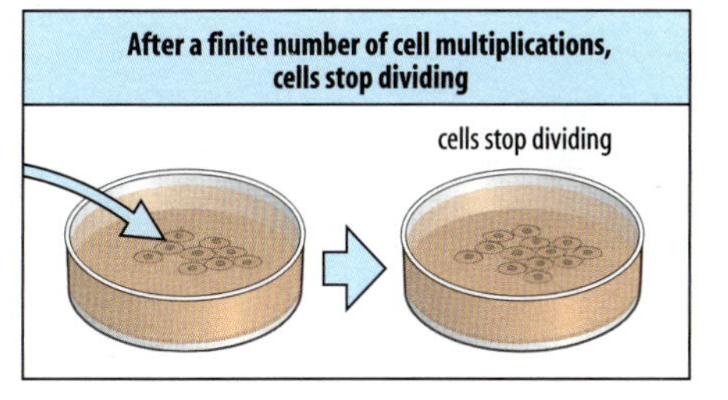

Galapagos tortoise (175 years)
Human (120 years)
Chicken (30 years)
Mouse (3.5 years)
0 25 50 75 100 125
number of population doublings of cells in culture

both on the species and the age of the animal from which they are taken. Human fibroblasts taken from a fetus go through about 60 doublings, those from an 80-year-old about 30, and those from an adult mouse about 12–15. When the cells stop dividing, they appear to be healthy, but are stuck at some point in the cell cycle, often G_0; this phenomenon is known as **cell senescence**. Cells taken from patients with Werner's syndrome, who show an acceleration of many features of normal aging, make significantly fewer divisions in culture than normal cells. However, it is far from clear how this behavior of cells in culture is significant for the aging of the organism and to what extent it reflects the way the cells are cultured.

A feature shared by senescent cells in culture and *in vivo* is shortening of the telomeres. These are the repetitive DNA sequences at the ends of the chromosomes that preserve chromosome integrity and ensure that chromosomes replicate themselves completely without loss of informational DNA at the ends. The length of the telomeres is reduced in older human cells. Telomere length is found to decrease at each DNA replication, suggesting that they are not completely replicated at each cell division and this may be related to senescence. If the enzyme telomerase, which maintains telomere length and is normally absent from cells in culture, is expressed in those cells, senescence in culture does not occur. However, it is not yet clear whether telomere shortening is a major cause of aging in somatic cells, as certain rodent cells such as Schwann cells can, under appropriate conditions, proliferate indefinitely and their telomeres do not control replication. ES cells also express telomerase and proliferate indefinitely.

Summary

Aging is largely caused by damage to cells, particularly by reactive oxygen, but is also under genetic control. Many normal cells in culture can only undergo a limited number of cell divisions, which correlate with their age at isolation and the normal lifespan of the animal from which they came. Genes that can increase lifespan have been identified in *C. elegans* and *Drosophila* and may act by increasing the animals' resistance to oxidative stress.

SUMMARY TO CHAPTER 12

The form of many animals is laid down in miniature during embryonic development, and they then grow in size, keeping the basic body form, although different regions grow at different rates. Growth may involve cell multiplication, cell enlargement, and the laying down of extracellular material. In vertebrates, structures have an intrinsic growth program, which is under hormonal control. Arthropod larvae grow by molting and both they and other animals, such as frogs, undergo metamorphosis, which is under hormonal control. Cancer can be viewed as an aberration of growth, as it usually results from mutations that lead to excessive cell proliferation and failure of cells to differentiate. The symptoms of aging appear mainly to be caused by damage to cells which accumulates over time, and aging is also under genetic control. In plants, all adult structures are derived by growth from specialized regions called meristems.

GENERAL FURTHER READING

Clarke, M.F.: **At the root of brain cancer.** *Nature* 2004, **432:** 281–282.

Nature Insight: **Cell division and cancer.** *Nature* 2004, **432:** 293–341.

SECTION FURTHER READING

12.1 Tissues can grow by cell proliferation, cell enlargement, or accretion

Goss, R.J.: *The Physiology of Growth.* New York: Academic Press, 1978.

12.2 Cell proliferation can be controlled by an intrinsic program

Edgar, B., Lehner, C.F.: **Developmental control of cell cycle regulators: a fly's perspective.** *Science* 1996, **274:** 1646–1652.

Follette, P.J., O'Farrell, PH.: **Connecting cell behavior to patterning: lessons from the cell cycle.** *Cell* 1997, **88:** 309–314.

12.3 Organ size can be controlled by external signals and intrinsic growth programs

Amthor, H., Huang, R., McKinnell, I., Christ, B., Kambadur, R., Sharma, M., Patel, K.: **The regulation and action of myostatin as a negative regulator of muscle development during avian embryogenesis.** *Dev. Biol.* 2002, **251:** 241–257.

Conlon, I., Raff, M.: **Size control in animal development.** *Cell* 1999, **96:** 235–244.

Shingleton, A.W.: **Body-size regulation: combining genetics and physiology.** *Curr. Biol.* 2005, **15:** R825–R827.

Taub, R.: **Liver regeneration: from myth to mechanism.** *Nat. Rev. Mol. Cell Biol.* 2004, **5:** 836–847.

12.4 Organ size may be determined by absolute dimension rather than cell number

Day, S., Lawrence, P.A.: **Measuring dimensions: the regulation of size and shape.** *Development* 2000, **127:** 2977–2987.

de la Cova, C., Abril, M., Bellosta, P., Gallant, P., Johnston, L.A.: ***Drosophila* myc regulates organ size by inducing cell competition.** *Cell* 2004, **117:** 107–116.

Schubiger, M., Tomita, S., Sung, C., Robinow, S., Truman, J.W.: **Isoform specific control of gene activity *in vivo* by the *Drosophila* ecdysone receptor.** *Mech. Dev.* 2003, **120:** 909–918.

Stern, D.: **Body-size control: how an insect knows it has grown enough.** *Curr. Biol.* 2003, **13:** R267–R269.

Su, T.T., O'Farrell, P.H.: **Size control: cell proliferation does not equal growth.** *Curr. Biol.* 1998, **8:** R687–R689.

12.5 Growth can be dependent on growth hormones

Barker, D.J.: **The Wellcome Foundation Lecture. 1994: The fetal origins of adult disease.** *Proc. R. Soc. Lond.* 1995, **262:** 37–43.

Colombani, J., Bianchini, L., Layalle, S., Pondeville, E., Dauphin-Villemant, C., Antoniewski, C., Carre, C., Noselli, S., Leopold, P.: **Antagonistic actions of ecdysone and insulins determine final size in *Drosophila*.** *Science* 2005, **310:** 667–670.

Kwong, W.Y., Wild, A.E., Roberts, P., Willis, A.C., Fleming, T.P.: **Maternal undernutrition in the postimplantation period of rat development causes blastocyst abnormalities and programming of postnatal hypertension.** *Development* 2000, **127:** 4195–4202.

Sanders, E.J., Harvey S.: **Growth hormone as an early embryonic growth and differentiation factor.** *Anat. Embryol.* 2004, **209:** 1–9.

12.6 Growth of the long bones occurs in the growth plates

Kember, N.F.: **Cell kinetics and the control of bone growth.** *Acta Paediatr. Suppl.* 1993, **391:** 61–65.

Kronenberg, H.M.: **Developmental regulation of the growth plate.** *Nature* 2003, **423:** 332–336.

Nilsson, O., Baron, J.: **Fundamental limits on longitudinal bone growth: growth plate senescence and epiphyseal function.** *Trends Endocrinol. Metab.* 2004, **8:** 370–374.

Roush, W.: **Putting the brakes on bone growth.** *Science* 1996, **273:** 579.

12.7 Growth of vertebrate striated muscle is dependent on tension

Schultz, E.: **Satellite cell proliferative compartments in growing skeletal muscles.** *Dev. Biol.* 1996, **175:** 84–94.

Williams, P.E., Goldspink, G.: **Changes in sarcomere length and physiological properties in immobilized muscle.** *J. Anat.* 1978, **127:** 450–468.

12.8 Cancer can result from mutations in genes that control cell multiplication and differentiation

Beachy, P.A., Karhadkar, S.S., Berman, D.M.: **Tissue repair and stem cell renewal in carcinogenesis.** *Nature* 2004, **432:** 324–331.

Hunter, T.: **Oncoprotein networks.** *Cell* 1997, **88:** 333–346.

Sawyers, C.L., Denny, C.T., Witte, O.N.: **Leukemia and the disruption of normal hematopoiesis.** *Cell* 1991, **64:** 337–350.

Shilo, B.Z.: **Tumor suppressors. Dispatches from patched.** *Nature* 1996, **382:** 115–116.

Taipale, J., Beachy, P.A.: **The hedgehog and Wnt signalling pathways in cancer.** *Nature* 2001, **422:** 349–353.

Vernon, A.E., LaBonne, C.: **Tumor metastasis: a new Twist on epithelial-mesenchymal transitions.** *Curr. Biol.* **14:** R719–R721.

12.9 Hormones control many features of plant growth

Cosgrove, D.J.: **Plant cell enlargement and the action of expansins.** *BioEssays* 1996, **18:** 533–540.

Hauser, M.T., Morikami, A., Benfey, P.N.: **Conditional root expansion mutants of *Arabidopsis*.** *Development* 1995, **121:** 1237–1252.

Raven, P.H., Evert, R.F., Eichhorn, S.E.: *The Biology of Plants* (7th edn). New York: Worth, 1998.

Sablowski, R.: **Root development: the embryo within.** *Curr. Biol.* 2004, **14:** R1054–R1055.

12.10 Arthropods have to molt in order to grow

Reynolds, S.E., Samuels, R.I.: **Physionomy and biochemistry of insect molting fluid.** *Adv. Insect Physiol.* 1996, **26:** 157–232.

12.11 Metamorphosis is under environmental and hormonal control

Ellinson, R.P., Remo, B., Brown, D.D.: **Novel structural element identified during tail resorption in *Xenopus laevis* metamorphosis: lessons from tailed frogs.** *Dev. Biol.* 1999, **215:** 243–252.

Huang, H., Brown, D.D.: **Prolactin is not juvenile hormone in *Xenopus laevis* metamorphosis.** *Proc. Natl Acad. Sci. USA* 2000, **97:** 195–199.

Tata, J.R.: *Hormonal Signaling and Postembryonic Development.* Heidelberg: Springer, 1998.

Thummel, C.S.: **Flies on steroids—*Drosophila* metamorphosis and the mechanisms of steroid hormone action.** *Trends Genet.* 1996, **12:** 306–310.

White, K.P., Rifkin, S.A., Hurban, P., Hogness, D.S.: **Microarray analysis of *Drosophila* development during metamorphosis.** *Science* 1999, **286:** 2179–2184.

Wu, H.H., Ivkovic, S., Murray, R.C., Jaramillo, S., Lyons, K.M., Johnson, J.E., Calof, A.L.: **Autoregulation of neurogenesis by GDF11.** *Neuron* 2003, **37:** 197–207.

12.12 Genes can alter the timing of senescence

Arantes-Oliviera, N., Berman, J.R., Kenyon, C.: **Healthy animals with extreme longevity.** *Science* 2003, **302:** 611.

Finch, C.E., Kirkwood, T.B.L.: *Chance, Development and Aging.* Oxford: Oxford University Press, 2000.

Harper, M.E., Bevilacqua, L., Hagopian, K., Weindruch, R., Ramsey, J.J.: **Ageing, oxidative stress, and mitochondrial uncoupling.** *Acta Physiol. Scand.* 2004, **182:** 321–331.

Kenyon, C.: **A conserved regulatory system for aging.** *Cell* 2001, **105:** 165–168.

Kipling, D., Davis, T., Ostler, E.L., Faragher, R.G.: **What can progeroid syndromes tell us about human aging?** *Science* 2004, **305:** 1426–1431.

Martinez, D.E.: **Mortality patterns suggest lack of senescence in hydra.** *Exp. Gerontol.* 1998, **33:** 217–225.

Murphy, C.T., Partridge, L., Gems, D.: **Mechanisms of ageing: public or private.** *Nat. Rev. Genet.* 2002, **3:** 165–175.

Murphy, C.T., McCarroll, S.A., Bargmann, C.I., Fraser, A., Kamath, R.S., Ahringer, J., Li, H., Kenyon, C.: **Genes that act downstream of DAF-16 to influence the lifespan of *Caenorhabditis elegans*.** *Nature* 2003, **424:** 277–283.

Tatar, M., Bartke, A., Antebi, A.: **The endocrine regulation of aging by insulin-like signals.** *Science* 2003, **299:** 1346–1351.

Weindruch, R.: **Caloric restriction and aging.** *Science* 1996, **274:** 46–52.

12.13 Cultured mammalian cells undergo cell senescence

Shay, J.W., Wright, W.E.: **When do telomeres matter?** *Science* 2001, **291:** 839–840.

Sherr, C.J., DePinho, R.A.: **Cellular senescence: mitotic clock or culture shock.** *Cell* 2000, **102:** 407–410.

13

Regeneration

- Limb and organ regeneration
- Regeneration in *Hydra*

Some animals such as newts can regenerate limbs, tails, and the eye lens. This process involves dedifferentiation of cells and subsequent growth, with correct development guided by the generation of new positional values. *Hydra*, a small coelenterate, regenerates its head without growth and this involves gradients in inhibition and positional information set up by the head. It provides insight into the mechanisms involved in embryonic development in ancestral animals.

Many of the cells in the adult body, such as heart muscle and most neurons of the central nervous system, are the same cells as were originally generated during embryonic development. But some tissues, such as blood and epithelia, are continually being replaced by stem cells, and others, such as skeletal muscle, can be regenerated from quiescent stem cells if they are damaged (see, for example, Section 8.11). We have also seen many examples of the capacity of the embryo to self-regulate when parts of it are removed or rearranged (see, for example, Section 3.9). Here we look at the related phenomenon of **regeneration** in adult organisms. Regeneration is the ability of the fully developed organism to replace tissues, organs and appendages by growth or repatterning of somatic tissue. Plants have remarkable powers of regeneration: a single somatic plant cell can give rise to a complete new plant (see Section 6.4). Some animals also show great ability to regenerate: small fragments of animals such as starfish, planarians (flatworms), and *Hydra* can give rise to a whole animal (Fig. 13.1). The ability of these animals to regenerate may be related to their ability to reproduce asexually; that is, to produce a complete new individual by budding or fission.

Among vertebrates, newts and other urodele amphibians show a remarkable capacity for regeneration, being able to regenerate complete new tails and limbs as well as some internal tissues (Fig. 13.2). The lens in the newt eye, for example, regenerates from the pigmented epithelium of the iris (Fig. 13.3), an example of transdifferentiation (see Section 8.16). A striking case of regeneration in vertebrates is the zebrafish, which can regenerate the heart after removal of part of the ventricle. Mammals cannot regenerate the heart, and so this ability in the genetically tractable zebrafish has implications for medical research. Some insects and other arthropods can also regenerate lost appendages, such as legs. The regenerative powers of mammals are much more restricted. The mammalian liver can regenerate if a part of it is removed, as hepatocytes retain the ability to divide, the antlers of male deer regrow each year, and fractured bones can mend by

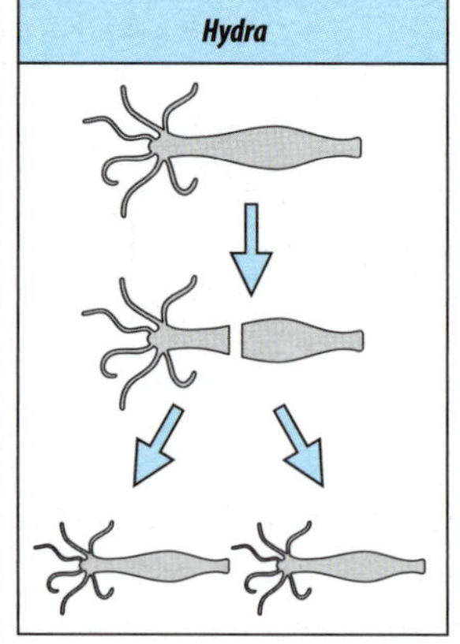

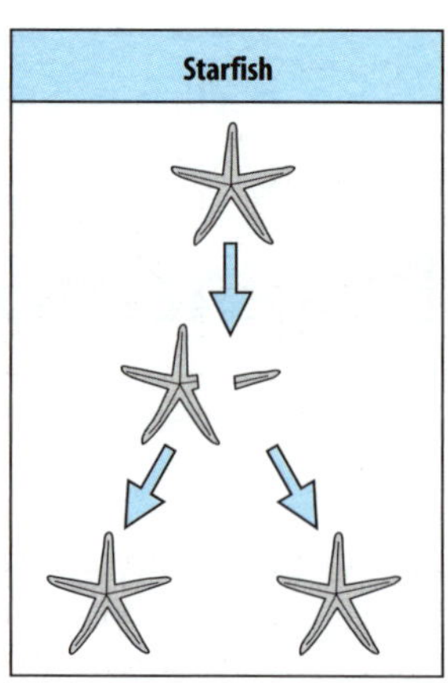

Fig. 13.1 Regeneration in some invertebrate animals. A planarian, *Hydra*, and a starfish all show remarkable powers of regeneration. When parts are removed or a small fragment isolated, a whole animal can be regenerated.

a regenerative process. But mammals cannot regenerate lost limbs, although they do have a limited capacity to replace the ends of digits.

The issue of regeneration raises several major questions. Why are some animals able to regenerate and others not? The lack of regenerative powers is not necessarily associated with increasing complexity: animals with no powers of regeneration at all include nematodes and rotifers. What is the origin of the cells that give rise to the regenerated structures? What mechanisms pattern the regenerated tissue and how are these related to the patterning processes that occur in embryonic development? We will focus on two systems in which regeneration has been intensively studied: regeneration of limbs in amphibians (newts and axolotls) and insects, and regeneration of a whole animal—the freshwater coelenterate *Hydra*. Understanding regeneration in these systems could lead to progress in stimulating regeneration in mammalian systems and thus help the development of regenerative medicine. We will also look briefly at regeneration of heart muscle in zebrafish and regeneration of nerves in mammals. Regeneration requires, at the minimum, the generation of a cohort of cells whose growth, differentiation, and patterning is controlled. There is also the question of the possible similarities between cells in a regenerating system and stem cells, but there is little evidence for this at present.

A distinction can be drawn at the outset between two types of regeneration. In one—**morphallaxis**—there is little new growth, and regeneration occurs mainly by the repatterning of existing tissues and the re-establishment of boundaries. Regeneration in *Hydra* is a good example of morphallaxis. By contrast, regeneration in the newt limb, for example, depends on the growth of new, correctly patterned structures, and this is known as **epimorphosis**. Both types of regeneration can be illustrated with reference to the French flag pattern (Fig. 13.4). In morphallaxis, new boundary regions are first established and new positional values are specified in relation to them; in epimorphosis, new positional values are linked to growth from the cut surface. In both cases, the origin of the progenitor cells for the regenerated tissue is a key issue.

Fig. 13.2 The capacity for regeneration in urodele amphibians. The emperor newt can regenerate its dorsal crest (1), limbs (2), retina and lens (3 and 4), jaw (5), and tail (not shown).

Limb and organ regeneration

Urodele amphibians such as newts and axolotls show a remarkable capacity for regenerating body structures such as tails, limbs, jaws, and the lens of the eye (see Fig. 13.2). Regeneration of all these structures involves new growth and is therefore of the epimorphic type. Regeneration of a structure such as an adult vertebrate

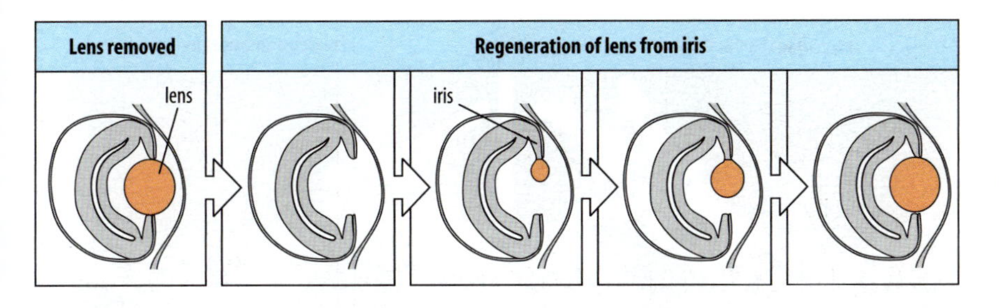

Fig. 13.3 Lens regeneration. Removal of the lens from the eye of a newt results in regeneration of a new lens from the dorsal pigmented epithelium of the iris.

limb, which contains a variety of fully differentiated cell types in a highly organized arrangement, raises a central question relating to the origin of the cells that give rise to the regenerated structure: are there special reserve cells or do cells dedifferentiate and change their character? As we shall see, in regeneration the fully differentiated cells of the mature vertebrate limb return to the cell cycle, dedifferentiate, and then re-differentiate into different cell types. Damaged insect legs and other appendages can also regenerate. This has been studied mainly in the larval cockroach, which has relatively large legs that are easy to manipulate, and we shall briefly discuss this system.

13.1 Vertebrate limb regeneration involves cell dedifferentiation and growth

Following amputation of a newt limb, there is a rapid migration of epidermal cells over the wound surface, which is essential for subsequent regeneration. A mass of undifferentiated cells called the **blastema** then begins to form under the epidermis and will give rise to the regenerated limb (Fig. 13.5). The blastema is formed from cells beneath the wound epidermis that lose their differentiated character and start to divide. As the limb regenerates over a period of weeks, these cells differentiate into cartilage, muscle, and connective tissue. The blastemal cells are derived locally from the mesenchymal tissues of the stump, close to the site of amputation. They particularly come from the dermis but also from the cartilage and muscle. Multinucleate post-mitotic muscle cells from the newt can be converted to mononuclear cells in culture in the presence of thrombin in the culture medium; thrombin is a proteolytic protein that is more familiar as part of the blood clotting cascade, but appears to be involved in providing an environment for dedifferentiation in some cases. Dedifferentiation of the muscle cells involves their expression of the homeodomain transcription factor Msx1. This is a multifunctional transcriptional factor that is known to prevent myogenic differentiation in mammalian cells, and its expression is characteristic of undifferentiated mesenchymal cells that can undergo regeneration.

This raises the question as to whether cells differentiating into cartilage and muscle in the regenerating limb are remaining true to type or whether once dedifferentiated, the cells can differentiate into a different cell type. Can dedifferentiated skeletal muscle cells in the stump then re-differentiate as cartilage, for example? In other words, is transdifferentiation occurring during limb regeneration, as in the regenerating eye lens of a newt (see Fig. 13.3)? The answer, at least in the newt, is yes. If cultured newt limb muscle myotubes, which are multinucleate and have stopped dividing, are labeled in culture with a retrovirus expressing alkaline phosphatase, and are introduced into regenerating limbs, strongly labeled mononucleate cells can be observed in blastemas after 1 week. The majority of the myotubes give rise to mononucleate cells. These mononucleate cells proliferate, and there is some evidence that they later give rise to cartilage as well as new muscle.

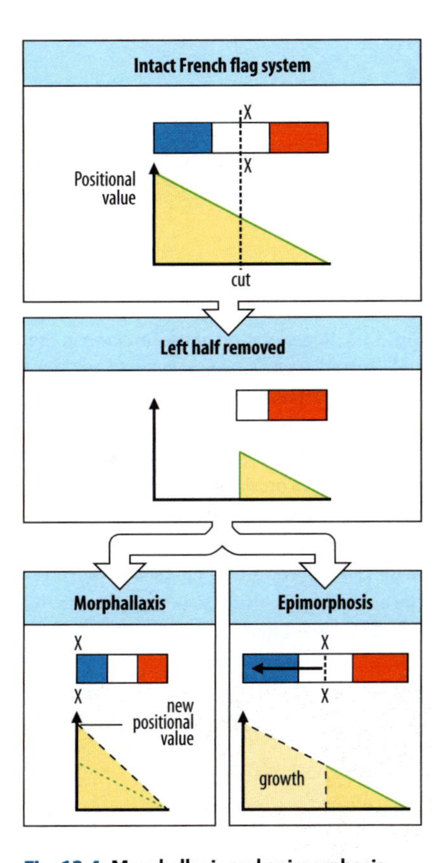

Fig. 13.4 Morphallaxis and epimorphosis. A pattern such as the French flag may be specified by a gradient in positional value (see Fig. 1.25). If the system is cut in half it can regenerate in one of two ways. In regeneration by morphallaxis, a new boundary is established at the cut and the positional values are changed throughout. In regeneration by epimorphosis, new positional values are linked to growth from the cut surface.

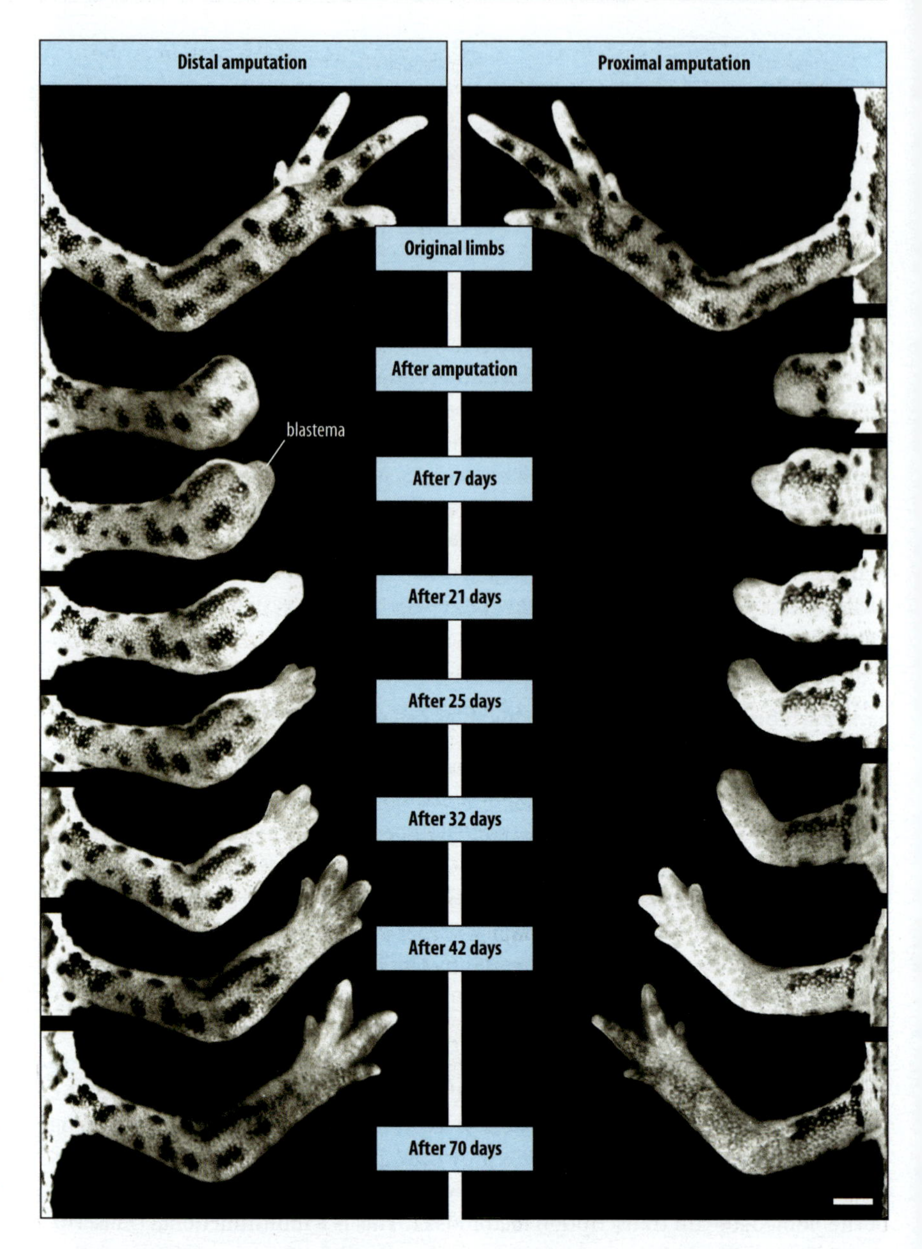

Fig. 13.5 Regeneration of the forelimb in the red-spotted newt *Notophthalmus viridescens*. The left panel shows the regeneration of a forelimb after amputation at a distal (mid-radius/ulna) site. The right panel shows regeneration after amputation at a proximal (mid-humerus) site. At the top, the limbs are shown before amputation. Successive photographs were taken at the times shown after amputation. Note that the blastema gives rise to structures distal to the cut. Scale bar = 1 mm.

These results reinforce the view that the early blastema is an environment that causes cells to dedifferentiate. The cells in this special environment then act as progenitor cells for the regenerating limb. This is further illustrated by the striking observation that if iris epithelium cells are grafted to different sites on the newt's body, they maintain their differentiated state. If they are grafted into a limb blastema, however, they transdifferentiate into lens cells. The blastema environment therefore contains the necessary cues for their dedifferentiation and re-differentiation as another cell type. Their re-differentiation as lens cells rather than a limb cell type reflects a restriction on developmental potential acquired during earlier development.

The ability of muscle cells to re-enter the cell cycle is a special feature of newt limb regeneration, since mature skeletal muscle cells normally never divide (Chapter 8). A general feature of vertebrate muscle differentiation is the withdrawal of the muscle precursor cells from the cell cycle after myoblast fusion to produce myotubes. This involves the dephosphorylation of the protein product of the retinoblastoma gene *Rb* (see Section 8.10) and cultured mouse muscle cells lacking the pRb protein can re-enter the cell cycle. The regenerating newt cells contain Rb protein but it is inactivated by phosphorylation, so the cells can re-enter the cell cycle and divide. The ability of newt muscle cells to enter the cell cycle is associated with the local activation of thrombin. As noted earlier, newt myotubes will dedifferentiate in culture in the prescence of thrombin. Thrombin may be a key factor that enables regeneration to occur in a variety of systems, as the regeneration of the newt lens from the dorsal margin of the iris also correlates with thrombin activity in that region. However, thrombin is not a universal requirement for dedifferentiation as mouse myotubes in culture are refractory to its action.

The regenerating axolotl tail provides an even more dramatic example of transdifferentiation. In this case, muscle and cartilage cells are produced by the transdifferentiation of glial cells (an ectodermal derivative) that migrate from the spinal cord. This is a rare example of transdifferentiation occurring from an ectodermal to a mesodermal cell type. In most cases, transdifferentiation occurs between related cell types (see Section 8.15).

Growth of the blastema is dependent both on its nerve supply (Fig. 13.6) and on the overlying wound epidermis, which may play a role similar to the apical ectodermal ridge in limb development (see Section 9.1). In limbs in which the nerves have been cut before amputation, a blastema forms but fails to grow. The nerves have no influence on the character or pattern of the regenerated structure; it is the amount of neural innervation, not the type of nerve that matters. Nerve cells, therefore, seem to be providing some essential growth factor. Members of the neuregulin family of growth factors are likely candidates and their local application can result in regeneration of denervated limbs.

An interesting phenomenon, as yet unexplained, is that if embryonic limbs are denervated very early in their development and so have never been exposed to the influence of nerves, they can regenerate in the complete absence of any nerve supply

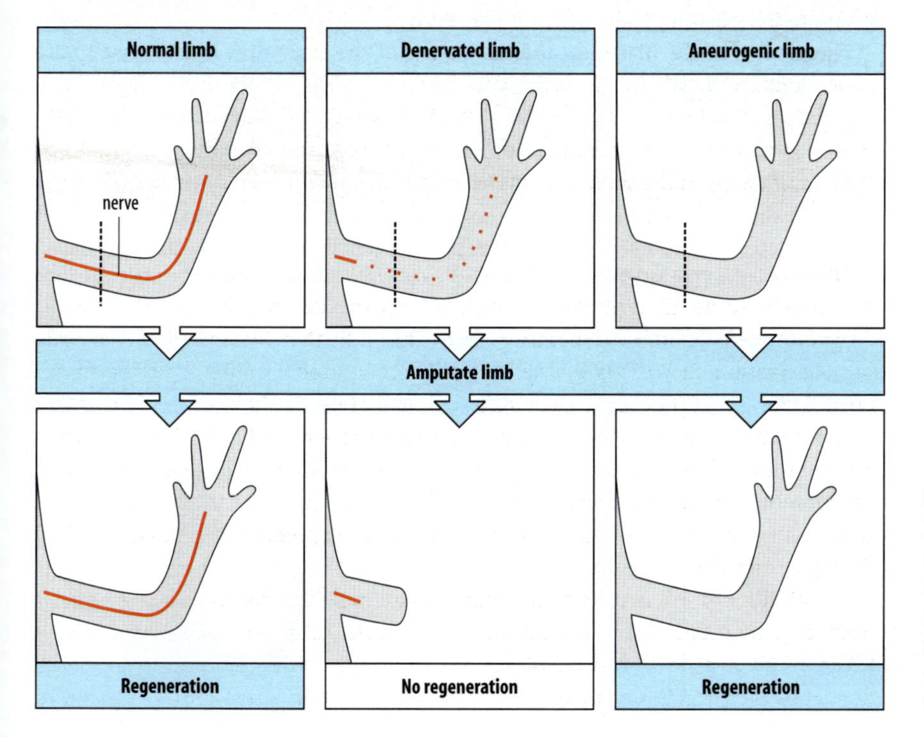

Fig. 13.6 Innervation and limb regeneration. Normal limbs require a nerve supply to regenerate (left panels). Limbs denervated prior to amputation will not regenerate (center panels). However, limbs that have never been innervated, because the nerve was removed during development, can regenerate normally in the absence of nerves (right panels).

(see Fig. 13.6, right panel). If such an aneurogenic limb is innervated, it rapidly becomes dependent on nerves for regeneration. This evidence suggests that the dependence on the nerve is imposed on the limb after the ingrowth of the nerve. Moreover, if a major peripheral nerve such as the sciatic nerve is cut and the branch inserted into a wound on the limb or an adjacent flank surface, a supernumerary limb develops.

Although mammals cannot regenerate whole limbs, many mammals, including young children, can regenerate the tips of their digits. In mice and children, the level from which digits are able to regenerate is limited to the base of the claw or nail, respectively. This probably reflects the presence of connective tissue cells under the nail organ that express *Msx1*, a homeobox-containing gene that regulates BMP-4 expression and is associated with undifferentiated mesenchymal cells. In mammalian limb development, the *Msx1* gene is expressed in the progress zone of the embryonic limb bud and in mice it continues to be expressed in the tips of the digits even after birth. The region in which digit regeneration in mice can occur corresponds with that of expression of *Msx1*.

13.2 The limb blastema gives rise to structures with positional values distal to the site of amputation

There is no evidence that limb regeneration is based on the same mechanisms that give rise to the limb in embryonic development, though there must be some relationship. Regeneration always proceeds in a direction distal to the cut surface, allowing replacement of the lost part of the limb. If the hand is amputated at the wrist, only the carpals and digits are regenerated, whereas if amputation is through the middle of the humerus, everything distal to the cut (including the distal humerus) is regenerated. Positional value along the axis is therefore of great importance. The blastema has considerable morphogenetic autonomy. If it is transplanted to a neutral location that permits growth, such as the dorsal crest of a newt larva or even the anterior chamber of the eye, it gives rise to a regenerate appropriate to the position from which it was taken.

The growth of the blastema and the nature of the structures it gives rise to are dependent on the site of the amputation and not on the nature of the more proximal tissues. The limb is not, however, simply 'trying' to replace missing parts. This was shown in a classic experiment in which the distal end of a newt limb that had been amputated at the wrist was inserted into the belly of the same animal, so as to establish a blood supply to it. The limb was then cut mid-humerus. Both surfaces regenerated distally, even though the part attached to the belly already had a radius and ulna (Fig. 13.7).

Fig. 13.7 Limb regeneration is always in the distal direction. The distal end of a limb is amputated and the limb inserted into the belly. Once vascular connections are established, a cut is made through the humerus. Both cut surfaces regenerate the same distal structures even though, in the case of one of the regenerating limbs, distal structures are already present.

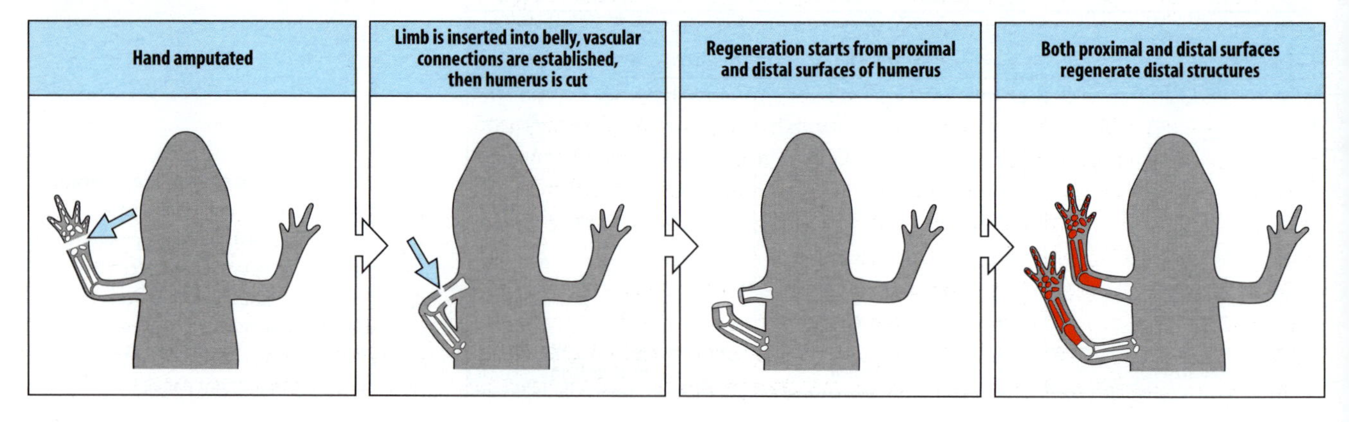

The blastema is much larger than the embryonic limb bud—it can be ten times larger in terms of cell numbers. This size makes it most unlikely that there are signals that diffuse right across the blastema. Instead, regeneration can best be understood in terms of an adult limb having a set of positional values along its proximo-distal axis, which are set up during embryonic development (see Section 9.6). The regenerating limb in some way reads the positional value at the site of the amputation and then regenerates all positional values distal to it. Epimorphic regeneration involves the retention of embryonic processes, like the ability to specify new positional values.

The ability of cells to recognize a discontinuity in positional values is illustrated by grafting a distal blastema to a proximal stump. In this experiment, the forelimb's stump and blastema have different positional values, corresponding to shoulder and wrist, respectively. The result is a normal limb in which structures between the shoulder and wrist have been generated by **intercalary growth**, predominantly from the proximal stump, while the cells from the wrist blastema mostly give rise to the hand (Fig. 13.8).

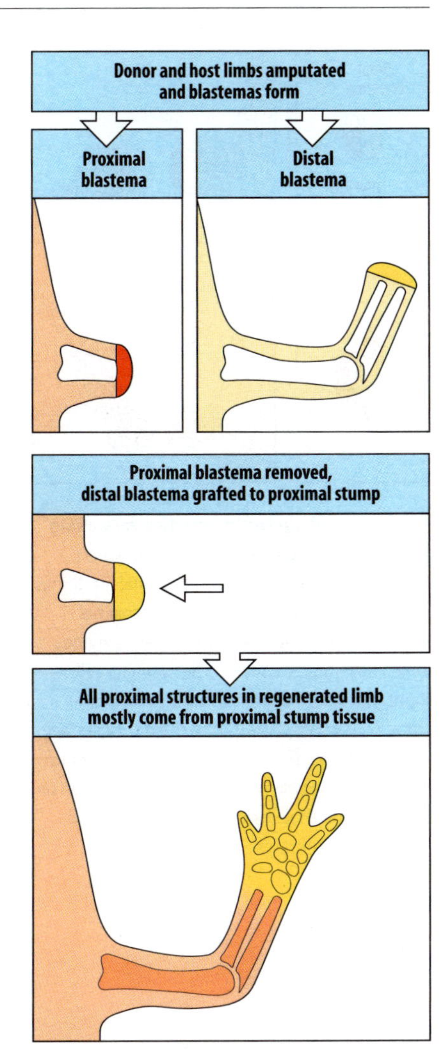

Fig. 13.8 Proximo-distal intercalation in limb regeneration. A distal blastema grafted to a proximal stump results in intercalation of all the structures proximal to the distal blastema. Almost all of the intercalated region comes from the proximal stump.

A fundamental question in any discussion of pattern formation is the molecular basis of the proposed positional information. A major advance in this system has been the identification of a cell-surface protein that is expressed in a graded manner along the proximo-distal axis of the newt blastema. Proximo-distal identity in the blastema can be altered by retinoic acid, as discussed in more detail in the next section, and this property was used to compare gene expression in blastemas whose proximo-distal identity had been altered. A cell-surface protein called Prod1, the newt equivalent of the mammalian cell-surface protein CD59, was identified that has an almost twofold difference in the level of gene expression between 'proximal' and 'distal' blastemas. This protein might therefore be used by the cells to determine positional identity.

The gradient in Prod1 fits very well with the observation that cell-surface properties are involved in regeneration. When mesenchyme from two blastemas from different proximo-distal sites are confronted in culture, the more proximal mesenchyme engulfs the distal (Fig. 13.9, left panels), whereas mesenchyme from two blastemas at similar sites maintains a stable boundary. This behavior is suggestive of a graded difference in cell adhesiveness along the axis, with adhesiveness being highest distally. The cells in the distal explant remain more tightly bound to each other than do the cells from the proximal blastema, which thus spread to a greater extent. This difference in adhesiveness is also suggested by the behavior of a distal blastema when grafted to the dorsal surface of a proximal blastema in such a way that their mesenchymal cells are in contact. Under these conditions, the distal blastema moves during limb regeneration to end up at the site from which it originated (see Fig. 13.9, right panels). This suggests that its cells adhere more strongly to the regenerated wrist region than they do to the proximal region to which it was transplanted. Transplantation of a shoulder-level blastema to a shoulder stump does not mobilize the stump tissue, but leads to a normal distal outgrowth from the shoulder blastema. These experiments again suggest that proximo-distal positional values in urodele limb regeneration are encoded as a graded property, probably in part at the cell surface, and that cell behavior relevant to axial specification—growth, movement, and adhesion—is a function of the expression of this property, relative to neighboring cells.

Maintaining the continuity of positional values by intercalation is a fundamental property of regenerating epimorphic systems and we consider it further in relation to the cockroach leg, below. Even normal regeneration by outgrowth from a blastema could be considered the result of intercalation between the cells at the level of

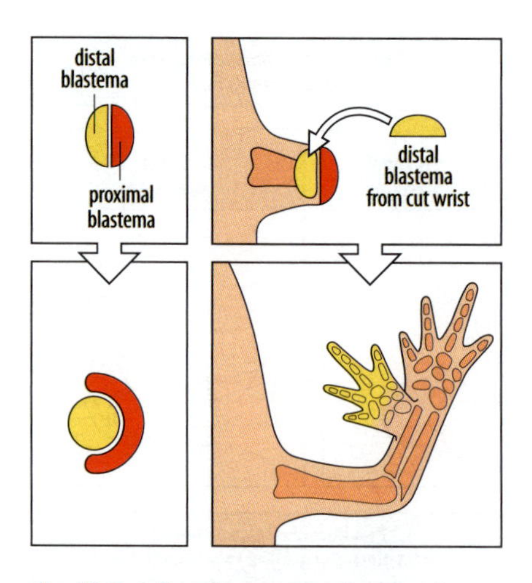

Fig. 13.9 Cell-surface properties vary along the proximo-distal axis. Left panels: when mesenchyme from distal and proximal blastemas is placed in contact in culture, the proximal mesenchyme engulfs the distal mesenchyme, which has greater adhesion between its cells. Right panels: if a distal blastema (in this case from a cut wrist) is grafted to the dorsal surface of a more proximal blastema, the regenerating wrist blastema will move distally to a position on the host limb that corresponds to its original level and regenerates a hand.

amputation and those with the most distal positional values, as specified by the wound epidermis. It is not clear to what extent blastemal cells inherit a particular positional value, for example, from their differentiated precursors, and to what extent they are subject to signals that induce the appropriate expression of positional value.

The precise relationship between Hox gene expression and positional identity is not understood, either for limb embryonic development or regeneration. During limb development in the embryo, the Hoxa genes are expressed with temporal and spatial co-linearity along the proximo-distal axis, as we saw in Chapter 9. In the regenerating axolotl limb, however, two Hoxa genes from the 3′ and 5′ ends of the complex are expressed in the stump cells at the same time—24 to 48 hours after amputation. This suggests that the most distal region of the blastema is specified first and that regeneration involves intercalation of positional values between this distal region and the stump. Grafting experiments also show that distal identity is established early, and by 4 days the blastema is already divided into distinct zones that will give rise to the proximo-distal regions of the limb.

13.3 Retinoic acid can change proximo-distal positional values in regenerating limbs

We have already seen that retinoic acid is present in developing vertebrate limbs and how experimental treatment with retinoic acid can alter positional values in the developing chick limb (see Section 9.5). It also has striking effects on regenerating amphibian limbs.

Exposing a regenerating limb to retinoic acid results in the blastema becoming proximalized; that is, the limb regenerates as if it had originally been amputated at a more proximal site. For example, if a limb is amputated through the radius and ulna, treatment with retinoic acid will result not only in the regeneration of the elements distal to the cut, but also in the production of an extra, complete radius and ulna. The effect of retinoic acid is dose dependent, and with a high dose it is possible to regenerate a whole extra limb, including part of the shoulder girdle, on a limb from which only the hand has been amputated (Fig. 13.10). Retinoic acid can therefore alter the proximo-distal positional value of the blastema, making it more proximal, probably by increasing the concentration of Prod1 (see Section 13.2). Overexpression of Prod1 in distal blastema cells causes them to translocate to more proximal positions. Retinoic acid can also, under some experimental conditions, shift positional values along the antero-posterior axis in a posterior direction. For the proximo-distal axis, retinoic acid shifts positional values in the blastema in a proximal direction through the activation of *meis* homeobox genes, which are involved in specifying proximal identity and are repressed in distal regions.

Fig. 13.10 Retinoic acid can proximalize positional values. A forelimb amputated at the level of the hand, as indicated by the dotted line, and then treated with retinoic acid, regenerates structures corresponding to a cut at the proximal end of the humerus. Scale bar = 1 mm.

In untreated regenerating limbs, endogenous retinoic acid is present in a distinct pattern, although there is no direct evidence that it is involved in regeneration. There is an antero-posterior gradient of retinoic acid in the blastema, and there is also a higher concentration in distal blastemas than in proximal blastemas, suggesting a proximo-distal gradient as well. The wound epidermis is a strong source of retinoic acid.

Retinoic acid is known to act through a variety of receptors. There are a number of different types of receptors in the limb but only one of them (δ_2) is involved in changes in positional values. By constructing a chimeric receptor from a δ_2 retinoic acid receptor and a thyroxine receptor, the retinoic acid receptor can be activated selectively by thyroxine, and the effects of its activation studied experimentally (Fig. 13.11). In these experiments, cells in the distal blastema are transfected with the chimeric receptor, which is then grafted to a proximal stump and treated with thyroxine. The transfected cells behave as if they have been treated with retinoic acid. The result is a movement of the transfected blastemal cells to more proximal regions in the intercalating regenerate. This shows that activation of the retinoic acid pathway can proximalize the positional values of the cells, which then respond by translocation to more proximal sites. Prod1 is a very strong candidate for this change as it too can cause cells to move proximally when its concentration is increased.

Another remarkable effect of retinoic acid is its ability to bring about a homeotic transformation of tails into limbs in tadpoles of the frog *Rana temporaria*. If the tail of a tadpole is removed it will regenerate. Treatment of regenerating tails with retinoic acid at the same time as hindlimbs are developing results in the appearance of additional hindlimbs in place of a regenerated tail (Fig. 13.12). We have, as yet, no satisfactory explanation for this result, but it has been speculated that the retinoic acid alters the antero-posterior positional value of the regenerating tail blastema to that of the site along the antero-posterior axis where hindlimbs would normally develop.

13.4 Insect limbs intercalate positional values by both proximo-distal and circumferential growth

The legs of some insects such as the cockroach can regenerate. The structure of insect legs is different from those of vertebrates as the key structural feature is the

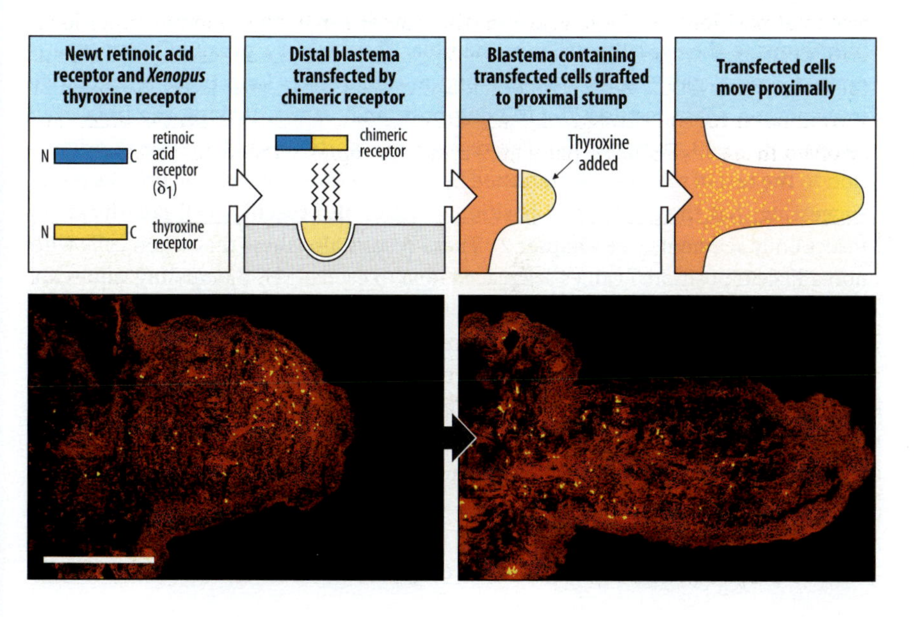

Fig. 13.11 Retinoic acid proximalizes the positional value of individual cells. Some of the cells of a newt distal blastema are transfected by a chimeric receptor, through which retinoic acid receptor function can be activated by thyroxine. This blastema is grafted to a proximal stump and treated with thyroxine. During intercalary growth, the transfected cells, which have been labeled, move proximally because their positional values have been proximalized by the activation of the retinoic acid receptor. The photographs illustrate proximalization of the transfected cells. Scale bar = 0.5 mm.

Photographs from Pecorino, L.T., et al.: 1996.

Fig. 13.12 Retinoic acid can induce additional limbs in regenerating tail of a frog, *Rana temporaria*. Treatment of the regenerating tail with retinoic acid at the time when hindlimbs are developing results in the appearance of additional hindlimbs in place of a regenerated tail. Scale bar = 5 mm.

Photograph courtesy of M. Maden.

ectodermally derived external cuticle. Nevertheless, intercalation of missing positional values, as already described for the proximo-distal axis of the amphibian limb (see Section 13.2), seems to be a general property of regenerating epimorphic systems. When cells with disparate positional values are placed next to one another, intercalary growth occurs in order to regenerate the missing positional values. Intercalation is particularly clearly illustrated by limb regeneration in the cockroach. Intercalation also occurs in *Drosophila* leg and wing imaginal discs.

A cockroach leg is made up of a number of distinct segments, arranged along the proximo-distal axis in the order coxa, femur, tibia, tarsus. Each segment seems to contain a similar set of proximo-distal and circumferential positional values, and will intercalate the missing positional values. When a distally amputated tibia is grafted onto a host tibia that has been cut at a more proximal site, localized growth occurs at the junction between graft and host, and the missing central regions of the tibia are intercalated (Fig. 13.13, left panels). In contrast to amphibian regeneration, there is a predominant contribution from the distal piece. As in the amphibian, however, regeneration is a local phenomenon and the cells are indifferent to the overall pattern of the tibia. Thus, when a proximally cut tibia is grafted onto a more distal site, making an abnormally long tibia, regenerative intercalation again restores the missing positional values, making the tibia even longer (Fig. 13.13, right panels). The regenerated portion is in the reverse orientation to the rest of the limb, as indicated by the direction in which the bristles point, suggesting that the gradient in positional values also specifies cell polarity, as in insect body segments (see Chapter 2). These results also show that when cells with non-adjacent positional values are placed next to each other, the missing values are intercalated by growth to provide a set of continuous positional values.

A similar set of positional values is present in each segment of the limb. Thus, a mid-tibia amputation, when grafted to the mid-femur of a host, will heal without intercalation. But grafting a distally amputated femur onto a proximally amputated host tibia results in intercalation, largely femur in type. There must be other factors making each segment different, rather like the segments of the insect larva.

Intercalary regeneration also occurs in a circumferential direction. When a longitudinal strip of epidermis is removed from the leg of a cockroach, normally

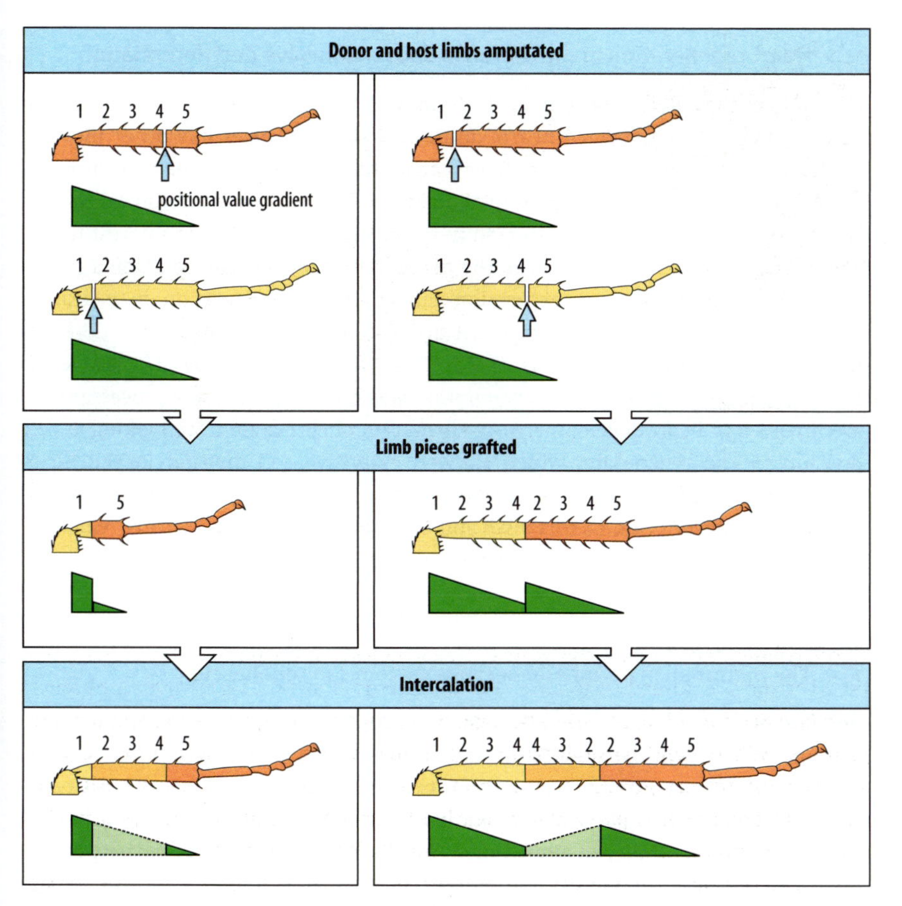

Fig. 13.13 Intercalation of positional values by growth in the regenerating cockroach leg. Left panels: when a distally amputated tibia (5) is grafted to a proximally amputated host (1), intercalation of the positional values 2–4 occurs, irrespective of the proximo-distal orientation of the grafts, and a normal tibia is regenerated. Right panels: when a proximally amputated tibia (1) is grafted to a distally amputated host (4), however, the regenerated tibia is longer than normal and the regenerated portion is in the reverse orientation to normal, as judged by the orientation of surface bristles. The reversed orientation of regeneration is due to the reversal in positional value gradient. The proposed gradient in positional value is shown under each figure.

After French, V., et al.: 1976.

non-adjacent cells come into contact with one another, and intercalation in a circumferential direction occurs after molting (Fig. 13.14). Cell division occurs preferentially at sites of mismatch around the circumference. One can treat positional values in the circumferential direction as a clock face, with values going continuously 12, 1, 2, 3 . . . 6 . . . 9 . . . 11. As in the proximo-distal axis, there is intercalation of the missing positional values.

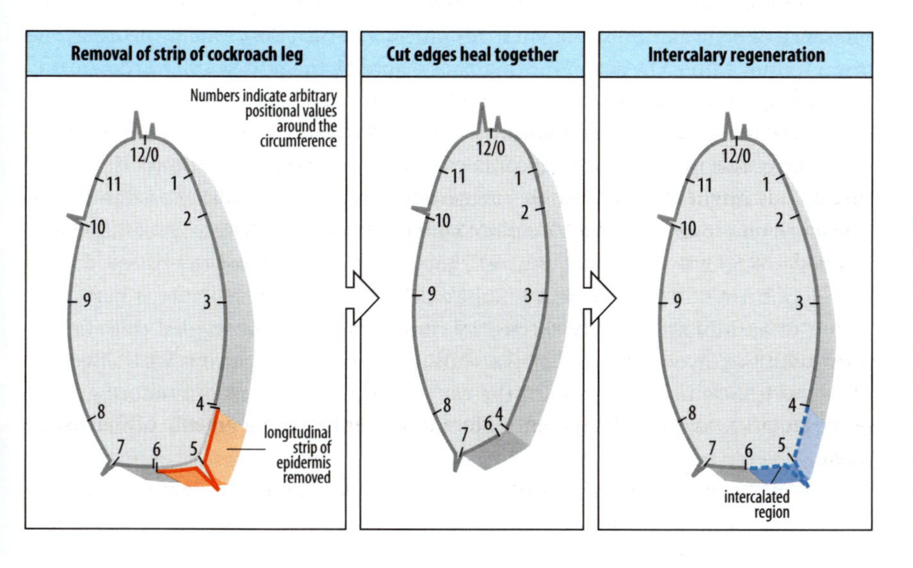

Fig. 13.14 Circumferential intercalation in the cockroach leg. The leg is seen in transverse section. When a piece of cockroach ventral epidermis is removed (left panel), the cut edges heal together (center panel). When the insect molts and the cuticle regrows, circumferential positional values are intercalated (right panel). The positional values are arranged around the circumference of the leg, rather like the hours on a clock face.

After French, V., et al.: 1976.

13.5 Heart regeneration in the zebrafish does not involve dedifferentiation

One tissue that does not renew itself in mammals is the cardiac muscle or myocardium. Thus, damaged hearts do not repair themselves like other muscles do. Some vertebrates, however, do have the ability to regenerate the myocardium. If a large portion of the ventricle of the adult zebrafish heart is removed, it will regrow. This phenomenon is also seen in newts, but is of particular interest in the zebrafish as in this model organism the regeneration process can be studied genetically. An analysis of gene expression during heart regeneration in zebrafish showed that genes were expressed that are also involved in the normal development of cardiac muscle, such as *nkx2.5*. This was expressed only at a low level, however, suggesting that heart regeneration was not using the same processes that are involved in heart development. Confirmation of this came from looking at cell division in the regenerate, which showed extensive cell division in the heart muscle cells that gave rise to the regenerated tissue. In addition the homeodomain-containing transcription factors *msxB* and *msxC* were upregulated during regeneration, whereas these genes are not expressed during the embryonic development of the heart.

13.6 The mammalian peripheral nervous system can regenerate

The nervous system can be divided into two distinct divisions, the central nervous system, which comprises the brain and spinal cord, and the peripheral nervous system, which contains the nerves, like motor nerves, that connect to different parts of the body. In spite of the dynamic nature of the maturing vertebrate nervous system and its ability to alter connectivity (see Chapter 10), the capacity of the mammalian central nervous system to replace lost connections after neurons have been destroyed is rather small. By contrast, the peripheral nervous system has considerable regenerative capability, even in adults. Even so, regeneration involves the regrowth of axons, and not the replacement of neurons themselves. If the cell body itself is destroyed, the neuron is not replaced.

Some axons of peripheral neurons in adult vertebrates, such as the motor and sensory axons between the spinal cord and the ends of the limbs, can be hundreds of centimeters long. When such axons are cut they can regenerate. A new growth cone forms at the cut surface and grows down the pathway of the original nerve trunk to make functional connections, leading to an almost complete recovery. The presence of Schwann cells promotes this process. In the case of motor neurons, the axon terminal finds the site of the original synapse on the muscle cell by recognizing the basal lamina that fills the synaptic cleft.

The adult central nervous system of birds and mammals is not able to regenerate correct axonal and dendritic connections after damage to neurons. This is not due to an intrinsic deficit in the neurons themselves but rather is due to the damaged environment, which does not support, and even actively prevents, regeneration. Although the remaining cell body produces outgrowths, these do not develop further. It seems that the failure of the axons to regenerate is partly the result of an inhibitory effect exerted by central nervous system glial cells, associated with the myelin of nerve fibers. Culturing immature neurons with oligodendrocytes (the source of myelin in the central nervous system) can induce growth cone collapse, and central nervous system astrocytes can be equally inhibitory in damaged grey matter. In contrast, the Schwann cells that are responsible for myelination in the peripheral nervous system can promote axon outgrowth, and when

Schwann cells are implanted into the central nervous system, axons migrate along them. Also, the cleaning up of debris in the damaged central nervous system is much slower than in the peripheral nervous system. It is not difficult to think of reasons for the general inhibition of new axonal sprouting in the mammalian central nervous system, with its highly complex architecture. If axons were allowed to sprout unchecked, this would be very likely to disrupt the elaborate pattern of neuronal connections, and destroy brain function.

Stem cells have great promise as sources for introducing new neurons and glia into damaged regions of the central nervous system. Embryonic stem cells (see Section 8.17), for example, can differentiate into neurons when transplanted into the mammalian brain. While such experiments are encouraging, it still needs to be shown that this approach can restore normal function after damage.

Summary

Urodele amphibians can regenerate amputated limbs and tails. Stump tissue at the site of amputation first dedifferentiates to form a blastema, which then grows and gives rise to a regenerated structure. The dedifferentiated cells of the blastema are the progenitors of the regenerate. Regeneration is usually dependent on the presence of nerves, but limbs that have never been innervated are capable of regeneration. Regeneration always gives rise to structures with positional values more distal than those at the site of amputation. When a blastema is grafted to a stump with different positional values, proximo-distal intercalation of the missing positional values occurs. Positional values may be related to a proximo-distal gradient in a cell-surface protein. Retinoic acid proximalizes the positional values of the cells of the blastema. Insect limbs can also regenerate, intercalation of positional values occurring in both proximo-distal and circumferential directions. Among other vertebrate organs, heart muscle does not regenerate in mammals but can do so in amphibians and fish. Central nervous system neurons do not regenerate in adults after damage, but peripheral nerves will regrow axons and reform connections, although if the cell body is damaged, the neuron cannot be replaced.

Summary: regeneration of an amphibian limb

amputation of a newt limb

⇓

local dedifferentiation of stump tissue to form a blastema

⇓

the blastema grows to form distal structures provided nerves are present

graft of distal blastema to proximal stump

intercalation of missing proximo-distal positional values from stump tissue growth

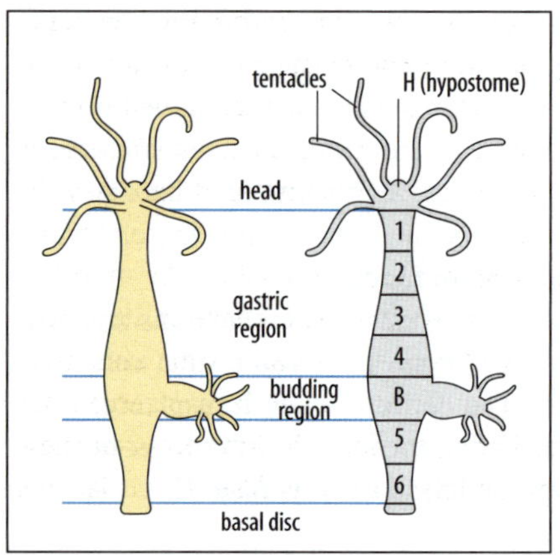

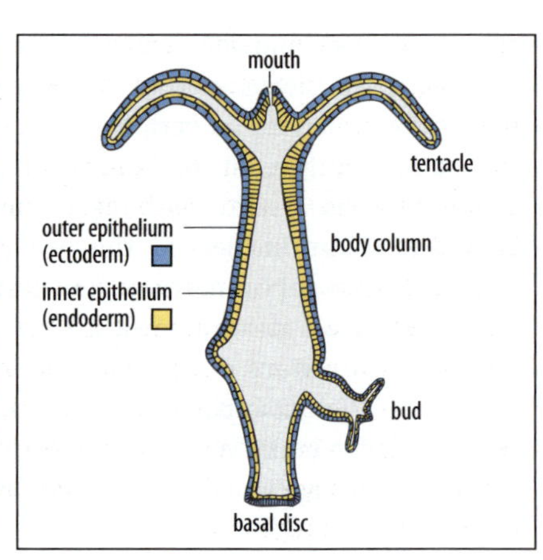

Fig. 13.15 ***Hydra.*** This freshwater coelenterate (photograph, left panel) has a head with tentacles and mouth at one end and a sticky foot at the other. It can reproduce by budding. For the purposes of grafting experiments, the body column is divided into a series of regions, as indicated in the center panel. The body wall is made up of two epithelial layers, corresponding to the ectoderm and endoderm found in other organisms featured in this book (right panel). Scale bar = 1 mm.

Photograph courtesy of W. Müller, from Müller, W.A.: 1989.

Regeneration in *Hydra*

Hydra is a freshwater coelenterate consisting of a hollow tubular body about 0.5 cm long, with a head region at one end (the distal end) and a basal region at the other (proximal) end, with which it can stick to surfaces (Fig. 13.15). The head consists of a small conical hypostome where the mouth opens, surrounded by a set of tentacles, which are used for catching the small animals on which *Hydra* feeds. Unlike most of the animals discussed so far in this book, which have three germ layers, *Hydra* has only two. The body wall is composed of an outer epithelium, which corresponds to the ectoderm, and an inner epithelium, which corresponds to the endoderm. These two layers are separated by a basement membrane. There are about 20 different cell types in *Hydra*, which include nerve cells, secretory cells, and nematocytes that are used to capture prey. Interstitial cells are stem cells that can give rise to nematocytes, nerve cells, and secretory cells, but are not required for regeneration. *Hydra* is of particular interest as it provides insights into mechanisms present in ancestral animals.

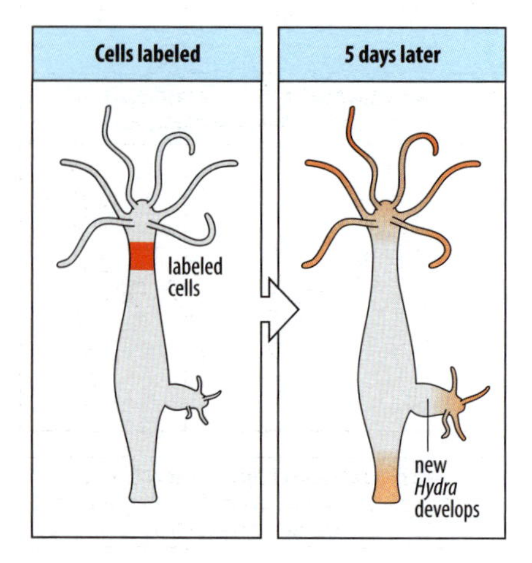

Fig. 13.16 Growth in *Hydra*. The cells in the body column of a *Hydra* are continually dividing and being displaced. If a group of cells in region 1 are labeled (left panel), 5 days later the labeled cells have been displaced to the tentacles and basal disc, where they are lost, and to the budding zone where they form a new *Hydra* (right panel).

13.7 *Hydra* grows continuously but regeneration does not require growth

Well-fed *Hydra* are in a dynamic state of continuous growth and pattern formation. The cells of both epithelial layers proliferate steadily and, as the tissues grow, cells are displaced along the body column toward the head or foot (Fig. 13.16). In order for an adult *Hydra* to maintain a constant size, excess cells must be continually lost. Cell loss occurs at the tips of the tentacles and at the basal disc of the foot bud. Most of the excess cell production is taken up by the asexual budding of new *Hydra* from the body column. Budding occurs about two-thirds of the way down the body column; the body wall evaginates by a morphogenetic change in cell shape, generated locally by the cells in the bud region, to form a new column that develops a head at the end and then detaches as a small new *Hydra*.

Continuous growth in *Hydra* means that cells are continually changing their relative positions and are forming new structures as they move up or down the body column. Moreover, new *Hydra* are generated asexually by budding from the body wall. There must, therefore, be mechanisms for repatterning cells during this dynamic process. It is these mechanisms that give *Hydra* its remarkable capacity for regeneration.

If the body column of a *Hydra* is cut transversely, the lower piece will regenerate a head but the upper piece will regenerate a foot. Thus, what structure the cells regenerate at a cut surface depends on their relative position within the regenerating piece. The cut surface nearest the original head end forms a head—this shows that *Hydra* has a well-defined overall polarity. This polarity is maintained even in small pieces of the body. This can be seen when a short piece is cut out of the body column; the distal end regenerates a head while the proximal end becomes the basal disc.

Regeneration in *Hydra* does not require growth and is thus said to be morphallactic. When a short fragment of the column regenerates, there is no initial increase in size and the regenerated animal will be a small *Hydra*. Only after feeding will the animal return to a normal size. The lack of a growth requirement for regeneration is shown in heavily irradiated *Hydra*; no cell divisions occur in these animals but they can still regenerate more or less normally. Also, *Hydra* lacking interstitial cells regenerate normally.

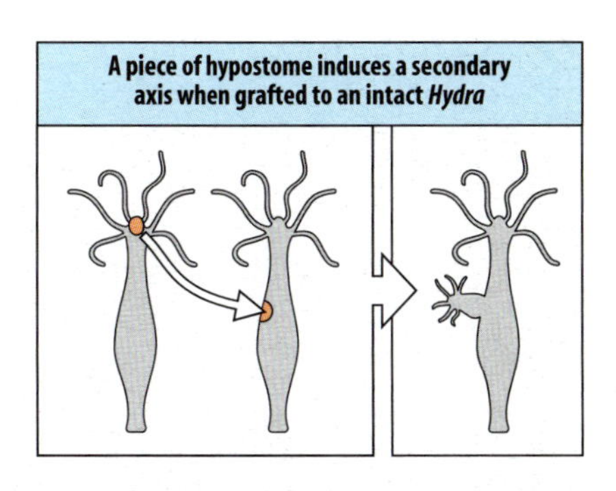

Fig. 13.17 The hypostome can induce a new head and body in *Hydra*. When an excised fragment of hypostome is grafted into the gastric region of another, intact *Hydra*, it can induce the formation of a complete new secondary axis with a head and tentacles.

13.8 The head region of *Hydra* acts both as an organizing region and as an inhibitor of inappropriate head formation

At the beginning of this century, it was shown that grafting a small fragment of the hypostome region of a *Hydra* into the gastric region of another *Hydra* induced a new head, complete with tentacles, and a body axis (Fig. 13.17). Similarly, transplantation of a fragment of the basal region induced a new body column with a basal disc at its end. *Hydra* therefore have two organizing regions, one at each end. The hypostome and the basal disc act as organizing regions (like the Spemann organizer in amphibians and the polarizing regions in vertebrate limb buds). These organizing regions at the ends give *Hydra* its overall polarity. The head region produces two signals down the body column, one that inhibits head formation and the other that specifies a gradient in positional value.

Grafting experiments also show that, as part of its organizing function, the hypostome produces an inhibitor of head formation whose effectiveness drops with distance from the head (Fig. 13.18). This inhibition normally prevents inappropriate head formation in the intact animal. When a body piece from just below the head (region 1 in Fig. 13.15) is grafted into the gastric region, it rarely induces a new head and is usually simply absorbed into the body. But if the head of the host is removed at the time of grafting, the graft can then induce a new axis and head. This suggests that the removal of the head results in the loss of some factor that is inhibiting head formation. This inhibitory effect falls off with distance from the head: when a region 1 is grafted to near the foot it can induce a head, even when the original head is still in place (see Fig. 13.18, bottom panel). These experiments suggest that the formation of extra heads is normally prevented in *Hydra* by a lateral inhibition mechanism (see Section 1.16) acting through a gradient of

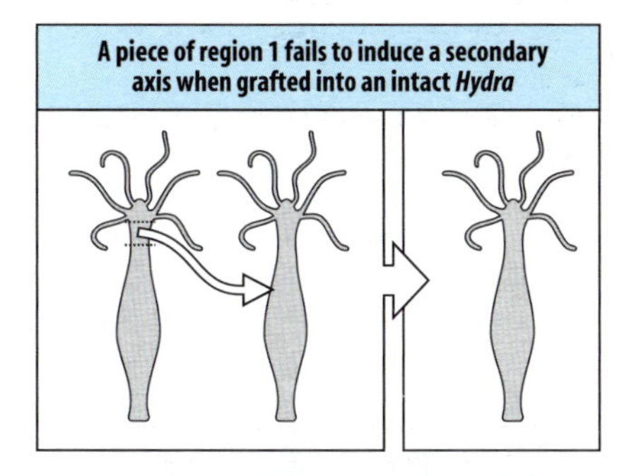

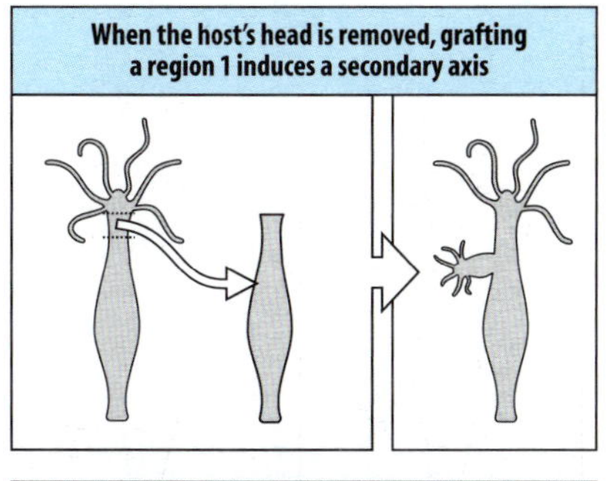

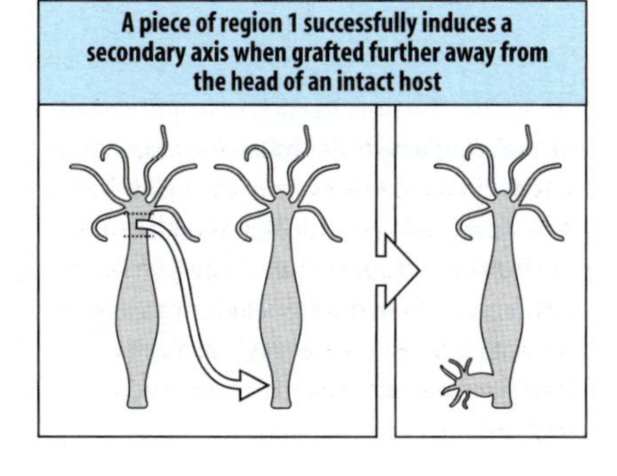

Fig. 13.18 The head region of *Hydra* produces an inhibitory signal that falls off with distance. Region 1 does not induce a head when grafted into the gastric region of an intact *Hydra* (top panel), indicating the presence of an inhibitory signal in the host. If the host's head is removed and a piece of region 1 is then grafted into the host, a secondary axis is induced (middle panel), indicating that the head region is the source of the inhibitory signal. Region 1 can induce a new axis in the foot region of an intact *Hydra* because the inhibitory signal grows weaker further away from the head (bottom panel).

inhibitory signal with its highest concentration at the head end. An opposing gradient inhibiting foot regeneration appears to be produced by the basal disc. These gradients are dynamic and when, for example, the head is removed, the concentration of the inhibitor falls.

13.9 Head regeneration in *Hydra* can be accounted for in terms of two gradients

One can account for the results of most of the regeneration experiments in *Hydra* in terms of two gradients set up along the body column by the head, one is a gradient in positional value, and the other is a gradient in head inhibitor. The gradient in positional value appears to determine both head-inducing ability and resistance to inhibition. A gradient in resistance to inhibition can be recognized by differences in the level of inhibitor required to suppress head formation by different regions of the body. This gradient in resistance decreases with distance from the head and is thought to represent a gradient of positional value with a high point at the head end. Thus, there is insufficient inhibitor near the foot to prevent a region 1 transplant forming a head when transplanted to this site, but sufficient inhibitor to prevent a region 5 from doing so.

The gradient in head-inducing ability also runs from a high point at the head end to a low point at the basal end. Evidence that this is determined by a gradient in positional value is provided by the difference in time required for different regions to acquire head-inducing properties after amputation. Region 1 from an intact *Hydra* will not induce an axis when transplanted into the gastric region of another *Hydra*. If, however, the head of the donor *Hydra* is amputated, the region 1 can induce a new axis if taken about 6 hours after head removal (Fig. 13.19). The further down the axis the amputation is made, the longer it takes for the remaining cells to acquire head-like inducing properties. A region 5 can take up to 30 hours.

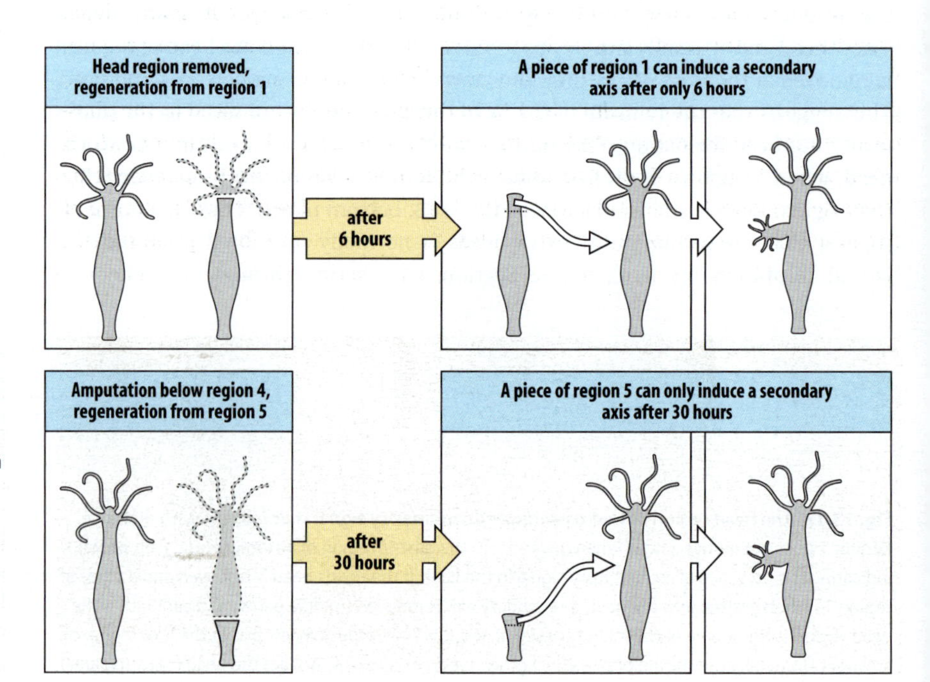

Fig. 13.19 The time needed to acquire head-like inducing properties following amputation increases with distance from the head. Top panels: a region 1 can induce a secondary axis if grafted into an intact *Hydra* 6 hours after head removal from the host. Bottom panels: if amputation is made lower down, it can take up to 30 hours for the region's cells to acquire head-like inducing properties.

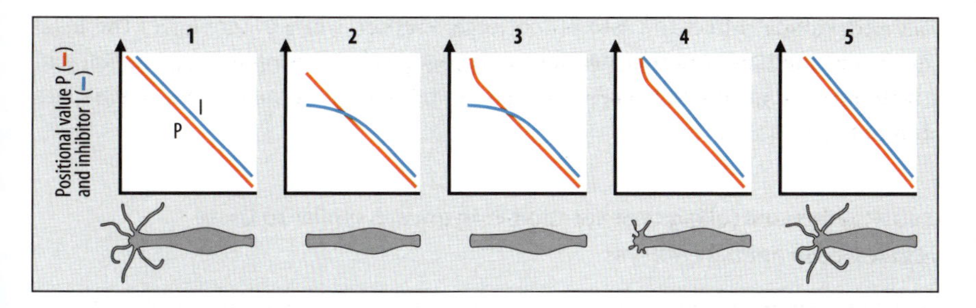

Fig. 13.20 A simplified model for head regeneration in *Hydra*. This model assumes that there are two gradients running from head to foot. One is a diffusible molecule (I) that inhibits head formation and is produced by the head. The other is a gradient of positional value (P) that is an intrinsic property of the cells. When the head is removed, the concentration of inhibitor falls at the cut surface (graph 2) until a threshold is reached at which the positional value increases to that of the head region (graph 3). This then re-establishes the inhibitor gradient (graph 4). The overall gradient in positional value takes a much longer time to return to normal (graph 5).

A simple model for these gradients assumes that the head inhibitor is a secreted factor, made by the head, that diffuses down the body column and is degraded at the basal end. The gradient in positional value is assumed to be an intrinsic property of cells. In this model, both gradients are linear—their values decrease at a constant rate with distance from the head. We further assume that, provided the level of inhibitor is greater than the threshold set by the positional value, head regeneration is inhibited. Removal of the head results in the concentration of inhibitor falling, as the inhibitor is degraded and cannot be replaced. The decrease in inhibitor concentration is greatest at the cut end, and when the inhibitor falls below the threshold concentration set by the local positional value, the positional value increases to that of the head end (Fig. 13.20). Thus, the first key step in this morphallactic regeneration, when the head region is removed, is the specification of a new head region at the cut surface.

When the positional value has increased to that of a normal head region, the cells start to make inhibitor and so prevent head formation in other body regions. The level of inhibitor will always first fall below the threshold for head inhibition where the positional value is highest, thus maintaining polarity. Once a new head has been specified and the inhibitory gradient re-established, the gradient in positional value also returns to normal, but this can take more than 24 hours. Morphallactic regeneration results in a smaller *Hydra* which, after feeding, will eventually grow back to a normal size.

Addition of diacylglycerol—an intracellular second messenger in many signal transduction systems—to the medium in which *Hydra* is growing, increases positional value throughout the body column and can cause ectopic head formation, producing a multiheaded *Hydra* (Fig. 13.21). Diacylglycerol is produced in the phosphatidylinositol signaling pathway; this pathway is affected by lithium, which when added to the medium can cause ectopic foot ends to form, apparently by lowering positional values throughout the body column. The molecular nature of the inhibitor remains to be discovered, but there is evidence for peptide signals.

Fig. 13.21 Diacylglycerol can induce formation of multiple heads in *Hydra*. Diacylglycerol is an intracellular second messenger in many signal transduction systems. When added to the medium in which a *Hydra* is growing, it increases the positional value of cells and can cause ectopic head formation, producing a multiheaded *Hydra*. Scale bar = 1 mm.

Photograph courtesy of W. Müller, from Müller, W.A.: 1989.

Several peptides affect the rate of foot regeneration, while others lower the head activation gradient. Another peptide is exposed in the developing head region and treatment of regenerating heads with this peptide increases the number of tentacles formed.

13.10 Genes controlling regeneration in *Hydra* are similar to those expressed in animal embryos

The Hox genes that control body pattern in many animals are also involved in patterning *Hydra*. Some Hox genes are expressed along the body axis of coelenterates in a regional pattern and may be involved in specifying head–foot positional values. A homeobox gene of the *paired* type related to the *Drosophila* gene *aristaless*, which is expressed in the head of the fly, is expressed in *Hydra* endoderm at the time of head determination. *Cnox-1* and *Cnox-2*, which are related to an anterior member of the Hox complex of *Drosophila*, are expressed early and late, respectively. The *Hydra* gene *budhead* is present in developing heads and is a homolog of the vertebrate *HNF3β* gene, which is expressed in the *Xenopus* organizer. A *Hydra* homolog of *goosecoid*, another organizer-specific gene, is expressed just above where tentacles will appear. When injected into an early *Xenopus* embryo, the *Hydra* version of *goosecoid* can induce a partial secondary axis. This suggests that the roles of those genes in tissue that can act as an organizer have been retained over many millions of years of evolution. In addition, the *Hydra* homolog of the T-box gene *Brachyury* is expressed in the hypostome and in buds is expressed at an early stage in the region of the future hypostomes. Its function is not known, but its presence has evolutionary implications as *Hydra* has no mesoderm, the usual tissue in which *Brachyury* is expressed, and suggests that the head corresponds to the proximal end, the blastopore, of other animals.

The canonical Wnt pathway leading to the activation of β-catenin (see Section 3.6) has an important role in *Hydra* development. *In situ* hybridization experiments revealed that expression of HyWnt, the *Hydra* Wnt homolog, is restricted to the apical tip of the body axis in adult polyps; this apical tissue is the *Hydra* head organizer (Fig. 13.22). During head regeneration, both *Hydra* β-catenin (Hyβ-Cat) and Wnt genes were expressed at the regenerating tip within an hour after

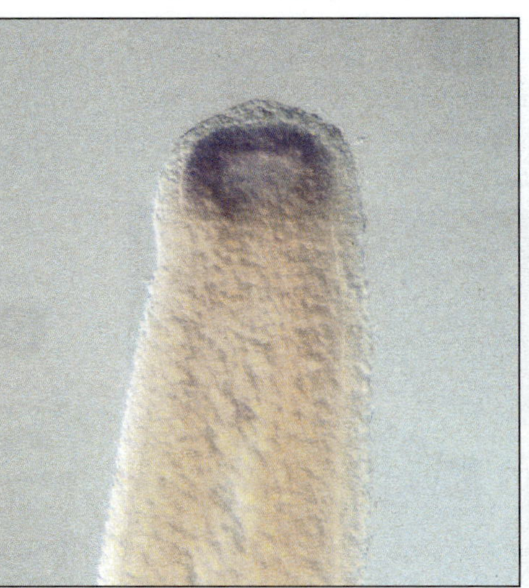
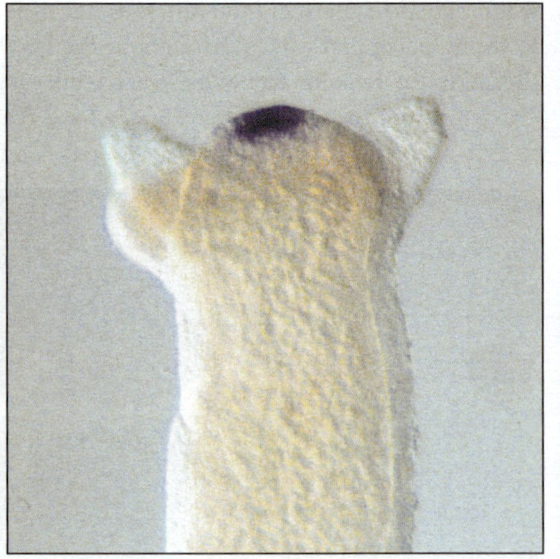

Fig. 13.22 Wnt expression in *Hydra* (left) and in a regenerating tip at 1 hour (center) and 48 hours (right).

Photographs courtesy of T.W. Holstein

head removal. During budding, a significant upregulation of Hyβ-Cat expression occurred in a ring-like domain in the prospective budding zone before tissue evagination started. This high expression level was maintained until bud detachment. A dramatic demonstration of the importance of the Wnt signaling pathway comes from blocking the activity of glycogen synthase kinase-3β, a component of the pathway (see Fig. 2.39). This caused a rise in the concentration of Hyβ-Cat along the body axis and all regions acquired the characteristics of the head organizer, as shown by grafting individual regions to another organism. These results demonstrate that Wnt signaling is involved in axis formation in *Hydra* and support the idea that it played a key part in the evolution of axial differentiation in early multicellular animals. Hyβ-Cat, shows a high degree of conservation, as injection of *Hyβ-Cat* mRNA into ventral *Xenopus* blastomeres at the eight-cell stage is able to induce a complete *Xenopus* secondary body axis. This contains the anterior-most structures, such as eyes and cement gland, and is indistinguishable from the secondary axis induced by *Xenopus* β-catenin.

Indeed, the importance of understanding patterning in *Hydra* is that it provides insight into an organizer and developmental gradients that arose early in the evolution of animal development. It is likely that the more complex body patterns of other animals evolved from a simple body plan like that of *Hydra*.

Summary

Hydra grows continually, losing cells from its ends and through the formation of buds. Two organizing regions, one at the head end and one at the basal end, pattern the body and maintain polarity. If transplanted to another site, a head region can induce a new body axis and head. If the body of an intact *Hydra* is severed in two, it regenerates by morphallaxis, which does not require new growth. Regeneration initially involves the respecification of cells at the cut end as 'head', leading to the establishment of an organizing region. The head region produces an inhibitor that prevents other regions forming a head. The concentration of inhibitor decreases with distance from the head. There is also a gradient in positional value specified by the head that determines the threshold at which head inhibition occurs. When the head is removed, the inhibitor level in the rest of the body falls, and a new head region develops where the positional value is highest, thus maintaining polarity. Genes similar to those acting in the development of higher animals are expressed.

Summary: regeneration in *Hydra*

Hydra grows continuously and loses cells from its ends and by budding

the head produces a gradient of inhibitor that prevents other heads forming;
the head also specifies a gradient in positional values

local reduction in concentration of inhibitor

local increase in positional value

formation of new head without new growth

SUMMARY TO CHAPTER 13

Regeneration is the ability of an adult organism to replace a lost part of its body. The capacity to regenerate varies greatly among different groups of organisms and even between different species within a group. Mammals have very limited powers of regeneration, whereas newts can regenerate limbs, jaws, and the lens of the eye. Regeneration of limbs involves dedifferentiation of cells and the formation of a blastema, which grows and forms the regenerate. There is evidence for a proximo-distal gradient in positional values. Intercalation of positional values can occur when normally non-adjacent values are placed next to each other in amphibian and insect limbs. The zebrafish heart can regenerate, as can the mammalian peripheral nervous system. Regeneration in the coelenterate *Hydra* does not require growth. The head end produces an inhibitory signal for head formation and a signal that specifies positional value along the body. When the head is removed, the inhibitor concentration falls and a new head forms. *Hydra* may provide a model for an evolutionarily early developmental system.

GENERAL FURTHER READING

Brockes, J.P., Kumar, A., Vellosoa, V.P.: **Regeneration as an evolutionary variable.** *J. Anat.* 2001, **199:** 3–11.

Brockes, J.P., Kumar, A.: **Principles of appendage regeneration in adult vertebrates and their implications for regenerative medicine.** *Science* 2005, **310:** 1919–1923.

Goss, R.J.: *Principles of Regeneration.* London and New York: Academic Press, 1969.

SECTION FURTHER READING

13.1 Vertebrate limb regeneration involves cell dedifferentiation and growth

Brockes, J.P.: **Amphibian limb regeneration: rebuilding a complex structure.** *Science* 1997, **276:** 81–87.

Brockes, J.P., Kumar, A.: **Plasticity and reprogramming of differentiated cells in a amphibian regeneration.** *Nat. Rev. Mol. Cell Biol.* 2002, **3:** 566–574.

Gardiner, D.M., Carlson, M.R., Roy, S.: **Towards a functional analysis of limb regeneration.** *Semin. Cell Dev. Biol.* 1999, **10:** 385–393.

Han, M., Yang, X., Farrington, J.E., Muneoka, K.: **Digit regeneration is regulated by *Msx1* and BMP4 in fetal mice.** *Development* 2003, **130:** 5123–5132.

Imokawa, Y., Simon, A., Brockes, J.P.: **A critical role for thrombin in vertebrate lens regeneration.** *Phil. Trans. R. Soc. Lond. B Biol. Sci.* 2004, **359:** 765–776.

Kumar, A., Velloso, C.P., Imokawa, Y., Brockes, J.P.: **The regenerative plasticity of isolated urodele myofibers and its dependence on MSX1.** *PLoS Biol.* 2004, **2:** E218.

Lee, H., Habas, R., Abate-Shen, C.: **Msx1 cooperates with histone H1b for inhibition of transcription and myogenesis.** *Science* 2004, **304:** 1675–1678.

Tanaka, E.M., Drechel, D.N., Brockes, J.P.: **Thrombin regulates S-phase re-entry by cultured newt myotubes.** *Curr. Biol.* 1999, **9:** 792–799.

Wang, L., Marchioni, M.A., Tassava, R.A.: **Cloning and neuronal expression of a type III newt neuregulin and rescue of denervated nerve-dependent newt limb blastemas by rhGGF2.** *J. Neurobiol.* 2000, **43:** 150–158.

13.2 The limb blastema gives rise to structures with positional values distal to the site of amputation

Da Silva, S., Gates, P.B., Brockes, J.P.: **New ortholog of CD59 is implicated in proximodistal identity during amphibian limb regeneration.** *Dev. Cell* 2002, **3:** 547–551.

Echeverri, K., Tanaka, E.M.: **Proximodistal patterning during limb regeneration.** *Dev. Biol.* 2005, **279:** 391–401.

Reginelli, A.D., Wang, Y.Q., Sassoon, D., Muneoka, K.: **Digit tip regeneration correlates with *Msx-1* (*Hox7*) expression in fetal and newborn mice.** *Devlopment* 1995, **121:** 1065–1076.

Torok, M.A., Gardiner, D.M., Shubin, N.H., Bryant, S.V.: **Expression of HoxD genes in developing and regenerating axolotl limbs.** *Dev. Biol.* 1998, **200:** 225–233.

13.3 Retinoic acid can change proximo-distal positional values in regenerating limbs

Bryant, S.V., Gardiner, D.M.: **Retinoic acid, local cell-cell interactions and pattern formation in vertebrate limbs.** *Dev. Biol.* 1992, **152:** 125.

Maden, M.: **The homeotic transformation of tails into limbs in *Rana temporaria* by retinoids.** *Dev. Biol.* 1993, **159:** 379–391.

Mercader, N., Tanaka, E.M., Torres, M.: **Proximodistal identity during limb regeneration is regulated by Meis homeodomain proteins.** *Development* 2005, **132:** 4131–4142.

Pecorino, L.T., Entwistle, A., Brockes, J.P.: **Activation of a single retinoic acid receptor isoform mediates proximo-distal respecification.** *Curr. Biol.* 1996, **6:** 563–569.

Scadding, S.R., Maden, M.: **Retinoic acid gradients during limb regeneration.** *Dev. Biol.* 1994, **162:** 608–617.

13.4 Insect limbs intercalate positional values by both proximo-distal and circumferential growth

Echeverri, K., Tanaka, E.M.: **Ectoderm to mesoderm lineage switching during axolotl tail regeneration.** *Science* 2002, **298:** 1993–1996.

French, V.: **Pattern regulation and regeneration.** *Phil. Trans. R. Soc. Lond. B Biol. Sci.* 1981, **295:** 601–617.

13.5 Heart regeneration in the zebrafish does not involve dedifferentiation

Raya, A., Consiglio, A., Kawakami, Y., Rodriguez-Esteban, C., Izpisúa-Belmonte, J.C.: **The zebrafish as a model of heart regeneration.** *Cloning Stem Cells* 2004, **6:** 345–351.

Ross, K.D., Wilson, L.G., Keating, M.T.: **Heart regeneration in zebrafish.** *Science* 2002, **298:** 2188–2190.

13.6 The mammalian peripheral nervous system can regenerate

Case, L.C., Tessier-Lavigne, M.: **Regeneration of the adult central nervous system.** *Curr. Biol.* 2005, **15:** R749–R753.

Johnson, E.O., Zoubos, A.B., Soucacos, P.N.: **Regeneration and repair of peripheral nerves.** *Injury* 2005, **36:** S24–S29.

13.7 *Hydra* grows continuously but regeneration does not require growth

Hicklin, J., Wolpert, L.: **Positional information and pattern regulation in *Hydra*: the effect of gamma-radiation.** *J. Embryol. Exp. Morph.* 1973, **30:** 741–752.

Otto, J.J., Campbell, R.D.: **Tissue economics of *Hydra*: regulation of cell cycle, animal size and development by controlled feeding rates.** *J. Cell Sci.* 1977, **28:** 117–132.

13.8 The head region of *Hydra* acts both as an organizing region and as an inhibitor of inappropriate head formation

Brown, M., Bode, H.R.: **Characterization of the head organizer in hydra.** *Development* 2002, **129:** 875–884.

Wolpert, L., Hornbruch, A., Clarke, M.R.B.: **Positional information and positional signaling in *Hydra*.** *Am. Zool.* 1974, **14:** 647–663.

13.9 Head regeneration in *Hydra* can be accounted for in terms of two gradients

Bosch, T.C.G., Fujisawa, T.: **Polyps, peptides and patterning.** *BioEssays* 2001, **23:** 420–427.

MacWilliams, W.: ***Hydra* transplantation phenomena and the mechanism of *Hydra* head regeneration. II. Properties of the head activation.** *Dev. Biol.* 1983, **96:** 239–257.

Müller, W.A.: **Pattern formation in the immortal *Hydra*.** *Trends Genet.* 1996, **12:** 91–96.

Takahashi, T., Hatta, M., Yum, S., Gee, L., Ohtani, M., Fujisawa, T., Bode, H.R.: **Hym-301, a novel peptide, regulates the number of tentacles formed in hydra.** *Development* 2005, **132:** 2225–2234.

13.10 Genes controlling regeneration in *Hydra* are similar to those expressed in animal embryos

Broun, M., Sokol, S., Bode, H.R.: ***Cngsc*, a homologue of *goosecoid*, participates in the patterning of the head, and is expressed in the organizer region of *Hydra*.** *Development* 1999, **126:** 5245–5254.

Broun, M., Gee, L., Reinhardt, B., Bode, H.R.: **Formation of the head organizer in Hydra involves the canonical Wnt pathway.** *Development* 2005, **132:** 2907–2916.

Galliot, B.: **Conserved and divergent genes in apex and axis development in cindiarians.** *Curr. Opin. Genet. Dev.* 2000, **19:** 629–637.

Hobmayer, B., Rentzsch, F., Kuhn, K., Happel, C.M., von Laue, C.C., Snyder, P., Rothbacher, U., Holstein, T.W.: **WNT signalling molecules act in axis formation in the diploblastic metazoan *Hydra*.** *Nature* 2000, **407:** 186–189.

Shenk, M.A., Gee, L., Steele R.E., Bode, H.R.: **Expression of *Cnox-2*, a Hom/Hox gene, is suppressed during head formation in *Hydra*.** *Dev. Biol.* 1993, **160:** 108–118.

Technau, U., Bode, H.R.: ***HyBra1*, a *Brachyury* homologue, acts during head formation in *Hydra*.** *Development* 1999, **126:** 999–1010.

14

Evolution and development

- The evolutionary modification of embryonic development
- Changes in the timing of developmental processes
- The evolution of development

The evolution of multicellular organisms is fundamentally linked to embryonic development, for it is through development that genetic changes are realized as changes in body form that can be passed on to future generations. This raises important questions as to how evolutionary and developmental processes are linked. How has embryonic development been modified during animal evolution? How important are changes in the timing of developmental processes? Finally, and most important, how did development itself evolve? In this chapter, we shall discuss these general questions using particular examples from animal evolution.

The disciplines of evolutionary biology and developmental biology have rather different histories and approaches. Whereas studies of development are based largely on model systems and experiments, evolutionary biology uses comparative methods to describe, study, and explain past and present patterns of diversity and change. All animals are thought to have evolved from a common ancestor that was a multicellular organism and that had, in turn, evolved from a unicellular organism. A simplified view of the evolution of animals and plants is shown in Figure 14.1. The history of evolution is a long one, taking place over thousands of millions of years, and it can only be accessed indirectly—through the study of fossils and by comparisons between living organisms. As Charles Darwin was the first to realize, evolution is the result of heritable changes in life forms and the selection of those that are best adapted to their environment.

Development is a fundamental process in evolution. The evolution of multicellular forms of life is the result of changes in embryonic development, and these in turn are entirely due to heritable changes in genes that control cell behavior in the embryo. It is also true, however, as the evolutionary biologist Theodosius Dobzhansky once said, that nothing in biology makes sense unless viewed in the light of evolution. Certainly, it would be very difficult to make sense of many aspects of development without an evolutionary perspective. For example, in our consideration of vertebrate development we have seen how, despite different modes of very early development, all vertebrate embryos develop through a rather similar stage, after which their development diverges again (see Fig. 3.1). This shared stage, which is the embryonic stage after neurulation and the formation of the somites, is a stage through which some distant ancestor of the vertebrates passed. It has persisted ever since, to become a fundamental characteristic of the

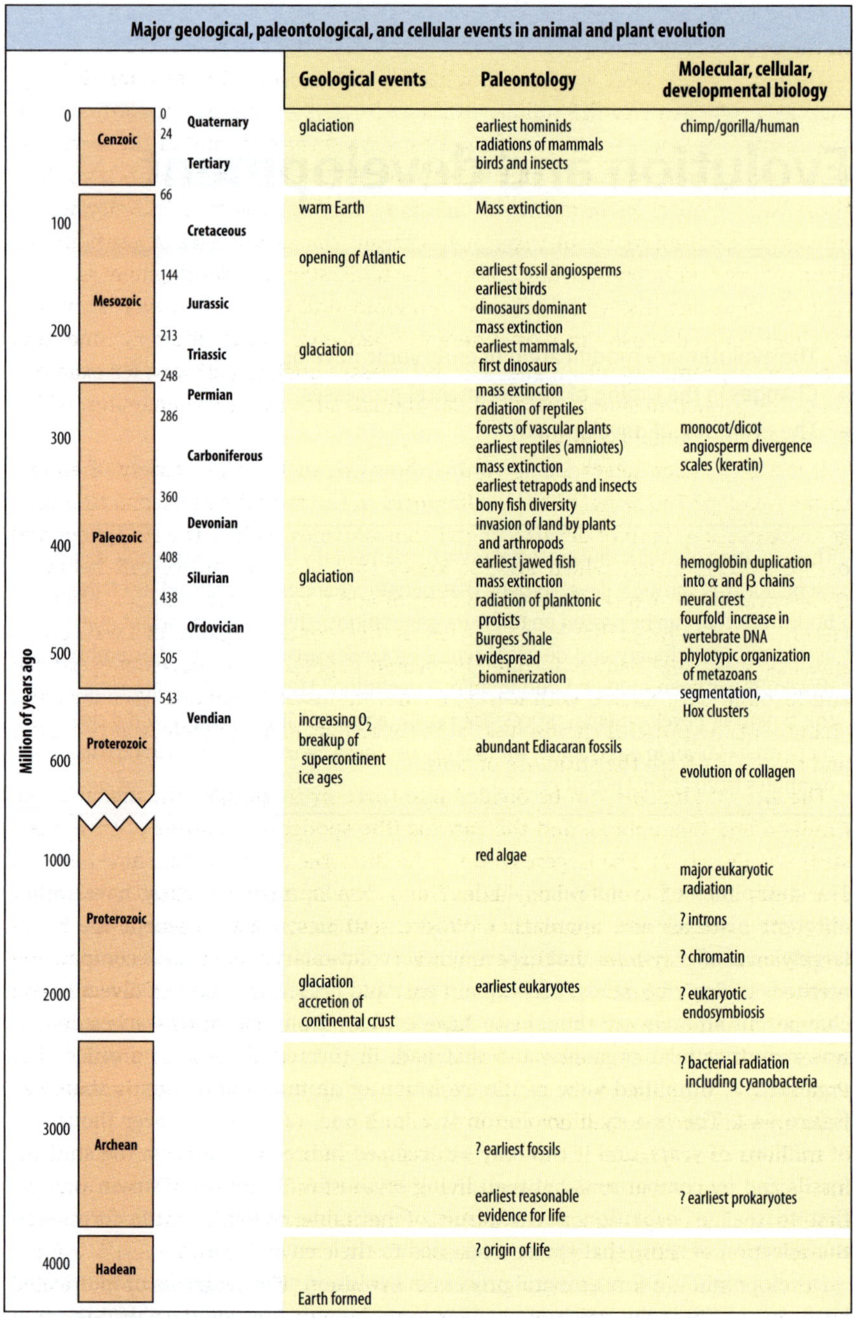

Major geological, paleontological, and cellular events in animal and plant evolution

Million of years ago	Era	Period	Geological events	Paleontology	Molecular, cellular, developmental biology
0–24	Cenzoic	Quaternary	glaciation	earliest hominids	chimp/gorilla/human
24–66		Tertiary		radiations of mammals birds and insects	
66–144	Mesozoic	Cretaceous	warm Earth opening of Atlantic	Mass extinction earliest fossil angiosperms	
144–213		Jurassic		earliest birds dinosaurs dominant mass extinction	
213–248		Triassic	glaciation	earliest mammals, first dinosaurs	
248–286	Paleozoic	Permian		mass extinction radiation of reptiles	
286–360		Carboniferous		forests of vascular plants earliest reptiles (amniotes) mass extinction	monocot/dicot angiosperm divergence scales (keratin)
360–408		Devonian		earliest tetrapods and insects bony fish diversity invasion of land by plants and arthropods	
408–438		Silurian	glaciation	earliest jawed fish mass extinction	hemoglobin duplication into α and β chains neural crest
438–505		Ordovician		radiation of planktonic protists Burgess Shale widespread biomineralization	fourfold increase in vertebrate DNA phylotypic organization of metazoans segmentation, Hox clusters
505–543					
543–600	Proterozoic	Vendian	increasing O_2 breakup of supercontinent ice ages	abundant Ediacaran fossils	evolution of collagen
1000–2000	Proterozoic		glaciation accretion of continental crust	red algae earliest eukaryotes	major eukaryotic radiation ? introns ? chromatin ? eukaryotic endosymbiosis
2500–3800	Archean			? earliest fossils earliest reasonable evidence for life	? bacterial radiation including cyanobacteria ? earliest prokaryotes
4000	Hadean		Earth formed	? origin of life	

Fig. 14.1 **Major geological, paleontological, and cellular events in the evolution of animals and plants.**

Modified from Gerhart, J. and Kirschner, M.: 1997.

development of all vertebrates, whereas the stages before and after this stage have evolved differently in different organisms.

Genetically based changes in development that generated more successful modes of reproduction or adult forms better adapted to their environment have been selected for during evolution. Changes in development occurring before neurulation are more often associated with changes in reproduction; those occurring after neurulation are more associated with the evolution of animal form. Genetic variability resulting from mutations, sexual reproduction, and genetic recombination is present in the populations of all the organisms we have looked at in this book, and provides new phenotypes upon which selection can act. Changes in the control regions of genes are particularly important in evolution.

Throughout this book we have emphasized the conservation of some developmental mechanisms at the cellular and molecular level among distantly related organisms. The widespread use of the Hox gene complexes and of the same few families of protein signaling molecules provide excellent examples of this. It is these basic similarities in molecular mechanisms that have made developmental biology so exciting in recent years; it has meant that discoveries of genes in one animal have had important implications for understanding development in other animals. It seems that when a useful developmental mechanism evolved, it was retained and redeployed in very different organisms, and at different times and places in the same organism. The Wnt (Wingless) signaling pathway, for example, is already present in simple multicellular animals such as the coelenterates, which arose early in animal evolution.

If the genes themselves are so similar, how can such a wide variety of animal forms develop? The answer lies in differences in the spatial pattern and timing of gene expression, both of developmental control genes such as the Hox genes and of their target genes, which are the genes that directly control cell behavior. Differences in the expression of both control and target genes among species are largely due to changes that have occurred in their control regions. In a way, evolution has been conservative; having 'discovered' a good way of developing pattern and form, it just tinkered with it to make new animals. There are, of course, also changes in the proteins themselves, both those that regulate biological processes and those that form the structure of cells.

The animal kingdom can be divided into three main groups—the Bilateria, the Cnidaria and Ctenophora, and the Parazoa (the sponges)—in terms of their basic structure (Fig. 14.2). The largest group is the Bilateria, animals that show bilateral symmetry across the main body axis. This group includes vertebrates and other chordates, arthropods, annelids, mollusks, and nematodes. These animals are **triploblasts**, as they have the three major germ layers—endoderm, mesoderm, and ectoderm. The Cnidaria and Ctenophora are **diploblasts**—they do not have bilateral symmetry and have only two germ layers, lacking mesoderm (there is a report, however, that the sea anemone *Nematostella* does have bilateral symmetry). The Parazoa are considered to be a quite separate group from the rest of the multicellular animals and we shall not consider them here.

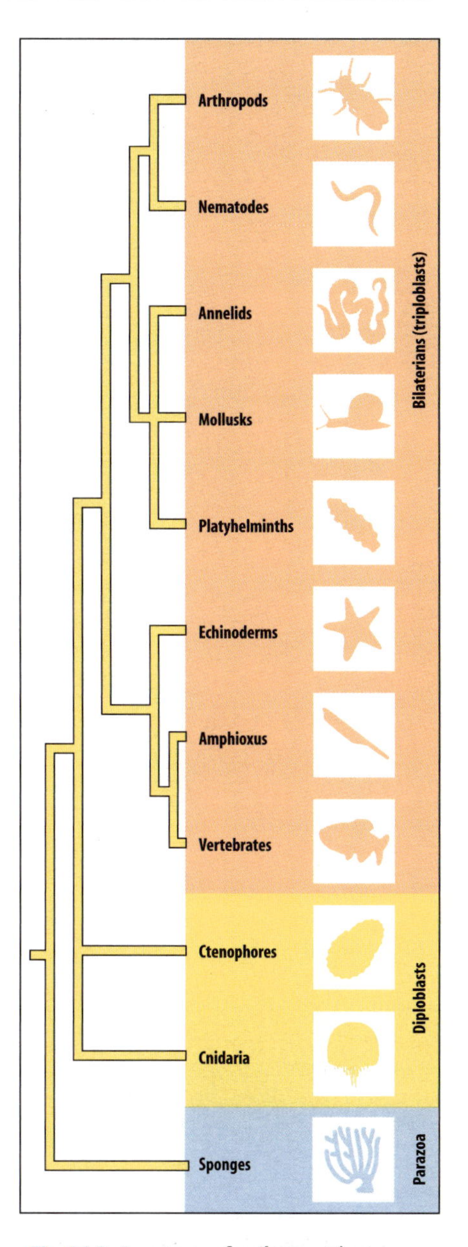

Fig. 14.2 A metazoan family tree. The tree shown here is based on ribosomal DNA and Hox gene sequences.

Adapted from Ferrier, D.E.K. and Holland, P.W.H.: 2001.

We have already looked at the early development of a wide variety of organisms and found some similarities, as well as a number of differences. In this chapter we mainly confine our attention to two phyla—the chordates (which include the vertebrates) and the arthropods (which include the insects and crustaceans). We focus on those differences that distinguish the members of a large group of related animals, such as the vertebrates or the insects, from each other. We will look at the relationship between the development of the individual organism (its **ontogeny**) and the evolutionary history of the species or group (the **phylogeny**): why, for example, do all vertebrate embryos pass through an apparently fish-like stage that has structures resembling gill slits? We then discuss the many variations that occur on the theme of a basic segmented body plan: what determines the different numbers and positions of paired appendages, such as legs and wings, in different groups of segmented organisms? We consider the timing of developmental events, and how variations in growth can have major effects on the shape and form of an organism. In all cases, we ultimately want to understand the changes in the developmental processes and the genes controlling them that have resulted in the extraordinary variety of multicellular animals. This is an exciting area of study in which many problems remain to be solved. Finally, we take a brief look at the evolution of development itself. But first we shall look at the importance of the unique interaction of a species with the environment—its **life history**.

14.1 The evolution of life histories has implications for development

Animals and plants have very diverse life histories: small birds breed in the spring following their birth, and continue to do so each year until their death; Pacific salmon breed in a suicidal burst at 3 years of age; and oak trees require 30 years of growth before producing acorns, which they eventually produce by the thousand. In order to understand the evolution of such life histories, evolutionary ecologists consider them in terms of probabilities of survival, rates of reproduction, and optimization of reproductive effort. These factors have important implications for the evolution of developmental strategies, particularly in relation to the rate of development. For example, a characteristic feature of many animal life histories is the presence of a free-living larval stage that is distinct from, and usually simpler in form, than the sexually mature adult, and which feeds in a different way. This strategy provides the developing individual with both a means of dispersal and a means of obtaining nutrition before it becomes adult, and also allows the organism to exploit different environments for feeding, and we discuss the evolution of larval forms later in the chapter. We will consider here two other issues in the evolution of life histories that impinge directly on development—selection for speed of development and selection for egg size.

Life histories help us to understand the evolution of long germ-band insects, such as *Drosophila*, which are of more recent origin than short germ-band insects, such as grasshoppers. Unlike *Drosophila*, short germ-band insects do not undergo metamorphosis and develop directly into small, immature adult-like forms. In contrast, *Drosophila* develops very rapidly into a feeding larva, taking only 24 hours to start feeding compared with the grasshopper's 5–6 days. It is easy to imagine conditions in which there would have been a selective advantage to insects whose larvae begin to feed as quickly as possible. It is also likely that embryos are more vulnerable than adults and so there would be selection for making the embryonic stage shorter. Thus it is likely that the complex developmental mechanisms of long-band insects—the system for setting up the complete antero-posterior axis in the egg (see Section 2.25)—evolved as a result of selection pressure for rapid development. In the case of fruit flies like *Drosophila*, this was perhaps related to the ability to feed on fast-disappearing fruit.

Egg size can also be best understood within the context of life histories. If we assume that the parent has limited energy resources to put into reproduction, the question is how should these resources best be invested in making gametes, particularly eggs; is it more advantageous to make lots of small eggs or a few large ones? In general, it seems that the larger the egg, and thus the larger the offspring at birth, the better are the chances of the offspring surviving. This would seem to suggest that in most circumstances an embryo needs to give rise to a hatchling as large as possible. Why then, do some species lay many small eggs capable of rapid development? One possible answer is that a parental investment in large eggs may reduce the parents' own chance of surviving and laying more eggs. This strategy may be especially successful in variable environmental conditions, where populations can suddenly crash. The rapid development of a larval stage enables early feeding and dispersal to new sites in such circumstances.

An interesting evolutionary change related to the environment is the loss of eyesight when animals come to live in perpetual darkness. Over the past 10,000 years, four different cavefish populations have independently evolved various degrees of eye degeneration. In one species that has been studied, small eye primordia are present but later degenerate. Changes in the expression of the secreted signaling protein Sonic hedgehog compared to sighted relatives is linked to this degeneration. Loss of eyesight may have resulted from mutations that were retained because they had no adaptive costs in the darkness.

The evolutionary modification of embryonic development

Comparisons of embryos of related species have suggested an important generalization about development: the more general characteristics of a group of animals (that is, those shared by all members of the group) usually appear earlier in their embryos than the more specialized ones, and arose earlier in evolution. In the vertebrates, a good example of a general characteristic would be the notochord, which is common to all vertebrates and is also found in other chordate embryos. The phylum Chordata comprises three subphyla—the vertebrates, the cephalochordates, such as amphioxus (Fig. 14.3), and the urochordates, such as the ascidians (see Chapter 5). They all have, at some embryonic stage, a notochord, flanked by muscle, and a dorsal neural tube. Paired appendages, such as limbs, which develop later, are special characters that develop only in the vertebrates, and differ in form among different vertebrates. All vertebrate embryos pass through a phylotypic stage, which then gives rise to the diverse forms of the different vertebrate classes. The phylotypic stage itself is somewhat variable, however, as can be seen in Figure 14.4, which shows that the number of somites can be very different, and that some organs such as limb buds are at different stages of development at this time. The development of the different vertebrate classes before the phylotypic stage is also highly divergent, because of their very different modes of reproduction; some developmental features that precede the phylotypic stage are evolutionarily highly advanced, such as the formation of a trophoblast and inner cell mass by mammals. This is an example of a special character that developed late in vertebrate evolution, and is related to the nutrition of the embryo through a placenta, rather than a yolky egg.

The reason that vertebrate embryos pass through a similar phylotypic stage may be related to that being the stage when the Hox genes that control pattern along

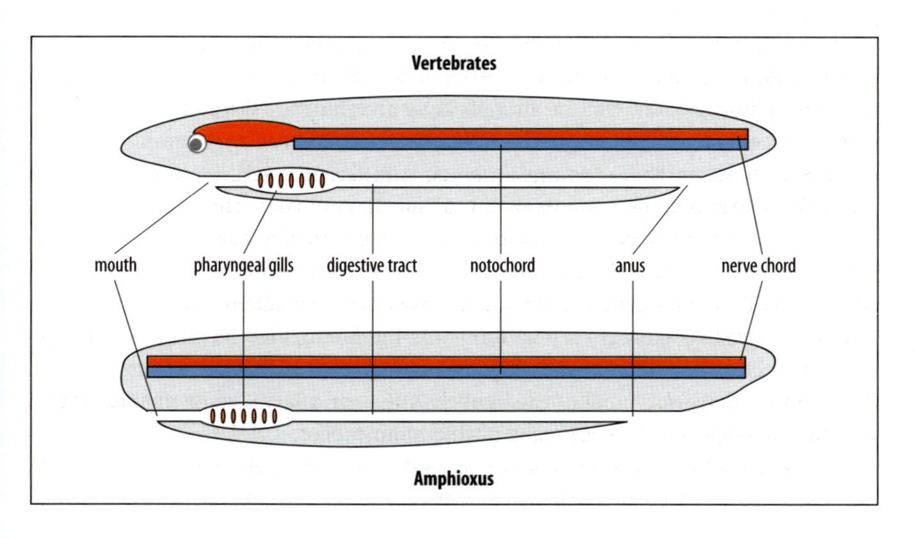

Fig. 14.3 The cephalochordate amphioxus compared to a hypothetical primitive vertebrate similar to a present-day lamprey. The overall construction of the organisms is very similar, with a dorsal nerve cord, a more ventral axial skeleton known as the notochord, and a ventral digestive tract. Both animals have gills in the pharyngeal region, structures that are designed to capture food or remove oxygen from the water. The notochord extends right to the anterior end of amphioxus, but the vertebrate has a prominent head at the anterior end, extending beyond the notochord. In more advanced vertebrates, the notochord is present only in the embryo, being replaced by the vertebrae.

Modified from Finnerty, J.R.: 2000.

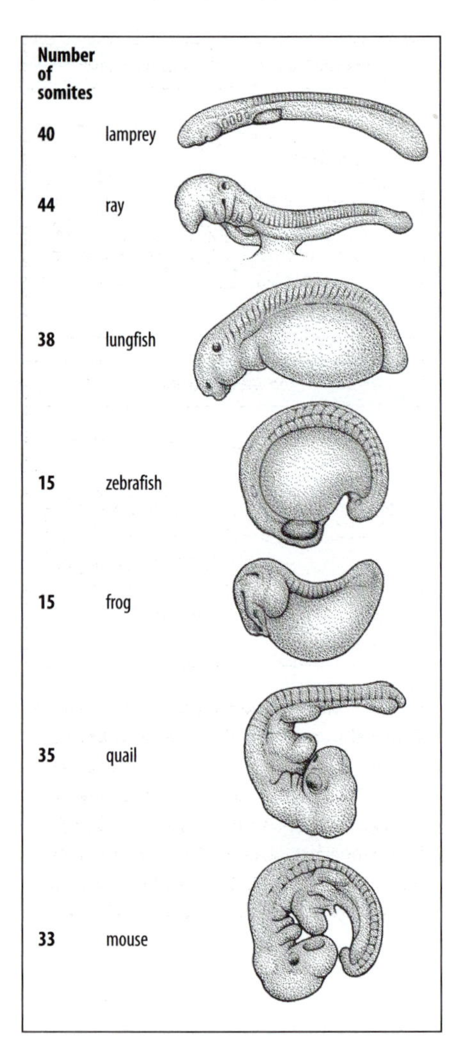

Fig. 14.4 Vertebrate embryos at the phylotypic (tailbud) stage show wide variation in somite number. Not to scale.

Modified from Richardson, M.K., et al.: 1997.

the main body axis are expressed. Also, it is the stage at which structures common to all vertebrates are developed, such as somites, notochord, and neural tube.

We will now consider the modifications that have occurred to a variety of embryonic structures during evolution, including the basic body plan and the limbs. We start by looking at the branchial arches in vertebrates.

14.2 Embryonic structures have acquired new functions during evolution

If two groups of animals that differ greatly in their adult structure and habits (such as fishes and mammals) pass through a very similar embryonic stage, this could indicate that they are descended from a common ancestor and, in evolutionary terms, are closely related. Thus, an embryo's development reflects the evolutionary history of its ancestors. Structures found at a particular embryonic stage have become modified during evolution into different forms in the different groups. In vertebrates, one good example of this is the evolution of limbs from the embryonic fin-like structures of a fish ancestor, which is discussed in the next section.

Division of the body into segments, which then diverge from each other in structure and function, is a common feature in the evolution of both vertebrates and arthropods. In the vertebrates one example is the branchial arches and clefts that are present in all vertebrate embryos (see Figs 3.8 and 4.28), including humans. These segmented structures are not the relics of the gill arches and gill slits of an adult fish-like ancestor, but of structures that would have been present in the embryo of the fish-like ancestor as developmental precursors to gill slits and arches. During evolution, the branchial arches have given rise both to the gills of the primitive jawless fishes and, in a later modification, to jaws (Fig. 14.5). In the lamprey, a present-day example of a jawless fish, jaw formation is inhibited by Hox genes, and it may have been the loss of Hox gene expression from the first arch and the associated neural crest that facilitated the evolution of jaws, which develop entirely from neural crest cells that have migrated into the first arch. With time the arches became further modified, and in mammals they now give rise to various structures in the face and neck (Fig. 14.6), many of which are derived from the neural crest cells that migrate into the branchial arches early in development (see Fig. 4.31). The cleft between the first and second branchial arches provides the opening for the eustachian tube, and endodermal cells in the clefts give rise to a variety of glands, such as the thyroid and thymus.

Evolution rarely, if ever, generates a completely novel structure out of the blue. New anatomical features usually arise from modification of an existing structure. One can therefore think of much of evolution as a tinkering with existing structures, which gradually fashions something different. This is possible because many structures are modular: that is, animals have anatomically distinct parts that can evolve independently of one another. Vertebrae are modules, for example, and can evolve independently; so too are limbs. At a different level, signal transduction pathways could represent modules. It is often changes in *cis*-regulatory regions of genes that allow such modules to evolve along separate pathways.

A nice example of a modification of an existing structure is provided by the evolution of the mammalian middle ear. This is made up of three bones that transmit sound from the eardrum (the tympanic membrane) to the inner ear. In the reptilian ancestors of mammals, the joint between the skull and the lower jaw was between the quadrate bone of the skull and the articular bone of the lower jaw, which were also involved in transmitting sound (Fig. 14.7). During mammalian evolution, the lower jaw became just one bone, the dentary, with the articular bone no longer attached to the lower jaw. By changes in their development, the articular and the quadrate bones in mammals were modified into two bones, the malleus and incus, whose function was now to transmit sound from the tympanic membrane to the inner ear. The quadrate is evolutionarily and developmentally homologous with the dorsal bone of the first arch, and the stapes with that of the second. **Homology** here refers to a morphological or structural similarity due to a common ancestry. Structures are homologous when they share a common ancestry, even if they no longer serve the same function. Structures are also considered homologous if they share a similar developmental program; thus, the vertebrae are homologous even though they vary in form.

A quite clear example of how change in shape could be due to change in expression of a single gene involves beak variation in finches. The variation in the beaks of Galapagos finches in relation to the food they eat is a classic example of evolutionary adaptation, which Darwin was the first to discover. More modern studies have shown that the secreted signaling molecule BMP-4, which we have encountered many times in previous chapters, may have played a key role in this evolution, as expression of BMP-4 in the developing beak of a chick embryo resulted in a beak similar to that of the large ground finch. Small differences in BMP dose and timing of expression may thus be the cause of the subtle differences in beak shape between the finches.

Another example of modification of a pre-existing structure is provided by the evolution of the vertebrate kidney. In birds and mammals, three kidney-like structures—metanephros, mesonephros, and pronephros—appear during development. The pronephros and mesonephros are transitory and the functional kidney develops from the metanephros. However, as discussed in Section 11.11, the mesonephros plays a key role in the development of the gonads, giving rise to the somatic cells of the testis and ovary. In lower vertebrates, such as fish and amphibians, the pronephros acts as the functional kidney in the immature juvenile stages, but the mesonephros is the functional kidney in the adult. Thus, in birds and mammals the embryonic kidneys of their ancestors have persisted as embryonic structures, but have been modified to provide structures essential to the development of the gonad.

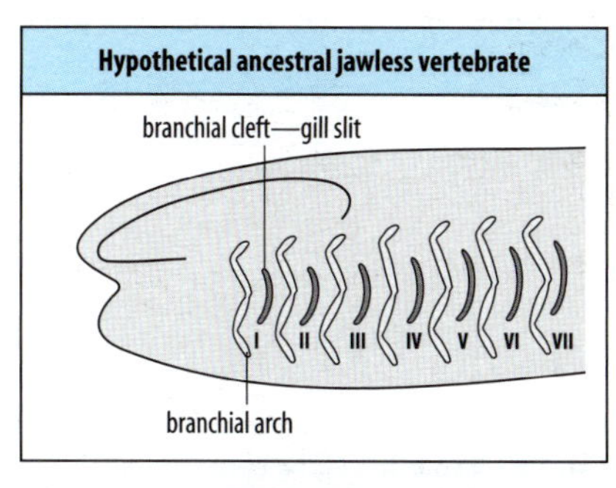

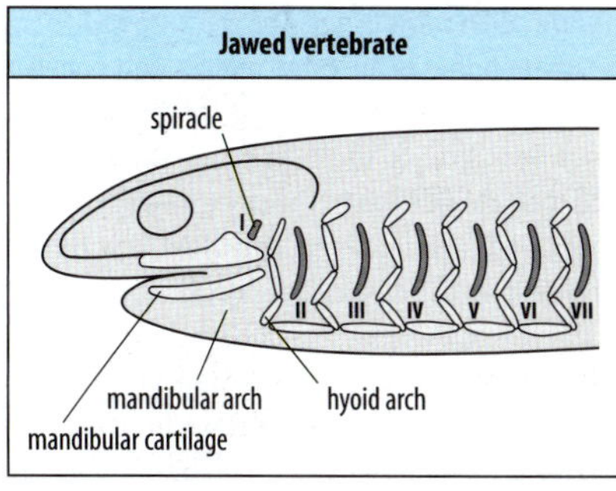

Fig. 14.5 Modification of the branchial arches during the evolution of jaws in vertebrates. The ancestral jawless fish had a series of at least seven gill slits—branchial clefts—supported by cartilaginous or bony arches. Jaws developed from a modification of the first arch to give the mandibular arch, with the mandibular cartilage of the lower jaw and the hyoid arch behind it.

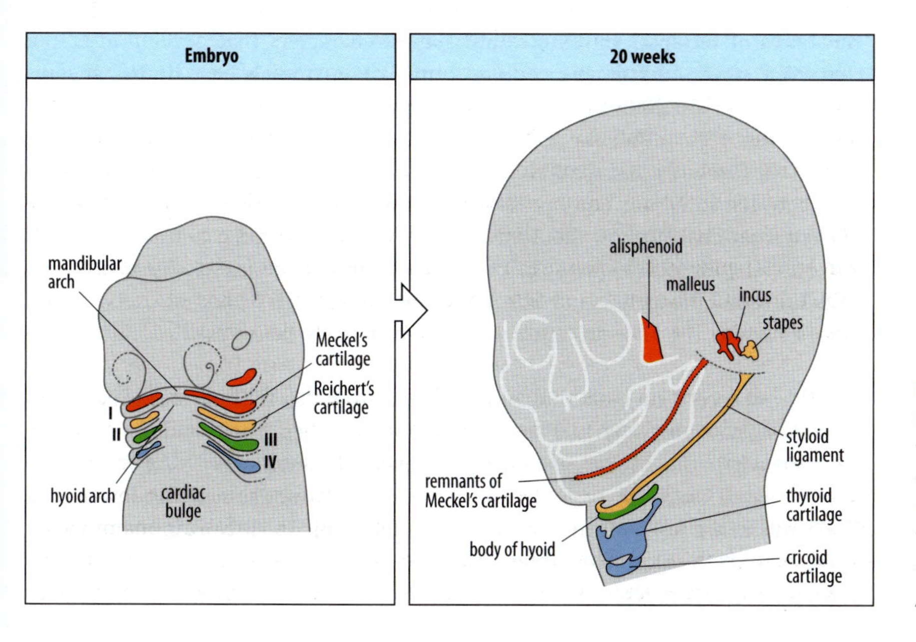

Fig. 14.6 Fate of branchial arch cartilage in humans. In the embryo, cartilage develops in the branchial arches, which gives rise to elements of the three auditory ossicles, the hyoid, and the pharyngeal skeleton. The fate of the various elements is shown by the color coding.

After Larsen, W.J.: 1993.

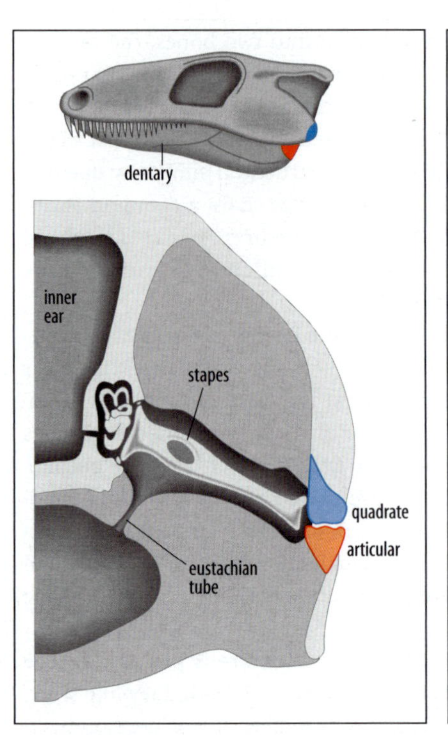

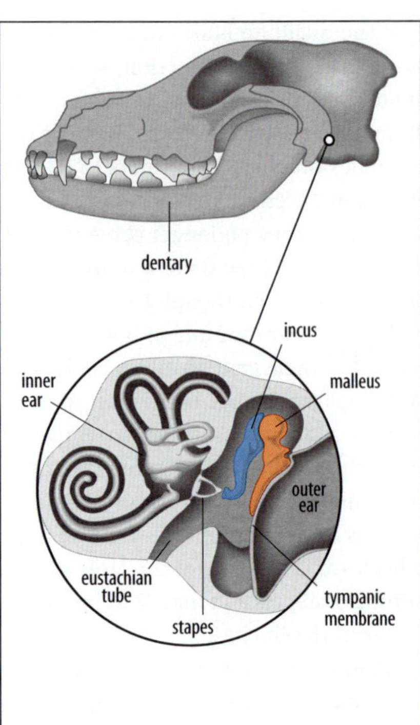

Fig. 14.7 Evolution of the bones of the mammalian middle ear. The articular and quadrate bones of ancestral reptiles (left panel) were part of the lower jaw articulation. Sound was transmitted to the inner ear via these bones and their connection to the stapes. When the lower jaw of mammals became a single bone (the dentary), the articular bone became the malleus, and the quadrate bone the incus of the middle ear, acquiring a new function in transmitting sound to the inner ear from the tympanic membrane (right panel). The eustachian tube forms between branchial arches I and II.

After Romer, A.S.: 1949.

These examples provide evidence of a key relationship between evolution and development, the gradual change of a structure into a different form. In many cases, however, we do not understand how intermediate forms were adaptive and gave a selective advantage to the animal. Consider, for example, the intermediate forms in the transition of the first branchial arch to jaws; what was the adaptive advantage? We do not know, and because of the passage of time and our current ignorance of the ecology of ancient organisms we may never know.

14.3 Limbs evolved from fins

The limbs of tetrapod vertebrates are special characters that develop after the phylotypic stage. Amphibians, reptiles, birds, and mammals have limbs, whereas fish have fins. The limbs of the first land vertebrates evolved from the pelvic and pectoral fins of their fish-like ancestors. The basic limb pattern is highly conserved in both the forelimbs and hindlimbs of all tetrapods, although there are differences both between forelimbs and hindlimbs, and between different vertebrates. Limbs evolved from fins, but how fins evolved is far less clear, and is a complex problem. Nevertheless, the development of these appendages made use of signaling molecules like Sonic hedgehog and fibroblast growth factor (FGF) and of transcription factors such as the Hox proteins, which were already being used to pattern the body.

The fossil record suggests that the transition from fins to limbs occurred in the Devonian period, between 400 and 360 million years ago. The transition probably occurred when the fish ancestors of the tetrapod vertebrates living in shallow waters moved onto the land. The fins of Devonian lobe-finned fishes, such as *Panderichthys*, are probably ancestral to tetrapod limbs, an early example of which is the limb of the Devonian tetrapod *Tulerpeton* (Fig. 14.8). The proximal skeletal elements corresponding to the humerus, radius, and ulna of the tetrapod limb are

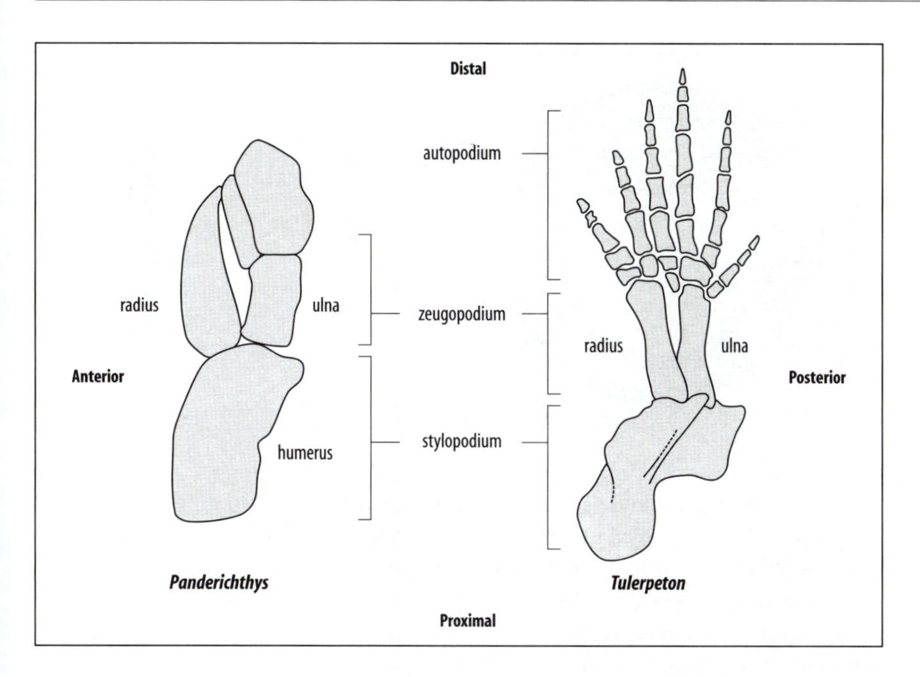

Fig. 14.8 The fin to limb transition. In the lobe-like fin of the Devonian fish *Panderichthys*, there were proximal elements corresponding to the humerus (the stylopodium), radius, and ulna (the zeugopodium), but no distal elements. The Devonian tetrapod *Tulerpeton* has similar proximal elements, but has also developed digits.

present in the ancestral fish, but there are no structures corresponding to digits. How did digits evolve? Some insights have been obtained by examining the development of fins in a modern fish, the zebrafish *Danio*.

The fin buds of the zebrafish embryo are initially similar to tetrapod limb buds, but important differences soon arise during development. The proximal part of the fin bud gives rise to skeletal elements, which may be homologous to the proximal skeletal elements of the tetrapod limb. There are four main proximal skeletal elements in a zebrafish fin, which arise from the subdivision of a cartilaginous sheet (Fig. 14.9). The essential difference between fin and limb development is in the distal skeletal elements. In the zebrafish fin bud, an ectodermal fin fold develops at the distal end of the bud and fine bony fin rays are formed within it. These rays have no relation to anything in the vertebrate limb.

As in the tetrapod limb bud (see Chapter 9), the key signaling gene *Sonic hedgehog* is expressed at the posterior margin of the zebrafish fins and the early expression pattern of Hoxd and Hoxa genes is similar to that in tetrapods. If zebrafish fin development reflects that of the primitive ancestor, then more distal cartilage elements may have evolved from the distal recruitment of the same developmental mechanisms and processes to generate the radius and ulna, wrist, and digits. There are, as discussed in Chapter 9, mechanisms in the limb for generating periodic cartilaginous structures such as digits. It is likely that such a mechanism was involved in the evolution of digits by an extension of the region in which the embryonic cartilaginous elements form, together with the establishment of a new pattern of Hox gene expression in the more distal region.

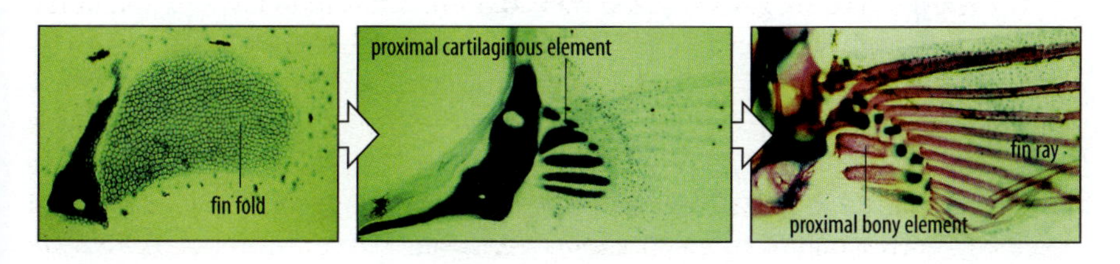

Fig. 14.9 The development of the pectoral fin of the zebrafish *Danio*. Left panel: the pectoral girdle and fin fold. Middle panel: four proximal cartilaginous elements and distal fin rays. Right panel: four proximal bony elements supporting the distal fin rays in the adult fish.

Photographs courtesy of D. Duboule, from Sordino, P., et al.: 1995.

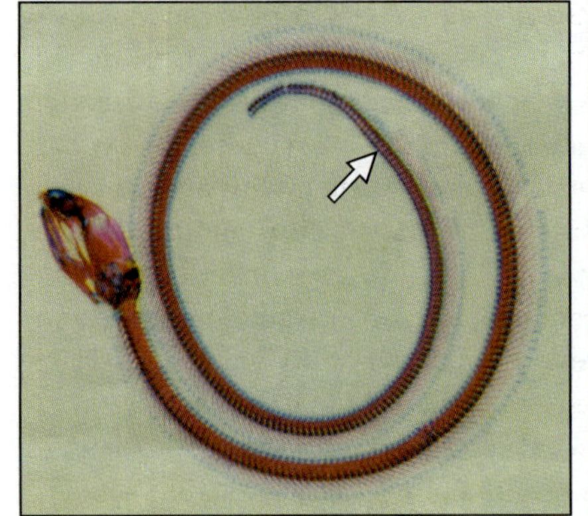

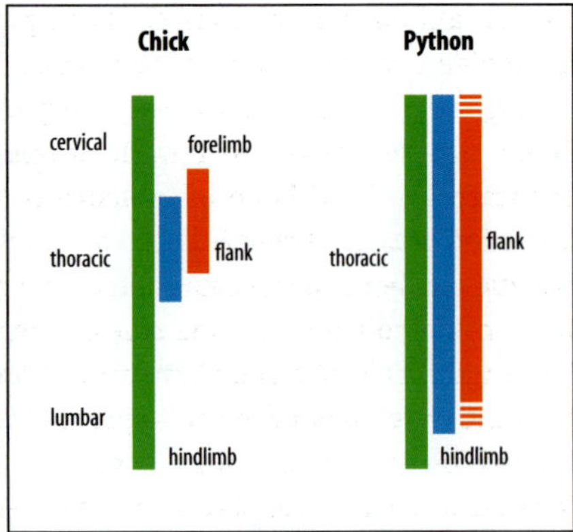

Fig. 14.10 Comparison of Hox gene expression in python and chick embryos. The photograph shows a skeleton of a python embryo at 24 days' incubation stained with Alcian blue and Alizarin red. The arrow marks the position of hindlimb rudiments, which have been removed in this preparation. Note the similarity of the vertebrae anterior to the arrow. The right panel shows a schematic comparison of domains of expression of *Hoxb5* (green), *Hoxc8* (blue) and *Hoxc6* (red) in chick and python embryos. The expansion of the *Hoxc8* and *Hoxc6* domains in the python correlates with the expansion of thoracic identity in the axial skeleton and flank identity in the lateral plate mesoderm.

Adapted from Cohn, M.J. and Tickle, C.: 1999.

Differences in fore- and hindlimb development are related to differences in Hox genes, which are involved in establishing positional differences in the lateral plate mesoderm from which the limb develops. Limb loss is quite common, and the python provides a good example of the developmental changes involved. Snakes have hundreds of similar vertebrae in their backbones, as can be seen in the skeleton of a python embryo in Figure 14.10 (left). Snakes have no forelimbs, but pythons have a pair of hindlimb rudiments at the junction between the rib-bearing thoracic vertebrae, and vertebrae with shorter, forked ribs. In four-limbed vertebrates, expression of *Hoxb5* and *Hoxc8* is confined to a short trunk region. In the python embryo, however, these genes are expressed along the whole body as far as the pelvic rudiment (Fig. 14.10). The expansion of these Hox-expression domains is thought to underlie the expansion of rib-bearing vertebrae and the loss of forelimbs in snake evolution. Hindlimb buds begin to develop, but the signals associated with apical ridge and polarizing region signaling (see Chapter 9) are not activated.

The great range of anatomical specializations in the limbs of mammals (Fig. 14.11) is due to changes both in limb patterning and in the differential growth of parts of the limbs during embryonic development, but the basic underlying pattern of skeletal elements is maintained. This is an excellent example of the modularity of the skeletal elements. If one compares the forelimb of a bat and a horse, one can

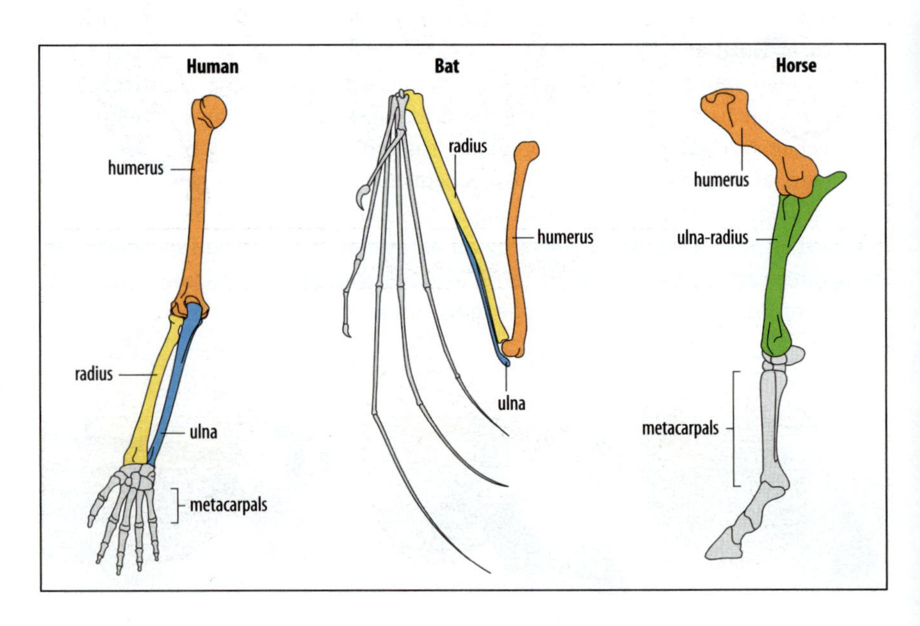

Fig. 14.11 Diversification of mammalian limbs. The basic pattern of bones in the forelimb is conserved throughout the mammals, but there are changes in the proportions of the different bones, as well as fusion and loss of bones. This is seen particularly in the horse limb, in which the radius and ulna have become fused into a single bone, and the central metacarpal (a hand bone in humans) is greatly elongated. In addition, there has been loss and reduction of the digits in the horse. In the bat wing, by contrast, the digits have become greatly elongated to support the membranous wing.

see that although both retain the basic pattern of limb bones, it has been modified to provide a specialized function in each. In the bat, the limb is adapted for flying: the digits are greatly lengthened to support a membranous wing. In the horse, the limb is adapted for running: in the forelimb, lateral digits are reduced, the central metacarpal (a hand bone in humans) is lengthened, and the radius and ulna are fused for greater strength. The role of differential growth rates and the loss of skeletal elements in the evolution of the horse's limb, and cases of limb reduction, are considered later. In some cases, differences in size of fore- and hindlimbs and the length of the skeletal elements are present from a very early embryonic stage, as in the development of the flightless kiwi and in bats (Fig. 14.12).

The limb can also provide examples of what can be considered a developmental constraint. That is, the nature of the developmental process does not allow certain changes. It is hard to imagine a limb in which the humerus-like element is interposed between radius/ulna and the wrist, because the mechanism of limb development would seem to make this impossible. It is even harder to imagine how a mouse could evolve a feathered wing—just too many control regions and modules are involved.

A feature of limb evolution is that while reduction in digit number is common (there are only three in the chick wing and reduction is common in lizards) species with more than five digits are very rare. It seems that there is a developmental constraint on evolving more than five different kinds of digits. This may be due to the Hox genes in the distal part of the limb providing only five discrete genetic programs for giving a digit an identity. In limbs with polydactyly, at least two of the digits are the same (see Fig. 9.12), such that there are still only five different kinds of digits. This may be the reason why in an animal with an additional distinctive digit-like element, such as the giant panda's 'thumb', this digit is in fact a modified wrist bone.

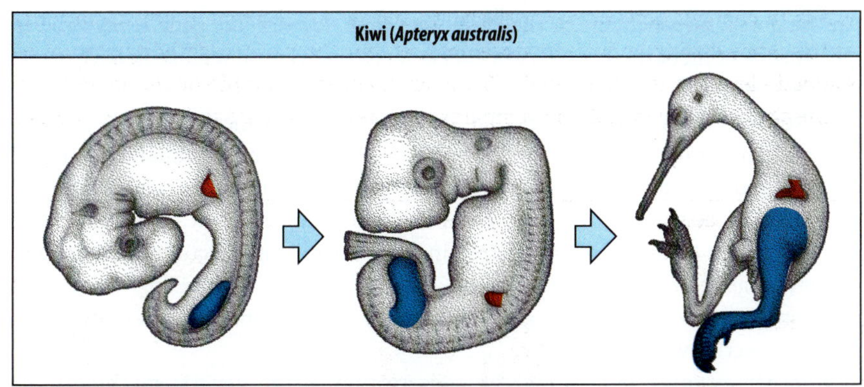

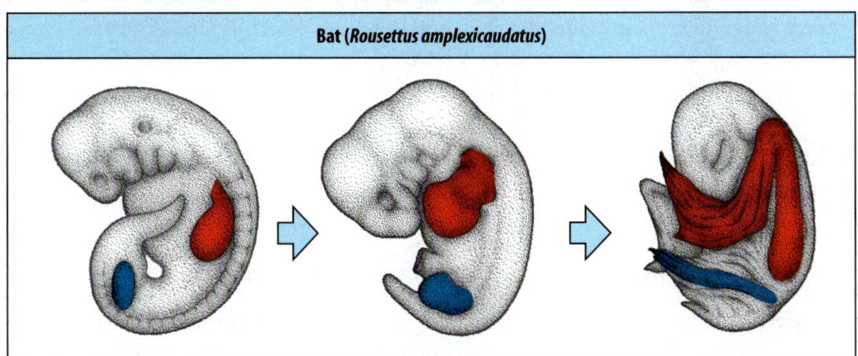

Fig. 14.12 Developmental expression of adult limb size. Top row: the flightless kiwi (*Apteryx australis*) has relatively short forelimbs and large hindlimbs at all stages of development, even in the early embryo. Bottom row: in the bat *Rousettus amplexicaudatus*, which has large forelimbs and smaller hindlimbs when adult, the forelimb bud is large at all stages, relative to the hindlimb.

Adapted from Richardson, M.K.: 1999.

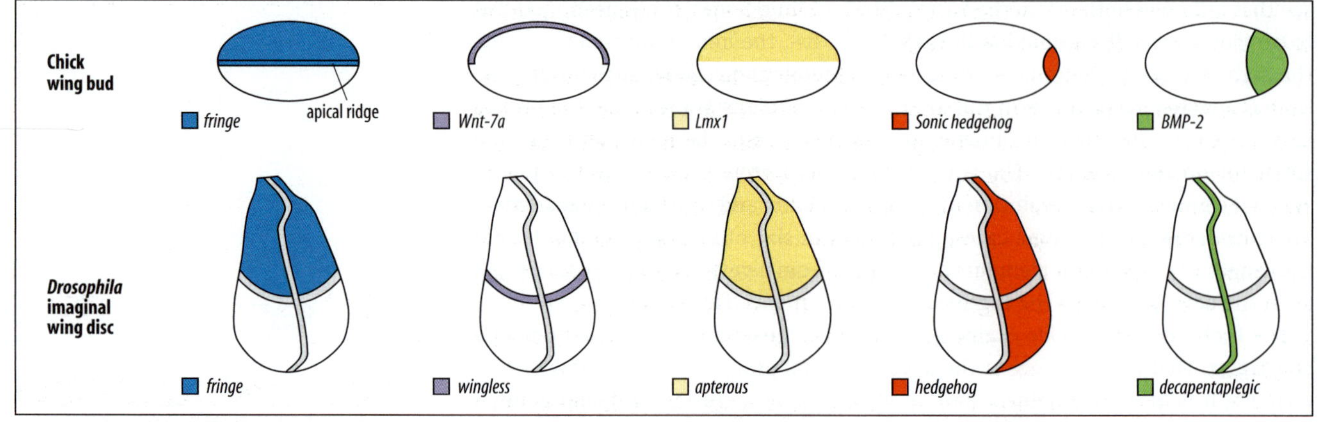

Fig. 14.13 Comparison of developmental signals in the chick wing bud and *Drosophila* wing imaginal disc. First column: the chick wing bud (top) is shown with the distal end facing. The double line bisecting it represents the apical ridge. The chick apical ectodermal ridge forms at the boundary between the dorsal cells, which express *radical fringe*, and the ventral cells, which do not. The dorso-ventral boundary in the insect wing disc also forms at the boundary between *fringe* and non-*fringe* cells (bottom). In the insect wing imaginal disc, the future dorsal and ventral regions are in the same plane. The vertical double lines represent the antero-posterior compartment border, and the horizontal double lines represent the dorso-ventral compartment border. Second column: the dorsal region of the chick wing is specified by *Wnt-7a* in the ectoderm, whereas *wingless* is expressed at the dorso-ventral insect wing margin. Third column: the gene *Lmx1* is expressed in the dorsal region of chick wing bud mesoderm, whereas the *Drosophila* gene *apterous*, to which it is structurally related, specifies the dorsal region of the insect wing. Fourth and fifth columns: *Sonic hedgehog* in the chick wing and *hedgehog* in *Drosophila* are expressed in posterior regions and both induce expression of genes of the TGF-β family—*BMP-2* and *decapentaplegic*, respectively.

14.4 Vertebrate and insect wings make use of evolutionarily conserved developmental mechanisms

Vertebrate and insect wings are not homologous but have some superficial similarities; they have similar functions yet are very different in structure. The insect wing is a double-layered epithelial structure, whereas the vertebrate limb develops mainly from a mesenchymal core surrounded by ectoderm. Despite these great anatomical differences, however, there are striking similarities in the genes and signaling molecules involved in setting up the axes and so patterning insect legs, insect wings, and vertebrate limbs (Fig. 14.13, and see Chapter 9). Patterning along the antero-posterior axis of all these appendages uses Hedgehog-related signals and members of the TGF-β family, such as Decapentaplegic (in insects) and BMP-2 (in vertebrates). It is remarkable that the dorsal surface of the insect wing is characterized by expression of the gene *apterous*, whereas the related gene *Lmx1* is expressed in the dorsal mesenchyme of the vertebrate limb. A *fringe*-like gene is involved in the specification of the boundary between dorsal and ventral regions in both insect and bird wings.

All these relationships suggest that, during evolution, a mechanism for patterning and setting up the axes of appendages appeared in some common ancestor of insects and vertebrates. Subsequently, the genes and signals involved acquired different downstream targets so that they could interact with different sets of genes, yet the same set of signals retain their organizing function in these very different appendages. The individual genes and their interacting circuits of proteins involved in specifying the limb axes are probably more ancient than either insect or vertebrate limbs.

Another example illustrating the conservation of the genetic machinery for insect appendages is provided by the gene *Distal-less*, which is also expressed along the proximo-distal axis of a wide variety of developing appendages in other animals, including annelid parapodia and the tube feet of sea urchins. It is also expressed during vertebrate limb outgrowth.

All this emphasizes how evolution has used a wonderfully modifiable system for setting up axes and specifying positional values.

14.5 Hox gene complexes have evolved through gene duplication

Hox genes play a key role in development in both vertebrates and insects—and in most other animals. By comparing the organization and structure of the Hox genes in insects and vertebrates, we can determine how one set of important developmental genes has changed during evolution.

A major general mechanism of evolutionary change has been gene duplication and divergence. Tandem duplication of a gene, which can occur by a variety of mechanisms during DNA replication, provides the embryo with an additional copy of the gene. This copy can diverge in both its nucleotide sequence and its regulatory region, thus changing its pattern of expression and its downstream targets without depriving the organism of the function of the original gene (Fig. 14.14). The process of gene duplication has been fundamental in the evolution of new proteins and new patterns of gene expression; it is clear, for example, that the different hemoglobins in humans (see Fig. 8.15) have arisen as a result of gene duplication.

The Hox gene complexes provide one of the clearest examples of the importance of gene duplication in developmental evolution. As we have seen, the Hox genes are members of the homeobox gene family, which is characterized by a short 180 base pair motif, the homeobox, which encodes a helix-turn-helix domain that is involved in transcriptional regulation (see Box 4A, p. 156). Two features characterize all known Hox genes: the individual genes are organized into one or more gene clusters or complexes, and the order of expression of individual genes along the antero-posterior axis is usually the same as their sequential order in the gene complex.

The pathway of Hox gene evolution, which is central to the laying down of the body plan, is relatively clear for the Bilateria, but the origin of these genes is less clear. Hox genes have been identified in the Cnidaria, such as *Hydra*, where the gene *Cnox2* is homologous to the Hox2 genes. One model suggests that the duplication and divergence of a 'ProtoHox' cluster of just five genes gave rise to all the later Hox gene complexes, and to the related 'ParaHox' genes, which are found outside the classical Hox clusters. Invertebrates have one Hox complex that specifies pattern along the antero-posterior axis, whereas vertebrates have four, suggesting two further rounds of duplication of the whole complex. In several fish there has been one further round of duplication: the zebrafish has seven Hox clusters, one having been lost after three rounds of duplication. The Japanese puffer fish *Fugu rubripes* has four Hox clusters, but many genes have been lost, and one cluster bears little relation to any in other vertebrates. It appears that in the ancestral chordates that gave rise to the vertebrates, the spatial co-linearity between the order of the Hox genes along the chromosome and their expression along the main body axis was confined to the ectoderm-derived neural tube, and possibly the epidermis, and that it was only later extended to other tissues, thus allowing increased complexity of body plan.

Comparing the Hox genes of a variety of species, it is possible to reconstruct the way in which they are likely to have evolved from a simple set of seven genes in a common ancestor of arthropods and vertebrates (Fig. 14.15). Amphioxus, which is a vertebrate-like chordate, has many features of a primitive vertebrate: it possesses a dorsal hollow nerve cord, a notochord, and segmental muscles that derive from somites. It has only one Hox gene cluster, and one can think of this cluster as most closely resembling the common ancestor of the four vertebrate Hox gene complexes—Hoxa, Hoxb, Hoxc, and Hoxd (see Fig. 14.15). It is possible that both the vertebrate and *Drosophila* Hox complexes evolved from a simpler ancestral complex by gene duplication. In *Drosophila*, the duplications could have given *abdominal-A* (*abd-A*), *Ubx*, and *Antennapedia* (*Antp*). In vertebrates, the Hox genes are arranged in four separate clusters, each of which is on a different chromosome, and are not linked to each other. These separate clusters probably arose from duplications

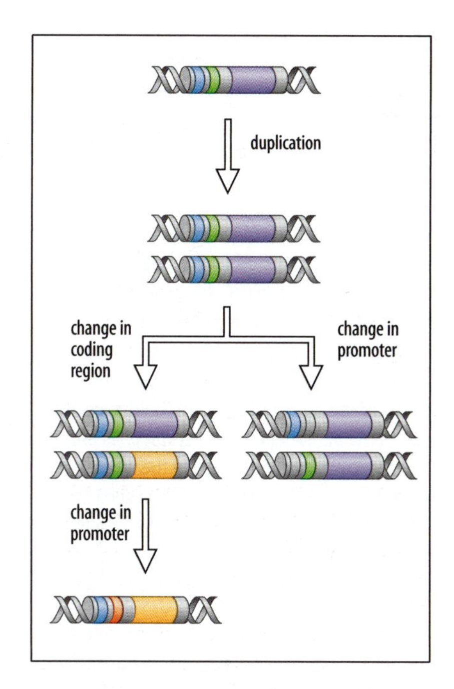

Fig. 14.14 Gene duplication and diversification. Once a gene has become duplicated, the second copy can evolve new functions and/or new expression patterns. Illustrated is a hypothetical example of a gene (purple) with two *cis*-regulatory modules (blue and green) that confer expression in two different tissues. Left pathway: after duplication of the complete gene and its regulatory regions, mutation in the coding region of one of the copies can generate a gene with a new and useful function that is selected for. Both copies will be retained in the genome as they both fulfill useful functions. Subsequent mutation in a regulatory module could lead to the new gene acquiring a new expression pattern. Right pathway: alternatively, a mutation in one regulatory region in one copy and the other regulatory region in the other copy can generate two copies of the gene that retain the original function but are now only expressed in a single tissue each. Both will be retained in the genome as they are both now needed to fulfill the function of the original gene. Examples of all these types of duplication and divergence are found in animal genomes.

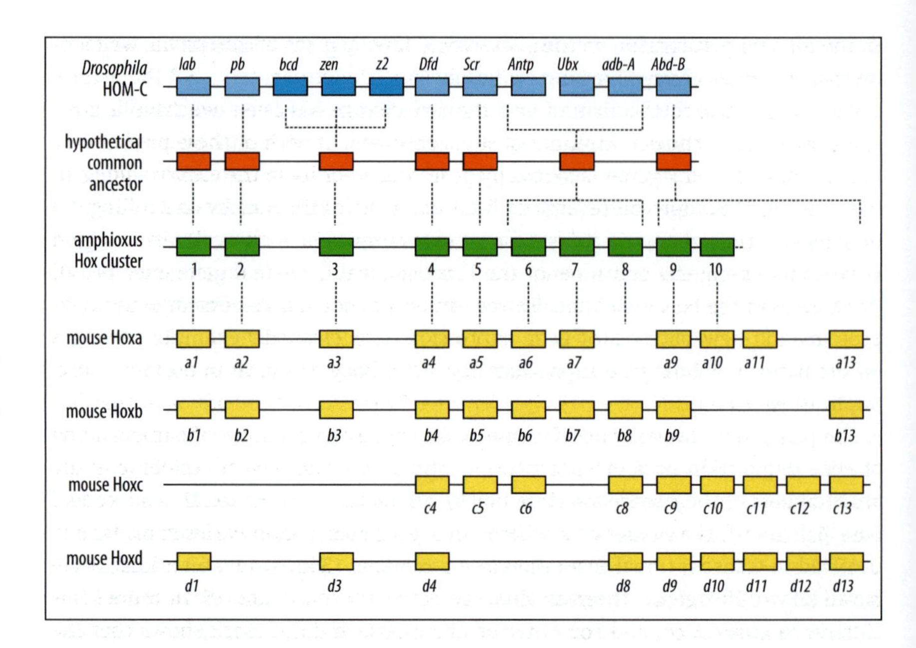

Fig. 14.15 Gene duplication and Hox gene evolution. A suggested evolutionary relationship between the Hox genes of a hypothetical common ancestor and *Drosophila* (an arthropod), amphioxus (a cephalochordate), and the mouse (a vertebrate). Duplications of genes of the ancestral set (red) could have given rise to the additional genes in *Drosophila* and amphioxus. Two duplications of the whole cluster in a chordate ancestor of the vertebrates could have given rise to the four separate Hox gene complexes in vertebrates. There has also been a loss of some of the duplicated genes in vertebrates.

of whole chromosomal regions, in which new Hox genes had already been generated by tandem duplication. For example, sequence comparisons suggest that the multiple mouse Hox genes related to the *Drosophila Abdominal-B* (*Abd-B*) gene do not have direct homologs within the *Drosophila* bithorax complex (see Fig. 14.15); they probably arose by tandem duplication from an ancestral gene after the split of the insect and vertebrate lineages, but before duplication of the whole cluster in vertebrates. The advantage of duplication was that the embryo had more Hox genes to control downstream targets and so could make a more complicated body. We now consider the role of these genes in evolution of the axial body plan.

14.6 Changes in Hox genes have generated the elaboration of vertebrate and arthropod body plans

From paleontological, molecular, and cellular evidence, all multicellular animals are presumed to descend from a common ancestor, and the broad similarities in the pattern of Hox gene expression in vertebrates and arthropods, whose evolution diverged hundreds of millions of years ago, are taken as good supporting evidence for this idea. Hox genes are key genes in the control of development and are expressed regionally along the antero-posterior axis of the embryo, as we have seen in earlier chapters. The apparently conserved nature of the Hox genes (and of certain other homologous genes) in animal development has led to the concept of the **zootype**. This defines the general pattern of expression of these key genes along the antero-posterior axis of the embryo, a pattern that is present in all animals.

Multicellular organisms are thought to have originated about 1500 million years ago, with the earliest generally accepted fossil of a multicellular animal dating from about 600 million years ago. There are currently about 35 extant animal phyla, each with its own distinctive basic body plan, and all of these had evolved by the end of the Cambrian period, around 500 million years ago. No new basic body plans with clear antero-posterior and dorso-ventral axes have evolved since then, although body plans within the different phyla have been modified and elaborated to give organisms as different as fish and mammals within the chordates.

In the following discussion on the elaboration of body plans within phyla, we focus on just two phyla—the chordates, which include the vertebrates, and the arthropods, which include the insects and the crustaceans—because we have a good understanding of the development of some members of both of these phyla.

The role of the Hox genes is to specify positional identity in the embryo along an axis, such as the antero-posterior axis. They exert their influence by controlling the activity of other genes that determine, for example, how the cells in a region develop into segments and appendages. Thus, changes in these target genes as well as changes in the Hox genes themselves can be a major source of change in evolution, and in general these targets are far from being identified. In addition, changes in the pattern of Hox gene expression along the body can have important consequences. An example is a relatively minor modification of the body plan that has taken place within vertebrates. One easily distinguishable feature of pattern along the antero-posterior axis in vertebrates is the number and type of vertebrae in the main anatomical regions—cervical (neck), thoracic, lumbar, sacral, and caudal (see Fig. 4.12). The number of vertebrae in a particular region varies considerably among the different vertebrate classes—mammals, with rare exceptions, have seven cervical vertebrae, whereas birds can have between 13 and 15. How does this difference arise? A comparison between the mouse and the chick shows that the domains of Hox gene expression have shifted in parallel with the change in number of vertebrae in the different regions (see Fig. 4.12). For example, the anterior boundary of *Hoxc6* expression in the mesoderm in mice and chicks is always at the boundary of the cervical and thoracic regions, irrespective of the number of cervical vertebrae. Moreover, the *Hoxc6* expression boundary is also at the cervical–thoracic boundary in geese, which have three more cervical vertebrae than chickens. The changes in the spatial expression of *Hoxc6* thus correlate specifically with the number of cervical vertebrae rather than with a fixed number of vertebrae. Other Hox genes are also involved in the patterning of the antero-posterior axis, and their boundaries also shift with a change in anatomy.

Changes in the targets of master regulatory genes such as Hox genes probably play a key role in determining the differences between animal species. There may be many—perhaps very many—downstream targets of the Hox genes. In *Drosophila*, for example, it is expression of the Hox gene *Ultrabithorax* that prevents the rudimentary haltere from growing and so makes it different from its neighboring appendage, the wing. Tests have shown that six of the twelve genes known to be involved in patterning the wing are under the control of *Ultrabithorax* and are repressed in the haltere. But because of changes in their *cis*-regulatory regions, only some of these genes are repressed by *Ultrabithorax* in butterflies, resulting in the development of a hind wing rather than a haltere.

Working out the mechanisms by which the Hox genes control their targets is further complicated by evidence that the amount of Hox protein present in a cell is important. Knock-outs of *Hoxa3* or *Hoxd3* in the mouse give quite different phenotypes, and it would seem reasonable to assume that this reflects the differences in the proteins encoded by these genes. However, when the protein-coding region of *Hoxa3* was replaced with that for *Hoxd3*, the mice developed normally, even though Hoxa3 protein was absent. There was, however, much more Hoxd3 protein present than normal because its expression was being controlled by *Hoxa3 cis*-regulatory regions. So it may simply be that a sufficient amount of Hox protein, such as Hoxd3, rather than a particular type, needs to be present, in this case at least.

An example of how subtle variations in body form can be controlled by Hox genes comes from insects. There are around 150,000 species of flies and all use roughly the same set of developmental genes. Diversity in form can reflect different

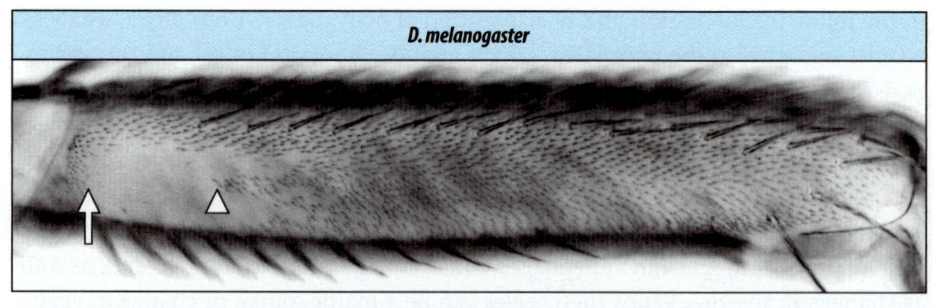

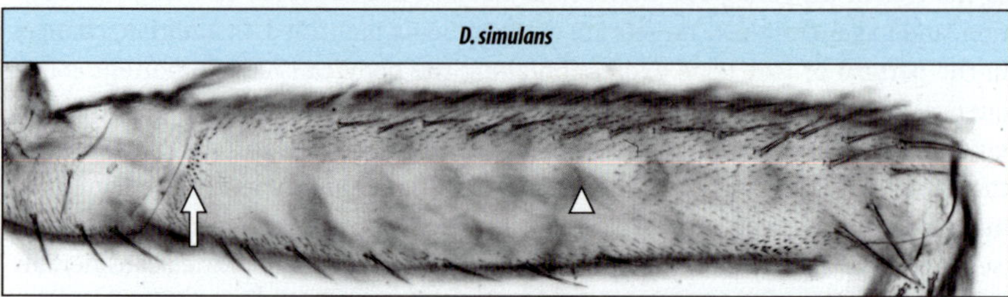

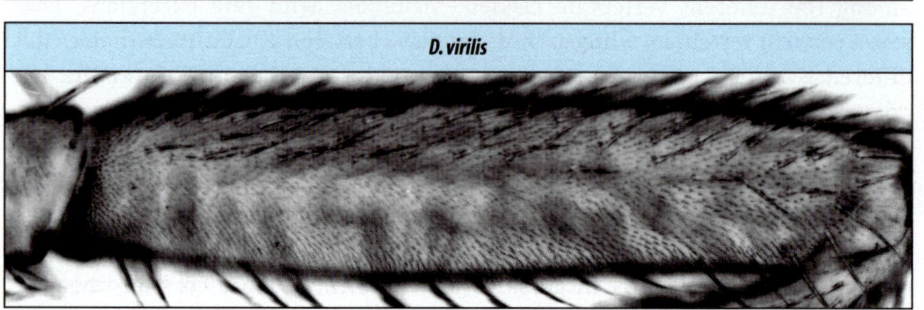

Fig. 14.16 Trichome patterns on the posterior second femur vary among *Drosophila* species. Top panel, *Drosophila melanogaster*. Middle panel, *D. simulans*. Bottom panel, *D. virilis*. The arrow and arrowhead in the top two panels indicate the extent of the patch of naked cuticle.

From Stern: 1998.

patterns of gene activity, but the fascinating question in this case is how the same sets of proteins can generate such subtle morphological diversity. In the genus *Drosophila*, different species have different patterns of non-sensory bristles—trichomes—on the femur of the second leg. *D. melanogaster* has a small naked patch, whereas *D. simulans* has a large one and *D. virilis* no naked patch at all (Fig. 14.16). *Ultrabithorax* represses trichome development, and the extent or presence of the naked patch is determined by subtle differences in the regulation of *Ultrabithorax* expression in this region. These differences are due to small differences in the regulatory region of *Ultrabithorax* in the different species, and this variation provides the evolutionary basis for the variation in pattern.

Changes in patterns of Hox gene expression can also help explain the evolution of arthropod body plans. Insects and crustaceans are distinct groups of arthropods that have evolved from a common arthropod ancestor, which probably had a body composed of more or less uniform segments. A comparison of Hox gene expression in an insect, the grasshopper, with that in a crustacean, the brine shrimp *Artemia*, shows how the different body plans might have evolved from the ancestral body plan. Such a comparison shows that both the pattern of Hox gene expression and the body regions to which particular Hox genes relate have changed during the evolution of these two groups (Fig. 14.17).

The grasshopper has a pattern of Hox gene expression similar to that of *Drosophila*. As in *Drosophila*, the Hox genes *Antennapedia*, *Ultrabithorax*, and *abdominal-A* specify distinct segment types in the thorax and abdomen, but are expressed in overlapping domains, and segment types are defined by combinatorial expression (see Section 2.27). In the brine shrimp, however, these genes are all expressed together in a thoracic region that is composed of uniform segments. This suggests

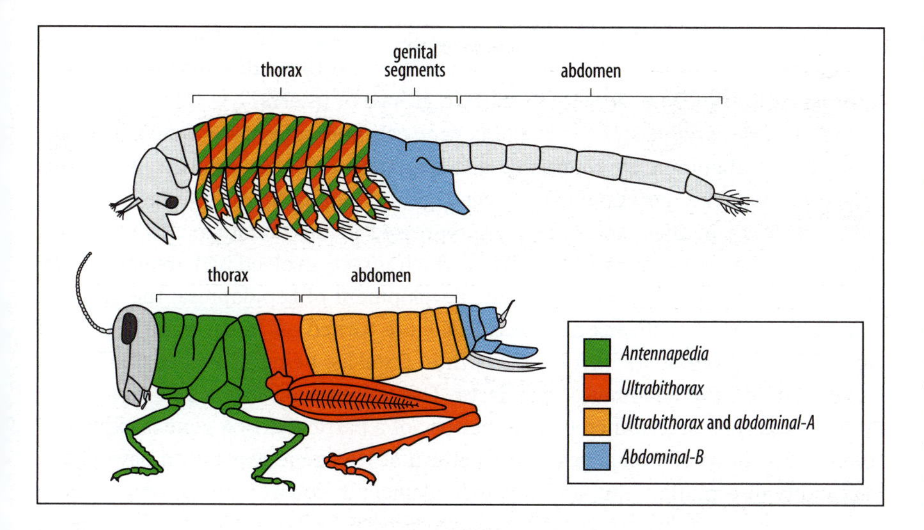

Fig. 14.17 A comparison of body plans and Hox gene expression in two arthropods. A comparison of Hox gene expression in an insect, the grasshopper (bottom), with that in a crustacean, the brine shrimp *Artemia* (top), shows that both the pattern of Hox gene expression and the body regions to which particular Hox genes relate have changed during the evolution of these two groups of arthropods from their common ancestor. In *Artemia*, the three Hox genes *Antennapedia*, *Ultrabithorax*, and *abdominal-A* are all expressed throughout the thorax, where most of the segments are similar. However, the expression of these genes in the thorax and abdomen of the grasshopper is different. They have overlapping and distinct patterns of expression that reflect the regional differences in the insect thorax. The gene *Abdominal-B* is expressed in the genital regions of both animals, indicating that these two regions are homologous.

After Akam, M.: 1995.

that the thorax of *Artemia* might be homologous to the whole insect thorax and much of the insect abdomen. Thus, in the evolution of the grasshopper and the brine shrimp from their common ancestor, the changes in body plan have resulted in part from spatial changes in the expression of particular Hox genes, but key changes in their downstream targets have also been involved.

These comparisons clearly show that Hox genes do not specify particular structures but simply provide a regional identity. How that identity is interpreted to produce a particular morphology is the role of genes acting downstream from the Hox genes, and there may be many of these. The head region of different arthropods, for example, can be very different, and indeed spiders do not seem to have a head at all. Their body is divided into a front and a back region, the front end having segments that bear the spider's fangs, pedipalps, and four pairs of walking legs; the front region thus acts as a head and thorax combined. In spite of these differences, the pattern of expression of the Hox genes in this region, including *orthodenticle* and *labial*, is very similar in spiders and insects, showing how much morphology can change as a result of changes in the downstream targets. Further examples of the role of Hox genes are provided by arthropod appendages, which we will look at next.

14.7 The position and number of paired appendages in insects is dependent on Hox gene expression

Insect fossils display a variety of patterns in the position and number of their paired appendages—principally the legs and wings. Some insect fossils have legs on every segment, whereas others only have legs in a distinct thoracic region. The number of abdominal segments bearing legs varies, as does the size and shape of the legs. Wings arose later than legs in insect evolution. Wing-like appendages are present on all the thoracic and abdominal segments of some insect fossils, but are restricted to the thorax in others. To understand how these different patterns of appendages arose during evolution we need to look at how the different patterns of appendages develop in two orders of modern insects, the Lepidoptera (butterflies and moths) and the Diptera (flies, including *Drosophila*).

The basic pattern of Hox gene expression along the antero-posterior axis is the same in all present-day insect species that have been studied. Yet the larvae of Lepidoptera have legs on the abdomen as well as on the thorax, and the adults have two pairs of wings, whereas the Diptera, which evolved later, have no legs on the abdomen in the larva or adult, and only one pair of wings, with the second pair of wings having been modified into halteres. How are these differences related to differences in Hox gene activity in the two groups of insects?

In *Drosophila*, products of the bithorax gene complex suppress appendage formation in the abdomen by repressing the expression of the *Distal-less* gene. This suggests that the potential for appendage development is present in every segment, even in flies, and is actively repressed in the abdomen. It thus seems likely that the ancestral arthropod from which insects evolved had appendages on all its segments. During the embryonic development of Lepidoptera, the bithorax complex genes *Ultrabithorax* and *Abdominal-B* are turned off in the ventral parts of the abdominal segments; this results both in *Distal-less* being expressed and in legs developing on the abdomen in the larva. The presence or absence of legs on the abdomen is thus determined by whether or not a particular Hox gene is expressed there. This shows that changes in the pattern of Hox gene expression have played a key role in evolution. But, and this is fundamental, downstream targets are even more important: in insects, *Ultrabithorax* and *abdominal-A* expressed together repress limb development in the abdomen, but in crustaceans these genes are expressed together in the thorax yet limbs do develop. The difference must lie in the downstream targets of these genes.

The Hox genes can also determine the nature of an appendage: we have seen how mutations can convert legs into antenna-like structures and an antenna into a leg (see Section 9.19). There is also good evidence that the coding sequences outside the DNA-binding homeodomain have evolved in Hox proteins and that this can have significant effects on segment identity and appendage development.

It has been proposed that insect wings originated as modifications of the gills of ancestral aquatic arthropods in their transition to life on land. However, Hox genes do not seem to have been involved in wing evolution, as *Antennapedia*, which is expressed in the wing-bearing second thoracic segment of *Drosophila*, is not required for wing development. It is likely that wings were originally present on all thoracic and abdominal segments and that Hox genes have repressed and modified their development in all segments except the second thoracic. For example, the differences between forewings and hindwings of insects with two pairs of wings, such as butterflies, are probably regulated by the *Ultrabithorax* gene (see Section 14.5).

It is clear that in the course of evolution, rather than new genes appearing, new regulatory interactions between the bithorax complex proteins and genes involved in leg and wing development have evolved.

14.8 The basic body plan of arthropods and vertebrates is similar, but the dorso-ventral axis is inverted

A comparison between the body plans of arthropods and chordates reveals an intriguing difference. In spite of many similarities in their basic body plan—both have an anterior head, a nerve cord running anterior to posterior, a gut, and appendages—the dorso-ventral axis of vertebrates is inverted when compared to that of arthropods. The most obvious manifestation of this is that the main nerve cord runs ventrally in arthropods and dorsally in vertebrates (Fig. 14.18, left panels).

One explanation for this, first proposed in the 19th century, is that during the evolution of the vertebrates from their common ancestor with the arthropods, the dorso-ventral axis was turned upside-down, so that the ventral nerve cord of the ancestor became dorsal. This startling idea has recently found some support from molecular evidence showing that the same genes are expressed along the dorso-ventral axis in both insects and vertebrates, but in inverse directions. This inversion

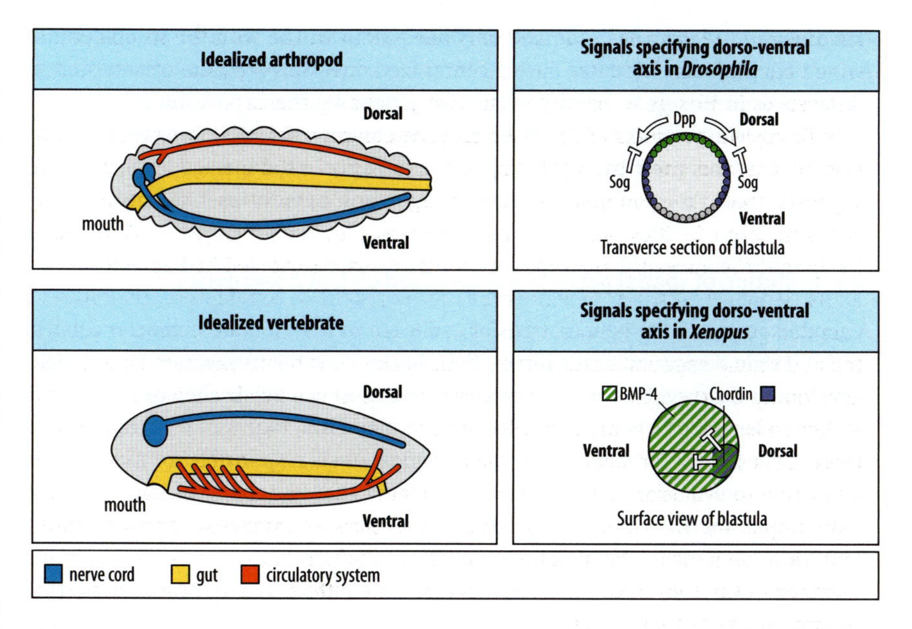

Fig. 14.18 The vertebrate and *Drosophila* dorso-ventral axes are related but inverted. In arthropods, the nerve cord is ventral whereas in vertebrates it is dorsal—dorsal and ventral being defined by the position of the mouth. In *Drosophila* and *Xenopus*, the signals specifying the dorso-ventral axis are similar, but are expressed in inverted positions. The protein Chordin, a dorsal specifier in vertebrates, is related to Sog, which is a ventral specifier in *Drosophila*, and the vertebrate ventral specifier BMP-4 is related to *Drosophila* Decapentaplegic (Dpp), which specifies dorsal.

After Ferguson, E.L.: 1996.

may have been dictated by the position of the mouth. The mouth defines the ventral side, and a change in the position of the mouth away from the side of the nerve cord would have resulted in the reversal of the dorso-ventral axis in relation to the mouth. The position of the mouth is specified during gastrulation, and it is not difficult to imagine how its position could have moved, but other changes in body structure and the nervous system were also involved.

We have seen in Chapters 2 and 3 that the patterning of the dorso-ventral axis in insects and vertebrates involves intercellular signaling. In *Xenopus*, the protein Chordin is one of the signals that specifies the dorsal region, whereas the growth factor BMP-4 specifies a ventral fate. In *Drosophila*, the pattern of gene expression is reversed: the protein Decapentaplegic, which is closely related to BMP-4, is the dorsal signal, and the protein short gastrulation (Sog), which is related to Chordin, is the ventral signal (see Fig. 14.18, right panels). These signaling molecules are experimentally interchangeable between insects and frogs. Chordin can promote ventral development in *Drosophila*, and Decapentaplegic protein promotes ventral development in *Xenopus*. The molecules and mechanisms that set up the dorso-ventral axes in the two groups of animals are thus homologous, strongly suggesting that the divergence in the body plans of the arthropods and vertebrates involved an inversion of this axis, by movement of the mouth during the evolution of the vertebrates. Thus our image of the common ancestor of chordates and arthropods is an animal in which the establishment of the two major body axes, antero-posterior and dorso-ventral, was similar to that in present-day animals.

But what did the ancestor of the chordates look like? Did it possess all the distinctive chordate features such as a hollow dorsal nerve cord, notochord, gill slits, and a post-anal tail? The nervous system is the key element in trying to identify the ancestor. The acorn worm is a member of the phylum Hemichordata, a small phylum of bilateral animals most closely related to the chordates, and studies on this worm have been important in this respect. Like chordates, the hemichordates are deuterostomes, and they also have gill slits that are homologous to those of chordates (Fig. 14.19). Orthologs of 22 chordate neural patterning genes have been mapped in the neural plate of the acorn worm and their antero-posterior pattern of expression is very similar to that of chordates. This is despite the fact

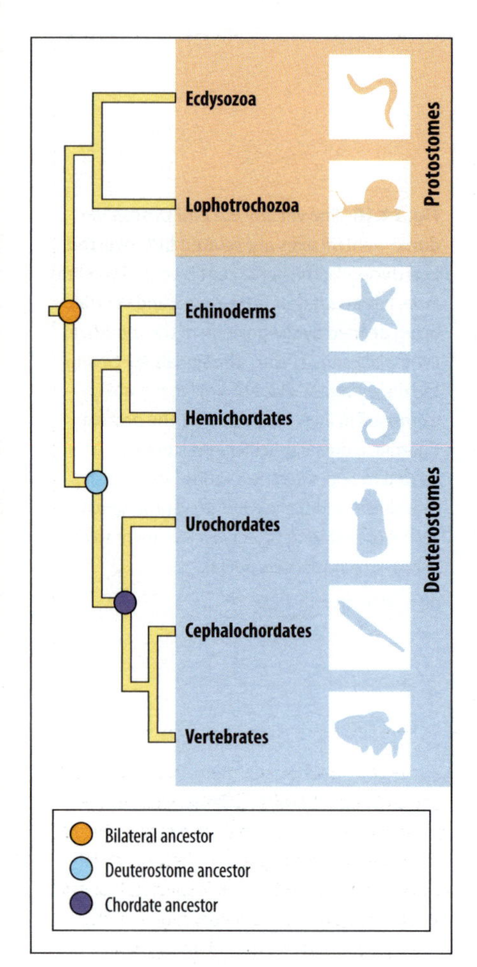

Fig. 14.19 The phylogeny of hemichordates and chordates. Hemichordates are the closest invertebrate relative to the chordates. The bilateral ancestor of both protostomes and deuterostomes would have had a dorso-ventral axis, mesoderm, a gut with mouth and anus, a distinct head, and the expression of Hox genes in domains in the trunk. The deuterostome ancestor had gill slits (among other features), while the chordate ancestor is distinguished from the lineage leading to the hemichordates and echinoderms by the possession of a notochord, post-anal tail, and a hollow dorsal nerve cord.

Adapted from Gerhart, J. et al.: 2005.

that the acorn worm has a diffuse nervous system in the form of an epidermal nerve net, whereas chordates have a centralized one. However, the orientation of the dorso-ventral axis in hemichordates, as judged by the expression of BMP and Chordin, is similar to that of protostomes. The inversion of the dorso-ventral axis seen in chordates must therefore have occurred only in the lineage leading to the chordates, as did the evolution of the notochord and a centralized nervous system.

14.9 Evolution of spatial pattern may be based on just a few genes

A good example of the genetic basis of evolution of a vertebrate character relates to the pelvic region of three-spine sticklebacks. Several freshwater stickleback populations have evolved complete or partial loss of the pelvic skeleton. This loss has been identified as being controlled by one major and four minor chromosomal regions. The gene *Pitx1* maps to the major region and is responsible for most of the variation in pelvic size. It is not the protein Pitx1 that has changed, but rather its *cis*-regulatory regions. This regulation of a major gene can underlie the mechanism for rapid evolutionary change in the vertebrate skeleton.

Another example of *cis*-regulatory evolution of a pattern is provided by the pigment pattern in the wing of male *Drosophila biarmipes*, which is closely related to our standard model organism, *D. melanogaster*, although separated by some 15 million years of evolution. *D. biarmipes* males have a conspicuous dark spot at the anterior tip of the wing that is lacking in male *D. melanogaster* (Fig. 14.20). This difference is due to differences in the expression of the gene *yellow*, which is required for production of the black pigment, in the wings of the two species. In *D. melanogaster*, *yellow* is expressed at a low level throughout the wing. The spot in *D. biarmipes* appears to have evolved as a result of changes in the *cis*-regulatory region of the *yellow* gene, such that it is now activated in a particular spatial pattern by transcription factors that are involved in patterning the wing. For example, the protein Engrailed is expressed in the posterior wing compartment (see Section 9.15) and it activates *yellow* in the distal region of this compartment. Other conserved gene regulatory proteins have also been co-opted to control the *yellow* gene by the evolution of binding sites for them within the gene's control region.

Summary

The development of an embryo provides insights into the evolutionary origin of the animal. Groups of animals that pass through a similar embryonic stage are descended from a common ancestor and have evolved modifications of gene-regulatory circuits. During evolution, the development of structures can be altered so that they acquire new functions, as has happened in the evolution of the mammalian middle ear from a reptilian jaw bone, which itself evolved from the branchial arch. The limbs of tetrapod vertebrates evolved from fins, with the digits as a novel feature. The development of vertebrate and insect limbs involves the same set of pattern-establishing genes, reflecting the evolution of limb development from an ancestral mechanism for specifying body appendages. The basic body plan of all animals is defined by patterns of Hox gene expression that provide positional identity, the interpretation of which has changed in evolution. The Hox genes themselves have undergone considerable evolution by gene duplication and divergence and are a good example of Darwin's description of evolution as "descent with modification". Comparison of patterns of dorso-ventral gene expression suggests that during the evolution of the vertebrates the dorso-ventral axis of an invertebrate ancestor was inverted.

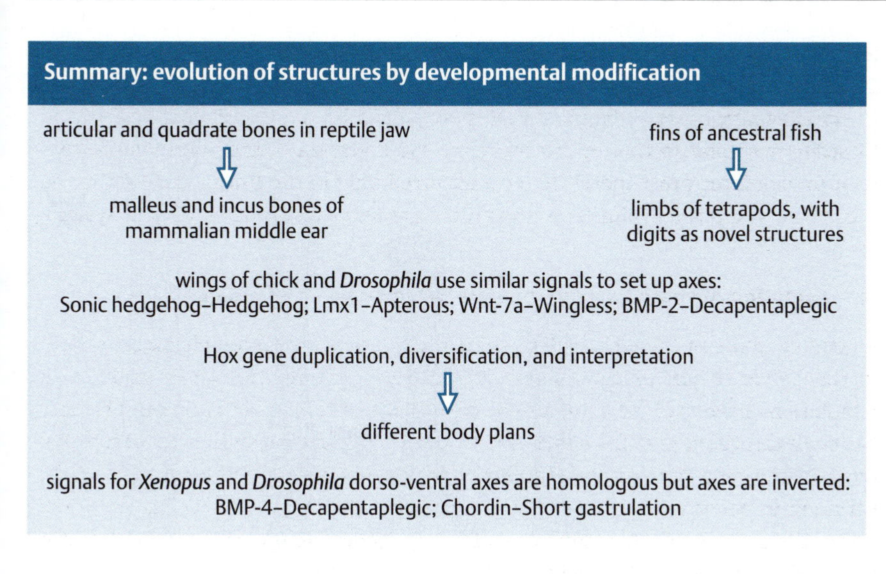

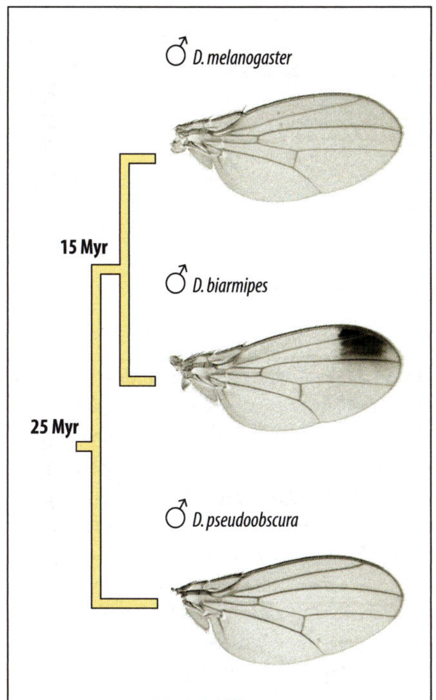

Fig. 14.20 Evolution of differences in pigment pattern in closely related *Drosophila* species. A spot of dark pigment on the tip of the wing of male *Drosophila biarmipes* is a relatively recently evolved trait among *Drosophila* species. It arises from a change in the regulatory region of the *yellow* gene, which is required for pigment production. Myr, millions of years ago.

From Gompel, N. et al.: 2005.

Changes in the timing of developmental processes

In the previous part of the chapter we focused on changes in spatial patterning that have occurred during evolution. But changes in the timing of developmental processes can also have major effects. In this part of the chapter, we look at some examples of how changes in the timing of growth and sexual maturation can affect animal form and behavior.

14.10 Changes in growth can alter the shapes of organisms

Many of the changes that occur during evolution reflect changes in the relative dimensions of parts of the body. We have seen how growth can alter the proportions of the human baby after birth, as the head grows much less than the rest of the body (see Fig. 12.6). The variety of face shapes in the different breeds of dog, which are all members of the same species, descended from the grey wolf, also provides a good example of the effects of differential growth after birth. All dogs are born with rounded faces; some keep this shape, but in others the nasal regions and jaws elongate during growth. The elongated face of the baboon is also the result of growth of this region after birth.

Because individual structures such as bones can grow at different rates, the overall shape of an organism can be changed substantially during evolution by heritable changes in the duration of growth that leads to an increase in the overall size of the organism. In the horse, for example, the central digit of the ancestral horse grew faster than the digits on either side, so that it ended up longer than the lateral digits (Fig. 14.21). As horses continued to increase in overall size during evolution, this discrepancy in growth rates resulted in the relatively smaller lateral digits no longer touching the ground because of the much greater length of the central digit. At a later stage in evolution, the now-redundant lateral digits became reduced even further in size because of a separate genetic change.

14.11 The timing of developmental events has changed during evolution

Differences among species in the time at which developmental processes occur relative to one another, and relative to their timing in an ancestor, can have

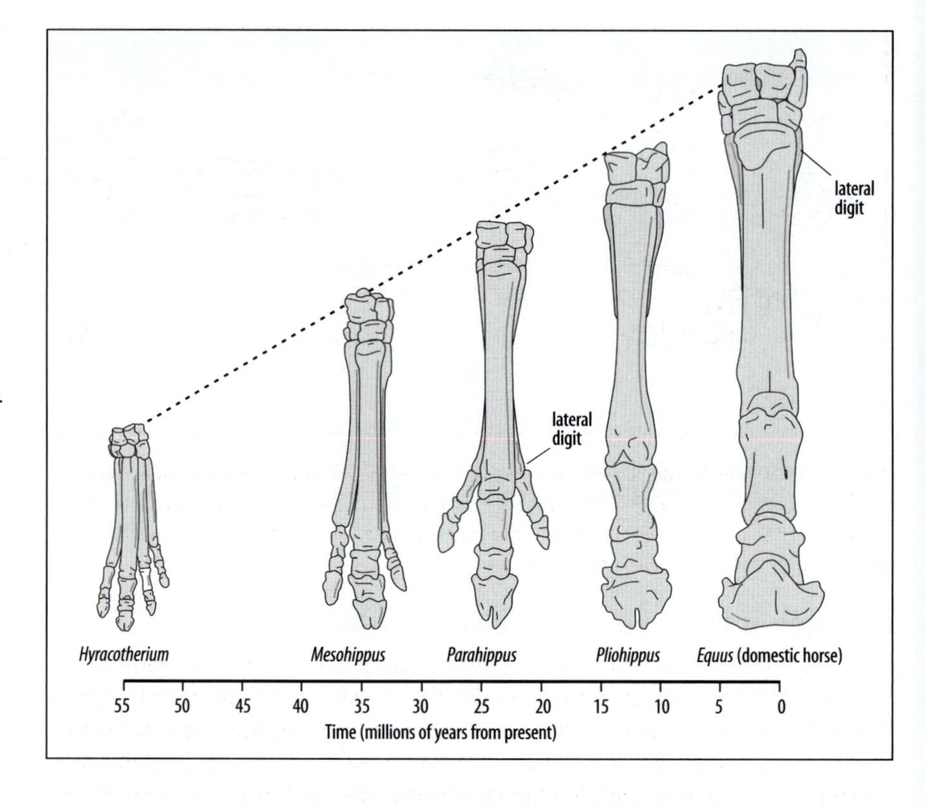

Fig. 14.21 Evolution of the forelimb in horses. *Hyracotherium*, the first true equid, was about the size of a large dog. Its forefeet had four digits, of which one (the third digit in anatomical terms) was slightly longer, as a result of a faster growth rate. All digits were in contact with the ground. As equids increased in size, the lateral digits lost contact with the ground, as the result of the relatively greater increase in length of metacarpal 3 (the hand bone of the third digit). At a later stage, the lateral digits became even shorter, because of a separate genetic change.

After Gregory, W.K.: 1957.

dramatic effects on both the structure and behavior of an organism. Differences in the feet of members of a genus of tropical salamanders illustrate the effect of a change in developmental timing on both the morphology and ecology of different species. Many species of the salamander genus *Bolitoglossa* are arboreal (tree living), rather than typically terrestrial, and their feet are modified for climbing on smooth surfaces. The feet of the arboreal species are smaller and more webbed than those of terrestrial species, and their digits are shorter (Fig. 14.22). These differences seem to be mainly the result of the development and growth of the foot ceasing at an earlier stage in the arboreal species than in the terrestrial species. The term used to describe such differences in timing is **heterochrony**.

Some of the clearest examples of heterochrony come from alterations in the timing of onset of sexual maturity in organisms with larval stages. The acquisition of sexual maturity by an animal while still in the larval stage is a process that goes under the name **neoteny**. The development of the animal, although not its growth, is retarded in relation to the maturation of the reproductive organs. This occurs in the Mexican axolotl, a type of salamander; the larva grows in size and matures sexually, but does not undergo metamorphosis. The sexually mature form remains aquatic and looks like an overgrown larva. However, the axolotl can be induced to undergo metamorphosis by treatment with the hormone thyroxine (Fig. 14.23).

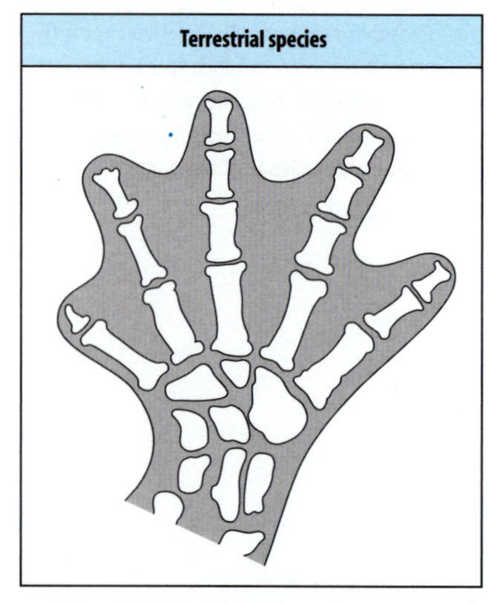

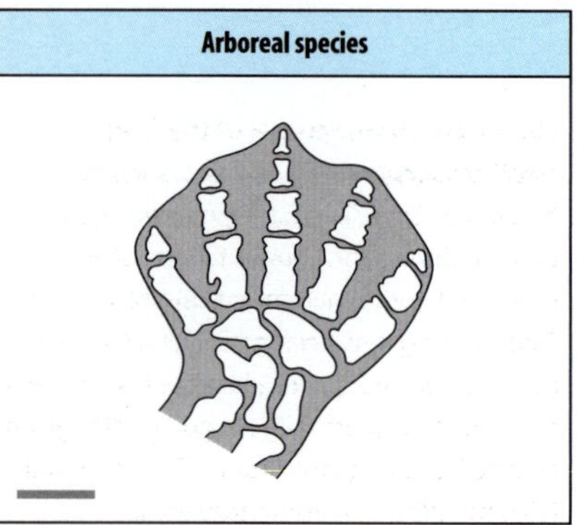

Fig. 14.22 Heterochrony in salamanders. In terrestrial species of the salamander *Bolitoglossa* (top panel), the foot is larger, has longer digits, and is less markedly webbed than in those that live in trees (bottom panel). This difference can be accounted for by foot growth ceasing at an earlier stage in the arboreal species. Scale bar = 1 mm.

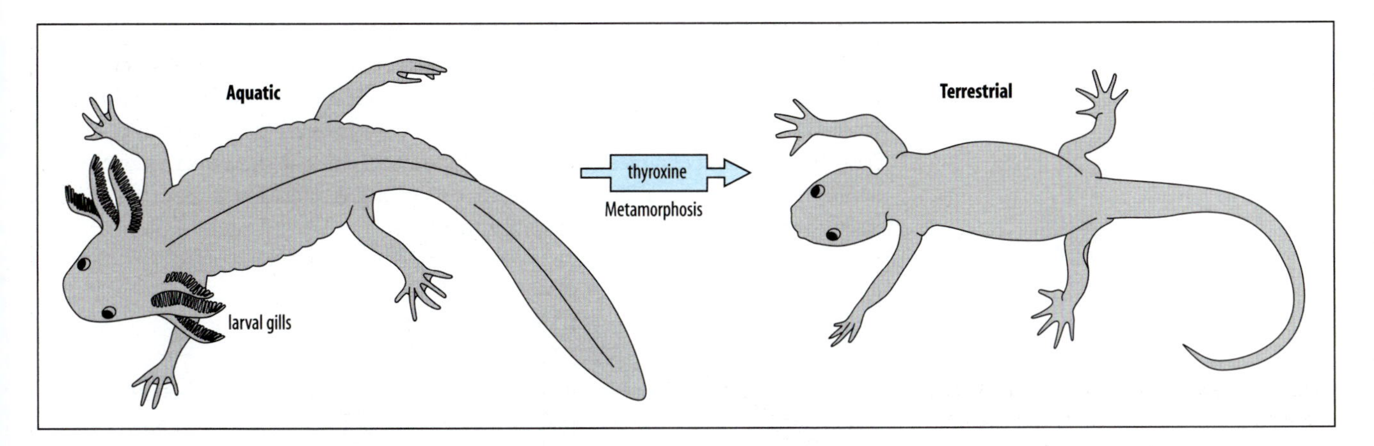

Fig. 14.23 Neoteny in salamanders. The sexually mature Mexican axolotl retains larval features, such as gills, and remains aquatic. This neotenic form metamorphoses into a typical terrestrial adult salamander if treated with thyroxine, which it does not produce, and which is the hormone that causes metamorphosis in other amphibians.

Larval stages themselves may have evolved as a result of heterochrony. Larval stages must have arisen as intercalations into a pre-existing developmental program before the sexually adult stage. They could not have been the original state because the metamorphosis that is necessary to bring larvae back to the developmental program into a mature adult is a highly complex process. If we assume, therefore, that frog ancestors developed directly into adults, a change in the timing of events in the post-neurula stages, together with structural modifications, could have led to the interposition of a feeding tadpole stage. One of the changes required would have been a delay in the development of limbs.

As well as being acquired during evolution, larval stages can be lost. Some modern frogs have evolved with direct development to the adult by a loss of the larval stage and the acceleration of the development of adult features. Frogs of the genus *Eleutherodactylus*, unlike the more typical amphibians *Rana* and *Xenopus*, develop directly into an adult frog and there is no aquatic tadpole stage, the eggs being laid on land. Typical tadpole features, such as gills and cement glands, do not develop, and prominent limb buds appear shortly after the formation of the neural tube (Fig. 14.24). In the embryo, the tail is modified into a respiratory organ. Such direct development requires a large supply of yolk to the egg in order to support development through to an adult without a tadpole feeding stage. This increase in yolk may itself be an example of heterochrony, involving a longer or more rapid period of yolk synthesis in the development of the egg.

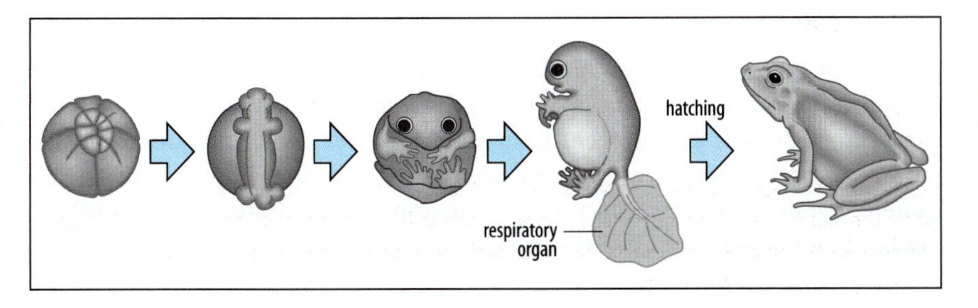

Fig. 14.24 Development of the frog *Eleutherodactylus*. Typical frogs, such as *Xenopus* or *Rana*, lay their eggs in water and develop through an aquatic tadpole stage, which undergoes metamorphosis into the adult. Frogs of the genus *Eleutherodactylus* lay their eggs on land and the frog hatches from the egg as a miniature adult, without going through an aquatic free-living larval stage. The embryonic tail is modified as a respiratory organ.

Most sea urchins have a larval stage that takes a month or more to become adult. But there are some species in which direct development has evolved, so that they no longer go through a functional larval stage. Such species have large eggs and, as a result of their very rapid development, become juvenile urchins within 4 days. This has involved changes in early development such that the directly developing embryo gives rise to a 'larval' stage that lacks a gut and cannot feed, and which metamorphoses rapidly into the adult form.

Summary

Changes in the timing of developmental processes that have occurred during evolution can alter the form of the body, for if different regions grow faster than others, their size is proportionately increased if the animal gets larger. The decrease in size of the lateral digits of horses is partly due to this type of developmental change. Speed of development and egg size also have important evolutionary implications. Changing the time at which an animal becomes sexually mature can result in adults with larval characteristics. Some animals, such as frogs and sea urchins, which usually have a larval form, have evolved species that develop directly into the adult without a larval stage.

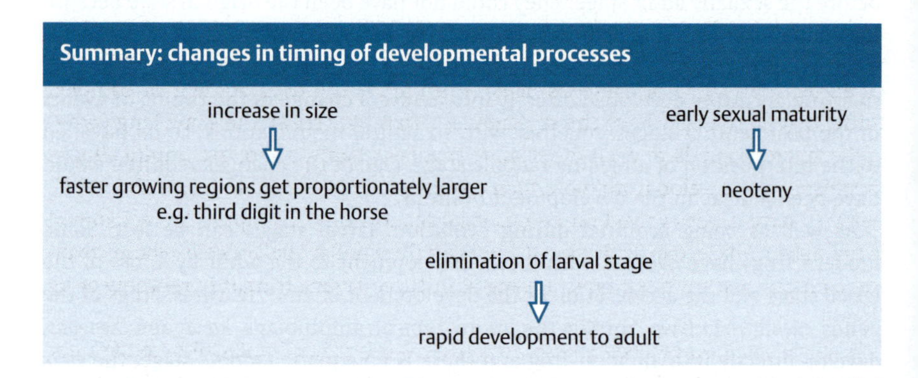

The evolution of development

How did development itself evolve? What is the origin of the egg and how did processes like pattern formation and gastrulation evolve? What follows are some possible answers, admittedly highly speculative.

14.12 How multicellular organisms evolved from single-celled ancestors is still highly speculative

Some 600 million years ago, multicellular animals evolved from a unicellular organism, possibly a flagellate. What had to be invented for this transition? And how did embryonic development from an egg evolve? The key requirements for embryonic development, as we have seen, are a program of gene activity, cell differentiation, signal transduction (so cells can communicate with each other), and cell motility and cohesion (so that overall form can change).

Judging by modern unicellular eukaryotes, the ancestral single-celled organism possessed all these features in primitive form, and little new would have had to be invented, although much had to evolve further and acquire different functions for multicellularity to evolve. At a minimum, the ancestor of multicellular organisms would have already possessed the characteristic eukaryotic cell cycle, which entails a complex program of gene activity, the ability to undergo cellular differentiation, signaling by receptor tyrosine kinases, and cadherin-mediated cell adhesion.

What then was the origin of multicellularity and the embryo? One possibility, and it is highly speculative, is that mutations resulted in the progeny of a single-celled organism not separating after cell division, leading to a loose colony of identical cells that occasionally fragmented to give new 'individuals'. One advantage of a colony might originally have been that when food was in short supply, the cells could feed off each other, and so the colony survived. This could have been the origin of both multicellularity and the requirement for cell death in multicellular organisms. The egg may subsequently have evolved as the cell fed by other cells; in sponges, the egg phagocytoses neighboring cells.

Once multicellularity evolved, it opened up all sorts of new possibilities, such as cell specialization for different functions. Some cells could specialize in providing motility, for example, and others in feeding, as we see in the endoderm of *Hydra*. The origin of pattern formation—that is, cell differentiation in a spatially organized arrangement—is not known, but may have depended on gradients set up by external influences such as inside–outside differences.

Development from a single cell may be essential for the evolution of complex organisms. Development from more than one cell presents problems, as mutations could occur in some of the cells. It is essential for the development of complex forms that all the cells have the same genes—that they speak the same language—and can send signals to each other and respond reliably. There are 'multicellular' organisms that develop from more than one cell, such as slime molds (see Chapter 5), but they have never evolved numerous complex forms. The simplest known animal is the marine placozoan *Trichoplax*. It is essentially a hollow ball of cells whose covering contains just four cell types arranged in three layers. It mainly reproduces by fission.

How gastrulation evolved is also unknown, but it is not implausible to consider a scenario in which a hollow sphere of cells, the common ancestor of all multicellular animals, changed its form to assist feeding (Fig. 14.25). This ancestor may, for example, have sat on the ocean floor, ingesting food particles by phagocytosis. A small invagination developing in the body wall could have promoted feeding by forming a primitive gut. Ciliary movement could have swept food particles more efficiently into this region, where they would be phagocytosed. Once the invagination formed, it is not too difficult to imagine how it could eventually extend right across the sphere, fuse with the other side, and form a continuous gut, which

Fig. 14.25 A possible scenario for the development of the gastrula. A colonial protozoan in the form of a hollow multicellular sphere could have settled on the sea bottom and developed a gut-like invagination to aid feeding.

Based on Jaegerstern, G.: 1956.

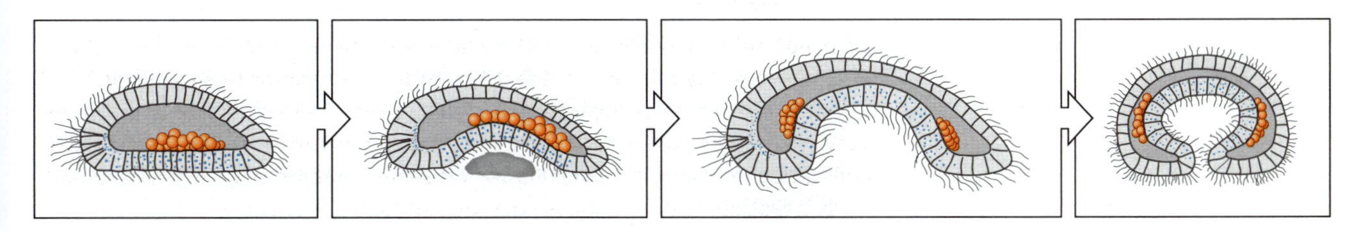

would be the endoderm. At a later stage in evolution, cells migrating inside between the gut and the outer epithelium would give rise to the mesoderm.

Gastrulation provides a good example of developmental change during evolution. While there is considerable similarity in the process of gastrulation in many different animals, there are also significant differences. But how these evolved and what could have been the adaptive nature of the intermediate forms is a hard problem.

Many animals have evolved larval forms that undergo a dramatic change in form to reach the adult state at metamorphosis. A larval stage has advantages when it comes to dispersal and feeding, but it is still difficult to understand how larval stages and metamorphosis could have evolved. The essence of development is gradual change; yet at metamorphosis there is no gradual continuity between larva and adult. This can be understood if it is assumed that all larval forms are due to the intercalation of the larval stage into the development of a directly developing animal. In frogs, for example, at some point in evolution the stage following formation of the body but before limb development became mobile, which aided dispersal, and later evolved into the feeding tadpole. The trick was to get back to the normal developmental program, and that is what metamorphosis is for.

Finally, the most important goal for the developing embryo is to reliably and consistently produce a fully functional adult. This is achieved in spite of environmental variations and small genetic changes, and this phenomenon, which buffers development against such variations, is called canalization. How this robustness is achieved is still a major question. Evolution selects for embryos whose development is reliable and consistent.

Summary

The embryo arose during the evolution of multicellular organisms from single-celled organisms, which have most of the cellular properties required by embryonic development. Multicellularity, having arisen, might have persisted originally because a colony of cells could provide a source of food for some of the members in times of scarcity. The egg might have evolved as the cell that is fed by other cells and provides the basis for the development of complex structures.

SUMMARY TO CHAPTER 14

Many developmental processes have been conserved during evolution. While many questions relating to evolution and development remain unanswered, it is clear that development reflects the evolutionary history of ancestral embryos. All vertebrate embryos pass through a conserved phylotypic developmental stage, although there can be considerable divergence both earlier and later in development. The signals involved in patterning of vertebrate and arthropod appendages, as well as of the dorso-ventral axis, show remarkable similarity and conservation. The pattern of Hox gene expression along the body axis of vertebrates and arthropods is conserved, and major changes in body plan reflect changes in both Hox gene expression and their downstream targets. Changes in gene-regulatory regions were crucial. Alterations in the timing of developmental events have played an important role in evolution. Such changes can alter the overall form of an organism as a result of differences in growth rates of different structures, and can also result in sexual maturation at larval stages. The origins of multicellularity and the embryo from a single-celled ancestor are still highly speculative.

GENERAL FURTHER READING

Carroll, S.B., Grenier, J.K., Weatherbee, S.D.: *From DNA to Diversity*, 2nd edn. Malden, MA: Blackwell Science, 2005.

Finnerty, J.R., Pang, K., Burton, P., Paulson, D., Martindale, M.Q.: **Origins of bilateral symmetry: Hox and *dpp* expression in a sea anemone.** *Science* 2004, **304:** 1335–1337.

Gerhart, J., Kirschner, M.: *Cells, Embryos and Evolution*. Malden, MA: Blackwell Science, 1997.

Kirschner, M., Gerhart, J.: **Evolvability.** *Proc. Natl Acad. Sci. USA* 1998, **95:** 8420–8427.

Raff, R.A.: *The Shape of Life*. University of Chicago Press, 1996.

Richardson, M.K.: **Vertebrate evolution: the developmental origins of adult variation.** *BioEssays* 1999, **21:** 604–613.

SECTION FURTHER READING

14.1 The evolution of life histories has implications for development

Partridge, L., Harvey, P.: **The ecological context of life history evolution.** *Science* 1988, **241:** 1449–1455.

Yamamoto, Y., Stock, D.W., Jeffery, W.R.: **Hedgehog signalling controls eye degeneration in blind cavefish.** *Nature* 2004, **431:** 844–847.

14.2 Embryonic structures have acquired new functions during evolution

Abzhanov, A., Protas, M, Grant, B.R., Grant, P.R., Tabin, C.J.: ***Bmp4*** **and morphological variation of beaks in Darwin's finches.** *Science* 2004, **305:** 1462–1465.

Cerny, R., Lwigale, P., Ericsson, R., Meulemans, D., Epperlein, H.H., Bronner-Fraser, M.: **Developmental origins and evolution of jaws: new interpretation of "maxillary" and "mandibular".** *Dev. Biol.* 2004, **276:** 225–236.

Cohn, M.J.: **Lamprey Hox genes and the origin of jaws.** *Nature* 2002, **416:** 386–387.

Romer, A.S.: *The Vertebrate Body*. Philadelphia: W.B. Saunders, 1949.

14.3 Limbs evolved from fins

Capdevila, J., Izpisua Belmonte, J.C.: **Perspectives on the evolutionary origin of tetrapod limbs.** *J. Exp. Zool. (Mol. Dev. Evol.)* 2000, **288:** 287–303.

Coates, M.I., Jeffery, J.E., Rut, M.: **Fins to limbs: what the fossils say.** *Evol. Dev.* 2002, **4:** 390–401.

Cohn, M.J., Tickle, C.: **Developmental basis of limblessness and axial patterning in snakes.** *Nature* 1999, **399:** 474–479.

Ruvinsky, I., Gibson-Brown, J.J.: **Genetic and developmental bases of serial homology in vertebrate limb evolution.** *Development* 2000, **127:** 5211–5244.

Sordino, P., van der Hoeven, F., Duboule, D.: **Hox gene expression in teleost fins and the origin of vertebrate digits.** *Nature* 1995, **375:** 678–681.

14.4 Vertebrate and insect wings make use of evolutionarily conserved developmental mechanisms

Panganiban, G., Irvine, S.M., Lowe, C., Roehl, H., Corley, L.S., Sherbon, B., Grenier, J.K., Fallon, J.F., Kimble, J., Walker, M., Wray, G.A., Swalla, B.J., Martindale, M.Q., Carroll, S.B.: **The origin and evolution of animal appendages.** *Proc. Natl Acad. Sci. USA* 1997, **94:** 5162–5166.

14.5 Hox gene complexes have evolved through gene duplication

Brooke, N.M., Gacia-Fernandez, J., Holland, P.W.H.: **The ParaHox gene cluster is an evolutionary sister of the Hox gene cluster.** *Nature* 1998, **392:** 920–922.

Ferrier, D.E.K., Holland, P.W.H: **Ancient origin of the Hox gene cluster.** *Nat. Rev. Genet.* 2001, **2:** 33–34.

Prince, V.E., Pickett, F.B.: **Splitting pairs: the diverging fates of duplicated genes.** *Nat. Rev. Genet.* 2002, **3:** 827–837.

Valentine, J.W., Erwin, D.H., Jablonski, D.: **Developmental evolution of metazoan bodyplans: the fossil evidence.** *Dev. Biol.* 1996, **173:** 373–381.

14.6 Changes in Hox genes have generated the elaboration of vertebrate and arthropod body plans

Akam, M.: **Hox genes and the evolution of diverse body plans.** *Phil. Trans. R. Soc. Lond. B* 1995, **349:** 313–319.

Averof, M., Akam, M.: **Hox genes and the diversification of insect and crustacean body plans.** *Nature* 1995, **376:** 420–423.

Averof, M.: **Origin of the spider's head.** *Nature* 1998, **395:** 436–437.

Duboule, D.: **A Hox by any other name.** *Nature* 2000, **403:** 607–610.

Galant, R., Carroll, S.B.: **Evolution of a transcriptional repression domain in an insect Hox protein.** *Nature* 2003, **415:** 910–913.

Pavlopoulos, A., Averof, M.: **Developmental evolution: Hox proteins ring the changes.** *Curr. Biol.* 2002, **12:** R291–R293.

Slack, J.M., Holland, P.W., Graham, C.F.: **The zootype and phylotypic stage.** *Nature* 1993, **361:** 490–492.

Stern, D.L.: **A role of *Ultrabithorax* in morphological differences between *Drosophila* species.** *Nature* 1998, **396:** 463–466.

14.7 The position and number of paired appendages in insects is dependent on Hox gene expression

Carroll, S.B., Weatherbee, S.D., Langeland, J.A.: **Homeotic genes and the regulation and evolution of insect wing number.** *Nature* 1995, **375:** 58–61.

Levine, M.: **How insects lose their wings.** *Nature* 2002, **415:** 848–849.

Weatherbee, S.D., Carroll, S.: **Selector genes and limb identity in arthropods and vertebrates.** *Cell* 1999, **97:** 283–286.

Weatherbee, S.D., Nijhout, H.F, Grunnert, L.W., Halder, G., Galant, R., Selegue, J., Carroll, S.: **Ultrabithorax function in butterfly wings and the evolution of insect wing patterns.** *Curr. Biol.* 1999, **9:** 109–115.

14.8 The basic body plan of arthropods and vertebrates is similar, but the dorso-ventral axis is inverted

Arendt, D., Nübler-Jung, K.: **Dorsal or ventral: similarities in fate maps and gastrulation patterns in annelids, arthropods and chordates.** *Mech. Dev.* 1997, **61:** 7–21.

Davis, G.K., Patel, N.H.: **The origin and evolution of segmentation.** *Trends Biochem. Sci.* 1999, **24:** M68–M72.

Gerhart, J., Lowe, C., Kirschner, M.: **Hemichordates and the origins of chordates.** *Curr. Opin. Genet. Dev.* 2005, **15:** 461–467.

Holley, S.A., Jackson, P.D., Sasai, Y., Lu, B., De Robertis, E., Hoffman, F.M., Ferguson, E.L.: **A conserved system for dorso-ventral patterning in insects and vertebrates involving *sog* and *chordin*.** *Nature* 1995, **376:** 249–253.

Lowe, C.J., Wu, M., Salic, A., Evans, L., Lander, E., Stange-Thomann, N., Gruber, C.E., Gerhart, J., Kirschner, M.: **Anteroposterior patterning in hemichordates and the origins of the chordate nervous system.** *Cell* 2003, **113:** 853–865.

14.9 Evolution of spatial pattern may be based on just a few genes

Gompel, N., Carrol, S.B.: **Genetic mechanisms and constraints governing the evolution of correlated traits in drosophilid flies.** *Nature* 2003, **424:** 931–935.

Gompel, N., Prud'homme, B., Wittkopp, P.J., Kassner, V.A., Carroll, S.B.: **Chance caught on the wing: cis-regulatory evolution and the origin of pigment patterns in *Drosophila*.** *Nature* 2005, **433:** 481–487.

Shapiro, M.D., Marks, M.E., Peichel, C.L., Blackman, B.K., Nereng, K.S., Jonsson, B., Schluter, D., Kingsley, D.M.: **Genetic and developmental basis of evolutionary pelvic reduction in threespine sticklebacks.** *Nature* 2004, **428:** 717–723.

14.10 Changes in growth can alter the shapes of organisms

Huxley, J.S.: *Problems of Relative Growth*. London: Methuen & Co. Ltd., 1932.

14.11 The timing of developmental events has changed during evolution

Alberch, P., Alberch, J.: **Heterochronic mechanisms of morphological diversification and evolutionary change in the neotropical salamander *Bolitoglossa occidentales* (Amphibia: Plethodontidae).** *J. Morphol.* 1981, **167:** 249–264.

Lande, R.: **Evolutionary mechanisms of limb loss in tetrapods.** *Evolution* 1978, **32:** 73–92.

Raynaud, A.: **Developmental mechanism involved in the embryonic reduction of limbs in reptiles.** *Int. J. Dev. Biol.* 1990, **34:** 233–243.

Wray, G.A., Raff, R.A.: **The evolution of developmental strategy in marine invertebrates.** *Trends Evol. Ecol.* 1991, **6:** 45–56.

14.12 How multicellular organisms evolved from single-celled ancestors is still highly speculative

Brooke, N.M., Holland, P.W.H.: **The evolution of multicellularity and early animal genomes.** *Curr. Opin. Genet. Dev.* 2003, **6:** 599–603.

Jaegerstern, G.: **The early phylogeny of the metazoa. The bilaterogastrea theory.** *Zool. Bidrag. (Uppsala)* 1956, **30:** 321–354.

King, N.: **The unicellular ancestry of animal development.** *Dev. Biol.* 2004, **7:** 313–325.

Miller, D.J., Ball, E.E.: **Animal evolution: the enigmatic phylum Placozoa revisited.** *Curr. Biol.* 2005, **15:** R26–R28.

Rudel, D., Sommer, R.J.: **The evolution of developmental mechanisms.** *Dev. Biol.* 2003, **264:** 15–37.

Siegal, M.L., Bergman, A.: **Waddington's canalization revisited: developmental stability and evolution.** *Proc. Natl Acad. Sci. USA* 2002, **99:** 10528–10532.

Szathmary, E., Wolpert, L.: **The transition from single cells to multicellularity.** in *Genetic and Cultural Evolution of Cooperation* (ed. Hammersteen, P.), 271–289. Cambridge, MA: MIT Press, 2004.

Wolpert L.: **Gastrulation and the evolution of development.** *Development (Suppl.)* 1992, 7–13.

Wolpert, L., Szathmary, E.: **Evolution and the egg.** *Nature* 2002, **420:** 745.

Glossary

In insects the body is divided into three distinct parts, the head at the anterior end, followed by the thorax and the posterior abdomen.

Abembryonic pole *see* embryonic–abembryonic axis.

The acron is a specialized structure associated with the most anterior region of the *Drosophila* embryo.

The acrosomal reaction is the release of enzymes and other proteins from the acrosomal vesicle of the sperm head that occurs once a sperm has bound to the outer surface of the egg. It helps the sperm to penetrate the outer layers of the egg.

Actin filaments or microfilaments are one of the three principal protein filaments of the cytoskeleton. They are involved in cell movement and changes in cell shape. Actin filaments are also part of the contractile apparatus of muscle cells.

Gene-regulatory proteins that turn genes on are known as activators.

Actomyosin is an assembly of actin filaments and the motor protein myosin that can undergo contraction.

In plants, the adaxial—abaxial axis runs from the center of the stem to the circumference. In plant leaves it runs from the upper surface to the lower surface (dorsal to ventral).

Adherens junctions are a type of adhesive cell junction in which the adhesion molecules linking the two cells together are cadherins that are linked intracellularly to the actin cytoskeleton.

The allantois is a set of extra-embryonic membranes that develops in some vertebrate embryos. In bird and reptile embryos it acts as a respiratory surface, while in mammals its blood vessels carry blood to and from the placenta.

An allele is a particular version of a gene. In diploid organisms two alleles of each gene are present, which may or may not be the same.

The amnion is an extra-embryonic membrane in birds, reptiles, and mammals, which forms a fluid-filled sac that encloses and protects the embryo. It is derived from extra-embryonic ectoderm and mesoderm.

The amnioserosa of the *Drosophila* embryo is an extra-embryonic membrane on the dorsal side of the embryo.

An androgenetic embryo is an embryo in which the two sets of homologous chromosomes are both paternal in origin.

Angioblasts are the mesodermal precursor cells that will give rise to blood vessels.

Angiogenesis is the process by which small blood vessels sprout from the larger vessels.

The animal region of an egg is the end of the egg where the nucleus resides, usually away from the yolk. The most terminal part of this region is the animal pole, which is directly opposite the vegetal pole at the other end of the egg. In *Xenopus* the pigmented animal half is called the animal cap.

The animal–vegetal axis runs from the animal pole to the vegetal pole in an egg or early embryo.

The anterior visceral ectoderm (AVE) is an extra-embryonic tissue in the early mouse embryo that is involved in inducing anterior regions of the embryo.

The antero-posterior axis defines which is the 'head' end and which is the 'tail' end of an animal. The head is anterior and the tail posterior. In the vertebrate limb, this axis runs from the thumb to little finger.

Anticlinal cell divisions are divisions in planes at right angles to the outer surface of a tissue.

Antisense RNA is an RNA complementary in sequence to an mRNA or gene coding sequence and is used to block the expression of a protein by binding to its mRNA.

The apical–basal axis of a plant is the axis running from shoot tip to root tip.

The apical ectodermal ridge or apical ridge is a thickening of the ectoderm at the distal end of the developing chick and mammalian limb bud. Signals from the ridge specify the progress zone in the underlying mesoderm.

An apical meristem is the region of dividing cells at the tip of a growing shoot or root.

Apoptosis or programmed cell death is a type of cell death that occurs widely during development. In programmed cell death, a cell is induced to commit 'suicide', which involves fragmentation of the DNA and shrinkage of the cell. These apoptotic cells are removed by the body's scavenger cells and, unlike necrosis, their death does not cause damage to surrounding cells.

The archenteron is the cavity formed inside the embryo when the endoderm and mesoderm invaginate during gastrulation. It forms the gut.

The area opaca is the outer dark area of the chick blastoderm.

The area pellucida is the central clear area of the chick blastoderm.

Asymmetric cell division or asymmetric division are cell divisions in which the daughter cells are different from each other because some cytoplasmic determinant(s) have been distributed unequally between them.

A developmental process is said to be proceeding **autonomously** when it can continue without a requirement for extracellular signals to be continuously present. *See also* **cell autonomous**.

AVE *see* **anterior visceral endoderm**.

Axial structures are those that form along the main axis of the body, such as the notochord, vertebral column, and neural tube in vertebrates.

Axons are long cell processes of neurons that conduct nerve impulses away from the cell body. The end of an axon, the **axon terminal**, forms contacts (synapses) with other neurons, muscle cells, or glandular cells.

The protein **β-catenin** functions both as a transcription factor and as one of the proteins present at cell junctions. In its role as a transcription factor it is activated in early development in many vertebrates as the end result of a Wnt signaling pathway.

The **basal lamina** is a sheet of extracellular matrix that separates an epithelial layer from the underlying tissues. For example, the epidermis of the skin is separated from the dermis by a basal lamina.

Animals in which the only axis of symmetry is the central axis running from head to tail are said to possess **bilateral symmetry**. The two sides of the body are mirror images of each other.

In amphibian limb regeneration, a **blastema** is formed from the dedifferentiation and proliferation of cells beneath the wound epidermis, and gives rise to the regenerated limb.

The **blastocoel** is the fluid-filled cavity that develops in the interior of a blastula.

The **blastocyst** stage of a mammalian embryo corresponds in form to the blastula stage of other animal embryos, and is the stage at which the embryo implants in the uterine wall.

A **blastoderm** is a post-cleavage embryo composed of a solid layer of cells rather than a spherical blastula, as in early chick and *Drosophila* embryos. The chick blastoderm is also known as the **blastodisc**.

Blastomeres are the cells derived from cleavage of the early embryo.

The **blastopore** is the slit-like or circular invagination on the surface of amphibian and sea urchin embryos where the mesoderm and endoderm move inside the embryo at gastrulation.

The **blastula** stage in animal development is the outcome of cleavage. The blastula is a hollow ball of cells, composed of an epithelial layer of small cells enclosing a fluid-filled cavity—the blastocoel.

The **body plan** describes the overall organization of an organism, for example the position of the head and tail, and the plane of bilateral symmetry, where it exists. The body plan of most animals is organized around two main axes, the antero-posterior axis and the dorso-ventral axis.

The **brachial** region of a vertebrate embryo is the region that includes the forelimb or that gives rise to the structures of the forelimb.

The **cadherins** are a family of cell-adhesion molecules with important roles in development.

The **cambium** in plants is a ring of meristem in the stem that gives rise to new stem tissue that increases the diameter of the stem.

The **canonical Wnt pathway** is an intracellular signaling pathway stimulated by members of the Wnt family of signal proteins that leads to the stabilization of β-catenin and its entry into nuclei, where it acts as a transcription factor.

The functional maturation of sperm after they have been deposited in the female reproductive tract is known as **capacitation**.

Caspases are intracellular proteases, some of which are involved in apoptosis.

Cell-adhesion molecules bind cells to each other and to the extracellular matrix. The main classes of adhesion molecules important in development are the cadherins, the immunoglobulin superfamily, and the integrins.

The effects of a gene are **cell autonomous** if they only affect the cell the gene is expressed in.

The **cell body** is the part of a neuron that contains the nucleus, and from which the axon and dendrites extend.

Cell–cell interaction and **cell–cell signaling** are general terms to describe many different types of intercellular communication by which one cell influences the behavior of another cell. Cells can communicate with each other via cell contact or by the secretion of signaling molecules that influence the behavior of other cells nearby or at a distance.

The **cell cycle** is the sequence of events by which a cell duplicates itself and divides in two.

During **cell differentiation**, cells become functionally and structurally different from one another and become distinct cell types, such as muscle or blood cells.

Cell-lineage restriction occurs when all the descendants of a particular group of cells remain within a 'boundary' and never mix with an adjacent group of cells. Compartment boundaries in insect development are boundaries of lineage restriction.

Cell senescence *see* **senescence**.

The **chemoaffinity hypothesis** proposes that each retinal neuron carries a chemical label that enables it to connect reliably with an appropriately labeled cell in the optic tectum.

A **chemoattractant** is a molecule that attracts cells to move towards it.

A **chemorepellant** is a molecule that repels cells, causing them to move away from it.

A **chimeric** organism or tissue (a chimera) is made up of cells from two or more different sources, and thus of different genetic constitutions.

Chondrocytes are differentiated cartilage cells.

The **chorion** is the outermost of the extra-embryonic membranes in birds, reptiles, and mammals. It is involved in respiratory gas exchange. In birds and reptiles it lies just beneath the shell. In mammals it is part of the placenta and is also involved in nutrition and waste removal. The chorion of insect eggs has a different structure.

Chromatin is the material of which chromosomes are made. It is composed of DNA and protein. Enzyme complexes called **chromatin-remodeling complexes** can act on chromatin to modify it and alter the ability of the DNA to be transcribed.

The **circadian clock** is an internal 24-hour timer present in living organisms that causes many metabolic and physiological processes, including the expression of some genes, to vary in a regular manner throughout the day.

The ***cis*-regulatory control region** of a gene comprises the sequences flanking the gene and containing sites at which the expression of that gene can be controlled. Many control regions contain a variety of different ***cis*-regulatory modules**, which are short regions containing multiple binding sites for different transcription factors; the combination of factors bound determines whether the gene is switched on or off.

Cleavage occurs after fertilization and is a series of rapid cell divisions without growth that divides the embryo up into a number of small cells.

A **clone** is a collection of genetically identical cells derived from a single cell by repeated cell division, or the genetically identical offspring of a single individual produced by asexual reproduction or artifical cloning techniques.

Cloning is the procedure by which an individual genetically identical to a 'parent' is produced by transplantation of a parental somatic cell nucleus into an unfertilized oocyte.

The **coding region** of a gene is that part of the DNA that encodes a polypeptide or functional RNA.

The correspondence between the order of Hox genes on a chromosome and their temporal and spatial order of expression in the embryo is known as **co-linearity**.

Induction of cell differentiation in some tissues depends on a **community effect**, in that there have to be a sufficient number of responding cells present for differentiation to occur.

Compaction of the mouse embryo occurs during early cleavage. The blastomeres flatten against each other and microvilli become confined to the outer surface of the ball of cells.

Compartments are discrete areas of an embryo that contain all the descendants of a small group of founder cells and which show cell-lineage restriction. Cells in compartments respect the compartment boundary and do not cross over into an adjacent compartment. Compartments tend to act as discrete developmental units.

Competence is the ability of a tissue to respond to an inducing signal. Embryonic tissues only remain **competent** to respond to a particular signal for a limited period of time.

The **control region** of a gene is the region to which regulatory proteins bind and so determine whether or not the gene is transcribed.

Convergent extension is the process by which a sheet of cells changes shape by extending in one direction and narrowing—converging—in a direction at right angles to the extension.

Cortical rotation occurs immediately after an amphibian egg is fertilized. The egg **cortex**, an actin-rich layer of cytoplasm lying immediately below the surface, rotates with respect to the underlying cytoplasm, toward the point of sperm entry.

A **cotyledon** is the part of the plant embryo that acts as a food storage organ.

Cyclins are proteins that periodically rise and fall in concentration during the cell cycle and are involved in controlling progression through the cycle. They act by binding to and activating **cyclin-dependent kinases (CDKs)**.

Cytoplasmic localization is the non-uniform distribution of some factor or determinant in a cell's cytoplasm, so that when the cell divides, the determinant is unequally distributed to the daughter cells.

The long-lived **dauer** larvae of *C. elegans* are a response to starvation conditions and neither eat nor grow until food is again available.

Dedifferentiation is loss of the structural characteristics of a differentiated cell, which may result in the cell then differentiating into a new cell type.

Delamination is the process in which epithelial cells leave an epithelium as individual cells. It occurs, for example, in the primitive streak and in the movement of neural crest cells out of the neural tube.

Dendrites are extensions from the body of a nerve cell that receive stimuli from other nerve cells.

Denticles are small tooth-like outgrowths of the cuticle on insect larvae.

The **dermatome** is the region of the somite that will give rise to the dermis.

The **dermis** of the skin is the connective tissue beneath the epidermis, from which it is separated by a basal lamina.

The **dermomyotome** is the region of the somite that will give rise to both muscle and dermis.

Determinants are cytoplasmic factors (e.g. proteins and RNAs) in the egg and in embryonic cells that can be asymmetrically distributed at cell division and so influence how the daughter cells develop.

Determination implies a stable change in the internal state of a cell such that its fate is now fixed, or **determined**. A determined cell will follow that fate when grafted into other regions of the embryo.

Deuterostomes are those animals, such as chordates and echinoderms, that have radial cleavage of the egg, and in which the primary invagination of the gut at gastrulation forms the anus, with the mouth developing independently.

Genes that specifically control a developmental process are known as **developmental genes**.

Diploblasts are animals with two germ layers (endoderm and ectoderm) only and include coelenterates such as *Hydra* and jellyfish.

Diploid cells contain two sets of homologous chromosomes, one from each parent, and thus two copies of each gene.

Directed dilation is the extension of a tube-like structure at each end due to hydrostatic pressure, the direction of extension reflecting greater circumferential resistance to expansion.

The **distal** end of a structure such as a limb is the end furthest away from the point of attachment to the body.

DNA microarrays, also known as **DNA chips**, are arrays of oligonucleotides that are used to detect and measure the expression of large numbers of genes simultaneously, by hybridization of cellular RNA or cDNA.

A **dominant** allele is one that determines the phenotype even when present in only a single copy.

A **dominant-negative** mutation inactivates a particular cellular function by the production of a defective RNA or protein molecule that blocks the normal function of the gene product.

Dorsalized embryos develop greatly increased dorsal regions at the expense of ventral regions.

Dorsalizing factors in vertebrate embryos are proteins that promote the formation of dorsal structures.

The **dorso-ventral** axis defines the relation of the upper surface or back (dorsal) to the under surface (ventral) of an organism or structure. The mouth is always on the ventral side.

Dosage compensation is the mechanism that ensures that although the number of X chromosomes in males and females is different, the level of expression of X-chromosome genes is the same in both sexes. Mammals, insects, and nematodes all have different dosage compensation mechanisms.

Ecdysis is a type of molting in arthropods in which the external cuticle is shed to allow for growth.

The **ectoderm** is the germ layer that gives rise to the epidermis and the nervous system.

The **egg cylinder** in early post-implantation mouse embryogenesis is the cylindrical structure comprising the epiblast covered by visceral endoderm.

Embryogenesis is the process of development of the embryo from the fertilized egg.

Embryology is the study of the development of an embryo.

The **embryonic-abembryonic axis** in the mammalian blastocyst runs from the site of attachment of the inner cell mass—the **embryonic pole**—to the opposite pole, the **abembryonic pole**.

The **embryonic ectoderm** is the name given to the mouse epiblast once it has developed into an epithelial sheet.

Embryonic stem cells (**ES cells**) are derived from the inner cell mass of a mammalian embryo, usually mouse, and can be indefinitely maintained in culture. When injected into another blastocyst, they combine with the inner cell mass and can potentially contribute to all the tissues of the embryo.

In the chick embryo, the hypoblast underlying the epiblast is replaced by a layer of cells called the **endoblast** that grows out from the posterior marginal zone prior to primitive streak formation.

The **endocardium** is the inner endothelial layer of the developing heart.

Endochondral ossification is the replacement of cartilage with bone in the growth plate of vertebrate embryonic skeletal elements, such as those that give rise to the long bones of the limbs.

The **endoderm** is the germ layer that gives rise to the gut and associated organs, such as the lungs and liver in vertebrates.

The **endodermis** is a tissue layer in plant roots interior to the cortex and outside the vascular tissue.

Endomesoderm is a tissue that can give rise to both mesoderm and endoderm.

The **endosperm** in higher plant seeds is a nutritive tissue that serves as a source of food for the embryo.

The **enhancer-trap** technique is used in *Drosophila* to turn on the expression of a specific gene in a particular tissue or stage in development.

Ephrins and their receptors, **Eph receptors**, are cell-surface molecules involved in delimiting compartments in rhombomeres and in axonal guidance.

The **epiblast** of mouse and chick embryos is a group of cells within the blastocyst or blastoderm, respectively, that gives rise to the embryo proper. In the mouse, it develops from cells of the inner cell mass.

Epiboly is the process during gastrulation in which the ectoderm extends to cover the whole of the embryo.

The **epidermis** in vertebrates, insects, and plants is the outer layer of cells that forms the interface between the organism and its environment. Its structure is quite different in the different organisms.

Epimorphosis is a type of regeneration in which the regenerated structures are formed by new growth.

ES cells *see* **embryonic stem cells**.

An epithelium is said to **evaginate** when it forms a tubular outgrowth from the surface.

Extra-embryonic ectoderm in mammals contributes to the formation of the placenta.

Extra-embryonic membranes are membranes external to the embryo proper that are involved in the protection and nutrition of the embryo.

The **fate** of cells describes what they will normally develop into. By marking cells in the embryo, a **fate map** of embryonic regions can be constructed. Having a particular normal fate does not, however, imply that a cell could not develop differently if placed in a different environment.

Fertilization is the fusion of sperm and egg to form the zygote.

A **file** of cells in a plant root is a vertical column of cells that originates from a single initial in the root meristem.

The **floor plate** is a small region of the developing neural tube at the ventral midline that is composed of non-neural cells. It is involved in patterning the ventral part of the neural tube.

A **floral meristem** is a region of dividing cells at the tip of a shoot that gives rise to a flower.

The individual parts of a flower develop from **floral organ primordia** generated by the floral meristem and are given their individual identities by the expression of **floral organ identity genes**.

Follicle cells are somatic cells that surround the oocyte and nurse cells during egg development in *Drosophila*.

Plant organs such as flowers and leaves each develop from a small number of **founder cells** that derive from the apical meristem.

The **gametes** are the cells that carry the genes to the next generation—in animals they are the eggs and sperm.

A **ganglion mother cell** is formed by division of a neuroblast in *Drosophila* and gives rise to neurons.

Gap genes are zygotic genes coding for transcription factors expressed in early *Drosophila* development that subdivide the embryo into regions along the antero-posterior axis.

The **gastrula** is the stage in animal development when the endoderm and mesoderm of the blastula move inside the embryo.

Gastrulation is the process in animal embryos in which the endoderm and mesoderm move from the outer surface of the embryo to the inside, where they give rise to internal organs.

Gene knock-out refers to the complete inactivation of a particular gene in an organism by means of genetic manipulation.

Gene-regulatory proteins are proteins that bind to control regions in DNA and help to switch genes on and off.

When genes are switched off by microRNAs, RNA interference, or changes to chromatin this is known as **gene silencing**.

General transcription factors are ubiquitous transcription factors that form a complex with sites in the gene promoter and with the RNA polymerase to enable the polymerase to start transcription.

The **genital ridge** in vertebrates is the region of mesoderm lining the abdominal cavity from which the gonads develop.

The **genotype** is a description of the exact genetic constitution of a cell or organism in terms of the alleles it possesses for any given gene.

The **germarium** in female *Drosophila* contains stem cells that give rise to a succession of egg chambers, each containing an oocyte.

The **germ band** is the name given to the ventral blastoderm of the early *Drosophila* embryo, from which most of the embryo will eventually develop.

Germ cells are those cells that give rise to eggs and sperm.

The **germ layers** refer to the regions of the early animal embryo that will give rise to distinct types of tissue. Most animals have three germ layers—ectoderm, mesoderm, and endoderm.

The **germline** cells give rise to the gametes.

Germplasm is the special cytoplasm in some animal eggs, such as those of *Drosophila*, that is involved in the specification of germ cells.

Glia are supporting cells of the nervous system, such as Schwann cells.

The **globular stage** of a plant embryo is a ball of around 32 cells.

The **gonads** are the reproductive organs of animals.

Growth is an increase in size, which occurs by cell multiplication, increase in cell size and deposition of extracellular material.

Axons of developing neurons extend by means of a **growth cone** at their tip. The growth cone both crawls forward on the substrate and senses its environment by means of filopodia.

Growth hormone is a protein hormone produced by the pituitary gland that is essential for the post-embryonic growth of humans and other mammals.

Growth of vertebrate long bones occurs at the cartilaginous **growth plates**. The cartilage grows and is eventually replaced by bone by the process of endochondral ossification.

A **gynogenetic** embryo is one in which the two sets of homologous chromosomes are both maternal in origin.

Haploid cells are derived from diploid cells by meiosis and contain only one set of chromosomes (half the diploid number of chromosomes), and thus contain only one copy of each gene. In most animals the only haploid cells are the gametes—the sperm or egg.

The **heart stage** is a stage in embryogenesis in dicotyledonous plants in which the cotyledons and embryonic root are starting to form, giving a heart-shaped embryo.

The **Hedgehog** signaling protein of *Drosophila* is a member of an important family of developmental signaling proteins that includes Sonic hedgehog in vertebrates.

Hematopoiesis is the process by which all the blood cells are derived from a pluripotent stem cell. This occurs mainly in the bone marrow.

A **hemisegment** in *Drosophila* is the lateral half of a segment, one on each side of the midline, and is the developmental unit for the nervous system.

Hensen's node is a condensation of cells at the anterior end of the primitive streak in chick and mouse embryos. Cells from the node give rise to the notochord. It corresponds to the Spemann organizer in amphibians.

A **hermaphrodite** is an organism that possesses both male and female gonads and produces both male and female gametes.

Heterochromatin is the state of chromatin in regions of the chromosome that are so condensed that transcription is not possible.

Heterochrony is an evolutionary change in the timing of developmental events. A mutation that changes the timing of a developmental event is called a **heterochronic** mutation.

A diploid individual is **heterozygous** for a given gene when it carries two different alleles of that gene.

The **homeobox** is a region of DNA in homeotic genes that encodes a DNA-binding domain called the **homeodomain**. Genes containing this motif are known generally as **homeobox genes**. The homeodomain is present in a large number of transcription factors that are important in development, such as the products of the Hox genes and the Pax genes.

Homeosis is the phenomenon in which one structure is transformed into another, homologous, structure. An example of a **homeotic transformation** is the development of legs in place of antennae in *Drosophila* as a result of mutation in a **homeotic gene**.

Homeotic selector genes in *Drosophila* are genes that specify the identity and developmental pathway of a group of cells. They encode homeodomain transcription factors and act by controlling the expression of other genes. Their expression is required throughout development. The *Drosophila* gene *engrailed* is an example of a homeotic selector gene.

Homologous genes share significant similarity in their nucleotide sequence and are derived from a common ancestral gene.

Homologous recombination is the recombination of two DNA molecules at a specific site of sequence similarity.

Homology refers to morphological or structural similarity due to common ancestry.

A diploid individual is **homozygous** for a given gene when it carries two identical alleles of that gene.

Hox genes are a particular family of homeobox-containing genes that are present in all animals (as far as is known) and are involved in patterning the antero-posterior axis. They are clustered on the chromosomes in one or more gene complexes.

The **hypoblast** in the early chick embryo is a sheet of cells that covers the yolk and gives rise to extra-embryonic structures such as the stalk of the yolk sac.

The **hypocotyl** is the seedling stem that develops from the region between the embryonic root and the future shoot.

The **hypophysis** is a cell in some plant embryos that is recruited from the suspensor and contributes to the embryonic root meristem and root cap.

IGF *see* **insulin-like growth factor**.

Imaginal discs are small sacs of epithelium present in the larva of *Drosophila* and other insects, which at metamorphosis give rise to adult structures such as wings, legs, antennae, eyes, and genitalia.

Some cell-adhesion molecules, such as N-CAM, are members of the **immunoglobulin superfamily** (which also contains many proteins that are not cell-adhesion molecules).

A gene is said to be **imprinted** when it is expressed differently (either active or inactive) in the embryo depending on whether it is derived from the mother or father. This genomic **imprinting** occurs during gamete formation.

Plants with **indeterminate** growth do not make a fixed number of leaves or flowers.

Induction is the process whereby one group of cells signals to another group of cells in the embryo and so affects how they will develop.

An **inflorescence** in plants is a flowerhead—a flowering shoot. Shoots that can bear flowers develop as a result of the conversion of a vegetative apical meristem into an **inflorescence meristem**.

Ingression is the movement of individual cells from the outside of the embryo into the blastocoel.

Initials are cells in the meristems of plants that are able to divide continuously, giving rise both to dividing cells that stay within the meristem and to cells that leave the meristem and go on to differentiate.

The **inner cell mass** of the early mammalian embryo is derived from the inner cells of the morula, which form a discrete mass of cells in the blastocyst. Some of the cells of the inner cell mass give rise to the embryo proper.

***In situ* hybridization** is a technique used to detect where in the embryo particular genes are being expressed. The mRNA that is being transcribed is detected by its hybridization to a labeled single-stranded complementary DNA probe.

In animals in which the larva goes through successive phases of growth and molting before developing into an adult, the phase between each molt is known as an **instar**.

In an **instructive** induction, the cells respond differently to different concentrations of the inducing signal.

Insulin-like growth factors (**IGF**) are polypeptide growth factors that mediate many of the effects of growth hormone and are essential for post-natal growth in mammals.

Integrins are a class of cell-adhesion molecules by which cells attach to the extracellular matrix.

Intercalary growth can occur in animals capable of epimorphic regeneration when two pieces of tissue with different positional values are placed next to each other. The intercalary growth replaces the intermediate positional values.

Intermediate filaments are one of the three principal protein filaments of the cytoskeleton. They are involved in strengthening tissues such as epithelia.

An **internode** is that portion of a plant stem between two nodes (sites at which a leaf or leaves form).

Invagination is the local inward deformation of a sheet of embryonic epithelial cells to form a bulge-like structure, as in early gastrulation in the sea urchin embryo.

Involution is a type of cell movement that occurs at the beginning of amphibian gastrulation, when a sheet of cells enters the interior of the embryo by rolling in under itself.

Keratinocytes are differentiated epidermal skin cells that produce keratin, eventually die, and are shed from the skin surface.

Knock-out *see* **gene knock-out**.

Koller's sickle is a crescent-shaped region of small cells lying at the front of the posterior marginal zone.

The **lateral geniculate nucleus** is the main region of the brain in mammals where the axons from the retina terminate.

Lateral inhibition is the mechanism by which cells inhibit neighboring cells from developing in a similar way to themselves.

The **lateral plate mesoderm** in vertebrate embryos lies lateral and ventral to the somites and gives rise to the tissues of the heart, kidney, gonads, and blood.

Lateral shoot meristems arise from the apical shoot meristem and give rise to lateral shoots.

A plant leaf develops from a small set of cells called a **leaf primordium** at the edge of the apical meristem.

The bilateral asymmetry of the arrangement and structure of most internal organs in vertebrates is known as **left–right asymmetry**. In mice and humans, for example, the heart is on the left side, the right lung has more lobes than the left, and the stomach and spleen lie to the left.

The **life history** of an organism or species is its life cycle viewed in terms of its reproductive strategy and its unique ecology or interaction with the environment.

The small embryonic structures that give rise to the limbs of vertebrates are called **limb buds**.

A cell's **lineage** is the sequence of cell divisions that give rise to that cell.

Lineage restriction *see* **cell-lineage restriction**.

In **long-germ development** the blastoderm gives rise to the whole of the future embryo, as in *Drosophila*.

Macromeres are the larger of the cells that result from unequal cleavage in certain embryos, such as those of sea urchins.

The **marginal zone** of an amphibian embryo is the belt-like region of presumptive mesoderm at the equator of the late blastula.

Maternal-effect mutations are mutations in genes of the mother that affect the development of the egg and later the embryo. Genes affected by such mutations are called maternal-effect genes.

Maternal factors are proteins and RNAs that are deposited in the egg by the mother during oogenesis. The production of these maternal proteins and RNAs is under the control of so-called **maternal genes**.

The **medio-lateral** axis in vertebrates runs from the midline to the periphery.

Medio-lateral intercalation of cells occurs during convergent extension in amphibian gastrulation. The sheet of cells narrows and elongates by cells pushing in sideways between their neighbors.

Meiosis is a special type of cell division that occurs during formation of sperm and eggs, and in which the number of chromosomes is halved from diploid to haploid.

In **mericlinal chimeras** a genetically marked cell gives rise to a sector of an organ or of a whole plant.

Meristems are groups of undifferentiated, dividing cells that persist at the growing tips of plants. They give rise to all the adult structures—shoots, leaves, flowers, and roots. The **meristem identity genes** specify whether a meristem is a vegetative or an inflorescence meristem.

Mesectoderm is composed of cells that may give rise to both ectoderm and mesoderm.

Mesenchyme describes loose connective tissue, usually of mesodermal origin, whose cells are capable of migration; some epithelia of ectodermal origin, such as the neural crest, undergo an epithelial to mesenchymal transition.

A **mesenchyme-to-epithelium transition** occurs when loose mesenchyme cells aggregate and then form an epithelium like a tube, as in kidney development.

Mesendoderm is composed of cells that may give rise to both endoderm and mesoderm.

The **mesoderm** is the germ layer that gives rise to the skeleto-muscular system, connective tissues, the blood, and internal organs such as the kidney and heart.

The **mesonephros** in mammals is an embryonic kidney that contributes to the male and female reproductive organs.

Messenger RNA (**mRNA**) is the RNA molecule that specifies the sequence of amino acids in a protein. It is produced by transcription from DNA.

Metamorphosis is the process by which a larva is transformed into an adult. It often involves a radical change in form, and the development of new organs, such as wings in butterflies and limbs in frogs.

Metastasis is the movement of cancer cells from their site of origin to invade underlying tissues and to spread to other parts of the body. Such cells are said to **metastasize**.

Microfilaments *see* **actin filaments**.

Micromeres are small cells that result from unequal cleavage during early animal development.

MicroRNAs (**miRNAs**) are small RNAs that suppress the expression of specific genes.

Microtubules are one of the three principal protein filaments of the cytoskeleton. They are involved in the transport of proteins and RNAs within cells.

The **mid-blastula transition** in amphibian embryos is when the embryo's own genes begin to be transcribed, cleavages become asynchronous, and the cells of the blastula become motile.

miRNAs *see* **microRNAs**.

Mitosis is the nuclear division that occurs during the proliferation of somatic diploid cells and results in both daughter cells having the same diploid complement of chromosomes as the parent cell.

The small number of species that are commonly studied in developmental biology are known as **model organisms**.

Molting is the shedding of an external cuticle when arthropods grow, and its replacement with a new one.

Morphallaxis is a type of regeneration that involves repatterning of existing tissues without growth.

A **morphogen** is any substance active in pattern formation whose spatial concentration varies and to which cells respond differently at different threshold concentrations.

Morphogenesis refers to the processes involved in bringing about changes in form in the developing embryo.

The **morphogenetic furrow** in *Drosophila* eye development moves across the eye disc and initiates the development of the ommatidia.

A **morula** is the very early stage in a mammalian embryo when cleavage has resulted in a solid ball of cells.

Mosaic development was a term used historically to describe the development of organisms that appeared to develop mainly by distribution of localized cytoplasmic determinants.

The **Müllerian duct** runs adjacent to the Wolffian duct in the mammalian embryo and becomes the oviduct in females.

Müllerian-inhibiting substance is secreted by the developing testis and induces regression of the Müllerian ducts in males.

A **multipotent** cell is one that can give rise to many different types of differentiated cell.

A **myoblast** is a committed but undifferentiated muscle cell. In developing skeletal muscle, it will first develop into a multinucleate **myotube** and then into a fully differentiated **muscle fiber**.

The **myocardium** is the outer contractile layer of the developing heart.

Myoplasm is special cytoplasm in ascidian eggs involved in the specification of muscle cells.

The **myotome** is that part of the somite that gives rise to muscle.

Necrosis is a type of cell death due to pathological damage in which cells break up, releasing their contents.

Negative feedback is a type of regulation in which the end-product of a pathway or process inhibits an earlier stage.

Neoteny is the phenomenon in which an animal acquires sexual maturity while still in larval form.

The **neural crest cells** in vertebrates are derived from the edge of the neural plate. They migrate to different regions of the body and give rise to a wide variety of tissues, including the autonomic nervous system, the sensory nervous system, pigment cells, and some cartilage of the head.

Neural folds, neural plate, neural tube, *see* **neurulation**.

The **neurectoderm** is embryonic ectoderm with the potential to form neural cells and epidermis.

A **neuroblast** is an embryonic cell that will give rise to neural tissue (neurons and glia).

The **neuromuscular junction** is the specialized area of contact between a motor neuron and a muscle fiber, where the neuron can stimulate muscle activity.

Neurons or nerve cells are the electrically excitable cells of the nervous system, which convey information in the form of electrical signals.

Neurotrophins are proteins that are necessary for neuronal survival, such as nerve growth factor.

A **neurula** is the stage of vertebrate development at the end of gastrulation when the neural tube is forming.

Neurulation in vertebrates is the process in which the ectoderm of the future brain and spinal cord—the **neural plate**—develops folds (**neural folds**) and forms the **neural tube**.

The **Nieuwkoop center** is a signaling center on the dorsal side of the early *Xenopus* embryo. It forms in the vegetal region as a result of cortical rotation.

Nodal and **Nodal-related proteins** comprise a subfamily of the TGF-β family of signaling proteins of vetebrates. They are involved in all stages of development, but particularly in early mesoderm induction and patterning.

A **node** in a plant is that part of the stem at which leaves and lateral buds form. In avian and mammalian embryos the term usually refers to Hensen's node or its equivalent.

The effects of a gene are **non cell autonomous** or non-autonomous if they affect cells other than the cell in which the gene is expressed.

The **notochord** in vertebrate embryos is a rod-like cellular structure that runs from head to tail and lies centrally beneath the future central nervous system. It is derived from mesoderm.

Nurse cells surround the developing oocyte in *Drosophila* and synthesize proteins and RNAs that are to be deposited in it.

The eight-cell stage of a plant embryo is called the **octant stage**.

Alternate **ocular dominance columns** in the visual cortex respond to the same visual stimulus from either the left or right eye.

Insect compound eyes are composed of hundreds of individual photoreceptor organs, the **ommatidia** (singular **ommatidium**).

Many of the genes involved in cell regulation can be mutated into **oncogenes**, which cause cells to become cancerous.

Ontogeny refers to the development of an individual organism.

An **oocyte** is an immature egg.

Oogenesis is the process of egg formation in the female.

The **optic tectum** is the region of the brain in amphibians and birds where the axons from the retina terminate.

The **oral–aboral axis** in sea urchins and other radially symmetrical organisms runs from the centrally situated mouth to the opposite side of the body.

An **organizer**, organizing region or organizing center is a signaling center that directs the development of the whole embryo or of part of the embryo, such as a limb. In amphibians, the organizer usually refers to the Spemann organizer.

Organogenesis is the development of specific organs such as limbs, eyes, and heart.

Osteoblasts are the precursors from which differentiated bone cells are formed.

The **oviduct** in female birds and mammals transports the eggs from the ovaries to the uterus.

An **ovule** is the structure in plants that contains an egg cell.

The **pair-rule genes** in *Drosophila* are involved in delimiting parasegments. They are expressed in transverse stripes in the blastoderm, each pair-rule gene being expressed in alternate parasegments.

Genes within a species that have arisen by duplication and divergence are called **paralogs**. Examples are the Hox genes in vertebrates, which comprise several **paralogous subgroups** made up of **paralogous genes**.

Parasegments in the developing *Drosophila* embryo are independent developmental units that give rise to the segments of the larva and adult.

The mesoderm lying on either side of the midline and forming the somites is sometimes called the **paraxial mesoderm**.

Pattern formation is the process by which cells in a developing embryo acquire identities that lead to a well ordered spatial pattern of cell activities.

Pax genes encode transcriptional regulatory proteins that contain both a homeodomain and another protein motif, the paired motif.

P elements are transposable DNA elements found in *Drosophila*. They are short sequences of DNA that can become inserted in different positions within a chromosome and can also move to other chromosomes. This property is exploited in the technique of P-element-mediated transformation for making transgenic flies.

Periclinal cell divisions are divisions in a plane parallel to the surface of the tissue.

In **periclinal chimeras** in plants, one of the three meristem layers has a genetic marker which distinguishes it from the other two.

Permissive inductions occur when a cell makes only one kind of response to an inducing signal, and makes it when a given level of signal is reached.

P granules are granules that become localized to the posterior end of the fertilized egg of *C. elegans*.

The **phenotype** is the observable or measurable characters and features of a cell or an organism.

The response of an organism to relative day length is known as **photoperiodism**, and in plants is responsible for promoting flowering as days become longer.

Phyllotaxy is the way the leaves are arranged along a shoot.

Phylogeny is the evolutionary history of a species or group.

Vertebrate embryos pass through a developmental stage known as the **phylotypic stage** at which the embryos of the different vertebrate groups closely resemble each other. This is the stage at which the embryo possesses a distinct head, a neural tube, and somites.

Planar cell polarity is the situation in which cells are polarized in the plane of the tissue, as in the epidermis of insect wings, in which wing hairs all point in the same direction.

Plasmodesmata (singular **plasmodesma**) are the threads of cytoplasm that run through the cell wall and interconnect adjacent plant cells.

A **pluripotent** stem cell is one that can give rise to many different types of differentiated cell.

The **pluteus** is the larval stage of the sea urchin.

Polar bodies are formed during meiosis in the developing egg. They are small cells containing a haploid nucleus and take no part in embryonic development.

When one end of a cell, structure, or organism is different from the other end it is said to have **polarity** or to be **polarized**.

In the developing chick and mouse limb buds, the **polarizing region** at the posterior margin of the bud produces a signal specifying position along the antero-posterior axis.

Pole cells give rise to the germ cells in *Drosophila* and are formed at the posterior end of the blastoderm.

Pole plasm is the cytoplasm at the posterior end of the *Drosophila* egg that is involved in specifying germ cells.

Polydactyly is the occurrence of extra digits on hands or feet.

Polyspermy is the entry of more than one sperm into the egg.

Positional information in the form, for example, of a gradient of an extracellular signaling molecule, can provide the basis for pattern formation. Cells acquire a **positional value** that is related to their position with respect to the boundaries of the given field of positional information. The cells then interpret this positional value according to their genetic constitution and developmental history, and develop accordingly.

Positive feedback is a type of regulation in which the end-product of a pathway or process can activate an earlier stage.

Posterior dominance or **posterior prevalence** is the process whereby the more posteriorly expressed Hox genes can inhibit the action of more anteriorly expressed Hox genes when they are expressed in the same region.

The **posterior marginal zone** of the chick embryo is a dense region of cells at the edge of the blastoderm that will give rise to the primitive streak.

Most neurons do not divide further once they are formed, when they are known as **post-mitotic neurons**.

The **post-synaptic** side of a synapse is the part that receives the signal.

Post-translational modification of a protein involves changes in the protein after it has been synthesized. The protein can, for example, be enzymatically cleaved, glycosylated, or acetylated.

A basic pattern generated automatically in a structure is known as a **prepattern**. It may subsequently be modified during development.

The **pre-somitic mesoderm** is the unsegmented mesoderm between the node (in chick and mouse) and the already formed somites. It will form somites from its anterior end.

The **pre-synaptic** side of a synapse is the part that generates the signal.

Primary embryonic induction is the induction of the whole body axis, as demonstrated by transplantation of the Spemann organizer in amphibians.

The **primitive ectoderm** or epiblast is the part of the inner cell mass in the mammalian blastocyst that gives rise to the embryo proper.

The **primitive endoderm** in mammalian embryos is that part of the inner cell mass that contributes to extra-embryonic membranes.

The **primitive streak** of the chick embryo is a strip of cells that extends inward from the posterior marginal zone and is the forerunner of the antero-posterior axis. During gastrulation, cells move through the streak into the interior of the blastoderm. The primitive streak in the mouse embryo has a similar function to that in the chick.

Minute undifferentiated outgrowths that will give rise to a structure such as a tooth, leaf, flower or floral organ are known as **primordia** (singular **primordium**).

The two-celled stage in plants is called the **proembryo**.

Programmed cell death *see* **apoptosis**.

In chick and mouse limb buds the cells in the **progress zone** at the tip of the bud proliferate and acquire positional values.

The **promeristem** is the central region of the meristem that contains cells capable of continued division—the initials.

The **promoter** is a region of DNA close to the coding sequence to which RNA polymerase binds to begin transcription of a gene.

Proneural clusters are small clusters of cells within the neurectoderm in which one cell will eventually become a neuroblast.

Genes that promote a neural fate in neurectoderm cells are called **proneural genes**.

A **pronucleus** is the haploid nucleus of sperm or egg after fertilization but before nuclear fusion and the first mitotic division.

A **proto-oncogene** is a gene that is involved in regulation of cell proliferation and that can cause cancer when mutated into an oncogene or expressed under abnormal control.

Protostomes are those animals, such as insects, in which cleavage of the zygote is not radial and in which gastrulation primarily forms the mouth.

The **proximo-distal axis** of a limb or other appendage runs from the point of attachment to the body (proximal) to the tip of the limb (distal).

The **pupa** in *Drosophila* and similar insects is a stage following the larval stages in which the organism can remain dormant for long periods and in which metamorphosis occurs.

The **quiescent center** in a plant root tip meristem is a central group of cells that divide rarely but are essential for meristem function.

The **radial axis** of a structure is the axis running from the center to the circumference. Cylindrical structures such as plant stems and roots that are completely symmetrical around a central axis are said to have **radial symmetry**.

Radial intercalation occurs in a multilayered ectoderm of an amphibian gastrula when cells intercalate in a direction perpendicular to the surface, so thinning and extending the cell sheet.

Reaction–diffusion mechanisms produce self-organizing patterns of chemical concentrations which could underlie periodic patterns.

A **recessive** mutation is a mutation in a gene that only changes the phenotype when both copies of the gene carry the mutation.

Redundancy refers to an apparent absence of an effect when a gene that is normally active during development is inactivated. It is assumed that other pathways exist which can substitute for the missing gene action.

Regeneration is the ability of a fully developed organism to replace lost parts.

Regulation is the ability of the embryo to develop normally even when parts are removed or rearranged. Embryos that can regulate are called **regulative**.

The **rhombomeres** are a sequence of compartments of cell-lineage restriction in the hindbrain of chick and mice embryos.

RNA interference (**RNAi**) is a means of suppressing gene expression by promoting the destruction of a specific mRNA by targeting them with short complementary RNAs called **short interfering RNAs** (**siRNAs**).

RNA processing is the process in eukaryotic cells in which newly transcribed RNAs are modified in various ways to make a functional messenger RNA or structural RNA. It includes **RNA splicing**, which removes introns from the transcript to leave a continuous coding messenger RNA or a functional structural RNA.

The **roof plate** of the developing neural tube is composed of non-neural cells and is involved in patterning the dorsal part of the tube.

Skeletal muscle can be renewed from undifferentiated stem cells called **satellite cells**.

The **sclerotome** is that part of a somite that will give rise to the cartilage of the vertebrae.

Segmentation is the division of the body of an organism into a series of morphologically similar units or **segments**.

Segmentation genes in *Drosophila* are involved in patterning the parasegments and segments.

Selector genes in *Drosophila* determine the activity of a group of cells, and their continued expression is required to maintain that activity.

A **semi-dominant** mutation is a mutation that affects the phenotype when just one allele carries the mutation but where the effect on the phenotype is much greater when both alleles carry the mutation.

Senescence is the impairment of function associated with aging.

A **sensory organ precursor** is an ectodermal cell that will give rise to a sensory bristle in the adult *Drosophila* epidermis.

Sex determination is the genetic and developmental process by which an organism's sex is specified. In many organisms, sexual phenotype is determined by specific chromosomes called **sex chromosomes**.

The **sex-determining region of the Y chromosome** (SRY) determines maleness by specifying the gonad as a testis.

The **shield stage** in zebrafish embryos is the stage in which the organizer, known as the shield, has been formed.

Short-germ development characterizes those insects in which most of the segments are formed sequentially by growth. The blastoderm itself only gives rise to the anterior segments of the embryo.

short interfering RNA (siRNA) *see* RNA interference.

A signaling center is a localized region of the embryo that exerts a special influence on surrounding cells and thus determines how they develop.

Signal transduction is the process by which a cell converts an extracellular signal, usually at the cell membrane, into a response, which is often a change in gene expression.

In the rare condition situs inversus in humans there is complete mirror-image reversal of the position of the internal organs.

Somatic cells are any cells other than germ cells. In most animals, the somatic cells are diploid.

Somites in vertebrate embryos are segmented blocks of mesoderm lying on either side of the notochord. They give rise to body and limb muscles, the vertebral column, and the dermis.

A specification map shows how the tissues of an embryo will develop when placed in a simple culture medium.

A group of cells is called specified if when isolated and cultured in a neutral medium they develop according to their normal fate.

The Spemann organizer or Spemann–Mangold organizer is a signaling center on the dorsal side of the amphibian embryo that acts as the main embryonic organizer. Signals from this center can organize new antero-posterior and dorso-ventral axes.

The sphere stage in zebrafish embryos comprises a blastoderm of around 1000 cells lying over a spherical yolk.

Spermatogenesis is the production of sperm.

A stem cell is a type of undifferentiated cell that is both self renewing and also gives rise to differentiated cell types. They are found in some adult tissues. They are maintained in microenvironments known as stem-cell niches. *See also* embryonic stem cells.

The subgerminal space is the cavity that develops under the area pellucida in the early chick blastoderm.

The superior colliculus in mammals is a region of the brain to which some retinal neurons project. It corresponds to the optic tectum of amphibians and birds.

The suspensor attaches the embryo to maternal tissue and is a source of nutrients.

A synapse is the specialized point of contact where a neuron communicates with another neuron or a muscle cell. Neurotransmitter is produced in synaptic vesicles in the presynaptic neuron and is released into the synaptic cleft separating the two cells.

A syncytium is a cell with many nuclei in a common cytoplasm. Cell walls do not develop during nuclear division within the very early *Drosophila* embryo. This gives rise to the syncytial blastoderm in which the nuclei are arranged around the periphery of the embryo.

The tailbud is the structure at the posterior end of vertebrate embryos that gives rise to the post-anal tail.

The telson is a distinctive structure at the posterior end of the *Drosophila* embryo.

Teratocarcinomas are solid tumors that arise from germ cells and which can contain a mixture of differentiated cell types.

A tetraploid cell is one that contains four sets of chromosomes.

Therapeutic cloning is the potential use of somatic cell transfer to alleviate disease, without affecting the germline.

In insects the body is divided into three distinct parts, the head at the anterior end, followed by the thorax and the posterior abdomen. The thoracic segments of *Drosophila* are the segments that carry the legs and wings in the adult. In vertebrates, the thorax is the chest region.

A threshold concentration is that concentration of a chemical signal or morphogen that can elicit a particular response from a cell. A specific response to a chemical signal that only occurs above or below a particular threshold concentration of the signal is known as a threshold effect.

Totipotency is the capacity of a cell to develop into any of the cell types found in that particular organism. Such a cell is called totipotent.

The tracheal system of insects is a system of fine tubules that deliver air (and thus oxygen) to the tissues.

When a gene is active its DNA sequence is copied, or transcribed, into a complementary RNA sequence, a process known as transcription.

A transcription factor is a regulatory protein required to initiate or regulate the transcription of a gene into RNA. Transcription factors act within the nucleus of a cell by binding to specific regulatory regions in the DNA.

Transdifferentiation is the process by which a differentiated cell can differentiate into a different cell type, such as pigment cell to lens.

Transfection is the technique by which mammalian and other animal cells are induced to take up foreign DNA molecules. The introduced DNA sometimes becomes inserted permanently into the host cell's DNA.

The process by which messenger RNA directs the order of amino acids in a protein during protein synthesis at ribosomes is known as translation. The messenger RNA is said to be translated into protein.

A transposon is a DNA sequence that can become inserted into a different site on the chromosome, either by the insertion of a copy of the original sequence or by excision and reinsertion of the original sequence.

A trichome is a hair-bearing cell in plant epidermis.

A triploblast is an animal with three germ layers—endoderm, mesoderm, and ectoderm.

The trophectoderm is the outer layer of cells of the early mammalian embryo. It gives rise to extra-embryonic structures such as the placenta.

Tumor suppressor genes are genes that can cause a cell to become cancerous when both copies of the gene have been inactivated.

Vasculogenesis is the formation of blood vessels.

The vegetal region of an amphibian egg is the most yolky region, and is the region from which the endoderm will develop. The most terminal part of this region is called the vegetal pole, and is directly opposite the animal pole.

Embryos that are **ventralized** are deficient in dorsal regions and have much increased ventral regions.

The **ventricular zone** is a layer of proliferating cells lining the lumen of the vertebrate neural tube, from which neurons and glia are formed.

The phenomenon by which flowering is accelerated after the plant has been exposed to a long period of cold temperature is known as **vernalization**.

The **vertebral column** is the backbone or spine of vertebrates, composed of a succession of vertebrae.

The **visceral endoderm** is derived from the primitive endoderm that develops on the surface of the egg cylinder in the mammalian blastocyst.

The **vitelline membrane** is an extracellular layer surrounding the eggs of the sea urchin and other animals. In the sea urchin it gives rise to the fertilization membrane.

The **Wnt family** of secreted signaling proteins are important in many aspects of development. It includes the Wingless protein in *Drosophila*. See also **canonical Wnt pathway** and **planar cell polarity**.

Wolffian ducts are ducts associated with the mesonephros in mammalian embryos, They become the vas deferens in males.

The **yolk sac** is an extra-embryonic membrane in birds and mammals. In the chick embryo it surrounds the yolk.

The **yolk syncytial layer** in zebrafish forms a continuous layer of multinucleate non-yolky cytoplasm underlying the blastoderm.

The **zona pellucida** is a layer of glycoprotein surrounding the mammalian egg that serves to prevent polyspermy.

The **zone of polarizing activity** is another name for the polarizing region at the posterior margin of the limb bud.

The **zootype** refers to a pattern of expression of Hox genes and certain other genes along the antero-posterior axis of the embryo that is characteristic of all animal embryos.

The **zygote** is the fertilized egg. It is diploid and contains the chromosomes of both the male and female parents.

Zygotic genes are those present in the fertilized egg and which are expressed in the embryo itself.

Index

see also glossary of terms

A

D

E

F

I

J

K

L

M

N

O

P

Q

R

S

W

X

Y

Z